Property of the
Electrical & Computer Engineering Department
360 Huntington Street
409 Dana Research Center
Boston, MA 02115
Track 5741

$110.00

PRENTICE HALL SIGNAL PROCESSING SERIES

Alan V. Oppenheim, Series Editor

ANDREWS AND HUNT *Digital Image Restoration*
BRIGHAM *The Fast Fourier Tranform*
BRIGHAM *The Fast Fourier Transform and Its Applications*
BURDIC *Underwater Acoustic System Analysis, 2/E*
CASTLEMAN *Digital Image Processing*
COWAN AND GRANT *Adaptive Filters*
CROCHIERE AND RABINER *Multirate Digital Signal Processing*
DUDGEON AND MERSEREAU *Mulidimensional Digital Signal Processing*
HAMMING *Digital Filters, 3/E*
HAYKIN, ED. *Advances in Spectrum Analysis and Array Processing, Vols. I & II*
HAYKIN, ED. *Array Signal Processing*
JAYANT AND NOLL *Digital Coding of Waveforms*
JOHNSON AND DUDGEON *Array Signal Processing: Concepts and Techniques*
KAY *Modern Spectral Estimation*
KINO *Acoustic Waves: Devices, Imaging, and Analog Signal Processing*
LEA, ED. *Trends in Speech Recognition*
LIM *Two-Dimensional Signal and Image Processing*
LIM, ED. *Speech Enhancement*
LIM AND OPPENHEIM, EDS. *Advanced Topics in Signal Processing*
MARPLE *Digital Spectral Analysis with Applications*
MCCLELLAN AND RADER *Number Theory in Digital Signal Processing*
MENDEL *Lessons in Digital Estimation Theory*
OPPENHEIM, ED. *Applictions of Digital Signal Processing*
OPPENHEIM AND NAWAB, ED. *Symbolic and Knowledge-Based Signal Processing*
OPPENHEIM, WILLSKY, WITH YOUNG *Signals and Systems*
OPPENHEIM AND SCHAFER *Digital Signal Processing*
OPPENHEIM AND SCHAFER *Discrete-Time Signal Processing*
QUACKENBUSH ET AL. *Objective Measures of Speech Quality*
RABINER AND GOLD *Theory and Applications of Digital Signal Processing*
RABINER AND SCHAFER *Digital Processing of Speech Signals*
ROBINSON AND TREITEL *Geophysical Signal Analysis*
STEARNS AND DAVID *Signal Processing Algorithms*
STEARNS AND DAVID *Signal Processing Algorithms in Fortran and C*
STEARNS AND HUSH *Digital Signal Analysis, 2/E*
TRIBOLET *Seismic Applications of Homomorphic Signal Processing*
VAIDYANATHAN *Multirate Systems and Filter Banks*
WIDROW AND STEARNS *Adaptive Signal Processing*

Multirate Systems and Filter Banks

P.P. Vaidyanathan

Department of Electrical Engineering
California Institute of Technology, Pasadena

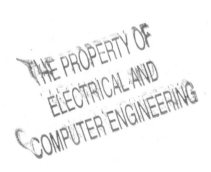

PRENTICE HALL P T R, Englewood Cliffs, New Jersey 07632

Library of Congress Cataloging-in-Publication Data

Vaidyanathan, P.P.
    Multirate systems and filter banks / P.P. Vaidyanathan.
       p.   cm.
    Includes bibliographical references (p.) and index.
    ISBN 0-13-605718-7
    1. Signal processiong--Digital techniques.   I. Title.
TK5102.5. V24   1993   92-6547
621.382'2--dc20   CIP

Editorial/production supervision: *Brendan M. Stewart*
Prepress buyer: *Mary McCartney*
Manufacturing buyer: *Susan Brunke*
Acquisitions editor: *Karen Gettman*

© 1993 by Prentice Hall P T R
Prentice-Hall, Inc.
A Simon & Schuster Company
Englewood Cliffs, New Jersey 07632

The publisher offers discounts on this book when ordered in bulk quantities. For more information, write:
Special Sales/Professional Marketing, Prentice-Hall, Professional & Technical Reference Division,
Englewood Cliffs, NJ 07632

All rights reserved. No part of this book may be
reproduced, in any form or by any means,
without permission in writing from the publisher.

Printed in the United States of America

10 9 8 7 6 5 4 3

ISBN 0-13-605718-7

Prentice-Hall International (UK) Limited, *London*
Prentice-Hall of Australia Pty. Limited, *Sydney*
Prentice-Hall Canada Inc., *Toronto*
Prentice-Hall Hispanoamericana, S.A., *Mexico*
Prentice-Hall of India Private Limited, *New Delhi*
Prentice-Hall of Japan, Inc., *Tokyo*
Simon & Schuster Asia Pte. Ltd., *Singapore*
Editora Prentice-Hall do Brasil, Ltda., *Rio de Janeiro*

*to
my Parents
and
my wife Usha*

Contents

Preface   xi

PART 1   INTRODUCTORY CHAPTERS

1 Introduction   1
    1.1   Major Developments   3
    1.2   Scope and Outline   8

2 Review of Discrete-Time Systems   12
    2.0   Introduction   12
    2.1   Discrete-Time Signals   12
    2.2   Multi-Input Multi-Output Systems   24
    2.3   Notations   28
    2.4   Discrete-Time Filters (Digital Filters)   31
    Problems   39

3 Review of Digital Filters   42
    3.0   Introduction   42
    3.1   Filter Design Specifications   42
    3.2   FIR Filter Design   45
    3.3   IIR Filter Design   60
    3.4   Allpass Filters   71
    3.5   Special Types of Filters   83
    3.6   IIR Filters Based on Two Allpass Filters   84
    3.7   Concluding Remarks   91
    Problems   93

4 Fundamentals of Multirate Systems   100
    4.0   Introduction   100
    4.1   Basic Multirate Operations   100
    4.2   Interconnection of Building Blocks   118
    4.3   The Polyphase Representation   120
    4.4   Multistage Implementations   134
    4.5   Some Applications of Multirate Systems   143
    4.6   Special Filters and Filter Banks   151
    4.7   Multigrid Methods   168
    Problems   178

PART 2   MULTIRATE FILTER BANKS

5   Maximally Decimated Filter Banks   188

- 5.0   Introduction   188
- 5.1   Errors Created in the QMF Bank   191
- 5.2   A Simple Alias-Free QMF System   196
- 5.3   Power Symmetric QMF Banks   204
- 5.4 M-channel Filter Banks   223
- 5.5   Polyphase Representation   230
- 5.6   Perfect Reconstruction (PR) Systems   234
- 5.7   Alias-Free Filter Banks   245
- 5.8   Tree Structured Filter Banks   254
- 5.9   Transmultiplexers   259
- 5.10 Summary and Tables   266
- *Problems*   272

6   Paraunitary Perfect Reconstruction (PR) Filter Banks   286

- 6.0   Introduction   286
- 6.1   Lossless Transfer Matrices   288
- 6.2   Filter Bank Properties Induced by Paraunitariness   294
- 6.3   Two Channel FIR Paraunitary QMF Banks   298
- 6.4   The Two Channel Paraunitary QMF Lattice   302
- 6.5 M-channel FIR Paraunitary Filter Banks   314
- 6.6   Transform Coding and the "LOT"   322
- 6.7   Summary, Comparisons, and Tables   326
- *Problems*   333

7   Linear Phase Perfect Reconstruction QMF Banks   337

- 7.0   Introduction   337
- 7.1   Some Necessary Conditions   337
- 7.2   Lattice Structures for Linear Phase FIR PR QMF Banks   339
- 7.3   Formal Synthesis of Linear Phase FIR PR QMF Lattice   347
- *Problems*   351

8   Cosine Modulated Filter Banks   353

- 8.0   Introduction   353
- 8.1   The Pseudo QMF Bank   354
- 8.2   Design of the Pseudo QMF Bank   363
- 8.3   Efficient Polyphase Structures   370
- 8.4   Deeper Properties of Cosine Matrices   373
- 8.5   Cosine Modulated Perfect Reconstruction Systems   377
- *Problems*   392

PART 3   SPECIAL TOPICS

9   Quantization Effects   394

- 9.0   Introduction   394
- 9.1   Types of Quantization Effects   394
- 9.2   Review of Standard Techniques   397

9.3 Noise Transmission in Multirate Systems 405
9.4 Noise in Filter Banks 408
9.5 Filter Bank Output Noise 412
9.6 Limit Cycles 416
9.7 Coefficient Quantization 418
Problems 424

10 Multirate Filter Bank Theory and Related Topics 427

10.0 Introduction 427
10.1 Block Filters, LPTV Systems and Multirate Filter Banks 427
10.2 Unconventional Sampling Theorems 436
Problems 454

11 The Wavelet Transform and its Relation to Multirate Filter Banks 457

11.0 Introduction 457
11.1 Background and Outline 458
11.2 The Short-Time Fourier Transform 463
11.3 The Wavelet Transform 481
11.4 Discrete-Time Orthonormal Wavelets 500
11.5 Continuous-Time Orthonormal Wavelet Basis 510
11.6 Concluding Remarks 536
Problems 539

12 Multidimensional Multirate Systems 545

12.0 Introduction 545
12.1 Multidimensional Signals 546
12.2 Sampling a Multidimensional Signal 555
12.3 Minimum Sampling Density 568
12.4 Multirate Fundamentals 572
12.5 Alias-Free Decimation 597
12.6 Cascade Connections 603
12.7 Multirate Filter Design 608
12.8 Special Filters and Filter Banks 623
12.9 Maximally Decimated Filter Banks 627
12.10 Concluding Remarks 641
Problems 650

PART 4   MULTIVARIABLE AND LOSSLESS SYSTEMS

13 Review of Discrete-Time Multi-Input Multi-Output LTI Systems 660

13.0 Introduction 660
13.1 Multi-Input Multi-Output Systems 661
13.2 Matrix Polynomials 661
13.3 Matrix Fraction Descriptions 665
13.4 State Space Descriptions 669
13.5 The Smith-McMillan Form 687
13.6 Poles of Transfer Matrices 699
13.7 Zeros of Transfer Matrices 703
13.8 Degree of a Transfer Matrix 707

13.9	FIR Transfer Matrices	708
13.10	Causal Inverses of Causal Systems	711
	Problems 715	

14 Paraunitary and Lossless Systems   722

14.0	Introduction	722
14.1	A Brief History	723
14.2	Fundamentals of Lossless Systems	724
14.3	Lossless Systems with Two Outputs	727
14.4	Structures for $M \times M$ and $M \times 1$ FIR Lossless Systems	731
14.5	State Space Manifestation of Lossless Property	740
14.6	Factorization of Unitary Matrices	745
14.7	Smith-McMillan Form and Pole-Zero Pattern	754
14.8	The Modulus Property	758
14.9	Structures for IIR Lossless Systems	759
14.10	Modified Lossless Structures	763
14.11	Preserving Lossless Property Under Quantization	768
14.12	Summary and Tables	771
	Problems 775	

APPENDICES

A Review of Matrices   782

A.0	Introduction	782
A.1	Definitions and Examples	782
A.2	Basic Operations	783
A.3	Determinants	786
A.4	Linear Independence, Rank, and Related Issues	787
A.5	Eigenvalues and Eigenvectors	790
A.6	Special Types of Matrices	793
A.7	Unitary Triangularization	798
A.8	Maximization and Minimization	798
A.9	Properties Preserved in matrix products	799
	Problems 800	

B Review of Random Processes   803

B.0	Introduction	803
B.1	Real Random Variables	803
B.2	Real Random Processes	806
B.3	Passage Through LTI Systems	810
B.4	The Complex Case	811
B.5	The Vector Case	812

C Quantization of Subband Signals   816

C.0	Introduction	816
C.1	Quantizer Noise Variance	816
C.2	The Ideal Subband Coder	818
C.3	The Orthogonal Transform Coder	826
C.4	Similarities and Differences	833

  C.5 Relation to Other Methods 839
      Problems 845

D Spectral Factorization Techniques 849
  D.0 Introduction 849
  D.1 The Complex Cepstrum 849
  D.2 A Cepstral Inversion Algorithm 853
  D.3 A Spectral Factorization Algorithm 854
      Problems 858

E Mason's Gain Formula 859

Glossary of Symbols 866

List of Important Summaries (Tables) 867

List of Important Summaries (Figures) 868

Bibliography 869
  Alphabetical List of References Cited in the Book 869
  Some References by Topic 888

Index 891

Preface

Multirate digital signal processing techniques have been practiced by engineers for more than a decade and a half. This discipline finds applications in speech and image compression, the digital audio industry, statistical and adaptive signal processing, numerical solution of differential equations, and in many other fields. It also fits naturally with certain special classes of time-frequency representations such as the short-time Fourier transform and the wavelet transform, which are useful in analyzing the time-varying nature of signal spectra.

Over the last decade, there has been a tremendous growth of activity in the area of multirate signal processing, perhaps triggered by the first book in this field [Crochiere and Rabiner, 1983]. Particularly impressive is the amount of new literature in digital filter banks, multidimensional multirate systems, and wavelet representations. The theoretical work in multirate filter banks appears to have reached a level of maturity which justifies a thorough, unified, and in-depth treatment of these topics. This book is intended to serve that purpose, and it presents the above mentioned topics under one cover. Research in the areas of multidimensional systems and wavelet transforms is still proceeding at a rapid rate. We have dedicated one chapter to each of these, in order to bring the reader up to a point where research can be begun.

I have always believed that it is important to appreciate the generality of principles and to obtain a solid theoretical foundation, and my presentation here reflects this philosophy. Several applications are discussed throughout the book, but the general principles are presented without bias towards specific application-oriented detail.

The writing style here is very much in the form of a *text*. Whenever possible I have included examples to demonstrate new principles. Many design examples and complete design rules for filter banks have been included. Each chapter includes a fairly extensive set of homework problems (totaling over 300). The solutions to these are available to instructors, from the publisher. Tables and summaries are inserted at many places to enable the reader to locate important results conveniently. I have also tried to simplify the reader's task by assigning separate chapters for more advanced material. For example, Chap. 11 is dedicated to wavelet transforms, and Chap. 14 contains detailed developments of many results on paraunitary systems. Whenever a result from an advanced chapter (for example, Chap. 14) is used in an earlier chapter, this result is first stated clearly within the context of use, and the reader is referred to the appropriate chapter for proof.

The text is self-contained for readers who have some prior exposure to digital signal processing. A one-term course which deals with sampling, discrete-time Fourier transforms, z-transforms, and digital filtering, is sufficient. In Chap. 2 and

3 a brief review of this material is provided. A thorough exposition can be found in a number of references, for example, [Oppenheim and Schafer, 1989]. Chapter 3 also contains some new material, for example, *eigenfilters,* and detailed discussions on allpass filters, which are very useful in multirate system design.

A detailed description of the text can be found in Chap. 1. Chapters 2 and 3 provide a brief review of signals, systems, and digital filtering. Chapter 4, which is the first one on multirate systems, covers the fundamentals of multirate building blocks and filter banks, and describes many applications. Chapter 5 introduces multirate filter banks, laying the theoretical foundation for alias cancelation, and elimination of other errors. The first two sections in Chap. 4 and 5 contain material overlapping with [Crochiere and Rabiner, 1983]. Most of the remaining material in these chapters, and in the majority of the chapters that follow, have not appeared in this form in text books.

Chapters 6 to 8 provide a deeper study of multirate filter banks, and present several design techniques, including those based on the so-called paraunitary matrices. (These matrices play a role in the design of many multirate systems, and are treated in full depth in Chap. 14.) Chapters 9 to 12 cover special topics in multirate signal processing. These include roundoff noise effects (Chap. 9), block filtering, periodically time varying systems and sampling theorems (Chap. 10), wavelet transforms (Chap. 11) and *multidimensional* multirate systems (Chap. 12). Chapters 13 and 14 give an in-depth coverage of multivariable linear systems and lossless (or paraunitary) systems, which are required for a deeper understanding of multirate filter banks and wavelet transforms.

There are five appendices which serve as references as well as supplementary reading. Three of these are review-material (matrix theory, random processes, and Mason's gain formula). Two of the appendices contain results directly related to filter bank systems. One of these is a technique for spectral factorization; the other one analyzes the effects of quantization of subband signals.

Many of these chapters have been taught at Caltech over the last three years. This text can be used for teaching a one, two, or three term (quarter or semester) course on one of many possible topics, for example, multirate fundamentals, multirate filter banks, wavelet representation, and so on. There are many homework problems. The instructor has a great deal of flexibility in choosing the topics, but I prefer not to bias him or her by providing specific course outlines here.

In summary, I have endeavored to produce a text which is useful for the classroom, as well as for self-study. It is also hoped that it will bring the reader to a point where he/she can start pursuing research in a vast range of multirate areas. Finally, I believe that the text can be comfortably used by the practicing engineer because of the inclusion of several design procedures, examples, tables, and summaries.

ACKNOWLEDGMENTS

The generous support and pleasant environment offered by the California Institute of Technology has been most crucial in the successful completion of this project. The funding provided by the National Science Foundation for our research in multirate signal processing has been very helpful in developing many of the results which are included in this text. I also take this opportunity to thank Prof. Alan Oppenheim for his enthusiasm about this project, and for including this text in his distinguished signal processing series. I am indebted to Brendan Stewart, Prentice Hall, for carefully supervising the production of the final book. Karen Gettman, Prentice Hall, was very helpful during all stages of the writing process.

Several professional colleagues and students have played a major role in the evolution and completion of this project. Professor Maurice Bellanger (TRT, France), and Drs. Ron Crochiere, N. S. Jayant, and Larry Rabiner of the AT&T Bell Laboratories, provided the valuable encouragement which I needed during the initial stages.

I am deeply thankful to Dr. Rabiner for his criticism of the first draft, and for providing many valuable suggestions on the style of presentation. His feedback has resulted in significant improvement of the presentation here. My heartfelt gratitude also goes to Prof. Martin Vetterli (Columbia University, New York) for his great enthusiasm, interaction, friendship, and valuable feedback over the years, and to Professors Mark Smith and Tom Barnwell (Georgia Institute of Technology), for their constant support. Prof. Smith had studied the first draft of the manuscript very carefully, and provided valuable suggestions. I am indebted to Dr. Ajay Luthra and his colleagues (Tektronix laboratories), for their enthusiasm and interest in the project. Some of the research work related to this book has been done with the support from Tektronix.

Dr. Rashid Ansari (Bellcore), Prof. Roberto Bamberger (Washington State University) and Dr. Jelena Kovačević (AT& T Bell laboratories) provided valuable comments on the chapter on multidimensional multirate systems. Dr. Ingrid Daubechies (AT& T Bell laboratories) provided very useful feedback, which improved parts of Chap. 11 (wavelets) substantially. Ramesh Gopinath (Rice University) also provided important comments on this chapter. I also wish to thank Prof. Tom Parks (Cornell University) for his valuable feedback on Chap. 1 to 6 and 11, Prof. P. K. Rajan (Tennesse Technological University) for his comments on Chap. 12, and Prof. A. Sideris (Caltech) for comments on Chap. 13.

I appreciate the interest shown by Prof. H. Malvar (University of Brasilia), Prof. George Moschytz (Swiss Federal Institute of Technology), and Prof. Tor Ramstad (Norwegian Institute of Technology). I also sincerely acknowledge the constant moral support and encouragement I received from Prof. Sanjit Mitra (University of California, Santa Barbara) during various stages.

Several graduate students at Caltech have participated in the research that eventually gave rise to this book. In this regard, my interactions with Tsuhan Chen, Zinnur Doğanata, Phuong-Quan Hoang, David Koilpillai, Vincent Liu, Truong Nguyen, Vinay Sathe, and Anand Soman have been most enjoyable. David's research at Caltech has had a major effect on Chap. 8 (cosine modulated filter banks). In addition to providing intellectual interactions at the deepest level, Tsuhan Chen, David Koilpillai, Truong Nguyen, and Anand Soman have also generated many of the multirate design examples in this text. They have also read various chapters of the manuscript and provided useful feedback. In this connection my special thanks go to Tsuhan Chen, Zinnur Doğanata, Ian Galton, David Koilpillai, Ramesh Rajaram, Ken Rose and Anand Soman for reading many of the chapters. Tsuhan was a very careful reader, and provided many valuable suggestions for Chap. 12. Ian Galton provided several comments on writing style as well as technical contents, which I found to be extremely useful. Ian's enthusiasm and friendship are gratefully acknowledged.

Debbie McGougan and Cynthia Stewart at Caltech were very helpful during various phases of the manuscript preparation. Solutions to all the homework problems have been prepared by Tsuhan Chen, Igor Djokovic, See-May Phoong, and Anand Soman. I appreciate their interest and patience.

My parents have been responsible for teaching me many valuable "theorems of life" which I could not find in text books and papers. These certainly were the principles which sustained me during all phases of this project.

My deepest gratitude goes to my wife Usha who showed infinite patience and understanding during my absorption in this long project, and offered the type of moral support and loving encouragement which only she can offer. Countless were the evenings, weekends, and holidays during which she provided me with the seclusion and peace of mind needed to pursue this ambitious goal. She is certainly my best blessing, and this book would have remained an idle dream without her support.

P. P. Vaidyanathan
California Institute of Technology

PART 1   *Introductory Chapters*

1

Introduction

A traditional single rate digital signal processing system can be schematically represented as shown in Fig. 1.1-1, which is an interconnection of computational building blocks such as multipliers, adders, and 'delay elements' (which store internal signals). Examples are digital filters, Fourier transformers, modulators, and so on. In a multirate signal processing system, there are two new building blocks, called the M-fold decimator and the L-fold expander (Fig. 1.1-2). These will be defined and illustrated in Chapter 4. For the purpose of the present discussion, the decimator is a device that reduces the sampling rate by an integer factor of M, whereas the expander is used to increase the rate by L. Such sampling rate alteration can be introduced at the input and/or output of the system or *internal* to the system, depending on the application.

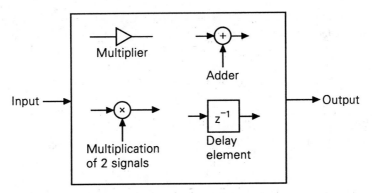

Figure 1.1-1   Schematic of a (single rate) digital signal processor.

Multirate techniques have been in use for many years. Some of the early references are Schafer and Rabiner [1973], Meyer and Burrus [1975], and Oetken, et al. [1975]. The use of multiple sampling rates offers many advantages, such as reduced computational complexity for a given task, reduced transmission rate (i.e., bits per second), and/or reduced storage requirement, depending on the application.

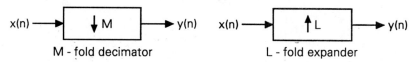

Figure 1.1-2 Multirate building blocks.

One of the earliest applications of multirate processing is in professional digital music [Digital audio, 1985]. Broadly speaking, the idea is as follows. Suppose we wish to digitize an analog signal $x_a(t)$. If the signal has significant energy only up to a frequency f_M, we can first bandlimit the signal to this range using an analog lowpass filter (antialiasing filter), and then sample and digitize it. The lowpass filter in this case has a sharp transition from passband to stopband.

A second technique proceeds in two stages: (a) First use an antialiasing filter with wider transition bandwidth, say by a factor of two. Then *oversample* by a factor of two before digitizing, so that aliasing due to the poor bandlimiting filter is avoided. (b) Pass the digitized signal through a linear phase *digital filter* and decimate by two, so that the sampling rate is reduced to the minimum rate. This two-stage process eliminates the need for sharp-cutoff antialiasing analog filters, which not only are expensive, but also introduce severe phase distortion. Details of this technique will be considered in Chap. 4.

A second application is in fractional sampling rate alteration, for example, converting a 48 kHz discrete-time signal to a 44.1 kHz discrete-time signal. Such requirements are common in the digital audio industry, where a number of sampling rates coexist [Bloom, 1985]. For example, the sampling rate for studio work is 48 kHz, whereas that for CD production is 44.1 kHz. These, in turn, are different from the broadcast rate (32 kHz). The obvious way to perform the rate conversion would be to first convert the discrete-time signal into a continuous-time signal and then resample it at the lower rate. This method is expensive and involves analog components, along with the associated inaccuracies. A direct digital (multirate) method is to perform the conversion directly in the discrete-time domain. Such fractional decimation (or interpolation) is done by combining integer decimators, expanders and filters appropriately. This is more accurate as well as convenient. Details of this technique will be described in Chap. 4.

There are many more applications of multirate processing, and several of them are based on the so-called subband decomposition, to be described next.

1.1 MAJOR DEVELOPMENTS

If a sequence $x(n)$ is bandlimited, then it is possible to decimate it either by an integer or by a fraction, by use of appropriate multirate techniques. The desire to reduce the sampling rate whenever possible is of course understandable, because it usually reduces the storage as well as the processing requirements.

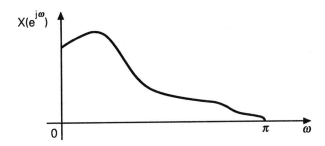

Figure 1.1-3 Example of Fourier transform of a sequence $x(n)$ which has most of the energy in the low frequency region.

Now suppose $x(n)$ is not bandlimited, but nevertheless has most of the energy in the low frequency region. Figure 1.1-3 demonstrates the Fourier transform of such a signal. Even though this cannot be decimated without aliasing, it seems only reasonable to expect that some kind of data rate reduction is still feasible. This is indeed made possible by a technique called *subband decomposition*, implemented with the so-called quadrature mirror filter bank. In this technique, the average number of bits per sample is reduced, even though the average number of samples per unit time is unchanged.

1.1.1 The Quadrature Mirror Filter (QMF) Bank

The quadrature mirror filter bank is shown in Fig. 1.1-4. Here a discrete-time signal $x(n)$ is passed through a pair of digital filters $H_k(z)$ called *analysis filters*, with frequency responses as demonstrated in the figure. The filtered signals $x_k(n)$ (subband signals) are thus approximately bandlimited (lowpass and highpass, respectively). They are then decimated by two, so that the number of samples per unit time [counting $v_0(n)$ as well as $v_1(n)$] is the same as that for $x(n)$. The decimated subband signals, $v_k(n)$, are then quantized and transmitted. At the receiver end, these are recombined by using expanders and synthesis filters $F_k(z)$. In this manner, an approximation $\widehat{x}(n)$ of the signal $x(n)$ is generated. This system will be studied in Chap. 5.

The above system can be regarded as a sophisticated quantizer. Thus, assume that we are allowed to transmit b bits per sample. In a direct method, we would quantize each sample of $x(n)$ independently to b bits. In the above filter-bank approach, we quantize the lower rate signals $v_0(n)$ and $v_1(n)$ to

b_0 bits and b_1 bits per sample, so that the average bit rate is $b = 0.5(b_0 + b_1)$. If the signal is dominantly lowpass, then we can make $b_0 > b$ and $b_1 < b$. An extreme case is to assign $b_0 = 2b$ and $b_1 = 0$. Thus, depending on the energy distribution in the frequency domain, we can allocate bits to the subbands appropriately, thereby increasing the accuracy of representation of $x(n)$, for a fixed bit rate b.

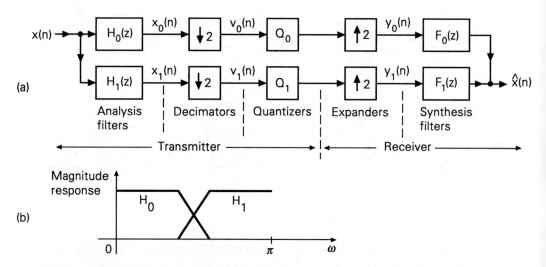

Figure 1.1-4 (a) The quadrature mirror filter (QMF) bank, and (b) typical frequency responses $|H_k(e^{j\omega})|$.

This scheme is called *subband coding* [Croisier, et al., 1976], [Crochiere, 1977], [Esteban and Galand, 1977], [Barnwell, 1982], [Galand and Nussbaumer, 1984]. This has been found to be very useful in speech coding [Crochiere, et al., 1976], where the perceptual properties of the human ear play a major role while assigning bits to $v_k(n)$.

More recently, the effectiveness of subband coding has been demonstrated for music signals. Digitized music normally uses 16 bits per sample (at a sampling rate of about 44 kHz). Using subband coding, it has been demonstrated that a major bit rate reduction can be obtained (compared to the traditional 16 bit repesentation), with little compromise of quality [Veldhuis, et al., 1989]. This has been used in the digital compact cassete (DCC). See also the papers in ICASSP, 1991, pp. 3597-3620, and [Fettweis, et al., 1990]. At the end of this section, more applications of subband splitting will be mentioned.

Reconstruction from subband signals. In many applications, the signals $v_k(n)$ (or, more properly, the quantized versions) are recombined to obtain an approximation $\hat{x}(n)$ of the original signal $x(n)$. This recombination is done by use of expanders (which restore the sampling rate) followed by digital filters $F_k(z)$ (whose purpose will be explained in Sec. 5.1). Such recombination is subject to several errors (apart from the error due to quantization). One of these is aliasing, created due to decimation of $x_k(n)$. Other distortions will be discussed in due course. One of the major developments in multirate signal processing is the recognition of the fact that all of these errors (except quantization error) can be eliminated *completely* at *finite* cost, by proper design of filters.

The QMF bank, introduced in the mid seventies, has since been extended to the case of more than two subbands. Thus, a system with M subbands would have M filters followed by M-fold decimators. The decimated (and quantized) signals would then be recombined using a synthesis bank (expanders and digital filters), to obtain an approximation $\hat{x}(n)$ of the signal $x(n)$. Such a system is called an M-channel maximally decimated filter bank or simply an M-channel QMF bank (even though QMF is a misnomer unless $M = 2$, as explained in Chap. 5).

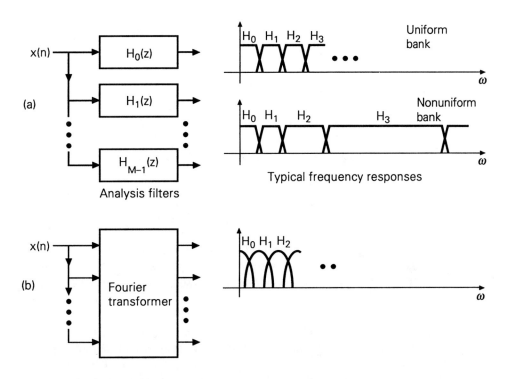

Figure 1.1-5 (a) An M-channel analysis bank, and typical frequency responses. (b) An M-point Fourier transformer, viewed as a filter bank.

Figure 1.1-5(a) shows the analysis filters of an M-channel system. Two sets of typical frequency responses are also sketched in the figure. One of these has uniform filter bandwidths and spacing, while the other has nonuniform (octave) spacing. The latter is particularly useful in the analysis and coding of speech and music.

A filter bank can be viewed as a sophisticated *spectrum analyzer* as elaborated in Chap. 4 and 11. For the moment, we note that a spectrum analyzer takes a signal $x(n)$ and computes the Fourier transform of short blocks, after some preliminary processing (such as windowing). Such a system can be interpreted as a filter bank [Fig. 1.1-5(b)]. The outputs of the 'filters' represent the discrete-time Fourier transform coefficients of the blocks of input data. While the details will be presented only in Chap. 4 and 11, the main point here is that the filter bank of Fig. 1.1-5(a) is a generalization of the Fourier transformer, with greater flexibility on the choice of frequency responses. While the Fourier transformer provides filters with overlapping responses, the generalized system can provide filters with arbitrarily sharp cutoff, better interband isolation and unequal bandwidths.

1.1.2 Polyphase Decomposition

One of the reasons why multirate processing became practically attractive is the invention of the polyphase decomposition [Bellanger, et al., 1976], [Vary, 1979]. This enables the designer to perform all computations at the "lowest rate permissible within the given context," and reduces the speed requirements on the processors. Polyphase decomposition is useful in virtually every application of multirate signal processing, and often results in dramatic computational efficiency. It is valuable in theoretical study, practical design and actual implementation of filter banks. This will be introduced in Chap. 4, and subsequently used throughout the text.

1.1.3 Perfect Reconstruction Systems

In a practical filter bank system, the filters $H_k(z)$ are not ideal, and decimation of the filter outputs results in aliasing errors. As will be seen in Chap. 5, the z-transform of the output signal $\widehat{x}(n)$ can be expressed as

$$\widehat{X}(z) = T(z)X(z) + \text{terms due to aliasing.}$$

It was shown in Croisier, et al. [1976] that aliasing can be completely eliminated in the two channel QMF bank, by proper choice of the synthesis filters $F_0(z)$ and $F_1(z)$.

If $T(z)$ can be forced to be a delay, that is, $T(z) = cz^{-K}$, then the alias-free system is said to have the perfect reconstruction (PR) property. If this is not the case, then the alias-free system still suffers from residual distortion. If the designer does not impose any specifications on the analysis filters such as large stopband attenuation, sharp cutoff rate, and so on, it is an easy matter to choose the filters $H_k(z)$ and $F_k(z)$ so as to satisfy the perfect reconstruction property. However, this is not very practical because, in order to utilize the benefits of subband coding, it is necessary to impose fairly stringent specifications on the attenuation characteristics of the filters.

For the two channel QMF bank a fundamental result was proved independently by Smith and Barnwell [1984 and 1986], and Mintzer [1985]. These papers showed that perfect reconstruction can be achieved even after imposing such practical attenuation requirements. This involves careful design of the four filters, as will be seen in Chap. 5.

1.1.4 Extension to M Channels

The above results from many authors stimulated further research, resulting in techniques to generalize the subband splitting ideas for the case of M-channel QMF banks [Nussbaumer, 1981], [Rothweiler, 1983], [Ramstad, 1984b], [Smith and Barnwell, 1985, 1987], [Masson and Picel, 1985], [Chu, 1985], [Cox, 1986], [Princen and Bradley, 1986], [Wackersreuther, 1986b] [Vetterli, 1986a], [Vaidyanathan, 1987a], [Malvar, 1990b], and [Akansu and Liu, 1991]. Nussbaumer's pioneering work on pseudo QMF banks provides approximate alias cancelation, which is sufficient in some applications. Smith and Barnwell as well as Ramstad independently showed how to formulate the perfect reconstruction conditions in matrix form. It was first recognized by Vetterli, and then independently by Vaidyanathan (in the two references mentioned above) that a polyphase component approach results in considerable simplification of the theory.

It has since been shown that, by using a class of filter banks called paraunitary filter banks, perfect reconstruction can be achieved quite easily. In these systems, the filter bank is constrained to have a paraunitary polyphase matrix (to be explained in Chap. 6). The designer can specify arbitrary filter attenuation, and at the same time obtain perfect reconstruction [Vaidyanathan, 1987a], [Nguyen and Vaidyanathan, 1988], [Vetterli and Le Gall, 1989].

Subsequent to this, a class of systems called the cosine modulated filter banks has been developed by some authors [Malvar, 1990b], [Ramstad, 1991], [Koilpillai and Vaidyanathan, 1991a, 1992]. These have the advantage that the cost of design as well as implementation is largely determined by the cost of *one prototype filter,* since all the other filters are derived from it.

The paraunitary property of filter banks offers many advantages, as elaborated in Chap. 6. Interestingly enough, paraunitary matrices have their origin in classical electrical network theory (see Sec. 14.1 and references therein). In the past, applications of these matrices have been confined mostly within the network theory and control theory communities. The use of paraunitary matrices in digital signal processing, especially filter bank theory, is relatively recent.

Filter bank theory has been extended to the case of nonuniform bandwidths and decimation ratios [Hoang and Vaidyanathan, 1989], [Kovačević and Vetterli, 1991a], and [Nayebi, Barnwell, and Smith, 1991a].

1.1.5 Other Applications and Interrelations

The success of subband coding encouraged researchers to extend the ideas to multidimensional signals. The extensions to two dimensional signals has application in image compression and coding. A systematic study of multidimensional filter banks was first undertaken by Vetterli [1984]. This idea has since been applied for image coding by Woods and O'Neil [1986]. Since then there has been major progress in multidimensional multirate systems [Ansari and Lau, 1987], [Viscito and Allebach, 1988b, 1991], [Smith and Eddins, 1990], [Chen and Vaidyanathan, 1991,1992], [Bamberger and Smith, 1992], and [Kovačević and Vetterli, 1992]. Further references will be cited in Chap. 12. Research results in multidimensional multirate systems are emerging at a rapid rate now.

Recently it has been observed that multirate/subband techniques are attractive in adaptive and statistical signal processing [Gilloire, 1987], [Sathe and Vaidyanathan, 1990,1991], and [Gilloire and Vetterli, 1992]. Research on these topics is still evolving; an excellent reference on the subject is provided by Shynk [1992]. Further applications in communications have been reported by some authors, for example, transmultiplexing [Vetterli, 1986b], high speed analog to digital conversion [Pertraglia and Mitra, 1990], and equalization [Ramesh, 1990].

In recent years, it has been recognized that there is a close connection between multirate filter banks and the so-called "wavelet transforms". This relation was revealed by the fundamental contributions by Daubechies [1988] and Mallat [1989a,b]. (See Sec. 11.0 for further references.) This work has opened up considerable amount of research activity in both the signal processing and mathematics communities. In Chap. 11 we will present this in considerable depth. It will be seen that wavelet analysis is closely related to the so called octave-band filter banks, introduced in the early seventies for analysis of sound signals. Research in wavelet transforms has grown very rapidly after the mid 1980s (and is still growing).

1.2 SCOPE AND OUTLINE

Contributions by many researchers, as outlined above, have resulted in a mature theory of multirate systems. In particular, the detailed aspects of filter bank theory were developed largely during the last decade, subsequent to (and in many cases triggered by) the publication of Crochiere and Rabiner [1983]. The theory of perfect reconstruction filter banks has now reached a state where such systems can be designed as well as implemented with ease. The underlying theory is somewhat complicated, but as a reward it has immense potential for further research and applications. For example, the theory can be applied directly to areas such as subband coding, voice privacy, image processing, multiresolution, and wavelet analysis.

The purpose of this text is to present an in-depth study of multirate systems and filter banks. We have assumed that the reader has some exposure to signal processing (e.g., a one-term course from Oppenheim and Schafer [1989], covering sampling, z-transforms, and digital filtering). Except for this requirement, the book is self-contained. However, this background material is reviewed in Chap. 2 and 3.

Each chapter is supplemented with several homework problems, making it suitable for classroom use. At the same time, our aim has also been to provide a useful reference for researchers. This is evidenced by the inclusion of several advanced topics. There are many examples, design methods, and tables which will aid the practicing professional as well. The chapters can be grouped naturally into the following four parts as elaborated.

Chapters 2 to 4: Introductory Material

A brief review of linear system fundamentals and digital filtering is provided in Chap. 2 and 3. More detailed presentation can be found in a number of references indicated in these chapters. In Chap. 3, IIR elliptic filters, FIR eigenfilters, and allpass filters have been treated in greater detail because of their special role in multirate systems.

Chapter 4 is a detailed study of multirate building blocks, and their interconnections with other systems (such as digital filters). This can be considered to be the 'foundation chapter' for this text. Some of the early sections overlap with the material covered in Crochiere and Rabiner [1983]. At the expense of this overlap, we have ensured that the chapter is self-contained.

A number of special types of digital filters, for example, Nyquist filters, power complementary filters and so on, which are frequently encountered in multirate systems, are also studied in Chap. 4. The polyphase decomposition is introduced, along with special types of filter banks, for example, the uniform-DFT bank.

Many applications of multirate processing are also described in Chap. 4. This includes subband coding, digital audio, and transmultiplexers, to name a few. A complete section of this chapter is devoted to "multigrid" techniques, which find application in the numerical solution of differential equations.

Chapters 5 to 8: Maximally Decimated Filter Banks

Chapter 5 is a study of the M-channel maximally decimated filter bank system (shown in Fig. 1.1-4(a) for the $M = 2$ case). Various distortions will be analyzed, foremost being *aliasing* caused by decimation. Conditions for alias cancelation and perfect reconstruction will be established. Transmultiplexers will also be studied.

Chapter 6 is dedicated to the design of M-channel QMF banks with the perfect reconstruction property. The method presented is based on a class of matrices called *paraunitary* or lossless transfer matrices. The presentation will use some of the results on paraunitary matrices, which will be proved only in Chap. 14. We have chosen to defer the proofs to Chapter 14 (which is devoted to paraunitary systems) in order to ensure an easy and smooth flow. (The results of Chap. 14 will also be stated and used in some other chapters, e.g., Chap. 8 and 11.)

Chapter 7 deals with linear-phase perfect reconstruction QMF banks. In these systems the analysis filters have linear phase, which is a requirement in some applications.

Chapter 8 describes a particular class of M-channel filter banks, in which all the analysis filters are derived from a single filter by use of *cosine modulation*. As a result, this system is very efficient both from the *design* and *implementation* points of view. It turns out that one can easily achieve perfect reconstruction in these systems, by further imposing the paraunitary property. We first describe cosine modulated systems with approximate reconstruction properties (pseudo QMF banks, Sec. 8.1–8.3), and then show how these can be modified to obtain perfect reconstruction (Sec. 8.4, 8.5). The cosine modulated perfect reconstruction system (Sec. 8.4, 8.5) can, however, be studied independently, with Sec. 8.1–8.3 used only as a reference.

Chapters 9 to 12: Special Topics on Multirate Systems

Chapter 9 studies the effects of finite precision in the implementations of multirate filter banks. This includes roundoff noise analysis and coefficient quantization analysis. The effect of quantization of subband signals is dealt with in Appendix C.

In Chap. 10 we study the connection between filter banks and a number of "peripheral" topics such as periodically time varying systems, block filtering, and unconventional sampling theorems.

Chapter 11 deals with a special type of time-frequency representation called the *short-time Fourier transformation,* and extends this to develop *wavelet transforms.* Wavelet transforms, in particular, have drawn considerable attention in recent years from a wide scientific community, including physicists, mathematicians, and signal processors. Many researchers in the signal processing community have taken the view that wavelet transforms are closely related to filter banks (see Chap. 11 for references). In Chap. 11, we will take this viewpoint; this makes it easier to understand, design, and implement wavelet transforms.

In Chap. 12 we study the multidimensional versions of many of the fundamental multirate concepts introduced in earlier chapters. These find applications in image and video signal processing.

Chapters 13 and 14: Multivariable and Paraunitary Systems

Many of the multirate (time-varying) systems discussed in the text can be represented in terms of multi-input multi-output (MIMO) linear time invariant (LTI) systems. This will be evident when we analyze filter banks using the polyphase approach. It turns out, therefore, that a deeper understanding of MIMO LTI systems is very useful in the study of filter banks. Chapter 13 is meant to serve this purpose. Even though the results of this chapter are not explicitly used in earlier ones, they are required to establish some of the deeper properties of paraunitary systems discussed in Chap. 14.

Chapter 14 is a complete treatment of paraunitary and lossless transfer matrices. These systems find application in perfect reconstruction filter banks (Chap. 6 and 8) as well as in wavelet transform theory (Chap. 11). As mentioned previously, some of the results in Chap. 14 are in turn *stated and used* in some of the earlier chapters. The detailed discussions in Chap. 13 and 14 ensure completeness of presentation, and also serve as research aids.

Appendices

There are five appendices. Appendix A provides a brief review of matrices, where we have summarized matrix concepts used throughout the text. Appendix B is on random processes, and its primary use is in the analysis of roundoff noise effects in filter banks (Chap. 9), and in the study of subband quantization (Appendic C). Appendix C deals with the effects of quantization in subbands, and summarizes theoretical results on bit-allocation strategies in subband and transform coding schemes. Appendix D is on 'spectral factorization' which is a frequently used tool in filter bank design. Appendix E is on Mason's gain formula, which is useful for the analytical evaluation of transfer functions.

Most of these appendices include several examples and homework problems. While the appendices are not substitutes for a good book or chapter on these topics, they serve to make the text self contained and complete.

2

Review of Discrete-Time Systems

2.0 INTRODUCTION

This chapter provides a brief review of the fundamentals of discrete-time systems. This serves as quick reference material throughout the text, and also familiarizes the reader with the notations we will use. In this chapter, we will present basic facts and results without detailed justification. Detailed treatments can be found in Oppenheim and Schafer [1989]. Other related references are Rabiner and Gold [1975], and Jackson [1991].

2.1 DISCRETE-TIME SIGNALS

Discrete-time signals are typically denoted as $u(n)$, $x(n)$, and so on, where n is an integer called the *time index*. We will use notations such as $x(n)$ to indicate the entire sequence (i.e., with $-\infty \leq n \leq \infty$) or, on occasions, just to denote the nth sample $x(n)$. The context will clarify the exact meaning. All sequences are taken to be complex valued unless mentioned otherwise. Figure 2.1-1 shows some typical sequences.

1. The *unit-pulse*, denoted $\delta(n)$, is defined according to

$$\delta(n) = \begin{cases} 1, & n = 0 \\ 0, & \text{otherwise.} \end{cases} \qquad (2.1.1)$$

This is sometimes called the *impulse function*, and should not be confused with the impulse function $\delta_a(\alpha)$ of the real continuous variable α. The function $\delta_a(\alpha)$ is usually called the *Dirac delta function*, and is defined to be zero everywhere except $\alpha = 0$, and such that $\int_p^q \delta_a(\alpha) d\alpha = 1$ if, and only if, $p < 0 < q$.

2. The *unit-step* sequence is defined as

$$\mathcal{U}(n) = \begin{cases} 1, & n \geq 0 \\ 0, & \text{otherwise.} \end{cases} \qquad (2.1.2)$$

3. *Exponentials.* A sequence of the form ca^n is said to be an exponential. Here c and a are arbitrary (possibly complex) constants. Sequences such as $ca^n\mathcal{U}(n)$ and $cb^n\mathcal{U}(-n)$ are called *one-sided exponentials* (or truncated exponentials). Thus, $ca^n\mathcal{U}(n)$ is right-sided, whereas $cb^n\mathcal{U}(-n)$ is left-sided.

4. *Single-frequency sequence.* The sequence $ce^{j\omega_0 n}$ is said to be a single-frequency sequence. This is an exponential sequence with $a = e^{j\omega_0}$. Here, ω_0 is real, but can have either sign. In less formal terms this is sometimes called a *sinusoid* with frequency ω_0. This is periodic if, and only if, the frequency ω_0 is a rational multiple of 2π, that is, $\omega_0 = 2\pi k/L$ for integer k and L.

5. A sequence of the form $A\cos(\omega_0 n + \theta)$ is a true *sinusoid*. Since we can write $\cos(\omega_0 n + \theta) = 0.5(e^{j(\omega_0 n+\theta)} + e^{-j(\omega_0 n+\theta)})$, it contains two frequencies, that is, ω_0 and $-\omega_0$. So it is not a single-frequency signal.

6. *Bounded sequence.* A sequence $u(n)$ is said to be bounded if there exists a finite B such that $|u(n)| \leq B$ for all n. Examples: (a) $a^n\mathcal{U}(n), |a| < 1$, (b) $\cos\omega_0 n$ (real ω_0). Note that an exponential a^n is not bounded unless $a = 0$ or $|a| = 1$.

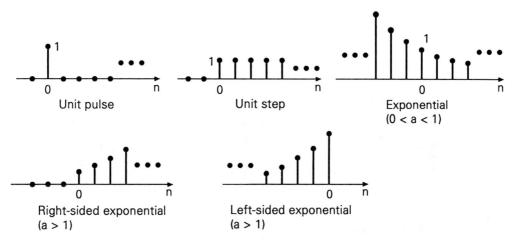

Figure 2.1-1 Demonstration of well known sequences.

2.1.1 Transform Domain Analysis

It is often convenient to work with transformed versions of signals such as the z-transform and Fourier transform. These are defined next.

The z-Transform

The z-transform of a sequence $x(n)$ is defined as

$$X(z) = \sum_{n=-\infty}^{\infty} x(n)z^{-n}. \qquad (2.1.3)$$

If this summation does not converge for any z, the z-transform does not exist; an example is the exponential sequence $a^n, a \neq 0$.

In general, the summation converges in an annulus defined as $R_1 < |z| < R_2$ in the z-plane. This is called the *region of convergence* (ROC). For example if $x(n) = a^n \mathcal{U}(n)$, then $X(z) = 1/(1 - az^{-1})$ with ROC given by $|z| > |a|$. This same $X(z)$ with ROC specified as $|z| < |a|$ would result in the inverse transform $x(n) = -a^n \mathcal{U}(-n-1)$. Given $X(z)$ and its associated ROC, $x(n)$ can be uniquely recovered from $X(z)$.

For a finite length sequence (with finite sample values) the z-transform converges *everywhere* except possibly at $z = 0$ and/or $z = \infty$.

The Fourier Transform (FT)

If the ROC of $X(z)$ includes the unit circle (i.e., points of the form $z = e^{j\omega}$ where ω is real), we say that $X(e^{j\omega})$ is the Fourier transform (FT) of $x(n)$. Thus,

$$X(e^{j\omega}) = \sum_{n=-\infty}^{\infty} x(n) e^{-j\omega n}. \qquad (2.1.4a)$$

The inverse transform relation is given by

$$x(n) = \frac{1}{2\pi} \int_0^{2\pi} X(e^{j\omega}) e^{j\omega n} d\omega. \qquad (2.1.4b)$$

Note that the frequency variable ω is in radians. The FT $X(e^{j\omega})$ is periodic in ω with period 2π, so that the region $\pi < \omega < 2\pi$ is considered to be the negative frequency region (equivalent to $-\pi < \omega < 0$).

Fourier transform of a single-frequency signal. Since the z-transform of a^n does not converge anywhere (unless $a = 0$), the Fourier transform of $e^{j\omega_0 n}$, in particular, does not exist in the usual sense. However, by using a Dirac delta function $\delta_a(\omega)$, one can write the FT of this sequence as $2\pi \delta_a(\omega - \omega_0)$ in the range $0 \leq \omega < 2\pi$ (and periodically repeating with period 2π).

Parseval's Relation

Let $U(e^{j\omega})$ and $V(e^{j\omega})$ be the Fourier transforms of $u(n)$ and $v(n)$. Parseval's relation says

$$\sum_{n=-\infty}^{\infty} u(n) v^*(n) = \frac{1}{2\pi} \int_0^{2\pi} U(e^{j\omega}) V^*(e^{j\omega}) d\omega. \qquad (2.1.5a)$$

In particular, if we set $u(n) = v(n)$, then

$$\sum_{n=-\infty}^{\infty} |u(n)|^2 = \frac{1}{2\pi} \int_0^{2\pi} |U(e^{j\omega})|^2 d\omega. \qquad (2.1.5b)$$

Energy of a sequence. The energy of a sequence $u(n)$ is defined as $E_u = \sum_{n=-\infty}^{\infty} |u(n)|^2$. If this summation does not converge, the energy is taken to be infinite. Eq. (2.1.5b) gives us two ways to express the energy.

Tables of z-transform and Fourier transform pairs, and Tables of their properties can be found in Chap. 2 and 4 of Oppenheim and Schafer [1989]. We will make use of these throughout the text.

2.1.2 Discrete-Time Systems

A discrete-time system operates on an input sequence $u(n)$ to produce an output sequence $y(n)$. It is assumed that the value of $u(n)$ in the range $-\infty \leq n \leq \infty$ uniquely determines the output $y(n)$ in the range $-\infty \leq n \leq \infty$. Of great interest to us in this text are linear systems and shift invariant (or time invariant) systems. The properties of linearity and shift invariance enable us to characterize the system using the notion of transfer functions.

Linearity. Suppose the input sequences $u_0(n)$ and $u_1(n)$ produce the output sequences $y_0(n)$ and $y_1(n)$ respectively. If the system output in response to the input $a_0 u_0(n) + a_1 u_1(n)$ is equal to $a_0 y_0(n) + a_1 y_1(n)$, and if this is true for every pair of constants a_0, a_1 and every possible $u_0(n)$ and $u_1(n)$, then we say that the system is linear.

Shift-invariance. Let $y(n)$ denote the output of a system in response to the input $u(n)$. If the output in response to the shifted version $u(n-N)$ is equal to $y(n-N)$, and if this holds for all integers N and all input sequences $u(n)$, we say that the system is shift-invariant or time-invariant.

LTI Systems

A system is said to be linear and shift-invariant (abbreviated LSI or LTI) if it is both linear and shift-invariant. Such a system can be completely characterized by the *impulse response* sequence $h(n)$ (also called the *unit-pulse response*) which is the output $y(n)$ in response to an unit-pulse input $\delta(n)$. For LTI systems, the input-output relation is given by

$$y(n) = \sum_{m=-\infty}^{\infty} h(m)u(n-m), \qquad (2.1.6)$$

which is called the *convolution summation*. This can be expressed in the transform domain as

$$Y(z) = H(z)U(z), \qquad (2.1.7)$$

where $H(z)$ is the z-transform of $h(n)$, that is,

$$H(z) = \sum_{n=-\infty}^{\infty} h(n)z^{-n}. \qquad (2.1.8)$$

$H(z)$ is called the *transfer function* of the LTI system. To physically visualize the meaning of $H(z)$, note that if we apply an exponential input a^n, the output is also an exponential, given by $y(n) = H(a)a^n$ [provided "a" belongs to the region of convergence of $H(z)$].

Eigenfunctions of LTI systems. If a nonzero input $f(n)$ to an LTI system $H(z)$ produces an output of the form $cf(n)$, where c is a constant, we say that $f(n)$ is an eigenfunction (and c an eigenvalue) of $H(z)$. Thus, *exponentials* are eigenfunctions of LTI systems.

Causality

A discrete-time system is said to be causal if the output $y(n)$ at time n does not depend on the future values of the input sequence, that is, does not depend on $u(m), m > n$. An LTI system is causal if and only if the impulse response satisfies the condition

$$h(n) = 0, \quad n < 0 \quad \text{(causality)}. \qquad (2.1.9)$$

This has given rise to the term 'causal sequence' for any sequence $x(n)$ which is zero for $n < 0$. We say that a sequence $x(n)$ is *anticausal* if $x(n) = 0$ for $n \geq 0$. An example is $\mathcal{U}(-n-1)$.

From (2.1.8), we see that a causal LTI system has $H(\infty) = h(0)$, and that an LTI system is causal if and only if $H(\infty)$ is finite. For convenience of language, we often use phrases such as '$H(z)$ is causal'. This means that the associated ROC has been so chosen that the inverse transform $h(n)$ is zero for $n < 0$.

Rational Transfer Functions

All transfer functions in this text are rational, that is, of the form

$$H(z) = \frac{A(z)}{B(z)}, \qquad (2.1.10)$$

with

$$A(z) = \sum_{n=0}^{N} a_n z^{-n}, \quad B(z) = \sum_{n=0}^{N} b_n z^{-n}. \qquad (2.1.11)$$

Here a_n and b_n are (possibly complex) finite numbers. If there are no common factors of the form $(\beta - \alpha z^{-1}), \alpha \neq 0$ between $A(z)$ and $B(z)$ (i.e., if $A(z)$ and $B(z)$ are relatively prime), we say that (2.1.10) is an *irreducible* rational form. Under this condition, N is called the order of the system (assuming that at least one of a_N or b_N is nonzero).

Zeros and poles. If $A(z)/B(z)$ is irreducible, the zeros of $A(z)$ and $B(z)$ are said to be the zeros and poles, respectively, of $H(z)$. The time-domain significance of poles and zeros is well known, and is discussed in Problems 2.4 to 2.6.

Realness. A system is said to be 'real' if the output $y(n)$ is real for real inputs $u(n)$. For LTI systems this is equivalent to the condition that $h(n)$ be real for all n. For rational LTI system in the irreducible form (2.1.10), this in turn is equivalent to the condition that a_n and b_n be real for all n. Real systems are also referred to as real coefficient systems.

FIR and IIR Systems

A finite impulse response (FIR) system is one for which b_n in (2.1.11) is nonzero for only one value of n. As an example, let $N = 3$, and let $b_2 = 1$. Then, $H(z) = a_0 z^2 + a_1 z + a_2 + a_3 z^{-1}$, and is FIR.

A *causal* Nth order FIR filter can be represented as

$$H(z) = \sum_{n=0}^{N} h(n) z^{-n}, \quad h(N) \neq 0. \qquad (2.1.12)$$

[This corresponds to $B(z) = 1$ and $A(z) = H(z)$.] The quantity $N+1$, which is the number of impulse response coefficients, is said to be the *length* of the filter (i.e., $H(z)$ is an $(N+1)$-point filter). In Sec. 2.4.2 we will see that FIR systems can be designed to have exactly linear phase response, which is required in some applications.

An LTI system which is not FIR is said to be an IIR (Infinite impulse response) system. An example is the system with impulse response $a^n \mathcal{U}(n)$, which has transfer function $H(z) = 1/(1 - az^{-1})$.

All-zero and all-pole systems. An FIR system is also said to be an all-zero system (because poles are located only at $z = 0$ and/or ∞). An IIR system of the form $H(z) = cz^{-K}/B(z)$ is said to be an all-pole system. The zeros for such a system are at $z = 0$ and/or ∞.

FIR sequences. A finite-duration (or finite-length) sequence $u(n)$ is often referred to as an FIR sequence. We often use the terms "FIR input", "FIR output" and so on, where the term FIR actually stands for "finite-length".

Stability

If a discrete-time system is such that every bounded input produces a bounded output, we say that the system is stable (more precisely *bounded input bounded output* stable or BIBO stable). For the case of LTI systems, it can be shown that BIBO stability is equivalent to the condition

$$\sum_{n=-\infty}^{\infty} |h(n)| < \infty. \qquad (2.1.13)$$

In other words, the impulse response must be *absolutely summable*.

Stability condition in terms of poles. If $H(z)$ is rational and $h(n)$ causal, then (2.1.13) is equivalent to the condition that all poles p_k of $H(z)$ be inside the unit circle, that is, $|p_k| < 1$. Unless mentioned otherwise, the statement "stable" in this text will imply this condition (i.e., $|p_k| < 1$).

Causality of $h(n)$ and rationality of $H(z)$ will be implicit. The only exception to this convention will be noncausal filters of the form $z^K H(z)$, where $K > 0$ and $H(z)$ is causal and stable. This system has K poles at $z = \infty$, yet it is stable.

2.1.3 Implementations of Rational Transfer Functions

It is well known that the system with transfer function (2.1.10) can be described in the time domain by a difference equation of the form

$$b_0 y(n) = -\sum_{m=1}^{N} b_m y(n-m) + \sum_{m=0}^{N} a_m u(n-m). \qquad (2.1.14)$$

Since the output $y(n)$ depends in general on past outputs $y(n-m)$, this is called a *recursive difference equation*. Without loss of generality, we can assume that at least one of a_0, b_0 in (2.1.11) is nonzero. If $b_0 = 0$ and $a_0 \neq 0$ this implies a noncausal system [since $H(\infty)$ is then not finite]. For causal systems $b_0 \neq 0$, and we can assume $b_0 = 1$ without loss of generality.

Direct form structure. With $b_0 = 1$ we obtain the structure of Fig. 2.1-2(a) (demonstrated for $N = 2$) for the implementation of this difference equation. This is called the *direct form structure,* and requires $2N+1$ multiplications and $2N$ additions for the computation of each output sample $y(n)$. The number of delays is N, which is the filter order.

Figure 2.1-2(b) shows the common notations and building blocks (multipliers, adders and delays) used to draw digital filter structures. Multipliers with values $\pm 2^{\pm K}$ are often not counted as multipliers, as these can be implemented with binary shifts on a digital machine.

FIR direct form. For the special case of FIR filters the structure reduces to the form shown in Fig. 2.1-3(a) (assuming causality), requiring $N+1$ multipliers, N adders, and N delays. An equivalent structure called the *transposed* direct form is shown in Fig. 2.1-3(b). FIR structures do not have any feedback paths (unlike Fig. 2.1-2(a)). Equivalently, the difference equation (2.1.14) has only the input terms $u(n-m)$ and no $y(n-m)$ terms on the right hand side, that is, the difference equation is nonrecursive.

Cascade form structures. Another popular structure used in digital filtering is the cascade form. This is obtained by expressing $A(z)$ and $B(z)$ in factored form as

$$A(z) = a_0 \prod_{n=1}^{N}(1 - z_n z^{-1}), \quad B(z) = \prod_{n=1}^{N}(1 - p_n z^{-1}), \qquad (2.1.15)$$

so that

$$H(z) = \frac{a_0 \prod_{n=1}^{N}(1 - z_n z^{-1})}{\prod_{n=1}^{N}(1 - p_n z^{-1})}. \qquad (2.1.16)$$

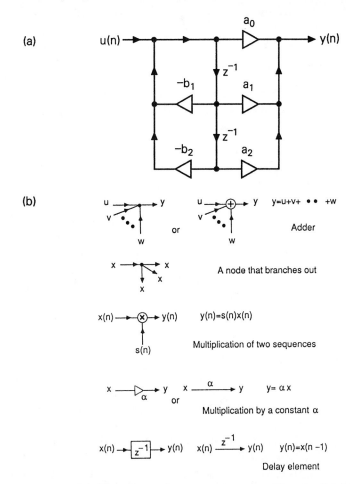

Figure 2.1-2 (a) The direct form structure for $N=2$. (b) Meanings of signal flow graph notations used.

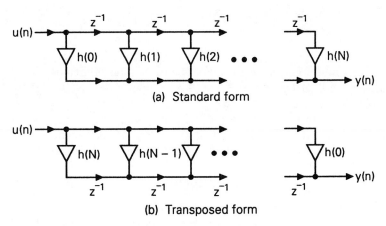

Figure 2.1-3 Direct form structures for FIR filters. (a) Standard form, and (b) transposed form.

Figure 2.1-4(a) shows the cascade form structure, where the building blocks are as in Fig. 2.1-4(b). Notice that p_n and z_n are, respectively, the poles and zeros of $H(z)$.

There exist more complicated structures for filters. A useful tool to compute the transfer functions of arbitrary structures is provided by *Mason's formula*, reviewed in Appendix E.

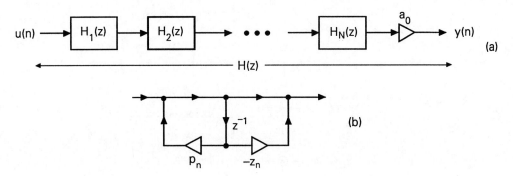

Figure 2.1-4 (a) The cascade form structure, and (b) the first order building blocks. Here z_n and p_n can be complex.

Real Coefficient Case

In the above cascade form structure, the multipliers can in general be complex even if a_n and b_n are real. For the special case where b_n are real, the poles are either real or occur in complex conjugate pairs. (The same is true of zeros if a_n are real.)

If p_k is complex, we can combine the factors $1 - p_k z^{-1}$ and $1 - p_k^* z^{-1}$ to produce the real factor $1 - c_k z^{-1} - d_k z^{-2}$. In this way, we can obtain a real coefficient cascade form. The transfer function can now be expressed as

$$H(z) = \frac{a_0(1 + az^{-1})^i \prod_{k=1}^{m}(1 + t_k z^{-1} + q_k z^{-2})}{(1 + bz^{-1})^\ell \prod_{k=1}^{m}(1 - c_k z^{-1} - d_k z^{-2})}, \qquad (2.1.17)$$

where i and ℓ can take the values 0 or 1. All coefficients in (2.1.17) are real. The second order sections can be implemented as in Fig. 2.1-5, using the direct form structure.

A complex conjugate pair of poles can also be represented as $R_k e^{\pm j\phi_k}$, as demonstrated in Fig. 2.1-6(a), where R_k is the pole radius and ϕ_k the pole angle. This gives rise to a factor in the denominator of $H(z)$, of the form

$$(1 - 2R_k \cos \phi_k z^{-1} + R_k^2 z^{-2}). \qquad (2.1.18)$$

The same is true of zeros. In particular, a complex conjugate pair of zeros on the unit circle [Fig. 2.1-6(b)] can be represented by the factor

$$(1 - 2\cos \omega_k z^{-1} + z^{-2}). \qquad (2.1.19)$$

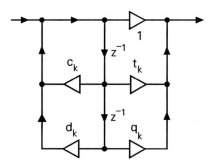

Figure 2.1-5 Implementation of a second order section in (2.1.17), using the direct form structure.

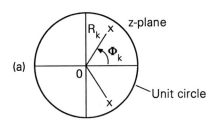

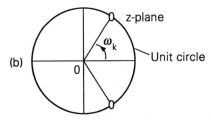

Figure 2.1-6 (a) A complex conjugate pair of poles inside the unit circle, and (b) a complex conjugate pair of zeros on the unit circle.

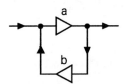

Figure 2.1-7 Example of a delay-free loop.

The number of delays in all the above structures is equal to N, which is the smallest possible. So these structures are minimal.†

Delay-free loops. A loop in which there is no delay is said to be a delay-free loop. Fig. 2.1-7 demonstrates the idea. Discrete-time structures with delay free loops cannot be implemented in practice. Such structures are, therefore, of no practical interest.

---

† Digital filter structures which use the smallest possible number of delays are said to be minimal in delays, or just *minimal* or *canonical*.

2.1.4 Continuous-Time Systems

Continuous-time functions are denoted as $x_a(t)$, $y_a(t)$ and so on. The subscript a, which stands for "analog", is deleted if the context makes it clear. The Fourier transform of $x_a(t)$, if it exists, is defined as

$$X_a(j\Omega) = \int_{-\infty}^{\infty} x_a(t)e^{-j\Omega t}dt, \qquad (2.1.20)$$

and the inverse transform relation is

$$x_a(t) = \frac{1}{2\pi}\int_{-\infty}^{\infty} X_a(j\Omega)e^{j\Omega t}d\Omega. \qquad (2.1.21)$$

Here, the *frequency variable* Ω has the dimension of *radians per second*.

Many of the concepts described earlier carry over to continuous-time systems in an obvious manner. A continuous-time LTI system is characterized by an impulse response $h_a(t)$ and transfer function $H_a(s)$. The transfer function is the Laplace's transform of $h_a(t)$, that is,

$$H_a(s) = \int_{-\infty}^{\infty} h_a(t)e^{-st}\,dt$$

One has to specify a region of convergence for this integral in the s-plane [Oppenheim, Willsky, and Young, 1983]. If the region of convergence includes the imaginary axis, then $H_a(j\Omega)$ [the Fourier transform of $h_a(t)$] is defined, and is called the frequency response of the system. The system is causal if, and only if, $h_a(t) = 0$ for $t < 0$. A causal system is (BIBO) stable if, and only if, all the poles of $H_a(s)$ are in the open left half plane (abbreviated LHP) which is the region characterized by Re$[s] < 0$. In this case the region of convergence includes the closed right half plane, that is, the region Re$[s] \geq 0$.

Sampling

We say that $x(n)$ is the sampled version of $x_a(t)$ if $x(n) = x_a(nT)$ for some $T > 0$. The quantity T is the sampling period (or sample spacing), and $2\pi/T$ the sampling frequency or sampling rate. Denote the Fourier transforms of $x(n)$ and $x_a(t)$ as $X(e^{j\omega})$ and $X_a(j\Omega)$. It can be shown that

$$X(e^{j\omega}) = \frac{1}{T}\sum_{k=-\infty}^{\infty} X_a\bigl(j(\Omega - \frac{2\pi k}{T})\bigr)\Big|_{\Omega=\omega/T} \qquad (2.1.22)$$

Thus, $X(e^{j\omega})$ is obtained as follows: (a) duplicate $X_a(j\Omega)$ at uniform intervals separated by $2\pi/T$, (b) add these copies and divide by T, and (c) replace Ω with ω/T. Figure 2.1-8(a) demonstrates this idea. In part (b) we demonstrate the physical dimensions of the frequency axis, by assuming that the sampling period T is one millisecond [i.e., $2\pi/T$ is 2π Kilo radians per second (Kr/s)]. Figure 2.1-8(c) shows the correspondence with the frequency variable ω associate with the sequence $x(n)$.

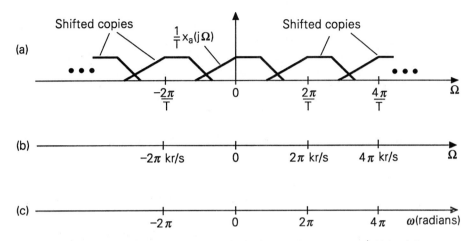

Figure 2.1-8 (a) Fourier transform of a sampled version of $x_a(t)$. (b) Example of frequency dimensions in kiloradians/second, assuming 1 kHz sampling rate, and (c) correspondence with discrete-time frequency variable ω (radians).

Aliasing. If there is no overlap between $X_a(j\Omega)$ and the shifted versions, we can recover $x_a(t)$ from the sampled version $x(n)$ by retaining only one copy. This is accomplished by filtering. If we have the apriori knowledge that the signal is lowpass, then an ideal lowpass filter is used. Otherwise a bandpass filter with appropriate center frequency is required.

The overlap-free condition can be ensured by requiring that $X_a(j\Omega)$ be zero for $|\Omega| \geq \pi/T$. (This is the lowpass case; see Problem 2.15 for other possibilities.) If there is overlap between $X_a(j\Omega)$ and any of its shifted versions, we say that there is *aliasing*.

Bandlimited signals and Nyquist rate. If $X_a(j\Omega)$ is zero for $|\Omega| \geq \sigma$, we say that $x_a(t)$ is σ-bandlimited or σ-BL. We see that if $x_a(t)$ is σ-BL, then we can avoid aliasing by sampling at the rate $\Theta \triangleq 2\sigma$ (*Shannon* or *Nyquist sampling theorem*). This is called the *Nyquist rate* for $x_a(t)$.

Samplers and A/D Converters

We often refer to a continuous-time signal as an "analog" signal, even though there is a fine distinction between these two [Oppenheim and Schafer, 1989]. The amplitude of an analog signal can take continuous values (possibly complex). A digital signal can take values only from a preassigned discrete set (e.g., the values represented by the binary number system). Either of these signals could be continuous-time or discrete-time. In this text, we often use the term "analog signal" to imply both analog and continuous-time signals. Similarly, terms such as "digital signals" and "digital filters" are used to imply "digital" as well as "discrete-time" versions. Whenever a finer distinction is appropriate, it is either mentioned or will be clear from the context.

We often refer to black boxes in figures, with the labels "A/D converters" and "D/A converters" (or just A/D and D/A). These stand for "analog-to-digital" and "digital-to-analog," respectively. In most cases, these black boxes really imply conversion between continuous-time and discrete time, and a notation such as C/D and D/C (as in Oppenheim and Schafer, [1989]) would have been appropriate. We will, however, use A/D and D/A everywhere, and the precise meaning will be clear from the context.

2.2 MULTI-INPUT MULTI-OUTPUT SYSTEMS

Consider a system with r inputs and p outputs, with a transfer function connected from every input to every output. Thus, let $H_{km}(z)$ denote the transfer function from the mth input to the kth output. This is demonstrated in Fig. 2.2-1 for $p = r = 2$. The kth output in response to all the inputs is given by

$$Y_k(z) = \sum_{m=0}^{r-1} H_{km}(z) U_m(z), \quad 0 \le k \le p-1. \qquad (2.2.1)$$

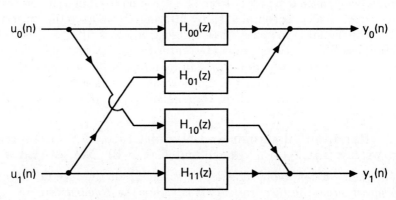

Figure 2.2-1 A two-input two-output system.

The entire system is said to be a multi-input multi-output (MIMO) LTI sytstem, and can be characterized by the set of pr transfer functions $H_{km}(z)$. In order to compactly represent the system, we define the input and output vectors

$$\begin{aligned} \mathbf{u}(n) &= [\, u_0(n) \quad u_1(n) \quad \ldots \quad u_{r-1}(n)\,]^T, \\ \mathbf{y}(n) &= [\, y_0(n) \quad y_1(n) \quad \ldots \quad y_{p-1}(n)\,]^T, \end{aligned} \qquad (2.2.2)$$

and their z-transforms

$$\mathbf{U}(z) = [U_0(z) \quad U_1(z) \quad \ldots \quad U_{r-1}(z)]^T = \sum_{n=-\infty}^{\infty} \mathbf{u}(n) z^{-n},$$

$$\mathbf{Y}(z) = [Y_0(z) \quad Y_1(z) \quad \ldots \quad Y_{p-1}(z)]^T = \sum_{n=-\infty}^{\infty} \mathbf{y}(n) z^{-n}.$$
(2.2.3)

Then the system can be described as

$$\mathbf{Y}(z) = \mathbf{H}(z)\mathbf{U}(z), \qquad (2.2.4)$$

where

$$\mathbf{H}(z) = [H_{km}(z)]. \qquad (2.2.5)$$

Note the use of bold letters to indicate matrices and vectors. The $p \times r$ matrix $\mathbf{H}(z)$ is called the *transfer matrix* of the system. We will use the terms "r-input p-output system" and "$p \times r$ system" interchangeably. A system with $p = r = 1$ is said to be a single-input single-output (SISO) system, or a *scalar* system.

Fig. 2.2-2 indicates two ways of representing the system. The input and output lines are indicated either by heavy arrows, or by double-line arrows according to convenience. The double lines do *not* imply that there are only two inputs or two outputs.

Figure 2.2-2 Two ways to represent a multi-input multi-output LTI system.

The impulse Response Matrix

Let $h_{km}(n)$ denote the impulse response of the transfer function $H_{km}(z)$. Define the $p \times r$ matrix of impulse response sequences as

$$\mathbf{h}(n) = [h_{km}(n)]. \qquad (2.2.6)$$

Then, the relation (2.2.4) can be expressed in the time-domain as

$$\mathbf{y}(n) = \sum_{i=-\infty}^{\infty} \mathbf{h}(i)\mathbf{u}(n-i) = \sum_{i=-\infty}^{\infty} \mathbf{h}(n-i)\mathbf{u}(i), \qquad (2.2.7)$$

which is the matrix version of the familiar convolution summation (2.1.6). From the definitions of $\mathbf{H}(z)$ and $\mathbf{h}(n)$ it is evident that they are related as

$$\mathbf{H}(z) = \sum_{n=-\infty}^{\infty} \mathbf{h}(n) z^{-n}. \qquad (2.2.8)$$

In general, the above infinite summation converges only in certain regions of the z-plane.

The matrix sequence $\mathbf{h}(n)$ is said to be the "impulse response" or "unit-pulse response" of the system $\mathbf{H}(z)$. For example, let

$$\mathbf{H}(z) = \begin{bmatrix} 1 + z^{-1} & 2 + z^{-2} \\ 1 + z^{-1} + z^{-2} & 2 \end{bmatrix}. \qquad (2.2.9)$$

This can be rewritten as

$$\mathbf{H}(z) = \underbrace{\begin{bmatrix} 1 & 2 \\ 1 & 2 \end{bmatrix}}_{\mathbf{h}(0)} + z^{-1} \underbrace{\begin{bmatrix} 1 & 0 \\ 1 & 0 \end{bmatrix}}_{\mathbf{h}(1)} + z^{-2} \underbrace{\begin{bmatrix} 0 & 1 \\ 1 & 0 \end{bmatrix}}_{\mathbf{h}(2)}, \qquad (2.2.10)$$

and the sequence $\mathbf{h}(n)$ can be readily identified, as indicated.

Transfer matrices which are row or column vectors. A system with one input and p outputs has a $p \times 1$ transfer matrix, that is, a column vector. A system with r inputs and one output has a $1 \times r$ transfer matrix, that is, a row vector. Figure 2.2-3 shows both types of examples, where

$$\mathbf{H}_1(z) = \begin{bmatrix} 1 + z^{-1} \\ 2 + 3z^{-1} \end{bmatrix}, \quad \mathbf{H}_2(z) = [\, 1 + z^{-1} \quad 2 + 3z^{-1} \,].$$

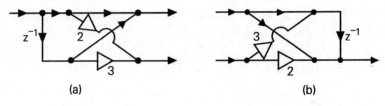

Figure 2.2-3 Examples of transfer matrices (a) a column vector, and (b) a row vector.

Such vector-transfer functions arise in the study of digital filter banks (e.g., Chap. 5).

The frequency response matrix. The discrete-time Fourier transform of a sequence is obtainable from the z-transform by setting $z = e^{j\omega}$. This is given by

$$\mathbf{H}(e^{j\omega}) = \sum_{n=-\infty}^{\infty} \mathbf{h}(n) e^{-j\omega n}, \qquad (2.2.11)$$

and is the frequency response matrix of the system.

Stability and causality. The system $\mathbf{H}(z)$ is said to be 'causal' if $\mathbf{h}(n)$ is causal [that is, $\mathbf{h}(n) = \mathbf{0}$ for $n < 0$]. This is equivalent to saying that each $h_{km}(n)$ is causal. We say that α is a pole of $\mathbf{H}(z)$ if it is a pole of some element $H_{km}(z)$. The system $\mathbf{H}(z)$ is stable (in the BIBO sense) if each of the functions $H_{km}(z)$ is stable. So, $\mathbf{H}(z)$ is causal and stable if, and only if, each of the systems $H_{km}(z)$ is causal with all poles strictly inside the unit circle. This is equivalent to the condition that the region of convergence of the summation (2.2.8) includes all points on and outside the unit circle of the z-plane. In particular, this means that the summation in (2.2.11) converges.

No poles at infinity. For a causal rational system $\mathbf{H}(z)$, the region of convergence is everywhere outside a certain circle in the z-plane. In particular, therefore, there are no poles at $z = \infty$, whether the system is stable or not. Since the ROC of $\mathbf{H}(z)$ includes $z = \infty$, the value $\mathbf{H}(\infty)$ can be obtained from the infinite power series $\sum_{n=0}^{\infty} \mathbf{h}(n) z^{-n}$. Thus $\mathbf{H}(\infty) = \mathbf{h}(0)$. In contrast, the value of $\mathbf{H}(0)$ cannot be found by using the infinite power series, since the ROC of the causal power series does not include the origin.

Parseval's relation for vector signals. Let $\mathbf{x}(n)$ and $\mathbf{y}(n)$ be vector sequences with Fourier transforms $\mathbf{X}(e^{j\omega})$ and $\mathbf{Y}(e^{j\omega})$. We then have (Problem 13.1)

$$\sum_{n=-\infty}^{\infty} \mathbf{y}^{\dagger}(n)\mathbf{x}(n) = \frac{1}{2\pi} \int_{0}^{2\pi} \mathbf{Y}^{\dagger}(e^{j\omega})\mathbf{X}(e^{j\omega}) d\omega. \quad (2.2.12)$$

Degree of a system. The degree or "McMillan degree" of an LTI system $\mathbf{H}(z)$ is defined to be the smallest number of delay elements (z^{-1} elements) required to implement the system. Unlike in the scalar case, the degree of an MIMO system cannot be determined just by inspection of $\mathbf{H}(z)$. In Chap. 13 we will study this topic carefully.

Exponential Inputs Produce Exponential Outputs

For a scalar LTI system $H(z)$, we know that an exponential input a^n produces the output $H(a)a^n$. Now consider a $p \times r$ system $\mathbf{H}(z)$, and apply the input $\mathbf{v}a^n$ where a is an arbitrary scalar and $\mathbf{v}$ an arbitrary vector. Using (2.2.7) we find

$$\mathbf{y}(n) = \sum_{i=-\infty}^{\infty} \mathbf{h}(i)\mathbf{v} a^{n-i} = a^n \sum_{i=-\infty}^{\infty} \mathbf{h}(i)\mathbf{v} a^{-i} = \mathbf{H}(a)\mathbf{v} a^n. \quad (2.2.13)$$

In other words, an exponential input $\mathbf{v}a^n$ aligned in the direction of the vector $\mathbf{v}$ produces the exponential output $\mathbf{H}(a)\mathbf{v}a^n$, which is aligned in the direction of the vector $\mathbf{H}(a)\mathbf{v}$. This gives us a beautiful 'physical' significance for the transfer matrix $\mathbf{H}(z)$.

Chapter 13 is dedicated to a thorough review of MIMO LTI systems, and is a preparation for some of the deeper results shown in the later sections of Chap. 14 on paraunitary system.

2.3 NOTATIONS

In what follows we summarize the notations used in the text. The reader may wish to glance through this section during first reading (a section on notations can hardly be entertaining!), and then use this primarily as a reference.

2.3.1 Preliminaries

1. The variables Ω and ω are the frequency variables for the continuous and discrete-time cases respectively.
2. $\mathcal{U}(n)$ denotes the unit step sequence and should not be confused with $u(n)$ which sometimes represents the input signal.
3. $\delta(n)$ is the unit-pulse (n is discrete) and $\delta_a(t)$ is the Dirac delta function (t is continuous). Both $\delta(n)$ and $\delta_a(t)$ are often termed as the "impulse functions," and the distinction is usually clear from the context.
4. The terms "inside the unit circle" and "outside the unit circle" are often abbreviated as "the region $|z| < 1$" and "region $|z| > 1$" respectively.
5. Superscript asterik, as in $H^*(z)$, denotes complex conjugation of $H(z)$, whereas subscript asterik, as in $H_*(z)$, means that only the coefficients are conjugated. For example, if $H(z) = a + bz^{-1}$ then, $H_*(z) = a^* + b^* z^{-1}$.

2.3.2 Polynomials

A polynomial in x has the form $\sum_{n=0}^{N} a_n x^n$, that is, has only nonnegative powers of x. Here N is a finite integer. If $a_N \neq 0$, N is said to be its order. Usually we do not use the word "degree," which is reserved for the number of delays required to implement a causal LTI system.

Let $H(z) = \sum_{n=0}^{N} h(n) z^{-n}$. This is a polynomial in z^{-1}, and represents a causal FIR filter. It is common to attach various adjectives to $H(z)$ depending upon its zero locations. Here are some commonly used ones.

1. $H(z)$ is strictly minimum-phase (or strictly Hurwitz, abbreviated SH) if all zeros are inside the unit circle.
2. $H(z)$ is strictly maximum-phase if all zeros are outside the unit circle.
3. $H(z)$ is minimum-phase if all the zeros satisfy $|z_k| \leq 1$ (i.e., are inside or on the unit circle).
4. $H(z)$ is maximum-phase if all the zeros satisfy $|z_k| \geq 1$.
5. $H(z)$ is a mixed phase polynomial if none of the above holds.

2.3.3 The "Tilde" Notation (Paraconjugation)

The notation $\widetilde{H}(z)$ plays a crucial role in our discussion. This is defined such that, on the unit circle, $\widetilde{H}(z) = [H(z)]^*$ (that is, complex conjugation). Examples:

Let $H(z) = 1 + 2z^{-1}$, then $\widetilde{H}(z) = 1 + 2z$.

Let $H(z) = (a + bz^{-1})/(c + dz^{-1})$, then $\widetilde{H}(z) = (a^* + b^*z)/(c^* + d^*z)$.

More generally, we define $\widetilde{H}(z)$ for a rational function $H(z)$ as follows: first conjugate the coefficients, and then replace z with z^{-1}. Using the subscript asterik notation defined earlier, we see that

$$\widetilde{H}(z) = H_*(z^{-1}).$$

As an application, if $H(z) = \sum_{n=0}^{N} h(n)z^{-n}$, then

$$z^{-N}\widetilde{H}(z) = h^*(N) + h^*(N-1)z^{-1} + \ldots + h^*(0)z^{-N},$$

that is, the coefficients are time-reversed and conjugated. A number of points about the "tilde notation" are worth noting.

1. The function $H_*(z^*)$ *also* reduces to $H^*(z)$ on the unit circle (in fact for any z), but it is not a rational function of z [unlike $H(z)$]. The function $\widetilde{H}(z)$, on the other hand, continues to be rational in z (hence analytic), and it is mathematically more convenient to use. $\widetilde{H}(z)$ is also called the *paraconjugate* of $H(z)$ and can be regarded as an analytic extension of unit-circle conjugation.

2. Given a function $H(z)$, the quantity $\widetilde{H}(z)H(z)$, evaluated on the unit circle, is the magnitude squared response $|H(e^{j\omega})|^2$.

3. We can write $\widetilde{H}(z) = H^*(1/z^*)$ for any z. This has been used, for example, in Oppenheim and Schafer [1989] when discussing magnitude squared responses of digital filters.

4. If $H(z) = 0$ for $z = \alpha$, then $\widetilde{H}(z) = 0$ for $z = 1/\alpha^*$ (the reciprocal conjugate point).

5. If $H(z)$ has (strictly) minimum phase, then $\widetilde{H}(z)$ has (strictly) maximum phase, and conversely.

6. If $H(z) = H_1(z)H_2(z)$, then $\widetilde{H}(z) = \widetilde{H}_1(z)\widetilde{H}_2(z)$. If $H(z) = H_1(z) + H_2(z)$, then $\widetilde{H}(z) = \widetilde{H}_1(z) + \widetilde{H}_2(z)$.

2.3.4 Matrices and Matrix Functions

Bold faced letters such as $\mathbf{A}, \mathbf{v}$ denote matrices and vectors. See Appendix A for a brief review of matrices, and matrix operations such as transpose, transpose conjugate, and so on. We often encounter matrix functions $\mathbf{H}(z)$. These are matrices in which each element is a rational function (e.g., polynomial) in z or z^{-1}. Once again the "tilde" notation plays a major role. Here is a summary of key matrix notations.

1. $\mathbf{A}^T$ denotes the transpose of $\mathbf{A}$, and $\mathbf{H}^T(z)$ stands for $[\mathbf{H}(z)]^T$.

2. $\mathbf{A}^\dagger$ denotes transpose-conjugate of $\mathbf{A}$, and $\mathbf{H}^\dagger(e^{j\omega})$ denotes $[\mathbf{H}(e^{j\omega})]^\dagger$.

3. $\mathbf{H}_*(z)$ denotes conjugation of *coefficients* without changing z (see example below).
4. $\tilde{\mathbf{H}}(z)$ denotes $\mathbf{H}_*^T(z^{-1})$.

As an example, let $\mathbf{H}(z) = \mathbf{h}(0) + \mathbf{h}(1)z^{-1}$. Then

$$\begin{aligned}
\mathbf{H}^T(z) &= \mathbf{h}^T(0) + \mathbf{h}^T(1)z^{-1}, \\
\mathbf{H}_*(z) &= \mathbf{h}^*(0) + \mathbf{h}^*(1)z^{-1}, \\
\tilde{\mathbf{H}}(z) &= \mathbf{h}^\dagger(0) + \mathbf{h}^\dagger(1)z, \\
\mathbf{H}(e^{j\omega}) &= \mathbf{h}(0) + \mathbf{h}(1)e^{-j\omega}, \\
\mathbf{H}^\dagger(e^{j\omega}) &= \mathbf{h}^\dagger(0) + \mathbf{h}^\dagger(1)e^{j\omega}.
\end{aligned} \quad (2.3.1)$$

The matrix $\tilde{\mathbf{H}}(z)$ is said to be the paraconjugate of $\mathbf{H}(z)$. For rational $\mathbf{H}(z)$ it continues to be rational. Notice that $\tilde{\mathbf{H}}(z) = \mathbf{H}^\dagger(z)$ for $z = e^{j\omega}$, that is, paraconjugation and transpose conjugation are identical on the unit circle. In general, one can write $\tilde{\mathbf{H}}(z) = \mathbf{H}^\dagger(1/z^*)$ for any z.

The word 'scalar' corresponds to a matrix with $p = r = 1$. Thus a 'scalar system' is a single-input single-output (SISO) system. All the notations introduced above apply to the scalar case; just remember that transposition leaves the scalar quantity unchanged.

2.3.5 Notations for FIR Functions

An FIR function is any function of the form

$$G(z) = \sum_{n=n_1}^{n_2} g(n)z^{-n}, \quad (2.3.2)$$

where $-\infty < n_1 \leq n_2 < \infty$. A causal FIR function (or a polynomial in z^{-1}) is of the form $H(z) = \sum_{n=0}^{N} h(n)z^{-n}$. Several types of causal FIR filters can be distinguished:

1. *Hermitian and skew-Hermitian polynomials.* We say that $H(z)$ is Hermitian [or $h(n)$ is Hermitian] if $h(n) = h^*(N-n)$ for all n, and skew-Hermitian if $h(n) = -h^*(N-n)$. In terms of $H(z)$ this means $H(z) = z^{-N}\tilde{H}(z)$ (Hermitian) and $H(z) = -z^{-N}\tilde{H}(z)$ (skew-Hermitian). For example, $1 + 2j + (1-2j)z^{-1}$ is Hermitian whereas $2 + j - (2-j)z^{-1}$ is skew-Hermitian.

2. *Generalized-Hermitian polynomial.* We say that $H(z)$ is generalized Hermitian if $h(n) = ch^*(N-n)$, [i.e., $H(z) = cz^{-N}\tilde{H}(z)$] for some c with $|c| = 1$. Examples: (a) $a + 2z^{-1} + a^*z^{-2}$, (b) $a + jz^{-1} + z^{-2} + ja^*z^{-3}$. Evidently, Hermitian and skew-Hermitian polynomials are special cases of this.

3. *Symmetric and antisymmetric polynomials.* $H(z)$ [or $h(n)$] is said to be symmetric if $h(n) = h(N-n)$ and antisymmetric if $h(n) = -h(N-n)$; [this definition is particularly useful if $h(n)$ is real]. These are equivalent, respectively, to $H(z) = z^{-N}H(z^{-1})$ and $H(z) = -z^{-N}H(z^{-1})$. Examples: $1 + 2z^{-1} + z^{-2}$ is symmetric; $1 + 2z^{-1} - 2z^{-2} - z^{-3}$ is antisymmetric.

4. *Hermitain image and mirror image.* Let $A(z)$ and $B(z)$ be two polynomials in z^{-1} with order N. We say that $B(z)$ is the generalized Hermitian image of $A(z)$ if $B(z) = cz^{-N}\tilde{A}(z)$ for some c with $|c| = 1$ (Hermitian image if $c = 1$, skew-Hermitian image if $c = -1$). Also $B(z)$ is the *mirror image* of $A(z)$ if $B(z) = z^{-N}A(z^{-1})$. Here are some examples, with $N = 1$.

   a) $1 + jz^{-1}$ and $j + z^{-1}$ (mirror images)
   b) $1 + jz^{-1}$ and $-j + z^{-1}$ (Hermitian images)
   c) $1 + jz^{-1}$ and $1 + jz^{-1}$ (Hermitian images; why?)
   d) $1 + jz^{-1}$ and $j - z^{-1}$ (Hermitian images)
   e) $1 + 2z^{-1}$ and $2 + z^{-1}$ (mirror *and* Hermitian images)

The above terminology can also be used for the more general form (2.3.2). For example, we say $G(z)$ is symmetric if $H(z)$ defined as $z^{-n_1}G(z) = \sum_{n=0}^{N} h(n)z^{-n}$ is symmetric. Thus, $z + 2 + z^{-1}$ is symmetric, and so is the transfer function $z^{-1} + 5z^{-2} + z^{-3}$.

2.3.6 Miscellaneous Mathematical Symbols

The following symbols are sometimes employed for economy: $\exists$ (there exists); $\iff$ (if and only if); $\forall$ (for all); $\approx$ (approximately equal to); $\Rightarrow$ (implies); $\in$ (belongs to); $\triangleq$ (defined as).

2.4 DISCRETE-TIME FILTERS (DIGITAL FILTERS)

We use the term 'digital filter' for discrete-time filters even though digitization (quantization) effects will be considered only in Chap. 9. A digital filter, then, is an LTI system with rational transfer function as in (2.1.10). The quantity $H(e^{j\omega})$ is called the *frequency response*. Its meaning is clear from (2.1.7) which yields

$$Y(e^{j\omega}) = H(e^{j\omega})U(e^{j\omega}). \tag{2.4.1}$$

From the time domain viewpoint, if we apply an input with frequency ω_0, that is, $u(n) = e^{j\omega_0 n}$, then the output is $y(n) = H(e^{j\omega_0})e^{j\omega_0 n}$. This follows because $e^{j\omega_0 n}$ is an eigenfunction of the system. $H(e^{j\omega_0})$ is the weighting function (or eigenvalue) or simply "the gain of the system" at frequency ω_0.

Transmission Zeros

If $H(e^{j\omega_0}) = 0$ then the frequency ω_0 is rejected by the filter. So the filter offers infinite attenuation at this frequency. We say that ω_0 is a transmission zero of $H(z)$.

Transmission zeros come from zeros of $H(z)$ on the unit circle. Let $H(z) = A(z)/B(z)$, with $A(z)$ and $B(z)$ as in (2.1.11). It is clear that ω_0 is a transmission zero if, and only if, $(1 - e^{j\omega_0}z^{-1})$ is a factor of $A(z)$. If all zeros of $H(z)$ are on the unit circle, then all the factors of the numerator $A(z)$ have the form $(1 - e^{j\omega_k}z^{-1})$.

For the real coefficient case, each factor $(1 - e^{j\omega_k}z^{-1})$ is paired with $(1 - e^{-j\omega_k}z^{-1})$ unless $\omega_k = 0$ or π. The factor of $A(z)$ which represents the complex conjugate pair of zeros is, therefore, $1 - 2\cos\omega_k z^{-1} + z^{-2}$. This is a symmetric polynomial so that the product of such factors is symmetric. Transmission zeros at $\omega = 0$ and π give rise to the factors $(1 - z^{-1})$ (antisymmetric) and $(1 + z^{-1})$ (symmetric). Thus, for a real filter with all zeros on the unit circle, the numerator $A(z)$ is either symmetric or antisymmetric. That is,

$$a_n = \begin{cases} a_{N-n} & \text{(even number of zeros at } \omega = 0) \\ -a_{N-n} & \text{(odd number of zeros at } \omega = 0). \end{cases} \quad (2.4.2)$$

For the more general case of complex systems, if all zeros are on the unit circle then $A(z)$ is generalized-Hermitian (Problem 2.3). Notice, however, that these conditions on $A(z)$ are not sufficient to ensure that all zeros are on the unit circle.

2.4.1 Magnitude Response and Phase Response

The frequency response which in general is a complex quantity, can be expressed as

$$H(e^{j\omega}) = |H(e^{j\omega})|e^{j\phi(\omega)}. \quad (2.4.3)$$

The real-valued quantities $|H(e^{j\omega})|$ and $\phi(\omega)$ are, respectively, called the magnitude response and the phase response of the filter. The quantity $\tau(\omega) = -d\phi(\omega)/d\omega$ is said to be the group delay of the system $H(z)$.

Depending on the nature of $|H(e^{j\omega})|$, filters are typically classified as lowpass, bandpass and so on. Figure 2.4-1 demonstrates this for real coefficient filters (see below).

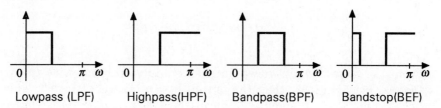

Figure 2.4-1 Various types of magnitude responses for real-coefficient filters.

Real Coefficient Filters

For real $h(n)$, the magnitude $|H(e^{j\omega})|$ is an even function of ω whereas the phase response $\phi(\omega)$ is an odd function. For example, let $H(z) = 1 - az^{-1}$ with real a. Then, $H(e^{j\omega}) = 1 - ae^{-j\omega}$, and

$$|H(e^{j\omega})| = (1 - 2a\cos\omega + a^2)^{0.5}, \quad \phi(\omega) = \tan^{-1}\left(\frac{a\sin\omega}{1 - a\cos\omega}\right),$$

so that $|H(e^{j\omega})|$ is even, and $\phi(\omega)$ odd indeed. So the response needs to be shown only for the region $0 \leq \omega \leq \pi$. If $H(z)$ has real coefficients, an input $\cos(\omega_0 n)$ produces the output $y(n) = |H(e^{j\omega_0})|\cos(\omega_0 n + \phi(\omega_0))$. [†]

Unwrapped and Wrapped Phase

Consider the filter $H(z) = (\frac{1+z^{-1}}{2})^6$. The frequency response is given by $H(e^{j\omega}) = e^{-j3\omega}\cos^6(\omega/2)$. So, the phase response is $\phi_u(\omega) = -3\omega$ and varies from 0 to -6π as ω changes from 0 to 2π. If we replace $\phi_u(\omega)$ with its principal value $\phi_w(\omega) = \phi_u(\omega) \mod 2\pi$, this does not change the value of $H(e^{j\omega})$, as seen from (2.4.3). The quantities $\phi_u(\omega)$ and $\phi_w(\omega)$ are called the *unwrapped* and *wrapped* phase responses, respectively (this also explains the introduction of subscripts for this discussion).

The unwrapped phase $\phi_u(\omega)$ can have any value whereas the wrapped phase $\phi_w(\omega)$ is always within a range of length 2π, for example, $-2\pi < \phi_w(\omega) \leq 0$ or $-\pi < \phi_w(\omega) \leq \pi$. The unwrapped phase $\phi_u(\omega)$ is related to the group delay $\tau(\omega)$ according to the integral

$$\phi_u(\omega) = -\int_0^\omega \tau(\theta)\,d\theta + \phi_u(0).$$

Most computer programs that evaluate the phase response actually return the wrapped phase $\phi_w(\omega)$. There exist good algorithms to obtain the unwrapped phase from the wrapped phase [Tribolet, 1977], [Oppenheim and Schafer, 1989].

If the distinction between the wrapped and unwrapped phases is not necessary (as in the majority of our discussions), the subscript will be omitted. The distinction is sometimes essential, for example, when dealing with the so-called *complex cepstrum*. We will have occasion to describe this in Appendix D, where we study the spectral factorization problem.

Decibel (dB) Plots

The plot of $20\log_{10}|H(e^{j\omega})|$ as a function of ω is particularly useful in revealing the stopband details of the response. This is often referred to as the dB plot of the (magnitude) response. Figure 2.4-2 demonstrates this for

---

[†] Notice that, in general, signals of the form $\cos(\omega_0 n)$ are *not* eigenfunctions of LTI systems.

a lowpass filter. It is helpful to remember that $|H(e^{j\omega})| = 10^{-k}$ implies a level of $-20k$ dB on this plot. For example, $|H(e^{j\omega_1})| = 0.01$ implies that the filter provides 40 dB attenuation at ω_1 (see Sec. 3.1 later).

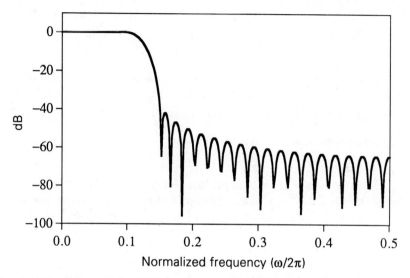

Figure 2.4-2 Demonstration of decibel (dB) plot of the magnitude response.

2.4.2 Linear Phase Filters

Strictly speaking, a digital filter is said to have linear phase if the phase response $\phi(\omega)$ is linear in ω (i.e., of the form $\alpha\omega$ where α is constant). In engineering practice, a less stringent definition is used, which we employ in this text. According to this definition, $H(z)$ has linear phase if

$$H(e^{j\omega}) = ce^{-jK\omega}H_R(\omega), \qquad (2.4.4)$$

where c is a possibly complex constant, K is real, and $H_R(\omega)$ is a real valued function of ω. Note that $H_R(\omega)$ does not necessarily have period 2π [for example, try $H(z) = 1 + z^{-1}$], and we have to avoid the notation $H_R(e^{j\omega})$.

The quantity $H_R(\omega)$ is called the *amplitude response* or *zero-phase response*. Linear phase filters for which c is real and $K = 0$ are called *zero-phase filters*. For a zero-phase filter $H(e^{j\omega})$ itself is real, but it can become negative (this typically happens in the stopband).

In a region where $H_R(\omega)$ has fixed sign, the group delay of a linear-phase filter is constant, that is, $\tau(\omega) = K$. For filters with nonlinear phase, it is common practice to plot the group delay $\tau(\omega)$ to show the nonlinearity. The degree to which $\tau(\omega)$ is a nonconstant reveals the degree of phase nonlinearity.

Four Types of Real Coefficient Linear Phase Filters

Let $H(z) = \sum_{n=0}^{N} h(n)z^{-n}$, with real $h(n)$. It is well known that the response has the form (2.4.4) if $h(n)$ is symmetric or antisymmetric. Depending on whether N is even or odd, and whether $h(n)$ is symmetric or antisymmetric, we obtain four types of real coefficient linear phase filters. These are summarized in Table 2.4.1.

Notice that some of these filters have transmission zeros at $\omega = 0$ and/or π, so that they cannot be used for certain applications. For example, Types 3 and 4 (antisymmetric cases) cannot be used for lowpass filter design, and Type 2 cannot be used for highpass design. If $H(z)$ is Type 1 or 3, then the filter $G(z) = z^M H(z)$ (where $M = N/2$) is a zero-phase filter. In this text whenever we refer to linear phase filters of Type 1–4, it is implicit that the coefficients are real so that the properties in Table 2.4.1 hold.

TABLE 2.4.1 Four types of real coefficient linear phase FIR filters.
Here $H(z) = \sum_{n=0}^{N} h(n)z^{-n}$, with $h(n)$ real

Type	1	2	3	4
Symmetry	$h(n) = h(N-n)$	$h(n) = h(N-n)$	$h(n) = -h(N-n)$	$h(n) = -h(N-n)$
Parity of N	N even	N odd	N even	N odd
Expression for frequency response $H(e^{j\omega})$	$e^{-j\omega N/2} H_R(\omega)$	$e^{-j\omega N/2} H_R(\omega)$	$je^{-j\omega N/2} H_R(\omega)$	$je^{-j\omega N/2} H_R(\omega)$
Amplitude response or zero-phase response $H_R(\omega)$	$\sum_{n=0}^{M} b_n \cos(\omega n)$ $M = N/2$	$\cos\frac{\omega}{2} \sum_{n=0}^{M} b_n \cos(\omega n)$ $M = (N-1)/2$	$\sin \omega \sum_{n=0}^{M} b_n \cos(\omega n)$ $M = (N-2)/2$	$\sin\frac{\omega}{2} \sum_{n=0}^{M} b_n \cos(\omega n)$ $M = (N-1)/2$
Special features		Zero at $\omega = \pi$	Zero at $\omega = 0$ and π	Zero at $\omega = 0$
Can be used for	Any type of bandpass response (LPF, HPF, etc.)	Any bandpass response except Highpass	Differentiators and Hilbert transformers[†]	Differentiators, Hilbert transformers, and high pass filters

[†]See Rabiner and Gold, 1975

For Type 1 filters, $H_R(\omega)$ does have period 2π, whereas for Type 2 filters the period is 4π. (This can be deduced from Table 2.4.1.) Notice that $H_R(\omega)$ changes sign at $\omega = 0$ in some cases. Note also that $[H_R(\omega)]^2$ has period 2π in all cases.

Efficient structures for linear phase filters. Because of the property $h(n) = \pm h(N-n)$, Type 1–4 linear phase FIR filters can be implemented with only about $(N+1)/2$ multipliers. For example, Fig. 2.4-3 shows how we can implement a fifth order (Type 2) filter with only 3 multipliers.

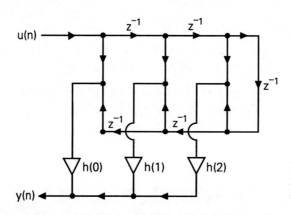

Figure 2.4-3 Efficient direct form implementation of a linear phase filter.

Advantage of Linear Phase Property

Suppose we have a signal $u(n)$ whose energy is dominantly in the region $|\omega| < \sigma$. Suppose this is passed through a real lowpass filter with passband edge at σ. (This is often done to attenuate the out-of-band noise.) The output signal is

$$Y(e^{j\omega}) = H(e^{j\omega})U(e^{j\omega}). \qquad (2.4.5)$$

If we assume that the filter $H(z)$ is a 'good' lowpass filter, then $|H(e^{j\omega})| \approx 1$ in the passband. Moreover, both $|Y(e^{j\omega})|$ and $|U(e^{j\omega})|$ are very small in the stopband. So we have

$$|Y(e^{j\omega})| \approx |U(e^{j\omega})|, \qquad (2.4.6)$$

for all ω. The approximate nature of this relation is due to the facts that (a) the filter is not ideal, and (b) $u(n)$ is not perfectly bandlimited. Nevertheless, (2.4.6) implies that the output signal $y(n)$ tends to resemble $u(n)$ provided that there is no phase distortion.

The phase distortion, in turn, is eliminated if $U(e^{j\omega})$ and $Y(e^{j\omega})$ have same phase (except for a linear offset term). This can be satisfied if $H(z)$ has linear phase. In this case (2.4.4) holds so that (2.4.5) can be replaced with

$$Y(e^{j\omega}) \approx ce^{-j\omega K}U(e^{j\omega}). \qquad (2.4.7)$$

For the case where c is real and K an integer, this implies $y(n) \approx cu(n-K)$, which is a (scaled) and delayed version of $u(n)$.

Summarizing, if the input has energy confined to the passband of the filter then the output signal is approximately equal to (a scaled and shifted version of) this input provided the filter has linear phase and 'good' passband and stopband. (The above discussion assumes K is an integer. But this is not always the case, e.g., when the filter order is odd.)

If the filter has nonlinear phase, then we still have (2.4.6), but due to phase distortion, the time domain relation $y(n) \approx cu(n - K)$ does not hold. Whether this distortion is acceptable or not depends largely on applications. For example, in speech processing a certain degree of phase distortion can be tolerated, but in image processing, phase distortion is often disastrous [Lim, 1990].

Most General Conditions for the Linear Phase Property

Let $H(z) = \sum_{n=0}^{N} h(n) z^{-n}$ with $h(0) \neq 0$ and $h(N) \neq 0$. Then it is a linear phase filter if and only if the impulse response is generalized-Hermitian, that is, satisfies the condition

$$h(n) = d h^*(N - n), \qquad (2.4.8)$$

for some d with $|d| = 1$ (Problem 2.12). The following points are worth noting:

1. For the real coefficient case, the above condition reduces to $h(n) = \pm h(N - n)$. So, the impulse response has to be symmetric or antisymmetric.
2. In the transform domain, (2.4.8) is equivalent to the condition

$$H(z) = d z^{-N} \widetilde{H}(z). \qquad (2.4.9)$$

3. *Zero-locations of linear-phase filters.* One consequence of (2.4.9) is that, if z_k is a zero then so is $1/z_k^*$. So, the zeros of a linear phase FIR filter $H(z)$ occur in reciprocal conjugate pairs. This also explains why it is not possible to design causal stable IIR linear phase filters (Problem 2.11 requests a more rigorous discussion.)

2.4.3 Analytic Continuation

Let $H_0(z)$ and $H_1(z)$ be two transfer functions such that $H_0(e^{j\omega}) = H_1(e^{j\omega})$ for all ω. This implies that their impulse responses are identical, that is, $h_0(n) = h_1(n)$, and, therefore, that the transfer functions are identical. Thus, two filters with identical frequency responses must have identical transfer functions. In other words, if

$$H_0(z) = H_1(z), \qquad z = e^{j\omega}, \qquad (2.4.10)$$

then

$$H_0(z) = H_1(z), \qquad \text{for all } z. \qquad (2.4.11)$$

This is called the analytic continuation property. If $H_0(z)$ and $H_1(z)$ in the above discussion are replaced with $p \times r$ transfer matrices $\mathbf{H}_0(z)$ and $\mathbf{H}_1(z)$, then the analytic continuation property still holds.

The above mentioned property might appear to be trivial but it holds because all practical transfer functions are rational (hence analytic), and the unit circle is in the region of analyticity. If $H_0(z)$ and $H_1(z)$ were arbitrary (nonanalytic) functions, then (2.4.10) would not imply (2.4.11). Consider, for example, a function $P(z)$, defined to be zero in an annulus around the unit-circle, but unity otherwise (Fig. 2.4-4). This satisfies $P(e^{j\omega}) = 0$, but $P(z)$ is not identically zero for all z.

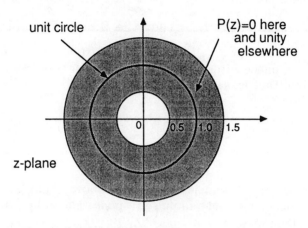

Figure 2.4-4 A function $P(z)$, defined to be zero everywhere on the unit circle, but not identically zero for all z.

PROBLEMS

2.1. In Sec. 2.1.2 we stated that exponential sequences are eigenfunctions of LTI systems. Conversely, suppose $s(n)$ is an eigenfunction of a rational IIR transfer function $H(z)$ (with order > 0 to avoid trivial answers). Does this necessarily mean that $s(n)$ is an exponential? Justify (that is, prove if *yes;* give counter example if *no*).

2.2. Suppose we apply the truncated exponential $a^n \mathcal{U}(-n)$ to a causal stable LTI system $H(z)$. Assume $|a| > 1$ so the input does not blow up for $n \to -\infty$. (a) What is the output $y(n)$ for $n \le 0$? (b) Suppose $y(n)$ is zero for $n > 0$, i.e., the output becomes zero as soon as the input becomes zero. This leads us to suspect that the system might be *memoryless,* [i.e., $H(z) = $ constant]. This is indeed true. Prove this.

2.3. Let $H(z) = \sum_{n=0}^{N} h(n) z^{-n}$ be a transfer function with all zeros on unit circle. Show that $h(n)$ is generalized-Hermitian. [In particular if $h(n)$ is real, this reduces to the fact that $h(n)$ is symmetric or antisymmetric.]

2.4. Let $H(z) = \sum_{n=0}^{\infty} h(n) z^{-n}, h(0) \ne 0$, be a rational transfer function representing a causal, stable LTI system. So, $H(z) = A(z)/B(z)$ where $A(z)$ and $B(z)$ are relatively prime polynomials in z^{-1}. We shall now develop a time domain interpretation of "poles", which is more appealing than the definition which says that $H(z)$ "blows up" at a pole.
   a) Suppose $p \ne 0$ is a pole of $H(z)$. Show that there exists a causal finite length input $x(n)$ and a finite integer L such that the output $y(n)$ has the form p^n for $n > L$.
   b) Conversely suppose there exists a causal finite length input $x(n)$ and a finite integer L such that the output $y(n)$ has the form p^n for $n > L$. Show that p is a pole of $H(z)$.

2.5. We know that if z_k is a zero of $H(z)$ then the output in response to the input z_k^n is zero for all n. This input is noncausal (and doubly infinitely long). In this problem we develop another engineering insight for the meaning of a zero of $H(z)$. This is based on causal inputs, and might be more appealing. Assume $H(z)$ is in irreducible rational form as in (2.1.10). Also, assume that it is causal so that it can be implemented with the difference equation (2.1.14) (with $b_0 = 1$). Assume $a_0 \ne 0$, and $b_N \ne 0$.
   a) Consider the first order case ($N = 1$). We know the system has a zero at $z_1 = -a_1/a_0$. Suppose, we apply an input of the form $z_1^n \mathcal{U}(n)$ where $\mathcal{U}(n)$ is the unit step. Find an initial value $y(-1)$ such that $y(n) = 0$ for all $n \ge 0$.
   b) More generally, for the Nth order system suppose z_1 is a zero and we apply the causal input $z_1^n \mathcal{U}(n)$. Show how you can find an initial state $y(-1), y(-2), \ldots, y(-N)$ such that $y(n) = 0$ for all $n \ge 0$.
   c) Conversely, suppose there exists an initial state $y(-1), y(-2), \ldots, y(-N)$ such that the input $z_1^n \mathcal{U}(n)$ produces zero output for all $n \ge 0$. Show that z_1 is a zero of $H(z)$.

2.6. Let $H(z) = \sum_{n=0}^{\infty} h(n) z^{-n}, h(0) \ne 0$, be a rational transfer function representing a causal, stable LTI system. So, we can write $H(z) = A(z)/B(z)$ where $A(z)$ and $B(z)$ are relatively prime polynomials in z^{-1}.

a) Let z_0 be a *zero* of the system. Then show that there exists a causal finite length sequence $s(n)$ such that the input

$$x(n) = z_0^n \mathcal{U}(n) + s(n) \qquad (P2.6)$$

produces a causal, finite-length output.

b) Conversely, let there exist an input of the form in (P2.6) where $s(n)$ is causal and finite-length, such that the output is of finite length. Then show that z_0 is indeed a zero, assuming $z_0 \neq 0$.

Note. This gives the following engineering interpretation of a *zero*: there exists a causal input such that, if you wait for finite time after applying the input, the input will look like an exponential z_0^n whereas the output will become zero and stay zero!

2.7. You are given a black box, which you can imagine to be a computer program. This black box takes an input sequence $x(n)$ and computes each sample of the output $y(n)$ in finite amount of time. You are given the additional information that any exponential input a^n produces an exponential output $H(a)a^n$. Can you conclude that the black box is an LTI system? Justify (i.e., prove if *yes;* give counter example if *no*).

2.8. Let $H(z) = \sum_{n=0}^{N} h(n) z^{-n}$ be a Type 1 linear phase FIR filter. Define a new filter $G(z)$ with $g(n) = h(n) \cos[\omega_0 (n - K)]$ where K is an integer. How would you choose K so that $G(z)$ also has linear phase?

2.9. Let $x(n) = \cos(\omega_0 n)$ be the input to a Type 1 linear phase FIR filter $H(z)$. Find an expression for the output $y(n)$ and simplify as best as you can. Can you say that $y(n) = cx(n - K)$ for some c and K? What if the filter were Type 3?

2.10. Let $H(z) = 1/[1 + \sum_{n=1}^{N} b_n z^{-n}]$ represent a causal filter with linear phase. Prove that it is unstable (unless $b_n = 0$ for all n).

2.11. Let $H(z)$ be causal stable with irreducible form $A(z)/B(z)$. Suppose this has linear phase, that is, satisfies (2.4.4). Show then that $H(z)$ is FIR! [This is a generalization of Problem 2.10. It is somewhat subtle because, you have to wonder whether $H(z)$ might have linear phase even if $A(z)$ and $B(z)$ are not, individually, linear phase functions.]

2.12. This is a continuation of (2.4.8) and the paragraph preceding it. We stated that (2.4.8) is necessary and sufficient for linear phase property. Prove this.

2.13. This pertains to stability.
a) Consider a causal IIR filter with transfer function $H(z) = 1/D(z)$ where

$$D(z) = 1 + d_1 z^{-1} + d_2 z^{-2} + \ldots + d_N z^{-N},$$

with d_n real for all n. Show that this is BIBO stable *only if* $D(1) > 0$ and $D(-1) > 0$.

b) As a special case, consider a second order causal IIR filter with transfer function $H(z) = 1/(1 + az^{-1} + bz^{-2})$, with a, b real. Show that this is BIBO stable only if the values of a and b are restricted to be *strictly* inside the triangular area in Fig. P2-13 (called the *stability triangle*).

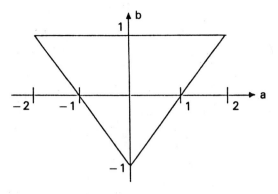

Figure P2-13

c) Finally, show that, in the above second-order case, the converse is also true, that is, if a, b are stritctly inside the triangular region then $H(z)$ is BIBO stable.

2.14. The deterministic cross correlation between two sequences $x(n)$ and $y(n)$ is defined as

$$R_{xy}(k) = \sum_{n=-\infty}^{\infty} x(n) y^*(n-k). \qquad (P2.14)$$

The integer k is called the lag variable. Let $S_{xy}(z)$ denote the z-transform of $R_{xy}(k)$. Show that $S_{xy}(z) = X(z)\widetilde{Y}(z)$. [Note: A special case of this result arises when $y(n) = x(n)$. The quantity $R_{xx}(k)$ is called the *deterministic autocorrelation* of $x(n)$. Its z-transform is $S_{xx}(z) = X(z)\widetilde{X}(z)$, so that $S_{xx}(e^{j\omega}) = |X(e^{j\omega})|^2$.]

2.15. Give example of a function $x_a(t)$ such that (a) it is not σ-BL, and (b) if it is sampled at the rate $2\pi/T = 2\sigma$, no two terms in (2.1.22) overlap.

3

Review of Digital Filters

3.0 INTRODUCTION

This chapter includes a brief review of digital filter design techniques. Many of these topics are treated in Oppenheim and Schafer [1989]. Other related texts are [Rabiner and Gold, 1975], [Antoniou, 1979], and [Jackson, 1989]. Because of the availability of these references, our review is brief and limited to those techniques that are directly relevant to multirate systems. Discussions on relatively recent developments, for example eigenfilters (Section 3.2.3), and allpass decomposition of IIR filters (Section 3.6), are not available in the above references.

Some design techniques and structures will be treated in greater detail here. These include (a) the window design (FIR), (b) eigenfilters approach (FIR), (c) elliptic filters (IIR), (d) properties of allpass functions, and (e) allpass lattice structures. This elaboration is motivated by the applications in multirate filter bank design, as we indicate at the appropriate places.

Section 3.1 describes the common filter design specifications. In Sec. 3.2 and 3.3 we consider the design of finite impulse response (FIR) and infinite impules response (IIR) filters. Section 3.4 discusses allpass filters, which play a key role in the design of filter banks. Section 3.5 summarizes several special filters. In Sec. 3.6 we will show that many IIR filters (e.g., elliptic) can be expressed as a sum of two allpass filters. A special case of this (called IIR power symmetric filters) will find application in two-channel QMF bank design (Chap. 5).

3.1 FILTER DESIGN SPECIFICATIONS

The specifications on the magnitude response of a digital filter are usually given in terms of certain tolerances as demonstrated in Fig. 3.1-1(a) for the lowpass case. We assume the coefficients to be real, so only the region $0 \leq \omega \leq \pi$ has to be specified. Between the passband and stopband, we have to specify a transition band region since the response cannot change abruptly from unity to zero. The region $0 \leq \omega \leq \omega_p$ is the passband and $\omega_S \leq \omega \leq \pi$ the stopband. The responses in the passband and stopband are required to lie within the tolerance regions.

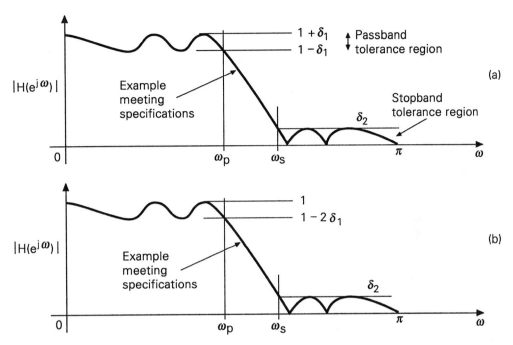

Figure 3.1-1 Magnitude response specifications for real-coefficient lowpass filters. (a) unnormalized magnitude, and (b) normalized magnitude.

The following terminology is standard.

$$\delta_1 = \text{peak passband ripple}$$
$$\delta_2 = \text{peak stopband ripple}$$
$$A_S = -20\log_{10}\delta_2 = \text{minimum stopband attenuation}$$
$$A_p = -20\log_{10}(1-\delta_1) = \text{peak passband ripple in dB}$$
$$\omega_p = \text{passband edge}$$
$$\omega_S = \text{stopband edge}$$
$$\Delta\omega = \omega_S - \omega_p = \text{transition BW (radians)}$$
$$\Delta f = \frac{\Delta\omega}{2\pi} = \text{normalized transition BW (dimensionless).}$$
(3.1.1)

(Note that BW is an abbreviation for bandwidth.) The variable $f \triangleq \omega/2\pi$ is said to be the normalized frequency. Frequency response plots in the range

Sec. 3.1 Design specifications 43

$0 \leq \omega \leq \pi$ correspond to the range $0 \leq f \leq 0.5$ in terms of f. Figure 3.1-1(a) also shows example of a response conforming to these specifications.

Normalized specifications. It is sometimes convenient to normalize the peak passband magnitude to unity. This can be done by dividing the response by $(1 + \delta_1)$. If $\delta_1 \ll 1$, this does not significantly affect the ripple sizes. Figure 3.1-1(b) shows this normalized set of specifications. One can verify that

$$A_p \approx 8.686\delta_1, \quad \text{for } \delta_1 \ll 1. \tag{3.1.2}$$

The quantity $\mathcal{A}(e^{j\omega}) = -20\log_{10}|H(e^{j\omega})|$ is said to be the attenuation characteristics for the filter. $-\mathcal{A}(e^{j\omega})$ is the magnitude response in dB. Fig. 3.1-2 shows how to specify the tolerances in terms of this quantity, for the case of normalized response. Note that the normalization of $|H(e^{j\omega})|$ corresponds to setting the minimum value of $\mathcal{A}(e^{j\omega})$ to 0 dB. The quantity A_{max} shown in this figure is called the *maximum passband attenuation*. With $\delta_1 \ll 1$, one can verify that

$$A_{max} = -20\log_{10}(1 - 2\delta_1) = 2A_p. \tag{3.1.3}$$

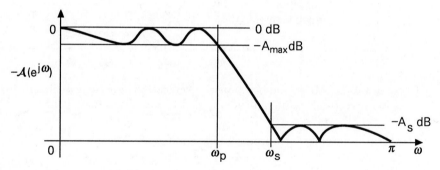

Figure 3.1-2 Specifications in terms of attenuation function, normalized to 0 dB. The attenuation goes to infinity at the transmission zeros.

Criteria for Optimality

We often talk about optimal filters, that is, filters that are "best" in some sense. The criterion of optimality has to be mentioned in order to make the meaning complete, as elaborated next.

Equiripple filters. For an equiripple filter, the extremal values of the error are the same throughout a given band. The examples in Fig. 3.1-1 are equiripple. To describe the optimality of such filters, let N denote the filter order, and let δ_1, δ_2 and Δf be defined as above. If any three of these four quantities are fixed, then the fourth parameter is minimum for an equiripple filter. Such a filter is said to be optimal in the *minimax* sense because, the *maximum* ripple sizes have been *minimized* for fixed N and Δf.

Least-squares filters. In the design of these filters, the square of the difference between the ideal and actual responses is integrated over the appropriate frequency bands and minimized. The simplest examples are FIR filters based on rectangular windows (Sec. 3.2). In Sec. 3.2.3 we will describe more useful variations, called *eigenfilters*.

Flatness constraints. In some applications it is desirable to have a high degree of flatness around zero frequency in the passband. Such flatness is usually specified in terms of the number of zeros of the derivative of $|H(e^{j\omega})|^2$ at $\omega = 0$. These specifications are called *flatness constraints*. In the IIR case, Butterworth filters (Sec. 3.3) serve this purpose. In the FIR case also it is possible to design filters with flatness constraints [Herrmann, 1971], and [Kaiser, 1979]. We will study this while discussing wavelet transforms in Chap. 11, where the flatness constraint is used to generate orthonormal basis functions with "regularity" properties.

3.2 FIR FILTER DESIGN

Consider Fig. 3.2-1(a) which shows an ideal lowpass response $H_i(e^{j\omega})$ with cutoff frequency ω_c, that is,

$$H_i(e^{j\omega}) = \begin{cases} 1 & |\omega| < \omega_c \\ 0 & \text{otherwise.} \end{cases} \quad (3.2.1)$$

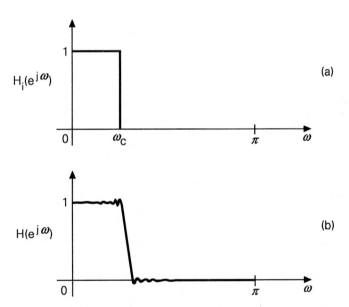

Figure 3.2-1 (a) The ideal lowpass response, and (b) truncated filter response.

So, this is a zero-phase filter with magnitude equal to unity in the passband and zero in the stopband (and no transition band whatsoever). Its impulse response [inverse transform of $H_i(e^{j\omega})$] is given by

$$h_i(n) = \frac{\omega_c}{\pi}\left(\frac{\sin \omega_c n}{\omega_c n}\right), \quad -\infty \leq n \leq \infty. \qquad (3.2.2)$$

It can be shown that this impulse response does not satisfy the BIBO stability requirement (Sec. 2.1.2). So, the ideal filter is not stable. In addition it is noncausal and IIR. No amount of delay would make the impulse response causal.

The simplest way to obtain an FIR lowpass filter from this would be to truncate the impulse response

$$h(n) = \begin{cases} h_i(n), & |n| \leq N/2, \\ 0, & \text{otherwise.} \end{cases} \qquad (3.2.3)$$

It can be shown (Problem 3.5) that the resulting response $H(e^{j\omega})$ approximates $H_i(e^{j\omega})$ in the least squares sense, that is, for a given N, $\int_0^{2\pi}[H_i(e^{j\omega}) - H(e^{j\omega})]^2 d\omega$ is minimized. (Note that both $H(e^{j\omega})$ and $H_i(e^{j\omega})$ are real.) However, the above truncation causes ripples in the passband and stopband [Fig. 3.2-1(b)], and the ripple size grows as we get closer to ω_c from either side. As N increases, the ripples get crowded closer to the cutoff frequency ω_c, but the size of the peak ripple does not decrease. This is called the *Gibbs phenomenon*. This is demonstrated in Fig. 3.2-2 where we show dB plots of the truncated response for $N = 20$ and $N = 50$. The minimum stopband attenuation A_S in both cases is only about 21 dB.

3.2.1 Window Design

An improvement over the truncation technique is offered by the use of windows. Here the impulse response is obtained as

$$h(n) = h_i(n)v(n), \qquad (3.2.4)$$

where $v(n)$ is a window function, which is zero for $|n| > N/2$. As long as $v(n)$ is symmetric, we obtain a zero-phase filter $h(n)$ [since $h_i(n)$ is already symmetric]. If we set $v(n) = 1$ for $|n| \leq N/2$ (rectangular window), windowing is equivalent to simple truncation.

In the frequency domain, the response $H(e^{j\omega})$ is a convolution of the ideal response $H_i(e^{j\omega})$ with $V(e^{j\omega})$. Two parameters of the window which control the quality of the filter response are: (a) the main lobe width ΔB of $V(e^{j\omega})$ which controls the filter transition bandwidth $\Delta\omega$, and (b) the peak sidelobe level of $V(e^{j\omega})$ which controls the peak passband and stopband ripples of $|H(e^{j\omega})|$. If the main lobe width of $V(e^{j\omega})$ is made smaller, the transition bandwidth of $H(e^{j\omega})$ is reduced. On the other hand, if the window has smaller side lobe ripples, the stopband attenuation provided by $H(e^{j\omega})$ is correspondingly improved. By appropriate choice of the window, it is, therefore, possible to control both the attenuation and the transition bandwidth of the response $H(e^{j\omega})$.

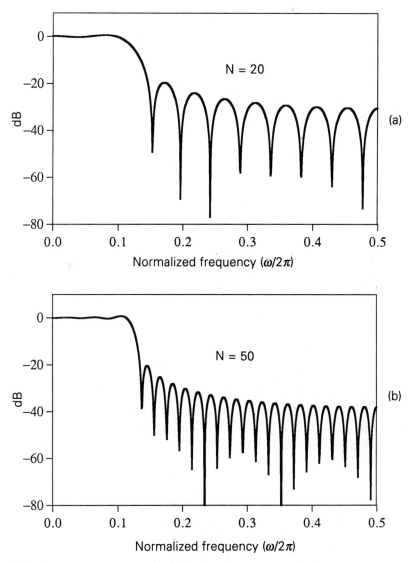

Figure 3.2-2 Magnitude responses obtained by truncation of the ideal impulse response.

Several windows have been invented (e.g., the Hamming window, Blackmann window, etc.), which offer various degrees of tradeoff between Δf and A_S [Oppenheim and Schafer, 1975]. A systematic way to obtain such a tradeoff is offered by the Kaiser window [Kaiser, 1974], which is actually a family of windows spanned by a parameter β. By adjusting β, one obtains any desired stopband attenuation for $|H(e^{j\omega})|$; the order N is adjusted to satisfy the requirement on Δf. In all window-based methods, the resulting filter has $\delta_1 \approx \delta_2$.

The Kaiser window is given by

$$v(n) = \begin{cases} I_0\left[\beta\sqrt{1-(n/0.5N)^2}\right]/I_0(\beta), & -\frac{N}{2} \leq n \leq \frac{N}{2}, \\ 0 & \text{otherwise,} \end{cases} \quad (3.2.5)$$

where $I_0(x)$ is the modified zeroth-order Bessel function, which can be computed from the power series

$$I_0(x) = 1 + \sum_{k=1}^{\infty}\left[\frac{(0.5x)^k}{k!}\right]^2. \quad (3.2.6)$$

Note that $I_0(x)$ is positive for all (real) x. In most practical designs, only about twenty terms in the above summation need to be retained. Once $v(n)$ is computed in this manner, the coefficients of the filter can be found from (3.2.4) where $h_i(n)$ is as in (3.2.2) (with $\omega_c = 0.5(\omega_p+\omega_S)$). The filter order is evidently equal to N. Since $v(n)$ and $h_i(n)$ are even functions of n, the resulting FIR filter $h(n)$ has zero phase.

The parameter β depends on the attenuation requirements of the lowpass filter. Kaiser has developed simple formulas for estimating the parameters β and N, for given A_S and Δf. The quantity β is found from

$$\beta = \begin{cases} 0.1102(A_S - 8.7) & \text{if } A_S > 50 \\ 0.5842(A_S - 21)^{0.4} + 0.07886(A_S - 21) & \text{if } 21 < A_S < 50 \\ 0 & \text{if } A_S < 21. \end{cases} \quad (3.2.7)$$

Given the quantities A_S and Δf, the filter order N is estimated from

$$N \approx \frac{A_S - 7.95}{14.36\Delta f}. \quad (3.2.8)$$

Notice that once β and N are determined, we do not have independent control over δ_1. In most designs, the resulting δ_1 comes out to be very close to δ_2 (which in turn is determined by A_S).

It has been demonstrated [Saramaki, 1989] that the Kaiser window can also be obtained from a rectangular window by means of a change of variables.

Design Example 3.2.1: FIR Lowpass Filter Using Kaiser Window

Suppose the design specifications are $\omega_p = 0.2\pi, \omega_S = 0.3\pi, A_S = 40$dB, and $\delta_1 = \delta_2$. (The value of A_S implies $\delta_2 = 0.01$.) The estimated values of β and N are $\beta = 3.395, N = 44.6$. The order can be rounded off to the next even integer, that is, $N = 46$. (This makes $N/2$ even in (3.2.5).) The cutoff frequency $\omega_c = 0.5(\omega_p + \omega_S) = 0.25\pi$.

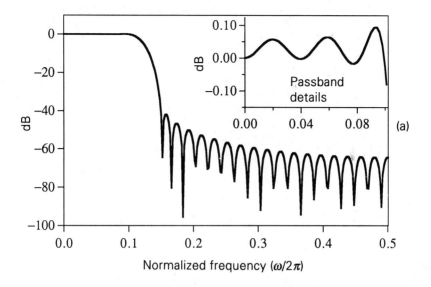

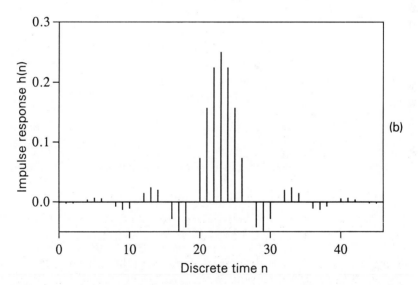

Figure 3.2-3 Design example 3.2.1. Lowpass filter based on Kaiser window. (a) Magnitude response, and (b) impulse response.

We can now compute the coefficients of $v(n)$ and $h_i(n)$ as above, and obtain $h(n)$ from (3.2.4). Fig. 3.2-3(a) shows the magnitude response plot (with $N = 46$) which meets the required specifications. For clarity, the passband details are shown separately in magnified form. One can verify

Sec. 3.2 FIR filter design

that the peak ripple δ_1 is very close to δ_2, that is, $\delta_1 \approx 0.01$. Part (b) of the figure shows the impulse response coefficients.

Summary. The Kaiser window technique offers a very simple means of designing linear phase FIR filters. No elaborate optimization steps are involved. The quantities ω_p, ω_S, and δ_2 can be specified independently, but not δ_1 (which turns out to be close to δ_2). The windowing technique is not suitable for design of filters with more sophisticated specifications (such as unequal ripple sizes, nonconstant passband responses, and so on).

The Dolph-Chebyshev Function: An Optimal Window

The Dolph-Chebyshev (DC) window $v(n)$ has the property that the *max*imum side lobe level of $|V(e^{j\omega})|$ is *min*imized in the region $\sigma \leq \omega \leq \pi$. Consequently, this is called a *minimax window*. The plot of $|V(e^{j\omega})|$ is equiripple in the region $\sigma \leq \omega \leq \pi$. For this window a closed form expression in the frequency domain, based on Chebyshev polynomials, is available. This can be found in Helms [1971]. The window coefficients in the time domain can be found by performing an inverse Fourier transform. Details are omitted.

3.2.2 The Prolate Sequence: Another Optimal Window

The Kaiser window is a good approximation to a class of optimal windows called *prolate spheroidal* (or just prolate) wave sequences $v(n)$ [Slepian, 1978]. A prolate sequence is a real sequence of finite length $N + 1$ and unit energy, with the energy in the frequency region $\sigma \leq \omega \leq \pi$ minimized. The quantities N and σ can be regarded as parameters of the prolate sequence family. In all our discussions we assume $0 < \sigma < \pi$.

We now show how the optimal window coefficients can be computed. This makes use of some results from matrix theory (especially Rayleigh's principle), reviewed in Appendix A. The derivation will show how we can pass from optimal *windows* to optimal (least squares) *filters* (Sec. 3.2.3).

Matrix-vector Formulation of the Optimization Problem

Assume $v(n)$ is causal with length $N+1$ so that $V(z) = \sum_{n=0}^{N} v(n)z^{-n}$. Define

$$\phi_S \triangleq \int_\sigma^\pi |V(e^{j\omega})|^2 \frac{d\omega}{\pi}. \tag{3.2.9}$$

This is the quantity to be minimized under the unit-energy constraint. Because of Parseval's theorem we can write the constraint either in the time domain or in the frequency domain:

$$\int_0^\pi |V(e^{j\omega})|^2 \frac{d\omega}{\pi} = 1 \quad \text{or} \quad \sum_{n=0}^{N} v^2(n) = 1. \tag{3.2.10}$$

[Remember $v(n)$ is real.] Fig. 3.2-4 demonstrates the idea: we minimize the area of the shaded region, for a fixed total area under the curve. Minimizing

ϕ_S under the constraint (3.2.10) is equivalent to maximizing

$$\phi \triangleq \int_0^\sigma |V(e^{j\omega})|^2 \frac{d\omega}{\pi}. \qquad (3.2.11)$$

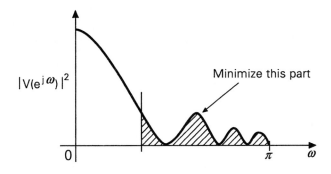

Figure 3.2-4 Design of the optimal window.

Defining the vectors

$$\mathbf{v} = [v(0) \quad v(1) \quad \ldots \quad v(N)]^T, \quad \mathbf{e}(z) = [1 \quad z^{-1} \quad \ldots \quad z^{-N}]^T, \qquad (3.2.12)$$

we have $V(e^{j\omega}) = \mathbf{v}^\dagger(n)\mathbf{e}(e^{j\omega})$ so that

$$|V(e^{j\omega})|^2 = V(e^{j\omega})V^*(e^{j\omega}) = \mathbf{v}^\dagger \mathbf{e}(e^{j\omega})\mathbf{e}^\dagger(e^{j\omega})\mathbf{v}. \qquad (3.2.13)$$

(We could have used $\mathbf{v}^T$ instead of $\mathbf{v}^\dagger$ since $\mathbf{v}$ is real; we will use $\mathbf{v}^\dagger$ for notational uniformity). We can rewrite the objective function ϕ as

$$\phi = \mathbf{v}^\dagger \left[\int_0^\sigma \mathbf{R}(\omega)\frac{d\omega}{\pi}\right]\mathbf{v}, \qquad (3.2.14)$$

where

$$\mathbf{R}(\omega) \triangleq \mathbf{e}(e^{j\omega})\mathbf{e}^\dagger(e^{j\omega}). \qquad (3.2.15)$$

The $(N+1) \times (N+1)$ matrix $\mathbf{R}(\omega)$ has (m,n) element

$$e^{-j(m-n)\omega} = \cos(m-n)\omega - j\sin(m-n)\omega, \qquad (3.2.16)$$

so that $\mathbf{R}(\omega)$ is Hermitian. Its imaginary part $\mathbf{Q}(\omega)$ is therefore antisymmetric. So, $\mathbf{v}^\dagger \mathbf{Q}(\omega)\mathbf{v} = 0$ (since $\mathbf{v}$ is real). Thus, ϕ can be simplified to

$$\phi = \mathbf{v}^\dagger \mathbf{P} \mathbf{v}, \qquad (3.2.17)$$

where $\mathbf{P}$ has (m,n)th entry

$$p_{mn} = \int_0^\sigma \cos(m-n)\omega \frac{d\omega}{\pi} = \frac{\sin(m-n)\sigma}{(m-n)\pi}, \quad 0 \le m, n \le N. \qquad (3.2.18)$$

The unit-energy constraint (3.2.10) can also be rewritten in terms of $\mathbf{v}$ as

$$\mathbf{v}^\dagger \mathbf{v} = 1, \qquad (3.2.19)$$

that is, $\mathbf{v}$ has unit norm. Summarizing, the problem of finding the unit-energy window function $v(n)$ with smallest energy in $\sigma \leq \omega \leq \pi$ has been converted to the problem of finding a unit-norm vector $\mathbf{v}$ which maximizes (3.2.17).

Solution to the Optimization Problem

Now the matrix $\mathbf{P}$ is evidently real and symmetric (so it is Hermitian). From Rayleigh's principle (Appendix A) we know that all the eigenvalues of $\mathbf{P}$ are real, and that ϕ is maximized under the constraint (3.2.19) if, and only if, $\mathbf{v}$ is an eigenvector corresponding to the largest eigenvalue λ_N. We can therefore compute the coefficients of the optimal window $v(n)$ simply by computing this eigenvector. There exist standard techniques such as the power-method (Appendix A) for this computation. The eigenvalue λ_N is the quantity (3.2.11) after maximization, and satisfies $\lambda_N < 1$ [in view of (3.2.10)].

Design Example 3.2.2: Optimal Window

As an example, let $\sigma = 0.1\pi$ and $N = 32$. The optimum window computed in the above manner has response shown in Fig. 3.2-5 which also shows the Kaiser window response with same N, and $\beta = 4.55$. The agreement between the plots demonstrates that the Kaiser window is an excellent approximation to the optimal window. The parameter σ of the optimal window has the same role as the tradeoff parameter β of the Kaiser window. It is intuitively clear that if we increase σ, then the optimum window has smaller peak side lobe level. This is indeed the case as one can verify by plotting $|V(e^{j\omega})|$ for different values of σ.

Notice that there are precisely 16 transmission zeros of $V(e^{j\omega})$ in the range $\sigma \leq \omega < \pi$, that is, a total of 32 on the unit circle. More generally, the optimal window always has all zeros on the unit circle. This is a consequence of the properties of $\mathbf{P}$, as will be elaborated.

Properties of P

The quantity ϕ is the energy of the window in the region $0 \leq |\omega| \leq \sigma$ and cannot be zero for $\sigma > 0$. This statement is true for any nonzero $v(n)$, hence any nonzero $\mathbf{v}$. In other words, $\mathbf{P}$ is positive definite for any $\sigma > 0$. It is also clear from (3.2.18) that $\mathbf{P}$ is Toeplitz (Appendix A). So, $\mathbf{P}$ is real, symmetric, positive definite and Toeplitz.

Denote the eigenvalues of $\mathbf{P}$ by λ_i, $0 \leq i \leq N$. We have $\lambda_i > 0$ due to positive definiteness. Combining with $\lambda_N < 1$, we get

$$0 < \lambda_0 \leq \lambda_1 \leq \ldots \leq \lambda_N < 1. \qquad (3.2.20)$$

Less obvious is the fact [Slepian, 1978] that the above inequalities are strict, that is, $\lambda_i < \lambda_{i+1}$. This means that the eigenvectors are unique (up to scale) so that, in particular, the optimal window **v** is unique.

Zeros of the window. Based on this uniqueness, one can show (Problem 3.6) that all the zeros of the optimal window $V(z)$ lie on the unit circle. This implies, in turn, that $v(n)$ is a symmetric sequence, that is, $v(n) = v(N-n)$. [It cannot be antisymmetric, as it would mean $V(e^{j0}) = 0$.] Redefining $v(n)$ to be $v(n+M)$, where $M = N/2$, we obtain the zero-phase optimal window which can now be used in (3.2.4) to design $h(n)$. The filter is guaranteed to have linear phase because of the symmetry of $v(n)$.

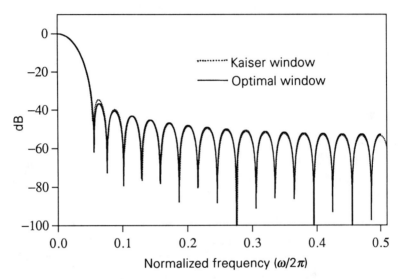

Figure 3.2-5 Design example 3.2.2. Responses of the optimal window and the Kaiser window.

3.2.3 Optimal Lowpass Eigenfilters

As discussed in Sec. 3.1, there are several classes of optimal filters, according to the choice of the performance measure (or objective function) to be minimized. In general, FIR filters based on the window approach do not yield filters which are optimal in any sense, even if the window is optimal in some sense.

An exception is the rectangular window which yields a filter $H(e^{j\omega})$ which is optimal in the least squares sense, that is, the integral

$$\int_0^{2\pi} |H_i(e^{j\omega}) - H(e^{j\omega})|^2 \, d\omega$$

is minimized, where $H_i(e^{j\omega})$ is the ideal response (3.2.1) (see Problem 3.5). These filters, however, suffer from Gibbs phenomenon as seen earlier. Furthermore, the above integral includes the transition band error, which should actually be excluded.

In this section, we introduce lowpass *eigenfilters* [Vaidyanathan and Nguyen, 1987a]. These are optimal in the least squares sense but the objective function itself is defined differently, by formulating it as a sum of the passband and stopband errors. The error of approximation in the transition band is not included. Such an objective function is obtained by adding a second term to (3.2.9), which was used to design an optimal window. The second term represents a 'squared measure' of the deviation from the ideal passband response. The formulation is such that we can obtain the optimal filter coefficients from an eigenvector of an appropriate matrix.

The eigenfilter approach is different from other types of least squares approaches for FIR design, which are obtainable by matrix inversion, for example, the one described in Roberts and Mullis [1987, Sec. 7.2].

Why eigenfilters? One of the advantages of eigenfilters over other FIR filters (such as equiripple filters) is that, they can be designed to incorporate a wide variety of time domain constraints such as the step response constraint, Nyquist constraint and so on, in addition to the usual frequency domain requirements. The filter coefficients are obtained simply by computing an eigenvector of a positive definite matrix, which is derived from the time and frequency domain specifications. Eigenfilters can be used for optimal design of the so-called Nyquist filters, which are ideally suited for interpolation filtering (Chap. 4). Nyquist filters also find use in filter bank design.

To introduce the basic idea of eigenfilters, consider Type 1 linear phase filters (Table 2.4.1). These have the form $H(z) = \sum_{n=0}^{N} h(n) z^{-n}$, where $h(n)$ is real and satisfies $h(n) = h(N-n)$. Moreover N is even. The amplitude response is

$$H_R(\omega) = \sum_{n=0}^{M} b_n \cos \omega n = \mathbf{b}^T \mathbf{c}(\omega), \qquad (3.2.21)$$

where $M = N/2$ and

$$\mathbf{b} = [\, b_0 \quad b_1 \quad \ldots \quad b_M \,]^T, \quad \mathbf{c}(\omega) = [\, 1 \quad \cos \omega \quad \ldots \quad \cos M\omega \,]^T \qquad (3.2.22)$$

The aim is to find the coefficients $\mathbf{b}$ such that an appropriate objective function is minimized. The objective function should reflect the stopband energy (energy in $\omega_S \leq \omega \leq \pi$) as well as the passband accuracy. We will formulate the minimization problem in such a way that the optimal $\mathbf{b}$ can be computed as an eigenvector of an appropriate positive definite matrix.

Since $H(e^{j\omega}) = e^{-j\omega M} H_R(\omega)$ we have

$$|H(e^{j\omega})|^2 = H_R^2(\omega) = \mathbf{b}^T \mathbf{c}(\omega) \mathbf{c}^T(\omega) \mathbf{b}. \qquad (3.2.23)$$

So the stopband energy is

$$E_S = \int_{\omega_S}^{\pi} |H(e^{j\omega})|^2 \frac{d\omega}{\pi} = \mathbf{b}^T \mathbf{P} \mathbf{b} \qquad (3.2.24)$$

where
$$\mathbf{P} = \int_{\omega_S}^{\pi} \mathbf{c}(\omega)\mathbf{c}^T(\omega)\frac{d\omega}{\pi}.$$

The (m,n) element of $\mathbf{P}$ is

$$p_{mn} = \int_{\omega_S}^{\pi} \cos(m\omega)\cos(n\omega)\frac{d\omega}{\pi} \qquad (3.2.25)$$

which can be evaluated in terms of ω_S, m, and n. The passband error can be included in the objective function as follows. The amplitude response at zero frequency is given by $H_R(0) = \mathbf{b}^T\mathbf{1}$, where $\mathbf{1}$ is the vector of all 1's. By taking this as a reference, the passband deviation at any frequency can be written as

$$\mathbf{b}^T\mathbf{1} - \mathbf{b}^T\mathbf{c}(\omega) = \mathbf{b}^T\left[\mathbf{1} - \mathbf{c}(\omega)\right], \qquad (3.2.26)$$

so that the quantity

$$E_p = \mathbf{b}^T\mathbf{Q}\mathbf{b} \qquad (3.2.27)$$

is a measure of mean square passband error, where

$$\mathbf{Q} = \int_0^{\omega_p} [\mathbf{1} - \mathbf{c}(\omega)][\mathbf{1} - \mathbf{c}(\omega)]^T \frac{d\omega}{\pi}. \qquad (3.2.28)$$

Now define the objective function

$$\phi = \alpha E_S + (1-\alpha)E_p, \qquad (3.2.29)$$

where $0 < \alpha < 1$. Here α is a tradeoff parameter between passband and stopband performances. We then have

$$\phi = \mathbf{b}^T\mathbf{R}\mathbf{b}, \qquad (3.2.30)$$

where

$$\mathbf{R} = \alpha\mathbf{P} + (1-\alpha)\mathbf{Q}. \qquad (3.2.31)$$

It is easily verified that $\mathbf{R}$ is a real, symmetric and positive definite matrix (Problem 3.8). The unit-norm vector $\mathbf{b}$ which mimimizes ϕ is the eigenvector corresponding to the minimum eigenvalue λ_0 of $\mathbf{R}$, and can be calculated using the 'power method' described in Sec. A.8, Appendix A.

Design Example 3.2.3. Linear Phase Eigenfilters.

Consider an example with bandedges $\omega_p = 0.3\pi, \omega_S = 0.5\pi$, and order $N = 30$. Fig. 3.2-6 shows the magnitude responses of the eigenfilters designed as above, for two values of α. The passband details are also shown separately. It is clear that as α is increased the peak stopband ripple is reduced at the expense of peak passband ripple.

Even though the role of the tradeoff parameter α is very clear, there is no known analytical relation between α and the relative peak ripple sizes δ_1 and δ_2.

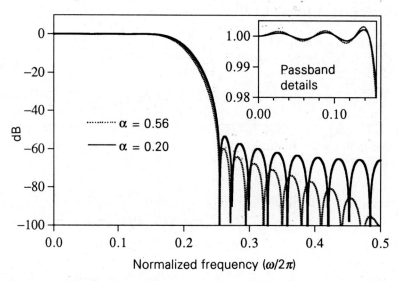

Figure 3.2-6 Design example 3.2.3. Magnitude response plots for linear phase eigenfilters.

Extensions of the eigenfilter approach. The approach is readily extended to the case of other types of filters such as highpass and bandpass filters, differentiators, and Hilbert transformers [Pei and Shyu, 1989]. It is also possible to extend the method to include certain *flatness* constraints in the passband. Finally, extensions to the case of two dimensional filters have also been made. These extensions permit faster design of two dimensional FIR filters than most other methods. See Nashashibi and Charalambous [1988], and Pei and Shyu [1990].

3.2.4 Equiripple FIR Filters

The filters which result from the Kaiser window approach, and the eigenfilter approach are such that the ripple size grows as we move closer to the band edge. Because of this, the filter performace exceeds (i.e., is better than the specifications) for most frequencies except around ω_p and ω_S. So, the filter has actually been overdesinged in this sense. If this can be avoided, it is possible to reduce N while meeting the same set of specifications. The way to achieve this is to distribute the approximation error uniformly in the passband (and also in the stopband). This leads to the idea of equiripple FIR filters (Fig. 3.2-7). Here all the local extrema of the approximation error in the passband are equal. The same is true in the stopband.

For a given set of specifications $\omega_p, \omega_s, \delta_1$, and δ_2 it turns out that an equiripple filter has the smallest possible order N. A more precise statement of optimality is given by the so-called 'alternation theorem' [Rabiner and Gold, 1975]. Based on this theorem, the problem of designing equiripple linear phase filters has been solved, by using a technique called the *Remez exchange algorithm*. The resulting algorithm [Parks and McClellan, 1972], often referred to as the McClellan-Parks algorithm, permits unequal ripple sizes in each of the frequency bands (unlike the window techniques). We skip details here. Suffice it to say that the method is very systematic, and permits one to design a large family of linear phase FIR filters (all four Types in Table 2.4.1.) including differentiators and Hilbert transformers. Special requirements such as time domain constraints and flatness constraints cannot, however, be incorporated in a straightforward manner.

Estimating the filter order. Several formulas have been proposed for estimating the order of a linear phase equiripple lowpass filter with specifications $\omega_p, \omega_S, \delta_1, \delta_2$. The most well known of these are:

$$N = \begin{cases} \dfrac{-20 \log_{10} \sqrt{\delta_1 \delta_2} - 13}{14.6 \Delta f} & \text{(Kaiser's formula)} \\ \dfrac{2 \log_{10}(\frac{1}{10 \delta_1 \delta_2})}{3 \Delta f} & \text{(Bellanger's formula)}, \end{cases} \quad (3.2.32)$$

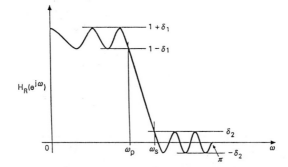

Figure 3.2-7 An equiripple amplitude response.

where $\Delta f = (\omega_S - \omega_p)/2\pi$. Notice the simplicity of these formulas. They also clearly reveal the nature of dependency of N on the ripple sizes and Δf. A more accurate (but not as simple) formula has been reported by Herrmann, and can be found in Rabiner, et al. [1975].

3.2.5 Spectral Factorization

In some filter design problems, one finds it necessary to compute a spectral factor (defined below) of a transfer function. An example is in the design of FIR QMF banks (Chap. 5). We now describe the basic idea.

Let $H(z) = \sum_{n=-M}^{M} h(n) z^{-n}$ be a zero-phase FIR transfer function so that $H(e^{j\omega})$ is real. If in addition $H(e^{j\omega}) \geq 0$ for all ω, we can factorize it as $H(e^{j\omega}) = |H_0(e^{j\omega})|^2$. That is, we can write

$$H(z) = \widetilde{H}_0(z) H_0(z), \quad (3.2.33)$$

where $H_0(z)$ is causal FIR with order M, that is, $H_0(z) = \sum_{n=0}^{M} h_0(n)z^{-n}$. The filter $H_0(z)$ is said to be a spectral factor of $H(z)$. [The notation $\widetilde{H}_0(z)$ is described in Sec. 2.3.]

To see how $H_0(z)$ can be identified, recall that the zero-phase property of $H(z)$ implies that, if z_k is a zero then so is $1/z_k^*$. The property $H(e^{j\omega}) \geq 0$, on the other hand, implies that if z_k is on the unit circle, then it is a zero of even multiplicity (e.g., a double zero). Fig. 3.2-8 shows a typical set of zeros of the function $H(z)$. Once these zeros are known, we can obtain $H_0(z)$ by assigning to it the zero located at either z_k or $1/z_k^*$, for each k. The figure demonstrates this. If z_k is assigned to $H_0(z)$, then $1/z_k^*$ is assigned to $\widetilde{H}_0(z)$. We can write $H_0(z)$ as

$$H_0(z) = c \prod_{k=1}^{M}(1 - z^{-1}z_k), \qquad (3.2.34)$$

so that

$$\widetilde{H}_0(z) = c^* \prod_{k=1}^{M}(1 - zz_k^*). \qquad (3.2.35)$$

Equation (3.2.33) can now be satisfied for appropriate constant c.

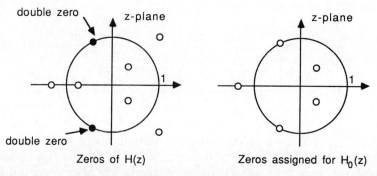

Figure 3.2-8 Obtaining a spectral factor of a transfer function $H(z)$.

Nonuniqueness. The spectral factor $H_0(z)$ is in general not unique because we can replace a particular factor $(1 - z^{-1}z_k)$ in $H_0(z)$ with $(1 - z^{-1}/z_k^*)$ (and readjust c) so that (3.2.33) continues to hold. In other words, if we replace a zero z_k of $H_0(z)$ with $1/z_k^*$, the result continues to be a spectral factor after scaling. If $H(z)$ happens to have all zeros on the unit circle, then the spectral factor is unique (up to a scale factor of unit-magnitude). If we choose all zeros such that they satisfy $|z_k| \leq 1$ (or $|z_k| \geq 1$) then we have a minimum (or maximum) phase spectral factor. Such a spectral factor is unique (up to a scale factor of unit magnitude). If $H(z)$ is free from unit-circle zeros, then these become strictly minimum (or maximum)

58   Chap. 3. Digital filters

phase factors. Finally, note that if the zero-phase function $H(e^{j\omega})$ is an even function of ω, then $h(n)$ is real. In this case we can find $H_0(z)$ with real coefficients; in particular we can find minimum or maximum phase spectral factors with real coefficients.

The most obvious technique to compute a spectral factor is to find the $2M$ zeros of $H(z)$ and pick an appropriate subset of M zeros to define $H_0(z)$. There exist more efficient procedures, which do not compute the zeros of $H(z)$ [Mian and Nainer, 1982], and [Friedlander, 1983]. One such procedure is described in Appendix D.

Application in Nonlinear Phase FIR Design

In some applications linearity of phase is not particularly important, even though FIR filters are still preferred for other reasons. This situation arises, for example, in one dimensional QMF banks and will be discussed in Chap. 5. In general, by relaxing the linear phase property, it is possible to reduce the filter order required for a given set of magnitude response specifications. We now describe a technique [Herrmann and Schüssler, 1970], for designing nonlinear phase FIR filters.

Let $G(z) = \sum_{n=-M}^{M} g(n)z^{-n}$ be a zero phase FIR filter designed using any one of the techniques described above. Let δ_2 denote the peak stopband ripple. Now consider the filter

$$H(z) = G(z) + \delta_2. \qquad (3.2.36)$$

The impulse response of $H(z)$ is given by

$$h(n) = \begin{cases} g(n) & n \neq 0 \\ g(n) + \delta_2 & n = 0. \end{cases} \qquad (3.2.37)$$

The frequency response of $H(z)$ is

$$H(e^{j\omega}) = G(e^{j\omega}) + \delta_2. \qquad (3.2.38)$$

Since $G(z)$ has zero phase, $G(e^{j\omega})$ is real, so that $H(e^{j\omega})$ is obtained just by lifting the response $G(e^{j\omega})$ by δ_2. This is demonstrated in Fig. 3.2-9 for the equiripple case. It is clear that $H(e^{j\omega}) \geq 0$ for all ω, so that we can find a spectral factor $H_0(z)$ of $H(z)$ (as demonstrated in Fig. 3.2-8). In particular if $G(z)$ has real coefficients, then so does $H_0(z)$. The spectral factor $H_0(z)$ in general does not have linear phase. As explained above it is possible to find a minimum or maximum phase (or even a mixed phase) spectral factor.

Suppose we wish to design a minimum phase equiripple FIR filter $H_0(z)$ with bandedges ω_p, ω_S, and peak ripples ϵ_1, and ϵ_2. We then design a zero phase filter $G(z)$ with same bandedges ω_p and ω_S but with ripples as follows:

$$\begin{aligned} \text{peak to peak passband ripple } 2\delta_1 &\approx 1 - (1 - 2\epsilon_1)^2 \\ \text{peak stopband ripple } \delta_2 &= \epsilon_2^2/2. \end{aligned} \qquad (3.2.39)$$

We can now obtain $H(z) = G(z) + \delta_2$ and compute the minimum-phase spectral factor $H_0(z)$.

Notice that the stopband attenuation of $G(z)$ is more than twice that of $H_0(z)$. For example suppose $H_0(z)$ requires a stopband attenuation of 60 dB. Then $H(z)$ has stopband attenuation 120 dB and $G(z)$ has 126 dB. Spectral factorization of systems with such large attenuation is typically subject to considerable numerical error, particularly if the procedure involves the computation of zeros of $H(z)$.

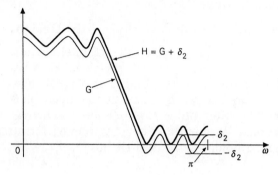

Figure 3.2-9 Lifting the amplitude response of a zero-phase filter $G(z)$ to obtain $H(z)$ with nonnegative amplitude response.

3.3 IIR FILTER DESIGN

The most striking advantage of FIR filters is that they can be designed to have exact linear phase. In situations where linearity of phase is not important, it is sometimes preferable to use IIR filters because an IIR filter usually requires a much lower order for the same set of magnitude response specifications. (See Design Example 3.3.2 later). This implies fewer multipliers and adders.

For various reasons, a comparison of IIR and FIR filters is more involved that the above remark appears to imply. First, there exist techniques (which are perhaps less readily available), for the design of nonlinear phase FIR filters. For a given magnitude response specification, such FIR filters are less expensive than the linear phase versions. Second, there are some commercial signal processing chips, specifically tailored for the implementation of FIR filters. In these chips, the implementation of IIR filters is not necessarily more efficient. Finally, there exist *multistage* design techniques for the design of narrowband FIR filters (Sec. 4.4) which are sometimes more efficient than IIR filters. It is, therefore, difficult to provide a comparison that is fair under all contexts. In this text, we will merely compare the number of multiplications and additons. It should be cautioned that in many cases these *do not* provide a good measure of complexity.

Working Principle of IIR Filters

An IIR filter has transfer function of the form $H(z) = P(z)/D(z)$, where

$P(z)$ and $D(z)$ are polynomials in z^{-1}. The zeros of $P(z)$ are typically located on the unit circle, and therefore, have the form $e^{j\omega_k}$. They can be seen in the magnitude response plots, since $H(e^{j\omega_k}) = 0$. These zeros are there to provide stopband attenuation. Figure 3.3-1(a) shows a typical plot of the numerator $|P(e^{j\omega})|$ with several zeros on the unit circle.

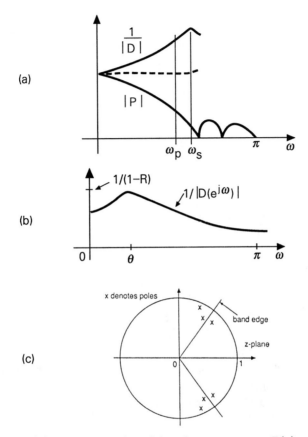

Figure 3.3-1 (a) The roles played by the numerator $P(z)$ and denominator $D(z)$, (b) typical behavior of $1/|D(e^{j\omega})|$ with a single pole, and (c) clustering of poles around the bandedge.

The plot of $|P(e^{j\omega})|$ has the appearance of a lowpass filter, but the passband response is very poor (i.e., not close to unity). The denominator $D(z)$ compensates for this. Figure 3.3-1(a) also indicates a typical response of $1/|D(e^{j\omega})|$, which grows as ω increases in the passband. Thus the magnitude $1/|D(e^{j\omega})|$ is large near the band edge. The product of the two solid curves tends to approximate unity (the broken curve) in the passband.

Figure 3.3-1(b) demonstrates the behavior of $1/|D(e^{j\omega})|$ for the case where $D(z) = 1 - Re^{j\theta}z^{-1}$. Here R and θ are the radius and angle of the pole of $1/D(z)$. We see that the plot has a peak near the pole angle θ. This

Sec. 3.3 IIR Filter design

peak gets steeper as the pole moves closer to the unit circle (i.e., as $R \to 1$). Since $1/|D(e^{j\omega})|$ is required to have large values near the passband edge, the zeros of $D(z)$ (i.e., poles of the filter) are typically crowded near the band edge [Fig. 3.3-1(c)].

Effect of narrow transition-bands. If the transition bandwidth Δf is small, then the quantity $|P(e^{j\omega_p})|$ gets smaller, so that $1/|D(e^{j\omega_p})|$ has to be 'large' in order for the product to be close to unity. For this reason, the zeros of $D(z)$ are placed *closer* to the unit circle for 'sharp cutoff' filters. Summarizing, the poles are typically crowded near the band edge, and for sharp cutoff filters they are also close to the unit circle.

3.3.1 The Bilinear Transformation

The most common technique to design an IIR filter is to first design an analog filter $H_a(s)$ and convert it into a digital filter using a transformation. Suppose we are given an analog filter with rational transfer function $H_a(s)$, having magnitude response $|H_a(j\Omega)|$ as shown in Fig. 3.3-2. This is lowpass with band edges Ω_p and Ω_S, peak passband ripple δ_1 and peak stopband ripple δ_2. (The frequency ∞ is shown at a finite point just for convenience.) Suppose we take the transfer function and replace the Laplace transform variable s as follows:

$$s = \frac{1 - z^{-1}}{1 + z^{-1}} \quad \text{(bilinear transformation)}. \quad (3.3.1)$$

The result is a rational function $H(z)$ in the variable z^{-1}. With $s = j\Omega$ and $z = e^{j\omega}$, we have from (3.3.1)

$$\Omega = \tan(\omega/2), \quad (3.3.2)$$

which shows that the mapping transforms $\Omega = 0$ into $\omega = 0$, and $\Omega = \infty$ into $\omega = \pi$ (Fig. 3.3-3). The above mapping is called the *bilinear transform* and is the most popular technique to convert analog filters into digital. It can be shown that the transformed version $H(z)$ is stable if and only if $H_a(s)$ is stable.

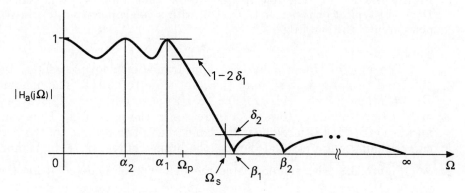

Figure 3.3-2 A typical magnitude response of an analog lowpass filter.

The digital filter response $|H(e^{j\omega})|$ corresponding to the analog response of Fig. 3.3-2 has the appearance shown earlier in Fig. 3.1-1(b). The band-edges ω_p and ω_S are determined by (3.3.2) as

$$\omega_p = 2\arctan\Omega_p, \quad \omega_S = 2\arctan\Omega_S. \qquad (3.3.3)$$

The sizes of the ripples δ_1 and δ_2 are unchanged.

If we are given the digital filter specifications $\omega_p, \omega_S, \delta_1,$ and δ_2, a design procedure based on bilinear transformation would run as follows: (a) find $\Omega_p = \tan(\omega_p/2)$ and $\Omega_S = \tan(\omega_S/2)$, (b) design the analog filter which meets the specifications $\Omega_p, \Omega_S, \delta_1,$ and δ_2, and (c) transform $H_a(s)$ into $H(z)$ using bilinear transformation. It remains only to provide details for the second step, that is, the design of classical analog filters.

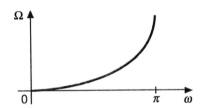

Figure 3.3-3 The frequency mapping property of bilinear transformation.

3.3.2 Analog Filters

The magnitude response for most of the standard analog filters takes the form

$$|H_a(j\Omega)|^2 = \frac{1}{1+F^2(\Omega)} \qquad (3.3.4)$$

where $F(\Omega)$ is a real-valued rational function of Ω. Clearly $|H_a(j\Omega)| \leq 1$. The extreme values of the response are

$$|H_a(j\Omega)| = \begin{cases} 1 & \text{if } F(\Omega) = 0 \\ 0 & \text{if } F(\Omega) = \infty. \end{cases} \qquad (3.3.5)$$

Butterworth (or Maximally Flat) Filters

The simplest and most illuminating example is the Butterworth filter for which $F(\Omega) = (\Omega/\Omega_c)^N$. So, the frequency response has the form

$$|H_a(j\Omega)|^2 = \frac{1}{1+(\Omega/\Omega_c)^{2N}}. \qquad (3.3.6)$$

This is monotone lowpass, and varies from unity (at $\Omega = 0$) to zero (at $\Omega = \infty$). See Fig. 3.3-4. The quantity N is the order of $H_a(s)$. Here is a summary of the main features:

1. We have $|H_a(j\Omega_c)|^2 = 1/2$ which corresponds to an attenuation of 3 dB. So Ω_c is called the *three dB point*. This is not necessarily the passband or stopband edge.
2. For $\Omega \gg \Omega_c$ we have

$$-20 \log_{10} |H_a(j\Omega)| \approx 20N \log_{10} \Omega - 20N \log_{10} \Omega_c. \qquad (3.3.7)$$

This shows that as Ω is increased by one decade (i.e., by a factor of ten) the attenuation increases by $20N$ dB. This is called the $20N$ *dB/decade property* (equivalent to $6.02N$ dB/octave).

3. The first $2N-1$ derivatives of $|H_a(j\Omega)|^2$ are equal to zero at $\Omega = 0$ (see Problem 3.11). This is the maximum possible number of derivatives that can be zero, since $H_a(j\Omega)$ has order N. So, the Butterworth filter is said to be *maximally* flat at $\Omega = 0$.

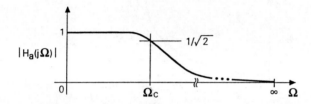

Figure 3.3-4 The magnitude response characteristics of a Butterworth filter.

Expression for the digital transfer function. Since the response is zero only at $\Omega = \pm\infty$, the transfer function $H_a(s)$ has the form

$$H_a(s) = \frac{1}{D_a(s)}, \qquad (3.3.8)$$

where

$$D_a(s) = 1 + d_1 s + \ldots + d_N s^N. \qquad (3.3.9)$$

(assuming $H_a(0) = 1$.) In other words, $H_a(s)$ is an all-pole filter. After bilinear transformation, the digital Butterworth filter therefore has the form

$$H(z) = \frac{c(1+z^{-1})^N}{B(z)}. \qquad (3.3.10)$$

All zeros are now at $z = -1$, that is, at $\omega = \pi$ which corresponds to $\Omega = \infty$. Note that $H(z)$ can be implemented with $2N$ adders and only $N+1$ multipliers (rather than $2N+1$) because of the special form of the numerator.

Location of poles. The N zeros of $D_a(s)$ [poles of the Butterworth filter $H_a(s)$] lie on a circle in the s plane, with center at the origin and radius Ω_c. The pole angles are given by

$$\frac{\pi}{2} + \frac{\pi}{2N} + \frac{k\pi}{N}, \quad 0 \le k \le N-1. \qquad (3.3.11)$$

Given Ω_c and N, one can compute the pole locations as above, and hence the coefficients of $D_a(s)$. The pole locations are demonstrated in Fig. 3.3-5 for $N = 3$. The pole angles are $\pm 2\pi/3$ and π so that, with $\Omega_c = 1$, we have

$$D_a(s) = (s+1)(s - e^{j2\pi/3})(s - e^{-j2\pi/3})$$
$$= (s+1)(s^2 + s + 1) = s^3 + 2s^2 + 2s + 1. \qquad (3.3.12)$$

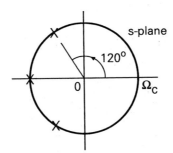

Figure 3.3-5 Pole locations of a third order Butterworth filter.

So, the third order Butterworth lowpass filter with $\Omega_c = 1$ is given by $H_a(s) = 1/(s^3 + 2s^2 + 2s + 1)$.

Note that the transfer function is determined completely by the two parameters N and Ω_c. We have only two degrees of freedom available. In the above example, we see that $D_a(s)$ is a symmetric polynomial; this is true for any N as long as $\Omega_c = 1$.

From the above demonstration, and more generally from (3.3.11) one can see that the poles are in the open left half plane (i.e, Re[s] < 0) so that the filter $H_a(s)$ is always stable.

Design Example 3.3.1: Butterworth Filters

Suppose we wish to design a Butterworth filter with $\Omega_p = 2\pi \times (10\text{kHz})$, $\Omega_S = 2\pi \times (20\text{kHz})$, $A_S = 60$ dB and $A_{max} = 0.1$ dB. This is illustrated in Fig. 3.3-6. Since $A_S = -10\log_{10}|H_a(j\Omega_S)|^2$ and so on, we obtain

$$(\Omega_p/\Omega_c)^{2N} = 10^{A_{max}/10} - 1,$$
$$(\Omega_S/\Omega_c)^{2N} = 10^{A_S/10} - 1. \qquad (3.3.13)$$

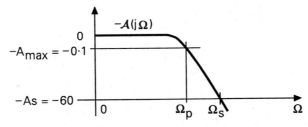

Figure 3.3-6 Specifications in dB for the Butterworth example.

Dividing one equation by the other we eliminate Ω_c and obtain $N = 12.677$. This is the estimated order which should be rounded to the nearest integer, that is, $N = 13$. Since Ω_p, Ω_S, A_S and A_{max} are known we can solve for Ω_c from either equation in (3.3.13). Suppose we use the second equation, then $\Omega_c = \Omega_S/1.701$. The resulting filter has the specified A_S, and the value of A_{max} is slightly better than specified (because N was rounded *up*).

Equiripple Filters

As in the digital FIR case, an analog filter with equiripple response requires smaller order for the same set of ripple sizes and transition bandwidth. A Chebyshev filter, for example, has an equiripple passband. This is obtained by choosing $F(\Omega) = \epsilon C_N(\Omega/\Omega_p)$ in (3.3.4) where $C_N(x)$ is the so-called Nth order Chebyshev polynomial. The transfer function $H_a(s)$ corresponding to this is again all-pole. The filter is optimal in the sense that among all all-pole filters of order N, this filter has the smallest A_{max} for fixed Ω_p, Ω_S, and A_S. We will not discuss Chebyshev filters (or *inverse* Chebyshev filters) further in this text. (However, Problems 3.13 and 3.14 cover some details.)

Elliptic filters

An elliptic filter is an improvement over Chebyshev in the sense that both the passband and stopband are equiripple, as in Fig. 3.3-2. From the figure, we see that there are transmission zeros at finite frequencies, so that $H_a(s)$ is not an all-pole filter. The filter is optimal in the sense that among all rational transfer functions of a given order, elliptic filters have the smallest A_{max} for fixed Ω_p, Ω_S, and A_S.

Design of Elliptic Filters

A simple algorithm for the design of analog elliptic filters is presented in pp. 125 to 128 of Antoniou [1979]. The algorithm designs the coefficients of $H_a(s)$ with magnitude response specified as in Fig. 3.3-2. It is assumed that the bandedges are related as

$$\Omega_p \Omega_S = 1. \qquad (3.3.14)$$

This is called the *frequency normalization condition*. Given the quantities δ_1, δ_2 (equivalently A_{max}, A_S) and $r \triangleq \Omega_p/\Omega_S$, the algorithm first estimates the required order which will meet these specifications. This estimate may turn out to be a noninteger. The coefficients of $H_a(s)$ with nearest integer N (or next higher integer N, if the user prefers it) are then calculated.

The complete procedure to design a digital elliptic filter is as follows: given the specifications $\omega_p, \omega_S, \delta_1$ and δ_2, compute

$$\Omega_p = \tan(\omega_p/2), \quad \Omega_S = \tan(\omega_S/2). \qquad (3.3.15)$$

If these bandedges do not satisfy (3.3.14) then define

$$\Omega'_p = \alpha \Omega_p, \quad \Omega'_S = \alpha \Omega_S, \qquad (3.3.16)$$

where $\alpha = (\Omega_p \Omega_S)^{-0.5} > 0$. This ensures that the frequency normalization $\Omega'_p \Omega'_S = 1$ holds. We can now design the analog elliptic filter $H'_a(s)$ whose specifications are $\Omega'_p, \Omega'_S, A_{max}$ and A_S. If we define $H_a(s) = H'_a(\alpha s)$, then $H_a(s)$ meets the specifications $\Omega_p, \Omega_S, A_{max}$ and A_S. We finally obtain the digital filter as

$$H(z) = H_a(s)\Big|_{s=(1-z^{-1})/(1+z^{-1})}. \qquad (3.3.17)$$

The filter can then be implemented using the direct form or cascade form structure (or, better still, using the structure to be derived in Section 3.6, which has least complexity).

For odd N the elliptic lowpass digital filter has one real pole, and a zero at $z = -1$. The remaining poles are complex conjugate pairs and so are the zeros. Furthermore all zeros are on the unit circle. Based on these facts we can express the transfer function as

$$H(z) = a_0 \left(\frac{1+z^{-1}}{1+bz^{-1}}\right)^{\ell} \prod_{k=1}^{m} \left(\frac{1 - 2\cos\omega_k z^{-1} + z^{-2}}{1 - 2R_k \cos\phi_k z^{-1} + R_k^2 z^{-2}}\right), \qquad (3.3.18)$$

where ω_k are the transmission zeros, ϕ_k are the pole angles and R_k the pole radii for the complex conjugate pairs. (See Chap. 2, discussions around eqs. (2.1.18), and (2.1.19).) The order is $N = 2m + \ell$, where $\ell = 0$ or 1 depending on N.

The numerator of the above $H(z)$ is a symmetric polynomial. The direct form as well as cascade form structures can be implemented with a total of $2N$ additions and about $1.5N$ (rather than $2N + 1$) multipliers because of numerator symmetry.

Design Example 3.3.2: Elliptic Filters

Suppose the digital lowpass filter specifications are $\omega_p = 0.15\pi, \omega_S = 0.20\pi$, $\delta_1 = 0.01$ and $\delta_2 = 0.001$. (This δ_2 implies 60dB attenuation.) By using the above procedure in conjunction with the algorithm in Antoniou [1979], the order of the elliptic filter is estimated as $N = 6.56$, which can be rounded to the integer seven. The resulting digital filter $H(z)$ has magnitude response as shown in Fig. 3.3-7(a), and satisfies the stated specifications. The group delay response is shown in Fig. 3.3-7(b). Since this is not constant, the system has a nonlinear phase response. The group delay shows a variation from about 10 samples to 60 samples.

By using (3.2.32) one can verify that a *linear phase* FIR equiripple filter with same specifications requires an order of 101. The 7th order elliptic IIR filter can be implemented with only 7 multiplications (as we will see in Sec. 3.6) whereas the FIR filter requires 51 multipliers!

Comparison with Butterworth filters. Because of the equiripple nature in both passband and stopband, the elliptic filter requires much smaller order that a Butterworth filter meeting same specifications. In the above design example, a Butterworth filter would require an order of 28.

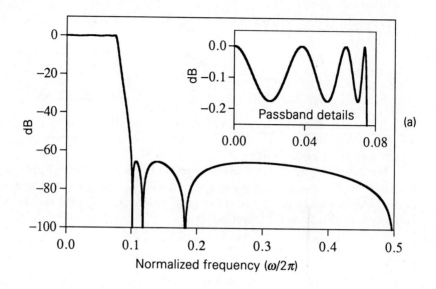

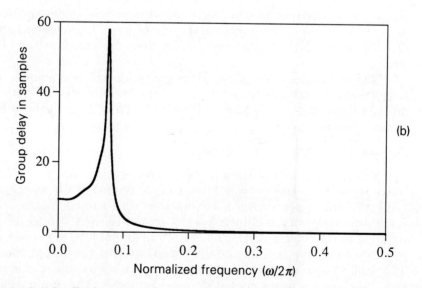

Figure 3.3-7 Design example 3.3.2. Responses of the elliptic filter. (a) Magnitude and (b) group delay.

3.3.3 Properties of Digital Elliptic Filters

Elliptic filters are very important in the design of multirate filter banks, as we will see in Chap. 5. For this reason we now describe their most important features. Since we will be using digital elliptic filters (i.e., bilinearly transformed analog elliptic filters), our discussions will directly address the digital transfer function $H(z)$.

For any real transfer function, we can write $|H(e^{j\omega})|^2$ as $H(z^{-1})H(z)$ (with $z = e^{j\omega}$). This is notationally more convenient. For an elliptic filter, $H(z^{-1})H(z)$ takes the form

$$H(z^{-1})H(z) = \frac{1}{1 + \epsilon^2 R(z)R(z^{-1})}, \qquad (3.3.19)$$

where $R(z)$ is a rational function of the form

$$R(z) = \left(\frac{1 - z^{-1}}{1 + z^{-1}}\right)^\ell \prod_{k=1}^{m} \frac{(1 - z^{-1}e^{j\theta_k})(1 - z^{-1}e^{-j\theta_k})}{(1 - z^{-1}e^{j\omega_k})(1 - z^{-1}e^{-j\omega_k})}. \qquad (3.3.20)$$

The order of $H(z)$ is given by $N = 2m + \ell$, with

$$\ell = \begin{cases} 1 & N \text{ odd} \\ 0 & N \text{ even}. \end{cases} \qquad (3.3.21)$$

If we set $z = e^{j\omega}$ then $R(z)R(z^{-1}) = |R(e^{j\omega})|^2 \geq 0$ so that

$$|H(e^{j\omega})|^2 = \frac{1}{1 + \epsilon^2 |R(e^{j\omega})|^2} \leq 1. \qquad (3.3.22)$$

It is easily verified that

$$|H(e^{j\omega})|^2 = \begin{cases} 0 & \text{for } \omega = \omega_k, \\ 1 & \text{for } \omega = \theta_k. \end{cases} \qquad (3.3.23)$$

This is illustrated in Fig. 3.3-8 for $N = 5$ and $N = 6$. There are precisely N frequencies (in the region $0 \leq \omega < 2\pi$) where $|H(e^{j\omega})| = 1$, and N frequencies where $|H(e^{j\omega})| = 0$. For odd N we have $\ell = 1$ so that $|H(e^{j0})| = 1$ and $|H(e^{j\pi})| = 0$. For even N this is not true. By inspecting plots of the form in Fig. 3.3-8 one can immediately identify the order of the elliptic filter.

The frequencies ω_k, as we know, are the transmission zeros of $H(z)$. The frequencies θ_k, where $|H(e^{j\omega})| = 1$ (maximum magnitude) are called the reflection zeros.[†] The values of θ_k and ω_k are such that the response $|H(e^{j\omega})|$ has equiripple behavior.

The elliptic family. It turns out that for a given ω_p, ω_S, and N, the quantities θ_k and ω_k are fixed. The parameter ϵ in (3.3.19) acts as a tradeoff

---

† This name has to do with doubly terminated LC realizations of electrical filters; the interested reader can see Chap. 12 in Antoniou [1979].

between δ_1 and δ_2 (Fig. 3.3-9). By varying ϵ, one spans a complete family of elliptic transfer functions. Each ϵ corresponds to a unique elliptic filter in the family.

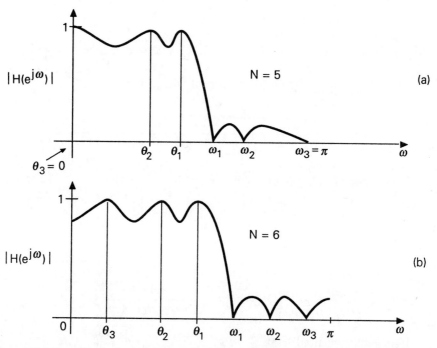

Figure 3.3-8 Typical responses of digital elliptic lowpass filters (a) odd order ($N = 5$), and (b) even order ($N = 6$).

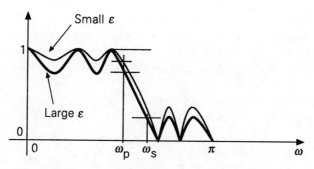

Figure 3.3-9 Two responses belonging to the same elliptic family characterized by w_p, w_S and N. The two curves differ only in terms of ϵ.

Uniqueness. Suppose $G(z)$ is an Nth order stable equiripple lowpass filter with bandedges ω_p, ω_S and the same number of ripples as an Nth order elliptic lowpass filter. Then, $G(z)$ is elliptic and belongs to the elliptic family characterized by N, ω_p, and ω_S.

3.4 ALLPASS FILTERS

Allpass filters play an important role in some multirate applications. Prominent among these is the two channel IIR QMF bank to be discussed in Sec. 5.3. In this section we study the fundamental properties of allpass functions. A tutorial on allpass filters can be found in Regalia, et al. [1988].

Definition and examples. A discrete-time transfer function $H(z)$ is said to be allpass if

$$|H(e^{j\omega})| = c, \quad c > 0, \quad \text{for all } \omega, \tag{3.4.1}$$

that is, the magnitude response is constant. As a result the frequency response has the form

$$H(e^{j\omega}) = ce^{j\phi(\omega)}, \tag{3.4.2}$$

where $\phi(\omega)$ is the phase response. If $|H(e^{j\omega})| = 1$ (i.e., $|c| = 1$) we say that $H(z)$ is unit-magnitude allpass.

Simple examples of allpass functions are: $H(z) = 1$ and $H(z) = z^{-K}$ where K is an integer. A nontrivial example is the first-order filter

$$H(z) = \frac{a^* + z^{-1}}{1 + az^{-1}}. \tag{3.4.3}$$

To verify that this is allpass, rewrite

$$H(z) = z^{-1} \frac{1 + a^* z}{1 + az^{-1}} \tag{3.4.4}$$

so that the frequency response is

$$H(e^{j\omega}) = e^{-j\omega} \left(\frac{1 + a^* e^{j\omega}}{1 + ae^{-j\omega}} \right). \tag{3.4.5}$$

Clearly, $|H(e^{j\omega})| = 1$ for all ω.

More complicated examples can be obtained by multiplying first order filters of the form (3.4.3) because the product of two allpass functions is allpass. The sum of two allpass functions is, in general, not allpass. For example $H_1(z) = 1$ and $H_2(z) = z^{-1}$ are allpass but their sum has magnitude response $2\cos(\omega/2)$ which is not constant.

3.4.1 Properties of Allpass Functions

We restrict attention only to allpass functions which can be expressed as rational functions (though not necessarily with real coefficients). In what follows, we will freely use notations and terms such as "tilde", "dagger", subscript asterik, and "generalized-Hermitian", which are summarized in Sec. 2.3.

It is often convenient to express the property (3.4.1) in terms of the z-transform variable. For this note that (3.4.1) is equivalent to the property

$$\widetilde{H}(z)H(z) = c^2, \quad z = e^{j\omega}. \tag{3.4.6}$$

Invoking analytic continuation (Sec. 2.4.3), we see that this implies

$$\widetilde{H}(z)H(z) = c^2, \quad \text{for all } z. \tag{3.4.7}$$

We can verify this for (3.4.3) as follows:

$$\widetilde{H}(z)H(z) = \left(\frac{a+z}{1+a^*z}\right)\left(\frac{a^*+z^{-1}}{1+az^{-1}}\right) = \left(\frac{1+az^{-1}}{a^*+z^{-1}}\right)\left(\frac{a^*+z^{-1}}{1+az^{-1}}\right) = 1. \tag{3.4.8}$$

The allpass property, expressed in the form (3.4.7), will be frequently used to derive deeper properties.

A. Poles and Zeros of Allpass Functions

The poles and zeros of an allpass function occur in *reciprocal conjugate* pairs. In other words, if α is a pole, then its reciprocal conjugate $1/\alpha^*$ is a zero. This is easily verified for (3.4.3), where the pole $= -a$ and the zero $= -1/a^*$ indeed.

For a general proof, note that (3.4.7) yields

$$\widetilde{H}(\alpha) = \frac{c^2}{H(\alpha)} = \frac{c^2}{\infty} = 0 \quad (\text{since } H(\alpha) = \infty). \tag{3.4.9}$$

In view of the definition of the 'tilde' notation this implies

$$H_*(1/\alpha) = 0. \tag{3.4.10}$$

Conjugating both sides and exploiting the meaning of "subscript asterisk" we see that this in turn implies $H(1/\alpha^*) = 0$. That is, $1/\alpha^*$ is a zero of $H(z)$.

B. Most General Form of Rational Allpass Functions

Suppose $H_N(z)$ is an Nth order rational allpass function with a pole at α_1. This implies that $H_N(z)$ has a zero at $1/\alpha_1^*$ so that $H_N(z)$ has the factor $(-\alpha_1^* + z^{-1})/(1 - \alpha_1 z^{-1})$. This factor is clearly a first order allpass function. We can then write

$$H_N(z) = \left(\frac{-\alpha_1^* + z^{-1}}{1 - \alpha_1 z^{-1}}\right) H_{N-1}(z). \tag{3.4.11}$$

By taking magnitudes on both sides, we see that $H_{N-1}(z)$ is allpass (with order $N-1$). Repeating the above factorization process, we arrive at

$$H_N(z) = \beta \prod_{k=1}^{N} \frac{-\alpha_k^* + z^{-1}}{1 - \alpha_k z^{-1}}, \quad \beta \neq 0, \qquad (3.4.12)$$

where β is a (possibly complex) constant. Summarizing, we have proved that an Nth order allpass function has the general form (3.4.12). Note that if $\alpha_k = 0$ for some k, the corresponding factor reduces to z^{-1}. Thus, the special case $H_N(z) = \beta z^{-N}$ is also covered by the above form.

Most general unfactored form. The form (3.4.12) is induced by the fact that the poles and zeros of an allpass function come in reciprocal conjugate pairs. It is often convenient to write an expression for an allpass function in unfactored form. This can be done by multiplying out the factors in (3.4.12). It can be shown that after such multiplication the result takes the form

$$H(z) = d \frac{b_N^* + b_{N-1}^* z^{-1} + \ldots + b_1^* z^{-(N-1)} + b_0^* z^{-N}}{b_0 + b_1 z^{-1} + \ldots + b_N z^{-N}}, \quad d > 0. \qquad (3.4.13)$$

We have restricted d to be real and positive because b_0 can be arbitrary.

Except for the scale factor d, the numerator coefficients are therefore obtainable by writing the denominator coefficients in reverse order and conjugating them. In other words if $H(z) = A(z)/B(z)$ with

$$A(z) = \sum_{n=0}^{N} a_n z^{-n}, \quad B(z) = \sum_{n=0}^{N} b_n z^{-n}, \qquad (3.4.14)$$

then $a_n = d b_{N-n}^*$. We can express this in the z-domain as

$$A(z) = d z^{-N} \widetilde{B}(z), \qquad (3.4.15)$$

so that (3.4.13) reduces to the form

$$H(z) = d z^{-N} \frac{\widetilde{B}(z)}{B(z)}. \qquad (3.4.16)$$

So any rational allpass filter can be expressed as above.

Conversely, the form (3.4.12) is allpass since each factor is allpass. Similarly any transfer function of the form (3.4.16) is allpass because

$$H(e^{j\omega}) = d e^{-j\omega N} \frac{B^*(e^{j\omega})}{B(e^{j\omega})}, \qquad (3.4.17)$$

which has magnitude d for all ω.

Summarizing, an Nth order rational function $H(z)$ is allpass if and only if it can be expressed as in (3.4.16) for $d > 0$ [or equivalently as in (3.4.12)]. Furthermore, for unit-magnitude allpass functions, we can always take $d = 1$.

C. Energy Balance Property (Losslessness)

Let $u(n)$ and $y(n)$ be the input and output of a stable allpass filter $H(z)$. In view of (3.4.1) we have $|Y(e^{j\omega})| = c|U(e^{j\omega})|$. So,

$$\int_0^{2\pi} |Y(e^{j\omega})|^2 \frac{d\omega}{2\pi} = c^2 \int_0^{2\pi} |U(e^{j\omega})|^2 \frac{d\omega}{2\pi}, \qquad (3.4.18)$$

for any input $u(n)$. By Parseval's theorem this implies

$$\underbrace{\sum_{n=-\infty}^{\infty} |y(n)|^2}_{\text{output energy } E_y} = c^2 \underbrace{\sum_{n=-\infty}^{\infty} |u(n)|^2}_{\text{input energy } E_u}. \qquad (3.4.19)$$

Thus $E_y = c^2 E_u$, that is, the energy-amplification factor c^2 is independent of the input. In particular if $c = 1$ in (3.4.1), then the output energy is equal to the input energy for all possible input sequences. For this reason allpass functions are also called *lossless functions*, (whether $c = 1$ or not).

D. Time Domain Meaning of Allpass Property

In Problem 2.14, we defined the autocorrelation $r(k)$ of a sequence $h(n)$ to be

$$r(k) = \sum_{n=-\infty}^{\infty} h(n)h^*(n-k).$$

From this problem we can conclude that the z-transform of $r(k)$ is given by $R(z) = \widetilde{H}(z)H(z)$. If $H(z)$ is allpass, $\widetilde{H}(z)H(z) = c^2$. This implies that $r(k)$ is an unit pulse function, that is,

$$r(k) = c^2 \delta(k). \qquad (3.4.20)$$

Conversely, if $r(k)$ is an unit pulse, then its z-transform $\widetilde{H}(z)H(z)$ is constant and $H(z)$ is allpass. Summarizing, $H(z)$ is allpass if and only if the autocorrelation of $h(n)$ is a unit pulse.

E. The Modulus Property of Allpass Functions

We now derive a property of causal stable allpass functions based on a well-known theorem in the theory of complex variables [Churchill and Brown, 1984] called the maximum modulus theorem. This property was observed in [Schüssler, 1976].

♠**The maximum modulus theorem.** Let $F(z)$ be a complex function of the complex variable z. Let $F(z)$ be analytic on and inside a closed contour $\mathcal{C}$ in the z-plane (Fig. 3.4-1). Let the maximum value of $|F(z)|$ on the contour $\mathcal{C}$ be denoted F_{max}. Then we have $|F(z)| \leq F_{max}$ for all z inside the contour $\mathcal{C}$. Equality holds somewhere inside the contour *if and only if* $F(z)$ is constant. ◊

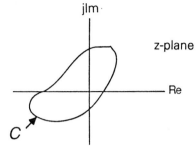

Figure 3.4-1 Pertaining to the maximum modulus theorem.

Now let $H(z)$ be a transfer function with all poles strictly inside the unit circle of the z-plane. Let $|H(e^{j\omega})|$ have maximum value (that is, maximum over all ω) equal to c. By defining $F(z) = H(1/z)$ and invoking the maximum modulus theorem we conclude that $|H(z)| \leq c$ for all z *outside* the unit circle. Equality holds for some z outside the unit circle if and only if $H(z)$ is a constant.

In particular, if the above $H(z)$ is allpass, then more is true. In this case we can also make a claim about the magnitude $|H(z)|$ *inside* the unit circle. For this note that (3.4.7) implies

$$\widetilde{H}(\alpha) = \frac{c^2}{H(\alpha)} \qquad (3.4.21)$$

for any α. Using the fact that $\widetilde{H}(\alpha) = H_*(1/\alpha)$ and conjugating both sides,

$$H(1/\alpha^*) = \frac{c^2}{H^*(\alpha)}. \qquad (3.4.22)$$

For $|\alpha| > 1$ we know $|H(\alpha)| < c$ so that by taking magnitudes we get

$$|H(1/\alpha^*)| > c \qquad (3.4.23)$$

for every α outside the unit circle, that is, $|H(\beta)| > c$ for every β inside the unit circle.

Summarizing we have proved this: if $H(z)$ is a causal stable allpass function with $|H(e^{j\omega})| = c$, then

$$|H(z)| \begin{cases} < c, & \text{for } |z| > 1 \\ > c, & \text{for } |z| < 1 \\ = c, & \text{for } |z| = 1, \end{cases} \qquad (3.4.24)$$

unless $H(z)$ is constant for all z. The example $H(z) = z^{-1}$ provides a simple way to remember the above inequalities.

F. The Monotone Phase-Response Property

Consider the delay function $H(z) = z^{-K}, K > 0$. This is allpass with $H(e^{j\omega}) = e^{-j\omega K}$. The phase response is $\phi(\omega) = -K\omega$, which is a monotone decreasing function spanning a total range of $2\pi K$ as ω increases from 0 to 2π. More generally, let $H(z)$ be any rational Nth order allpass function. If $H(z)$ has all poles inside the unit circle, we will prove that $\phi(\omega)$ is monotone decreasing, and spans a range of $2\pi N$ as ω increases from 0 to 2π.

First order case. First consider $H(z) = (a^* + z^{-1})/(1 + az^{-1})$. The pole is at $z = -a$. Let R and θ represent the radius and angle of the pole so that $a = -Re^{j\theta}$. Then

$$H(e^{j\omega}) = e^{-j\omega}\frac{1 - Re^{j(\omega-\theta)}}{1 - Re^{-j(\omega-\theta)}}. \quad (3.4.25)$$

The phase response $\phi(\omega)$ can be obtained from this as

$$\phi(\omega) = -\omega + 2\tan^{-1}\frac{R\sin(\omega - \theta)}{R\cos(\omega - \theta) - 1}. \quad (3.4.26)$$

Differentiating with respect to ω we arrive at

$$\frac{d\phi(\omega)}{d\omega} = \frac{-(1 - R^2)}{(1 - R)^2 + 2R[1 - \cos(\omega - \theta)]}. \quad (3.4.27)$$

If the pole is inside the unit circle, we have $0 \leq R < 1$. So $d\phi(\omega)/d\omega < 0$, that is, $\phi(\omega)$ is monotone decreasing.

Figure 3.4-2 demonstrates this for $\theta = 0$ (real pole at $-R$). As ω varies from 0 to 2π, the range spanned by the phase is 2π. For arbitrary θ we can simply shift this curve by θ (and add a constant) to obtain $\phi(\omega)$. If the pole is outside the unit circle, then all discussions remain the same except that the phase is monotone *increasing*.

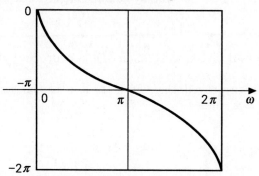

Figure 3.4-2 The monotone phase response of a first order allpass filter ($\theta = 0$).

Since an Nth order stable allpass function is a product of N first order stable allpass functions, its (unwrapped) phase response is the sum of the N individual phase responses, and is thus monotone. The range spanned by $\phi(\omega)$ is the sum of individual ranges, that is, $2\pi N$.

The converse result. Suppose $H(z)$ is a causal Nth order allpass function with monotone decreasing phase response spanning a range of $2\pi N$ as ω varies from 0 to 2π. This is possible only if each of the N first order factors has a monotone decreasing phase response. So, each factor is stable, showing that $H(z)$ is stable.

Some of the above discussions turn out to be conceptually simpler if we think in terms of *continuous time* allpass functions. See Problems 3.17 and 3.18.

3.4.2 Simple Structures for Allpass Filters

Figure 3.4-3 shows the direct form structure for the first order allpass function $H(z) = (a^* + z^{-1})/(1 + az^{-1})$. It has one (complex) delay, two (complex) multipliers and two (complex) adders. For the real coefficient case we have $a = a^*$ and the equivalent structure of Fig. 3.4-4 can be obtained (since a and z^{-1} are interchangeable in this case).

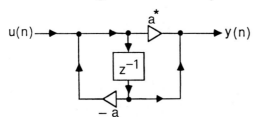

Figure 3.4-3 The direct form structure for a first order allpass function.

An arbitrary Nth order allpass function can be implemented by cascading first order sections (Fig. 3.4-5). For the real-coefficient case we know that poles (and zeros) are either real or occur in complex conjugate pairs so that the allpass function is a product of first order sections of the form

$$H_{1,k}(z) = \frac{a_k + z^{-1}}{1 + a_k z^{-1}}, \quad a_k \text{ real}, \qquad (3.4.28)$$

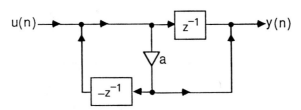

Figure 3.4-4 A one-multiplier implementation of a real coefficient first order allpass function.

second order sections of the form

$$H_{2,k}(z) = \frac{R_k^2 - 2R_k \cos\theta_k z^{-1} + z^{-2}}{1 - 2R_k \cos\theta_k z^{-1} + R_k^2 z^{-2}} \qquad (3.4.29)$$

and a scale factor β.

Figure 3.4-5 Cascade form implementation of an Nth order allpass function. Each section is first-order allpass.

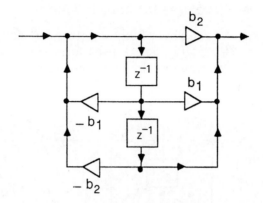

(a) 4 multiplier, 2 delay version

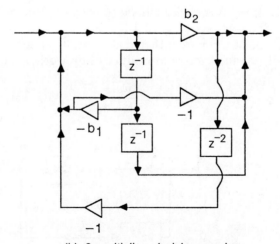

(b) 2 multiplier, 4 delay version

Figure 3.4-6 Direct form structures for second order real coefficient allpass fuctions (a) 4 multiplier, 2 delay version, and (b) 2 multiplier, 4 delay version.

A real-coefficient second order section can be implemented as in Fig. 3.4-6(a) requiring four multipliers, four adders and two delays. A second implementation requiring two multipliers, four adders and four delays is shown in Fig. 3.4-6(b). It is possible to obtain more efficient implementations having the smallest possible number of multipliers and delays. One of these is the one-multiplier lattice structure to be derived in the next section. Several other interesting allpass structures have been derived in Mitra and Hirano [1974], by use of a systematic technique called the multiplier extraction approach. Also see Szczupak, et al. [1988].

3.4.3 Lattice Structures for Allpass Filters

We now derive an allpass structure called the *cascaded lattice structure*, also known as the *Gray and Markel structure* [Gray and Markel, 1973]. Such a structure can be derived for any Nth order stable unit-magnitude allpass filter, and has the property that all multipliers have magnitude less than unity. Its importance lies in the fact that the transfer function remains stable (and allpass) inspite of multiplier quantization, as long as the multiplier magnitudes remain less than unity. The derivation of this structure depends on the following result.

♠**Theorem 3.4.1. The order reduction step.** Let $G_m(z)$ be an mth order causal, stable, unit-magnitude allpass function. Then it can be implemented as in Fig. 3.4-7 where (a) $|k_m| < 1$, and (b) $G_{m-1}(z)$ is a causal, stable, unit-magnitude allpass function with order $m - 1$. ◊

Proof. $G_m(z)$ has the form $G_m(z) = z^{-m}\widetilde{B}_m(z)/B_m(z)$ where

$$B_m(z) = b_{m,0} + b_{m,1}z^{-1} + \ldots + b_{m,m}z^{-m}. \qquad (3.4.30)$$

$B_m(z)$ has all zeros inside the unit circle, since $G_m(z)$ is stable. Now Fig. 3.4-7 implies

$$G_m(z) = \frac{k_m^* + z^{-1}G_{m-1}(z)}{1 + k_m z^{-1}G_{m-1}(z)}. \qquad (3.4.31)$$

Equivalently, by inversion of this, we have

$$z^{-1}G_{m-1}(z) = \frac{G_m(z) - k_m^*}{1 - k_m G_m(z)}, \qquad (3.4.32)$$

that is,

$$z^{-1}G_{m-1}(z) = \frac{z^{-m}\widetilde{B}_m(z) - k_m^* B_m(z)}{B_m(z) - k_m z^{-m}\widetilde{B}_m(z)}. \qquad (3.4.33)$$

Our aim is to show that there exists k_m with $|k_m| < 1$, such that the right-hand side above does indeed have the form $z^{-1}G_{m-1}(z)$, with $G_{m-1}(z)$ having the stated properties. It is clear that k_m must be such that the polynomial $B_m(z) - k_m z^{-m}\widetilde{B}_m(z)$ has order $m - 1$ (so that it can be taken as the denominator of $G_{m-1}(z)$.) By using the definition of *tilde* we see that this polynomial has highest term

$$(b_{m,m} - k_m b_{m,0}^*)z^{-m}, \qquad (3.4.34)$$

so that the only possible choice of k_m is

$$k_m = b_{m,m}/b_{m,0}^*. \qquad (3.4.35)$$

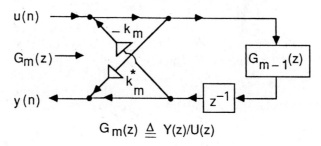

Figure 3.4-7 Generation of an allpass function $G_m(z)$ from a lower order allpass function $G_{m-1}(z)$.

Now the constant term in the numerator polynomial $z^{-m}\widetilde{B}_m(z) - k_m^* B_m(z)$ is $b_{m,m}^* - k_m^* b_{m,0}$ which automatically reduces to zero by the above choice of k_m. As a result, the ratio on the righthand side of (3.4.33) has the form $z^{-1}A_{m-1}(z)/B_{m-1}(z)$, where $A_{m-1}(z)$ and $B_{m-1}(z)$ are polynomials in z^{-1} with order $\leq m-1$. From the relation $z^{-1}A_{m-1}(z) = z^{-m}\widetilde{B}_m(z) - k_m^* B_m(z)$ it is easy to verify that $A_{m-1}(z) = z^{-(m-1)}\widetilde{B}_{m-1}(z)$ so that $G_{m-1}(z) = z^{-(m-1)}\widetilde{B}_{m-1}(z)/B_{m-1}(z)$.

Summarizing, the above choice of k_m ensures that $G_{m-1}(z)$ in (3.4.32) is indeed a causal allpass function, with order $m-1$. [It cannot be less than $m-1$, as $G_m(z)$ in (3.4.31) has order m.] To prove that $|k_m| < 1$, note that the magnitude of the product of all roots of $B_m(z)$ is equal to $|b_{m,m}/b_{m,0}|$. Since all poles of $G_m(z)$ are inside the unit circle, this implies $|k_m| < 1$ indeed.

Next, let α be a pole of $G_{m-1}(z)$. From (3.4.32) we then have $1 - k_m G_m(\alpha) = 0$ so that $|G_m(\alpha)| = 1/|k_m| > 1$. In view of the modulus property (3.4.24), this implies $|\alpha| < 1$ proving that all poles of $G_{m-1}(z)$ are inside the unit circle, that is, $G_{m-1}(z)$ is stable. ▽▽▽

Repeated Application of the Order Reduction Step

We can repeat the order reduction step, and express $G_{m-1}(z)$ in terms of a reduced order allpass function $G_{m-2}(z)$. If we continue this we finally obtain the constant function G_0 with $|G_0| = 1$. This proves that an Nth order unit-magnitude allpass function $G_N(z)$ with all poles inside the unit circle can be implemented with the cascaded lattice structure of Fig. 3.4-8, with $|k_m| < 1$ for all m. The N quantities k_m are called the *lattice coefficients* of $G_N(z)$. All multipliers (k_m and k_m^*) in the structure have magnitude less than unity.

The real-coefficient case. If $G_N(z)$ has real coefficients, then k_N is real [see (3.4.35)], so that $G_{N-1}(z)$ also has real coefficients. So all the lattice coefficients k_m are real. Since G_0 has the form $\widetilde{B}_0/B_0$, we have $G_0 = 1$.

The Lattice Guarantees Stability and Allpass Property

We saw above that any causal, stable, unit magnitude allpass function $G_N(z)$ can be implemented as in Fig. 3.4-8, where $|k_m| < 1$ and $|G_0| = 1$. Conversely, the transfer functions $G_m(z)$ indicated in the figure are stable unit-magnitude allpass filters, as long as $|k_m| < 1$ and $|G_0| = 1$. This can be proved by a minor variation of the above reasonings (Problem 3.19). One consequence of this property is that, stability and allpass property are preserved inspite of quantization of k_m, as long as the quantized multipliers satisfy $|k_m| < 1$ and $|G_0| = 1$.

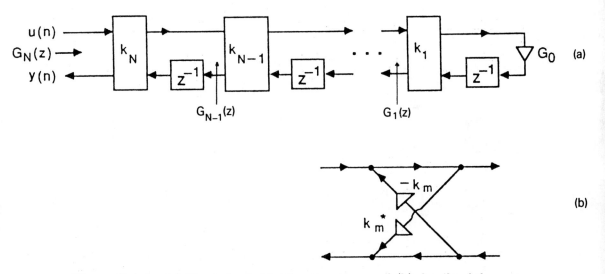

Figure 3.4-8 (a) The cascaded lattice structure, and (b) details of the rectangular boxes labelled k_m.

Variations of the Lattice Structure

Many variations of the lattice structure are known. We now present two of these, which are particularly attractive in practice. To derive these, notice that the structure of Fig. 3.4-7 can be schematically represented as in Fig. 3.4-9, where the quantities $T_{ij}(z)$ are

$$T_{00}(z) = k_m^*, \; T_{01}(z) = \left(1 - |k_m|^2\right)z^{-1}, \; T_{10}(z) = 1, \; T_{11}(z) = -k_m z^{-1}. \tag{3.4.36}$$

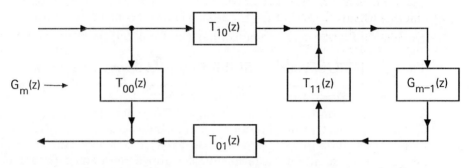

Figure 3.4-9 Schematic redrawing of Fig. 3.4-7.

More generally, for arbitrary $T_{ij}(z)$, the relation between $G_m(z)$ and $G_{m-1}(z)$ is given by

$$G_m(z) = T_{00}(z) + \frac{T_{01}(z)T_{10}(z)G_{m-1}(z)}{1 - T_{11}(z)G_{m-1}(z)}. \tag{3.4.37}$$

Thus, $G_m(z)$ is unchanged if we change $T_{01}(z)$ and $T_{10}(z)$ in such a way that the product $T_{10}(z)T_{01}(z)$ is unchanged. For example if

$$T_{00}(z) = k_m^*, \; T_{01}(z) = \widehat{k}_m z^{-1}, \; T_{10}(z) = \widehat{k}_m, \; T_{11}(z) = -k_m z^{-1}, \tag{3.4.38}$$

where

$$\widehat{k}_m = \sqrt{1 - |k_m|^2}, \tag{3.4.39}$$

then $G_m(z)$ is unchanged, for a given $G_{m-1}(z)$. The resulting lattice section, shown in Fig. 3.4-10, is called the *normalized lattice*. An advantage of this structure is that the internal signals are automatically scaled in a certain sense [Gray and Markel, 1975].

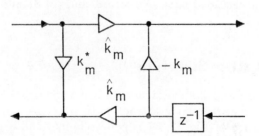

Figure 3.4-10 The normalized lattice section.

For the special case of real coefficient filters, the choice

$$T_{00}(z) = k_m, \ T_{01}(z) = (1+k_m)z^{-1}, \ T_{10}(z) = (1-k_m), \ T_{11}(z) = -k_m z^{-1}, \tag{3.4.40}$$

results in a useful structure, requiring only one multiplier per lattice section! This structure is shown in Fig. 3.4-11, and requires an extra adder. The complete allpass lattice structure, therefore, requires N real multipliers, N delays and $3N$ adders.

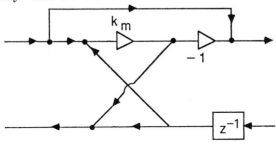

Figure 3.4-11 The one-multiplier lattice section for real coefficient allpass filters.

3.5 SPECIAL TYPES OF FILTERS

We now summarize a number of special transfer functions, that arise frequently in this text.

1. Linear phase transfer functions (Sec. 2.4.2).
2. Allpass transfer functions (Sec. 3.4).
3. *Bounded and BR transfer functions.* If $H(z)$ is stable and such that $|H(e^{j\omega})| \leq 1$, then we say that $H(z)$ is bounded. A bounded transfer function with real coefficients is said to be bounded real (BR).
4. *Lossless transfer functions.* A transfer function is said to be lossless if it is stable and allpass. The name arises from the fact that for such a system the input and output energies are related as $E_y = c^2 E_u$, for all finite-energy inputs. (If $c^2 = 1$, the name "lossless" is particularly appealing, but this condition is not there in the definition.)
5. *LBR transfer functions.* A lossless transfer function with real coefficients is said to be (LBR). So an LBR function is a real-coefficient stable allpass function.
6. *Power complementary transfer functions.* Two transfer functions $H_0(z)$ and $H_1(z)$ are said to be power complementary if

$$|H_0(e^{j\omega})|^2 + |H_1(e^{j\omega})|^2 = c^2 > 0, \quad \text{for all } \omega. \tag{3.5.1a}$$

This can also be rewritten as

$$\widetilde{H}_0(z)H_0(z) + \widetilde{H}_1(z)H_1(z) = c^2 > 0, \tag{3.5.1b}$$

with $z = e^{j\omega}$. Since our filters are always rational functions, this condition holds *for all z* (Sec. 2.4.3). In practice the transfer functions are scaled so that $c^2 = 1$. Thus, if $H_0(z)$ is a good lowpass filter, then $H_1(z)$ is a good highpass filter. As a generalization, a set of M transfer functions $H_k(z)$ is said to be power complementary if

$$\sum_{k=0}^{M-1} |H_k(e^{j\omega})|^2 = c^2 > 0, \quad \text{for all } \omega. \tag{3.5.2}$$

This concept will be used in many chapters.

7. *Mth band or Nyquist(M) filters.* These will be described in Section 4.6.1.

3.6 IIR FILTERS BASED ON TWO ALLPASS FILTERS

3.6.1 The Allpass Decomposition Theorem

A wide family of practical transfer functions including Butterworth, Chebyshev, and elliptic filters can be represented as

$$H_0(z) = \frac{A_0(z) + A_1(z)}{2}$$

where $A_0(z)$ and $A_1(z)$ are stable unit-magnitude allpass filters. This has been observed by a number of authors, for example, Fettweis [1974], Constantinides and Valenzuela [1982], Ansari and Liu [1985], Saramäki [1985], and Vaidyanathan, et al. [1986].

The following special case is particularly noteworthy: Let the transfer function $H_0(z)$ be Butterworth, Chebyshev or elliptic *lowpass*, with order N. Let n_0 and n_1 denote the orders of $A_0(z)$ and $A_1(z)$. Then the following things are true.

1. If N is odd, $A_0(z)$ and $A_1(z)$ have real coefficients, and $N = n_0 + n_1$.
2. If N is even, $A_0(z)$ and $A_1(z)$ have complex coefficients and $n_0 = n_1 = N/2$. In this case, the coefficients of $A_1(z)$ are conjugates of those of $A_0(z)$.

The proof of the first statement (odd N) follows from the theorem to be proved next. In this text, only odd N will be of interest, and will be used in Sec. 5.3 (alias-free IIR QMF banks). See Vaidyanathan et al. [1987] for details of even N, which will not be considered here. Also see Problem 3.20.

The fact that a sum of two allpass filters can give rise to good lowpass behavior might occassion an initial surprise. To appreciate the basic idea, recall that the allpass functions have frequency responses $A_0(e^{j\omega}) = e^{j\phi_0(\omega)}$ and $A_1(e^{j\omega}) = e^{j\phi_1(\omega)}$. Now the behavior of the magnitude of

$$H_0(e^{j\omega}) = \frac{e^{j\phi_0(\omega)} + e^{j\phi_1(\omega)}}{2}. \tag{3.6.1}$$

is governed by the phase difference $\phi_0(\omega) - \phi_1(\omega)$. From Sec. 3.4.1 we know that the phase responses of stable allpass filters are monotone decreasing functions. Figure 3.6-1 shows typical sketches of $\phi_0(\omega)$ and $\phi_1(\omega)$ which will ensure that $H_0(z)$ is a good lowpass filter. In the passband $\phi_0(\omega) \approx \phi_1(\omega)$ so that $|H_0(e^{j\omega})| \approx 1$. In the stopband $\phi_0(\omega) - \phi_1(\omega) \approx \pi$ so that $|H_0(e^{j\omega})| \approx 0$.

Thus, an appropriate behavior of relative phases of the two allpass filters can give rise to a good lowpass response. More generally, we will now state and prove the following result.

♠**Theorem 3.6.1. Allpass decomposition.** Let $H_0(z)$ and $H_1(z)$ be two Nth order bounded real (BR) transfer functions (Sec. 3.5) with irreducible rational forms $H_0(z) = P_0(z)/D(z)$ and $H_1(z) = P_1(z)/D(z)$ where,

$$P_0(z) = \sum_{n=0}^{N} p_{0,n} z^{-n}, \quad P_1(z) = \sum_{n=0}^{N} p_{1,n} z^{-n}, \quad D(z) = \sum_{n=0}^{N} d_n z^{-n}. \quad (3.6.2)$$

Suppose the following conditions are satisfied:
1. $P_0(z)$ is symmetric and $P_1(z)$ antisymmetric, that is,

$$p_{0,n} = p_{0,N-n}, \quad p_{1,n} = -p_{1,N-n}. \quad (3.6.3)$$

2. $H_0(z)$ and $H_1(z)$ are power complementary, satisfying (3.5.1b) with $c = 1$.

Then $H_0(z)$ and $H_1(z)$ can be expressed as

$$H_0(z) = \frac{A_0(z) + A_1(z)}{2}, \quad (3.6.4)$$

$$H_1(z) = \frac{A_0(z) - A_1(z)}{2}, \quad (3.6.5)$$

where $A_0(z)$ and $A_1(z)$ are stable real coefficient allpass functions

$$A_0(z) = \frac{z^{-n_0} \widetilde{D}_0(z)}{D_0(z)}, \quad A_1(z) = \frac{z^{-n_1} \widetilde{D}_1(z)}{D_1(z)}, \quad (3.6.6)$$

with orders n_0 and n_1, respectively. Moreover $N = n_0 + n_1$. ◇

Comments
1. The BR nature of $H_0(z)$ and $H_1(z)$ means that these are stable, that the coefficients $p_{0,n}, p_{1,n}$ and d_n are real and that the magnitudes on the unit circle are bounded by unity.
2. The allpass functions $A_0(z)$ and $A_1(z)$ have unit magnitude on the unit circle and their orders $A_0(z)$ and $A_1(z)$ add up to N.

3. Out of the N poles of $H_0(z)$, a subset of n_0 poles are assigned to $A_0(z)$ and the remaining n_1 poles assigned to $A_1(z)$. This partitioning of the poles of $H_0(z)$ *completely* determines its numerator. The zeros of $H_0(z)$ are, therefore, *not* independent parameters any more. The transfer function has only N degrees of freedom.

4. Figure 3.6-2 indicates a structure which implements the two transfer functions.

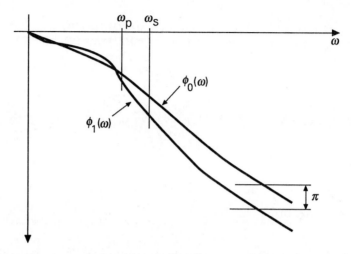

Figure 3.6-1 Demonstrating the phase responses of the two allpass functions.

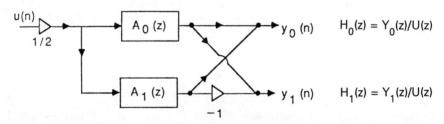

Figure 3.6-2 Implementing two transfer functions by adding and subtracting two allpass filters.

Proof of Theorem 3.6.1. First notice that (3.5.1b) can be rearranged as
$$\widetilde{P}_0(z)P_0(z) + \widetilde{P}_1(z)P_1(z) = \widetilde{D}(z)D(z), \qquad (3.6.7)$$
since $c^2 = 1$. In view of (3.6.3) we have
$$\widetilde{P}_0(z) = z^N P_0(z), \quad \widetilde{P}_1(z) = -z^N P_1(z). \qquad (3.6.8)$$
Substituting into (3.6.7) we obtain $P_0^2(z) - P_1^2(z) = z^{-N}\widetilde{D}(z)D(z)$, which can be rewritten as
$$\bigl(P_0(z) + P_1(z)\bigr)\bigl(P_0(z) - P_1(z)\bigr) = z^{-N}\widetilde{D}(z)D(z). \qquad (3.6.9)$$

Notice that $P_0(z) - P_1(z) = z^{-N}\left(\tilde{P}_0(z) + \tilde{P}_1(z)\right)$ so that the zeros of $P_0(z) - P_1(z)$ are the reciprocal conjugates of those of $P_0(z) + P_1(z)$.

We know that the zeros of $D(z)$ are inside the unit circle so that those of $\tilde{D}(z)$ are outside. So, none of the zeros of $P_0(z) + P_1(z)$ can be on the unit circle (from (3.6.9)). Let n_1 be the number of zeros of $P_0(z) + P_1(z)$ inside the unit circle. Then, there is a factor of $D(z)$, denote it $D_1(z)$, of order n_1 which is also a factor of $P_0(z) + P_1(z)$. Clearly, $P_0(z) + P_1(z)$ has $n_0 \triangleq N - n_1$ zeros outside the unit circle. As seen from (3.6.9) there is then a factor of $\tilde{D}(z)$, say $\tilde{D}_0(z)$, of order n_0 which is also a factor of $P_0(z) + P_1(z)$. Clearly $D_0(z)$ is a n_0th order factor of $D(z)$. Summarizing we can always write

$$P_0(z) + P_1(z) = \alpha D_1(z) z^{-n_0} \tilde{D}_0(z), \quad (3.6.10)$$

where

$$D_0(z) = 1 + \sum_{n=1}^{n_0} d_{0,n} z^{-n}, \quad \text{and} \quad D_1(z) = 1 + \sum_{n=1}^{n_1} d_{1,n} z^{-n}, \quad (3.6.11)$$

are factors of $D(z)$, and α is a real nonzero constant.

As the orders of $D_0(z)$ and $D_1(z)$ add up to the order of $D(z)$, we get

$$D(z) = D_0(z) D_1(z). \quad (3.6.12)$$

By using (3.6.10) and (3.6.12) in (3.6.9) we obtain

$$P_0(z) - P_1(z) = \frac{1}{\alpha} D_0(z) z^{-n_1} \tilde{D}_1(z). \quad (3.6.13)$$

But the symmetry relation (3.6.8) along with (3.6.10) also leads to this equation, with α in place of $1/\alpha$. This implies $\alpha = \pm 1$. We take $\alpha = 1$ (because the other choice $\alpha = -1$ does not change the magnitude responses of $H_0(z)$ and $H_1(z)$ anyway). Dividing both sides of (3.6.10) and (3.6.13) by $D(z)$ we finally arrive at

$$H_0(z) + H_1(z) = A_0(z), \quad H_0(z) - H_1(z) = A_1(z). \quad (3.6.14)$$

Rearranging (3.6.14), we therefore obtain (3.6.4) and (3.6.5). ▽▽▽

3.6.2 Elliptic, Butterworth, and Chebyshev Filters

Figure 3.6-3 shows the typical magnitude response of a fifth-order elliptic lowpass filter $H_0(z) = P_0(z)/D(z)$. The coefficients are known to be real, and the magnitude is bounded by unity so that $H_0(z)$ is BR. We know that all the zeros are on the unit circle. The zero at $\omega = \pi$ contributes to the factor $(1 + z^{-1})$ and the complex conjugate pairs of zeros contribute to factors of the form $(1 - 2z^{-1}\cos\omega_k + z^{-2})$. So, the numerator $P_0(z)$ is indeed a symmetric polynomial.

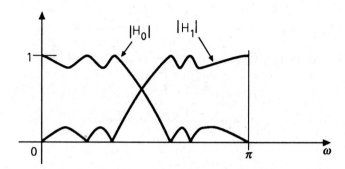

Figure 3.6-3 A fifth order elliptic lowpass filter, and its power complementary response.

The figure also shows the magnitude of the power complementary filter $H_1(z)$. Clearly, $|H_1(e^{j\omega})|$ is equal to zero at frequencies where, $|H_0(e^{j\omega})|$ takes the maximum value of unity. $|H_1(e^{j\omega})|$ has one zero at $\omega = 0$ and two complex conjugate pairs of zeros on the unit circle so that all the zeros are on the unit circle again. The zero at $\omega = 0$, however, contributes to an *antisymmetric factor* $(1 - z^{-1})$. As a result, the numerator $P_1(z)$ of $H_1(z)$ is antisymmetric. Summarizing, $H_0(z)$ has a symmetric numerator and $H_1(z)$ has an antisymmetric numerator.

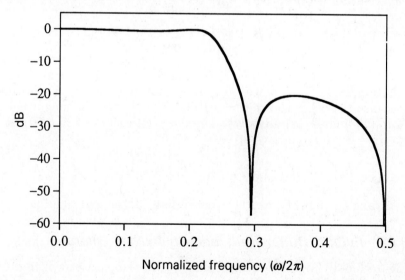

Figure 3.6-4 Example 3.6.1. Magnitude response of a 3rd order elliptic filter.

More generally, if $H_0(z)$ represents an *odd order lowpass* Butterworth, Chebyshev or elliptic filter, the above conclusions remain valid. That is, the numerator of $H_0(z)$ is symmetric and that of $H_1(z)$ is antisymmetric. We can, therefore, apply Theorem 3.6.1 to conclude that $H_0(z)$ and $H_1(z)$ can be expressed as in (3.6.4) and (3.6.5).

Example 3.6.1

Consider the third order elliptic lowpass filter

$$H_0(z) = \frac{0.23179 + 0.36021z^{-1} + 0.36021z^{-2} + 0.23179z^{-3}}{1 - 0.38409z^{-1} + 0.70390z^{-2} - 0.13581z^{-3}} \quad (3.6.15)$$

whose magnitude response is shown in Fig. 3.6-4. The reader can verify that $H_0(z)$ can be expressed as

$$H_0(z) = 0.5 \left[\underbrace{\frac{-0.20356 + z^{-1}}{1 - 0.20356z^{-1}}}_{A_0(z)} + \underbrace{\frac{0.66715 - 0.18053z^{-1} + z^{-2}}{1 - 0.18053z^{-1} + 0.66715z^{-2}}}_{A_1(z)} \right].$$

Evidently $A_0(z)$ and $A_1(z)$ indicated above are unit-magnitude allpass.

Efficiency of the Allpass Based Structure

The cost of the implementation of Fig. 3.6-2 (say an elliptic filter) is equal to the cost of the two allpass filters plus the two adders. We know from Sec. 3.4.3 that a real coefficient allpass filter of order n_k can be implemented with n_k multipliers. So, the structure requires only $n_0 + n_1 = N$ multipliers. For this cost, we get two filters $H_0(z)$ and $H_1(z)$, that is, we require $N/2$ multipliers per filter! In contrast, a direct form implementation of a single elliptic filter would require as many as $1.5N$ multipliers (even after taking numerator symmetry into account)!

The Pole Interlace Property

Given an odd order elliptic transfer function $H_0(z) = P_0(z)/D(z)$, what is the procedure to identify the allpass functions $A_0(z)$ and $A_1(z)$? One method would be to identify $P_1(z)$ using (3.6.7), and compute the zeros of $P_0(z) + P_1(z)$. The zeros inside the unit circle determine $D_1(z)$, and those outside are used to determine $D_0(z)$. The allpass functions can then be found from (3.6.6).

There exists a simpler procedure, whenever the zeros of $D(z)$ [poles of $H_0(z)$] are known. Let the poles of $H_0(z)$ be $z_0, z_1, \ldots$, with pole angles $\theta_0, \theta_1, \ldots$ Let the numbering of poles be such that $\theta_0 < \theta_1 < \ldots$. Then the poles of $A_0(z)$ are given by z_{2k} and those of $A_1(z)$ by z_{2k+1}. This is called the *pole interlace property* [Gazsi, 1985]. Using this we can identify the allpass functions as demonstrated in Fig. 3.6-5.

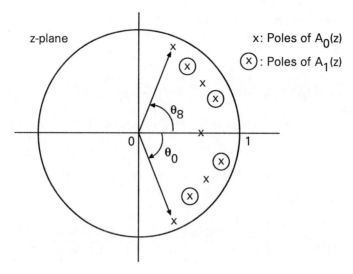

Figure 3.6-5 Demonstration of interlace property. The nine poles of $H_0(z)$ are split into those of $A_0(z)$ and $A_1(z)$ as indicated.

Case When N is Even

What happens if the filter has even order? Consider a sixth order elliptic lowpass filter $H_0(z) = P_0(z)/D(z)$ with response as shown in Fig. 3.6-6. In the region $0 \le \omega \le \pi$, there are three zeros. Thus we have three complex conjugate pairs of zeros, giving rise to three factors of the form $(1 - 2z^{-1}\cos\omega_k + z^{-2})$ for the numerator $P_0(z)$. This numerator, therefore, is symmetric.

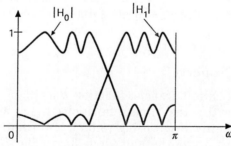

Figure 3.6-6 A sixth order elliptic lowpass filter and its power complementary response.

Now consider the power complementary response $|H_1(e^{j\omega})|$ which is also shown in the figure. This is zero whenever $|H_0(e^{j\omega})|$ is unity. Since $|H_0(e^{j\omega})|$ does *not* have a maximum at $\omega = 0$, we conclude that $|H_1(e^{j\omega})| \ne 0$ at $\omega = 0$. So the numerator $P_1(z)$ of $H_1(z)$ does not have the factor $(1 - z^{-1})$. In fact $P_1(z)$ has three factors of the form $(1 - 2z^{-1}\cos\theta_k + z^{-2})$ because it also has three pairs of complex conjugate zeros on the unit circle. As a result $P_1(z)$ is symmetric rather than antisymmetric. More generally whenever $H_0(z)$ is a Butterworth, Chebyshev or elliptic lowpass filter of even order, the above conclusion remains true. That is, the numerators of $H_0(z)$ and $H_1(z)$ are both symmetric. So, the conditions of Theorem 3.6.1 are not satisfied.

In this case it can be shown [Vaidyanathan, et. al., 1987] that we can still express $H_0(z)$ as $0.5[A_0(z)+A_1(z)]$, where $A_0(z)$ and $A_1(z)$ are complex-coefficient allpass filters, and the coefficients of $A_1(z)$ are conjugates of those of $A_0(z)$. Finally, note that if $H_0(z)$ is bandpass or bandstop, then it can often be implemented in terms of real allpass filters, even if its order is even. (Example: start from an odd order elliptic lowpass filter and replace z with z^2 or $-z^2$.)

TABLE 3.7.1 Comparison of four techniques for lowpass filter design. The specifications are $\omega_p = 0.15\pi$, $\omega_S = 0.20\pi$, $\delta_1 = 0.01$ and $\delta_2 = 0.001$.

Method	IIR elliptic	IIR Butterworth	FIR equiripple	FIR Kaiser window
Special features	Optimal in minimax sense	Maximally flat at $\omega = 0, \pi$	Linear phase. Also optimal in minimax sense	Linear phase. Very easy to design
Required order N	7	28	101	146
Complexity of Implementation	11 mul* 14 add (direct form)	28 mul, 56 add (direct form)	51 mul, 101 add	74 mul, 146 add

* 7 mul and 22 add, if the allpass based structure is used (with one-multiplier lattice sections).

3.7 CONCLUDING REMARKS

In Sec. 3.1 to 3.3 we reviewed many techniques for digital filter design. A summary and comparison of many of the earlier methods can be found in Rabiner and Gold [1975]. In Table 3.7.1 we have compared the filter orders and computational complexities of several methods, for a given set of specifications on the magnitude response. It is clear that the IIR elliptic design is the least expensive, but it introduces phase distortion. The FIR filters, on the other hand, have exact linear phase, but are more expensive. As explained at the beginning of Sec. 3.3, the complexity in terms of multiplications and additions is not always a fair measure of comparison. One should take into account the architecture of the implementation and, if possible, use more efficient FIR implementations (e.g., multistage implementations, Sec. 4.4).

The following Chapters will show that in the context of multirate signal processing, some of the methods we described are particularly suitable, for example, window techniques, eigenfilter techniques and IIR elliptic designs. For this reason, we have elaborated them in this chapter. We will see later (Sec. 5.3) that IIR filters based on a sum of two allpass functions (Sec. 3.6) are particulary useful in filter bank designs.

PROBLEMS

3.1. Let $H(z) = \sum_{n=0}^{N} h(n)z^{-n}$ be a Type 1 linear phase FIR filter and let $G(z) = \sum_{n=0}^{N} g(n)z^{-n}$ with

$$g(n) = (-1)^M \delta(n - M) - (-1)^n h(n), \quad M = N/2. \qquad (P3.1)$$

Prove that $G(z)$ is also a linear phase filter. Assuming that the amplitude response of $H(z)$ is as in Fig. 3.2-7, give a qualitative plot of the amplitude response of $G(z)$. Clearly indicate the bandedges and ripple sizes in terms of the known quantities $\omega_p, \omega_S, \delta_1, \delta_2$. What sort of filter is $G(z)$ (i.e., lowpass or highpass or …)?

3.2. Let $G(z)$ be an ideal transfer function such that $G(e^{j\omega}) = 1$ for $0 \leq |\omega| \leq \pi/4$ and zero elsewhere.

a) Consider the new system $H(z) = G(z^2)$. Plot $|H(e^{j\omega})|$ for $0 \leq \omega \leq \pi$.
b) Consider the system constructed according to the following flowgraph:

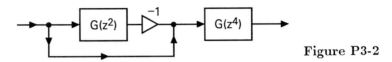

Figure P3-2

Plot the magnitude response of this new system for $0 \leq \omega \leq \pi$. What kind of a filter is this (i.e., lowpass or highpass or …)?

3.3. Let $H(z)$ and $\widehat{H}(z)$ be two lowpass filters. Let δ_1 and δ_2 be the peak passband and stopband ripples for $H(z)$, and $\widehat{\delta}_1$ and $\widehat{\delta}_2$ the corresponding ripples for $\widehat{H}(z)$. Assume that all these ripples are very small compared to unity. The cascaded filter $H(z)\widehat{H}(z)$ is clearly lowpass. Show that its peak passband ripple $\leq \delta_1 + \widehat{\delta}_1$, and peak stopband ripple $\leq \max(\delta_2, \widehat{\delta}_2)$.

3.4. Consider a zero phase FIR lowpass filter $H(z)$ with the frequency response shown in Fig. P3-4(a). Here δ_1 and δ_2 represent the peak ripple sizes. Our aim is to generate a better filter by making multiple use of the filter $H(z)$. In Fig. P3-4(b)–(d) we have shown three structures which attempt to do this.

a) In each case give a qualitative plot of the amplitude response and verify that the resulting filter continues to be lowpass with (nearly) same bandedges as $H(z)$.
b) In each case find the peak to peak passband and stopband ripple. Assuming $\delta_1 = 0.0025$ and $\delta_2 = 0.002$, compute all these ripple sizes, and present them in the form of a neat table.
c) If you wish to design a filter which is better than $H(z)$ in the passband, which of the three methods would you choose?
d) If you wish to design a filter which is better than $H(z)$ in the stopband, which of the three methods would you choose?
e) If you wish to design a filter which is better than $H(z)$ in the passband *and* stopband, which of the three methods would you choose?

f) Show how these three structures should be modified if $H(z)$ is not zero-phase, but Type 1 linear-phase with order N (with Fig. P3-4(a) representing the amplitude response).

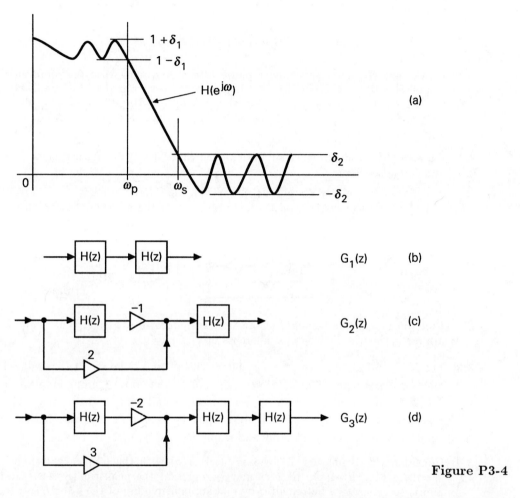

Figure P3-4

Note. You can assume that δ_1 and δ_2 are 'sufficiently small' in order to make *wise* engineering simplification of your expressions.

3.5. Let

$$F(\omega) = \sum_{k=-\infty}^{\infty} f(k)\Phi_k(\omega), \qquad (P3.5a)$$

where $\Phi_k(\omega)$ is a set of orthonormal functions in the range $a \leq \omega < b$, that is, $\int_a^b \Phi_k(\omega)\Phi_m^*(\omega)d\omega = \delta(k-m)$. In other words, $F(\omega)$ is a linear combination of an orthonormal set of basis functions $\Phi_k(\omega)$ in the interval $a \leq \omega < b$. The most common example is when

$$\Phi_k(\omega) = e^{jk\omega}/\sqrt{2\pi}, \quad a = 0, \quad b = 2\pi. \qquad (P3.5b)$$

In this case, the above summation reduces to the familiar Fourier transform of the sequence $f(k)$.

a) Suppose we wish to approximate $F(\omega)$ with the finite summation

$$F_M(\omega) = \sum_{k=-M}^{M} \widehat{f}(k)\Phi_k(\omega) \qquad (P3.5c)$$

in the region $a \leq \omega < b$. We wish the approximation to be "best in the least squares sense", that is, $e \triangleq \int_a^b |F(\omega) - F_M(\omega)|^2 d\omega$ must be minimized. Show that the choice $\widehat{f}(k) = f(k), -M \leq k \leq M$ achieves this.

b) With $\widehat{f}(k) = f(k)$, what is the minimized error e? Simplify as best as you can.

c) Consider the window design procedure for zero-phase FIR lowpass filters. Show that, if we use the rectangular window, the resulting filter is optimal in the least square sense.

3.6. In Sec. 3.2.2 we stated that the matrix **P** in the real-coefficient optimal window design problem is real, symmetric, positive definite and Toeplitz, with a unique eigenvector (up to scale) for each eigenvalue. In this problem we use some or all of these properties to derive two useful conclusions.

a) Prove that the optimal window is symmetric, that is, $v(n) = v(N - n)$. (*Note:* You can ignore the possibility of an antisymmetric window, as that would imply a zero at $\omega = 0$, which in turn conflicts the energy maximization requirement.)

b) Prove that the z-transform $V(z)$ of the optimal window $v(n)$ has all zeros on the unit circle.

Note: Part (b) was stated in text without proof. Evidently (b) implies (a). However, an independent proof of (a) is easier than (b).

3.7. Shown in the following figure is an example of a typical bandpass response.

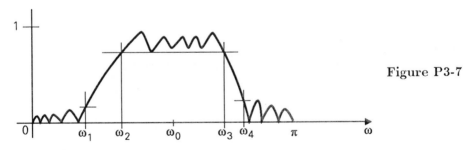

Figure P3-7

Here ω_0 is the center (or reference) frequency and $\omega_1, \omega_2, \omega_3, \omega_4$ represent the bandedges. In Sec. 3.2.3 we described how to design lowpass eigenfilters, which are optimal in the sense of minimizing the stopband and passband errors in a certain least square sense. Describe how this can be extended for the design of bandpass filters of the above form. Assume the filter to be Type 1 linear phase FIR for simplicity.

3.8. Consider the method of Sec. 3.2.3 for design of lowpass eigenfilters. It is clear from the definition of E_S that it cannot be zero as long as $\omega_S \neq \pi$. Similarly E_p cannot be negative. Based on these physical considerations prove that the Hermitian matrices $\mathbf{P}$ and $\mathbf{R}$ are positive definite (as long as α is restricted to $0 < \alpha < 1$).

3.9. Let $H(z)$ be a Type 3 linear phase FIR filter, with passband in the range $0 < \omega_1 \leq \omega \leq \omega_2 < \pi$. Suppose $x(n)$ is a real signal with no energy outside the passband of $H(z)$. It is obvious that the output $y(n)$ is real because $h(n)$ and $x(n)$ are real. However, we also know that $H(e^{j\omega}) = ce^{-j\omega N/2} H_R(\omega)$, where c is a *complex* constant ($c = j$). Assuming that $H(e^{j\omega})$ has a "good passband", that is, very small passband ripple, do you still think that $y(n)$ has the form $y(n) \approx \alpha x(n - M)$ where α is a real constant? (*Hint:* Try $x(n) = \cos(\omega_0 n + \theta)$ where ω_0 belongs to the passband and θ is real.)

3.10. Find the coefficients of a second order digital lowpass Butterworth filter with 3 dB point at $\omega = 0.2\pi$.

3.11. Consider the Butterworth response (3.3.6). Show that the first $2N - 1$ derivatives are zero at $\Omega = 0$.

3.12. Suppose we wish to transform an analog filter $H_a(s)$ into digital filter $H(z)$. Assume that the following transformation has been used: $s = 1 - z^{-1}$. This is called the backward difference approach. (The motivation for this substitution is that s represents differentiation and $1 - z^{-1}$ represents a first difference.)
 a) Suppose $H_a(s)$ has all poles in $\text{Re}[s] < 0$. Does it necessarily mean that $H(z)$ has all poles inside the unit circle?
 b) Suppose $H(z)$ has all poles inside the unit circle. Does it mean that $H_a(s)$ has all poles in $\text{Re}[s] < 0$?
 c) Instead of the above mapping assume that we use the mapping $s = z - 1$ (forward difference approach). Repeat parts (a) and (b).

3.13. In this problem we shall give an overview of Chebyshev polynomials. Recall that the hyperbolic cosine function is defined as $\cosh \theta = (e^\theta + e^{-\theta})/2$. Let us denote this as x, that is,

$$\cosh \theta = x = \frac{e^\theta + e^{-\theta}}{2}. \quad (P3.13a)$$

Here θ could be real or complex. If x is real then either θ is real or e^θ is the conjugate of $e^{-\theta}$.
 a) For real θ show that $x \geq 1$. Also justify that $-1 \leq x \leq 1$ if, and only if, $\theta = j\omega$ where ω is real.
 b) Show that

$$\cosh((N \pm 1)\theta) = \cosh(N\theta)\cosh\theta \pm \sinh(N\theta)\sinh\theta. \quad (P3.13b)$$

Hence, prove the recursion

$$C_{N+1}(x) = 2xC_N(x) - C_{N-1}(x) \quad (P3.13c)$$

where $C_N(x) = \cosh(N\theta) = \cosh(N\cosh^{-1} x)$. For example $C_1(x) = \cosh\theta = x$.

c) Evidently, $C_0(x) = 1$ and $C_1(x) = x$. Prove by use of the recursion (P3.13c) that $C_N(x)$ is a polynomial for any x. $C_N(x)$ is called the Nth order Chebyshev polynomial. Give qualitative plots of $C_N(x)$ for $N = 0, 1, 2, 3, 4$.

d) Prove that $C_N(x)$ is an even polynomial (i.e., has only even powers of x) for even N, and odd polynomial for odd N. Also show that $C_N(1) = 1$ and that the highest power x^N has coefficient 2^{N-1} for $N \geq 1$.

e) Show that all the N zeros of $C_N(x)$ are real and lie in the range $-1 < x < 1$. (Hint. $C_N(x) = \cosh(N\theta) = \cos(N\omega)$ for $\theta = j\omega$, ω real.) So, in the region $|x| > 1$, the behavior is monotone as demonstrated below.

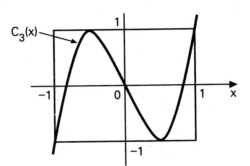

Figure P3-13

f) Let $C_N(x) = c_N(0) + c_N(1)x + \ldots c_N(N)x^N$. Let $P(x) = p(0) + p(1)x + \ldots + p(N)x^N$ be some real coefficient polynomial with $p(N) = c_N(N)$. Prove that

$$\max_{-1 \leq x \leq 1} |P(x)|^2 \geq \max_{-1 \leq x \leq 1} |C_N(x)|^2. \qquad (P3.13d)$$

This shows that among all polynomials of order N with highest coefficient equal to that of $C_N(x)$, the Chebyshev polynomial has the smallest peak value in $-1 \leq x \leq 1$. So, the polynomial has the minimax property (i.e., *max*imum magnitude in $-1 \leq x \leq 1$ is *min*imized).

3.14. Consider the response

$$|H_a(j\Omega)|^2 = \frac{1}{1 + \epsilon^2 C_N^2(\Omega/\Omega_p)}, \qquad (P3.14a)$$

where $\Omega_p > 0$. This is called the *Chebyshev response* and a stable tranfer function $H_a(s)$ with this response is called a *Chebyshev filter*.

a) Justify that the magnitude response has the behavior shown in Fig. P3-14, when $N = 7$. The quantity Ω_p is the passband edge, and ϵ directly controls the passband ripple size. The passband is equiripple. All the $2N$ zeros of (P3.14a) are at $\Omega = \infty$.

b) By making wise engineering assumptions show that the required order N (for a given set of specifications) can be estimated from

$$N = \frac{A_S + 20 \log_{10}(1/\epsilon) + 6.02}{8.686 \cosh^{-1}(\Omega_S/\Omega_p)}. \qquad (P3.14b)$$

c) Plot the response $|G_a(j\Omega)|^2 = 1 - |H_a(j/\Omega)|^2$. This should be lowpass with equiripple stopband and monotone passband. A stable transfer function $G_a(s)$ with this behavior is called an *inverse Chebyshev filter*.

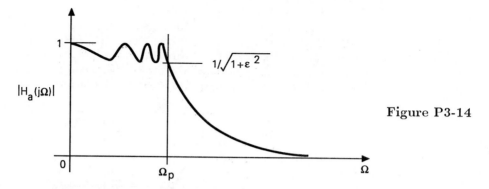

Figure P3-14

d) Suppose $H(z)$ is obtained by use of bilinear transformation on $H_a(s)$. Specify any special feature that the numerator of $H(z)$ might have.

3.15. For an analog elliptic lowpass filter $H_a(s)$, assume that the bandedges are related as $\Omega_p \Omega_S = 1$. Let $G_a(s)$ be a stable filter such that $|G_a(j\Omega)|^2 = 1 - |H_a(j/\Omega)|^2$.
   a) Qualitatively plot the responses $|H_a(j\Omega)|^2$ and $|G_a(j\Omega)|^2$ for $N = 5$.
   b) Give a simple argument to justify that the reflection and transmission zeros (α_k's and β_k's in Fig. 3.3-2) satisfy $\alpha_k = 1/\beta_k$.

3.16. Suppose we wish to design a digital lowpass filter with specifications $\omega_p, \omega_S, \delta_1$ and δ_2 as in Table 3.7.1. Verify that the required order N is as in the table, for the two FIR cases and for the IIR Butterworth case.

3.17. In the continuous-time world a rational transfer function $G(s)$ of degree $N > 0$ with real coefficients is said to be a reactance if it satisfies the following two properties: (a) $\text{Re}[G(j\Omega)] = 0$ for all frequencies Ω and (b) $\text{Re}[G(s)] > 0$ for all s in the right half plane, that is, for all s such that $\text{Re}[s] > 0$. Now define a discrete-time transfer function $H(z)$ as follows:

$$H(z) = \left. \frac{1 - G(s)}{1 + G(s)} \right|_{s = (1-z^{-1})/(1+z^{-1})} \quad (P3.17)$$

Show then that $H(z)$ represents an allpass function with all poles strictly inside the unit circle. (*Note.* It turns out that $G(s)$ is a reactance if and only if it is the input impedance of a lossless electrical network (LC network with positive elements). This establishes the link between digital allpass functions and continuous-time LC networks.)

3.18. Allpass functions have played an important role in continuous-time filter theory also. Consider a causal system with transfer function

$$H(s) = \frac{-s - a^*}{s - a}. \quad (P3.18)$$

Here a and $-a^*$ are the pole and zero respectively.
a) Prove that this is allpass, that is, $|H(j\Omega)| = 1$.
b) Assuming $\text{Re}[a] < 0$ (that is, $H(s)$ stable) prove by explicitly writing down $|H(s)|$ that $|H(s)| < 1$ for $\text{Re}[s] > 0$ and $|H(s)| > 1$ for $\text{Re}[s] < 0$.
c) Consider the following pole-zero diagram for $H(s)$.

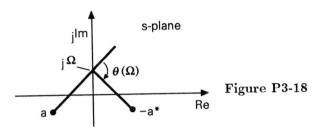

Figure P3-18

Show that the phase response at frequency Ω is the angle $\theta(\Omega)$ indicated in the figure. This shows, essentially by inspection, that this is a monotone decreasing function of Ω.

An Nth order unit-magnitude allpass function is a product of N first order functions of the form (P3.18). So if all the poles are in the left half plane, properties (b) and (c) continue to hold. Since any discrete-time stable allpass function can be derived from a continuous-time counterpart using bilinear transform, this gives a second proof of these same properties for the discrete-time case.

3.19. Consider Fig. 3.4-7, and assume $|k_m| < 1$. Assume $G_{m-1}(z)$ has all poles inside the unit circle, and let $|G_{m-1}(e^{j\omega})| = 1$ for all ω.
  a) Show that $|G_m(e^{j\omega})| = 1$ for all ω, and that all its poles are inside the unit circle.
  b) Consider the lattice structure of Fig. 3.4-8. Let $|G_0| = 1$, and $|k_m| < 1$ for all m. Show that the transfer function $G_N(z)$ has all poles inside the unit circle, and that $|G_N(e^{j\omega})| = 1$.

3.20. *Generalization of allpass decomposition.* Let $H_0(z) = P_0(z)/D(z)$ and $H_1(z) = P_1(z)/D(z)$ be two stable transfer functions (with possibly complex coefficients) of order N with

$$P_0(z) = \sum_{n=0}^{N} p_{0,n} z^{-n}, \quad P_1(z) = \sum_{n=0}^{N} p_{1,n} z^{-n}, \quad D(z) = 1 + \sum_{n=1}^{N} d_n z^{-n}.$$

Assume that the following properties are true. (a) $P_0(z)$ is Hermitian, (b) $P_1(z)$ is generalized-Hermitian (i.e., $\widetilde{P}_1(z) = cz^N P_1(z)$ for some c with $|c| = 1$) and (c) $|H_0(e^{j\omega})|^2 + |H_1(e^{j\omega})|^2 = 1$, (i.e., power complementarity). Prove that $H_0(z)$ and $H_1(z)$ can be expressed as

$$H_0(z) = \frac{\beta A_0(z) + \beta^* A_1(z)}{2}, \quad H_1(z) = d\frac{\beta A_0(z) - \beta^* A_1(z)}{2},$$

where $A_0(z)$ and $A_1(z)$ are stable unit-magnitude allpass of orders n_0 and n_1 with $n_0 + n_1 = N$ and where $|\beta| = |d| = 1$.

3.21. Let $G(z) = A_0(z) + A_1(z)$, where $A_0(z)$ and $A_1(z)$ are allpass. Show that $G(z)$ is allpass if, and only if, $A_1(z) = cA_0(z)$ for some constant c.

4

Fundamentals of Multirate Systems

4.0 INTRODUCTION

This chapter is basic to the study of multirate systems and filter banks. Section 4.1 introduces decimation, interpolation, and filter bank systems, and Sec. 4.2 discusses interconnections of building blocks. The polyphase decomposition is introduced in Sec. 4.3, along with some applications. Multistage filter design is discussed in Sec. 4.4. Several applications of multirate systems are described in Sec. 4.5. Many special types of filters such as halfband filters and Nyquist filters, and complementary filter banks are discussed in Sec. 4.6. Finally, Sec. 4.7 introduces *multigrid techniques* which are well known in the literature on numerical computation.

Some of these topics have also been covered in various chapters of Crochiere and Rabiner [1983]. However, a number of new topics are also introduced here, for example, complementary filters (power complementary, Euclidean complementary, etc.), and multigrid methods.

4.1 BASIC MULTIRATE OPERATIONS

4.1.1 Decimation and interpolation

The most basic operations in multirate digital signal processing are decimation and interpolation. In order to describe these, two new building blocks are introduced, called the *decimator* and the *expander*.

The M-fold decimator. Figure 4.1-1(a) shows the M-fold decimator, which takes an input sequence $x(n)$ and produces the output sequence

$$y_D(n) = x(Mn), \qquad (4.1.1)$$

where M is an integer. Only those samples of $x(n)$ which occur at time equal to multiples of M are retained by the decimator. Figure 4.1-2 demonstrates the idea for $M = 2$. The decimator is also called a *downsampler*, *subsampler, sampling rate compressor*, or merely a *compressor*. We will use the term "decimator" consistently. As will be mathematically substantiated, decimation results in aliasing unless $x(n)$ is bandlimited in a certain way. In general, therefore, it may not be possible to recover $x(n)$ from $y_D(n)$ because of loss of information.

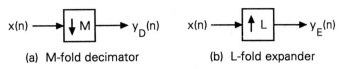

(a) M-fold decimator   (b) L-fold expander

Figure 4.1-1 The decimator and expander.

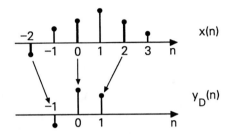

Figure 4.1-2 Demonstration of decimation for $M = 2$. The samples of $x(n)$ shown by heavy lines are retained.

The L-fold expander. Figure 4.1-1(b) shows a building block which is commonly called an *L-fold expander*. This device takes an input $x(n)$ and produces an output sequence

$$y_E(n) = \begin{cases} x(n/L), & \text{if } n \text{ is integer-multiple of } L \\ 0, & \text{otherwise.} \end{cases} \quad (4.1.2)$$

Here L is an integer. Figure 4.1-3 is a demonstration of this operation for $L = 2$. It is evident that the expander does not cause loss of information. We can recover the input $x(n)$ from $y_E(n)$ by L-fold decimation.

Other names for the expander are: sampling rate expander, upsampler, and interpolator. Of these, the term 'interpolator' is really a misnomer. We will consistently use the term 'expander' in this text. The expander is used in interpolation, but a filter is required to complete the process; we will see how the zero-valued samples are converted into interpolated samples by using a lowpass filter at the output of the expander.

Transform Domain Analysis of Decimators and Expanders

First consider the expander which is easier to analyze. We have

$$Y_E(z) = \sum_{n=-\infty}^{\infty} y_E(n)z^{-n} = \sum_{n=\text{mul. of } L} y_E(n)z^{-n}$$

$$= \sum_{k=-\infty}^{\infty} y_E(kL)z^{-kL} = \sum_{k=-\infty}^{\infty} x(k)z^{-kL} \quad \text{from (4.1.2)} \quad (4.1.3)$$

$$= X(z^L).$$

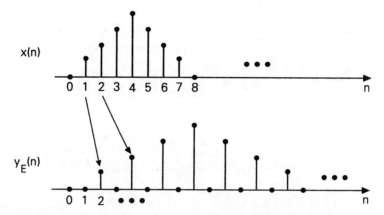

Figure 4.1-3 Demonstration of the expander for $L = 2$.

So $Y_E(e^{j\omega}) = X(e^{j\omega L})$. This means that $Y_E(e^{j\omega})$ is an L-fold compressed version of $X(e^{j\omega})$ as demonstrated in Figs. 4.1-4(a),(b). The multiple copies of the compressed spectrum are called *images*, and we say that the expander creates an imaging effect. [The quantity $X(e^{j\omega})$ in the figure is taken to be nonsymmetric with respect to $\omega = 0$, to improve clarity and generality.]

For the M-fold decimator (4.1.1), we now derive an expression for the output $Y_D(e^{j\omega})$ in terms of $X(e^{j\omega})$. We will show that

$$Y_D(e^{j\omega}) = \frac{1}{M} \sum_{k=0}^{M-1} X(e^{j(\omega - 2\pi k)/M}). \quad (4.1.4)$$

This can be graphically interpreted as follows: (a) stretch $X(e^{j\omega})$ by a factor M to obtain $X(e^{j\omega/M})$, (b) create $M - 1$ copies of this stretched version by shifting it uniformly in successive amounts of 2π, and (c) add all these shifted stretched versions to the unshifted stretched version $X(e^{j\omega/M})$, and divide by M. The stretched quantity $X(e^{j\omega/M})$ does not have period 2π, but after adding the shifted versions the result is periodic with period 2π (which is a requirement for the Fourier transform of a sequence). See Figs. 4.1-4(c) and 4.1-5 which demonstrate these for $M = 2$, and 3.

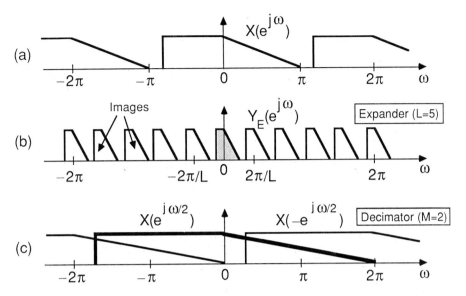

Figure 4.1-4 Transform-domain effects of the expander and decimator. The Fourier transforms of (a) the input signal $x(n)$, (b) the expanded signal ($L = 5$), and (c) the decimated signal ($M = 2$).

Proof of (4.1.4). The z-transform of $y_D(n)$ can be written as

$$Y_D(z) = \sum_{n=-\infty}^{\infty} y_D(n) z^{-n} = \sum_{n=-\infty}^{\infty} x(Mn) z^{-n}.$$

Define an intermediate sequence

$$x_1(n) = \begin{cases} x(n), & n = \text{mul. of } M \\ 0, & \text{otherwise} \end{cases} \qquad (4.1.5)$$

so that $y_D(n) = x(Mn) = x_1(Mn)$. Now

$$Y_D(z) = \sum_{n=-\infty}^{\infty} x_1(Mn) z^{-n} = \sum_{k=-\infty}^{\infty} x_1(k) z^{-k/M}. \qquad (4.1.6)$$

This step is valid because $x_1(k)$ is zero unless k is a multiple of M. So

$$Y_D(z) = X_1(z^{1/M}). \qquad (4.1.7)$$

It only remains to express $X_1(z)$ in terms of $X(z)$. For this note that (4.1.5) can be written as
$$x_1(n) = \mathcal{C}_M(n)x(n) \tag{4.1.8}$$
where $\mathcal{C}_M(n)$ is the 'comb' sequence defined as
$$\mathcal{C}_M(n) = \begin{cases} 1, & n = \text{mul. of } M, \\ 0, & \text{otherwise.} \end{cases} \tag{4.1.9}$$
We can express the comb sequence as
$$\mathcal{C}_M(n) = \frac{1}{M} \sum_{k=0}^{M-1} W_M^{-kn}, \tag{4.1.10}$$
where W_M is the Mth root of unity defined as
$$W_M = e^{-j2\pi/M}. \tag{4.1.11}$$

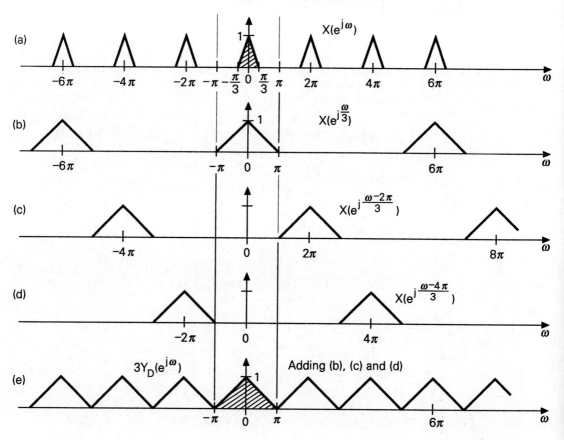

Figure 4.1-5 Demonstrating the frequency-domain effect of decimation with $M = 3$.

The subscript M on W is usually deleted, unless there is room for confusion. We can now obtain

$$X_1(z) = \frac{1}{M} \sum_{k=0}^{M-1} \sum_{n=-\infty}^{\infty} x(n) W^{-kn} z^{-n} = \frac{1}{M} \sum_{k=0}^{M-1} \sum_{n=-\infty}^{\infty} x(n)(zW^k)^{-n}. \quad (4.1.12)$$

The inner summation above is equal to $X(zW^k)$ so that from (4.1.7)

$$Y_D(z) = \frac{1}{M} \sum_{k=0}^{M-1} X(z^{1/M} W^k). \quad (4.1.13)$$

In terms of the frequency variable ω this becomes (4.1.4) indeed.

We often use the following notation to indicate the relation (4.1.13):

$$Y_D(z) = X(z)\Big|_{\downarrow M} \quad (4.1.14)$$

This notation means that $y_D(n)$ is the M-fold decimated version of $x(n)$.

Aliasing Created by Decimation

From Fig. 4.1-4(c), which demonstrates the effect of decimation for $M = 2$, we see that the stretched version $X(e^{j\omega/M})$ can in general overlap with its shifted replicas. If this happens, we cannot recover $x(n)$ from the decimated version $y_D(n)$. This overlap effect is called *aliasing*.

Avoiding aliasing. It is clear that aliasing can be avoided if $x(n)$ is a lowpass signal bandlimited to the region $|\omega| < \pi/M$. The example in Fig. 4.1-5 demonstrates this for $M = 3$. In this case we can recover $x(n)$ from the decimated version by use of an expander, followed by filtering, as demonstrated in Fig. 4.1-6. This recovery scheme works as follows: in the frequency domain, the output $V(e^{j\omega})$ of the expander is a compressed version of $Y_D(e^{j\omega})$ [part (c)]. By using a lowpass filter $H(e^{j\omega})$ [part(d)] we can therefore eliminate the images and extract the original spectrum $X(e^{j\omega})$ [part (e)].

The above condition on bandwidth is, however, not *necessary* to avoid aliasing. For example if $X(e^{j\omega})$ is zero everywhere in $0 \le \omega < 2\pi$ except in $\omega_1 < \omega < \omega_1 + 2\pi/M$ for some ω_1, then there is no overlap between any pair of terms in (4.1.4). Also see Problem 4.3. The most general condition for alias-free decimation can be found in [Sathe and Vaidyanathan, 1993].

It can be verified (Problem 4.4) that the decimator and expander are linear but *time-varying* (LTV) systems.

Decimation Filters and Interpolation Filters.

In most applications, the decimator is preceded by a lowpass digital filter called the *decimation filter* [Fig. 4.1-7(a)]. The filter ensures that the signal being decimated is bandlimited. The exact bandedges of the filter depend on how much aliasing is permitted. For example, in QMF banks (Chapters 5–8), a certain degree of aliasing is usually permitted because this can eventually be canceled off. The simplest form of lowpass decimation filter has magnitude response as sketched in Fig. 4.1-7(b).

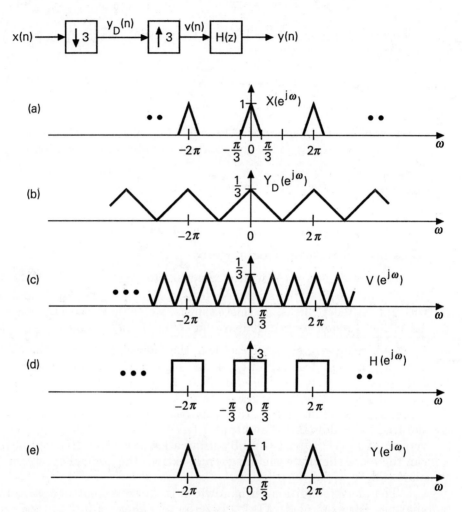

Figure 4.1-6 Recovering bandlimited $x(n)$ from its decimated version.

Next, an interpolation filter (Fig. 4.1-8) is a digital filter that *follows* an expander. The typical purpose is to suppress all the images. Thus, it retains only the shaded portion of the compressed spectrum $Y_E(e^{j\omega})$ in Fig. 4.1-4(b). Typically the interpolation filter is lowpass with cutoff frequency π/L. In the time domain, $y(n)$ is a convolution of $y_E(n)$ with the impulse response $h(n)$. The effect is that the zero-valued samples introduced by the expander are filled with 'interpolated' values [Fig. 4.1-8(c)].

106   Chap. 4. Multirate system fundamentals

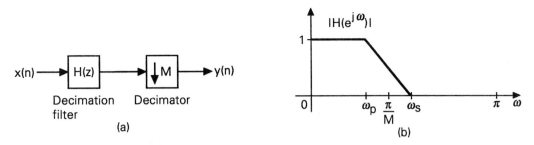

Figure 4.1-7 (a) The complete decimation circuit, and (b) typical response of the decimation filter.

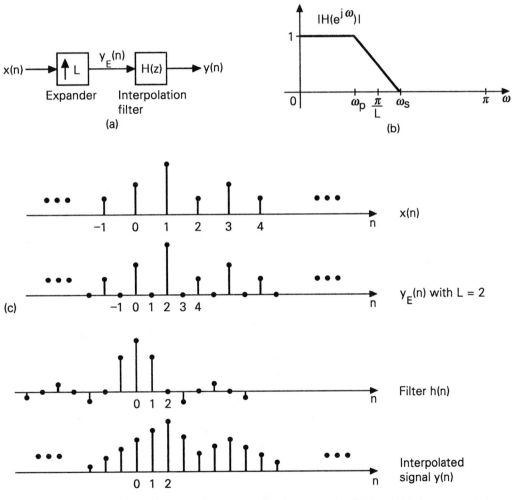

Figure 4.1-8 (a) The complete interpolation circuit, (b) typical response of the interpolation filter, and (c) examples of the sequence $x(n)$, the filter $h(n)$, and the interpolated signal $y(n)$.

Sec. 4.1 Basic multirate operations   107

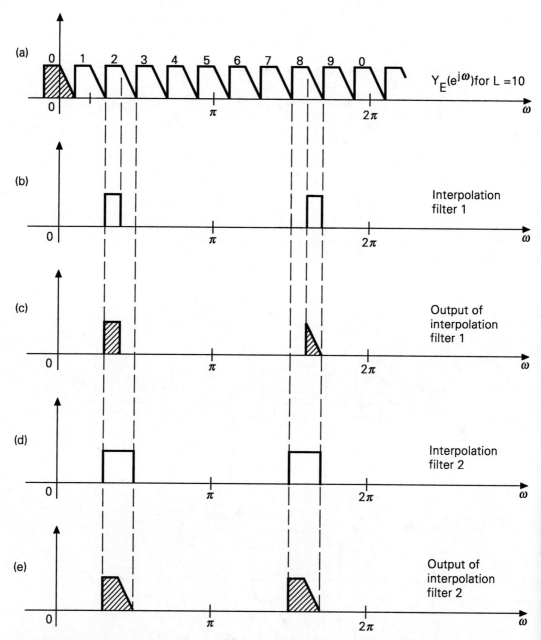

Figure 4.1-9 Demonstrating several possible choices for the interpolation filter.

More generally, it is possible to make other choices of the interpolation filter, as demonstrated in Fig. 4.1-9 for $L = 10$. Here $Y_E(e^{j\omega})$ has nine images (unshaded copies in Fig. 4.1-9(a)). If the filter is chosen as in part (b), the filter output is as in part (c), and contains all information about $X(e^{j\omega})$ [which is the 10-fold stretched version of the shaded portion in part (a)]. If the filter is as in part (d), then two images are retained. This is analogous to cosine modulation of the shaded portion in part (a). See Problem 4.5 for precise relation between cosine modulation and interpolation-filtering. Both filtering schemes in this figure are such that the filter coefficients are real [so that the filter output is real if $x(n)$ is].

Fractional Sampling Rate Alteration

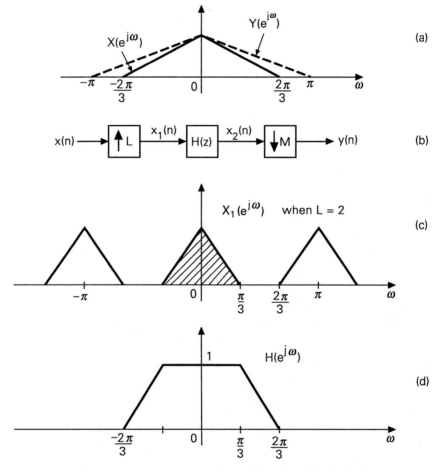

Figure 4.1-10 Pertaining to fractional decimation.

The above techniques permit us to alter the sampling rate of a signal by an integer factor (such as L or M). In some applications, however, it is necessary to change the rate by a rational fraction (such as L/M). For example consider Fig. 4.1-10(a) which shows the transform $X(e^{j\omega})$ of a signal bandlimited to $|\omega| < 2\pi/3$. We cannot decimate the signal by two because that would create aliasing error. It appears to be possible to decimate by the factor 1.5 (so that the Fourier Transform gets stretched as shown by broken lines). One procedure for this would be to convert the signal into a continuous-time signal and resample at the lower rate. It is however simpler to perform the fractional rate-alterations directly in the digital domain, by judicious combination of interpolation and decimation.

Figure 4.1-10(b) shows a simple technique which can be used for this purpose. For the example under consideration, we will take $L = 2, M = 3$ so that the overall reduction of sampling rate is by the factor $M/L = 3/2$. The quantity $X_1(e^{j\omega}) = X(e^{j2\omega})$ is shown in part (c). If we design $H(z)$ to be a zero-phase lowpass filter with response as in part (d), then the filter output $X_2(e^{j\omega})$ is the shaded part in part (c). Decimation by 3 finally results in $y(n)$ whose transform $Y(e^{j\omega})$ is as shown by broken lines in part (a)

It is clear that this technique can be generalized to reduce the sampling rate by any rational number M/L. In practice the quality of the filter $H(z)$ [i.e., passband and stopband ripples] determines the quality of the result [i.e., the degree to which $Y(e^{j\omega})$ agrees with the stretched version $X(e^{j\omega L/M})$ in the region $0 \leq \omega \leq 2\pi$.] Fig. 4.1-11 demonstrates the time-domain meaning of decimation by a factor of $3/2$. The samples numbered $1, 3, \ldots$ are the newly generated (interpolated) samples.

Notice that the permissible transition bandwidth of $H(z)$ is not unduly narrow. In the above example it can be as large as $\pi/3$ (equal to the passband width!). As a quick example, suppose we wish $H(z)$ to be a linear phase equiripple filter with peak ripples $\delta_1 = \delta_2 = 0.01$ (so that $A_S = 40$ dB). The normalized transition bandwidth is $\Delta f = (\pi/3)/2\pi = 1/6$. The estimated order from Sec. 3.2.4 is then $N \approx 11$ which requires only 6 multipliers and 11 adders for implementation.

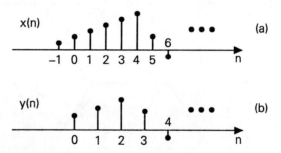

Figure 4.1-11 A signal $x(n)$ and the fractionally decimated version $y(n)$. The decimation factor is $3/2$.

More generally the transition bandwidth of the filter is $\Delta f = (\pi - \sigma)/\pi L$ [where $x(n)$ is bandlimited to $|\omega| < \sigma$]. So, for a given σ, the filter order is proportional to L. In Sec. 4.3 we see how this system can be implemented with even fewer computations per input sample, using a technique called the *polyphase approach*. We will show that even though the filter order goes up as L increases, the number of computations per output sample is independent of L.

The scheme of Fig. 4.1-10(b) also works when $L > M$. In this case, there is an overall *increase* of sampling rate by L/M. There exist other methods for fractional sampling rate alteration. These combine filtering techniques with polynomial fitting, and are more suitable when L/M is a ratio of very large integers. See Lagadec et al. [1982] and Ramstad [1984a].

The Physical Time Scale.

In the above we defined decimators and expanders purely as devices which work on sequences of numbers $x(n)$ to produce a related sequence of numbers. Upon reflection, these definitions are somewhat strange. For example, an M-fold decimator is a noncausal device, that is, output sample $y_D(n)$ in general depends on $x(m), m > n$. (Thus, with $M = 2$, the sample $y_D(1) = x(2)$.) The L-fold expander is also a noncausal device. To see this, let $L = 2$ and think of negative values of n. The sample $y_E(-2) = x(-1)$ which implies noncausality.

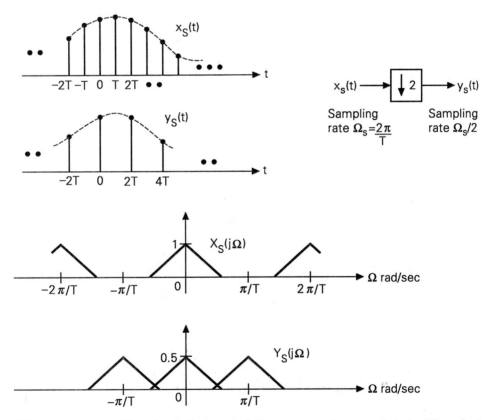

Figure 4.1-12 The physical time and frequency scales associated with a decimator.

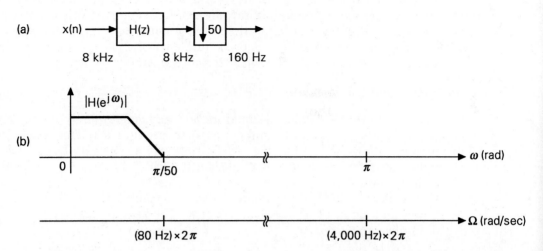

Figure 4.1-13 Demonstration of various real-time dimensions that go with a decimator.

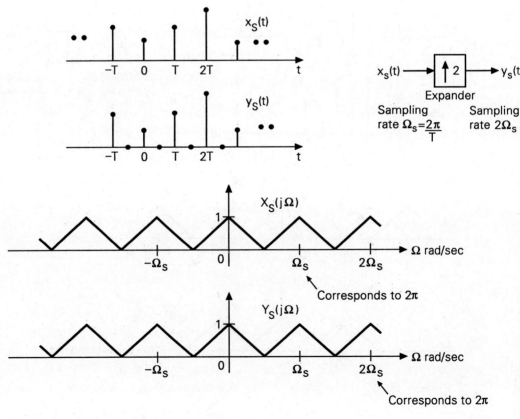

Figure 4.1-14 Demonstration of real-time dimensions that go with an expander.

However, when these devices are operated in real time, noncausality does not arise. Figure 4.1-12 shows a sampled analog waveform $x_S(t)$ with sampling instants labeled in multiples of T (sampling interval). The output signal $y_S(t)$ is an undersampled version with samples spaced apart by $2T$. No physical noncausality is really involved. Typical Fourier transforms are also shown in the figure.† Here Ω is the real-time frequency variable in radians/second, and $\Omega_S = 2\pi/T$ is the sampling rate for $x_S(t)$. There is no physical stretching of the Ω variable. Only the repetition rate of the basic spectrum is changed by a factor of 2 (consistent with decimation factor $= 2$).

In Fig. 4.1-13 we indicate the various real-time frequencies for the example of a 50-fold decimator with input sampling frequency of 8 kHz and output sampling frequency 160 Hz. So the 'digital' frequency π for the input sequence corresponds to 4 kHz. The lowpass filter $H(z)$ now has stopband edge at $\pi/50$ ($= 2\pi/100$) which corresponds to $8,000/100 = 80$ Hz. In other words, the signal has to be bandlimited to 80 Hz before the sampling rate can be cut down to 160 Hz.

The operation of the expander in real time is demonstrated in Fig. 4.1-14 for $L = 2$, along with frequency domain quantities. In the frequency domain there is really no fundamental difference between the input $X_S(j\Omega)$ and the output $Y_S(j\Omega)$. If we think of the plot of $X_S(j\Omega)$ as a discrete-time Fourier transform, we would label Ω_S as 2π whereas for $Y_S(j\Omega)$ we would label $2\Omega_S$ as 2π.

4.1.2 Digital Filter Banks

A digital filter bank is a collection of digital filters, with a common input or a common output. Both of these cases are shown in Fig. 4.1-15. The system in Fig. 4.1-15(a) is called an *analysis bank,* and the filters $H_k(z)$ the analysis filters. The system splits a signal $x(n)$ into M signals $x_k(n)$ typically called *subband signals.* The system in Fig. 4.1-15(b) is called a *synthesis bank,* and $F_k(z)$ are the synthesis filters. These filters combine the M subband signals into a single signal $\widehat{x}(n)$. Figure 4.1-15(c)–(e) shows typical frequency responses for the analysis filters. These could be marginally overlapping, non overlapping, or very much overlapping, depending on the application.

Example 4.1.1: The DFT Filter Bank

We will present a filter bank based on the DFT matrix (Sec. A.6, Appendix A). The $M \times M$ DFT matrix $\mathbf{W}$ has elements $[\mathbf{W}]_{km} = W^{km}$ where $W = e^{-j2\pi/M}$. Now consider Fig. 4.1-16(a). Here $x(n)$ is a sequence from which we generate M sequences $s_i(n)$ by passing $x(n)$ through a delay chain, so that $s_i(n) = x(n-i)$. The matrix $\mathbf{W}^*$ repre-

---

† Frequency domain effects of sampling were reviewed in Sec. 2.1.4.

sents the conjugate of **W**. ‡ From the definition of **W** we, therefore, have

$$x_k(n) = \sum_{i=0}^{M-1} s_i(n) W^{-ki}. \qquad (4.1.15)$$

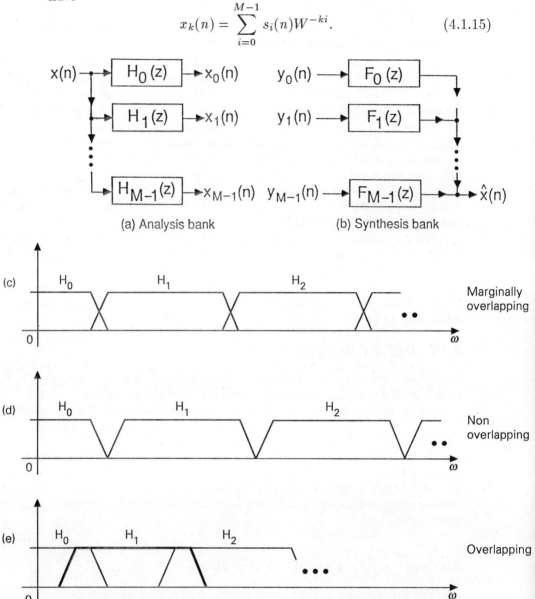

Figure 4.1-15 Digital filter banks and typical filter responses.

In other words, for every value of time index n, we compute the set of M signals $x_k(n)$ from the set of M signals $s_i(n)$ according to the above

---

‡ Since $\mathbf{W}^T = \mathbf{W}$, the quantity $\mathbf{W}^*$ is same as $\mathbf{W}^\dagger$.

equation which is exactly the inverse DFT (IDFT) relation given in Sec. A.6 (Appendix A), except for a scale factor $1/M$. From (4.1.15) we have

$$X_k(z) = \sum_{i=0}^{M-1} S_i(z)W^{-ki} = \sum_{i=0}^{M-1} z^{-i}W^{-ki}X(z) = \sum_{i=0}^{M-1} (zW^k)^{-i}X(z). \quad (4.1.16)$$

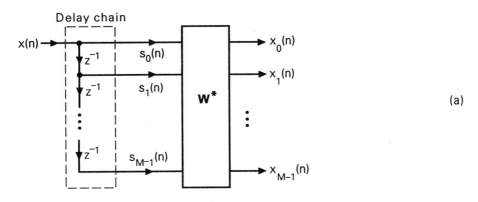

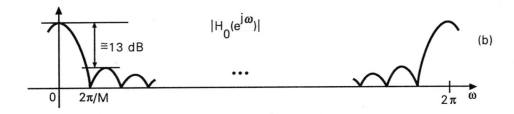

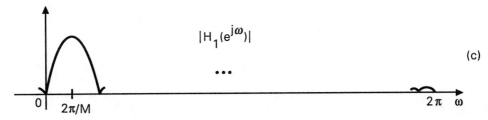

Figure 4.1-16 The simplest example of a uniform-DFT filter bank.

So we can write $X_k(z) = H_k(z)X(z)$ where

$$H_k(z) \triangleq H_0(zW^k), \quad (4.1.17)$$

with

$$H_0(z) = 1 + z^{-1} + \ldots + z^{-(M-1)}. \quad (4.1.18)$$

Summarizing, the system is equivalent to an analysis bank with analysis filters $H_k(z)$ as above. Let us take a closer look at these filters. Equation

Sec. 4.1 Basic multirate operations

(4.1.18) implies $|H_0(e^{j\omega})| = |\sin(M\omega/2)/\sin(\omega/2)|$, which is plotted in part (b) of the figure. The filter $H_k(z)$ has response

$$H_k(e^{j\omega}) = H_0(e^{j(\omega - (2\pi k/M))}), \qquad (4.1.19)$$

which is a shifted version of $H_0(e^{j\omega})$. So we have a bank of M filters, and these are uniformly shifted versions of $H_0(z)$ (part(c) in the figure). The responses evidently have large amount of overlap. Each analysis filter in this example has order M, and offers about 13 dB of minimum stopband attenuation (with respect to zero-frequency gain).

Uniform DFT banks. A filter bank in which the filters are related as in (4.1.17) is called a uniform DFT filter bank, even though the matrix appearing in Fig. 4.1-16(a) is $\mathbf{W}^*$ rather than $\mathbf{W}$.

Can we attach a physical meaning to $x_k(n)$ which are outputs of the analysis filters? Like $x(n)$, these are time-domain quantities (assuming that n represents time). For convenience let us talk about the shifted version $x_k(n + M - 1)$. For fixed n we have

$$x_k(n+M-1) = \sum_{i=0}^{M-1} x(n+M-1-i)W^{-ki} = W^k \sum_{\ell=0}^{M-1} x(n+\ell)W^{k\ell}, \quad (4.1.20)$$

after making a change of variables $\ell = M - 1 - i$ and using $W^M = 1$. So, $x_k(n+M-1)$ is W^k times the kth point of the DFT of the M-point sequence

$$x(n), x(n+1), \ldots, x(n+M-1). \qquad (4.1.21)$$

This sequence is nothing but a M-point segment of the input sequence $x(n)$, starting from time n. So the magnitude of $x_k(n + M - 1)$ represents the magnitude of the (kth point of the) DFT of the sequence (4.1.21). As time advances (that is, as n increases), this quantity gets updated that is, recomputed for the next segment of M samples. And this goes on for ever.

Summarizing, we can think of the filter bank of Fig. 4.1-16(a) as a spectrum analyzer. The kth output $x_k(n)$ is the 'spectrum' (i.e., kth point of the DFT except for scale factor W^k) computed based on the most recent M samples of the sequence $x(n)$. Since $x_k(n)$ is the output of $H_k(e^{j\omega})$, it dominantly represents the portion of $X(e^{j\omega})$ around the region $\omega = 2\pi k/M$. Actually $x_k(n)$ represents some kind of averaged version of the exact spectrum $X(e^{j\omega})$ at this frequency, because the filter actually permits a range of frequencies to pass. The resolution of the spectrum analyzer can be improved by increasing M. In any case, the overlapping nature of the filter responses ensures that $x_k(n)$ tends to represent an averaged effect around $2\pi k/M$.

Figure 4.1-17 explains the operation in a pictorial way. In practice one can multiply $x(m)$ with a window as indicated, in order to reduce the sidelobe level of the frequency response. This is equivalent to inserting the multipliers α_i just after the delay chain in Fig. 4.1-16(a). As time advances, the window in Fig. 4.1-17 merely slides past the data, computing the M DFT coefficients afresh for each increment of n. Thus, the window helps to localize the time domain data, before computation of the Fourier transform.

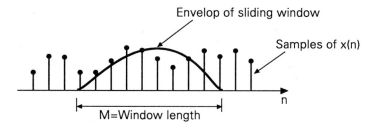

Figure 4.1-17 The sliding window interpretation of the uniform DFT bank.

This sliding window mechanism gives rise to the idea of short-time Fourier transform (STFT), discussed in Chap. 11. A generalization of this, called the *wavelet transform* is also discussed in that chapter.

Reducing Overlap by Improving the Filters

In Example 4.1.1, the M analysis filters are obtained from a single filter $H_0(z)$ (which we call the prototype filter) by uniformly shifting the response according to the relation $H_k(z) = H_0(zW^k)$. The filters themselves are not very good (they had wide transition bands, and stopband attenuation of only 13 dB) because the prototype (4.1.18) itself is a very simple filter.

Now suppose that we use a higher order prototype $H_0(z)$ with a sharper response, for example, as in Fig. 4.1-15(c). Then, the shifted versions have reduced amount of overlap. Such systems with marginal overlap are used in quadrature mirror filter banks (Chap. 5). In Sec. 4.3.2 we will show how the DFT bank can be modified to implement these M filters at the cost of (almost) one filter.

4.1.3 Time Domain Descriptions of Multirate Filters

So far we have seen three types of multirate filters: decimation filters, interpolation filters, and fractional decimation filters [Figs. 4.1-7, 4.1-8, and 4.1-10(b)]. For each of these, we give below the input-output relation in the time domain:

$$y(n) = \begin{cases} \sum_{k=-\infty}^{\infty} x(k)h(nM - k), & M\text{-fold decimation filter} \\ \sum_{k=-\infty}^{\infty} x(k)h(n - kL), & L\text{-fold interpolation filter} \\ \sum_{k=-\infty}^{\infty} x(k)h(nM - kL), & \frac{M}{L}\text{-fold decimation filter} \end{cases} \quad (4.1.22)$$

Note that the relation for decimation filter can be written in two ways:

$$y(n) = \sum_{k=-\infty}^{\infty} x(k)h(nM - k) = \sum_{k=-\infty}^{\infty} h(k)x(nM - k). \qquad (4.1.23)$$

4.2 INTERCONNECTION OF BUILDING BLOCKS

We now consider some interconnections of building blocks which occur commonly in multirate systems. Figure 4.2-1 shows a number of interconnections and equivalences which can be easily verified.

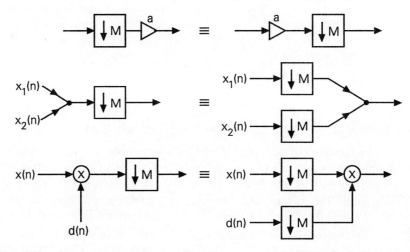

Figure 4.2-1 Simple identities for interconnected systems. All of these hold if the decimators are replaced with expanders.

Decimator-expander cascades. Figure 4.2-2 shows two common interconnections of a decimator with an expander. The two structures in the figure are in general not equivalent (i.e., the decimator and expander do not commute). For example, if $L = M$ then we can verify that $y_2(n) = x(n)$ whereas $y_1(n) = x(n)\mathcal{C}_M(n)$ where $\mathcal{C}_M(n)$ is the comb sequence (4.1.9).

The two systems in Fig. 4.2-2 are equivalent [i.e., $y_1(n) = y_2(n)$ for every possible input $x(n)$] if, and only if, L and M are relatively prime integers (i.e., greatest common divisor = 1). To prove this, note first that

$$Y_1(z) = \frac{1}{M} \sum_{k=0}^{M-1} X(z^{L/M} W_M^k), \text{ and } Y_2(z) = \frac{1}{M} \sum_{k=0}^{M-1} X(z^{L/M} W_M^{kL}). \quad (4.2.1)$$

We know that the set $\mathcal{S}_1$ of numbers W_M^k, $0 \leq k \leq M-1$ are the M distinct Mth roots of unity. The set $\mathcal{S}_2$ of numbers W_M^{kL}, $0 \leq k \leq M-1$ is equal to the set S_1 if and only if L and M are relatively prime (Problem 4.7). As a result, for arbitrary $X(z)$, the set of M terms in $Y_1(z)$ is same as the set of M terms in $Y_2(z)$ if, and only if, L and M are relatively prime.

A time-domain proof of the above result (which is perhaps more appealing) is requested in Problem 4.8.

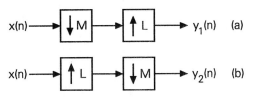

Figure 4.2-2 Two popular interconnections of decimators with expanders. These are equivalent if and only if L and M are relatively prime.

The Noble Identities

We have already seen cascades of decimators and expanders with LTI systems [e.g., Figs. 4.1-7(a) and 4.1-8(a)]. A different type of cascade is shown in Fig. 4.2-3(a) where a filter $G(z)$ *follows* a decimator, and in Fig. 4.2-3(c) where a filter $G(z)$ *precedes* an expander. Such interconnections arise when we try to use the polyphase representation (Sec. 4.3) for decimation and interpolation filters. If the function $G(z)$ is rational (i.e., a ratio of polynomials in z or z^{-1}) then we can redraw Fig. 4.2-3(a) as in Fig. 4.2-3(b) and Fig. 4.2-3(c) as in Fig. 4.2-3(d). These are called *noble identities* and are very useful in the theory and implementation of multirate systems.

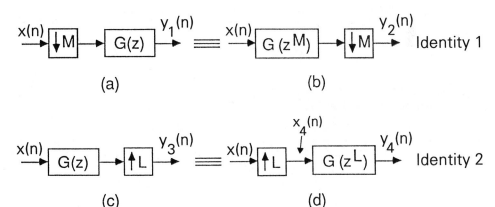

Figure 4.2-3 The noble identitites for multirate systems.

Before proving these identities, note that they may not work if $G(z)$ is irrational, for example, $G(z) = z^{-1/2}$. Thus, consider the system of Fig. 4.2-4(a). If the identities were applicable, this could be redrawn as in Fig. 4.2-4(c). But it is easy to verify that these are not equivalent. For example,

Sec. 4.2 Interconnections 119

if the input $x(n)$ is such that $x(2n) = 0$ for all n, then $y_2(n)$ is zero for all n, but $y_1(n)$ is not necessarily so.

To prove the noble identities note that

$$Y_2(z) = \frac{1}{M}\sum_{k=0}^{M-1} X(z^{1/M}W^k)G((z^{1/M}W^k)^M) = \frac{1}{M}\sum_{k=0}^{M-1} X(z^{1/M}W^k)G(z) \qquad (4.2.2)$$

which is the same as $Y_1(z)$. Also

$$Y_4(z) = G(z^L)X_4(z) = G(z^L)X(z^L), \qquad (4.2.3)$$

which agrees with $Y_3(z)$, completing the proof.

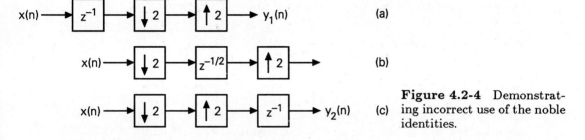

Figure 4.2-4 Demonstrating incorrect use of the noble identities.

4.3 THE POLYPHASE REPRESENTATION

An important advancement in multirate signal processing is the invention of the polyphase representation [Bellanger, et al., 1976], [Vary, 1979]. This permits great simplification of theoretical results and also leads to computationally efficient implementations of decimation/interpolation filters, as well as filter banks (both single and multirate). These applications will be elaborated in Sec. 4.3.1–4.3.3, and used throughout the book.

To explain the basic idea, consider a filter $H(z) = \sum_{n=-\infty}^{\infty} h(n)z^{-n}$. By separating the even numbered coefficients of $h(n)$ from the odd numbered ones, we can write

$$H(z) = \sum_{n=-\infty}^{\infty} h(2n)z^{-2n} + z^{-1}\sum_{n=-\infty}^{\infty} h(2n+1)z^{-2n}. \qquad (4.3.1)$$

Defining

$$E_0(z) = \sum_{n=-\infty}^{\infty} h(2n)z^{-n}, \quad E_1(z) = \sum_{n=-\infty}^{\infty} h(2n+1)z^{-n}, \qquad (4.3.2)$$

we can, therefore, write $H(z)$ as

$$H(z) = E_0(z^2) + z^{-1}E_1(z^2). \qquad (4.3.3)$$

Note that these representations hold whether $H(z)$ is FIR or IIR; causal or noncausal. As an example if $H(z) = 1 + 2z^{-1} + 3z^{-2} + 4z^{-3}$ then

$$E_0(z) = 1 + 3z^{-1}, \quad E_1(z) = 2 + 4z^{-1}. \qquad (4.3.4)$$

For an IIR example, let $H(z) = 1/(1 - \alpha z^{-1})$. By using the identity $(1-x) = (1-x^2)/(1+x)$ we can write

$$H(z) = \frac{1}{1 - \alpha z^{-1}} = \frac{1}{1 - \alpha^2 z^{-2}} + \frac{\alpha z^{-1}}{1 - \alpha^2 z^{-2}}, \qquad (4.3.5)$$

so that $E_0(z) = 1/(1 - \alpha^2 z^{-1})$ and $E_1(z) = \alpha/(1 - \alpha^2 z^{-1})$.

Extending this idea further, suppose we are given any integer M. We can always decompose $H(z)$ as

$$\begin{aligned}H(z) = &\sum_{n=-\infty}^{\infty} h(nM) z^{-nM} \\ &+ z^{-1} \sum_{n=-\infty}^{\infty} h(nM+1) z^{-nM} \\ &\vdots \\ &+ z^{-(M-1)} \sum_{n=-\infty}^{\infty} h(nM+M-1) z^{-nM}.\end{aligned} \qquad (4.3.6)$$

This can be compactly written as

$$H(z) = \sum_{\ell=0}^{M-1} z^{-\ell} E_\ell(z^M) \quad \text{(Type 1 polyphase)}, \qquad (4.3.7)$$

where

$$E_\ell(z) = \sum_{n=-\infty}^{\infty} e_\ell(n) z^{-n}, \qquad (4.3.8a)$$

with

$$e_\ell(n) \stackrel{\Delta}{=} h(Mn + \ell), \quad 0 \le \ell \le M - 1. \qquad (4.3.8b)$$

Equation (4.3.7) is called the Type 1 polyphase representation (with respect to M) and $E_\ell(z)$ the polyphase components of $H(z)$. Figure 4.3-1 summarizes the generation of $e_\ell(n)$ from $h(n)$. Notice that $E_\ell(z)$ depends on choice of M. So a notation such as $E_\ell^{(M)}(z)$ would have been more logical, but is avoided here for simplicity. Normally the value of M is clear from the context.

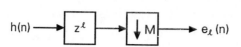

Figure 4.3-1 Schematic of the relation between $h(n)$ and its ℓth polyphase component.

A variation of (4.3.7) is given by

$$H(z) = \sum_{\ell=0}^{M-1} z^{-(M-1-\ell)} R_\ell(z^M) \quad \text{(Type 2 polyphase)}. \tag{4.3.9}$$

The Type 2 polyphase components $R_\ell(z)$ are permutations of $E_\ell(z)$, that is, $R_\ell(z) = E_{M-1-\ell}(z)$.

The reader who is curious about the origin of the term 'polyphase' should see Sec. 4.6.5.

4.3.1 Efficient Structures for Decimation and Interpolation Filters

Consider the decimation filter (Fig. 4.1-7) with $M = 2$. If we represent $H(z)$ as in (4.3.3) then we can redraw the system as in Fig. 4.3-2(a). By invoking noble identity 1, this can be redrawn as in Fig. 4.3-2(b). This implementation is more efficient than a direct implementation of $H(z)$ as explained next.

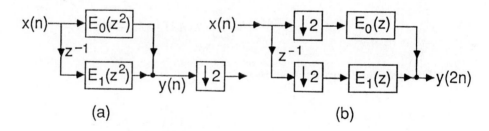

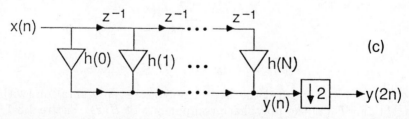

Figure 4.3-2 The decimation filter. (a) Polyphase implementation, (b) moving the polyphase components, and (c) direct implementation.

Let $H(z)$ be Nth order FIR. Its traditional direct form implementation is shown in Fig. 4.3-2(c).‡ Only the even numbered output samples $y(2n)$ are computed. This computation requires $N+1$ multiplications and N additions. As time changes from $2n$ to $2n+1$, the stored signals in the delays change, so that the above computation must be completed in *one* unit of time. The speed of operation should therefore correspond to $N+1$ multiplications and N additions *per unit time*. However, during the odd instants of time, the structure is merely resting. This is inefficient resource utilization.

Next consider the polyphase implementation of Fig. 4.3-2(b). Let n_0 and n_1 be the orders of $E_0(z)$ and $E_1(z)$ (so that $N+1 = n_0 + n_1 + 2$). So $E_\ell(z)$ requires $n_\ell + 1$ multiplications and n_ℓ additions. The total cost [including the extra adder in Fig. 4.3-2(b)] is again $N+1$ multipliers and N adders. However, since $E_\ell(z)$ operates at the lower rate, only a total of $(N+1)/2$ multiplications per *unit time* (abbreviated MPUs), and $N/2$ additions per unit time (APUs) are required. The multipliers and adders in each of the filters $E_0(z)$ and $E_1(z)$ now have two units of time available for doing their work, and they are continually operative (i.e., no resting time).

Interpolation Filters

Now consider an interpolation filter (Fig. 4.1-8) with $L=2$. A direct-form implementation of $H(z)$ is again inefficient because, at most 50% of the input samples to $H(z)$ are nonzero, which means that at any point in time, only 50% of the multipliers $h(n)$ have nonzero input. So the remaining multipliers are resting. And those multipliers which are not resting are expected to complete their job in *half unit* of time because the outputs of the delay elements will change by that time. A more efficient structure can again be obtained by using the Type 2 polyphase decomposition

$$H(z) = R_1(z^2) + z^{-1} R_0(z^2). \tag{4.3.10}$$

This is shown in Fig. 4.3-3. Here $R_\ell(z)$ are operating at the input rate, and none of the multipliers is resting. Each multiplier gets one unit of time to finish its task. The complexity of the system is $(N+1)$ MPUs and $N-1$ APUs. Note that the extra adder following the expander is *not counted* because, the signal $y(n)$ is obtained merely by interlacing $y_0(n)$ and $y_1(n)$, that is, $y(n)$ is

$$\ldots y_1(0)\ y_0(0)\ y_1(1)\ y_0(1)\ y_1(2)\ y_0(2) \ldots$$

which is a time-multiplexed version of the outputs of $R_0(z)$ and $R_1(z)$.

More generally, an M-fold decimation filter can be implemented with approximately M-fold reduction in the number of MPUs and APUs by using the polyphase structure of Fig. 4.3-4(a). To see this note that the number of

---

‡ In all discussions, the input sample spacing is taken to be one unit of time.

multipliers and adders required to implement the M polyphase components independently is equal to $N+1$ and $N-M+1$ respectively. This is followed by $M-1$ additions (to combine the outputs of polyphase components) so that we have a total of $N+1$ multiplications and N additions. All of these are performed at $1/M$th of the input rate. In other words, M units of time are available to perform this computation once. The polyphase structure has complexity $(N+1)/M$ MPUs and N/M APUs.

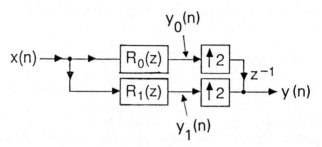

Figure 4.3-3 The polyphase implementation of an interpolation filter.

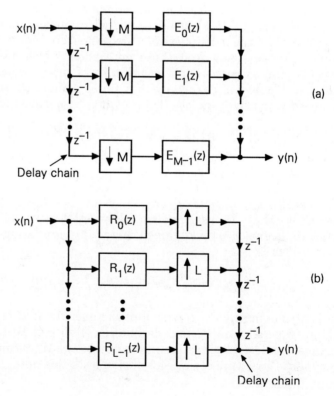

Figure 4.3-4 Polyphase implementations of (a) M-fold decimation filter and (b) L-fold interpolation filter.

Similar comments hold for L-fold interpolation filters [Fig. 4.3-4(b)], which require $(N+1)$ MPUs and $(N-L+1)$ APUs.

Case of Linear Phase FIR Decimation Filters

Suppose $H(z) = \sum_{n=0}^{N} h(n)z^{-n}$ where $h(n) = h(N-n)$. Let us see how the symmetry of $h(n)$ reflects into the polyphase components $E_0(z)$ and $E_1(z)$. For example, let $N = 4$ and

$$H(z) = 1 + 2z^{-1} + 4z^{-2} + 2z^{-3} + z^{-4}.$$

Then,
$$E_0(z) = 1 + 4z^{-1} + z^{-2}, \quad E_1(z) = 2 + 2z^{-1}.$$

So each of the filters $E_0(z)$ and $E_1(z)$ has symmetric impulse response. Now consider odd N, say $N = 5$. Let

$$H(z) = 1 + 2z^{-1} + 4z^{-2} + 4z^{-3} + 2z^{-4} + z^{-5}.$$

Now
$$E_0(z) = 1 + 4z^{-1} + 2z^{-2}, \quad E_1(z) = 2 + 4z^{-1} + z^{-2}.$$

So $E_0(z)$ and $E_1(z)$ do not have symmetric impulse responses, but the impulse response $e_1(n)$ is the mirror image of $e_0(n)$.

These facts can be generalized as follows: Let $H(z) = \sum_{n=0}^{N} h(n)z^{-n}$ with $h(n) = h(N-n)$. Let $E_0(z)$ and $E_1(z)$ be the Type 1 polyphase components. If N is even, then $e_0(n)$ and $e_1(n)$ are symmetric sequences. If N is odd then $e_0(n)$ is the mirror image of $e_1(n)$.

Impact on computational complexity. Consider Fig. 4.3-2(b). When N is even (so that $e_0(n)$ and $e_1(n)$ are symmetric), we obtain a factor of two saving in multiplication rate (in addition to the factor of two saving due to decimation). On the other hand if N is odd, then $e_0(n)$ and $e_1(n)$ are not symmetric, but $e_1(n)$ is the mirror image of $e_0(n)$. This fact can be exploited to obtain a factor of two saving in multiplication rate (see Problem 4.17).

Summarizing, the polyphase structure for a two fold decimation filter (symmetric impulse response with order N) requires about $N/4$ MPUs whether N is even or odd. See Problem 4.22 for the case of linear phase M-fold decimation filters.

4.3.2 Polyphase Implementation of Uniform DFT Filter Banks

Recall that a set of M filters is said to be a uniform-DFT filter bank if they are related as in (4.1.17), where $W = e^{-j2\pi/M}$. The polyphase decomposition can be used to implement such a filter bank in a very efficient manner. Assume that the prototype $H_0(z)$ has been expressed as in (4.3.7). The kth filter can now be expressed as

$$H_k(z) = H_0(zW^k) = \sum_{\ell=0}^{M-1} (z^{-1}W^{-k})^\ell E_\ell(z^M), \qquad (4.3.11a)$$

because $(zW^k)^M = z^M$. With $X_k(z)$ denoting the output of $H_k(z)$, we obtain

$$X_k(z) = \sum_{\ell=0}^{M-1} W^{-k\ell}\Big(z^{-\ell}E_\ell(z^M)X(z)\Big). \quad (4.3.11b)$$

This shows that the M filters can be implemented by using the structure shown in Fig. 4.3-5(a). If $H_0(z)$ is FIR with order N, we require $N+1$ multiplications and $N-M+1$ additions to implement the polyphase components. Add to this the DFT cost, and this gives the total cost for implementing the M filters.

If we set $E_\ell(z) = 1$ for all ℓ, Fig. 4.3-5(a) gives rise to the special case of Fig. 4.1-16(a). The presence of $E_\ell(z)$ permits the use of a prototype $H_0(z)$ with larger length as compared to Fig. 4.1-16(a), where the filter length was limited to M. Thus the prototype (and hence all the M filters) can have sharper cutoff and higher stopband attenuation [Fig. 4.3-5(b)]. As in Fig. 4.1-17, we can once again interpret the filter bank system as a DFT computation with a sliding window. But now the window is longer [with length equal to the length of the filter $H_0(z)$], even though the DFT is still $M \times M$. A more elaborate discussion of this can be found in Sec. 11.2 (short-time Fourier transform).

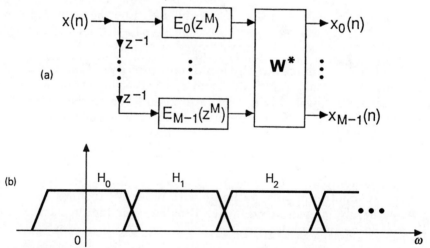

Figure 4.3-5 (a) Implementation of the uniform DFT bank using polyphase decomposition, and (b) typical magnitude responses of filters. Here $H_k(z) = X_k(z)/X(z)$.

For composite M (for example $M = 2^m$) the DFT can be performed efficiently using fast Fourier Transform (FFT) techniques [Rabiner and Gold, 1975]. For example, let $M = 32$ and $N = 50$. A standard 32-point radix-2 FFT requires (approximately) 136 real multiplications. So the total number of multiplications is $136 + 51 = 187$. If each of the 32 filters were implemented indepedently, the number of multiplications would have been $32 \times 51 = 1{,}632$ which is about nine times larger! [In fact it is even worse because most of the filters $H_k(z)$ have complex multipliers even if $H_0(z)$ has real coefficients.] The polyphase implementation of Fig. 4.3-5(a) is, therefore, very efficient indeed.

Decimated uniform DFT banks. In many applications (such as QMF banks, Chap. 8,9), we are interested in decimating the outputs of $H_k(z)$ by the factor M. This is logical because each of these outputs has a bandwidth which is approximately M times narrower than that of $x(n)$. By using noble identity 1, the polyphase uniform-DFT filter bank structure with decimators can be redrawn as in Fig. 4.3-6. This structure requires M times fewer MPUs and APUs than Fig. 4.3-5(a) so that it is even more efficient.

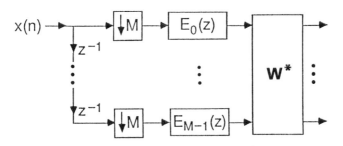

Figure 4.3-6 Redrawing Fig. 4.3-5(a) when $x_k(n)$ have to be decimated by M.

For the above numerical example, the total number of real MPUs is only about 6. So for 6 multiplications per unit time (i.e., per input sample) we are able to implement 32 filters, each of order 50.

4.3.3 Efficient Structures for Fractional Decimation

Recall that Fig. 4.1-10(b) is a technique for decimating a sequence by a rational number M/L. Implementing $H(z)$ directly (i.e., no polyphase) is inefficient because of two reasons: first, at any point in time, $L - 1$ out of L multipliers have input equal to zero. Second, only one out of M output samples is being retained. Neither of these facts has been exploited in the implementation.

To obtain a more efficient implementation, we begin by considering the $M = 3, L = 2$ example again. By using the Type 1 polyphase representation and rearranging we obtain the implementation of Fig. 4.3-7(a). On the other hand if we use the Type 2 decomposition and rearrange, we get Fig. 4.3-7(b). Clearly these are more efficient than a direct implementation by factors of M and L respectively. Notice that Fig. 4.3-7(a) exploits the presence of the decimator, whereas Fig. 4.3-7(b) exploits the presence of the expander. But we have not exploited *both* yet. For example in Fig. 4.3-7(a) the inputs to $E_k(z)$ still have some zero-valued samples. In Fig. 4.3-7(b), some of the outputs computed by $R_k(z)$ are discarded by the decimator. Can we rearrange the structure so that we can take full advantage of both the decimator and the expander?

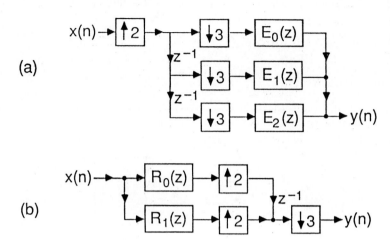

Figure 4.3-7 Two ways to improve the efficiency of the fractional-decimation filter. (© Adopted from 1990 IEEE.)

The answer is in the affirmative [Crochiere and Rabiner, 1981, p. 311], and [Hsiao, 1987]. We now outline the technique reported by Hsiao, which is based on the z-domain formulation and polyphase decomposition. First notice that we are stuck in Fig. 4.3-7 mainly because we cannot move the expander any more to the right (and decimator any more to the left). This is because, the noble identities simply cannot be applied anymore!

However, here is the nice trick which comes to our rescue: we can write $z^{-1} = z^{-3} z^2$ so that Fig. 4.3-7(b) can be redrawn as in Fig. 4.3-8(a). With the help of the noble identities this becomes Fig. 4.3-8(b). Next we can interchange the decimator with the expander (which is valid because 2 and 3 are relatively prime; see Sec. 4.2) to obtain Fig. 4.3-8(c). Finally we can perform a Type 1 polyphase decomposition of the polyphase components $R_0(z)$ and $R_1(z)$ as follows:

$$R_0(z) = R_{00}(z^3) + z^{-1} R_{01}(z^3) + z^{-2} R_{02}(z^3),$$
$$R_1(z) = R_{10}(z^3) + z^{-1} R_{11}(z^3) + z^{-2} R_{12}(z^3),$$
(4.3.12)

so that Fig. 4.3-8(c) can be redrawn as in Fig. 4.3-8(d). So Fig. 4.3-8(d) is equivalent to Fig. 4.3-8(a)!

If $H(z) = \sum_{n=0}^{N} h(n) z^{-n}$, then we still have only $N+1$ multipliers in Fig. 4.3-8(d). However, each multiplier operates at the lowest possible rate, which is one-third of the input rate. This trick works for arbitrary M, L as long as they are relatively prime because in that case, two things are true: (a) there exist integers n_0 and n_1 satisfying $-n_0 L + n_1 M = 1$

(Euclid's theorem, [see Sec. 2.3 in Bose, 1985]) so that we can replace each z^{-1} with $z^{n_0 L} z^{-n_1 M}$, and (b) the decimator and expander in cascade can be interchanged.

The above polyphase implementation requires only $(N+1)/M$ MPUs and $(N+1-L)/M$ APUs. Note that the 'additions' which follow expanders are not counted, as these represent time-domain interlace operations. (The reduction of MPUs obtainable due to linear-phase symmetry has not been accounted here.) The structure is most efficient because, the decimators have been moved to the left of all the computational units, and the interpolators moved to the right of these computational units.

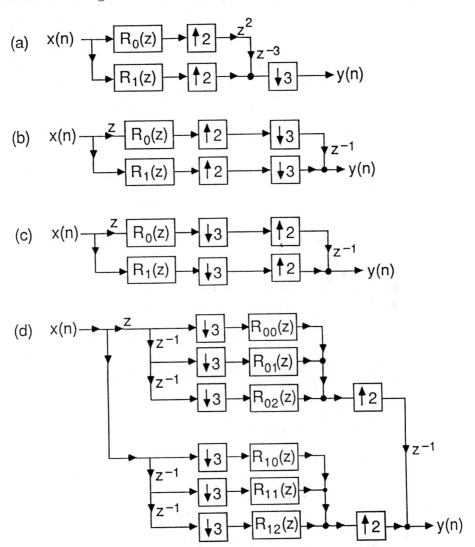

Figure 4.3-8 Successive redrawings of the fractional decimation circuit, to maximize computational efficiency. (© Adopted from 1990 IEEE.)

Computations per output sample. It is instructive to estimate the complexity per computed output sample $y(n)$. Since there are L/M output samples per unit time, we have $(M/L) \times (N+1)/M = (N+1)/L$ multiplications per output sample. Similarly, there are $(N+1-L)/L$ additions per output sample. In Sec. 4.1 we saw that for a given set of specifications on in-band ripples, and for a given σ, the order N of $H(z)$ is proportional to L. So the number of multiplications and additions per *output sample* are essentially independent of L!

4.3.4 Commutator models

A very appealing conceptual tool to visualize the operation of polyphase implementations is the commutator model. To understand this note that any polyphase implementation is characterized either by a delay chain followed by a set of decimators, or by a set of expanders followed by a delay chain (e.g., Fig. 4.3-4).

Figure 4.3-9 demonstrates a commutator-model equivalent circuit (with $M = 3$) which is useful for Type 1 polyphase implementations. The model is almost self-explanatory but the following comments are in order. We have a switch (shown by a heavy line) which can assume one of three possible positions. The switch position shown in the figure corresponds to $n = 0$ (because of the label '$n = 0$' attached to the switch). The switch rotates at uniform speed and takes on the three postions in the manner indicated. Thus whenever $n \bmod 3 = -2$, the switch is at the bottom position; when $n \bmod 3 = -1$, the switch is at the middle position; and when $n \bmod 3 = 0$, the switch returns to the top position. The switch keeps rotating in this way in a counter clockwise direction. Figure 4.3-10 shows the polyphase implementation of the decimation filter, redrawn using the counterclockwise commutator model. It is explicitly clear that each polyphase component has only to operate at three-fold lower rate.

A second type of commutator model, useful for Type 2 polyphase implementations, is shown in Fig. 4.3-11. This is called the *clockwise model*, as the switch rotates in a clockwise direction. The label '$n = 0$' in the figure means that the switch is in this particular position (i.e., bottom position) for $n \bmod 3 = 0$. The switch is in the middle position for $n \bmod 3 = 1$ and in the top position for $n \bmod 3 = 2$.

4.3.5 Further Applications of the Polyphase Concept

Periodically Time Varying Property

Some of the multirate systems introduced in this chapter can be considered to be linear periodically time varying (LPTV) systems. The simplest example is the cascade shown in Fig. 4.2-2(a) with $L = M$. We then have the input output relation

$$y_1(n) = \begin{cases} x(n) & \text{if } n \text{ is mul. of } M, \\ 0 & \text{otherwise.} \end{cases} \quad (4.3.13)$$

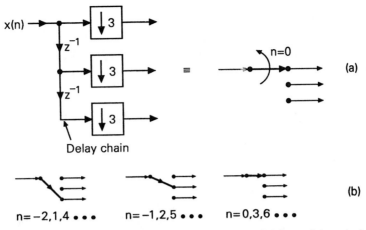

Figure 4.3-9 The counterclockwise commutator model for a delay chain followed by decimators. (a) Example with $M = 3$, and (b) operation of the commutator switch.

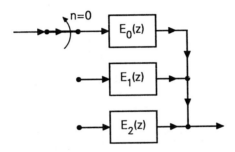

Figure 4.3-10 The polyphase implementation of a decimation filter ($M = 3$) using a counterclockwise commutator model.

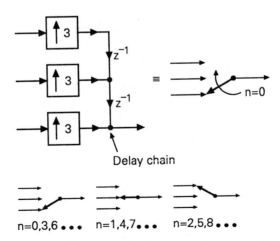

Figure 4.3-11 The clockwise commutator model for a set of expanders followed by a delay chain.

This system is equivalent to a multiplier, whose value is unity when n is a multiple of M and zero otherwise. So, this is a linear system, with periodically time varying coefficient (period M).

A second example is the interpolation filter (Fig. 4.3-3). Denoting the impulse responses of the polyphase filters $R_k(z)$ by $r_k(n)$, the output is given by

$$y(2n) = \sum_{m=-\infty}^{\infty} r_1(m)x(n-m),$$
$$y(2n+1) = \sum_{m=-\infty}^{\infty} r_0(m)x(n-m). \qquad (4.3.14)$$

In other words, the even numbered samples of $y(n)$ are obtained by filtering $x(n)$ with $R_1(z)$, and odd numbered samples obtained by filtering $x(n)$ with $R_0(z)$. The rate of $y(n)$ is two times that of $x(n)$, unlike the previous example. Strictly speaking, this system does not fall under the category of LPTV systems (which should have the same input and output rates).

A third example is the quadrature mirror filter (QMF) bank, (shown in Fig. 4.5-3 later). In Chap. 5 we will see that this is an LPTV system (with same input and output rates), and reduces to an LTI system when aliasing is completely eliminated.

Perfect Reconstruction (PR) Systems

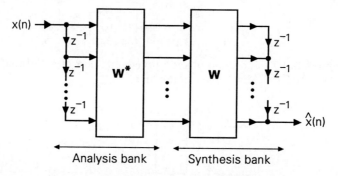

Figure 4.3-12 An analysis/synthesis system with the perfect reconstruction property.

In this text we will see many applications where the analysis bank and synthesis bank of Fig. 4.1-15 are connected back to back, that is, $y_k(n)$ is taken to be equal to $x_k(n)$. [Most of these applications are, in fact, such that there are decimators following $x_k(n)$ and expanders preceding $y_k(n)$, but let us ignore them for this discussion.] Such a system is said to be a perfect reconstruction (PR) system if $\hat{x}(n) = cx(n - n_0)$ for some $c \neq 0$ and integer n_0.

A simple PR system is shown in Fig. 4.3-12 where the analysis bank is the same as the uniform DFT bank of Fig. 4.1-16(a). The analysis filters are $H_k(z) = \sum_{\ell=0}^{M-1} z^{-\ell} W^{-k\ell}$. The synthesis filters $F_k(z)$ for this system are given by (Problem 4.20)

$$F_k(z) = W^{-k} H_0(zW^k). \qquad (4.3.15)$$

So the synthesis filters form a uniform DFT bank except for the scale factors W^{-k}. Since $\mathbf{WW}^* = M\mathbf{I}$, the system of Fig. 4.3-12 has the perfect reconstruction property, that is, $\widehat{x}(n) = M^2 x(n - M + 1)$.

The Polyphase Identity.

Consider the structure of Fig. 4.3-13(a) which is a cascade of an expander followed by a filter $H(z)$, which in turn is followed by a decimator. This is similar to the fractional decimation scheme of Fig. 4.1-10(b), but with $L = M$. Such a strange interconnection arises in many applications (e.g., transmultiplexers to be discussed in Sec. 4.5.4).

Even though the decimator and expander are time-varying building blocks, the above cascaded system happens to be time-invariant. To see this note that the input to the decimator has z-transform $X(z^M)H(z)$. The decimated version therefore has the z-transform

$$[X(z^M)H(z)]\Big|_{\downarrow M} = X(z)[H(z)|_{\downarrow M}] = X(z)E_0(z), \qquad (4.3.16)$$

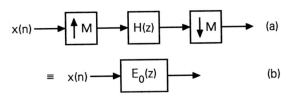

Figure 4.3-13 The polyphase identity. (a) An unusual cascade, and (b) its equivalent circuit.

where $E_0(z)$ is the 0th polyphase component of $H(z)$. The remaining polyphase components of $H(z)$ are irrelevant here! Thus Fig. 4.3-13(a) represents a linear time invariant system with transfer function $E_0(z)$.

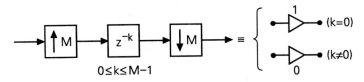

Figure 4.3-14 An application of the polyphase identity.

As an interesting application of this, suppose $H(z) = z^{-k}$, with $0 \le k \le M - 1$. We then have

$$E_0(z) = \begin{cases} 1 & \text{if } k = 0 \\ 0 & \text{if } 1 \le k \le M - 1. \end{cases} \qquad (4.3.17)$$

Fig. 4.3-14 shows this equivalence.

4.4 MULTISTAGE IMPLEMENTATIONS

In many applications, it is necessary to decimate by a large integer factor. Even though this can be done by designing a filter $H(z)$ and using polyphase structures, it is even more efficient (in terms of number of computations per unit time) to design the decimation filter in multiple stages. Figure 4.4-1 shows an example where $M = 16$. Since $16 = 4 \times 2 \times 2$, it is possible to implement the system in three stages as shown.

This proposal raises several questions: What is the best way to split M into factors? In what order should these factors be arranged? These questions are not easy to answer. However, the fact that multistage implementations result in more efficient systems can be demonstrated very easily.

Our discussion relies on the fact that a linear phase FIR lowpass filter meeting specifications as in Fig. 3.1-1 has order

$$N \approx \frac{D(\delta_1, \delta_2)}{\Delta f}, \qquad (4.4.1)$$

where $D(\delta_1, \delta_2)$ is a function of the peak passband ripple δ_1 and peak stop ripple δ_2. A formula of this type holds for equiripple designs [Eq. (3.2.32)] as well as Kaiser-window based designs [Eq. (3.2.8)], even though $D(\delta_1, \delta_2)$ depends on the method. Note that for fixed ripple size, the order varies as $1/\Delta f$.

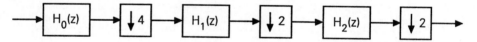

Figure 4.4-1 A multistage implementation of a 16-fold decimator.

4.4.1 The Interpolated FIR (IFIR) Approach

Multistage techniques for decimation and interpolation filters are best motivated by first presenting an efficient technique for the design (and implementation) of narrow band lowpass filters called the *Interpolated FIR* (IFIR) technique [Neuvo, et al., 1984]. This technique, by itself, has nothing to do with decimators and expanders, and is applicable for any narrowband filter. In the context of decimation and interpolation filtering, it also shows how multistage structures arise naturally. Independent treatement of multistage structures can also be found in Crochiere and Rabiner [1983].

To explain the basic idea, consider Fig. 4.4-2(a) which shows a lowpass specification (with ripples not shown). Let N denote the required filter order. Now, instead of meeting these specifications, suppose we try to meet a two-fold stretched specification (Fig. 4.4-2(b)). The stretched filter $G(z)$ has transition bandwidth $2\Delta f$ so that its order is $N/2$. This means that the number of multiplications and additions are reduced by a factor of two.

Figure 4.4-2(c) shows the magnitude response of $G(z^2)$. This filter has two passbands. One of these is similar to the desired passband, whereas the other passband (centered around π) is unwanted. The unwanted passband can be suppressed by cascading $G(z^2)$ with a new filter $I(z)$ [Fig. 4.4-2(d)]. For small ω_S, this filter has a very wide transition band so that it requires very low order.

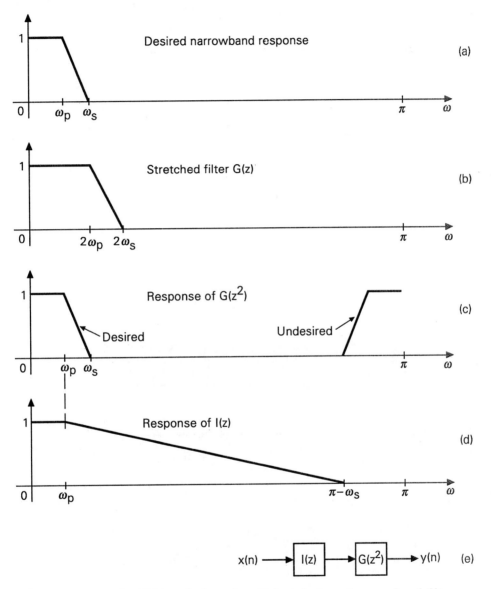

Figure 4.4-2 The IFIR technique for efficient design of narrowband filters.

Figure 4.4-2(e) shows the complete system. Denoting the orders of $G(z)$

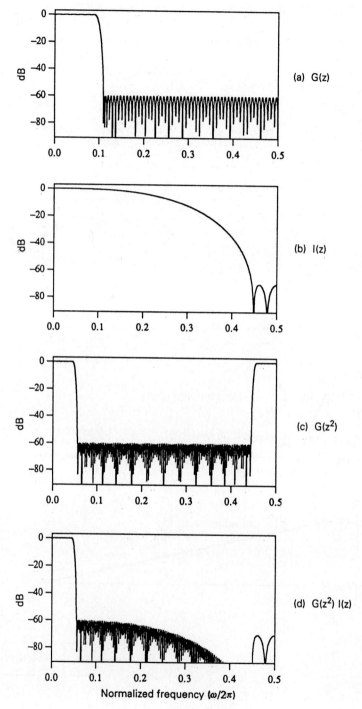

Figure 4.4-3 Design example 4.4.1. Magnitude response plots for various filters.

and $I(z)$ by N_g and N_i, the system requires a total of $(N_g + 1) + (N_i + 1)$ multipliers and $N_g + N_i$ adders. This is much less than the complexity required to implement a filter designed directly to meet the specifications of Fig. 4.4-2(a).

Adjusting ripple sizes. Notice that the passband of the overall response $G(z^2)I(z)$ can have ripples larger than the ripples of $G(z^2)$ and $I(z)$. To meet the desired specifications, we can take the peak passband ripples of $G(z)$ and $I(z)$ to be $\delta_1/2$ so that the peak passband ripple of the overall system $G(z^2)I(z)$ is no greater than δ_1. (Review Problem 3.3). If the peak stopband ripples of $G(z)$ and $I(z)$ are δ_2, then the system $G(z^2)I(z)$ has stopband ripple no larger than δ_2. The order of $G(z)$ is somewhat larger than $N/2$ because of the more stringent passband requirement.

Terminology. The filter $G(z)$ is called the *model filter*. The purpose of $I(z)$ is to suppress the extra copy (image) of the basic passband. $I(z)$ was originally called an interpolator [Neuvo, et al., 1984] because the odd-numbered impulse response coefficients of $G(z^2)$, which are zero-valued, get "filled in" when we cascade $G(z^2)$ with $I(z)$. In this text, since an interpolator has a different meaning, we will call $I(z)$ an *image suppressor*.

Design Example 4.4.1: IFIR Design

As a specific example, let $\omega_p = 0.09\pi, \omega_S = 0.11\pi, \delta_1 = 0.02$, and $\delta_2 = 0.001$. Then a direct equiripple design using the McClellan-Parks algorithm would require an order $N = 233$.

If we use the IFIR method, the filter $G(z)$ has specifications $\omega_p = 0.18\pi, \omega_S = 0.22\pi, \delta_1 = 0.01$ and $\delta_2 = 0.001$. The filter $I(z)$ has specifications $\omega_p = 0.09\pi, \omega_S = 0.89\pi, \delta_1 = 0.01$ and $\delta_2 = 0.001$. If $G(z)$ and $I(z)$ are designed using the McClellan-Parks algorithm, then the filter orders $N_g = 131$ and $N_i = 6$ are found to be sufficient. So, $G(z)$ [hence $G(z^2)$] requires 66 multipliers and 131 adders, whereas $I(z)$ requires 4 multipliers and 6 adders. The total complexities of the IFIR method and the direct method are summarized in Table 4.4.1. The filter $I(z)$ is very inexpensive, whereas the cost of $G(z^2)$ is little more than half the cost of the direct design. Notice finally that the system $G(z^2)I(z)$ has linear phase since $G(z)$ and $I(z)$ have this property. Fig. 4.4-3 shows plots of appropriate frequency responses.

In addition to the computational advantage, the IFIR approach also offers advantages during the filter design phase: a direct equiripple design of a filter with as high an order as 233 can result in numerical inaccuracies, and convergence-difficulties. The IFIR technique allows us to design two filters $G(z)$ and $I(z)$ of much lower order, which can be used to eventually meet the same specifications.

Extensions. Several extensions of this idea can be found in Neuvo, et al. [1984]. For example, instead of stretching the specifications by two, it is possible to stretch by an amount $M_1 > 2$. In principle M_1 can be as large as the integer-part of π/ω_S. (So, in the above example, a much larger M_1 is possible.) If $G(z)$ is the filter which meets the stretched specifications, then the system which meets the original specifications is given by $G(z^{M_1})I(z)$. The filter $G(z^{M_1})$ has $M_1 - 1$ unwanted passbands (images) in addition to the desired passband centered around $\omega = 0$. The image-suppressor $I(z)$ eliminates these unwanted passbands.

The transition bandwidth of $I(z)$ depends on ω_S (the original desired stopband edge) and M_1 (see Fig. 4.4-4). If M_1 is too large (i.e., if the given specifications are stretched too much), then the transition band of $I(z)$ becomes very narrow [so that $I(z)$ dominates the cost], and we begin to get decreasing returns. (Essentially we have "stretched our luck" too far!) Summarizing, as M_1 increases the cost of $G(z)$ decreases and that of $I(z)$ increases. There is some M_1 which minimizes the cost.

Extensions of the IFIR technique for wideband lowpass filters, and general bandpass filters have been made. Also see Problem 4.25.

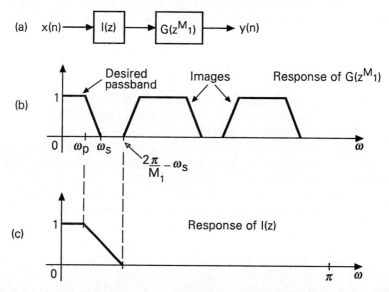

Figure 4.4-4 Extension of the IFIR technique for stretching factor $M_1 > 2$.

4.4.2 Multistage Design of Decimation and Interpolation Filters

Suppose the lowpass filter $H(z)$ in Fig. 4.1-7 has stopband edge near π/M. If M is large, then $H(z)$ is a narrow-band lowpass and can be implemented using the IFIR technique, to reduce computations. Suppose we take the stretch factor M_1 to be a factor of M, that is, $M = M_1 M_2$. Then, the system is as in Fig. 4.4-5(a). This can be rearranged as in Fig. 4.4-5(b) using a noble identity. Thus, we perform the decimation in two stages: first by M_1 and then by M_2. So, the IFIR approach naturally leads to the two-stage implementation. We can, in fact, repeat this process by factorizing M_1 and M_2 further.

TABLE 4.4.1 Design example 4.4.1. Comparison of direct method with the IFIR method.

Quantity Compared	Conventional Method	IFIR Method		
		$G(z)$	$I(z)$	Total
Filter order	233	131	6	268
Number of Mul.	117	66	4	70
Number of Add.	233	131	6	137

TABLE 4.4.2 Design example 4.4.2. Complexity comparison between direct design and multistage design.

	Direct design $H(z)$	Multistage Design		
		$G(z)$	$I(z)$	Total
Filter order	2,028	90	139	2,389
MPUs	≈ 21	0.92	2.8	3.72
APUs	≈ 41	1.8	5.56	7.36
Mul per sec (8 kHz)	168,000			29,760
Add per sec (8 kHz)	328,000			58,880

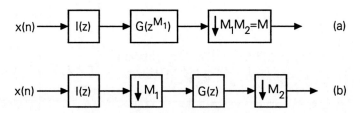

Figure 4.4-5 The two-stage decimator, developed from the IFIR decimation filter.

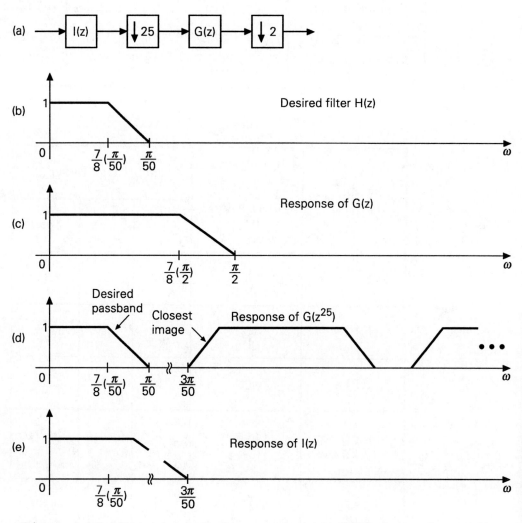

Figure 4.4-6 Pertaining to Design example 4.4.2. (IFIR design of a 50-fold decimation filter.)

Design Example 4.4.2: Multistage Decimation Filter

Consider the 50-fold decimation of an 8 KHz signal as in Fig. 4.1-13. Assume the passband and stopband ripples for the filter $H(z)$ are $\delta_1 = 0.01, \delta_2 = 0.001$. If the decimation filter has band edges at 70 Hz and 80 Hz, then the normalized transition bandwidth for the digital filter is $\Delta f = 1/800$. A direct equiripple design using the McClellan-Parks algorithm would require an order $N = 2,028$.

Now consider using the IFIR approach to design the same filter with $M_1 = 25$. Figure 4.4-6(a) shows the multistage system. The desired filter response is as in Fig. 4.4-6(b). The response of $G(z)$ is as in Fig. 4.4-6(c), with $\Delta f = 25 \times (1/800)$. The specifications are: $\omega_p = 0.4375\pi$, $\omega_S = 0.5\pi$, $\delta_1 = 0.005, \delta_2 = 0.001$ as explained earlier, so that the order of $G(z)$ is 90. The image suppressor $I(z)$ has appearance as in Fig. 4.4-6(e). This has $\Delta f = 17/800$. The specifications are $\omega_p = 0.0175\pi$, $\omega_S = 0.06\pi$, $\delta_1 = 0.005$, and $\delta_2 = 0.001$. An equiripple filter of order $N_i = 139$ is sufficient to meet these specifications. Notice that $I(z)$ has higher order than $G(z)$. Fig. 4.4-7 gives various frequency response plots for the resulting design.

Table 4.4.2 summarizes the details, comparing the conventional approach to the IFIR approach. By using a polyphase structure [Fig. 4.3-4(a) with $M = 50$] and exploiting the linear-phase symmetry, we obtain the MPU and APU count shown in Table 4.4.2. For the multistage design, the APU count is obtained as follows: $I(z)$ requires 139 additions. By using the polyphase structure, we can reduce the complexity to 139/25 APU. Next, $G(z)$ which has order 90 requires 90 additions. By using appropriate polyphase decomposition, we can peform this at 1/50th of the input rate. So we have $90/50 = 1.8$ APUs only. The MPU counts shown in the Table can be verified similarly.

The improvement obtainable using the multistage approach is therefore by a factor of almost six. The table also indicates the computations per second with 8 KHz sampling rate prior to decimation.

From the above discussions, we see that the order of $G(z)$ in terms of the specifications $\delta_1, \delta_2, \omega_p$, and ω_S can be written as

$$N_g = \frac{2\pi D(0.5\delta_1, \delta_2)}{M_1(\omega_S - \omega_p)}. \qquad (4.4.2)$$

The image suppressor has order

$$N_i = \frac{2\pi D(0.5\delta_1, \delta_2)M_1}{2\pi - (\omega_S + \omega_p)M_1}. \qquad (4.4.3)$$

The number of MPUs is approximately

$$\frac{N_g}{2M} + \frac{N_i}{2M_1} = \frac{\pi D(0.5\delta_1, \delta_2)}{MM_1(\omega_S - \omega_p)} + \frac{\pi D(0.5\delta_1, \delta_2)}{\left(2\pi - (\omega_S + \omega_p)M_1\right)}. \qquad (4.4.4)$$

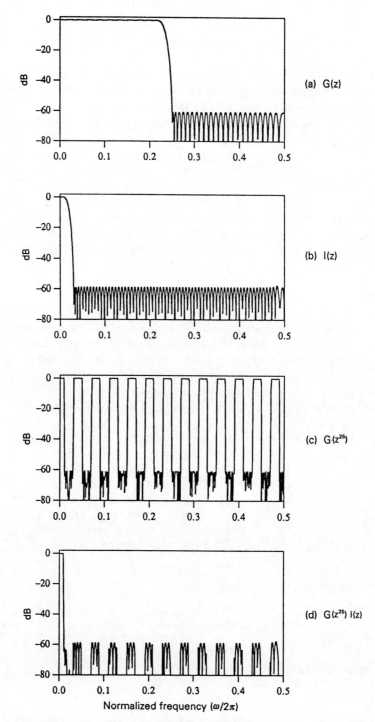

Figure 4.4-7 Design example 4.4.2. Magnitude response plots for various filters.

The factor '2' appears in the denominator because we can save a factor of two due to linear-phase symmetry. For fixed M, the tradeoff is, therefore, clear. As M_1 increases, the first term decreases whereas the second term increases. Among all integer factors of M, there is an optimal factor M_1 which minimizes (4.4.4).

For interpolation filters, the multistage idea still works. Figure 4.4-8 demonstrates the required manipulations with an example. We now see that interpolation by 25 *follows* [rather than precedes, as in Fig. 4.4-6(a)] interpolation by 2.

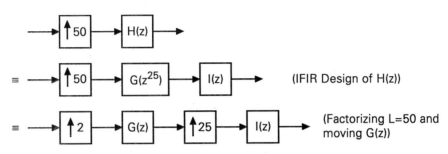

Figure 4.4-8 Multistage design of interpolation filters.

4.5 SOME APPLICATIONS OF MULTIRATE SYSTEMS

We now review a number of applications of multirate filters and filter banks. More applications are outlined in Sec. 4.6.4.

4.5.1 Digital Audio Systems

In the digital audio industry, it is a common requirement to change the sampling rates of bandlimited sequences [Digital audio, 1985]. This arises for example when an analog music waveform $x_a(t)$ is to be digitized [Bloom, 1985]. Assuming that the significant information is in the band $0 \leq |\Omega|/2\pi \leq$ 22kHz, a minimum sampling rate of 44kHz is suggested [Fig. 4.5-1(a)]. It is, however, necessary to perform analog filtering before sampling to eliminate aliasing of out-of-band noise. Now the requirements on the analog filter $H_a(j\Omega)$ [Fig. 4.5-1(b)] are strigent: it should have a fairly flat passband [so that $X_a(j\Omega)$ is not distorted] and a narrow transition band. Optimal filters for this purpose (such as elliptic filters which are optimal in the minimax sense) have a very nonlinear phase response around the bandedge, that is, around 22kHz. [See Design Example 3.3.2.] In high quality music this is considered to be objectionable.

A common strategy to solve this problem is to oversample $x_a(t)$ by a factor of two (and often four). The filter $H_a(j\Omega)$ now has a much wider transition band (Fig. 4.5-1(c)), so that the phase-response nonlinearity is acceptably low. In fact it is possible to use an analog filter with approximately linear phase in the passband (such as the Bessel filter, [p. 129, Antoniou, 1979]). Such filters are sufficient to provide the required stopband attenuation to avoid aliasing.

The sequence $x_1(n)$ obtained by the above oversampling method is then lowpass filtered (Fig. 4.5-1(d)) by a digital filter $H(z)$ and then decimated by the same factor of two, to obtain the final digital signal $x(n)$. The crucial point is that since $H(z)$ is digital, it can be designed to have linear phase while at the same time providing the desired degree of sharpness.

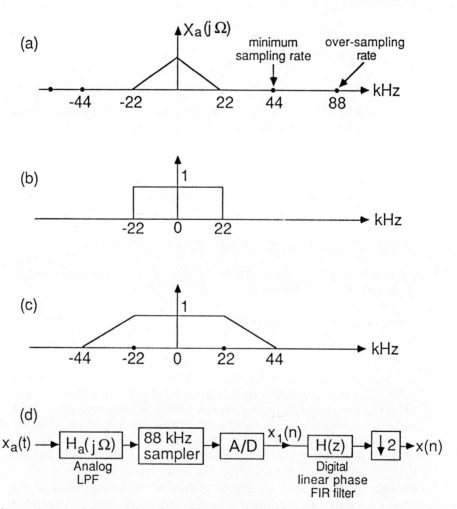

Figure 4.5-1 (a) Spectrum of $x_a(t)$; (b) Antialiasing filter response for sampling at 44 kHz; (c) Antialiasing filter response for sampling at 88 kHz; (d) Improved scheme for A/D stage of a digital audio system.

A similar problem arises after the D/A conversion stage, where the digital music signal $y(n)$ should be converted to an analog signal by lowpass filtering. To eliminate the images of $Y(e^{j\omega})$ in the region outside 22kHz, a sharp cutoff (hence nonlinear phase) analog lowpass filter is required. This problem is avoided by using an expander and a digital interpolation filter. After this D/A conversion is performed, and is followed by analog filtering. The interpolation filter $H(z)$ is once again a linear-phase FIR lowpass filter, and introduces no phase distortion.

The obvious price paid in these systems is the increased internal rate of computation. However, by using the polyphase framework (Sec. 4.3) the efficiency of decimation and interpolation filters can be significantly improved.

Fractional Sampling Rate Alterations

In digital audio, there are several applications which require fractional sampling rate alterations. This is because at least three sampling rates coexist. Thus for most studio work the sampling rate is 48 kHz, whereas for CD mastering (both digital tape and compact discs) the rate is 44.1 kHz. For broadcasting of digital audio, a sampling rate of 32 kHz is expected to become the standard. To convert from studio frequency to CD mastering standards, one would, therefore, use the arrangement of Fig. 4.1-10(b) with $L = 441$ and $M = 480$. Such large values of L normally imply that $H(z)$ has very high order. A multistage design (Sec. 4.4) is more convenient (as well as computationally efficient) in such cases.

Further applications of multirate *filter banks* in digital audio can be found in Sec. 4.6.4.

4.5.2 Subband Coding of Speech and Image Signals

In practice one often encounters signals with energy dominantly concentrated in a particular region of frequency. An extreme example was considered earlier where *all the energy* is in $0 \le |\omega| < 2\pi/3$. In that case it was possible to compress the signal simply by decimating it by a factor of 3/2, or less.

It is more common, however, to encounter signals that are not band limited, but still have dominant frequency bands. An example is shown in Fig. 4.5-2(a). The information in $|\omega| > \pi/2$ is not small enough to be discarded. And we cannot decimate $x(n)$ without causing aliasing either. It does seem unfortunate that a small (but not negligible) fraction of energy in the high frequency region should prevent us from obtaining any kind of signal compression at all.

But there is a way to get around this difficulty: we can split the signal into two frequency bands by using an analysis bank with responses as in Fig. 4.5-2(b). The subband signal $x_1(n)$ has less energy than $x_0(n)$ and so can be encoded with fewer bits than $x_0(n)$. As an example, let $x(n)$ be a 10kHz signal (10,000 samples/sec) normally requiring 16 bits per sample so that the data rate is 160 kilo bits/sec. Let us assume that the subband signals $x_0(n)$ and $x_1(n)$ can be represented with 16 bits and 8 bits per sample, respectively. Because these signals are also decimated by two, the data rate now works out to be $80 + 40 = 120$ kilo bits/sec, which is a compression by 4/3. This is the basic principle of subband coding: split the signal into two or more subbands, decimate each subband signal, and allocate bits for samples in each subband depending on the energy content. This strategy is clearly a generalization of the simple decimation process (which works only for strictly bandlimited signals). In Appendix C the theoretical basis for the bit allocation strategy is studied, by making certain simplifying assumptions on the statistics of the input $x(n)$. In speech coding practice the number of subbands, filter bandwidths and bit allocations are chosen to further exploit the *perceptual properties* of human hearing.

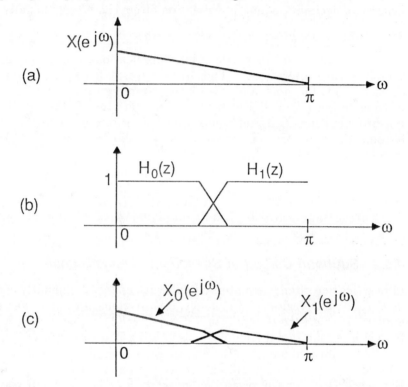

Figure 4.5-2 Splitting a signal into subband signals $x_0(n)$ and $x_1(n)$.

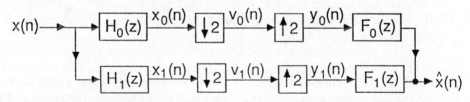

Figure 4.5-3 The analysis/synthesis system for subband coding. (Also called the two band QMF bank; see text.)

The reconstruction of the full band signal is done using the expanders and synthesis filters as in Fig. 4.5-3. The expanders restore the original sampling rate, and the filters $F_k(z)$ eliminate the images. Further generalizations follow immediately: the signal can be split into M subbands with each subband signal decimated by M and independently quantized. The complete analysis synthesis system in Fig. 4.5-3 is called the *quadrature mirror filter* (QMF) bank, and is the topic of Chap. 5 to 8.

For details on subband coding, see Crochiere, et al. [1976], Crochiere [1977 and 1981], and Woods [1991]. The coding in each subband is typically more sophisticated than just quantization. For example, techniques such as DPCM and ADPCM are commonly used [Jayant and Noll, 1984]. The specific properties of speech signals and their relation to human perception are carefully exploited in the coding process; the appropriate number of subbands and the coding accuracy in each subband are judged based on the *articulation index*. The quality of subband coders is usually judged by what is called the *mean opinion score* (MOS). This score is obtained by performing listening tests with the help of a wide variety of unbiased listeners, and asking them to assign a score for the quality of the reproduced signal $\hat{x}(n)$ [in comparison to $x(n)$]. The maximum score is normalized to 5 by convention. Subband-coded speech with an average bit rate of 16 kilo bits/sec can typically achieve a MOS of 3.1, whereas at 32 kilo bits/sec a score of 4.3 has been achieved [Chap. 11, Jayant and Noll, 1984]. Note that, if $x(n)$ is obtained by 8 kHz sampling (typical for speech), then 16 kilo bits/sec corresponds to 2 bits per sample, and 32 kilo bits/sec corresponds to 4 bits per sample. These are much lower than the typical precisions used in digital filtering (e.g., in the implementation of the subband filters). Thus, any new technique which further reduces the bit rate by a small amount (such as one bit per sample) could still qualify as a "significant contribution".

Image compression. Two dimensional multirate filter banks for image processing were first introduced by Vetterli [1984]. Subband coding has been applied for image compression by several authors [Safranek, et al., 1988], [Woods and O'Neil, 1986], [Smith and Eddins, 1990]. (See Sec. 12.0 for further references.) Coded images with only 0.48 bits per pixel (the original uncoded picture being a 8 bits per pixel image), have been found to give "acceptable" quality. Again, this statement does not give complete information, without comparing the detailed picture qualities before and after coding. Further details can be found in Woods [1991], and references therein. Multirate filter banks have been used in image processing applications in various other forms, for example, multiresolution systems, and the Laplacian pyramid (Sec. 5.8).

Music signals. Subband coding has also been applied for compression of digitized music, e.g., the discrete compact cassete (DCC) [Veldhuis, et al., 1989], [pp. 3597–3620, ICASSP, 1991], [Fettweis, et al., 1990]. A judgement on quality of the 'compressed' music depends on the nature of the waveform, as well as the threshold of acceptability imposed by the listener. It has been shown that a great deal of compression can be achieved in many cases.

General remarks on subband coding. Several comments are now in order: first, in order for subband coding to work, it is necessary to have some knowledge about the energy distribution of $X(e^{j\omega})$. Such knowledge is usually obtainable by means of adaptive estimation of energy in various subbands [Jayant and Noll, 1984]. The bit allocation strategy depends crucially upon this energy information (Sec. C.2.1, Appendix C). Second, the band splitting and decimation operation inevitably results in *aliasing* because the filters $H_k(z)$ are not ideal. The filters $F_k(z)$ should be chosen carefully in such a way that aliasing is actually canceled. These details are the topic of Chap. 5.

4.5.3 Analog Voice Privacy Systems

These systems are intended to communicate speech over standard analog telephone links, while at the same time ensuring voice privacy. The main idea is to split a signal $x(n)$ into M subband signals $x_k(n)$ and then divide each subband signal into segments in the time domain. The time segments as well as the subbands are then permuted and recombined to generate an encrypted signal $y(n)$, which can then be transmitted (after D/A conversion). Unless an eavesdropper is aware of the details of the permutation (i.e., unless he has the 'key'), he will be unable to eavesdrop. The aims of the designer of such a privacy system are: the encrypted message should be unintelligible, decryption without a key should be very difficult, and the decrypted signal should be of good quality retaining naturalness and voice characteristics. Some of these goals have been accomplished [Cox, et al., 1987].

At the receiver end, $y(n)$ is again split into subbands, and the time segments of the subbands unshuffled to get $x_k(n)$ which can then be interpolated and recombined through the synthesis filters.

4.5.4 Transmultiplexers

In digital telephone networks, it is sometimes necessary to convert between two formats called the *time-division multiplexed* (TDM) format and the *frequency-division multiplexed* (FDM) format.

To describe the TDM format consider Fig. 4.5-4(a) where three signals are passed through 3-fold expanders and added through a delay chain. It can be shown that $y(n)$ is an interleaved version of the three signals, that is, it has the form

$$\ldots x_0(0)\ x_1(0)\ x_2(0)\ x_0(1)\ x_1(1)\ x_2(1) \ldots$$

This is the TDM version of the three signals. We can recover the three signals from $y(n)$ by using the time-domain demultiplexer shown in Fig. 4.5-4(b). Note that $X_0(z), X_1(z)$ and $X_2(z)$ are polyphase components of $Y(z)$.

To explain the FDM operation, consider Fig. 4.5-5 where transforms of three signals $x_0(n), x_1(n)$, and $x_2(n)$ are shown. The FDM signal $y(n)$ is a single composite signal, whose transform $Y(e^{j\omega})$ is obtained by "pasting" the transforms of the individual signals next to each other. Note that each individual spectrum has to be compressed by 3 to make enough room for all three signals in the range $0 \leq \omega < 2\pi$. The FDM operation can be performed using the circuit shown in the figure. Each individual signal is first passed through an expander to obtain a 3-fold compression of the transform. The interpolation filter $F_k(z)$ (assumed ideal for this discussion) retains one out of the three images which appear in $X_k(e^{j3\omega})$. The shaded portions in Fig. 4.5-5 (d), (e) and (f) are the retained images from each signal. The filter responses (which pass the shaded regions of the respective signals) are shifted with respect to each other so that the retained images from $X_k(e^{j3\omega})$ do not overlap with the retained images from $X_m(e^{j3\omega})$, $m \neq k$. If we add the outputs of the three filters, the result is the FDM signal $Y(e^{j\omega})$.

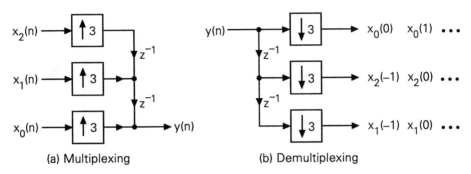

Figure 4.5-4 (a) Time domain multiplexing and (b) time domain demultiplexing, represented using multirate building blocks.

Since the shaded areas in the figure are not symmetric with respect to $\omega = 0$, the filters have complex coefficients, and $y(n)$ is complex (even if the individual signals might be real). If $x_k(n)$ are real, we can avoid this by judicious choice of responses $|F_k(e^{j\omega})|$ so that there is symmetry with respect to $\omega = 0$. Each filter then has one passband in the region $0 \leq \omega < \pi$ and one in the conjugate region $\pi \leq \omega < 2\pi$.

Figure 4.5-6 shows the complete transmultiplexer. The components $x_k(n)$ of the TDM version can be recovered by separating the consecutive regions of $Y(e^{j\omega})$ (which are the M message signals) with the help of an analysis bank and then decimating the outputs. Now, if the synthesis filters $F_k(z)$ are non ideal, the adjacent spectra in Fig. 4.5-5(g) will actually tend to overlap. Similarly if the analysis filters $H_k(z)$ are non ideal then the outputs of $H_k(z)$ have contributions from $X_k(e^{j\omega})$ as well as $X_\ell(e^{j\omega})$, $\ell \neq k$. So in general each of the reconstructed signals $\widehat{x}_k(n)$ has contribution from the desired signal $x_k(n)$ as well as the 'cross-talk' terms $x_\ell(n)$, $\ell \neq k$. An obvious approach to reduce the extent of cross talk is to design $H_k(z)$ and $F_k(z)$ to be very sharp cutoff filters, that is, practically *non overlapping* frequency responses. To obtain acceptable cross talk reduction this requires filters of very high order (e.g., exceeding 2,000; see [Bellanger, 1982]).

A novel approach to transmultiplexing was proposed by Vetterli [1986b]. In this approach, cross-talk is permitted in TDM→FDM converter and then *canceled* by the FDM→TDM converter stage. It can be shown that the cross-talk terms can be completely eliminated by careful choice of the relation

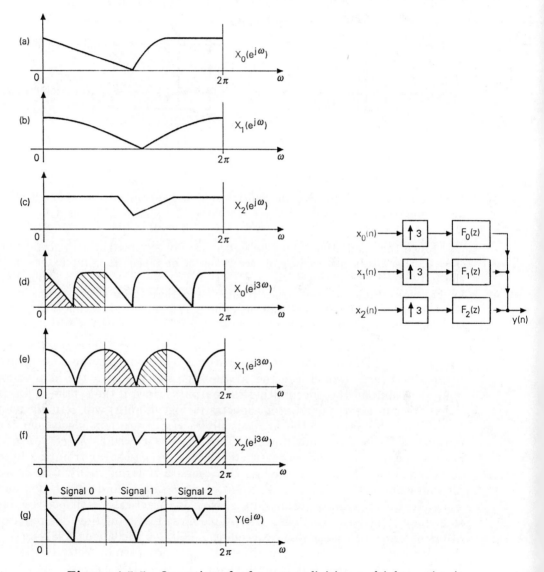

Figure 4.5-5 Operation of a frequency-division multiplexer circuit.

150 Chap. 4. Multirate system fundamentals

between the analysis and synthesis filters. In Sec. 5.9 we will derive the conditions for this and show that the set of M original signals $x_k(n)$ can be recovered by the TDM→FDM→TDM system with no distortion. Since cross talk is permitted (and then canceled), the filters $H_k(z)$ and $F_k(z)$ are more economical than in conventional designs which tend to suppress cross talk altogether.

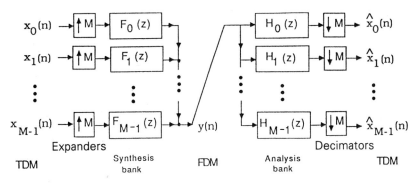

Figure 4.5-6 The complete transmultiplexer structure.

4.5.5 Multirate adaptive filters

The subject of adaptive filtering has been treated well in many texts, for example, Widrow and Stearns [1985], and Haykin [1991]. We can not get into the details here without spending several pages. We merely wish to draw attention to the fact that the concepts of decimation, interpolation, and subband decomposition have been used to obtain 'improved' adaptive filters. The 'improvement' is typically in terms of reduced filter lengths, faster convergence, and/or lower implementation complexity. Some of the references in this context are: Gilloire [1987], Gilloire and Vetterli [1992], and [Sathe and Vaidyanathan, 1990,1991 and 1993]. The reference Shynk [1992] provides an excellent introduction and overview of this topic.

4.6 SPECIAL FILTERS AND FILTER BANKS

A number of multirate applications employ special types of systems such as half-band filters, Mth band filters, and power complementary filter-banks. We now review these systems, and indicate some applications.

4.6.1 M-th Band Filters or Nyquist(M) Filters

For an M-fold interpolation filter (Fig. 4.1-8 with $L = M$), the output $y(n)$ has z-transform $Y(z) = X(z^M)H(z)$. Consider the polyphase decomposition (4.3.7) for $H(z)$. Suppose the 0th polyphase component $E_0(z)$ is a constant c, that is,

$$H(z) = c + z^{-1}E_1(z^M) + \ldots + z^{-(M-1)}E_{M-1}(z^M). \qquad (4.6.1)$$

Then
$$Y(z) = cX(z^M) + \sum_{\ell=1}^{M-1} z^{-\ell} E_\ell(z^M) X(z^M). \qquad (4.6.2)$$

This means that $y(Mn) = cx(n)$. In practice we can scale the filter such that $c = 1$. Thus, even though the interpolation filter inserts new samples, the *existing* samples themselves are communicated to the output without distortion.

A filter with the above property is said to be a Nyquist [or Nyquist(M)] filter [Mueller, 1973], or an Mth band filter [Mintzer, 1982]. We see that the impulse response $h(n)$ satisfies

$$h(Mn) = \begin{cases} c, & n = 0 \\ 0, & \text{otherwise.} \end{cases} \qquad (4.6.3)$$

In other words, $h(n)$ has periodic zero-crossings separated by M samples [except that $h(0) = c$]. Figure 4.6-1(a) demonstrates this for $M = 3$. In Fig. 4.6-1(b) we show typical appearance of $y_E(n)$. If these two sequences are convolved, then the nonzero samples of $y_E(n)$ are unaffected (except for a scale factor c).

Generalized definition. We generalize the above definition so that any delayed version of an Mth band filter is also an Mth band filter. We will say that $H(z)$ is a Mth band filter or Nyquist(M) filter, if *any one* of the M polyphase components, say $E_k(z)$, has the form $E_k(z) = cz^{-n_k}$. Thus

$$H(z) = E_0(z^M) + z^{-1} E_1(z^M) + \ldots + cz^{-k} z^{-Mn_k} + \ldots + z^{-(M-1)} E_{M-1}(z^M). \qquad (4.6.4a)$$

In terms of the impulse response, the above property is equivalent to

$$h(Mn + k) = \begin{cases} c & n = n_k \\ 0 & \text{otherwise.} \end{cases} \qquad (4.6.4b)$$

Fig. 4.6-2 demonstrates this for $M = 3, k = 2$ and $n_k = 1$. [If $k = n_k = 0$, we obtain the special case (4.6.3).]

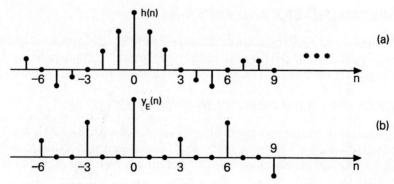

Figure 4.6-1 Mth band or Nyquist(M) filters. (a) Example of impulse response ($M = 3$), and (b) typical input to the filter if used as a 3-fold interpolation filter (see Fig. 4.1-8(a)).

The output of the above Mth band interpolation filter is

$$Y(z) = cz^{-k}z^{-Mn_k}X(z^M) + \sum_{\ell \neq k} z^{-\ell} E_\ell(z^M) X(z^M). \qquad (4.6.5a)$$

This means that the input samples $x(n)$ are communicated, without distortion, to the output according to the rule

$$y(Mn + Mn_k + k) = cx(n). \qquad (4.6.5b)$$

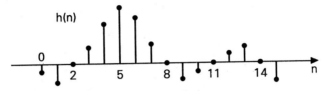

Figure 4.6-2 Example of a Mth band filter with $M = 3$, $k = 2$, and $n_k = 1$.

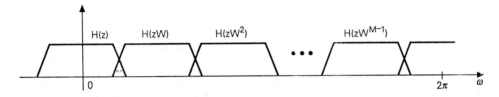

Figure 4.6-3 For an Mth band filter $H(z)$, the responses $H(e^{j(\omega - \frac{2\pi k}{M})})$ add up to a constant.

Manifestation in frequency domain. If $H(z)$ satisfies (4.6.1), we can show that

$$\sum_{k=0}^{M-1} H(zW^k) = Mc = 1 \quad \text{(assuming } c = 1/M\text{)}, \qquad (4.6.6)$$

where $W = e^{-j2\pi/M}$. The frequency response of $H(zW^k)$ is the shifted version $H(e^{j(\omega - (2\pi k/M))})$ of $H(e^{j\omega})$. So we conclude that all these M uniformly shifted versions of $H(e^{j\omega})$ add up to a constant (see Fig. 4.6-3).

The definition of Nyquist filters can be extended to the continuous-time case as well. If the impulse response $h_a(t)$ is such that $h_a(\tau n) = 0$ for any integer $n \neq 0$, we say that $h_a(t)$ is Nyquist(τ). Here τ need not be an integer.

Half-Band Filters

A half-band filter $H(z)$ is an Mth band filter with $M = 2$. For the simple case where $k = n_k = 0$, we, therefore, have

$$H(z) = c + z^{-1} E_1(z^2). \qquad (4.6.7a)$$

In terms of the impulse response $h(n)$ this means

$$h(2n) = \begin{cases} c & n = 0 \\ 0 & \text{otherwise.} \end{cases} \quad (4.6.7b)$$

The condition (4.6.6) reduces to

$$H(z) + H(-z) = 1 \quad \text{(assuming } c = 0.5\text{).} \quad (4.6.7c)$$

Here are some examples, with $c = 1$ and various choices of $E_1(z)$.

$$H(z) = \begin{cases} 1 + z^{-3}, & \text{with } E_1(z) = z^{-1} \\ z + 1 + z^{-1}, & \text{with } E_1(z) = 1 + z \\ 1 + z^{-1} + z^{-3}, & \text{with } E_1(z) = 1 + z^{-1} \\ 1 + jz^{-1}, & \text{with } E_1(z) = j. \end{cases}$$

Notice from these examples that a half-band filter may or may not be causal; it may or may not have real coefficients or linear-phase.

If $H(z)$ has real coefficients, then $H(-e^{j\omega}) = H(e^{j(\pi-\omega)})$, and this becomes

$$H(e^{j\omega}) + H(e^{j(\pi-\omega)}) = 1. \quad (4.6.7d)$$

This shows that $H(e^{j(\pi/2-\theta)})$ and $H(e^{j(\pi/2+\theta)})$ add up to unity for all θ. In other words, we have a symmetry with respect to the half-band frequency $\pi/2$, justifying the name "half-band filters". Figure 4.6-4 demonstrates the effect of this symmetry for a lowpass filter: the peak passband and stopband errors δ_1 and δ_2 are equal, and the bandedges ω_p and ω_S are *equally away from* $\pi/2$.

Zero-phase FIR half-band filters. Half-band (more generally Mth band) filters can be FIR or IIR. In the FIR case, it is possible to design them to have linear phase property, and this is the most commonly used case. In particular suppose a half-band filter is zero-phase so that $h(n) = ch^*(-n)$, with $|c| = 1$. Let the highest nonzero coefficient be $h(K)$. Then, K is odd [except in the trivial case $K = 0$], in view of (4.6.7b). So $K = 2J + 1$ for some integer J. Thus, the length of the impulse response is restricted to be of the form $4J + 2 + 1$ (unless $H(z)$ is a constant).

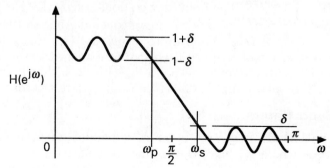

Figure 4.6-4   Frequency response of a zero-phase half band filter.

Design of Zero-Phase FIR Half-Band and M-th Band Filters

Techniques for the design of zero-phase FIR half-band and Mth band filters are discussed in Mintzer [1982] and Vaidyanathan and Nguyen [1987b]. Problem 4.30 takes the reader through all the detailed steps of an efficient method for the half-band case. The problem also shows how the response can be made to satisfy $H(e^{j\omega}) \geq 0$, which is required in perfect reconstruction filter-bank design (Sec. 5.3.6).

Window designs. The simplest way to design a good lowpass FIR Nyquist(M) filter with cutoff $\omega_c = \pi/M$ would be to use the windowing technique, that is, design $h(n)$ as

$$h(n) = \frac{\sin(\pi n/M)}{\pi n} v(n),$$

where $v(n)$ is a suitable window, say the Kaiser window. Since $\sin(\pi n/M) = 0$ for $n =$ nonzero multiple of M, (4.6.3) is indeed satisfied. Design example 3.2.1, which was presented in Chap. 3, already satisfies the Nyquist(4) property, since $\omega_c = \pi/4$. As explained in Sec. 3.2.1, any desired stopband attenuation and transition bandwidth can be obtained by using an appropriate Kaiser window. We know, however, that window methods do not result in filters that are optimal in any way (except rectangular windows; Problem 3.5).

Eigenfilter designs. A technique to design Mth band filters which are optimal in the least squares sense is provided by the eigenfilter approach described in Sec. 3.2.3. Here the design problem is formulated in terms of the coefficients b_n [Eq. (3.2.21)]. The Mth band condition can be satisfied by forcing

$$b_{Mn} = \begin{cases} c & n = 0 \\ 0 & n \neq 0. \end{cases}$$

In the eigenfilter approach we minimize an objective function ϕ [Eq. (3.2.30)] which represents a linear combination of passband and stopband accuracies. The optimum filter vector $\mathbf{b}$ is equal to the eigenvector of the matrix $\mathbf{R}$ corresponding to its smallest eigenvalue λ_0. Now if we want to impose the Mth band constraint, we can do so very easily by modifying the objective function as follows: define a new vector $\hat{\mathbf{b}}$ by deleting the components b_{Mn} ($n \neq 0$). For example with $M = 3$ we have

$$\hat{\mathbf{b}} = [b_0 \quad b_1 \quad b_2 \quad b_4 \quad b_5 \quad b_7 \quad b_8 \quad b_{10} \quad \ldots]^T.$$

Having defined $\hat{\mathbf{b}}$ like this, we replace the matrix $\mathbf{R}$ with a reduced matrix $\hat{\mathbf{R}}$, obtained by deleting from $\mathbf{R}$ all the rows and columns whose indices are multiples of M (except the 0th row and column).

We now compute $\hat{\mathbf{b}}$ such that the quantity $\hat{\mathbf{b}}^\dagger \hat{\mathbf{R}} \hat{\mathbf{b}}$ is minimized under the constraint $\hat{\mathbf{b}}^\dagger \hat{\mathbf{b}} = 1$. (The matrix $\hat{\mathbf{R}}$ continues to be positive definite.) The solution $\hat{\mathbf{b}}$ is the eigenvector of $\hat{\mathbf{R}}$ corresponding to the smallest eigenvalue. This represents the optimal [in the sense of minimizing (3.2.30)] Mth band linear-phase filter.

Design Example 4.6.1: Nyquist(5) Eigenfilters

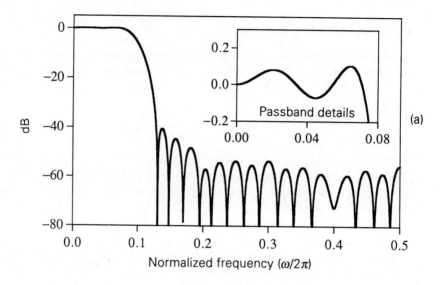

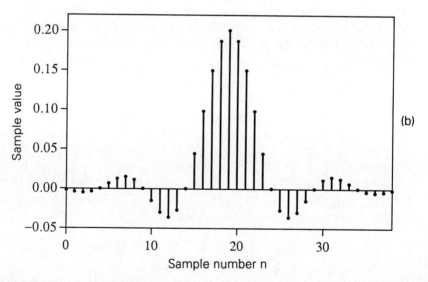

Figure 4.6-5 Design example 4.6.1. Design of 5th band (i.e., Nyquist(5)) lowpass filter. Order $N = 38$. (a) Frequency response magnitude, and (b) impulse response.

Figure 4.6-5(a) shows the magnitude response plot of a 5th band lowpass eigenfilter of order 38 designed in this manner. The band edges are $\omega_p = 0.15\pi$ and $\omega_S = 0.25\pi$. The impulse response is shown in Fig. 4.6-5(b). Design procedures which seek to obtain approximately equiripple (i.e., optimal in the minimax sense) FIR Mth band filters can also be found in the above references.

4.6.2 Complementary Transfer Functions

A. Strictly Complementary (SC) Funtions

A set of transfer functions $[H_0(z), H_1(z), \ldots, H_{M-1}(z)]$ is said to be strictly complementary (abbreviated SC) if they add up to a delay, that is,

$$\sum_{k=0}^{M-1} H_k(z) = cz^{-n_o}, \quad c \neq 0. \tag{4.6.8}$$

If we split a signal $x(n)$ into M subband signals using the SC analysis filters $H_k(z)$, then we can just add the subband signals to get back the original signal $x(n)$ with *no distortion*, except a delay.

When $M = 2$, we can design an SC pair easily as follows: let $H_0(z)$ be a Type 1 linear phase FIR filter. Then $H_0(e^{j\omega}) = e^{-j\omega N/2} H_R(\omega)$, where $H_R(\omega)$ is real. Here $N/2$ is an integer (since the order N is even for Type 1). Define $H_1(z) = z^{-N/2} - H_0(z)$. Then

$$H_1(e^{j\omega}) = e^{-j\omega N/2}\Big(1 - H_R(\omega)\Big)$$

Figure 4.6-6 shows a typical response $H_R(\omega)$ which can be obtained using the McClellan-Parks algorithm. The response $1 - H_R(\omega)$ is also shown. Thus $H_0(z)$ and $H_1(z)$ are lowpass and highpass filters, and satisfy the SC property $H_0(z) + H_1(z) = z^{-N/2}$ by construction.

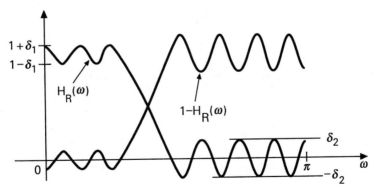

Figure 4.6-6 Example of a strictly complementary (SC) pair.

For arbitrary M we can generate an example as follows: define

$$H_k(z) = H(zW^k), \quad 0 \leq k \leq M - 1$$

where $H(z)$ is an Mth band filter satisfying (4.6.3). Then (4.6.6) holds, which means that (4.6.8) is satisfied! If the filter $H(z)$ is lowpass with appropriate passband width, then the set of M filters has response resembling Fig. 4.1-15(c).

B. Power Complementary (PC) Functions

A set of M transfer functions is said to be power-complementary (abbreviated PC) if

$$\sum_{k=0}^{M-1} |H_k(e^{j\omega})|^2 = c \quad \text{for all } \omega, \tag{4.6.9a}$$

where $c > 0$ is constant. This property is equivalent to $\sum_{k=0}^{M-1} \widetilde{H}_k(z) H_k(z) = c$ for all z, by analytic continuation. [The notation $\widetilde{H}_k(z)$ was explained in Section 2.3.] Such a property is useful in analysis/synthesis systems [that is, in systems where Fig. 4.1-15(b) is cascaded to Fig. 4.1-15(a)]. If the synthesis filters are chosen as $F_k(z) = \widetilde{H}_k(z)$, then we have $\widehat{x}(n) = cx(n)$ which implies perfect recovery of $x(n)$. In practice, the noncausality of $\widetilde{H}_k(z)$ is avoided by insertion of a delay z^{-n_0} so that $\widehat{x}(n) = cx(n - n_0)$.

Given an FIR $H_0(z)$ with $|H_0(e^{j\omega})| \leq 1$, it is easy to find an FIR filter $H_1(z)$ such that $[H_0(z), H_1(z)]$ is PC. For this note that PC property is equivalent to

$$|H_1(e^{j\omega})|^2 = 1 - |H_0(e^{j\omega})|^2. \tag{4.6.9b}$$

In other words, $H_1(z)$ is a spectral factor of the quantity $1 - |H_0(e^{j\omega})|^2$. The coefficients of such a spectral factor can be calculated as described earlier in Sec. 3.2.5.

C. Allpass Complementary (AC) Functions

A set of transfer functions is said to be allpass-complementary (abbreviated AC) if

$$\sum_{k=0}^{M-1} H_k(z) = A(z) \tag{4.6.10}$$

where $A(z)$ is allpass (Sec. 3.4). If such a set is used in an analysis bank, and if the subband signals $x_k(n)$ are recombined by adding, the result $\widehat{x}(n)$ satisfies $|\widehat{X}(e^{j\omega})| = |X(e^{j\omega})|$. So the reconstructed signal is free from magnitude distortion. Note that strictly complementary functions are also allpass complementary but not necessarily power complementary.

D. Doubly Complementary (DC) Functions

A set of transfer functions is said to be doubly complementary (DC) if it is allpass complementary as well as power complementary. There are several applications of this (including digital audio; see below).

From the results of Sec. 3.6 we obtain a simple technique to design doubly complementary filters. Recall that many standard IIR filters (including odd order Butterworth, Chebyshev and elliptic lowpass filters) can be written as in (3.6.4) where $A_0(z)$ and $A_1(z)$ are real-coefficient allpass. Recall also that $H_1(z)$ defined by (3.6.5) is power complementary to $H_0(z)$. Clearly $H_0(z) + H_1(z) = A_0(z)$ which is allpass, so that the pair $[H_0(z), H_1(z)]$ is both AC and PC, that is, doubly complementry!

E. Euclidean Complementary (EC) Functions

A pair of FIR transfer functions $[H_0(z), H_1(z)]$ is said to be Euclidean complementary (abbreviated EC) if the polynomials $H_0(z)$ and $H_1(z)$ are relatively prime [that is, do not share a common factor of the form $(\beta - \alpha z^{-1})$ with $0 < |\alpha| < \infty$]. It is well known (Euclid's theorem [see Sec. 2.3, Bose, 1985]) that if $H_0(z)$ and $H_1(z)$ are relatively prime, there exist polynomials $F_0(z)$ and $F_1(z)$ such that

$$H_0(z)F_0(z) + H_1(z)F_1(z) = c, \quad c \neq 0. \tag{4.6.11}$$

This means that we can combine the outputs of $H_0(z)$ and $H_1(z)$ to reproduce $x(n)$ with *no delay* even when the filters $H_0(z), H_1(z), F_0(z), F_1(z)$ are causal FIR!

Here is a simple example which helps to remove the initial surprise which accompanies this result: let $H_0(z) = 1 + z^{-1}$ and $H_1(z) = 1 - z^{-1}$. Then the choice $F_0(z) = F_1(z) = 0.5$ results in $\hat{x}(n) = x(n)$. Given the relatively prime pair $[H_0(z), H_1(z)]$, there exists a unique pair $[F_0(z), F_1(z)]$ (up to a scale factor) of lowest degree which can be solved using Euclid's algorithm.

An example is shown in Figure 4.6-7. The magnitude response plots of the four filters satisfying (4.6.11) are shown in the figure. The filter orders are 39 for $H_k(z)$ and 38 for $F_k(z)$.

A warning is appropriate in this context. Suppose the relatively prime polynomials $H_0(z)$ and $H_1(z)$ are causal lowpass and highpass filters with $|H_0(e^{j\omega})|, |H_1(e^{j\omega})| \leq 1$. Then the unique lowest degree pair $[F_0(z), F_1(z)]$ turns out to be *highpass* and *lowpass* respectively! So with $F_0(z)$ and $F_1(z)$ normalized so that $|F_0(e^{j\omega})|, |F_1(e^{j\omega})| \leq 1$, the constant c in (4.6.11) can be very small. This means, quantization errors in the design and implementation of the filters can dominate c, rendering this scheme impractical.

More generally, a set of M FIR functions $[H_0(z), H_1(z), \ldots, H_{M-1}(z)]$ is said to be EC if there is no factor $(\beta - \alpha z^{-1})$ with $0 < |\alpha| < \infty$ common to all of these. Under this condition, there exists a set of M FIR filters $[F_0(z), F_1(z), \ldots, F_{M-1}(z)]$ such that $\sum_{k=0}^{M-1} H_k(z)F_k(z) = 1$. In particular, if the filters $H_k(z)$ are causal then so are $F_k(z)$.

4.6.3 Relation Between Nyquist(M) Filters and Power Complementary Filters

There is a relation between Mth band filters and power-complementary filters, which can be stated as follows: Consider a transfer function $H(z)$ represented in the M-component polyphase form (4.3.7). Define the new transfer function $G(z) = \tilde{H}(z)H(z)$. Then the set $[E_0(z), E_1(z), \ldots, E_{M-1}(z)]$ is power complementary *if, and only if*, $G(z)$ is an Mth band filter.

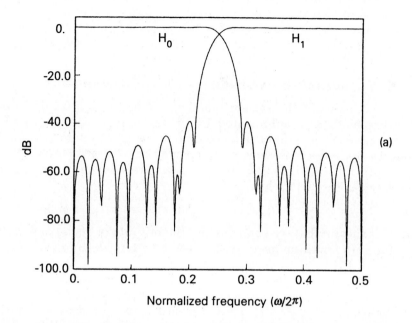

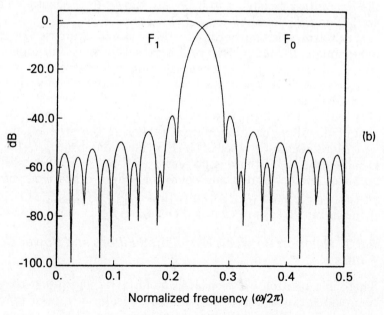

Figure 4.6-7 Example of a Eucidean complementary (EC) pair $[H_0(z), H_1(z)]$ and its synthesis counterpart $[F_0(z), F_1(z)]$. Note that $F_0(z)$ is highpass, and $F_1(z)$ lowpass!

To prove this, define

$$H_n(z) \triangleq H(zW^{-n}) = \sum_{k=0}^{M-1} z^{-k} W^{kn} E_k(z^M), \qquad (4.6.12)$$

for $0 \leq n \leq M-1$, where $W = e^{-j2\pi/M}$. This set of M transfer functions can be represented in vector form as

$$\underbrace{\begin{bmatrix} H_0(z) \\ H_1(z) \\ \vdots \\ H_{M-1}(z) \end{bmatrix}}_{\mathbf{H}(z)} = \mathbf{W} \underbrace{\begin{bmatrix} 1 & 0 & \cdots & 0 \\ 0 & z^{-1} & \cdots & 0 \\ \vdots & \vdots & \ddots & \vdots \\ 0 & 0 & \cdots & z^{-(M-1)} \end{bmatrix}}_{\mathbf{\Lambda}(z)} \underbrace{\begin{bmatrix} E_0(z^M) \\ E_1(z^M) \\ \vdots \\ E_{M-1}(z^M) \end{bmatrix}}_{\mathbf{E}(z)} \qquad (4.6.13)$$

where $\mathbf{W}$ is the $M \times M$ DFT matrix satisfying $\mathbf{W}^\dagger \mathbf{W} = M\mathbf{I}$. If the set $[E_0(z), E_1(z), \ldots, E_{M-1}(z)]$ is power complementary then $\widetilde{\mathbf{E}}(z)\mathbf{E}(z) = c$ so that

$$\widetilde{\mathbf{H}}(z)\mathbf{H}(z) = \widetilde{\mathbf{E}}(z)\widetilde{\mathbf{\Lambda}}(z)\mathbf{W}^\dagger \mathbf{W} \mathbf{\Lambda}(z)\mathbf{E}(z) = Mc. \qquad (4.6.14)$$

This in turn means that the set $H_k(z)$, $0 \leq k \leq M-1$ is power complementary. In other words, if we define $G(z) \triangleq \widetilde{H}(z)H(z)$, then $G(z)$ satisfies $\sum_{k=0}^{M-1} G(zW^{-k}) = Mc$, so that it is an Mth band filter.

Conversely, assuming that $G(z)$ is an Mth band filter, we can prove that the set of polyphase components $[E_0(z), E_1(z), \ldots, E_{M-1}(z)]$ is power complementary simply by inverting the matrices in (4.6.13) and carrying out a similar argument as above.

4.6.4 Applications of Special Transfer Functions

From the above discussions it is clear that Mth band filters and complementary filters have applications in the exact reconstruction of a signal $x(n)$ after it has been split into M subbands (provided the subband signals are not decimated; decimation would cause aliasing error, which is a major issue discussed in Chap. 5). A second application of Mth band filters is in the design of interpolation filters, as explained earlier. Some applications are also described in Mitra, et al. [1985] and Regalia and Mitra, [1987].

Doubly complementary Filters in Digital Audio

The loudspeaker system in most audio equipment typically has different subspeakers for different frequency ranges such as the tweeter (high frequency) and woofer (low frequency). In a digital audio system it is desirable to split the audio signal $y(n)$ (before D/A conversion) into the lowpass signal $y_0(n)$ and highpass signal $y_1(n)$ using an anlaysis bank $[H_0(z), H_1(z)]$. The analysis bank is more commonly called a *crossover network* in the audio industry [Bullock, III, 1986]. These subband signals can then be D/A converted and fed into the speakers (Fig. 4.6-8). Assuming that the loudspeaker introduces negligible distortion (which in general is not true), the human ear eventually perceives an analog version of $y_0(n) + y_1(n)$. In the transform domain, this is $[H_0(z) + H_1(z)]Y(z)$. To avoid any distortion in the reconstruction, it is desirable to design $H_0(z)$ and $H_1(z)$ to be a strictly complementary pair. As elaborated earlier this can be done by using a Type 1 FIR linear phase filter for $H_0(z)$, but this is more expensive than IIR filters.

With IIR filters it is possible to force the allpass complementarity. This means that $[H_0(z) + H_1(z)]Y(z) = A(z)Y(z)$ where $A(z)$ is allpass, so that the reconstructed signal represents $Y(z)$ faithfully except for phase distortion. If necessary, phase distortion can be equalized using an allpass filter.

It is desirable to design $H_0(z)$ and $H_1(z)$ to be good lowpass and highpass filters so that the speakers are not damaged by out-of-band energy. Notice, however, that if $H_0(z)$ is a good lowpass filter and if the pair $[H_0(z), H_1(z)]$ is allpass complementary, this does not necessarily mean that $H_1(z)$ is a good highpass filter. This is because the responses $H_0(e^{j\omega})$ and $H_1(e^{j\omega})$ are in general complex. For example, it is possible at some frequency ω_0 to have $H_0(e^{j\omega_0}) = e^{j\pi/3}$ and $H_1(e^{j\omega_0}) = e^{-j\pi/3}$ so that the sum is $2\cos(\pi/3) = 1$ consistent with the AC requirement $|H_0(e^{j\omega}) + H_1(e^{j\omega})| = 1$.

For this reason a *doubly complementary pair* $[H_0(z), H_1(z)]$ is most suitable: the PC property ensures that $H_1(z)$ is a good highpass filter (if $H_0(z)$ is lowpass) and the AC property eliminates amplitude distortion. Such IIR filters can be implemented much more efficiently than FIR filters, as elaborated earlier in Sec. 3.6. For systems with several subband speakers, an M-band AC filter bank can be used; see Renfors and Saramäki [1987] for design of such filters.

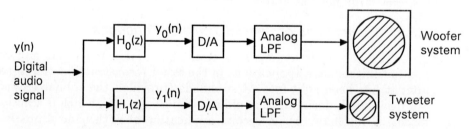

Figure 4.6-8 Splitting the digital audio signal into woofer and tweeter components.

Digital/Analog Hybrid QMF Banks in Digital Audio

A second potential approach to split the audio signal for loudspeaker driving is shown in Fig. 4.6-9. Here, the digital audio signal $y(n)$ is first split into lowpass and highpass versions by using digital analysis filters. Then D/A conversion is performed at the lower rate, on the *decimated* subband signals $v_0(n)$ and $v_1(n)$. The analog subband signals are then passed through analog synthesis filters $F_{a,0}(s)$ and $F_{a,1}(s)$ before feeding the speakers. The aim here is to choose the filters such that aliasing is canceled and amplitude and phase distortion are reduced to the desired extent. The frequency response characteristics of the speakers should be taken into consideration in such a design. The advantage of this hybrid digital/analog QMF bank is that the D/A conversion is performed at half the sampling rate. At this point in time, no results are available on the design of such hybrid QMF banks, but the idea appears promising.

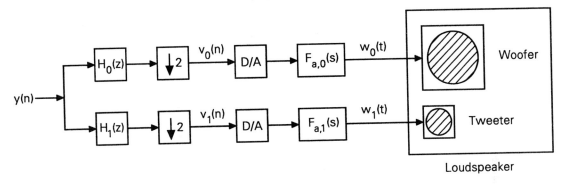

Figure 4.6-9 A digital/analog hybrid QMF bank with potential application in digital audio.

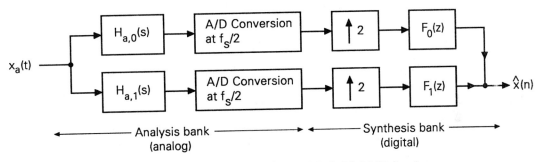

Figure 4.6-10 An analog/digital hybrid QMF bank.

Analog/Digital Hybrid QMF Banks in A/D Conversion

An immediate dual of the above idea is the use of analog analysis filters and digital synthesis filters. Such systems can have applications in A/D conversion, where a high sampling rate A/D converter can be designed by using M converters operating at M fold lower rate. Figure 4.6-10 depicts the basic idea where the analog signal $x_a(t)$ is split into subband signals by the analog filters $H_{a,0}(s)$ and $H_{a,1}(s)$. These signals are then sampled at half the intended rate f_S, and converted into digital format. The digitized subband signals are then passed through expanders (to get back the desired sampling rate) and recombined through the synthesis filters $F_0(z)$ and $F_1(z)$ to obtain $\hat{x}(n)$. The aim here is to design the filters such that $\hat{x}(n)$ represents, as closely as possible, the signal $x(n)$ which would have been obtained by direct A/D conversion of $x_a(t)$ at the rate f_S. Again, since the analog filters are not ideal, there is aliasing at the output of the analysis bank. The synthesis filters should be chosen to minimize the effect of aliasing, as well as amplitude and phase distortions.

High speed A/D conversion using filter banks is also discussed in Petraglia and Mitra [1990]. More applications of special transfer functions are scattered throughout this text in various forms.

4.6.5 Adjustable Multilevel Filters and Tunable Filters

A *multilevel* filter has typical response as shown in Fig. 4.6-11. Basically, the frequency axis is divided into a number of regions, and the response has some constant value in each region. There are transition bands between these regions so that the filter is realizable. The levels β_k are real or complex numbers. The multilevel response is a generalization of lowpass and bandpass responses.

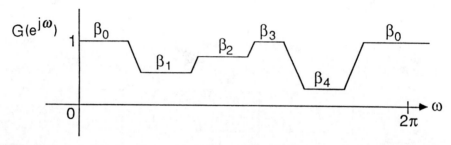

Figure 4.6-11 A typical response of a multilevel filter. (© Adopted from 1990 IEEE.)

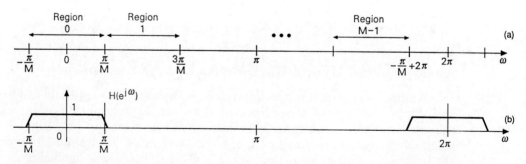

Figure 4.6-12 (a) A typical splitting of the frequency axis into M divisions and (b) a prototype lowpass response, which can be used to generate a multilevel response.

We will consider the example where the multilevel filter $G(z)$ has M

regions with equal width $2\pi/M$, as shown in Fig. 4.6-12(a).† Suppose $H(z)$ is a Mth band zero-phase filter with response as shown in Fig. 4.6-12(b). It is clear that we can obtain the multilevel filter $G(z)$ as

$$G(z) = \sum_{k=0}^{M-1} \beta_k H(zW^k), \qquad (4.6.15)$$

because the response of $H(zW^k)$ is obtained by shifting $H(e^{j\omega})$ to the right by $2\pi k/M$. We know that the M responses $H(zW^k)$ can be realized in terms of the polyphase components of $H(z)$ as in Fig. 4.3-5(a). This means that the multilevel filter can be implemented as in Fig. 4.6-13. The levels β_k appear in the structure as independent multipliers, and can be separately tuned. Similarly by changing $H(z)$ [i.e., the polyphase components $E_\ell(z)$] we can adjust the sharpness of the level transitions without affecting the levels.

In a practical implementation of this idea, we have to be more careful. When the M responses are added as in (4.6.15), we have no difficulty in obtaining the in-band levels, but the transition bands may exhibit dips or bumps depending on the degree of overlap between adjacent responses [such as $H(zW^k)$ and $H(zW^{k+1})$]. As explained below, a very simple way to avoid these dips and bumps is to take $H(z)$ to be an Mth band [i.e., Nyquist(M)] filter satisfying (4.6.6).

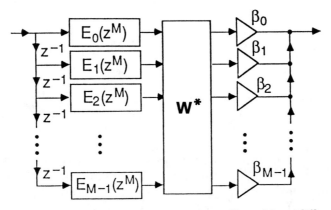

Figure 4.6-13 Polyphase implementation of an adjustable multilevel filter. Here β_k represents the 'response level' in the kth band. (© Adopted from 1990 IEEE.)

If we assume that in the region of overlap of $H(zW^k)$ and $H(zW^{k+1})$ the remaining terms of (4.6.6) are negligible, we have $H(zW^k) + H(zW^{k+1}) \approx 1$.

---

† By permitting adjacent levels to be equal if necessary, and taking M to be sufficiently large, we can in practice cover less restricted responses as in Fig. 4.6-11 also.

This implies

$$\beta_k H(zW^k) + \beta_{k+1} H(zW^{k+1}) \approx \beta_k + (\beta_{k+1} - \beta_k) H(zW^{k+1}). \quad (4.6.16)$$

This is plotted in Fig. 4.6-14 in the neighbourhood of the transition from β_k to β_{k+1} and is monotone (i.e., free from bumps and dips).

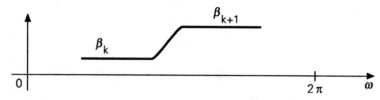

Figure 4.6-14 Behavior of (4.6.16) in the neighborhood of transition.

Summarizing, we can design multilevel filters very efficiently by the structure of Fig. 4.6-13 where $E_\ell(z)$ are the polyphase components of a zero phase Mth band filter. The Mth band property ensures smooth transition from band to band.

Design Example 4.6.2: Multilevel Filters

Figure 4.6-15 shows the response of a 5th band lowpass filter (order 58) designed using the Kaiser window approach, and two examples of multilevel responses derived from it. The values of the parameters β_k are $(0.4, 1.0, 0.7, 0.1, 0.9)$ and $(1.0, 0.5, 0.0, 0.7, 0.7)$, respectively.

The reader will notice that in the above example, the multilevel filter does not have real coefficients even though the prototype $H(z)$ does.

Tunable filters.

We can use the structure of Fig. 4.6-13 to obtain a lowpass filter whose cutoff frequency is tunable. Consider, for example, real coefficient filters so that the magnitude response is symmetric with respect to zero frequency. If we set $\beta_0 = 1$, and $\beta_k = 0$ otherwise, the 'cutoff' is π/M. If we set $\beta_0 = \beta_1 = \beta_{M-1} = 1$ and $\beta_k = 0$ otherwise, then the cutoff frequency is $3\pi/M$. [Refer to Fig. 4.6-12(a).] By making M sufficiently large, we can thus tune the cutoff frequency in very fine (discrete) steps.

Why the Name "Polyphase" Decomposition?

This seems to be the best place to explain the reason for use of the term "polyphase" decomposition. Suppose we have a Mth band filter with response as in Fig. 4.6-12(b). We know that the impulse response $e_\ell(n)$ of the polyphase component $E_\ell(z)$ is obtained by decimating $h(n + \ell)$ (Fig. 4.3-1). This means that the polyphase component $E_\ell(e^{j\omega})$ is an aliased version of $e^{j\omega\ell} H(e^{j\omega})$, so that it has the appearance of an allpass function with magnitude $1/M$ (except around $\omega = \pm\pi$ because of aliasing). This is demonstrated in Fig. 4.6-16. Now let us see how the summation

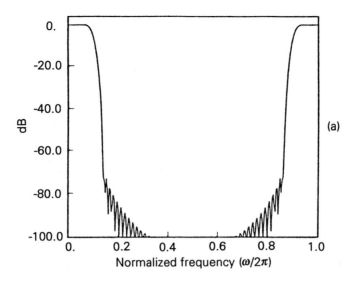

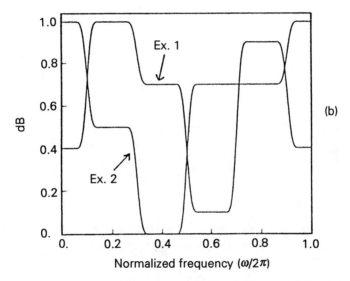

Figure 4.6-15 Design example 4.6.2. Multilevel filters. (a) Prototype 5th band filter, and (b) two multilevel examples. (© Adopted from 1990 IEEE.)

$$H(z) = \sum_{\ell=0}^{M-1} z^{-\ell} E_\ell(z^M)$$

works in the passband region in Fig. 4.6-12(b). There are M terms, each with magnitude $\approx 1/M$. These add up to nearly unity which shows that the

Sec. 4.6 Special filters and filter Banks 167

M terms $z^{-\ell}E_\ell(z^M)$ are almost in phase. But for an Mth band filter, $E_0(z)$ is constant. This shows that the phase responses of $z^{-\ell}E_\ell(z^M)$ are nearly zero in the passband. In other words, $E_\ell(e^{j\omega})$ tries to approximate $e^{j\omega\ell/M}$. So the phase response $\phi_\ell(\omega)$ of the ℓth polyphase component is trying to approximate $\omega\ell/M$, for each ℓ. This is the motivation for use of the term "polyphase" [Bellanger, et al., 1976].

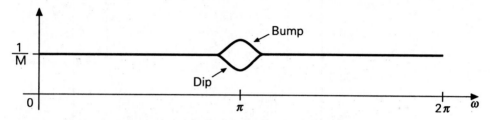

Figure 4.6-16 Typical behaviors of $|E_\ell(e^{j\omega})|$ when $|H(e^{j\omega})|$ is as in Fig. 4.6-12(b).

4.7 MULTIGRID METHODS

The term "multigrid methods" represents a wide range of techniques used in iterative numerical computations. These are used to solve large sets of linear or nonlinear equations (thousands of unknowns) which in turn may be discretized versions of (partial) differential equations. Multigrid techniques improve the speed/accuracy of solutions, in some cases dramatically. This is a discipline with vast amount of literature, and the reader wishing to pursue the literature should begin with Briggs [1987], and Brandt [1977]. Our goal here is to describe the philosophy with simple examples (ordinary linear differential equations), and show the connection to multirate signal processing. The discussion in this section is intended to convey the idea with emphasis on concepts rather than rigor.

4.7.1 Discretizing a Continuous-Time Problem

Matrix equations arise either directly, or by discretization of continuous problems. To demonstrate, consider the second order differential equation

$$\frac{d^2 y_a(t)}{dt^2} + \mu y_a(t) = u_a(t), \qquad (4.7.1)$$

to be solved in the time range $0 < t < 1$, with boundary conditions $y_a(0) = y_a(1) = 0$. This is a linear constant-coefficient differential equation. Here $u_a(t)$ is the continuous-time input to the differential equation, and $y_a(t)$ is the output.

When attempting to solve this [i.e., find $y_a(t)$] on a computer, we discretize the problem. We first define a uniform "grid" $\mathcal{G}^{(\Delta)}$ of the domain $0 \leq t \leq 1$ (Fig. 4.7-1), and try to find (approximate value of) $y_a(t)$ on the grid, that is, at points $t = n\Delta$, where n is an integer with $1 \leq n \leq N$. If we define $y(n) = y_a(n\Delta)$, and $u(n) = u_a(n\Delta)$ and approximate the second derivative with the second difference, we obtain from (4.7.1)

$$\frac{y(n-1) - 2y(n) + y(n+1)}{\Delta^2} + \mu y(n) = u(n). \tag{4.7.2}$$

Figure 4.7-1 A uniform grid defined on $0 \leq t \leq 1$. The spacing Δ is called fineness.

The aim is to solve this for $1 \leq n \leq N$, under the boundary conditions $y(0) = y(N+1) = 0$. We can write the equations (4.7.2) in matrix form

$$\mathbf{Ay = u}, \tag{4.7.3}$$

where

$$\mathbf{u} = -\Delta^2 \begin{bmatrix} u_a(\Delta) \\ \vdots \\ u_a(N\Delta) \end{bmatrix}, \quad \mathbf{y} = \begin{bmatrix} y_a(\Delta) \\ \vdots \\ y_a(N\Delta) \end{bmatrix}, \tag{4.7.4}$$

and $\mathbf{A}$ is $N \times N$. For example with $N = 5$,

$$\mathbf{A} = \begin{bmatrix} 2-\mu\Delta^2 & -1 & 0 & 0 & 0 \\ -1 & 2-\mu\Delta^2 & -1 & 0 & 0 \\ 0 & -1 & 2-\mu\Delta^2 & -1 & 0 \\ 0 & 0 & -1 & 2-\mu\Delta^2 & -1 \\ 0 & 0 & 0 & -1 & 2-\mu\Delta^2 \end{bmatrix}. \tag{4.7.5}$$

Summarizing, we have converted our problem into a problem in linear algebra. The original problem (4.7.1) has been reformulated on a grid $\mathcal{G}^{(\Delta)}$, where the superscript Δ denotes the "fineness" of the grid. The solution $\mathbf{y}$ of the equation (4.7.3) is an approximation to the sampled version of the solution $y_a(t)$ of (4.7.1).

Notice that $\mathbf{A}$ is symmetric and Toeplitz (Appendix A). Furthermore, it is "banded." For a banded matrix all elements sufficiently away from the main diagonal are zero. The banded nature is a consequence of discretizing a differential equation.

4.7.2 Traditional Techniques to Solve Ay=u

Assume that **A** is nonsingular. There exist many techniques to find the exact solution **y** to (4.7.3) (subject only to finite precision errors), in a finite amount of time. This includes Gauss-elimination and its variations [Golub and Van Loan, 1989]. These have $O(N^3)$ complexity [i.e., the computational time is cN^3 where c is independent of N]. If **A** has special structure (symmetry, positive definiteness and so on, Appendix A) the methods can be made more efficient, but are still $O(N^3)$. If **A** is Toeplitz then there exist $O(N^2)$ techniques (see Blahut [1985] and references therein).

Iterative Techniques

In many applications, however, iterative techniques are used. Instead of finding the exact solution, these methods attempt to find an approximate solution $\mathbf{y}_k$, whose accuracy improves with iteration number k. These have some computational advantages when the matrix is banded, with the matrix size N much larger than the number of nonzero elements per row. Iterative techniques can also be applied to more general situations (such as nonlinear equations). Finally there are some applications where the matrix **A** grows in real time (such as in real time recursive estimation); iterative techniques can incorporate a mechanism to update the approximations efficiently.

A typical iteration has the form

$$\mathbf{y}_{k+1} = \mathbf{P}\mathbf{y}_k + \mathbf{q}, \qquad (4.7.6)$$

where **P** is related to **A** (details depending on the particular iterative technique), and **q** depends on **A** as well as **u**. Ideally this converges to the exact solution **y** for any initial estimate $\mathbf{y}_0$ as long as **P** is stable (i.e., has eigenvalues strictly inside unit circle; Sec. 13.4.3).

For fixed **A**, the choice of **P** is not unique. To see how **P** is related to **A**, note that upon convergence (4.7.6) implies

$$\mathbf{y} = \mathbf{P}\mathbf{y} + \mathbf{q}, \qquad (4.7.7)$$

that is, $(\mathbf{I} - \mathbf{P})\mathbf{y} = \mathbf{q}$. So we can choose $\mathbf{I} - \mathbf{P} = \mathbf{TA}, \mathbf{q} = \mathbf{Tu}$, for any nonsingular **T**. For our discussion the choice of **T** is irrelevant, as long as **P** is stable. We shall assume $\mathbf{T} = \alpha\mathbf{I}$ where α is a nonzero scalar. So $\mathbf{P} = \mathbf{I} - \alpha\mathbf{A}$.

The error vector and the residual vector. Suppose $\hat{\mathbf{y}}$ is an approximation to the solution **y**. Then $\mathbf{A}\hat{\mathbf{y}} = \hat{\mathbf{u}} \neq \mathbf{u}$. Define

$$\begin{aligned}\mathbf{e} &= \mathbf{y} - \hat{\mathbf{y}}, \quad \text{(error vector)} \\ \mathbf{r} &= \mathbf{u} - \hat{\mathbf{u}} = \mathbf{u} - \mathbf{A}\hat{\mathbf{y}}, \quad \text{(residual vector)}.\end{aligned} \qquad (4.7.8)$$

As the exact solution **y** is not known, we do not know the error **e**. On the contrary, the residual **r** *can* be computed from the approximate solution $\hat{\mathbf{y}}$.

It is easily verified that the error and residual are related as $\mathbf{Ae} = \mathbf{r}$, so that, at the kth iteration we have the relation
$$\mathbf{Ae}_k = \mathbf{r}_k, \tag{4.7.9}$$
where $\mathbf{e}_k$ and $\mathbf{r}_k$ are the error and residual at the kth iteration, that is,
$$\mathbf{e}_k = \mathbf{y} - \mathbf{y}_k, \quad \mathbf{r}_k = \mathbf{u} - \mathbf{u}_k, \tag{4.7.10}$$
where $\mathbf{u}_k = \mathbf{Ay}_k$. The iteration (4.7.6) can be rewritten in terms of the residual $\mathbf{r}_k$ as
$$\mathbf{y}_{k+1} = \mathbf{y}_k + \alpha \mathbf{r}_k. \tag{4.7.11}$$
Thus, the solution $\mathbf{y}_{k+1}$ is obtained by adding a correction term to $\mathbf{y}_k$, proportional to the residual $\mathbf{r}_k$.

More insight can be gained by subtracting (4.7.6) from (4.7.7) and rewriting in terms of $\mathbf{e}_k$. This gives
$$\mathbf{e}_{k+1} = \mathbf{Pe}_k, \quad \mathbf{r}_{k+1} = \mathbf{Pr}_k, \tag{4.7.12}$$
by using $\mathbf{P} = \mathbf{I} - \alpha \mathbf{A}$.

Thus, as long as $\mathbf{P}$ is stable, $\mathbf{r}_k$ and $\mathbf{e}_k$ eventually become zero. We see below that the speed of convergence has to do with the eigenvalues of $\mathbf{P}$. Choice of α (which affects eigenvalues of $\mathbf{P}$) is therefore crucial.

4.7.3 The Stalling Phenomenon

We see that the convergence of the residual is governed by
$$\mathbf{r}_{k+1} = \mathbf{Pr}_k. \tag{4.7.13}$$
We can judge the rate of convergence by considering the behavior of the residual energy
$$\theta_k = \mathbf{r}_k^\dagger \mathbf{r}_k. \tag{4.7.14}$$
Figure 4.7-2 shows a typical behavior of this quantity, which arises in many practical situations: initially the energy decreases rapidly, and then there is a phase of slower decrease, and then even slower, and so on. (The iteration gets into "slower and slower" modes.) At some stage the rate of decrease becomes very small. This phenomenon is called *stalling*. Under this condition, it takes several iterations (i.e., prohibitive amount of computations) to reduce the energy θ_k by a significant amount.

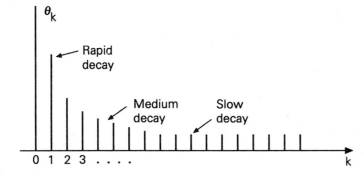

Figure 4.7-2 Reduction of residual energy as k increases.

An "obvious" way to overcome the stalling problem is suggested by (4.7.13). Since $\mathbf{r}_{k+L} = \mathbf{P}^L \mathbf{r}_k$, we can replace $\mathbf{P}$ with $\mathbf{P}^L$ for some large integer L and perform the iteration. The convergence is speeded up in proportion to L, that is, the slow modes will appear to be faster. The disadvantage is that the evaluation of $\mathbf{P}^L$ itself requires $O(LN^3)$ computations, far in excess of Gaussian elimination!

In this situation, *multigrid processing* comes to our rescue. Broadly speaking, as soon as the iteration begins to stall, we reformulate a new problem by "decimating" the original problem into a problem of smaller dimension (as described below). The iterations for the problem proceed at a much faster rate. We then use an interpolation scheme on the result, to obtain a correction term for the original (larger) problem. Before describing the details, we study the reason for the stalling behavior (Fig. 4.7-2) more quantitatively.

Why Does Stalling Occur?

To explain stalling, first note that $\mathbf{r}_k = \mathbf{P}^k \mathbf{r}_0$. Assume for simplicity that $\mathbf{P}$ is diagonalizable with a unitary matrix (Sec. A.6, Appendix A), that is, can be written as $\mathbf{P} = \mathbf{U}\boldsymbol{\Lambda}\mathbf{U}^\dagger$, where (i) $\boldsymbol{\Lambda}$ is a diagonal matrix with diagonal elements λ_i equal to the eigenvalues of $\mathbf{P}$, and (ii) $\mathbf{U}^\dagger \mathbf{U} = \mathbf{I}$. We then have $\mathbf{P}^k = \mathbf{U}\boldsymbol{\Lambda}^k\mathbf{U}^\dagger$ so that $\mathbf{r}_k = \mathbf{U}\boldsymbol{\Lambda}^k\mathbf{U}^\dagger \mathbf{r}_0$. Denoting the elements of the column vector $\mathbf{U}^\dagger \mathbf{r}_0$ as $v_0, \ldots v_{N-1}$, we can write

$$\mathbf{r}_k = v_0 \lambda_0^k \mathbf{u}_0 + v_1 \lambda_1^k \mathbf{u}_1 + \ldots + v_{N-1} \lambda_{N-1}^k \mathbf{u}_{N-1}, \tag{4.7.15}$$

where $\mathbf{u}_i$ are the columns of $\mathbf{U}$. Since $\mathbf{u}_i^\dagger \mathbf{u}_\ell = \delta(i - \ell)$, we get

$$\theta_k = \mathbf{r}_k^\dagger \mathbf{r}_k = |v_0|^2 |\lambda_0|^{2k} + |v_1|^2 |\lambda_1|^{2k} + \ldots + |v_{N-1}|^2 |\lambda_{N-1}|^{2k}. \tag{4.7.16}$$

To visualize the idea clearly, assume $|\lambda_0| < |\lambda_1| < \ldots < |\lambda_{N-1}|$. Referring now to Fig. 4.7-2, it is clear that the initial steep decrease of θ_k is due to $|v_0|^2 |\lambda_0|^{2k}$ (assuming this term is nonzero) and that stalling is created by the slowest-decaying term $|v_{N-1}|^2 |\lambda_{N-1}|^{2k}$ (assuming $v_{N-1} \neq 0$). This slowest mode corresponds to the eigenvalue λ_{N-1} with largest magnitude. As long as $|\lambda_{N-1}| < 1$, the residual $\mathbf{r}_k$ (hence the error $\mathbf{e}_k$) eventually decreases to zero as $k \to \infty$.

In Problem 4.31 we consider some additional details for the case where $\mathbf{A}$ is positive definite. We derive the condition on α for stability of $\mathbf{P}$, and also the value of α that maximizes convergence rate.

4.7.4 Basic Idea of Multigrid Approach

In what follows, the residual $\mathbf{r}_k$ and error $\mathbf{e}_k$ upon stalling will be denoted as $\mathbf{r}^{(\Delta)}$ and $\mathbf{e}^{(\Delta)}$. Also, we use $\mathbf{A}^{(\Delta)}$ in place of $\mathbf{A}$. Thus (4.7.9) is equivalent to

$$\mathbf{A}^{(\Delta)} \mathbf{e}^{(\Delta)} = \mathbf{r}^{(\Delta)}. \tag{4.7.17a}$$

The superscript indicates that these quantities pertain to the grid $\mathcal{G}^{(\Delta)}$ of fineness Δ (Fig. 4.7-1). In problems where the equation (4.7.3) is obtained by discretization of differential equations, the iteration (4.7.6) often exhibits an extremely interesting behavior upon stalling, viz., the residual vector $\mathbf{r}^{(\Delta)}$ has a smooth appearance. In other words, if $r^{(\Delta)}(n)$ denotes the nth component of $\mathbf{r}^{(\Delta)}$, then the plot of $r^{(\Delta)}(n)$ as a function of n is 'smooth'. Figure 4.7-3 demonstrates the meaning of "smooth". For comparison, we have also shown a "nonsmooth" or "oscillatory" function.

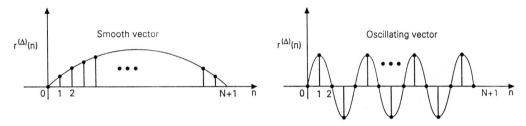

Figure 4.7-3 Demonstrating smooth and oscillating vectors.

Before explaining the reason for this behavior, let us explore the consequences. At the end of the kth iteration we know $\mathbf{r}^{(\Delta)}$. If we can solve for $\mathbf{e}^{(\Delta)}$ from (4.7.17a), we can obtain the exact solution $\mathbf{y}$ since

$$\mathbf{y} = \mathbf{y}_k + \mathbf{e}^{(\Delta)}.$$

This might appear to be as involved as the original problem (4.7.3), but the smooth nature of $\mathbf{r}^{(\Delta)}$ can be exploited now. Since the sequence $r^{(\Delta)}(n)$ appears to have only low-frequency components, we can "decimate it", say by a factor of 2, and consider the following equation

$$\mathbf{A}^{(2\Delta)} \mathbf{e}^{(2\Delta)} = \mathbf{r}^{(2\Delta)}. \tag{4.7.17b}$$

In this equation $\mathbf{r}^{(2\Delta)}$ is a decimated version of the residual $\mathbf{r}^{(\Delta)}$, and has about half as many elements. The matrix $\mathbf{A}^{(2\Delta)}$ is a "decimated version" of $\mathbf{A}$, and is approximately $\frac{N}{2} \times \frac{N}{2}$. Essentially, it is obtainable by discretizing the original differential equation with a coarser grid, but we shall fill these mathematical details later.

Equation (4.7.17b) is the analog of (4.7.17a), but on a coarser grid. The basic idea is that we can now solve for the error $\mathbf{e}^{(2\Delta)}$ using this smaller system. We then perform an 'interpolation operation' to find an approximation to the error $\mathbf{e}^{(\Delta)}$ on the original fine grid. The solution $\mathbf{y}_k$ on the fine grid can now be corrected by adding this estimate $\mathbf{e}^{(\Delta)}$ to it. This idea can of course be carried to deeper levels. Thus, if the iterative process for (4.7.17b) stalls, we can repeat the decimation process, and so on. We now turn to quantitative details of the various components of this idea.

Why Does the Residual Look "Smooth" Upon Stalling?

There is no law which says that this will always happen. However, with proper choice of α, this can usually be made to happen. First refer to the example where (4.7.1) was discretized, and consider the case where $\mu = 0$. The differentiator $d^2 y(t)/dt^2$ was replaced by a second order difference operator $c(z - 2 + z^{-1}) = cz(1 - z^{-1})^2$, which is a zero-phase highpass filter. If we now look at the matrix $\mathbf{A}$ given in (4.7.5), we see that all rows have the form

$$[0 \quad \ldots \quad -1 \quad 2 \quad -1 \quad 0 \quad \ldots \quad 0], \tag{4.7.18}$$

except the 0th and last rows. (We assume that the matrix size N is large, so that these border effects can be ignored.) Moreover, each row is a shifted version of the preceding row. As a result, the product $\mathbf{Av}$ approximates a convolution of the sequence $v(n)$ with the highpass filter $(-z + 2 - z^{-1})$ (where $v(n)$ is the nth element of $\mathbf{v}$).

Now consider the iteration matrix $\mathbf{P} = \mathbf{I} - \alpha\mathbf{A}$. This has all the properties of $\mathbf{A}$ except that it represents a convolution with the filter

$$H(z) = \alpha z + (1 - 2\alpha) + \alpha z^{-1}. \tag{4.7.19}$$

For proper choice of α this filter has *lowpass* behavior. For example let $\alpha = 1/3$, then $H(z) = (z + 1 + z^{-1})/3$. This has response

$$H(e^{j\omega}) = \frac{1 + 2\cos\omega}{3}, \tag{4.7.20}$$

which is plotted in Fig. 4.7-4(a). Thus the operation $\mathbf{Pv}$ resembles lowpass filtering of the sequence $v(n)$. Summarizing, the effect of the iteration (4.7.13) is that of a lowpass filter. For large k therefore, this has a smoothing effect on the residual as mentioned above.

If we choose $\alpha = 1/2$, then the filter has response $H(e^{j\omega}) = \cos\omega$ [Fig. 4.7-4(b)] which is not lowpass. In this case, the iteration does *not* result in smoothing.

Relation to eigenvector viewpoint. In Sec. 4.7.3 we saw that the iteration stalls when $\mathbf{r}_k$ is close to an eigenvector corresponding to the eigenvalue λ_{N-1} with largest magnitude. In the above discussion, however, we see that stalling typically occurs when $\mathbf{r}_k$ has been smoothed out. The latter observation is a special case of the former, when $\mathbf{P}$ can be interpreted as a lowpass filtering operator.

Thus when $\mathbf{P}$ approximates a filter operator, vectors of the form

$$\begin{bmatrix} 1 \\ e^{j\omega} \\ \vdots \\ e^{j\omega(N-1)} \end{bmatrix} \tag{4.7.21}$$

are approximately eigenvectors, with eigenvalues $H(e^{j\omega})$. (This is because $e^{j\omega n}$, and more generally a^n, are eigenfunctions of discrete time LTI systems; see Sec. 2.1.2.) If $H(e^{j\omega})$ is a lowpass response, then the eigenvalues with large magnitude correspond to low frequencies ω. This means that, after several iterations, only the *low frequency components* of the initial vector $\mathbf{r}_0$ survive. So the residual vector $\mathbf{r}_k$ appears to be smooth upon stalling.

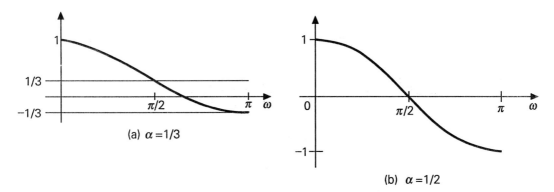

Figure 4.7-4 Explaining the smoothing effect created by the iteration matrix. The response $H(e^{j\omega})$ is shown. (a) $\alpha = 1/3$ causes smoothing and (b) $\alpha = 1/2$ does not cause smoothing.

The fact that the response $H(e^{j\omega})$ is real and symmetric in the above example implies that if we replace ω with $-\omega$ in (4.7.21), the eigenvalue does not change. This means that we can find a smooth eigenvector with real elements $\cos \omega n$.

Decimation of the Residual, and Interpolation of the Error

Decimation operation is needed to transfer the residual $\mathbf{r}^{(\Delta)}$ from the grid $\mathcal{G}^{(\Delta)}$ to the grid $\mathcal{G}^{(2\Delta)}$. Denote the nth component of $\mathbf{r}^{(2\Delta)}$ as $r^{(2\Delta)}(n)$. The simplest decimation scheme is to take $r^{(2\Delta)}(n) = r^{(\Delta)}(2n)$. A more sophisticated scheme would be to define $r^{(2\Delta)}(n)$ as the output of a decimation filter (Fig. 4.7-5). The most commonly used decimation filter in multigrid literature is $G_D(z) = (z+2+z^{-1})/4$. This is lowpass with frequency response $G_D(e^{j\omega}) = \cos^2(\omega/2)$.

Similarly, when we transfer the error $\mathbf{e}^{(2\Delta)}$ to the finer grid to obtain an approximation of $\mathbf{e}^{(\Delta)}$, we use an interpolation filter. A common example is $G_I(z) = (z + 2 + z^{-1})/2$.

Matrix representation. To be consistent with the rest of the problem, it is convenient to express the decimation and interpolation filters in matrix form. We do this by taking the above examples for $G_D(z)$ and $G_I(z)$. Figure 4.7-6 shows the grids $\mathcal{G}^{(\Delta)}$ and $\mathcal{G}^{2\Delta}$ for $N = 7$. Assuming the boundary conditions

$$v^{(\Delta)}(0) = v^{(\Delta)}(8) = 0 \qquad v^{(2\Delta)}(0) = v^{(2\Delta)}(4) = 0, \qquad (4.7.22)$$

we compute the decimated samples as

$$\underbrace{\begin{bmatrix} v^{(2\Delta)}(1) \\ v^{(2\Delta)}(2) \\ v^{(2\Delta)}(3) \end{bmatrix}}_{\mathbf{v}^{(2\Delta)}} = \frac{1}{4} \underbrace{\begin{bmatrix} 1 & 2 & 1 & 0 & 0 & 0 & 0 \\ 0 & 0 & 1 & 2 & 1 & 0 & 0 \\ 0 & 0 & 0 & 0 & 1 & 2 & 1 \end{bmatrix}}_{\mathbf{M}_D} \underbrace{\begin{bmatrix} v^{(\Delta)}(1) \\ v^{(\Delta)}(2) \\ v^{(\Delta)}(3) \\ v^{(\Delta)}(4) \\ v^{(\Delta)}(5) \\ v^{(\Delta)}(6) \\ v^{(\Delta)}(7) \end{bmatrix}}_{\mathbf{v}^{(\Delta)}} \qquad (4.7.23)$$

For example, we use a transformation of this type to obtain $\mathbf{r}^{(2\Delta)}$ from $\mathbf{r}^{(\Delta)}$. Similarly when we convert from the coarse to the dense grid (interpolation) we use the matrix transformation

$$\underbrace{\begin{bmatrix} v^{(\Delta)}(1) \\ v^{(\Delta)}(2) \\ v^{(\Delta)}(3) \\ v^{(\Delta)}(4) \\ v^{(\Delta)}(5) \\ v^{(\Delta)}(6) \\ v^{(\Delta)}(7) \end{bmatrix}}_{\mathbf{v}^{(\Delta)}} = \frac{1}{2} \underbrace{\begin{bmatrix} 1 & 0 & 0 \\ 2 & 0 & 0 \\ 1 & 1 & 0 \\ 0 & 2 & 0 \\ 0 & 1 & 1 \\ 0 & 0 & 2 \\ 0 & 0 & 1 \end{bmatrix}}_{\mathbf{M}_I} \underbrace{\begin{bmatrix} v^{(2\Delta)}(1) \\ v^{(2\Delta)}(2) \\ v^{(2\Delta)}(3) \end{bmatrix}}_{\mathbf{v}^{(2\Delta)}}. \qquad (4.7.24)$$

We use this to obtain the approximation of $\mathbf{e}^{(\Delta)}$ from $\mathbf{e}^{(2\Delta)}$.

Finding the Matrix $\mathbf{A}^{(2\Delta)}$ for the Coarse Grid

In the example of Fig. 4.7-6, the matrix $\mathbf{A}^{(\Delta)}$ is 7×7, whereas $\mathbf{A}^{(2\Delta)}$ would be 3×3. It remains to show how to find the elements of this smaller matrix. In multigrid literature, the following formula is used:

$$\mathbf{A}^{(2\Delta)} = \mathbf{M}_D \mathbf{A}^{(\Delta)} \mathbf{M}_I, \qquad (4.7.25)$$

where $\mathbf{M}_D$ and $\mathbf{M}_I$ are the decimator and interpolator matrices indicated above. The reason for the above choice of $\mathbf{A}^{(2\Delta)}$ can be understood as follows: suppose the error vector $\mathbf{e}^{(\Delta)}$ can be obtained *exactly* by interpolation of $\mathbf{e}^{(2\Delta)}$. That is suppose

$$\mathbf{A}^{(\Delta)} \mathbf{M}_I \mathbf{e}^{(2\Delta)} = \mathbf{r}^{(\Delta)}. \qquad (4.7.26)$$

Since $\mathbf{r}^{(2\Delta)} = \mathbf{M}_D \mathbf{r}^{(\Delta)}$, the above equation implies

$$\mathbf{M}_D \mathbf{A}^{(\Delta)} \mathbf{M}_I \mathbf{e}^{(2\Delta)} = \mathbf{r}^{(2\Delta)}. \qquad (4.7.27)$$

Comparing with (4.7.17b) we see that the above choice of $\mathbf{A}^{(2\Delta)}$ is well-motivated. Detailed numerical examples of the use of multigrid techniques can be found in Brandt [1977].

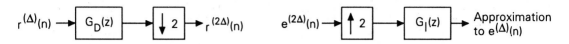

Figure 4.7-5 Transfer of information between two grids, with the help of decimation and interpolation filters.

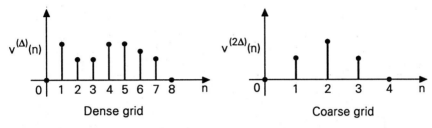

Figure 4.7-6 Demonstration of dense grid and coarse grid ($N = 7$ and 3 respectively).

PROBLEMS

4.1. Consider the structures shown in Fig. P4-1, with input transforms and filter responses as indicated.

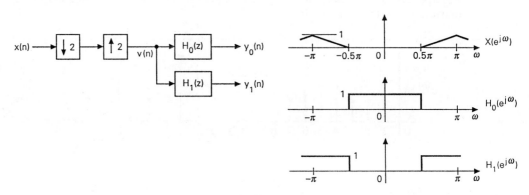

Figure P4-1

Sketch the quantities $Y_0(e^{j\omega})$ and $Y_1(e^{j\omega})$.

4.2. For the system in Fig. P4-2, find an expression for $y(n)$ in terms of $x(n)$.

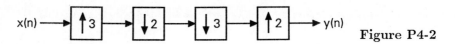

Figure P4-2

Simplify the expression as best as you can.

4.3. Consider a sequence $x(n)$ with $X(e^{j\omega})$ as shown in Fig. P4-3.

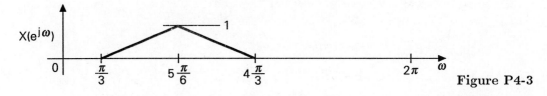

Figure P4-3

Let $y(n) = x(2n)$. Show how we can recover $x(n)$ from $y(n)$ using filters and multirate building blocks.

4.4. Show that the decimator and expander are linear time varying systems.

4.5. Show that the two systems shown in Fig. P4-5(a) (where k is some integer) are equivalent (that is, $y_0(n) = y_1(n)$) when $h_k(n) = h_0(n)\cos(2\pi kn/L)$.

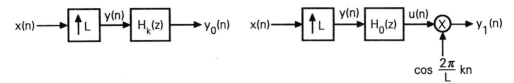

Figure P4-5(a)

This is a structure where filtering followed by cosine modulation has the same effect as filtering with the cosine modulated impulse response. (This is not true in all situations; see next problem). Now consider the example where $L = 5$, and $k = 1$. Let $X(e^{j\omega})$ and $H_0(e^{j\omega})$ be as sketched in Fig. P4-5(b).

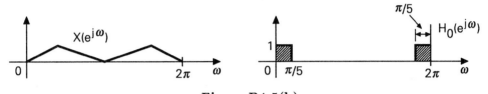

Figure P4-5(b)

Give sketches of $Y(e^{j\omega})$, $Y_0(e^{j\omega})$ and $U(e^{j\omega})$.

4.6. Show that the two systems shown in Fig. P4-6 are not equivalent, that is, $y_0(n)$ and $y_1(n)$ are not necessarily the same, even if $h_k(n) = h_0(n)\cos(2\pi kn/L)$.

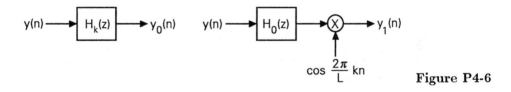

Figure P4-6

4.7. Consider the two sets of M numbers given by W^k, $0 \le k \le M-1$ and W^{kL}, $0 \le k \le M-1$ where $W = e^{-j2\pi/M}$. Show that these sets are identical if and only if L and M are relatively prime.

4.8. For the two systems in Fig. 4.2-2 we can write down $y_1(n)$ and $y_2(n)$ in terms of $x(n)$, M and L. For example

$$y_1(n) = \begin{cases} x(\frac{Mn}{L}), & n = \text{mul. of } L \\ 0, & \text{otherwise.} \end{cases}$$

a) Similarly write an expression for $y_2(n)$.
b) Verify that these two expressions yield the same result (i.e., $y_1(n) = y_2(n)$ for any sequence $x(n)$), if, and only if, L and M are relatively prime.

4.9. *The jumping painter problem.* Consider a circular arrangement of objects as shown in Fig. P4-9.

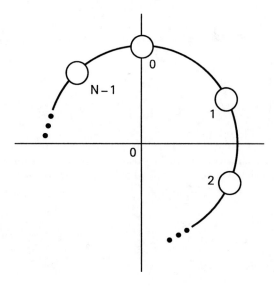

Figure P4-9

There are N objects numbered as indicated. A painter is requested to paint these one at a time. To avoid a boring job, he decides to paint the objects in nonconsecutive order as: $0, M, 2M, \ldots$ Now there are two possibilities:
   a) either all the objects get painted,
   b) or the painter returns to an object already painted, before all objects are covered. This means that he will never be able to paint a subset of objects.

Find a set of necessary and sufficient conditions so that the first possibility takes place.

4.10. Let $x(n)$ be periodic with period N, that is, N is the smallest integer such that $x(n) = x(n + N)$ for all n. Let $y(n)$ be the M-fold decimated version, that is, $y(n) = x(Mn)$. Show that $y(n)$ is periodic, that is, there exists $L < \infty$ such that $y(n) = y(n + L)$ for all n. Assuming no further knowledge about the input, what is the smallest value of L in terms of M and N?

4.11. Consider a sequence $x(n)$ with $X(e^{j\omega})$ as shown in Fig. P4-11(a).

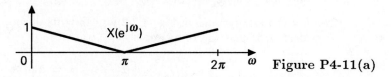

Figure P4-11(a)

Suppose we generate the sequences $y(n)$ and $s(n)$ from $x(n)$ as in Fig. P4-11(b)

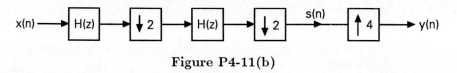

Figure P4-11(b)

180   Chap. 4. Multirate system fundamentals

where
$$H(e^{j\omega}) = \begin{cases} 1 & \text{for } |\omega| < \pi/2 \\ 0 & \text{for } \pi/2 \leq |\omega| \leq \pi. \end{cases}$$

Plot the quantities $Y(e^{j\omega})$ and $S(e^{j\omega})$.

4.12. In this problem, the term 'polyphase components' stands for the Type 1 components with $M = 2$.
   a) Let $H(z)$ represent an FIR filter of length 10 with impulse response coefficients $h(n) = (1/2)^n$ for $0 \leq n \leq 9$ and zero otherwise. Find the polyphase components $E_0(z)$ and $E_1(z)$.
   b) Let $H(z)$ be IIR with $h(n) = (1/2)^n \mathcal{U}(n) + (1/3)^n \mathcal{U}(n-3)$. Find the polyphase components $E_0(z)$ and $E_1(z)$. Give simplified, closed form expressions.

4.13. Let $H(z) = (a + z^{-1})/(1 + az^{-1})$. Write down expressions for the Type 1 polyphase components (with $M = 2$). For real a, notice that $H(z)$ is allpass. Are the polyphase components allpass as well?

4.14. Let $H(z) = 1/(1 - 2R\cos\theta z^{-1} + R^2 z^{-2})$, with $R > 0$ and θ real. This is a system with a pair of complex conjugate poles at $Re^{\pm j\theta}$. Find the Type 1 polyphase components for $M = 2$.

4.15. Consider the fractional decimation circuit of Fig. 4.1-10(b) with $L = 3, M = 4$. Suppose $H(z)$ is a linear phase FIR filter with length 60. Assume that $x(n)$ has a sampling rate of 100 KHz. (a) If $H(z)$ is implemented directly (i.e., no polyphase forms) what is the time available for each multiplier to perform one multiplication? (b) Suppose the structure is implemented in the best possible way (i.e., using polyphase form similar to Fig. 4.3-8). Then what is the time available for each multiplier to perform one multiplication? (c) Find the number of multiplications and additions *per second* in part (b).

4.16. Is the following statement true or false? Justify. "Let $h(n)$ be the impulse response of an allpass filter. Let $g(n) \triangleq h(2n)$. Suppose the filter $G(z)$ [whose impulse response is $g(n)$] is allpass as well. Then $h(n)$ *must be* an impulse (i.e., $h(n) = c\delta(n - n_0)$ where n_0 is some integer, and c is some constant)."

4.17. Consider the systems shown in Fig. P4-17

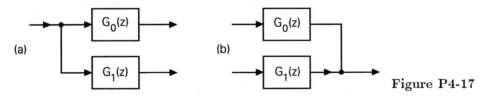

Figure P4-17

where $G_0(z) = \sum_{n=0}^{N} g(n) z^{-n}$ and $G_1(z) = \sum_{n=0}^{N} g(N - n) z^{-n}$. The impulse response of $G_1(z)$ is the mirror image of that of $G_0(z)$. Draw a structure for the system in Fig. P4-17(a), using only $N + 1$ multipliers. (*Hint.* Draw one set of multipliers, and two sets of delay chains running in opposite directions). Repeat for the system in Fig. P4-17(b).

4.18. Consider the uniform DFT analysis bank [Fig. 4.3-5(a)] with $M = 4$. Assume $E_0(z) = 1+z^{-1}, E_1(z) = 1+2z^{-1}, E_2(z) = 2+z^{-2}$ and $E_3(z) = 0.5+z^{-1}$. Find explicit expressions for $H_k(z)$, $0 \le k \le 3$, working out the numerical values of these filter coefficients.

4.19. Consider the structure shown in Fig. P4-19(a), where $\mathbf{W}$ is the 3×3 DFT matrix.

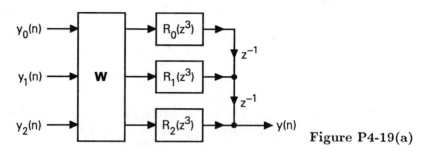

Figure P4-19(a)

This is a three channel synthesis bank with three filters $F_0(z), F_1(z)$ and $F_2(z)$. (For example $F_0(z) = Y(z)/Y_0(z)$ with $y_1(n)$ and $y_2(n)$ set to zero.)
  a) Assuming $R_0(z) = 1 + z^{-1}, R_1(z) = 1 - z^{-2}$ and $R_2(z) = 2 + 3z^{-1}$, find expressions for the three synthesis filters $F_0(z), F_1(z), F_2(z)$.
  b) Let the magnitude response of $F_1(z)$ be as shown in Fig. 4-19(b).

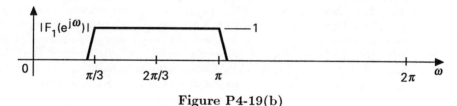

Figure P4-19(b)

Plot the responses $|F_0(e^{j\omega})|$ and $|F_2(e^{j\omega})|$.

4.20. For the structure of Fig. 4.3-12, prove that the synthesis filters are indeed given by (4.3.15).

4.21. Let $H_0(z) = 1 + 2z^{-1} + 4z^{-2} + 2z^{-3} + z^{-4}$ and let $H_1(z) = H_0(-z)$. Draw an implementation for the pair $[H_0(z), H_1(z)]$ in the form of a uniform DFT analysis bank, explicitly showing the polyphase components, the 2×2 IDFT box, and other relevant details.

4.22. Let $H(z) = \sum_{n=0}^{N} h(n)z^{-n}$ with $h(n) = h(N - n)$. Consider the polyphase decomposition (4.3.7). The symmetry of $h(n)$ reflects into the coefficients of $E_\ell(z)$ in some way. To be more specific, we can make the following statement: there exists an integer m_0 (with $0 \le m_0 \le M - 1$) such that $e_k(n)$ is the image of $e_{m_0-k}(n)$ for $0 \le k \le m_0$, and $e_k(n)$ is the image of $e_{M+m_0-k}(n)$ for $m_0 + 1 \le k \le M - 1$.
  a) Take an example of 7th order $H(z)$, and *verify* the above statement for $M = 3$. What is m_0? Repeat for $M = 4$.
  b) *Prove* the above statement. How is m_0 related to N and M?

This problem shows that if $h(n)$ is symmetric (as in linear phase filters), then we can implement pairs of polyphase components [such as $E_k(z)$ and $E_{m_0-k}(z)$] using one set of multipliers. This gives rise to an additional saving (by two) on the number of MPUs in the implementation of a decimation filter.

4.23. Consider the design of real coefficient narrow band lowpass filters using the IFIR method. We know that the savings depends on the stretching factor M_1 used. We assume that the image suppressor $I(z)$ and model filter $G(z)$ are designed using McClellan-Parks equiripple technique. For a given set of specifications $\delta_1, \delta_2, \omega_p$ and ω_S, we will design $G(z)$ and $I(z)$ to have peak passband ripples $0.5\delta_1$ and stopband ripples δ_2 (for reasons explained in text). Let N_g and N_i denote the orders of $G(z)$ and $I(z)$.

   a) Write down the orders N_g and N_i in terms of $\delta_1, \delta_2, \omega_p, \omega_S$, and M_1. Show that as M_1 increases, N_g decreases whereas N_i increases. Evidently there is some 'best' M_1 for a given set of specifications.

   b) For $\omega_p = 0.18\pi, \omega_S = 0.2\pi$, $\delta_1 = 0.01$ and $\delta_2 = 0.001$, estimate the orders N_g and N_i for all permissible values of M_1, and plot the multiplier count against M_1. What is the value of M_1 that minimizes the multiplier count?

4.24. Suppose we wish to design a linear phase FIR filter with specifications $\delta_1 = \delta_2 = 0.001$, $\omega_p = 0.015\pi$ and $\omega_S = 0.02\pi$. (a) For direct equiripple design, estimate the filter order, and number of multiplications and additions required. (b) If we use the IFIR approach with stretching factor of 25, what are the orders of equiripple $G(z)$ and $I(z)$? What is the total number of multiplications and additions? What is the order of the overall filter? (c) Repeat part (b) with stretching factor $= 45$.

4.25. We know that the IFIR technique can be used to design narrow band filters in two stages, thereby improving the computational efficiency. Now suppose that we are interested in designing a real-coefficient *wide* band lowpass filter with magnitude response as in Fig. P4-25(a).

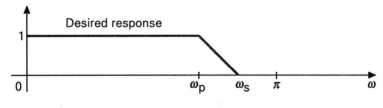

Figure P4-25(a)

Evidently, we cannot stretch the response by an integer factor to obtain a model response $G(e^{j\omega})$ as in the IFIR approach. Consider the modified lowpass specification shown in Fig. P4-25(b).

   a) Let $H_1(z)$ be a Type 1 Nth order linear phase filter (Table 2.4.1) meeting this specification. Define $H_2(z) = z^{-N/2} - H_1(z)$ and $H(z) = H_2(-z)$. Sketch the magnitude responses of $H_2(e^{j\omega})$ and $H(e^{j\omega})$. Note that $H(z)$ is lowpass, and has same band edges as the desired wideband filter.

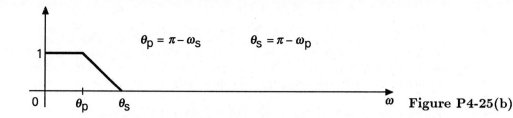

Figure P4-25(b)

b) Suppose the specifications for $H(z)$ are $\omega_p = 0.85\pi, \omega_S = 0.9\pi, \delta_1 = 0.01$, $\delta_2 = 0.001$. What is the total number of multipliers required if we design linear-phase equiripple $H(z)$ directly?

c) Suppose we meet the specifications of part (b) by proceeding as in part (a) where $H_1(z)$ is designed using the IFIR approach (stretching factor 2). What are the specifications for $H_1(z)$? What is the required number of multipliers for implementing the wideband filter $H(z)$? (Take the model filter and image suppressor to be equiripple.)

4.26. Suppose we wish to design a 25-fold lowpass linear-phase interpolation filter (i.e., $L = 25$). Let the input signal $x(n)$ be bandlimited to $|\omega| < 0.95\pi$. Assume that the ripple specifications are $\delta_1 = 0.01, \delta_2 = 0.002$. (a) Find reasonable band edges ω_p and ω_S for $H(z)$. (b) What is the filter order if a direct design is used? (c) Suppose the filter is designed using a two stage approach. What are the orders of $G(z)$ and $I(z)$? (d) What are the total number of multiplications and additions in the direct design and how do these compare with the two-stage design? (e) Assuming an input sampling rate of 8 KHz, what is the number of multiplications and additions *per second* in the two-stage design?

4.27. For a uniform DFT analysis bank, we know that the filters are related by $H_k(z) = H_0(zW^k), 0 \le k \le M - 1$, with $W = e^{-j2\pi/M}$. Let $M = 5$ and define two new transfer functions $G_1(z) = H_1(z) + H_4(z)$ and $G_2(z) = H_2(z) + H_3(z)$. Let $h_0(n)$ denote the impulse response of $H_0(z)$, assumed to be real for all n.

a) Are $h_k(n), 1 \le k \le 4$ real for all n?

b) Express the impulse responses $g_1(n)$ and $g_2(n)$ of $G_1(z)$ and $G_2(z)$ in terms of $h_0(n)$. Are $g_1(n)$ and $g_2(n)$ real for all n?

c) Let $|H_0(e^{j\omega})|$ be as shown in Fig. P4-27.

Figure P4-27

Plot the responses $|G_1(e^{j\omega})|$ and $|G_2(e^{j\omega})|$, for $0 \le \omega \le 2\pi$. Does $|G_2(e^{j\omega})|$ necessarily look 'good' in its passband?

4.28. Consider the analysis/synthesis system in Fig. P4-28.

a) Let the analysis filters be $H_0(z) = 1 + 3z^{-1} + 0.5z^{-2} + z^{-3}$ and $H_1(z) = H_0(-z)$. Find causal stable IIR filters $F_0(z)$ and $F_1(z)$ such that $\hat{x}(n)$

agrees with $x(n)$ except for a possible delay and (nonzero) scale factor.

b) Let $H_0(z) = 1 + 2z^{-1} + 3z^{-2} + 2z^{-3} + z^{-4}$, and $H_1(z) = H_0(-z)$. Find causal FIR filters $F_0(z)$ and $F_1(z)$ such that $\hat{x}(n)$ agrees with $x(n)$ except for a possible delay and (nonzero) scale factor.

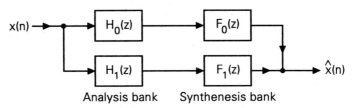

Figure P4-28

(*Hint.* This is perhaps tricky, but not tedious or difficult. It helps to use polyphase decomposition. Review of complementary filters might help.)

4.29. Let $H_0(z) = (1 + z^{-1})/2$. Find real-coefficient causal FIR $H_1(z)$ such that the pair $H_0(z), H_1(z)$ is power complementary. Are these filters also allpass complementary? Euclidean complementary? Doubly complementary?

4.30. *A trick for the design of zero-phase FIR equiripple half-band filters.* Suppose $G(z) = \sum_{n=0}^{N} g(n) z^{-n}$ is a Type 2 linear phase filter (Sec. 2.4.2). This means that N is odd and $g(n)$ is real, with $g(n) = g(N - n)$. This also means that there is a zero at $\omega = \pi$. We know we can write $G(e^{j\omega}) = e^{-j\omega N/2} G_R(\omega)$ where $G_R(\omega)$ real. Suppose we have designed $G(z)$ such that the response $G_R(\omega)$ is as shown in Fig. P4-30.

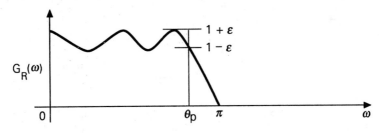

Figure P4-30

This design can be done by defining the passband to be $0 \leq \omega \leq \theta_p$ and transition band to be $\theta_p \leq \omega \leq \pi$. There is no stopband. Such filters with one equiripple passband and no stopband can indeed be designed using the McClellan-Parks algorithm (Section 3.2.4). Now define the transfer function $F(z) = [z^{-N} + G(z^2)]/2$. This is a Type 1 linear phase filter.

a) Show that $F(e^{j\omega}) = e^{-j\omega N} F_R(\omega)$, where $F_R(\omega)$ is real. Express $F_R(\omega)$ in terms of $G_R(\omega)$.

b) Plot the amplitude response $F_R(\omega)$ in $0 \leq \omega \leq \pi$. Verify that it resembles Fig. 4.6-4. What are the values of δ, ω_p, and ω_S in terms of ϵ and θ_p?

c) Let $f(n)$ and $g(n)$ denote the impulse responses of $F(z)$ and $G(z)$. Show that

$$f(n) = \begin{cases} 0.5g(\frac{n}{2}), & n \text{ even} \\ 0.5, & n = N \\ 0, & n \text{ odd}, n \neq N. \end{cases} \quad (P4.30a)$$

This also verifies that $F(z)$ is a half-band filter.

d) Suppose we define $\widehat{H}(z) = z^N F(z)$. Then $\widehat{H}(z)$ is a zero-phase half-band filter. Its length is $2N+1$, and since N is odd, we have $2N+1 = 4J+2+1$ for some integer J. Clearly $\widehat{H}(e^{j\omega})$ is real. Define $H(z) = 0.5\epsilon + \widehat{H}(z)$, i.e.,

$$h(n) = \begin{cases} \widehat{h}(n), & n \neq 0 \\ 0.5\epsilon + \widehat{h}(0) & n = 0. \end{cases} \quad (P4.30b)$$

Show that $H(e^{j\omega}) \geq 0$ for all ω.

This shows how we can design a zero-phase equiripple half-band filter $H(z)$ (with $H(e^{j\omega}) \geq 0$) just by designing a Type 2 one-band filter $G(z)$ of half the order, and making minor changes to its coefficients!

4.31. Let $\mathbf{A}$ be $N \times N$ Hermitian positive definite with eigenvalues $0 < \lambda_0 < \lambda_1 \ldots < \lambda_{N-1}$. Define $\mathbf{P} = \mathbf{I} - \alpha\mathbf{A}$, with $\alpha > 0$.
  a) Show that $\mathbf{P}$ is stable (i.e., eigenvalues strictly inside the unit circle) if and only if $\alpha < 2/\lambda_{N-1}$.
  b) In general $\mathbf{P}$ is not positive definite even though Hermitian. Show that the maximum eigenvalue-magnitude is minimized by the choice $\alpha = 2/(\lambda_0 + \lambda_{N-1})$. This choice therefore gives the fastest decrease of α_k after 'stalling' has set in (Fig. 4.7-2).

4.32. The matrix in (4.7.5) is a demonstration, for $N = 5$, of the $N \times N$ matrix $\mathbf{A}$ in (4.7.3). In this problem we consider this matrix and assume N is arbitrary. Let $\mu = 0$ for simplicity.
  a) Show that any vector of the form

$$\mathbf{v}_k = \begin{bmatrix} \sin(\frac{k\pi}{N+1}) \\ \sin(\frac{2k\pi}{N+1}) \\ \vdots \\ \sin(\frac{Nk\pi}{N+1}) \end{bmatrix}, \quad (P4.32)$$

is an eigenvector, with eigenvalue $\lambda_k = 4\sin^2\frac{k\pi}{2(N+1)}$. Here k is an integer in the range $1 \leq k \leq N$.
  b) Thus the elements $v_k(n)$ of the vector $\mathbf{v}_k$ can be written as $v_k(n) = \sin(\frac{k\pi n}{N+1})$. This sequence is "smooth" for small k and "oscillatory" for large k. To demonstrate this, let $N = 4$, and plot this sequence for $1 \leq k \leq 4$. Also plot the eigenvalue λ_k, as a function of k for $1 \leq k \leq 4$.
  c) The above exercise demonstrates that the eigenvectors corresponding to large eigenvalues are more "oscillatory". In other words, the matrix $\mathbf{A}$ acts like a highpass filtering operator (Sec. 4.7.4). Suppose we define $\mathbf{P} = \mathbf{I} - \alpha\mathbf{A}$. The eigenvalues of this matrix are $\mu_k \stackrel{\Delta}{=} 1 - \alpha\lambda_k$, and the eigenvectors continue to be $\mathbf{v}_k$. Give example of a choice of α such that

μ_k^2 decreases as k increases in the range $1 \leq k \leq N$. The matrix $\mathbf{P}$ would now resemble a lowpass filtering operator.

4.33. Recall the matrix $\mathbf{A}$ in (4.7.3), demonstrated in (4.7.5) for $N = 5$. In this problem, $\mathbf{A}^{(\Delta)}$ stands for this matrix, with $\mu = 0$ and $N = 7$. With $\mathbf{M}_D$ and $\mathbf{M}_I$ as in (4.7.23) and (4.7.24), compute the elements of $\mathbf{A}^{(2\Delta)}$ defined in (4.7.25).

PART 2 Multirate Filter Banks

5

Maximally Decimated Filter Banks

5.0 INTRODUCTION

The basic philosophy of subband coding was explained in Chap. 4. The analysis/synthesis system used for this purpose is the maximally decimated filter bank. Figure 5.1-1(a) shows the two channel version, popularly called the Quadrature Mirror Filter (QMF) bank. This system was introduced in the mid seventies [Croisier, et al., 1976], and has since then been studied by many other researchers, as we cite at the appropriate sections. The input signal $x(n)$ is first filtered by two filters $H_0(z)$ and $H_1(z)$, typically lowpass and highpass as shown in part (b). Each signal $x_k(n)$ (subband signal) is therefore approximately bandlimited to a total width of π (in the frequency region $0 \le \omega < 2\pi$). The subband signals are decimated by a factor of 2 to produce $v_k(n)$.

Each decimated signal $v_k(n)$ is then coded in such a way that the special properties of the subband (such as energy level, perceptive importance and so on) are exploited. At the receiver end, the received signals are decoded to produce (approximations of) the signals $v_0(n)$ and $v_1(n)$ which are then passed through two-fold expanders. The output signals $y_0(n)$ and $y_1(n)$ are then passed through the filters $F_0(z)$ and $F_1(z)$ (whose purpose we will explain) to produce the output signal $\hat{x}(n)$.

$H_0(z)$ and $H_1(z)$ are called analysis filters, and the pair $[H_0(z), H_1(z)]$ the analysis bank. This pair followed by the two decimators is the *decimated analysis bank*. Similarly $F_0(z)$ and $F_1(z)$ are the synthesis (or reconstruction) filters, and the pair $[F_0(z), F_1(z)]$ the synthesis bank. In this chapter we will see that the reconstructed signal $\hat{x}(n)$ differs from $x(n)$ due to *three* reasons: aliasing, amplitude distortion, phase distortion, It will be shown that the filters can be designed in such a way that some or all of these distortions are eliminated. These results will then be extended to the case of M channel filter banks.

There is a fourth reason why the reconstructed signal differs from $x(n)$. This is due to the coding or quantization of the subband signals. The effect of this cannot be corrected, but can only be analyzed. This is done in Appendix C.

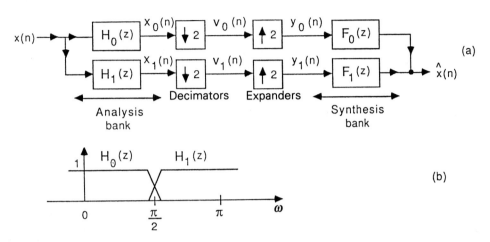

Figure 5.1-1 (a) The quadrature mirror filter bank and (b) typical magnitude responses.

5.0.1 A Brief History

For the two channel case, it was shown in Croisier, et al. [1976] that aliasing can be completely eliminated by a simple choice of the synthesis filters. Design techniques were later developed by other authors to minimize the remaining distortions [Johnston, 1980], [Jain and Crochiere, 1984], [Fettweis, et al., 1985] and efficient structures developed [Galand and Nussbaumer, 1984]. It was shown by Smith and Barnwell [1984] and Mintzer [1985] that all the three distortions mentioned above can be eliminated (i.e., perfect reconstruction achieved) in a two channel QMF bank with properly designed FIR filters, and further optimization techniques were developed [Grenez, 1988].

For the case of M channel filter banks, the conditions for alias cancelation and perfect reconstruction are much more complicated. The pseudo QMF technique was introduced [Nussbaumer, 1981], as a means of obtaining approximate alias cancelation in this case, and has since been developed by a number of authors [Rothweiler, 1983], [Chu, 1985], [Masson and Picel, 1985], and [Cox, 1986]. The general theory of perfect reconstruction in the M channel case was developed by a number of authors [Ramstad, 1984b], [Smith and Barnwell, 1985], [Vetterli, 1985], [Princen and Bradley, 1986], [Wackershruther, 1986b], [Vaidyanathan, 1987a,b], [Nguyen and Vaidyanathan,

1988], and [Viscito and Allebach, 1988a]. Vetterli and Vaidyanathan showed independently that the use of polyphase components leads to considerable simplification of the theory. A technique for the design of M channel perfect reconstruction systems was developed [Vaidyanathan, 1987a,b], based on polyphase matrices with the so-called paraunitary property. It has since been shown that the two channel perfect reconstruction system developed in Smith and Barnwell [1984] and Mintzer [1985] satisfy the paraunitary property. (This same property also finds application in the theory of orthonormal wavelet transforms, which we will study in Chap. 11.)

A particular class of M-channel perfect reconstruction systems was subsequently developed, with the property that all the analysis filters are derived starting from a prototype, by modulation. This has the advantage of economy during the design as well as implementation phases. The theory was developed by Malvar [1990b], Koilpillai and Vaidyanathan [1991 and 1992], and Ramstad [1991] independently. It turns out that these systems can be regarded as the generalization of the so-called lapped orthogonal transforms independently devloped by Cassreau [1985] and studied in Malvar and Staelin [1989], and Malvar [1990a].

Further advancement in the theory and design of filter banks have been made by several authors, but these will not be covered in our limited exposure here. This includes time domain design techniques, [Nayebi, et al., 1990], nonuniform filter banks, and filter banks with noninteger decimation ratios [Hoang and Vaidyanathan, 1989], [Kovačević and Vetterli, 1991a], [Nayebi, et al., 1991a], and filter banks with minimum reconstruction delay [Nayebi, et al., 1991b]. Also see Padmanabhan and Martin [1992], Horng, Samueli, and Willson [1991], and Horng and Willson [1992]. In Problems 5.25 and 5.32 we will consider some issues pertaining to nonuniform QMF banks.

5.0.2 Chapter Outline

In this chapter we present a detailed study of the QMF bank and its M channel extensions. Section 5.1 analyzes the various errors (aliasing, amplitude and phase distortions) created by the two channel QMF bank, and develops conditions for alias cancelation. Section 5.2 describes an alias-free system in greater detail. Section 5.3 considers a special class of alias free systems called the power symmetric QMF banks. These systems have very low complexity, and yet provide freedom from aliasing and amplitude distortion.

In Sec. 5.4 to 5.6 we extend these ideas for the case of M channel filter banks, and develop the theory of perfect reconstruction based on polyphase matrices. Section 5.7 develops the general theory of alias-free systems. Tree-structured filter banks are considered in Sec. 5.8, and Sec. 5.9 develops the theory of transmultiplexers.

The study of filter banks will be continued in the next few chapters. Paraunitary perfect reconstruction systems will be introduced in Chap. 6, along with several structures for implementing these systems. Some of the structures have the property that the perfect reconstruction property is retained in spite of coefficient quantization. Pseudo QMF banks and cosine modulated perfect reconstruction banks will be studied in Chap. 8.

5.1 ERRORS CREATED IN THE QMF BANK

The decimated signals $v_k(n)$ are encoded using one of many possible coding techniques [Jayant and Noll, 1984], and the resulting signals are actually transmitted. The receiver reconstructs an approximation $v'_k(n)$ of $v_k(n)$ from these encoded signals. The decoding error $v_k(n) - v'_k(n)$ represents a nonlinear distortion (like quantization error). This is called the *subband quantization error*. It cannot be corrected, that is, there is no way to exactly reconstruct $v_k(n)$ from $v'_k(n)$.

The subband quantization error will be treated in greater detail in Appendix C. In this chapter will ignore this error, that is, assume $v'_k(n) = v_k(n)$. The QMF bank still suffers from three fundamental errors, viz., aliasing, amplitude distortion, and phase distortion to be described next.

5.1.1 Aliasing and Imaging

In practice, the analysis filters have nonzero transition bandwidth and stopband gain. The signals $x_k(n)$ are, therefore, not bandlimited, and their decimation results in aliasing. To study this effect further, consider Fig. 5.1-2 where two situations are shown. In Fig. 5.1-2(a), the responses $|H_0(e^{j\omega})|$ and $|H_1(e^{j\omega})|$ do not overlap. Assuming that the stopband attenuations are sufficiently large, the effect of aliasing is not serious. In Fig. 5.1-2(b), however, the responses overlap, and each subband signal can in general have substantial energy for a bandwidth exceeding the ideal passband region. Decimation of these signals therefore results in aliasing regardless of how good the stopbands of the filters are.

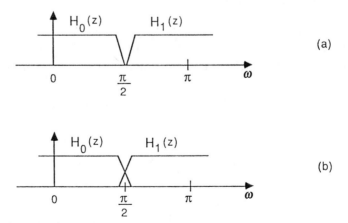

Figure 5.1-2 Two possible magnitude responses for the analysis filters. (a) Nonoverlapping, and (b) overlapping.

In principle it is true that the choice of filters as in Fig. 5.1-2(a) takes care of this problem. However, non overlapping responses imply severe attenuation of the input signal around $\omega = \pi/2$. Even though this can be compensenated, in principle, by appropriately boosting this frequency region [by proper design of $F_0(z)$ and $F_1(z)$], it will result in severe amplification of noise (such as coding noise, channel noise and filter roundoff noise). A second solution would be to make the transition widths of the responses very narrow but this requires very expensive filters. The overlapping response in Fig. 5.1-2(b) is therefore the more practical choice. Even though this results in aliasing, this effect can be canceled by careful choice of the *synthesis filters* as we will see.

Expression for the Reconstructed Signal

Using the results developed in Sec. 4.1.1 it is easy to find an expression for $\widehat{X}(z)$. From Fig. 5.1-1(a) we have

$$X_k(z) = H_k(z)X(z), \quad k = 0, 1. \tag{5.1.1}$$

The z-tranforms of the decimated signals $v_k(n)$ are [from (4.1.13) with $M = 2$]

$$V_k(z) = \frac{1}{2}[X_k(z^{1/2}) + X_k(-z^{1/2})], \quad k = 0, 1. \tag{5.1.2}$$

The second term above represents aliasing. The z-transform of $Y_k(z)$ is $V_k(z^2)$ so that

$$\begin{aligned} Y_k(z) = V_k(z^2) &= \frac{1}{2}[X_k(z) + X_k(-z)] \\ &= \frac{1}{2}[H_k(z)X(z) + H_k(-z)X(-z)], \quad k = 0, 1. \end{aligned} \tag{5.1.3}$$

The reconstructed signal is

$$\widehat{X}(z) = F_0(z)Y_0(z) + F_1(z)Y_1(z). \tag{5.1.4}$$

Substituting from (5.1.3) and rearranging, we finally obtain

$$\begin{aligned} \widehat{X}(z) =& \frac{1}{2}[H_0(z)F_0(z) + H_1(z)F_1(z)]X(z) \\ &+ \frac{1}{2}[H_0(-z)F_0(z) + H_1(-z)F_1(z)]X(-z). \end{aligned} \tag{5.1.5}$$

Or, in matrix-vector notation,

$$2\widehat{X}(z) = [\,X(z) \quad X(-z)\,] \underbrace{\begin{bmatrix} H_0(z) & H_1(z) \\ H_0(-z) & H_1(-z) \end{bmatrix}}_{\mathbf{H}(z)} \begin{bmatrix} F_0(z) \\ F_1(z) \end{bmatrix}. \tag{5.1.6}$$

The matrix $\mathbf{H}(z)$ is called the alias component (AC) matrix. The term which contains $X(-z)$ originates because of the decimation. On the unit circle, $X(-z) = X(e^{j(\omega-\pi)})$ which is a right-shifted version of $X(e^{j\omega})$ by an amount π. This term takes into account aliasing due to the decimators and imaging due to the expanders. We refer to this just as the alias term or alias component.

Alias Cancelation

From (5.1.5) it is clear that we can cancel aliasing by choosing the filters such that the quantity $H_0(-z)F_0(z) + H_1(-z)F_1(z)$ is zero. Thus the following choice cancels aliasing:

$$F_0(z) = H_1(-z), \quad F_1(z) = -H_0(-z). \tag{5.1.7}$$

Given $H_0(z)$ and $H_1(z)$, it is thus possible to *completely* cancel aliasing by this choice of synthesis filters. If the analysis filters have large transition bandwidths and low stopband attenuations, this implies large aliasing errors, but yet these errors are canceled by the choice (5.1.7).

So, the basic philosophy in the QMF bank is that we *permit* aliasing in the analysis bank instead of trying to avoid it. We then choose the synthesis filters so that the alias component in the upper branch is canceled by that in the lower branch.

Pictorial viewpoint. It helps to visualize the mechanism of alias cancelation in terms of frequency response plots. For this refer to Fig. 5.1-3 which shows an arbitrary input spectrum $X(e^{j\omega})$, the lowpass subband signal $X_0(e^{j\omega})$, and the decimated signal $V_0(e^{j\omega})$. The alias component $0.5X_0(-e^{j\omega/2})$ overlaps with $0.5X_0(e^{j\omega/2})$. The signal $Y_0(e^{j\omega})$ has contributions from $X(z)$ as well as $X(-z)$. The contribution which arises from $X(-z)$ (shaded region) is the alias component, and in general overlaps with the unshaded area.

In a similar way if we trace through the bottom channel, we can obtain qualitative plots of $X_1(e^{j\omega}), V_1(e^{j\omega})$ and $Y_1(e^{j\omega})$. The shaded areas in $Y_0(e^{j\omega})$ and $Y_1(e^{j\omega})$ represent aliasing (and imaging) effect(s), and dominantly occupy the *highpass* and *lowpass* regions, respectively. The filters $F_0(z)$ and $F_1(z)$, which are *lowpass* and *highpass* respectively, tend to eliminate these shaded portions. Because of the nonideal nature of these practical filters, the output of $F_0(z)$ still contains some residual shaded area (Fig. 5.1-3(h)), and so does the output of $F_1(z)$ (Fig. 5.1-3(i)). These two residual alias components can be made to cancel each other, and the choice (5.1.7) does precisely that.

The LPTV property. From Chap. 4 we know that the decimator and expander are linear and time varying (LTV) building blocks. So the QMF bank is a LTV system. Now (5.1.5) can be written as

$$\widehat{X}(z) = T(z)X(z) + A(z)X(-z). \tag{5.1.8}$$

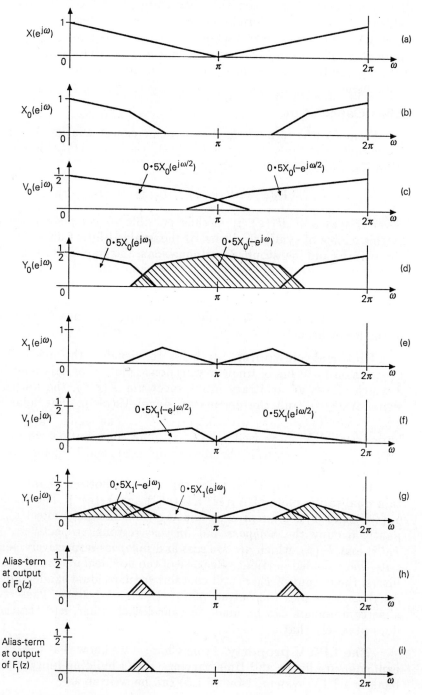

Figure 5.1-3 Various internal signals, and alias cancelation mechanism in the QMF bank. (© Adopted from 1990 IEEE.)

Denoting the impulse responses of $T(z)$ and $A(z)$ as $t(n)$ and $a(n)$, we can rewrite the above as

$$\hat{x}(n) = \sum_{k}\Bigl(t(k) + (-1)^{k-n}a(k)\Bigr)x(n-k). \qquad (5.1.9)$$

Defining $g_0(k) = t(k) + (-1)^k a(k)$ and $g_1(k) = t(k) - (-1)^k a(k)$, we then have

$$\hat{x}(n) = \begin{cases} \sum_k g_0(k) x(n-k) & n \text{ even} \\ \sum_k g_1(k) x(n-k) & n \text{ odd}, \end{cases} \qquad (5.1.10)$$

which proves that $\hat{x}(n)$ is produced by passing $x(n)$ through the systems $G_0(z)$ and $G_1(z)$ in parallel, and taking the output of $G_0(z)$ for even n and that of $G_1(z)$ for odd n (Fig. 5.1-4). So the QMF bank is a linear periodically time varying (LPTV) system with period two. If aliasing is canceled (i.e., $A(z) = 0$), the system becomes LTI, and has transfer function $T(z)$.

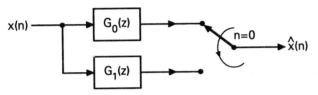

Figure 5.1-4 The QMF bank viewed as a LPTV system.

5.1.2 Amplitude and Phase Distortions

Suppose the choice (5.1.7) is made so that the QMF bank is free from aliasing. We then have

$$\hat{X}(z) = T(z) X(z). \qquad (5.1.11)$$

Thus even after aliasing is canceled, the signal $\hat{x}(n)$ suffers from a linear shift-invariant distortion $T(z)$. Here

$$T(z) = \frac{1}{2}[H_0(z) F_0(z) + H_1(z) F_1(z)], \qquad (5.1.12)$$

and is called the *distortion transfer function*, or "overall" transfer function of the alias-free system. Using (5.1.7) we get

$$T(z) = \frac{1}{2}[H_0(z) H_1(-z) - H_1(z) H_0(-z)]. \qquad (5.1.13)$$

Letting $T(e^{j\omega}) = |T(e^{j\omega})| e^{j\phi(\omega)}$, we have

$$\hat{X}(e^{j\omega}) = |T(e^{j\omega})| e^{j\phi(\omega)} X(e^{j\omega}). \qquad (5.1.14)$$

Unless $T(z)$ is allpass (i.e., $|T(e^{j\omega})| = d \neq 0$ for all ω), we say that $\hat{X}(e^{j\omega})$ suffers from "amplitude distortion." Similarly unless $T(z)$ has linear phase (that is, $\phi(\omega) = a + b\omega$ for constant a, b), $\widehat{X}(e^{j\omega})$ suffers from phase distortion.

We will use the following abbreviations for convenience: ALD (aliasing distortion), AMD (amplitude distortion), PHD (phase distortion).

Periodicity of $|T(e^{j\omega})|$. From (5.1.13) we see that $T(z)$ has the form $V(z) - V(-z)$. This means $T(z)$ has only odd powers of z, that is, $T(z) = z^{-1}S(z^2)$. So $|T(e^{j\omega})|$ has period π rather than 2π. For the real coefficient case this implies that $|T(e^{j\omega})|$ is symmetric with respect to $\pi/2$.

The Perfect Reconstruction (PR) QMF Bank

If a QMF bank is free from aliasing, amplitude distortion, and phase distortion, it is said to have the perfect reconstruction (abbreviated PR) property. This is equivalent to the condition $T(z) = cz^{-n_0}$. For a PR QMF bank we have

$$\widehat{X}(z) = cz^{-n_0}X(z), \quad \text{i.e.,} \quad \widehat{x}(n) = cx(n - n_0), \quad c \neq 0, \quad (5.1.15)$$

for all possible inputs $x(n)$. In other words, $\widehat{x}(n)$ is merely a scaled and delayed version of $x(n)$. This, of course, ignores the coding/decoding error and filter roundoff noise.

5.2 A SIMPLE ALIAS-FREE QMF SYSTEM

In the earliest known QMF banks the analysis filters were related as

$$H_1(z) = H_0(-z). \quad (5.2.1)$$

For the real coefficient case this means $|H_1(e^{j\omega})| = |H_0(e^{j(\pi-\omega)})|$. This ensures that $H_1(z)$ is a good highpass filter if $H_0(z)$ is a good lowpass filter. In fact $|H_1(e^{j\omega})|$ is a mirror image of $|H_0(e^{j\omega})|$ with respect to the quadrature frequency $2\pi/4$, justifying the name *quadrature mirror filters*.

With the choice (5.2.1) the alias cancelation constraint (5.1.7) becomes

$$F_0(z) = H_0(z), \quad F_1(z) = -H_1(z). \quad (5.2.2)$$

Thus all the four filters are completely determined by a single filter $H_0(z)$. The designer has to concentrate on the design of only this filter. According to the earliest nomenclatures, the system with the four filters related as above was known as the 'QMF' bank. But as a matter of convenience, the term 'QMF' has since been used to indicate generalized versions, for example, M-channel systems.

From (5.2.2) we see that $F_0(z)$ and $F_1(z)$ are lowpass and highpass respectively [consistent with the fact that $F_0(z)$ attenuates the 'highpass image' and $F_1(z)$ attenuates the 'lowpass image' created by the expanders]. With filters chosen as above, the distortion function is

$$T(z) = \frac{1}{2}\left(H_0^2(z) - H_1^2(z)\right) = \frac{1}{2}\left(H_0^2(z) - H_0^2(-z)\right). \quad (5.2.3)$$

5.2.1 Polyphase Representation

It is often beneficial, both conceptually and computationally, to represent the analysis and synthesis banks in terms of polyphase components (Section 4.3). Thus let

$$H_0(z) = E_0(z^2) + z^{-1}E_1(z^2) \quad \text{(Type 1 polyphase)}. \tag{5.2.4}$$

Since $H_1(z) = H_0(-z)$, we have $H_1(z) = E_0(z^2) - z^{-1}E_1(z^2)$, that is,

$$\begin{bmatrix} H_0(z) \\ H_1(z) \end{bmatrix} = \begin{bmatrix} 1 & 1 \\ 1 & -1 \end{bmatrix} \begin{bmatrix} E_0(z^2) \\ z^{-1}E_1(z^2) \end{bmatrix}. \tag{5.2.5}$$

The synthesis filters $F_0(z)$ and $F_1(z)$, which satisfy (5.2.2), can also be represented in terms of $E_0(z)$ and $E_1(z)$ as follows:

$$[F_0(z) \quad F_1(z)] = [z^{-1}E_1(z^2) \quad E_0(z^2)] \begin{bmatrix} 1 & 1 \\ 1 & -1 \end{bmatrix}. \tag{5.2.6}$$

By using (5.2.5) and (5.2.6) we can draw the analysis and synthesis banks as in Fig. 5.2-1(a) and (b) respectively, and the complete QMF bank as in Fig. 5.2-2(a). By using the noble identities (Fig. 4.2-3) we can redraw this as in Fig. 5.2-2(b). The polyphase components are now operating at the lowest possible rate, so that the number of multiplications and additions per unit time (MPUs and APUs) is minimized[†].

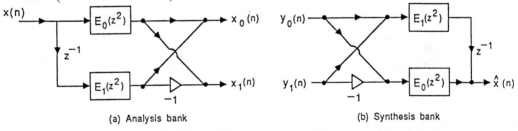

(a) Analysis bank   (b) Synthesis bank

Figure 5.2-1 The analysis and synthesis banks in polyphase form.

Limitations Imposed by the Constraint $H_1(z) = H_0(-z)$.

With the analysis filters related as $H_1(z) = H_0(-z)$ and synthesis filters chosen to cancel aliasing (eqn. (5.1.7)), the distortion function has the form (5.2.3). This can be written in terms of the polyphase components as

$$T(z) = 2z^{-1}E_0(z^2)E_1(z^2). \tag{5.2.7}$$

[†] As in Chap. 4, a unit of time is the separation between adjacent samples of the input $x(n)$.

This expression holds for any QMF bank (FIR or IIR; linear-phase or non-linear phase) for which the filters are related by (5.2.1) and (5.2.2). From this expression we can draw a number of important conclusions.

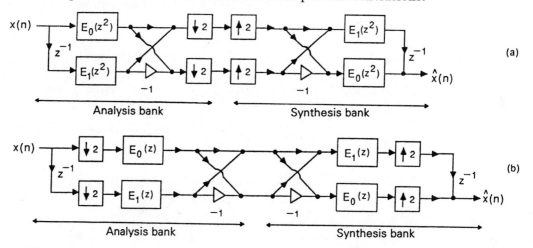

Figure 5.2-2 (a) The complete QMF bank in polyphase form. (b) Rearrangement using noble identities.

For example, let $H_0(z)$ be FIR so that $E_0(z)$, $E_1(z)$ and $T(z)$ are FIR as well. From (5.2.7) we note that amplitude distortion can be eliminated in this case if and only if each of the FIR functions $E_0(z)$ and $E_1(z)$ is a delay, that is, $E_0(z) = c_0 z^{-n_0}$ and $E_1(z) = c_1 z^{-n_1}$. This means

$$H_0(z) = c_0 z^{-2n_0} + c_1 z^{-(2n_1+1)}, \quad H_1(z) = c_0 z^{-2n_0} - c_1 z^{-(2n_1+1)}. \quad (5.2.8)$$

This conclusion holds whether or not $H_0(z)$ has linear phase.

Summarizing, if the analysis filters are related as $H_1(z) = H_0(-z)$ and $H_0(z)$ is FIR, we can eliminate amplitude distortion only if $H_0(z)$ and $H_1(z)$ have the above form! That is, the filters cannot have sharp cutoff and good stopband attenuations. We cannot, therefore, obtain useful FIR perfect reconstruction systems under the constraint $H_1(z) = H_0(-z)$.

If we choose $E_1(z) = 1/E_0(z)$, then (5.2.7) becomes a delay, thereby resulting in perfect reconstruction. But the filters become IIR.

5.2.2 Eliminating Phase Distortion with FIR Filters

A QMF bank in which the analysis *and* synthesis filters are FIR is said to be an FIR QMF bank. From Chap. 2 we know that FIR filters with exactly linear phase can be designed. If $H_0(z)$ has linear phase, then $T(z)$ given by (5.2.3) also has linear phase, thereby eliminating phase distortion.

The residual amplitude distortion $|T(e^{j\omega})|$ can now be analyzed with the help of (5.2.3). Let $H_0(z) = \sum_{n=0}^{N} h_0(n) z^{-n}$, with $h_0(n)$ real. The linear phase constraint requires $h_0(n) = \pm h_0(N-n)$. Since $H_0(z)$ has to be lowpass, the only possibility is $h_0(n) = h_0(N-n)$ (Section 2.4.2). With this choice

$$H_0(e^{j\omega}) = e^{-j\omega N/2} R(\omega), \qquad (5.2.9)$$

where $R(\omega)$ is real for all ω. Substituting (5.2.9) into (5.2.3) and using the fact that $|H(e^{j\omega})|$ is an even function, we get

$$T(e^{j\omega}) = \frac{e^{-jN\omega}}{2}\Big(|H_0(e^{j\omega})|^2 - (-1)^N |H_0(e^{j(\pi-\omega)})|^2\Big). \qquad (5.2.10)$$

Constraint on the order N. If N is even, then the above expression reduces to zero at $\omega = \pi/2$, resulting in severe amplitude distortion. So we have to chose N to be odd so that

$$\begin{aligned}T(e^{j\omega}) &= \frac{e^{-jN\omega}}{2}\Big(|H_0(e^{j\omega})|^2 + |H_0(e^{j(\pi-\omega)})|^2\Big) \\ &= \frac{e^{-jN\omega}}{2}\Big(|H_0(e^{j\omega})|^2 + |H_1(e^{j\omega})|^2\Big). \quad \text{(from (5.2.1))}\end{aligned} \qquad (5.2.11)$$

Minimization of Residual Amplitude Distortion

From the previous section we know that if $H_0(z)$ is FIR, then the constraint $H_1(z) = H_0(-z)$ rules out perfect reconstruction, unless the filters have the simple form (5.2.8). Having eliminated aliasing and phase distortion, we can therefore only *minimize* amplitude distortion, that is, we can make (5.2.11) only approximately constant.

If $H_0(z)$ has good passband and stopband responses, then $|T(e^{j\omega})|$ is almost constant in the passbands of $H_0(z)$ and $H_1(z)$. The main difficulty comes in the transition band region. The degree of overlap of $H_0(z)$ and $H_1(z)$ is very crucial in determining this distortion. To demonstrate this, Fig. 5.2-3 shows responses of three linear phase designs of $H_0(z)$. If the passband edge is too large as in curve 1 (i.e., $H_0(z)$ and $H_1(z)$ have too much overlap), $|T(e^{j\omega})|$ exhibits a peaking effect around $\pi/2$. If the passband edge is too small (curve 2), then $|T(e^{j\omega})|$ exhibits a dip around $\pi/2$. The third curve, where the passband edge is carefully chosen by trial and error, produces a much better response of $|T(e^{j\omega})|$.

The aim, therefore, is to adjust the coefficients of $H_0(z)$ so that the filters satisfy the condition

$$|H_0(e^{j\omega})|^2 + |H_1(e^{j\omega})|^2 = 1, \qquad (5.2.12)$$

approximately. Systematic computer-aided optimization techniques for this have been developed [Johnston, 1980], [Jain and Crochiere, 1984]. In Johnston's technique, an objective function is formulated which reflects two things: (a) the stopband attenuation of the filter $H_0(z)$, and (b) the extent to which (5.2.12) is satisfied. For example the objective function could be

$$\phi = \alpha \phi_1 + (1-\alpha)\phi_2, \qquad (5.2.13)$$

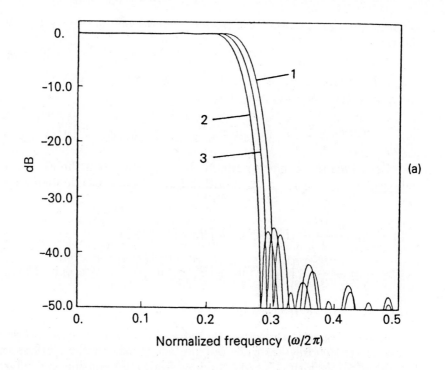

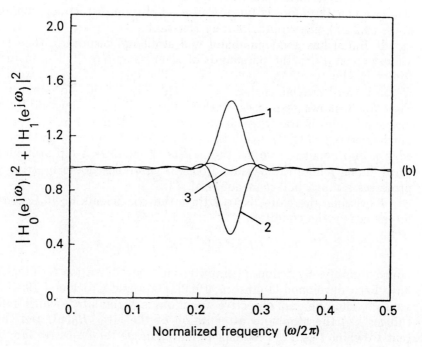

Figure 5.2-3 Amplitude distortion as a function of the degree of overlap between analysis filters. (© Adopted from 1990 IEEE.)

where
$$\phi_1 = \int_{\omega_S}^{\pi} |H_0(e^{j\omega})|^2 d\omega, \quad \phi_2 = \int_0^{\pi} \left(1 - |H_0(e^{j\omega})|^2 - |H_0(e^{j(\pi-\omega)})|^2\right)^2 d\omega,$$
(5.2.14)

and $0 < \alpha < 1$. The coefficients $h_0(n)$ of $H_0(z)$ are then optimized in order to minimize ϕ. Since $|T(e^{j\omega})|$ has symmetry with respect to $\pi/2$, we can replace $\int_0^{\pi}$ with $2\int_0^{\pi/2}$. The quantity ω_S is typically chosen as $\pi/2 + \epsilon$ for some small $\epsilon > 0$.

What controls the passband shape? If the optimized response is satisfactory, the quantities ϕ_1 and ϕ_2 will be very small, and (5.2.12) will hold approximately. This means that $|H_1(e^{j\omega})|$ (i.e., $|H_0(-e^{j\omega})|$) is close to unity in the stopband of $H_0(z)$. This is equivalent to saying that $|H_0(e^{j\omega})|$ is close to unity in its own passband. Summarizing, minimization of ϕ ensures that $H_0(z)$ has good stopband as well as passband responses.

Design Example 5.2.1: Johnston's Filters

Filters with a wide range of specifications have been designed, and impulse response coefficients tabulated in Johnston [1980]. These tables can also be found in Crochiere and Rabiner [1983]. Fig. 5.2-4(a) shows the magnitude response plots of the analysis filters for Johnston's 32D filter. For this design the filter order $N = 31$, $\omega_S = 0.586\pi$, and the minimum stopband attenuation is 38 dB. The quantity $|H_0(e^{j\omega})|^2 + |H_1(e^{j\omega})|^2$ (which is twice the amplitude distortion), is shown in Fig. 5.2-4(b). On a dB scale, this is close to 0 dB for all ω, with peak distortion equal to ± 0.025 dB.

Computational complexity. With N representing the order of $H_0(z)$, there are $N+1$ coefficients $h_0(n)$. There are $(N+1)/2$ coefficients in each of $E_0(z)$ and $E_1(z)$. So from Fig. 5.2-2(b) we see that the analysis bank requires a total of $N+1$ multiplications and additions, that is,

$$\frac{N+1}{2} \text{ MPUs} \quad \text{and} \quad \frac{N+1}{2} \text{ APUs}$$

(since these are performed after decimation). The synthesis bank has the same complexity.

For our design example, we have $N = 31$ so that the analysis bank can be implemented using 16 MPUs and 16 APUs. This is an efficient implementation exploiting two facts: (a) the presence of decimators and expanders, and (b) the relation $H_1(z) = H_0(-z)$. Once these are exploited the symmetry of $h_0(n)$ (due to linear phase) cannot, unfortunately, be exploited (Problem 5.3).

5.2.3 Eliminating Amplitude Distortion with IIR Filters

The question that arises now is this: is it possible to completely eliminate amplitude distortion, rather than just minimize it using a computer program? We address this now. In order to eliminate amplitude distortion,

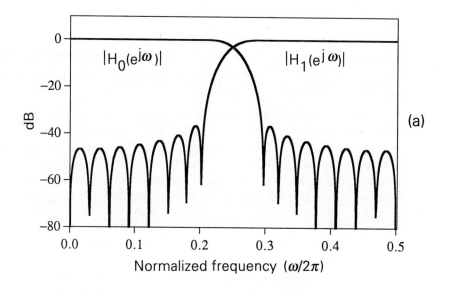

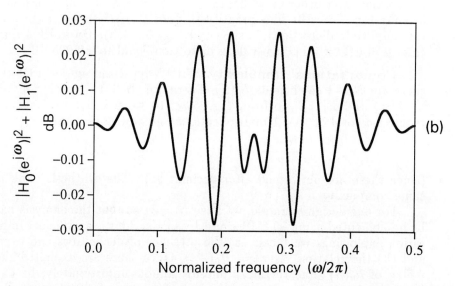

Figure 5.2-4 Design example 5.2.1 (Johnston's method). (a) Magnitude responses of the analysis filters, and (b) amplitude distortion measure.

we have to force $T(z)$ to be allpass. From (5.2.7) we see that this can be done by forcing $E_0(z)$ and $E_1(z)$ to be IIR and allpass [Vaidyanathan, et al, 1987], [Ramstad, 1988]. This also results in filters with a more general form than (5.2.8). Phase distortion still remains, and is governed by the phase responses of $E_0(z)$ and $E_1(z)$.†

To pursue this idea further, let us write the polyphase components as

$$E_0(z) = \frac{a_0(z)}{2}, \quad E_1(z) = \frac{a_1(z)}{2}, \qquad (5.2.15)$$

where $a_0(z)$ and $a_1(z)$ are allpass, with $|a_0(e^{j\omega})| = |a_1(e^{j\omega})| = 1$. The analysis filter $H_0(z)$ now takes the form

$$H_0(z) = \frac{a_0(z^2) + z^{-1} a_1(z^2)}{2}. \qquad (5.2.16)$$

Since $H_1(z) = H_0(-z)$, we have

$$\underbrace{\begin{bmatrix} H_0(z) \\ H_1(z) \end{bmatrix}}_{\mathbf{h}(z)} = \underbrace{\frac{1}{2} \begin{bmatrix} 1 & 1 \\ 1 & -1 \end{bmatrix}}_{\mathbf{R}} \underbrace{\begin{bmatrix} a_0(z^2) \\ z^{-1} a_1(z^2) \end{bmatrix}}_{\mathbf{a}(z)}. \qquad (5.2.17)$$

The synthesis filters, which are given by (5.2.2), can be expressed as

$$[F_0(z) \quad F_1(z)] = \frac{1}{2} [z^{-1} a_1(z^2) \quad a_0(z^2)] \begin{bmatrix} 1 & 1 \\ 1 & -1 \end{bmatrix}. \qquad (5.2.18)$$

The distortion function, which is allpass, is now given by

$$T(z) = \frac{z^{-1}}{2} a_0(z^2) a_1(z^2). \qquad (5.2.19)$$

Figure 5.1-1(a) can now be redrawn as in Fig. 5.2-5, showing the complete QMF bank. This is free from aliasing and amplitude distortion, regardless of the details of the allpass functions $a_0(z)$ and $a_1(z)$!

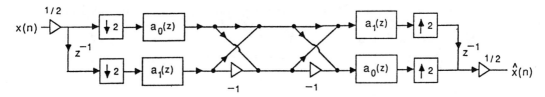

Figure 5.2-5 QMF bank with allpass polyphase components.

---

† The allpass constraint on $E_k(z)$ is, however, not *necessary*. For example, if $E_0(z) = 0.5 + z^{-1}$ and $E_1(z) = 1/(1 + 0.5z^{-1})$, then also $T(z)$ is allpass, that is, amplitude distortion is eliminated. However, since $H_0(z)$ has band edge around $\pi/2$ and since the coefficients $E_i(z)$ are decimated versions of $h_0(n+i)$, it is not counter-intuitive that $E_i(z)$ should be constrained to be allpass.

Can We Get Good Filter Responses with (5.2.16)?

The next question is, if we constrain the IIR analysis filter $H_0(z)$ to be of the form (5.2.16), is it still possible to have good attenuation characteristics? The answer is in the affirmative. For example, elliptic lowpass filters are of this form, if the bandedges and ripple sizes are chosen with appropriate symmetry (Fig. 5.2-6, to be explained next). With an elliptic filter so designed, we can easily identify the components $a_0(z)$ and $a_1(z)$ and then implement the structure of Fig. 5.2-5. It turns out that this technique is one of the most efficient ways (Sec. 5.3.5) to implement QMF banks free from aliasing and amplitude distortion. For example, we will see that if $H_0(z)$ is a fifth order elliptic filter, the entire analysis bank requires only one multiplication and three additions per input sample! In the next section we justify these claims, and also show how the filter $H_0(z)$ can be designed with the above constraint.

5.3 POWER SYMMETRIC QMF BANKS

We begin this section by summarizing the outcome of Sec. 5.2.3, concerning the IIR QMF bank. We assumed that the four filters are related as

$$H_1(z) = H_0(-z), \quad F_0(z) = H_0(z), \quad F_1(z) = -H_1(z).$$

This ensures that aliasing is canceled, and the distortion function is $T(z) = 2z^{-1}E_0(z^2)E_1(z^2)$, where $E_i(z)$ are polyphase components of $H_0(z)$, that is, $H_0(z) = E_0(z^2) + z^{-1}E_1(z^2)$. If $H_0(z)$ is IIR and the polyphase components $E_i(z)$ are allpass, then $T(z)$ becomes allpass. This, then, is a simple way to eliminate aliasing *and* amplitude distortion. The phase responses of $E_0(z)$ and $E_1(z)$ determine the remaining phase distortion.

♠**Main points of this section.** In this section we first study the properties of filters $H_0(z)$ for which $E_0(z)$ and $E_1(z)$ are allpass (i.e., filters which have the form (5.2.16) where $|a_0(e^{j\omega})| = |a_1(e^{j\omega})| = 1$) and then show how to design them.

1. We first show that if $H_0(z)$ is of the form (5.2.16), then it satisfies two symmetry-properties viz., numerator symmetry, and power symmetry (to be defined).
2. Conversely, we will show that if a transfer function satisfies these symmetry properties, it can be expressed as in (5.2.16). A more precise statement is given in Theorem 5.3.1.
3. As a consequence of the preceding result we will show the following: Let $H_0(z)$ be an odd order elliptic lowpass filter with ripple sizes δ_1, δ_2 and band edges ω_p, ω_S defined as usual [Figure 3.1-1(b)]. Suppose the response $|H_0(e^{j\omega})|^2$ exhibits symmetry with respect to $\pi/2$ as shown in Fig. 5.2-6. In other words, the ripple curve in the passband is a mirror image of the ripple curve in the stopband, with respect to the half-band frequency $\pi/2$. Mathematically this means $1 - (1 - 2\delta_1)^2 = \delta_2^2$, that is,

$$\delta_2^2 = 4\delta_1(1 - \delta_1), \qquad (5.3.1a)$$

and also
$$\omega_p + \omega_S = \pi. \qquad (5.3.1b)$$
(So if ω_S and δ_2 are specified then ω_p and δ_1 are determined, and the filter specifications are complete.) Under this symmetry condition, we can indeed express $H_0(z)$ as in (5.2.16), where $a_0(z)$ and $a_1(z)$ are unit-magnitude allpass filters. In other words the constraints (5.3.1) on the specifications ensure that the polyphase components of $H_0(z)$ are all-pass! We will present a modification of the standard elliptic filter design algorithm [Antoniou, 1979] to obtain the coefficients of $a_0(z)$ and $a_1(z)$, starting from the specifications ω_S and δ_2. ◇

The reader interested only in the design procedure can proceed directly to Sec. 5.3.4.

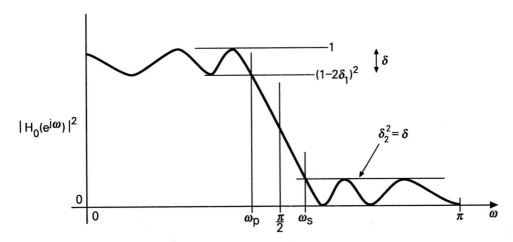

Figure 5.2-6 Square of the magnitude response function for a power symmetric filter.

5.3.1 Properties Induced by (5.2.16)

Power Symmetric Property

The quantities in (5.2.17) satisfy
$$\widetilde{\mathbf{a}}(z)\mathbf{a}(z) = 2, \quad \mathbf{R}^\dagger \mathbf{R} = 0.5\mathbf{I}, \qquad (5.3.2)$$
so that $\widetilde{\mathbf{h}}(z)\mathbf{h}(z) = 1$. In terms of ω this means
$$|H_0(e^{j\omega})|^2 + |H_1(e^{j\omega})|^2 = 1. \qquad (5.3.3)$$
So $H_1(z)$ is related to $H_0(z)$ in two ways: first by the constraint $H_1(z) = H_0(-z)$, and secondly by the power complementary property (5.3.3). Combining these we obtain the constraint
$$\widetilde{H}_0(z)H_0(z) + \widetilde{H}_0(-z)H_0(-z) = 1 \quad \text{(power symmetry condition)}. \qquad (5.3.4)$$

Now, on the unit circle we have $|H_0(-e^{j\omega})| = |H_0(e^{j(\omega-\pi)})|$. For the real coefficient case this is the same as $|H_0(e^{j(\pi-\omega)})|$ so that (5.3.4) implies

$$|H_0(e^{j(\frac{\pi}{2}+\theta)})|^2 + |H_0(e^{j(\frac{\pi}{2}-\theta)})|^2 = 1, \qquad (5.3.5)$$

for any real θ. This shows that the magnitude-squared function exhibits symmetry with respect to $\pi/2$, as demonstrated in Fig. 5.2-6. For this reason, (5.3.4) is called the *power symmetric property* and $H_0(z)$ is said to be power symmetric, even though (5.3.5) holds only for the real coefficient case. Also the right hand side of (5.3.4) is often permitted to be different from unity. We can restate (5.3.4) in any one of the following equivalent ways:

1. $\widetilde{H}_0(z)H_0(z)$ is a *half-band* filter [i.e., it satisfies (4.6.7c)].
2. $\widetilde{H}_0(z)H_0(z)\big|_{\downarrow 2} = 0.5$. Here the notation $A(z)\big|_{\downarrow 2}$ is as defined in Section 4.1.1. See, for example, (4.1.14).
3. $H_0(z)$ is power-symmetric.

Symmetry of Numerator of $H_0(z)$

From Sec. 3.4 we know that the allpass functions can be expressed as

$$a_0(z) = c_0 z^{-k_0} \frac{\widetilde{d}_0(z)}{d_0(z)}, \quad a_1(z) = c_1 z^{-k_1} \frac{\widetilde{d}_1(z)}{d_1(z)}, \qquad (5.3.6)$$

where $|c_0| = |c_1| = 1$, and $k_i \geq$ order of $d_i(z)$. [By convention $d_i(z)$ is a polynomial in z^{-1}.] Substituting into (5.2.16) we obtain

$$H_0(z) = \frac{0.5\left(c_0 z^{-2k_0}\widetilde{d}_0(z^2)d_1(z^2) + z^{-1}c_1 z^{-2k_1}\widetilde{d}_1(z^2)d_0(z^2)\right)}{d_0(z^2)d_1(z^2)} \qquad (5.3.7)$$

Thus, $H_0(z) = P_0(z)/d_0(z^2)d_1(z^2)$, that is, the denominator has only even powers of z^{-1}. It is easy to verify that the numerator $P_0(z)$ is generalized Hermitian (Sec. 2.3). For the most common case where $d_0(z)$ and $d_1(z)$ have real coefficients and $c_0 = c_1 = 1$, this means that $P_0(z)$ is *symmetric*. More specifically, $P_0(z^{-1}) = z^N P_0(z)$ where $N = 2(k_0 + k_1) + 1$. If $p_0(n)$ denotes the coefficients of $P_0(z)$, this property means $p_0(n) = p_0(N-n)$.

Irreducibility. It can be shown (Problem 5.7) that there are no common factors between $P_0(z)$ and the denominator $d_0(z^2)d_1(z^2)$, under the reasonable assumptions that (a) $d_0(z)$ and $d_1(z)$ have all zeros inside the unit circle, and (b) $d_0(z)$ and $d_1(z)$ do not have common factors. In practical examples such as Butterworth and elliptic filters, these two assumptions are true. The second assumption is reasonable because, if $(1-\alpha z^{-1})$ is a common factor between $d_0(z)$ and $d_1(z)$ then the allpass factor $(-\alpha^* + z^{-2})/(1-\alpha z^{-2})$ can be extracted from the right hand side of (5.2.16), and does not contribute to the magnitude response $|H_0(e^{j\omega})|$.

5.3.2 Power Symmetry and Numerator Symmetry Imply (5.2.16)

Assuming that (5.2.16) holds, we showed that $H_0(z)$ satisfies two symmetry properties. We now consider the converse, restricting our discussion to real coefficient filters. We show that if $H_0(z)$ is power symmetric and $P_0(z)$ symmetric, then $H_0(z)$ can be expressed as $(a_0(z^2) + z^{-1}a_1(z^2))/2$. The theorem below is a more precise statement of this. At this time, recall that $H_0(z)$ is said to be bounded real (BR) if it (a) has real coefficients, (b) is stable, and (c) satisfies $|H_0(e^{j\omega})| \leq 1$.

♠**Theorem 5.3.1.** Let $H_0(z) = P_0(z)/D(z)$ be the irreducible representation of a BR function with symmetric (or antisymmetric) numerator of odd order N. If $H_0(z)$ satisfies the power symmetric condition (5.3.4) then the following are true:

1. $H_0(z)$ can be expressed as in (5.2.16) where $a_0(z)$ and $a_1(z)$ are stable, real-coefficient, unit-magnitude allpass functions.
2. Moreover the order of $H_0(z)$ is $N = 2(k_0 + k_1) + 1$ where k_i is the order of $a_i(z)$. So there are no cancelations in (5.2.16). ◇

Some practical examples. As a special case, suppose $H_0(z)$ is an odd order elliptic lowpass filter satisfying (5.3.1a) and (5.3.1b). Then, the conditions of the theorem are satisfied. Notice, however, that power symmetric filters are not restricted to be elliptic. For example, odd order Butterworth filters can be designed to satisfy (5.3.4). Chebyshev filters, on the other hand, are not suitable because they are inherently nonsymmetric (the passband is equiripple and stopband monotone, or vice versa).

Proof of Theorem 5.3.1. Substituting $H_0(z) = P_0(z)/D(z)$ into the power symmetric condition (5.3.4) and rearranging, we obtain

$$\frac{\widetilde{P}_0(-z)P_0(-z)}{\widetilde{D}(-z)D(-z)} = \frac{\widetilde{D}(z)D(z) - \widetilde{P}_0(z)P_0(z)}{\widetilde{D}(z)D(z)}$$

Since $P_0(z)/D(z)$ is irreducible, there is no common factor of the form $(1 - \beta z^{-1})$, $\beta \neq 0$ between $P_0(z)$ and $D(z)$. Also since $\widetilde{P}_0(z) = z^N P_0(z)$ (by symmetry of $P_0(z)$), there are no such common factors between $\widetilde{P}_0(z)$ and $D(z)$ either. As a result, there are no common factors between the numerator and denominator of the left hand side of the above equation. The denominators on the two sides should therefore be equal except for a scale factor. Equating, in particular, the factors of these denominators which have zeros inside the unit circle, we obtain $D(-z) = cD(z)$. Assuming that $D(z)$ is normalized such that its constant coefficient is unity, we have $D(z) = D(-z)$. From this we also see that $D(z)$ has only even powers of z^{-1}, that is $D(z) = d(z^2)$.

Let $H_1(z) \stackrel{\Delta}{=} H_0(-z)$. Then

$$H_0(z) = \frac{P_0(z)}{d(z^2)}, \quad H_1(z) = \frac{P_0(-z)}{d(z^2)}$$

The power symmetric condition, (5.3.4) means that the function $H_1(z)$ is power complementary to $H_0(z)$. Since $P_0(z)$ is a odd order symmetric (antisymmetric) polynomial, $P_0(-z)$ is, therefore, antisymmetric (symmetric). Summarizing, $H_0(z)$ and $H_1(z)$ are a set of stable, real coefficient power complementary functions with symmetric and antisymmetric numerators. Moreover they have the same denominator. We can therefore apply Theorem 3.6.1 to conclude

$$\begin{bmatrix} H_0(z) \\ H_1(z) \end{bmatrix} = \frac{1}{2} \begin{bmatrix} 1 & 1 \\ 1 & -1 \end{bmatrix} \begin{bmatrix} A_0(z) \\ A_1(z) \end{bmatrix}, \qquad (5.3.8)$$

where $A_0(z)$ and $A_1(z)$ are stable unit-magnitude allpass, with orders n_0 and n_1 such that $N = n_0 + n_1$. Notice now that since $H_1(z) = H_0(-z)$ we can always write the pair as in (5.2.5). Since the 2×2 matrix in (5.3.8) is nonsingular, we conclude by comparing (5.2.5) with (5.3.8) that $E_0(z^2) = A_0(z)/2$ and $z^{-1}E_1(z^2) = A_1(z)/2$. This proves that the polyphase components $E_0(z)$ and $E_1(z)$ are allpass with magnitude 0.5. Thus $H_0(z)$ can be expressed as in (5.2.16), where $a_0(z)$ and $a_1(z)$ are stable unit-magnitude allpass. Since $N = n_0 + n_1$ we have $N = 2(k_0 + k_1) + 1$. $\qquad \triangledown\triangledown\triangledown$

Even-order filters. The above result is restricted to odd order filters. Recall from Sec. 3.6 that if $H_0(z)$ is an even order elliptic lowpass filter, then the allpass decomposition can still be done but the allpass filters now have complex coefficients [even though $H_0(z)$ has real coefficients]. Since the polyphase components evidently have real coefficients, these allpass filters cannot, therefore, be the polyphase components. In the even order case it is possible to use a modified IIR QMF bank which overcomes this difficulty (Problem 5.6).

5.3.3 Poles of Power Symmetric Elliptic Filters

Let $G(z)$ be a lowpass (or highpass) power symmetric elliptic filter. Then all its poles are located on the imaginary axis. Thus the poles have the form $j\beta_k$ (with $-1 < \beta_k < 1$ due to stability). The rest of this section is devoted to proving this, and can be skipped without loss of continuity.

Proof of the above claim. From Sec. 3.3.3 we know that for an Nth order elliptic filter $G(z)$ we can write

$$G(z^{-1})G(z) = \frac{1}{1 + \epsilon^2 R(z)R(z^{-1})}, \qquad (5.3.9)$$

where $R(z)$ is a rational function of the form

$$R(z) = \left(\frac{1 - z^{-1}}{1 + z^{-1}} \right)^\ell \prod_{k=1}^m \frac{(1 - z^{-1}e^{j\theta_k})(1 - z^{-1}e^{-j\theta_k})}{(1 - z^{-1}e^{j\omega_k})(1 - z^{-1}e^{-j\omega_k})}, \qquad (5.3.10)$$

with $\ell = 1$ for odd N and $\ell = 0$ for even N. Here m is such that $N = 2m+\ell$. The frequencies θ_k are the reflection zeros (i.e., points where $|G(e^{j\omega})|$ attains the maximum of unity) and ω_k the transmission zeros [Fig. 5.3-1(a)]. From (5.3.10) we have the relation

$$R(z) = (-1)^\ell R(z^{-1}). \tag{5.3.11}$$

The power-symmetric property means

$$G(z^{-1})G(z) = 1 - G(-z^{-1})G(-z). \tag{5.3.12}$$

Evidently the right hand side of (5.3.12) has reflection zeros at $\pi - \omega_k$ and transmission zeros at $\pi - \theta_k$. These, therefore, should agree with θ_k and ω_k respectively, that is, $\pi - \theta_k = \omega_k$ as demonstrated in Fig. 5.3-1. Substituting this into (5.3.10) we can show $R(-z) = 1/R(z)$, that is,

$$R(z)R(-z) = 1. \tag{5.3.13}$$

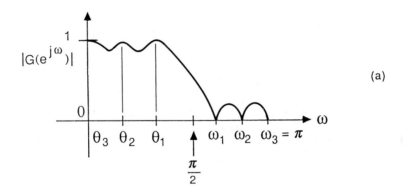

(a)

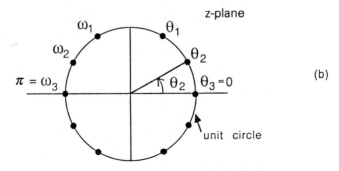

(b)

Figure 5.3-1 For a power symmetric elliptic lowpass filter $G(z)$, the relation $\omega_k + \theta_k = \pi$ holds.

Now by substituting (5.3.9) into the right hand side of (5.3.12) we get

$$G(z^{-1})G(z) = \frac{1}{1 + \frac{1}{\epsilon^2 R(-z)R(-z^{-1})}} = \frac{1}{1 + \frac{1}{\epsilon^2}R(z)R(z^{-1})} \qquad (5.3.14)$$

using (5.3.13). By comparing (5.3.9) and (5.3.14) we conclude $\epsilon^2 = 1$. For a power symmetric elliptic filter $G(z)$ we thus have

$$G(z^{-1})G(z) = \frac{1}{1 + R(z)R(z^{-1})}, \qquad (5.3.15)$$

where $R(z)$ is as in (5.3.10), with $\omega_k + \theta_k = \pi$. If p is a pole then $1 + R(p)R(p^{-1}) = 0$. In view of (5.3.11) and (5.3.13) this implies

$$\left|\frac{R(p)}{R(-p)}\right| = 1, \quad \text{if } p \text{ is a pole of } G(z). \qquad (5.3.16)$$

From Fig. 5.3-1 we see that $\omega_k > \pi/2$ and $\theta_k < \pi/2$ for all k. So the poles of the rational function $R(z)/R(-z)$ are restricted to the open left-half of the z-plane. Moreover, $R(z)/R(-z)$ has unit magnitude on the imaginary axis so that by maximum modulus theorem (Sec. 3.4.1) we have $|R(z)/R(-z)| < 1$ for Re $z > 0$. By replacing z with $-z$ and repeating this argument we find that $|R(-z)/R(z)| < 1$ for Re $z < 0$. Summarizing, we have

$$\left|\frac{R(z)}{R(-z)}\right| \begin{cases} < 1, & \text{Re } z > 0, \\ = 1, & \text{Re } z = 0, \\ > 1, & \text{Re } z < 0. \end{cases} \qquad (5.3.17)$$

which proves that all the poles of $G(z)$ are indeed on the imaginary axis of the z plane. $\triangledown\triangledown\triangledown$

5.3.4 Design of Power Symmetric Filters

By Theorem 5.3.1, elliptic lowpass filters whose specifications satisfy (5.3.1) have the form (5.2.16). For example, with $N = 5$, we have $n_0 = 2, n_1 = 3$, that is, $k_0 = k_1 = 1$ so that a fifth order power symmetric elliptic filter can be expressed as

$$H_0(z) = 0.5\frac{\alpha_0 + z^{-2}}{1 + \alpha_0 z^{-2}} + 0.5z^{-1}\frac{\alpha_1 + z^{-2}}{1 + \alpha_1 z^{-2}} \qquad (5.3.18)$$

where $0 < \alpha_0, \alpha_1 < 1$. Similarly a third order power symmetric elliptic filter can be expressed as

$$H_0(z) = 0.5\left(\frac{\alpha + z^{-2}}{1 + \alpha z^{-2}} + z^{-1}\right), \qquad (5.3.19)$$

where $0 < \alpha < 1$. [Since the zeros of an elliptic filter are on the unit circle, we have $(1/3) < \alpha$ as well.] The constants c_0 and c_1 in (5.3.6) which are obviously real in this case, are taken to be unity so that $H_0(1) = 1$ as required.

Design Procedure (Butterworth and Elliptic Cases)

Our discussions are made easier in terms of analog filters, reviewed in Sec. 3.3. If we design a Butterworth filter with 3 dB point $\Omega_c = 1$ and obtain $H_0(z)$ using the bilinear transformation (3.3.1), then $H_0(z)$ automatically satisfies the power symmetric property (Problem 5.8).

We now consider the elliptic case. The design specifications are δ_1, δ_2, Ω_p and Ω_S. The parameter

$$r \triangleq \Omega_p/\Omega_S \qquad (5.3.20)$$

governs the filter sharpness. The analog domain equivalent of the condition $\omega_p + \omega_S = \pi$ is, in view of the bilinear transform,

$$\Omega_p \Omega_S = 1. \qquad (5.3.21)$$

Since the ripples are constrained as in (5.3.1a), we have only two degrees of freedom, viz., δ_2 and r. Given these specifications, if we compute δ_1 using (5.3.1a) and then use standard techniques to design the analog elliptic filter, the estimated filter order may not turn out to be an integer. If it is rounded to the nearest (or next higher) odd integer N, the resulting filter does not in general satisfy the desired ripple constraint.

It is, however, possible to modify the standard elliptic filter algorithm such that the condition (5.3.1a) is exactly satisfied after the order has been rounded up. In this process the value of δ_2 is readjusted (reduced), such that N is exactly an integer under the constraint (5.3.1a). Table 5.3.1 shows the design-algorithm for this, obtained by modifying the procedure given by Antoniou [1979]. The resulting analog elliptic filter has the desired r, and usually has smaller ripple size δ_2 than specified. In any case it satisfies the power symmetric condition (5.3.1a) exactly. If this is transformed into a digital filter by use of the bilinear transform, the resulting $H_0(z)$ satisfies (5.2.16). In particular the denominator has only even powers of z^{-1}, and the poles are all on the imaginary axis of the z-plane.

Identifying the two allpass filters. Since the above algorithm gives $H_0(z)$ in factored form, the poles are already known, so the method described in Sec. 3.6 can be used to identify $a_0(z)$ and $a_1(z)$. In the elliptic filter case, the pole interlace property can be used to simplify this identification (recall Fig. 3.6-5 and associated comments). Fig. 5.3-2 demonstrates this for $N = 7$. Once the poles of $a_0(z^2)$ and $z^{-1}a_1(z^2)$ are identified, the polynomials $d_0(z^2)$ and $d_1(z^2)$ in (5.3.6) are known. By setting $c_0 = c_1 = 1$ and taking $k_i = $ order of $d_i(z)$, we can identify $a_i(z)$. These are summarized in Table 5.3.1.

A second way to identify the allpass filters is as follows. We have $H_0(z) = P_0(z)/d(z^2)$, with the coefficients of $P_0(z)$ and $d(z)$ known from

TABLE 5.3.1 Design of power symmetric elliptic filters

We summarize the procedure to design an odd order, lowpass, power-symmetric elliptic filter $H_0(z)$. Let the filter order be $N = 2m + 1$.

Specifications.

The given specifications are ω_S and δ_2, i.e., the stopband edge and peak stopband ripple, as in Fig. 3.1-1(b). The minimum stopband attenuation is then $A_S = -20\log_{10} \delta_2$. The passband edge ω_p and peak passband ripple δ_1 are determined according to the halfband symmetry conditions

$$\omega_p = \pi - \omega_S, \quad \text{and} \quad 4\delta_1(1 - \delta_1) = \delta_2^2.$$

Also recall $A_{max} = -20\log_{10}(1 - 2\delta_1)$.

Order estimation.

Define the quantities $r = \tan(0.5\omega_p)/\tan(0.5\omega_S)$, $\hat{r} = \sqrt{1 - r^2}$, $q_0 = 0.5(1 - \sqrt{\hat{r}})/(1 + \sqrt{\hat{r}})$,

$$q = q_0 + 2q_0^5 + 15q_0^9 + 150q_0^{13}, \quad \text{and} \quad D = \left(\frac{1 - \delta_2^2}{\delta_2^2}\right)^2.$$

Estimate the order N to be the smallest odd integer such that

$$N \geq \frac{\log_{10} 16D}{\log_{10}(1/q)}$$

Readjusting ripple size.

Since N is obtained by rounding-up the right hand side above, the resulting peak ripple δ_2 is smaller than specified. To recompute this ripple first recompute D from

$$D = \frac{10^{N \log_{10}(1/q)}}{16}.$$

Then find readjusted δ_2 from $D = (1 - \delta_2^2)^2/\delta_2^4$, and δ_1 from $4\delta_1(1 - \delta_1) = \delta_2^2$. Values of A_S and A_{max} are also readjusted accordingly.

Computing the filter coefficients.

Let

$$\Lambda = \frac{1}{2N} \log_e \left(\frac{10^{0.05 A_{max}} + 1}{10^{0.05 A_{max}} - 1}\right) \quad \text{(use readjusted } A_{max}\text{),}$$

and $w = (1 + r)/\sqrt{r}$. For $1 \leq k \leq m$ (where $m = (N - 1)/2$) compute

$$\Omega_k = \frac{2q^{0.25} \sum_{i=0}^{\infty}(-1)^i q^{i(i+1)} \sin\bigl((2i + 1)k\pi/N\bigr)}{1 + 2\sum_{i=1}^{\infty}(-1)^i q^{i^2} \cos(2\pi ki/N)},$$

Table 5.3.1 continued ...

$$v_k = \sqrt{\left(1 - r\Omega_k^2\right)\left(1 - \frac{\Omega_k^2}{r}\right)},$$

$$b_k = \frac{2v_k}{1 + \Omega_k^2}, \quad \text{and} \quad \alpha_{k-1} = \frac{2 - b_k}{2 + b_k},$$

in the order mentioned. Usually, it is sufficient to retain five or six terms of the infinite summations above. The quantities α_k computed above are distinct and satisfy $0 < \alpha_k < 1$. Renumber them so that

$$0 < \alpha_0 < \alpha_1 < \ldots < \alpha_{m-1} < 1.$$

Define the polynomials

$$d_0(z) = \prod_{k \text{ even}} (1 + \alpha_k z^{-1}), \quad d_1(z) = \prod_{k \text{ odd}} (1 + \alpha_k z^{-1}).$$

Let k_0 and k_1 denote the orders of $d_0(z)$ and $d_1(z)$. Define the allpass functions

$$a_0(z) = \frac{z^{-k_0} \widetilde{d}_0(z)}{d_0(z)}, \quad a_1(z) = \frac{z^{-k_1} \widetilde{d}_1(z)}{d_1(z)}.$$

Then the lowpass power symmetric elliptic filter is $H_0(z) = 0.5\left[a_0(z^2) + z^{-1} a_1(z^2)\right]$. Its order is $N = 2(k_0 + k_1) + 1$.

---

the above design. From (5.2.16) we have $H_0(z) + H_0(-z) = a_0(z^2)$, i.e., $a_0(z^2) = [P_0(z) + P_0(-z)]/d(z^2)$. After reducing this rational function to its irreducible form, we can identify $d_0(z)$. Thus, $a_0(z)$ given by $z^{-k_0}\widetilde{d}_0(z)/d_0(z)$ is found. Similarly, $a_1(z)$ can be identified from $[P_0(z) - P_0(-z)]/d(z^2)$.

Design Example 5.3.1: Power Symmetric Elliptic Filter

Suppose we wish to design a power symmetric elliptic filter $H_0(z)$ with stop band edge $\omega_S = 0.608\pi$ and stopband attenuation $A_S = 35$dB. This A_S corresponds to $\delta_2 \approx 0.0178$. From ω_S we determine $\omega_p = \pi - \omega_S$. The quantities Ω_p and Ω_S can now be identified using (3.3.15). From these we obtain $r = \Omega_p/\Omega_S \approx 0.5$. If we compute δ_1 using (5.3.1a), then the required filter order N for this combination of δ_1, δ_2 and r is $N = 4.7$, which is not an integer. If this is readjusted to $N = 5$, the ripples will not satisfy (5.3.1a) any more.

By using the values of δ_2 and r in the algorithm of Table 5.3.1, we can obtain the readjusted ripple size $\delta_2 = 0.0132$ (i.e., $A_S = 37.58$dB). If this is used in (5.3.1a), we get $\delta_1 = 4.36 \times 10^{-5}$. These values of δ_1 and δ_2, together with the specified r (i.e., $r = 0.5$) imply a filter order $N = 5$, which is exactly an integer. This filter, therefore is power symmetric.

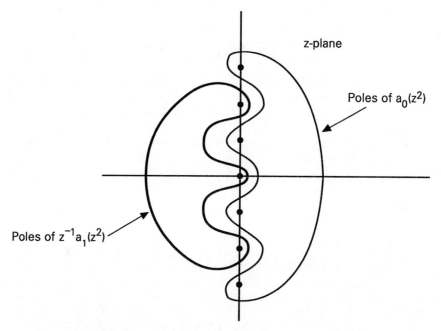

Figure 5.3-2 Grouping of poles into those of $a_0(z^2)$ and $z^{-1}a_1(z^2)$. Here $N = 7$.

The poles of this power symmetric elliptic filter are at the locations

$$z = 0, \quad z = \pm j\sqrt{\alpha_0}, \quad z = \pm j\sqrt{\alpha_1}, \qquad (5.3.22)$$

where $\alpha_0 = 0.226634$ and $\alpha_1 = 0.703653$. See Fig. 5.3-3. So we associate the poles $z = \pm j\sqrt{\alpha_0}$ with $a_0(z^2)$, and the poles $z = 0$ and $z = \pm j\sqrt{\alpha_1}$ with $z^{-1}a_1(z^2)$. Thus the elliptic filter $H_0(z)$ has the form (5.3.18), with α_0 and α_1 as above. Fig. 5.3-4(a) shows the magnitude response of $H_0(z)$.

Phase distortion. The distortion function $T(z)$ is given by

$$T(z) = \frac{z^{-1}a_0(z^2)a_1(z^2)}{2} = 0.5z^{-1}\left(\frac{\alpha_0 + z^{-2}}{1 + \alpha_0 z^{-2}}\right)\left(\frac{\alpha_1 + z^{-2}}{1 + \alpha_1 z^{-2}}\right)$$

This is allpass with nonlinear phase response (i.e., nonconstant group delay). [The phase response is linear only if $a_0(z)$ and $a_1(z)$ are pure delays, which is uninteresting]. Figure 5.3-4(b) shows a plot of the group delay of $T(z)$ for the above example. This exhibits a variation from 3 samples to about 16 samples. Whether this is acceptable or not depends on the application in hand, and several subjective considerations come into play. For example, some amount of phase distortion is acceptable in speech processing, but not in image processing [Lim, 1990].

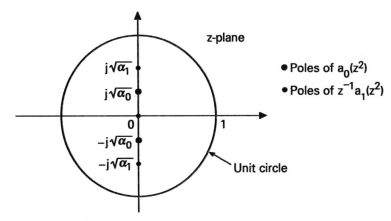

Figure 5.3-3 Identifying poles of $a_0(z^2)$ and $z^{-1}a_1(z^2)$ in Design example 5.3.1.

A Direct Optimization Approach (Non-Elliptic Design)

The fact that the poles of elliptic power symmetric filters are located on the imaginary axis implies that the denominators $d_0(z)$ and $d_1(z)$ of the allpass functions in (5.2.16) are of the form

$$d_0(z) = \prod_i (1 + \alpha_{0,i} z^{-1}), \quad d_1(z) = \prod_i (1 + \alpha_{1,i} z^{-1}), \qquad (5.3.23)$$

with $0 < \alpha_{j,i} < 1$. This gives us the hint that if we wish to optimize the coefficients of $a_0(z)$ and $a_1(z)$ directly (rather than by designing an elliptic filter), then we can restrict $d_0(z)$ and $d_1(z)$ to be of this form. For example, we can optimize the parameters $\alpha_{j,i}$ in (5.3.23) in order to minimize the stopband energy

$$\phi = \frac{1}{\pi} \int_{\omega_S}^{\pi} |H_0(e^{j\omega})|^2 d\omega. \qquad (5.3.24)$$

Such an optimization is generally fast because in practice we have very few parameters. A fifth order filter of the form (5.3.18) has only two parameters to optimize! Note that even though the passband error is not included in the objective function, it automatically turns out to be small because of the power symmetric condition ensured by (5.2.16).

Design example 5.3.1. Power symmetric elliptic filter(*continuation*).

Fig. 5.3-4(c) shows the plot of $|H_0(e^{j\omega})|$ designed by optimizing the function ϕ. Again the filter order is taken to be $N = 5$, i.e., the power symmetric filter is as in (5.3.18). In this example $\omega_S = 0.6\pi$ (the lower

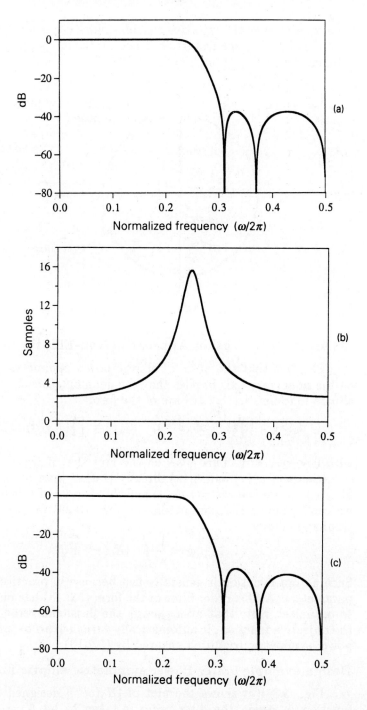

Figure 5.3-4 Design example 5.3.1. (a) Magnitude of elliptic power symmetric filter (b) its group delay response, and (c) magnitude of minimum energy power symmetric filter.

limit of the integral in (5.3.24)). The first peak ripple in the stopband is $A_S \approx 38$ dB. With (5.3.24) used as the objective function, the attenuation at ω_S is not equal to A_S, but typically less. In our example, the lowest frequency with attenuation equal to A_S is 0.614π. The optimized system has $\alpha_0 = 0.2121846$, and $\alpha_1 = 0.689796$. The optimized filter response is not equiripple, since this is not a minimax design. The peak ripple decreases as ω increases. This is desirable in some applications.

Table 5.3.2 gives the values of α_0 and α_1 in (5.3.18) which minimize (5.3.24), for various choices of ω_S appearing in (5.3.24). These are fifth order filters ($N = 5$), and cover a wide range of requirements. The table also shows the attenuation A_S at the location of the first peak-ripple in the stopband. The table serves as a quick design aid for IIR power symmetric filters which can be used to design alias-free QMF banks with freedom from amplitude distortion. For other combinations of N, ω_S and A_S, the reader can obtain designs by direct optimization of (5.3.24), or by using the algorithm of Table 5.3.1.

TABLE 5.3.2 Optimal IIR power symmetric filters with N=5

ω_S	α_0	α_1	A_S
0.550π	0.2790	0.7652	28.6
0.575π	0.2401	0.7231	33.2
0.600π	0.2122	0.6898	37.6
0.625π	0.1910	0.6626	41.7
0.650π	0.1744	0.6399	46.5
0.675π	0.1611	0.6206	50.0
0.700π	0.1502	0.6042	54.5

5.3.5 Low Complexity of the IIR Power Symmetric QMF Bank

We know that the allpass filters $a_j(z)$ have denominators of the form (5.3.23). So $a_j(z)$ is a product of k_j first order sections of the form

$$\frac{\alpha_{j,i} + z^{-1}}{1 + \alpha_{j,i} z^{-1}}, \qquad (5.3.25)$$

with $0 < \alpha_{j,i} < 1$. Each of these sections can be implemented with one multiplier, two adders and two delays as shown in Fig. 3.4-4. So $a_j(z)$ can be implemented by cascading k_j such sections, requiring a total of k_j

multipliers, $2k_j$ adders, and $2k_j$ delays. In fact, it is possible to share a delay between adjacent sections, as demonstrated in Fig. 5.3-5.

The total complexity to implement $a_0(z)$ and $a_1(z)$ is equal to $k_0 + k_1 = 0.5(N-1)$ multiplications and $(N-1)$ additions. The outputs of $a_i(z)$ are added and subtracted, which costs two more adders. These multipliers and adders operate at the lower rate (see Fig. 5.2-5) so that the analysis bank requires

$$0.25(N-1) \text{ MPUs} \quad \text{and} \quad 0.5(N+1) \text{ APUs}. \tag{5.3.26}$$

The complexity of the synthesis bank is the same.

In our design example $N = 5$, so that the analysis bank requires one MPU and three APUs. For this cost, the analysis filters provide 37.6 dB stopband attenuation, and the QMF bank is entirely free from aliasing and amplitude distortion. This system, therefore, is very efficient indeed!

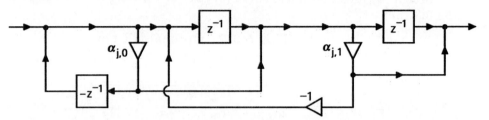

Figure 5.3-5 A cascade of two sections of the form (5.3.25). Each section is implemented as in Fig. 3.4-4, but a delay has been shared so that only three delays are required.

Robustness to Quantization

In any practical implementation, the multiplier coefficients are quantized (Chap. 9). In general this can result in the loss of some or all of the desirable properties (e.g., alias-cancelation, freedom from amplitude distortion, etc.). It is easy to verify that the allpass based structure of Fig. 5.2-5 is free from aliasing, as long as $a_i(z)$ in the analysis bank is quantized the same way as $a_i(z)$ in the synthesis bank. [This is because the alias cancelation condition (5.1.7) continues to hold.]

Furthermore, suppose the allpass filters are implemented such that they remain allpass inspite of multiplier quantization. This is easily ensured since $a_i(z)$ is a product of first order allpass functions which can be implemented as in Fig. 5.3-5 with real multiplier coefficients $\alpha_{j,i}$. Under this condition, the distortion function (5.2.19) continues to be allpass. Summarizing, the structure can be made free from aliasing as well as amplitude distortion, in spite of multiplier quantization.

5.3.6 FIR PR System with Power Symmetric Filters

We will now present an FIR perfect reconstruction system by modification of the above ideas. This system was introduced independently by Smith and

Barnwell [1984] and Mintzer [1985]. Let the synthesis filters be chosen in the usual way to cancel aliasing [i.e., as in (5.1.7)]. We have

$$\widehat{X}(z) = \frac{1}{2}[H_0(z)H_1(-z) - H_1(z)H_0(-z)]X(z), \qquad (5.3.27)$$

as shown in Section 5.1.2. For perfect reconstruction, we require this to be a delay. Note that we have not made the assumption $H_1(z) = H_0(-z)$ here. In particular, therefore, the alias-free system need not satisfy (5.2.2).

Assume now that $H_0(z)$ is power symmetric, that is, (5.3.4) holds. By comparing this with (5.3.27), we see that if the filter $H_1(z)$ is chosen as

$$H_1(z) = -z^{-N}\widetilde{H}_0(-z), \qquad (5.3.28)$$

for some odd N, then (5.3.27) reduces to $\widehat{X}(z) = 0.5z^{-N}X(z)$, that is, we have a perfect reconstruction system! In order for this system to be practical, $H_0(z)$ has to be FIR. (Otherwise $H_1(z)$ would be unstable for stable $H_0(z)$). By using (5.3.28) in (5.1.7) we see that the synthesis filters are given by

$$F_0(z) = z^{-N}\widetilde{H}_0(z), \quad F_1(z) = z^{-N}\widetilde{H}_1(z). \qquad (5.3.29)$$

The above choices of filters can be rewritten in the time domain as

$$h_1(n) = (-1)^n h_0^*(N-n), \; f_0(n) = h_0^*(N-n), \text{ and } f_1(n) = h_1^*(N-n). \qquad (5.3.30)$$

Assuming that $H_0(z)$ is causal, we see that the remaining filters are causal as long as $N \geq$ order of $H_0(z)$.

We can summarize these results as follows. Let

$$H_0(z) = \sum_{n=0}^{N} h_0(n)z^{-n} \qquad (5.3.31)$$

be power symmetric [i.e., satisfies (5.3.4)]. Then the choice of the remaining three filters according to (5.3.30) results in a perfect reconstruction system satisfying $\widehat{x}(n) = 0.5x(n-N)$.

Other properties. It is easily verified that the filters chosen as above satisfy these properties: (a) $|F_k(e^{j\omega})| = |H_k(e^{j\omega})|$, that is, the synthesis filters have the same magnitude responses as the analysis filters, and (b) $|H_1(e^{j\omega})| = |H_0(-e^{j\omega})|$. In the real coefficient case, the second property means that if $H_0(z)$ is lowpass then $H_1(z)$ is highpass, with same ripple sizes. Note that the relation $H_1(z) = H_0(-z)$ is in general *not* satisfied by this system.

Design Procedure

Only the filter $H_0(z)$ remains to be designed. The power symmetric property means that the zero-phase filter $H(z) = \widetilde{H}_0(z)H_0(z)$ is a half-band filter. Note that $H(e^{j\omega})$ has to be nonnegative. The design steps for the real coefficient case ($h_0(n)$ real) are as follows:

1. First design a zero-phase FIR half band filter $G(z) = \sum_{n=-N}^{N} g(n) z^{-n}$ of order $2N$ (e.g., by using the McClellan-Parks algorithm). The half-band property can be achieved by constraining the bandedges to be such that $\omega_p + \omega_S = \pi$, and the peak ripples in the passband and stopband to be identical as shown in Fig. 5.3-6(a).
2. Then define $H(z) = G(z) + \delta$, where δ is the peak stopband ripple of $G(e^{j\omega})$. This ensures that $H(e^{j\omega}) \geq 0$, as seen from Fig. 5.3-6(a).
3. Finally compute a spectral factor $H_0(z)$ of the filter $H(z)$. In principle, this can be done by computing the zeros of $H(z)$ and assigning an appropriate subset to $H_0(z)$ (Sec. 3.2.5). However there exist more efficient techniques which do not require the computation of zeros. One of these, due to Mian and Nainer [1982], is described in Appendix D. Once $H_0(z)$ has been computed, the remaining three filters are obtained using (5.3.30).

Comments.

1. *Order is odd.* As shown in Sec. 4.6.1, the order of $G(z)$ in the above design is of the form $4J + 2$ so that the order of $H_0(z)$ is $2J + 1$, that is, odd. Since the integer N in (5.3.28) is also required to be odd, we can take N to be same as the order of $H_0(z)$. This also ensures that the filters defined as in (5.3.30) are causal.
2. *Choosing the specifications.* Let ω_S and A_S be the stopband edge and minimum stopband attenuation specified for $H_0(z)$. Then the filter $G(z)$ has the same stopband edge ω_S, and stopband attenuation $\approx 2A_S + 6.02$ dB (why?). The passband specifications of $G(z)$ are automatically determined by the half-band constraint as follows: (a) peak passband ripple is identical to peak stopband ripple, and (b) $\omega_p + \omega_S = \pi$.
3. *Efficient design of $G(z)$.* The half-band filter $G(z)$ can also be designed using a more efficient trick, which was outlined in Problem 4.30 (using slightly different notations for the filters).
4. *Phase of $H_0(z)$.* As explained in Sec. 3.2.5, the spectral factor $H_0(z)$ is not unique because of the many ways in which the zeros of $H(z)$ can be grouped into those of $H_0(z)$ and $\widetilde{H}_0(z)$. The efficient technique described in Appendix D gives a minimum-phase spectral factor (i.e., the zeros are on and inside the unit circle). If one desires to have a spectral factor with nearly *linear* phase response, it can be done by other groupings of the zeros [Smith and Barnwell, 1984]. However, $H_0(z)$ cannot have exactly linear phase, unless it has the form $az^{-K} + bz^{-L}$. This is because, if $H_0(z)$ has linear phase, then so does $H_1(z)$ defined according to (5.3.28). But $H_0(z)$ and $H_1(z)$ are *also* power complementary, and cannot therefore have more than two nonzero coefficients (as proved later in Sec. 7.1).

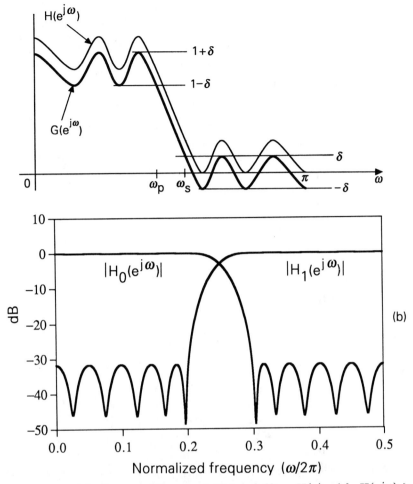

Figure 5.3-6 (a) Construction of a half-band filter $H(z)$ with $H(e^{j\omega}) \geq 0$. (b) Design example 5.3.2. Magnitude responses of the analysis filters of the perfect reconstruction system.

Design Example 5.3.2. FIR Power Symmetric Filter Bank

Suppose $H_0(z)$ is required to be a real-coefficient, equiripple, power symmetric FIR lowpass filter with specifications: $\omega_S = 0.6\pi$ and $A_S = 32$ dB. This means that the half-band filter $G(z)$ has stopband attenuation 70 dB (and stopband edge 0.6π). The required order of $G(z)$ (hence $H(z)$) turns out to be 38. So the power symmetric analysis filter $H_0(z)$ has order $N = 19$. The coefficients of the spectral factor $H_0(z)$ are found using the technique due to Mian and Nainer [1982], described in Appendix D. Table 5.3.3 shows the coefficients $h_0(n)$. The magnitude responses of the analysis filters are shown in Fig. 5.3-6(b).

Computational complexity. If implemented independently, each analysis filter would require $(N+1)$ multiplications and N additions. However, since the impulse responses are related as in (5.3.30), we can implement the analysis bank as shown in Fig. 5.3-7, requiring a total of $(N+1)$ multiplications and $2N$ additions. The total complexity of the direct form implementation is therefore $(N+1)$ MPUs and $2N$ APUs for the analysis bank (and the same for the synthesis bank).

TABLE 5.3.3 Filter coefficients in Design example 5.3.2

n	$h_0(n)$
0	0.1605476 e+00
1	0.4156381 e+00
2	0.4591917 e+00
3	0.1487153 e+00
4	−0.1642893 e+00
5	−0.1245206 e+00
6	0.8252419 e−01
7	0.8875733 e−01
8	−0.5080163 e−01
9	−0.6084593 e−01
10	0.3518087 e−01
11	0.3989182 e−01
12	−0.2561513 e−01
13	−0.2440664 e−01
14	0.1860065 e−01
15	0.1354778 e−01
16	−0.1308061 e−01
17	−0.7449561 e−02
18	0.1293440 e−01
19	−0.4995356 e−02

Instead of using the structure of Fig. 5.3-7 which exploits the relation between $H_1(z)$ and $H_0(z)$, we can also implement $H_0(z)$ and $H_1(z)$ individually in polyphase form. We then require only $(N+1)$ MPUs and N APUs for the entire analysis bank.

The above MPU and APU counts are higher than the cost for Johnston's designs ($0.5(N + 1)$ MPUs and $0.5(N + 1)$ APUs for the analysis bank). The increased complexity above is partly due to the fact that we have not *simultaneously* exploited the relation (5.3.30) and the decimation operations.

In Sec. 6.4 we will present a lattice structure for the QMF bank which overcomes this, and has the smallest possible complexity (same number of MPUs and APUs as Johnston's filters). This lattice has the additional advantage that the perfect reconstruction property is preserved in spite of multiplier quantization. Such a feature is not offered by the direct form structure (Fig. 5.3-7); for example, quantization of $h_0(n)$ results in the loss of power symmetric property (hence loss of perfect reconstruction).

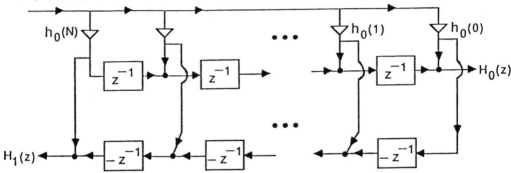

Figure 5.3-7 An $(N+1)$-multiplier implementation of the real-coefficient analysis bank satisfying $h_1(n) = (-1)^n h_0(N-n)$.

5.4 M-CHANNEL FILTER BANKS

For the two channel QMF bank, we considered a specific case where the analysis filters are related as $H_1(z) = H_0(-z)$, and studied it in detail. It is important to analyze the more general case [where restrictions such as $H_1(z) = H_0(-z)$ are not imposed a priori], so that we can understand the general conditions for alias cancelation and perfect reconstruction.

However, in attempting to study the general theory of alias cancelation and perfect reconstruction, it turns out to be more efficient to deal directly with the M-channel maximally decimated filter bank shown in Fig. 5.4-1. We, therefore, study this system in the next few sections. The special properties which arise for the two channel case ($M = 2$) will be pointed out at appropriate places, along with several examples.

In Fig. 5.4-1 the signal $x(n)$ is split into M subband signals $x_k(n)$ by the M analysis filters $H_k(z)$. Fig. 4.1-15(c) in Chap. 4 shows typical frequency responses of the analysis filters. Each signal $x_k(n)$ is then decimated by M to obtain $v_k(n)$. The decimated signals are eventually passed through M-fold expanders, and recombined via the synthesis filters $F_k(z)$ to produce $\widehat{x}(n)$. For convenience, and to be consistent with the literature, we sometimes refer to this system as the (M-channel) QMF bank, even though the name "QMF" is not justified any more. Many applications of this system were outlined in Chap. 4. More can be found in Chap. 10 and 11, where this system is used as a unifying tool for a number of diverse topics such as block filtering, nonuniform sampling, periodically time varying systems, and wavelet transform theory.

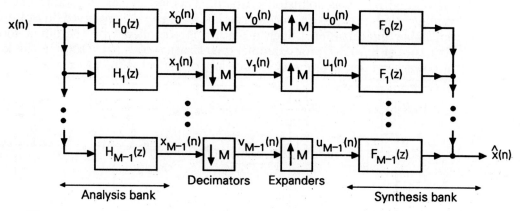

Figure 5.4-1 The M-channel (or M-band) maximally decimated filter bank. Also called M-channel QMF bank.

In this section we introduce the fundamentals of alias cancelation and perfect reconstruction. These results will be used in other chapters frequently. For notational convenience we define the vectors

$$\mathbf{h}(z) = \underbrace{\begin{bmatrix} H_0(z) \\ H_1(z) \\ \vdots \\ H_{M-1}(z) \end{bmatrix}}_{\text{Analysis bank}}, \quad \mathbf{f}(z) = \underbrace{\begin{bmatrix} F_0(z) \\ F_1(z) \\ \vdots \\ F_{M-1}(z) \end{bmatrix}}_{\text{Transposed synthesis bank}}, \quad \mathbf{e}(z) = \underbrace{\begin{bmatrix} 1 \\ z^{-1} \\ \vdots \\ z^{-(M-1)} \end{bmatrix}}_{\text{Delay chain}}$$

(5.4.1)

Notice that the analysis bank is a one-input M-output system with transfer matrix $\mathbf{h}(z)$; the synthesis bank is an M-input one-output system with transfer matrix $\mathbf{f}^T(z)$. The delay chain vector will be used in polyphase representations; this was already encountered in Chap. 4 [e.g., see Fig. 4.1-16(a)].

5.4.1 Expression for the Reconstructed Signal

We first obtain an expression for $\widehat{X}(z)$ in terms of $X(z)$, by ignoring the presence of coding and quantization errors. Each subband signal is given by

$$X_k(z) = H_k(z)X(z) \tag{5.4.2}$$

so that the decimated signals $v_k(n)$ have z-transform (Sec. 4.1.1)

$$V_k(z) = \frac{1}{M} \sum_{\ell=0}^{M-1} H_k(z^{1/M} W^\ell) X(z^{1/M} W^\ell), \qquad (5.4.3)$$

where $W = W_M = e^{-j2\pi/M}$. The outputs of the expanders are therefore given by

$$U_k(z) = V_k(z^M) = \frac{1}{M} \sum_{\ell=0}^{M-1} H_k(zW^\ell) X(zW^\ell), \qquad (5.4.4)$$

so that the reconstructed signal is

$$\widehat{X}(z) = \sum_{k=0}^{M-1} F_k(z) U_k(z) = \frac{1}{M} \sum_{\ell=0}^{M-1} X(zW^\ell) \sum_{k=0}^{M-1} H_k(zW^\ell) F_k(z). \qquad (5.4.5)$$

We can rewrite this in the more convenient form

$$\widehat{X}(z) = \sum_{\ell=0}^{M-1} A_\ell(z) X(zW^\ell), \qquad (5.4.6)$$

where

$$A_\ell(z) = \frac{1}{M} \sum_{k=0}^{M-1} H_k(zW^\ell) F_k(z), \quad 0 \le \ell \le M-1. \qquad (5.4.7)$$

The quantity $X(zW^\ell)$ can be written for $z = e^{j\omega}$ as

$$X(e^{j\omega} W^\ell) = X(e^{j(\omega - \frac{2\pi\ell}{M})}). \qquad (5.4.8)$$

For $\ell \neq 0$, this represents a shifted version of the spectrum $X(e^{j\omega})$. So the reconstructed spectrum $\widehat{X}(e^{j\omega})$ is a linear combination of $X(e^{j\omega})$ and its $M-1$ uniformly shifted versions.

5.4.2 Errors Created by the Filter Bank System

In a manner analogous to the two-channel QMF bank, the reconstructed signal $\widehat{x}(n)$ differs from $x(n)$ due to several reasons such as aliasing, imaging, amplitude distortion, and phase distortion as explained next.

Aliasing and Imaging

The presence of shifted versions $X(zW^\ell), \ell > 0$ is due to the decimation and interpolation operations. We say that $X(zW^\ell)$ is the ℓth aliasing term, and $A_\ell(z)$ is the gain for this aliasing term. It is clear that aliasing can be eliminated for *every possible* input $x(n)$, if, and only if,

$$A_\ell(z) = 0, \quad 1 \le \ell \le M - 1. \tag{5.4.9}$$

We now demonstrate alias cancelation ideas graphically for $M = 3$. We have two alias cancelation conditions to satisfy, namely

$$H_0(zW)F_0(z) + H_1(zW)F_1(z) + H_2(zW)F_2(z) = 0, \tag{5.4.10}$$

$$H_0(zW^2)F_0(z) + H_1(zW^2)F_1(z) + H_2(zW^2)F_2(z) = 0. \tag{5.4.11}$$

Figure 5.4-2(a)–(c) show the magnitude responses of the three analysis filters, along with their shifted versions. It is assumed that $|H_k(e^{j\omega})|$ is symmetric with respect to zero-frequency, which is consistent with the common situation where the filter coefficients are real.

The signal which enters the filter $F_0(z)$ contains the terms

$$H_0(z)X(z), \quad H_0(zW)X(zW), \quad \text{and } H_0(zW^2)X(zW^2). \tag{5.4.12}$$

The purpose of the filter $F_0(z)$, broadly speaking, is to eliminate the terms involving $X(zW)$ and $X(zW^2)$. This is done if $F_0(z)$ attenuates the replicas $H_0(zW)$ and $H_0(zW^2)$, and retains only $H_0(z)$. For this reason, the response $|F_0(e^{j\omega})|$ resembles $|H_0(e^{j\omega})|$, as shown in Fig. 5.4-2(d). The responses of $F_1(z)$ and $F_2(z)$, based on same reasoning, are also indicated in the same figure.

Thus, the output of $F_0(z)$ is a lowpass filtered version of $x(n)$, plus some alias terms. Similarly, the output of $F_1(z)$ is a bandpass filtered version of $x(n)$ plus alias terms. The relation between these outputs and the so-called multiresolution components will be discussed in Section 5.8.

Note that if the filters were ideal, with responses given by

$$H_k(e^{j\omega}) = F_k(e^{j\omega}) = \begin{cases} \sqrt{M} & \text{(passband)} \\ 0 & \text{(stopband)} \end{cases}$$

then there is perfect reconstruction, that is, $\widehat{x}(n) = x(n)$. Since the filters $F_k(z)$ are not ideal in practice, they do not completely eliminate the shifted replicas $H_k(zW)$ and $H_k(zW^2)$. For instance, the three terms in (5.4.10) are not individually equal to zero. The *residual alias terms* are demonstrated in Fig. 5.4-2(e)-(g). The responses of $H_0(zW)F_0(z)$ and $H_1(zW)F_1(z)$ have an overlap, and so do the responses of $H_1(zW)F_1(z)$ and $H_2(zW)F_2(z)$. The basic idea behind alias cancelation is to choose the synthesis filters such that these overlapping terms cancel out.

Amplitude and Phase Distortions

Unless aliasing is canceled, the M-channel QMF bank is a periodically time varying system (LPTV) with period M. (This was shown in Section

5.1.1, by taking $M = 2$; also see Sec. 10.1.2.) If the aliasing terms are somehow eliminated by forcing $A_\ell(z) = 0$ for $\ell > 0$, we have

$$\widehat{X}(z) = T(z)X(z). \qquad (5.4.13)$$

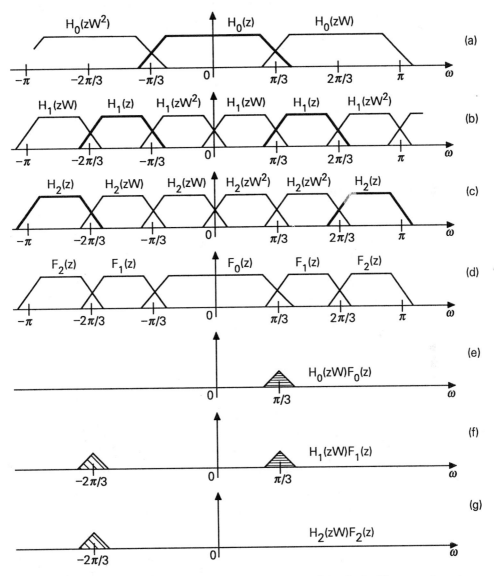

Figure 5.4-2 (a), (b), (c) Magnitude respones of analysis filters and various shifted versions. (d) Magnitude responses of synthesis filters. (e), (f), (g) Residual alias terms with $\ell = 1$, indicating overlap between adjacent-channel alias terms which can be canceled with each other.

Here $T(z)$ is the distortion function (or overall transfer function)

$$T(z) \triangleq A_0(z) = \frac{1}{M} \sum_{k=0}^{M-1} H_k(z) F_k(z). \qquad (5.4.14)$$

Thus, when aliasing is canceled, the QMF bank is an LTI system with transfer function $T(z)$. If $|T(e^{j\omega})|$ is not a constant (i.e., $T(z)$ not allpass) we say that there is amplitude distortion, and if $T(z)$ has nonlinear phase we say that there is phase distortion.

Perfect reconstruction (PR) systems. If $H_k(z)$ and $F_k(z)$ are such that (a) aliasing is completely canceled and (b) $T(z)$ is a pure delay (i.e., $T(z) = cz^{-n_0}, c \neq 0$), then the system is free from aliasing, amplitude distortion and phase distortion. Such a system satisfies $\hat{x}(n) = cx(n - n_0)$, and is called a perfect reconstruction system.

5.4.3 The Alias Component (AC) Matrix

We can rewrite (5.4.7) in matrix-vector form as

$$M \underbrace{\begin{bmatrix} A_0(z) \\ A_1(z) \\ \vdots \\ A_{M-1}(z) \end{bmatrix}}_{\mathbf{A}(z)} = \underbrace{\begin{bmatrix} H_0(z) & H_1(z) & \cdots & H_{M-1}(z) \\ H_0(zW) & H_1(zW) & \cdots & H_{M-1}(zW) \\ \vdots & \vdots & \ddots & \vdots \\ H_0(zW^{M-1}) & H_1(zW^{M-1}) & \cdots & H_{M-1}(zW^{M-1}) \end{bmatrix}}_{\mathbf{H}(z)} \underbrace{\begin{bmatrix} F_0(z) \\ F_1(z) \\ \vdots \\ F_{M-1}(z) \end{bmatrix}}_{\mathbf{f}(z)},$$

$$(5.4.15)$$

To cancel aliasing, we have to force all elements on the left side to zero (except the top element). So, the conditions for alias cancelation can be written as

$$\mathbf{H}(z)\mathbf{f}(z) = \mathbf{t}(z) \qquad (5.4.16)$$

where

$$\mathbf{t}(z) = \begin{bmatrix} MA_0(z) \\ 0 \\ \vdots \\ 0 \end{bmatrix} = \begin{bmatrix} MT(z) \\ 0 \\ \vdots \\ 0 \end{bmatrix}. \qquad (5.4.17)$$

The $M \times M$ matrix $\mathbf{H}(z)$ is called the *Alias Component* (AC) matrix.

By combining (5.4.6) with (5.4.15) we can express

$$\widehat{X}(z) = \mathbf{A}^T(z)\mathbf{x}(z) = \frac{1}{M}\mathbf{f}^T(z)\mathbf{H}^T(z)\mathbf{x}(z), \qquad (5.4.18)$$

where

$$\mathbf{x}(z) = \begin{bmatrix} X(z) \\ X(zW) \\ \vdots \\ X(zW^{M-1}) \end{bmatrix}. \qquad (5.4.19)$$

It is clear that, given a set of analysis filters $H_k(z)$, we can in principle cancel aliasing by solving for the synthesis filters from (5.4.16) as

$$\mathbf{f}(z) = \mathbf{H}^{-1}(z)\mathbf{t}(z). \qquad (5.4.20)$$

This works as long as [det $\mathbf{H}(z)$] is not identically zero. We can go a step further and obtain perfect reconstruction, simply by requiring that $\mathbf{t}(z)$ be of the form

$$\mathbf{t}(z) = \begin{bmatrix} z^{-n_0} \\ 0 \\ \vdots \\ 0 \end{bmatrix}. \qquad (5.4.21)$$

Practical Difficulties with the AC Matrix Inversion

If we attempt to solve the alias cancelation or perfect reconstuction problem by use of (5.4.20), then we would have to invert $\mathbf{H}(z)$. This is in principle possible, unless the determinant of $\mathbf{H}(z)$ is identically zero for all z. However the resulting filters $F_k(z)$ may not be practical. To elaborate this point, let us write (5.4.20) explicitly as (Appendix A)

$$\mathbf{f}(z) = \frac{\text{Adj } \mathbf{H}(z)}{\det \mathbf{H}(z)}\mathbf{t}(z). \qquad (5.4.22)$$

Notice from here that $F_k(z)$ could be IIR even if each analysis filter $H_k(z)$ is FIR. The zeros of the quantity [det $\mathbf{H}(z)$] are related to the analysis filters $H_k(z)$ in a very complicated manner, and it is difficult to ensure that they are inside the unit circle [which is necessary for stability of $F_k(z)$].

If we are willing to give up perfect reconstruction, and be satisfied with alias cancelation, then we can replace (5.4.22) with

$$\mathbf{f}(z) = [\text{Adj } \mathbf{H}(z)]\mathbf{t}(z), \qquad (5.4.23)$$

so that the distortion function after alias cancelation is

$$T(z) = cz^{-n_0}[\det \mathbf{H}(z)], \qquad (5.4.24)$$

for some $c \neq 0$. The synthesis filters $F_k(z)$ are now FIR (whenever the analysis filters are FIR). But the entries of the matrix [Adj $\mathbf{H}(z)$] are determinants of $(M-1) \times (M-1)$ submatrices of $\mathbf{H}(z)$ and can represent FIR filters of very large order even if $H_k(z)$ have moderate order. Another difficulty with this approach is that if [det $\mathbf{H}(z)$] has zeros on the unit circle, say at $z = e^{j\omega_0}$, then $|T(e^{j\omega_0})| = 0$, that is, there is severe amplitude distortion around ω_0.

In the next section we will outline a different technique for perfect reconstruction, in which all the above difficulties 'go away'. This is based on the polyphase representation.

Singularity of $\mathbf{H}(e^{j\omega})$ versus Amplitude Distortion

Consider a QMF bank in which the filters have been chosen to cancel aliasing completely. This means that (5.4.16) holds with $\mathbf{t}(z)$ as in (5.4.17). If $T(z)$ has a zero at $z = e^{j\omega_0}$, then $\mathbf{t}(e^{j\omega_0}) = \mathbf{0}$ so that

$$\mathbf{H}(e^{j\omega_0})\mathbf{f}(e^{j\omega_0}) = \mathbf{0}. \qquad (5.4.25)$$

Unless all synthesis filters $F_k(z)$ have a zero at ω_0, this implies that $\mathbf{H}(e^{j\omega_0})$ is singular. Summarizing, the situation $T(e^{j\omega_0}) = 0$ in a alias-free system implies singularity of the AC matrix at the frequency ω_0. We can restate this as follows: if the AC matrix is nonsingular for all ω, then the alias-free system cannot satisfy $T(e^{j\omega_0}) = 0$ for any ω_0 (unless $F_k(e^{j\omega_0}) = 0$ for all k, which does not happen in a good design).

In Section 5.2 we designed a class of two channel alias-free systems satisfying the constraint $H_1(z) = H_0(-z)$. In these systems the analysis filters had linear phase. The filter order was required to be odd, in order to avoid the situation $T(e^{j\pi/2}) = 0$. In Problem 5.19 we request the reader to verify the connection between that issue and the singularity of $\mathbf{H}(e^{j\pi/2})$.

5.5 POLYPHASE REPRESENTATION

In Sec. 4.3 we studied the polyphase representation, and found it to be very useful, both theoretically and in engineering practice. This representation finds application in filter bank theory as well [Vetterli, 1986], [Swaminathan and Vaidyanathan, 1986], [Vaidyanathan, 1987a,b].

We know from Sec. 4.3 that any transfer function $H_k(z)$ can be expressed in the form

$$H_k(z) = \sum_{\ell=0}^{M-1} z^{-\ell} E_{k\ell}(z^M) \quad \text{(Type 1 polyphase)}. \qquad (5.5.1)$$

We can rewrite this as

$$\begin{bmatrix} H_0(z) \\ \vdots \\ H_{M-1}(z) \end{bmatrix} = \begin{bmatrix} E_{00}(z^M) & E_{01}(z^M) & \cdots & E_{0,M-1}(z^M) \\ \vdots & \vdots & \ddots & \vdots \\ E_{M-1,0}(z^M) & E_{M-1,1}(z^M) & \cdots & E_{M-1,M-1}(z^M) \end{bmatrix} \begin{bmatrix} 1 \\ z^{-1} \\ \vdots \\ z^{-(M-1)} \end{bmatrix},$$

(5.5.2a)

that is, as
$$\mathbf{h}(z) = \mathbf{E}(z^M)\mathbf{e}(z), \qquad (5.5.2b)$$

where

$$\mathbf{E}(z) = \begin{bmatrix} E_{00}(z) & E_{01}(z) & \cdots & E_{0,M-1}(z) \\ E_{10}(z) & E_{11}(z) & \cdots & E_{1,M-1}(z) \\ \vdots & \vdots & \ddots & \vdots \\ E_{M-1,0}(z) & E_{M-1,1}(z) & \cdots & E_{M-1,M-1}(z) \end{bmatrix}, \qquad (5.5.3)$$

and $\mathbf{h}(z)$ and $\mathbf{e}(z)$ are as in (5.4.1). Fig. 5.5-1 shows this idea pictorially. The matrix $\mathbf{E}(z)$ is the $M \times M$ Type 1 polyphase component matrix (or polyphase matrix) for the analysis bank.

We can express the set of synthesis filters also in an identical manner. Thus

$$F_k(z) = \sum_{\ell=0}^{M-1} z^{-(M-1-\ell)} R_{\ell k}(z^M) \quad \text{(Type 2 polyphase)}. \qquad (5.5.4)$$

Using matrix notations we have

$$[F_0(z) \quad \cdots \quad F_{M-1}(z)] =$$

$$[z^{-(M-1)} \quad z^{-(M-2)} \quad \cdots \quad 1] \begin{bmatrix} R_{00}(z^M) & \cdots & R_{0,M-1}(z^M) \\ R_{10}(z^M) & \cdots & R_{1,M-1}(z^M) \\ \vdots & \ddots & \vdots \\ R_{M-1,0}(z^M) & \cdots & R_{M-1,M-1}(z^M) \end{bmatrix}.$$

(5.5.5a)

In terms of $\mathbf{e}(z)$ and the synthesis-bank vector $\mathbf{f}^T(z)$, this becomes

$$\mathbf{f}^T(z) = z^{-(M-1)}\widetilde{\mathbf{e}}(z)\mathbf{R}(z^M), \qquad (5.5.5b)$$

where

$$\mathbf{R}(z) = \begin{bmatrix} R_{00}(z) & R_{01}(z) & \cdots & R_{0,M-1}(z) \\ R_{10}(z) & R_{11}(z) & \cdots & R_{1,M-1}(z) \\ \vdots & \vdots & \ddots & \vdots \\ R_{M-1,0}(z) & R_{M-1,1}(z) & \cdots & R_{M-1,M-1}(z) \end{bmatrix}. \qquad (5.5.6)$$

The matrix $\mathbf{R}(z)$ is the Type 2 polyphase matrix for the synthesis bank. Fig. 5.5-2 shows this representation. In Sec. 5.6.3 we provide many examples.

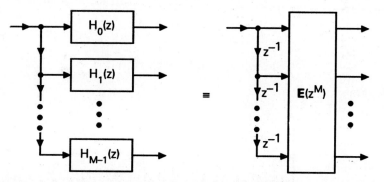

Figure 5.5-1 Type 1 polyphase representation of an analysis bank. $\mathbf{E}(z)$ is called the polyphase component matrix for the analysis bank.

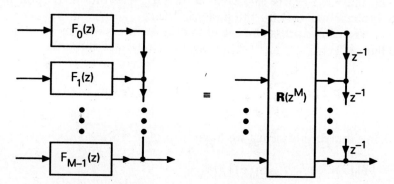

Figure 5.5-2 Type 2 polyphase representation of a synthesis bank. $\mathbf{R}(z)$ is the polyphase component matrix for the synthesis bank.

Using these two representations in the filter bank of Fig. 5.4-1, we obtain the equivalent representation shown in Fig. 5.5-3(a), which we refer to as the polyphase representation of the M-channel QMF bank.

By using noble identities (Fig. 4.2.3), we can redraw this in the equivalent form shown in Fig. 5.5-3(b). This simplified structure can even be used in practical implementations, and has the advantage that the filter coefficients (coefficients of $\mathbf{E}(z)$ and $\mathbf{R}(z)$) are operating at the *lower* rate.

Finally, we can combine the matrices and redraw the system as in Fig. 5.5-3(c), where the $M \times M$ matrix $\mathbf{P}(z)$ is defined as

$$\mathbf{P}(z) = \mathbf{R}(z)\mathbf{E}(z). \tag{5.5.7}$$

As we will see, these equivalent circuits are extremely useful for analytical study as well as in the design and efficient implementation of QMF banks.

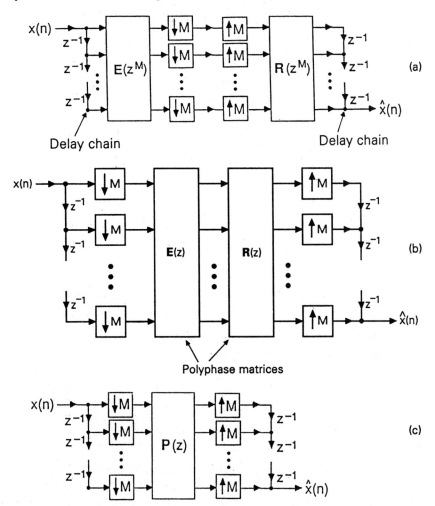

Figure 5.5-3 (a) Polyphase representation of an M-channel maximally decimated filter bank. (b) Rearragement using noble identitites. (c) Further simplification, where $\mathbf{P}(z) = \mathbf{R}(z)\mathbf{E}(z)$.

Causality

Unless mentioned otherwise, the analysis filters $H_k(z)$ will be assumed to be causal so that $\mathbf{E}(z)$ is causal. The synthesis filters $F_k(z)$, which are normally chosen to satisfy certain conditions (such as alias cancelation, perfect reconstruction and so on) can be made causal by insertion of appropriate delays.

Relation Between Polyphase Matrix and AC Matrix

The study of filter banks can be done using either the alias component matrix $\mathbf{H}(z)$ [Eq. (5.4.15)] or the polyphase matrix $\mathbf{E}(z)$. The later approach has the advantage that $\mathbf{E}(z)$ is a physical matrix which makes appearance in the polyphase implementation [Figs. 5.5-3(a),(b)]. However, all theoretical conclusions obtained from use of one of these matrices can also be obtained from the other.

We shall prove that the AC matrix $\mathbf{H}(z)$ and the polyphase component matrix $\mathbf{E}(z)$ of *any* M-channel analysis bank are related as

$$\mathbf{H}(z) = \mathbf{W}^\dagger \mathcal{D}(z) \mathbf{E}^T(z^M), \qquad (5.5.8)$$

where

$$\mathcal{D}(z) = \text{diag}\begin{bmatrix} 1 & z^{-1} & \ldots & z^{-(M-1)} \end{bmatrix}. \qquad (5.5.9)$$

and $\mathbf{W}$ is the $M \times M$ DFT matrix.

To see this, note that the definitions of $\mathbf{H}(z)$ and $\mathbf{h}(z)$ give us

$$\begin{aligned}\mathbf{H}^T(z) &= [\mathbf{h}(z) \quad \mathbf{h}(zW) \quad \ldots \quad \mathbf{h}(zW^{M-1})] \\ &= \mathbf{E}(z^M)[\mathbf{e}(z) \quad \mathbf{e}(zW) \quad \ldots \quad \mathbf{e}(zW^{M-1})],\end{aligned} \qquad (5.5.10)$$

using $\mathbf{h}(z) = \mathbf{E}(z^M)\mathbf{e}(z)$. From the definition of $\mathbf{e}(z)$ we find

$$\mathbf{e}(zW^k) = \mathcal{D}(z) \begin{bmatrix} 1 \\ W^{-k} \\ \vdots \\ W^{-(M-1)k} \end{bmatrix}. \qquad (5.5.11)$$

By using this in (5.5.10) (and remembering $\mathbf{W} = \mathbf{W}^T$), we obtain (5.5.8).

5.6 PERFECT RECONSTRUCTION (PR) SYSTEMS

Recall that a perfect reconstruction (PR) system satisfies $\widehat{x}(n) = cx(n - n_0)$. This means that aliasing has been canceled, and that $T(z)$ has been forced to be a delay. Such systems can indeed be designed. We will show that FIR PR systems can be built for arbitrary M. Moreover, these can be designed such that $H_k(z)$ provides as much attenuation as the user specifies. If designed properly, the implementation cost of such a system is quite competitive with the cost of well-known *approximate* reconstruction systems (Chap. 8).

5.6.1 The Delay Chain Perfect Reconstruction System

We begin with a very simple FIR perfect reconstuction system, and use it to build more useful systems. Consider Fig. 5.6-1(a) which is a two-channel system ($M = 2$) with analysis and synthesis filters

$$H_0(z) = 1, \quad H_1(z) = z^{-1}, \quad F_0(z) = z^{-1}, \quad F_1(z) = 1. \tag{5.6.1}$$

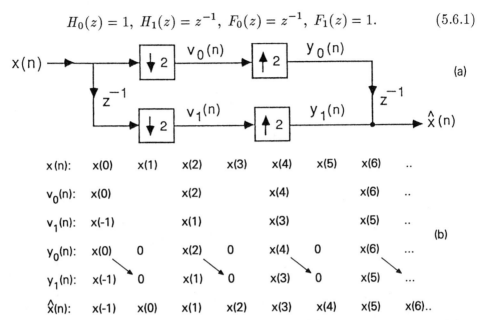

Figure 5.6-1 (a) The delay chain perfect reconstruction QMF bank, and (b) its operation explained in the time domain.

By substituting in (5.4.5) we obtain $\widehat{X}(z) = z^{-1}X(z)$. The distortion function simplifies to $T(z) = z^{-1}$ so that this is a PR system indeed. It is instructive to see how the system works in the time domain. This is demonstrated in Fig. 5.6-1(b). The output of the upper decimator permits the even numbered samples $x(0), x(2), x(4)\ldots$, whereas the lower decimator permits odd numbered samples $x(-1), x(1), x(3), \ldots$ The expanders insert zero-valued samples as shown. The signals in these two branches are beautifully interlaced by the synthesis bank as indicated by the oblique arrows. So the reconstructed signal is precisely $x(n)$ except for one unit of delay.

Figure 5.6-2 shows the M-channel generalization of this. This is a filterbank with analysis and synthesis filters

$$H_k(z) = z^{-k}, \quad F_k(z) = z^{-(M-1-k)}, \quad 0 \le k \le M-1. \tag{5.6.2}$$

By substituting into (5.4.5), one can verify that this is a perfect reconstruction system, with

$$\widehat{X}(z) = z^{-(M-1)}X(z), \quad \text{i.e.,} \quad \widehat{x}(n) = x(n - M + 1). \tag{5.6.3}$$

Sec. 5.6 Perfect reconstruction (PR) systems

So the overall system is an LTI system [with transfer function $T(z) = z^{-(M-1)}$] even though there are multirate building blocks in it.

Viewed in the time domain, we see that the kth channel passes the subset of input samples $x(nM - k)$. In other words, the analysis bank merely splits the input $x(n)$ into M subsequences

$$x(nM - k), \quad 0 \leq k \leq M - 1. \tag{5.6.4}$$

These subsequences are then interlaced by the synthesis bank, in order to resynthesize $x(n)$.

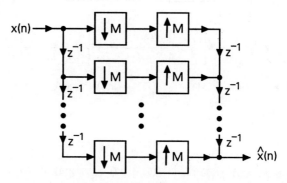

Figure 5.6-2 The delay chain perfect reconstruction system. Here $\widehat{x}(n) = x(n - M + 1)$. Number of channels $= M$.

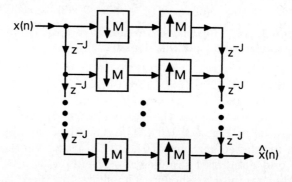

Figure 5.6-3 A generalization of Fig. 5.6-2. This is a perfect recronstruction system if and only if M and J are relatively prime. Again, M is the number of channels.

A further generalization is shown in Fig. 5.6-3. This is obtained by replacing each delay in Fig. 5.6-2 with z^{-J} where J is some integer. This is a perfect reconstruction system if and only if the integers M and J are relatively prime (Problem 5.15).

5.6.2 More General Perfect Reconstruction Systems

The above PR system has allpass analysis filters, which are not useful in practice. Our aim is to use this simple system to develop more useful and practical PR systems. For this imagine that we insert two matrices $\mathbf{E}(z)$ and $\mathbf{R}(z)$ in this system to obtain Fig. 5.5-3(b). It is clear that if

$$\mathbf{R}(z)\mathbf{E}(z) = \mathbf{I}, \qquad (5.6.5)$$

then the output $\widehat{x}(n)$ is unchanged. Next, suppose we move $\mathbf{E}(z)$ and $\mathbf{R}(z)$ (using noble identities) to obtain Fig. 5.5-3(a). This system continues to be equivalent to Fig. 5.6-2, so that an observer who measures $\widehat{x}(n)$ [in response to $x(n)$] does not even notice our manipulations! In particular, Fig. 5.5-3(a) continues to have the perfect reconstruction property, except that the analysis filters can now be nontrivial.

We can now do our thinking backwards: suppose we are given a set of analysis filters $H_k(z), 0 \le k \le M-1$. This completely determines $\mathbf{E}(z)$ (Sec. 5.5). Assuming that $\mathbf{E}(z)$ can be inverted, we can then obtain a PR system by choosing $\mathbf{R}(z)$ to be $\mathbf{E}^{-1}(z)$ and then computing the synthesis filter coefficients from (5.5.5).

Matrix inversion again? The first thought that crosses the mind now is that this will bring home the same difficulties (including instability) we encountered in the inversion of the AC matrix $\mathbf{H}(z)$ (Sec. 5.4.3). As we will substantiate in Chap. 6, this alarm is unwarranted. We can avoid direct inversion of $\mathbf{E}(z)$ in many ways; one of these is to constrain it to be paraunitary (Sec. 6.1). Notice also that, unlike the AC matrix, $\mathbf{E}(z)$ is a *physical* matrix which will be used in implementation as well as in filter design.

Necessary and Sufficient Conditions for Perfect Reconstruction

The condition (5.6.5) is sufficient for perfect reconstruction, whether the system is FIR or IIR. It is clear that if we replace this with

$$\mathbf{R}(z)\mathbf{E}(z) = cz^{-m_0}\mathbf{I}, \qquad (5.6.6)$$

we still have perfect reconstruction but now $T(z) = cz^{-(Mm_0+M-1)}$. More generally it can be shown that, the system has perfect reconstruction if and only if the product $\mathbf{R}(z)\mathbf{E}(z)$ has the form

$$\underbrace{\mathbf{R}(z)\mathbf{E}(z)}_{\mathbf{P}(z)} = cz^{-m_0} \begin{bmatrix} 0 & \mathbf{I}_{M-r} \\ z^{-1}\mathbf{I}_r & 0 \end{bmatrix} \quad \text{(most general PR condition)},$$

$$(5.6.7)$$

for some integer r with $0 \le r \le M-1$, some integer m_0, and some constant $c \ne 0$. Under this condition the reconstructed signal is $\widehat{x}(n) = cx(n-n_0)$, where $n_0 = Mm_0 + r + M - 1$. This result is a consequence of a general result which we will prove later in Sec. 5.7.2. It holds whether the system is FIR or IIR.

As a special case consider the two channel QMF bank. The matrices $\mathbf{E}(z), \mathbf{R}(z)$ and $\mathbf{P}(z)$ are now 2×2. This system has perfect reconstruction if and only if $\mathbf{P}(z)$ has the form

$$cz^{-m_0} \begin{bmatrix} 1 & 0 \\ 0 & 1 \end{bmatrix}, \quad \text{or} \quad cz^{-m_0} \begin{bmatrix} 0 & 1 \\ z^{-1} & 0 \end{bmatrix}. \qquad (5.6.8)$$

Every QMF bank satisfying (5.6.7) for some r can be obtained by starting from a QMF bank satisfying (5.6.5) and inserting a delay z^{-r} in front of each synthesis filter. (This will be shown in Sec. 5.7.2.) As a result c, r, and m_0 are not fundamental quantities. We sometimes use the term 'perfect reconstruction' to imply the simpler condition (5.6.5).

Condition on determinant. The reader can verify (Problem 5.16) that under the condition (5.6.7) we have

$$\det \mathbf{R}(z) \det \mathbf{E}(z) = c_0 z^{-k_0}. \qquad (5.6.9)$$

for some $c_0 \neq 0$ and some integer k_0. So any perfect reconstruction system (FIR or IIR) *has* to satisfy this determinant condition.

FIR Perfect Reconstruction Systems

Perfect reconstruction QMF banks with FIR filters $H_k(z)$ and $F_k(z)$ are of great interest in practice. For these systems the elements of $\mathbf{E}(z)$ and $\mathbf{R}(z)$ are FIR. The FIR nature of $\mathbf{E}(z)$ and $\mathbf{R}(z)$ implies that their determinants are FIR. If the product of these FIR functions has to be a delay [see (5.6.9)], then we must have

$$\det \mathbf{E}(z) = \alpha z^{-K}, \quad \alpha \neq 0, \quad K = \text{integer}. \qquad (5.6.10)$$

Thus every FIR perfect reconstruction system must satisfy the above condition; and [det $\mathbf{R}(z)$] must have similar form.

Characterization using paraunitary and unimodular systems. In Chap. 6 we will study a particular family of causal FIR matrices called paraunitary matrices, which satisfy the condition (5.6.10) with $K =$ McMillan degree of $\mathbf{E}(z)$. In Chap. 13 we will encounter *another* family of causal FIR matrices called unimodular matrices, which, by definition, satisfy (5.6.10) with $K = 0$. It is shown in Vaidyanathan [1990b] that any causal FIR matrix satisfying (5.6.10) is a product of a paraunitary matrix and a unimodular matrix, motivating us to study these two classes of matrices in the chapters mentioned above.

5.6.3 Examples of Perfect Reconstruction Systems

Using the above principles we now generate a number of examples which demonstrate the idea of perfect reconstruction.

Example 5.6.1

Consider the two channel system in Fig. 5.6-4(a). By comparing with Fig. 5.6-1 we see that $\mathbf{E}(z) = \mathbf{T}$ and $\mathbf{R}(z) = c\mathbf{T}^{-1}$ so that the perfect reconstruction condition is satisfied, and $\widehat{x}(n) = cx(n-1)$. We can find the analysis and synthesis filters using (5.5.2) and (5.5.5), that is,

$$\begin{bmatrix} H_0(z) \\ H_1(z) \end{bmatrix} = \mathbf{E}(z^2) \begin{bmatrix} 1 \\ z^{-1} \end{bmatrix}, \quad [F_0(z) \; F_1(z)] = [z^{-1} \; 1] \mathbf{R}(z^2). \quad (5.6.11)$$

Take an example with $c = 2$ and

$$\mathbf{T} = \begin{bmatrix} 1 & 1 \\ 1 & -1 \end{bmatrix}, \quad \text{so that} \quad c\mathbf{T}^{-1} = \begin{bmatrix} 1 & 1 \\ 1 & -1 \end{bmatrix} = \mathbf{T}. \quad (5.6.12)$$

This is shown in Fig. 5.6-4(b), and can be redrawn in the form of the usual QMF bank as in Fig. 5.6-4(c). So the filters are

$$\begin{aligned} H_0(z) = 1 + z^{-1}, & \quad H_1(z) = 1 - z^{-1}, \\ F_0(z) = 1 + z^{-1}, & \quad F_1(z) = -1 + z^{-1}. \end{aligned} \quad (5.6.13)$$

This PR system is less trivial than Fig. 5.6-1(a) because the filters $H_0(z)$ and $H_1(z)$ are lowpass and highpass (rather than just allpass). We can generate endless examples like this. For example let

$$\mathbf{T} = \begin{bmatrix} 2 & 1 \\ 3 & 2 \end{bmatrix}, \quad \text{so that} \quad \mathbf{T}^{-1} = \begin{bmatrix} 2 & -1 \\ -3 & 2 \end{bmatrix}. \quad (5.6.14)$$

We then have with $c = 1$,

$$\begin{aligned} H_0(z) = 2 + z^{-1}, & \quad H_1(z) = 3 + 2z^{-1}, \\ F_0(z) = -3 + 2z^{-1}, & \quad F_1(z) = 2 - z^{-1}. \end{aligned} \quad (5.6.15)$$

In this case $\widehat{x}(n) = x(n-1)$. Notice that the condition $H_1(z) = H_0(-z)$ is not satisfied by this perfect reconstruction example.

Example 5.6.2.

Let

$$\mathbf{E}(z) = \begin{bmatrix} 1 + z^{-1} & 1 - z^{-1} \\ 1 - z^{-1} & 1 + z^{-1} \end{bmatrix}, \quad (5.6.16)$$

which is FIR. Notice that the determinant of this matrix is a delay, as required by (5.6.10). We choose $\mathbf{R}(z)$ to satisfy (5.6.6), that is,

$$\mathbf{R}(z) = cz^{-m_0} \mathbf{E}^{-1}(z) = \frac{cz^{-m_0}}{4} \begin{bmatrix} 1 + z & 1 - z \\ 1 - z & 1 + z \end{bmatrix}, \quad (5.6.17)$$

so that the perfect reconstruction condition holds. Choosing $c = 4$ and $m_0 = 1$, this becomes

$$\mathbf{R}(z) = \begin{bmatrix} 1 + z^{-1} & -1 + z^{-1} \\ -1 + z^{-1} & 1 + z^{-1} \end{bmatrix}. \qquad (5.6.18)$$

The only purpose of m_0 has been to avoid the positive powers of z (non causal terms). The analysis and synthesis filters corresponding to the above $\mathbf{E}(z)$ and $\mathbf{R}(z)$ are

$$\begin{aligned} H_0(z) &= 1 + z^{-1} + z^{-2} - z^{-3}, \quad H_1(z) = 1 + z^{-1} - z^{-2} + z^{-3}, \\ F_0(z) &= -1 + z^{-1} + z^{-2} + z^{-3}, \quad F_1(z) = 1 - z^{-1} + z^{-2} + z^{-3}. \end{aligned} \qquad (5.6.19)$$

(a)
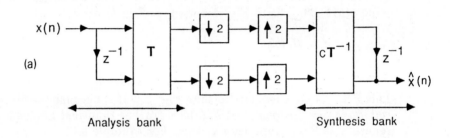

(b)
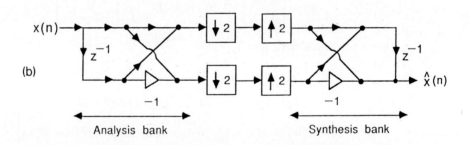

(c)
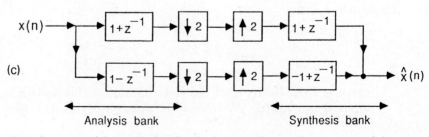

Figure 5.6-4 (a) Example of a perfect reconstruction system, (b) a specific choice of **T** and (c) redrawing in conventional form.

Example 5.6.3: The Uniform-DFT Filter Bank

A simple FIR perfect reconstruction system can be constructed by referring to Example 4.1.1 (the DFT filter bank) in Chap. 4. In that example, the analysis bank is as in Fig 4.1-16(a), so that the filters are related as

$$H_k(z) = H_0(zW^k), \qquad (5.6.20)$$

where

$$H_0(z) = 1 + z^{-1} + \ldots + z^{-(M-1)}. \qquad (5.6.21)$$

Notice that the filters have length M, which is equal to the number of channels. The frequency responses $H_k(e^{j\omega})$ are shifted versions of the lowpass response $H_0(e^{j\omega})$, as shown in Fig. 4.1-16. In this example, we clearly have $\mathbf{E}(z) = \mathbf{W}^*$, so that we can obtain a perfect reconstruction system by taking $\mathbf{R}(z) = \mathbf{W}$. Under this condition $\mathbf{R}(z)\mathbf{E}(z) = \mathbf{W}\mathbf{W}^* = M\mathbf{I}$ so that the reconstructed signal satisfies the perfect reconstruction property $\widehat{x}(n) = Mx(n - M + 1)$. It can be shown that the synthesis filters are related as

$$F_k(z) = W^{-k} F_0(zW^k), \qquad (5.6.22)$$

and that $F_0(z) = H_0(z)$. So each synthesis filter has precisely the same magnitude response as the corresponding analysis filter. Fig. 5.6-5 shows the complete analysis/synthesis system.

Recall from Fig. 4.1-16 that each analysis filter has about 13 dB attenuation, and adjacent responses have substantial overlap. This shows that there is substantial amount of aliasing error at the output of each decimator. However, the filters $F_k(z)$ and $H_k(z)$ are related in such a delicate manner that the aliasing has canceled off.

Higher Order FIR Perfect Reconstruction Systems

Even though (5.6.10) can be trivially satisfied by taking $\mathbf{E}(z)$ to be a constant nonsingular matrix (as we did in the above example), it is of greater practical interest to employ $\mathbf{E}(z)$ having higher degree, so that the filters $H_k(z)$ have higher order. In this way $|H_k(e^{j\omega})|$ can have higher stopband attenuation and sharper cutoff rate.

One way to obtain FIR $\mathbf{E}(z)$ of higher degree while at the same time satisfying (5.6.10) is shown in Fig. 5.6-6. Here $\mathbf{R}_m$ are constant $M \times M$ nonsingular matrices. Clearly

$$\mathbf{E}(z) = \mathbf{R}_J \mathbf{\Lambda}(z) \mathbf{R}_{J-1} \ldots \mathbf{\Lambda}(z) \mathbf{R}_0, \qquad (5.6.23)$$

where

$$\mathbf{\Lambda}(z) = \begin{bmatrix} \mathbf{I}_{M-1} & 0 \\ 0 & z^{-1} \end{bmatrix}. \qquad (5.6.24)$$

Evidently $[\det \mathbf{E}(z)] = \alpha z^{-J}, \alpha \neq 0$. We can choose $\mathbf{R}(z) = z^{-J}\mathbf{E}^{-1}(z)$ so that it is causal. We then have

$$\mathbf{R}(z) = \mathbf{R}_0^{-1}\mathbf{\Gamma}(z)\mathbf{R}_1^{-1}\ldots\mathbf{\Gamma}(z)\mathbf{R}_J^{-1}, \qquad (5.6.25)$$

where

$$\mathbf{\Gamma}(z) = \begin{bmatrix} z^{-1}\mathbf{I}_{M-1} & \mathbf{0} \\ \mathbf{0} & 1 \end{bmatrix}. \qquad (5.6.26)$$

Figure 5.6-7 shows the synthesis bank obtained in this manner. The filters $H_k(z)$ and $F_k(z)$ can be found using (5.5.2) and (5.5.5).

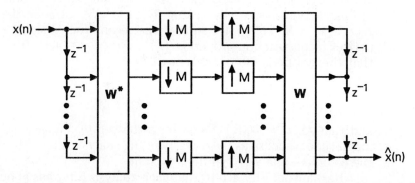

Figure 5.6-5 An FIR perfect reconstruction system with $\mathbf{E}(z) = \mathbf{W}^*$ and $\mathbf{R}(z) = \mathbf{W}$, where $\mathbf{W}$ = DFT matrix. Here $\widehat{x}(n) = Mx(n - M + 1)$.

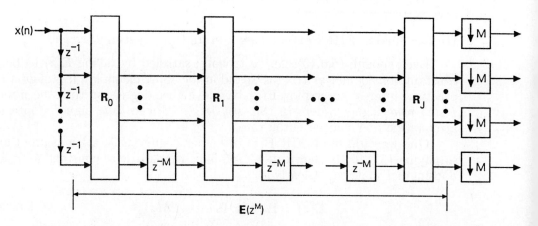

Figure 5.6-6 Analysis bank in which $\mathbf{E}(z)$ is a cascade of nonsingular matrices $\mathbf{R}_m$ separated by delays. Clearly $[\det \mathbf{E}(z)]$ = delay.

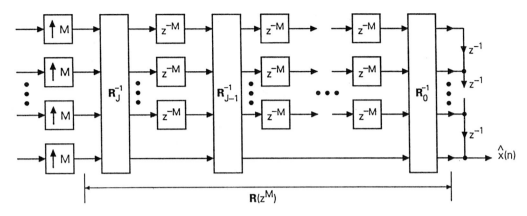

Figure 5.6-7 The synthesis-bank corresponding to Fig. 5.6-6, which would result in a perfect reconstruction system.

Example 5.6.4

Consider a special case with $J = 1$, and the matrices $\mathbf{R}_0$ and $\mathbf{R}_1$ chosen (rather arbitrarily) as

$$\mathbf{R}_0 = \begin{bmatrix} 1 & 0 & 0 \\ 2 & 1 & 0 \\ 1 & 2 & 1 \end{bmatrix}, \quad \mathbf{R}_1 = \mathbf{R}_0^T. \qquad (5.6.27)$$

Since $\mathbf{R}_0$ is triangular, its determinant is the product of its diagonal elements, and is nonzero. So $\mathbf{R}_0$ is nonsingular, and

$$\mathbf{R}_0^{-1} = \begin{bmatrix} 1 & 0 & 0 \\ -2 & 1 & 0 \\ 3 & -2 & 1 \end{bmatrix}. \qquad (5.6.28)$$

Also $\mathbf{R}_1^{-1}$ is the transpose of $\mathbf{R}_0^{-1}$. The matrix $\mathbf{E}(z)$ is

$$\mathbf{E}(z) = \mathbf{R}_1 \mathbf{\Lambda}(z) \mathbf{R}_0 = \begin{bmatrix} 5 + z^{-1} & 2 + 2z^{-1} & z^{-1} \\ 2 + 2z^{-1} & 1 + 4z^{-1} & 2z^{-1} \\ z^{-1} & 2z^{-1} & z^{-1} \end{bmatrix}. \qquad (5.6.29)$$

The analysis filters obtained using (5.5.2) are given by

$$\begin{aligned} H_0(z) &= 5 + 2z^{-1} + z^{-3} + 2z^{-4} + z^{-5} \\ H_1(z) &= 2 + z^{-1} + 2z^{-3} + 4z^{-4} + 2z^{-5} \\ H_2(z) &= z^{-3} + 2z^{-4} + z^{-5} \end{aligned} \qquad (5.6.30)$$

The synthesis filters for perfect reconstruction are obtained by taking

$$\mathbf{R}(z) = \mathbf{R}_0^{-1}\boldsymbol{\Gamma}(z)\mathbf{R}_1^{-1} = \begin{bmatrix} z^{-1} & -2z^{-1} & 3z^{-1} \\ -2z^{-1} & 5z^{-1} & -8z^{-1} \\ 3z^{-1} & -8z^{-1} & 1+13z^{-1} \end{bmatrix}. \quad (5.6.31)$$

By using (5.5.5) we obtain

$$\begin{aligned} F_0(z) &= 3z^{-3} - 2z^{-4} + z^{-5} \\ F_1(z) &= -8z^{-3} + 5z^{-4} - 2z^{-5} \\ F_2(z) &= 1 + 13z^{-3} - 8z^{-4} + 3z^{-5}. \end{aligned} \quad (5.6.32)$$

This example demonstrates that we can construct FIR perfect reconstruction systems of arbitrarily high order, by structuring $\mathbf{E}(z)$ and $\mathbf{R}(z)$ as in Figs. 5.6-6 and 5.6-7. Since the matrices $\mathbf{R}_m$ in Fig. 5.6-6 can be chosen arbitrarily (subject only to nonsingularity requirement), we can optimize the elements of $\mathbf{R}_m$ to obtain good filter responses $|H_k(e^{j\omega})|$. The resulting system is guaranteed to have perfect reconstruction. Practical design examples of this nature can be found in Chap. 6 to 8.

Example 5.6.5

Let $H_0(z)$ and $H_1(z)$ be related as $H_1(z) = H_0(-z)$ so that the analysis bank has the form (5.2.5). We then have

$$\mathbf{E}(z) = \begin{bmatrix} 1 & 1 \\ 1 & -1 \end{bmatrix} \begin{bmatrix} E_0(z) & 0 \\ 0 & E_1(z) \end{bmatrix}. \quad (5.6.33)$$

Using (5.6.6) with $c = 2$ and $m_0 = 0$ results in

$$\mathbf{R}(z) = \begin{bmatrix} \frac{1}{E_0(z)} & 0 \\ 0 & \frac{1}{E_1(z)} \end{bmatrix} \begin{bmatrix} 1 & 1 \\ 1 & -1 \end{bmatrix}. \quad (5.6.34)$$

The analysis and synthesis banks can now be drawn as in Fig. 5.6-8. So in this case the PR system is obtained merely by using, on the synthesis bank side, the *reciprocals* of the polyphase components of $H_0(z)$. The synthesis filters are

$$F_0(z) = \frac{1}{E_1(z^2)} + z^{-1}\frac{1}{E_0(z^2)}, \quad F_1(z) = \frac{-1}{E_1(z^2)} + z^{-1}\frac{1}{E_0(z^2)}, \quad (5.6.35)$$

and are stable as long as the zeros of $E_i(z)$ are strictly inside the unit circle. In this case [i.e., with $H_1(z) = H_0(-z)$] there is no way to obtain perfect reconstruction if all the filters are required to be FIR (unless the filters have trivial responses). This is consistent with the observation made in Sec. 5.2, where we studied this case in detail. As a numerical example, let $E_0(z) = E_1(z) = 2 + z^{-1}$. Then

$$H_0(z) = 2 + 2z^{-1} + z^{-2} + z^{-3}, \quad H_1(z) = H_0(-z),$$
$$F_0(z) = \frac{0.5(1 + z^{-1})}{1 + 0.5z^{-2}}, \quad F_1(z) = \frac{0.5(-1 + z^{-1})}{1 + 0.5z^{-2}}. \tag{5.6.36}$$

In the above example, the requirement that the zeros of $E_i(z)$ be inside the unit circle, is severe. It puts severe constraints on the frequency response of $H_0(z)$. So this is not a very practical system.

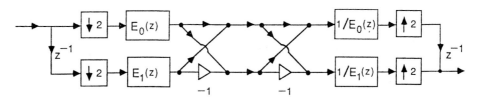

Figure 5.6-8 Another example of a PR QMF bank. The synthesis bank is IIR if analysis bank is FIR.

5.7 ALIAS-FREE FILTER BANKS

Alias cancelation is evidently a less stringent requirement than perfect reconstruction. Even though it is possible to achieve perfect reconstruction as explained in the previous section, it is important to study the most general conditions under which aliasing is canceled. We first demonstrate some useful M-channel alias-free QMF banks. We then study the general theory for alias cancelation.

5.7.1 Examples of Alias-Free Systems

Starting from the conceptually simple perfect reconstruction system of Fig. 5.6-2, we now obtain some examples of alias-free systems.

Example 5.7.1

Consider Fig. 5.7-1(a) in which we have M transfer functions $S_k(z)$ 'sandwiched' between the decimators and expanders. Evidently $\widehat{X}(z)$ is a linear combination of $X(z)$ and the alias components $X(zW^\ell)$. What is the set of necessary and sufficient conditions on $S_k(z)$ so that aliasing terms are canceled?

First, we claim that aliasing is absent if

$$S_k(z) = S(z), \quad \text{for all } k. \tag{5.7.1}$$

To see this, simply move $S(z)$ all the way to the right using the appropriate noble identity (Fig. 4.2-3). The result [Fig. 5.7-1(b)] is identical to the perfect reconstruction structure of Fig. 5.6-2, in cascade with $S(z^M)$. Under this condition we have

$$\widehat{X}(z) = z^{-(M-1)} S(z^M) X(z). \tag{5.7.2}$$

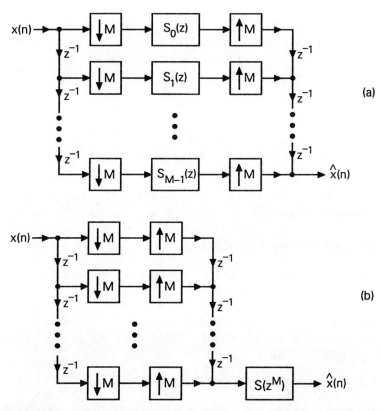

Figure 5.7-1 (a) Pertaining to Example 5.7.1 and (b) simplification when $S_k(z) = S(z)$ for all k.

So the system is alias free, and has distortion $T(z) = z^{-(M-1)} S(z^M)$. It turns out that (5.7.1) is also a *necessary* condition for alias cancelation. To see this, we first express $\widehat{X}(z)$ in terms of $X(z)$:

$$\widehat{X}(z) = \frac{z^{-(M-1)}}{M} \sum_{\ell=0}^{M-1} X(zW^\ell) \sum_{k=0}^{M-1} S_k(z^M) W^{-k\ell}. \tag{5.7.3}$$

This is free from the alias components $X(zW^\ell), \ell > 0$ [for all possible inputs $x(n)$] if, and only if,

$$\sum_{k=0}^{M-1} S_k(z) W^{-k\ell} = 0, \quad 1 \le \ell \le M-1. \qquad (5.7.4)$$

i.e.,
$$\mathbf{W}^\dagger \begin{bmatrix} S_0(z) \\ S_1(z) \\ \vdots \\ S_{M-1}(z) \end{bmatrix} = \begin{bmatrix} \times \\ 0 \\ \vdots \\ 0 \end{bmatrix}. \qquad (5.7.5)$$

where $\times$ denotes a possibly nonzero entry. Since $\mathbf{WW}^\dagger = M\mathbf{I}$, this implies

$$\begin{bmatrix} S_0(z) \\ S_1(z) \\ \vdots \\ S_{M-1}(z) \end{bmatrix} = \mathbf{W} \begin{bmatrix} S(z) \\ 0 \\ \vdots \\ 0 \end{bmatrix}, \qquad (5.7.6)$$

for some $S(z)$, from which (5.7.1) follows. This result can be used to generate some useful alias-free systems, as demonstrated next.

Example 5.7.2.

Suppose each transfer function $S_k(z)$ in Fig. 5.7-1(a) is factorized into $S_k(z) = E_k(z)R_k(z)$ (Fig. 5.7-2(a)). By use of the noble identities we can move $E_k(z)$ all the way to the left and $R_k(z)$ all the way to the right (Fig. 5.7-2(b)). If we now insert a nonsingular matrix $\mathbf{T}$ and its inverse as shown, the input-output behavior of the system is still unchanged. In particular if the product

$$S_k(z) = E_k(z)R_k(z), \quad 0 \le k \le M-1, \qquad (5.7.7)$$

is the same ($= S(z)$) for all k, then the system is free from aliasing, and $\widehat{X}(z)$ is given by (5.7.2), *regardless of* the choice of $\mathbf{T}$!

For example, imagine that $\mathbf{T} = \mathbf{W}^*$. Then the analysis bank is the familiar uniform-DFT bank. In this case, $E_k(z)$ and $R_k(z)$ are, respectively, the Type 1 and Type 2 polyphase components of the prototype filters $H_0(z)$ and $F_0(z)$. The filters are related by uniform shifts (precisely as in (5.6.20) and (5.6.22)). This little exercise shows that we can eliminate aliasing in a uniform-DFT filter bank by enforcing the condition that $R_k(z)E_k(z)$ be the same for all k, that is,

$$R_k(z)E_k(z) = S(z), \quad \text{for all } k.$$

One way to do so would be to take $R_k(z) = 1/E_k(z)$, which also yields perfect reconstruction. This choice, however, makes $R_k(z)$ (and hence

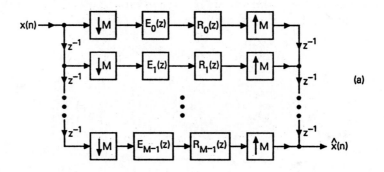

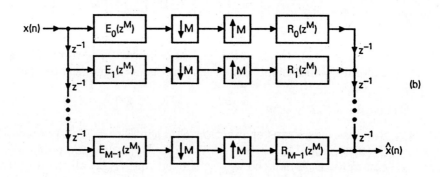

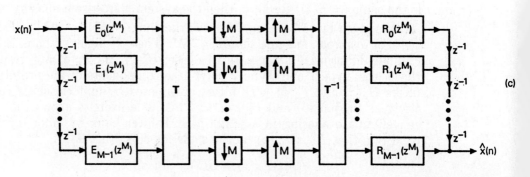

Figure 5.7-2 Step by step development of a fairly general alias-free system. All three systems have the same input/output behavior. Here $S_k(z) = E_k(z)R_k(z)$.

the synthesis filters) unstable unless $E_k(z)$ has all zeros inside the unit circle. A second way to enforce the above condition would be to take

$$R_k(z) = \prod_{\ell \neq k} E_\ell(z). \tag{5.7.8}$$

Then

$$R_k(z)E_k(z) = S(z) = \prod_{\ell=0}^{M-1} E_\ell(z), \quad 0 \leq k \leq M-1. \tag{5.7.9}$$

For large M, (5.7.8) implies that the synthesis filters have much higher order than the analysis filters. For $M = 2$, (5.7.8) means

$$R_0(z) = E_1(z), \quad R_1(z) = E_0(z).$$

This is consistent with the special cases we saw in Section 5.2. For example see Fig. 5.2-2 where the synthesis bank has $E_1(z)$ in the top branch and $E_0(z)$ in the bottom branch. Also in Fig. 5.2-5, the synthesis bank has $a_1(z)$ in the top branch and $a_0(z)$ in the bottom branch.

The ideas introduced above can also be used to compensate for channel distortion in QMF systems, as well as to design M-channel IIR systems free from amplitude distortion. We will skip these details (many of which are covered in Problems 5.21–5.23), and return to the general problem.

5.7.2 The Most General Alias-Free System

What is the most general set of *necessary and sufficient* conditions so that aliasing is canceled? One way to answer this question is to refer to (5.4.16), where $\mathbf{H}(z)$ is the alias component matrix (determined completely by the analysis bank) and $\mathbf{f}(z)$ is the synthesis filter bank. The filter bank is alias free if and only if the product $\mathbf{H}(z)\mathbf{f}(z)$ has the form (5.4.17).

We now obtain an equivalent set of necessary and sufficient conditions based on the polyphase matrices $\mathbf{E}(z)$ and $\mathbf{R}(z)$ [Vaidyanathan and Mitra, 1988]. We will show that the filter bank is alias free if and only if $\mathbf{P}(z)$, defined as the product $\mathbf{R}(z)\mathbf{E}(z)$, is a *pseudocirculant* matrix (defined below). Under alias free condition, additional properties of the distortion function $T(z)$ can be expressed entirely in terms of this matrix very conveniently.

Pseudocirculant Matrices

First, a matrix is said to be *circulant* if every row is obtained using a right-shift (by one position) of the previous row with the added requirement that the righmost element which 'spills over' in the process be 'circulated back' to become the leftmost element. Here is an example:

$$\begin{bmatrix} P_0(z) & P_1(z) & P_2(z) \\ P_2(z) & P_0(z) & P_1(z) \\ P_1(z) & P_2(z) & P_0(z) \end{bmatrix} \quad \text{(circulant matrix)}. \tag{5.7.10}$$

Actually it is more appropriate to call this a *right*-circulant, because the definition involves *right* shifts. In a similar way one can define left circulants. In this book, 'circulant' stands for 'right circulant', unless mentioned otherwise.

A pseudocirculant matrix is essentially a circulant matrix with the additional feature that the elements below the main diagonal are multiplied with z^{-1}. An example is:

$$\begin{bmatrix} P_0(z) & P_1(z) & P_2(z) \\ z^{-1}P_2(z) & P_0(z) & P_1(z) \\ z^{-1}P_1(z) & z^{-1}P_2(z) & P_0(z) \end{bmatrix} \quad \text{(pseudocirculant matrix)}. \quad (5.7.11a)$$

In other words, the element that spills over during the right shift is circulated after multiplying with z^{-1}. In the time domain, the above matrix has the form

$$\begin{bmatrix} p_0(n) & p_1(n) & p_2(n) \\ p_2(n-1) & p_0(n) & p_1(n) \\ p_1(n-1) & p_2(n-1) & p_0(n) \end{bmatrix} \quad \text{(pseudocirculant matrix)}. \quad (5.7.11b)$$

Evidently, all the rows of a $M \times M$ pesudocirculant matrix $\mathbf{P}(z)$ are determined by the 0th row which is

$$[\, P_0(z) \quad P_1(z) \quad \ldots \quad P_{M-1}(z) \,]. \quad (5.7.12)$$

For a pseudocirculant, the kth column is obtainable from the $(k+1)$st column as follows: (a) shift the $(k+1)$st column upwards by one element, (b) circulate the element that spills over so that it becomes the bottom most element, and (c) multiply the circulated element with z^{-1}. The result is equal to the kth column. [The reader can verify this for (5.7.11a).] This can in fact be taken as an equivalent definition for pseudocirculants.

The occurence of pseudocirculant matrices in the context of multirate filter banks was noticed by Marshall [1982]. It was studied later in Vaidyanathan and Mitra [1988]. These matrices have also been found to arise in the context of block digital filtering [Barnes and Shinnaka, 1980]; Sec. 10.1 provides a more complete discussion. The following result was proved in Vaidyanathan and Mitra [1988].

♠**Theorem 5.7.1. Necessary and sufficient condition for alias cancelation.** The M-channel maximally decimated filter bank (Fig. 5.4-1) is free from aliasing if and only if the $M \times M$ matrix $\mathbf{P}(z)$ (defined as the product $\mathbf{R}(z)\mathbf{E}(z)$) is pseudocirculant. Under this condition $\widehat{X}(z) = T(z)X(z)$, and the distortion function $T(z)$ can be expressed as

$$T(z) = z^{-(M-1)}\Big(P_0(z^M) + z^{-1}P_1(z^M) + \ldots + z^{-(M-1)}P_{M-1}(z^M)\Big), \quad (5.7.13)$$

where $P_m(z)$ are the elements of the 0th row of $\mathbf{P}(z)$. ◊

Proof. Consider Fig. 5.7-3 which is the familiar equivalent circuit for the QMF bank in terms of $\mathbf{P}(z)$. We will express $\widehat{X}(z)$ in terms of $X(z)$ and the elements $P_{s,\ell}(z)$ of $\mathbf{P}(z)$. First, using standard decimation formulas (Sec. 4.1.1) we have

$$C_\ell(z) = \frac{1}{M} \sum_{k=0}^{M-1} \left(z^{1/M} W^k\right)^{-\ell} X(z^{1/M} W^k), \quad 0 \le \ell \le M-1, \quad (5.7.14)$$

with $W = e^{-j2\pi/M}$. The outputs of $\mathbf{P}(z)$ are given by

$$B_s(z) = \sum_{\ell=0}^{M-1} P_{s,\ell}(z) C_\ell(z). \quad (5.7.15)$$

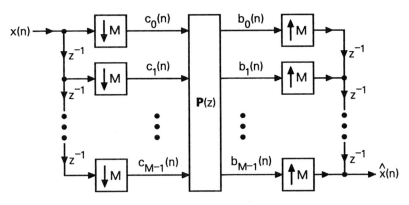

Figure 5.7-3 The equivalent circuit for the maximally decimated filter bank.

The reconstructed signal is

$$\widehat{X}(z) = \sum_{s=0}^{M-1} z^{-(M-1-s)} B_s(z^M)$$

$$= \sum_{s=0}^{M-1} z^{-(M-1-s)} \sum_{\ell=0}^{M-1} P_{s,\ell}(z^M) C_\ell(z^M) \quad (5.7.16)$$

$$= \frac{1}{M} \sum_{s=0}^{M-1} z^{-(M-1-s)} \sum_{\ell=0}^{M-1} P_{s,\ell}(z^M) \sum_{k=0}^{M-1} \left(zW^k\right)^{-\ell} X(zW^k).$$

This can be rearranged as

$$\widehat{X}(z) = \frac{1}{M} \sum_{k=0}^{M-1} X(zW^k) \sum_{\ell=0}^{M-1} W^{-k\ell} \sum_{s=0}^{M-1} z^{-\ell} z^{-(M-1-s)} P_{s,\ell}(z^M). \quad (5.7.17)$$

The terms of the form $X(zW^k), k \neq 0$ represent aliasing. The above expression is free from these aliasing terms [for all input signals $x(n)$] if and only if

$$\sum_{\ell=0}^{M-1} W^{-k\ell} \underbrace{\sum_{s=0}^{M-1} z^{-\ell} z^{-(M-1-s)} P_{s,\ell}(z^M)}_{V_\ell(z)} = 0, \quad k \neq 0. \tag{5.7.18}$$

This can be written using matrix notation as

$$\mathbf{W}^\dagger \begin{bmatrix} V_0(z) \\ V_1(z) \\ \vdots \\ V_{M-1}(z) \end{bmatrix} = \begin{bmatrix} \times \\ 0 \\ \vdots \\ 0 \end{bmatrix}. \tag{5.7.19}$$

where $\mathbf{W}$ is the $M \times M$ DFT matrix, and $\times$ indicates a possibly nonzero entry. Using the fact that $\mathbf{WW}^\dagger = M\mathbf{I}$, we can rewrite this as

$$\begin{bmatrix} V_0(z) \\ V_1(z) \\ \vdots \\ V_{M-1}(z) \end{bmatrix} = \mathbf{W} \begin{bmatrix} \times \\ 0 \\ \vdots \\ 0 \end{bmatrix}. \tag{5.7.20}$$

This implies

$$V_\ell(z) = V(z), \quad 0 \le \ell \le M-1, \tag{5.7.21}$$

since the 0th column of $\mathbf{W}$ has all entries equal to unity. Thus the QMF bank is alias-free if and only $V_\ell(z)$ defined in (5.7.18) is the same for all ℓ.

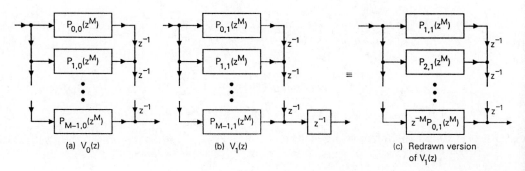

Figure 5.7-4 Comparing the Type 2 polyphase implementations of $V_0(z)$ and $V_1(z)$.

In Figs. 5.7-4(a) and (b) we demonstrate polyphase structures for $V_0(z)$ and $V_1(z)$. The structure for $V_1(z)$ can be rearranged as shown in Fig. 5.7-4(c). Because of the requirement $V_0(z) = V_1(z)$, the polyphase components in Figs. 5.7-4(a) and (c) should be the same. This shows that the 0th column of $\mathbf{P}(z)$ is an upwards-shifted version of the 1st column, with the top most element recirculated with a z^{-1} attached to it. Similarly we can verify that the ℓth column is obtained from the $(\ell + 1)$st column in this manner. This proves that $\mathbf{P}(z)$ is pseudocirculant.

Having canceled aliasing, (5.7.18) holds so that $\widehat{X}(z) = T(z)X(z)$ with $T(z)$ obtained from (5.7.17) as

$$T(z) = \frac{1}{M} \sum_{s=0}^{M-1} z^{-(M-1-s)} \sum_{\ell=0}^{M-1} z^{-\ell} P_{s,\ell}(z^M). \qquad (5.7.22)$$

Since the elements $P_{s,\ell}(z)$ are completely determined by the 0th row elements $P_{0,\ell}(z)$, we can rearrange this (Problem 5.29) into the form (5.7.13). This completes the proof. ▽▽▽

The Special Case of Perfect Reconstruction (PR) Systems

A PR system is an alias free system with $T(z) = $ delay. The alias-free nature implies that $\mathbf{P}(z)$ is pseudocirculant. With the 0th row of $\mathbf{P}(z)$ as in (5.7.12), $T(z)$ has the form (5.7.13). This is a delay only if $P_m(z) = 0$ for all but one value of m in the range $0 \leq m \leq M - 1$. And this nonzero $P_m(z)$ must have the form cz^{-m_0}. Summarizing, an alias free system has perfect reconstruction if and only if the pseudocirculant $\mathbf{P}(z)$ has 0th row equal to

$$[0 \quad \ldots \quad 0 \quad cz^{-m_0} \quad 0 \quad \ldots 0]. \qquad (5.7.23a)$$

In other words $\mathbf{P}(z)$ (i.e., $\mathbf{R}(z)\mathbf{E}(z)$) has the form

$$\mathbf{R}(z)\mathbf{E}(z) = cz^{-m_0} \begin{bmatrix} 0 & \mathbf{I}_{M-r} \\ z^{-1}\mathbf{I}_r & 0 \end{bmatrix}, \qquad (5.7.23b)$$

for some r in $0 \leq r \leq M - 1$. This was stated earlier in (5.6.7) without proof. Under this condition (5.7.13) reduces to

$$T(z) = cz^{-r}z^{-(M-1)}z^{-m_0 M}. \qquad (5.7.23c)$$

Some Practical Special Cases of Alias Free Systems

1. Consider the special case when $\mathbf{P}(z)$ is diagonal. This means that the structure is as in Fig. 5.7-1(a). The pseudocirculant condition on $\mathbf{P}(z)$ now means that all diagonal elements are identical, so that

$$\mathbf{P}(z) = S(z)\mathbf{I}. \qquad (5.7.24)$$

This result agrees with the alias-cancelation condition obtained earlier in Example 5.7.1. In this case $T(z)$ reduces to

$$T(z) = z^{-(M-1)} S(z^M). \qquad (5.7.25)$$

2. A generalization of the above is the case where $\mathbf{P}(z)$ has one nonzero entry per row. In this case, the pseudocirculant property means

$$\mathbf{P}(z) = S(z) \begin{bmatrix} \mathbf{0} & \mathbf{I}_{M-r} \\ z^{-1}\mathbf{I}_r & \mathbf{0} \end{bmatrix}, \qquad (5.7.26)$$

where $0 \leq r \leq M - 1$. The 0th row of $\mathbf{P}(z)$ has all zeros except $P_r(z) = S(z)$, so that (5.7.13) yields

$$T(z) = z^{-(M-1)} z^{-r} S(z^M). \qquad (5.7.27)$$

The presence of r merely introduces additional delay. We can obtain this from the first special case simply by replacing each synthesis filter $F_k(z)$ with $z^{-r} F_k(z)$. Thus every PR system satisfying (5.7.26) can be obtained from a PR system satisfying (5.7.24) simply by replacing each synthesis filter $F_k(z)$ with $z^{-r} F_k(z)$. In this sense, the form (5.7.26) is only "trivially" more general than (5.7.24).

If $S(z)$ is a delay, then (5.7.26) reduces to the form (5.7.23b) implying perfect reconstruction. Practically all the alias-free systems we consider belong to simple special cases of the form (5.7.26), that is, $\mathbf{P}(z)$ is a pseudocirculant with one nonzero entry per row.

Further results on amplitude and phase distortion in alias-free systems can be found in Sec. 10.1. In particular, it will be shown that $T(z)$ is allpass (i.e., there is no amplitude distortion) if and only if the pseudocirculant $\mathbf{P}(z)$ satisfies a property called paraunitariness.

5.8 TREE STRUCTURED FILTER BANKS

Consider the structure shown in Fig. 5.8-1(a). Here a signal is split into two subbands, and after decimation, each subband is again split into two and decimated. The subbands are then recombined, two at a time, by use of two-channel synthesis banks. This system is said to be a maximally decimated (binary) tree structured filter bank. The complete system can be redrawn in the equivalent nontree form of Fig. 5.4-1, with $M = 4$. The resulting filters $H_m(z)$ and $F_m(z)$ ($0 \leq m \leq M - 1$) can be expressed in terms of the filters $H_i^{(k)}(z)$ and $F_i^{(k)}(z)$ (Problem 5.24).

Fig. 5.8-1(b) shows an example of the magnitude responses of the four analysis filters $H_m(z)$ for the two level tree. In this example, the tree filters $H_0^{(k)}(z)$ have the power symmetric response shown earlier in Fig. 5.3-4(a), and $H_1^{(k)}(z) = H_0^{(k)}(-z)$. Note that the four analysis filters are not equiripple, even though $H_0^{(k)}(z)$ and $H_1^{(k)}(z)$ are.

Suppose the filters $H_0^{(k)}(z), H_1^{(k)}(z), F_0^{(k)}(z)$ and $F_1^{(k)}(z)$ are such that the two-channel QMF bank with these filters is alias-free. Then the complete system is also alias-free. Similarly, if the two channel system has perfect reconstruction, then so does the complete system. (Problems 5.24 and 5.25). These results can also be extended to more than two levels of splitting. The results also extend to trees other than binary (e.g., split a signal into two subbands, then split one subband into three and the other into four, etc.).

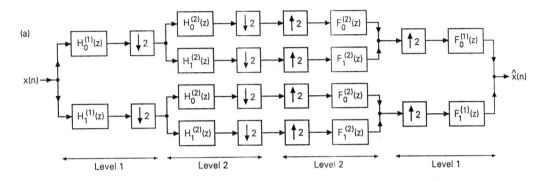

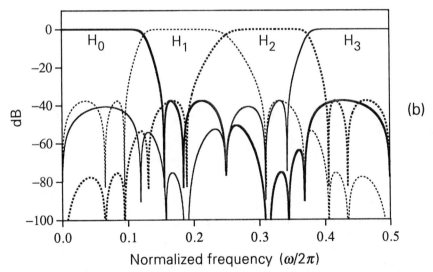

Figure 5.8-1 (a) A two-level maximally decimated tree structured filter bank, and (b) example of magnitude responses.

Assume that all the two-channel systems in Fig. 5.8-1(a) have perfect reconstruction. Suppose, however, that the upper two-channel QMF bank and the lower two-channel QMF bank at the second level do *not* have the same set of analysis and synthesis filters. Then it may be necessary to introduce appropriate scale factors and delays at proper places so that the complete system still has perfect reconstruction (why?).

Tree structured filter banks are used in a number of applications both in one and two-dimensional signal processing. We now mention two of these, which were originally intended for image processing. The presentation here is brief.

Multiresolution Analysis Algorithm

Consider the variation of the analysis bank shown in Fig. 5.8-2(a). This is equivalent to the system shown in Fig. 5.8-2(b). This is a four-channel system with unequal decimation ratios. (It is still a maximally decimated system.) Each $[G(z), H(z)]$ is typically a lowpass/highpass pair, as in a two channel QMF bank.

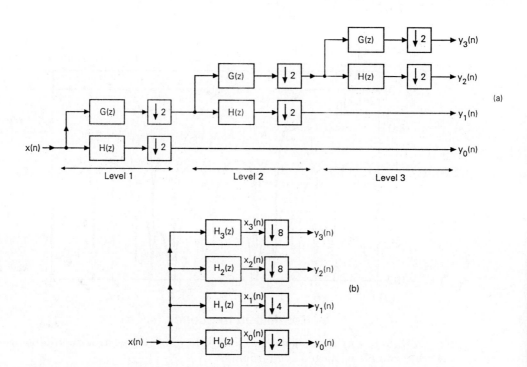

Figure 5.8-2 (a) A 3-level binary tree structured QMF bank, and (b) the equivalent four-channel system.

Figure 5.8-3(a) shows the synthesis bank that goes with this system, and Fig. 5.8-3(b) shows the non-tree equivalent structure. Assume that $G_s(z), H_s(z)$ are chosen so that the two channel QMF bank with filters

$G(z), H(z), G_s(z)$ and $H_s(z)$ has perfect reconstruction, with unit-gain and no delay. We then have $\hat{x}(n) = x(n)$.

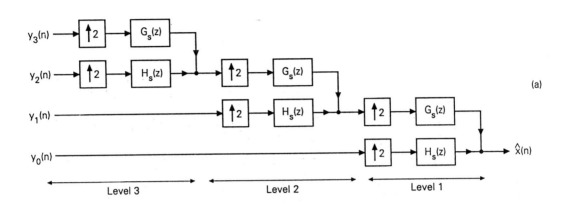

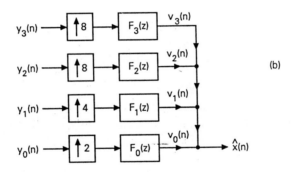

Figure 5.8-3 (a) The synthesis bank corresponding to Fig. 5.8-2, and (b) equivalent four-channel system.

Figure 5.8-4 shows typical frequency responses of the analysis and synthesis filters (assuming that $G(z)$ and $H(z)$ form a lowpass and highpass pair, with cutoff around $\pi/2$.) The signals $v_k(n)$ which are outputs of $F_k(z)$, are called *multiresolution* components. For example the signal $v_3(n)$ represents a lowpass version (or a 'coarse' approximation) of $x(n)$, subject to aliasing and other errors. (Note that $v_k(n)$ has the same 'sampling rate' as $x(n)$.) The signal $v_2(n)$ adds some high frequency (bandpass) details, so that $v_3(n) + v_2(n)$ is a finer approximation of $x(n)$. The signal $v_0(n)$ adds the finest ultimate (high-frequency) detail, so that $\hat{x}(n) = x(n)$ (by perfect reconstruction property). An obvious generalization of the tree structure uses different filter pairs at different levels of the tree.

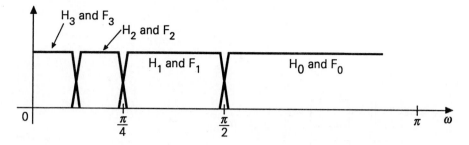

Figure 5.8-4 Typical appearances of magnitude responses of filters in the 3-level tree.

There are several ways in which this structure can be used to obtain image compression. For example, one can choose to retain only $v_3(n)$; or one can add a quantized version of $v_2(n)$ to $v_3(n)$. More generally we can attach decreasing weights (bits) to the finer and finer detail signals $v_k(n)$. This technique is the ingredient of Mallat's multiresolution algorithm for image compression [Mallat, 1989a,b]. The above observation can also be used to transmit finer and finer versions of video data (e.g., in teleconferencing).

The above algorithm is extremely appealing even from an intuitive and philosophical view point: any kind of 'learning' or 'understanding' in life always occurs at various levels of resolutions, which get finer and finer as we improve our skills. Think of the way we mature in any of these: baseball, music, scientific skills, writing skills ...

The Laplacian Pyramid

This is a well-known scheme for image coding [Burt and Adelson, 1983], and is demonstrated in Fig. 5.8-5. Here $G(z)$ is an FIR lowpass filter. The notation $\widetilde{G}(z)$ is defined as usual, so that $\widetilde{G}(e^{j\omega}) = G^*(e^{j\omega})$. Thus $\widehat{x}_0(n)$ is a coarse lowpass approximation of the input $x(n)$. This approximation introduces no phase distortion [since $\widetilde{G}(z)G(z)$ has zero phase]. We can subtract $\widehat{x}_0(n)$ from $x(n)$ to recover the high frequency details, denoted $d_0(n)$.

This process is now repeated on the decimated signal $x_1(n)$. The analysis bank, therefore, produces the highpass signals $d_0(n), \ldots d_{L-1}(n)$, and the lowpass signal $\widehat{x}_{L-1}(n)$. (In the figure $L = 2$.) These signals can then be recombined using a synthesis bank as demonstrated in the figure, to recover $x(n)$.

Notice that the perfect reconstruction property is trivially satisfied, regardless of the design of $G(z)$. This is not surprising because the difference signals (highpass signals) $d_k(n)$ are not maximally decimated. For example, $d_0(n)$ is not decimated at all. This results in increased data rate (nearly by a factor of two). In order for the scheme to be beneficial, this must be compensated by the compression obtainable by the quantization of the signals $d_k(n)$ and $\hat{x}_{L-1}(n)$. Traditional QMF banks (such as Fig. 5.8-2) are, on the other hand, maximally decimated, and do not have this problem (but require special design procedures).

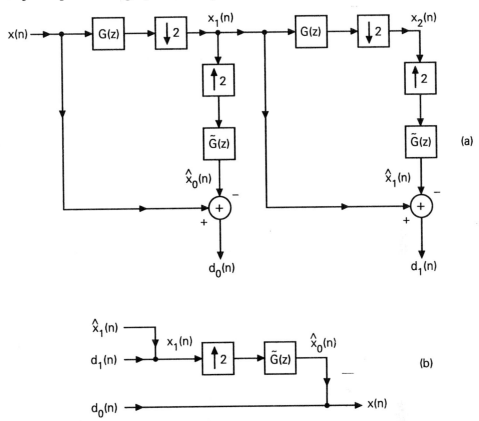

Figure 5.8-5 Burt and Adelson's algorithm. (a) analysis, and (b) synthesis.

5.9 TRANSMULTIPLEXERS

An introduction to transmultiplexers was given in Section 4.5.4, which the reader should review at this time. Figs. 4.5-4 and 4.5-5 demonstrate the time domain and frequency domain multiplexing operations, and Fig. 4.5-6 shows the complete TDM → FDM → TDM converter (transmultiplexer), also reproduced in Fig. 5.9-1.

Fig. 5.9-2 demonstrates how the signal $V_1(e^{j\omega})$ is generated starting from $X_1(e^{j\omega})$. If all the signals $x_k(n)$ are bandlimited to $|\omega| < \sigma_k$ with $\sigma_k < \pi$, there is no overlap between adjacent signals in the FDM format, that is, there exists a guard band between adjacent frequency bins, as demonstrated in Fig. 5.9-3. In this case the FDM signals can be separated by filtering operations (followed by M-fold decimation to stretch the signal back to the full band $-\pi \leq \omega \leq \pi$). The presence of guard bands ensures that there is no cross talk between adjacent signals, even though the filters have nonzero

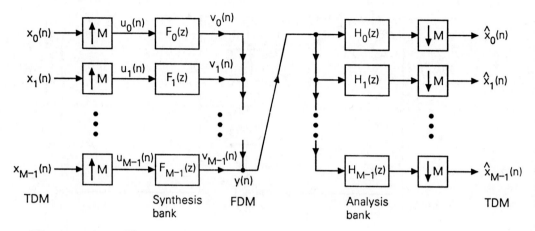

Figure 5.9-1 The transmultiplexer circuit, drawn in terms of filter bank notations.

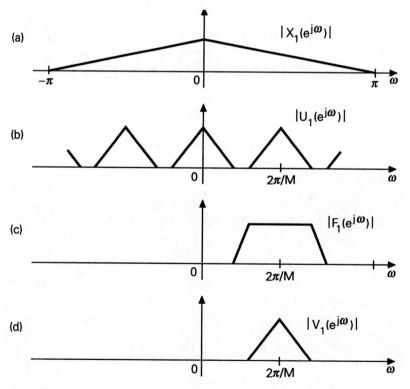

Figure 5.9-2 Generation of the signal $v_1(n)$ by use of interpolation and filtering.

260 Chap. 5. Maximally decimated filter banks

transition band. A larger guard band implies larger permissible transition band (hence lower cost) for the filters $H_k(z)$, which attempt to recover the signals $x_k(n)$ from the FDM version. However, the existence of guard bands also means that the full channel bandwidth is *not* utilized in the transmission process.

The following observation was made in Vetterli [1986]: even if there are no guard bands (thereby permitting cross talk), we can subsequenty eliminate the cross talk in a manner analogous to alias cancelation in QMF banks. This idea makes judicious use of the relation between the mathematics of QMF banks and transmultiplexers as we will elaborate. We remind the reader that the term 'QMF', which is used for convenience, really stands for 'maximally decimated analysis synthesis systems'.

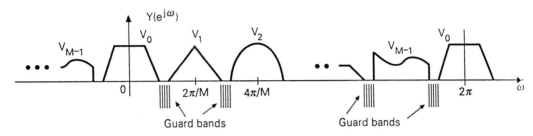

Figure 5.9-3 Stacking up the M signals $V_k(e^{j\omega})$ in the frequency domain, to obtain the FDM version $y(n)$.

5.9.1 Input-Output Relations for Transmultiplexers

We show that it is possible to achieve perfect cross talk elimination as well as perfect recovery of each TDM component $x_k(n)$ with finite-cost (in fact FIR) filters $H_k(z)$ and $F_k(z)$. In analogy with the QMF bank, we continue to use terms such as "analysis" and "synthesis" filters, and "filter banks" as indicated in Fig. 5.9-1. Notice the conceptual duality between the QMF bank and the transmultiplexer. In the former, we first "analyze" and then "synthesize"; this is in reverse order as compared to the transmultiplexer. (The QMF bank can also be conceptually looked upon as a FDM → TDM → FDM convertor.) We will see that the problem of designing filters for 'perfect reconstruction transmultiplexers' is same as the design of perfect reconstruction (PR) QMF banks.

The relation between $\widehat{x}_k(n)$ and $x_m(n)$ can be schematically represented as in Fig. 5.9-4. By using the polyphase identity (Fig. 4.3-13) we see that each branch in this figure is in reality an LTI system. We can therefore express

$$\widehat{X}_k(z) = \sum_{m=0}^{M-1} S_{km}(z) X_m(z), \quad 0 \le k \le M-1, \quad (5.9.1)$$

where $S_{km}(z)$ is the 0th polyphase component of $H_k(z)F_m(z)$. By defining

$$\mathbf{x}(n) = \begin{bmatrix} x_0(n) \\ \vdots \\ x_{M-1}(n) \end{bmatrix}, \quad \widehat{\mathbf{x}}(n) = \begin{bmatrix} \widehat{x}_0(n) \\ \vdots \\ \widehat{x}_{M-1}(n) \end{bmatrix}, \quad (5.9.2)$$

we can express (5.9.1) more compactly as

$$\widehat{\mathbf{X}}(z) = \mathbf{S}(z)\mathbf{X}(z). \quad (5.9.3)$$

So the transmultiplexer is an LTI system with transfer matrix $\mathbf{S}(z)$. The system is free from cross talk if and only if $\mathbf{S}(z)$ is diagonal. (This is the same as the requirement that the 0th polyphase component of $H_k(z)F_m(z)$ be zero unless $k = m$.) Under this condition, each reconstructed TDM signal $\widehat{x}_k(n)$ is related to the original signal $x_k(n)$ according to

$$\widehat{X}_k(z) = S_{kk}(z)X_k(z). \quad (5.9.4)$$

The transfer functions $S_{kk}(z)$ represent the distortions that remain after cross talk elimination. If $S_{kk}(z)$ is allpass for all k, there is no amplitude distortion; if $S_{kk}(z)$ has linear phase, there is no phase distortion. Finally, a perfect reconstruction (PR) transmultiplexer is one for which

$$S_{kk}(z) = c_k z^{-n_k}, \quad \text{for all } k, \quad (5.9.5)$$

for some nonzero c_k and integer n_k. The TDM signals are then recovered without error, that is, $\widehat{x}_k(n) = c_k x_k(n - n_k)$.

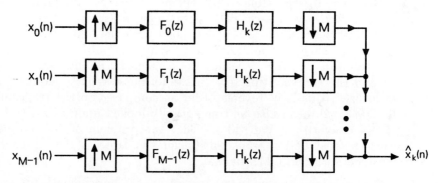

Figure 5.9-4 Equivalent circuit for generation of $\widehat{x}_k(n)$.

5.9.2 Study Based on Polyphase Matrices

The use of polyphase decomposition adds further insight into the operation of the transmultiplexer [Koilpillai et al., 1991]. As in Sec. 5.5, we can redraw the analysis and synthesis banks in terms of the polyphase matrices $\mathbf{E}(z)$ and

$\mathbf{R}(z)$. The resulting equivalent transmultiplexer circuit is shown in Fig. 5.9-5(a), which simplifies to Fig. 5.9-5(b) after invoking the noble identities.†
This structure can be further simplified into the equivalent form shown in Fig. 5.9-5(c), by using the equivalence of Fig. 4.3-14. It is, therefore, clear that the transfer matrix $\mathbf{S}(z)$ can be expressed as

$$\mathbf{S}(z) = \mathbf{E}(z)\mathbf{\Gamma}(z)\mathbf{R}(z), \qquad (5.9.6)$$

where

$$\mathbf{\Gamma}(z) = \begin{bmatrix} 0 & 1 \\ z^{-1}\mathbf{I}_{M-1} & 0 \end{bmatrix}. \qquad (5.9.7)$$

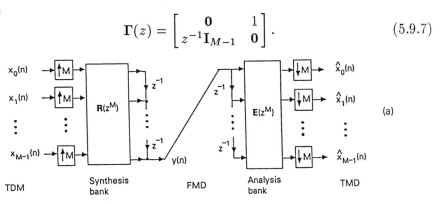

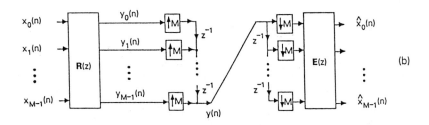

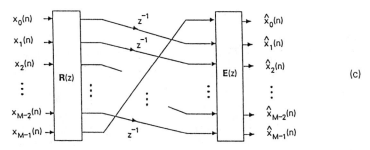

Figure 5.9-5 (a) Equivalent structures for the transmultiplexer in terms of polyphase matrices, (b) rearrangement using noble identitites, and (c) simplification using the equivalence of Fig. 4.3-14.

---

† Note that if we set $\mathbf{E}(z) = \mathbf{I}$ as a special case, then $y(n)$ becomes the TDM (rather than FDM) signal!

So the set of reconstructed signals $\hat{\mathbf{x}}(n)$ is related to $\mathbf{x}(n)$ by the transfer matrix (5.9.6). From this expression we can explore the conditions for cross talk elimination and perfect reconstruction.

Perfect Reconstruction

A *sufficient condition* for perfect reconstruction is obtained by setting $\mathbf{S}(z) = cz^{-n_0}\mathbf{I}$. Now

$$\begin{aligned} \mathbf{S}(z) &= cz^{-n_0}\mathbf{I} \\ \Longleftrightarrow \quad \mathbf{E}(z)\boldsymbol{\Gamma}(z)\mathbf{R}(z) &= cz^{-n_0}\mathbf{I} \\ \Longleftrightarrow \quad c\mathbf{E}^{-1}(z)\mathbf{R}^{-1}(z) &= z^{n_0}\boldsymbol{\Gamma}(z) \\ \Longleftrightarrow \quad \mathbf{R}(z)\mathbf{E}(z) &= cz^{-n_0}\boldsymbol{\Gamma}^{-1}(z). \end{aligned} \qquad (5.9.8)$$

Substituting for $\boldsymbol{\Gamma}(z)$, this becomes

$$\mathbf{R}(z)\mathbf{E}(z) = cz^{-m_0}\begin{bmatrix} \mathbf{0} & \mathbf{I}_{M-1} \\ z^{-1} & \mathbf{0} \end{bmatrix}, \qquad (5.9.9)$$

for appropriate integer m_0.

Relation to perfect reconstruction (PR) QMF banks. From the previous section we know that the product $\mathbf{R}(z)\mathbf{E}(z)$ of a PR QMF bank satisfies (5.7.23b) for some integer r in $0 \leq r \leq M - 1$. If the QMF bank is such that $r = 1$, then this condition is same as (5.9.9). On the other hand, if $r \neq 1$, then we can insert appropriate amount of delay in front of the filters $F_k(z)$ to force $r = 1$.

The amount of delay to be introduced can be judged as follows: for arbitrary r the PR QMF bank has overall transfer function (5.7.23c). This has the form $cz^{-\ell M}$ for integer ℓ if and only if $r = 1$. So the amount of delay to be inserted is such that $T(z)$ takes this form. For example, suppose the PR QMF bank has $T(z) = cz^{-2}z^{-iM}$. If we insert the delay $z^{-(M-2)}$ in front of each $F_k(z)$, then $T(z)$ becomes $cz^{-(i+1)M}$. So insertion of this delay results in a PR QMF bank with $r = 1$. Its analysis and synthesis filters can then be used in the transmultiplexer to obtain perfect reconstruction!

Summary of perfect-reconstruction condition. This important conclusion can be summarized as follows: Let $H_k(z)$ and $F_k(z)$ be the analysis and synthesis filters of a perfect reconstruction QMF bank, with overall transfer function $T(z) = cz^{-L}$ for some $c \neq 0$ and integer L. Then the transmultiplexer with analysis filters $H_k(z)$ and synthesis filters $z^{-J}F_k(z)$ has perfect reconstruction property for some integer J in the range $0 \leq J \leq M - 1$. The appropriate value of J is such that $L + J$ is a multiple of M. (That is, J is such that a QMF bank with filters $H_k(z)$ and $z^{-J}F_k(z)$ would have $T(z) = cz^{-\ell M}$ for some integer ℓ.)

Cross Talk Free Transmultiplexers

The next natural question is this: suppose we are not interested in perfect reconstruction, but only in perfect cross talk elimination, and minimization of other distortions. (This will cut the cost of filters to some extent.) Can we obtain such a system starting from a QMF bank? Since $\widehat{\mathbf{X}}(z) = \mathbf{S}(z)\mathbf{X}(z)$, the transmultiplexer is cross talk free if $\mathbf{S}(z)$ is diagonal.

We now show that this can be accomplished by starting from a suitable alias-free QMF bank. The most common alias-free QMF bank satisfies (5.7.26), where $\mathbf{P}(z) = \mathbf{R}(z)\mathbf{E}(z)$. We will assume $r = 1$, as this can be ensured by inserting the right amount of delay z^{-J} in front of $F_k(z)$. So we have

$$\mathbf{R}(z)\mathbf{E}(z) = S(z) \underbrace{\begin{bmatrix} \mathbf{0} & \mathbf{I}_{M-1} \\ z^{-1} & \mathbf{0} \end{bmatrix}}_{z^{-1}\boldsymbol{\Gamma}^{-1}(z)}, \qquad (5.9.10)$$

where $\boldsymbol{\Gamma}(z)$ is as in (5.9.7). The QMF bank satisfying (5.9.10) has distortion function (5.7.27), with $r = 1$. That is,

$$T(z) = z^{-M} S(z^M). \qquad (5.9.11)$$

In other words, $T(z)$ is a function of z^M, i.e., z appears only in the form z^M. Now the condition (5.9.10) implies

$$z^{-1} S(z) \mathbf{R}^{-1}(z) \boldsymbol{\Gamma}^{-1}(z) \mathbf{E}^{-1}(z) = \mathbf{I}, \qquad (5.9.12)$$

which in turn implies

$$\mathbf{E}(z)\boldsymbol{\Gamma}(z)\mathbf{R}(z) = z^{-1} S(z) \mathbf{I}. \qquad (5.9.13)$$

The quantity on the left is the transfer matrix $\mathbf{S}(z)$ of the transmultiplexer with same analysis and synthesis filters as the QMF bank. So (5.9.13) is equivalent to

$$\mathbf{S}(z) = z^{-1} S(z) \mathbf{I}. \qquad (5.9.14)$$

Since this is a diagonal matrix, cross talk has been eliminated, and the reconstructed signals satisfy

$$\frac{\widehat{X}_k(z)}{X_k(z)} = z^{-1} S(z). \qquad (5.9.15)$$

Summary of cross talk cancelation condition. This result can be summarized as follows: Let $H_k(z)$ and $F_k(z)$ be the analysis and synthesis filters in a QMF bank satisfying (5.9.10). This QMF bank is therefore alias-free with distortion function $T(z) = z^{-M} S(z^M)$. If we now design a transmultiplexer with analysis filters $H_k(z)$ and synthesis filters $F_k(z)$, then it is free from cross talk. Moreover, the reconstructed signals satisfy $\widehat{X}_k(z) = z^{-1}S(z)X_k(z)$.

The cross talk free transmultiplexer in general suffers from amplitude and phase distortions since $S(z)$ in (5.9.15) is an arbitrary transfer function. This same $S(z)$ appears in (5.9.11) which represents the distortions in the alias-free QMF bank. If $S(z)$ is allpass, then both systems are free from amplitude distortion. If $S(z)$ has linear phase, then both systems are free from phase distortion.

Notice that the distortion functions $S_{kk}(z)$ in the transmultiplexer are not required to be same for all k. This freedom is not exploited above, because $S_{kk}(z) = z^{-1}S(z)$ for all k.

5.10 SUMMARY AND TABLES

In this chapter we studied the quadrature mirror filter bank. Both two-channel and M channel cases were considered.

For the two channel case we also presented design techniques for alias-free QMF banks; in the FIR case we showed how to eliminate phase distortion and minimize amplitude distortion. For the IIR case we showed that if the analysis filters are constrained to be power symmetric, we can design alias-free QMF banks free from amplitude distortion. The very low computational complexity of this IIR system was also demonstrated. With FIR filters, the same power symmetric condition was then used to obtain perfect reconstruction.

For the M channel case we developed the theory of alias cancelation and perfect reconstruction, and demonstrated the ideas with several examples. These results were extended to the study of transmultiplexers. We also considered tree structured filter banks.

Tables 5.10.1–5.10.4 summarize the main results of this chapter. Table 5.10.5 presents a summary of important matrix quantities, and the relations between them. In the next few chapters, we will present design techniques for M channel QMF banks.

TABLE 5.10.1 Two-channel QMF bank at a glance

1. **Basic facts (Section 5.1)**

    Reconstructed signal: $\widehat{X}(z) = T(z)X(z) + A(z)X(-z)$.
    $T(z) = $ Distortion function $= \frac{1}{2}[H_0(z)F_0(z) + H_1(z)F_1(z)]$.
    $A(z) = $ Aliasing gain $= \frac{1}{2}[H_0(-z)F_0(z) + H_1(-z)F_1(z)]$.
    Suff. cond. for alias cancelation: $F_0(z) = H_1(-z)$, $F_1(z) = -H_0(-z)$.
    After aliasing is canceled $\widehat{X}(z) = T(z)X(z)$.
    $T(z)$ not allpass $\Rightarrow$ amplitude distortion (AMD)
    $T(z)$ not linear-phase $\Rightarrow$ phase distortion (PHD).
    FIR QMF bank: $H_0(z), H_1(z), F_0(z), F_1(z)$ are FIR.
    Linear-phase QMF bank: $H_0(z), H_1(z)$ have linear phase.

2. **A simple choice of filters for alias cancelation (Section 5.2)**

    Choose $H_1(z) = H_0(-z)$, $F_0(z) = H_0(z)$, $F_1(z) = -H_1(z)$. Then
    a) This is alias-free with $T(z) = \frac{1}{2}[H_0^2(z) - H_0^2(-z)]$.
    b) Let $H_0(z) = E_0(z^2) + z^{-1}E_1(z^2)$, then $T(z) = 2z^{-1}E_0(z^2)E_1(z^2)$.
    c) This expression for $T(z)$ (a consequence of the constraint $H_1(z) = H_0(-z)$) shows that perfect reconstruction is obtained if and only if $E_1(z) = az^{-b}/E_0(z)$, imposing severe restrictions on analysis filters. For example, in the FIR case $H_0(z)$ has to be a sum of two delays.

    A polyphase implementation:

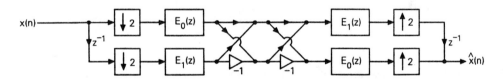

    FIR case (see Table 5.10.2 for IIR case).
    a) If $H_0(z)$ is linear-phase FIR with order N, then $T(z)$ has linear phase, and the system has only AMD. N must be odd, or else $T(e^{j\pi/2}) = 0$. Perfect reconstruction is not possible unless $E_0(z)$ and $E_1(z)$ are delays, which would make $H_0(z)$ trivial.
    b) If N denotes the order (odd) of $H_0(z)$, the analysis bank requires $0.5(N+1)$ MPUs and $0.5(N+1)$ APUs (using polyphase form). This is true whether $H_0(z)$ has linear phase or not.

TABLE 5.10.2 IIR allpass based QMF banks

In what follows, the analysis and synthesis filters are related as $H_1(z) = H_0(-z), F_0(z) = H_0(z), F_1(z) = -H_1(z)$, so that aliasing is canceled.

1. **Power symmetric filters.**

   a) $H_0(z)$ is said to be power symmetric if $\widetilde{H}_0(z)H_0(z)$ is a half-band filter, i.e., $\widetilde{H}_0(z)H_0(z) + \widetilde{H}_0(-z)H_0(-z) = \beta$ for some nonzero constant β.

   b) Under some mild conditions (Theorem 5.3.1), an IIR power symmetric filter can be written as $H_0(z) = 0.5[a_0(z^2) + z^{-1}a_1(z^2)]$, where $a_0(z), a_1(z)$ are real-coefficient allpass. Then the QMF bank can be implemented as shown below.

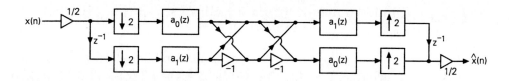

The distortion function is $T(z) = \frac{1}{2}z^{-1}a_0(z^2)a_1(z^2) = $ allpass, so that the QMF bank is free from AMD. Only PHD is still present, since aliasing has already been canceled.

2. **Power symmetric elliptic filters (Table 5.3.1 has design algorithm).**

   a) *Major fact.* If $H_0(z)$ is elliptic lowpass with ripples related as $\delta_2^2 = 4\delta_1(1-\delta_1)$ and band edges related as $\omega_p + \omega_S = \pi$, then it is power symmetric.

   b) *Low complexity.* If in addition the order N is odd, it can be expressed as $H_0(z) = [a_0(z^2) + z^{-1}a_1(z^2)]/2$, and the QMF bank implemented as above. Here $a_0(z), a_1(z)$ are real-coefficient allpass. The analysis bank requires only $0.25(N-1)$ MPUs and $0.5(N+1)$ APUs.

   c) *Pole locations.* Power symmetric elliptic filters have all poles on the imaginary axis. So the denominator has the form $D(z) = d(z^2)$.

TABLE 5.10.3 FIR power symmetric QMF banks

Basic result. (Section 5.3.6). Let $H_0(z) = \sum_{n=0}^{N} h_0(n) z^{-n}$ be power symmetric, that is, $\widetilde{H}_0(z) H_0(z)$ is a half-band filter, that is,

$$\widetilde{H}_0(z) H_0(z) + \widetilde{H}_0(-z) H_0(-z) = \beta$$

for some nonzero constant β. Then N is automatically odd (assuming that $h_0(0) \neq 0$ and $h_0(N) \neq 0$). Let the filters $H_1(z), F_0(z)$ and $F_1(z)$ be chosen as

$$H_1(z) = -z^{-N} \widetilde{H}_0(-z), \ F_0(z) = z^{-N} \widetilde{H}_0(z), \ F_1(z) = z^{-N} \widetilde{H}_1(z).$$

Then the two channel QMF bank has *perfect reconstruction*. All the filters are FIR and have same order N. Efficient lattice structures for this system will be presented in Section 6.4.

Design procedure. It only remains to design $H_0(z)$. This can be done by first designing a zero-phase FIR half-band filter $H(z)$ with $H(e^{j\omega}) \geq 0$ and taking $H_0(z)$ to be a spectral factor. See Section 5.3.6 for more details.

TABLE 5.10.4 Facts about M-channel QMF banks

Fig. 5.4-1 represents an M channel QMF bank. The reconstructed signal $\widehat{X}(z)$ is given by $\widehat{X}(z) = T(z)X(z) + \sum_{\ell=1}^{M-1} A_\ell(z)X(zW^\ell)$. This is a linear and time varying system. The terms $X(zW^\ell), \ell > 0$ are the alias terms. The system is free from aliasing if $A_\ell(z) = 0$ for $\ell > 0$. Under such condition, the QMF bank becomes a linear time invariant (LTI) system with transfer function $T(z) = \sum_{k=0}^{M-1} H_k(z)F_k(z)/M$, called the *distortion* function.

Any M-channel QMF bank can be redrawn in terms of the polyphase component matrices $\mathbf{E}(z)$ and $\mathbf{R}(z)$ (Fig. 5.5-3(a),(b)). This in turn can be redrawn in terms of a $M \times M$ matrix $\mathbf{P}(z) = \mathbf{R}(z)\mathbf{E}(z)$ (Fig. 5.5-3(c)).

1. The QMF bank is alias-free if and only if $\mathbf{P}(z)$ is a pseudocirculant (demonstrated in (5.7.11) for $M = 3$.)

2. Under this alias-free condition, the QMF bank is an LTI system with transfer function $T(z) = z^{-(M-1)} \sum_{k=0}^{M-1} z^{-k} P_k(z^M)$.

3. An alias-free system is free from amplitude distortion (i.e., $T(z)$ is stable allpass) if and only if $\mathbf{P}(z)$ is a lossless matrix. (To be proved later in Section 10.1.)

4. An alias-free system has perfect reconstruction if $T(z)$ is a delay, i.e., $T(z) = cz^{-n_0}$. This happens if and only if the pseudocirculant $\mathbf{P}(z)$ has the special form (5.7.23b). The most common special case has $r = 0$ so that $\mathbf{R}(z)\mathbf{E}(z) = cz^{-m_0}\mathbf{I}$, i.e.,

$$\mathbf{R}(z) = cz^{-m_0}\mathbf{E}^{-1}(z).$$

Given a perfect reconstrucion (PR) QMF bank satisfying (5.7.23b) for some r in the range $0 \leq r \leq M - 1$, we can obtain a PR QMF bank with a different value of r just by replacing the synthesis filters $F_k(z)$ with $z^{-m}F_k(z)$ for appropriate integer m.

5. An FIR QMF bank is one for which $H_k(z)$ as well as $F_k(z)$ are FIR. If such a system has PR property then

$$\det \mathbf{E}(z) = \alpha_0 z^{-K_0}, \quad \text{and} \quad \det \mathbf{R}(z) = \alpha_1 z^{-K_1}$$

6. *Special case where* $\mathbf{P}(z)$ *is diagonal.* A multirate system of the form shown in Fig. 5.7-1(a) is alias-free if and only if $S_k(z)$ is same for all k. Letting $S_k(z) = S(z)$ we then have $T(z) = z^{-(M-1)}S(z^M)$.

TABLE 5.10.5 Matrix notations in filter bank theory

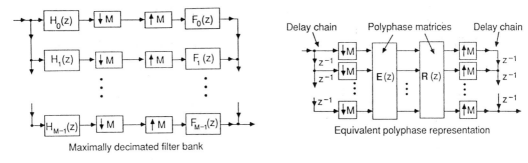

Maximally decimated filter bank

Equivalent polyphase representation

$$\underbrace{\begin{bmatrix} H_0(z) \\ \vdots \\ H_{M-1}(z) \end{bmatrix}}_{\mathbf{h}(z)} = \underbrace{\begin{bmatrix} E_{00}(z^M) & E_{01}(z^M) & \cdots & E_{0,M-1}(z^M) \\ \vdots & \vdots & \ddots & \vdots \\ E_{M-1,0}(z^M) & E_{M-1,1}(z^M) & \cdots & E_{M-1,M-1}(z^M) \end{bmatrix}}_{\mathbf{E}(z^M)} \underbrace{\begin{bmatrix} 1 \\ z^{-1} \\ \vdots \\ z^{-(M-1)} \end{bmatrix}}_{\mathbf{e}(z)}$$

$$\underbrace{[F_0(z) \ \cdots \ F_{M-1}(z)]}_{\mathbf{f}^T(z)} =$$

$$\underbrace{[z^{-(M-1)} \ z^{-(M-2)} \ \cdots \ 1]}_{z^{-(M-1)}\widetilde{\mathbf{e}}(z)} \underbrace{\begin{bmatrix} R_{00}(z^M) & \cdots & R_{0,M-1}(z^M) \\ R_{10}(z^M) & \cdots & R_{1,M-1}(z^M) \\ \vdots & \ddots & \vdots \\ R_{M-1,0}(z^M) & \cdots & R_{M-1,M-1}(z^M) \end{bmatrix}}_{\mathbf{R}(z^M)}$$

AC matrix $\mathbf{H}(z) = \begin{bmatrix} H_0(z) & H_1(z) & \cdots & H_{M-1}(z) \\ H_0(zW) & H_1(zW) & \cdots & H_{M-1}(zW) \\ \vdots & \vdots & \ddots & \vdots \\ H_0(zW^{M-1}) & H_1(zW^{M-1}) & \cdots & H_{M-1}(zW^{M-1}) \end{bmatrix}$

The matrix $\mathbf{F}(z) = \begin{bmatrix} F_0(z) & F_1(z) & \cdots & F_{M-1}(z) \\ F_0(zW) & F_1(zW) & \cdots & F_{M-1}(zW) \\ \vdots & \vdots & \ddots & \vdots \\ F_0(zW^{M-1}) & F_1(zW^{M-1}) & \cdots & F_{M-1}(zW^{M-1}) \end{bmatrix}$

$$\mathbf{H}(z) = \mathbf{W}^\dagger \mathcal{D}(z) \mathbf{E}^T(z^M) \quad \text{(Sec. 5.5)}$$

$$\mathbf{F}(z) = \mathbf{\Gamma} \mathbf{W} \mathbf{\Lambda}(z) \mathbf{R}(z^M) \quad \text{(see Sec. 11.4.3 later)}$$

PROBLEMS

5.1. Suppose the analysis filters in a two-channel QMF bank (Fig. 5.1-1(a)) are given by

$$H_0(z) = 2 + 6z^{-1} + z^{-2} + 5z^{-3} + z^{-5}, \quad H_1(z) = H_0(-z).$$

Find a set of stable synthesis filters that result in perfect reconstruction.

5.2. In Sec. 5.2 we considered QMF banks in which the filters are related as in (5.2.1) and (5.2.2). We saw that with $H_0(z)$ chosen to have real coefficients and linear phase, the distortion function is given by (5.2.10). If N is even this implies $T(e^{j\pi/2}) = 0$ so that the filter order N has to be odd. Now consider the modified QMF bank shown below where the filters could be FIR or IIR [Galand and Nussbaumer, 1984].

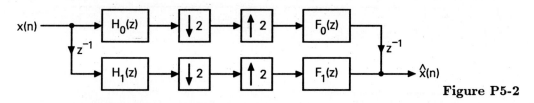

Figure P5-2

Express $\widehat{X}(z)$ in terms of $X(z)$. With $H_1(z) = H_0(-z)$, show that the choice $F_0(z) = H_0(z)$ and $F_1(z) = H_1(z)$ cancels aliasing. With this choice write down the distortion $T(z)$ in terms of $H_0(z)$.

a) Now let $H_0(z)$ be a real coefficient linear phase FIR lowpass filter of order N. Simplify $T(z)$ and show that there is no phase distortion. Also show that N has to be *even*, in order to avoid the condition $T(e^{j\pi/2}) = 0$.

b) For the system in part (a) with N even, what is the number of MPUs required to implement the analysis bank? (Try to exploit as many of the following facts as you can: (i) the relation $H_1(z) = H_0(-z)$, (ii) the linear phase property, and (iii) the presence of decimators). How does this compare with the numbers we obtained for the case of Fig. 5.1-1(a) with odd N?

5.3. Consider Fig. 5.2-2(b). Here the analysis filters are related as $H_1(z) = H_0(-z)$. Assuming that $H_0(z)$ is a real coefficient Nth order filter (N odd), we know that the analysis bank requires $0.5(N+1)$ MPUs. This implementation uses two facts, namely that the coefficients of $H_1(z)$ are related to those of $H_0(z)$, and that we decimate the filter outputs. Curiously enough, we have not used the fact that $H_0(z)$ has linear phase, i.e., $h_0(n) = h_0(N-n)$. At first sight it appears that there should be some way to reduce complexity further by exploiting this relation. This, however, is not true.

a) Prove that linear-phase of $H_0(z)$ implies that $E_0(z)$ is the Hermitian image of $E_1(z)$.

b) In view of Problem 4.17, it appears therefore that we can share the multipliers between $E_0(z)$ and $E_1(z)$. This, however, is not true. To see this, consider the decimated outputs $v_0(n)$ and $v_1(n)$ (Fig. 5.1-1). Show that if the product $x(i)h_0(j)$ is computed in the process of evaluating $v_0(n)$, this same product is never computed in the evaluation of $v_1(m)$ for any choice of n, m (unless the input sequence $x(n)$ is restricted to have special values). (*Note.* Another way to look at this is as follows. The analysis bank requires the implementation of the two systems $E_0(z) + E_1(z)$ and $E_0(z) - E_1(z)$. Each of these resembles Fig. P4-17(b), and can therefore be implemented with $(N+1)/2$ multipliers each. If each of these systems is implemented this way, we cannot share the multipliers in $E_0(z) + E_1(z)$ with those in $E_0(z) - E_1(z)$. This is because the multiplier α cannot be shared between $(x+y)\alpha$ and $(x-y)\alpha$.)

c) What is the story if N is even? Can we exploit linear phase property of $H_0(z)$ to implement the analysis bank with only about $0.25(N+1)$ MPUs? Explain.

5.4. For the case of an odd order power symmetric BR transfer function we presented a result which showed that it can be expressed as in (5.2.16) so that the polyphase components are allpass (Theorem 5.3.1). For the even order case, the situation is somewhat different. Suppose for example that $H_0(z)$ is an IIR elliptic lowpass filter with even order $N > 0$. It can be shown [Vaidyanathan, et al., 1987] that this can be expressed as

$$H_0(z) = \frac{A(z) + A_*(z)}{2} \qquad (P5.4a)$$

where $A(z)$ is a unit-magnitude allpass function of order $N/2$ with complex coefficients, and $A_*(z)$ is obtained by conjugating the coefficients of $A(z)$. (You can accept this as a fact for this Problem). Suppose now that $H_0(z)$ is, in addition, power symmetric. Define the new real coefficient transfer function

$$H_1(z) = \frac{A(z) - A_*(z)}{2j} \qquad (P5.4b)$$

a) Show that $\widetilde{H}_0(z)H_0(z) + \widetilde{H}_1(z)H_1(z) = 1$.
b) Show that $H_1(z) = \widehat{c}H_0(-z)$ where $\widehat{c} = \pm 1$.
c) Let $E_0(z)$ and $E_1(z)$ be the polyphase components of $H_0(z)$, i.e., $H_0(z) = E_0(z^2) + z^{-1}E_1(z^2)$. Show that $z^{-i}E_i(z^2) = c_i A(z) + c_i^* A_*(z)$ for some c_i.
d) Hence show that $E_i(z)$ cannot be allpass (*Hint.* Use Problem 3.21.)

The application of this problem in QMF banks is considered in Problem 5.6.

5.5. Let $H_0(z)$ be as in Problem 5.4.
  a) Show that $A_*(z) = \pm jA(-z)$.
  b) Show that $A(z)$ has the form

$$A(z) = c \prod_{k=1}^{M} \left(\frac{-j\beta_k + z^{-1}}{1 + j\beta_k z^{-1}} \right) \qquad (P5.5)$$

where $c = e^{\pm j\pi/4}$ and $-1 < \beta_k < 1$.

5.6. Consider the system shown in Fig. P5-2 again. Suppose that the analysis filters are related as $H_1(z) = H_0(-z)$ and let the synthesis filters be chosen as in Problem 5.2 to eliminate aliasing. Assume, however, that $H_0(z)$ is an elliptic power symmetric filter with even order $N > 0$. We know from Problem 5.4 that the polyphase components are not allpass.

  a) Find the distortion function $T(z)$ and show that amplitude distortion has actually been eliminated.
  b) Draw the complete QMF bank in terms of $A(z)$. In your scheme, what is the number of MPUs required to implement the analysis bank? (*Note*. When the multipliers are complex, you must carefully count the number of real MPUs).

5.7. Let $H_0(z)$ be as in (5.2.16) where $a_0(z)$ and $a_1(z)$ are allpass filters as in (5.3.6). We can then write $H_0(z) = P_0(z)/d_0(z^2)d_1(z^2)$ where $P_0(z)$ is a polynomial in z^{-1}. Assume that (i) $d_0(z)$ and $d_1(z)$ have all poles inside the unit circle, and (ii) $d_0(z)$ and $d_1(z)$ have no common factors of order ≥ 1. Show that there are no common factors (of order ≥ 1) between $P_0(z)$ and the denominator $d_0(z^2)d_1(z^2)$.

5.8. Let $H_0(z)$ be a digital filter obtained from an analog Butterworth filter $H_a(s)$ (Sec. 3.3.2), using the bilinear transform (3.3.1). Suppose the Butterworth filter has 3dB point $\Omega_c = 1$. Show then that $H_0(z)$ is power symmetric.

5.9. Consider a QMF bank with analysis filters related by $H_1(z) = H_0(-z)$ so that (5.2.5) holds. $H_0(z)$ and $H_1(z)$ could be FIR or IIR, but assume that they are stable.

  a) Assume that the polyphase components $E_0(z)$ and $E_1(z)$ have all zeros *outside* the unit circle (the poles, of course, are inside). Find a set of stable synthesis filters so that aliasing as well as amplitude distortion are eliminated.
  b) Repeat (a) under the condition that $E_0(z)$ and $E_1(z)$ have some zeros inside and some outside (but none *on*) the unit circle.

5.10. Let $H_0(z) = P_0(z)/D(z)$ and $H_1(z) = P_1(z)/D(z)$ with

$$P_k(z) = \sum_{n=0}^{N} p_k(n)z^{-n}, \quad D(z) = 1 + \sum_{n=1}^{N} d(n)z^{-n}. \qquad (P5.10)$$

Assume $D(z)$ has all zeros inside the unit circle. Suppose the following conditions are true: (i) $|H_0(e^{j\omega})|^2 + |H_1(e^{j\omega})|^2 = 1$ for all ω. (ii) $P_0(z)$ is Hermitian and $P_1(z)$ is skew Hermitian. Show that we can express these transfer functions as $H_0(z) = [A_0(z) + A_1(z)]/2$ and $H_1(z) = [A_0(z) - A_1(z)]/2$ where $A_0(z)$ and $A_1(z)$ are stable unit-magnitude allpass.

5.11. Let $H_0(z)$ be a causal stable rational transfer function with $|H_0(e^{j\omega})| \leq 1$, with irreducible representation $H_0(z) = P_0(z)/D_0(z)$. Assume that if α is a zero of $D_0(z)$ then $1/\alpha^*$ cannot be a zero of $P_0(z)$. This means that there are no nontrivial allpass factors in $H_0(z)$.

  a) Let $H_1(z)$ be a causal stable system such that $|H_1(e^{j\omega})|^2 + |H_0(e^{j\omega})|^2 = 1$, and let $H_1(z) = P_1(z)/D_1(z)$ be an irreducible representation. Assume that $H_1(z)$ has no nontrivial allpass factors. Show that $D_1(z) = cD_0(z)$ for some constant c.
  b) Assume now that $H_0(z)$ above is power symmetric. Show that its denominator can be written in the form $D_0(z) = 1 + d(2)z^{-2} + \ldots d(2K)z^{-2K}$. That is, $D_0(z) = G(z^2)$ for some FIR $G(z)$.

5.12. Let $H_0(z)$ be a stable IIR power symmetric transfer function with possibly complex coefficients. Let $H_0(z) = P_0(z)/D(z)$ be an irreducible representation, where

$$P_0(z) = \sum_{n=0}^{N} p_0(n) z^{-n}, \quad D(z) = 1 + \sum_{n=1}^{N} d(n) z^{-n}, \qquad (P5.12a)$$

with none of $d(N), p_0(0), p_0(N)$ equal to zero. Assume that there are no allpass factors (of order > 0) in $H_0(z)$. Suppose N is odd and $P_0(z)$ Hermitian. Prove that $H_0(z)$ can be expressed as

$$H_0(z) = \frac{a_0(z^2) + z^{-1} a_1(z^2)}{2} \qquad (P5.12b)$$

where $a_0(z)$ and $a_1(z)$ are stable (rational) unit-magnitude allpass functions. Thus, the polyphase components are allpass. (*Hint.* use Problem 5.10.)

5.13. *Analog QMF bank.* In this problem we consider an extension of the maximally decimated filter bank system for the case where the input is a continuous-time signal. Consider the following 'two-channel filter bank' system, where $x_a(t)$ is a continuous-time signal with Laplace transform $X_a(s)$.

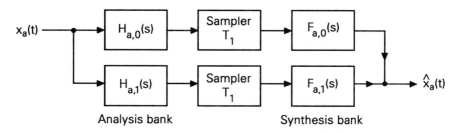

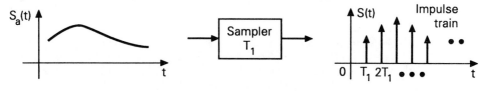

Figure P5-13(a),(b)

The device labeled "sampler" operates as follows: in response to a continuous time input $s_a(t)$, it produces the sampled version

$$s(t) = \sum_{n=-\infty}^{\infty} s_a(nT_1) \delta_a(t - nT_1),$$

where $\delta_a(.)$ is the Dirac delta function (Sec. 2.3). This is illustrated in Fig. P5-13(b).

a) Express $\widehat{X}_a(j\Omega)$ in terms of $X_a(j\Omega)$, and the various filter transfer functions. You will see that this expression is a linear combination of an infinite number of terms of the form

$$X_a(j\Omega - j\frac{2\pi m}{T_1}), \quad -\infty \leq m \leq \infty.$$

b) Suppose $x_a(t)$ is a σ-bandlimited signal, that is, $X_a(j\Omega) = 0$ for $|\Omega| \geq \sigma$. (Assume $0 < \sigma < \pi$.) We know that the Nyquist sampling rate is $\Theta = 2\sigma$. (If we sample $x_a(t)$ at this rate, we can recover $x_a(t)$ from the samples with the help of a ideal lowpass filter with passband $-\sigma < \Omega < \sigma$.) Define the corresponding Nyquist sampling period $T = 2\pi/\Theta$. Suppose $T_1 = 2T$, that is, each channel in the figure performs sampling at *half* the Nyquist rate (so that the total number of samples per unit time, counting both channels, corresponds to Nyquist sampling). Show that, for any fixed frequency Ω, only *two* out of the infinite number of terms in the expression for $\widehat{X}_a(j\Omega)$ can be nonzero.

c) The aim is to choose the synthesis filters such that aliasing and other distortions are eliminated. Continuing with part (b), assume that the synthesis filters satisfy $F_{a,k}(j\Omega) = 0$ for $|\Omega| \geq \sigma$. Show that we can obtain perfect reconstruction (that is, $\widehat{x}_a(t) = x_a(t)$) by solving for $F_{a,k}(j\Omega)$ from the equations

$$\begin{bmatrix} H_{a,0}(j\Omega) & H_{a,1}(j\Omega) \\ H_{a,0}(j\Omega + j\sigma) & H_{a,1}(j\Omega + j\sigma) \end{bmatrix} \begin{bmatrix} F_{a,0}(j\Omega) \\ F_{a,1}(j\Omega) \end{bmatrix} = \begin{bmatrix} T_1 \\ 0 \end{bmatrix}, \quad (P5.13a)$$

for $-\sigma < \Omega \leq 0$, and

$$\begin{bmatrix} H_{a,0}(j\Omega) & H_{a,1}(j\Omega) \\ H_{a,0}(j\Omega - j\sigma) & H_{a,1}(j\Omega - j\sigma) \end{bmatrix} \begin{bmatrix} F_{a,0}(j\Omega) \\ F_{a,1}(j\Omega) \end{bmatrix} = \begin{bmatrix} T_1 \\ 0 \end{bmatrix}, \quad (P5.13b)$$

for $0 \leq \Omega < \sigma$. In other words, given the analysis filters $H_{a,0}(s)$ and $H_{a,1}(s)$, we can solve for the frequency responses of the synthesis filters from the above equations. The sets of equations to be used depends on the frequency region as indicated. (Outside this frequency region we just take $F_{a,k}(j\Omega) = 0$.) (*Note.* This idea works as long as the 2×2 matrices in the equations above are nonsingular, but the resulting synthesis filters, in general, are not guaranteed to be stable or realizable!)

d) Continuing with part (c), let $H_{a,0}(s) = 1$, and let $H_{a,1}(s) = s$ (i.e., a differentiator). Verify that the matrices above are nonsingular. So we can indeed find synthesis filters for perfect reconstruction. Find expressions for $F_{a,0}(j\Omega)$ and $F_{a,1}(j\Omega)$ in the above frequency regions. Show that these synthesis filters have impulse responses

$$f_{a,0}(t) = 4\sin^2(\sigma t/2)/\sigma^2 t^2, \quad f_{a,1}(t) = 4\sin^2(\sigma t/2)/\sigma^2 t. \quad (P5.13c)$$

Evidently these are noncausal (unrealizable) continuous-time filters.

Note: The above scheme gives rise to a number of generalizations to Nyquist sampling theorem. If $x_a(t)$ is σ-bandlimited, Nyquist theorem says that we can reconstruct it (by lowpass filtering) from its samples uniformly spaced apart by T seconds. According to above scheme, we can split $x_a(t)$ into two signals and sample each at *half the rate*, and still reconstruct the original version. In part (d) we are essentially sampling $x_a(t)$ and its derivative (output of $H_{a,1}(s)$) at half the Nyquist rate. We can recover $x_a(t)$ from these two undersampled signals by using the filters in (P5.13c). This gives a proof of the *derivative sampling theorem,* originally proposed in [Shannon, 1949] four decades ago. More generally, if we consider the M-channel version of this problem we will find that we can recover a bandlimited signal by sampling it and its $M - 1$ derivatives M times slower than the Nyquist rate. As part (d) shows, the filters required to do this reconstruction are, as such, unrealizable. (In fact the ideal lowpass filter which is used to reconstruct a bandlimited signal from its 'traditional Nyquist-rate samples' is also unstable and noncausal.) These filters should therefore be replaced with practical approximations.

5.14. Given below are three sets of FIR analysis banks for a 3 channel maximally decimated QMF bank (Fig. 5.4-1, with $M = 3$). In each case, answer the following: (i) Is it possible to obtain a set of FIR synthesis filters for perfect reconstruction? If so find them. (ii) If not, find a set of IIR synthesis filters for perfect reconstruction. (iii) In the latter event, are the synthesis filters stable?

a) $H_0(z) = 1$, $H_1(z) = 2 + z^{-1}$, $H_2(z) = 3 + 2z^{-1} + z^{-2}$.
b) $H_0(z) = 1$, $H_1(z) = 2 + z^{-1} + z^{-5}$, $H_2(z) = 3 + 2z^{-1} + z^{-2}$.
c) $H_0(z) = 1$, $H_1(z) = 2 + z^{-1} + z^{-5}$, $H_2(z) = 3 + z^{-1} + 2z^{-2}$.

5.15. Prove that the structure of Fig. 5.6-3 has perfect reconstruction property if and only if the integers M and J are relatively prime. Under this condition, find $\widehat{x}(n)$ in terms of $x(n), M,$ and J.

5.16. Consider the $M \times M$ matrix

$$\begin{bmatrix} 0 & \mathbf{I}_{M-r} \\ z^{-1}\mathbf{I}_r & 0 \end{bmatrix}. \qquad (P5.16)$$

Show that its determinant is of the form $\pm z^{-r}$. This shows that if the product $\mathbf{P}(z) = \mathbf{R}(z)\mathbf{E}(z)$ takes the form (5.6.7), then its determinant is a delay, that is, has the form (5.6.9).

5.17. Suppose the filter bank of Fig. 5.4-1 is alias-free, and let $T(z)$ be the distortion function. Suppose we define a new filter bank in which the analysis and synthesis filters are interchanged, that is, $F_k(z)$ are the analysis filters and $H_k(z)$ the synthesis filters. Show that the resulting system is free from aliasing and has the same distortion function $T(z)$. So we can swap each $F_k(z)$ with corresponding $H_k(z)$, without changing these input/output properties! (*Hint.* Use AC matrix formulation cleverly.)

5.18. Consider the M channel maximally decimated system of Fig. 5.4-1. Let the choice of filters be such that this is a perfect reconstruction system. Suppose we replace each synthesis filters $F_k(z)$ with $F_k(zW^\ell)$, where $W = e^{-j2\pi/M}$, and ℓ is an integer (independent of k) with $0 \le \ell \le M - 1$. Let $\widehat{x}_1(n)$ be the new output of the QMF bank. How is it related to the input $x(n)$? Given $\widehat{x}_1(n)$, would you be able to recover $x(n)$? If so, how?

5.19. Consider the two channel QMF bank with analysis filters related as $H_1(z) = H_0(-z)$. Suppose the synthesis filters are chosen as $F_0(z) = H_0(z)$ and $F_1(z) = -H_1(z)$, so that aliasing is canceled.
   a) Write down the AC matrix $\mathbf{H}(z)$, and express its determinant in terms of $H_0(z)$.
   b) Show that the distortion function $T(z)$ is zero for some z if and only if $\mathbf{H}(z)$ is singular (i.e., the determinant equals zero) for this value of z.
   c) Suppose $H_0(z) = \sum_{n=0}^{N} h(n) z^{-n}$. Let N be even and let $h(n)$ be real with $h(n) = h(N-n)$ (Type 1 linear phase). Show that $\mathbf{H}(z)$ is singular for $z = e^{j\pi/2}$. In view of part (b) this proves that $T(e^{j\pi/2}) = 0$, a conclusion we already know from Sec. 5.2.2.

5.20. Consider the following uniform DFT analysis bank,

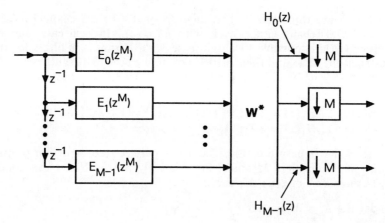

Figure P5-20

where $E_k(z)$ are stable allpass functions with $|E_k(e^{j\omega})| = 1$ for all ω. (Evidently, these are the polyphase components of $H_0(z)$.)
   a) Show that the anlaysis filters are power complementary.
   b) Show that each analysis filter is a spectral factor of an Mth band filter, that is, show that $\widetilde{H}_k(z) H_k(z)$ satisfies the Mth band property.
   c) Draw a (stable) synthesis bank structure so that (i) aliasing is canceled, and (ii) $T(z)$ becomes allpass thereby eliminating amplitude distortion.

5.21. Consider Fig. 5.7-2(c) with $\mathbf{T} = \mathbf{W}^*$ (uniform DFT analysis bank). Suppose $R_k(z)$ are chosen as in (5.7.8), so that the product $R_k(z) E_k(z)$ is independent of k. This ensures that aliasing has been canceled.
   a) Just as a review, verify that the uniform-shift relations $H_k(z) = H_0(zW^k)$ and $F_k(z) = W^{-k} F_0(zW^k)$ hold, where $W = e^{-j2\pi/M}$.
   b) Express the distortion function $T(z)$ in terms of $E_k(z)$, $0 \le k \le M-1$.
   c) Show that the AC matrix $\mathbf{H}(z)$ is a left circulant.
   d) Find the determinant of $\mathbf{H}(z)$ in terms of $E_k(z)$. (Review of Sec. 5.5 helps here.) Show that this determinant is equal to $cz^{-K} T(z)$ where $c \ne 0$, and K is some integer. Thus $\mathbf{H}(e^{j\theta})$ is singular if and only if $T(e^{j\theta}) = 0$.
   e) Suppose $H_0(z) = \sum_{n=0}^{N} h(n) z^{-n}$. Assume $h(n)$ is real with $h(n) = h(N-n)$ (linear phase FIR). This property imposes certain constraints on the

polyphase components $E_k(z)$ (Problem 4.22). In particular, for some combinations of N and M, it is possible that some polyphase component $E_L(z)$ is an *odd* order filter with *symmetric* impulse response. This implies $E_L(e^{j\pi}) = 0$. Using part (b) prove that this implies $T(e^{j\omega}) = 0$ for $\omega = \pi/M, 3\pi/M, \ldots$ etc. To avoid this situation, the relative values of M and N must be carefully chosen. Explain how. (You can do it using the notation 'm_0' from Problem 4.22) (*Hint.* For $M = 2$, this should reduce to the requirement that N be odd, as seen in Sec. 5.2.2.)

5.22. In the above problem we took $R_k(z)$ according to (5.7.8). This has a disadvantage: each filter $R_k(z)$, which is a product of $M - 1$ of the $E_\ell(z)$'s, can have very high order (for large M), so that the synthesis filters have high order. We can partially rectify this situation if we take a closer look at the form of $E_k(z)$. Thus, let

$$E_k(z) = \frac{N_{k,1}(z) N_{k,2}(z)}{D_k(z)}. \quad (P5.22a)$$

where $N_{k,1}(z), N_{k,2}(z)$ and $D_k(z)$ are polynomials in z^{-1}. Here $N_{k,2}(z)$ is the part with all zeros inside the unit circle, (and $N_{k,1}(z)$ has zeros on and/or outside).

a) Show that the choice

$$R_k(z) = \frac{D_k(z) \prod_{\ell \neq k} N_{\ell,1}(z)}{N_{k,2}(z)}. \quad (P5.22b)$$

cancels aliasing. (Note that this choice gives stable synthesis filters.)

b) With such choice of $R_k(z)$, what is the distortion function $T(z)$?

c) Making the further assumption that $N_{k,1}(z)$ has no zeros *on* the unit circle for any k, how would you modify $R_k(z)$ [without destroying stability of $R_k(z)$, and the alias-free property] so that $T(z)$ now becomes allpass (thereby eliminating amplitude distortion)?

5.23. Consider the following M channel multirate system, which is essentially a QMF bank with the additional transfer functions $C_k(z)$ inserted.

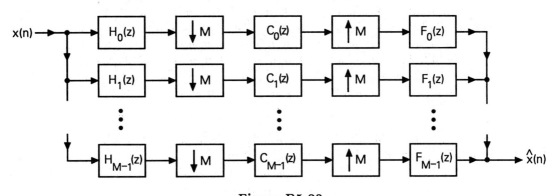

Figure P5-23

We can imagine that $C_k(z)$ represents the amplitude and phase distortions introduced by the kth channel. Assume throughout that the functions $F_k(z)$,

$H_k(z)$ and $C_k(z)$ are rational and stable; unless stated otherwise, do not make specific assumptions about *zeros* of these transfer functions.

a) Suppose $H_k(z)$ and $F_k(z)$ are such that the system is alias-free in absence of channel distortion $C_k(z)$ (i.e., with $C_k(z)$ replaced with unity for all k). Now with $C_k(z)$ present, find a modified set of (stable) synthesis filters $G_k(z)$ to retain alias-free property.

b) Repeat part (a) by replacing "alias-free" with "alias-free and free from amplitude distortion" everywhere. Assume, however, that $C_k(z)$ has no zeros on the unit circle. (*Hint.* First write the numerator of $C_k(z)$ as $A_k(z)B_k(z)$ where $A_k(z)$ has all zeros inside the unit circle and $B_k(z)$ has them outside.)

c) Repeat part (a) by replacing "alias-free" with "perfect reconstruction" everywhere. Assume now that $C_k(z)$ has no zeros on or outside the unit circle.

Hint. This is not tedious, but you have to think straight!

5.24. Consider the tree structure shown in Fig. 5.8-1(a).

a) Show that the complete system is equivalent to Fig. 5.4-1 with $M = 4$ (four-band QMF bank) and identify $H_k(z)$ and $F_k(z)$ for $0 \le k \le 3$, in terms of the filters in Figure 5.8-1(a).

b) Assume that the two-channel QMF bank with filters $H_0^{(1)}(z)$, $H_1^{(1)}(z)$, $F_0^{(1)}(z)$, and $F_1^{(1)}(z)$ is alias-free with distortion function $T^{(1)}(z)$. Let the same be true of $H_0^{(2)}(z)$, $H_1^{(2)}(z)$, $F_0^{(2)}(z)$, and $F_1^{(2)}(z)$, with distortion function $T^{(2)}(z)$. Prove that the equivalent four-band QMF bank is alias-free, and find its distortion function $T(z)$ in terms of $T^{(1)}(z)$ and $T^{(2)}(z)$.

c) Continuing with part (b), prove that $T(z)$ is allpass if $T^{(1)}(z)$ and $T^{(2)}(z)$ are allpass. Thus, if each two-channel QMF bank is free from amplitude distortion, then so is the overall four-channel system. Similarly verify that $T(z)$ has linear phase if $T^{(1)}(z)$ and $T^{(2)}(z)$ have linear phase.

d) If each two-channel QMF bank

$$H_0^{(k)}(z),\ H_1^{(k)}(z),\ F_0^{(k)}(z),\ F_1^{(k)}(z),\quad k = 1, 2,$$

is a perfect reconstruction system, verify that the same is true for the equivalent four-channel system.

Note. These results can be extended to tree structures with more than two (say m) levels. We can thus build QMF banks with $M = 2^m$, with any set of desired properties (such as freedom from selected set of distortions etc.), including perfect reconstruction. For composite M which is not a power of two, the idea can be extended. Thus if $M = 3 \times 2$, we can build the QMF bank in terms of two-channel systems and three channel systems. So tree structures cover a wide class of useful filter banks.

5.25. Tree structures can be used to obtain QMF banks in which the decimation ratio is not the same for all channels (called nonuniform filter banks). Consider the system shown in Fig. P5-25(a).

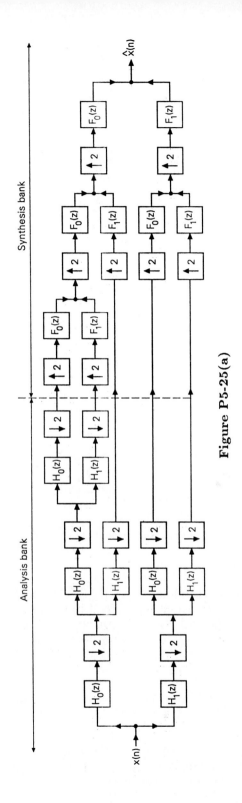

Figure P5-25(a)

This is equivalent to the following five channel system.

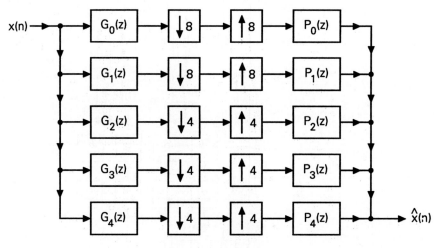

Figure P5-25(b)

a) Identify the filters $G_k(z)$ and $P_k(z)$ in terms of the filters $H_0(z)$, $H_1(z)$, $F_0(z)$ and $F_1(z)$. Suppose $H_0(z)$ and $H_1(z)$ are real coefficient filters with magnitude responses as shown below.

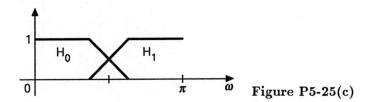

Figure P5-25(c)

Sketch the magnitude responses $|G_k(e^{j\omega})|$ for $0 \leq k \leq 4$. Thus, the filters have unequal bandwidths, and the decimation ratios are inversely proportional to these bandwidths. This is called a nonuniform (maximally decimated) QMF bank.

b) Suppose the filters $H_0(z), H_1(z), F_0(z)$ and $F_1(z)$ are such that the traditional two channel QMF bank (Fig. 5.1-1(a)), has perfect reconstruction property, with distortion function $T(z) = 1$. Show that the five-channel nonuniform system also has perfect reconstruction property.

c) Suppose the filters $H_0(z), H_1(z), F_0(z)$ and $F_1(z)$ are such that the traditional two channel QMF bank is alias-free, with distortion function $T(z)$. Does the above five channel nonuniform system remain alias-free? If not, how would you modify the structure of Fig. P5-25(a) to obtain this property, and what is the resulting distortion function?

5.26. Consider a transmultiplexer with $M = 2$.
  a) Let the analysis filters be $H_0(z) = 1 + z^{-1}$ and $H_1(z) = 1 - z^{-1}$. Find a set of FIR synthesis filters $F_0(z)$ and $F_1(z)$ such that the system has perfect reconstruction property.

b) Let the analysis filters be $H_0(z) = 1 + z^{-1}$ and $H_1(z) = 1 + z^{-1} + z^{-2}$. Can we find FIR synthesis filters such that there is perfect reconstruction? If so find them.

5.27. Consider a three-channel transmultiplexer with *synthesis* filters

$$F_0(z) = 1, \quad F_1(z) = 2 + z^{-1}, \quad F_2(z) = 3 + 2z^{-1} + z^{-2}.$$

Find a set of FIR analysis filters such that perfect reconstruction property is satisfied.

5.28. Suppose we have a three channel alias-free QMF bank with distortion function $T(z) = z^{-2}/(1 - az^{-1})$. Find closed form expressions for the elements of the 3×3 matrix $\mathbf{P}(z)$ [Fig. 5.5-3(c)] in terms of a, z.

5.29. Assuming that the matrix $\mathbf{P}(z)$ is pseudocirculant, verify that (5.7.22) indeed reduces to (5.7.13) (with $P_m(z)$ denoting $P_{0,m}(z)$).

5.30. Consider the following multirate system.

Figure P5-30

In each of the following cases, what can you say about the input output relation of the system? Give as much information as possible, based on given data.
 a) $M = 2$ and $H(z)$ is an IIR power symmetric elliptic filter of odd order.
 b) M is arbitrary, and $H(z)$ is a zero-phase Mth band lowpass filter.

5.31. Consider the following M-channel analysis/synthesis system.

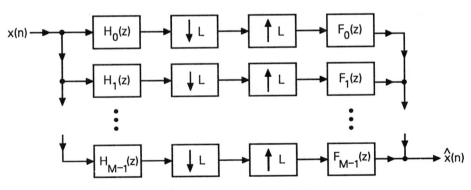

Figure P5-31(a)

This reduces to the QMF bank of Fig. 5.4-1 if $L = M$. If $L < M$, this is called a *nonmaximally* decimated QMF bank. With such a system, elimination of aliasing turns out to be relatively easy (as this exercise will demonstrate).
 a) Find an expression for $\widehat{X}(z)$.

b) Suppose $M = 4$ and $L = 3$. Suppose the analysis bank is the uniform DFT bank, i.e., $H_k(z) = H_0(zW_4^k), 0 \leq k \leq 3$. Assume that $H_0(z)$ has response shown below.

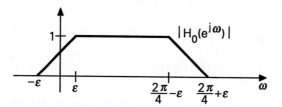

Figure P5-31(b)

How small would ϵ have to be so that $H_0(z)$ does not overlap with the aliased versions $H_0(zW_3^n), n = 1, 2$? With such ϵ, show typical responses of $F_k(z), 0 \leq k \leq 3$ such that aliasing terms are eliminated. (The trivial choice $F_k(z) = 0$ is forbidden, of course!). What is the distortion transfer function after such elimimation of aliasing?

Note. In nonmaximally decimated systems, the total number of samples per unit time at the output of the decimated analysis bank is evidently more than for $x(n)$. This is the price paid to obtain the simplicity of alias elimination.

5.32. Consider the following system which is a general M channel nonuniform filter bank.

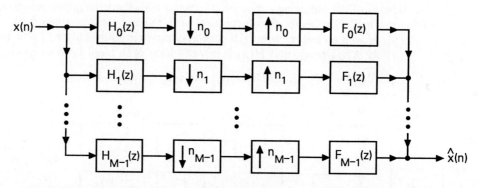

Figure P5-32

If the integers n_k are such that

$$\sum_{k=0}^{M-1} \frac{1}{n_k} = 1, \qquad (P5.32a)$$

the system is said to be maximally decimated. (Note that this condition holds for Fig. 5.4-1, where $n_k = M$ for all k. Also the tree structured system in Problem 5.25 is a special case of this nonuniform filter bank, with decimation ratios $8, 8, 4, 4, 4$.) The kth analysis filter $H_k(z)$ has total passband width $\approx 2\pi/n_k$, so that it makes sense to decimate its output by n_k. This system suffers from the usual set of errors (aliasing, amplitude distortion, and phase distortion) as does

Fig. 5.4-1. Given an arbitrary set $\{n_i\}$ of M integers n_k satisfying (P5.32a), it is not in general possible to find nonideal filters $H_k(z)$ and $F_k(z)$ to eliminate aliasing completely. The alias terms at the output of the kth expander are

$$X(zW_{n_k}^n), \quad 1 \leq n \leq n_k - 1. \qquad (P5.32b)$$

(W_m stands for $e^{-j2\pi/m}$.) So, unlike in Fig. 5.4-1, the "shifted versions" created by different channels have different amounts of shift. For example, the 0th channel generates

$$X(zW_{n_0}), X(zW_{n_0}^2) \ldots \qquad (P5.32c)$$

Unless each of these is also generated by at least one other channel, we cannot "cancel" all alias terms with nonideal filters.

If a set of integers $\{n_i\}$ is such that every shifted copy of $X(e^{j\omega})$ appears at the output of at least two expanders, we say that $\{n_i\}$ is a *compatible set*. Compatibility of $\{n_i\}$ is thus a *necessary* condition for complete alias cancelation in Fig. P5-32. If the filter bank is derived from a tree structure, (i.e., by starting from a system similar to Fig. P5-25(a) where uniform filter banks, not necessarily two-channel each, are used repeatedly), then the compatibility property is satisfied automatically because we know the system can be designed to be alias free (Problems 5.24, 5.25).

Which of the following sets are compatible?

a) $(2, 3, 6)$

b) $(2, 6, 6, 6)$

c) For large sets of integers [e.g., part e) below], it is tedious to directly check compatibility. Devise an efficient test for compatibility of a given set $\{n_i\}$.

d) Using the test developed above, show that the set $(2, 6, 10, 12, 12, 30, 30)$ is compatible.

e) Show that the set in part d) cannot be derived from a tree structure. Thus, there exist compatible sets which are not derived from tree structures.

5.33. Consider Fig. 5.4-1. Suppose the filters are chosen such that the system has the perfect reconstruction (PR) property. Now suppose that we replace each of the analysis and synthesis filters with $H_k(z^2)$ and $F_k(z^2)$, for $0 \leq k \leq M - 1$. Does the resulting system still have the PR property?

5.34. In Problem 5.33, suppose we replace $H_k(z)$ and $F_k(z)$ with $H_k(z^L)$ and $F_k(z^L)$ for some integer $L > 0$. (So with $L = 2$, we obtain Problem 5.33). Find a necessary and sufficient condition on L such that the resulting system continues to have the PR property.

6
Paraunitary Perfect Reconstruction (PR) Filter Banks

6.0 INTRODUCTION

From Sec. 5.5 we know that the analysis and synthesis filters of the M channel maximally decimated filter bank can be expressed in terms of the polyphase matrices $\mathbf{E}(z)$ and $\mathbf{R}(z)$. If the filters are FIR and the filter bank has the perfect reconstruction (PR) property, then the polyphase matrix $\mathbf{E}(z)$ has to satisfy the property (5.6.10). That is, the determinant of $\mathbf{E}(z)$ must be a delay. This ensures that its inverse is FIR so that we can find FIR $\mathbf{R}(z)$ satisfying the PR requirement.

In this chapter we study PR filter banks in which $\mathbf{E}(z)$ satisfies a special property called the *lossless* or *paraunitary* property. This property automatically ensures (5.6.10) [even though paraunitariness is *not* a necessary condition for (5.6.10)]. In addition to perfect reconstruction, the FIR filter bank based on paraunitary $\mathbf{E}(z)$ satisfies many other useful properties. These are summarized at the end of Sec. 6.7.1. Here is a short preview of some of the benefits.

1. The synthesis filter $F_k(z)$ has the same length as the analysis filter $H_k(z)$.
2. $F_k(z)$ can be found from $H_k(z)$ by inspection.
3. There exist good design techniques with fast convergence.
4. The paraunitary property is basic to the design of cosine modulated perfect reconstruction systems, described in Sec. 8.5. We will see that these systems combine the perfect reconstruction property with very low design as well as implementation complexity.
5. The paraunitary property is also basic to the generation of the so called 'orthonormal wavelet basis' to be studied in Chap. 11.

Brief historical notes

The ideas of passivity, paraunitariness, and losslessness originate from classical electrical network theory [Brune, 1931], [Darlington, 1939], [Guillemin, 1957], [Potapov, 1960], [Belevitch, 1968], [Balabanian and Bickart, 1969], and [Anderson and Vongpanitlerd, 1973]. These have been applied for the design of robust digital filter structures [Fettweis, 1971], [Swamy and Thyagarajan, 1975], [Bruton and Vaughan-Pope, 1976], [Antoniou, 1979], [Deprettere and Dewilde, 1980], [Rao and Kailath, 1984], [Vaidyanathan and Mitra, 1984], and [Vaidyanathan, 1985a,b]. Also see Sec. 14.1.

Paraunitary transfer matrices were applied to the design of perfect reconstruction systems in Vaidyanathan [1987a,b]. It turns out that the two channel power symmetric PR QMF bank (Sec. 5.3.6) has paraunitary $E(z)$ (Sec. 6.3.2), even though the authors [Smith and Barnwell, 1984] and [Mintzer,1985] used a different approach in their derivation. It has been shown [Vetterli and Le Gall, 1989] that some of the earlier filter bank designs [Princen and Bradley, 1986] also have the paraunitary property. The lapped orthogonal transform (LOT) [Cassereau, 1985], [Malvar and Staelin, 1989], has been shown later to have the paraunitary property (see Sec. 6.6). Paraunitary matrices have also been considered in the context of multidimensional multirate filter banks [Karlsson and Vetterli, 1990] (see Chap. 12). Subsequently, paraunitary systems have been used in the design of cosine modulated filter banks, which offer great simplicity of design as well as implementation [Malvar, 1990b], [Koilpillai and Vaidyanathan, 1991a], [Ramstad, 1991].

We will see that the paraunitary condition is a very natural choice. For example, if the anaysis filters have ideal brick-wall responses, then $E(z)$ is paraunitary [see comments after eqn. (6.2.13) later]. Second, some of the approximate reconstruction designs (the pseudo QMF design, Chap. 8), developed prior to the introduction of paraunitary filter banks, are such that $E(z)$ is "approximately" paraunitary.

Outline

The presentation in this chapter is in terms of discrete-time language, and will not require the electrical network theoretic background mentioned above. In Sec. 6.1 we introduce the lossless and paraunitary properties. Section 6.2 studies the properties of filter banks with paraunitary $E(z)$. In Sec. 6.3 and 6.4 the two channel case is studied in depth. We present design techniques as well as robust lattice structures for FIR PR QMF banks with paraunitary $E(z)$. These results are extended to the M channel case in Sec. 6.5. In Sec. 6.6 we introduce transform coding and the lapped orthogonal transform (LOT). Section 6.7 provides a summary and comparison of the many design techniques introduced in this and the previous chapters.

In Sec. 8.5 we will return to the study of cosine modulated paraunitary filter banks. A detailed study of paraunitary systems is presented in Chap. 14. As in Chap. 5, we will sometimes use the term 'QMF' even for the M

channel case. This is for simplicity, and is a misnomer unless $M = 2$.

6.1 LOSSLESS TRANSFER MATRICES

In Sec. 2.2 we introduced r-input p-output linear time invariant (LTI) systems. These are characterized by $p \times r$ 'transfer matrices' $\mathbf{H}(z)$. Such matrices were used in Chap. 5 to characterize filter banks. For example, the analysis bank was described by an $M \times 1$ transfer matrix $\mathbf{h}(z)$, and synthesis bank by a $1 \times M$ matrix $\mathbf{f}^T(z)$ [see Eq. (5.4.1)]. These two vectors were expressed in terms of the polyphase matrices $\mathbf{E}(z)$ and $\mathbf{R}(z)$ according to

$$\mathbf{h}(z) = \mathbf{E}(z^M)\mathbf{e}(z), \quad \mathbf{f}^T(z) = z^{-(M-1)}\widetilde{\mathbf{e}}(z)\mathbf{R}(z^M). \qquad (6.1.1)$$

The matrices $\mathbf{E}(z)$ and $\mathbf{R}(z)$ were defined in Sec. 5.5.

In this chapter we will impose the "paraunitary" or "lossless" property on $\mathbf{E}(z)$, and thereby obtain the perfect reconstruction property. Towards this end, we give a brief review of lossless systems, which will serve the purpose of this chapter.† For the interested reader, a detailed treatment can be found in Chap. 14.

6.1.1 Definition and properties

A $p \times r$ causal transfer matrix $\mathbf{H}(z)$ is said to be lossless if (a) each entry $H_{km}(z)$ is stable and (b) $\mathbf{H}(e^{j\omega})$ is unitary, that is,

$$\mathbf{H}^\dagger(e^{j\omega})\mathbf{H}(e^{j\omega}) = d\mathbf{I}_r, \quad \text{for all } \omega, \qquad (6.1.2)$$

for some $d > 0$. If in addition the coefficients of $\mathbf{H}(z)$ are real (i.e., $\mathbf{H}(z)$ real for real z), we say that $\mathbf{H}(z)$ is *lossless bounded real* (abbreviated LBR). Note that the phrase "$\mathbf{H}(z)$ is lossless" is equivalent to the phrase "the LTI system with transfer function $\mathbf{H}(z)$ is lossless."

The property (6.1.2) is the unitary property. Thus, $\mathbf{H}(z)$ is unitary on the unit circle of the z-plane. For $p = r = 1$ this reduces to the allpass property (Sec. 3.4). In order to satisfy (6.1.2) we require $p \geq r$. (If $p < r$ the rank of the left hand side in (6.1.2) is less than r, and the right hand side cannot be $d\mathbf{I}_r$.) The subscript r on $\mathbf{I}_r$, which is a reminder that it is an $r \times r$ matrix, will be deleted unless there is room for confusion.

Paraunitary property. For rational transfer functions, it can be verified (Problem 14.1) that (6.1.2) implies

$$\widetilde{\mathbf{H}}(z)\mathbf{H}(z) = d\mathbf{I}, \quad \text{for all } z, \qquad (6.1.3)$$

which is termed the *paraunitary* property. Conversely, (6.1.3) implies (6.1.2). We can, therefore, define a lossless system to be a causal, stable paraunitary

---

† At this point, it is useful to review Sec. 2.3 on matrix notations, particularly $\mathbf{H}^\dagger, \widetilde{\mathbf{H}}(z), \mathbf{H}_*(z)$ and so on.

system. So, in order to prove that a causal system is lossless, it is sufficient to prove (a) stability and (b) paraunitariness. If $\mathbf{H}(z)$ is square and lossless, then $\widetilde{\mathbf{H}}(z)$ is paraunitary but not lossless (unless it is a constant). Whenever causality and stability are obvious from the context, we do not mention them; we then use "lossless" and "paraunitary" interchangeably. Notice that any constant unitary matrix is (trivially) paraunitary as well as lossless.

Columnwise orthogonality. Letting $\mathbf{H}_k(z)$ and $\mathbf{H}_m(z)$ denote the kth and mth columns of $\mathbf{H}(z)$, we see that these columns are mutually orthogonal, that is, $\widetilde{\mathbf{H}}_k(z)\mathbf{H}_m(z) = 0$ for $k \neq m$. Moreover, each column represents a set of p power complementary filters, i.e., $\widetilde{\mathbf{H}}_k(z)\mathbf{H}_k(z) = d$. We say "each column is power complementary."

Normalized systems. If a lossless system has $d = 1$ in (6.1.2) we say that it is *normalized-lossless*. Correspondingly the properties (6.1.2) and (6.1.3) are termed *normalized-unitary* and *normalized-paraunitary*.

Square Matrices

For the case of square matrices, (6.1.3) implies

$$\mathbf{H}^{-1}(z) = \widetilde{\mathbf{H}}(z)/d, \quad \text{for all } z, \qquad (6.1.4)$$

so that the inverse is obtained essentially by use of 'tilde' operation. Moreover, in this case we have $\widetilde{\mathbf{H}}(z)\mathbf{H}(z) = \mathbf{H}(z)\widetilde{\mathbf{H}}(z) = d\mathbf{I}$. So every *row* is power complementary, and any pair of rows is orthogonal.

For the special case where $\mathbf{H}(z)$ is 2×2, that is,

$$\mathbf{H}(z) = \begin{bmatrix} H_{00}(z) & H_{01}(z) \\ H_{10}(z) & H_{11}(z) \end{bmatrix}, \qquad (6.1.5)$$

the paraunitary property $\widetilde{\mathbf{H}}(z)\mathbf{H}(z) = d\mathbf{I}$ can be written explicitly in terms of the elements $H_{km}(z)$ as:

$$\begin{aligned}
\widetilde{H}_{00}(z)H_{00}(z) + \widetilde{H}_{10}(z)H_{10}(z) &= d, \\
\widetilde{H}_{01}(z)H_{01}(z) + \widetilde{H}_{11}(z)H_{11}(z) &= d, \\
\widetilde{H}_{00}(z)H_{01}(z) + \widetilde{H}_{10}(z)H_{11}(z) &= 0.
\end{aligned} \qquad (6.1.6)$$

Some properties of Paraunitary Systems

1. *Determinant is allpass.* Assume $p = r$, and let $A(z)$ denote the determinant of $\mathbf{H}(z)$. From (6.1.3) we get $\widetilde{A}(z)A(z) = d^r$ for all z, proving that $A(z)$ is allpass. In particular, if $\mathbf{H}(z)$ is FIR then $A(z)$ is a delay, that is,

$$\det \mathbf{H}(z) = az^{-K}, \quad K \geq 0, a \neq 0 \quad \text{(for FIR paraunitary } \mathbf{H}(z)\text{)}. \qquad (6.1.7)$$

2. *Power complementary (PC) property.* For an $M \times 1$ transfer matrix $\mathbf{h}(z) = [\, H_0(z) \; \ldots \; H_{M-1}(z)\,]^T$, the paraunitary property implies the power complementary property, that is,

$$\sum_{k=0}^{M-1} |H_k(e^{j\omega})|^2 = c, \quad \text{for all } \omega. \qquad (6.1.8)$$

This follows directly from $\widetilde{\mathbf{h}}(z)\mathbf{h}(z) = c$.

3. *Submatrices of paraunitary* $\mathbf{H}(z)$. From the definition it is clear that every column of a paraunitary transfer matrix is itself paraunitary (i.e., PC). In fact any $p \times L$ submatrix of $\mathbf{H}(z)$ is paraunitary (Problem 14.2).

These three properties also hold if we replace "paraunitary" with "lossless" everywhere.

6.1.2 Interconnections and Examples

We now consider examples of lossless systems, with particular emphasis on filter banks. We begin by noting a number of operations and interconnections which preserve the lossless property.

Operations Preserving Paraunitary and Lossless Properties

We can verify that if $\mathbf{H}(z)$ is square and paraunitary, then so are the following matrices: (a) $\mathbf{H}(z^M)$ for any integer M (b) $\mathbf{H}^T(z)$, and (c) $\widetilde{\mathbf{H}}(z)$. If $\mathbf{H}(z)$ is lossless, then the first two are lossless as well.

Consider next the cascaded structure of Fig. 6.1-1. The overall transfer matrix is $\mathbf{H}(z) = \mathbf{H}_1(z)\mathbf{H}_0(z)$. (The sizes of $\mathbf{H}_0(z)$ and $\mathbf{H}_1(z)$ can be different as long as the product makes sense.) This product is paraunitary if $\mathbf{H}_0(z)$ and $\mathbf{H}_1(z)$ are. (*Proof:* $\widetilde{\mathbf{H}}(z)\mathbf{H}(z) = \widetilde{\mathbf{H}}_0(z)\widetilde{\mathbf{H}}_1(z)\mathbf{H}_1(z)\mathbf{H}_0(z) = d\mathbf{I}$, since $\mathbf{H}_0(z)$ and $\mathbf{H}_1(z)$ are paraunitary.) Furthermore if $\mathbf{H}_0(z)$ and $\mathbf{H}_1(z)$ are lossless, so is the product (as it does not have *new* poles). Thus, the operation of cascading (or product) preserves losslessness.

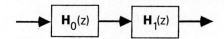

Figure 6.1-1 A cascade of two paraunitary systems.

Example 6.1.1: Cascaded Paraunitary Systems

Consider the transfer matrix

$$\mathbf{R}_m = \begin{bmatrix} \cos\theta_m & \sin\theta_m \\ -\sin\theta_m & \cos\theta_m \end{bmatrix}, \quad \theta_m \text{ real} \quad \text{(Givens rotation)}. \qquad (6.1.9)$$

Fig. 6.1-2(a) shows a flowgraph of this system. If $\mathbf{y} = \mathbf{R}_m\mathbf{x}$, then $\mathbf{y}$ is obtainable by rotating $\mathbf{x}$ by θ_m, clockwise. This can be seen from Fig.

6.1-2(b)), which shows the components of **x** and **y** in terms of θ_m. In this figure r is the 'length' of **x**, that is, $r = \sqrt{\mathbf{x}^T\mathbf{x}}$.

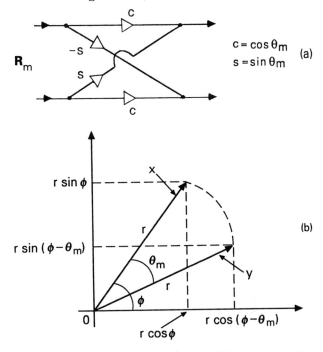

Figure 6.1-2 (a) The Givens rotation, and (b) demonstration of rotation.

The operator $\mathbf{R}_m$ is known as the Givens rotation [Givens, 1958], [Golub and Van Loan, 1989], planar rotation or simply *rotation*. It is easily verified that $\mathbf{R}_m$ is unitary (with $\mathbf{R}_m^T \mathbf{R}_m = \mathbf{I}$). Next consider the 2×2 system of Fig. 6.1-3 with transfer matrix

$$\mathbf{\Lambda}(z) = \begin{bmatrix} 1 & 0 \\ 0 & z^{-1} \end{bmatrix}. \tag{6.1.10}$$

We have

$$\widetilde{\mathbf{\Lambda}}(z)\mathbf{\Lambda}(z) = \begin{bmatrix} 1 & 0 \\ 0 & z \end{bmatrix} \begin{bmatrix} 1 & 0 \\ 0 & z^{-1} \end{bmatrix} = \mathbf{I}, \tag{6.1.11}$$

so that $\mathbf{\Lambda}(z)$ is paraunitary.

Figure 6.1-4 shows a cascade of the above paraunitary systems, which is therefore paraunitary. Its transfer matrix is

$$\mathbf{H}_N(z) = \mathbf{R}_N \mathbf{\Lambda}(z) \mathbf{R}_{N-1} \mathbf{\Lambda}(z) \ldots \mathbf{R}_1 \mathbf{\Lambda}(z) \mathbf{R}_0. \tag{6.1.12}$$

As an example let $N = 1$ so that $\mathbf{H}_N(z) = \mathbf{R}_1 \mathbf{\Lambda}(z) \mathbf{R}_0$. With $\theta_0 = \theta_1 = \pi/4$ the transfer function of the cascaded system is

$$\mathbf{H}(z) = \frac{1}{2} \begin{bmatrix} 1 - z^{-1} & 1 + z^{-1} \\ -(1 + z^{-1}) & -(1 - z^{-1}) \end{bmatrix}. \tag{6.1.13}$$

A second verification of the fact that (6.1.13) is paraunitary is obtained by noting that the matrix

$$\mathbf{H}(e^{j\omega}) = e^{-j\omega/2} \begin{bmatrix} j\sin(\omega/2) & \cos(\omega/2) \\ -\cos(\omega/2) & -j\sin(\omega/2) \end{bmatrix} \qquad (6.1.14)$$

is unitary. In this example, the building blocks are also causal and FIR so that $\mathbf{H}_N(z)$ is lossless.

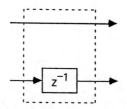

Figure 6.1-3 A simple, yet fundamental paraunitary system.

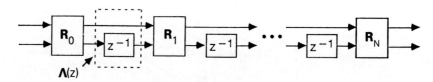

Figure 6.1-4 A cascade of paraunitary building blocks. $\mathbf{R}_m$ is the planar or Givens rotation shown in Fig. 6.1-2(a).

Example 6.1.2: Paraunitary Vectors

The system of Fig. 6.1-5(a) has transfer matrix

$$\mathbf{e}(z) = \begin{bmatrix} 1 \\ z^{-1} \end{bmatrix}. \qquad (6.1.15)$$

We have

$$\widetilde{\mathbf{e}}(z)\mathbf{e}(z) = \begin{bmatrix} 1 & z \end{bmatrix} \begin{bmatrix} 1 \\ z^{-1} \end{bmatrix} = 2, \qquad (6.1.16)$$

so that $\mathbf{e}(z)$ is paraunitary. Next consider the system of Fig. 6.1-5(b) with transfer matrix

$$\mathbf{P}_0 = \begin{bmatrix} \cos\theta \\ \sin\theta \end{bmatrix}. \qquad (6.1.17)$$

We have $\mathbf{P}_0^T \mathbf{P}_0 = 1$ so that $\mathbf{P}_0$ is normalized lossless.

Finally consider the cascade of Fig. 6.1-6. Here the leftmost building block is as in (6.1.17) and the other building blocks have transfer functions of the forms (6.1.9) or (6.1.10). Since all the building blocks are paraunitary, the cascaded system is paraunitary. It has transfer matrix

$$\mathbf{P}_N(z) = \mathbf{R}_N \mathbf{\Lambda}(z) \mathbf{R}_{N-1} \mathbf{\Lambda}(z) \ldots \mathbf{R}_1 \mathbf{\Lambda}(z) \mathbf{P}_0. \qquad (6.1.18)$$

This system is essentially a power complementary analysis bank.

Figure 6.1-5 Examples of 2 × 1 paraunitary systems.

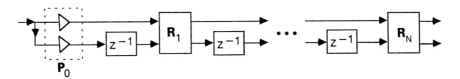

Figure 6.1-6 Example of more general 2 × 1 FIR paraunitary system. Each $\mathbf{R}_m$ is as in Figure 6.1-2(a).

Example 6.1.3: More on Paraunitary Filter Banks

Consider the system of Fig. 6.1-7 which is a cascade of two systems with transfer matrices $\mathbf{e}(z)$ and $\mathbf{W}^*$ respectively. Note that $\mathbf{W}$ represents the $M \times M$ DFT matrix so that $\mathbf{W}^*$ is unitary. Moreover $\widetilde{\mathbf{e}}(z)\mathbf{e}(z) = M$ so that $\mathbf{e}(z)$ is paraunitary. The overall transfer matrix

$$\underbrace{\begin{bmatrix} H_0(z) \\ H_1(z) \\ \vdots \\ H_{M-1}(z) \end{bmatrix}}_{\mathbf{h}(z)} = \mathbf{W}^* \mathbf{e}(z) \qquad (6.1.19)$$

is thus paraunitary. This implies in particular that $\mathbf{h}^\dagger(e^{j\omega})\mathbf{h}(e^{j\omega}) = M^2$, so that the set $H_k(z)$ is power complementary. Recall (example 4.1.1) that $H_0(z)$ is lowpass with approximately 13 dB stopband attenuation, and that the filters $H_k(z)$ form a uniform-DFT analysis bank.

IIR lossless systems. One can obtain examples of IIR lossless systems simply by replacing each delay z^{-1} in the above examples with a stable unit-magnitude allpass function $A_k(z)$. For instance, in Example 6.1.1 if the building block $\mathbf{\Lambda}(z)$ is replaced with the IIR lossless system

$$\mathbf{\Lambda}_k(z) = \begin{bmatrix} 1 & 0 \\ 0 & A_k(z) \end{bmatrix}, \qquad (6.1.20)$$

we obtain an IIR lossless system.

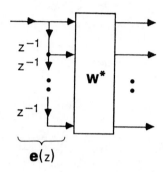

Figure 6.1-7 An M-channel filter bank. This is $M \times 1$ paraunitary.

6.2 FILTER BANK PROPERTIES INDUCED BY PARAUNITARINESS

We now study some consequences of constraining the polyphase matrix $\mathbf{E}(z)$ to be paraunitary. Whenever $\mathbf{E}(z)$ is paraunitary, we often express it by saying that "the analysis filters form a paraunitary set," or "the analysis bank is paraunitary", or "paraunitary QMF bank" (whenever $\mathbf{R}(z)$ is also paraunitary). In the two channel case, we often say "$[H_0(z), H_1(z)]$ is a paraunitary pair".

The paraunitary property implies

$$\widetilde{\mathbf{E}}(z)\mathbf{E}(z) = d\mathbf{I}, \quad \text{that is,} \quad \mathbf{E}^{-1}(z) = \widetilde{\mathbf{E}}(z)/d, \quad \text{for all } z \quad d > 0. \quad (6.2.1)$$

So we choose $\mathbf{R}(z)$ as

$$\mathbf{R}(z) = cz^{-K}\widetilde{\mathbf{E}}(z), \qquad (6.2.2)$$

for some $c \neq 0$, to satisfy the perfect reconstruction condition (5.6.6). Neither c nor K is fundamental, but choice of a positive K serves to ensure that $\mathbf{R}(z)$ [hence $F_k(z)$] is causal. For example, we know from the previous section that if $\mathbf{E}(z)$ is a cascade as in Fig. 6.1-4, and if the matrices $\mathbf{R}_m$ satisfy

$$\mathbf{R}_m^\dagger \mathbf{R}_m = \mathbf{I} \qquad (6.2.3)$$

then $\mathbf{E}(z)$ is paraunitary. With $\mathbf{E}(z)$ so chosen, we can satisfy (6.2.2) by taking $\mathbf{R}(z)$ to be

$$\mathbf{R}(z) = \mathbf{R}_0^\dagger \mathbf{\Gamma}(z) \mathbf{R}_1^\dagger \ldots \mathbf{\Gamma}(z) \mathbf{R}_N^\dagger, \qquad (6.2.4)$$

where
$$\mathbf{\Gamma}(z) = z^{-1}\widetilde{\mathbf{\Lambda}}(z) = \begin{bmatrix} z^{-1} & 0 \\ 0 & 1 \end{bmatrix}. \qquad (6.2.5)$$

This corresponds to (6.2.2) with $c = 1$ and $K = N$.

Stability. If the analysis filters (hence $\mathbf{E}(z)$) are stable and IIR, then the choice (6.2.2) results in unstable filters. This is because the poles of the (k, m) element of $\widetilde{\mathbf{E}}(z)$ are reciprocal conjugates of those of $E_{m,k}(z)$, and therefore are outside the unit circle. So $\mathbf{R}(z)$ in (6.2.2) is unstable. We cannot therefore build useful perfect reconstruction systems with IIR lossless $\mathbf{E}(z)$. We therefore restrict attention to the FIR case. (However, see Problem 6.12 where the adjugate of $\mathbf{E}(z)$ is used to define $\mathbf{R}(z)$, thereby eliminating stability problems).

6.2.1 Relation Between Analysis and Synthesis Filters

The condition $\mathbf{R}(z) = cz^{-K}\widetilde{\mathbf{E}}(z)$ between the analysis and synthesis banks implies a very important relation between the analysis and synthesis filters. This relation enables us to find $f_k(n)$ simply by 'flipping' and conjugating the coefficients $h_k(n)$. More precisely we will show that the relation $\mathbf{R}(z) = cz^{-K}\widetilde{\mathbf{E}}(z)$ implies

$$f_k(n) = ch_k^*(L - n), \quad 0 \le k \le M - 1, \qquad (6.2.6)$$

where $L = M - 1 + MK$. For the FIR case this means, in particular, that the synthesis filters have same length as the analysis filters! In the z-domain, (6.2.6) is equivalent to

$$F_k(z) = cz^{-L}\widetilde{H}_k(z), \quad 0 \le k \le M - 1, \qquad (6.2.7)$$

that is, $F_k(z)$ is the Hermitian image of $H_k(z)$ (up to scale). To prove this, just substitute $\mathbf{R}(z) = cz^{-K}\widetilde{\mathbf{E}}(z)$ into (6.1.1) to obtain

$$\mathbf{f}^T(z) = cz^{-(M-1+MK)}\widetilde{\mathbf{e}}(z)\widetilde{\mathbf{E}}(z^M) = cz^{-(M-1+MK)}\widetilde{\mathbf{h}}(z), \qquad (6.2.8)$$

from which the desired result follows. As the proof shows, the relation (6.2.6) is induced by (6.2.2), and really has nothing to do with paraunitariness of $\mathbf{E}(z)$. Moreover, it holds whether the filters are FIR or IIR. If $H_k(z)$ are IIR with poles inside the unit circle, then the synthesis filters given by (6.2.7) have poles outside the unit circle.

Frequency domain implication. Eq. (6.2.7) implies

$$|F_k(e^{j\omega})| = |c| \times |H_k(e^{j\omega})|, \qquad (6.2.9)$$

which means that the magnitude response of $F_k(z)$ is exactly the same as that of $H_k(z)$ (up to scale).

Converse of the above property. As mentioned above, the relation (6.2.6) follows entirely from (6.2.2), and not from paraunitariness. In fact if (6.2.6) holds for all k, the polyphase matrices are related as in (6.2.2). To prove this converse, note that (6.2.6) is equivalent to

$$\mathbf{f}^T(z) = cz^{-(M-1+MK)}\widetilde{\mathbf{h}}(z). \qquad (6.2.10)$$

In terms of $\mathbf{E}(z)$ and $\mathbf{R}(z)$ this becomes

$$\widetilde{\mathbf{e}}(z)\mathbf{R}(z^M) = cz^{-MK}\widetilde{\mathbf{e}}(z)\widetilde{\mathbf{E}}(z^M). \qquad (6.2.11)$$

This implies (6.2.2) indeed (Problem 6.6).

The following theorem [Vaidyanathan, 1987a] summarizes the crucial relations between paraunitariness, perfect reconstruction, and the condition (6.2.6).

♠**Theorem 6.2.1.** Consider the maximally decimated QMF bank with causal FIR analysis filters $H_k(z)$, and let $\mathbf{E}(z)$ be the polyphase matrix for the analysis filters. Consider the following three statements.

1. $\mathbf{E}(z)$ is lossless (that is, paraunitary).
2. The synthesis filters are given by $f_k(n) = ch_k^*(L-n), 0 \leq k \leq M-1$, for some $c \neq 0$ and some integer L.
3. The system has perfect reconstruction property.

If any two of these statements are true, then the remaining statement also holds.   ◇

Proof. (Not very entertaining!) Suppose the first two statements are true. Paraunitariness implies $\widetilde{\mathbf{E}}(z)\mathbf{E}(z) = d\mathbf{I}$, whereas the second statement implies $\mathbf{R}(z) = cz^{-K}\widetilde{\mathbf{E}}(z)$ (as proved above). Combining these we arrive at $\mathbf{R}(z)\mathbf{E}(z) = c_0 z^{-m_0}\mathbf{I}$, implying statement 3.

Next let statements 2 and 3 be true. Statement 2 implies the relation $\mathbf{R}(z) = cz^{-K}\widetilde{\mathbf{E}}(z)$ whereas statement 3 implies $\mathbf{R}(z)\mathbf{E}(z) = c_0 z^{-m_0}\mathbf{I}$. Combinining these we arrive at $\widetilde{\mathbf{E}}(z)\mathbf{E}(z) = d\mathbf{I}$, that is, $\mathbf{E}(z)$ is paraunitary. We can prove that statements 1 and 3 imply 2 in a similar manner.   ▽▽▽

6.2.2 Other Properties Induced by Paraunitary E(z)

Power Complementary Property

Consider the vector of analysis filters given by $\mathbf{h}(z) = \mathbf{E}(z^M)\mathbf{e}(z)$. Paraunitariness of $\mathbf{E}(z)$ implies that $\mathbf{E}(z^M)$ is paraunitary. Moreover, the delay chain $\mathbf{e}(z)$ is also paraunitary. So, the product $\mathbf{h}(z)$ is paraunitary as well. This in turn implies that the analysis filters $H_k(z)$ are power complementary, that is,

$$\sum_{k=0}^{M-1} |H_k(e^{j\omega})|^2 = \text{positive constant.}$$

With $\mathbf{E}(z)$ chosen to be paraunitary, the only choice of synthesis filters $F_k(z)$ to obtain perfect reconstruction is given by (6.2.6) (use Theorem 6.2.1). Since this implies (6.2.9), the M synthesis filters are also power complementary.

The AC Matrix is Paraunitary if, and only if, E(z) is

Recall from Section 5.5 that the alias component (AC) matrix $\mathbf{H}(z)$ and polyphase matrix $\mathbf{E}(z)$ are related as

$$\mathbf{H}(z) = \mathbf{W}^\dagger \mathcal{D}(z) \mathbf{E}^T(z^M), \qquad (6.2.12)$$

where

$$\mathcal{D}(z) = \text{diag}\begin{bmatrix} 1 & z^{-1} & \ldots & z^{-(M-1)} \end{bmatrix}, \qquad (6.2.13)$$

and $\mathbf{W}$ is the $M \times M$ DFT matrix. By using the facts that $\mathbf{W}\mathbf{W}^\dagger = M\mathbf{I}$ and $\widetilde{\mathcal{D}}(z)\mathcal{D}(z) = \mathbf{I}$, we conclude that $\widetilde{\mathbf{H}}(z)\mathbf{H}(z) = \beta\mathbf{I}$ if and only if $\widetilde{\mathbf{E}}(z)\mathbf{E}(z) = \beta\mathbf{I}/M$. Summarizing, the AC matrix $\mathbf{H}(z)$ is paraunitary if and only $\mathbf{E}(z)$ is paraunitary.

As an application, we can show that if $\widetilde{\mathbf{E}}(z)\mathbf{E}(z) = \mathbf{I}$, then each analysis filter has unit energy, that is, $\int_0^{2\pi} |H_k(e^{j\omega})|^2 d\omega/2\pi = 1$. This is because $\widetilde{\mathbf{E}}(z)\mathbf{E}(z) = \mathbf{I}$ implies $\widetilde{\mathbf{H}}(z)\mathbf{H}(z) = M\mathbf{I}$, that is, in particular,

$$\sum_{\ell=0}^{M-1} |H_k(e^{j\omega}W^\ell)|^2 = M.$$

If we integrate both sides over the range $[0, 2\pi)$, each term on the left hand side yields the same answer. From this we arrive at the desired result.

Another simple application of the above result is this: suppose the analysis filters are ideal and nonoverlapping, with

$$|H_k(e^{j\omega})| = \begin{cases} 1 & \text{for } 2\pi k/M \leq \omega < 2\pi(k+1)/M \\ 0 & \text{otherwise.} \end{cases}$$

Then the AC matrix $\mathbf{H}(z)$ is paraunitary (why?). So $\mathbf{E}(z)$ is paraunitary as well.

Relation to M-th band [or Nyquist(M)] filters

From the above property, we obtain a very interesting relation between paraunitary filter banks and Mth band filters (or Nyquist(M) filters; these were defined in Sec. 4.6.1): if $\mathbf{E}(z)$ is paraunitary, then each analysis filter $H_k(z)$ is a spectral factor of a (zero-phase) Mth band filter. In other words, the filter $G_k(z)$ defined as

$$G_k(z) \stackrel{\Delta}{=} \widetilde{H}_k(z)H_k(z), \qquad (6.2.14)$$

is an Mth band filter.

To prove this, recall that the Mth band property is essentially defined by the property (4.6.6) (for zero phase systems). Thus it is sufficient to prove

$$\sum_{\ell=0}^{M-1} G_k(zW^\ell) = \text{nonzero constant}. \qquad (6.2.15)$$

Since paraunitariness of $\mathbf{E}(z)$ implies that of $\mathbf{H}(z)$, each column of $\mathbf{H}(z)$ is paraunitary, that is,

$$\sum_{\ell=0}^{M-1} \widetilde{H}_k(zW^\ell) H_k(zW^\ell) = \text{nonzero constant}. \qquad (6.2.16)$$

Using the definition of $G_k(z)$, the desired property (6.2.15) follows.

6.3 TWO CHANNEL FIR PARAUNITARY QMF BANKS

In this section we consider the two channel QMF bank (Fig. 5.1-1(a)) in which the analysis filters are causal and FIR, that is,

$$H_0(z) = \sum_{n=0}^{N} h_0(n) z^{-n}, \quad H_1(z) = \sum_{n=0}^{N} h_1(n) z^{-n}. \qquad (6.3.1)$$

We know that if the 2×2 polyphase matrix $\mathbf{E}(z)$ is paraunitary, then all properties stated in Sec. 6.2 are true. For the two channel case, some additional properties are satisfied, which we study next.

6.3.1 Further properties

Power Symmetric Property

We know the alias-component (AC) matrix (defined in Sec. 5.4) is given by

$$\mathbf{H}(z) = \begin{bmatrix} H_0(z) & H_1(z) \\ H_0(-z) & H_1(-z) \end{bmatrix}. \qquad (6.3.2)$$

From Sec. 6.2.2 we know that paraunitariness of $\mathbf{E}(z)$ (i.e., $\widetilde{\mathbf{E}}(z)\mathbf{E}(z) = d\mathbf{I}$) implies that of $\mathbf{H}(z)$, i.e., $\widetilde{\mathbf{H}}(z)\mathbf{H}(z) = \beta \mathbf{I}$, where $\beta = 2d$. From this we obtain the three equations

$$\widetilde{H}_0(z) H_0(z) + \widetilde{H}_0(-z) H_0(-z) = \beta, \qquad (6.3.3a)$$
$$\widetilde{H}_1(z) H_1(z) + \widetilde{H}_1(-z) H_1(-z) = \beta, \qquad (6.3.3b)$$
$$\widetilde{H}_0(z) H_1(z) + \widetilde{H}_0(-z) H_1(-z) = 0. \qquad (6.3.3c)$$

Using the decimation notation $A(z)\big|_{\downarrow 2}$ defined in (4.1.14), we can rewrite the above three equations as

$$\widetilde{H}_0(z)H_0(z)\big|_{\downarrow 2} = 0.5\beta, \quad \widetilde{H}_1(z)H_1(z)\big|_{\downarrow 2} = 0.5\beta, \quad \widetilde{H}_0(z)H_1(z)\big|_{\downarrow 2} = 0, \tag{6.3.4}$$

Thus paraunitariness of $\mathbf{E}(z)$ is equivalent to (6.3.4). From the first of the above equations we see that $\widetilde{H}_0(z)H_0(z)$ is a half-band filter (Sec. 4.6.1). In other words, $H_0(z)$ is *power symmetric*. (This property was defined in Sec. 5.3. The original definition had $\beta = 1$, but it is convenient to allow arbitrary $\beta > 0$.) Same comment holds for $H_1(z)$.

Order of $H_0(z)$ is necessarily odd. Assume $h_0(0), h_0(N)$ and N are nonzero (as in any useful design). Then the order N is necessarily odd, that is, $N = 2J + 1$. This is because $\widetilde{H}_0(z)H_0(z)$ is a zero-phase half band filter and has order of the form $4J + 2$ (Sec. 4.6.1).

Relation Between the Two Analysis Filters

From Sec. 6.2.2 we already know that the two analysis filters are power complementary if $\mathbf{E}(z)$ is paraunitary. We will show that the analysis filters are also related as

$$H_1(z) = cz^{-L}\widetilde{H}_0(-z), \quad |c| = 1, \quad L = \text{odd}. \tag{6.3.5a}$$

In other words, in the time domain,

$$h_1(n) = -c(-1)^n h_0^*(L - n). \tag{6.3.5b}$$

In view of this, we can find $h_1(n)$ from $h_0(n)$ by inspection. (In practice we do not lose anything by setting $c = 1$.)

Derivation of (6.3.5a). If $\mathbf{E}(z)$ is paraunitary then the AC matrix $\mathbf{H}(z)$ is paraunitary, so that (6.3.3) holds. Furthermore $\mathbf{H}(z)\widetilde{\mathbf{H}}(z) = \beta\mathbf{I}$ as well. From this we obtain three equations similar to (6.3.3), two of which are

$$H_0(z)\widetilde{H}_0(z) + H_1(z)\widetilde{H}_1(z) = \beta, \tag{6.3.6a}$$
$$H_0(z)\widetilde{H}_0(-z) + H_1(z)\widetilde{H}_1(-z) = 0. \tag{6.3.6b}$$

From (6.3.6b) we have

$$\frac{H_1(z)}{H_0(z)} = \frac{-\widetilde{H}_0(-z)}{\widetilde{H}_1(-z)}. \tag{6.3.6c}$$

Equation (6.3.6a) implies that there are no common factors between $H_0(z)$ and $H_1(z)$ (since the right hand side is constant). So $\widetilde{H}_0(-z)$ and $\widetilde{H}_1(-z)$ have no common factors either. From (6.3.6c) we therefore conclude that

$H_1(z) = cz^{-L}\widetilde{H}_0(-z)$. By substituting this into (6.3.6a) and using (6.3.3a) we obtain $|c| = 1$. By substituting $H_1(z) = cz^{-L}\widetilde{H}_0(-z)$ into (6.3.6b), it can further be verified that L has to be odd.

Frequency domain implication. The relation (6.3.5a) implies that the two analysis filters are such that

$$|H_1(e^{j\omega})| = |H_0(-e^{j\omega})| = |H_0(e^{j(\omega-\pi)})|. \qquad (6.3.7)$$

So the magnitude response of $H_1(z)$ is obtained by shifting that of $H_0(z)$ by π. For the real coefficient case this means that if $H_0(z)$ is lowpass then $H_1(z)$ is highpass; both filters have the same ripple sizes, and same transition band widths.

Example 6.3.1

Let $\mathbf{E}(z) = \mathbf{I}$, which is paraunitary. The analysis filters are $H_0(z) = 1$, and $H_1(z) = z^{-1}$, and can be verified to satisfy all the above properties. For example, (a) each of these is power symmetric, (b) they are related as in (6.3.5a) for $L = 1$, and (c) they form a power complementary pair. The synthesis filters for perfect reconstruction can be found from (6.2.7), where $c \neq 0$ and L is odd. Choosing $c = 1$ and $L = 1$ we get $F_0(z) = z^{-1}$ and $F_1(z) = 1$.

Example 6.3.2

As a second example let $\mathbf{E}(z) = \begin{bmatrix} 1 & 1 \\ 1 & -1 \end{bmatrix}$ which is paraunitary. Then, $H_0(z) = 1 + z^{-1}$ and $H_1(z) = 1 - z^{-1}$, and again the three properties listed in the previous example are satisfied. We can find a set of synthesis filters for perfect reconstruction, by setting $c = 1$ and $L = 1$ in (6.2.7). Thus $F_0(z) = 1 + z^{-1}$ and $F_1(z) = -1 + z^{-1}$.

Power Symmetry of $H_0(z)$ Implies $E(z)$ is Paraunitary

We now consider the converse of some of the above results. We will show that, given any power symmetric $H_0(z)$, we can always force $\mathbf{E}(z)$ to be paraunitary by defining $H_1(z)$ as in (6.3.5a). For this, note that power symmetry of $H_0(z)$ implies that the 0th column of the AC matrix $\mathbf{H}(z)$ is paraunitary. In view of the relation $H_1(z) = cz^{-L}\widetilde{H}_0(-z)$, the 1st column of $\mathbf{H}(z)$ is also paraunitary. By using this relation we can also verify that the two columns of $\mathbf{H}(z)$ are mutually orthogonal, that is,

$$\widetilde{H}_0(z)H_1(z) + \widetilde{H}_0(-z)H_1(-z) = 0. \quad \text{(using } L \text{ odd)}. \qquad (6.3.8)$$

In other words, $\mathbf{H}(z)$ satisfies the three properties (6.3.4), that is, we have $\widetilde{\mathbf{H}}(z)\mathbf{H}(z) = \beta \mathbf{I}$. This implies that $\mathbf{E}(z)$ is paraunitary.

We summarize all the above results as follows:

◆**Theorem 6.3.1.** Let $H_0(z)$ and $H_1(z)$ be causal FIR and let $\mathbf{E}(z)$ be the 2×2 polyphase matrix of the analysis bank $[H_0(z), H_1(z)]$. Then the following statements are equivalent:

1. $H_0(z)$ is power symmetric (i.e., $\widetilde{H}_0(z)H_0(z)\big|_{\downarrow 2} = 0.5\beta, \beta > 0$), and $H_1(z) = cz^{-L}\widetilde{H}_0(-z)$ for some $|c| = 1$ and odd L.
2. $\widetilde{\mathbf{E}}(z)\mathbf{E}(z) = 0.5\beta\mathbf{I}$, that is, $\mathbf{E}(z)$ is paraunitary, (same as 'lossless' since $\mathbf{E}(z)$ is causal FIR).
3. $\mathbf{H}(z)$ is paraunitary, that is, $H_0(z)$ and $H_1(z)$ together satisfy (6.3.4). ◊

Comments. It is also true that, for a given power-symmetric $H_0(z)$, the function $H_1(z)$ which forces $\mathbf{E}(z)$ to be paraunitary *must* have the form (6.3.5a). This follows from the steps of the derivation of (6.3.5a).

Summary of Properties Induced by Paraunitary E(z)

Summarizing, the two-channel causal FIR QMF bank with paraunitary $\mathbf{E}(z)$ has the following properties.

1. $H_0(z)$ is power symmetric. This statement is equivalent to any one of the following:
    a) $H_0(z)$ is a spectral factor of a half band filter, that is, $\widetilde{H}_0(z)H_0(z)$ is a half band filter.
    b) $\widetilde{H}_0(z)H_0(z)\big|_{\downarrow 2} = 0.5\beta$.
2. $H_1(z)$ has all the properties of $H_0(z)$. Together they satisfy (6.3.4), that is, the AC matrix $\mathbf{H}(z)$ is paraunitary.
3. $H_1(z) = cz^{-L}\widetilde{H}_0(-z)$ where $|c| = 1$, and L is odd.
4. $H_0(z)$ and $H_1(z)$ form a power complementary pair.
5. $H_0(z)$ has odd order N (as long as $h_0(0), h_0(N)$ and N are nonzero).

6.3.2 Design of Perfect Reconstruction QMF Bank

The above results place in evidence the following procedure for the design of a two channel FIR perfect reconstruction QMF bank: first design a zero-phase half-band filter $H(z)$ with $H(e^{j\omega}) \geq 0$, by using any standard technique (Sec. 4.6.1). Then compute a spectral factor $H_0(z)$ by using any method mentioned in Sec. 3.2.5 or Appendix D. This gives one of the analysis filters $H_0(z) = \sum_{n=0}^{N} h_0(n)z^{-n}$ (a power-symmetric function), with order $N = 2J + 1$. Obtain the other analysis filter $H_1(z)$ and the two synthesis filters $F_0(z)$ and $F_1(z)$ as

$$H_1(z) = -z^{-N}\widetilde{H}_0(-z), \quad F_0(z) = z^{-N}\widetilde{H}_0(z), \quad \text{and} \quad F_1(z) = z^{-N}\widetilde{H}_1(z). \tag{6.3.9a}$$

or, equivalently, in terms of impulse response coefficients as:

$$h_1(n) = (-1)^n h_0^*(N-n), \quad f_0(n) = h_0^*(N-n), \text{ and } f_1(n) = h_1^*(N-n). \tag{6.3.9b}$$

This system is identical to the one presented in Section 5.3.6! In this section we have obtained an independent derivation starting from the paraunitary property. This derivation enables us to generalize the technique for M channels, as we will see. In addition, it gives rise to lattice structures which preserve perfect reconstruction in spite of quantization (see below).

Properties of the Above Filter Bank

If the above four filters (6.3.9) are used in the QMF bank of Fig. 5.1-1(a) then the following are true:

1. There is perfect reconstruction, and $\hat{x}(n) = 0.5\beta x(n-N)$.
2. The analysis filters are power complementary, and furthermore satisfy the relation $|H_1(e^{j\omega})| = |H_0(-e^{j\omega})|$.
3. The synthesis filters $F_k(z)$ satisfy $|F_k(e^{j\omega})| = |H_k(e^{j\omega})|$, and are also power compementary.
4. All filters have order $N = 2J+1$ which is automatically odd.
5. The polyphase matrix $\mathbf{E}(z)$ is paraunitary. Since the synthesis filters are chosen as in (6.3.9), the polyphase matrix $\mathbf{R}(z)$ of the synthesis bank is given by $\mathbf{R}(z) = z^{-J}\widetilde{\mathbf{E}}(z)$, and is paraunitary as well.

Completeness. It is worth emphasizing that, every filter bank designed according to the design procedure in Sec. 5.3.6 has paraunitary $\mathbf{E}(z)$. Conversely, whenever $\mathbf{E}(z)$ is FIR and paraunitary, the four filters in the QMF bank are such that they can, in principle, be obtained by the above design procedure (Theorem 6.3.1).

6.4 THE TWO CHANNEL PARAUNITARY QMF LATTICE

We now consider the specical case of the above FIR two channel QMF bank, with real coefficient filters. In this case the paraunitary matrix $\mathbf{E}(z)$ has real coefficients. From Example 6.1.1 we know that a cascaded structure of the form in Fig. 6.1-4 is paraunitary whenever $\mathbf{R}_m$ has the form given in (6.1.9) (Givens rotations), and $\mathbf{\Lambda}(z)$ is as in (6.1.10).

It turns out that the above cascade is very general in the sense that every paraunitary system can be implemented like this! More precisely, we will show in Sec. 14.3.1 that any 2×2 real coefficient (causal, FIR) paraunitary matrix can be factorized as

$$\mathbf{E}(z) = \alpha \mathbf{R}_J \mathbf{\Lambda}(z) \mathbf{R}_{J-1} \ldots \mathbf{\Lambda}(z) \mathbf{R}_0 \begin{bmatrix} 1 & 0 \\ 0 & \pm 1 \end{bmatrix} \tag{6.4.1}$$

where α is a positive scalar. To obtain the synthesis bank which would result in perfect reconstruction, we have to use (6.2.2). Let us choose $K = J$ (so that $\mathbf{R}(z)$ is causal), and $c = 1$. We then have

$$\mathbf{R}(z) = \alpha \begin{bmatrix} 1 & 0 \\ 0 & \pm 1 \end{bmatrix} \mathbf{R}_0^T \mathbf{\Gamma}(z) \ldots \mathbf{R}_{J-1}^T \mathbf{\Gamma}(z) \mathbf{R}_J^T, \qquad (6.4.2)$$

where $\mathbf{\Gamma}(z)$ is as in (6.2.5). Fig. 6.4-1 shows the complete lattice structure for the QMF bank. This is called the QMF lattice [Vaidyanathan and Hoang, 1988]. The analysis and synthesis filters have order $N = 2J + 1$.

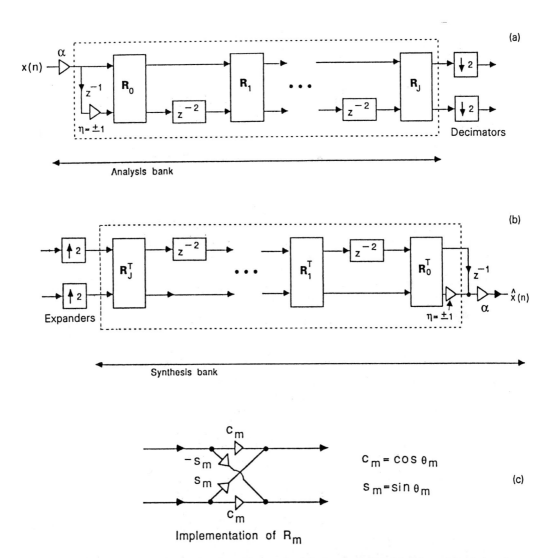

Figure 6.4-1 The QMF lattice structure. (a) Analysis bank, (b) synthesis bank, and (c) details of the unitary matrix $\mathbf{R}_m$.

Sec. 6.4 Two channel QMF lattice

The Two Multiplier QMF Lattice

In practice a more efficient version of the lattice structure can be used to implement this system. For this note that the rotation matrix $\mathbf{R}_m$ can be written as

$$\mathbf{R}_m = \cos\theta_m \begin{bmatrix} 1 & \alpha_m \\ -\alpha_m & 1 \end{bmatrix}, \qquad (6.4.3)$$

if $\cos\theta_m \neq 0$, and as

$$\mathbf{R}_m = \pm \begin{bmatrix} 0 & 1 \\ -1 & 0 \end{bmatrix}, \qquad (6.4.4)$$

otherwise. Assume for simplicity of discussion that $\cos\theta_m \neq 0$ for any m. We can then redraw the lattice structure as in Fig. 6.4-2 with $S = \alpha(\prod_m \cos\theta_m)$. Notice that we have also moved the decimators and expanders in accordance with the noble identities. The quantity S can be expressed directly in terms of α_m as

$$S = \alpha/\gamma, \quad \gamma = \prod_m (1 + \alpha_m^2)^{0.5}. \qquad (6.4.5)$$

Notice that the element ± 1 in (6.4.2) has been replaced with '-1.' The reason is that, the other choice of sign would be equivalent to replacing z with $-z$. This means that $H_1(z)$ becomes lowpass and $H_0(z)$ highpass. This does not add generality, as it can also be covered by renumbering the filters. Also see Problem 6.5.

6.4.1 Properties of the Paraunitary QMF Lattice

Most of the properties of the lattice structure follow immediately from the fact that $\mathbf{E}(z)$ is paraunitary and that $\mathbf{R}(z) = z^{-J}\widetilde{\mathbf{E}}(z)$. These are summarized below, with notations adapted to the real coefficient case.

1. $|H_0(e^{j\omega})|^2 + |H_1(e^{j\omega})|^2 = 2\alpha^2$.
2. $H_1(z) = -z^{-N}H_0(-z^{-1})$, that is, $h_1(n) = (-1)^n h_0(N-n)$, with $N = 2J+1$.
3. $F_k(z) = z^{-N}H_k(z^{-1})$, that is, $f_k(n) = h_k(N-n)$ for $k = 0, 1$. So the synthesis filters have same lengths as the analysis filters.
4. $|H_1(e^{j\omega})| = |H_0(-e^{j\omega})|$. Also $|F_k(e^{j\omega})| = |H_k(e^{j\omega})|, k = 0, 1$.
5. $H_0(z)$ is power symmetric, that is, $\widetilde{H}_0(z)H_0(z)$ is a half-band filter.
6. The system satisfies $\widehat{x}(n) = \alpha^2 x(n - N)$, that is, it is an FIR PR QMF bank.

Completeness

It is worth emphasizing that every two channel (real-coefficient, FIR) paraunitary QMF bank can be represented using the above lattice structure. It is also important to note these points:

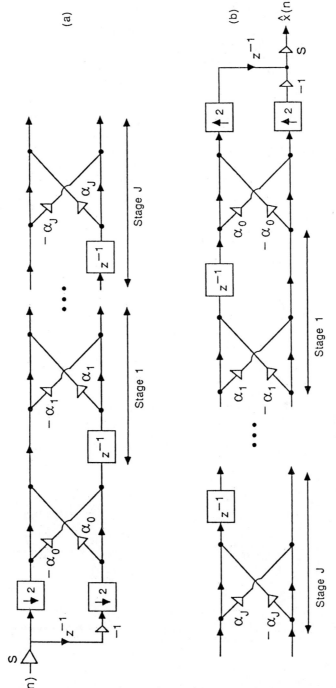

Figure 6.4-2 More efficient implementation of the real coefficient perfect reconstruction lattice structure. (a) Analysis bank, and (b) synthesis bank.

1. Given a real coefficient power symmetric FIR filter $H_0(z)$, we can always define $H_1(z) = -z^{-N}H_0(-z^{-1})$, and implement the analysis bank using the above lattice; the polyphase matrix $\mathbf{E}(z)$ is guaranteed to be paraunitary.
2. Any system designed as in Sec. 5.3.6 (by spectral factorization of a half-band filter) can be represented using the above lattice structure.

Hierarchical Property

Figure 6.4-3(a) is a redrawing of the analysis bank of Fig. 6.4-2(a), with decimators moved to the right for convenience of discussion. Suppose we eliminate the Jth stage, that is, just delete the lattice section which has the multiplier α_J, along with its delay element. We now have a new system with analysis and synthesis filters of reduced order $N-2$. The polyphase matrix for the analysis filter is still paraunitary. Similar comment holds for the synthesis bank. So the reduced system is a perfect reconstruction system and satisfies all the properties listed above, with N replaced by $N-2$.

So if we "cut" the lattice structure after m stages, we still have a FIR PR QMF bank with filters of order $2m+1$. In practice the only thing that happens as m increases is that the filters $H_0(z)$ and $H_1(z)$ have better and better attenuation characteristics (i.e., sharper cutoff and higher attenuation). Referring to Fig. 6.4-3(b), the analysis filters for the m-stage lattice are given by

$$H_0^{(m)}(z) = H_0^{(m-1)}(z) + \alpha_m z^{-2} H_1^{(m-1)}(z),$$
$$H_1^{(m)}(z) = -\alpha_m H_0^{(m-1)}(z) + z^{-2} H_1^{(m-1)}(z). \quad (6.4.6)$$

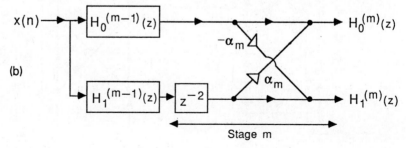

Figure 6.4-3 (a) The analysis bank of the QMF lattice, and (b) schematic for the m-th stage.

So the new lowpass filter $H_0^{(m)}(z)$ (order $2m+1$) is obtained from the old lowpass filter $H_0^{(m-1)}(z)$ (order $2m-1$) by adding a "correction" proportional to α_m. The same holds for the highpass filter $H_1^{(m)}(z)$. Summarizing, the lattice structure represents a whole sequence of perfect reconstruction pairs $[H_0^{(m)}(z), H_1^{(m)}(z)]$ with improved frequency responses as m grows. (However, given a lattice structure with good frequency responses, if we delete the rightmost section, the resulting responses are not necessarily good, even though the perfect reconstruction conditions are preserved.)

Such a hierarchical property is not possible if we implement the filters $H_0(z), H_1(z), F_0(z), F_1(z)$ using a direct-form structure. For example, if we merely replace the highest impulse response coefficient $h_0(N)$ with zero [and adjust the remaining filters to satisfy (6.3.9b)], the result is not a perfect reconstruction system.

Robustness to Coefficient Quantization

Suppose that we implement the lattice using finite precision arithmetic. So the coefficients α_m have to be quantized to some value $Q[\alpha_m]$. Now the matrix

$$\mathbf{Q}_m = \begin{bmatrix} 1 & Q[\alpha_m] \\ -Q[\alpha_m] & 1 \end{bmatrix} \quad (6.4.7)$$

remains unitary, that is, $\mathbf{Q}_m^T \mathbf{Q}_m = c_1 \mathbf{I}, c_1 > 0$. So the cascaded lattice structures for $\mathbf{E}(z)$ and $\mathbf{R}(z)$ remain paraunitary, and all six properties in the preceding list continue to hold including (a) the power symmetric property of $H_0(z)$ and (b) the perfect reconstruction property of the QMF bank.

Such robustness to quantization is however not offered by the direct-form implementation of the filters. Direct quantization of the coefficients $h_0(n)$ [with other filters re-adjusted to satisfy (6.3.9b)] results in loss of power symmetric property of $H_0(z)$, which is crucial to perfect reconstruction. Properties 2–4 in the preceding list are retained by the direct form under quantization, but these are only of secondary importance.

6.4.2 Calculating Lattice Coefficients from the Impulse Response

Suppose the power symmetric filter $H_0(z)$ has been designed somehow, say as in Sec. 6.3.2. We can now define $h_1(n)$ as in (6.3.9b) and write down $\mathbf{E}(z)$. From Theorem 6.3.1 we know that $\mathbf{E}(z)$ is paraunitary. So we are assured of the existence of a lattice structure as in Fig. 6.4-2(a) for the analysis bank (as we show in Sec. 14.3.1).

One can find the coefficients α_m by inverting the recursion (6.4.6) to obtain

$$\begin{aligned} (1+\alpha_m^2)H_0^{(m-1)}(z) &= H_0^{(m)}(z) - \alpha_m H_1^{(m)}(z), \\ (1+\alpha_m^2)z^{-2}H_1^{(m-1)}(z) &= \alpha_m H_0^{(m)}(z) + H_1^{(m)}(z). \end{aligned} \quad (6.4.8)$$

We initialize this recursion by setting $H_0^{(J)}(z) = H_0(z)$ and $H_1^{(J)}(z) = H_1(z)$. The coefficient α_m is chosen so that the highest power of z^{-1} in $H_0^{(m)}(z) - \alpha_m H_1^{(m)}(z)$ is canceled. The fact that the lattice exists in this case assures the following things: (a) the next highest power of z^{-1} is also canceled (so that the order is reduced by two), and (b) the coefficients of z^0 and z^{-1} in $\alpha_m H_0^{(m)}(z) + H_1^{(m)}(z)$ are reduced to zero so that $H_1^{(m-1)}(z)$ in (6.4.8) is indeed causal. Problem 6.7 develops a direct proof that the recursion (6.4.8) works for any power symmetric filter $H_0(z)$.

Table 6.4.1 shows the lattice coefficients calculated in this manner for Design example 5.3.2 presented earlier. Notice that $|\alpha_m|$ gets smaller and smaller as m gets large. Also note that the sign of α_m alternates. This alternation property is consistently observed in all good designs with minimum phase $H_0(z)$, but has not yet been theoretically explained!

TABLE 6.4.1 Lattice coefficients for the perfect reconstruction analysis bank example

m	α_m
0	-0.2588883 e$+01$
1	0.8410785 e$+00$
2	-0.4787637 e$+00$
3	0.3148984 e$+00$
4	-0.2179341 e$+00$
5	0.1522899 e$+00$
6	-0.1046526 e$+00$
7	0.6906427 e-01
8	-0.4258295 e-01
9	0.3111448 e-01

Numerical Accuracy

The lattice structure of the above form exists only if $H_0(z)$ is the spectral factor of a half-band filter. In practice, due to numerical errors (e.g., accuracy of spectral factorization, degree of quantization of $h_0(n)$ etc), this property does not hold. So the lattice generated by the above recursion represents $H_0(z)$ only approximately. The numerical accuracy can be improved considerably as follows: since $[H_0(z), H_1(z)]$ is a power complementary pair, we can synthesize a paraunitary lattice as shown later in Fig 14.3-3. And since $H_0(z)$ is almost power symmetric, the even numbered coefficients satisfy $c_m \approx 1$ and $s_m \approx 0$ (why?). If we replace these with $c_m = 1$ and $s_m = 0$, then the resulting structure resembles Fig. 6.4-1(a) (and can then be denormalized as in Fig. 6.4-2(a)). In our experience, this lattice represents $H_0(z)$ more accurately than the lattice obtained by direct use of (6.4.8).

6.4.3 Direct Design Technique Based on Lattice

Since the lattice guarantees perfect reconstruction regardless of values of α_m, we can use it to *design* the transfer function $H_0(z)$. Thus, we optimize α_m in order to minimize

$$\phi = \int_{\omega_S}^{\pi} |H_0(e^{j\omega})|^2 d\omega, \qquad (6.4.9)$$

which is proportional to the stopband energy of $H_0(z)$. The remaining three filters are completely determined by $H_0(z)$ because of (6.3.9b). We do not have to worry about the passband of $H_0(z)$ because the power symmetric property, which is guaranteed by the structure, ensures that the passband is good.

There are many standard optimization programs which can be used to minimize a specified nonnegative function of several parameters [Press, et al., 1989]. These programs require the designer to supply a routine which can calculate the objective function ϕ for a given set of coefficients α_m. The first step in this calculation would be to compute the coefficients of $H_0(z)$ using the recursion (6.4.6). The recursion is initialized with

$$H_0^{(0)}(z) = (1 - \alpha_0 z^{-1})S, \quad H_1^{(0)}(z) = (-\alpha_0 - z^{-1})S.$$

After J steps we get $H_0^{(J)}$ which is the desired $H_0(z)$. The quantity ϕ can now be written as

$$\phi = (\pi - \omega_S)r(0) + 2\int_{\omega_S}^{\pi} \sum_{k=1}^{N} r(k)\cos(k\omega)d\omega, \qquad (6.4.10)$$

where $r(k)$ is the autocorrelation of $h_0(n)$, that is,

$$r(k) = \sum_{n=0}^{N} h_0(n)h_0(n-k). \qquad (6.4.11)$$

We interchange the summation with the integral to rewrite (6.4.10) as

$$\phi = (\pi - \omega_S)r(0) - 2\sum_{k=1}^{N} r(k)\frac{\sin(k\omega_S)}{k}. \qquad (6.4.12)$$

So for a given set of α_m, we can evaluate the objective function ϕ without having to perform numerical integration. Standard optimization techniques

[Press, et al., 1989] can now be employed to minimize ϕ with respect to the coefficients α_m. The resulting $H_0(z)$ is a power symmetric FIR filter with smallest possible stopband energy. Note that in this design procedure no computation of spectral factors is required.

TABLE 6.4.2 Design example 6.4.1. Optimized lattice coefficients for the perfect reconstruction analysis bank

m	α_m
0	-0.3836487 e+01
1	0.1247866 e+01
2	-0.7220668 e+00
3	0.4951553 e+00
4	-0.3688423 e+00
5	0.2885146 e+00
6	-0.2327588 e+00
7	0.1913137 e+00
8	-0.1598938 e+00
9	0.1348106 e+00
10	-0.1140321 e+00
11	0.9681786 e−01
12	-0.8223478 e−01
13	0.6963367 e−01
14	-0.5867790 e−01
15	0.4913793 e−01
16	-0.4081778 e−01
17	0.3353566 e−01
18	-0.2713113 e−01
19	0.2149517 e−01
20	-0.1658255 e−01
21	0.1238607 e−01
22	-0.8895189 e−02
23	0.6072120 e−02

Design example 6.4.1. Perfect Reconstruction QMF Lattice

In order to demonstrate the above ideas, consider a lattice structure with 24 sections ($J = 23$) so that the filters have order $N = 2J + 1 = 47$.

Let the stopband edge be at $\omega_S = 0.54\pi$. Table 6.4.2 shows the coefficients α_m, optimized in order to minimize the stopband energy (6.4.12). Notice that $|\alpha_m|$ gets smaller as m grows, and that the sign of α_m alternates. The impulse response $h_0(n)$ can be calculated from α_m by using (6.4.6). Figure 6.4-4(a) shows the analysis filter responses, which have a minimum stopband attenuation of 32 dB.

Figure 6.4-4(b) shows plots of $|H_0(e^{j\omega})|^2 + |H_1(e^{j\omega})|^2$ (which is twice the amplitude distortion, that is, $2|T(e^{j\omega})|$), with quantized coefficients. For this example, the quantization was done by rounding the mantissa part to two decimal digits. For example, in Table 6.4.2, α_{20} was replaced with $[-0.17\ e-01]$. The solid curve is for the direct form implementation of the analysis and synthesis filters, whereas the broken curve is for the lattice structure. This demonstrates the perfect reconstruction property of the lattice inspite of coefficient quantization. The only effect of quantization in the lattice structure is a deterioration of the attenuation characteristics of the filters, as demonstrated in Fig. 6.4-4(c).

Figure 6.4-4(d) shows the response of another analysis bank designed using the same technique. The lattice has $J = 31$ (i.e., 32 sections) so that the filter order is $N = 63$. The value $\omega_S = 0.58\pi$ was used in the optimization. The minimum stopband attenuation of the resulting system is 74 dB. This example demonstrates that we can design perfect reconstruction systems with very large attenuation. If this filter were designed by starting from a half-band filter $H(z)$ (as in Design example 5.3.2), the required attenuation would be 148 dB. Finding a spectral factor of such a half-band filter is subject to severe numerical errors, and the resulting analysis filter will not satisfy the PR condition. The lattice structure on the other hand avoids these steps, and provides the designer with filter coefficients which are guaranteed to have the PR property. The lattice coefficients for this and many other designs can be found in Vaidyanathan and Hoang [1988]; in this reference, the above two designs have been numbered as 48E and 64D, respectively.

6.4.4 Complexity of the Paraunitary QMF Lattice

The total number of multipliers required to implement the lattice sections in the analysis bank is equal to $2(J+1) + 2$. Each of these operates at half the input sampling rate so that we have an average of $J + 2$ MPUs, which simplifies to $0.5(N+3)$ for the analysis bank. With each lattice section requiring two additions, the number of APUs can be verified to be $0.5(N+1)$. The synthesis bank has same complexity. So the lattice structure is much more efficient, requiring half as many MPUs as the direct form. For a given filter order N, the lattice has nearly the same complexity as the polyphase implementation of Johnston's QMF bank (Sec. 5.2.2).

For a given set of specifications (e.g., stopband attenuation, transition bandwidth etc.) Johnston's filters may have higher or lower order (as compared to the above perfect reconstruction system) depending on the acceptable level of amplitude distortion. See Sec. 6.7.2 for more comparisons.

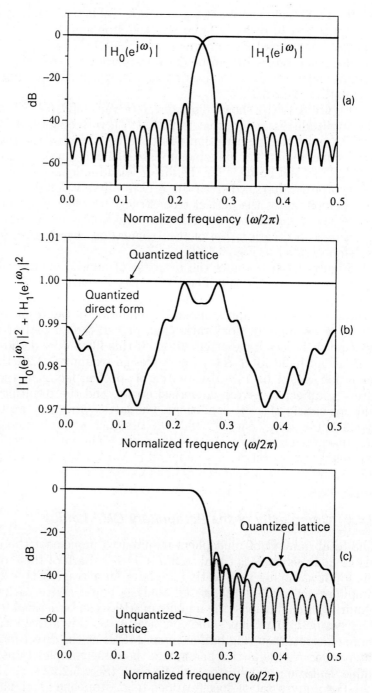

Figure 6.4-4 Design example 6.4.1. (a) Magnitude responses of analysis filters, (b) amplitude distortion after quantization, and (c) response of $H_0(z)$ after lattice quantization.

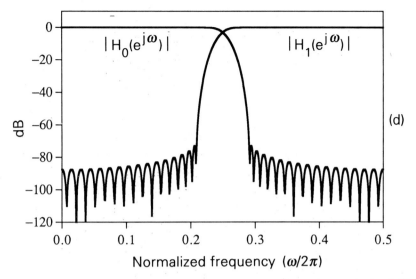

Figure 6.4-4 (*continued*) (d) Another example of lattice-optimized $H_0(z)$.

6.4.5 Summary of Advantages of the QMF Lattice

1. The lattice has the lowest implementation complexity (in terms of MPUs and APUs) among all known (real coefficient FIR PR QMF) structures with paraunitary $\mathbf{E}(z)$.
2. All six properties listed at the beginning of Sec. 6.4.1 are retained in spite of the values of coefficients α_m.
3. In particular, perfect reconstruction property is preserved inspite of coefficient quantization.
4. Moreover the lattice can be used as a design tool. We can optimize the lattice coefficients to minimize the stopband energy. This method has the advantages that (a) there is no need to compute spectral factors of high order half-band filters, (b) the resulting filter $H_0(z)$ is automatically power symmetric with smallest stopband energy, and (c) the resulting filter bank is guaranteed to have perfect reconstruction regardless of the quality of convergence of optimization.
5. In fact, if we want the power symmetric filter $H_0(z)$ to have minimum energy (rather than equiripple as in Design example 5.3.2), then there is no other known technique to design it, other than optimization of the above lattice.
6. *Hierarchical property.* If we delete an arbitrary number of lattice sections, the resulting structure still satisfies all the six properties listed earlier, including perfect reconstruction. Thus, as we add more stages to the lattice, the frequency responses of $H_k(z)$ improve, while retaining perfect reconstruction property.

For an arbitrary FIR filter $H_0(z)$ of order N, it is possible for all the N zeros to be on the unit circle. However, if $H_0(z)$ is power-symmetric, then the maximum number (n_{max}) of zeros on the unit circle is restricted. The number n_{max} can be easily found. For example, if $N = 47$ (as in Design example 6.4.1), then $n_{max} = 24$. See Appendix A of Vaidyanathan and Hoang [1988] for details. From the plot of Fig. 6.4-4(a) we see that there are indeed 24 zeros on the unit circle (as there are twelve in the range $0 \leq \omega \leq \pi$). Experience shows that, in all the examples obtained by lattice optimization, the filter $H_0(z)$ has this maximum permissible number (n_{max}) of zeros on the unit circle. The same is true if $H_0(z)$ is generated as a spectral factor of a optimal equiripple half band filter.

6.5 M-CHANNEL FIR PARAUNITARY FILTER BANKS

We now consider the case of M-channel filter banks. If the analysis filters $H_k(z)$ are constrained to be FIR with paraunitary $\mathbf{E}(z)$, then the choice of synthesis filters as in (6.2.7) results in perfect reconstruction. In this section we outline some methods for finding the analysis filter coefficients $h_k(n)$, so that they have good bandpass responses, under the constraint that $\mathbf{E}(z)$ be paraunitary.

6.5.1 The Basic Optimization Problem

In Chap. 14 we will develop several systematic techniques for representing (or implementing) paraunitary systems in terms of simple building blocks. In this section, we state and use one of these results.

A Characterization of Paraunitary Matrices

Consider the transfer matrix

$$\mathbf{V}_m(z) = \mathbf{I} - \mathbf{v}_m \mathbf{v}_m^\dagger + z^{-1} \mathbf{v}_m \mathbf{v}_m^\dagger, \tag{6.5.1}$$

where $\mathbf{v}_m$ are column vectors (size $M \times 1$), with unit norm, that is, $\mathbf{v}_m^\dagger \mathbf{v}_m = 1$. Using this unit norm property, it is easy to verify that $\widetilde{\mathbf{V}}_m(z)\mathbf{V}_m(z) = \mathbf{I}$ so that $\mathbf{V}_m(z)$ is paraunitary. This matrix can be implemented as in Fig. 6.5-1 using one delay, and therefore has degree equal to one. †

It can be shown that any causal degree-J FIR paraunitary $\mathbf{E}(z)$ can be expressed as

$$\mathbf{E}(z) = \mathbf{V}_J(z)\mathbf{V}_{J-1}(z)\ldots\mathbf{V}_1(z)\mathbf{U}, \tag{6.5.2}$$

where $\mathbf{V}_m(z)$ are paraunitary systems as above, and $\mathbf{U}$ is a constant unitary matrix, that is, $\mathbf{U}^\dagger \mathbf{U} = d\mathbf{I}$. This factorization result [Vaidyanathan, et al., 1989] will be proved in Sec. 14.4.2. Fig. 6.5-2 shows the cascaded structure

---

† The degree of a transfer matrix is the minimum number of delays required to implement it.

corresponding to this. We will also show (Sec. 14.6.2) that any constant unitary matrix $\mathbf{U}$ can be expressed as

$$\mathbf{U} = \sqrt{d}\,\mathbf{U}_1 \mathbf{U}_2 \ldots \mathbf{U}_{M-1} \mathbf{D}, \tag{6.5.3}$$

where $\mathbf{D}$ is a diagonal matrix with diagonal elements $D_{ii} = e^{j\theta_i}$, and

$$\mathbf{U}_i = \mathbf{I} - 2\mathbf{u}_i\mathbf{u}_i^\dagger \qquad \text{(Householder matrix)}. \tag{6.5.4}$$

Here $\mathbf{u}_i$ are unit-norm column vectors. Matrices of the form (6.5.4) are unitary with $\mathbf{U}_i^\dagger \mathbf{U}_i = \mathbf{I}$, and are called Householder matrices. Fig. 6.5-3 shows a structure implementing the building block $\mathbf{U}_i$.

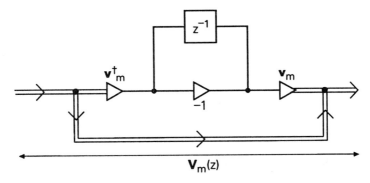

Figure 6.5-1 Implementation of paraunitary $\mathbf{V}_m(z)$ using one delay.

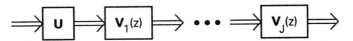

Figure 6.5-2 Factorization of paraunitary $\mathbf{E}(z)$. Building blocks are as in Fig. 6.5-1.

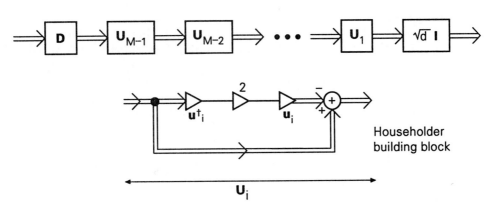

Figure 6.5-3 Cascaded structure for unitary $\mathbf{U}$ with Householder building blocks.

These representations hold whether the coefficients of $\mathbf{E}(z)$ (i.e., coefficients of analysis filters) are real or complex. For the important *real coefficient case*, the vectors $\mathbf{v}_k$ and $\mathbf{u}_i$ are real.

Optimization of the Coefficients $\mathbf{v}_m$ and $\mathbf{u}_i$

It now remains only to optimize the components of $\mathbf{v}_m$ and $\mathbf{u}_i$ such that the responses $|H_k(e^{j\omega})|$ represent 'good' filters. For this we formulate an objective function ϕ as in the two channel case, representing the attenuation characteristics of the filters. We optimize $\mathbf{v}_m$ and $\mathbf{u}_i$ using nonlinear optimization techniques (e.g., see [Press, et al, 1989]), so as to minimize ϕ. The resulting vectors $\mathbf{v}_m$ and $\mathbf{u}_i$ completely determine $\mathbf{E}(z)$, thereby determining all the analysis filters $H_k(z)$. The synthesis filters are then found from (6.2.7), resulting in a perfect reconstruction system.

Completeness. The characterization (6.5.2) is complete in the sense that all (causal FIR) paraunitary $\mathbf{E}(z)$ of degree J are covered. Moreover, the matrix $\mathbf{E}(z)$ is guaranteed to be paraunitary regardless of the values of the quantities $\mathbf{v}_m$ and $\mathbf{u}_i$ as long as they have unit-norm. As a result, the filter-bank system is guaranteed to have the perfect reconstruction property regardless of the values of these unit norm vectors, as long as the synthesis filters are chosen as in (6.2.7). So when the vectors $\mathbf{v}_m$ and $\mathbf{u}_i$ are being optimized, we are searching precisely over the *complete class* of FIR filter banks with paraunitary $\mathbf{E}(z)$.

Objective Function to be Minimized

For simplicity assume that the filters have real coefficients, so that $|H_k(e^{j\omega})|$ is an even function of ω. (Extension to complex coefficient case is easy.) Figure 6.5-4 demonstrates for $M = 3$, a typical set of desired magnitude responses for the analysis filters. The passbands of the filters are nonoverlapping. In the frequency region designated as "passband" for the filter $H_k(z)$, all other filters have their stopbands.

The paraunitary property of $\mathbf{E}(z)$ ensures that the analysis filters are power complementary (Sec. 6.2.2), that is, satisfy (6.1.8). Consequently, if the stopband responses are sufficiently small, then the passband responses of $|H_k(e^{j\omega})|$ are sufficiently close to unity (assuming $c = 1$ in (6.1.8) for simplicity). This means that, if we minimize an objective function ϕ which reflects the stopband energies of $H_k(e^{j\omega})$, then the passbands of all the filters will automatically be "good". Based on this logic we conclude that it is sufficient to formulate an objective function of the form

$$\phi = \sum_{k=0}^{M-1} \int_{k\text{th stopband}} |H_k(e^{j\omega})|^2 d\omega. \qquad (6.5.5)$$

Minimizing ϕ (by optimization of the parameters $\mathbf{u}_i$ and $\mathbf{v}_m$) results in filters which have good stopbands as well as passbands.

Design example 6.5.1

Figure 6.5-5 shows the magnitude responses of the analysis filters for a three channel system designed in this manner. This is obtained by minimizing

$$\phi = \int_{\frac{\pi}{3}+\epsilon}^{\pi} |H_0(e^{j\omega})|^2 d\omega + \int_{0}^{\frac{2\pi}{3}-\epsilon} |H_2(e^{j\omega})|^2 d\omega \qquad (6.5.6)$$
$$+ \int_{0}^{\frac{\pi}{3}-\epsilon} |H_1(e^{j\omega})|^2 d\omega + \int_{\frac{2\pi}{3}+\epsilon}^{\pi} |H_1(e^{j\omega})|^2 d\omega.$$

The impulse responses of the optimized analysis filters $h_k(n)$ are given in Table 6.5.1, with filter order $N = 14$. The coefficients of the synthesis filters for perfect reconstruction are given by $f_k(n) = h_k(14-n)$. Notice that about 20 dB stopband attenuation has been obtained for each filter.

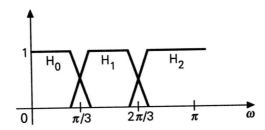

Figure 6.5-4 Typical magnitude responses for an analysis bank with $M = 3$.

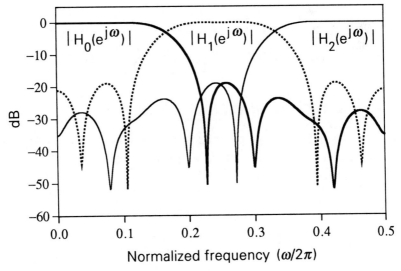

Figure 6.5-5 Design example 6.5.1. Magnitude responses of optimized analysis filters for a 3 channel FIR perfect reconstruction system. Filter order $N = 14$.

TABLE 6.5.1 Design example 6.5.1. The optimized filter coefficients for the FIR perfect reconstruction analysis bank. Here $N = 14$ and $M = 3$.

n	$h_0(n)$	$h_1(n)$	$h_2(n)$
0	-0.0429753	-0.0927704	0.0429888
1	0.0000139	0.0000008	-0.0000139
2	0.1489104	0.0087654	-0.1489217
3	0.2971954	0.0000226	0.2972354
4	0.3537539	0.1864025	-0.3537496
5	0.2672266	-0.0000020	0.2672007
6	0.0870758	-0.3543303	-0.0870508
7	-0.0521155	-0.0000363	-0.0520909
8	-0.0875973	0.3564594	0.0875756
9	-0.0427096	-0.0000049	-0.0427067
10	0.0474530	-0.1931082	-0.0474452
11	0.0429618	0.0000230	0.0429677
12	0.0	0.0	0.0
13	-0.0232765	-0.0000026	-0.0232749
14	0.0000022	0.0	0.0000022

6.5.2 Incorporation of Symmetry

In most practical designs the response $|H_{M-1-k}(e^{j\omega})|$ can be taken to be the image of $|H_k(e^{j\omega})|$ with respect to $\pi/2$, that is,

$$|H_{M-1-k}(e^{j\omega})| = |H_k(e^{j(\pi-\omega)})|. \quad (6.5.7)$$

This is demonstrated in Fig. 6.5-6 for $M = 5$. (We assume the filter coefficients to be real so that the magnitude responses are automatically symmetric with respect to π.) Indeed, in Design example 6.5.1 the responses $|H_0(e^{j\omega})|$ and $|H_2(e^{j\omega})|$ do satisfy this property approximately, (and moreover $|H_1(e^{j\omega})|$ is self-symmetric with respect to $\pi/2$) because the stopband regions [regions of integration in (6.5.6)] were chosen with such symmetry.

This opens up an idea. Suppose we modify the polyphase structure such that the symmetric constraint is built into the structure. Because of this constraint, the number of degrees of freedom (i.e., $\mathbf{v}_m$'s) will be reduced by almost a factor of 2, which reduces the number of parameters to be optimized for a given filter order. This in turn will result in drastic reduction of optimization time [Nguyen and Vaidyanathan, 1988].

For the three channel real coefficient case, we can incorporate the symmetry condition (6.5.7) by the constraint

$$H_2(z) = H_0(-z), \quad H_1(z) = \alpha_1(z^2). \tag{6.5.8}$$

[This is not the only way to achieve (6.5.7), but it works.] In particular $H_1(z)$ is constrained to be a function of z^2. These imply $|H_2(e^{j\omega})| = |H_0(e^{j(\pi-\omega)})|$ and $|H_1(e^{j\omega})| = |H_1(e^{j(\pi-\omega)})|$, so that (6.5.7) is satisfied. Notice from Table 6.5.1 that the filter coefficients of the previous design example almost satisfy (6.5.8)! It will, therefore, be judicious to force this symmetry prior to optimization.

Figure 6.5-7 shows a structure for imposing (6.5.8). Here the transfer functions are given by

$$H_0(z) = \alpha_0(z^2) + z^{-3}\alpha_2(z^2), \quad H_1(z) = \sqrt{2}\alpha_1(z^2), \quad H_2(z) = H_0(-z), \tag{6.5.9}$$

so that (6.5.8) is satisfied.

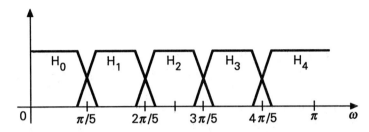

Figure 6.5-6 Demonstrating the symmetry of responses with respect to $\pi/2$.

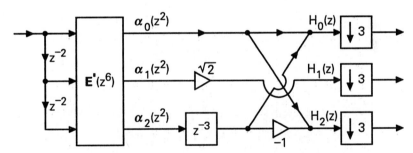

Figure 6.5-7 The three-channel analysis bank, with $H_k(z) = H_{2-k}(-z)$.

Retaining the Perfect Reconstruction Property

It only remains to 'worry about' the perfect reconstruction property of this modified structure. Note that the analysis bank can be written as

$$\mathbf{h}(z) = \underbrace{\begin{bmatrix} 1 & 0 & 1 \\ 0 & \sqrt{2} & 0 \\ 1 & 0 & -1 \end{bmatrix}}_{\mathbf{A}} \begin{bmatrix} 1 & 0 & 0 \\ 0 & 1 & 0 \\ 0 & 0 & z^{-3} \end{bmatrix} \mathbf{E}'(z^6) \begin{bmatrix} 1 \\ z^{-2} \\ z^{-4} \end{bmatrix}. \tag{6.5.10}$$

The real matrix $\mathbf{A}$ is unitary, with $\mathbf{A}^T\mathbf{A} = 2\mathbf{I}$. So if we force $\mathbf{E}'(z)$ to be paraunitary in the usual way, and choose the synthesis bank as in Fig. 6.5-8, then the complete analysis/synthesis system is equivalent to Fig. 6.5-9. This, indeed, is a perfect reconstruction system, since the integers 2 and 3 are relatively prime (see Fig. 5.6.3 and associated comments). The synthesis filter vector is given by

$$\mathbf{f}^T(z) = [\,z^{-4}\ \ z^{-2}\ \ 1\,]\,\widetilde{\mathbf{E}}'(z^6)\begin{bmatrix} z^{-3} & 0 & 0 \\ 0 & z^{-3} & 0 \\ 0 & 0 & 1 \end{bmatrix}\mathbf{A}, \qquad (6.5.11)$$

from which we verify that the relation

$$f_k(n) = c_0 h_k(L-n), \quad 0 \le k \le M-1, \qquad (6.5.12)$$

holds for appropriate c_0 and L.

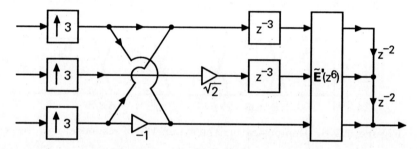

Figure 6.5-8 The synthesis bank corresponding to Fig. 6.5-7. This results in a perfect reconstruction system when $\mathbf{E}'(z)$ is paraunitary. (See text.)

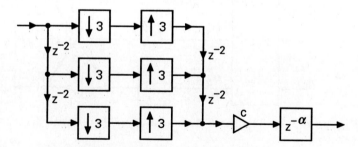

Figure 6.5-9 The analysis bank of Fig. 6.5-7 followed by the synthesis bank of Fig. 6.5-8 is equivalent to the above structure when $\mathbf{E}'(z)$ is paraunitary. Here c is a nonzero constant, and α = integer.

Paraunitariness of E(z). Notice that the polyphase component matrix $\mathbf{E}(z)$ is *not* the same as $\mathbf{E}'(z)$. In the above design problem, it is not necessary to know what $\mathbf{E}(z)$ is, as it does not directly enter the optimization. It turns out, however, that $\mathbf{E}(z)$ is paraunitary. [This follows from

Theorem 6.2.1 because the system under consideration is a perfect reconstruction system satisfying (6.5.12)].

An extension of this idea for arbitrary M is possible [Nguyen and Vaidyanathan, 1988]. The details depend on whether M is even or odd. Figure 6.5-10 shows the structure for odd M (of which Fig. 6.5-7 is a special case), which forces the condition

$$H_k(z) = H_{M-1-k}(-z), \qquad (6.5.13)$$

and hence (6.5.7). Here **A** is a generalization of the matrix **A** in (6.5.10). For example, with $M = 5$, we have

$$\mathbf{A} = \begin{bmatrix} 1 & 0 & 0 & 0 & -1 \\ 0 & 1 & 0 & -1 & 0 \\ 0 & 0 & \sqrt{2} & 0 & 0 \\ 0 & 1 & 0 & -1 & 0 \\ 1 & 0 & 0 & 0 & -1 \end{bmatrix}. \qquad (6.5.14)$$

Clearly **A** is unitary with $\mathbf{A}^T\mathbf{A} = 2\mathbf{I}$. If $\mathbf{E}'(z)$ is constrained to be paraunitary, we have to choose the synthesis filters as $f_k(n) = c_0 h_k(L - n)$ for perfect reconstruction. Notice that for a given filter order N, the number of parameters which enter the optimization problem is nearly halved, when compared with the direct approach [which uses $\mathbf{E}(z^M)$ rather than $\mathbf{E}'(z^{2M})$].

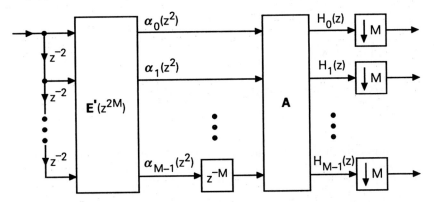

Figure 6.5-10 Extension of the symmetric analysis bank for arbitrary odd M.

Design example 6.5.2

Figure 6.5-11 shows the frequency response of an FIR perfect reconstruction system designed in the above manner, using an optimization program from [IMSL, 1987]. The analysis filters have order ≤ 55, and the impulse response coefficients can be found in Table I of Vaidyanathan, et al. [1989].

In order for the optimization to converge to an acceptable solution in reasonable time, it is very important to 'initialize' the unknown parameters in a judicious way. This initialization can be done by designing approximate reconstruction systems (called pseudo QMF designs, Sec. 8.1). In Sec. 8.5 we return to a more systematic design procedure (cosine modulated perfect reconstruction systems), and provide further details.

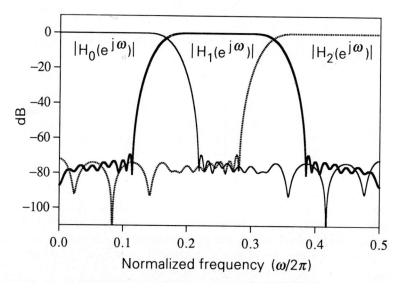

Figure 6.5-11 Design example 6.5-2. Magnitude responses of analysis filters for a three channel FIR perfect reconstruction system. Analysis filters are related as in (6.5.8), and filter order $N = 55$. (© Adopted from 1989 IEEE.)

6.6 TRANSFORM CODING AND THE "LOT"

Before the introduction of FIR QMF banks with paraunitary $\mathbf{E}(z)$, some authors have independently reported other techniques for perfect recovery systems, which work for the case where the filter order is $N = 2M - 1$. One of these is the *lapped orthogonal transform* (LOT) studied in Cassereau [1985], Malvar and Staelin [1989], and Malvar [1990a]. The second is a special case (introduced in [Princen and Bradley, 1986]) of the pseudo QMF bank to be discussed in Chap. 8. The polyphase matrices in these methods have the form $\mathbf{E}(z) = \mathbf{e}_0 + \mathbf{e}_1 z^{-1}$ so that the analysis filters have order $N = 2M - 1$.

It was observed in Vetterli and Le Gall [1989] that the above two systems have paraunitary $\mathbf{E}(z)$, thereby offering a very simple explanation of the perfect reconstruction property. We now elaborate this point.

6.6.1 Review of Transform Coding

In the area of waveform quantization and coding, there exists a popular technique called *transform coding* [Jayant and Noll, 1984]. In this technique, a signal $x(n)$ is divided into blocks of length M, and each block transformed into a new block of length M by a linear (nonsingular) transformation $\mathbf{T}$. This can be schematically represented as in Fig. 6.6-1, using multirate notation. Let us denote a block of length M as

$$\mathbf{x}(n) = [\,x(nM) \quad x(nM-1) \quad \ldots \quad x(nM-M+1)\,]^T. \qquad (6.6.1)$$

Then the transformed block is obtained as

$$\mathbf{y}(n) = \mathbf{T}\mathbf{x}(n). \qquad (6.6.2)$$

Notice that successive blocks of input do not overlap, that is, $\mathbf{x}(n)$ and $\mathbf{x}(n+1)$ do not have overlapping samples. If we take $\mathbf{T} = \mathbf{W}^*$, the above system becomes the familiar uniform-DFT bank.

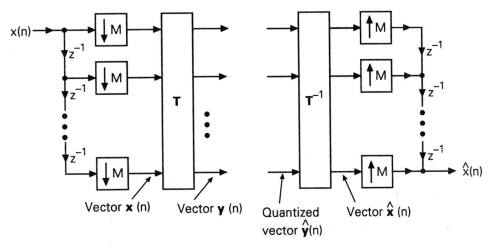

Figure 6.6-1 The transform-coder/decoder schematic.

The components of the transformed data $\mathbf{y}(n)$ are typically quantized and transmitted. With $\widehat{\mathbf{y}}(n)$ denoting the quantized $\mathbf{y}(n)$, the reconstructed signal is obtained by using the inverse transformation $\mathbf{T}^{-1}$, and unblocking the resulting vector $\widehat{\mathbf{x}}(n)$ by use of expanders and delay chain as shown in the figure. In the absence of quantization of $\mathbf{y}(n)$, the system has the perfect reconstruction property, i.e., $\widehat{x}(n) = x(n-M+1)$.

The aim is to quantize the components of $\mathbf{y}(n)$ in such a way that the quantization error in the reconstructed signal $\widehat{x}(n)$ is minimized. In practically all applications, the transformation matrix $\mathbf{T}$ satisfies $\mathbf{T}^\dagger \mathbf{T} = \mathbf{I}$ (orthogonal transform coding). Under this condition, and with suitable statistical assumptions one can solve for the best set of quantizers, resulting in the so called *optimal bit-allocation* schemes (Appendix C).

The main advantage of transform coding is that, an appropriate choice of $\mathbf{T}$ results in reduced number of bits per second, after quantization. (The extent of this reduction is quantitatively measured by the so-called coding gain.[†] A related problem in transform coding is the choice of best unitary **T**. This has been solved for the case of wide sense stationary $x(n)$: the best **T** is the one whose rows are equal to the eigenvectors of the $M \times M$ autocorrelation matrix $E[\mathbf{x}(n)\mathbf{x}^\dagger(n)]$. This **T** is called the *Karhunen-Loeve Transform* (KLT) (see Appendix C for details).

A commonly observed disadvantage of transform coding is the *blocking effect,* caused by the encoding of $x(n)$ block-by-block without overlap. In image coding, this manifests in the form of visible discontinuities across block boundaries; in speech coding it is perceived as extraneous tones. Elegant techniques for reducing the blocking effect have been proposed based on the so-called lapped orthogonal transforms (LOT) [Cassereau, 1985], [Malvar and Staelin, 1989]. We shall now define the LOT in the framework of multirate systems, and observe that it is a filter bank with paraunitary polyphase matrix $\mathbf{E}(z)$.

6.6.2 Transforms with Overlap

Figure 6.6-2(a) shows a modification of the transform coder, where the matix **T** is $M \times L$ rather than $M \times M$, with $L \geq M$. This means that the input is partitioned into *overlapping* blocks

$$\mathbf{x}(n) = [\, x(nM) \quad x(nM-1) \quad \ldots \quad x(nM-L+1)\,]^T.$$

of length L, and each block transformed into a block of (smaller) length M. Fig. 6.6-2(b) demonstrates this overlap for $L = 5, M = 3$. The existence of overlap between blocks has been shown to reduce the 'blocking effect' in speech and image processing.

The 'inverse' transform operation is also indicated in the figure (where **Q** has to be chosen appropriately; see below). Note that the process of obtaining $\mathbf{y}(n)$ from $x(n)$ is equivalent to the use of an analysis bank, with the analysis filter vector $\mathbf{h}(z)$ [eqn. (5.4.1)] given by

$$\mathbf{h}(z) = \mathbf{T}\mathbf{e}_1(z), \qquad (6.6.3)$$

with $\mathbf{e}_1(z) = [\,1 \quad z^{-1} \quad \ldots \quad z^{-(L-1)}\,]^T$. In all known applications we have $L \leq 2M$. We shall let $L = 2M$, (with the provision that some columns of **T** are allowed to be zero, permitting the $L < 2M$ case). Evidently the analysis filters have lengths $\leq 2M$. By partitioning **T** as

$$\mathbf{T} = [\,\mathbf{T}_0 \quad \mathbf{T}_1\,], \qquad (6.6.4)$$

[†] Notice the similarity of this philosophy to subband coding. We can think of subband coding as a generalization of this transform coding idea, with **T** replaced by $\mathbf{E}(z)$ [Fig. 5.5-3(b)].

where $\mathbf{T}_0$ and $\mathbf{T}_1$ are $M \times M$, we can write

$$\mathbf{h}(z) = [\mathbf{T}_0 + z^{-M}\mathbf{T}_1]\mathbf{e}(z), \qquad (6.6.5)$$

where $\mathbf{e}(z)$ is the delay chain vector $[1 \quad z^{-1} \quad \ldots \quad z^{-(M-1)}]^T$. By comparing with (5.5.2b), we see that the analysis bank has $M \times M$ polyphase component matrix

$$\mathbf{E}(z) = \mathbf{T}_0 + z^{-1}\mathbf{T}_1. \qquad (6.6.6)$$

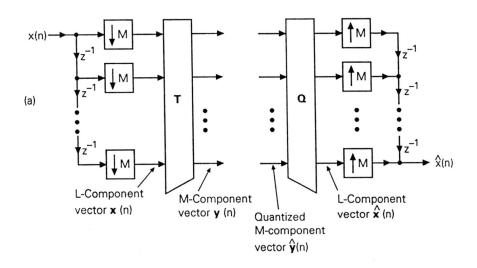

Figure 6.6-2 (a) The transform-coder/decoder with successive blocks overlapping. For appropriate choice of $\mathbf{T}$, this leads to the Lapped Orthogonal Transform (LOT) technique. (b) Demonstration of overlap, with $L = 5$ and $M = 3$.

We know that if $\mathbf{E}(z)$ is paraunitary, we can obtain perfect reconstruction (in absence of quantization) by taking the synthesis filters to be

$$f_k(n) = h_k^*(2M - 1 - n).$$

Paraunitariness of $\mathbf{E}(z)$ implies $\mathbf{E}(z)\widetilde{\mathbf{E}}(z) = c\mathbf{I}$. In terms of $\mathbf{T}_0$ and $\mathbf{T}_1$ this simplifies to

$$\mathbf{T}_0\mathbf{T}_0^\dagger + \mathbf{T}_1\mathbf{T}_1^\dagger = c\mathbf{I}, \qquad (6.6.7)$$

and
$$\mathbf{T}_0 \mathbf{T}_1^\dagger = \mathbf{0}. \tag{6.6.8}$$

After making appropriate notational changes, we find that these are precisely the conditions (4) and (5) given by Malvar and Staelin, [1989]. So the LOT structure is a subclass of filter banks with paraunitary $\mathbf{E}(z)$, with analysis filter length $= 2M$.

We know that the above choice of $f_k(n)$, which results in perfect reconstruction, corresponds to the choice $\mathbf{R}(z) = z^{-1}\widetilde{\mathbf{E}}(z)$ [Fig. 5.5-3(b)]. This in turn helps us to identify the matrix $\mathbf{Q}$ in Fig. 6.6-2(a) (Problem 6.17).

The LOT has been extended by Malvar to obtain the extended LOT (abbreviated ELT). This system, again, is a paraunitary filter bank with some additional structure on the filters, namely the cosine modulation property. We will return to this in Sec. 8.5.

6.7 SUMMARY, COMPARISONS, AND TABLES

In this and the previous chapters several types of filter banks have been presented. In Chap. 5 a number of techniques for the design of two-channel QMF bank were studied, and the general theory of M-channel maximally decimated filter banks was developed. In Tables 5.10.1–5.10.4 we already summarized these results.

In this chapter we have concentrated on filter banks with paraunitary polyphase matrix $\mathbf{E}(z)$. The main points are summarized in Table 6.7.1. Special properties of two-channel paraunitary filter banks were studied in Sec. 6.3, and various results summarized at appropriate places in that section. In Sec. 6.4 we presented a lattice structure for these systems. This structure is such that, perfect reconstruction (PR) is preserved inspite of multiplier quantization.

6.7.1 Venn Diagram for Perfect Reconstruction Systems

Paraunitariness of $\mathbf{E}(z)$ is not a necessary condition for obtaining the PR property in FIR QMF systems. From Chap. 5 we know that (5.6.10) is really necessary and sufficient. Indeed, in Chap. 7 we will show for the two channel case, that if the analysis filters are required to have linear phase, it is necessary to *give up* the paraunitary property.

In Fig. 6.7-1 we show a Venn diagram which summarizes various possibilities with FIR $\mathbf{E}(z)$. Set 2 here is the set of $\mathbf{E}(z)$ for which the determinant is a strictly minimum phase polynomial, that is, has all zeros inside the unit circle (except possibly some zeros at $z = \infty$, as in the function z^{-1}). In this case if we take $\mathbf{R}(z) = \mathbf{E}^{-1}(z)$, we obtain perfect reconstruction with stable synthesis filters. But the synthesis filters are IIR, and typically have very high order (for large M). Set 3 is further constrained by the requirement that the determinant of $\mathbf{E}(z)$ be a delay. In this case the synthesis filters are also FIR but still can have very high order for large M. Set 4 represents

TABLE 6.7.1 Perfect reconstruction filter banks with paraunitary $\mathbf{E}(z)$

Any M-channel QMF bank (Fig. 5.4-1) can always be redrawn as shown in Fig. 5.5-3(a), where $\mathbf{E}(z)$ and $\mathbf{R}(z)$ are the polyphase component matrices of the analysis and synthesis banks.

1. We say that $\mathbf{E}(z)$ is paraunitary if $\widetilde{\mathbf{E}}(z)\mathbf{E}(z) = d\mathbf{I}$ for some $d > 0$. If in addition $\mathbf{E}(z)$ is causal and stable (as we assume in following summary), we say that it is lossless. We use 'lossless' and 'paraunitary' interchangeably whenever causality and stability are clear from context.

2. Paraunitariness of $\mathbf{E}(z)$ is not *necessary* for perfect reconstruction. If $\mathbf{E}(z)$ is paraunitary, the choice $\mathbf{R}(z) = cz^{-K}\widetilde{\mathbf{E}}(z)$ results in perfect reconstruction. Assuming $\mathbf{E}(z)$ is FIR, this choice of $\mathbf{R}(z)$ is also FIR and results in causal synthesis filters $F_k(z)$ for proper choice of K.

3. The condition $\mathbf{R}(z) = cz^{-K}\widetilde{\mathbf{E}}(z)$ is equivalent to

$$f_k(n) = ch_k^*(L - n), \quad \text{i.e.,} \quad F_k(z) = cz^{-L}\widetilde{H}_k(z),$$

   with $L = MK + M - 1$. That is, coefficients of $F_k(z)$ are essentially obtained by writing the coefficients of $H_k(z)$ in reverse order and conjugating; c is just a scale factor.) This choice of $F_k(z)$ also implies $|F_k(e^{j\omega})| = |cH_k(e^{j\omega})|$.

4. If $\mathbf{E}(z)$ is FIR and paraunitary, we obtain perfect reconstruction by choosing the synthesis filters as above, i.e., $f_k(n) = ch_k^*(L - n)$. See Theorem 6.2.1 for further complete summary.

5. **Factorization.** An FIR paraunitary system $\mathbf{E}(z)$ of degree J can always be factorized as in (6.5.2), in terms of unit norm vectors $\mathbf{v}_m$ and $\mathbf{u}_i$ (see (6.5.1)–(6.5.4)). Conversely, (6.5.2) always represents a paraunitary system as long as $\mathbf{v}_m$ and $\mathbf{u}_i$ have unit norm. The vectors $\mathbf{v}_m$ and $\mathbf{u}_i$ can be optimized to obtain good analysis filters $H_k(z)$ for perfect reconstruction.

6. Paraunitariness of $\mathbf{E}(z)$ induces further properties (Section 6.2.2).
   b) The analysis filters are power complementary: $\sum_{k=0}^{M-1} |H_k(e^{j\omega})|^2 =$ nonzero constant.
   c) The function $\widetilde{H}_k(z)H_k(z)$ is an Mth band (Nyquist(M)) filter.
   d) The alias component matrix $\mathbf{H}(z)$ is paraunitary. In fact $\mathbf{H}(z)$ is paraunitary if and only if $\mathbf{E}(z)$ is paraunitary because of the relation (6.2.12).

Results on two-channel paraunitary QMF banks.

These were derived in Sections 6.3 and 6.4, and the main results are summarized at appropriate places in these sections.

$\mathbf{E}(z)$ of the form

$$\mathbf{E}(z) = \mathbf{R}_J \mathbf{\Lambda}(z) \mathbf{R}_{J-1} \mathbf{\Lambda}(z) \ldots \mathbf{R}_1 \mathbf{\Lambda}(z) \mathbf{R}_0 \qquad (6.7.1)$$

where $\mathbf{R}_m$ are nonsingular (not necessarily unitary) matrices, and $\mathbf{\Lambda}(z)$ is a diagonal matrix of delay elements [e.g., as in (6.1.10)]. This is a convenient subset of all matrices satisfing [det $\mathbf{E}(z)$] = delay.

Finally set 5 is the paraunitary set, and has many advantages explained in this chapter. These advantages are summarized below, and hold for two- as well as M-channel cases.

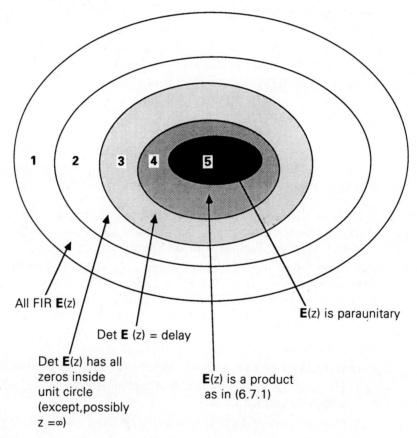

Figure 6.7-1 Summary of various ways to force the perfect reconstruction property in FIR QMF banks.

Advantages of Paraunitary E(z) in FIR Filter Banks

1. No matrix inversions are involved in the design.
2. The synthesis filters are FIR, have the same length as the analysis filters, and can be obtained by time-reversal and conjugation of the analysis filter coefficients.

3. If the paraunitary matrix $\mathbf{E}(z)$ is implemented as a cascaded structure (Fig. 6.5-2), the perfect reconstruction property can be retained in spite of multiplier quantization. For the two-channel case, this was justified in Sec. 6.4.1. For the M channel case, see Sec. 14.11.2.
4. The cascaded paraunitary structure also ensures that the computational complexity (for implementing the analysis bank) is low. This was justified in Sec. 6.4.4 for the two channel case. For the M channel case, see below.
5. The objective function to be optimized during design is simple, and does not have to explicitly include passband error. (It is implicitly there because of power complementary property of analysis filters, induced by paraunitariness of $\mathbf{E}(z)$.)
6. Filter banks with paraunitary $\mathbf{E}(z)$ can be used to generate an orthonormal basis for the so-called *wavelet transforms.* See Chap. 11.

6.7.2 Complexity comparisions

For each of the methods studied in this and the previous chapters, we also presented design examples, and counted the number of multiplications and additions per unit time (MPU and APU). Table 6.7.2 gives a summary of the major features of various two-channel QMF banks, along with computational complexities. Table 6.7.3 provides a comparison for a chosen set of specifications for the response of $H_0(z)$. It is clear that the perfect reconstruction QMF bank implemented with lattice structures (Method 2) is quite competetive with the approximate reconstruction systems (Method 1). Finally Table 6.7.4 compares the FIR PR lattice with the IIR power-symmetric method (Sec. 5.3). The IIR system has the lowest complexity among all the methods with given specifications on $H_0(z)$.

Complexity of M-channel Paraunitary QMF Bank

First suppose that we implement the filters $H_k(z)$ in direct form. Even though the filters are constrained by the fact that $\mathbf{E}(z)$ is paraunitary, there is no obvious relation among the coefficients of these filters (except the optional relation (6.5.13), if symmetry has been imposed through the structure). So the cost of a direct form implementation is roughly proportional to MN where N is the filter order and M is the number of channels. Since each $H_k(z)$ is followed by a decimator, we can implement it in polyphase form (Sec. 4.3). So the analysis bank requires N MPUs and N APUs.

The second method is to implement the system using polyphase matrices, that is, exactly as in Fig. 5.5-3(b). In this manner, the implicit relation between the analysis filters, induced by the paraunitariness of $\mathbf{E}(z)$, can be exploited. We will show below that the complexity of this system is only

$$\frac{2N}{M} + M \text{ MPUs} \quad \text{and} \quad \frac{2N}{M} + \frac{(M-1)^2}{M} \text{ APUs}. \qquad (6.7.2)$$

For example if $J = 10$ and $M = 5$, then $N = 54$ and the analysis bank requires 26.6 MPUs and 24.8 APUs. For the first method where each $H_k(z)$

TABLE 6.7.2 Comparison of three types of two-channel QMF banks.
Note: ALD = Aliasing distortion; AMD = Amplitude distortion; PHD = Phase distortion.
N = Order of $H_0(z)$

	Method 1 (FIR) Section 5.2.2	Method 2 (FIR) Perfect Reconstruction with lossless $E(z)$ Section 6.4	Method 3 IIR Allpass based. Section 5.3.4
Relation between filters	$H_1(z) = H_0(-z)$ $F_0(z) = H_0(z), F_1(z) = -H_1(z)$	$H_1(z) = -z^{-N}\widetilde{H}_0(-z)$ $F_0(z) = z^{-N}\widetilde{H}_0(z),$ $F_1(z) = z^{-N}\widetilde{H}_1(z)$ H_0, H_1 power comp.	Same as method 1
Phase response of $H_0(z)$	linear	nonlinear	nonlinear since $H_0(z)$ IIR
Distortions in QMF bank	ALD canceled AMD minimized PHD eliminated	ALD canceled AMD eliminated PHD eliminated	ALD canceled AMD eliminated PHD remains
Features of $H_0(z)$		$\widetilde{H}_0(z)H_0(z)$ is a (zero-phase FIR) half-band filter, i.e., $H_0(z)$ is power symmetric	$2H_0(z) = a_0(z^2)$ $+z^{-1}a_1(z^2),$ $a_0(z), a_1(z)$ IIR allpass. $H_0(z)$ is power symmetric.
Complexity, i.e., No., of (MPU, APU) for analysis bank[‡]	$0.5(N+1), 0.5(N+1),$ using polyphase	$(N+1), N$ using direct form polyphase. $0.5(N+3), 0.5(N+1)$ using paraunitary lattice	$0.25(N-1),$ $0.5(N+1)$
Group delay of entire analysis/synthesis system	N samples	N samples	nonconstant

[‡] In each case, complexity of the sythesis bank is essentially same as that of the analysis bank.
MPU = multiplications per unit time. APU = additions per unit time.
One unit of time = separation between samples of input $x(n)$.

TABLE 6.7.3 Comparison of Design examples for two FIR two-channel QMF banks. In both methods, $H_0(z)$ has $A_S \simeq 40$dB and $\omega_S \simeq 0.586\pi$.

	Method 1 (Johnston's 32D filter) i.e., imperfect reconstruction with linear phase analysis and sythesis filters	Method 2 (Vaidyanathan and Hoang) Perfect reconstruction lattice with nonlinear phase analysis and synthesis filters
Required order of $H_0(z)$	$N = 31$ for peak amplitude distortion $= 0.025$ dB	$N = 31$
Group delay of entire analysis/synthesis system	31	31
No. of (MPU, APU) for analysis bank	(16, 16)	(17, 16)
Price paid by the method	Amplitude distortion not equal to zero	$H_0(z)$ and $H_1(z)$ do not have linear-phase

TABLE 6.7.4 Comparison of Methods 2 and 3. In both cases, $A_S \simeq 40$dB and $\omega_S \simeq 0.62\pi$.

	Method 2 FIR Perfect reconstruction QMF lattice	Method 3 IIR allpass based system with nonzero phase distortion
Required order N for $H_0(z)$	21	5
No. of (MPU, APU) for analysis bank	(12, 11)	(1, 3)
Group delay distortion	None	Difference between largest and smallest delays $=14$ samples
Other distortions	None	None

is implemented independently in polyphase form, we require 54 MPUs and APUs. This shows that the implementation based on $\mathbf{E}(z)$ is more efficient.

To justify (6.7.2), recall that any FIR paraunitary $\mathbf{E}(z)$ of degree J can be implemented in cascade form as in Fig. 6.5-2. Here $\mathbf{V}_m(z)$ and $\mathbf{U}$ are $M \times M$ matrices. Each of the building blocks $\mathbf{V}_m(z)$ requires $2M$ real multipliers (since $\mathbf{v}_m$ is real for real-coefficient case). The unitary matrix $\mathbf{U}$ can be implemented as a cascade of $M - 1$ Householder matrices [eqn. (6.5.3)]. Each Householder matrix is implemented as in Fig. 6.5-3. The unit-norm vectors $\mathbf{u}_i$ appearing in these Householder matrices have a restricted number of nonzero entries as elaborated in Sec. 14.6.2. From these details, we can verify that $\mathbf{E}(z)$ (hence the analysis bank) requires $2N + M^2$ multipliers. The number of additions is $2N + (M - 1)^2$. Since $\mathbf{E}(z)$ is operating at M times lower rate, the matrix $\mathbf{E}(z)$ (hence the analysis bank) has the complexity given in (6.7.2). If the filters are constrained by symmetry conditions such as (6.5.13), we obtain a further saving by about a factor of 2 (since the degree of $\mathbf{E}'(z)$ in Fig. 6.5-10 is nearly halved for fixed filter order N).

The implementation in terms of the cascaded structure for $\mathbf{E}(z)$ has the additional advantage that when the multipliers are quantized, the perfect reconstruction property is unaffected (Sec. 14.11). The same is not true for the direct form implementation of $H_k(z)$.

Cosine modulated FIR PR systems. In Chap. 8 we will study a class of FIR perfect reconstruction systems in which all analysis filters are derived from a single prototype by cosine modulation. This has the advantage that during the design (optimization) phase the number of parameters to be optimized is very small, even for large M. Another advantage is that the complexity of the implementation is very small. Indeed, among all techniques for designing (perfect or approximate) QMF banks for arbitrary M, this method appears to have least complexity (both during design and implementation). The perfect reconstruction property in this scheme is again acheived by exploiting paraunitariness of $\mathbf{E}(z)$, as we shall see in Sec. 8.5.

PROBLEMS

6.1. Consider the following FIR analysis bank.

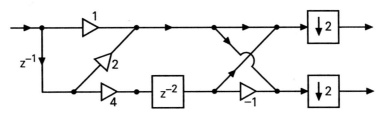

Figure P6-1

a) Find $\mathbf{E}(z)$, and the analysis filters $H_0(z), H_1(z)$. Is $\mathbf{E}(z)$ paraunitary? What is its determinant?

b) Find a set of causal FIR synthesis filters which result in perfect reconstruction.

6.2. Find an example of a real-coefficient two-channel FIR perfect reconstruction QMF bank, with the following features: (i) $\mathbf{E}(z)$ is paraunitary, and (ii) $H_0(z)$ is causal, with order ≥ 3. You must explicitly indicate the values of the filter coefficients $H_0(z), H_1(z), F_0(z)$ and $F_1(z)$.

6.3. Consider the example generated in Problem 6.2.

a) Verify that $H_0(z)$ is power symmetric.

b) Draw the lattice structures for the analysis and synthesis banks, and indicate the values of the lattice coefficients α_m.

6.4. Let $\mathbf{E}(z)$ denote the polyphase matrix of an M-channel analysis bank.

a) Suppose $\mathbf{E}(z)$ is normalized lossless. Show that all the analysis filters $H_k(z)$ have unit energy, i.e., $\sum_n |h_k(n)|^2 = 1$.

b) Conversely, let all the filters $H_k(z)$ have unit energy, and furthermore let them be power complementary. This does not necessarily mean that $\mathbf{E}(z)$ is paraunitary. Prove this by counter example.

6.5. Consider the lattice structure of Fig. 6.4-3(a). Suppose we perform the following changes: (a) replace the multiplier "-1" (just prior to stage 0) with a multiplier equal to "$+1$," and (b) exchange each α_k with $-\alpha_k$. Show then that the analysis filter $H_0(z)$ remains unchanged, and that $H_1(z)$ gets replaced with $-H_1(z)$. This shows that the multiplier with value "-1" in the figure is not of major importance.

6.6. Let $\mathbf{R}(z)$ and $\mathbf{E}(z)$ be matrices with rational entries, and let $\mathbf{e}(z)$ be the delay chain vector [as in (5.4.1)]. Prove that the relation (6.2.11) implies $\mathbf{R}(z) = cz^{-K}\widetilde{\mathbf{E}}(z)$.

6.7. In this problem we obtain a second, independent, proof that the recursion (6.4.8) for synthesizing the perfect reconstruction lattice works. Let $H_0^{(m)}(z) =$

$\sum_{n=0}^{2m+1} h^{(m)}(n) z^{-n}$ be a transfer function with real coefficients, satisfying

$$|H_0^{(m)}(e^{j\omega})|^2 + |H_0^{(m)}(-e^{j\omega})|^2 = c_m, \qquad (P6.7a)$$

where c_m is a nonzero constant. This is nothing but the power symmetric property, except that c_m is not necessarily unity. Define

$$H_1^{(m)}(z) = -z^{-(2m+1)} H_0^{(m)}(-z^{-1}). \qquad (P6.7b)$$

Assume $h^{(m)}(0) \neq 0$. Suppose we define $H_0^{(m-1)}(z)$ as in (6.4.8) where $\alpha_m = -h^{(m)}(2m+1)/h^{(m)}(0)$. Prove the following.

a) $H_1^{(m-1)}(z)$ defined in (6.4.8) satisfies a relation similar to (P6.7b) with m replaced by $m-1$.
b) $H_0^{(m-1)}(z)$ satisfies a property similar to (P6.7a) with m replaced by $m-1$.
c) $h^{(m-1)}(0) \neq 0$.
d) $H_0^{(m-1)}(z)$ and $H_1^{(m-1)}(z)$ are causal with degree $\leq 2m-1$.

Summarizing, the pair $[H_0^{(m-1)}(z), H_1^{(m-1)}(z)]$ satisfies all the properties of the higher order pair $[H_0^{(m)}(z), H_1^{(m)}(z)]$. So the recursion can be repeated. After a finite number of recursions we obtain $H_0^{(0)}(z)$ and $H_1^{(0)}(z)$. Write down the forms of these. Hence show that the pair $[H_0^{(m)}(z), H_1^{(m)}(z)]$ can be implemented in the form of the following lattice structure.

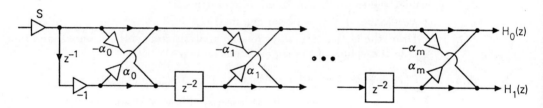

Figure P6-7

6.8. The system in (6.5.1) can be written as $\mathbf{h}(0) + \mathbf{h}(1)z^{-1}$ where $\mathbf{h}(n)$ are the 'impulse response' coefficients of $\mathbf{V}_m(z)$. Let

$$\mathbf{v}_m = \frac{1}{3} \begin{bmatrix} 1 \\ 2 \\ 2 \end{bmatrix}.$$

This is a unit norm vector. By using this in (6.5.1), evaluate the coefficients $\mathbf{h}(n)$. Hence fill all the nine entries of $\mathbf{V}_m(z)$.

6.9. Consider the paraunitary system $\mathbf{V}_m(z)$ of Problem 6.8. Suppose we take this as the polyphase matrix $\mathbf{E}(z)$ of a three channel maximally decimated filter bank.
  a) Find the coefficients of the analysis filters.
  b) Find the coefficients of a set of FIR synthesis filters, which would result in perfect reconstruction.

6.10. Consider the product $\mathbf{V}_M(z)\ldots\mathbf{V}_1(z)$ where $\mathbf{V}_m(z)$ are $M \times M$ matrices as in (6.5.1). Let $\mathbf{v}_m$ be chosen so that $\mathbf{v}_m^\dagger \mathbf{v}_k = \delta(m-k)$. Show that the product reduces to $z^{-1}\mathbf{I}$.

6.11. Consider the M-channel maximally decimated system of Fig. 5.4-1. Let $\mathbf{H}(z)$ be the alias component (AC) matrix for the analysis bank, and let $\mathbf{E}(z)$ be the polyphase component matrix for the analysis bank. Show the following:

  a) $[\det \mathbf{H}(z)] = cz^{-K}$ for some $c \neq 0$ and integer K if and only if $[\det \mathbf{E}(z)] = dz^{-L}$ for some $d \neq 0$ and integer L.
  b) $[\det \mathbf{H}(z)]$ is allpass if and only if $[\det \mathbf{E}(z)]$ is allpass.

6.12. In a certain QMF bank suppose the $M \times M$ polyphase matrix $\mathbf{E}(z)$ is lossless. Let the polyphase matrix for the synthesis bank be given by $\mathbf{R}(z) = \mathrm{Adj}\,[\mathbf{E}(z)]$, where 'Adj' denotes the adjugate matrix (Appendix A). Show that the system is alias-free. Find an expression for the distortion function $T(z)$ and hence show that amplitude distortion has been eliminated.

6.13. Consider an analysis bank $H_k(z), 0 \leq k \leq M-1$, with paraunitary polyphase matrix $\mathbf{E}(z)$. Define a new analysis bank $H'_k(z), 0 \leq k \leq M-1$, by replacing z with $ze^{j\theta}$ where θ is an arbitrary real number. In other words, $H'_k(z) = H_k(ze^{j\theta})$. Show that the polyphase matrix $\mathbf{E}'(z)$ of the resulting system is paraunitary. [Note. In general $\mathbf{E}'(z) \neq \mathbf{E}(ze^{j\theta})$.]

6.14. Consider Fig. 5.2-5, where $a_i(z)$ are causal stable with $|a_i(e^{j\omega})| = 1$. This represents a two channel QMF bank. Let $\mathbf{E}(z)$ and $\mathbf{R}(z)$ be the polyphase matrices of the analysis and synthesis banks.

  a) Identify $\mathbf{E}(z)$ and $\mathbf{R}(z)$.
  b) Is $\mathbf{E}(z)$ lossless?
  c) Write down $\mathbf{R}(z)$ in terms of $\mathbf{E}(z)$.
  d) What is the product $\mathbf{R}(z)\mathbf{E}(z)$?

6.15. We now look into some deeper properties of tree structured QMF banks. Consider Fig. 5.8-1 again. Let $H_0^{(k)}(z) = \sum_{n=0}^{N_k} h^{(k)}(n)z^{-n}$ and let the filters be related as

$$H_1^{(k)}(z) = z^{-N_k}\widetilde{H}_0^{(k)}(-z), \qquad (P6.15a)$$

and

$$\widetilde{H}_0^{(k)}(z)H_0^{(k)}(z) + \widetilde{H}_1^{(k)}(z)H_1^{(k)}(z) = 1, \qquad (P6.15b)$$

for $k=1,2$. Assume N_k to be odd. Let the synthesis filters be chosen as

$$F_0^{(k)}(z) = z^{-N_k}\widetilde{H}_0^{(k)}(z), \quad F_1^{(k)}(z) = z^{-N_k}\widetilde{H}_1^{(k)}(z), \quad k=1,2. \qquad (P6.15c)$$

  a) Prove that the system has perfect reconstruction property.
  b) Consider the equivalent four channel system of the form in Fig. 5.4-1. Verify that $F_m(z)$ and $H_m(z)$ are related as $F_m(z) = cz^{-L}\widetilde{H}_m(z)$ for $0 \leq m \leq 3$, for some c and L.
  c) Show that the 4×4 polyphase matrix $\mathbf{E}(z)$ of the equivalent four channel system is paraunitary.

  Note. This problem is not tedious, if the properties in Sec. 6.2 are used judiciously. These results can also be generalized to tree structures with arbitrary number of levels, and arbitrary number of channels in a given level.

6.16. Consider Fig. 5.4-1. Find an example of a FIR perfect reconstruction system for which $F_k(z) = H_k(z)$ for each k. To avoid trivialities, make sure $M \geq 2$ in the example.

6.17. Consider the LOT structure of Fig. 6.6-2(a). Let $\mathbf{T} = [\mathbf{T}_0 \quad \mathbf{T}_1]$, where $\mathbf{T}_0$ and $\mathbf{T}_1$ are $M \times M$, and assume that (6.6.7) and (6.6.8) hold. Find $\mathbf{Q}$ in terms of $\mathbf{T}_0$ and $\mathbf{T}_1$, in order to have perfect reconstruction.

7

Linear Phase Perfect Reconstruction QMF Banks

7.0 INTRODUCTION

In some applications it is desirable to have a filter bank in which the analysis filters $H_k(z)$ are constrained to have linear phase. Such systems are called linear phase filter banks. These should not be confused with filter banks free from phase distortion, that is, filter banks for which the distortion function $T(z)$ has linear phase. For example the system in Design example 5.3.2 is not a linear phase filter bank (since the impulse response coefficients in Table 5.3.3 do not exhibit any symmetry), yet it is a perfect reconstruction system. On the other hand, the system in Design example 5.2.1 is a linear phase QMF bank (all filters have linear phase), but it is not a PR system since there is residual amplitude distortion (Fig. 5.2-4(b)).

In this chapter we show how to design linear phase filter banks which at the same time satisfy the perfect reconstruction (PR) property. The basic results were independently reported in Nguyen and Vaidyanathan [1989] and Vetterli and Le Gall [1988, 1989].

7.1 SOME NECESSARY CONDITIONS

In both the design examples mentioned above, the analysis filters $H_1(z)$ and $H_0(z)$ are constrained in some manner. In Design example 5.3.2 they are power complementary, whereas in Design example 5.2.1, $H_1(z) = H_0(-z)$. It turns out that, in order to design FIR linear phase QMF banks which at the same time enjoy the PR property, it is *necessary* to give up the power complementary property, as well as the relation $H_1(z) = H_0(-z)$. We begin the chapter by explaining why.

Power Complementary Constraint Must be Avoided

Suppose $H(z)$ and $G(z)$ are linear phase FIR filters, which at the same time satisfy the power complementary property. We will show that $H(z)$ is a sum of two delays, that is, $H(z) = az^{-K} + bz^{-L}$, where K and L are integers. $G(z)$ has similar form. As a result, the responses $|H(e^{j\omega})|$ and $|G(e^{j\omega})|$ are very restricted.

To prove this we assume that $H(z)$ and $G(z)$ are causal with the impulse response coefficients $h(0) \neq 0, g(0) \neq 0$. If this is not the case, we can redefine $H(z)$ and $G(z)$ by shifting the impulse responses, which does not affect either the linear phase property or the power complementary property. Let N denote the order of $H(z)$, so $h(N) \neq 0$. Let $N > 0$. (Otherwise there is nothing to prove.) The power complementary property implies

$$\widetilde{H}(z)H(z) + \widetilde{G}(z)G(z) = c^2 > 0. \qquad (7.1.1)$$

Equating like powers on both sides we see that $G(z)$ also has order N. So $H(z) = \sum_{n=0}^{N} h(n)z^{-n}$ and $G(z) = \sum_{n=0}^{N} g(n)z^{-n}$, with $g(N) \neq 0$. From Sec. 2.4.2 we know that the linear phase property of $H(z)$ and $G(z)$ implies

$$H(z) = e^{-j\alpha}z^{-N}\widetilde{H}(z), \quad G(z) = e^{-j\beta}z^{-N}\widetilde{G}(z), \qquad (7.1.2)$$

for real α, β. Substituting into (7.1.1) and simplifying we get

$$\left(e^{j\alpha/2}H(z) + je^{j\beta/2}G(z)\right)\left(e^{j\alpha/2}H(z) - je^{j\beta/2}G(z)\right) = c^2 z^{-N}. \qquad (7.1.3)$$

Since all quantities on the left hand side are FIR, this implies

$$e^{j\alpha/2}H(z) + je^{j\beta/2}G(z) = pz^{-K}, \quad e^{j\alpha/2}H(z) - je^{j\beta/2}G(z) = qz^{-L}, \qquad (7.1.4)$$

where $pq = c^2$, and $K + L = N$. Adding and subtracting these two equations we get

$$H(z) = az^{-K} + bz^{-L}, \quad G(z) = \gamma\left(az^{-K} - bz^{-L}\right) \qquad (7.1.5)$$

for appropriate a, b and γ, with $|\gamma| = 1$.

Consequences. As a consequence of this result, we have to remove the power complementary restriction on the analysis filters in order to obtain good responses. Since paraunitariness of the polyphase matrix $\mathbf{E}(z)$ (Sec. 6.2.2) implies that $H_0(z), H_1(z)$ are power complementary, it is necessary to *give up* paraunitariness of $\mathbf{E}(z)$ as well.

The Relation $H_1(z) = H_0(-z)$ Must be Avoided

In Sec. 5.2 we studied alias-free FIR QMF banks with analysis filters related as $H_1(z) = H_0(-z)$. The overall distortion function is $T(z) = 0.5[H_0^2(z) - H_0^2(-z)]$. If $H_0(z)$ has linear phase, then $T(z)$ has linear phase, and phase distortion is eliminated. This system however suffers from amplitude distortion, that is, $|T(e^{j\omega})|$ is not perfectly flat. The residual amplitude distortion can be made very small using Johnston's procedure (Design example 5.2.1). This means that we have already seen examples of linear phase FIR QMF banks which "almost" satisfy the PR property. By increasing the order of $H_0(z)$ we can decrease the amplitude distortion to any desired degree [while maintaining the attenuation requirements of $H_0(z)$], so that the system gets as close to PR as we wish.

With this system, however, we can never achieve PR property *exactly*! We proved this in Sec. 5.2.1 by showing that the distortion function of the alias-free system has the form $T(z) = 2z^{-1}E_0(z^2)E_1(z^2)$, where $H_0(z) = E_0(z^2) + z^{-1}E_1(z^2)$. For perfect reconstruction $T(z)$ has to be a delay, that is, $H(z)$ has to be a sum of two delays, which is not useful. As a result, it is necessary to give up the condition $H_1(z) = H_0(-z)$ as well.

7.2 LATTICE STRUCTURES FOR LINEAR PHASE FIR PR QMF BANKS

Recall that neither the relation $H_1(z) = H_0(-z)$ nor the power complementary property is necessary for perfect reconstruction in FIR QMF banks. The condition (5.6.10) is really (necessary and) sufficient. It turns out that we can design very good linear phase analysis filters which at the same time satisfy this condition. As a first step, we generate an example with nontrivial analysis filters, such that neither the power complementary property nor the relation $H_1(z) = H_0(-z)$ is satisfied.

Example 7.2.1 An FIR Linear-Phase PR QMF Bank

Consider the analysis bank of Fig. 7.2-1(a). Here the polyphase matrix $\mathbf{E}(z) = \mathbf{T}_1 \mathbf{\Lambda}(z) \mathbf{T}_0$, where

$$\mathbf{T}_0 = \begin{bmatrix} 1 & k \\ k & 1 \end{bmatrix}, \quad \mathbf{\Lambda}(z) = \begin{bmatrix} 1 & 0 \\ 0 & z^{-1} \end{bmatrix}, \quad \mathbf{T}_1 = \begin{bmatrix} 1 & 1 \\ 1 & -1 \end{bmatrix}. \quad (7.2.1)$$

Assume k is real and $k \neq \pm 1$. This ensures that $\mathbf{T}_0$ is nonsingular. A corresponding synthesis bank which gives rise to perfect reconstruction is shown in Fig. 7.2-1(b). This is obtained by taking

$$\mathbf{R}(z) = cz^{-1}\mathbf{E}^{-1}(z) = c\mathbf{T}_0^{-1} \begin{bmatrix} z^{-1} & 0 \\ 0 & 1 \end{bmatrix} \mathbf{T}_1^{-1}$$

for appropriate c. The analysis and synthesis filters are verified to be

$$H_0(z) = 1 + kz^{-1} + kz^{-2} + z^{-3}, \quad H_1(z) = 1 + kz^{-1} - kz^{-2} - z^{-3},$$
$$F_0(z) = 1 - kz^{-1} - kz^{-2} + z^{-3}, \quad F_1(z) = -1 + kz^{-1} - kz^{-2} + z^{-3}.$$
$$(7.2.2)$$

Many points are worth noting here. The synthesis filters satisfy $F_0(z) = H_1(-z)$ and $F_1(z) = -H_0(-z)$, consistent with the alias cancelation condition (5.1.7). The analysis filters evidently have linear phase, and are nontrivial in the sense that they are not just sums of two delays. However, they are not power complementary, nor is the relation $H_1(z) = H_0(-z)$ satisfied. Finally, the synthesis filters are *not* given by $F_k(z) = z^{-3}H_k(z^{-1})$ as in a paraunitary perfect reconstruction system.

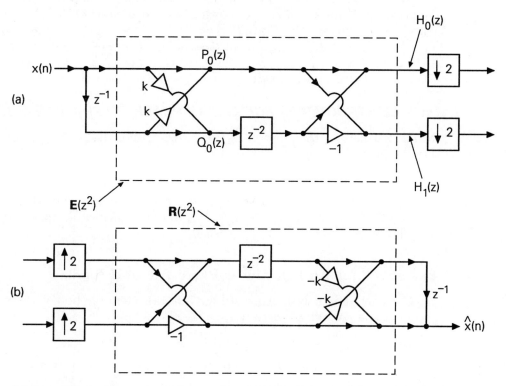

Figure 7.2-1 Example of linear phase PR QMF bank. (a) Analysis bank, and (b) synthesis bank.

What is the trick behind the success of this example? The matrices $\mathbf{T}_0$ and $\mathbf{T}_1$ have, no doubt, been 'carefully' chosen. The choice of $\mathbf{T}_0$ is such that $Q_0(z)$ is the Hermitian image of $P_0(z)$ (see Fig. 7.2-1(a)). The choice of $\mathbf{T}_1$ is such that $H_0(z)$ and $H_1(z)$ are the sum and difference of the image pair $P_0(z), z^{-2}Q_0(z)$, so that $h_0(n)$ is symmetric and $h_1(n)$ is antisymmetric!

Example 7.2.2

We can in fact generate similar examples of arbitrary order. To demonstrate this, consider Fig. 7.2-2(a) in which

(a)
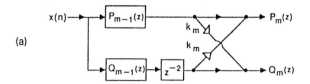

(b)
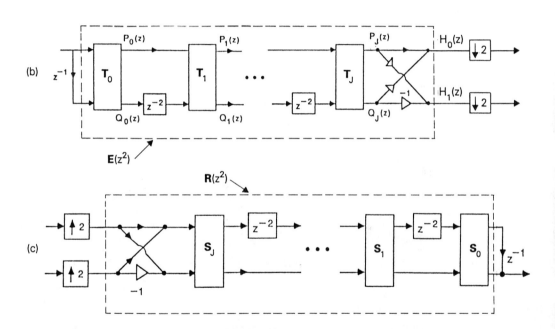

(c)

(d)
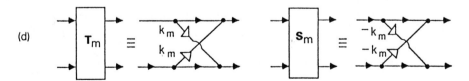

Figure 7.2-2 More general linear phase FIR PR QMF bank. (a) Basic generation technique for analysis bank. (b) Complete analysis bank. (c) Complete synthesis bank. (d) Details of the building blocks.

$$P_{m-1}(z) = \sum_{n=0}^{2m-1} p_{m-1}(n)z^{-n}, \quad Q_{m-1}(z) = \sum_{n=0}^{2m-1} q_{m-1}(n)z^{-n}, \quad (7.2.3)$$

are real coefficient polynomials and k_m is real. Let $Q_{m-1}(z)$ be the Hermitian image of $P_{m-1}(z)$, i.e., $Q_{m-1}(z) = z^{-(2m-1)}P_{m-1}(z^{-1})$. Then the transfer functions

$$\begin{aligned} P_m(z) &= P_{m-1}(z) + k_m z^{-2} Q_{m-1}(z), \\ Q_m(z) &= k_m P_{m-1}(z) + z^{-2} Q_{m-1}(z), \end{aligned} \quad (7.2.4)$$

are also Hermitian images, i.e., satisfy $Q_m(z) = z^{-(2m+1)}P_m(z^{-1})$. (This can be verified by substitution.) By repeated application of this, we see that the cascaded lattice structure shown in Fig. 7.2-2(b) has the property

$$Q_J(z) = z^{-N} P_J(z^{-1}), \quad N = 2J + 1. \quad (7.2.5)$$

The analysis filters in this figure are given by

$$\begin{bmatrix} H_0(z) \\ H_1(z) \end{bmatrix} = \begin{bmatrix} 1 & 1 \\ 1 & -1 \end{bmatrix} \begin{bmatrix} P_J(z) \\ Q_J(z) \end{bmatrix}. \quad (7.2.6)$$

From this it is easily verified that these filters satisfy

$$H_0(z) = z^{-N} H_0(z^{-1}), \quad H_1(z) = -z^{-N} H_1(z^{-1}), \quad (7.2.7)$$

that is, in terms of impulse response coefficients,

$$h_0(n) = h_0(N-n), \quad h_1(n) = -h_1(N-n), \quad (7.2.8)$$

so that they have linear phase. Figure 7.2-2(c) shows a synthesis bank which results in perfect reconstruction. This is obtained by choosing the polyphase matrix of the synthesis bank to be $\mathbf{R}(z) = cz^{-J}\mathbf{E}^{-1}(z)$. Here $\mathbf{S}_m = (1 - k_m^2)\mathbf{T}_m^{-1}$. The synthesis filters satisfy (5.1.7) within a scale factor (Problem 7.1).

The main point of the example is that we can generate linear phase FIR PR QMF banks in which the analysis filters are nontrivial (i.e., not restricted to be a sum of two delays). The next design example shows that these analysis filters can provide excellent attenuation as well.

Design Example 7.2.1. Linear Phase FIR PR Lattice

Consider a lattice with $J = 31$ so that the filters have order $N = 63$. The lattice coefficients should now be optimized in order to minimize an appropriate objective function. The simple function (6.4.9) is not suitable any more because $H_0(z)$ is not power symmetric, and moreover there is no power complementary relation between $H_0(z)$ and $H_1(z)$. It is necessary to define an objective function which reflects the passbands *and* stopbands of *both* filters. For example we can take

$$\phi = \int_0^{\omega_p} \left[1 - |H_0(e^{j\omega})|\right]^2 d\omega + \int_{\omega_s}^{\pi} |H_0(e^{j\omega})|^2 d\omega \\ + \int_{\omega_s}^{\pi} \left[1 - |H_1(e^{j\omega})|\right]^2 d\omega + \int_0^{\omega_p} |H_1(e^{j\omega})|^2 d\omega. \quad (7.2.9)$$

Figure 7.2-3(a) shows the analysis filter responses of the optimized design. The filter coefficients are tabulated in Nguyen and Vaidyanathan [1989]. The transition bandwidth is about 0.172π. For comparison, we show in Fig. 7.2-3(b) the responses of Johnston's 64D filters (which also have order 63, and the same transition bandwidth). Johnston's filters offer a minimum stopband attenuation of 65 dB, in comparison to only 42.5 dB offered by the perfect reconstruction system.[†] The peak amplitude distortion of Johnston's 64D QMF bank is about 0.002 dB. Johnston's 32D filter, on the other hand, has nearly the same attenuation as the PR system.

Even though the PR system has higher order for a given attenuation, it can be implemented more efficiently than Johnston's filters, because of the lattice structure (see *computational complexity* below).

Recall that in order to obtain perfect reconstruction using linear phase filters, we had to give up the relation $H_1(z) = H_0(-z)$ as well as the power complementary property. Also the plot of $|H_0(e^{j\omega})|^2 + |H_1(e^{j\omega})|^2$ is very flat for Johnston's design but not for the linear phase PR pair (see Fig. 7.2-4). In spite of this the linear phase lattice structure enjoys perfect reconstruction because the quantity $|H_0(e^{j\omega})|^2 + |H_1(e^{j\omega})|^2$ is *not* proportional to the amplitude distortion unlike in Johnston's design!

Computational Complexity of Linear Phase QMF Lattice

The lattice structure of Fig. 7.2-2(b) has $J + 1$ sections, with two multipliers per section. However, each section can be rearranged, permitting an implementation with only one multiplier (and three adders) per section (Problem 7.2)[‡]. From this we deduce that the analysis bank requires $0.25(N+1) + 1$ MPUs and $0.75(N+1) + 1$ APUs (where N = filter order). In our example $N = 63$ so this reduces to 17 MPUs and 49 APUs. For comparison suppose we consider Johnston's 32D filter, which has nearly the same

---

[†] Improved optimization has recently been reported by Nguyen [1992a], whereby the perfect reconstruction system can provide almost as good attenuation as Johnston's filters of the same order.

[‡] This is unlike in the paraunitary lattice (Fig. 6.4-2), which required a minumum of two multipliers per section.

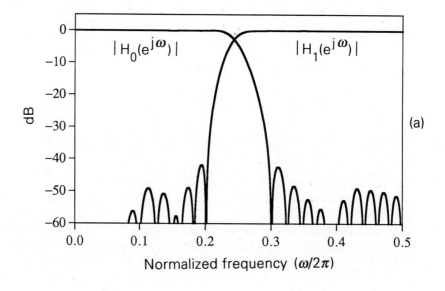

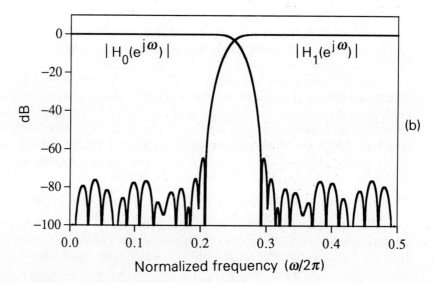

Figure 7.2-3 Design example 7.2.1 (Linear phase QMF banks). Magnitude responses of analysis filters for (a) perfect reconstruction system, and (b) Johnston's 64D system. Both systems have analysis filters of length 64. (© Adopted from 1989 IEEE.)

specifications (including minimum stopband attenuation) as the linear-phase lattice filters, and has a peak amplitude distortion of 0.025 dB. This can be implemented with a total of 16 MPUs and 16 APUs for the analysis bank. Summarizing, the linear phase PR QMF lattice has about the same number of MPUs as Johnston's filters with same specifications (and amplitude distortion of 0.025 dB). The number of APUs, however, is higher.

Table 7.2.1 summarizes the comparison between the linear phase PR QMF bank and Johnston's design.

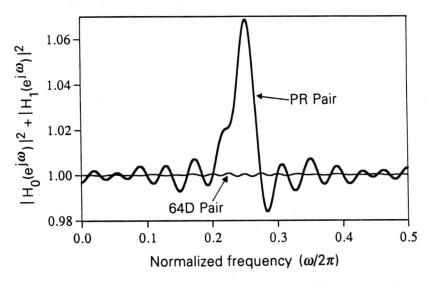

Figure 7.2-4 Pertaining to Design example 7.2.1. (© Adopted from 1989 IEEE.)

Initialization of the Lattice Parameters for Optimization

Since Johnston's filters have linear phase and "almost" satisfy the PR property, it is possible to obtain a lattice structure for these, which "almost" resembles Fig. 7.2-2(b). This can be used to initialize the parameters k_m.

Let us make the above statement more precise. Given the Nth order pair $H_0(z), H_1(z)$ from Johnston's design, suppose we define $P_J(z)$ and $Q_J(z)$ according to (7.2.6). Then (7.2.5) is satisfied because (7.2.7) holds. Define $P'_N(z) = P_J(z), Q'_N(z) = Q_J(z)$ for convenience. We can now construct a pair of lower order transfer functions as follows:

$$\begin{aligned} P'_{N-1}(z) &= P'_N(z) - \ell_N Q'_N(z), \\ z^{-1}Q'_{N-1}(z) &= -\ell_N P'_N(z) + Q'_N(z). \end{aligned} \qquad (7.2.10)$$

We choose $\ell_N = p'_N(N)/q'_N(N)$ so that $P'_{N-1}(z)$ has degree $N-1$. Because of the relation (7.2.5), this same value of ℓ_N ensures that $Q'_{N-1}(z)$ defined in (7.2.10) is causal with degree $N-1$ or less. Assuming $\ell_N^2 \ne 1$ we can invert (7.2.10) to get

$$\begin{aligned} (1-\ell_N^2)P'_N(z) &= P'_{N-1}(z) + \ell_N z^{-1}Q'_{N-1}(z), \\ (1-\ell_N^2)Q'_N(z) &= \ell_N P'_{N-1}(z) + z^{-1}Q'_{N-1}(z), \end{aligned} \qquad (7.2.11)$$

This gives rise to the lattice representation of Fig. 7.2-5(a). It can be verified from above that $Q'_{N-1}(z) = z^{-(N-1)} P'_{N-1}(z^{-1})$, so that we can repeat the process, resulting in Fig. 7.2-5(b). ‡ Here the scale factors $(1 - \ell_m^2)$ have all been lumped together into α.

TABLE 7.2.1 Comparison between the linear-phase perfect reconstruction design and Johnston's 32D design.

Feature	Johnston's 32D Pair of Filters	Linear phase QMF lattice
Phase response	Linear	Linear
Filter order	31	63
Stopband Attenuation	38 dB	42.5 dB
Peak amplitude distortion	0.025 dB	No error
Number of MPU for analysis bank	16	17
Number of APU for analysis bank	16	49
Power Complementarity	Approximately holds	Does not hold
Relation between Analysis Filters	$H_1(z) = H_0(-z)$	Not explicit. Implicitly such that det E(z)=delay
Overall Group Delay of QMF bank	31	63
Aliasing	Canceled	Canceled
Phase distortion	Eliminated	Eliminated
Amplitude distortion	Minimized	Eliminated

---

‡ Readers familiar with linear predictive coding will notice the resemblance to the LPC lattice [Markel, and Gray, Jr., 1976]. However, there are two differences. In the LPC lattice the coefficients ℓ_m (called *reflection coefficients*) are typically bounded as $\ell_m^2 < 1$. Also the rightmost section, which generates $H_0(z), H_1(z)$, is absent.

So Johnston's analysis bank can be represented in this manner, provided $\ell_m^2 \neq 1$ for any m (and this is the case in all practical examples). This structure, however, is not in polyphase form because the system inside broken lines in Fig. 7.2-5(b) is not a function of z^{-2} (so that it is not equal to $\mathbf{E}(z^2)$). However, since the filters have linear phase and almost satisfy the PR property, the coefficients $\ell_2, \ell_4, \ldots$ and so on, turn out to be very close to zero. By setting these to zero, the remaining coefficients ℓ_{2m+1} can be used to initialize the coefficients k_n in the linear-phase lattice of Fig. 7.2-2(b). Such initialization leads to significantly faster convergence of optimization, as compared to random initialization. This method was used in the above design example.

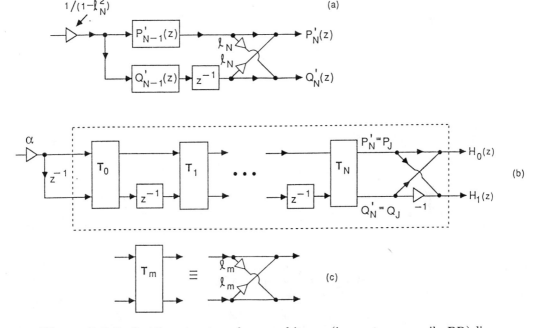

Figure 7.2-5 Lattice structure for an arbitrary (i.e. not necessarily PR) linear phase pair $[H_0(z), H_1(z)]$.

7.3 FORMAL SYNTHESIS OF LINEAR PHASE FIR PR QMF LATTICE

In Example 7.2.2 $H_0(z)$ and $H_1(z)$ are filters with odd order ($N = 2J + 1$) satisfying (7.2.7) and the PR condition $[\det \mathbf{E}(z)] = $ delay. The next question is, given such a pair of FIR filters, is it always possible to find a structure like Fig. 7.2-2(b)? In other words, does the lattice cover every analysis bank satisfying the said properties? The answer is *yes* with some minor exceptions which will be made clear soon.

The given set of filters $H_0(z), H_1(z)$ can always be expressed as in (7.2.6) by defining $P_J(z)$ and $Q_J(z)$ as

$$\begin{bmatrix} P_J(z) \\ Q_J(z) \end{bmatrix} = \frac{1}{2} \begin{bmatrix} 1 & 1 \\ 1 & -1 \end{bmatrix} \begin{bmatrix} H_0(z) \\ H_1(z) \end{bmatrix}. \qquad (7.3.1)$$

With analysis filters expressed in polyphase form (5.6.11), we have

$$\begin{bmatrix} P_J(z) \\ Q_J(z) \end{bmatrix} = \underbrace{\frac{1}{2} \begin{bmatrix} 1 & 1 \\ 1 & -1 \end{bmatrix} \mathbf{E}(z^2)}_{\mathbf{F}_J(z^2)} \begin{bmatrix} 1 \\ z^{-1} \end{bmatrix}. \qquad (7.3.2)$$

In other words

$$\begin{bmatrix} P_J(z) \\ Q_J(z) \end{bmatrix} = \mathbf{F}_J(z^2) \begin{bmatrix} 1 \\ z^{-1} \end{bmatrix}, \qquad (7.3.3)$$

and the PR condition (5.6.10) is equivalent to the condition

$$\det \mathbf{F}_J(z) = -c_0 z^{-m_0} \qquad \text{(PR condition)}. \qquad (7.3.4)$$

The linear phase condition (7.2.7) is, of course, equivalent to (7.2.5).

So the problem of designing linear phase FIR PR QMF banks can be transformed to that of finding a lattice structure for $[P_J(z)\, Q_J(z)]^T$ which satisfies the properties (7.3.4) and (7.2.5). For convenience of discussion let us write

$$P_J(z) = \sum_{n=0}^{N} p_J(n) z^{-n}, \quad Q_J(z) = \sum_{n=0}^{N} q_J(n) z^{-n} \qquad (7.3.5)$$

so that (7.2.5) is equivalent to

$$p_J(n) = q_J(N - n) \qquad \text{(linear-phase condition)}. \qquad (7.3.6)$$

The following Lemma is crucial to our discussion.

♦**Lemma 7.3.1.** Let $P_J(z)$ and $Q_J(z)$ be as in (7.3.5), and satisfy (7.3.6) and (7.3.4), where $\mathbf{F}_J(z)$ is defined as in (7.3.3). Let N be odd, that is, $N = 2J + 1$. Assume $p_J(n), q_J(n)$ are real, and $0 \neq p_J(0)$ and $p_J(0) \neq \pm p_J(N)$. Then we can find two FIR filters $P_{J-1}(z) = \sum_{n=0}^{N-2} p_{J-1}(n) z^{-n}$ and $Q_{J-1}(z) = \sum_{n=0}^{N-2} q_{J-1}(n) z^{-n}$ and a real $k_J \neq \pm 1$ such that

$$\begin{bmatrix} P_{J-1}(z) \\ z^{-2} Q_{J-1}(z) \end{bmatrix} = \frac{1}{1 - k_J^2} \begin{bmatrix} 1 & -k_J \\ -k_J & 1 \end{bmatrix} \begin{bmatrix} P_J(z) \\ Q_J(z) \end{bmatrix}. \qquad (7.3.7)$$

Moreover $Q_{J-1}(z) = z^{-(N-2)}P_{J-1}(z^{-1})$ and $p_{J-1}(0) \neq 0$. ◊

Remark. The condition $k_J^2 \neq 1$ automatically ensures that the 2×2 matrix in (7.3.7) is nonsingular. By inverting it, we can obtain the lattice structure of Fig. 7.2-2(a) with J in place of m. The remainder $[P_{J-1}(z) \ Q_{J-1}(z)]^T$ has all the properties of $[P_J(z) \ Q_J(z)]^T$ so that we can repeat this process provided $p_{J-1}(0) \neq \pm p_{J-1}(N-2)$. So we can obtain the lattice structure shown in Fig. 7.2-2(b), for the system $[P_J(z) \ Q_J(z)]^T$ *provided that* each one of the remainders satisfies

$$p_m(0) \neq \pm p_m(2m+1). \tag{7.3.8}$$

This means that we can implement the analysis bank $[H_0(z) \ H_1(z)]^T$ as in Fig. 7.2-2(b).

Proof of Lemma 7.3.1. From (7.3.7) it is clear that k_J has to satisfy $p_J(N) - k_J q_J(N) = 0$ so that

$$k_J = \frac{p_J(N)}{q_J(N)} = \frac{p_J(N)}{p_J(0)} \neq \pm 1. \tag{7.3.9}$$

With this choice of k_J, the coefficient of z^{-N} in $P_J(z) - k_J Q_J(z)$ drops off.

We now show that the coefficient of $z^{-(N-1)}$ in $P_J(z) - k_J Q_J(z)$ is also zero, so that $P_{J-1}(z)$ has order $N-2$ as claimed. For this note that this coefficient is

$$p_J(N-1) - k_J q_J(N-1) = [p_J(N-1)q_J(N) - q_J(N-1)p_J(N)]/q_J(N), \tag{7.3.10}$$

by (7.3.9). With

$$\mathbf{F}_J(z) = \begin{bmatrix} A(z) & B(z) \\ C(z) & D(z) \end{bmatrix}, \tag{7.3.11a}$$

the condition (7.3.4) implies

$$A(z)D(z) - B(z)C(z) = -c_0 z^{-m_0}. \tag{7.3.11b}$$

But we have

$$\begin{aligned} A(z) &= p_J(0) + p_J(2)z^{-1} + \ldots + p_J(N-1)z^{-M}, \\ B(z) &= p_J(1) + p_J(3)z^{-1} + \ldots + p_J(N)z^{-M}, \\ C(z) &= q_J(0) + q_J(2)z^{-1} + \ldots + q_J(N-1)z^{-M}, \\ D(z) &= q_J(1) + q_J(3)z^{-1} + \ldots + q_J(N)z^{-M}, \end{aligned} \tag{7.3.12}$$

where $M = (N-1)/2$. The coefficient of z^0 in the LHS of (7.3.11b) is

$$p_J(0)q_J(1) - q_J(0)p_J(1). \tag{7.3.13a}$$

The coefficient of z^{-2M} is, on the other hand,

$$p_J(N-1)q_J(N) - p_J(N)q_J(N-1). \qquad (7.3.13b)$$

Since $N \geq 3$, we have $2M > 0$. So (7.3.11b) implies that at least one of (7.3.13a),(7.3.13b) is zero. By using the image property (7.3.6) it is verified that (7.3.13a) and (7.3.13b) have the same value. Setting this to zero we see that (7.3.10) is indeed zero. So $P_{J-1}(z)$ has the stated form.

We can verify that $Q_{J-1}(z) = z^{-(N-2)}P_{J-1}(z^{-1})$ by substituting (7.3.6) into (7.3.7). So $Q_{J-1}(z)$ has the state form too. Inverting (7.3.7) one obtains

$$\begin{bmatrix} P_J(z) \\ Q_J(z) \end{bmatrix} = \begin{bmatrix} 1 & k_J \\ k_J & 1 \end{bmatrix} \begin{bmatrix} P_{J-1}(z) \\ z^{-2}Q_{J-1}(z) \end{bmatrix}, \qquad (7.3.14)$$

from which it follows that $p_J(0) = p_{J-1}(0)$, so $p_{J-1}(0) \neq 0$ indeed. ▽▽▽

Condition on $H_0(z), H_1(z)$ for Lattice Realization

The analysis filters have the form $H_0(z) = \sum_{n=0}^{N} h_0(n)z^{-n}$, $H_1(z) = \sum_{n=0}^{N} h_1(n)z^{-n}$, $N \geq 3$. The condition $p_J(0) \neq \pm p_J(N)$ in Lemma 7.3.1 can be satisfied by assuming that neither of $h_0(N), h_1(N)$ is zero. Now what does $p_J(0) = 0$ mean? Since (7.3.13a) equals zero, this means $q_J(0) = 0$ or $p_J(1) = 0$. This means that either $h_0(0) = h_1(0) = 0$ or that $P_J(z)$ and $Q_J(z)$ have the form $P_J(z) = z^{-2}P_{J-1}(z)$ and $Q_J(z) = Q_{J-1}(z)$. The former case is trivial and can be avoided by shifting. The latter case implies that we can interchange the roles of $P_J(z)$ and $Q_J(z)$ and then take $k_J = 0$. So the situations created by violation of the condition '$0 \neq p_J(0) \neq \pm p_J(N)$', can be handled easily. It is, however, still possible to find examples where (7.3.8) is violated for some value of $m < J$. In such cases the lattice cannot be synthesized.

It nevertheless remains a significant fact that the lattice of Fig. 7.2-2 can be used to generate a wide class of linear phase FIR PR systems. More general study of two-channel linear phase FIR PR systems can be found in Nguyen and Vaidyanathan [1989]. For instance, it can be shown that one can take both $h_0(n)$ and $h_1(n)$ to be symmetric (provided their orders are even and unequal). Such systems have a different type of lattice structure. Also see [Vetterli and Le Gall, 1989] and Nguyen and Vaidyanathan [1990] for M-channel linear phase FIR perfect reconstruction systems. For a general theory of M-channel linear phase FIR paraunitary filter banks see [Soman, Vaidyanathan, and Nguyen, 1992].

PROBLEMS

7.1. The lattice structure shown in Fig. 7.2-2 (b) and (c) represents a perfect reconstruction QMF bank. Express the synthesis filters $F_k(z)$ in terms of the analysis filters $H_k(z)$.

7.2. Consider the two multiplier lattice section shown in Fig. P7-2(a).

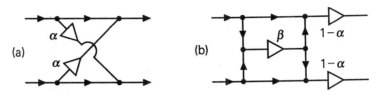

Figure P7-2

Show that this can be redrawn as shown in Fig. P7-2(b) where $\beta = \alpha/(1-\alpha)$, assuming $\alpha \neq 1$. Hence show that the linear phase perfect reconstruction system (Fig. 7.2-2) can be rearranged so that the analysis bank requires a total of $0.25(N+1)+1$ MPUs and $0.75(N+1)+1$ APUs.

7.3. Consider the analysis bank structure given below, where α_m, β_n are real.

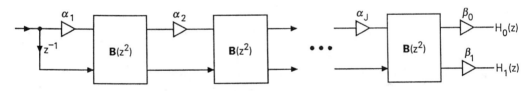

Figure P7-3

The building block $\mathbf{B}(z)$ has the form

$$\mathbf{B}(z) = \begin{bmatrix} 1+z^{-1} & 1 \\ 1+z^{-1}+z^{-2} & 1+z^{-1} \end{bmatrix}. \qquad (P7.3a)$$

Evidently the analysis filters $H_0(z)$ and $H_1(z)$ are causal and FIR. Show that they have linear phase. More specifically show that the impulse responses satisfy

$$h_0(n) = h_0(N_0 - n), \quad h_1(n) = h_1(N_1 - n) \qquad (P7.3b)$$

where $N_0 = 2J, N_1 = 2J + 2$. Find a set of FIR synthesis filters [in terms of $H_0(z)$ and $H_1(z)$] which result in perfect reconstruction. (Note. this problem shows that we can obtain an FIR perfect reconstruction QMF bank in which both analysis filters are symmetric unlike in Sec. 7.2 where $H_1(z)$ was antisymmetric.)

7.4. Let $H(z)$ and $G(z)$ be two real coefficient linear phase FIR filters satisfying $|H(e^{j\omega})|^2 + |G(e^{j\omega})|^2 = 1$ for all ω. Prove that $|H(e^{j\omega})|^2 = \alpha \cos^2(a\omega + b)$ for some real α, a, b.

7.5. Let $H(z)$ be a linear phase FIR filter, and let $\widetilde{H}(z)H(z)$ satisfy the Mth band property.

a) For $M = 2$ (i.e., $\widetilde{H}(z)H(z)$ is half-band) show that $H(z)$ can have at most two nonzero coefficients.

b) For $M > 2$, a similar statement is not true. Show, by construction, that the number of nonzero coefficients of $H(z)$ can exceed L, for any arbitrary integer L.

Hints. Use the results of Sec. 4.6.3 and 7.1.

8

Cosine Modulated Filter Banks

8.0 INTRODUCTION

In Chap. 5 and 6 we considered M channel maximally decimated analysis/synthesis systems, and studied various errors, as well as techniques to eliminate these. In particular, we studied the concept of perfect reconstruction (PR) in detail, and presented techniques to design FIR PR systems.

In this Chapter we will present filter banks based on cosine modulation. In these systems, all the M analysis filters are derived from a prototype filter $P_0(z)$ by cosine modulation. Two outstanding advantages of these systems are:

1. The cost of the analysis bank is equal to that of one filter, plus modulation overhead. The modulation itself can be done by fast techniques such as the fast discrete cosine transform (DCT). See, for example, Yip and Rao [1987]. The synthesis filters have the same cost as the analysis filters.
2. During the design phase, where we optimize the filter coefficients, the number of parameters to be optimized is very small because only the prototype has to be optimized.

Two classes of such systems will be studied — approximate reconstruction systems (pseudo QMF) and perfect reconstruction systems.

A. Cosine Modulated Pseudo QMF Banks (Sec. 8.1–8.3)

Prior to the development of perfect reconstruction systems, several authors have developed techniques for designing *approximate reconstruction* systems. These are called the pseudo QMF systems, introduced by Nussbaumer [1981] and developed further by Rothweiler [1983], Chu [1985], Masson and Picel [1985], and Cox [1986]. In these systems the analysis and synthesis filters $H_k(z)$ and $F_k(z)$ are chosen so that only "adjacent-channel aliasing" (to be explained below) is canceled, and the distortion function

$T(z)$ is only approximately a delay. Such approximate systems, called *pseudo* QMF banks, are acceptable in some practical applications.

B. Cosine Modulated Perfect Reconstruction Systems (Sec. 8.5)

More recently, cosine modulated systems with the perfect reconstruction property have been developed independently by Malvar [1990b and 1991], Ramstad [1991], and Koilpillai and Vaidyanathan [1991a and 1992]. These are paraunitary systems. They retain all the simplicity and economy of the pseudo QMF system, and yet have the perfect reconstruction property. In Sec. 8.4 we study some properties of cosine modulation matrices. These are then used in Sec. 8.5 to derive cosine modulated perfect reconstruction systems. *These two sections can be studied independently of the pseudo QMF derivations, with Sections 8.1–8.3 serving only as references.*

8.1 THE PSEUDO QMF BANK

In this section we present the theory of pseudo QMF banks. In Sec. 8.2 and 8.3 we will outline design procedures and structures for these. Readers interested only in perfect reconstruction systems can go directly to Sec. 8.4 (and use sections 8.1–8.3 only as a reference).

8.1.1 Generation of M Real Coefficient Analysis Filters

In Sec. 4.3.2 we saw how a set of M filters can be derived from one prototype filter by use of the structure of Fig. 4.3-5(a). In that structure, the filters $E_\ell(z)$ represent the Type 1 polyphase components of the prototype filter $H_0(z)$, and the filters $H_k(z)$ are related to $H_0(z)$ as $H_k(z) = H_0(zW_M^k)$, where $W_M = e^{-j2\pi/M}$. This means that the frequency responses $H_k(e^{j\omega})$ are uniformly shifted versions of the prototype, as demonstrated in Fig. 4.3-5(b). Since $h_k(n)$ is obtained by exponential modulation of $h_0(n)$, (that is, $h_k(n) = h_0(n)e^{j2\pi kn/M}$), the coefficients $h_k(n)$ are in general complex even if $h_0(n)$ is real. This means that the output of $H_k(z)$ could be a complex signal even if the input $x(n)$ is real.

We now derive a class of filters with real coefficients, by using *cosine* modulation rather than exponential modulation. This can be done by first obtaining $2M$ complex filters using exponential modulation, and then combining appropriate pairs of filters.

Consider Fig. 8.1-1 which is a modification of Fig. 4.3-5(a) (replace M with $2M$). This is a uniform-DFT analysis bank, with the $2M$ filters related as

$$P_k(z) = P_0(zW^k), \quad \text{that is,} \quad p_k(n) = p_0(n)W^{-kn}. \tag{8.1.1}$$

In this section, unsubscripted W stands for W_{2M}, that is,

$$W = W_{2M} = e^{-j2\pi/2M} = e^{-j\pi/M}. \tag{8.1.2}$$

Also, **W** is the $2M \times 2M$ DFT matrix.

$P_0(z)$ is called the prototype filter. Throughout this chapter, its impulse response $p_0(n)$ is real so that $|P_0(e^{j\omega})|$ is symmetric with respect to $\omega = 0$. This filter is typically lowpass, with cutoff frequency $\pi/2M$ [Fig. 8.1-2(a)]. The polyphase components of $P_0(z)$ are $G_k(z), 0 \le k \le 2M - 1$.

From (8.1.1) we have

$$P_k(e^{j\omega}) = P_0\left(e^{j\left(\omega - \frac{k\pi}{M}\right)}\right), \qquad (8.1.3)$$

that is, the response $P_k(e^{j\omega})$ is the right-shifted version of $P_0(e^{j\omega})$ by an amount $k\pi/M$ [Fig. 8.1-2(b)]. From the figure we see that the responses $|P_k(e^{j\omega})|$ and $|P_{2M-k}(e^{j\omega})|$ are images of each other with respect to zero-frequency, so that they are suitable candidates to be combined, to get a real coefficient filter. The typical passband width of such a 'combined filter' is equal to $2\pi/M$, which is twice that of $P_0(z)$ (which is not combined with any other filter).

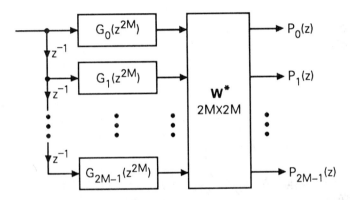

Figure 8.1-1 Generation of $2M$ uniformly shifted filters from prototype $P_0(z)$. Here $G_m(z)$, $0 \le m \le 2M - 1$, are the polyphase components of $P_0(z)$.

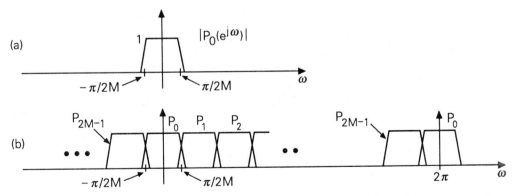

Figure 8.1-2 (a) Magnitude response of the prototype $P_0(z)$, and (b) Magnitude responses of shifted versions.

In order to make all the filter bandwidths equal after combining pairs, we use a *right-shifted version* of the original set of $2M$ responses [Fig. 8.1-3(a)], the amount of right-shift being $\pi/2M$. This is accomplished by replacing z with $zW^{0.5}$ as indicated in Fig. 8.1-3(b). (The quantity z^{2M} is replaced with $-z^{2M}$ since $W^M = W_{2M}^M = -1$.) The complex filters $Q_k(z)$ are given in terms of the prototype $P_0(z)$ by

$$Q_k(z) = P_0(zW^{k+0.5}), \quad 0 \le k \le 2M - 1. \tag{8.1.4}$$

The magnitude responses of $Q_k(z)$ and $Q_{2M-1-k}(z)$ are now images of each other with respect to zero-frequency, that is, $|Q_k(e^{j\omega})| = |Q_{2M-1-k}(e^{-j\omega})|$. The impulse response coefficients of $Q_k(z)$ and $Q_{2M-1-k}(z)$ are conjugates of each other, that is,

$$Q_{2M-1-k}(z) = Q_{k,*}(z)$$

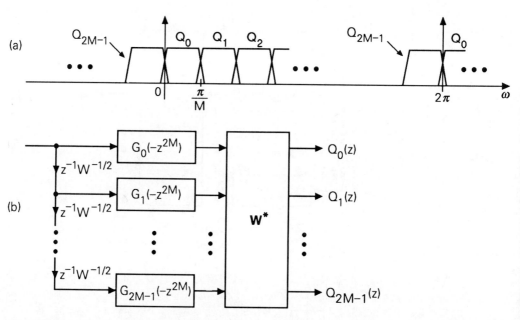

Figure 8.1-3 Shifting the responses by $\pi/2M$, by replacing z with $zW^{1/2}$.

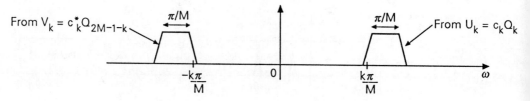

Figure 8.1-4 Magnitude response of the kth analysis filter $H_k(z)$. Synthesis filter $F_k(z)$ is chosen to have similar magnitude response. See text.

Definition of the Real Coefficient Analysis Filters

Define
$$U_k(z) = c_k P_0(zW^{k+0.5}) = c_k Q_k(z) \qquad (8.1.5)$$

and
$$V_k(z) = c_k^* P_0(zW^{-(k+0.5)}) = c_k^* Q_{2M-1-k}(z). \qquad (8.1.6)$$

We then generate the M analysis filters as follows:
$$H_k(z) = a_k U_k(z) + a_k^* V_k(z), \quad 0 \le k \le M-1. \qquad (8.1.7)$$

Here c_k and a_k are unit-magnitude constants, whose purpose will be clarified soon. (Actually, we could have done away with c_k by absorbing it in a_k, but the above form is more convenient for discussion.) Fig. 8.1-4 summarizes the situation.

We will assume the prototype to be of the form
$$P_0(z) = \sum_{n=0}^{N} p_0(n) z^{-n}, \qquad (8.1.8)$$

that is, Nth order FIR. All analysis filters are then FIR with order $\le N$, that is,
$$H_k(z) = \sum_{n=0}^{N} h_k(n) z^{-n}, \quad 0 \le k \le M-1. \qquad (8.1.9)$$

Since the coefficients of $P_0(z)$ are real, the coefficients of $V_k(z)$ and $U_k(z)$ are conjugates of each other. So $h_k(n)$ are real.

8.1.2 Alias Cancelation

Recall the intricacies of alias cancelation (Sec. 5.4.2). The decimated output of $H_k(z)$ gives rise to the alias components $H_k(zW_M^\ell)X(zW_M^\ell)$, [i.e., frequency-shifted versions of $H_k(z)X(z)$]. The synthesis filter $F_k(z)$, whose passband coincides with that of $H_k(z)$, retains the unshifted version $H_k(z)X(z)$, and also permits a small leakage of the shifted versions. When we add the outputs of all M synthesis filters, these leakages should "somehow" be canceled.

Remembering that the passbands of $F_k(z)$ should coincide with those of $H_k(z)$, we generate $F_k(z)$ as
$$F_k(z) = b_k U_k(z) + b_k^* V_k(z), \quad 0 \le k \le M-1, \qquad (8.1.10)$$

where b_k are unit-magnitude constants. (The choice of a_k, b_k and c_k will soon be settled.)

In general, the output of $F_k(z)$ has the components $H_k(zW_M^\ell)X(zW_M^\ell)$ for all values of ℓ, i.e., $0 \le \ell \le M-1$. However, if the stopband attenuation of $F_k(z)$ is sufficiently high, only some of these components are of practical significance. The other components, though not exactly equal to zero, will be ignored, giving rise to the term "approximate alias cancelation."

Figure 8.1-5 shows some of the shifted versions $U_k(zW_M^\ell)$ and $V_k(zW_M^\ell)$. Notice that, the response of $U_k(zW_M)$ does *not* overlap with that of $U_k(z)$. However, the responses of $U_k(zW_M^{-k})$ and $U_k(zW_M^{-(k+1)})$ overlap with the response of $V_k(z)$. Similarly, the responses of $V_k(zW_M^k)$ and $V_k(zW_M^{(k+1)})$ have overlap with that of $U_k(z)$. This means that the alias-components $X(zW_M^\ell)$, which are significant at the output of $F_k(z)$, correspond to

$$\ell = -(k+1), -k, k, k+1 \quad [\text{output of } F_k(z)] \qquad (8.1.11)$$

Similarly, at the output of $F_{k-1}(z)$, the significant alias components are for

$$\ell = -k, -(k-1), (k-1), k \quad [\text{output of } F_{k-1}(z)]. \qquad (8.1.12)$$

Note that negative values of ℓ should be interpreted modulo M. For example, $\ell = -1$ is equivalent to $\ell = M - 1$.

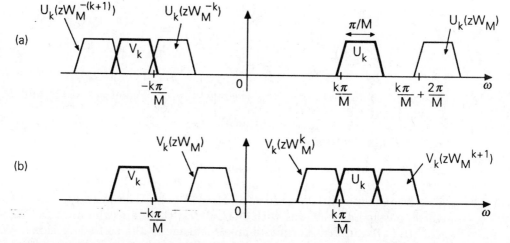

Figure 8.1-5 Demonstration of alias components which overlap with main synthesis filter response $|F_k(e^{j\omega})|$.

Constraint on a_k and b_k to Cancel Aliasing

Here, then, is the fundamental principle behind the approximate alias cancelation scheme: since the outputs of $F_k(z)$ as well as $F_{k-1}(z)$ have the common alias components $X(zW_M^{\pm k})$, we try to choose $F_{k-1}(z)$ and $F_k(z)$ such that this component is canceled when these outputs are added. In fact, such cancelation can be accomplished just by appropriately constraining a_k and b_k, as we show next.

The negative-frequency part of $F_k(z)$, [i.e., $b_k^* V_k(z)$] has the following significant alias components:

$$\left(a_k b_k^* U_k(zW_M^{-k}) V_k(z)\right) X(zW_M^{-k})$$
$$+ \left(a_k b_k^* U_k(zW_M^{-(k+1)}) V_k(z)\right) X(zW_M^{-(k+1)}), \tag{8.1.13}$$

and the negative-frequency part of $F_{k-1}(z)$ has

$$\left(a_{k-1} b_{k-1}^* U_{k-1}(zW_M^{-(k-1)}) V_{k-1}(z)\right) X(zW_M^{-(k-1)})$$
$$+ \left(a_{k-1} b_{k-1}^* U_{k-1}(zW_M^{-k}) V_{k-1}(z)\right) X(zW_M^{-k}). \tag{8.1.14}$$

The alias component $X(zW_M^{-k})$ can therefore be eliminated if

$$a_k b_k^* U_k(zW_M^{-k}) V_k(z) + a_{k-1} b_{k-1}^* U_{k-1}(zW_M^{-k}) V_{k-1}(z) = 0. \tag{8.1.15}$$

By using the definitions for $U_k(z)$ and $V_k(z)$, and the condition $|c_k| = |c_{k-1}| = 1$, we can rewrite (8.1.15) entirely in terms of $V_i(z)$'s as

$$\left(a_k b_k^* + a_{k-1} b_{k-1}^*\right) V_{k-1}(z) V_k(z) = 0. \tag{8.1.16}$$

This condition can be satisfied by constraining a_i and b_i such that

$$a_k b_k^* = -a_{k-1} b_{k-1}^*, \quad 1 \leq k \leq M - 1. \tag{8.1.17}$$

By considering the signal at the output of the positive frequency component $b_k U_k(z)$ of $F_k(z)$, we again obtain the same condition for alias cancelation. More specific choice of a_k and b_k will be given soon.

8.1.3 Eliminating Phase Distortion

Having canceled aliasing, we now turn to the distortion function $T(z)$. From Sec. 5.4.2 we know that $T(z)$ can always be expressed as

$$T(z) = \frac{1}{M} \sum_{k=0}^{M-1} F_k(z) H_k(z). \tag{8.1.18}$$

The QMF bank is free from phase distortion if $T(z)$ has linear phase. We can ensure this if the synthesis filters are chosen according to the mirror image condition

$$f_k(n) = h_k(N - n), \tag{8.1.19a}$$

or equivalently as

$$F_k(z) = z^{-N} H_k(z^{-1}) = z^{-N} \widetilde{H}_k(z) \quad (\text{as } h_k(n) \text{ is real}). \tag{8.1.19b}$$

In this case (8.1.18) becomes

$$T(z) = \frac{z^{-N}}{M} \sum_{k=0}^{M-1} H_k(z^{-1}) H_k(z). \tag{8.1.20}$$

Clearly

$$MT(e^{j\omega}) = e^{-j\omega N} \sum_{k=0}^{M-1} |H_k(e^{j\omega})|^2, \tag{8.1.21}$$

which shows that $T(z)$ has linear phase.

We know that $H_k(z)$ and $F_k(z)$ are already related in some way because of their definitions in terms of the same set of components $U_k(z)$ and $V_k(z)$. By careful choice of the constants a_k, b_k, c_k, we can satisfy the additional relation (8.1.19) as well. We do this in two steps as follows.

Choice of c_k to Ensure Linear Phase of $U_k(z)$ and $V_k(z)$

The phase response of $P_0(z)$ has not entered our discussion so far. We will now restrict $P_0(z)$ to be a linear phase filter with symmetric $p_0(n)$, that is, $p_0(n) = p_0(N-n)$, so that

$$\widetilde{P}_0(z) = z^N P_0(z). \tag{8.1.22}$$

[Antisymmetric $p_0(n)$ would be inconsistent, since $P_0(z)$ is lowpass.] We then have

$$P_0(e^{j\omega}) = e^{-j\omega N/2} P_R(\omega), \tag{8.1.23}$$

where $P_R(\omega)$ is real-valued (Sec. 2.4.2). We will choose c_k so that the complex-coefficient filters $U_k(z)$ and $V_k(z)$ have same (linear) phase as $P_0(z)$. (This will make it easier to determine the appropriate choice of a_k and b_k later.) From (8.1.5) we have

$$U_k(e^{j\omega}) = c_k W^{-(k+0.5)N/2} e^{-j\omega N/2} P_R\left(\omega - \frac{\pi(k+0.5)}{M}\right). \tag{8.1.24}$$

If we choose

$$c_k = W^{(k+0.5)N/2}, \tag{8.1.25}$$

then

$$U_k(e^{j\omega}) = e^{-j\omega N/2} P_R\left(\omega - \frac{\pi(k+0.5)}{M}\right). \tag{8.1.26}$$

Since $P_R(\omega)$ is real, $U_k(z)$ is a linear phase filter with phase response $\phi(\omega) = -\omega N/2$. Thus the phase responses of the modulated filters $U_k(z)$ are identical to that of the prototype $P_0(e^{j\omega})$. Same is true of $V_k(z)$, with c_k chosen as above.

Choice of b_k to Ensure the Relation $F_k(z) = z^{-N}H_k(z^{-1})$

The linear phase nature of $U_k(z)$ and $V_k(z)$ permits us to write

$$\widetilde{U}_k(z) = z^N U_k(z), \quad \widetilde{V}_k(z) = z^N V_k(z), \tag{8.1.27}$$

analogous to (8.1.22). By using these relations in (8.1.7) we can verify

$$z^{-N}\widetilde{H}_k(z) = a_k^* U_k(z) + a_k V_k(z). \tag{8.1.28}$$

If we now choose

$$b_k = a_k^*, \tag{8.1.29}$$

then the RHS of (8.1.28) reduces to $F_k(z)$ [defined as in (8.1.10)]. This proves that the mirror image condition (8.1.19b) can indeed be satisfied by enforcing the constraint $b_k = a_k^*$. The distortion $T(z)$ now takes the form (8.1.20), and hence has linear phase.

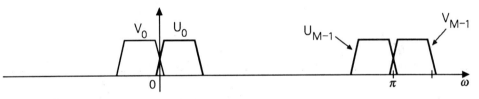

Figure 8.1-6 Demonstration of overlap of $V_0(e^{j\omega})$ with $U_0(e^{j\omega})$, and overlap of $V_{M-1}(e^{j\omega})$ with $U_{M-1}(e^{j\omega})$.

Choice of a_k

It only remains to choose a_k. The alias cancelation constraint (8.1.17) can be simplified by using the further relation (8.1.29) to obtain $a_k^2 = -a_{k-1}^2$, that is,

$$a_k = \pm j a_{k-1}. \tag{8.1.30}$$

This can be used to determine all a_k's, provided a_0 is somehow determined. To make this final choice, we note that the components $U_k(z)$ and $V_k(z)$ do not overlap significantly except when $k = 0$ or $M-1$ (Fig. 8.1-6). So the expression (8.1.18) can be simplified into

$$\begin{aligned} MT(z) \approx\ & (a_0^2 + a^{*2}_{\ 0})U_0(z)V_0(z) \\ & + (a_{M-1}^2 + a^{*2}_{\ M-1})U_{M-1}(z)V_{M-1}(z) \\ & + \sum_{k=0}^{M-1}\left(U_k^2(z) + V_k^2(z)\right), \end{aligned} \tag{8.1.31}$$

by using the condition $a_k b_k = a_k a_k^* = 1$. The cross-terms $U_0(z)V_0(z)$ and $U_{M-1}(z)V_{M-1}(z)$ can create significant distortions around the frequencies $\omega = 0$ and $\omega = \pi$, respectively. By constraining a_0 and a_{M-1} such that

$$a_0^4 = a_{M-1}^4 = -1, \tag{8.1.32}$$

Sec. 8.1 Pseudo QMF bank

we can eliminate these cross-terms, yielding

$$T(z) \approx \frac{1}{M}\sum_{k=0}^{M-1}\left(U_k^2(z) + V_k^2(z)\right). \qquad (8.1.33)$$

Based on these considerations we choose

$$a_k = e^{j\theta_k}, \quad \theta_k = (-1)^k \frac{\pi}{4}, \quad 0 \le k \le M-1. \qquad (8.1.34)$$

This implies $a_k = (-1)^k j a_{k-1}$, satisfying (8.1.30). Evidently (8.1.32) is also satisfied. We also choose b_k and c_k as stated above. All constants are now determined.

Summary

1. The condition for alias cancelation is given by $a_k b_k^* = -a_{k-1} b_{k-1}^*$.
2. The choice $c_k = W^{(k+0.5)N/2}$ (where $W = e^{-j\pi/M}$) ensures that $U_k(z)$ and $V_k(z)$ have the same (linear) phase response as the propotype $P_0(z)$. (This is a convenience which simplifies further design rules.)
3. The further constraint $b_k = a_k^*$ forces the relation $F_k(z) = z^{-N} H_k(z^{-1})$. This in turn leads to the linear phase form (8.1.20) for $T(z)$.
4. The constraint $a_k = (-1)^k j a_{k-1}$, together with $b_k = a_k^*$ ensures that the above alias cancelation condition is satisfied. Consistent with this constraint on a_k, we choose $a_k = e^{j\theta_k}$, $\theta_k = (-1)^k \pi/4$. This also ensures (8.1.32) so that $T(z)$ is further simplified to (8.1.33) (i.e., the two cross-terms given by $U_0(z)V_0(z)$ and $U_{M-1}(z)V_{M-1}(z)$, which can cause amplitude distortion around $\omega = 0$ and π, are eliminated).
5. Summarizing, the M analysis filters are given by (8.1.37) (see below), with $\theta_k = (-1)^k \pi/4$. With the synthesis filters chosen as in (8.1.19a), we have approximate alias cancellation and complete elimination of phase distortion. Amplitude distortion still remains, and should be minimized as shown in the next section. Notice finally that all the analysis and synthesis filters have real coefficients.

8.1.4 Closed Form Expressions for the Filters

We now find expressions for the analysis filters $h_k(n)$. The first term in (8.1.7) is

$$\begin{aligned}a_k U_k(z) &= e^{j\theta_k} c_k P_0(zW^{k+0.5}) \\ &= e^{j\theta_k} W^{(k+0.5)N/2} \sum_{n=0}^{N} p_0(n) W^{-(k+0.5)n} z^{-n},\end{aligned} \qquad (8.1.35)$$

so that its impulse response coefficients are

$$e^{j\theta_k} W^{(k+0.5)N/2} p_0(n) W^{-(k+0.5)n}. \qquad (8.1.36)$$

The coefficients of the second term in (8.1.7) are obtained by conjugating this. So $h_k(n)$ equals two times the real part of the first term, that is,

$$h_k(n) = 2p_0(n)\cos\left(\frac{\pi}{M}(k+0.5)(n-\frac{N}{2})+\theta_k\right), \qquad (8.1.37)$$

(since $p_0(n)$ is real). The synthesis filters $f_k(n)$ are obtained by replacing a_k with b_k. Since $b_k = a_k^*$, this is equivalent to replacing θ_k with $-\theta_k$, that is,

$$f_k(n) = 2p_0(n)\cos\left(\frac{\pi}{M}(k+0.5)(n-\frac{N}{2})-\theta_k\right). \qquad (8.1.38)$$

We can obtain this same $f_k(n)$ by using the mirror image relation (8.1.19), which we imposed in the above derivation. The analysis and synthesis filters, in general, do not have linear phase (even though the prototype $P_0(z)$ has linear phase). The distortion function $T(z)$, however, has linear phase.

As all the analysis and synthesis filters are related to the prototype $p_0(n)$ by cosine modulation, the only design freedom for the QMF bank is in the choice of $p_0(n)$. This design issue will be addressed in the next section.

8.2 DESIGN OF THE PSEUDO QMF BANK

In the previous section we considered the pseudo QMF bank, and eliminated phase distortion and (approximately) eliminated aliasing. It only remains to reduce amplitude distortion. Recall that amplitude distortion arises if $|T(e^{j\omega})|$ is not exactly flat. The prototype $P_0(z)$ should now be designed in such a way that $|T(e^{j\omega})|$ is acceptably flat.

8.2.1 Reducing Amplitude Distortion

We begin by pointing out some subtleties about the behavior of the distortion function $T(z)$.

Unit-Circle Zeros of T(z)

We know that it is undesirable for $T(z)$ to have zeros on the unit circle, as this would imply severe amplitude distortion. From the expression (8.1.21) we see that $T(e^{j\omega_0})$ is nonzero unless all the M filters satisfy $H_k(e^{j\omega_0}) = 0$. This unfortunate situation will not arise, unless the passband width of the prototpe $P_0(e^{j\omega})$ is unreasonably narrow.

Periodicity of $|T(e^{j\omega})|$

Consider the linear phase prototype (8.1.23). From Sec. 2.4.2 we know that $P_R^2(\omega)$ has period 2π for any N. Define $F(e^{j\omega}) = P_R^2(\omega)$ and $G(z) = F(zW^{0.5})$. We can then express (8.1.33) as

$$T(z) \approx \frac{z^{-N}}{M}\sum_{k=0}^{2M-1}G(zW^k), \qquad (8.2.1)$$

by using the simplified expression for $U_k(z)$ [i.e. eqn. (8.1.26)] and the fact that $v_k(n) = u_k^*(n)$. With $G(z) = \sum g(n)z^{-n}$, the summation in the above equation simplifies to

$$\sum_{k=0}^{2M-1} G(zW^k) = \sum_n g(n)z^{-n} \sum_{k=0}^{2M-1} W^{-kn} \qquad (8.2.2)$$
$$= 2M \sum_n g(2Mn)z^{-2Mn},$$

showing that the variable z appears only in powers such as z^{2M}. This shows that $T(z)$ has the form

$$T(z) \approx z^{-N} f(z^{2M}), \qquad (8.2.3)$$

for some FIR $f(z)$. In particular, therefore, $|T(e^{j\omega})|$ has period $2\pi/2M$.

Origin of Amplitude Distortion

Consider the expression (8.1.33). If ω is a frequency belonging to passband of some filter $U_k(z)$, then $T(e^{j\omega}) \approx U_k^2(e^{j\omega})/M$. This shows that $|T(e^{j\omega})|$ is nearly the same at all frequencies which belong to the passbands of $U_k(z)$'s (or $V_k(z)$'s). However, if ω is at the transition between $U_k(e^{j\omega})$ and $U_{k+1}(e^{j\omega})$ [see Fig. 8.2-1(a)], then

$$T(e^{j\omega}) \approx \frac{1}{M}\left(U_k^2(e^{j\omega}) + U_{k+1}^2(e^{j\omega})\right). \qquad (8.2.4)$$

Substituting from (8.1.26), this reduces to

$$\frac{e^{-j\omega N}}{M}\left(\left|P_0(e^{j(\omega - \frac{k\pi}{M} - \frac{\pi}{2M})})\right|^2 + \left|P_0(e^{j(\omega - \frac{(k+1)\pi}{M} - \frac{\pi}{2M})})\right|^2\right).$$
$$(8.2.5)$$

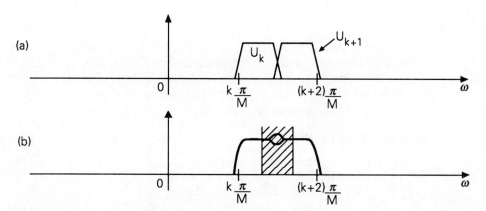

Figure 8.2-1 (a) Overlapping responses $|U_k(e^{j\omega})|$ and $|U_{k+1}(e^{j\omega})|$, and (b) two possible behaviors of $|U_k^2(e^{j\omega}) + U_{k+1}^2(e^{j\omega})|$, explaining the origin of amplitude to distortion.

Typical behaviors of this quantity are illustrated in Fig. 8.2-1(b). Assuming the prototype $P_0(z)$ to have 'good' stopband attenuation, this quantity is significant only in the frequency interval

$$\frac{k\pi}{M} \leq \omega \leq \frac{(k+2)\pi}{M}.$$

It can exhibit a 'bump' or 'dip' around the transition frequency $(k+1)\pi/M$. This is the source of amplitude distortion, that is, nonflatness of $|T(e^{j\omega})|$.

An Objective Function Representing the Flatness Requirement

Notice that the quantity in paranthesis in (8.2.5) is nothing but a frequency-shifted version of

$$|P_0(e^{j\omega})|^2 + |P_0(e^{j(\omega-\frac{\pi}{M})})|^2. \tag{8.2.6}$$

It follows that if we force this to be sufficiently 'flat,' then $|T(e^{j\omega})|$ will be "sufficiently flat" for all frequencies. This can be accomplished during the design of the prototype $p_0(n)$ by including a term in the objective function, to reflect the nonflatness of (8.2.6). Such an objective function is given by

$$\phi_1 = \int_0^{\pi/M} \left(|P_0(e^{j\omega})|^2 + |P_0(e^{j(\omega-\frac{\pi}{M})})|^2 - 1\right)^2 d\omega. \tag{8.2.7}$$

The above limits of integration are justified because $|T(e^{j\omega})|$ has period π/M as shown above.

8.2.2 Design of the Prototype Filter

The prototype $P_0(z)$ is a real coefficient linear phase FIR lowpass filter with cutoff $\pi/2M$ [Fig. 8.1-2(a)]. By designing it to have good stopband attenuation, we improve the attenuation characteristics of all the filters $H_k(z)$ and $F_k(z)$. Our choice of constants a_k, b_k, c_k above already ensures that aliasing and phase distortion are eliminated. By designing $P_0(z)$ such that (8.2.7) is small, one can reduce the amplitude distortion as well.

An appropriate measure of stopband attenuation of $P_0(z)$ is given by

$$\phi_2 = \int_{\frac{\pi}{2M}+\epsilon}^{\pi} |P_0(e^{j\omega})|^2 d\omega. \tag{8.2.8}$$

(The choice of $\epsilon > 0$ depends on the acceptable transition bandwidth.) So we optimize the coefficients $p_0(n)$ of $P_0(z)$ to minimize

$$\phi = \alpha\phi_1 + (1-\alpha)\phi_2 \quad \text{(composite objective function)}, \tag{8.2.9}$$

where α is a tradeoff parameter with $0 < \alpha < 1$. Standard nonlinear optimization packages [Press, et al, 1989] can be used for this.

TABLE 8.2.1 Design example 8.2.1. Impulse response of the FIR prorotype filter for pseudo QMF design.

n	$p_0(n)$
0	-2.9592103 e-03
1	-4.0188527 e-03
2	-4.9104756 e-03
3	-5.4331753 e-03
4	-5.3730961 e-03
5	-4.5222385 e-03
6	-2.6990818 e-03
7	2.3096829 e-04
8	4.3373153 e-03
9	9.6099830 e-03
10	1.5951440 e-02
11	2.3175400 e-02
12	3.1013020 e-02
13	3.9127130 e-02
14	4.7132594 e-02
15	5.4622061 e-02
16	6.1194772 e-02
17	6.6485873 e-02
18	7.0193888 e-02
19	7.2103807 e-02

Design example 8.2.1: Pseudo QMF Bank

We now show details of an 8-channel system ($M = 8$), with prototype filter order $N = 39$. The coefficients of the prototye $P_0(z)$, designed as described above, are shown in Table 8.2.1. (Only the first half is shown due to linear phase). Fig. 8.2-2 shows the prototype magnitude response $|P_0(e^{j\omega})|$, whereas Fig. 8.2-3 shows the magnitude responses of the analysis filters. Adjacent filter responses intersect approximately at the 3 dB level. This is consistent with the expression

$$|T(e^{j\omega})| = \frac{1}{M} \sum_{k=0}^{M-1} |H_k(e^{j\omega})|^2,$$

because, at the transition between two filters, only two of the M terms in the above summation are significant, and these are required to add up to unity.

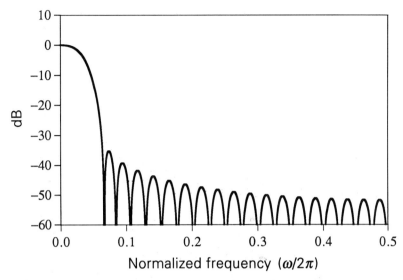

Figure 8.2-2 Design example 8.2.1. Pseudo QMF design. Magnitude response of the FIR linear phase prototype $P_0(z)$. Filter order $N = 39$.

Peak distortions E_a and E_{pp}. Let us now look at various distortions. Recall that (5.4.7) represents the gain for the ℓth alias component $X(zW^\ell), \ell > 0$. Fig. 8.2-4 shows a plot of

$$\sqrt{\sum_{\ell=1}^{M-1} |A_\ell(e^{j\omega})|^2}, \qquad (8.2.10)$$

which demonstrates that each of the terms $|A_\ell(e^{j\omega})|$ is very small for all ω. This shows that aliasing has been reduced satisfactorily. The quantity E_a, which is the maximum value of (8.2.10) over all ω, is the worst possible *peak aliasing distortion*.

Next, Fig. 8.2-5 shows a plot of $M|T(e^{j\omega})|$. This is very close to unity for all ω, verifying that amplitude distortion has been reduced satisfactorily. As argued earlier, $|T(e^{j\omega})|$ is seen to have period $2\pi/2M = \pi/8$. By design, $T(z)$ has linear phase, so we need not worry about phase distortion. The maximum peak to peak ripple of $M|T(e^{j\omega})|$, denoted E_{pp}, is usually taken to be a measure of worst possible amplitude distortion.

From (8.1.20) we see that $T(z)$ has order 78, i.e., $T(z) = \sum_{n=0}^{78} t(n)z^{-n}$. Because of the form (8.2.3), only a subset of the coefficients $t(n)$ are nonzero. The coefficients $Mt(n)$ are shown in Table 8.2.2, which also verifies the linear phase nature of $T(z)$. In fact, we see that $T(z)$ is nearly a delay.

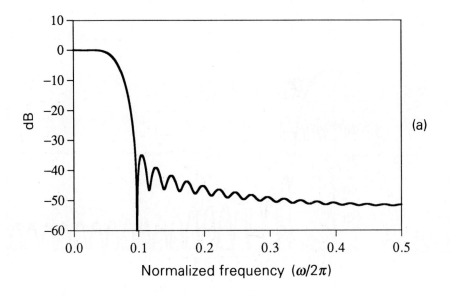

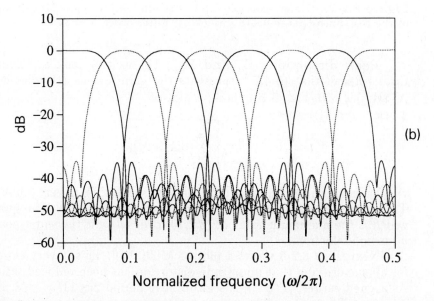

Figure 8.2-3 Design example 8.2.1. Pseudo QMF design. Magnitude responses of the analysis filters. (a) $H_0(z)$ only, and (b) all eight filters. Filter order $N = 39$; number of channels $M = 8$.

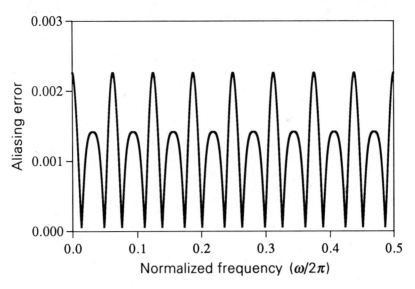

Figure 8.2-4 Design example 8.2.1. Plot of aliasing error in pseudo QMF design. The quantity (8.2.10) is shown above.

TABLE 8.2.2 Design example 8.2.1. Set of nonzero coefficients of $MT(z)$ where $T(z)$ is the distortion function for the pseudo QMF design.

n	$Mt(n)$
7	0.0022752
23	0.0008191
39	0.9988325
55	0.0008191
71	0.0022752

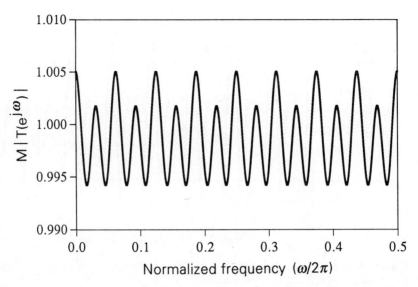

Figure 8.2-5 Design example 8.2.1. Plot of amplitude distortion function in pseudo QMF bank.

8.3 EFFICIENT POLYPHASE STRUCTURES

With the constants a_k, b_k and c_k constrained as summarized at the end of Sec. 8.1.3, we can rewrite the expression for the analysis filters as

$$H_k(z) = d_k Q_k(z) + d_k^* Q_{2M-1-k}(z) \qquad (8.3.1)$$

with $0 \le k \le M - 1$. Here

$$d_k = a_k c_k = e^{j\theta_k} W^{(k+0.5)N/2}, \quad 0 \le k \le M - 1. \qquad (8.3.2)$$

Since the coefficients of $Q_{2M-1-k}(z)$ are conjugates of those of $Q_k(z)$, $h_k(n)$ are real as intended. As the filters $Q_k(z)$ can be represented by the structure of Fig. 8.1-3, we can implement the M analysis filters as in Fig. 8.3-1(a). From this figure we can write

$$H_k(z) = \sum_{n=0}^{2M-1} t_{kn} z^{-n} G_n(-z^{2M}), \quad 0 \le k \le M - 1, \qquad (8.3.3)$$

where

$$t_{kn} = W^{-(k+0.5)(n-\frac{N}{2})} e^{j\theta_k} + W^{(k+0.5)(n-\frac{N}{2})} e^{-j\theta_k}. \qquad (8.3.4)$$

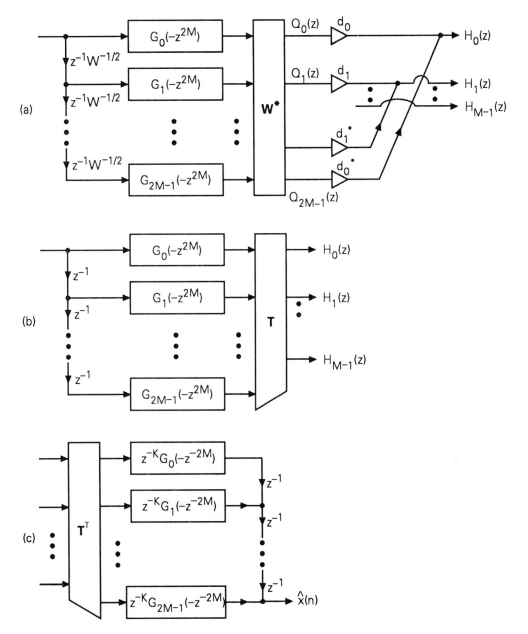

Figure 8.3-1 (a) Polyphase implementation of the M-channel cosine modulated analysis bank, (b) simplified drawing, where **T** is a real matrix, and (c) corresponding synthesis bank. Here $K = N - 2M + 1$.

The elements t_{kn} simplify to

$$t_{kn} = 2\cos\left(\frac{\pi}{M}(k+0.5)(n-\frac{N}{2}) + \theta_k\right), \quad (8.3.5)$$

where $\theta_k = (-1)^k \pi/4$. Equation (8.3.3) permits us to draw the analysis bank as in Fig. 8.3-1(b), where $\mathbf{T}$ is $M \times 2M$ with elements t_{kn}. Note that the coefficients t_{kn} are precisely the elements which modulate $p_0(n)$ in (8.1.37) to obtain $h_k(n)$. The M cosine modualted filters $H_k(z)$ are therefore obtained by implementing the polyphase components $G_n(-z^{2M})$ which come from the single prototype $P_0(z)$, and then using the cosine modulation matrix $\mathbf{T}$.

In terms of matrix notation, the analysis bank vector $\mathbf{h}(z)$ defined in (5.4.1) becomes

$$\mathbf{h}(z) = \mathbf{T}\mathbf{g}(z), \quad (8.3.6)$$

where

$$\mathbf{g}(z) \stackrel{\Delta}{=} \begin{bmatrix} G_0(-z^{2M}) \\ z^{-1}G_1(-z^{2M}) \\ \vdots \\ z^{-(2M-1)}G_{2M-1}(-z^{2M}) \end{bmatrix}. \quad (8.3.7)$$

To obtain the structure for the synthesis bank we use the relation $F_k(z) = z^{-N}H_k(z^{-1})$, and write the synthesis filter vector $\mathbf{f}(z)$ in terms of the analysis filter vector $\mathbf{h}(z)$ as

$$\mathbf{f}^T(z) = z^{-N}\mathbf{h}^T(z^{-1}) = z^{-N}\mathbf{g}^T(z^{-1})\mathbf{T}^T. \quad (8.3.8)$$

This system can be implemented as shown in Fig. 8.3-1(c), where the quantity $K = N - 2M + 1$.

In practice, we have decimators following the analysis filters, and expanders preceding the synthesis filters (Fig. 5.4.1). These devices can be moved by employing the noble identities (Fig. 4.2.3) to obtain more efficient polyphase structures. Figure 8.3-2 shows this scheme for the analysis bank, where the filters $G_n(-z^2)$ operate at the lowest possible rate. Similar arrangement can be obtained for the synthesis bank.

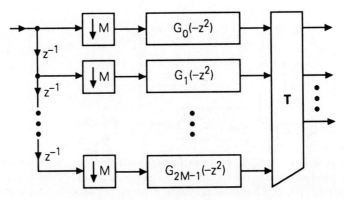

Figure 8.3-2 Improved polyphase implementation of the pseudo QMF analysis bank, with decimators moved "all the way to the left."

Implementation Using the Discrete Cosine Transform (DCT)

A special case of interest arises when the filter length $N+1$ is restricted to be $N+1 = 2mM$ for some integer m. In this case, the polyphase structure can be redrawn in such a way that the main computational load is represented by a $M \times M$ matrix called the discrete cosine transform (DCT). Moreover, this matrix can be implemented using fast transform techniques [Yip and Rao, 1987]. Details of this special case are developed in the next few sections, where we also show how to modify the results of the present section to achieve *perfect reconstruction* in cosine modulated QMF banks.

Computational Complexity of Pseudo QMF Systems

We can implement the analysis bank as in Fig. 8.3-1(b), where $G_k(z)$ are the $2M$ polyphase components of $P_0(z)$. The total number of multiplications and additions required for these components is nearly equal to the order N of the filter $P_0(z)$. So the complexity of the analysis bank is equal to about N multipliers and adders, plus the overhead required to implement the modulation matrix. The exact cost of this overhead depends on the value of M and the details of the fast DCT. Assuming this cost is negligible for simplicity, the complexity of the analysis bank is about N/M MPUs (and the same number of APUs). Recall from Sec. 6.7 that, for this *same filter order* N if $N \gg M$, the perfect reconstruction system is only about two times more expensive.

8.4 DEEPER PROPERTIES OF COSINE MATRICES

The cosine modulation matrix **T** which appears in the pseudo QMF structure of Fig. 8.3-2 satisfies some very useful mathematical properties. These properties, while not obvious, are important in the design of *perfect reconstruction* cosine modulated filter banks, as we see in Sec. 8.5. The purpose of this section is to state and prove these properties.

8.4.1 The DCT and DST matrices

We first introduce the discrete cosine transform (DCT) matrix **C**, and the the discrete sine transform (DST) matrix **S**. These will play a crucial role in the theory as well as *fast* implementation of the cosine-modulated perfect reconstruction systems to be studied in Sec. 8.5.

The discrete cosine transform has been known since the early seventies [Ahmed, et al., 1974]. Four types of DCT and DST matrices have been documented in the literature [Yip and Rao, 1987]. Of these only Type 4 matrices are relevant to our discussion. These are $M \times M$ matrices with elements

$$c_{kn} = \sqrt{\frac{2}{M}} \cos \frac{\pi}{M}(k+0.5)(n+0.5), \quad s_{kn} = \sqrt{\frac{2}{M}} \sin \frac{\pi}{M}(k+0.5)(n+0.5). \quad (8.4.1)$$

We omit the adjective 'Type 4' in all further discussions. Evidently $\mathbf{C}$ and $\mathbf{S}$ are real. They satisfy the following properties.

1. Symmetry, that is, $\mathbf{C}^T = \mathbf{C}$, and $\mathbf{S}^T = \mathbf{S}$.
2. $\mathbf{C}$ and $\mathbf{S}$ are related as
$$\mathbf{C} = \mathbf{\Gamma S J}, \qquad (8.4.2)$$

where

$$\mathbf{\Gamma} = \begin{bmatrix} 1 & 0 & 0 & \cdots & 0 \\ 0 & -1 & 0 & \cdots & 0 \\ \vdots & \vdots & \vdots & \ddots & \vdots \\ 0 & 0 & 0 & \cdots & (-1)^{M-1} \end{bmatrix}, \qquad (8.4.3)$$

and $\mathbf{J}$ is the reversal (or anti-diagonal) matrix defined in Appendix A (Sec. A.2). In words, if we renumber the nth column of $\mathbf{S}$ as the $(M-1-n)$th column (for each n), and insert a minus sign on all elements of every odd numbered row, the result is the $\mathbf{C}$ matrix. We can also rewrite (8.4.2) as $\mathbf{S} = \mathbf{\Gamma C J}$, since $\mathbf{\Gamma}^{-1} = \mathbf{\Gamma}$ and $\mathbf{J}^{-1} = \mathbf{J}$.

3. $\mathbf{C}$ and $\mathbf{S}$ are orthonormal, that is, $\mathbf{C}^T\mathbf{C} = \mathbf{S}^T\mathbf{S} = \mathbf{I}$. By combining with symmetry, we have $\mathbf{C}^2 = \mathbf{S}^2 = \mathbf{I}$ so that $\mathbf{C}^{-1} = \mathbf{C}$ and $\mathbf{S}^{-1} = \mathbf{S}$.

Proofs

The first property is obvious. A proof of the second property is requested in Problem 8.6. We will now prove the third property. It is sufficient to prove orthonormality of $\mathbf{C}$. Orthonormality of $\mathbf{S}$ then follows from $\mathbf{S} = \mathbf{\Gamma C J}$.

Orthonormality of C. Consider Fig. 8.4-1 which is a system with $2M$ inputs and M outputs. Here, $\mathbf{W}$ is the $2M \times 2M$ DFT matrix, $W = e^{-j2\pi/2M}$, and $\beta_n = W^{-(n+0.5)/2}$. This system has a $M \times 2M$ transfer matrix, which we denote as $\mathbf{V}$. This matrix has elements

$$\begin{aligned} V_{kn} &= \beta_k W^{-(k+0.5)n} + \beta_k^* W^{(k+0.5)n} \\ &= W^{-(k+0.5)(n+0.5)} + W^{(k+0.5)(n+0.5)} \\ &= 2\cos\frac{\pi}{M}(k+0.5)(n+0.5). \end{aligned}$$

So we can write
$$\mathbf{V} = \sqrt{2M}\,[\,\mathbf{C}\quad \times\,]. \qquad (8.4.4)$$

In other words, except for the scale factor $\sqrt{2M}$, the first M columns of $\mathbf{V}$ are the same as those of $\mathbf{C}$. (The $\times$ denotes an $M \times M$ matrix whose details are not relevant here.) By using the structure of Fig. 8.4-1 one verifies that

$$\mathbf{C} = \frac{1}{\sqrt{2M}}\underbrace{[\,\mathbf{I}\quad -\mathbf{J}\,]}_{M \times 2M}\mathbf{\Lambda}_\beta \mathbf{U}\mathbf{\Lambda}_w, \qquad (8.4.5)$$

where Λ_β and Λ_w are diagonal matrices of sizes $2M \times 2M$ and $M \times M$ respectively, with diagonal elements

$$[\Lambda_\beta]_{kk} = \beta_k, \quad [\Lambda_w]_{kk} = W^{-0.5k}, \qquad (8.4.6)$$

and $\mathbf{U}$ is the left $2M \times M$ submatrix of $\mathbf{W}^*$. By using the fact that

$$\begin{bmatrix} \mathbf{I} \\ -\mathbf{J} \end{bmatrix} [\mathbf{I} \ -\mathbf{J}] = \mathbf{I}_{2M} - \mathbf{J}_{2M},$$

we can verify

$$\mathbf{C}^T \mathbf{C} = \mathbf{C}^\dagger \mathbf{C} = \left(\Lambda_w^\dagger \mathbf{U}^\dagger \Lambda_\beta^\dagger \Lambda_\beta \mathbf{U} \Lambda_w - \Lambda_w^\dagger \mathbf{U}^\dagger \Lambda_\beta^\dagger \mathbf{J}_{2M} \Lambda_\beta \mathbf{U} \Lambda_w \right)/2M, \quad (8.4.7)$$

In Problem 8.7 we verify that the first term above is $2M\mathbf{I}$ and the second term is zero. This proves that $\mathbf{C}^T \mathbf{C} = \mathbf{I}$. ▽▽▽

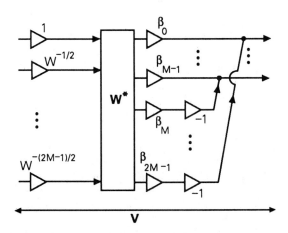

Figure 8.4-1 Pertaining to the proof that the DCT matrix is orthonormal.

8.4.2 Cosine Modulation Expressed Using DCT and DST

Consider the cosine modulation matrix $\mathbf{T}$ in the pseudo QMF structure of Fig. 8.3-2. We now show how this can be expressed in terms of the DCT and DST matrices. Recall that $\mathbf{T}$ is $M \times 2M$ with elements t_{kn} as in (8.3.5), where N is the order of the prototype filter.

We consider only the special case where the filter length $N + 1$ and the number of channels M are related as

$$N + 1 = 2mM, \qquad (8.4.8)$$

for some integer m. Let us partition $\mathbf{T}$ as

$$\mathbf{T} = [\mathbf{A}_0 \ \mathbf{A}_1], \qquad (8.4.9)$$

Sec. 8.4 Properties of cosine matrices

where $\mathbf{A}_0$ and $\mathbf{A}_1$ are $M \times M$. We will show that these matrices can be expressed in terms of the DCT and DST matrices as follows:

$$\mathbf{A}_0 = \sqrt{M}\mathbf{\Lambda}_c(\mathbf{C} - \mathbf{\Gamma}\mathbf{S}), \quad \mathbf{A}_1 = -\sqrt{M}\mathbf{\Lambda}_c(\mathbf{C} + \mathbf{\Gamma}\mathbf{S}) \quad (m \text{ even}), \quad (8.4.10)$$

$$\mathbf{A}_0 = \sqrt{M}\mathbf{\Lambda}_s\mathbf{\Gamma}(\mathbf{C} + \mathbf{\Gamma}\mathbf{S}), \quad \mathbf{A}_1 = \sqrt{M}\mathbf{\Lambda}_s\mathbf{\Gamma}(\mathbf{C} - \mathbf{\Gamma}\mathbf{S}) \quad (m \text{ odd}). \quad (8.4.11)$$

Here $\mathbf{\Lambda}_c$ and $\mathbf{\Lambda}_s$ are $M \times M$ diagonal matrices with diagonal elements

$$[\mathbf{\Lambda}_c]_{kk} = \cos(\pi(k+0.5)m), \quad [\mathbf{\Lambda}_s]_{kk} = \sin(\pi(k+0.5)m). \quad (8.4.12)$$

Notice that for fixed m, one of these two diagonal matrices is null, and the other has diagonal elements ± 1. Using the above expressions for $\mathbf{A}_0$ and $\mathbf{A}_1$ we will also show that they satisfy

$$\mathbf{A}_0^T \mathbf{A}_0 = 2M(\mathbf{I} - (-1)^m \mathbf{J}), \quad (8.4.13)$$

$$\mathbf{A}_1^T \mathbf{A}_1 = 2M(\mathbf{I} + (-1)^m \mathbf{J}), \quad (8.4.14)$$

$$\mathbf{A}_0^T \mathbf{A}_1 = \mathbf{0}. \quad (8.4.15)$$

Readers interested only in the consequences of these relations can skip the following proof, and proceed to Sec. 8.5.

Proof of the Relations (8.4.10)–(8.4.15)

For $N = 2mM - 1$, the elements t_{kn} in (8.3.5) become

$$t_{kn} = 2\Big(\widehat{c}_{kn} \cos \phi_k - \widehat{s}_{kn} \sin \phi_k\Big), \quad 0 \le k \le M-1, \ 0 \le n \le 2M-1, \quad (8.4.16)$$

where

$$\widehat{c}_{kn} = \cos\Big(\frac{\pi}{M}(k+0.5)(n+0.5)\Big), \quad \widehat{s}_{kn} = \sin\Big(\frac{\pi}{M}(k+0.5)(n+0.5)\Big), \quad (8.4.17)$$

and

$$\phi_k = -\pi(k+0.5)m + (-1)^k \frac{\pi}{4}. \quad (8.4.18)$$

The elements of the $M \times M$ matrix $\mathbf{A}_0$ are therefore

$$[\mathbf{A}_0]_{kn} = 2\Big(\widehat{c}_{kn} \cos \phi_k - \widehat{s}_{kn} \sin \phi_k\Big), \quad 0 \le k, n \le M-1. \quad (8.4.19)$$

The elements of $\mathbf{A}_1$ are found by replacing n with $n + M$ in (8.4.16), and simplifying. Thus

$$[\mathbf{A}_1]_{kn} = -2(-1)^k \Big(\widehat{s}_{kn} \cos \phi_k + \widehat{c}_{kn} \sin \phi_k\Big) \quad 0 \le k, n \le M-1. \quad (8.4.20)$$

The quantities $\cos\phi_k$ and $\sin\phi_k$ can be simplified into

$$\cos\phi_k = \frac{1}{\sqrt{2}}\Big(\cos\big(\pi(k+0.5)m\big) + (-1)^k \sin\big(\pi(k+0.5)m\big)\Big),$$
$$\sin\phi_k = \frac{1}{\sqrt{2}}\Big(-\sin\big(\pi(k+0.5)m\big) + (-1)^k \cos\big(\pi(k+0.5)m\big)\Big). \quad (8.4.21)$$

Using the diagonal matrices $\mathbf{\Lambda}_c$ and $\mathbf{\Lambda}_s$ we can then express

$$\mathbf{A}_0 = \sqrt{M}\Big((\mathbf{\Lambda}_c + \mathbf{\Gamma}\mathbf{\Lambda}_s)\mathbf{C} - (-\mathbf{\Lambda}_s + \mathbf{\Gamma}\mathbf{\Lambda}_c)\mathbf{S}\Big),$$
$$\mathbf{A}_1 = -\sqrt{M}\mathbf{\Gamma}\Big((\mathbf{\Lambda}_c + \mathbf{\Gamma}\mathbf{\Lambda}_s)\mathbf{S} + (-\mathbf{\Lambda}_s + \mathbf{\Gamma}\mathbf{\Lambda}_c)\mathbf{C}\Big). \quad (8.4.22)$$

Depending on the value of m, this expression can be simplified further. If m is even then $\mathbf{\Lambda}_s = \mathbf{0}$ and $[\mathbf{\Lambda}_c]_{kk} = \pm 1$. If m is odd, then the opposite situation prevails. This leads to the simplified relations claimed in (8.4.10) and (8.4.11).

If m is even, we find from (8.4.10)

$$\mathbf{A}_0^T \mathbf{A}_0 = M(\mathbf{C}^2 + \mathbf{S}^2 - \mathbf{C}\mathbf{\Gamma}\mathbf{S} - \mathbf{S}\mathbf{\Gamma}\mathbf{C}), \quad (8.4.23)$$
$$\mathbf{A}_1^T \mathbf{A}_1 = M(\mathbf{C}^2 + \mathbf{S}^2 + \mathbf{C}\mathbf{\Gamma}\mathbf{S} + \mathbf{S}\mathbf{\Gamma}\mathbf{C}), \quad (8.4.24)$$
$$\mathbf{A}_0^T \mathbf{A}_1 = -M(\mathbf{C}^2 - \mathbf{S}^2 + \mathbf{C}\mathbf{\Gamma}\mathbf{S} - \mathbf{S}\mathbf{\Gamma}\mathbf{C}), \quad (8.4.25)$$

The three properties of $\mathbf{C}$ and $\mathbf{S}$ listed at the beginning of Sec. 8.4.1 imply, in particular, $\mathbf{C}^2 = \mathbf{S}^2 = \mathbf{I}$, and $\mathbf{C}\mathbf{\Gamma}\mathbf{S} = \mathbf{S}\mathbf{\Gamma}\mathbf{C} = \mathbf{J}$. As a result the above relations reduce to (8.4.13)–(8.4.15) indeed. For odd m the proof can be carried out similarly. ▽▽▽

8.5 COSINE MODULATED PERFECT RECONSTRUCTION SYSTEMS

By using the results of the previous sections, it is now very easy to obtain a maximally decimated FIR perfect reconstruction system in which the analysis filters are related by cosine modulation [as in (8.1.37)], and the synthesis filters are as in (8.1.19a). This observation was made independently by Malvar [1990b], Ramstad [1991] and Koilpillai and Vaidyanathan, [1991 and 1992]. Among all FIR perfect reconstruction systems known today for arbitrary filter lengths, this system is perhaps the simplest (both in terms of design and implementation complexities). It inherits all the simplicity and elegance of the cosine modulated pseudo QMF system and yet offers perfect reconstruction property. We now proceed to derive this. Historically, a special case of this result for $N = 2M - 1$ was first reported [Princen and Bradley, 1986], [Malvar and Staelin, 1989], and is related to the concept of lapped orthogonal transforms (LOT, Sec. 6.6). The result of this section can be considered to be a generalization of the LOT, and is presented in Malvar [1990b] in that light. Our presentation here is based on Koilpillai and Vaidyanathan [1991a, 1992].

Expression for the Polyphase Matrix E(z)

For the cosine modulated system the analysis bank has the structure shown in Fig. 8.3-1(b), where $G_k(z)$ are the $2M$ polyphase components of the prototype $P_0(z)$ (see Fig. 8.1-1). Thus,

$$\mathbf{h}(z) = \mathbf{T} \begin{bmatrix} \mathbf{g}_0(z^{2M}) & 0 \\ 0 & \mathbf{g}_1(z^{2M}) \end{bmatrix} \begin{bmatrix} \mathbf{e}(z) \\ z^{-M}\mathbf{e}(z) \end{bmatrix},$$

(8.5.1)

$$= \mathbf{T} \begin{bmatrix} \mathbf{g}_0(z^{2M}) \\ z^{-M}\mathbf{g}_1(z^{2M}) \end{bmatrix} \mathbf{e}(z)$$

where $\mathbf{e}(z)$ is the delay chain vector [eqn. (5.4.1)] and $\mathbf{g}_i(z)$ are diagonal matrices with

$$[\mathbf{g}_0(z)]_{kk} = G_k(-z), \quad [\mathbf{g}_1(z)]_{kk} = G_{M+k}(-z). \tag{8.5.2}$$

Comparing with $\mathbf{h}(z) = \mathbf{E}(z^M)\mathbf{e}(z)$, we identify the polyphase matrix $\mathbf{E}(z)$ of the analysis bank as

$$\mathbf{E}(z) = \mathbf{T} \begin{bmatrix} \mathbf{g}_0(z^2) \\ z^{-1}\mathbf{g}_1(z^2) \end{bmatrix}. \tag{8.5.3}$$

Using the partition $\mathbf{T} = [\,\mathbf{A}_0 \quad \mathbf{A}_1\,]$ as before, we have

$$\mathbf{E}(z) = [\,\mathbf{A}_0 \quad \mathbf{A}_1\,] \begin{bmatrix} \mathbf{g}_0(z^2) \\ z^{-1}\mathbf{g}_1(z^2) \end{bmatrix}. \tag{8.5.4}$$

8.5.1 Forcing E(z) to be Paraunitary when N + 1 = 2mM

From Chapter 6 we know that we can achieve perfect reconstruction by constraining $\mathbf{E}(z)$ to be paraunitary (i.e., $\widetilde{\mathbf{E}}(z)\mathbf{E}(z) = d\mathbf{I}$) and taking the synthesis filter coeffcients to be the time reversed conjugates as in (6.2.6). Recall that the paraunitary property is the same as losslessness, since we are discussing only causal FIR systems. The main result is summarized as follows:

♦**Theorem 8.5.1.** Let the prototype $P_0(z)$ be a real-coefficient FIR filter with length $N+1 = 2mM$ for some integer m. Assume $p_0(n) = p_0(N-n)$ (linear phase constraint). Let $G_k(z)$, $0 \le k \le 2M-1$, be the $2M$ polyphase components of $P_0(z)$. Suppose the M analysis filters $H_k(z)$ are generated by cosine modulation as in (8.1.37) with $\theta_k = (-1)^k \pi/4$. Then the $M \times M$ polyphase component matrix $\mathbf{E}(z)$ is paraunitary if, and only if, $G_k(z)$ satisfy the pairwise power complementary conditions

$$\widetilde{G}_k(z)G_k(z) + \widetilde{G}_{M+k}(z)G_{M+k}(z) = \alpha, \quad 0 \le k \le M-1, \qquad (8.5.5)$$

for some $\alpha > 0$. $\diamond$

Proof. From (8.5.4) we have

$$\begin{aligned}\widetilde{\mathbf{E}}(z)\mathbf{E}(z) =& \widetilde{\mathbf{g}}_0(z^2)\mathbf{A}_0^T\mathbf{A}_0\mathbf{g}_0(z^2) + \widetilde{\mathbf{g}}_1(z^2)\mathbf{A}_1^T\mathbf{A}_1\mathbf{g}_1(z^2) \\ &+ z\widetilde{\mathbf{g}}_1(z^2)\mathbf{A}_1^T\mathbf{A}_0\mathbf{g}_0(z^2) + z^{-1}\widetilde{\mathbf{g}}_0(z^2)\mathbf{A}_0^T\mathbf{A}_1\mathbf{g}_1(z^2). \end{aligned} \qquad (8.5.6)$$

Since $\mathbf{A}_0$ and $\mathbf{A}_1$ satisfy (8.4.13)–(8.4.15), this becomes

$$\begin{aligned}\widetilde{\mathbf{E}}(z)\mathbf{E}(z) =& 2M\Big(\widetilde{\mathbf{g}}_0(z^2)\mathbf{g}_0(z^2) + \widetilde{\mathbf{g}}_1(z^2)\mathbf{g}_1(z^2)\Big) \\ & - 2M(-1)^m\Big(\widetilde{\mathbf{g}}_0(z^2)\mathbf{J}\mathbf{g}_0(z^2) - \widetilde{\mathbf{g}}_1(z^2)\mathbf{J}\mathbf{g}_1(z^2)\Big). \end{aligned} \qquad (8.5.7)$$

Since $N+1 = 2mM$, each polyphase component $G_k(z)$ has length m, that is, order $m-1$. The relation $p_0(n) = p_0(N-n)$ imposes the following constraint on these polyphase components (Problem 8.8):

$$G_k(z) = z^{-(m-1)}\widetilde{G}_{2M-1-k}(z) \quad \text{(linear phase constraint).} \qquad (8.5.8a)$$

In other words the diagonal matrices $\mathbf{g}_0(z)$ and $\mathbf{g}_1(z)$ are related as

$$\mathbf{g}_1(z) = z^{-(m-1)}(-1)^{m-1}\mathbf{J}\widetilde{\mathbf{g}}_0(z)\mathbf{J} \quad \text{(linear phase constraint).} \qquad (8.5.8b)$$

(Recall from Appendix A (Sec. A.2) that $\mathbf{JDJ}$ has the effect of reversing the order of the diagonal entries of a diagonal matrix $\mathbf{D}$.) If this relation is used in (8.5.7), the second term vanishes, and

$$\widetilde{\mathbf{E}}(z)\mathbf{E}(z) = 2M\Big(\widetilde{\mathbf{g}}_0(z^2)\mathbf{g}_0(z^2) + \widetilde{\mathbf{g}}_1(z^2)\mathbf{g}_1(z^2)\Big). \qquad (8.5.9)$$

It is now clear that $\mathbf{E}(z)$ is paraunitary if, and only if,

$$\widetilde{\mathbf{g}}_0(z)\mathbf{g}_0(z) + \widetilde{\mathbf{g}}_1(z)\mathbf{g}_1(z) = \alpha\mathbf{I}, \quad \alpha > 0. \qquad (8.5.10)$$

This condition is equivalent to saying that the polyphase components $G_k(z)$ and $G_{M+k}(z)$ are power complementary, that is, that (8.5.5) holds. $\triangledown\triangledown\triangledown$

Remark. The cosine modulated pseudo QMF system developed in Sec. 8.1 and 8.2 satisfies the relation (8.1.19a). Furthermore if the filters are well-designed as described in Sec. 8.2, the system is almost like a perfect reconstruction system (as demonstrated by Design example 8.2.1). This suggests, in view of Theorem 6.2.1, that $\mathbf{E}(z)$ is "almost paraunitary". This leads us to expect that $\widetilde{\mathbf{E}}(z)\mathbf{E}(z)$ is "almost a diagonal matrix," and the diagonal elements are "almost constant." This, indeed, has been verified with the help of several design examples. This also shows that, if we force the matrix $\mathbf{E}(z)$ to be paraunitary *apriori*, that is, before optimizing the filter coefficients, then the resulting filters are almost the same as the pseudo QMF filters, except that they satisfy the perfect reconstruction property "perfectly"!

8.5.2 The Design Procedure

We know that all the analysis filter responses are controlled by the prototype response $|P_0(e^{j\omega})|$. As in Sec. 8.2.2 we have to optimize the coefficients of $P_0(z)$ to minimize an objective function. For pseudo QMF design we minimized a linear combination of ϕ_1 and ϕ_2 [defined in (8.2.7) and (8.2.8)]. But in the present case, it is sufficient to minimize only the stopband energy ϕ_2. The quantity ϕ_1 which represents the degree of nonflatness of $|T(e^{j\omega})|$ is automatically zero, because of the perfect reconstruction property guaranteed by paraunitariness of $\mathbf{E}(z)$.

Instead of minimizing the stopband energy ϕ_2, it is also possible to minimize the maximum magnitude of $|P_0(e^{j\omega})|$ in its stopband region. Either of these minimizations can be done using standard optimization routines [Press, et al., 1989], [IMSL, 1987].

Imposing Paraunitary Constraint Using Two-Channel Lattice

During optimization it is however necessary to impose the paraunitary constraint on $\mathbf{E}(z)$, which we have shown to be equivalent to the power complementary constraint (8.5.5). Now the power complementary property is equivalent to the condition that the FIR vector $\begin{bmatrix} G_k(z) \\ G_{M+k}(z) \end{bmatrix}$ be paraunitary. In a manner similar to Sec. 6.4, this paraunitary vector can be implemented with the cascaded lattice structure of Fig. 8.5-1. (This will be proved in Sec. 14.3.2). Conversely, the transfer functions $G_k(z)$ and $G_{M+k}(z)$ in this structure remain power complementary [i.e., satisfies (8.5.5) with $\alpha = 1$] regardless of the values of the angular parameters $\theta_{k,\ell}$. This follows because the matrices $\mathbf{R}_{k,\ell}$ are unitary; see Sec. 6.1.2.

The cosine modulated analysis bank, shown earlier in Fig. 8.3-2., now takes the appearance shown in Fig. 8.5-2. We now optimize the angles $\theta_{k,\ell}$ so as to minimize ϕ_2. During optimization, each lattice section remains paraunitary regardless of the values of $\theta_{k,\ell}$ so that the pair $(G_k(z), G_{M+k}(z))$ remains power complementary (i.e., satisfies (8.5.5) with $\alpha = 1$). Thus, at the end of optimization, the matrix $\mathbf{E}(z)$ remains paraunitary, guaranteeing perfect reconstruction.

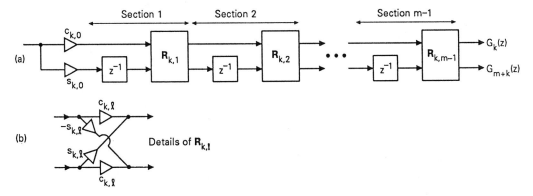

Figure 8.5-1 (a) Representation of the power complementary pair of functions $[G_k(z), G_{M+k}(z)]$ using a lossless lattice. (b) Details of $\mathbf{R}_{k,\ell}$. Here $c_{k,\ell} = \cos\theta_{k,\ell}$ and $s_{k,\ell} = \sin\theta_{k,\ell}$.

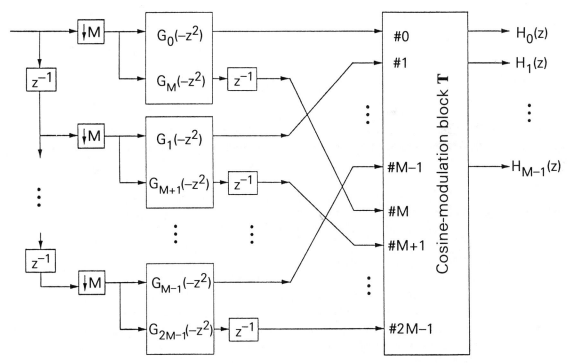

Figure 8.5-2 Implementation of cosine modulated PR analysis filter bank. Each polyphase component pair $[G_k(-z^2), G_{M+k}(-z^2)]$ is implemented by a two-channel lossless lattice.

Number of parameters to be optimized. In view of the linear-phase relation (8.5.8a), only $M/2$ lattice sections [$(M-1)/2$ for odd M; see below] need to be optimized. For example, let $M = 17$. Since $2M = 34$, there are 34 polyphase components $G_k(z)$. The pair $[G_0(z), G_{17}(z)]$ is generated using one lattice structure, the pair $[G_1(z), G_{18}(z)]$ using another lattice structure, and so on. Thus the first eight lattice structures generate the sixteen polyphase components

$$G_0(z), \ldots G_7(z), \quad \text{and} \quad G_{17}(z), \ldots G_{24}(z). \tag{8.5.11}$$

From these we can find the sixteen polyphase components

$$G_{33}(z), \ldots G_{26}(z), \quad \text{and} \quad G_{16}(z), \ldots G_9(z), \tag{8.5.12}$$

by use of the linear phase constraint (8.5.8a). There are two more components $G_8(z)$ and $G_{25}(z)$ to be determined. But the linear phase constraint implies

$$G_8(z) = z^{-(m-1)} \widetilde{G}_{25}(z), \tag{8.5.13}$$

so that (8.5.5) reduces to

$$\widetilde{G}_8(z) G_8(z) + \widetilde{G}_8(z) G_8(z) = 1. \tag{8.5.14}$$

Thus we have to choose $G_8(z) = \sqrt{0.5\alpha} z^{-K}$ and $G_{25}(z) = \sqrt{0.5\alpha} z^{-(m-1-K)}$.

For odd M, we can generalize this discussion and show (Problem 8.9) that

$$G_{\frac{M-1}{2}}(z) = \sqrt{0.5\alpha} z^{-K}, \quad \text{and} \quad G_{\frac{3M-1}{2}}(z) = \sqrt{0.5\alpha} z^{-(m-1-K)}.$$

Using the fact that $P_0(z)$ is a lowpass filter with cutoff $\pi/2M$ [Fig. 8.1-2(a)], it can be shown (Problem 8.9) that the only acceptable choice of K is given by

$$K = \begin{cases} \dfrac{m-1}{2} & \text{for odd } m, \\ \dfrac{m}{2} & \text{for even } m. \end{cases} \tag{8.5.15}$$

Thus, K is completely determined. Moreover α does not affect frequency responses except for a scale factor. Thus, the only parameters to be optimized are the parameters $\theta_{k,\ell}$ of the eight lattice structures.

More generally, the number of parameters to be optimized is nearly equal to $mM/2 \approx N/4$, which is half the number required for the pseudo QMF approach! This technique for design of perfect reconstruction systems is, therefore, simpler than the pseudo QMF design, and dramatically simpler than the (more general) perfect reconstruction design described in Sec. 6.5.

Hierarchial property. If we wish to increase the prototype length, we have to do it in integer multiples of $2M$ (because of the constraint $N+1=$

$2mM$). This can be done as shown in Fig. 8.5-3, where one new section is added to each lattice structure. This hierarchial approach can be used in the design process, to recursively intialize the parameters to be optimized. Thus we optimize the angles $\theta_{k,\ell}$ for small m, and then use these as initial values with m replaced by $m+1$. (The newly introduced parameters $\theta_{k,m}$ have to be initialized rather arbitrarily.) ‡

Obtaining the analysis and synthesis filters. Once the prototype coefficients $p_0(n)$ are obtained as above, the M analysis filters are found from (8.1.37). The synthesis filters are then obtained as $f_k(n) = h_k(N-n)$. In general these do not have linear phase, even though $P_0(z)$ does.

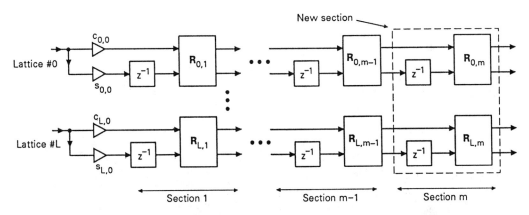

Figure 8.5-3 Explaining the hierarchial property of lattice-based design. $L+1$ is the number of lattice sections to be optimized.

Design Example 8.5.1: Cosine Modulated PR Systems

Let the number of channels be $M = 17$. For this choice of M we showed above that only eight lattice structures have to be optimized. Suppose the prototype filter $P_0(z)$ has order $N = 101$, so that $m = 3$. The kth lattice structure now has three angular parameters

$$\theta_{k,\ell}, \quad 0 \le k \le 7, 0 \le \ell \le 2.$$

These 24 parameters are optimized to minimize the peak stopband error of $P_0(z)$. Figure 8.5-4 shows the magnitude responses of $P_0(z)$ and all the

---

‡ See Koilpillai and Vaidyanathan [1992], for further details about initialization. A computer program, along with documentation, is available upon request.

17 analysis filters. Each analysis filter offers a stopband attenuation of about 40 dB. The impulse response $p_0(n)$ of the optimized prototype $P_0(z)$ is tabulated in Koilpillai and Vaidyanathan [1992].

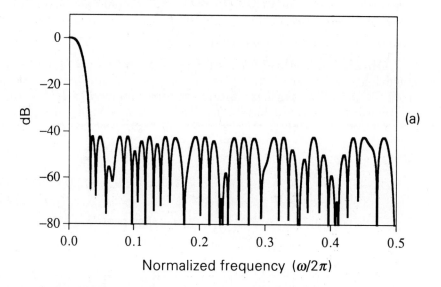

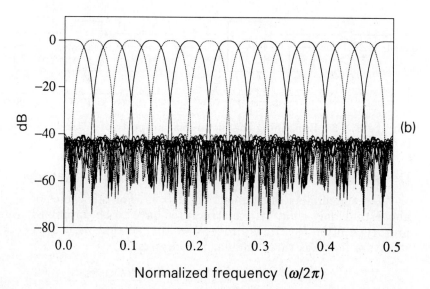

Figure 8.5-4 Design Example 8.5-1. Magnitude responses for the 17-channel cosine modulated perfect reconstruction system. (a) Prototype of order $N = 101$ and (b) the seventeen analysis filters. (© Adopted from 1992 IEEE.)

In this example the number of parameters to be optimized is equal to 24. For the same filter length and number of channels, the pseudo QMF system

(Sec. 8.2) has 51 parameters to be optimized, whereas the more general perfect reconstruction system (Sec. 6.5) has as many as 216 parameters! The method described in this section therefore has the fewest parameters, resulting in much faster design time. A more thorough comparison is given in the next section.

8.5.3 Complexity Comparison

We will now compare three types of M-channel maximally decimated filter banks, in terms of design complexity as well as implementation complexity. Recall that the filter coefficients are real. The quantity N denotes the order of the analysis filters, and M is the number of channels. The three types are:

Type 1. The general perfect reconstruction system with paraunitary $\mathbf{E}(z)$ described in Sec. 6.5, where $\mathbf{E}(z)$ was represented as a cascade of paraunitary building blocks of the form (6.5.1).

Type 2. The cosine modulated pseudo QMF (approximate reconstruction) system of Sec. 8.2.

Type 3. The cosine modulated perfect reconstruction (PR) system derived in this section.

Design Complexity

The number of parameters to be optimized during the design of the analysis filters depends on N, M, and the type of filter bank. In Table 8.5.1 we have listed these, for the three types of filter banks. Table 8.5.2 shows this number for various choices of M and N. We see that for fixed N and M, the cosine modulated PR system has significantly fewer parameters to be optimized than either of the other methods.

Next, in Table 8.5.3 we compare the two cosine modulated systems for the specific case where $N = 101$ and $M = 17$. To describe this table, first recall that the pseudo QMF system suffers from reconstruction errors, that is, residual aliasing and amplitude distortions. In Sec. 8.2.2 we defined quantitative measures for the aliasing error E_a and the peak-to-peak amplitude distortion E_{pp}. By varying the parameter α in the composite objective function (8.2.9), we can obtain a tradeoff between A_s and these reconstruction errors. In Table 8.5.3 we have shown a number of such tradeoffs. (The error $\approx 10^{-15}$ in the PR case is due to machine precision.) For the same N and M, the table also shows the attentuation A_s obtainable for the perfect reconstruction system. It is clear that, when we pass from the cosine modulated pseudo QMF system to the perfect reconstruction system, we pay a price in terms of the stopband attenuation A_s. This price however is not severe; it is less than 5 dB in most practical examples.

TABLE 8.5.1 The number of real valued parameters to be optimized during the design phase, for the three types of FIR M-channel maximally decimated QMF banks.

General FIR paraunitary perfect reconstruction system (section 6.5) (Type 1)	Cosine modulated pseudo QMF (Type 2)	Cosine modulated perfect reconstruction (Type 3)
$(M-1)\left(\frac{N+1-M}{M} + \frac{M}{2}\right)$	$\left(\frac{N+1}{2}\right)$	$\left(\frac{N+1}{2}\right)\left(\frac{M-1}{2M}\right), M$ odd

$N+1$ = filter length, M = number of channels

TABLE 8.5.2 Number of real valued parameters to be optimized for three types of FIR QMF banks.

M	N+1	Number of parameters to be optimized for		
		General FIR paraunitary perfect reconstruction (Type 1)	Cosine modulated pseudo QMF (Type 2)	Cosine modulated perfect reconstruction (Type 3)
3	48	33	24	8
	60	41	30	10
5	40	38	20	8
	60	54	30	12
7	42	51	21	9
	84	87	42	18
16	64	165	32	15
	96	195	48	23
17	68	184	34	16
	102	216	51	24

$N+1$ = Analysis filter length, M = Number of channels

Implementation Complexity

In Sec. 6.7 we summarized the cost of the Type 1 filter bank in terms of the number of multiplications and additions per unit time (MPUs and APUs). Both Type 2 and Type 3 systems are cosine modulated systems with polyphase implementation as in Fig. 8.3-2. If these are implemented like this, the analysis bank requires nearly $(N+1)/M$ MPUs and N/M APUs in both cases, plus the cost of implementing the modulation matrix. This additional cost is independent of the filter order N, and depends only on M. Table 8.5.4 is a summary of the implementation costs. Once again, the cosine modulated pseudo QMF and PR systems have significantly lower complexity than the Type 1 perfect reconstruction system.

TABLE 8.5.3 Comparison of the two cosine modulated systems (pseudo QMF versus perfect reconstruction). $N = 101$ and $M = 17$.

	Prototype		Reconstruction Error (E_{pp})	Aliasing Error (E_a)
	A_S (dB)	ω_S		
Pseudo-QMF bank	40.65 38.68 38.42	0.0590π 0.0585π 0.0581π	6.790 e−03 2.139 e−04 8.749 e−05	3.794 e−04 3.193 e−04 8.113 e−04
Cosine-modulated PR bank	35.72	0.0586π	8.216 e−15	1.041 e−15

TABLE 8.5.4 Computational complexity of the analysis bank for three types of FIR M-channel maximally decimated QMF banks. For cosine modulated system, cost of modulation must be added to the above numbers.

Complexity	General FIR paraunitary perfect reconstruction system (section 6.5) (Paraunitary cascade implementation) (Type 1)	Cosine modulated pseudo QMF (Type 2)	Cosine modulated perfect reconstruction system (Type 3)
MPU	$\frac{2N}{M} + M$ $\cong \frac{2N}{M}(N \gg M)$	$\frac{N+1}{M}$	$\frac{N+1}{M}$
APU	$\frac{2N}{M} + \frac{(M-1)^2}{M}$ $\cong \frac{2N}{M}(N \gg M)$	$\frac{N}{M}$	$\frac{N}{M}$

$N + 1 =$ filter length, $M =$ number of channels

Implementation using the lattice. The cosine modulated PR system can be implemented directly using the lattice structures which generate the pairs of polyphase components (Fig. 8.5-1). The schematic for this is shown in Fig. 8.5-2. From Chapter 6 we know that the two-channel lattice structure can be redrawn with two-multiplier sections (Fig. 6.4-2), by extracting some scale factors. If this is done, the complexity of the analysis bank (i.e., the number of MPUs and APUs) remains nearly the same as for the direct polyphase implementation of Fig. 8.3-2.

8.5.4 Implementing Cosine Modulation with DCT and DST

In Sec. 8.4 we established a relation between the cosine modulation matrix $\mathbf{T} = [\mathbf{A}_0 \quad \mathbf{A}_1]$ and the DCT and DST matrices. These relations are given in (8.4.10), (8.4.11), and hold when $N + 1 = 2mM$. Based on this we can redraw the analysis bank entirely in terms of the DCT matrix $\mathbf{C}$. This holds for both types of cosine modulated systems, that is, Type 2 (pseudo QMF) and Type 3 (perfect reconstruction).

Case When m is Even

The details depend on whether m is even or odd. We assume that m is even. (We leave it to the reader to work out the odd m case.) Since $\mathbf{S} = \mathbf{\Gamma} \mathbf{C} \mathbf{J}$ we can rewrite these entirely in terms of $\mathbf{C}$ to obtain

$$\mathbf{A}_0 = \sqrt{M}\mathbf{\Lambda}_c \mathbf{C}(\mathbf{I} - \mathbf{J}), \quad \mathbf{A}_1 = -\sqrt{M}\mathbf{\Lambda}_c \mathbf{C}(\mathbf{I} + \mathbf{J}). \qquad (8.5.16)$$

The set of M analysis filters can be expressed as in (8.5.1), where $\mathbf{e}(z)$ is the delay chain vector defined in (5.4.1). By using the above $\mathbf{A}_0$ and $\mathbf{A}_1$ we obtain

$$\mathbf{h}(z) = \sqrt{M}\mathbf{\Lambda}_c \mathbf{C}\left[\mathbf{I} - \mathbf{J} \quad -(\mathbf{I}+\mathbf{J})\right] \begin{bmatrix} \mathbf{g}_0(z^{2M}) & \mathbf{0} \\ \mathbf{0} & \mathbf{g}_1(z^{2M}) \end{bmatrix} \begin{bmatrix} \mathbf{e}(z) \\ z^{-M}\mathbf{e}(z) \end{bmatrix}. \qquad (8.5.17)$$

Here $\mathbf{g}_i(z)$ are the diagonal matrices of polyphase components, defined as in (8.5.2). Using (8.5.17) we can draw the analysis bank as in Fig. 8.5-5(a). Fig. 8.5-5(b) shows the more explicit structure in terms of $G_k(z)$. (The decimators can be moved to the left as we did earlier in Fig. 8.3-2.)

Recall that the synthesis filters are given by $f_k(n) = h_k(N - n)$. From this we obtain the synthesis bank structure of Fig. 8.5-6, which the reader is requested to justify in Problem 8.10.

Fast implementation of the DCT. The DCT matrix $\mathbf{C}$ in the above figures can itself be implemented using fast techniques. A quick way to see this is to note that $\mathbf{C}$ can be embedded into the matrix $\mathbf{V}$ as shown in (8.4.4). This matrix can, in turn, be implemented as in Fig. 8.4-1. The main cost here is the implementaion of $\mathbf{W}^*$, where $\mathbf{W}$ is the DFT matrix. $\mathbf{W}^*$ can be implemented efficiently by use of the Fast Fourier Transform (FFT) [Oppenheim and Schafer, 1989]. For more efficient and direct 'fast DCT algorithms', see Yip and Rao [1987] and references therein.

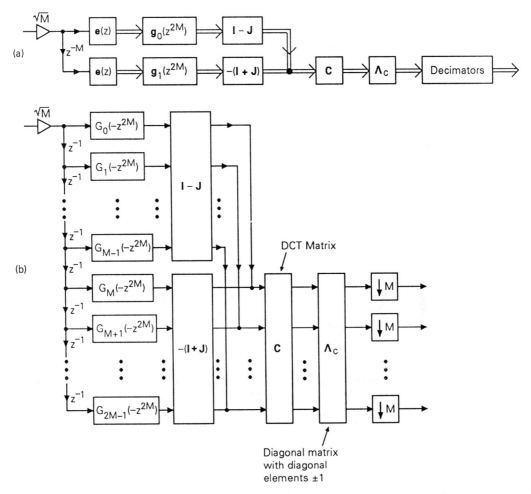

Figure 8.5-5 The cosine modulated analysis bank in terms of DCT. (a) Using matrix notations and (b) using more explicit notations. Here $N + 1 = 2mM$, and $m = $ even.

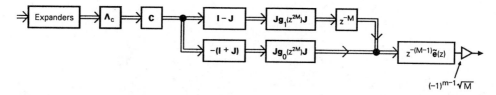

Figure 8.5-6 The cosine modulated synthesis bank, when $N+1 = 2mM$, with $m =$ even.

8.5.5 Advantages of the Cosine Modulated PR System

We now summarize the advantages of the FIR cosine modulated perfect reconstruction system.

1. All analysis filters $H_k(z)$ are obtained from a real coefficient prototype $P_0(z)$, by cosine modulation as in (8.1.37). Only this prototype has to be optimized during the design, so that the design complexity is low. Due to the paraunitary constraint on the polyphase matrix, the number of parameters to be optimized is in fact only about half the number used in pseudo QMF design. Tables 8.5.1 and 8.5.2 give quantitative details.

2. With the synthesis filters chosen as $f_k(n) = h_k(N-n)$, we have perfect reconstruction. (In particular the analysis and synthesis filters have same order N). The objective function to be minimized during optimization of the coefficients of $P_0(z)$ is therefore very simple, and has to reflect only the stopband attenuation of $P_0(z)$.

3. If we optimize the lattice coefficients $\theta_{k,\ell}$ as in Sec. 8.5.2, then the paraunitary constraint is automatically imposed during the design of the prototype $P_0(z)$. So we can use an unconstrained optimization routine to compute $\theta_{k,\ell}$.

4. The *implementation* complexity for the entire analysis bank is equal to the cost of the prototype $P_0(z)$ plus the modulation cost (which depends on the number of channels M but not on the filter order N). This is same as that of the pseudo QMF system.

5. The modulation cost can be reduced by expressing the analysis and synthesis banks in terms of the DCT matrix (Figs. 8.5-5 and 8.5-6), for which there exist fast implementations.

Summarizing, the system has all the advantages and simplicity of the cosine modulated pseudo QMF system of Sec. 8.1, and in addition offers perfect reconstruction. The price paid for this is in terms of reduced stopband attenuation of the prototype $P_0(z)$, but this is a minor loss in practice (Table 8.5.3).

It should be emphasized that, even though the prototype filter has linear phase, the cosine modulated analysis filters do not, in general, have linear phase. In fact, if we give up the linear phase property of the prototype, there are some advantages [Nguyen, 1992b]. Also see [Mau, 1992] for further generalizations.

PROBLEMS

8.1. In the pseudo QMF system discussed in Sec. 8.1 and 8.2, there is some residual aliasing distortion, which is measured by the quantity (8.2.10). Suppose we construct a new filter bank in which each $F_k(z)$ is interchanged with the corresponding $H_k(z)$. How is the measure (8.2.10) affected? How is the distortion function $T(z)$ affected?

8.2. For the pseudo QMF system we can find the synthesis filters either from (8.1.38) or from the relation $f_k(n) = h_k(N - n)$ where $h_k(n)$ is as in (8.1.37). Verify that these two yield the same synthesis filter coefficients.

8.3. Let $H_k(z)$ be a transfer function of the form

$$H_k(z) = \sum_{\ell=0}^{r-1} z^{-\ell} G_\ell(z^r) \cos(2\pi/r)(\ell + \ell_1)(k + k_0) \qquad (P8.3a)$$

where k_0 is a half-integer (i.e., $k_0 - 0.5$ is an integer) and ℓ_1 is arbitrary. Show that the impulse response of $H_k(z)$ has the form

$$h_k(n) = h(n) \cos(2\pi/r)(\ell_1 + n)(k + k_0), \qquad (P8.3b)$$

where $h(n)$ is the impulse response of $H(z)$, which is defined by

$$H(z) = \sum_{\ell=0}^{r-1} z^{-\ell} G_\ell(-z^r). \qquad (P8.3c)$$

8.4. Suppose we wish to design two-channel real coefficient FIR perfect reconstruction QMF banks using the method described in Sec. 8.5. Does this cover every design that can be generated using the two channel lattice structure of Sec. 6.4.3?

8.5. Consider the three types of FIR filter banks summarized in Sec. 8.5.3. Suppose $M = 15$ and filter lengths $N + 1 = 60$.
  a) For each type, what is the number of parameters to be optimized during the design phase?
  b) For each type, what are the number of MPUs and APUs required to implement the analysis bank?

8.6. Show that the DCT matrix **C** and DST matrix **S** are related as in (8.4.2).

8.7. Consider the expression inside the brackets in (8.4.7). Show that

$$\mathbf{\Lambda}_w^\dagger \mathbf{U}^\dagger \mathbf{\Lambda}_\beta^\dagger \mathbf{\Lambda}_\beta \mathbf{U} \mathbf{\Lambda}_w = 2M\mathbf{I}, \quad \text{and} \quad \mathbf{\Lambda}_w^\dagger \mathbf{U}^\dagger \mathbf{\Lambda}_\beta^\dagger \mathbf{J}_{2M} \mathbf{\Lambda}_\beta \mathbf{U} \mathbf{\Lambda}_w = 0. \qquad (P8.7)$$

Note. the second equality requires more work.

8.8. Assuming that the prototype satisfies the linear phase condition, $p_0(n) = p_0(N - n)$, establish the relation (8.5.8a) among the $2M$ polyphase components. (*Note.* $N + 1 = 2mM$.)

8.9. Let the $2M$ polyphase components of $P_0(z)$ satisfy (8.5.5) as well as (8.5.8a). Assuming $N+1 = 2mM$ and that M is odd, verify that

$$G_{\frac{M-1}{2}}(z) = \sqrt{0.5\alpha}z^{-K}, \quad \text{and} \quad G_{\frac{3M-1}{2}}(z) = \sqrt{0.5\alpha}z^{-(m-1-K)}. \quad (P8.9)$$

Show further that K satisfies (8.5.15). You can use the fact that $P_0(z)$ is a linear phase lowpass filter with cutoff $\pi/2M$.

8.10. In the text we saw that the analysis bank represented by (8.5.17) can be implemented as in Fig. 8.5.5. Let the M synthesis filters be chosen as $f_k(n) = h_k(N-n)$. Then show that the synthesis filter bank can be realized as in Fig. 8.5-6. Also draw the structure more explicitly in terms of polyphase components (i.e., as we did in Fig. 8.5-5(b) for the analysis bank).

8.11. Let $P_0(z)$ be the FIR prototype described in Theorem 8.5.1. Let the polyphase components of this prototype satisfy (8.5.5). Show then that $\widetilde{P}_0(z)P_0(z)$ is a Nyquist($2M$) filter.

PART 3 Special Topics

9

Quantization effects

9.0 INTRODUCTION

In any digital filter bank implementation, the multipliers as well as internal signals have to be represented in quantized form. The effect of this quantization is that the filter output is different from the ideal one. Broadly speaking, we can classify the quantization effects into three categories, viz., coefficient sensitivity effects, roundoff noise and limit cycles. In this chapter we will analyze these effects quantitatively. In Appendix C we will deal with another important effect in filter bank systems, arising due to quantization of subband signals.

9.1 TYPES OF QUANTIZATION EFFECTS

Consider Fig. 9.1-1(a) which shows the implementation of a first order digital filter with transfer function $H(z) = 1/(1 - az^{-1})$. If the signal $y_i(n-1)$ and multiplier a are represented with a certain precision, then the product $ay_i(n-1)$ in general requires a higher precision. So the signal $y_i(n)$ requires higher precision than $y_i(n-1)$. Since $y_i(n)$ is circulated back during the next cycle, this process continues indefinitely, implying infinite bit accumulation.

In a practical system this is not feasible, and the result of a computation has to be quantized before recirculation. This is indicated schematically as shown in Fig. 9.1-1(b), where the box labeled Q represents a quantizer. The signal $y(n)$ which is recirculated is the quantized version of an intermediate signal $w(n)$, and we write $y(n) = Q[w(n)]$. In general there could be more than one quantizer in a system, but in order to avoid infinite bit accumulation it is *sufficient* to make sure that there are no *loops* without quantizers.

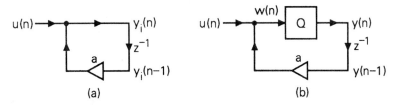

Figure 9.1-1 (a) A first order filter, and (b) implementation with a quantizer in the loop.

Effects of Multiplier (or Coefficient) Quantization

Quantization of the multiplier coefficients, for example, a in the above figure, results in a change of the transfer function from $H(z)$ to $H_q(z)$. Thus, the passband and stopband ripples after quantization can be significantly larger than the specified values. As an extreme case a stable filter may become unstable after coefficient quantization.

In a filter bank system, coefficient quantization can result in deeper consequences. For example, a QMF bank may lose the alias-free property, or perfect reconstruction (PR) property, because of multiplier quantization. It turns out, however, that in any perfect reconstruction system with paraunitary polyphase matrix, the paraunitary property (and hence the PR property) can be retained in spite of multiplier quantization. (In this sense the structure is 'robust' to quantization.) This will be justified only in Sec. 14.11 where we show how the paraunitary property of an $M \times M$ matrix can be retained in spite of coefficient quantization. A special case of this has already been noticed in Sec. 6.4.1 (two channel QMF lattice). In Sec. 5.3.5 we also studied a two channel IIR QMF bank which is free from aliasing as well as amplitude distortion in spite of coefficient quantization.

Effects of Signal Quantization

The effect of quantizing internal signals is more involved. Consider again Fig. 9.1-1(b). The quantity

$$q(n) = Q[w(n)] - w(n) \qquad (9.1.1)$$

is called the *quantizer error,* and is a function of time n. We can model the structure as shown in Fig. 9.1-2. We say that $q(n)$ is the *noise source* associated with the quantizer. Notice that the output $y(n)$ in Fig. 9.1-2 is different from the ideal output $y_i(n)$ in Fig. 9.1-1(a). The difference $y(n) - y_i(n)$ is *not* equal to the quantizer error $q(n)$. This is because the effect of quantizer accumulates with time as explained below.

We can think of $q(n)$ as an input to the filter (just like $u(n)$ is). Its effect on the output is governed by the transfer function between $q(n)$ and

$y(n)$, called the *noise transfer function*. This is given by

$$G(z) = 1/(1 - az^{-1}), \qquad (9.1.2)$$

which, in this example, happens to be the same as the filter transfer function $H(z)$. Let $e(n)$ denote the output of the system $G(z)$, in response to the input $q(n)$. Then the signal $y(n)$ can be written as $y(n) = y_i(n) + e(n)$, where $y_i(n)$ is the output of the ideal system of Fig. 9.1-1(a). So the noise source affects the filter output in a manner which depends on the noise transfer function.

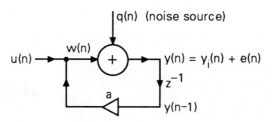

Figure 9.1-2 The roundoff noise model for the structure of Fig. 9.1-1(b).

In order to understand the effect of $q(n)$ on the filter output more quantitatively, it is common practice to model $q(n)$ as a random process (Appendix B), satisfying a set of simplifying assumptions. This allows us to estimate the variance of the noise which actually reaches the filter output, using simple and elegant techniques. The main point to note here is that the effect of signal quantization is to contribute a random component $e(n)$ to the filter output. Under some conditions, quantization also results in nonrandom components, called limit cycles. We will return to this in Sec. 9.6.

Subband Quantization

In subband coding applications, a third source of noise exists, due to quantization of the subband signals. This tends to dominate the total noise whenever it is present, but its effect is difficult to analyze. We will study this in Appendix C. In this chapter we will concentrate only on quantization noise due to filter implementation. Such a study is useful in many applications, for example, in voice privacy systems (Sec. 4.5.3) and transmultiplexers, where subband quantization effects do not dominate.

Chapter Outline

In Sec. 9.2 we present a brief summary of well known techniques for roundoff noise analysis. In Sec. 9.3–9.5 we use this to present a roundoff noise analysis for multirate filter banks. In Sec. 9.6 we consider limit cycles. We return to coefficient quantization effects in Sec. 9.7. It will be seen that many filter bank structures exhibit low passband sensitivity to coefficient quantization, particularly if the polyphase matrix $\mathbf{E}(z)$ is paraunitary.

Special prerequisites. We review the standard noise analysis techniques in Sec. 9.2. It is, however, assumed that the reader has some familiarity with fixed-point binary number representation, and random process representation of noise waveforms. Thus, we make free use of such terms as (a) b-bit fixed point arithmetic, (b) uncorrelated white noise source, (c) noise source with variance σ_q^2, and so on. There exist excellent treatements of this material [Oppenheim and Schafer, 1989], [Jackson, 1989], and [Rabiner and Gold, 1975]. Appendix B includes a brief review of random process, and we will freely use the definitions and properties in that appendix (e.g., uniform random variables, wide sense stationary random process, autocorrelation, power spectrum, white noise, and so on).

9.2 REVIEW OF STANDARD TECHNIQUES

9.2.1 Quantizers and Noise Models

All signals are represented as fixed point binary fractions, as shown in Fig. 9.2-1. This is a b-bit binary representation, with s representing the sign bit. We say that b is the *wordlength*. If $s = 0$ the number is nonnegative, whereas with $s = 1$ the number is nonpositive, and its decimal value depends on the convention chosen (e.g., two's complement convention, sign magnitude convention, etc.). All the numbers representable in this form are in the range $-1 \le x < 1$ (with $x = -1$ permitted only in some conventions, e.g., two's complement). This is said to be the permissible *dynamic range*. The quantity

$$\Delta \triangleq 2^{-b}, \qquad (9.2.1)$$

is the smallest positive number permitted, and is also the smallest possible increment. It is said to be the *quantization step* or stepsize.

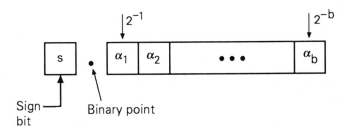

Figure 9.2-1 Format for the b-bit fixed-point binary fraction.

Quantizers

A quantizer is a device which takes an arbitrary real number and converts it into a b-bit fraction using some arithmetic rules. Thus, the quantizer input [e.g., $w(n)$ in Fig. 9.1-1(b)] need not be a b-bit fraction, but its output is. In this process, some error is introduced, and is denoted as $q(n)$

[see (9.1.1) and Fig. 9.1-2]. We say that $q(n)$ is the noise source due to the quantizer. If $w(n)$ does not belong to the permitted dynamic range, we say that a computational overflow has occured. The quantizer brings the number back to the permitted dynamic range by using certain rules (called *overflow handling rules*). So $y(n)$ still belongs to the dynamic range, but the error $q(n)$ is large.

Assume that there is no overflow, that is, $w(n)$ belongs to the dynamic range. In general $w(n)$ may still have more than b bits in its representation, that is, there could be some extra bits to the right of the bth bit in Fig. 9.2-1. When this is converted to a b bit number, the error $q(n)$ is 'small' and is of the order of the quantization step Δ. The exact details depend on the type of quantizer, that is, the rule used for quantization. Some rules are: (a) roundoff arithmetic where $y(n)$ is the quantized number closest to $w(n)$, (b) magnitude truncation, where the magnitude of quantized number $y(n)$ is no larger than that of the unquantized number $w(n)$, and (c) truncation arithmetic, where the extra bits to the right of the b bits are merely discarded.

The Noise Model Assumptions

Unless mentioned otherwise, we will assume roundoff arithmetic. In this case, we have

$$\frac{-\Delta}{2} \le q(n) \le \frac{\Delta}{2}. \tag{9.2.2}$$

We will assume that $q(n)$ is a random variable, uniformly distributed in the above range. Under this condition, it has zero mean and variance

$$\sigma_q^2 = \frac{2^{-2b}}{12}. \tag{9.2.3}$$

We make the further assumption that the sequence $q(n)$ is a white, wide sense stationary (WSS) random process. Summarizing, the quantizer noise source $q(n)$ is zero-mean white, with variance σ_q^2.

Multiple noise sources. In most practical structures, there are many quantizers. Fig. 9.2-2 shows the example of a cascade form structure (Sec. 2.1.3) with two quantizers. In such situations, each quantizer is replaced with a noise source, as shown by broken lines. We assume that each noise source satisfies the above model (i.e., white, etc.). To study the total effect of these at the filter output, we assume that any two noise sources are uncorrelated, and that each of them in turn is uncorrelated to the input $u(n)$. These assumptions will enable us to add the noise variances due to various sources, in order to obtain the total output noise variance.

Examples which violate these assumptions are not hard to generate (e.g., when the filter input is a sinusoid). However, in a large number of situations, the above assumptions have been verified to be reasonable [Barnes, et al., 1985]. In any case, noise analysis under these assumptions gives a very good qualitative idea of the nature of noise propagation. For example, one of the useful conclusions obtainable is that, in a direct form structure, the output noise variace increases as the poles get closer to the unit circle [see discussions following (9.2.6) later].

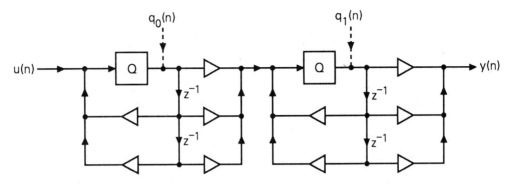

Figure 9.2-2 A cascade form structure, with two quantizers.

9.2.2 Noise Gain of a Filter

In Fig. 9.1-2 we associated a noise transfer function $G(z)$ with the noise source $q(n)$. This transfer function governs the extent to which $q(n)$ affects the output. More generally let there be many quantizers in the structure, each modeled by a noise source $q_k(n)$. Let $G_k(z)$ denote the transfer function from the noise source $q_k(n)$ to the filter output. We say that $G_k(z)$ is the noise transfer function for $q_k(n)$.

Let $e_k(n)$ denote the output of $G_k(z)$ in response to the input $q_k(n)$. Under the above noise model assumptions, $e_k(n)$ is a zero mean WSS random process with variance

$$\sigma_{e_k}^2 = \sigma_q^2 \underbrace{\sum_n |g_k(n)|^2}_{noise\ gain}, \qquad (9.2.4)$$

where $g_k(n)$ is the impulse response of $G_k(z)$. The summation in (9.2.4) is the energy of $G_k(z)$. So the output noise variance is the quantizer noise variance amplified by the *energy* of the noise transfer function (which is therefore called the *noise gain*).

Each of the quantizer noise sources $q_k(n)$ contributes a noise component at the filter output. In view of the uncorrelated assumption the output noise $e(n)$ has total variance

$$\begin{aligned}\sigma_e^2 &= \sigma_q^2 \sum_{k=0}^{N-1} \sum_n |g_k(n)|^2 \\ &= \frac{2^{-2b}}{12} \sum_{k=0}^{N-1} \sum_n |g_k(n)|^2 \quad \text{(total output noise variance)}.\end{aligned} \qquad (9.2.5)$$

Returning to the example of Fig. 9.1-2, the impulse response $g(n)$ of the noise transfer function is $g(n) = a^n \mathcal{U}(n)$, so that the noise gain is

$$\sum_n |g(n)|^2 = \frac{1}{1 - |a|^2}. \qquad (9.2.6)$$

This gain increases as the pole 'a' (which is inside the unit circle) gets closer and closer to the unit circle. As an example, if $a = 0.99$ then the noise gain ≈ 50. This demonstrates that the noise gain can be quite large indeed.

Effect of increasing the wordlength. If we increase the number of bits from b to $b+1$, this results in a four fold reduction in the noise variance (using (9.2.5)). On a dB scale, this is equivalent to $10 \log_{10} 4.0 = 6.02$ dB. So the output noise variance decreases by about 6 dB per every additional bit of internal precision.

9.2.3 Dynamic Range and Scaling

In a practical implementation we have to ensure that the signals do not overflow the dynamic range permitted by the number system. In order to study this issue, it is useful to find upper bounds on the magnitudes of various signals. In such an analysis, the presence of quantizers can be ignored, as they do not affect these bounds significantly.

Thus consider Fig. 9.1-1(b), and ignore the quantizer for this discussion. If the input is in the range $-1 \leq u(n) < 1$ (consistent with fixed point fractional representation), this does not imply that the signal $w(n)$ is in this range for all n. It can, however, be shown that $w(n)$ is bounded as

$$|w(n)| \leq \sum_n |f(n)| \qquad (9.2.7)$$

where $f(n)$ is the impulse response from $u(n)$ to the node $w(n)$. In our example, this impulse response happens to be the same as $h(n)$, that is, $f(n) = a^n \mathcal{U}(n)$. So the right hand side of (9.2.7) reduces to $1/(1-|a|)$. For example, if $a = 0.99$, this quantity equals 100. In other words, $|w(n)|$ can get as large as a hundred!

A simple way to ensure that $w(n)$ does not overflow (i.e., does not exceed the range $-1 \leq w(n) < 1$), is to insert a multiplier $1/L$ as shown in Fig. 9.2-3, with

$$L = \sum_n |f(n)|. \qquad (9.2.8)$$

We then say that the structure has been scaled. The price we pay for this freedom from overflow is that the output *signal level* goes down. Since the roundoff noise level is unaffected by scaling, the *signal to noise ratio* is reduced. This is an example of interaction between roundoff noise and dynamic range in digital filter implementations.

It is in principle possible to reduce the roundoff noise level simply by inserting a scale factor at the output node, but this results in reduced signal level as well. If we try to restore the signal level by insertion of another multiplier at the filter input, this will affect the probability of internal overflow. So, 'finite word length' will have its effect one way or the other.

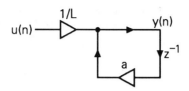

Figure 9.2-3 Scaling a first order digital filter.

Scaling a Structure

In practice, digital filter structures are more complicated than Fig. 9.1-1. There are several internal nodes, and one has to ensure that none of these suffers from computational overflow. Let $F_k(z)$ be the transfer function from the filter input to the kth internal node, and let $f_k(n)$ be its impulse response. We say that $F_k(z)$ is the scaling transfer function for the kth node. If

$$\sum_n |f_k(n)| < 1, \qquad (9.2.9)$$

then the kth node [or the transfer function $F_k(z)$] is scaled to be free from overflow. If all nodes satisfy this property, then the entire structure is said to be scaled. Scaling can be accomplished by rearrangement of the internal structure, which may or may not involve explicit insertion of multipliers (as in Fig. 9.2-3).

Which nodes to scale? With certain types of arithmetic conventions (e.g., two's complement), it can be shown [Jackson, 1970] that only those nodes which are *inputs to multipliers* have to be scaled. For example, consider Fig. 9.2-4. Here every multiplier input is a delayed version of the signal $s(n)$. So it is sufficient to scale this node and, of course, the output node $y(n)$. Even if there is an overflow at any of the other nodes, it will not affect the final output $y(n)$. (This has to do with the fact that two's complement arithmetic has similarities to modulo arithmetic).

Types of Scaling

If each of the scaling transfer functions $F_k(z)$ satisfies (9.2.9), we say that the strcuture is *sum-scaled*. The structure is completely free from overflow, but the price paid is in terms of the signal to roundoff noise ratio at the filter output. There exist less stringent scaling rules (called $\mathcal{L}_p$ *scaling rules*) which are sufficient under some conditions. We will not go into these details (which can be found in [Jackson, 1970] and [Rabiner and Gold, 1975]) but merely mention a scheme called the $\mathcal{L}_2$ *scaling policy.* In Problem 9.4 we cover some of the other scaling rules.

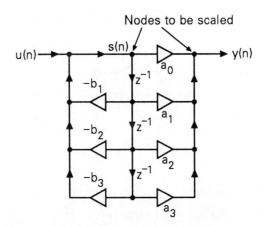

Figure 9.2-4 Pertaining to the choice of nodes to be scaled.

$\mathcal{L}_2$ **scaling.** In this scheme, instead of ensuring the condition (9.2.9), we ensure that

$$\sum_n |f_k(n)|^2 \le 1, \qquad (9.2.10)$$

(usually with equality), where $f_k(n)$ is the impulse response from the input to the kth node to be scaled. This is called $\mathcal{L}_2$ scaling because the above summation is the (square of) the $\mathcal{L}_2$ norm of $F_k(z)$.† If a node $x_k(n)$ is scaled (i.e., $F_k(z)$ is scaled) in the $\mathcal{L}_2$ sense, then it can be shown that $|x_k(n)| < 1$ as long as the filter input $u(n)$ has energy bounded by unity, that is,

$$\sum_n |u(n)|^2 < 1. \qquad (9.2.11)$$

Use of $\mathcal{L}_2$ scaling. $\mathcal{L}_2$ scaling is less stringent than sum-scaling (which guarantees complete freedom from overflow), and therefore results in increased signal to roundoff noise ratio in absence of overflow. However, the chances of overflow are higher; note that the condition (9.2.11) is rather unrealistic. (For example if $u(n)$ is a sinusoid, its energy is infinite.) However, $\mathcal{L}_2$ scaling is still useful for several reasons.

First, in most practical cases, the sequence $f_k(n)$ is significant only over a finite duration. If the energy of $u(n)$ over such a duration is properly bounded, we can still control the possibility of overflow of $x_k(n)$.

Second, if we view the input as a wide sense stationary random process (which is sometimes a realistic assumption, at least over short segments of

---

† For integer p the $\mathcal{L}_p$ norm of $F(z)$ is defined as $[\int_0^{2\pi} |F(e^{j\omega})|^p \frac{d\omega}{2\pi}]^{1/p}$.

time), we can obtain some useful conclusions. In this case, the energy of $u(n)$ is not finite, but only the power spectrum of $u(n)$ is of relevance. It is possible to obtain a bound on the variance of $x_k(n)$ as follows:

$$\sigma_{x_k}^2 \leq \sum_n |f_k(n)|^2 \times \max_\omega \left(S_{uu}(e^{j\omega}) \right).$$

This variance can in turn be used to bound the *probability* of overflow at the node $x_k(n)$ [Jackson, 1970]. If all the scaling transfer functions satisfy (9.2.10) with equality, then we can reduce the probability of overflow at all the internal nodes to the *same value*, simply by inserting a common scale factor (as we did in Fig. 9.2-3), in front of the input. For the rest of the chapter, we consider only $\mathcal{L}_2$ scaling.

9.2.4 Some Useful Special Cases

FIR Direct Form Structures

Many of the filter banks we studied are FIR systems, for which noise analysis is fairly simple. Consider the FIR direct form structure shown in Fig. 9.2-5(a). The transfer function is $H(z) = \sum_{n=0}^{N} h(n) z^{-n}$.

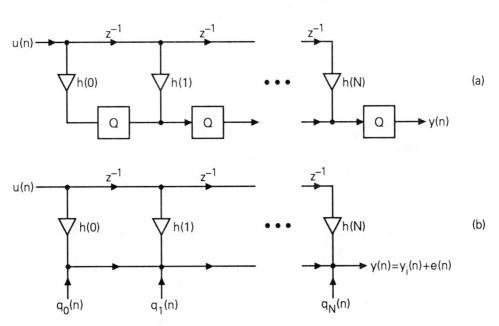

Figure 9.2-5 (a) The direct form FIR structure with quantizers, and (b) the noise model.

Here the output of every multiply/add operation is quantized, and the noise model is shown in Fig. 9.2-5(b). All noise sources $q_k(n)$ have the same noise transfer function, that is, $G_k(z) = 1$ for all k. Under the usual (white, uncorrelated) assumptions, the output noise variance is thus $(N+1)\sigma_q^2$ where σ_q^2 is the quantizer noise variance (9.2.3). For the case of linear phase FIR filters where we require only half as many multiplications (e.g., see Fig. 2.4-3), the output noise variance is approximately half the above value.

Since $G_k(z) = 1$ and since $q_k(n)$ are white as well as uncorrelated, the output noise $e(n)$ is also white! This is true regardless of the transfer function $H(z)$ (which does not affect the noise transfer function). This is a somewhat unusual situation, which is not common with IIR filters. For example, in Fig. 9.1-2 the noise transfer function $G(z)$ is not constant, and the output noise is not white.

A second quantization scheme for the FIR case would be to carry higher internal precision and quantize only the output $y(n)$. This scheme has less output noise variance ($= \sigma_q^2$ only), at the expense of higher internal wordlength. The extra internal wordlength depends on the number of multipliers as well as the multiplier precisions, and complicates things in general. We do not consider it here.

Scaling. Since the inputs to multipliers are derived by delaying $u(n)$, these are already scaled. The only extra scaling necessary is to ensure that the output $y(n)$ does not overflow. This can be done by insertion of a scale factor $1/L$ as we did in Fig. 9.2-3. The value of L depends on the scaling policy chosen.

Allpass Cascade form

Consider Fig. 9.2-6(a) which represents a cascade of L first order filters $H_k(z)$, each implemented in direct form. Assuming that α_k are real, $H_k(z)$ are allpass so that the overall filter $H(z)$ is allpass (with real poles only). Such real-pole allpass functions find application in power symmetric IIR QMF banks, as seen in Sec. 5.3. In that section, a two channel IIR QMF bank was introduced with analysis filters

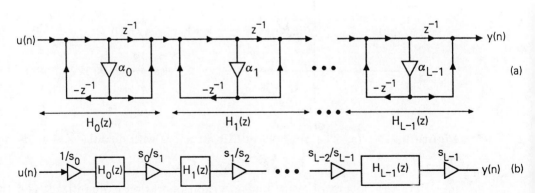

Figure 9.2-6 (a) A cascade of L first order allpass filters, and (b) insertion of scaling multipliers.

$$H_0(z) = \frac{a_0(z^2) + z^{-1}a_1(z^2)}{2}, \quad H_1(z) = H_0(-z).$$

The allpass functions $a_i(z)$ have only real poles, and can be implemented using the above cascade form.

Scaling. In this structure, the only nodes to be scaled are the inputs to the multipliers α_k. If we wish to scale these nodes in the $\mathcal{L}_2$ sense, then we define $s_k = 1/\sqrt{(1 - \alpha_k^2)}$ and insert $1/s_k$ at the input of the kth section. We also insert s_k at the output of the section to ensure that $H_k(z)$ is unaffected. Simplifying, we obtain the scaled structure of Fig. 9.2-6(b).

Noise variance. The noise model of the scaled structure is shown in Fig. 9.2-7, assuming that a quantizer is inserted in each section (exactly as we did in Fig. 9.2-2). The noise transfer function for the noise source $q_k(n)$ is

$$G_k(z) = s_k \prod_{\ell=k}^{L-1} H_\ell(z) \qquad (9.2.12)$$

which is allpass. Under the usual noise model assumptions, the output noise component $e(n)$ is therefore zero-mean and white, with total variance

$$\sigma_e^2 = \sigma_q^2 \sum_{k=0}^{L-1} s_k^2. \qquad (9.2.13)$$

Complex case. These discussions can be generalized to the case where filter coeffcients and inputs are complex. In this case we have to define a complex quantizer (with b-bit real part and b-bit imaginary part). Under proper assumptions, many of the above results can be extended.

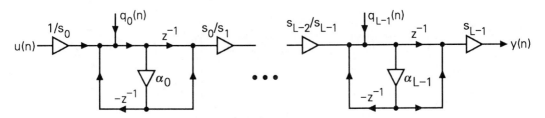

Figure 9.2-7 Noise model for the cascaded allpass structure.

9.3 NOISE TRANSMISSION IN MULTIRATE SYSTEMS

The study of noise generation and propagation in multirate systems is facilitated if we first note a number of useful properties exhibited by random processes in the presence of some familiar building blocks.

Decimators and Expanders

Let $x(n)$ be a wide sense stationary (WSS) random process with mean m and variance $\sigma^2 > 0$. The following properties are easily verified (Problem 9.5).

1. If $x(n)$ is input to an M-fold decimator, then the output $y(n) = x(Mn)$ is also WSS, with mean m and variance σ^2. In fact the autocorrelations of $y(n)$ and $x(n)$ are related as $R_{yy}(k) = R_{xx}(Mk)$ so that the power spectrum $S_{yy}(e^{j\omega})$ of $y(n)$ is related to the power spectrum $S_{xx}(e^{j\omega})$ of the input $x(n)$ in terms of the familiar aliasing relation (4.1.4) (i.e., simply replace $Y_D(e^{j\omega})$ and $X(e^{j\omega})$ in (4.1.4) with $S_{yy}(e^{j\omega})$ and $S_{xx}(e^{j\omega})$ respectively.)

2. If $x(n)$ is input to an M-fold expander ($M > 0$), then the output $y(n)$ is *not* WSS. For example, the random variable $y(0) [= x(0)]$ has variance σ^2, whereas $y(1) = 0$ (which is a 'random variable' with variance $= 0$). Since the variance is not constant with time, this rules out wide sense stationarity.

Expander/Delay-Chain Combination

Consider Fig. 9.3-1 where $x_k(n), 0 \leq k \leq M-1$ are WSS random processes. The signal $y(n)$ is an interlaced version of $x_k(n)$. (This is similar to the time-domain multiplexer in Fig. 4.5-4(a)). In general $y(n)$ is not WSS. For example, if $x_0(n)$ and $x_1(n)$ have unequal variances, then $y(n)$ has variance changing with time, and it cannot therefore be WSS.

A special example of interest arises when $x_k(n)$ are zero-mean, white-noise sources with variance σ^2 for all k. Assume further that $x_k(n)$ and $x_m(n)$ are uncorrelated for $k \neq m$. In this case, the output $y(n)$ is zero-mean and white (since $y(n_0)$ and $y(n_1)$ are uncorrelated for $n_0 \neq n_1$) with variance σ^2.

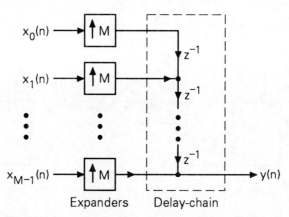

Figure 9.3-1 The time domain multiplexing circuit.

Paraunitary Systems

Some of the filter banks we have studied contain lossless (i.e., stable paraunitary) building blocks. For example, consider Fig. 9.3-2 which is the polyphase implementation of a synthesis bank (Section 5.5). In many examples $\mathbf{R}(z)$ is paraunitary. We will derive a result which applicable in such situations.

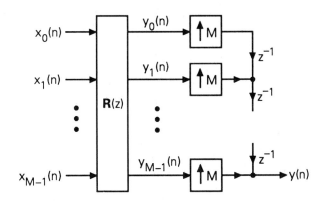

Figure 9.3-2 A synthesis bank in polyphase form.

Let $x_k(n), 0 \le k \le M-1$ be WSS random processes. These may, for example, represent the noise generated in the analysis bank. Suppose the following assumptions are true:

1. Each sequence $x_k(n)$ is white.
2. Any two of these sequences are uncorrelated, that is, $x_k(n_0)$ and $x_m(n_1)$ are uncorrelated unless $k = m$ and $n_0 = n_1$.
3. $x_k(n)$ have zero mean.
4. All the M sequences $x_k(n)$ have equal variance, that is, $\sigma_k^2 = \sigma^2$ for all k.

We then say that the vector

$$\mathbf{x}(n) \triangleq [\, x_0(n) \quad \ldots \quad x_{M-1}(n) \,]^T, \qquad (9.3.1a)$$

which is a vector-random process (Section B.5, Appendix B), is WUZE(σ^2). This is an abbreviation for *white, uncorrelated, zero-mean,* and *equal* variance σ^2. A zero-mean WSS random (vector) process $\mathbf{x}(n)$ is WUZE(σ^2) if, and only if, $E[\mathbf{x}(m)\mathbf{x}^\dagger(k)] = \sigma^2 \delta(m-k)\mathbf{I}$ (Problem 9.6).

Now suppose that $\mathbf{R}(z)$ is stable, and $\widetilde{\mathbf{R}}(z)\mathbf{R}(z) = d\mathbf{I}$ (i.e., lossless). Then the vector

$$\mathbf{y}(n) \triangleq [\, y_0(n) \quad \ldots \quad y_{M-1}(n) \,]^T \qquad (9.3.1b)$$

has all the four properties of $\mathbf{x}(n)$. More precisely, $\mathbf{y}(n)$ is WUZE($d\sigma^2$). (This is a consequence of the theorem to be proved below.) In view of this, the output signal $y(n)$ [which is a time multiplexed version of $y_k(n)$] is a white, zero-mean process, with variance $d\sigma^2$. We now state and prove a more general result, which is useful in the study of filter banks.

♠**Theorem 9.3.1.** Let $\mathbf{T}(z)$ be a $p \times r$ transfer matrix and let $\mathbf{T}^T(z)$ be lossless, that is, stable with $\mathbf{T}(z)\widetilde{\mathbf{T}}(z) = d\mathbf{I}_p$. Let $\mathbf{x}(n)$ and $\mathbf{y}(n)$ denote the input and output (vector-)sequences. If $\mathbf{x}(n)$ is WUZE(σ^2), then the output is WUZE($d\sigma^2$). ◊

Proof. Let $\mathbf{S}_{\mathbf{xx}}(e^{j\omega})$ and $\mathbf{S}_{\mathbf{yy}}(e^{j\omega})$ denote the power spectral density matrices of the vector WSS processes $\mathbf{x}(n)$ and $\mathbf{y}(n)$. We then have

$$\mathbf{S}_{\mathbf{yy}}(e^{j\omega}) = \mathbf{T}(e^{j\omega})\mathbf{S}_{\mathbf{xx}}(e^{j\omega})\mathbf{T}^\dagger(e^{j\omega}) \qquad (9.3.2)$$

In view of the WUZE property of $\mathbf{x}(n)$, its autocorrelation sequence is

$$\mathbf{R}_{\mathbf{xx}}(k) = \sigma^2 \delta(k) \mathbf{I}_r \qquad (9.3.3)$$

so that $\mathbf{S}_{\mathbf{xx}}(e^{j\omega}) = \sigma^2 \mathbf{I}_r$ for all ω. Substituting in (9.3.2), we get $\mathbf{S}_{\mathbf{yy}}(e^{j\omega}) = d\sigma^2 \mathbf{I}_p$. This shows that $\mathbf{y}(n)$ is WUZE($d\sigma^2$) indeed. ▽▽▽

Here are some applications of this result in filter-banks. More can be found in the next two sections.

1. When $p = r$, losslessness of $\mathbf{T}^T(z)$ also implies that of $\mathbf{T}(z)$, and we can apply this to Fig. 9.3-2 with $\mathbf{R}(z) = \mathbf{T}(z)$. The special case where $\mathbf{T}(z)$ is a constant $M \times M$ unitary matrix also arises in filter bank theory (orthogonal transform coding, Appendix C).
2. Another useful example arises when $p = 1$ and $r = M$. In this case $\mathbf{T}(z)$ is an M channel *synthesis bank*, with power complementary property. If its input is WUZE(σ^2), then the output is zero-mean white with variance $d\sigma^2$.

9.4 NOISE IN FILTER BANKS

In a complete analysis/synthesis system (as in Fig. 5.4-1), roundoff noise is generated both by the analysis bank and the synthesis bank. In addition, the noise due to analysis bank propagates through the synthesis bank. We therefore have to consider not only the noise *generated* by individual filters, but also the way in which the synthesis bank transmits the noise entering its inputs. In transmultiplexers, where the analysis bank *follows* the synthesis bank (Fig. 5.9-1), the reverse situation prevails (Problem 9.7).

The effect of quantization of subband signals will be studied in Appendix C. In this section, we will concentrate only on quantization noise due to filter implementation. We will study the noise generated by some popular analysis banks introduced in Chap. 5 and 6. In the next section, the total noise due to analysis and synthesis filters will be considered.

Consider the QMF bank of Fig. 5.4-1, and let the analysis filters $H_k(z)$ be FIR with order N. Then each output has noise component which is white, with variance $(N+1)\sigma_q^2$ (Sec. 9.2.4). Since the decimated version of white noise is white, the noise $\epsilon_k(n)$ contaminating the decimated signal $v_k(n)$ is also white. The noise components $\epsilon_k(n)$ and $\epsilon_m(n)$ ($k \neq m$) are in general not uncorrelated (unless the filters $H_k(e^{j\omega})$ and $H_m(e^{j\omega})$ have completely non overlapping frequency responses, which is not the case in most filter banks; see Problem 9.12). Surprisingly however, in most filter banks, these noise components turn out to be uncorrelated for various other reasons (as we will elaborate).

Summary of Notations and Assumptions

a) σ_q^2 denotes the b-bit quantizer noise variance (9.2.3), and all noise components have zero-mean.

b) $H_k(z)$ denotes the kth analysis filter, and N denotes analysis filter order (which equals the synthesis filter order in all cases considered).

c) As in Fig. 5.4-1, $v_k(n)$ denotes the M-fold decimated version of the output of $H_k(z)$. Also, $\epsilon_k(n)$ is the noise component affecting $v_k(n)$ due to roundoff noise in the implementation of the analysis bank. In other words, $v_k(n) = v_{k,ideal}(n) + \epsilon_k(n)$.

d) All filter coefficients are assumed to be real for simplicity.

♠**Main points of this section.** We will justify the following conclusions pertaining to the noise generated by some of the popular analysis bank systems.

Case 1. *Two channel FIR system of Fig. 5.1-1(a), with $H_1(z) = H_0(-z)$.* This was considered in Sec. 5.2.2 (and listed as Method 1 in Table 6.7.2). The filter $H_0(z)$ is FIR with odd order N, and $h_0(n) = h_0(N - n)$. We assume that this is implemented in polyphase form [Fig. 5.2-2(b)], with $E_0(z)$ and $E_1(z)$ implemented in direct form. Then $\epsilon_0(n)$ and $\epsilon_1(n)$ can be assumed to be white and uncorrelated with each other, and have equal variance $(N + 1)\sigma_q^2$.

Case 2. *Two-channel FIR perfect reconstruction (PR) system (direct-form).* (Sec. 5.3.6.) The filters are related by (5.3.28), and the order N is odd. Here $\epsilon_0(n)$ and $\epsilon_1(n)$ can be assumed to be white and uncorrelated with each other, with equal variance $(N + 1)\sigma_q^2$.

Case 3. *Lattice implementation of the PR QMF bank.* (Sec. 6.4). We know that the above perfect reconstruction system can be implemented using the lattice structure of Fig. 6.4-1. The analyis bank is reproduced in Fig. 9.4-1 (with quantizers), by setting $\alpha = 1$ and $\eta = -1$ in Fig. 6.4-1. In this system $\epsilon_0(n)$ and $\epsilon_1(n)$ can again be assumed to be white and uncorrelated, with equal variance. But now the variance is $0.5(N + 1)\sigma_q^2$.

Case 4. *M channel FIR PR system with paraunitary $\mathbf{E}(z)$.* (Sec. 6.5.) Let $\mathbf{E}(z)$ be implemented as a cascade of simpler paraunitary building blocks (e.g., as in Fig. 6.5-2). Assume that there is no quantization inside a building block, and that there are M b-bit quantizers at the output of each building block (Fig. 9.4-2). (This can be arranged by employing higher precision for all arithmetic inside the building block; the extra precision is finite, since there are no loops. The reader can modify the analysis for the case where there are more quantizers.). Then, $\epsilon_k(n)$ can again be assumed to be white (and $\epsilon_k(n)$ uncorrelated with $\epsilon_\ell(m)$ for $k \neq \ell$), with equal variance $(N+1)\sigma_q^2/M$ for all k.

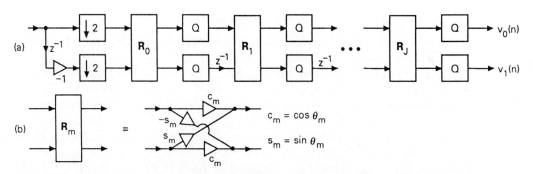

Figure 9.4-1 (a) Lattice structure for the analysis bank of the two channel PR QMF bnak. (b) Details of $\mathbf{R}_m$.

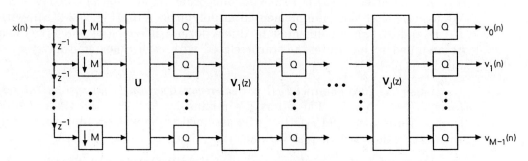

Figure 9.4-2 The anlaysis bank of a FIR PR QMF bank, with paraunitary polyphase matrix $\mathbf{E}(z)$. The paraunitary matrix is implemented as a cascade of simpler building blocks $\mathbf{U}$ and $\mathbf{V}_m(z)$.

Case 5. *Two channel IIR QMF bank with power symmetric analysis filters.* (Sec. 5.3). We know that this can be implemented as in Fig. 5.2-5 where $a_0(z)$ and $a_1(z)$ are unit-magnitude allpass. Each allpass filter can be implemented as in Fig. 9.2-7, with slight change of notations. Thus, let $s_{0,m}$ and $s_{1,m}$ be the scale factors used to scale the internal nodes. Also let k_i stand for the order of $a_i(z)$. In this case, $\epsilon_0(n)$ and $\epsilon_1(n)$ can be assumed to be white, and have equal variance given by

$$(\beta_0 + \beta_1)\sigma_q^2, \tag{9.4.1}$$

where
$$\beta_i = \sum_{m=0}^{k_i-1} s_{i,m}^2, \qquad i = 0, 1. \qquad (9.4.2)$$

However $\epsilon_0(n)$ and $\epsilon_1(n)$ are not uncorrelated with each other.

Justifications

Case 1. Consider the polyphase implementation of Fig. 5.2-2(b). Since N is odd, the FIR filters $E_0(z)$ and $E_1(z)$ have $(N+1)/2$ coefficients each. So their outputs have noise components which are white, with equal variance $K\sigma_q^2$, with $K = 0.5(N+1)$. Now, the linear phase condition $h_0(n) = h_0(N-n)$ implies that the coefficients of $E_1(z)$ are time reversed versions of those of $E_0(z)$. Inspite of this, the noise components at the outputs of $E_0(z)$ and $E_1(z)$ can be assumed to be uncorrelated because, the samples entering $E_0(z)$ and $E_1(z)$ are even and odd numbered subsets of $x(n)$, respectively. The multipliers in $E_0(z)$ and $E_1(z)$ cannot be shared (as seen in Problem 5.3), and we cannot obtain a fifty-percent noise reduction (which is normally obtainable in linear-phase filters).

The 2×2 matrix $\mathbf{T}$ which follows $E_0(z)$ and $E_1(z)$ in Fig. 5.2-2(b) can be written as
$$\mathbf{T} = \begin{bmatrix} 1 & 1 \\ 1 & -1 \end{bmatrix}.$$

This satisfies $\mathbf{TT}^T = 2\mathbf{I}$, and we can invoke Theorem 9.3.1 to conclude that the noise components $\epsilon_0(n)$ and $\epsilon_1(n)$ are white and uncorrelated, with variance $2K\sigma_q^2 = (N+1)\sigma_q^2$.

Case 2. We know that the filter coefficients are related according to $h_1(n) = (-1)^n h_0(N-n)$, and since the same input $x(n)$ enters both filters, terms of the form $x(i)h_0(m)$ are shared by the filter outputs. So we cannot claim that the roundoff noise components at the outputs of the filters are uncorrelated. However, consider the decimated outputs

$$v_0(n_0) = \sum_{m_0=0}^{N} h_0(m_0)x(2n_0 - m_0), \quad v_1(n_1) = \sum_{m_1=0}^{N} h_1(m_1)x(2n_1 - m_1). \qquad (9.4.3)$$

We will show that [for arbitrary $x(n)$] the same product $h_0(k)x(i)$ will not be shared by the two summations. Suppose this is not true. Then we must have
$$h_0(m_0) = \pm h_1(m_1), \quad \text{and} \quad x(2n_0 - m_0) = x(2n_1 - m_1), \qquad (9.4.4)$$

for some n_0, n_1, m_0 and m_1. Since $h_1(n) = (-1)^n h_0(N-n)$, this implies $m_0 = N - m_1$ and $2n_0 - m_0 = 2n_1 - m_1$. This in turn means $2m_0 = 2(n_0 - n_1) + N$, which contradicts the fact that N is odd. Summarizing, we can assume that $v_0(n)$ and $v_1(n)$ are uncorrelated. Furthermore, we already know (Sec. 9.2.4) that $v_0(n)$ and $v_1(n)$ are white with variance $(N+1)\sigma_q^2$.

Case 4. Since Case 3 follows from Case 4, we now proceed directly to Case 4. The analysis bank is shown in Fig. 9.4-2, and there are $J+1$ paraunitary building blocks in cascade. The 0th building block is a constant unitary matrix $\mathbf{U}$, whereas the remaining ones are degree-one paraunitary systems. This cascade covers the situations in Fig. 9.4-1 as well, if we replace $\mathbf{U}$ and $\mathbf{V}_m(z)$ appropriately. At the output of the mth building block, we have the noise source vector $\mathbf{e}_m(n)$, generated by the M quantizers. According to our noise model assumptions, each component of this noise vector is white with variance σ_q^2, and any two components are uncorrelated. So $\mathbf{e}_m(n)$ is WUZE(σ_q^2). The transfer matrix from $\mathbf{e}_m(n)$ to the output terminal is a cascade of paraunitary systems and is therefore paraunitary. As a result, the noise vector $\mathbf{e}_{m,out}(n)$ which contaminates the output vector $\mathbf{v}(n) = [v_0(n) \; \ldots \; v_{M-1}(n)]^T$ is WUZE(σ_q^2). Since any two noise vectors $\mathbf{e}_m(n)$ and $\mathbf{e}_\ell(n)$ are uncorrelated, the total noise vector contaminating $\mathbf{v}(n)$ is WUZE($(J+1)\sigma_q^2$). Using the relation $N = MJ + M - 1$ this can be written as WUZE($(N+1)\sigma_q^2/M$). Summarizing, the noise components $\epsilon_k(n)$ at the analysis bank output are white and uncorrelated, with variance $(N+1)\sigma_q^2/M$.

Case 5. The allpass filters $a_i(z)$ are products of first order real coefficient allpass filters, and can be implemented as in Fig. 9.2-7. From Sec. 9.2.4 we therefore conclude that the roundoff noise at the output of $a_i(z)$ is white with variance $\beta_i\sigma_q^2$, where β_i is as in (9.4.2). Using standard assumptions, it follows that the noise components at the two allpass filter outputs are uncorrelated. As a result, the noise components at the locations of $v_0(n)$ and $v_1(n)$ are white, with variance $(\beta_0 + \beta_1)\sigma_q^2$.

9.5 FILTER BANK OUTPUT NOISE

In the QMF bank of Fig. 5.4-1, the roundoff noise components $\epsilon_k(n)$ generated by the analysis bank are propagated through the synthesis bank. This contributes a noise component $e_a(n)$ at the output node [i.e., node labeled $\widehat{x}(n)$]. In addition to this, the roundoff operations in the implementation of the synthesis bank contribute a noise component $e_S(n)$. So the total noise component $e(n)$ affecting $\widehat{x}(n)$ can be written as

$$e(n) = e_a(n) + e_S(n). \qquad (9.5.1)$$

In other words, we can write $\widehat{x}(n) = \widehat{x}_i(n) + e(n)$ where $\widehat{x}_i(n)$ is the filter bank output under ideal conditions (i.e., no quantizers). Under normal conditions, we can assume that $e_a(n)$ and $e_S(n)$ are uncorrelated, with zero mean. We will now study further properties of $e_a(n)$ and $e_S(n)$ for each of the five cases listed in Sec. 9.4. In order to compare various structures on a common ground, we will adopt some conventions:

9.5.1 Conventions and Assumptions

1. *Scaling.* We will insert scale factors at appropriate places to satisfy the following requirement: signals that are inputs to appropriate computa-

tional building blocks should be scaled in the $\mathcal{L}_2$ sense. This requirement means that if the filter-bank input $x(n)$ is white with unit variance, then the variance at the scaled node is also unity.

2. Whenever necessary, we will insert a scale factor in front of $\widehat{x}(n)$ so that there is no discrepancy between $x(n)$ and $\widehat{x}(n)$ (except for possible amplitude and/or phase distortions, etc.). For example, in a perfect reconstruction system we will have $\widehat{x}(n) = cx(n - n_0)$, with $c = 1.0$.

3. We will neglect the noise generated by the above scale factors, as their contribution to total noise is small.

Figure 9.5-1 shows all the QMF banks of interest, with scale factors inserted according to these conventions. Quantizers, which are inserted as explained in earlier sections, are not shown just to keep the figures simple. We now make some observations and leave it to the reader to verify them.

1. **Fig. 9.5-1(a).** In this system, the inputs to the filters $E_0(z)$ and $E_1(z)$ in the analysis bank (in fact any nodes connected directly to $x(n)$) are automatically scaled (in the $\mathcal{L}_2$ sense). The same is approximately true of the filters $E_1(z)$ and $E_0(z)$ in the synthesis bank, under the assumption that the analysis filter $H_0(z)$ has energy ≈ 0.5. (This holds to the extent that $|H_0(e^{j\omega})| \approx 1$ in the passband and $|H_0(e^{j\omega})| \approx 0$ in the stopband). So we do not require any scale factor except the '2' inserted in front of $\widehat{x}(n)$, to satisfy convention 2.

2. **Fig. 9.5-1(b).** Next consider Fig. 9.5-1(b). If we insert the three scale factors $\sqrt{2}$ as shown, then the inputs to $F_k(z)$ are scaled in the $\mathcal{L}_2$ sense, and furthermore convention 2 is satisfied.

3. **Fig. 9.5-1(c).** In Fig. 9.5-1(c), $\mathbf{E}(z)$ is implemented as in Fig. 6.5-2, where $\mathbf{U}$ is unitary and $\mathbf{V}_m(z)\widetilde{\mathbf{V}}_m(z) = \mathbf{I}$. We assume $\mathbf{U}$ is normalized such that $\mathbf{U}^\dagger \mathbf{U} = \mathbf{I}$. So $\mathbf{E}(z)\widetilde{\mathbf{E}}(z) = \mathbf{I}$. The same comments hold for the paraunitary system $\mathbf{R}(z)$. As a result we have $\widehat{x}(n) = x(n - n_0)$, and no scale factors are necessary to satisfy convention 2. Also, the inputs to each of the paraunitary building blocks ($\mathbf{V}_m(z)$ and $\mathbf{U}$) are scaled in the $\mathcal{L}_2$ sense automatically.

4. **Fig. 9.5-1(d).** Finally, in Fig. 9.5-1(d), the allpass filters satisfy $|a_i(e^{j\omega})| = 1$. Insertion of $1/2$ as indicated ensures that the inputs to the allpass filters in the synthesis bank are scaled in the $\mathcal{L}_2$ sense. In addition, each allpass filter (implemented as in Fig. 9.2-7) has its own internal scale factors. For this figure, we have $\widehat{X}(e^{j\omega}) = e^{j\phi(\omega)}X(e^{j\omega})$, consistent with the fact that aliasing and amplitude distortion have been eliminated.

9.5.2 Output Roundoff Noise e(n)

For each of the above cases, we can compute the output noise variance as follows. We will freely use the standard noise model assumptions, as well as the results in Sec. 9.2.4 (FIR roundoff noise, and allpass roundoff noise) and Sec. 9.3.

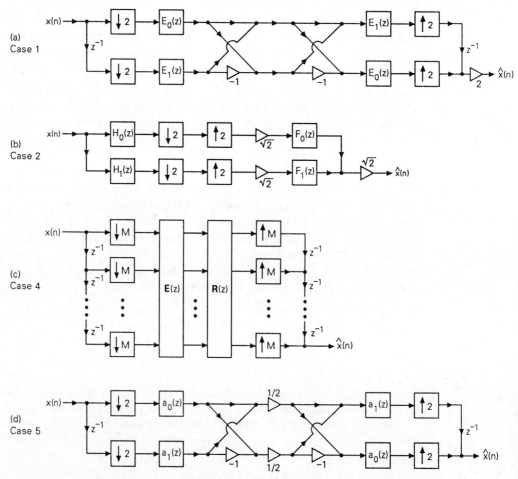

Figure 9.5-1 Examples of QMF banks, with scale factors inserted (a) case 1, (b) case 2, (c) case 4 and (d) case 5.

Case 1

Each polyphase component $E_k(z)$ in the synthesis bank, which is FIR with length $0.5(N+1)$, generates white noise $\delta_k(n)$ with variance $0.5(N+1)\sigma_q^2$. Under normal assumptions, $\delta_0(n)$ and $\delta_1(n)$ are uncorrelated. The output noise $e_S(n)$, which is the interlaced version of $\delta_0(n)$ and $\delta_1(n)$ (scaled by two) is, therefore, white with variance $2(N+1)\sigma_q^2$.

The noises generated by $E_0(z)$ and $E_1(z)$ in the analysis bank are also white and uncorrelated, with variance $0.5(N+1)\sigma_q^2$. If $|H_0(e^{j\omega})| \approx 1$ in the passband, then $|E_k(e^{j\omega})|$ are 'approximately' constant (≈ 0.5). We can then verify that this contributes a noise component $e_a(n)$ ('almost' white) at the output of the filter bank, with variance $2(N+1)\sigma_q^2$. The total noise $e(n)$ is therefore essentially white, with variance $4(N+1)\sigma_q^2$.

Case 2

Each synthesis filter $F_k(z)$ is FIR with order N, and its input is the output of an expander. So only $0.5(N+1)$ multiplication are involved per computed output. Thus, the roundoff noise at the outputs of the two synthesis filters $F_k(z)$ are uncorrelated and white, with variance $0.5(N+1)\sigma_q^2$ each. So the output noise component $e_s(n)$ is white, with variance $2(N+1)\sigma_q^2$.

To study the effect due to analysis filter noise, consider Fig. 9.5-2, which is the synthesis bank in polyphase form. The noise entering the paraunitary system $\mathbf{R}(z)$ from the analysis bank is WUZE($2(N+1)\sigma_q^2$). By Theorem 9.3.1 the noise at the output of $\mathbf{R}(z)$ is WUZE($(N+1)\sigma_q^2$). (This is because $\widetilde{\mathbf{R}}(z)\mathbf{R}(z) = c\mathbf{I}$, with $c = 0.5$, which is consistent with $|F_k(e^{j\omega})| \leq 1$). So the noise component $e_a(n)$ is white with variance $2(N+1)\sigma_q^2$. Summarizing, the total output noise $e(n)$ is white with variance $4(N+1)\sigma_q^2$.

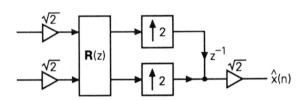

Figure 9.5-2 The synthesis bank of the PR QMF system, drawn in polyphase form.

Case 4

We will proceed to Case 4, since Case 3 is covered by this. The synthesis bank is implemented in a manner similar to Fig. 9.4-2 (i.e., as a cascade of paraunitary building blocks). Here $\mathbf{E}(z)\widetilde{\mathbf{E}}(z) = \mathbf{R}(z)\widetilde{\mathbf{R}}(z) = \mathbf{I}$. Proceeding as in Sec. 9.4 we conclude that the noise vector at the output of $\mathbf{R}(z)$, due to roundoff in synthesis bank, is WUZE($(N+1)\sigma_q^2/M$). So the noise component $e_S(n)$ which is the interlaced version of these, is white with variance $(N+1)\sigma_q^2/M$.

The noise entering $\mathbf{R}(z)$ from the analysis bank is also WUZE($(N+1)\sigma_q^2/M$). Using Theorem 9.3.1 as before, this noise vector propagates to the output of $\mathbf{R}(z)$ as WUZE($(N+1)\sigma_q^2/M$). The interlaced version $e_a(n)$ is again white with variance $(N+1)\sigma_q^2/M$. So the total noise $e(n)$ is white, with variance $2(N+1)\sigma_q^2/M$. For the special case of the two channel lattice this reduces to $(N+1)\sigma_q^2$.

Case 5.

The allpass filters $a_i(z)$ in Fig. 9.5-1(d) are implemented in cascade form (Fig. 9.2-7), with scale factors $s_{i,m}$. The noise generated in the implementation of $a_i(z)$ contributes a white noise component at its output, with variance $\beta_i \sigma_q^2$, where $\beta_i = \sum_{m=0}^{k_i-1} s_{i,m}^2$. In the synthesis bank, the noises at the outputs of $a_i(z)$ get interlaced. Since $\beta_0 \neq \beta_1$, the interlaced version $e_S(n)$ is not stationary. However, we will see that the total noise $e(n)$ is stationary.

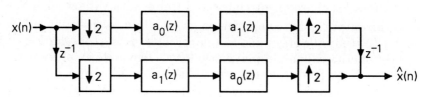

Figure 9.5-3 The power symmetric QMF bank, redrawn for the purpose of study.

Recall that the noise components $\epsilon_0(n)$ and $\epsilon_1(n)$ generated by the analysis bank are white but not uncorrelated. In order to study the properties of $e_a(n)$ it therefore turns out to be convenient to use the equivalent structure of Fig. 9.5-3. The noise from $a_0(z)$ in the analysis bank enters the filter $a_1(z)$ in the synthesis bank, and vice versa. So the noise at the output of $a_i(z)$ in the synthesis bank is a white noise component with variance $\beta_{1-i}\sigma_q^2$. If these are interlaced to obtain $e_a(n)$, the result is again non stationary. However, the total noise $e(n) = e_a(n) + e_S(n)$ has variance $(\beta_0 + \beta_1)\sigma_q^2$, which is same for all n. Summarizing, $e(n)$ is white with variance $(\beta_0 + \beta_1)\sigma_q^2$.

Summary. Table 9.5.1 summarizes the output noise variance for all the cases. It is interesting to note that the total noise $e(n)$ at the output of the QMF bank is white in each case. For $M=2$ the variance for case 4 is ony $(N+1)\sigma_q^2$ which is four times smaller than for cases 1 and 2. This has to do with the choice of scale factors, and is not a very significant difference. It corresponds to about 6.02 dB improvement in noise (which is equivalent to one additional bit of internal word length).

9.6 LIMIT CYCLES

Signal quantization in a digital filter usually generates a random error at the filter output, as we have seen in the previous sections. Under some conditions, however, signal quantization causes periodic oscillations called *limit cycles*. The most well understood type of limit cycles are zero-input limit cycles. As the name implies, these are self sustained oscillations which remain after the input $u(n)$ has been turned off.

TABLE 9.5.1 Summary of properties of output noise e(n) in various QMF banks.

Case considered	Case 1 2-channel FIR linear phase QMF (direct form polyphase)	Case 2 2-channel FIR PRQMF (direct form)	Case 4 M-channel FIR PRQMF with paraunitary $\mathbf{E}(z)$ (cascaded structure for $\mathbf{E}(z)$)	Case 5 2-channel IIR power symmetric QMF
Variance of output noise $e(n)$	$4(N+1)\sigma_q^2$	$4(N+1)\sigma_q^2$	$2(N+1)\sigma_q^2/M$	$(\beta_0 + \beta_1)\sigma_q^2$ (see text)

In all cases $e(n)$ has zero mean. For cases 2,4 and 5 $e(n)$ is white.
For case 1, $e(n)$ is 'approximately' white (see text). N denotes filter order,
and σ_q^2 is the basic quantizer noise variance.

Limit cycles arise because the quantizers, which are nonlinear elements, are inserted in feedback loops. Even though the linear system (i.e., structure without quantizer) is stable and therefore does not suffer from zero-input limit cycles, the system with quantizers can support such oscillations. Two types of limit cycles have been distinguished. The first is the granular or "roundoff" type, which is due to the "small" error introduced by the quantizer. The magnitude of this oscillation is proportional to the step size Δ, and can be reduced by adding more bits of precision. The second type, called overflow oscillations, can arise when the quantizer input exceeds the dynamic range. These are "large" oscillations, (with magnitude close to unity!) and cannot be reduced by adding more bits of precision. Examples of both types can be found in Oppenheim and Schafer [1989].

It is clear that FIR filter structures are free from limit cycles, since they have no feedback loops. Limit cycles arise only in IIR structures. In multirate filter bank systems, the only significant IIR filters we have seen are power symmetric filters (Sec. 5.3). These systems can be implemented in terms of allpass filters $a_0(z)$ and $a_1(z)$ as shown in Fig. 5.2-5. In these applications, $a_i(z)$ are real coefficient filters and furthermore can be factorized into first order allpass sections with *real coefficients* (Sec. 5.3.5). If these first order sections are free from limit cycles, then so is the complete structure. We are therefore interested in suppressing limit cycles in first order real coefficient sections. We will conclude this section by showing how.

First Order Sections

Consider Fig. 9.1-1(b) again, which shows a first order section with a quantizer Q. Under zero-input conditions the behavior of the closed loop system is governed by the equations

$$y(n) = Q[w(n)], \qquad (9.6.1)$$

and
$$w(n) = ay(n-1). \qquad (9.6.2)$$

Assume $|a| < 1$ (i.e., the system without quantizer is stable). We then have

$$|w(n)| < |y(n-1)|, \qquad (9.6.3)$$

unless $y(n-1) = 0$. Suppose now that the quantizer has the property

$$|Q[x]| \leq |x|, \qquad (9.6.4)$$

for any number x. This is easily accomplished in practice. For example, if the quantizer is of the magnitude truncation type, this is satisfied for any x within the dynamic range. If x is outside the dynamic range (i.e., overflow situation), then (9.6.4) is still satisfied because $Q[x]$ is within the dynamic range.

Quantizers satisfying (9.6.4) are said to be passive. With such quantizers, we have

$$|y(n)| \leq |w(n)|. \qquad (9.6.5)$$

Combining with (9.6.3) we see that $|y(n)| < |y(n-1)|$. But since $y(n)$ is a b-bit fraction with step size Δ, this implies

$$|y(n)| \leq |y(n-1)| - \Delta, \qquad (9.6.6)$$

for any $y(n-1) \neq 0$. This means that as n increases, the magnitude of $y(n)$ keeps decreasing (at least by Δ each time) until it becomes zero (in a finite amount of time).

Summarizing, the first order section of Fig. 9.1-1(b) does not support zero-input limit cycles of either type as long as the quantizer is passive and $|a| < 1$.

9.7 COEFFICIENT QUANTIZATION

Detailed presentations of coefficient quantization effects in digital filters can be found in a number of references, for example, Oppenheim and Schafer [1975], and Rabiner and Gold [1975]. So our presentation is brief, and we will discuss only some special issues particularly relevant to multirate filter banks. When the multiplier coefficients in a filter structure are quantized, the transfer function changes, say from $H(z)$ to $H_q(z)$. This means that the magnitude as well as phase responses have changed. In some extreme cases, some of the poles, which are close to the unit circle, may move outside, resulting in unstable filters. The IIR direct-form structure (demonstrated in Fig. 2.1-5 for order $N = 2$) is known to be very sensitive to coefficient quantization, particularly for large N. The effect is less severe for FIR direct form structures (Fig. 2.1-3), even though improved structures are available.

Magnitude response sensitivity. For linear phase FIR filter structures, the coefficient symmetry (hence linearity of phase) can be preserved in spite of quantization (e.g., see Fig. 2.4-3). So only the magnitude response $|H(e^{j\omega})|$ changes due to quantization. For filters which do not have linear phase (e.g., IIR), the phase response also changes, but this is usually not of concern since it is not linear anyway. It is therefore important to discuss only the sensitivity of the magnitude response $|H(e^{j\omega})|$.

Improved Structures

There exist structures for which the effects of quantization are less severe. In general cascade form structures (Sec. 2.1.3) are less sensitive to quantization. In these structures, quantization of a denominator coefficient affects only one complex-conjugate pole pair (or real pole, as the case may be). Similar comment holds for zeros. In Sec. 3.4.3 we introduced lattice structures which have some other advantages. One of these is that the transfer function remains stable as long as the quantized lattice coefficients k_m (Fig. 3.4-8) satisfy $|k_m| < 1$. From Sec. 3.6 we know that many IIR filters (including elliptic) can be implemented as a sum of two allpass filters (Fig. 3.6-2). Each allpass filter in turn can be implemented using the lattice. Such structures are therefore *stable even under quantization*. It is also known [Gray, Jr., 1980] that lattice structures are free from zero-input limit cycles. Other structures with improved finite-wordlength behavior are wave digital filters [Fettweis, 1971] and orthogonal filters [Deprettere and Dewilde, 1980].

In this section, we will show that a number of filter bank structures which we have presented in Chap. 5 and 6 exhibit low passband sensitivity. This means that the passband response of the quantized system is 'acceptably close' to the specified response. This is a consequence of a property called structural passivity, which we will elaborate. In two-channel PR QMF banks, since the analysis filters are power symmetric, this also implies that the stopband response is well-controlled under quantization, provided the structure retains power symmetry in spite of quantization.

9.7.1 Structural Passivity

Let m_i denote the multiplier coefficients in the structure, and assume $-1 < m_i < 1$. (This can always be arranged, since we can write $m_i =$ integer plus fraction, and eliminate the integer part by using adders.) Suppose the structure is such that $|H(e^{j\omega})| \leq 1$ for all values of the coefficients in the range $-1 < m_i < 1$. (Assume further that the transfer function remains stable for $-1 < m_i < 1$.) We then say that the implementation is structurally *passive* (or *bounded*) [Vaidyanathan and Mitra, 1984]. This means, in particular, that the response is bounded by unity even if the multipliers are quantized, as long as the quantized value does not exceed unity.

Consequence of Structural Passivity

Consider Fig. 9.7-1(a). This represents the ideal (unquantized) response of a digital elliptic filter. The magnitude attains a maximum of unity at the frequencies θ_k in the passband, that is, $|H(e^{j\theta_k})| = 1$. If we now quantize a coefficient m_i in the structure, the response $|H(e^{j\theta_k})|$ can only decrease, as demonstrated in Fig. 9.7-1(b). In other words, we have

$$\frac{\partial |H(e^{j\theta_k})|}{\partial m_i} = 0. \tag{9.7.1}$$

This means that the sensitivity of the magnitude response with respect to the coefficients, evaluated at the nominal coefficient values, is equal to zero. This is true with respect to every coefficient, and at every extremal frequency θ_k in the pass band. If there are several extrema in the passband, we can expect the sensitivity of $|H(e^{j\omega})|$ with respect to the coefficients m_i to be low for all frequencies in the passband. (This has also been verified by simulation, as we will demonstrate below.) This is the key behind the low passband sensitivity of many structures, for example, wave filters and orthogonal filters mentioned above.

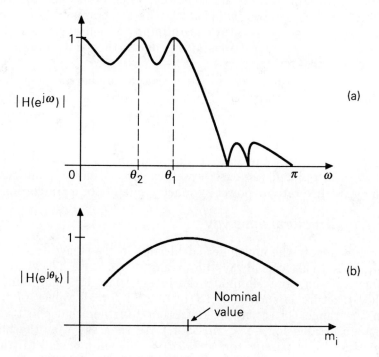

Figure 9.7-1 (a) Example of an elliptic filter response, and (b) demonstrating the effect of structural passivity.

9.7.2 Application in QMF Banks

Power Symmetric IIR QMF Bank

Consider the power symmetric IIR QMF bank discussed in Sec. 5.3. This system can be implemented as in Fig. 5.2-5 where $a_0(z)$ and $a_1(z)$ are unit magnitude allpass filters. In this system each allpass filter can be implemented as in Fig. 9.2-7, where the coefficients α_i satisfy $0 < \alpha_i < 1$. When these coefficients are quantized, the filters $a_0(z)$ and $a_1(z)$ still remain stable and satisfy $|a_i(e^{j\omega})| < 1$ (as long as $|\alpha_i| < 1$ continues to hold). Since

$$H_i(z) = \frac{a_0(z^2) \pm z^{-1} a_1(z^2)}{2}, \qquad (9.7.2)$$

we have $|H_i(e^{j\omega})| \leq 1$, that is, the implementation is structurally passive.

To demonstrate the low sensitivity property, consider the case where $a_i(z)$ are first order filters so that $H_0(z)$ is as in (5.3.18). We now implement this system (i.e., Fig. 5.2-5), with the allpass filters $a_i(z)$ implemented in cascade form (i.e., as in Fig. 9.2-7). Fig. 9.7-2(a) shows $|H_0(e^{j\omega})|$ for the quantized as well as ideal systems. The quantization level is 6 bits per multiplier (i.e., $b = 6$ in Fig. 9.2-1). For comparison, Fig. 9.7-2(b) shows the response of a direct form structure, with multipliers quantized to the same level (for convenience all plots are normalized to have a peak value of unity). It is clear that the passband response of the quantized structurally-passive implementation is far superior to the direct-form structure. It is also worth noting that the direct form structure does not preserve the power symmetric property under quantization, unlike the structure of Fig. 5.2-5.

FIR Perfect-Reconstruction QMF Lattice

Consider now the two channel QMF lattice of Fig. 6.4-1. The analysis filters are determined by the $J + 1$ angles θ_m. The scale factor α does not affect the sensitivity of the response and can be assumed to be $1/\sqrt{2}$ for the purpose of discussion. We know from Sec. 6.4 that the analysis filters are power complementary and satisfy $|H_k(e^{j\omega})|^2 \leq 1$, regardless of the values of the angles θ_m. This structure therefore exhibits low passband sensitivity.

We now demonstrate this, using the more economic structure of Fig. 6.4-3(a), and quantizing the coefficients α_m. For this we consider a system designed using the technique described in Sec. 6.3.2, with filter order $N = 23$. Figure 9.7-3(a) shows $|H_0(e^{j\omega})|$, for the quantized as well as unquantized lattice structures. For the quantized lattice we use 8 bits per coefficient α_m. For comparison, Fig. 9.7-3(b) shows the response when the impulse response coefficients $h_0(n)$ are directly quantized (i.e., direct-form implementation). From the passband details it is clear that the lattice structure has much lower passband sensitivity compared to the direct form, demonstrating the effect of structural passivity. Once again, the direct form structure does not preserve the power symmetric property, unlike the lattice structure.

Further examples of structurally passive implementations can be found in Problem 14.29.

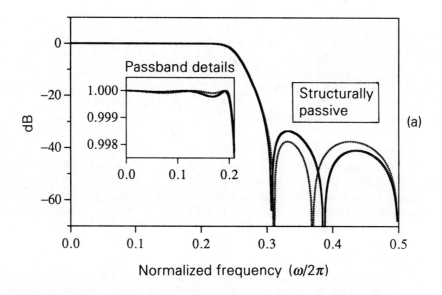

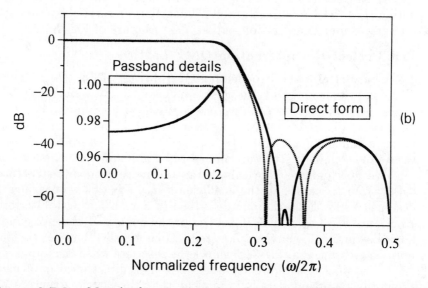

Figure 9.7-2 Magnitude response plots for quantized IIR power symmetric elliptic filters. (a) allpass-based structure (structurally passive) and (b) direct form structure. Broken lines indicate unquantized responses.

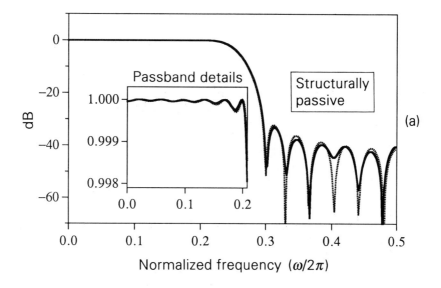
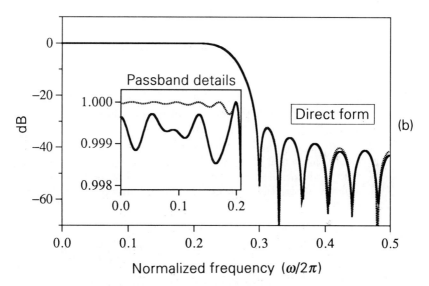

Figure 9.7-3 Magnitude responses for quantized FIR power symmetric filters. (a) QMF-lattice structure (structurally passive), and (b) direct form structure. Broken lines indicate unquantized responses.

PROBLEMS

Note. Familiarity with the material in Appendix B will be helpful while solving some of the following problems.

9.1. Consider the following lattice structure, where k is real with $k^2 < 1$, and $\hat{k} = \sqrt{1 - k^2}$.

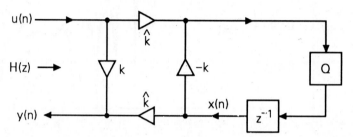

Figure P9-1

From Sec. 3.4 we know that the transfer function $H(z)$ is stable allpass. With a quantizer inserted in the feedback loop as shown, draw the noise model, and estimate the output noise variance under the usual assumptions.

9.2. Consider again the lattice structure in Fig. P9-1. There are four multipliers, two of which have input $u(n)$, and two of which have input $x(n)$. So in order to scale the structure, it is sufficient to scale the node $x(n)$. Show that this node is in fact already scaled in the $\mathcal{L}_2$ sense!

9.3. Consider Fig. 9.2-3, where a multiplier $1/L$ is inserted to scale the node which represents the output of the adder. For sum-scaling, we know we have to choose $L = 1/(1 - |a|)$.

   a) For $\mathcal{L}_2$ scaling, how would you choose L?

   b) Now assume that the input $u(n)$ is a zero-mean white WSS random process with unit variance, and let $a = 0.99$. Estimate the variance of the output *signal* $y(n)$ for the two cases (i) sum-scaling and (ii) $\mathcal{L}_2$ scaling.

9.4. Let p be a positive integer. The $\mathcal{L}_p$ norm of a transfer function $F(z)$ is defined as

$$\|F\|_p = \left(\int_0^{2\pi} |F(e^{j\omega})|^p \frac{d\omega}{2\pi} \right)^{1/p}. \qquad (P9.4a)$$

Let $y(n)$ be the output of $F(z)$ in response to an input $u(n)$. Let $U(e^{j\omega})$ be the Fourier Transform of $u(n)$. It can then be shown that

$$|y(n)| \leq \|F\|_p \|U\|_q, \qquad (P9.4b)$$

for any pair of positive integers p, q such that $p^{-1} + q^{-1} = 1$. Examples are $(p = 1, q = \infty)$, $(q = 1, p = \infty)$, and $(p = q = 2)$. If $\|F\|_p = 1$, we see that $|y(n)| \leq \|U\|_q$ for all n. Under this condition we say that the node $y(n)$ is scaled in the $\mathcal{L}_p$ sense. If $\|U\|_q \leq 1$ as well, then $|y(n)| \leq 1$, that is, there is no overflow at node $y(n)$. (For simplicity, assume that $y(n) = \pm 1$ is not considered as overflow). Summarizing, $\mathcal{L}_p$ scaling prevents overflow if the input is such that $\|U\|_q \leq 1$.

a) Show that $||F||_\infty$ is equal to the maximum value of $|F(e^{j\omega})|$.

b) In each of the following cases, what kind of $\mathcal{L}_p$ scaling will be appropriate (i.e., what p should be chosen) to avoid overflow? (i) $u(n)$ is a sequence with energy $\sum_n |u(n)|^2 = 1$, (ii) $u(n) = e^{j\omega_0 n}$ for some real ω_0, and (iii) $u(n)$ is such that $|U(e^{j\omega})| \le 1$. (*Note.* In each of the above cases, sum-scaling (Sec. 9.2) could also have avoided overflow, but is more stringent than necessary.)

9.5. Let $x(n)$ be WSS with autocorrelation $R_{xx}(k)$, and let $y(n) = x(Mn)$. Show that the autocorrelation of $y(n)$ is given by $R_{yy}(k) = R_{xx}(Mk)$.

9.6. Let $\mathbf{x}(n)$ be a zero-mean WSS process. Show that it is WUZE(σ^2) if and only if $E[\mathbf{x}(n)\mathbf{x}^\dagger(n+i)] = \sigma^2 \delta(i)\mathbf{I}$.

9.7. Consider the transmultiplexer structure of Fig. 5.9-1. Assume that the filters $H_k(z)$ and $F_k(z)$ are FIR with length $N+1 = MK$ for some K. Let each analysis filter $H_k(z)$ have energy $1/M$. Let $e_k(n)$ denote the roundoff noise component affecting the node labeled $\widehat{x}_k(n)$. Using the usual fixed-point b-bit roundoff noise model, estimate the variance of $e_k(n)$. With no further assumptions, can you say that $e_k(n)$ and $e_\ell(m)$ are uncorrelated for $k \ne \ell$?

9.8. Consider the two-stage decimation filter of Fig. 4.4-5(b). (This was the topic of Sec. 4.4.2 which should be reviewed at this time.) Here N_g and N_i are the orders of the FIR filters $G(z)$ and $I(z)$. Using the usual fixed-point roundoff noise model of Sec. 9.2, we wish to compute some noise variances, in terms of N_g, N_i, the quantizer noise variance σ_q^2, and the energy of $G(e^{j\omega})$. For simplicity, ignore the noise reduction obtainable by exploiting linear-phase symmetry.

a) Estimate the variance σ_1^2 of the roundoff noise at the final output, that is, output of M_2.

b) Instead of the above system suppose we use a single stage decimation filter for this problem, that is, Fig. 4.1-7 with $M = M_1 M_2$. Let N denote the order of $H(z)$. Estimate the noise variance σ_2^2 at the output of M.

c) In Design example 4.4.2, we started with some specifications, and arrived at specific values for M_1, M_2, N_g, N_i and N. Using these values, find the improvement in noise variance due to multistage implementation, i.e., find σ_2^2/σ_1^2. You can make the assumption that the energy of $G(e^{j\omega})$ is 0.5, which is consistent with the specifications of this design.

9.9. Consider the fractional decimation circuit of Fig. 4.1-10(b). Assume $L = 2$ and $M = 3$, and let $H(z)$ be FIR with order $N = 59$. For this problem, ignore any simplicity offered by linear-phase symmetry. Assume the usual fixed-point roundoff noise model of Sec. 9.2.

a) Estimate the roundoff noise variance at the output node [labeled $y(n)$].

b) Now consider Fig. 4.3-8(d). This represents an efficient polyphase implementation of the above fractional decimation circuit. Estimate the roundoff noise variance at the output node.

9.10. Consider Fig. 8.5-2 which represents the cosine modulated analysis filter bank. Here each pair $G_k(z), G_{M+k}(z)$ is power complementary, and is implemented using the lattice Fig. 8.5-1. The cosine modulation matrix $\mathbf{T}$ is as in (8.4.9). (At this time you must review Section 8.4, in particular the meanings of $\mathbf{A}_0, \mathbf{A}_1, \mathbf{C}, \mathbf{S}$, and so on). Assume that (i) each lattice has quantizer inserted similar to Fig. 9.4-1(a), (ii) there are no quantizers inside **T**, and (iii) there are M quantizers for the M outputs of **T**. Let $\epsilon_k(n)$ denote the total roundoff noise affecting the M outputs of the decimated analysis bank (i.e., outputs of **T**). Using the standard fixed-point roundoff noise model of Sec. 9.2, estimate the variance of $\epsilon_k(n)$.

9.11. Let $\mathbf{x}(n)$ be a vector WSS random process with autocorrelation matrix $\mathbf{R}(k)$ and power spectral density matrix $\mathbf{S}(e^{j\omega})$. (These were defined in Appendix B. Note that if $\mathbf{x}(n)$ is $M \times 1$ then $\mathbf{R}(k)$ and $\mathbf{S}(e^{j\omega})$ are $M \times M$ matrices.) Show that $\mathbf{S}(e^{j\omega})$ is a positive semidefinite matrix for all ω. (*Hint*. Somehow try to relate this to a scalar WSS process $t(n)$, and use the fact that its power spectrum is nonnegative).

9.12. In this problem we consider the joint behavior of two WSS random process $x(n)$ and $y(n)$. Assume that they have zero mean.

   a) Let the power spectra $S_{xx}(e^{j\omega})$ and $S_{yy}(e^{j\omega})$ be non overlapping, i.e., $S_{yy}(e^{j\omega})S_{xx}(e^{j\omega}) = 0$ for all ω. Show that this does not in general imply that the two random processes $x(n)$ and $y(n)$ are uncorrelated.
   b) Suppose $x(n)$ and $y(n)$ are jointly WSS. Show then that the condition $S_{yy}(e^{j\omega})S_{xx}(e^{j\omega}) = 0$ does imply that the two processes are uncorrelated. (*Hint*. The result of Problem 9.11 might help!)
   c) Suppose $x(n)$ and $y(n)$ are generated as follows,

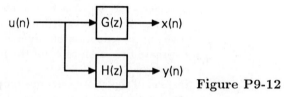

Figure P9-12

where $u(n)$ is a WSS process. Show that $x(n)$ and $y(n)$ are joinlty WSS. Hence show that if the filters $H(e^{j\omega})$ and $G(e^{j\omega})$ are nonoverlapping [that is, $H(e^{j\omega})G(e^{j\omega}) = 0$ for all ω], and $u(n)$ has zero-mean, then $x(n)$ and $y(n)$ are uncorrelated.

Note. If the zero-mean assumption is not true, the above statements should be modified by replacing "uncorrelated" with "orthogonal" everywhere.

9.13. Consider an M channel analysis bank $H_k(z), 0 \le k \le M-1$. Let the polyphase matrix $\mathbf{E}(z)$ be lossless with $\widetilde{\mathbf{E}}(z)\mathbf{E}(z) = \mathbf{I}$. In Sec. 6.2.2 we showed that this implies $\sum_n |h_k(n)|^2 = 1$ for each k, that is, each analysis filter has unit energy. Give a second proof of this using Theorem 9.3.1.

10

Multirate filter bank theory and related topics

10.0 INTRODUCTION

In this chapter we study the interrelation between multirate filter bank theory, and several "neighbouring" topics in signal processing. In Sec. 10.1 we consider the connection between alias-free (maximally decimated) filter banks, block digital filtering, and linear periodically time varying (LPTV) systems. We will see that the pseudocirculant matrix defined in Sec. 5.7.2 unifies these three topics in a nice way. In Sec. 10.2 we study a number of unconventional sampling theorems (such as "difference-sampling," and nonuniform sampling) using the framework of multirate filter banks. Readers who have looked at Problem 5.13 will recall that Shannon's well-known derivative sampling theorem can be derived based on an analog filter bank [Papoulis, 1977b], and [Brown, 1981]. Such a viewpoint not only simplifies the understanding of these sampling techniques, but also opens up new digital ways to reconstruct signals from unconventionally sampled data.

Further applications can be found in Vetterli [1988], where a multirate filter bank framework is used for the efficient implementation of FIR and IIR filters. Also see Sathe and Vaidyanathan [1993] where the role of pseudocirculat matrices in random process theory is discussed.

10.1 BLOCK FILTERS, LPTV SYSTEMS, AND MULTIRATE FILTER BANKS

10.1.1 Block Filtering

The processing of a scalar signal in blocks is a common approach in many applications. Block processing has been studied by a number of authors [Burrus, 1971], [Mitra and Gnanasekaran, 1978], [Barnes and Shinnaka, 1980], and [Clark, et al, 1981]. One example of block processing was indicated in Sec. 6.6 (transform-coding and LOT). Block digital filtering, in particular, is a technique to implement a scalar filter $H(z)$ in such a way as to increase the parallelism in the computations. This finds application in high speed digital filtering, that is, where the sampling rate is very high.

Definition of Block Digital Filters

Let $x(n)$ and $y(n)$ denote, respectively, the input and output of the scalar filter $H(z)$. Consider two vector sequences $\mathbf{x}_B(n)$ and $\mathbf{y}_B(n)$ defined by

$$\mathbf{x}_B(n) = \begin{bmatrix} x(nM + M - 1) \\ x(nM + M - 2) \\ \vdots \\ x(nM) \end{bmatrix}, \quad \mathbf{y}_B(n) = \begin{bmatrix} y(nM + M - 1) \\ y(nM + M - 2) \\ \vdots \\ y(nM) \end{bmatrix}. \quad (10.1.1)$$

We say that the vector sequences $\mathbf{x}_B(n)$ and $\mathbf{y}_B(n)$ are *blocked versions* of (or blocked sequences corresponding to) the scalar sequences $x(n)$ and $y(n)$. The block-length (or size) is M.

Now imagine that we have a system which generates the sequence $\mathbf{y}_B(n)$ in response to $\mathbf{x}_B(n)$. Evidently this is an M-input M-output system. Not surprisingly this is an LTI system (Problem 13.24), and can be characterized by a $M \times M$ transfer matrix $\mathbf{H}(z)$. In other words we have $\mathbf{Y}_B(z) = \mathbf{H}(z)\mathbf{X}_B(z)$, where

$$\mathbf{X}_B(z) = \sum_n \mathbf{x}_B(n) z^{-n}, \quad \mathbf{Y}_B(z) = \sum_n \mathbf{y}_B(n) z^{-n}. \quad (10.1.2)$$

The matrix $\mathbf{H}(z)$ is called the *blocked version* of $H(z)$. From its definition it is clear that $\mathbf{H}(z)$ is completely determined by the scalar system $H(z)$. Figure 10.1-1 is a summary of the situation. The "blocking mechanism" can be considered to be a serial to parallel converter of data, and the "unblocking mechanism" a parallel to serial converter.

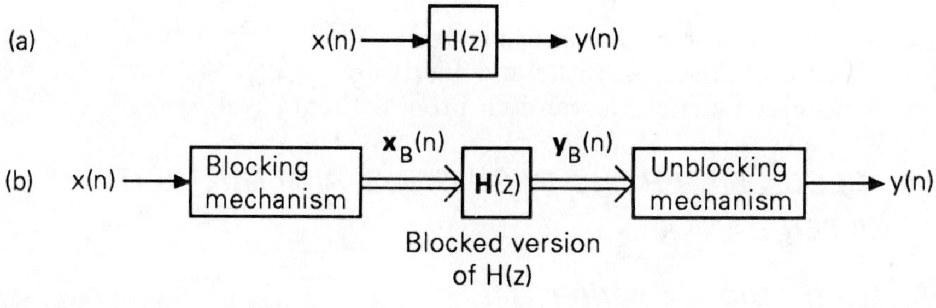

Figure 10.1-1 (a) A scalar transfer function, and (b) its blocked implementation.

Multirate Filter-Bank Notation

Figure 10.1-2 shows a schematic diagram of block digital filtering in terms of multirate notation. Here the signals $x_k(n)$ are given by $x_k(n) =$

$x(nM+k)$. Similarly $y_k(n) = y(nM+k)$, so that the blocked versions (10.1.1) can also be represented as

$$\mathbf{x}_B(n) = \begin{bmatrix} x_{M-1}(n) \\ \vdots \\ x_1(n) \\ x_0(n) \end{bmatrix}, \quad \mathbf{y}_B(n) = \begin{bmatrix} y_{M-1}(n) \\ \vdots \\ y_1(n) \\ y_0(n) \end{bmatrix}. \quad (10.1.3)$$

The transfer matrix $\mathbf{H}(z)$ in this figure produces $\mathbf{y}_B(n)$ in response to $\mathbf{x}_B(n)$, and is therefore the blocked version of $H(z)$.

Let $X_\ell(z)$ and $Y_\ell(z)$ denote the z-transforms of $x_\ell(n)$ and $y_\ell(n)$. Then the z-transforms of $x(n)$ and $y(n)$ can be expressed as

$$X(z) = \sum_{\ell=0}^{M-1} z^{-\ell} X_\ell(z^M), \quad Y(z) = \sum_{\ell=0}^{M-1} z^{-\ell} Y_\ell(z^M). \quad (10.1.4)$$

In other words, the components of $\mathbf{x}_B(n)$ and $\mathbf{y}_B(n)$ are the polyphase compononents of $x(n)$ and $y(n)$, respectively.

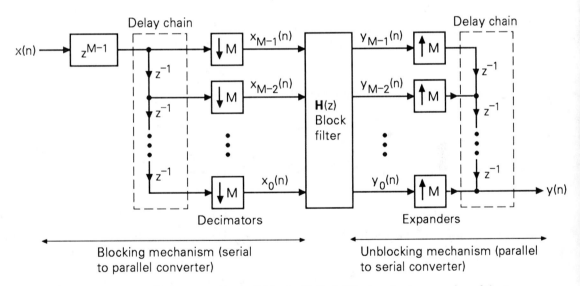

Figure 10.1-2 Representation of block digital filtering in terms of multirate building blocks.

Increased Parallelism Offered by the Block Filter

The structure of Fig. 10.1-2 also shows the "speed advantage" of blocking. The system $\mathbf{H}(z)$ is operating at a rate which is M times lower than the input rate. So the sampling rate of the input signal $x(n)$ can be M times larger than the speed of the basic computational unit. This advantage, which depends on the block size M, can be made arbitrarily large by increasing M. However, there is a price paid for this: since $\mathbf{H}(z)$ is an $M \times M$ system, it requires larger number of computational units (multipliers and adders) than the original scalar system. Summarizing, we have obtained increased computational parallelism in the blocked implementation, by increasing the number of computational units. As a result we are able to process signals which arrive at M times higher rate (than the rate that can normally be handled by *one* computational unit).

Relation to Alias-Free Filter Banks

The decimators and expanders in the representation of Fig. 10.1-2 produce the alias components $X(zW_M^k)$, just as in a filter bank. However, magically, $Y(z)$ is free from these alias-components because, by definition of $\mathbf{H}(z)$, the overall system in Fig. 10.1-2 is still a linear time invariant system with transfer function $H(z)$. The explanation of this is that, alias components are somehow canceled.

Returning now to filter banks, we know that any M-channel maximally decimated filter bank (Fig. 5.4-1) can be redrawn as in Fig. 5.5-3(c), where $\mathbf{P}(z)$ is the product $\mathbf{R}(z)\mathbf{E}(z)$ of polyphase matrices. The structures of Figs. 5.5-3(c) and 10.1-2 are identical (except for the advance operator z^{M-1}, which will not affect any significant conclusions.) Since the structure of Fig. 10.1-2 is indeed alias-free by definition of $\mathbf{H}(z)$, we conclude that this structure is equivalent to an M-channel alias-free maximally decimated filter bank.

We know from Sec. 5.7.2 that the filter bank is alias-free *if, and only if*, $\mathbf{P}(z)$ is a pseudocirculant. This shows that the blocked version $\mathbf{H}(z)$ of a scalar transfer function $H(z)$ is *necessarily* a pseudocirculant. The pseudocirculant property has been observed implicitly in Barnes and Shinnaka [1980], and also in Marshall [1982]. It has been further studied in Vaidyanathan and Mitra [1988].

Next, how can we determine the elements of $\mathbf{H}(z)$? The pseudocirculant property means that all rows are determined by the elements $H_{0,k}(z)$ of the 0th row. We also know that the transfer function of the alias-free filter bank is given by (5.7.13) where $P_\ell(z)$ are the elements of the 0th row of $\mathbf{P}(z)$. From this we conclude that the scalar filter $H(z)$ is related to the blocked version as

$$H(z) = \sum_{\ell=0}^{M-1} z^{-\ell} H_{0,\ell}(z^M). \qquad (10.1.5)$$

So the elements of the 0th row of $\mathbf{H}(z)$ are the Type 1 polyphase components (usually denoted $E_\ell(z)$) of $H(z)$:

$$H_{0,\ell}(z) = E_\ell(z). \qquad (10.1.6)$$

Example 10.1.1

Let $H(z) = 1 + 2z^{-1} + 3z^{-2} + 4z^{-3}$. We can rewrite

$$H(z) = [1 + 4z^{-3}] + 2z^{-1} + 3z^{-2} \quad \text{(scalar filter)}, \qquad (10.1.7)$$

from which we identify the Type 1 polyphase components (for $M = 3$) as $E_0(z) = 1 + 4z^{-1}$, $E_1(z) = 2$, and $E_2(z) = 3$. So the 3×3 blocked version is the pseudocirculant

$$\begin{bmatrix} 1 + 4z^{-1} & 2 & 3 \\ 3z^{-1} & 1 + 4z^{-1} & 2 \\ 2z^{-1} & 3z^{-1} & 1 + 4z^{-1} \end{bmatrix} \quad \text{(blocked filter)}. \qquad (10.1.8)$$

Next consider an IIR example; let $H(z) = (-a + z^{-1})/(1 - az^{-1})$. This can be written as

$$H(z) = \frac{-a(1 - z^{-2})}{1 - a^2 z^{-2}} + \frac{(1 - a^2)z^{-1}}{1 - a^2 z^{-2}} \quad \text{(scalar filter)}, \qquad (10.1.9)$$

so that the Type 1 polyphase components (for $M = 2$) are $E_0(z) = -a(1 - z^{-1})/(1 - a^2 z^{-1})$, and $E_1(z) = (1 - a^2)/(1 - a^2 z^{-1})$. So the 2×2 blocked version is the pseudocirculant

$$\mathbf{H}(z) = \frac{1}{1 - a^2 z^{-1}} \begin{bmatrix} -a(1 - z^{-1}) & 1 - a^2 \\ (1 - a^2)z^{-1} & -a(1 - z^{-1}) \end{bmatrix} \quad \text{(blocked filter)}. \qquad (10.1.10)$$

For real a, the scalar IIR filter $H(z)$ is allpass, that is, $\widetilde{H}(z)H(z) = 1$. How did this allpass property reflect into the blocked version? In Problem 10.2 we request the reader to verify the interesting fact that $\widetilde{\mathbf{H}}(z)\mathbf{H}(z) = \mathbf{I}$. In other words, $\mathbf{H}(z)$ is paraunitary!

More generally, we can summarize the above results as follows.

♦**Theorem 10.1.1. On blocked version of a scalar filter.** Let $\mathbf{H}(z)$ represent the $M \times M$ blocked version of a scalar transfer function $H(z)$. Then $\mathbf{H}(z)$ is a pseudocirculant, and its 0th row is given by

$$[\, E_0(z) \quad E_1(z) \quad \ldots \quad E_{M-1}(z)\,], \qquad (10.1.11)$$

where $E_\ell(z)$ are the Type 1 polyphase components of $H(z)$ [i.e., $H(z) = \sum_{\ell=0}^{M-1} z^{-\ell} E_\ell(z^M)$]. Moreover $\mathbf{H}(z)$ is paraunitary if and only if $H(z)$ is allpass, that is, $\widetilde{H}(z)H(z) = c$ if and only if $\widetilde{\mathbf{H}}(z)\mathbf{H}(z) = c\mathbf{I}$. ◊

Proof. It only remains to prove the part of the statement involving the paraunitary property. With the 0th row of pseudocirculant $\mathbf{H}(z)$ given by (10.1.11), the kth row ($k > 0$) is

$$[z^{-1}E_{M-k}(z) \quad \ldots \quad z^{-1}E_{M-1}(z) \quad E_0(z) \quad \ldots \quad E_{M-k-1}(z)]. \quad (10.1.12)$$

Since $E_\ell(z)$ are the Type 1 polyphase components of $H(z)$, it is easily verified (Problem 10.4) that the polyphase components of $z^{-k}H(z)$ are the elements of the kth row above. So we can express

$$z^{-k}H(z) = \sum_{\ell=0}^{M-1} z^{-\ell} H_{k\ell}(z^M). \quad (10.1.13)$$

By writing this for all values of k ($0 \leq k \leq M - 1$) we obtain the matrix equation

$$\begin{bmatrix} H(z) \\ z^{-1}H(z) \\ \vdots \\ z^{-(M-1)}H(z) \end{bmatrix} = \mathbf{H}(z^M)\mathbf{e}(z), \quad (10.1.14)$$

where $\mathbf{e}(z) = [1 \quad z^{-1} \quad \ldots \quad z^{-(M-1)}]^T$. Since (10.1.14) holds for all z, it holds if we replace z with zW^{-k}, where $W = e^{-j2\pi/M}$. By doing this for $k = 0, \ldots M - 1$, we arrive at M equations which can be collected together as follows:

$$\mathbf{\Lambda}(z)\mathbf{W}\mathbf{Q}(z) = \mathbf{H}(z^M)\mathbf{\Lambda}(z)\mathbf{W}, \quad (10.1.15)$$

where

$$\begin{aligned}\mathbf{\Lambda}(z) &= \text{diag}\,[\,1 \quad z^{-1} \quad \ldots \quad z^{-(M-1)}\,], \\ \mathbf{Q}(z) &= \text{diag}\,[\,H(z) \quad H(zW^{-1}) \quad \ldots \quad H(zW^{-(M-1)})\,],\end{aligned} \quad (10.1.16)$$

and $\mathbf{W}$ is the $M \times M$ DFT matrix. Clearly $\mathbf{W}$ and $\mathbf{\Lambda}(z)$ are paraunitary. So $\mathbf{Q}(z)$ is paraunitary if and only if $\mathbf{H}(z)$ is paraunitary. But since $\mathbf{Q}(z)$ is diagonal with elements $H(zW^{-k})$, it is paraunitary if and only if $H(z)$ is allpass. This completes the proof. $\triangledown\triangledown\triangledown$

Application to alias-free filter banks. For an alias-free filter bank, $\mathbf{P}(z)$ is pseudocirculant and the distortion function $T(z)$ is given by (5.7.13). From the above theorem we conclude that $T(z)$ is allpass (i.e., the filter bank is free from amplitude distortion) if and only if $\mathbf{P}(z)$ is paraunitary.

10.1.2 Linear Periodically Time Varying (LPTV) Systems

In this text, we have seen linear time varying (LTV) systems on many occassions. The decimator and expander, defined in Chap. 4, are examples of such systems. The fractional sampling rate changer (Fig. 4.1-10(b)) is another such example. In these examples the input and output signals have different rates.

For a more sophisticated example, consider the M-channel maximally decimated filter bank (Fig. 5.4-1). In Chap. 5 we found that this is an LTV system, characterized by the input output relation (5.4.5). This relation reduces to $\widehat{X}(z) = T(z)X(z)$, i.e., the filter bank becomes an LTI system, if and only if it is alias-free.

Recall that an LTI system is characterized by an impulse response $h(n)$ such that the output $y(n)$ is computed by convolution:

$$y(n) = \sum_m h(m)x(n-m). \qquad (10.1.17)$$

For an LTV system (with input rate = output rate), $y(n)$ is still a linear combination of the samples $x(n-m)$ as above, but $h(m)$ is not fixed; it depends on the output time index n. The relation is of the form

$$y(n) = \sum_m a_n(m)x(n-m). \qquad (10.1.18)$$

This idea is best understood by drawing a schematic structure, as shown in Fig. 10.1-3. Here we have an Nth order FIR filter, whose impulse response coefficients are not fixed (as for LTI systems) but varies with output time index n.

An LPTV system (with period M) has the further property that $a_n(m)$ is a periodic function (period M) of the output time index n [which is also the subscript on $a(m)$ in (10.1.18)]. In other words,

$$a_n(m) = a_{n+M}(m), \quad \text{for all } n, m. \qquad (10.1.19)$$

Figure 10.1-4 demonstrates an LPTV system with period = 2. Whenever n is even, the output is taken to be that of the filter with impulse response $a_0(m)$. When n is odd, the ouput is taken to be that of $a_1(m)$. This behavior can be compactly represented using multirate notation as shown in Fig. 10.1-5(a). Here the filter $A_n(z)$ is given by

$$A_n(z) \triangleq \sum_m a_n(m) z^{-m}. \qquad (10.1.20)$$

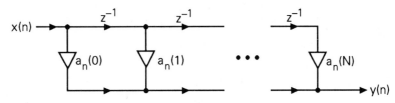

Figure 10.1-3 An LTV FIR filter. Here $a_n(m)$ is a function of the output time index n.

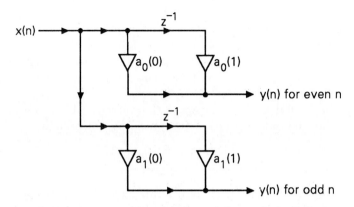

Figure 10.1-4 An FIR LPTV system with period two.

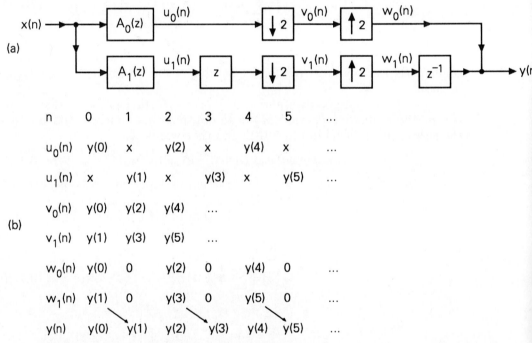

Figure 10.1-5 (a) Representation of an LPTV system in terms of multirate notations. (b) Explaining the operation.

Relation to Filter Banks

Extending the above discussion, the more general case where the LPTV system has period M can similarly be represented by the structure of Fig. 10.1-6. (This representation is restricted to systems where the input and output have equal rates; but the system can be FIR or IIR.) The system output $y(n)$ at time n is equal to the output of $A_k(z)$ at time n, where $k = n \mod M$.

We can think of this system as a M channel filter bank with analysis and synthesis filters

$$H_n(z) = z^{M-1-n}A_{M-1-n}(z), \quad F_n(z) = z^{-(M-1-n)}. \tag{10.1.21}$$

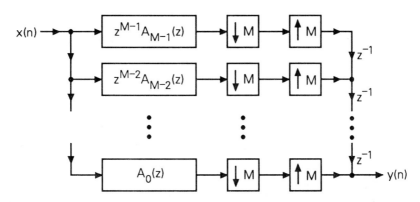

Figure 10.1-6 Representation of an arbitrary (FIR or IIR) LPTV system of period M, using multirate building blocks. (Input rate = output rate).

By using the polyphase decomposition on the filters $A_n(z)$, we gain further insight. Thus, let the filters $A_n(z)$ be represented as

$$A_n(z) = \sum_{\ell=0}^{M-1} z^{-\ell} G_{n,\ell}(z^M), \quad 0 \leq n \leq M-1. \tag{10.1.22}$$

Let $\mathbf{E}(z)$ be the Type 1 polyphase component of the analysis bank, and $\mathbf{R}(z)$ the Type 2 polyphase matrix of synthesis bank. Clearly $\mathbf{R}(z) = \mathbf{I}$ in this case. The quantity $\mathbf{E}(z)$, on the other hand depends on $G_{n,\ell}(z)$. As a demonstration, for $M = 3$ we can verify that

$$\mathbf{E}(z) = \begin{bmatrix} G_{2,2}(z) & zG_{2,0}(z) & zG_{2,1}(z) \\ G_{1,1}(z) & G_{1,2}(z) & zG_{1,0}(z) \\ G_{0,0}(z) & G_{0,1}(z) & G_{0,2}(z) \end{bmatrix}. \tag{10.1.23}$$

For arbitrary M, the form of $\mathbf{E}(z)$ can be written in a similar way (Problem 10.5). Thus the LPTV system is equivalent to Fig. 10.1-7 where $\mathbf{P}(z) = \mathbf{R}(z)\mathbf{E}(z)$.

We know from filter bank theory that this system is alias-free (hence time invariant) if and only if $\mathbf{P}(z)$ is pseudocirculant. Now let us see what happens to $G_{n,\ell}(z)$ when $\mathbf{P}(z)$ [i.e., $\mathbf{E}(z)$] is pseudocirculant. By insepection of (10.1.23) we conclude that under this condition $G_{n,\ell}(z)$ is independent of

n. This means that $A_n(z)$ is same for all n in Fig. 10.1-6, so that the original LPTV system becomes *time invariant!* This is consistent with the fact that the maximally decimated filter bank is time invariant if and only if it is alias free.

The main points of this section are summarized in Table 10.1.1.

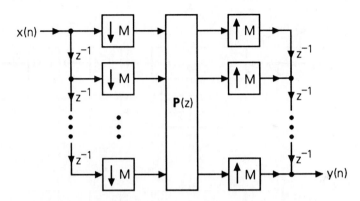

Figure 10.1-7 Equivalent representation of an arbitrary LPTV system of period M (input rate = output rate). The matrix $\mathbf{P}(z)$ is a pseudocirculant if and only if the LPTV system degenerates into an LTI system.

10.2 UNCONVENTIONAL SAMPLING THEOREMS

Let $x_a(t)$ be a continuous-time signal and define

$$x_a^{(s)}(t) = \sum_{n=-\infty}^{\infty} x_a(nT)\delta_a(t-nT), \qquad (10.2.1)$$

where $\delta_a(t)$ is the impulse function defined in Chap. 2. $x_a^{(s)}(t)$ is the uniformly sampled version of $x_a(t)$, with sample spacing equal to T (Fig. 10.2-1). Equivalently, the sampling frequency (or rate) is $2\pi/T$. The Fourier transform of $x_a^{(s)}(t)$ is given by the relation

$$X_a^{(s)}(j\Omega) = \frac{1}{T} \sum_{k=-\infty}^{\infty} X_a\left(j(\Omega - \frac{2\pi k}{T})\right), \qquad (10.2.2)$$

provided this summation converges [Oppenheim and Schafer, 1989]. Thus $X_a^{(s)}(j\Omega)$ is obtained by adding to $X_a(j\Omega)$ an infinite number of shifted copies (images), the shift being in uniform integer multiples of $2\pi/T$.

Figure 10.2-2 is a demonstration of this effect. In this figure we have assumed that $x_a(t)$ is σ-BL (defined in Sec. 2.1.4), that is, $|X_a(j\Omega)| = 0$ for $|\Omega| \geq \sigma$. From Sec. 2.1.4 we know that if the sampling rate $2\pi/T$ exceeds the Nyquist rate $\Theta = 2\sigma$, then none of the images has an overlap with the

TABLE 10.1.1 Block filtering, LPTV systems, and filter banks.

---

1. **Block digital filters.** Given a scalar transfer function $H(z)$ with input $x(n)$ and output $y(n)$, define the vector sequences $\mathbf{x}_B(n)$ and $\mathbf{y}_B(n)$ as in (10.1.3). Then these are related by an $M \times M$ transfer function $\mathbf{H}(z)$, called the blocked version of $H(z)$. Fig. 10.1-2 shows this block implementation of $H(z)$.

    a) The blocked version $\mathbf{H}(z)$ is pseudocirculant. The scalar function $H(z)$ can be obtained from the 0th row of $\mathbf{H}(z)$ as

    $$H(z) = \sum_{\ell=0}^{M-1} z^{-\ell} H_{0,\ell}(z^M). \qquad (T10.1)$$

    b) Conversely, if $\mathbf{H}(z)$ is pseudocirculant, the structure of Fig. 10.1-2 is an LTI system (this being not true for arbitrary $\mathbf{H}(z)$) and $\mathbf{H}(z)$ represents the blocked version of the scalar transfer function given by (T10.1).

2. **Linear periodically time varying systems.** A linear periodically time varying (LPTV) system with period M (and same input and output rates) is characterized by a set of M transfer functions $A_n(z)$. The system can be represented by the structure of Fig. 10.1-6.

    a) The output at time n is equal to the output of $A_k(z)$ at time n, where $k = n \bmod M$.
    b) An LPTV system with period M (with equal input and output rates) can *always* be represented by the equivalent structure of Fig. 10.1-7, where $\mathbf{P}(z)$ is an $M \times M$ transfer matrix.
    c) Conversely, for arbitrary transfer matrix $\mathbf{P}(z)$, Fig. 10.1-7 represents an LPTV system (with equal input and output rates) of period M.

3. **Relation to filter banks.** An M-channel maximally decimated filter bank (Fig. 5.4-1) can always be represented by the structure of Fig. 10.1-7 where $\mathbf{P}(z)$ is the product $\mathbf{R}(z)\mathbf{E}(z)$ of the polyphase matrices of the analysis and synthesis banks.

    a) This representation closely resembles the block implementation of a scalar transfer function $H(z)$ (Fig. 10.1-2).
    b) The representation also resembles the general representation of an LPTV system (with equal input and output rates).
    c) The blocked version $\mathbf{H}(z)$ of a scalar $H(z)$ is always pseudocirculant; the filter bank is alias-free if and only if $\mathbf{P}(z)$ in Fig. 10.1-7 is pseudocirculant; the LPTV system is actually time *invariant* if and only if $\mathbf{P}(z)$ in Fig. 10.1-7 is pseudocirculant.
    d) Let $\mathbf{H}(z)$ be $M \times M$ pseudocirculant. Consider the transfer function $H(z)$ defined in (T10.1) above. $H(z)$ is allpass if and only if $\mathbf{H}(z)$ is paraunitary. This means two things: (i) the distortion function $T(z)$ of an alias-free filter bank is allpass if and only if $\mathbf{P}(z)$ is paraunitary, and (ii) a scalar transfer function is allpass if and only if its blocked version is paraunitary.

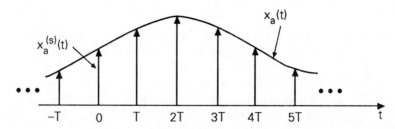

Figure 10.2-1 A continuous-time signal $x_a(t)$ and the uniformly sampled version $x_a^{(s)}(t)$.

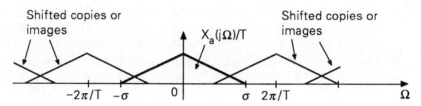

Figure 10.2-2 Fourier transform of the sampled version $x_a^{(s)}(t)$ of a bandlimited signal $x_a(t)$. Sampling rate $= 2\pi/T$.

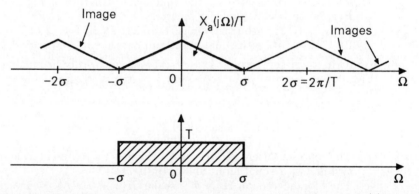

Figure 10.2-3 (a) Fourier transform of the sampled version $x_a^{(s)}(t)$ of a σ-bandlimited signal $x_a(t)$ sampled at Nyquist rate 2σ (i.e. $2\pi/T = 2\sigma$). (b) The ideal lowpass filter which reconstructs $x_a(t)$ from $x_a^{(s)}(t)$.

original version $X_a(j\Omega)$. In this case we can reconstruct $x_a(t)$ from $x_a^{(s)}(t)$ by removing the images with an ideal lowpass filter (Fig. 10.2-3). This filter has impulse response

$$h(t) = \frac{\sin(\pi t/T)}{(\pi t/T)}. \qquad (10.2.3)$$

In the time domain, the recovered signal $x_a(t)$ is therefore the convolution of

$x_a^{(s)}(t)$ with $h(t)$. This simplifies to the well-known reconstruction formula,

$$x_a(t) = \sum_{n=-\infty}^{\infty} x_a(nT) \frac{\sin(\pi/T)(t-nT)}{(\pi/T)(t-nT)}, \qquad (10.2.4)$$

which is also called the "interpolation" formula. The Nyquist frequency $\Theta = 2\sigma$ is the minimum rate at which $x_a(t)$ should be sampled so that it can be recovered from these samples.[†] Sampling at the rate Θ is called Nyquist sampling.

The above result is the *uniform sampling* theorem [Nyquist, 1928], [Whittaker, 1929], and [Shannon, 1949]: if we "uniformly sample" the σ-BL signal $x_a(t)$ at the Nyquist rate Θ, then we do not lose any information, and can reconstruct $x_a(t)$ from these samples using (10.2.4). It should be noticed, however, that (10.2.4) is equivalent to passing $x_a^{(s)}(t)$ through an ideal lowpass filter (10.2.3). This filter is noncausal and unstable [since $h(t)$ is not absolutely integrable]. In practice, we have to live with an approximation of $h(t)$, and the reconstruction is not exact. If the sampling rate $2\pi/T$ exceeds the minimum required rate Θ by a significant margin, then the images in Fig. 10.2-3 are more widely separated from the main term. So the lowpass filter can have a wider transition bandwidth and the reconstruction can be done more accurately with a practical filter.

Unconventional Sampling

Instead of sampling the σ-BL signal $x_a(t)$ at the Nyquist rate Θ, suppose we sample $x_a(t)$ and its derivative $\dot{x}_a(t)$ at *half* the Nyquist rate. It is possible to recover $x_a(t)$ from these two undersampled signals. This was actually shown in Problem 5.13, by formulating this as a two-channel analog QMF bank problem. By using an M-channel analog QMF formulation, it is possible to derive other extensions of the sampling theorem. For example, if we sample $x_a(t)$ and its $M-1$ derivatives at the rate Θ/M, we can recover $x_a(t)$ from this information (Problem 10.7). As seen from these Problems, the reconstruction filters are unrealizable, and should be replaced with practical approximations. (To be fair, the lowpass reconstruction filter (10.2.3) used in the case of uniform sampling is also unrealizable.) In the next subsection, we will obtain the discrete-time analog of this result, called the difference-sampling theorem. In contrast to the continuous-time case, this theorem involves practical (in fact FIR) reconstruction filters.

Another generalization of sampling is the so-called nonuniform sampling, demonstrated in Fig. 10.2-4. Here the samples are spaced 'too far

---

[†] This assumes, of course, that no further information is available about $x_a(t)$ except that it is σ-BL. If this is not true, then the situation is different. For example, if $x_a(t)$ is *known* to be a sinusoid $A\sin(\omega_0 t + \beta)$, then it can be recovered from a *finite number* of samples since we need to extract only three pieces of information $(A, \omega_0, \text{ and } \beta)$ from the samples! Similar comment holds if $x_a(t)$ is known to be a sum of finite number of sinusoids.

apart' (compared to Nyquist rate Θ) in some regions and 'too close' in some regions. Yet, theory has it [Jerri, 1977] that we can recover $x_a(t)$ from such samples as long as the 'average sampling rate' $\geq \Theta$. (A special case is the situation when only the *past values* of $x_a(t)$ are sampled at the rate 2Θ!) We will not prove this general result, as it does not place in evidence practical reconstruction techniques. In Sec. 10.2.2 we prove more practical special cases, for which reconstruction techniques can be found based on a FIR digital filter bank approach.

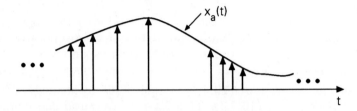

Figure 10.2-4 Nonuniform sampling of $x_a(t)$.

10.2.1 Difference Sampling Theorems for Sequences

We will now discuss difference sampling theorems, which can be considered as discrete time counterparts of derivative sampling theorems. We begin with an example.

Example 10.2.1

Let $x(n)$ be an arbitrary sequence, and let $x_1(n)$ denote its first difference, i.e.,

$$x_1(n) = x(n) - x(n-1). \qquad (10.2.5)$$

Consider the two sequences

$$y_0(n) = x(2n), \quad y_1(n) = x_1(2n). \qquad (10.2.6)$$

These are the two-fold decimated versions of $x(n)$ and its first-difference. Can we recover $x(n)$ from these two undersampled sequences? [Evidently, the number of samples per unit time, counting both the signals $y_0(n)$ and $y_1(n)$, is the same as that for $x(n)$.] The even numbered samples of $x(n)$ are already available in $y_0(n)$. It only remains to see if the odd numbered samples can be recovered from $y_1(n)$. Now $y_1(n)$ has samples

$$\ldots x(-2) - x(-3), \quad x(0) - x(-1), \quad x(2) - x(1), \quad x(4) - x(3) \ldots \quad (10.2.7)$$

From this it is clear that we can recover all odd-numbered samples of $x(n)$ by subtracting out the even-numbered samples from these differences.

A more systematic approach will help us to extend this idea to the case of higher order differences. For this we view the problem as a two channel QMF problem, as shown in Fig. 10.2-5. The analysis filters are $H_0(z) = 1$, and $H_1(z) = 1 - z^{-1}$ (representing the first-difference operation). The aim is to find synthesis filters $F_0(z), F_1(z)$ such that we have perfect reconstruction. This is an easy problem and is made easier by use of the polyphase approach. Thus, we can redraw the system as in Fig. 10.2-6, where

$$\mathbf{E} = \begin{bmatrix} 1 & 0 \\ 1 & -1 \end{bmatrix}. \qquad (10.2.8)$$

It is clear that the choice $\mathbf{R} = \mathbf{E}^{-1}$ results in perfect recovery, that is, $\widehat{x}(n) = x(n-1)$. It is readily verified that the matrix $\mathbf{E}$ is its own inverse, so we take

$$\mathbf{R} = \begin{bmatrix} 1 & 0 \\ 1 & -1 \end{bmatrix}. \qquad (10.2.9)$$

The synthesis filters are now computed according to

$$[F_0(z) \quad F_1(z)] = [z^{-1} \quad 1]\mathbf{R} = [z^{-1} \quad 1]\begin{bmatrix} 1 & 0 \\ 1 & -1 \end{bmatrix}. \qquad (10.2.10)$$

This simplifies to $F_0(z) = 1 + z^{-1}$ and $F_1(z) = -1$. If these filters are used in Fig. 10.2-5, we have perfect reconstruction that is, $\widehat{x}(n) = x(n-1)$.

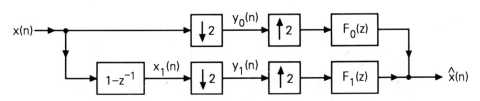

Figure 10.2-5 The difference sampling and reconstruction, viewed as a QMF bank problem.

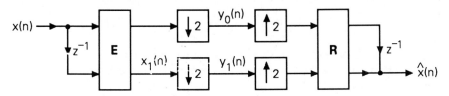

Figure 10.2-6 Redrawing Fig. 10.2-5 using the polyphase notation.

The above example can be considered to be the discrete-time equivalent of the derivative sampling theorem discussed above for continuous-time signals. (Instead of derivatives, we have differences.) Notice however, that the reconstruction scheme is very simple (and realizable), involving only FIR filters.

One motivation for thinking about such 'sampling theorems' is demonstrated in Fig. 10.2-7, where $x(n)$ is a slowly varying sequence, i.e., the adjacent samples differ by a 'very small' amount. Assume that each sample $x(n)$ requires 16 bits for its representation. Let us say that the differences $x(n) - x(n-1)$, being very small, require only 8 bits for their representation. Now instead of 'storing' or 'transmitting' all samples of $x(n)$ with 16 bits per sample, we can store (two-fold) decimated versions of $x(n)$ (16 bits per sample) and the first difference (8 bits per sample). This reduces the data rate to an average of 12 bits per sample. This is similar in principle to subband coding (Sec. 4.5.2).

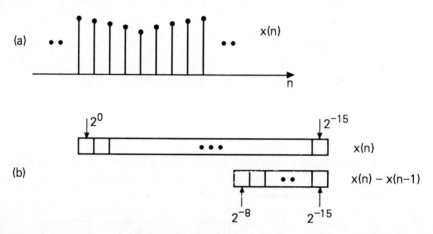

Figure 10.2-7 (a) A slowly varying signal $x(n)$, and (b) binary representations for $x(n)$ and its first difference.

Extension to Higher Order Differences

Can we extend this idea for higher differences? We can define the second difference in terms of the first difference $x_1(n)$ as $x_2(n) = x_1(n) - x_1(n-1)$, and so on. Thus, let $x_k(n), 1 \leq k \leq M-1$ denote the first $M-1$ differences of the signal $x(n)$. We wish to recover $x(n)$ from the M-fold decimated versions

$$y_k(n) = x_k(Mn), \quad 0 \leq k \leq M - 1. \qquad (10.2.11)$$

[For $k = 0$ we define $x_0(n) = x(n)$, i.e., the original sequence.] Once again, this problem can be handled using the filter-bank approach. For this note that the kth difference $x_k(n)$ has the z-transform

$$X_k(z) = (1 - z^{-1})X_{k-1}(z) = (1 - z^{-1})^k X(z), \qquad (10.2.12)$$

so that the kth difference operator is the transfer function

$$H_k(z) = (1 - z^{-1})^k. \qquad (10.2.13)$$

Thus the difference sampling scheme can be represented by a maximally decimated analysis bank (Fig. 10.2-8). If this is redrawn using the polyphase notation (Sec. 5.5), the $\mathbf{E}(z)$ matrix is a constant given by

$$\mathbf{E} = \begin{bmatrix} 1 & 0 & 0 & \cdots & 0 \\ 1 & -1 & 0 & \cdots & 0 \\ 1 & -2 & 1 & \cdots & 0 \\ \vdots & \vdots & \vdots & \ddots & \vdots \end{bmatrix} \quad (10.2.14)$$

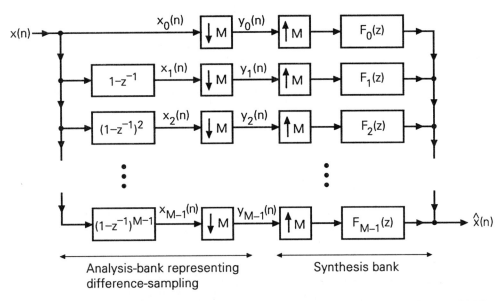

Figure 10.2-8 The difference sampling and reconstruction, posed as a QMF problem.

The rows of this $M \times M$ matrix are coefficients of $(1 - z^{-1})^k$, which are the binomial coefficients with alternating sign. It can be shown (see below) that this matrix is its own inverse so that we can take $\mathbf{R}(z) = \mathbf{E}$ in Fig. 5.5-3(b), for perfect reconstruction. In other words, if the synthesis filters are chosen as

$$[F_0(z) \quad F_1(z) \quad \cdots \quad F_{M-1}(z)] = [1 \quad z^{-1} \quad \cdots \quad z^{-(M-1)}] \mathbf{E} \quad (10.2.15)$$

the reconstructed signal is given by $\widehat{x}(n) = x(n - M + 1)$.

Proof that E is its own inverse. By definition, $\mathbf{E}$ has the following property:

$$\mathbf{E}\mathbf{v}(x) = \mathbf{v}(1 - x), \quad (10.2.16)$$

where $\mathbf{v}(x) \triangleq [1 \quad x \quad \cdots \quad x^{M-1}]^T$. From this we obtain

$$\mathbf{E}^2 \mathbf{v}(x) = \mathbf{v}(x). \quad (10.2.17)$$

Since this holds for all x, we conclude, in particular, that

$$\mathbf{E}^2 \underbrace{[\mathbf{v}(x_0) \quad \mathbf{v}(x_1) \ldots \quad \mathbf{v}(x_{M-1})]}_{\mathbf{V}} = [\mathbf{v}(x_0) \quad \mathbf{v}(x_1) \ldots \quad \mathbf{v}(x_{M-1})].$$

(10.2.18)

The matrix indicated as $\mathbf{V}$ is an $M \times M$ Vandermonde matrix and is non-singular if x_i are distinct (Appendix A). So it can be canceled in the above equation, yielding $\mathbf{E}^2 = \mathbf{I}$, i.e., $\mathbf{E}^{-1} = \mathbf{E}$. ▽▽▽

Example 10.2.2

Consider the case when $M = 3$. The difference signals are

$$x_1(n) = x(n) - x(n-1), \quad x_2(n) = x_1(n) - x_1(n-1). \tag{10.2.19}$$

The decimated signals from which we wish to recover $x(n)$ are $y_0(n) = x(3n), y_1(n) = x_1(3n), y_2(n) = x_2(3n)$. The reconstruction is done using the synthesis bank

$$[F_0(z) \quad F_1(z) \quad F_2(z)] = [z^{-2} \quad z^{-1} \quad 1] \begin{bmatrix} 1 & 0 & 0 \\ 1 & -1 & 0 \\ 1 & -2 & 1 \end{bmatrix}. \tag{10.2.20}$$

so that $F_0(z) = 1 + z^{-1} + z^{-2}, F_1(z) = -2 - z^{-1}$, and $F_2(z) = 1$.

Note that the synthesis filters are FIR, and that the highest required order is equal to $M - 1$. Compare this with the case of derivative sampling [e.g., Problem 5.13(d)] of continuous-time signals, where the synthesis filters are unrealizable. To be fair, it should be mentioned that $(1 - z^{-1})$ is only an approximate equivalent of the derivative operation; in fact if we perform bilinear transformation of $H(s) = s$ (which is a differentiator), we obtain $(1 - z^{-1})/(1 + z^{-1})$, which (is unstable and) represents the exact discrete-time equivalent of differentiation.

10.2.2 Nonuniform Sampling Theorems for Sequences

We now explain the concept of nonuniform "sampling" of sequences with an example. Consider a σ-BL sequence (i.e., a sequence $x(n)$ such that $X(e^{j\omega}) = 0$ for $\sigma \le |\omega| \le \pi$) with $\sigma = 2\pi/3$. Fig. 10.2-9(a) shows an example. In Sec. 4.1.1 we showed how we can decimate such a sequence by the noninteger quantity 3/2 to obtain the full band signal $Y(e^{j\omega})$ (Fig. 4.1-10). Fig. 4.1.-11 also shows the two signals $x(n)$ and $y(n)$ in the time domain. This can be considered to be uniform decimation by a factor of 3/2, since the samples $y(n)$ are still uniformly spaced in time.

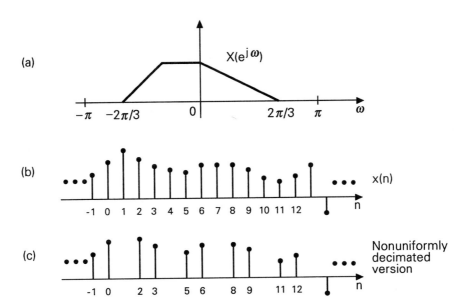

Figure 10.2-9 (a) Fourier transform of a bandlimited sequence, (b) the sequence $x(n)$, and (c) a nonuniformly decimated version.

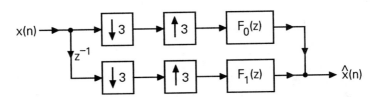

Figure 10.2-10 The nonuniform decimation and reconstruction, posed as a QMF problem.

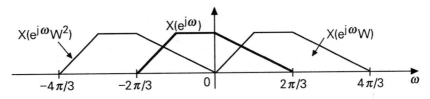

Figure 10.2-11 Demonstration of alias components of $X(e^{j\omega})$, which appear in nonuniform decimation.

But there is another (simpler) way to 'decimate' $x(n)$ by a factor 3/2, which can be described as follows: (i) divide the time axis into intervals of length three, (ii) retain the first two samples in each interval, and discard the third. This is demonstrated in Fig. 10.2-9. The resulting sequence is

Sec. 10.2 Unconventional sampling theorems 445

a *nonuniformly* decimated version of $x(n)$. Can we recover $x(n)$ from this version?

The answer is in the affirmative. We will show that this problem can be formulated as a multirate digital filter bank problem. The reconstruction of $x(n)$ from the nonuniformly decimated version is equivalent to finding a set of *synthesis filters* for perfect reconstruction. (The analysis filters are predetermined and are not under our control; see below). These synthesis filters are ideal (unrealizable) filters, but can actually be approximated using linear phase FIR filters, as we will demonstrate with practical designs.

Filter Bank Model for Nonuniform Sampling

Consider Fig. 10.2-10 which is a 3-channel maximally decimated filter bank, in which only two of the analysis filters are nonzero. More precisely we have $H_0(z) = 1, H_1(z) = z^{-1}$, and $H_2(z) = 0$. The analysis bank can be considered to be a nonuniform decimator, retaining only the samples indicated in Fig. 10-2-9(c). Our aim is to find synthesis filters such that $\widehat{x}(n) = x(n)$, under the assumption that $x(n)$ is bandlimited to $|\omega| < 2\pi/3$. (Without the bandlimited constraint we cannot do this because we have only two-thirds of the original number of samples per unit time.)

Solving for Synthesis Filters

First recall that the most general equations for perfect reconstruction are given by the AC matrix formulation (Sec. 5.4.3). From this we obtain

$$H_0(z)F_0(z) + H_1(z)F_1(z) = 3,$$
$$H_0(zW)F_0(z) + H_1(zW)F_1(z) = 0 \quad \text{[to eliminate } X(zW)\text{]}, \quad (10.2.21)$$
$$H_0(zW^2)F_0(z) + H_1(zW^2)F_1(z) = 0 \quad \text{[to eliminate } X(zW^2)\text{]},$$

with $W = e^{-j2\pi/3}$. Substituting $H_0(z) = 1, H_1(z) = z^{-1}$, and $z = e^{j\omega}$, this gives rise to

$$\begin{bmatrix} 1 & 1 \\ 1 & W^{-1} \\ 1 & W^{-2} \end{bmatrix} \begin{bmatrix} F_0(e^{j\omega}) \\ e^{-j\omega}F_1(e^{j\omega}) \end{bmatrix} = \begin{bmatrix} 3 \\ 0 \\ 0 \end{bmatrix}. \quad (10.2.22)$$

In general we cannot solve for the two filters $F_0(z), F_1(z)$ to satisfy the *three* conditions (10.2.22). We can, however, make further progress by using the bandlimited property of $x(n)$. First, we constrain $F_0(z)$ and $F_1(z)$ such that

$$F_0(e^{j\omega}) = 0, \quad F_1(e^{j\omega}) = 0, \quad \frac{2\pi}{3} \leq |\omega| \leq \pi. \quad (10.2.23)$$

(We are discussing only ideal filters for the moment). This eliminates any alias components which occupy the region outside the band of $X(e^{j\omega})$. It only remains to cancel aliasing in the region $|\omega| < 2\pi/3$.

Figure 10.2-11 demonstrates typical plots of $X(e^{j\omega}W^k)$, for $k = 0, 1$ and 2. In the frequency region $0 \leq \omega < 4\pi/3$, the quantity $X(e^{j\omega}W^2)$ is zero, so that the third equation in (10.2.21) need not be satisfied. So we have to choose $F_0(z)$ and $F_1(z)$ such that

$$\begin{bmatrix} 1 & 1 \\ 1 & W^{-1} \end{bmatrix} \begin{bmatrix} F_0(e^{j\omega}) \\ e^{-j\omega}F_1(e^{j\omega}) \end{bmatrix} = \begin{bmatrix} 3 \\ 0 \end{bmatrix}, \qquad (10.2.24)$$

for $0 \le \omega < 2\pi/3$. Similarly the alias component $X(e^{j\omega}W)$ is zero in the region $-4\pi/3 < \omega \le 0$ so that the middle equation in (10.2.21) need not be considered in this region. So $F_0(z)$ and $F_1(z)$ have to satisfy

$$\begin{bmatrix} 1 & 1 \\ 1 & W^{-2} \end{bmatrix} \begin{bmatrix} F_0(e^{j\omega}) \\ e^{-j\omega}F_1(e^{j\omega}) \end{bmatrix} = \begin{bmatrix} 3 \\ 0 \end{bmatrix}, \qquad (10.2.25)$$

for $-2\pi/3 < \omega \le 0$. Summarizing, we choose $F_0(z)$ and $F_1(z)$ to satisfy the two equations (10.2.24) if $0 \le \omega < 2\pi/3$, and the two equations (10.2.25) if $-2\pi/3 < \omega \le 0$. For ω outside either of these regions, we set $F_0(e^{j\omega}) = 0$, and $F_1(e^{j\omega}) = 0$ as stated earlier.

After solving the above two sets of equations we arrive at the following results:

$$F_0(e^{j\omega}) = \begin{cases} e^{-j\omega}(1 - c + js), & 0 \le \omega < 2\pi/3 \\ e^{-j\omega}(1 - c - js), & -2\pi/3 < \omega \le 0 \\ 0 & \text{otherwise,} \end{cases} \qquad (10.2.26)$$

and

$$F_1(e^{j\omega}) = \begin{cases} (1 - c - js), & 0 \le \omega < 2\pi/3 \\ (1 - c + js), & -2\pi/3 < \omega \le 0 \\ 0 & \text{otherwise.} \end{cases} \qquad (10.2.27)$$

Here c and s are defined as $c = \cos(2\pi/3)$, $s = -\sin(2\pi/3)$. Notice that the responses $F_0(e^{j\omega})$ and $F_1(e^{j\omega})$ are piecewise constants. Essentially, the frequency axis has been divided into three regions of equal widths (Fig. 10.2-12), and $F_k(e^{j\omega})$ takes on a fixed (complex) value in each of these regions. We say that $F_k(z)$ is a *multilevel* filter (Sec. 4.6.5.)

Implementing the Multilevel Synthesis Filters

The above solutions can be expressed neatly in terms of an ideal lowpass filter $G_L(e^{j\omega})$ and an ideal Hilbert transformer $G_H(e^{j\omega})$ (defined below). The ideal lowpass filter is

$$G_L(e^{j\omega}) = \begin{cases} 1 & |\omega| < 2\pi/3 \\ 0 & \text{otherwise,} \end{cases} \qquad (10.2.28)$$

and the ideal Hilbert transformer [Rabiner and Gold, 1975] is

$$G_H(e^{j\omega}) = \begin{cases} j & 0 < \omega < \pi \\ -j & -\pi < \omega < 0 \end{cases} \qquad (10.2.29)$$

(see Fig. 10.2-13). It is then clear that the above synthesis filters can be expressed as

$$F_0(z) = z^{-1}\Big(1 - c + sG_H(z)\Big)G_L(z),$$
$$F_1(z) = \Big(1 - c - sG_H(z)\Big)G_L(z). \tag{10.2.30}$$

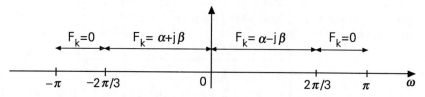

Figure 10.2-12 The interval $-\pi \leq \omega \leq \pi$ is divided into three equal regions. $F_k(e^{j\omega})$ is constant in each region. Also it has conjugate symmetry with respect to zero-frequency.

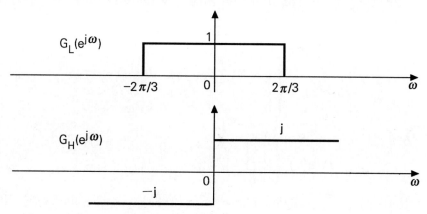

Figure 10.2-13 Definition of the ideal filters $G_L(e^{j\omega})$ and $G_H(e^{j\omega})$.

In practice, we can approximate the zero-phase filters $G_L(z)$ and $G_H(z)$ with real coefficient linear phase FIR filters $\widehat{G}_L(z)$ and $\widehat{G}_H(z)$ by using the McClellan-Parks algorithm as elaborated in Rabiner and Gold [1975]. Assuming that $\widehat{G}_H(z)$ has order $2K$, we have $\widehat{G}_H(e^{j\omega}) \approx e^{-j\omega K} \times G_H(e^{j\omega})$. So we can implement the synthesis bank as in Fig. 10.2-14. The extra delay z^{-K} in the top branch compensates for the group delay due to $\widehat{G}_H(z)$. The synthesis filters in this practical structure are

$$F_0(z) = z^{-1}\Big((1-c)z^{-K} + s\widehat{G}_H(z)\Big)\widehat{G}_L(z),$$
$$F_1(z) = \Big((1-c)z^{-K} - s\widehat{G}_H(z)\Big)\widehat{G}_L(z). \tag{10.2.31}$$

Reconstruction Error Created by Filter Approximation

The practical approximations (10.2.31) to the ideal solution evidently result in reconstruction error, so that the filter bank in Fig. 10.2-14 is not a perfect reconstruction system. For any maximally decimated filter bank, we know that

$$\widehat{X}(z) = T(z)X(z) + \text{alias terms.} \qquad (10.2.32)$$

If aliasing terms have been sufficiently attenuated we have $\widehat{X}(z) \approx T(z)X(z)$, and the distortion function $T(z)$ reduces to

$$T(z) = (1-c)z^{-(K+1)}\widehat{G}_L(z). \qquad (10.2.33)$$

This distortion is free from $\widehat{G}_H(z)$! This shows that the approximation error involved in the design of the Hilbert transformer does not affect $T(z)$; it affects only the extent to which alias-terms have been canceled. The lowpass filter $\widehat{G}_L(z)$ completely determines the amplitude and phase distortions in the reconstructed signal $\widehat{x}(n)$.

Summarizing, the Hilbert transformer $\widehat{G}_H(z)$ controls the extent to which aliasing has been canceled in the range $|\omega| < 2\pi/3$, whereas the lowpass filter $\widehat{G}_L(z)$ suppresses aliasing components in the range $|\omega| > 2\pi/3$. The passband ripple of $\widehat{G}_L(z)$ completely determines the amplitude distortion in $\widehat{x}(n)$. There is no phase distortion if $\widehat{G}_L(z)$ (hence $T(z)$) has linear phase.

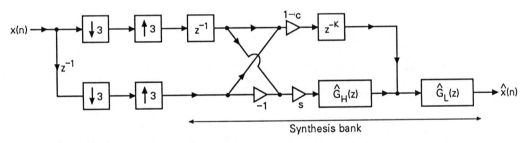

Figure 10.2-14 The complete analysis/synthesis system representing nonuniform decimation and reconstruction.

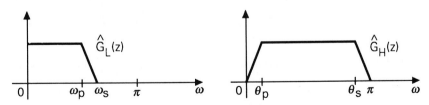

Figure 10.2-15 Defining bandedges for the real coefficient linear phase filters $\widehat{G}_L(z)$ and $\widehat{G}_H(z)$.

Design Example 10.2.1

We now demonstrate these ideas with an example. Figure 10.2-15 shows the definitions of the bandedges for $\widehat{G}_L(z)$ and $\widehat{G}_H(z)$, both of which are real coefficient linear phase filters. In Fig. 10.2-16(a) we show the magnitude $|X(e^{j\omega})|$ for our test sequence $x(n)$, which is a real finite length sequence of length 71. The plot shows that $x(n)$ is (approximately) bandlimited to $|\omega| < 2\pi/3$. We will reconstruct $x(n)$ from the nonuniform subset of samples indicated in Fig. 10.2-9(c), by using the synthesis bank in Fig. 10.2-14. $\widehat{G}_L(z)$ is taken to be of order 72, and has following features: $\omega_p = 0.58\pi, \omega_S = 0.70\pi$, and stopband attenuation > 55dB. The Hilbert transformer $\widehat{G}_H(z)$ has order 50, with $\theta_p = 0.04\pi$ and $\theta_S = 0.96\pi$. The magnitude responses of $\widehat{G}_L(z)$ and $\widehat{G}_H(z)$ are shown in the figure. With these filters used in the structure of Fig. 10.2-14, the quantity $|\widehat{X}(e^{j\omega})|$ for the reconstructed signal is shown in Fig. 10.2-16(c), which is in good agreement with $|X(e^{j\omega})|$. Since $T(z)$ has linear phase, there is no phase distortion, so we conclude that $\widehat{x}(n)$ is indeed a good approximation of $x(n)$.

As explained above, the Hilbert transformer serves to eliminate aliasing in the signal band. But a practical Hilbert transformer has to have a transition bandwidth around zero-frequency (because $G_H(1) = 0$; see Rabiner and Gold [1975]). If this bandwidth around zero-frequency is large, it results in poor reconstruction.

Efficieny of Reconstruction

The above reconstruction scheme is not the most efficient technique for the purpose, and is meant only to demonstrate the fundamental principles underlying the recovery of a signal from nonuniform samples. For more efficient (polyphase) techniques, and for further detailed comparison between various techniques, the reader is referred to Vaidyanathan and Liu [1990].

Generalizations

Several generalizations of the above approach are available. Thus consider the system of Fig. 10.2-17. Here we have an M-channel filter bank in which only the first L analysis filters are nonzero. And these nonzero filters are delay elements of the form $z^{-n_k}, 0 \leq k \leq L-1$, with

$$0 \leq n_0 < n_1 \ldots < n_{L-1} \leq M - 1. \qquad (10.2.34)$$

The effect of the analysis bank is merely to retain the subset of samples

$$x(nM - n_k), \quad 0 \leq k \leq L - 1. \qquad (10.2.35)$$

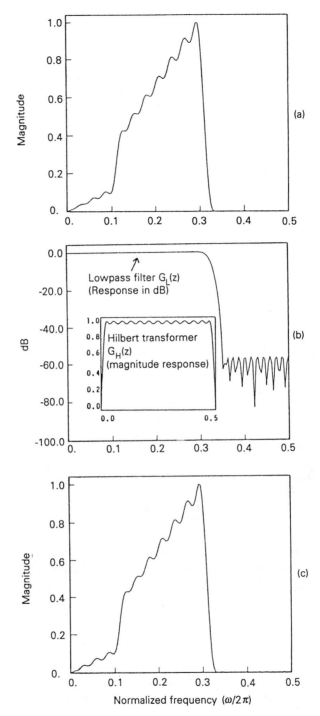

Figure 10.2-16 Design example 10.2.1. (a) $|X(e^{j\omega})|$, (b) filter responses, and (c) $|\widehat{X}(e^{j\omega})|$.

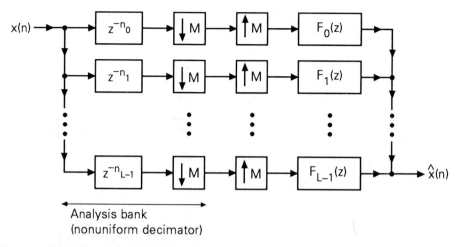

Figure 10.2-17 Generalization of nonuniform decimation and reconstruction.

This means that the time-index is divided into intervals of length M, and L samples are retained in each interval. (The time indices for these L samples are equal to $-n_k$ modulo M.) The analysis bank, therefore, is a nonuniform decimator. If the signal $x(n)$ is bandlimited to the region $|\omega| < L\pi/M$, we can find the L synthesis filters $F_k(z)$ such that there is perfect recovery of $x(n)$ from the nonuniformly decimated version! These synthesis filters, once again, turn out to be multilevel filters. (More precisely, if we imagine the frequency region $0 \le \omega < 2\pi$ to be divided into M contiguous intervals, then $F_k(e^{j\omega})$ is a (complex) constant in each interval.) So these are ideal filters and must be approximated by practical designs.

A further generalization arises when we consider multiband signals, that is, signals that are not bandlimited, but limited to a union of bands in the frequency domain. (For example, $X(e^{j\omega})$ could be zero everywhere except in $0 \le \omega \le 0.1\pi$ and $0.4\pi \le \omega \le 1.3\pi$.) The nonuniform decimation/reconstruction process works in this case as well. Many of these ideas also generalize to the case of two dimensional signals [Vaidyanathan and Liu, 1990].

Relation to Sampling of a Continuous-Time Signal

The signal $x(n)$, which is bandlimited to $L\pi/M$, can be considered to be the oversampled version of a continuous-time signal $x_a(t)$. Here $x_a(t)$ is sampled at the rate $2\pi/T$, which is M/L times the Nyquist rate Θ. The set of nonuniformly decimated samples (10.2.35) can be considered to be a set of nonuniformly sampled values of $x_a(t)$, (which is a subset of the original oversampled values). The nonuniformity is such that the *average* number of samples per unit time is reduced by the factor M/L, so that it becomes equal to the Nyquist rate Θ.

Evidently the nonuniform pattern repeats periodically after every L

samples (see Fig. 10.2-9(c)). In this sense, the above results address only a special case of nonuniform sampling, namely the case of *recurring* or *periodic* nonuniformity. It should also be noted that, in our special case, the locations of the nonuniformly spaced samples are not permitted to be arbitrary; if $x_a(t_1)$ and $x_a(t_2)$ are two samples in the nonuniformly spaced system, then t_1/t_2 is required to be rational.

PROBLEMS

10.1. Consider the scalar system $H(z) = 1/(1 - az^{-1})$. Write down the general form of the blocked version for arbitrary block size M.

10.2. Verify that the 2×2 matrix in (10.1.10) is paraunitary.

10.3. Let $\mathbf{P}_0(z)$ and $\mathbf{P}_1(z)$ be $M \times M$ pseudocirculants. Prove that the product $\mathbf{P}_0(z)\mathbf{P}_1(z)$ is pseudocirculant, and that $\mathbf{P}_0(z)\mathbf{P}_1(z) = \mathbf{P}_1(z)\mathbf{P}_0(z)$. (*Hint.* A pseudocirculant is related to the blocked version of an LTI system.)

10.4. Let $E_\ell(z), 0 \le \ell \le M-1$ be the Type 1 polyphase components of $H(z)$. Derive an expression for the polyphase components of $z^{-k}H(z)$. Assume $0 \le k \le M-1$ for simplicity.

10.5. Show that the polyphase matrix $\mathbf{E}(z)$ for the analysis bank of Fig. 10.1-6 has the form (10.1.23) for $M = 3$, where $G_{n,\ell}(z)$ are the polyphase components defined in (10.1.22). Also, obtain the form of $\mathbf{E}(z)$ for arbitrary M.

10.6. A sequence $x(n)$ is said to be bandlimited if $X(e^{j\omega}) = 0$ for $\sigma \le |\omega| \le \pi$ for some $\sigma < \pi$. Show that $x(n)$ cannot be bandlimited if it is causal (unless $x(n) = 0$ for all n). *Hint.* Assume $x(n)$ is causal and bandlimited. Construct a sequence $y(n)$ from $x(n)$ such that it is also causal, but bandlimited to a *narrower* band. Keep repeating till ...

Another hint. First try with the assumption $\sigma < \pi/2$ if you wish!

10.7. Consider the following system which is the continuous-time analog of the M-channel filter bank.

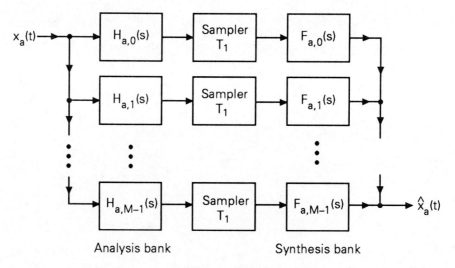

Figure P10-7

The continuous-time signal $x_a(t)$, which we assume to be σ-BL, is passed through M analog filters $H_{a,k}(s)$, and then sampled at the rate $2\pi/T_1$. (The sampler is defined precisely as in Problem 5.13.) The rate $2\pi/T_1$ is equal to

Θ/M, where $\Theta \triangleq 2\sigma$ is the Nyquist rate. Each of the sampled signals is in general subject to aliasing since it is not necessarily bandlimited. We assume $F_{a,k}(j\Omega) = 0$ for $|\Omega| \geq \sigma$ so that aliasing terms which fall outside the band of $x_a(t)$ are automatically eliminated.

a) Express $\widehat{X}_a(j\Omega)$ in terms of $X_a(j\Omega)$ and the filters in the structure.

b) Suppose we are given the set of M analysis filters $H_{a,k}(j\Omega)$, and wish to find the M synthesis filters for perfect reconstruction. Show that the frequency region $-\sigma \leq \Omega < \sigma$ can be divided into M intervals such that in each interval we have to solve M equations for the M unknowns $F_{a,k}(j\Omega)$.

c) Consider the special case where $H_{a,k}(s) = s^k$. This means that the output of $H_{a,k}(s)$ is the kth derivative of $x_a(t)$. Show that the $M \times M$ matrix which should (in principle) be inverted to obtain the M synthesis filters for perfect reconstruction is nonsingular. This proves the existence (but does not assert realizability!) of these synthesis filters, and gives a proof of the generalized derivative sampling theorem. (*Hint.* Review Vandermonde matrices from Appendix A).

10.8. Consider a sequence $x(n)$ with $|X(e^{j\omega})| = 0$ for $3\pi/4 \leq |\omega| \leq \pi$. It is clear that we can decimate it by $4/3$ without losing information. Suppose we wish to perform *nonuniform* decimation as in Sec. 10.2.2. We can do so by using the analysis bank shown in Fig. P10-8(a).

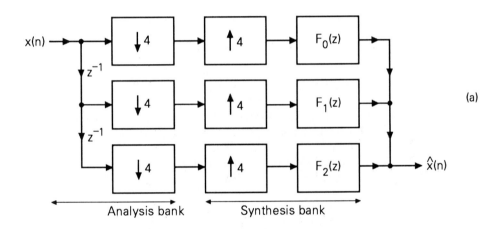

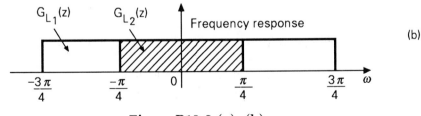

Figure P10-8 (a), (b)

We wish to reconstruct $x(n)$ by using the synthesis bank system shown. Assume

$F_k(e^{j\omega}) = 0$ for $3\pi/4 \le |\omega| \le \pi$ so that alias components outside the signal band are automatically removed.

a) Show how $F_0(z), F_1(z)$ and $F_2(z)$ can be found, in order to provide perfect reconstruction. (Divide the frequency region into appropriate number of intervals, and solve 3×3 equations in each interval. *Hint.* even though there are several sets of 3×3 equations, you need to invert *only one* constant matrix, if you do things right). Show that $F_k(e^{j\omega})$ can be expressed as

$$e^{-jk\omega} F_k(e^{j\omega}) = \begin{cases} a_k & \frac{-3\pi}{4} < \omega \le \frac{-\pi}{4} \\ b_k & \frac{-\pi}{4} < \omega \le \frac{\pi}{4} \\ c_k & \frac{\pi}{4} \le \omega < \frac{3\pi}{4} \end{cases} \qquad (P10.8a)$$

where a_k, b_k, c_k are (possibly complex) constants. Identify the constants a_k, b_k, c_k for $k = 0, 1, 2$.

b) Show that the filters $F_k(z)$ have real-valued impulse response.

c) Show that these filters can be expressed as

$$F_0(z) = \Big(1 + G_{L_2}(z) - (1 - G_{L_2}(z))G_H(z)\Big)G_{L_1}(z),$$
$$F_1(z) = 2z\Big(1 - G_{L_2}(z)\Big)G_{L_1}(z), \qquad (P10.8b)$$
$$F_2(z) = z^2\Big(1 + G_{L_2}(z) + (1 - G_{L_2}(z))G_H(z)\Big)G_{L_1}(z),$$

where $G_{L_1}(z), G_{L_2}(z)$ are ideal lowpass filters with response shown in Fig. P10-8(b), and $G_H(z)$ is an ideal Hilbert transformer. Draw a structure for the synthesis bank using the filters $G_{L_1}(z), G_{L_2}(z)$ and $G_H(z)$ as building blocks. In practice we would like to replace $G_{L_1}(z), G_{L_2}(z)$ and $G_H(z)$ with causal FIR linear phase approximations $\widehat{G}_{L_1}(z), \widehat{G}_{L_2}(z)$ and $\widehat{G}_H(z)$ of orders N_{L_1}, N_{L_2} and N_H. Show how the expressions given in (P10.8b) should be modified to incorporate the group delays of these filters. (Assume filter orders are even where necessary.)

11

The wavelet transform and its relation to multirate filter banks

11.0 INTRODUCTION

In this chapter we study the wavelet transform, which has received a great deal of attention since the middle eighties. The mathematical aspects of wavelet transforms were introduced in Grossmann and Morlet [1984]. The topic has since been treated in considerable detail by several authors in the mathematics literature [Meyer, 1986], [Daubechies, 1988], [Mallat, 1989a,b], and [Strang, 1989], and a number of books have appeared [Coifman, et al., 1990], [Chui, 1991], [Daubechies, 1992]. The fundamental papers by Daubechies and by Mallat were influential in generating an unprecedented amount of activity in this area. Daubechies developed a systematic technique for generating finite-duration orthonormal wavelets, and also established the connection between continuous-time 'orthonormal wavelets' and the digital filter bank studied in Sec. 5.3.6 (FIR power symmetric filter bank). This result on finite-duration orthonormal wavelets triggered considerable interest in the mathematics as well as the signal processing communities.

Wavelet transforms are closely related to tree structured digital filter banks, and hence to multiresolution analysis described in Sec. 5.8. We know that tree structured filter banks give rise to nonuniform filter bandwidths (Fig. 5.8-4) and nonuniform decimation ratios in the subbands. These two nonuniformities can be considered to be the fundamental ingredients of the wavelet transform.

Even before the wavelet transform was formally introduced, such non uniform filter banks were already employed in the speech processing literature. See Nelson, et al. [1972], Schafer, et al. [1975], and pp. 301–303 of the text by Rabiner and Schafer [1978]. Also see McGee and Zhang [1990] for the design and use of such filter banks in music. Using nonuniform filter banks, one can exploit the decreasing resolution of the human ear at higher frequencies [Flanagan, 1972]. This nonuniform nature of the ear also explains the evolution of the musical scale. Figure 11.1-1 demonstrates this, by showing the locations of the notes c and g in the major diatonic scale, for several octaves. On a logarithmic scale, these would appear to be nearly equispaced. The notes in between c and g are not shown, but it is clear that they become sparser and sparser as the frequency increases.

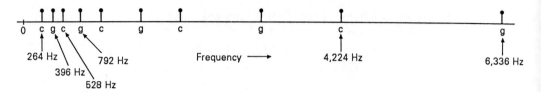

Figure 11.1-1 Pitch-frequencies corresponding to the keys 'c' and 'g' in a piano. These correspond to the major diatonic scale of western music. The spacing is very nonuniform, and will appear to be almost uniform on a logarithmic scale.

The literature on wavelet transforms is extensive, but most of it requires a level of mathematical preparation which is perhaps unsuitable for many signal-processing experts. In the signal processing literature, a number of authors have explored the relation between wavelets and multirate filter banks. Tutorial treatments can be found in the magazine articles by Rioul and Vetterli [1991], and by Hlawatsch and Boudreaux-Bartels [1992]. Further references in the signal processing literature are Evangelista [1989], Wornell [1990], Gopinath and Burrus [1991], Vaidyanathan [1991c], Soman and Vaidyanathan [1991 and 1992a,b], Tewfik and Kim [1992], Akansu and Liu [1991], Akansu, et al. [1992], Wornell and Oppenheim [1992], and Vetterli and Herley [1992]. In this chapter we will develop the basic ideas of wavelet transforms in a manner suitable for the signal processing person who understands the traditional Fourier transform, and has some familiarity with filter banks.

11.1 BACKGROUND AND OUTLINE

The conventional discrete-time Fourier transform pair is given by

$$X(e^{j\omega}) = \sum_{n=-\infty}^{\infty} x(n)e^{-j\omega n}, \quad (11.1.1)$$

$$x(n) = \frac{1}{2\pi} \int_0^{2\pi} X(e^{j\omega})e^{j\omega n} d\omega. \quad (11.1.2)$$

If $x(n)$ is a single frequency signal,[†] that is, $x(n) = e^{j\omega_0 n}$ then

$$X(e^{j\omega}) = 2\pi\delta_a(\omega - \omega_0), \quad 0 \leq \omega < 2\pi. \quad (11.1.3)$$

[†] A signal of the form $\cos(\omega_0 n + \theta)$ is sometimes referred to as a signal

We say that the transform is completely localized at ω_0. In contrast, the time domain plot of $e^{j\omega_0 n}$ is infinite in extent (in fact its magnitude is unity for all n). This is consistent with the *uncertainty principle* which says (heuristically) that if $x(n)$ has a 'wide' support, then $X(e^{j\omega})$ has 'narrow' support in $-\pi \le \omega < \pi$.

In Sec. 11.2.4 a more quantitative statement of the uncertainty principle will be presented. The principle is most easily demonstrated for the Fourier transform pair in Fig. 11.1-2. As N increases, the signal $x(n)$ becomes less localized, but the main lobe of the Fourier transform gets narrower. As N approaches infinity, the transform looks more and more like the impulse (Dirac delta) function.

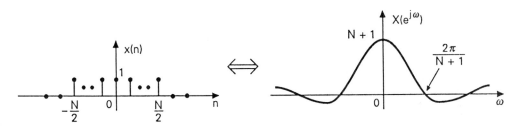

Figure 11.1-2 A Fourier transform pair, demonstrating the uncertainty principle.

The above localization property of the Fourier transform rejects the notion of "frequency that varies with time." But such a notion is often useful. For example, as the musician passes from a low to a high note, the 'frequency' (more accurately the 'pitch') is said to change in real time. According to Fourier transform theory this is meaningless because a single frequency is always associated with *infinite* time duration. If we apply Fourier analysis to the signal shown in Fig. 11.1-3(a) (where "frequency" makes an abrupt transition), we find that it is composed of an infinite number of frequencies. Another example is shown in Fig. 11.1-3(b) (a rising signal). In the regions $t < 0$ and $t > 1$ we naturally wish to think of this as a 'zero frequency signal', whereas in $0 < t < 1$ the signal has high frequency components. For signals of the above kind, it is desirable to find a time-frequency representation where the notion of 'frequency changing with time' can be formally accomodated.

The short-time Fourier transform (or time dependent Fourier transform), abbreviated as STFT, is a tool that fills this need [Gabor, 1946], [Rabiner and Schafer, 1978], [Portnoff, 1980], [Oppenheim and Schafer, 1989]. Here the signal $x(n)$ is multiplied with a window (typically of finite duration), and then the Fourier transform computed. The window is then shifted in uniform amounts, and the above computation repeated. We will see that

---

with frequency ω_0. However, this is a superposition of two single-frequency signals, viz., $e^{j\omega_0 n}$ and $e^{-j\omega_0 n}$.

the computation of the STFT is equivalent to the implementation of a filter bank where all the filters have the same bandwidth, and each filter is followed by a decimator. The duration of the window governs the time localization of the analysis, the bandwidths of the filters govern the frequency resolution, and the decimator governs the stepsize for window movement.

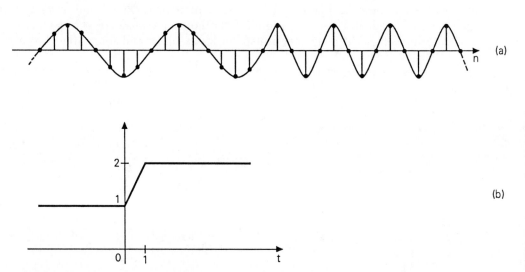

Figure 11.1-3 (a) A signal whose 'frequency' changes abruptly, and (b) a rising signal.

The wavelet transform, which is a more recent advance, generalizes the STFT by incorporating two novel features into its definition. First it permits nonuniform bandwidths (as in a tree structured filter bank system; see Fig. 5.8-4), so that the resolution is higher (i.e., bandwidth smaller) at lower frequencies. This makes the "fractional resolution" identical for all center-frequencies. Second, the nonuniform bandwidths automatically lead us to use different decimators for the different filter outputs. Such nonuniform systems are well-suited for the processing of sound signals, because of the decreasing resolution of the human ear at higher frequencies.

The traditional Fourier transform representation (2.1.21) can be regarded as an expansion of $x_a(t)$ in terms of the basis functions $e^{j\Omega t}$. (It will be easier to switch our discussion to continuous-time for a while.) The basis functions (which are functions of time) are parameterized by the frequency variable Ω. We will see that the short-time Fourier transform is a representation of a signal in terms of a different class of basis functions, indexed by *two variables*, namely *time* as well as *frequency*.

The wavelet transform is a further modification of this, which allows nonuniform frequency resolution. We will see that, in the wavelet transformation, the basis functions have a very unusual property, namely, all the basis functions are generated by dilation and shift of a single function $\psi(t)$ (called a mother wavelet). Thus, instead of representing $x_a(t)$ as a linear combination of the functions $e^{j\Omega t}$ (as does the Fourier transform representation), the wavelet transform attempts to represent $x_a(t)$ as a linear combination of

$$\psi_{k\ell}(t) \triangleq 2^{-k/2}\psi(2^{-k}t - \ell) \quad \text{(wavelet basis functions)}$$

where k and ℓ are integers. (This is only an example.) That is,

$$x_a(t) = \sum_k \sum_\ell X_{WT}(k,\ell) \underbrace{2^{-k/2}\psi(2^{-k}t - \ell)}_{\text{basis } \psi_{k\ell}(t)} \quad \text{(wavelet expansion)}.$$

This *double* summation should be compared with the traditional Fourier transform representation

$$x_a(t) = \frac{1}{2\pi}\int_{-\infty}^{\infty} X_a(j\Omega)e^{j\Omega t}d\Omega,$$

which is a *single* integral, in the frequency variabe Ω. In the wavelet expansion, k plays the role of 'frequency' and ℓ plays the role of 'time'. The index variables (k, ℓ) are integers, unlike 'Ω' which is a continuous variable in the Fourier expansion.

In a manner analogous to the orthonormality of the basis functions $\{e^{j\Omega t}\}$ in the Fourier representation, we will later discuss orthonormality of the basis functions $\{\psi_{k\ell}(t)\}$. As a preview example, the functions in Fig. 11.5-7 represent an orthonormal basis (the Haar basis) for the set of finite energy functions.

Time-frequency representations. The tools we develop in this chapter come under the general class of "time-frequency representations." In these representations, the signal is represented in a domain which is a hybrid between time and frequency, for example, a time-localized Fourier transform with the center of localization shifted uniformly. We will see later that the double index "$k\ell$" in $\psi_{k\ell}(t)$ above represents time-frequency.

The use of time-frequency representations reflects the philosophy that some aspects of a signal are most conveniently represented in the time domain, whereas there are certain other aspects which are best represented in terms of frequency. Consider, for example, the two signals shown in Fig. 11.1-4. Both of these have an underlying periodic waveform $p(t)$, for example a musical note of fixed pitch. The envelopes of the signals [$e_1(t)$ and $e_2(t)$] are, however, different. The first signal has a rapid rise followed by a steady state, and then a slow decay. The second envelope has a totally different behavior. (The envelope for a given note is typically determined by the source, for example, the musical instrument chosen). While it is useful to regard the pitch of the note in terms of impulses in the Fourier transform domain, it is more convenient to describe the envelope in the time domain, as directly preceived by the ear.

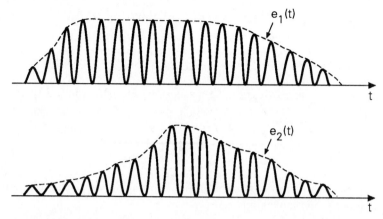

Figure 11.1-4 Two signal pulses, with nearly the same pitch, but different envelopes.

Chapter Outline

In view of the philosophy outlined above, we will first develop the short-time Fourier transform (STFT) and then the wavelet transform. In Sec. 11.2 we introduce the discrete time STFT and inverse STFT, and then develop the filter bank interpretation. In Sec. 11.2.4 we will briefly summarize the main features of the continuous-time STFT. We will show in Sec. 11.3.1 that (continuous-time) wavelet transforms can be obtained by performing two simple modifications to the STFT. This will also place in evidence the relation to filter banks.

We then introduce, in Sec. 11.3.3, the discrete-time wavelet transform (which *cannot* be obtained merely by sampling the continuous time version) and establish the connection to tree-structured, maximally decimated filter banks. The fundamental relation between paraunitary filter banks (Chap. 6) and the so-called "orthonormal wavelet decomposition" will be developed in Sec. 11.4.

In Sec. 11.5 we return to continuous time wavelets, and establish the relation between *continuous* time orthonormal wavelets and *discrete* time paraunitary filter banks. This section will show how to systematically generate finite duration orthonormal basis functions $\psi_{k\ell}(t)$ for the representation of finite energy functions. Finally, in Sec. 11.5.4 we will describe a technique for generating orthonormal wavelets with deeper properties, such as *regularity*.

We will see that the continuous time wavelet functions satisfy some very interesting mathematical properties (e.g., self-similarity) which are not relevant in the discrete time case. For such reasons, the discrete time counterpart is sometimes not regarded as "wavelets" but merely as "filter banks." In any case, both the continuous- and discrete-time versions share several common properties (sufficient for many applications), when viewed in terms of the frequency domain quantities and subsampling operations.

Notice that our presentation in this chapter has the following order: (a) discrete-time STFT, (b) continuous-time STFT, (c) continuous-time wavelets, (d) discrete-time wavelets, and (e) connection between continuous time wavelets and discrete-time filter banks. We have opted for this ordering because there appears to be a natural "flow" between two successive items on this list. Historically, however, items (b) and (a) should have been switched, as the continuous-time STFT has been known since the middle forties [Gabor, 1946].

11.2 THE SHORT-TIME FOURIER TRANSFORM

In Sec. 4.1.2 we introduced the DFT filter bank (Fig. 4.1-16). This system merely computes the DFT of M successive samples of the input, and then repeats the operation for the next set of M samples. We saw that this could be interpreted in terms of a window that slides past the data. This could also be interpreted in terms of a bank of bandpass filters with 13 dB stopband attenuation. We also generalized this to the case of a uniform DFT filter bank (Fig. 4.3-5), with the advantage that the filters could now offer sharper cutoff and higher attenuation. These systems are essentially implementations of the so called short-time Fourier transform (STFT) to be described now.

In short-time Fourier transformation, a signal $x(n)$ is multiplied with a window $v(n)$ (typically finite in duration). See Fig. 11.2-1. The Fourier transform of the product $x(n)v(n)$ is computed, and then the window $v(n)$ is shifted in time, and the Fourier transform of the product computed again. This operation results in a separate Fourier transformation for each location m of the center of the window. In other words, we obtain a function $X_{STFT}(e^{j\omega}, m)$ of two variables ω and m. The frequency variable ω is continuous, and takes the usual range $-\pi \leq \omega < \pi$. The shift-variable m is typically an integer multiple of some fixed integer K. Figure 11.2-2 shows a two dimensional plot which represents the idea. These are often called *spectrograms*. See Oppenheim and Schafer [1989] for real-time examples of spectrograms. (Also notice the cover picture in that reference!)

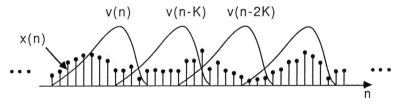

Figure 11.2-1 Pertaining to the short-time Fourier transform.

Essentially, for any fixed m, the window captures the features of the signal $x(n)$ in the local region around m. The window therefore helps to localize the time domain data, before obtaining the frequency domain information. From the above discussions it is clear that the short time Fourier transform can be written mathematically as

$$X_{STFT}(e^{j\omega}, m) = \sum_{n=-\infty}^{\infty} x(n)v(n-m)e^{-j\omega n}. \qquad (11.2.1)$$

If we set $v(n) = 1$ for all n, this reduces to the traditional Fourier transform for any choice of m.

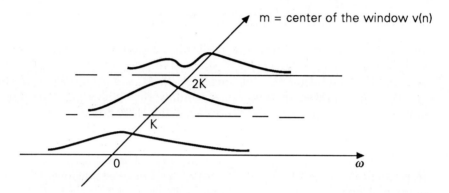

Figure 11.2-2 Demonstrating $X_{STFT}(e^{j\omega}, m)$ for $m = 0, K, 2K \ldots$

Existence of the STFT. The traditional Fourier transform (11.1.1) exists only if the signal $x(n)$ satisfies some subtle mathematical conditions [Oppenheim, Willsky and Young, 1983]. In a practical STFT system, however, $v(n)$ has finite duration, so that the above summation always converges (i.e., the STFT exists). Thus many of the subtle mathematical questions which are raised in connection with the Fourier transform do not arise in the STFT regime.

11.2.1 Interpretation Using Bandpass Filters

For a variety of reasons, it is convenient to interpret the STFT using the notion of filter banks. In addition to enhancing insight, this also gives a practical scheme for implementation. Furthermore this interpretation helps us to generalize the STFT to obtain more flexibility (Sec. 11.2.3). Finally, the theory of perfect reconstruction filter banks can be used to obtain practical "inversion" formulas for STFT.

Traditional Fourier Transform as a Bank of Filters

We will begin by presenting a filter bank interpretation for the traditional Fourier transform (11.1.1). The evaluation of $X(e^{j\omega})$ at a fixed frequency ω_0 can be pictorially represented as in Fig. 11.2-3(a). This is a cascade of two systems.

1. *The modulator* $e^{-j\omega_0 n}$. This performs a frequency-shift. More specifically, it shifts the Fourier transform towards the left by the amount ω_0, so that the zero-frequency value of $S(e^{j\omega})$ is equal to $X(e^{j\omega_0})$ (Fig. 11.2-3(b)).
2. *The LTI system* $H(e^{j\omega})$. This has impulse response $h(n) = 1$ for all n. This system is evidently unstable (Sec. 2.1.2), but let us ignore these fine details for the moment. Its "frequency response" is

$$H(e^{j\omega}) = \sum_{n=-\infty}^{\infty} h(n)e^{-j\omega n} = 2\pi \delta_a(\omega), \quad -\pi \leq \omega < \pi,$$

where $\delta_a(.)$ is the Dirac delta function which was defined at the beginning of Sec. 2.1.† Thus, $H(e^{j\omega})$ is an ideal "lowpass" filter, which "passes" only the zero-frequency signal. Every other frequency is completely suppressed. This filter can be regarded as the limit, as $\Delta\omega \to 0$, of an ideal filter with response $2\pi/\Delta\omega$ for $|\omega| \leq \Delta\omega/2$ and zero elsewhere.

The output $y(n)$ of the system is therefore a zero-frequency signal with $Y(e^{j\omega}) = 2\pi X(e^{j\omega_0})\delta_a(\omega)$ for $-\pi \leq \omega < \pi$, that is,

$$y(n) = X(e^{j\omega_0}), \quad \text{for all } n.$$

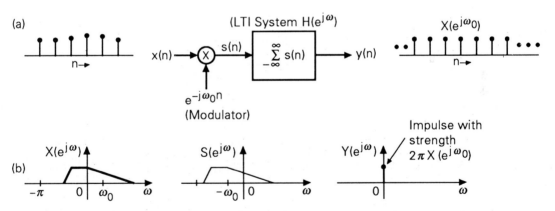

Figure 11.2-3 (a) Representation of Fourier transformation in terms of linear systems. (b) Frequency domain quantitites, sketched for an arbitrary example.

Summarizing, the process of evaluating $X(e^{j\omega_0})$ can be looked upon as a linear system, which takes the input $x(n)$ and produces a *constant* output

---

† Of course, the Fourier transform $H(e^{j\omega})$ repeats periodically with period 2π, but we will not explicity show it in the formulas.

$y(n)$ whose value is equal to $X(e^{j\omega_0})$ for all time n. Thus, any sample of $y(n)$ can be taken to be the value of $X(e^{j\omega_0})$. The Fourier transform operator which evaluates $X(e^{j\omega})$ for *all* ω is, therefore, a *bank* of modulators followed by filters. This system has an uncountably infinite number of channels.

The STFT as a Bank of Filters

From its definition it is clear that the STFT can be represented as in Fig. 11.2-4(a). In this figure, ω_0 and m are constants. So $y(n)$ is constant for all n, with $y(n) = X_{STFT}(e^{j\omega_0}, m)$. To gain further insight, let us rearrange the definition to obtain

$$X_{STFT}(e^{j\omega}, m) = e^{-j\omega m} \sum_{n=-\infty}^{\infty} x(n)v(n-m)e^{j\omega(m-n)}. \qquad (11.2.2)$$

Figure 11.2-4(b) shows this interpretation, where the indices k and m have been replaced with n, to be consistent with standard notations (Sec. 2.1.2). This is a linear system with two parts. The first is an LTI filter with impulse response $v(-n)e^{j\omega_0 n}$. This is followed by the modulator $e^{-j\omega_0 n}$ (linear time varying device). The output $y_0(n)$ of this system is now a function of n [unlike in Fig. 11.2-4(a)]. For any specific value of n, say $n = m$, this output represents the Fourier transform of $x(.)$ "in the neighbourhood of m," because m represents the location of the window $v(k)$ in the time domain. For the special case where $v(k) = 1$ for all k, this output becomes a constant [equal to the traditional Fourier transform $X(e^{j\omega_0})$] for all n.

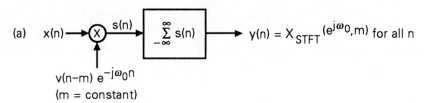

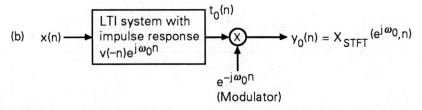

Figure 11.2-4 (a) The STFT represented in terms of a linear system and (b) a rearrangement.

In most applications, $v(n)$ has a lowpass transform $V(e^{j\omega})$. So $v(-n)$ has the lowpass transform $V(e^{-j\omega})$. The modulated version $v(-n)e^{j\omega_0 n}$ represents a bandpass filter $V(e^{-j(\omega-\omega_0)})$. See Fig. 11.2-5. The output sequence

$t_0(n)$ in Fig. 11.2-4(b) is, therefore, the output of a bandpass filter, whose passband is centered around ω_0. The effect of the modulator $e^{-j\omega_0 n}$ is merely to re-center this around zero frequency.

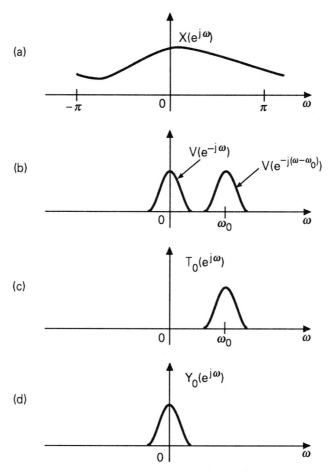

Figure 11.2-5 Demonstration of how the STFT works. (a) $X(e^{j\omega})$, (b) the window-transform and its shifted version, (c) output of LTI filter and (d) traditional Fourier transform of $X_{STFT}(e^{j\omega_0}, n)$.

For every frequency ω_0 the STFT performs the filtering operation of Fig. 11.2-4(b) to produce an output sequence $X_{STFT}(e^{j\omega_0}, n)$. So the STFT can be looked upon as a *filter bank*, with infinite number of filters (one 'per frequency'). In practice, we are interested in computing the Fourier transform at a discrete set of frequencies

$$0 \leq \omega_0 < \omega_1 < \ldots < \omega_{M-1} < 2\pi. \qquad (11.2.3)$$

In this case the STFT reduces to a filter bank with M bandpass filters $H_k(z)$ with responses $H_k(e^{j\omega}) = V(e^{-j(\omega-\omega_k)})$ (and followed by modulators). This is shown in Fig. 11.2-6. The passband of $H_k(e^{j\omega})$ is centered around ω_k. The output signals $y_k(n)$ represent the STFT coefficients.

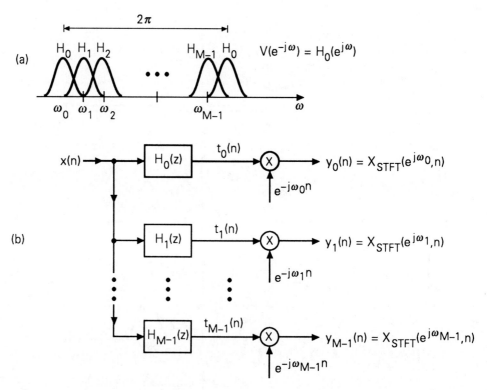

Figure 11.2-6 The STFT operation viewed as a filter bank. (a) $V(e^{-j\omega})$ and shifted versions, and (b) the filter bank.

The uniform DFT bank. If the frequencies ω_k are uniformly spaced, then the above system becomes the uniform DFT bank (Sec. 4.1.2). In this system the M filters are related as

$$H_k(z) = H_0(zW^k), \quad 0 \leq k \leq M-1, \qquad (11.2.4)$$

where $W = e^{-j2\pi/M}$. This means that the frequency responses are uniformly shifted versions of $H_0(e^{j\omega})$, i.e.,

$$H_k(e^{j\omega}) = H_0(e^{j(\omega - \frac{2\pi k}{M})}), \quad 0 \leq k \leq M-1. \qquad (11.2.5)$$

The unshifted filter is $H_0(e^{j\omega}) = V(e^{-j\omega})$. In Fig. 4.3-5 we saw an implementation of this set of filters, in terms of the polyphase components

$E_\ell(z)$ (Sec. 4.3) and the DFT matrix **W**. The window $v(n)$ is completely determined by $E_\ell(z)$ because

$$V(z^{-1}) = H_0(z) = E_0(z^M) + z^{-1}E_1(z^M) + \ldots + z^{-(M-1)}E_{M-1}(z^M). \tag{11.2.6}$$

It is now clear that the uniform DFT bank is a device to compute the short-time Fourier transform at uniformly spaced frequencies. In particular if the polyphase components are replaced with unity [i.e., $E_k(z) = 1$ for all k], then the system merely computes the DFT of a block of M samples (Fig. 4.1-16). In this case $v(n)$ is a rectangular window of length M.

Choice of v(n), and Time-Frequency Tradeoff

Unlike the Fourier transform, the STFT is not uniquely defined unless we specify the window $v(n)$. The fact that $v(n)$ can be chosen by the user offers a great deal of flexibility. The choice of $v(n)$ essentially governs the tradeoff between 'time localization' and 'frequency resolution' as explained next.

The signal $y_0(m)$ can be considered to represent the change or 'evolution' of the Fourier transform of $x(n)$, evaluated around frequency ω_0. Thus $y_0(m)$ represents the local information, *around* time m (since m represents the location of the window $v(k)$ in the time domain) *around* frequency ω_0. It is clear that as $V(e^{j\omega})$ becomes narrower, the bandpass filters in Fig. 11.2-6(b) get narrower, and $y_k(n)$ gets closer to $X(e^{j\omega_k})$. This means that the information in the frequency domain is becoming more and more localized. However, as $V(e^{j\omega})$ gets narrower, the window $v(n)$ gets wider (uncertainty principle) so that the localization of information in the time domain is compromised. Fig. 11.2-7 demonstrates the tradeoff between time localization and frequency resolution.

The fact that time localization and frequency resolution are conflicting requirements has given rise to interesting theoretical questions. For example, what is the choice of window that will give the best frequency resolution for a given time localization? To answer this question, one has to define the term "best" using a mathematical measure. To take a specific case, suppose we constrain $v(n)$ to be a symmetric window of finite length $N + 1$. What is the best choice of the coefficients $v(n)$ so that the energy of $V(e^{j\omega})$ is most concentrated in a specified region $|\omega| \leq \alpha$? This problem was in fact addressed in Sec. 3.2.2, where we found the solution to be the *prolate spheroidal sequence*, obtainable from an eigenvector of a positive definite matrix. See Sec. 11.2.4 for another measure of "best localization."

Time-Frequency Representation, and Decimation

It is often stated that $X_{STFT}(e^{j\omega}, m)$ is a time-frequency representation, because it is a function of time m as well as frequency ω. If the passband width of $V(e^{j\omega})$ [hence that of $V(e^{-j\omega})$] is narrow, then the signal $y_0(n)$ in Fig. 11.2-4(b) is a narrowband lowpass signal. This means that $y_0(n)$ varies slowly with the time index n. An extreme case is when $V(e^{j\omega})$ is an impulse (traditional Fourier transform) so that $y_0(n) = X(e^{j\omega_0})$ for all n.

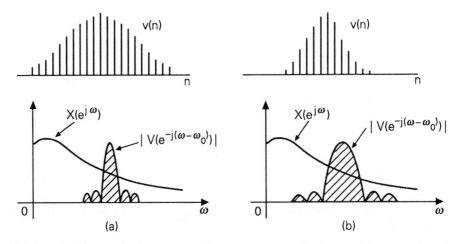

Figure 11.2-7 Demonstrating the tradeoff between time localization and frequency resolution. (a) Wide window $v(n)$; poor time localization and good frequency resolution. (b) Narrow window $v(n)$; good time localization and poor frequency resolution.

The slowly-varying nature of $X_{STFT}(e^{j\omega}, n)$, [i.e., $y_0(n)$ in Fig. 11.2-4(b)] can be exploited to decimate it, thereby resulting in a more economical time-frequency representation. If the decimation ratio is M, then this is equivalent to moving the window $v(k)$ by M samples at a time. (That is, M is the 'step size' for window movement). If $y_0(n)$ were not varying at all (as with traditional Fourier transform), then we would have to retain only one sample, and its value is $X(e^{j\omega_0})$.

Figure 11.2-8 shows a decimated STFT system, where the modulators have been moved past the decimators. Since the filters have equal bandwidth, the decimation ratios n_k can be taken to be equal. With $n_k = M$ this represents a maximally decimated analysis bank. In a more general system n_k could be different for different k, and moreover $H_k(z)$ may not be derived from one prototype by modulation. Such a system, however, does not represent the STFT obtainable by moving a single window $v(k)$ across the data $x(n)$. When we introduce the wavelet transforms in Sec. 11.3, we will admit such generalized systems.

The time-frequency representation offers a whole family of tradeoffs ('time localization' versus 'frequency resolution' tradeoff) between the two extremes, viz., $x(n)$ (time domain representation), and $X(e^{j\omega})$ (frequency domain representation). The filter bank system performs an operation analogous to Fourier transformation, yet the outputs of the transform are time varying. After performing the maximal decimation, the time-frequency representation has the same number of samples per unit time as does $x(n)$. There is no redundancy in the representation.

The time-frequency grid. Figure 11.2-9 demonstrates a grid in the two-dimensional time-frequency plane. The vertical lines represent the 'frequencies' where the STFT is computed (i.e., center frequencies of the filters). The horizontal lines represent the sample locations in the time domain, for the decimated filter outputs. The intersections of the lines represent the location of a sample of $X_{STFT}(e^{j\omega_k}, m)$. This grid represents uniform sampling of both 'frequency' ω and 'time' n. The fact that the time spacing is M corresponds to the fact that the window is moved in steps of M units at a time. The frequency spacing (spacing between center frequencies of adjacent filters) is $2\pi/M$ because there are M filters of identical bandwidths.

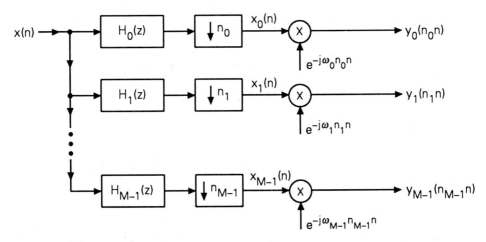

Figure 11.2-8 An analysis bank with decimators and modulators. The signal $y_k(n_k n)$ represents the decimated version of $X_{STFT}(e^{j\omega_k}, n)$ where ω_k is the center frequency of $H_k(e^{j\omega})$. Usually $n_k = M$ for all k.

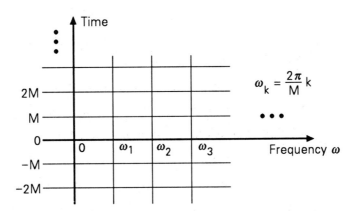

Figure 11.2-9 The two-dimensional time-frequency grid for evaluating the short-time Fourier transform.

11.2.2 Inversion of the STFT

Starting from the definition (11.2.1) it is easy to derive a number of inversion formulas that recover $x(n)$ from $X_{STFT}(e^{j\omega}, m)$. As $X_{STFT}(e^{j\omega}, m)$ is the traditional Fourier transform of $x(n)v(n-m)$, it is clear that

$$x(n)v(n-m) = \frac{1}{2\pi} \int_0^{2\pi} X_{STFT}(e^{j\omega}, m) e^{j\omega n} d\omega. \qquad (11.2.7)$$

For example if we set $n = m$ we obtain the STFT inversion formula

$$x(m)v(0) = \frac{1}{2\pi} \int_0^{2\pi} X_{STFT}(e^{j\omega}, m) e^{j\omega m} d\omega, \qquad (11.2.8)$$

so that we can recover $x(m)$ for all m as long as $v(0) \neq 0$. [If $v(0) = 0$, pick some other value of m in (11.2.7)]. Notice that it is not necessary to know $v(n)$ for all n, in order to recover $x(n)$ from its STFT.

A second inversion formula is given by

$$x(n) = \frac{1}{2\pi} \int_0^{2\pi} \left(\sum_{m=-\infty}^{\infty} X_{STFT}(e^{j\omega}, m) v^*(n-m) \right) e^{j\omega n} d\omega, \qquad (11.2.9)$$

provided $\sum_m |v(m)|^2 = 1$. To prove this we merely substitute (11.2.1) into the RHS of (11.2.9), which then reduces to $x(n) \sum_m |v(m)|^2$. There are some subtleties about the formula, which we mention here. Problem 11.3 covers details.

1. If $\sum_m |v(m)|^2 \neq 1$ but finite, we can divide the right side of (11.2.9) by $\sum_m |v(m)|^2$ and obtain the inversion. However, if the window $v(n)$ has infinite energy (as in the important special case of $v(m) = 1$), this inversion formula cannot be applied.
2. Suppose we replace $v^*(n-m)$ in (11.2.9) with $w^*(n-m)$ where $w(n)$ is an arbitrary sequence with the restriction that $\sum_n v(n) w^*(n) = 1$. Then the inversion formula still works!
3. The function $G(e^{j\omega}, m)$ satisfying

$$x(n) = \frac{1}{2\pi} \int_0^{2\pi} \left(\sum_{m=-\infty}^{\infty} G(e^{j\omega}, m) v^*(n-m) \right) e^{j\omega n} d\omega, \qquad (11.2.10)$$

is not unique. For example, suppose z_0 is a zero of the z-transform $\sum_k v^*(k) z^{-k}$. Then $G(e^{j\omega}, m) \stackrel{\Delta}{=} X_{STFT}(e^{j\omega}, m) + z_0^m$ satisfies (11.2.10) for the same sequence $x(n)$. This is unlike the case of traditional Fourier transform, where (11.1.2) is not satisfied if we replace $X(e^{j\omega})$ with something else.

Filter Bank Interpretation of the Inverse Transform

It is valuable to express the inverse transform using filter bank notation. Recall that Fig. 11.2-6(b) offers a practical means of implementing the STFT. Viewed like this, the STFT is a transformation of a one dimensional sequence $x(n)$ into a two dimensional sequence $y_k(m)$ (i.e., function of two integer variables k and m). The number and locations of the frequencies ω_k might appear to be arbitrary, and so might the shapes of the filters. In fact they *are* somewhat arbitrary, subject primarily to the requirement that the signal $x(n)$ be reconstructible from the STFT coefficients $y_k(n)$ with reasonable accuracy in reasonable amount of time. It turns out that, as long as the filters $H_k(z)$ are chosen properly, we can find stable synthesis filters $F_k(z)$ to recover $x(n)$ perfectly. On the other hand, inversion of the traditional Fourier transformation (11.1.1) requires the computation of an *integral*.

With the STFT implemented as in Fig. 11.2-8, the reconstruction is done by using a synthesis bank as shown in Fig. 11.2-10. Typically $n_k = M$ for all k, but we will use n_k for generality. The z-transform of $\widehat{x}(n)$ is given by

$$\widehat{X}(z) = \sum_{k=0}^{M-1} X_k(z^{n_k}) F_k(z). \qquad (11.2.11)$$

In the time domain this is equivalent to

$$\widehat{x}(n) = \sum_{k=0}^{M-1} \sum_{m=-\infty}^{\infty} x_k(m) f_k(n - n_k m)$$

$$= \sum_{k=0}^{M-1} \sum_{m=-\infty}^{\infty} y_k(n_k m) e^{j\omega_k(n_k m)} f_k(n - n_k m). \qquad (11.2.12)$$

If the synthesis filters $F_k(z)$ are such that $\widehat{x}(n) = x(n)$, we can say that (11.2.12) is the representation of $x(n)$ in terms of the decimated STFT coefficients $y_k(n_k m)$ just as (11.1.2) is the representation of $x(n)$ in terms of the traditional Fourier transform 'coefficients' $X(e^{j\omega})$. While (11.1.2) is an integral in terms of the single variable ω, the new representation is a *double* summation (in the integer variables k and m.) The reconstruction is stable if the filters $F_k(z)$ are stable.

Example 11.2.1: Reconstructing x(n) from STFT Coefficients

Assume that there are no decimators, that is, the STFT is as in Fig. 11.2-6. Let the window $v(n)$ satisfy the Nyquist property, viz., $v(Mn) = 0$ for $n \neq 0$. Then, the M filters $H_k(z)$ defined in (11.2.4) satisfy the property $\sum_{k=0}^{M-1} H_k(z) = c_0$ for constant $c_0 \neq 0$ (Sec. 4.6.1). This means that we can reconstruct $x(n)$ from the STFT coefficients $y_k(n)$ simply by adding them after demodulation, that is,

$$x(n) = c_1 \sum_{k=0}^{M-1} y_k(n) e^{j\omega_k n} \qquad (11.2.13)$$

for some constant c_1.

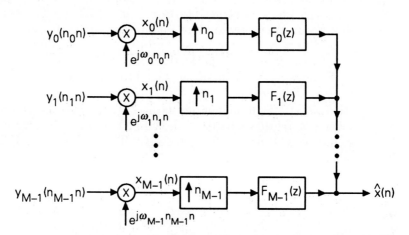

Figure 11.2-10 The synthesis bank used to reconstruct $x(n)$ from its STFT coefficients. Usually $n_k = M$ for all k.

11.2.3 Generalizations of the STFT

We know that we can recover an arbitrary signal $x(n)$ from its (decimated) STFT coefficients, provided we can design a synthesis bank with perfect reconstruction (PR) property. However, since the analysis filters are derived from a single prototype by modulation, the PR requirement will in turn restrict the coefficients of $v(n)$ severely (Example 5.7.2).

If we relax the requirement that all analysis filters be derived from one prototype $v(n)$, we can obtain more flexibility. For this we generalize the STFT idea by viewing any analysis filter bank as a *generalized Fourier transformer*. The outputs of the filters are narrowband signals, and represent the localized Fourier transform as described above. This generalized system, however, is not derivable from a traditional single sliding window system as in Fig. 11.2-1, i.e., the simple description (11.2.1) does not hold. These are more appropriately called "spectrum analyzers" rather than short time Fourier transformers. From Chap. 5–8 we know that there exist many techniques to perfectly reconstruct $x(n)$ from the (possibly decimated) filter outputs.

We will say that the quantities

$$x_k(n), \quad 0 \le k \le M-1, \quad -\infty \le n \le \infty, \qquad (11.2.14)$$

in Fig. 11.2-8 are the generalized STFT coefficients of the signal $x(n)$. The transform domain is characterized by *two* integer variables k and n. The STFT pair can be written as

$$x_k(n) = \sum_{m=-\infty}^{\infty} x(m) h_k(n_k n - m) \quad \text{(STFT)}, \qquad (11.2.15)$$

and

$$x(n) = \sum_{k=0}^{M-1} \sum_{m=-\infty}^{\infty} x_k(m) f_k(n - n_k m) \quad \text{(Inverse STFT)}. \qquad (11.2.16)$$

The decimators n_k should be chosen to be inversely proportional to the passband widths of the filters $H_k(z)$. If these bandwidths are equal, then $n_k = M$ for all k. Notice that the modulators and demodulators have not been included in the definition for simplicity; they cancel each other anyway, and do not provide further insights. We can regard (11.2.15) and (11.2.16) as the generalized STFT pair or the "spectrum analyzer/synthesizer pair". It is also called the "filter-bank transform pair." For the special case where the analysis filters are as in (11.2.5) with $H_0(z) = V(z^{-1})$, this reduces to the traditional STFT which is computed from a single window $v(n)$.

Comments.

1. The above STFT/inverse STFT definition assumes that the filters $H_k(z)$ and $F_k(z)$ are such that the filter bank system (Fig. 11.2-8 followed by Fig. 11.2-10) has perfect reconstruction property. It is easy to ensure that $H_k(z)$ are stable. If $F_k(z)$ are also stable, we have a stable reconstruction scheme for performing the inverse transform.
2. *Nonuniqueness.* In Fig. 11.2-6(b), suppose α is a zero of all the analysis filters, that is, $H_k(\alpha) = 0$ for all k. This means that if we replace $x(n)$ with $x(n) + \alpha^n$, the STFT coefficients do not change. In other words, the sequence $x(n)$ producing the transform domain coefficients is not unique. Such a situation is easily avoided in practice. For example, in a perfect reconstruction system this situation will not arise. (Because, it would imply that the input $x(n) = \alpha^n$ produces zero output).

Basis Functions and Orthonormality

Consider the conventional Fourier transform representation (11.1.2). Here $x(n)$ is a linear combination of the sequences $e^{j\omega n}$, and the set $\{e^{j\omega n}\}$ is said to be a basis for the space of sequences representable as in (11.1.2). The basis functions $e^{j\omega n}$, are orthonormal in following sense:

$$\sum_{n=-\infty}^{\infty} (e^{j\omega_1 n})^* e^{j\omega_2 n} = 2\pi \delta_a(\omega_1 - \omega_2), \qquad (11.2.17)$$

that is, the "inner-product" of $e^{j\omega_1 n}$ and $e^{j\omega_2 n}$ is equal to zero unless $\omega_1 = \omega_2$.

By the above analogy we see that the quantities

$$\eta_{km}(n) \stackrel{\Delta}{=} f_k(n - n_k m) \qquad (11.2.18)$$

play the role of "basis functions" in (11.2.16). Notice that this is a doubly indexed family of functions. The first index is the filter number k, and the second index determines the time shift. Such basis functions will be called the "filter-bank like basis".

It is of interest to impose the orthonormality property on the basis functions $\{\eta_{km}(n)\}$. This property means that

$$\sum_{n=-\infty}^{\infty} f_{k_1}^*(n - n_{k_1} m_1) f_{k_2}(n - n_{k_2} m_2) = \delta(k_1 - k_2)\delta(m_1 - m_2). \qquad (11.2.19)$$

In other words, the above summation should be zero except for the case where $k_1 = k_2$ and $m_1 = m_2$ (and reduces to unity in that case). How should we design the filters $F_k(z)$ in order to ensure this? We will return to this very interesting issue in Sec. 11.4, and show that the paraunitary property of the polyphase matrix (Chap. 6) is sufficient!

Relation between $h_k(n)$ and $f_k(n)$. In the traditional Fourier transform pair, the basis function $e^{j\omega n}$ appears in (11.1.2) whereas its conjugate appears in (11.1.1). Inspection of the STFT pair reveals no obvious analogy of this relation. The only requirement is that the functions $h_k(n_k n - m)$ and $f_k(n - n_k m)$ be related in such a way as to ensure perfect reconstruction. We will return to this later, and show that if the basis functions are orthonormal then $f_k(n) = h_k^*(-n)$. This is very similar to the relation between analysis and synthesis filters in a paraunitary perfect reconstruction system (Sec. 6.2.1).

Table 11.2.1 provides a summary of the discrete-time STFT.

11.2.4 The Continuous-Time Case

Historically, the STFT idea was first developed for the continuous-time case even though our presentation here started with the discrete case. In 1946 Dennis Gabor [†] considered windowed versions of the continuous-time Fourier transform. Gabor used a Gaussian window, that is, a function of the form $v(t) = v(0)e^{-bt^2}$, $b > 0$. The corresponding continuous-time STFT is called the *Gabor transform* [Gabor, 1946]. Notice that this window does not have finite duration.

[†] He received the Nobel Prize in 1971 for contributions to the principles of holography.

TABLE 11.2.1 Short-Time Fourier Transform (STFT), discrete-time.

Here are the key equations governing the discrete-time traditional and short-time Fourier transforms. Note that the subscript on $\eta_\omega(n)$ is the real-valued continuous variable ω. For the STFT, there are two subscripts, as in $\eta_{\omega,m}(n)$ and $\eta_{km}(n)$. In the former, ω is a real (continuous) variable. In the latter, k is integer-valued (*center-frequency number* or *filter number*). The double subscripts arise because the transform domain is *time-frequency* rather than *frequency*, as would be the case for the traditional Fourier transform. Read the text, particularly in the neighbourhood of the following equations, to fully appreciate assumptions and conditions.

Traditional Discrete-time Fourier Transform

$$X(e^{j\omega}) = \sum_{n=-\infty}^{\infty} x(n)e^{-j\omega n} \qquad \text{(transform)} \qquad (11.1.1)$$

$$x(n) = \frac{1}{2\pi} \int_0^{2\pi} X(e^{j\omega}) \underbrace{e^{j\omega n}}_{\text{basis } \eta_\omega(n)} d\omega \qquad \text{(inverse transform)} \qquad (11.1.2)$$

Discrete-Time STFT

$$X_{STFT}(e^{j\omega},m) = \sum_{n=-\infty}^{\infty} x(n)v(n-m)e^{-j\omega n} \qquad \text{(STFT)} \qquad (11.2.1)$$

$$x(m)v(0) = \frac{1}{2\pi} \int_0^{2\pi} X_{STFT}(e^{j\omega},m)e^{j\omega m} d\omega \qquad \text{(inverse STFT)} \qquad (11.2.8)$$

$$x(n) = \frac{1}{2\pi} \int_0^{2\pi} \sum_{m=-\infty}^{\infty} X_{STFT}(e^{j\omega},m) \underbrace{v^*(n-m)e^{j\omega n}}_{\text{basis } \eta_{\omega,m}(n)} d\omega \qquad \text{(inverse STFT)}$$

$$(11.2.9)$$

Generalized Discrete-Time STFT (Filter-Bank Transformer)

$$x_k(n) = \sum_{m=-\infty}^{\infty} x(m)h_k(n_k n - m), \quad 0 \le k \le M-1 \qquad \text{(STFT)}, (11.2.15)$$

$$x(n) = \sum_{k=0}^{M-1} \sum_{m=-\infty}^{\infty} x_k(m) \underbrace{f_k(n - n_k m)}_{\text{basis } \eta_{km}(n)} \qquad \text{(inverse STFT)}. \qquad (11.2.16)$$

It is assumed that the filter bank with analysis filters $h_k(n)$, synthesis filters $f_k(n)$ and decimation ratios n_k (Fig. 11.2-8, 11.2-10) has perfect reconstruction ($\hat{x}(n) = x(n)$). For orthonormal basis, $f_k(n) = h_k^*(-n)$.

Because of the close resemblance to the discrete-time case, we only summarize the main points. Given a signal $x(t)$ we define the STFT as

$$X_{STFT}(j\Omega, \tau) = \int_{-\infty}^{\infty} x(t)v(t-\tau)e^{-j\Omega t}dt \quad \text{(STFT)}, \qquad (11.2.20)$$

where $v(t)$ is an appropriate window function, typically with lowpass Fourier transform $V(j\Omega)$. (Proper choice of $v(t)$ ensures existence of the integral). Once again we can find an inversion formula similar to (11.2.8), and obtain

$$x(\tau)v(0) = \frac{1}{2\pi} \int_{-\infty}^{\infty} X_{STFT}(j\Omega, \tau)e^{j\Omega\tau}d\Omega \quad \text{(inv. STFT)}, \qquad (11.2.21)$$

which works as long as $v(0) \neq 0$. The inverse transform analogous to (11.2.9) is given by the double integral

$$x(t) = \frac{1}{2\pi} \int_{-\infty}^{\infty} \left(\int_{-\infty}^{\infty} X_{STFT}(j\Omega, \tau)v^*(t-\tau)d\tau \right) e^{j\Omega t}d\Omega \quad \text{(inv. STFT)}, \qquad (11.2.22)$$

and this is associated with similar subtleties as itemized after (11.2.9).

Choice of "Best Window" to Optimize Localization

We know that if the window $v(t)$ is 'narrow' in the time domain, its Fourier transform is 'broad' and vice versa. This means that there is a tradeoff between time localization and frequency resolution. To make this idea more precise, the rms (root-mean squared) duration of a signal is introduced in the literature ([Gabor, 1946], [Papoulis, 1977a]). Thus consider the two nonnegative quantities D_t and D_f defined by

$$D_t^2 = \frac{1}{E} \int_{-\infty}^{\infty} t^2 v^2(t)dt, \quad D_f^2 = \frac{1}{2\pi E} \int_{-\infty}^{\infty} \Omega^2 |V(j\Omega)|^2 d\Omega. \qquad (11.2.23)$$

where E is the window energy, that is, $E = \int v^2(t)dt$. (For this discussion $v(t)$ is real.) We say that D_t is the rms time domain duration and D_f the rms frequency domain duration of $v(t)$. Figure 11.2-11 shows the rms duration D_t for a number of signals (the reader is requested to verify these in Problem 11.5). It is interesting that a triangular waveform has a smaller rms duration than a rectangular waveform, even though they have identical 'traditional duration'! This is because the factor t^2 in the definition of D_t^2 increases the penalty on nonzero values of $v(t)$, as t increases.

Uncertainty principle. It turns out that the product $D_t D_f$ cannot be arbitrarily small. Here is a quantitative statement of uncertainty principle: $D_t D_f \geq 0.5$, with equality if and only if $v(t) = Ae^{-\alpha t^2}, \alpha > 0$ (Problem 11.6). Thus the optimal window is a Gaussian waveform, and its 'traditional' duration is infinite.

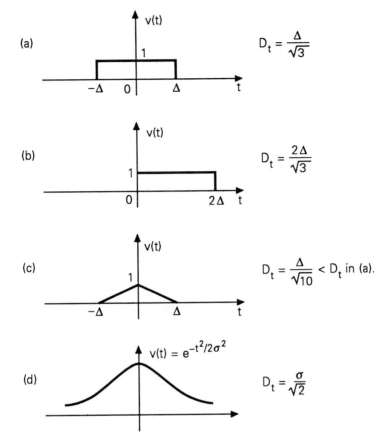

Figure 11.2-11 RMS duration of some typical signals. (a) Rectangular window, (b) one-sided rectangular window, (c) triangular window, and (d) Gaussian window.

Filter Bank Interpretation

To obtain further insight we rewrite the STFT for fixed frequency Ω_k as

$$X_{STFT}(j\Omega_k, \tau) = e^{-j\Omega_k \tau} \int_{-\infty}^{\infty} x(t)v(t-\tau)e^{j\Omega_k(\tau-t)}dt. \qquad (11.2.24)$$

Defining $h(t) = v(-t)$, we see that the integral above is a convolution of $x(t)$ with the filter having impulse response

$$h_k(t) = h(t)e^{j\Omega_k t} = v(-t)e^{j\Omega_k t}. \qquad (11.2.25)$$

Thus

$$X_{STFT}(j\Omega_k, \tau) = e^{-j\Omega_k \tau} \int_{-\infty}^{\infty} x(t)h_k(\tau-t)dt. \qquad (11.2.26)$$

Sec. 11.2 The short-time Fourier transform

Figure 11.2-12 shows this "filtering" interpretation. We have replaced τ with t everywhere to conform with traditional notations. The output $y_k(t)$ represents the STFT of $x(t)$, evaluated with window $v(t)$, at the frequency Ω_k. Assuming $v(-t)$ is narrowband lowpass, the output of the filter in the figure is narrowband bandpass, centered at Ω_k. So the signal $y_k(\tau)$ is narrowband lowpass, that is, it is "slowly varing in τ." It represents an estimate of the Fourier transform of $x(t)$ 'localized around time τ' and around the frequency Ω_k. With the system of Fig. 11.2-12 repeated for several values of Ω_k, the complete system is equivalent to a bank of bandpass filters (Fig. 11.2-13). All filter responses are shifted versions of the prototype response $H(j\Omega)$. Finally, if $v(t) = 1$ for all t, then $y_k(t)$ is constant for all t, that is, $y_k(t) = X(j\Omega_k)$ (traditional Fourier transform).

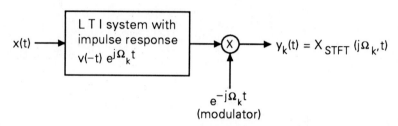

Figure 11.2-12 The continuous-time STFT as an LTI filter followed by modulator. For each frequency of interest Ω_k, we have one such filter, resulting in a filter bank.

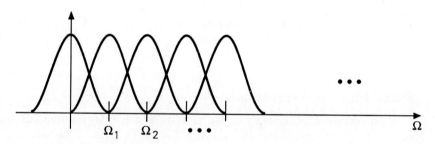

Figure 11.2-13 Performing the STFT at a discrete set of frequencies is equivalent to the use of a bank of bandpass filters.

Uniformly Sampled Version of the STFT

Since $y_k(t)$ is narrowband lowpass, we can sample it with appropriate sampling period, say T, to obtain $X_{STFT}(j\Omega_k, nT)$. See Fig. 11.2-14. (With nonideal filters, aliasing due to sampling is unavoidable, and must somehow be canceled later). One special choice of Ω_k is particularly illuminating, viz.,

$\Omega_k = k\Omega_0$, for integer k and fixed Ω_0. In this case we have

$$\mathcal{X}(k,n) \triangleq X_{STFT}(jk\Omega_0, nT) = e^{-j(\Omega_0 T)kn} \int_{-\infty}^{\infty} x(t)v(t-nT)e^{jk\Omega_0(nT-t)} dt. \quad (11.2.27)$$

Essentially, $\mathcal{X}(k,n)$ is a uniformly sampled version of the two-dimensional function $X_{STFT}(j\Omega, \tau)$. The set of sample points described by $(k\Omega_0, nT)$ is called the Gabor lattice (or Von Neumann lattice in quantum mechanics literature; [Von Neumann, 1955]). The question now is, can we reconstruct $x(t)$ from this sampled transform? If the product $\Omega_0 T$ is sufficiently small, the answer is in the affirmative. It is shown in the literature (also see Problem 11.4) that if $\Omega_0 T = 2\pi$, the function $x(t)$ can indeed be written explicitly in terms of the samples $\mathcal{X}(k,n)$. However, the reconstruction procedure itself is unstable; one requires the condition $\Omega_0 T < 2\pi$ for stability. See Daubechies [1990] for elaboration on this point.

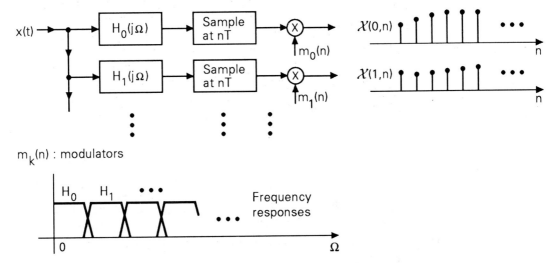

Figure 11.2-14 Sampled version of the continuous-time STFT.

The fact that we can reconstruct $x(t)$ from the two dimensional sampled version might occassion an initial surprise, since we have not assumed any bandlimited property of $x(t)$. Consider, however, the situation where we increase Ω_0 for a fixed value of the product $\Omega_0 T < 2\pi$. In the extreme case where $\Omega_0 \to \infty$, we have $T \to 0$. This is equivalent to having a single filter with infinite bandwidth, whose output is $x(t)$ itself; this output is sampled with samples spaced infinitesimally close together. In other words, the transformed version is essentially $x(t)$ itself!

11.3 THE WAVELET TRANSFORM

While the short-time Fourier transform is a convenient generalization of the Fourier transform, it still has some disadvantages. To appreciate this, consider Fig. 11.3-1 which shows two cases. For the first case $x(t)$ is a high-frequency signal, and many cycles are captured by the window. For the second, $x(t)$ is of low frequency, so that very few cycles are within the window. Thus the accuracy of the estimate of the Fourier transform is poor at low frequencies, and improves as the frequency increases. This can also be seen from the fact that the bandpass filters in Fig. 11.2-13 have equal bandwidths, rather than bandwidth increasing with center-frequency.

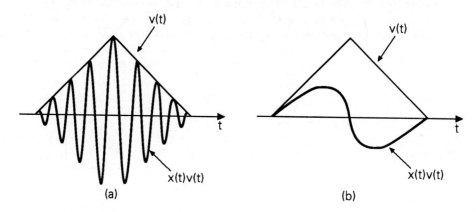

Figure 11.3-1 The windowed function $x(t)v(t)$ for (a) high-frequency signal $x(t)$, and (b) low-frequency signal $x(t)$.

Another issue is revealed by considering the rising signal of Fig. 11.1-3(b). We see that if the window is narrow, it helps to localize the rising portion very well, as compared to a wide window. With a narrow window, however, the information in the steady part of the signal changes very slowly. It will be appropriate here to have a window whose width adjusts itself with 'frequency'. This can be accomplished by using a filter bank where the lowpass filter has a narrower bandwidth (wider time-width) than the bandpass and highpass filters.

One (conceptual) way to do this is to replace the window $v(t)$ with a function of both frequency and time, so that the time domain plot of the window gets wider (i.e., bandwidth gets narrower) as frequency decreases. In this way, the window captures nearly the same number of zero-crossings for any sinusoidal input with arbitrary frequency. Furthermore, as the window gets wider, it is also desirable to have wider step sizes for moving the window (equivalently larger decimation ratio n_k in Fig. 11.2-8).

These goals are nicely accomplished by the wavelet transform. We begin by developing the continuous-time wavelet transform which is conceptually easier.

11.3.1 Passing from STFT to Wavelets

Step 1. Nonuniform filter banks

The bandpass filters in Fig. 11.2-6 have equal bandwidth because they are obtained by modulation of a single filter. As a first step, we give up this modulation scheme, and obtain the filters $h_k(t)$ as

$$h_k(t) = a^{-k/2} h(a^{-k} t), \quad a > 1, \ k = \text{integer}. \quad (11.3.1)$$

Equivalently, in the frequency domain,

$$H_k(j\Omega) = a^{k/2} H(ja^k \Omega). \quad (11.3.2)$$

Thus all the responses are obtained by *frequency-scaling* of a prototype response $H(j\Omega)$. This is unlike the case of STFT, where all filters were obtained by *frequency-shift* of a prototype.

The scale factor $a^{-k/2}$ in (11.3.1) is meant to ensure that the energy $\int_{-\infty}^{\infty} |h_k(t)|^2 dt$ is independent of k. This can be regarded as a normalizing convention or height convention.

Example 11.3.1

Assuming that $H(j\Omega)$ is bandpass with cutoff frequencies α and β, we obtain the responses shown in Fig. 11.3-2(a). Note that $H_0(j\Omega) = H(j\Omega)$. We have assumed $a = 2$, and $\beta = 2\alpha$. The bandedges of adjacent filters overlap, as indicated. The passband gets narrower as the center frequency decreases. Quantitatively, we define the center frequency to be the geometric mean of the two cutoff edges, that is,

$$\Omega_k = 2^{-k} \sqrt{\alpha \beta} = \alpha 2^{-k} \sqrt{2}. \quad (11.3.3)$$

These are nonuniformly located, and appear to be uniform if the frequency axis is represented on a logarithmic scale. Notice that $H(j\Omega)$ is bandpass rather than highpass. One often restricts k to be nonnegative, so that there are no filters to the right of the bandpass filter $H_0(j\Omega)$. This is acceptable if the input signal has no information beyond this filter, that is, if it is bandlimited. From Fig. 11.3-2(a) we see that the ratio

$$\frac{\text{bandwidth}}{\text{center-frequency } \Omega_k} = \frac{2^{-k}(\beta - \alpha)}{2^{-k}\sqrt{\alpha\beta}} = \frac{\beta - \alpha}{\sqrt{\alpha\beta}} = \frac{1}{\sqrt{2}} \quad (11.3.4)$$

is independent of the filter number k.† Notice a slight change in convention here: as k increases, the center frequency *decreases*. This happens to be more convenient.

---

† In the language of electrical filter theory [Sedra and Brackett, 1978], such a system is often said to be a 'constant Q' system. The quantity 'Q' (Quality factor) is usually defined as (center-frequency/bandwidth), i.e., the reciprocal of (11.3.4).

Nonuniform bandwidths as in the above example are very useful for the analysis of sound signals. See, for example, [Nelson, et al., 1972], and pp. 301-303 of Rabiner and Schafer [1978]. This is because of the decreasing frequency-resolution of the human ear with increasing frequency [Flanagan, 1972].

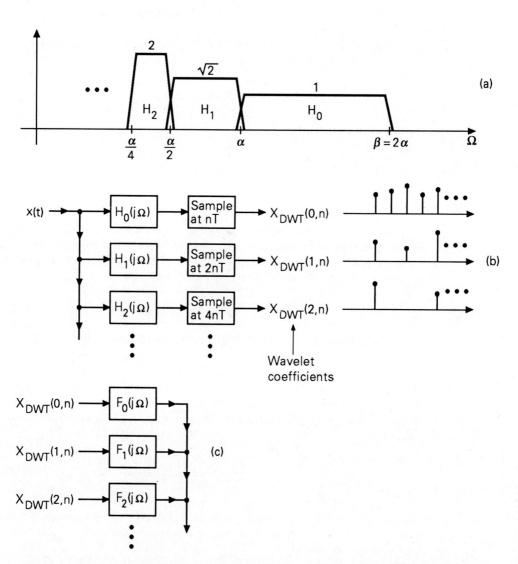

Figure 11.3-2 (a) Frequency responses obtained by the scaling process (11.3.2) with $a = 2$. (b) Analysis bank representation of discrete wavelet transform. (c) Synthesis bank which would reconstruct $x(t)$ from the set of wavelet coefficients $X_{DWT}(k,n)$. In this figure, the signal $X_{DWT}(k,n)$ indicates a continuous-time impulse train $\sum_{n=-\infty}^{\infty} X_{DWT}(k,n)\delta(t - 2^k nT)$.

With the filters redefined as in (11.3.1), the filter outputs can be obtained by modifying the right hand side of (11.2.26) with

$$a^{-k/2}e^{-j\Omega_k\tau}\int_{-\infty}^{\infty} x(t)h\left(a^{-k}(\tau - t)\right)dt. \qquad (11.3.5)$$

This is the first of the two modifications of the STFT, which will lead to wavelet transforms.

Step 2. Nonuniform decimation

Since the bandwidth of $H_k(j\Omega)$ is smaller for large k, we can sample its output at a correspondingly lower rate. Equivalently, viewed in the time domain, the width of $h_k(t)$ is larger so that we can afford to move the window by a larger step size. We will do this by replacing the continuous variable τ with na^kT in (11.3.5), where n is an integer. This means that the step size for window movement is a^kT and it increases with k, that is, increases as the center-frequency Ω_k (hence bandwidth) of the filter *decreases*. Thus the quantity $h[a^{-k}(\tau - t)]$ in (11.3.5) is now replaced with

$$h\left(a^{-k}(na^kT - t)\right) = h(nT - a^{-k}t). \qquad (11.3.6)$$

Summarizing, we are computing

$$X_{DWT}(k,n) = a^{-k/2}\int_{-\infty}^{\infty} x(t)h(nT - a^{-k}t)dt, \quad k,n \text{ integers}, \qquad (11.3.7a)$$

i.e.,

$$X_{DWT}(k,n) = \int_{-\infty}^{\infty} x(t)h_k(na^kT - t)dt, \quad k,n \text{ integers}. \qquad (11.3.7b)$$

We have omitted the inconsequential factor $e^{-j\Omega_k\tau}$ which appeared earlier in (11.3.5). The above integral represents the convolution between $x(t)$ and $h_k(t)$, evaluated at a discrete set of points na^kT. In other words, the output of the convolution (a continuous-time function) is sampled with spacing a^kT. Fig. 11.3-2(b) is a schematic of this for $a = 2$. The kth sampler merely retains the samples at the locations $(2^kT)n$, where $n =$ integer.

The subscript DWT above stands for *discrete wavelet transform*. Also, the dependence on a and $h(t)$ is not explicitly indicated in the notation $X_{DWT}(k,n)$.

Time-frequency grid. Figure 11.3-3 shows the time-frequency diagrams for the STFT and the wavelet transform, and this summarizes the fundamental difference between these two. In the former, the frequency spacing and time spacing are uniform. In the latter, the frequency spacing

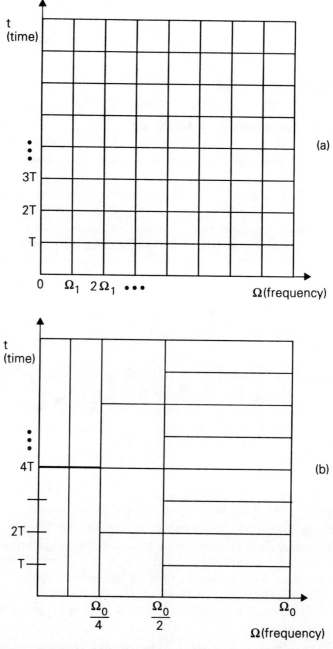

Figure 11.3-3 Fundamental differences between the STFT and the wavelet transform. (a) In the STFT, time and frequency axes are typically uniformly divided. (b) In the wavelet transform, the frequency samples are spaced closer together at lower frequencies, and the corresponding time samples are spaced wider apart.

is smaller at lower frequencies, and the corresponding time-spacing is larger. Notice that 'frequency' spacing refers to the spacing between adjacent filters, and time-spacing refers to the sampling period used for the filtered outputs. The wavelet transform is not explicitly implemented by a moving window because there is in reality no unique window here. The system is in essence a filter bank, and is somewhat analogous to a family of windows (wider for low frequencies, etc.) as explained above.

General Definition of the Wavelet Transform

Equation (11.3.7a) is a special case of the more general definition of the continuous wavelet transformation (CWT) given in the literature, viz.,

$$X_{CWT}(p,q) = \frac{1}{\sqrt{|p|}} \int_{-\infty}^{\infty} x(t) f\left(\frac{t-q}{p}\right) dt, \qquad (11.3.8)$$

where p and q are real-valued continuous variables. This reduces to (11.3.7a) if we identify

$$p = a^k, \quad q = a^k T n, \quad \text{and} \quad f(t) = h(-t). \qquad (11.3.9)$$

This choice is equivalent to evaluating (11.3.8) at a discrete set of points in the (p,q) plane, hence the name DWT for (11.3.7). DWT is different from discrete-time wavelet transforms (DTWT) to be discussed in Sec. 11.3.3. Quantities such as $X_{CWT}(p,q)$ and $X_{DWT}(k,n)$ are also called *wavelet coefficients*.

The CWT is a mapping of the function $x(t)$ into a two dimensional function $X_{CWT}(p,q)$ of the continuous variables p,q. The DWT is a mapping of $x(t)$ (t being still continuous-time) into a two dimensional *sequence* $X_{DWT}(k,n)$. The computation of $X_{DWT}(k,n)$ is equivalent to the implementation of the bank of filters $H_k(j\Omega)$, followed by sampling of their outputs at rates proportional to the filter bandwidths.

11.3.2 Inversion of the Wavelet Transform

The "inverse wavelet transform", if it exists, reconstructs the signal $x(t)$ from the wavelet coefficients. A direct inversion formula for (11.3.8) can be found in Daubechies [1990].

We will consider only the discretized case. Whether we can reconstruct $x(t)$ from the discretized version $X_{DWT}(k,n)$ depends on the prototype filter $h(t)$, and the discretizing parameters a and T which completely characterize the transformation. If the inverse transform exists, it has the appearance

$$x(t) = \sum_k \sum_n X_{DWT}(k,n) \psi_{kn}(t) \quad \text{(inverse DWT)}, \qquad (11.3.10)$$

where $\psi_{kn}(t)$ are the basis functions.

Filter Bank Interpretation of Inversion

Suppose the wavelet coefficients $X_{DWT}(k,n)$ are generated using the analysis bank in Fig. 11.3-2(b). The reconstruction of $x(t)$ from these coefficients can be visualized as a problem of designing the synthesis filters $F_k(j\Omega)$ shown in Fig. 11.3-2(c). If the analysis/synthesis system has the perfect reconstruction property, then the recovery is perfect.

We have to be careful with the interpretation of Fig. 11.3-2(c). Since $X_{DWT}(k,n)$ is a *sequence*, the signal which is input to the *continuous-time* filter $F_k(j\Omega)$ is actually an impulse train of the form

$$v_k(t) = \sum_n X_{DWT}(k,n)\delta_a(t - a^k nT). \qquad (11.3.11)$$

The output of the synthesis filter bank is therefore

$$\widehat{x}(t) = \sum_k \sum_n X_{DWT}(k,n) f_k(t - a^k nT). \qquad (11.3.12a)$$

Since the synthesis filters $F_k(j\Omega)$ have to retain only the frequency region passed by the analysis filters, it is reasonable that they be all generated from a fixed prototype synthesis filter $f(t)$, similar to (11.3.1). That is

$$f_k(t) = a^{-k/2} f(a^{-k}t). \qquad (11.3.12b)$$

Substituting this into the preceding equation, and assuming perfect reconstruction, we get

$$x(t) = \sum_k \sum_n X_{DWT}(k,n) \underbrace{a^{-k/2} f(a^{-k}t - nT)}_{\text{basis } \psi_{kn}(t)}. \qquad (11.3.12c)$$

Thus

$$\psi_{kn}(t) = a^{-k/2}\psi(a^{-k}t - nT) = a^{-k/2}\psi\left[a^{-k}(t - na^k T)\right], \qquad (11.3.12d)$$

where we have defined

$$\psi(t) = f(t). \qquad (11.3.12e)$$

Thus, Eqn. (11.3.12c) expresses $x(t)$ as a linear combination of a set of basis functions $\psi_{kn}(t)$ which are obtained by *dilations* (i.e., $t \to a^{-k}t$) and *shifts* (i.e., $t \to t - na^k T$) of a single *wavelet function* $\psi(t)$ or *mother wavelet*. The DWT coefficients $X_{DWT}(k,n)$ are the weights of these basis functions.

Using the relations $\psi(t) = f(t)$ and $f_k(t) = a^{-k/2}f(a^{-k}t)$, we can express each basis function $\psi_{kn}(t)$ in terms of the filter $f_k(t)$. Thus,

$$\begin{aligned}
\psi_{kn}(t) &= a^{-k/2} f(a^{-k}t - nT) \\
&= a^{-k/2} f(a^{-k}(t - na^k T)) \\
&= f_k(t - na^k T),
\end{aligned} \qquad (11.3.12f)$$

which is a shifted version of the synthesis filter $f_k(t)$.

Orthonormal Basis

Of particular interest is the case where $\{\psi_{kn}(t)\}$ is a set of orthonormal functions. Such functions satisfy

$$\int_{-\infty}^{\infty} \psi_{kn}^*(t)\psi_{\ell m}(t)dt = \delta(k-\ell)\delta(n-m). \qquad (11.3.13a)$$

Using Parseval's theorem, this becomes

$$\frac{1}{2\pi}\int_{-\infty}^{\infty} \Psi_{kn}^*(j\Omega)\Psi_{\ell m}(j\Omega)d\Omega = \delta(k-\ell)\delta(n-m). \qquad (11.3.13b)$$

By using the orthonormality property in (11.3.10) we obtain

$$X_{DWT}(k,n) = \int_{-\infty}^{\infty} x(t)\psi_{kn}^*(t)dt. \qquad (11.3.13c)$$

Comparing with (11.3.7a) we conclude

$$\begin{aligned}\psi_{kn}(t) &= a^{-k/2}h^*(nT - a^{-k}t) \\ &= a^{-k/2}h^*\big(a^{-k}(na^kT - t)\big) \\ &= h_k^*(a^k nT - t).\end{aligned} \qquad (11.3.14a)$$

In particular, $\psi_{00}(t) = \psi(t) = h^*(-t)$. But we have $\psi(t) = f(t)$ so that, in the orthonormal case, $f(t) = h^*(-t)$. Thus,

$$f_k(t) = h_k^*(-t) \qquad \text{(orthonormal case)}. \qquad (11.3.14b)$$

This is very similar to the relation (6.2.6) for the perfect reconstruction paraunitary QMF banks! (We could let $c = 1$ and $L = 0$ in (6.2.6) without affecting any significant conclusions of Sec. 6.2).

Table 11.3.1 summarizes the definition and the main features of the wavelet transform.

Completeness, uniqueness, and so forth

Given an arbitrary function $x(t)$, suppose we compute $X_{DWT}(k,n)$ using (11.3.7a). Can we then express $x(t)$ as in (11.3.10)? That is, can we invert the transformation? The answer depends on $\psi(t)$, and the discretization parameters T and a. If this is possible for a specified class of functions $\{x(t)\}$, then the wavelet basis $\{\psi_{kn}(t)\}$ is said to be *complete* over this class. In Sec. 11.5 we will see how to generate a complete orthonormal basis for the class of finite energy functions.

Another practical requirement in addition to completeness is that, the transformation and reconstruction formulas (11.3.7a) and (11.3.10) should be 'stable' so that the computations do not "blow up." In the next section where we study discrete time wavelets, we will see that the issues of completeness and stability are much easier to address.

TABLE 11.3.1 The continuous-time wavelet transform

Traditional Continuous-Time Fourier Transform

$$X(j\Omega) = \int_{-\infty}^{\infty} x(t)e^{-j\Omega t} dt \quad \text{(Fourier transform)} \qquad (2.1.20)$$

$$x(t) = \frac{1}{2\pi} \int_{-\infty}^{\infty} X(j\Omega)e^{j\Omega t} d\Omega \quad \text{(inverse transform)} \qquad (2.1.21)$$

Wavelet transform, general

$$X_{CWT}(p,q) = \frac{1}{\sqrt{|p|}} \int_{-\infty}^{\infty} x(t) f\left(\frac{t-q}{p}\right) dt \qquad (11.3.8)$$

Discrete Wavelet Transform, DWT (still continuous-time)

Obtained by setting $p = a^k$, $q = a^k Tn$, $f(t) = h(-t)$ in (11.3.8), for integer k, n. (We can take $T = 1$ for simplicity.)

$$X_{DWT}(k,n) = a^{-k/2} \int_{-\infty}^{\infty} x(t) h(nT - a^{-k}t) dt \quad \text{(DWT)}, \qquad (11.3.7a)$$

$$x(t) = \sum_k \sum_n X_{DWT}(k,n) \underbrace{a^{-k/2} f(a^{-k}t - nT)}_{\text{basis } \psi_{kn}(t)} \quad \text{(inverse DWT)}.$$

$$(11.3.12c)$$

The inversion formula assumes that the filter bank of Fig. 11.3-2 has the perfect reconstruction property, with the filters chosen as

$$\underbrace{h_k(t) = a^{-k/2} h(a^{-k}t)}_{\text{analysis filters}}, \qquad \underbrace{f_k(t) = a^{-k/2} f(a^{-k}t)}_{\text{synthesis filters}}$$

Thus, the functions $h(t)$ and $f(t)$ in (11.3.7a) and (11.3.12c) play the role of prototype filters in a filter bank where all the filters are derived by *dilation* of a single filter. The basis functions in (11.3.12c) are dilated ($t \to a^{-k}t$) and shifted ($t \to t - na^k T$) versions of $f(t)$, that is,

$$\psi_{kn}(t) = a^{-k/2} \psi(a^{-k}t - nT), \qquad (11.3.12d)$$

where $\psi(t) = f(t) =$ wavelet function or mother wavelet.

Continued →

Table 11.3.1 continued

Special case of orthonormal basis functions

$$\int_{-\infty}^{\infty} \psi_{kn}^*(t)\psi_{\ell m}(t)dt = \delta(k-\ell)\delta(n-m) \quad \text{(orthonormality)}. \quad (11.3.13a)$$

Under this condition,

$$X_{DWT}(k,n) = \int_{-\infty}^{\infty} x(t)\underbrace{a^{-k/2}f^*(a^{-k}t - nT)}_{\psi_{kn}^*(t)}dt, \quad (11.3.13c)$$

$$x(t) = \sum_k \sum_n X_{DWT}(k,n)\underbrace{a^{-k/2}f(a^{-k}t - nT)}_{\text{basis } \psi_{kn}(t)}. \quad (11.3.12c)$$

Again, $\psi_{00}(t) = f(t) = \psi(t) =$ wavelet function.

Filter Bank Properties in the Orthonormal Case
1. Synthesis filters for perfect reconstruction: $f_k(t) = h_k^*(-t)$.
2. Relation between prototype filters: $f(t) = h^*(-t)$.

11.3.3 Discrete-Time Wavelet Transforms

We now extend the wavelet transformation to the case of discrete-time signals. The starting point again is a set of filters with frequency responses having an appearance similar to Fig. 11.3-2(a). In the continuous-time case these filters were related as in (11.3.1). If we attempt to mimic this by replacing t with the discrete-time index n, the quantity $a^{-k}n$ does not in general remain an integer. Let us, therefore, try to imitate the frequency domain relation (11.3.2) rather than the time domain relation.

Consider the example $a = 2$. The equivalent of (11.3.2) for digital filters would be

$$H_k(e^{j\omega}) = H(e^{j2^k\omega}), \quad (11.3.15)$$

that is, $H_k(z) = H(z^{2^k})$ where k is a nonnegative integer. For highpass $H(e^{j\omega})$, the responses of $H_k(e^{j\omega})$ for $k = 1$ and $k = 2$ are shown in Fig. 11.3-4. This shows that in general $H_k(z)$ is a multiband (rather than bandpass) filter, and further modification is required to obtain bandpass responses. For this we cascade $H_k(z)$ with appropriate filters. Thus, let $G(z)$ be a lowpass filter with response as in Fig. 11.3-5(a). Then the responses of

$$H(z), \; G(z)H(z^2), \; G(z)G(z^2)H(z^4), \; \ldots \qquad (11.3.16)$$

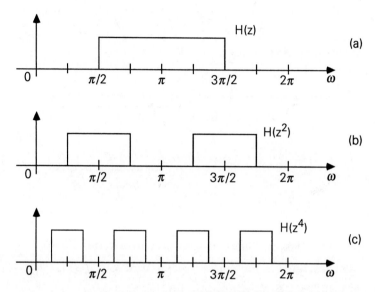

Figure 11.3-4 Magnitude response plots of (a) $H(z)$, (b) $H(z^2)$, and (c) $H(z^4)$.

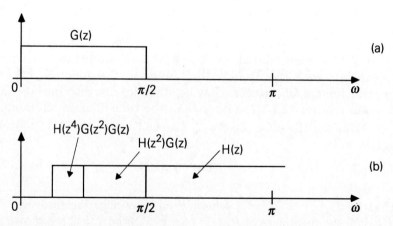

Figure 11.3-5 Magnitude responses of (a) lowpass $G(z)$, and (b) combinations of $H(z)$ and $G(z)$.

492   Chap. 11. Wavelet transforms and filter banks

are as shown in Fig. 11.3-5(b). These resemble Fig. 11.3-2(a) (except for the heights, which can be adjusted to make all the filter energies equal; see Fig. 11.3-8 later). The plots are shown only for $0 \leq \omega < \pi$ as we assume, for simplicity, magnitude symmetry with respect to π. The filters are bandpass, with center frequencies

$$\omega_k = c \times 2^{-k}\pi, \quad 0 \leq k \leq M-1, \qquad (11.3.17)$$

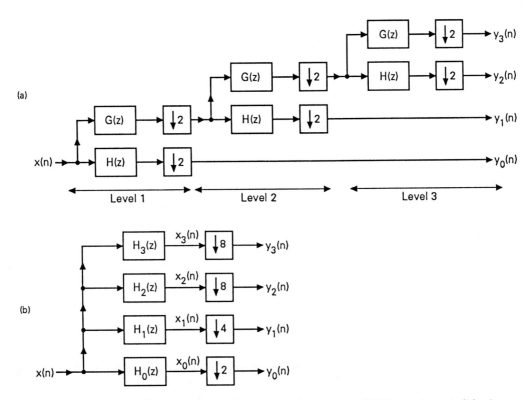

Figure 11.3-6 (a) A 3-level binary tree-structured QMF bank and (b) the equivalent four-channel system.

for appropriate c, and passband widths $BW_k = 2^{-k}\pi/2$ (measured only in the range $[0, \pi]$). Thus, the ratio BW_k/ω_k is independent of k.

From our experience with QMF banks (Sec. 5.8) we already know that filters of the form (11.3.16) can be generated with the help of a binary tree structure. Fig. 11.3-6 shows a three-level tree structure along with the equivalent nontree form, which has four filters. More generally if the tree has L levels, the number of channels is $M = L + 1$. The signals $x_k(n)$ can be decimated by the amounts shown, in order to obtain a maximally decimated system precisely as in a QMF bank. The decimation ratios are also consistent with the bandwidths of the signals $x_k(n)$. As in any multirate system with nonideal filters, decimation introduces aliasing. By using the techniques of Chap. 5 this can be canceled with an appropriate synthesis bank.

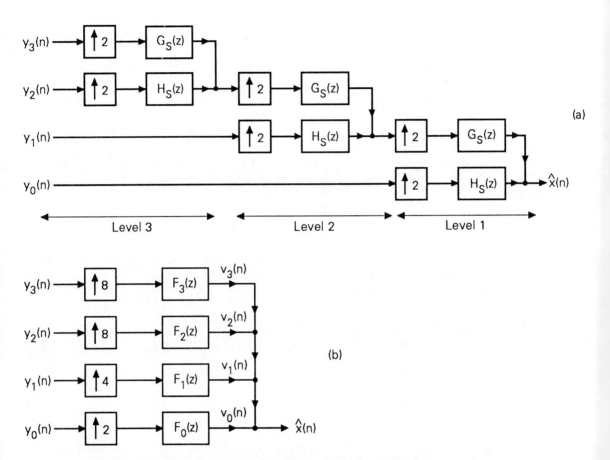

Figure 11.3-7 (a) The synthesis bank corresponding to Fig. 11.3-6 and (b) equivalent four-channel system.

Defining the Discrete-Time Wavelet Transform

In the z-domain we have $X_k(z) = H_k(z)X(z)$ so that

$$x_k(n) = \sum_m x(m) h_k(n-m). \qquad (11.3.18)$$

The decimated signals $y_k(n)$ are the *wavelet coefficients*, and the wavelet transform is given by

$$y_k(n) = \sum_{m=-\infty}^{\infty} x(m)h_k(2^{k+1}n - m), \quad 0 \le k \le M - 2,$$

$$y_{M-1}(n) = \sum_{m=-\infty}^{\infty} x(m)h_{M-1}(2^{M-1}n - m) \quad \text{(DTWT)}.$$

(11.3.19)

This is analogous to the situation in Fig. 11.2-8 where we obtained the STFT coefficients using a multirate filter bank. Equation (11.3.19) is the discrete-time wavelet transform (DTWT).

The inverse transform. The inversion of the above transform can be performed by designing an appropriate synthesis bank. Consider the synthesis filter bank of Fig. 11.3-7(a). This is equivalent to Fig. 11.3-7(b) with filters expressible entirely in terms of $G_s(z)$ and $H_s(z)$. For example,

$$\begin{aligned} F_0(z) &= H_s(z), \\ F_1(z) &= H_s(z^2)G_s(z), \\ F_2(z) &= H_s(z^4)G_s(z^2)G_s(z), \end{aligned}$$

(11.3.20)

and so on. We know from Sec. 5.8 that if the filters $G(z), H(z), G_s(z)$ and $H_s(z)$ are appropriately designed, the tree structured system produces perfect reconstruction, that is, $\widehat{x}(n) = x(n)$. Under this condition we can express

$$X(z) = F_0(z)Y_0(z^2) + F_1(z)Y_1(z^4) + \ldots + F_{M-2}(z)Y_{M-2}(z^{2^{M-1}})$$
$$+ F_{M-1}(z)Y_{M-1}(z^{2^{M-1}}) \quad \text{(inverse DTWT)}.$$

(11.3.21)

In Eqn. (4.1.22) we showed how to express the output of an interpolation filter in the time domain. Using similar principles, the above expression for inverse DTWT can be written in the time domain as

$$x(n) = \sum_{k=0}^{M-2} \sum_{m=-\infty}^{\infty} y_k(m) \underbrace{f_k(n - 2^{k+1}m)}_{\eta_{k\,m}(n)}$$
$$+ \sum_{m=-\infty}^{\infty} y_{M-1}(m) \underbrace{f_{M-1}(n - 2^{M-1}m)}_{\eta_{M-1,m}(n)},$$

(11.3.22a)

that is,

$$x(n) = \sum_{k=0}^{M-1} \sum_{m=-\infty}^{\infty} y_k(m)\eta_{km}(n) \quad \text{(inverse DTWT)}.$$

(11.3.22b)

Here $\eta_{km}(n)$ are the wavelet basis functions, and the weights $y_k(m)$ are the wavelet coefficients of $x(n)$ with respect to the above basis. Notice that $\eta_{k0}(n) = f_k(n) =$ synthesis filters. The basis function $\eta_{km}(n)$ is the impulse response $f_k(n)$ *shifted* by an appropriate amount.

Relation to Multiresolution Components

Refer to the analysis/synthesis system in Figs. 11.3-6 and 11.3-7. Assuming that $H_3(z)$ and $F_3(z)$ are good lowpass filters, the signal $v_3(n)$ is a lowpass filtered version of $x(n)$. (This is only approximately so, because of aliasing and imaging caused by the decimation and interpolation operations). The signal $v_2(n)$, on the other hand is a bandpass filtered version, and adds finer high frequency details. Thus, we can regard $v_3(n)$ to be a lowpass (i.e., smoothed) approximation of $x(n)$, whereas the sum $v_3(n) + v_2(n)$ is a 'higher resolution' approximation. By adding the component $v_1(n)$ we get a further refined approximation. Finally when $v_0(n)$ (the finest 'detail signal') is added, we obtain perfect recovery of $x(n)$. The tree structure (or wavelet decomposition) can therefore be used to transmit information (e.g., a picture in video conferencing) in various installments, with successively improved fine details.

Some Practical Requirements

1. *Stability.* In practice we require the filters $H_k(z)$ and $F_k(z)$ to be stable (equivalently $H(z), G(z), H_s(z)$ and $G_s(z)$ stable). This ensures that the procedure (11.3.19) to construct the wavelet coefficients, as well as the inversion procedure (11.3.22b) are stable.
2. *Orthonormality.* It is also desirable to have an orthonormal set of basis functions $\eta_{km}(n)$. This means

$$\sum_{n=-\infty}^{\infty} \eta_{km}(n)\eta_{\ell i}^*(n) = \delta(k-\ell)\delta(m-i). \qquad (11.3.23)$$

By using this in (11.3.22) we verify that orthonormality implies

$$y_k(m) = \sum_{n=-\infty}^{\infty} x(n)\eta_{km}^*(n). \qquad (11.3.24)$$

By comparing (11.3.19), (11.3.22a), and (11.3.24) we can eliminate $\eta_{km}(n)$ and obtain the relation

$$f_k(n) = h_k^*(-n). \qquad (11.3.25)$$

Thus, the analysis and synthesis filters are related as above when the basis is orthonormal. This is similar to the relation between filters in a *paraunitary* PR QMF bank! (Sec. 6.2). In the next section, we will present the precise relation between paraunitary QMF banks and orthonormal wavelets.

3. *Height conventions.* The increasing heights in Fig. 11.3-8 are chosen such that the energies of the filters are equal. This, however, is only a convention. It should be realized that the output $y_k(n)$ of the filter $H_k(z)$ is an estimate of $X(e^{j\omega_k})$ (with no scale-factor discrepancies) only if the heights of $|H_k(e^{j\omega})|$ in their passbands are inversely proportional to the bandwidths (i.e., heights are $c, 2c, 4c, \ldots$). See Problem 11.1. This requirement is *not consistent* with Fig. 11.3-8 (where the heights are $c, \sqrt{2}c, 2c \ldots$). In this chapter, conventions for heights will be flexible, depending on the particular context.

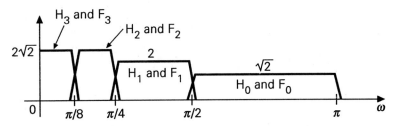

Figure 11.3-8 Typical appearances of magnitude responses of filters in the 3-level tree.

11.3.4 Summary

The STFT

The short time Fourier transform system is reproduced in Fig. 11.3-9. This is an M channel filter bank with equal-bandwidth filters having equispaced center frequencies. All the filters are generated from a prototype $H_0(z)$ as in (11.2.5). $E_k(z)$ are the polyphase components of this prototype as shown by (11.2.6). The system can be viewed as the sliding window system (Fig. 11.2-1) with $H_0(e^{j\omega}) = V(e^{-j\omega})$.

The quantity $|x_k(n)|$ represents the estimate of the magnitude of the transform of $x(i)$, *around* the center-frequency of $H_k(e^{j\omega})$, with the data $x(i)$ 'localized' *around* time n (which is the window position). As shown in earlier figures, each of the outputs $x_k(n)$ can be decimated by a factor $\leq M$. For any fixed n, the set of values

$$|x_k(n)|, \quad 0 \leq k \leq M - 1, \qquad (11.3.26)$$

i.e., the vector $[x_0(n) \quad \ldots \quad x_{M-1}(n)]^T$, provides a 'snapshot' of the magnitude of the Fourier transform of $x(i)$, localized around time n. The snapshot is delivered as a *uniformly* sampled version (in the frequency domain). The figure demonstrates this sampling for $n = -1, 3,$ and 7. In this demonstration, we see that the signal changes slowly from lowpass to highpass. Thus, the STFT keeps track of the evolution of the Fourier transform. As shown in the figure the set of filters can be either sharp-cutoff, or highly overlapping. The latter happens, for example, when $E_k(z) = 1$ for all k; this corresponds to the computation of DFT of blocks of the input.

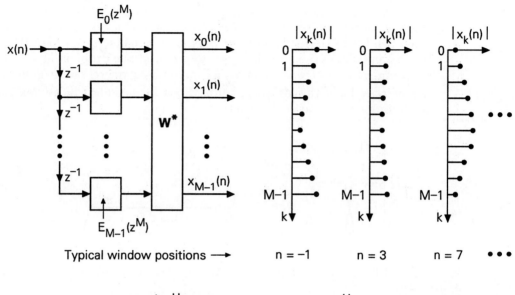

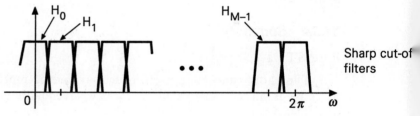

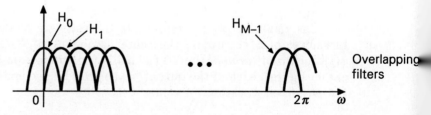

Figure 11.3-9 Uniform DFT filter banks as short-time Fourier transformers. For each window position n, we obtain a 'snapshot' of localized Fourier transform.

The Discrete-Time Wavelet Transform

The wavelet system (nonuniform filter bank) is reproduced in Fig. 11.3-10. This is an M-channel filter bank with nonuniform bandwidths for the filters. All the filters are generated from a tree structure as in Fig. 11.3-6.

The system cannot be viewed as the sliding window system, but one can imagine that there is an underlying window whose width is adjusted according to frequency. The outputs of the filters (bandpass signals) are modulated to obtain $x_k(n)$ which are *lowpass* signals, with *increasing* bandwidths as k decreases.

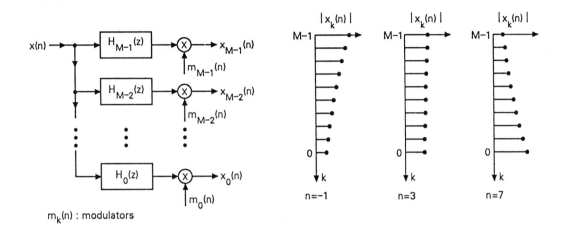

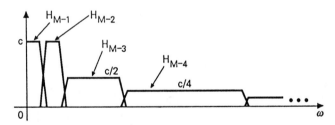

Figure 11.3-10 Summarizing the operation of a wavelet filter bank.

Figure 11.3-10 also shows the responses of the nonuniformly spaced filters. The quantity $x_k(n)$ represents the estimate of the transform of $x(i)$, *around* the center-frequency of $H_k(e^{j\omega})$, with the data $x(i)$ "localized" around time n. Even though all the signals $x_k(n)$ are lowpass, their bandwidths are different. Thus $x_1(n)$ varies "more slowly" than $x_0(n)$, and so forth. So the signals are decimated by unequal amounts (in fact by a factor inversely proportional to the filter bandwidth).

For any fixed n, the set of values

$$x_k(n), \quad 0 \leq k \leq M-1, \qquad (11.3.27)$$

provides a 'snapshot' of the nonuniformly sampled version of the Fourier transform of $x(i)$, localized around n. (This assumes that the heights of the filters are inversely proportional to the bandwidths, see Problem 11.1). The figure demonstrates this for $n = -1, 3$, and 7.

General Comments

1. If the analysis/synthesis system (i.e., Fig. 11.3-6 followed by Fig. 11.3-7) has the perfect reconstruction property $\hat{x}(n) = x(n)$, then (11.3.22b) represents the expansion of $x(n)$ in terms of the wavelet basis functions $\eta_{km}(n)$. If (11.3.23) holds, then the basis is orthonormal.

2. *FIR wavelet basis.* If the filters $H_s(z)$ and $G_s(z)$ are FIR, then the basis functions $\eta_{km}(n)$ are finite-length sequences.

3. When the QMF system satisfies the perfect reconstruction property, the basis is *complete* in the sense that any $x(n)$ can be expressed in this manner. Since we know that there exist FIR perfect reconstruction systems, this shows how to obtain a complete FIR wavelet basis (though $x(n)$ may not be FIR).

4. *The disk-partition diagram.* Figure 11.3-11(a) explains how the frequency domain is partitioned when performing the wavelet analysis. For simplicity we have shown only the upper half of the unit circle in the z-plane. (If the filters have real coefficients, the lower half need not be shown). The first level of the tree partitions the half-circle into two quarter circles. The second level splits the low-frequency quarter circle into two equal halves. This process is repeated L times in an L-level tree. For comparison, Fig. 11.3-11(b) shows the frequency partition for the case of uniform bandwidth filter banks, (e.g., STFT). Here the circle is divided into wedges of equal size. In both methods, the output of each filter is decimated in inverse proportion to the angular width of the wedge.

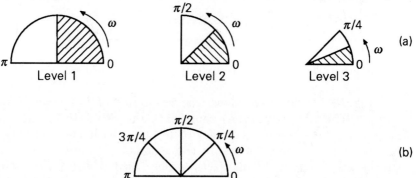

Figure 11.3-11 The disk partitioning diagram. (a) Wavelet transform, and (b) short-time Fourier transform.

11.4 DISCRETE-TIME ORTHONORMAL WAVELETS

From the previous sections we know that we can obtain a wavelet decomposition of a sequence $x(n)$ by using a tree structured perfect reconstruction

QMF bank of Figs. 11.3-6, 11.3-7. The coefficients $f_k(n)$ of the synthesis filters $F_k(z)$ govern the basis functions (see eqn. (11.3.22a)) whereas the decimated outputs $y_k(n)$ are the wavelet coefficients. Since the perfect reconstruction property holds for any $x(n)$, the expansion (11.3.22a) holds for any $x(n)$.

♠**Main points of this section.** We will show that if the filters $G_s(z)$ and $H_s(z)$ in the synthesis bank have the paraunitary property, then the basis functions $\{\eta_{km}(n)\}$ are orthonormal. Our development is in two steps. In Sec. 11.4.1 we will prove this result for a one-level tree (i.e., just a two channel QMF bank). In Sec. 11.4.2 this will be extended to an arbitrary number of levels. Finally in Sec. 11.4.3 a similar result will be proved for an M channel system with identical decimation ratio in all channels.

11.4.1 Two Channel Paraunitary QMF Banks

The fundamental building block in Figs. 11.3-6, 11.3-7 is the two-channel QMF bank reproduced in Fig. 11.4-1. Here $H_0(z)$ and $H_1(z)$ are typically lowpass and highpass, respectively, like $G(z)$ and $H(z)$ in Fig. 11.3-5. [In Fig. 11.3-6(a) we used the notations $G(z)$ and $H(z)$ instead of $H_0(z)$ and $H_1(z)$ in order to avoid confusion with Fig. 11.3-6(b)]. Here the two filters have equal bandwidth. We can think of this system as a simple special case of wavelet decomposition. Thus the wavelet coefficients are

$$y_k(n) = \sum_{m=-\infty}^{\infty} x(m) h_k(2n-m), \quad k = 0, 1. \qquad (11.4.1)$$

Assuming that the synthesis bank gives perfect reconstruction, we can express $x(n)$ as

$$x(n) = \sum_{m=-\infty}^{\infty} y_0(m) \underbrace{f_0(n-2m)}_{\eta_{0m}(n)} + \sum_{m=-\infty}^{\infty} y_1(m) \underbrace{f_1(n-2m)}_{\eta_{1m}(n)}. \qquad (11.4.2)$$

The wavelet bases $\eta_{km}(m)$ are indicated. These are stable if $F_k(z)$ are stable, and FIR if $F_k(z)$ are FIR.

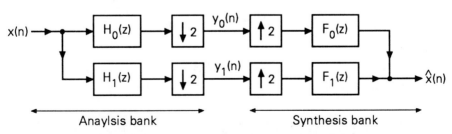

Figure 11.4-1 The two-channel QMF bank.

Orthonormality of Wavelet Basis

The wavelet basis is said to be orthonormal if (11.3.23) holds. Substituting for $\eta_{km}(n)$ from (11.4.2), this can be rewritten as

$$\sum_{n=-\infty}^{\infty} f_k(n-2m)f_\ell^*(n-2i) = \delta(k-\ell)\delta(m-i), \qquad (11.4.3)$$

that is, after a change of variables as

$$\sum_{n=-\infty}^{\infty} f_k(n)f_\ell^*(n-2m) = \delta(k-\ell)\delta(m). \qquad (11.4.4)$$

This means that $f_k(n)$ is orthogonal to the even-shifted versions of $f_\ell(n)$. Now the left side of this equation is the cross-correlation between $f_k(n)$ and $f_\ell(n)$ evaluated for even-lag $2m$ (Problem 2.14). Since the z-transform of the cross-correlation function is $F_k(z)\widetilde{F}_\ell(z)$, we can rewrite (11.4.4) in the z-domain as follows:

$$\left. F_k(z)\widetilde{F}_\ell(z) \right|_{\downarrow 2} = \delta(k-\ell). \qquad (11.4.5)$$

As explained in Sec. 4.1, the notation $\left. A(z) \right|_{\downarrow M}$ is an abbreviation which indicates decimation by M, for example,

$$\left. A(z) \right|_{\downarrow 2} = \frac{1}{2}[A(z^{1/2}) + A(-z^{1/2})]. \qquad (11.4.6)$$

We can rewrite (11.4.5) as

$$\widetilde{F}_k(z)F_\ell(z) + \widetilde{F}_k(-z)F_\ell(-z) = 2\delta(k-\ell), \quad 0 \le k,\ell \le 1, \qquad (11.4.7)$$

that is,

$$\widetilde{\mathbf{F}}(z)\mathbf{F}(z) = 2\mathbf{I} \quad \text{(wavelet orthonormality condition)}, \qquad (11.4.8)$$

where

$$\mathbf{F}(z) \triangleq \begin{bmatrix} F_0(z) & F_1(z) \\ F_0(-z) & F_1(-z) \end{bmatrix}. \qquad (11.4.9)$$

Thus orthonormality of the wavelet basis is equivalent to the paraunitary condition (11.4.8). Notice, in particular, that this implies the power complementary property $|F_0(e^{j\omega})|^2 + |F_1(e^{j\omega})|^2 = 2$.

Paraunitariness of R(z) and Wavelet Orthonormality

Recall (Sec. 5.5) that the two synthesis filters $F_0(z)$ and $F_1(z)$ can be expressed in terms of their 2×2 polyphase matrix $\mathbf{R}(z)$ in the form

$$[F_0(z) \quad F_1(z)] = [z^{-1} \quad 1]\mathbf{R}(z^2). \qquad (11.4.10a)$$

Using this relation, we can express the matrix $\mathbf{F}(z)$ as

$$\mathbf{F}(z) = \begin{bmatrix} 1 & 1 \\ -1 & 1 \end{bmatrix} \begin{bmatrix} z^{-1} & 0 \\ 0 & 1 \end{bmatrix} \mathbf{R}(z^2). \qquad (11.4.10b)$$

The above equation implies

$$\widetilde{\mathbf{F}}(z)\mathbf{F}(z) = 2\widetilde{\mathbf{R}}(z^2)\mathbf{R}(z^2). \qquad (11.4.11)$$

Thus, the discrete-time wavelet orthonormality condition holds if and only if

$$\widetilde{\mathbf{R}}(z)\mathbf{R}(z) = \mathbf{I}, \qquad (11.4.12)$$

which of course is the well-known (normalized-) paraunitary condition (Sec. 6.1.1)!

Designs Which Satisfy Orthonormality

In Sec. 5.3.6 we introduced an FIR perfect reconstruction QMF bank (invented independently in Smith and Barnwell [1984] and Mintzer [1985]). In this system, the FIR filter $H_0(z)$ is power symmetric, that is, satisfies

$$\widetilde{H}_0(z)H_0(z) + \widetilde{H}_0(-z)H_0(-z) = 2.$$

Under this condition we showed that if the remaining filters are chosen according to

1. $H_1(z) = -z^{-N}\widetilde{H}_0(-z)$, where $N = $ order of $H_0(z)$, and
2. $F_0(z) = \widetilde{H}_0(z)$ and $F_1(z) = \widetilde{H}_1(z)$,

then we have perfect reconstruction, with $\widehat{x}(n) = x(n)$. In Sec. 6.3.2 we saw that the above choice of filters ensures that the polyphase matrices $\mathbf{E}(z)$ and $\mathbf{R}(z)$ are paraunitary. This, therefore, ensures that the wavelet basis is orthonormal! Summarizing, the procedure to obtain a finite duration (FIR) orthonormal wavelet basis is as follows.

1. Design the Nth order FIR power symmetric filter $H_0(z)$. This is done either by starting from a zero-phase half band filter and computing a spectral factor $H_0(z)$ (Sec. 5.3.6), or equivalently by optimizing the lattice structure of Fig. 6.4-1. (Sec. 6.4.3).
2. Define the second analysis filter $H_1(z)$ and the synthesis filters $F_0(z)$ and $F_1(z)$ as above. Then the polyphase matrix $\mathbf{R}(z)$ satisfies the paraunitary condition $\widetilde{\mathbf{R}}(z)\mathbf{R}(z) = \mathbf{I}$. Notice that the synthesis filters are noncausal, but this is consistent with the delay-free reconstruction.

3. Define the wavelet basis as indicated in (11.4.2). Then the wavelet orthonormality condition is satisfied.
4. The analysis/synthesis system has the perfect reconstruction property, that is, $\widehat{x}(n) = x(n)$. So, (11.4.2) holds, and represents the expansion of the arbitrary input sequence $x(n)$ in terms of the orthonormal wavelet basis functions $\eta_{km}(n)$.

Completeness. Since any FIR power symmetric filter can be generated using the lattice mentioned in Step 1, any two-channel FIR orthonormal basis can be generated using this lattice.

11.4.2 Orthonormal Wavelets from Tree-Structured Paraunitary QMF Banks

Now consider an L-level tree-structured QMF bank (as demonstrated in Figs. 11.3-6, 11.3-7 for $L = 3$). The wavelet basis functions $\eta_{km}(n)$ are indicated in (11.3.22a) in terms of the filter coefficients $f_k(i)$. We now show that these basis functions are orthonormal if each of the two channel systems $[G_s(z), H_s(z)]$ has polyphase matrix $\mathbf{R}(z)$ satisfying $\widetilde{\mathbf{R}}(z)\mathbf{R}(z) = \mathbf{I}$.

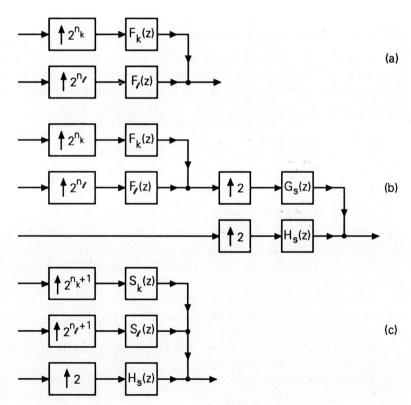

Figure 11.4-2 (a) Two of the $L+1$ branches in the synthesis bank of an L-level tree; (b) adding the $(L+1)$th level; and (c) redrawing the three branches.

From the previous section we know this to be true for $L = 1$. For arbitrary L we use an inductive reasoning. The tree structure with L levels has $L+1$ branches. Fig. 11.4-2(a) shows two of these branches, with $n_k \geq n_\ell$. Suppose we add another level to the tree. This adds a new branch, and modifies the existing branches as shown in Fig. 11.4-2(b). Assuming that

1. the wavelet bases are orthonormal for the L-level tree, and that
2. the new set of filters $[G_s(z), H_s(z)]$ has polyphase matrix $\mathbf{R}(z)$ satisfying $\widetilde{\mathbf{R}}(z)\mathbf{R}(z) = \mathbf{I}$,

we prove that the wavelet bases for the $(L+1)$-level tree are orthonormal.

From Sec. 6.3.1 recall that the paraunitary relation $\widetilde{\mathbf{R}}(z)\mathbf{R}(z) = \mathbf{I}$ is equivalent to the conditions

$$G_s(z)\widetilde{G}_s(z)\Big|_{\downarrow 2} = 1, \quad H_s(z)\widetilde{H}_s(z)\Big|_{\downarrow 2} = 1, \quad G_s(z)\widetilde{H}_s(z)\Big|_{\downarrow 2} = 0. \quad (11.4.13)$$

Instead of saying that $\mathbf{R}(z)$ is paraunitary, we will often say that the filter pair $[G_s(z), H_s(z)]$ is paraunitary (i.e., (11.4.13) holds).

Expressing Wavelet Orthonormality in z-Domain

Orthonormality of wavelets for the L-level tree implies

$$\sum_{n=-\infty}^{\infty} f_k(n - 2^{n_k}m)f_\ell^*(n - 2^{n_\ell}i) = \delta(k - \ell)\delta(m - i). \quad (11.4.14)$$

After a change of variables this can be rearranged as

$$\sum_{n=-\infty}^{\infty} f_k(n)f_\ell^*(n - 2^{n_\ell}m) = \delta(k - \ell)\delta(m), \quad (11.4.15)$$

using $2^{n_k} \geq 2^{n_\ell}$ (see Problem 11.13).

The summation on the left hand side of (11.4.15) is the cross-correlation between $f_k(n)$ and $f_\ell(n)$, evaluated at lags $2^{n_\ell}m$. Since the z-transform of the cross-correlation sequence is $F_k(z)\widetilde{F}_\ell(z)$, we can rephrase the above as

$$F_k(z)\widetilde{F}_\ell(z)\Big|_{\downarrow 2^{n_\ell}} = \delta(k - \ell), \quad 0 \leq k, \ell \leq L. \quad (11.4.16)$$

This is therefore another way of saying that the wavelet basis obtained from the L-level tree is orthonormal.

The inductive reasoning. The three branches of the $(L+1)$-level tree, shown in Fig. 11.4-2(b), can be redrawn as in Fig. 11.4-2(c) where

$$S_k(z) \triangleq F_k(z^2)G_s(z), \quad S_\ell(z) \triangleq F_\ell(z^2)G_s(z). \quad (11.4.17)$$

By using the identity

$$\left. \left(A(z^2)B(z) \right) \right|_{\downarrow 2^{K+1}} = \left. \left(A(z)(B(z)|_{\downarrow 2}) \right) \right|_{\downarrow 2^K} \qquad (11.4.18)$$

(Problem 11.9), we can prove that (11.4.13) and (11.4.16) imply

$$\left. S_k(z)\widetilde{S}_\ell(z) \right|_{\downarrow 2^{(n_\ell+1)}} = \delta(k-\ell), \ \left. S_k(z)\widetilde{H}_s(z) \right|_{\downarrow 2} = 0, \ \left. S_\ell(z)\widetilde{H}_s(z) \right|_{\downarrow 2} = 0,$$
(11.4.19)

which is sufficient to prove that the wavelet basis generated at the $(L+1)$th level remains orthonormal! The above reasoning does not assume that the pair $[G_s(z), H_s(z)]$ has to be the same for all levels of the tree. Notice finally that the FIR nature of the filters $G_s(z), H_s(z)$ ensures that the wavelet basis functions are FIR as well. The FIR nature also means, in particular, that the wavelet transformation as well as the inverse transformation are stable. Summarizing, we have proved:

♠**Theorem 11.4.1. Wavelet orthonormality.** Consider the L-level tree structure demonstrated in Figs. 11.3-6 and 11.3-7 for $L = 3$. Let the filters $G(z), H(z), G_s(z)$, and $H_s(z)$ be such that this is a perfect reconstruction system, that is, $\hat{x}(n) = x(n)$ so that $x(n)$ has the wavelet expansion (11.3.22a), where $M = L + 1$. Let $\mathbf{R}(z)$ be the 2×2 polyphase matrix of $[G_s(z), H_s(z)]$. Then the discrete-time wavelet basis $\{\eta_{km}(n)\}$ is orthonormal if and only if $\mathbf{R}(z)$ is paraunitary, that is, if and only if $[G_s(z), H_s(z)]$ forms a paraunitary pair. ◊

Unit energy property. Equation (11.4.15) implies, in particular, that $\sum_n |f_k(n)|^2 = 1$, that is, all the filters have unit energy. This is consistent with the increasing heights shown in Fig. 11.3-8. Note that the unit energy property holds *regardless* of the quality of the frequency responses (e.g., stopband attenuation, sharpness of cutoff, etc.), and is a direct consequence of the paraunitary property of the pair $[G_s(z), H_s(z)]$.

Use of lattice structure. As a converse, it can be shown that essentially all orthonormal wavelet bases can be generated using the lattice. See [Soman and Vaidyanathan, 1993] for precise statements.

Design example 11.4.1: STFT and Wavelet Filter Banks

A. Generalized STFT with orthonormal basis. Consider a four-channel maximally decimated filter bank system, designed using the tree structure of Fig. 5.8-1. For simplicity we take the filters at various levels to be the same, that is,

$$H_0^{(k)}(z) = G(z), H_1^{(k)}(z) = H(z), F_0^{(k)}(z) = G_s(z), F_1^{(k)}(z) = H_s(z).$$

We design this to be a perfect reconstruction system by designing $G(z)$ to be a causal Nth order FIR power symmetric filter (Sec. 5.3.6) and taking the remaining filters to be

$$H(z) = -z^{-N}\widetilde{G}(-z), \ G_s(z) = \widetilde{G}(z), \ H_s(z) = \widetilde{H}(z).$$

$G_s(z)$ and $H_s(z)$ are noncausal FIR filters. Assuming that $G(z)$ has been properly scaled, the power symmetric property implies

$$\widetilde{G}_s(z)G_s(z) + \widetilde{G}_s(-z)G_s(-z) = 2,$$

which in turn ensures $\widehat{x}(n) = x(n)$. The above choice of filters ensures the paraunitary condition $\widetilde{\mathbf{R}}(z)\mathbf{R}(z) = \mathbf{I}$.

For our demonstration we take $G(z)$ to be a 23rd order filter, designed as in Sec. 6.3.2. Fig. 11.4-3(a) shows the responses of $G_s(z)/\sqrt{2}$ and $H_s(z)/\sqrt{2}$. Fig. 11.4-3(b) shows the responses of the four analysis filters $H_k(z)$. These filters have equal passband bandwidths. Note that $|G_s(e^{j\omega})| = |G(e^{j\omega})|$ and $|H_s(e^{j\omega})| = |H(e^{j\omega})|$. Also $|F_k(e^{j\omega})| = |H_k(e^{j\omega})|$ for each k, according to the construction of these filters. Unlike in traditional STFT, the filters are not obtained by modulation of a prototype, hence the name "generalized" STFT.

B. *Wavelet filter bank with orthonormal basis.* We take a three level tree as in Fig. 11.3-6. Let $G(z), H(z), G_s(z)$ and $H_s(z)$ be as above so that we again have a perfect reconstruction system. The responses of the four analysis filters are now as in Fig. 11.4-3(c). In these plots, we have normalized $|H_k(e^{j\omega})|_{max}$ to be unity for convenience. Notice again that $|F_k(e^{j\omega})| = |H_k(e^{j\omega})|$ according to construction.

Since $\mathbf{R}(z)$ is paraunitary, the basis functions for the above generalized STFT as well as the wavelet filter bank are orthonormal.

11.4.3 Orthonormal Wavelets From M-channel Uniform-Decimation Paraunitary QMF Banks

We now generalize the result of Sec. 11.4.1 for the case of an M channel maximally decimated QMF bank with uniform decimation ratio for all channels (Fig. 5.4-1). If this system has the perfect reconstruction property, then $\widehat{x}(n) = x(n)$ so that we can express $x(n)$ as

$$x(n) = \sum_{k=0}^{M-1} \sum_{m=-\infty}^{\infty} y_k(m) \underbrace{f_k(n - Mm)}_{\eta_{km}(n)}. \qquad (11.4.20)$$

We have used the notation $y_k(n)$, instead of $v_k(n)$ as in Fig. 5.4-1, to be consistent with the rest of this chapter. We can regard $y_k(m)$ as the wavelet coefficients of $x(n)$ with respect to the wavelet basis defined by $\eta_{km}(n)$ indicated above. Note that $\eta_{k0}(n) = f_k(n)$, whereas $\eta_{km}(n)$ for arbitrary m is obtained by shifting $f_k(n)$ by *multiples of M*.

In the binary tree structure (Sec. 11.3.3) the bandwidths of the filters become smaller and smaller as the center frequency decreases (Fig. 11.3-8). But in the QMF bank under consideration the analysis filters typically tend to have equal bandwidths because the channels have equal decimation

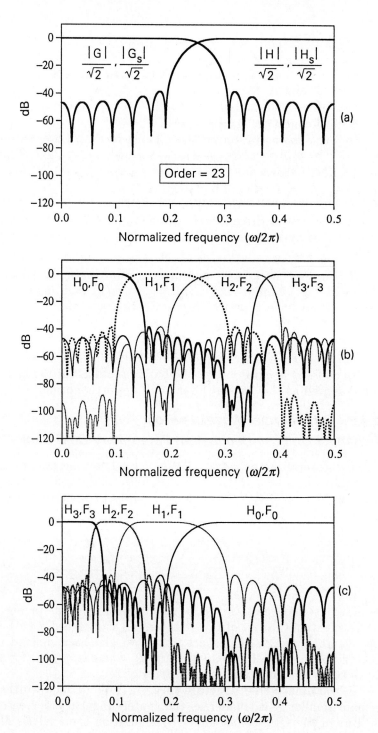

Figure 11.4-3 Design example 11.4-1. Magnitude responses of various filters. (a) $G_s(z)$, $G(z)$, $H_s(z)$ and $H(z)$, (b) STFT filters and (c) wavelet filters.

ratio M. This system is therefore closer to traditional (uniform bandwidth) spectrum analyzers rather than wavelet transformers.

Review of perfect reconstruction. Recall that the above QMF bank can always be redrawn as in Fig. 5.5-3(b), where $\mathbf{E}(z)$ and $\mathbf{R}(z)$ are the $M \times M$ polyphase matrices defined in Sec. 5.5. We also know that if $\mathbf{E}(z)$ is FIR and paraunitary then the choice of synthesis filters according to $f_k(n) = h_k^*(-n)$ gives rise to an FIR perfect reconstruction system. In this case, the expansion (11.4.20) holds. The above choice of synthesis filters implies, in particular, that $\mathbf{R}(z)$ is also paraunitary.

Paraunitariness of R(z) Implies Wavelet Orthonormality

We now show that the discrete time wavelet basis functions $\{\eta_{km}(n)\}$ are orthonormal if, and only if, $\widetilde{\mathbf{R}}(z)\mathbf{R}(z) = \mathbf{I}$.

Proof. Proceeding as in the previous subsection, we see that orthonormality of the basis implies

$$\sum_{n=-\infty}^{\infty} f_k(n) f_\ell^*(n - Mm) = \delta(k-\ell)\delta(m). \qquad (11.4.21)$$

Notice that this implies, in particular, that all filters have unit energy [regardless of the quality of the responses $F_k(e^{j\omega})$]. The above equation can be re-expressed in the z-domain as

$$\widetilde{F}_\ell(z) F_k(z) \Big|_{\downarrow M} = \delta(k-\ell) \quad \text{(wavelet orthonormality condition).} \qquad (11.4.22)$$

Using the z-domain expression for decimation (Sec. 4.1) this becomes

$$\sum_{m=0}^{M-1} \widetilde{F}_\ell(zW^m) F_k(zW^m) = M\delta(k-\ell), \quad 0 \le k, l \le M-1, \qquad (11.4.23)$$

with $W = e^{-j2\pi/M}$. This can be expressed compactly by defining

$$\mathbf{F}(z) \triangleq \begin{bmatrix} F_0(z) & F_1(z) & \cdots & F_{M-1}(z) \\ F_0(zW) & F_1(zW) & \cdots & F_{M-1}(zW) \\ \vdots & \vdots & \ddots & \vdots \\ F_0(zW^{M-1}) & F_1(zW^{M-1}) & \cdots & F_{M-1}(zW^{M-1}) \end{bmatrix}. \qquad (11.4.24)$$

Condition (11.4.23) can be expressed in terms of $\mathbf{F}(z)$ as

$$\widetilde{\mathbf{F}}(z)\mathbf{F}(z) = M\mathbf{I} \quad \text{(wavelet orthonormality condition).}$$

In a manner analogous to (11.4.10b), we can show (Problem 11.10) that $\mathbf{F}(z)$ is related to the $M \times M$ polyphase matrix $\mathbf{R}(z)$ according to

$$\mathbf{F}(z) = \mathbf{\Gamma}\mathbf{W}\mathbf{\Lambda}(z)\mathbf{R}(z^M). \qquad (11.4.25)$$

Here Γ and $\Lambda(z)$ are diagonal matrices with $[\Gamma]_{ii} = e^{-j2\pi i/M}$, $[\Lambda(z)]_{ii} = z^{-(M-1-i)}$, and $\mathbf{W}$ is the $M \times M$ DFT matrix. Using this we can show that the condition $\widetilde{\mathbf{F}}(z)\mathbf{F}(z) = M\mathbf{I}$ is equivalent to $\widetilde{\mathbf{R}}(z)\mathbf{R}(z) = \mathbf{I}$. Thus the wavelet basis functions are orthonormal if, and only if, the matrix $\mathbf{R}(z)$ is normalized paraunitary. ▽▽▽

This is an important result. From Sec. 6.4 and 6.5 we know how to generate the complete class of paraunitary matrices (of given degree and size). Using this we can, therefore, generate all discrete-time finite duration orthonormal wavelets.

11.4.4 Generalizations

The tree structured system of Fig. 11.3-6 can be generalized in several ways. First, instead of splitting a signal into two bands at a time, we can split into several bands. Second, signals such as $y_0(n)$ which are not split any further can themselves be decomposed into further subbands. In this way we obtain a very general tree structure. By modifying the synthesis bank appropriately, we can retain the perfect reconstruction property. The wavelet expansion equation (11.3.22a) can be modified for this case. This is called the *wavelet packet expansion* and was introduced in Coifman et al. [1990]. It can be shown that the paraunitary property of the filters ensures orthonormality of the basis functions of this expansion [Soman and Vaidyanathan, 1992].

The maximally decimated filter bank with arbitrary decimators n_k in the subbands (Fig. P5-32, Chap. 5 Problems section) is clearly a generalization of the filter banks discussed above. In this case, the transform domain coefficients are given by (11.2.15) and the inverse transform by (11.2.16). This is called the generalized STFT pair, the "filter bank transform" pair, or the "general discrete-time wavelet transform pair." Some of the properties of this system are summarized in Table 11.4.1. The nonuniform filter bank has been studied in a number of references, for example, Hoang and Vaidyanathan [1989], Kovačević and Vetterli [1991a], Nayebi et. al. [1991a], and Soman and Vaidyanathan [1993]. Also see Problem 11.22

11.5 CONTINUOUS-TIME ORTHONORMAL WAVELET BASIS

Consider the wavelet decomposition (11.3.7a) for a continuous time signal $x(t)$. Since k and n are integers, this is the discrete-wavelet transform (DWT) (but not discrete-time, as t is continuous). The case where $a = 2$ is commonly known as the diadic (or binary) wavelet decomposition. We will consider this special case, and further assume $T = 1$. For this case we show how, under certain conditions, an orthonormal basis can be generated starting from the discrete-time QMF bank.

For the discrete-time case (Figs. 11.3-6, 11.3-7) all the synthesis filters $F_k(z)$ (hence the wavelet basis functions $\eta_{km}(n)$) were constructed in terms of the basic filters $G_s(z)$ and $H_s(z)$. In a similar way we wish to construct a continuous-time wavelet basis starting from two basic functions. For this

TABLE 11.4.1 The discrete-time wavelet transform (DTWT)

The most general form is a nonuniform maximally decimated perfect reconstruction filter bank (Fig. P5-32, in Chap. 5 Problems section). Here $\sum_i 1/n_i = 1$ (maximal decimation). Let $n_i =$ integers for simplicity.

$$x_k(n) = \sum_{m=-\infty}^{\infty} x(m) h_k(n_k n - m), \quad 0 \le k \le M-1 \quad \text{(DTWT)} \quad (11.2.15)$$

$$x(n) = \sum_{k=0}^{M-1} \sum_{m=-\infty}^{\infty} x_k(m) \underbrace{f_k(n - n_k m)}_{\text{basis } \eta_{km}(n)} \quad \text{(inverse DTWT)} \quad (11.2.16)$$

This is identical to the *generalized* STFT, i.e., the 'filter bank' transform.

Orthonormal basis

1. *Definition:* $\sum_{n=-\infty}^{\infty} \eta_{km}(n) \eta_{\ell i}^*(n) = \delta(k-\ell)\delta(m-i)$.
2. *Equivalently* $\sum_n f_k(n) f_m^*(n + n_{k,m} p) = \delta(k-m)\delta(p)$. That is, $\left[F_k(z) \widetilde{F}_m(z) \right]_{\downarrow n_{k,m}} = \delta(k-m)$. Here $n_{k,m} = \gcd(n_k, n_m)$ (Problem 11.22).
3. *Perfect reconstruction condition:* $f_k(n) = h_k^*(-n)$.
4. *Paraunitariness.* A nonuniform filter bank can be converted into an equivalent uniform system with larger number of channels. If the larger system is paraunitary, the original system is orthonormal.

Special cases

1. *Uniform filter bank.* Here $n_k = M$ for all k. (Uniform bandwidth filters.) Now orthonormality is equivalent to the paraunitary property of the $M \times M$ polyphase matrix $\mathbf{R}(z)$.
2. *Octave filter bank.* Here $n_k = 2^{k+1}$ for $0 \le k \le M-2$ and $n_{M-1} = n_{M-2}$. Typical frequency responses have octave spacing and bandwidth, as in Fig. 11.3-8. Can be designed using the binary tree structure (Fig. 11.3-6, 11.3-7). $[G_s(z), H_s(z)]$ need not be the same at all levels of the tree. Orthonormality can be achieved by forcing $[G_s(z), H_s(z)]$ to be paraunitary at each level [i.e., forcing (11.4.13)]. Equivalently,
   a) $|G_s(e^{j\omega})|^2 + |H_s(e^{j\omega})|^2 = 2$ and
   b) $H_s(z) = z^{-N} \widetilde{G}_s(-z)$ for some odd N.
3. *Wave packets.* If the binary tree (Fig. 11.3-6, 11.3-7) is replaced with a more general version (with maximal decimation at each level), we obtain wavelet packets. The basis functions are orthonormal if the filter bank at each level has the paraunitary property.

we introduce the two infinite products

$$\Phi(\omega) = \frac{1}{\sqrt{2}} G_s(e^{j\omega/2}) \prod_{m=2}^{\infty} \frac{1}{\sqrt{2}} G_s(e^{j2^{-m}\omega}), \qquad (11.5.1a)$$

$$\Psi(\omega) = \frac{1}{\sqrt{2}} H_s(e^{j\omega/2}) \prod_{m=2}^{\infty} \frac{1}{\sqrt{2}} G_s(e^{j2^{-m}\omega}), \qquad (11.5.1b)$$

where

$$G_s(z) = \sum_n g_s(n) z^{-n}, \quad H_s(z) = \sum_n h_s(n) z^{-n}. \qquad (11.5.2)$$

We denote the inverse Fourier tranforms of $\Phi(\omega)$ and $\Psi(\omega)$ as $\phi(t)$ and $\psi(t)$ respectively. These are functions of the continuous time variable t, since $\Phi(\omega)$ and $\Psi(\omega)$ are not periodic in ω.[†] The function $\psi(t)$ (called the *wavelet function*) will play a crucial role in our discussions. The function $\phi(t)$ (called the *scaling function*) will enter many equations involving $\psi(t)$ [e.g., (11.5.15b) and (11.5.18b)].

To demonstrate how the infinite products work, assume that $G_s(e^{j\omega})$ and $H_s(e^{j\omega})$ are ideal lowpass and highpass filters with cutoff $\pi/2$. Fig. 11.5-1 shows a number of stretched (or dilated) versions of $G_s(e^{j\omega})$. From these we can judge that the infinite products $\Phi(\omega)$ and $\Psi(\omega)$ are lowpass and bandpass as shown in Fig. 11.5-2.

♠**Main points of this section.** Our aim is to generate wavelet basis functions $\psi_{k\ell}(t)$ by dilations and shifts of the wavelet function $\psi(t)$, that is,

$$\psi_{k\ell}(t) = 2^{-k/2} \psi(2^{-k} t - \ell). \qquad (11.5.3)$$

1. We will see that if $G_s(z)$ and $H_s(z)$ are FIR, then $\phi(t)$ and $\psi(t)$ are of finite duration. Thus, for each finite k and ℓ the function $\psi_{k\ell}(t)$ has finite duration.
2. Suppose $G_s(z)$ and $H_s(z)$ are FIR and form a paraunitary pair, that is, their polyphase matrix $\mathbf{R}(z)$ is paraunitary (i.e., (11.4.13) holds). Under some further mild conditions (to be made precise in Theorem 11.5.1), we will see that the set of functions $\{\psi_{k\ell}(t)\}$ is orthonormal. In fact, the set $\{\psi_{k\ell}(t)\}$ can be used to represent any finite energy function $x(t)$ in the form (11.3.10). That is, $\{\psi_{k\ell}(t)\}$ is complete over the so-called $L^2(R)$ class of functions.
3. In general, the function $\psi(t)$ obtained from the digital filters $G_s(z)$ and $H_s(z)$ as above is not "smooth". We will see that by constraining

---

[†] For convenience we continue to use ω rather than Ω in this section, even though $\phi(t)$ and $\psi(t)$ are continuous-time functions.

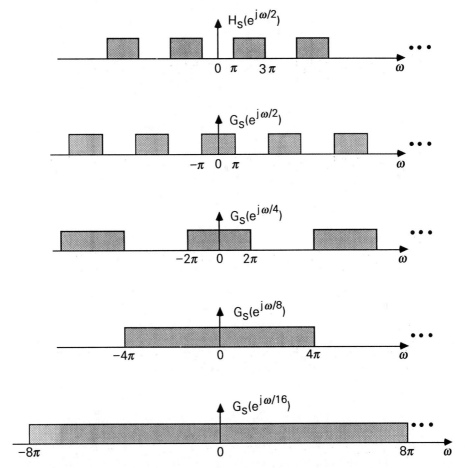

Figure 11.5-1 Various dilated versions of the basic filters, which take part in the infinite product.

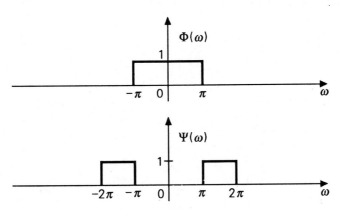

Figure 11.5-2 The magnitudes of $\Phi(\omega)$ and $\Psi(\omega)$ when $G_s(e^{j\omega})$ and $H_s(e^{j\omega})$ are ideal filters as in Fig. 11.5-1.

Sec. 11.5 Continuous-time orthonormal wavelets   513

$G_s(e^{j\omega})$ to have multiple zeros at $\omega = \pi$, the function $\psi(t)$ can be made "smooth" or "regular." We will also see how to design the Nth order FIR filter $G_s(z)$ in such a way that $G_s(z)$ has "as many zeros as possible" at $\omega = \pi$, under the additional constraint that $[G_s(z), H_s(z)]$ be a paraunitary pair. This maximizes "regularity" under the FIR orthonormal constraint.

The goal of this section is to provide a detailed derivation of the above results. We will also consider several illustrative examples to demonstrate these ideas. Some of the deeper mathematical issues (such as "completeness") will not be proved, but references will be provided.

11.5.1 Study of the Functions $\Phi(\omega)$ and $\Psi(\omega)$

As a first step towards the above goal, we now study some of the key properties of the infinite products $\Phi(\omega)$ and $\Psi(\omega)$. We begin with some examples.

Example 11.5.1: Infinite Products

Convergence of infinite products is tricky [Apostol, 1974]. Let $S = \prod_{m=1}^{\infty} s_m$. If there is an $\epsilon > 0$ such that $|s_m| < 1 - \epsilon$ for all m exceeding some integer m_0, then this product vanishes. On the other hand if $|s_m| > 1 + \epsilon$ for all $m > m_0$, the product does not converge to a finite value at all.

There do exist infinite products that converge to finite nonzero values. For example, let $s_m = a^{b^m}$, with $|b| < 1$. Then

$$S = \prod_{m=1}^{\infty} s_m = \prod_{m=1}^{\infty} a^{b^m} = a^{(\sum_{m=1}^{\infty} b^m)} = a^{b/(1-b)}. \qquad (11.5.4)$$

The above result is clearly finite. Readers who do not feel comfortable with the transition from infinite products to infinite sums in (11.5.4) must see Problem 11.14. For theorems on convergence of infinite products, see [Apostol, 1974] and pp.11-12 of Gradshteyn and Ryzhik [1980].

As a second example, let $s_m = \cos(2^{-m}\omega)$. We will show that the product converges:

$$\prod_{m=1}^{\infty} \cos(2^{-m}\omega) = \frac{\sin \omega}{\omega}. \qquad (11.5.5)$$

Note that $|\cos(2^{-m}\omega)| < 1$ for almost all ω, and yet the product converges to a nonzero value for almost all ω. This does not violate the statement in the preceding paragraph because there is no $\epsilon > 0$ such that $|\cos(2^{-m}\omega)| < 1 - \epsilon$ for all m greater than some m_0.

A simple proof of (11.5.5) is as follows: by using the identity $\cos \alpha = (\sin 2\alpha)/(2 \sin \alpha)$ we can simplify the partial product of the first K factors as

$$\prod_{m=1}^{K} \cos(2^{-m}\omega) = \frac{\sin \omega}{2^K \sin(\omega/2^K)}$$

As $K \to \infty$ we have $\sin(\omega/2^K) \to \omega/2^K$, so that

$$\lim_{K \to \infty} \prod_{m=1}^{K} \cos(2^{-m}\omega) = \frac{\sin \omega}{\omega}$$

indeed. Thus, the partial product converges to $(\sin \omega/\omega)$ pointwise, for each ω.

To obtain a 'system theoretic feeling' for the identity (11.5.5), recall that a product in the ω-domain translates to a convolution in the time domain. The inverse transform of $\cos(\omega/2)$ is a sum of two impulses, and that of $\cos(\omega/4)$ is a sum of two impulses placed closer together, and so on [Fig. 11.5-3(a)]. If these impulse trains are convolved, we obtain the rectangular window function [Fig. 11.5-3(b)] which indeed is the inverse transform of $(\sin \omega/\omega)$.

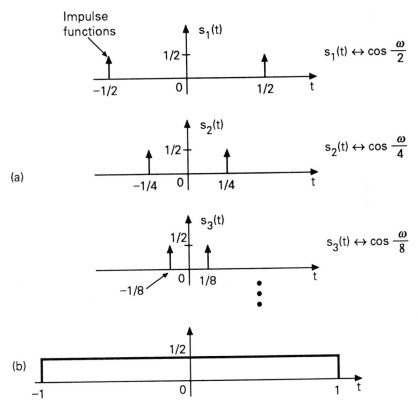

Figure 11.5-3 (a) Impulse functions whose Fourier transforms have the form $\cos(\omega/2^m)$ and (b) results of convolution of an infinite number of these.

Convergence of the Infinite Products

Consider the partial products

$$\Phi_L(\omega) = \frac{1}{\sqrt{2}}G_s(e^{j\omega/2})\prod_{m=2}^{L}\frac{1}{\sqrt{2}}G_s(e^{j2^{-m}\omega}), \qquad (11.5.6a)$$

$$\Psi_L(\omega) = \frac{1}{\sqrt{2}}H_s(e^{j\omega/2})\prod_{m=2}^{L}\frac{1}{\sqrt{2}}G_s(e^{j2^{-m}\omega}), \qquad (11.5.6b)$$

with

$$\Phi_1(\omega) = \frac{1}{\sqrt{2}}G_s(e^{j\omega/2}), \quad \Psi_1(\omega) = \frac{1}{\sqrt{2}}H_s(e^{j\omega/2}). \qquad (11.5.6c)$$

The infinite products $\Phi(\omega)$ and $\Psi(\omega)$ are, by definition, limits of these partial products as $L \to \infty$. Since $G_s(e^{j\omega})$ and $H_s(e^{j\omega})$ have period 2π, $\Phi_L(2^L\omega)$ and $\Psi_L(2^L\omega)$ have period 2π. In other words, the partial products $\Phi_L(\omega)$ and $\Psi_L(\omega)$ have period $2^{L+1}\pi$, and the infinite products $\Phi(\omega)$ and $\Psi(\omega)$ are nonperiodic. The inverse transforms therefore represent continuous-time functions $\phi(t)$ and $\psi(t)$. Notice that the quantities

$$\begin{aligned}\Phi_L(2^L\omega) = (\frac{1}{\sqrt{2}})^L G_s(e^{j\omega})\ldots G_s(e^{j2^{L-2}\omega})G_s(e^{j2^{L-1}\omega}),\\ \Psi_L(2^L\omega) = (\frac{1}{\sqrt{2}})^L G_s(e^{j\omega})\ldots G_s(e^{j2^{L-2}\omega})H_s(e^{j2^{L-1}\omega}),\end{aligned} \qquad (11.5.7)$$

are related to the synthesis filters $F_L(z)$ and $F_{L-1}(z)$ of the L-level tree structure (Fig. 11.3-7) as follows:

$$\Phi_L(2^L\omega) = \frac{F_L(e^{j\omega})}{(\sqrt{2})^L}, \quad \Psi_L(2^L\omega) = \frac{F_{L-1}(e^{j\omega})}{(\sqrt{2})^L}. \qquad (11.5.8)$$

Types of convergence. It is said that $\Psi_L(\omega)$ converges to $\Psi(\omega)$ *pointwise*, if for any ω in the range $-2^L\pi \le \omega < 2^L\pi$, the quantity $|\Psi_L(\omega) - \Psi(\omega)|$ tends to zero as $L \to \infty$. We say that $\Psi_L(\omega)$ converges to $\Psi(\omega)$ in the mean square sense if $\int_{-2^L\pi}^{2^L\pi}|\Psi_L(\omega) - \Psi(\omega)|^2 d\omega$ tends to zero as $L \to \infty$. Finally the convergence is said to be 'uniform' in an interval if some deeper requirements are satisfied [Kreyszig, 1972]. In this chapter, 'convergence' stands for pointwise convergence.

As mentioned above, pointwise convergence is not always guaranteed. However, if $|G_s(\omega)|/\sqrt{2} \le 1$ for all ω, then we have pointwise convergence. To see this note that when $|G_s(\omega)|/\sqrt{2} \le 1$ we have

$$|\Phi_{L+1}(\omega)| \le |\Phi_L(\omega)|. \qquad (11.5.9)$$

Thus $|\Phi_L(\omega)|$ is a monotone nonincreasing sequence in L, with a lower bound equal to zero. As a result, $\Phi_L(\omega)$ converges pointwise to a limit. The infinite-product $\Phi(\omega)$ denotes this limiting function. Similar comments hold for $\Psi(\omega)$. Notice, in particular, that if $[G_s(z), H_s(z)]$ is a paraunitary pair satisfying (11.4.13), we have

$$|G_s(e^{j\omega})|^2 + |H_s(e^{j\omega})|^2 = 2, \qquad (11.5.10)$$

so that $|G_s(e^{j\omega})|/\sqrt{2} \leq 1$ indeed, and pointwise convergence holds.

Example 11.5.2

Let $G_s(e^{j\omega})$ and $H_s(e^{j\omega})$ be ideal lowpass and highpass filters with cutoff $\pi/2$. From Fig. 11.5-2 we know that $\Phi(\omega)$ and $\Psi(\omega)$ are ideal lowpass and bandpass filters respectively. Their impulse responses are

$$\phi(t) = \frac{\sin(\pi t)}{\pi t}, \quad \psi(t) = \frac{\sin(\pi t/2)}{\pi t/2} \cos(3\pi t/2). \qquad (11.5.11)$$

Example 11.5.3.

Next assume that $G_s(z)$ and $H_s(z)$ are FIR with

$$G_s(z) = \frac{1 + z^{-1}}{\sqrt{2}}, \quad H_s(z) = \frac{1 - z^{-1}}{\sqrt{2}}. \qquad (11.5.12)$$

Then

$$G_s(e^{j\omega}) = \sqrt{2} e^{-j\omega/2} \cos(\omega/2), \quad H_s(e^{j\omega}) = j\sqrt{2} e^{-j\omega/2} \sin(\omega/2). \qquad (11.5.13)$$

By using these in the definitions of $\Phi(\omega)$ and $\Psi(\omega)$ and applying the identity (11.5.5), we obtain

$$\Phi(\omega) = e^{-j\omega/2} \frac{\sin(\omega/2)}{\omega/2}, \quad \Psi(\omega) = je^{-j\omega/2} \frac{\sin^2(\omega/4)}{\omega/4}. \qquad (11.5.14)$$

The inverse transforms $\psi(t)$ and $\phi(t)$ are shown in Fig. 11.5-4. These are also causal because of our causal choice of $G_s(z)$ and $H_s(z)$. We will use the above function $\psi(t)$ later to generate the so-called Haar-basis of wavelets (Example 11.5.5).

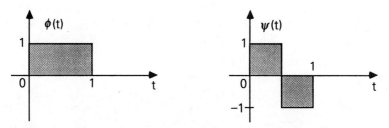

Figure 11.5-4 Example 11.5-3. The functions $\phi(t)$ and $\psi(t)$ which result from the choice of $G_s(z)$ and $H_s(z)$ as in (11.5.12).

Time Domain Interpretation of the Infinite Products

From the definition (11.5.1) we see that $\Phi(\omega)$ and $\Psi(\omega)$ can be expressed as

$$\Phi(\omega) = \frac{1}{\sqrt{2}} G_s(e^{j\omega/2}) \Phi(\omega/2) \qquad (11.5.15a)$$

$$\Psi(\omega) = \frac{1}{\sqrt{2}} H_s(e^{j\omega/2}) \Phi(\omega/2). \qquad (11.5.15b)$$

The above products can be expressed as convolutions in the time domain. To do this, note that the inverse transform of $G_s(e^{j\omega})$, viewed as a function of continuous time argument t, can be written as an impulse train:

$$\sum_{n=-\infty}^{\infty} g_s(n) \delta_a(t-n). \qquad (11.5.16)$$

Coupled with the fact that the inverse transform of $\frac{1}{2}F(\omega/2)$ is $f(2t)$, we see that

$$\phi(t) = 2\sqrt{2} \left(\sum_n g_s(n) \delta_a(2t-n) \right) * \phi(2t), \qquad (11.5.17)$$

where $*$ denotes convolution. Simplifying this we obtain the recursion

$$\phi(t) = \sqrt{2} \sum_{n=-\infty}^{\infty} g_s(n) \phi(2t-n). \qquad (11.5.18a)$$

Summarizing, the infinite product (11.5.1a) is equivalent to the above recursive relation, where $g_s(n)$ is related to $G_s(z)$ as in (11.5.2). Similarly, from (11.5.15b) we obtain the recursion

$$\psi(t) = \sqrt{2} \sum_{n=-\infty}^{\infty} h_s(n) \phi(2t-n). \qquad (11.5.18b)$$

Usually we take the filters $G_s(z)$ and $H_s(z)$ to be FIR, so that (11.5.18a) and (11.5.18b) are finite summations. For example if $G_s(z)$ and $H_s(z)$ are as in (11.5.12), the summations have only two terms. In this case $\phi(t)$ and $\psi(t)$ are as in Fig. 11.5-4, and the above recursions are easily verified.

Self-similarity. Figure 11.5-5 demonstrates the recursion (11.5.18a) for $N = 3$. Part (a) shows an example of $\phi(t)$. Part (b) shows the shifted and scaled versions $\sqrt{2}g_s(n)\phi(2t - n)$. Since $N = 3$, there are four curves, and these add up to $\phi(t)$ for all t. Notice that in the region $0 \le t \le 0.5$, the signal $\phi(t)$ is identical to a scaled version of $\phi(2t)$, which indicates the self-similar behavior of $\phi(t)$. Similar comments holds for $2.5 \le t \le 3.0$. It should be noticed that there is no simple equivalent of this elegant property, for the case of discrete-time wavelets.

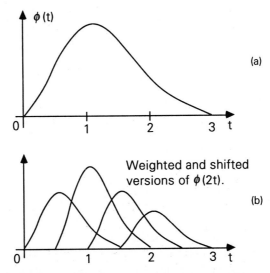

Figure 11.5-5 Demonstrating how a function is formed from superposition of shifted versions of a compressed version.

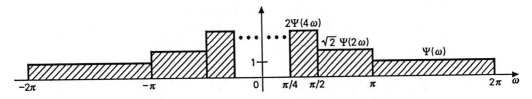

Figure 11.5-6 Various compressed versions of $\Psi(\omega)$. These are used to generate an orthonormal basis of wavelets.

Generation of the Wavelet Basis

Suppose $\Psi(\omega)$ is a bandpass response (e.g., approximating the one in

Fig. 11.5-2). Then the functions $2^{k/2}\Psi(2^k\omega)$ are bandpass with center frequencies as well as bandwidths reduced by the factor 2^k (e.g., see Fig. 11.5-6). (The scale factor $2^{k/2}$ serves to keep the energy of the filter same for all k.) Let these functions be used in the filter bank arrangement of Fig. 11.3-2(c), that is, $F_k(j\omega) = 2^{k/2}\Psi(2^k\omega)$. If this system has perfect reconstruction, then the reconstructed signal is

$$x(t) = \sum_k 2^{-k/2} \sum_\ell X_{DWT}(k,\ell)\psi(2^{-k}t - \ell). \qquad (11.5.19)$$

Thus, the wavelet basis functions are

$$\psi_{k\ell}(t) \triangleq 2^{-k/2}\psi(2^{-k}t - \ell), \qquad (11.5.20)$$

where k and ℓ are integers. The basis functions have Fourier transforms

$$\Psi_{k\ell}(\omega) = 2^{k/2} e^{-j2^k \ell \omega} \Psi(2^k\omega). \qquad (11.5.21)$$

If $x(t)$ is appropriately bandlimited, we can restrict k to be nonnegative.

For appropriate choice of the FIR filters $G_s(z)$ and $H_s(z)$, the functions $\psi_{k\ell}(t)$ can be made orthonormal. The following examples will demonstrate this. We will see later that the orthonormality property can be induced by designing $[G_s(z), H_s(z)]$ to be a paraunitary pair.

Example 11.5.4

Recall that when $G_s(e^{j\omega})$ and $H_s(e^{j\omega})$ are ideal filters (as in Fig. 11.5-1), $\Psi(\omega)$ is the ideal bandpass filter shown in Fig. 11.5-2. The quantities $\Psi(2^k\omega)$ do not overlap for two different values of k (Fig. 11.5-6). From this it is clear (use Parseval's relation) that the functions $\psi(2^{-k_1}t)$ and $\psi(2^{-k_2}t)$ are orthogonal for $k_1 \neq k_2$. As a further step, it is easy to verify that any two functions $\psi_{k\ell}(t)$ and $\psi_{mn}(t)$ are orthonormal (Problem 11.15).

Example 11.5.5: The Haar Basis

In the previous example, orthonormality was easy to see by considering the frequency domain, whereas a time domain approach is more suited in some cases. We know that if $G_s(z)$ and $H_s(z)$ are as in (11.5.12), $\psi(t)$ is the function shown in Fig. 11.5-4. Some of the wavelet basis functions $\psi_{k\ell}(t)$ are sketched in Fig. 11.5-7, and are evidently orthonormal. These are called the Haar basis functions.

It is easily verified that any two members of the family $\{\psi_{k\ell}(t)\}$ are orthonormal. It can be shown (see references in [Daubechies, 1988]) that these functions form a basis for the class of all real finite energy functions [commonly called the class $L^2(R)$].

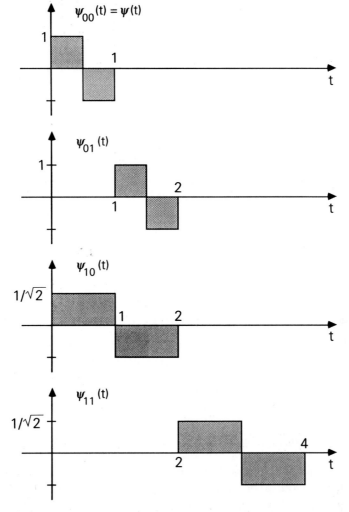

Figure 11.5-7 Example 11.5.5. Some of the basis functions belonging to the Haar basis.

Basis Functions with Finite Duration

We now show that if $G_s(z)$ and $H_s(z)$ are FIR, the functions $\phi(t)$, $\psi(t)$, and $\psi_{k\ell}(t)$ have finite duration. The infinite products in (11.5.1), are equivalent to infinite convolutions in the time domain. The inverse transform of $G_s(e^{j\omega})$, viewed as a function of the continuous variable t, can be expressed as the impulse-train function (11.5.16). The same comment holds for $H_s(e^{j\omega})$. Using this we obtain

$$\phi(t) = \frac{2}{\sqrt{2}} \sum_n g_s(n)\delta_a(2t-n) * \frac{4}{\sqrt{2}} \sum_n g_s(n)\delta_a(4t-n) * \ldots,$$

$$\psi(t) = \frac{2}{\sqrt{2}} \sum_n h_s(n)\delta_a(2t-n) * \frac{4}{\sqrt{2}} \sum_n g_s(n)\delta_a(4t-n) * \ldots,$$

(11.5.22)

Sec. 11.5 Continuous-time orthonormal wavelets 521

where $*$ indicates convolution. Assume that $G_s(z)$ and $H_s(z)$ are causal FIR with order N, that is,

$$G_s(z) = \sum_{n=0}^{N} g_s(n)z^{-n}, \quad H_s(z) = \sum_{n=0}^{N} h_s(n)z^{-n}. \qquad (11.5.23)$$

The region where the inverse transform of $G_s(e^{j2^{-m}\omega})$ can be nonzero is given by $0 \le t \le 2^{-m}N$. As m increases, this quantity has smaller and smaller duration, and the impulses are squeezed tighter. The convolution of infinite number of these will result in a function whose duration is

$$\frac{N}{2} + \frac{N}{4} + \frac{N}{8} + \ldots = N. \qquad (11.5.24)$$

Thus, $\phi(t)$ and $\psi(t)$ are causal functions with duration N. The wavelet basis functions $\psi_{k\ell}(t)$ are therefore of finite duration $2^k N$.

11.5.2 Generating Orthonormal Wavelets

From Example 11.5.4 we know that if the filters $G_s(e^{j\omega})$ and $H_s(e^{j\omega})$ are ideal then the wavelet basis $\psi_{k\ell}(t)$ is orthonormal. We also know from Example 11.5.5, that non ideal filters $G_s(z)$ and $H_s(z)$ can sometimes be used to get an orthonormal basis. In Sec. 11.5.3 we will derive a more general result, viz., if $[G_s(z), H_s(z)]$ is a paraunitary pair then $\{\psi_{k\ell}(t)\}$ is an orthonormal family, subject to some further mild conditions (Theorem 11.5.1). In this section we will prove a preliminary result required for this purpose.

Recall that the orthonormality of the wavelet basis is defined by the equation

$$\int_{-\infty}^{\infty} \psi_{k_1\ell_1}(t)\psi^*_{k_2\ell_2}(t)dt = \delta(k_1 - k_2)\delta(\ell_1 - \ell_2). \qquad (11.5.25)$$

In view of Parseval's relation, this is equivalent to

$$\int_{-\infty}^{\infty} \Psi_{k_1\ell_1}(\omega)\Psi^*_{k_2\ell_2}(\omega) \frac{d\omega}{2\pi} = \delta(k_1 - k_2)\delta(\ell_1 - \ell_2). \qquad (11.5.26)$$

Using (11.5.21) and making appropriate change of variables, this becomes

$$2^{i/2} \int_{-\infty}^{\infty} \Psi(\omega)\Psi^*(2^i\omega)e^{-j\omega(\ell_1 - 2^i\ell_2)} \frac{d\omega}{2\pi} = \delta(i)\delta(\ell_1 - \ell_2), \qquad (11.5.27)$$

where $i = k_2 - k_1$.

Our aim is to establish the connection between the above orthonormality and the paraunitary property of $[G_s(z), H_s(z)]$. Recall that $[G_s(z), H_s(z)]$ is a paraunitary pair if (11.4.13) holds. This condition means that if we define the polyphase matrix $\mathbf{R}(z)$ according to

$$[G_s(z) \quad H_s(z)] = [z^{-1} \quad 1] \mathbf{R}(z^2), \qquad (11.5.28)$$

then $\mathbf{R}(z)$ is paraunitary, i.e., $\widetilde{\mathbf{R}}(z)\mathbf{R}(z) = \mathbf{I}$. In Sec. 11.4 we already used the paraunitary property to derive discrete-time orthonormal wavelets.

We will now show that the above paraunitariness implies the relation

$$2^{i/2} \int_{-2^L \pi}^{2^L \pi} \Psi_L(\omega) \Psi_{L+i}^*(2^i \omega) e^{-j\omega(\ell_1 - 2^i \ell_2)} \frac{d\omega}{2\pi} = \delta(i)\delta(\ell_1 - \ell_2), \qquad (11.5.29)$$

for all integers $L > 0$, where $\Psi_L(\omega)$ is the partial product defined earlier. In the next section we will derive further conditions under which we can let $L \to \infty$ and obtain (11.5.27) from this.

Proof that Paraunitary Property Implies (11.5.29)

For convenience we first make a change of variables and rewrite (11.5.29) as

$$2^{i/2} 2^L \int_{-\pi}^{\pi} \Psi_L(2^L \omega) \Psi_{L+i}^*(2^{L+i} \omega) e^{-j 2^L \omega(\ell_1 - 2^i \ell_2)} \frac{d\omega}{2\pi} = \delta(i)\delta(\ell_1 - \ell_2). \qquad (11.5.30)$$

Case when $i > 0$. From the definition of $\Psi_L(\omega)$ we have (11.5.8), and we can rewrite (11.5.30) as

$$\int_{-\pi}^{\pi} F_{L-1}(e^{j\omega}) F_L^*(e^{j\omega}) A(e^{j 2^L \omega}) e^{-j 2^L \omega(\ell_1 - 2^i \ell_2)} \frac{d\omega}{2\pi} = 0, \qquad (11.5.31)$$

where $F_L(z)$ and $F_{L-1}(z)$ are the top two filters in the L-level tree-structured synthesis bank (Sec. 11.3). Here $A(z)$ is a discrete-time transfer function whose details are irrelevant for the proof.

If the pair $[G_s(z), H_s(z)]$ satisfies (11.4.13), the L-level tree structure generates orthonormal wavelets (Sec. 11.4.2). This implies, in particular,

$$\left(F_{L-1}(z)\widetilde{F}_L(z)\right)\Big|_{\downarrow 2^L} A(z) = 0, \qquad (11.5.32)$$

that is,

$$\left(F_{L-1}(z)\widetilde{F}_L(z) A(z^{2^L})\right)\Big|_{\downarrow 2^L} = 0. \qquad (11.5.33)$$

In other words, the inverse transform of $F_{L-1}(z)\widetilde{F}_L(z) A(z^{2^L})$ is zero at locations of the form $2^L m$, where $m = $ integer. This proves (11.5.31).

Case when $i = 0$. Now (11.5.30) reduces to

$$\int_{-\pi}^{\pi} F_{L-1}(e^{j\omega}) F_{L-1}^{*}(e^{j\omega}) e^{-j2^L \omega(\ell_1 - \ell_2)} \frac{d\omega}{2\pi} = \delta(\ell_1 - \ell_2). \quad (11.5.34)$$

But $F_{L-1}(z)$ satisfies (11.4.16), that is,

$$\left(F_{L-1}(z)\widetilde{F}_{L-1}(z)\right)\Big|_{\downarrow 2^L} = 1. \quad (11.5.35)$$

In other words, if we evaluate the inverse transform of $F_{L-1}(z)\widetilde{F}_{L-1}(z)$ at the locations $2^L n$, the result is zero for $n \neq 0$ and unity for $n = 0$. This establishes (11.5.34) indeed. ▽▽▽

Summarizing, we have proved that the paraunitary property of the filter bank $[G_s(z), H_s(z)]$ implies the orthonormality (11.5.29) for finite L. It has been shown [Mallat, 1989b] that this also implies (11.5.27) (i.e., it holds for $L \to \infty$) as long as $G_s(e^{j\omega}) \neq 0$ in $|\omega| \leq \pi/2$. This is only a mild requirement, since a lowpass filter in a QMF bank usually satisfies this. In Sec. 11.5.3 we will deal with the details of Mallat's conditions, and state the result more formally (Lemma 11.5.1 and Theorem 11.5.1).

Class of functions covered by the basis. Notice that $\psi_{k\ell}(t)$ has finite duration since $G_s(z)$ and $H_s(z)$ are FIR. The finite duration orthonormal wavelet basis $\psi_{k\ell}(t)$ generated in the above manner can be used to represent any *finite energy* signal $x(t)$ as in (11.3.10). See Daubechies [1988], Mallat [1989a,b], and references therein. Real signals with finite energy are said to belong to the class $L^2(R)$. So, we say that $\{\psi_{k\ell}(t)\}$ is complete over $L^2(R)$. With further restrictions on the FIR filters $G_s(z)$, the basis can be used to represent a wider class of functions, as elaborated in the above references. Proofs of these 'completeness statements' are beyond the scope of this chapter.

Design Example 11.5.1

Consider again the filters $G(z)$ and $H(z)$ used in Design example 11.4.1. With $G_s(z) = \widetilde{G}(z)$ and $H_s(z) = \widetilde{H}(z)$, the magnitude responses are as in Fig. 11.4-3(a). Since $[G_s(z), H_s(z)]$ is paraunitary, we can use these filters to obtain the orthonormal wavelet basis described above. Figure 11.5-8 shows the functions $\phi(t)$ and $\psi(t)$ as well as their Fourier transform magnitudes, verifying that they are lowpass and bandpass, respectively. Since the order of $G_s(z)$ is $N = 23$, the durations of $\phi(t)$ and $\psi(t)$ are also 23 as seen from the plots.

11.5.3 Orthonormality as L Approaches Infinity

The fact that (11.5.29) holds for all L is *not sufficient* to imply the orthonormality condition (11.5.27) of the continuous time wavelets, as demonstrated in the following example, shown to the author by Ingrid Daubechies.

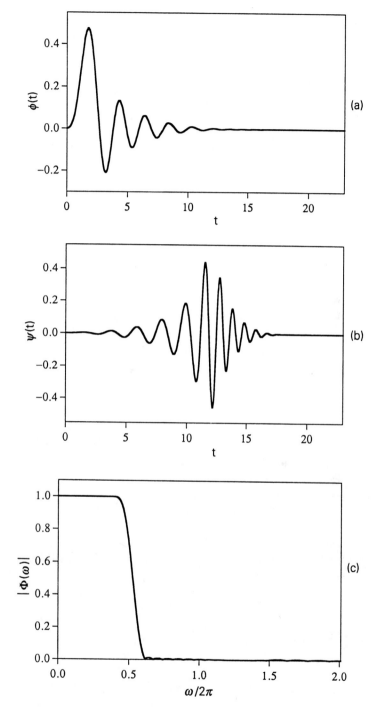

Figure 11.5-8 Design example 11.5.1. *Continued* →

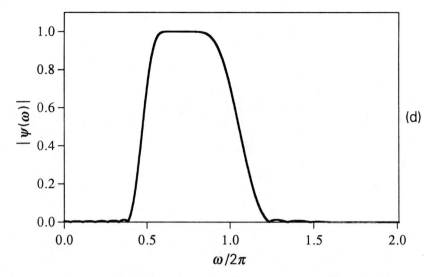

Figure 11.5-8 Design example 11.5.1. The functions $\phi(t)$ and $\psi(t)$, obtained by starting from $G_s(z)$ and $H_s(z)$. The magnitudes of $G_s(e^{j\omega})$ and $H_s(e^{j\omega})$ are as in Fig. 11.4-3(a).

Example 11.5.6

Consider the filter bank with filters

$$G_s(z) = \frac{1+z^{-3}}{\sqrt{2}}, \quad H_s(z) = \frac{1-z^{-3}}{\sqrt{2}}. \qquad (11.5.36)$$

This is a modification of the Haar basis example, with z replaced by z^3. It can be verified that the polyphase matrix $\mathbf{R}(z)$ is still paraunitary [i.e., (11.4.13) holds]. The functions $\Psi(\omega)$ and $\Phi(\omega)$ are obtained by replacing ω with 3ω in (11.5.14). The inverse transforms $\phi(t)$ and $\psi(t)$ are shown in Fig. 11.5-9. From the plots of Fig. 11.5-9 we can verify, in particular, that $\psi_{00}(t)$ and $\psi_{02}(t)$ are not mutually orthogonal.

We will now derive the conditions under which the paraunitary property of $[G_s(z), H_s(z)]$ implies the orthonormality condition (11.5.27).

Case when $i = 0$. Eqn. (11.5.27) reduces to

$$\int_{-\infty}^{\infty} |\Psi(\omega)|^2 e^{-j\omega n} \frac{d\omega}{2\pi} = \delta(n) \qquad (11.5.37)$$

for all integers n. We will express this in terms of $\Phi(\omega)$ for convenience of future discussion. The functions $\Phi(\omega)$ and $\Psi(\omega)$ are expressible as in (11.5.15).

Using this and the power complementary relation (11.5.10) (induced by paraunitariness), it can be shown that

$$|\Phi(\omega)|^2 + |\Psi(\omega)|^2 = |\Phi(\omega/2)|^2. \tag{11.5.38}$$

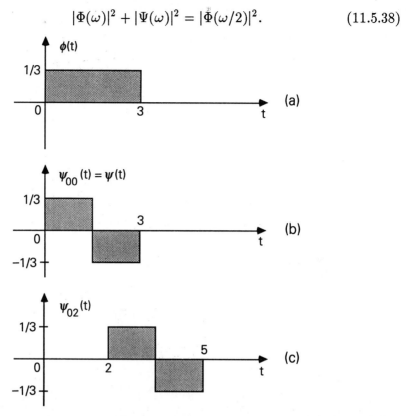

Figure 11.5-9 Example 11.5.6. Generation of basis functions using $G_s(z) = (1 + z^{-3})/\sqrt{2}$, $H_s(z) = (1 - z^{-3})/\sqrt{2}$. (a) Scaling function $\phi(t)$, (b) $\psi(t)$ and $\psi_{00}(t)$ and (c) $\Psi_{02}(t)$.

By using this we can verify that (11.5.37) is satisfied if

$$\int_{-\infty}^{\infty} |\Phi(\omega)|^2 e^{-j\omega n} \frac{d\omega}{2\pi} = \delta(n) \tag{11.5.39}$$

for all integers n. This equation has an interesting time domain interpretation. Thus, define $r_\phi(\tau)$ to be the deterministic autocorrelation function

$$r_\phi(\tau) = \int_{-\infty}^{\infty} \phi(t)\phi^*(t-\tau)\,dt \tag{11.5.40}$$

(as we did in Problem 2.14). Then, $|\Phi(\omega)|^2$ is the Fourier transform of $r_\phi(\tau)$. So (11.5.39) is equivalent to

$$r_\phi(n) = \delta(n), \quad \text{for all integers } n. \tag{11.5.41}$$

That is, if we 'sample' the function $r_\phi(\tau)$ with sampling period $T = 1$, the result is the unit pulse function $\delta(n)$. In other words, $r_\phi(\tau)$ has periodic zero-crossings, at (nonzero) integer values of the argument τ. This is the continuous-time analog of the *Nyquist(1)* property (Sec. 4.6.1), and we say that $r_\phi(\tau)$ is a Nyquist(1) function.

Using this interpretation and the standard expression (2.1.22) for the Fourier transform of a sampled signal, we can re-express (11.5.39) as

$$\sum_{k=-\infty}^{\infty} |\Phi(\omega + 2\pi k)|^2 = 1. \qquad (11.5.42)$$

Summarizing, the equations (11.5.39), (11.5.41) and (11.5.42) describe the same condition. If this condition is satisfied then (11.5.27) holds for $i = 0$.

Case when $i \neq 0$. Now (11.5.27) can be rewritten as

$$\int_{-\infty}^{\infty} \Psi(\omega)\Psi^*(2^i\omega)e^{-j\omega n}\,d\omega = 0, \quad \text{for all integers } n. \qquad (11.5.43)$$

By using the definitions of the partial products (11.5.6), we can simplify this to the equivalent form

$$\int_{-\infty}^{\infty} \Psi_1(\omega)\Psi_{i+1}^*(2^i\omega)|\Phi(\omega/2)|^2 e^{-j\omega n}\,d\omega = 0, \quad \text{for all integers } n. \qquad (11.5.44)$$

Using the fact that $\Psi_1(\omega)$ and $\Psi_{i+1}^*(2^i\omega)$ have a common period of 4π, this can be rewritten as

$$\sum_{k=-\infty}^{\infty} \int_{-2\pi}^{2\pi} \Psi_1(\omega)\Psi_{i+1}^*(2^i\omega)e^{-j\omega n}\left|\Phi(\frac{\omega}{2} + 2\pi k)\right|^2 d\omega = 0, \quad \text{for all integers } n. \qquad (11.5.45)$$

Interchanging the integral with the summation, we can rewrite this as

$$\int_{-2\pi}^{2\pi} \Psi_1(\omega)\Psi_{i+1}^*(2^i\omega)e^{-j\omega n} \sum_{k=-\infty}^{\infty}\left|\Phi(\frac{\omega}{2} + 2\pi k)\right|^2 d\omega = 0, \quad \text{for all integers } n. \qquad (11.5.46)$$

If $r_\phi(\tau)$ is Nyquist(1) that is, if (11.5.42) holds, then this is equivalent to

$$\int_{-2\pi}^{2\pi} \Psi_1(\omega)\Psi_{i+1}^*(2^i\omega)e^{-j\omega n}\,d\omega = 0, \quad \text{for all integers } n. \qquad (11.5.47)$$

But we have already shown this to be true in the paraunitary case [set $L = 1$ and $i \neq 0$ in (11.5.29)]. Summarizing, we have proved the following result.

♠**Lemma 11.5.1. Continuous time orthonormal wavelets.** Let $\Psi(\omega)$ and $\Phi(\omega)$ be defined as in (11.5.1) where $G_s(z)$ and $H_s(z)$ satisfy the paraunitary conditions (11.4.13) [i.e., the polyphase matrix $\mathbf{R}(z)$ is paraunitary]. If the scaling function $\phi(t)$ is such that its autocorrelation $r_\phi(\tau)$ is Nyquist(1), [i.e., if any one of (11.5.39), (11.5.41) or (11.5.42) holds], then the wavelet basis functions $\psi_{k\ell}(t) \triangleq 2^{-k/2}\psi(2^{-k}t - \ell)$ are orthonormal. ◊

Example 11.5.7

Consider the Haar basis example again. The function $\phi(t)$ is as in Fig. 11.5-4, and its autocorrelation is a triangular waveform [Fig. 11.5-10(a)], which is zero for $|\tau| \geq 1$. So the Nyqusit(1) property is automatically satisfied, and the wavelet basis $\psi_{k\ell}(t)$ is orthonormal as seen earlier.

For Example 11.5.6 on the other hand, the function $\psi(t)$ [Fig. 11.5-9(a)] has the triangular autocorrelation $r_\phi(\tau)$, shown in Fig. 11.5-10(b). This does not satisfy the Nyquist(1) condition [e.g., $r_\phi(1) \neq 0$]. So we do not expect the basis functions $\psi_{k\ell}(t)$ to be orthonormal. Thus, as demonstrated earlier, $\psi_{00}(t)$ and $\psi_{02}(t)$ are not orthonormal.

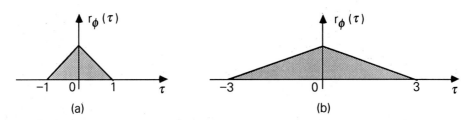

(a) (b)

Figure 11.5-10 The autocorrealtion $r_\phi(t)$ of $\phi(t)$, generated from (a) $G_s(z) = (1 + z^{-1})/\sqrt{2}$ and (b) $G_s(z) = (1 + z^{-3})/\sqrt{2}$.

Satisfying the Nyquist(1) Condition

The natural question now is this: how should we design $G_s(z)$ so that the above Nyquist(1) condition is satisfied? The answer to this has been provided by Mallat. We will state this in a slightly modified form here: if $[G_s(z), H_s(z)]$ is a FIR paraunitary pair [i.e., if (11.4.13) holds] and if the following two conditions hold:

1. $|G_s(e^{j0})| = \sqrt{2}$ and
2. $G_s(e^{j\omega}) \neq 0$ for $|\omega| \leq \pi/2$,

then the Nyquist(1) property (11.5.42) is indeed satisfied. The proof is beyond the scope of this chapter, and can be found in Mallat [1989b]. Note that the second condition is trivially satisfied in most QMF designs.

To demonstrate this result, consider again the Haar-basis (example 11.5.3). We have $|G_s(e^{j\omega})| = \sqrt{2}|\cos(\omega/2)|$, and the above two conditions are satisfied; so the Nyquist(1) condition holds, and the basis $\psi_{k\ell}(t)$ is orthonormal. In example 11.5.6, however, we have $|G_s(e^{j\omega})| = \sqrt{2}|\cos(3\omega/2)|$, so that $G_s(e^{\pm j\pi/3}) = 0$. This violates the second condition above. The Nyquist(1) property is not satisfied in this case, and the basis $\psi_{k\ell}(t)$ is not orthonormal as demonstrated earlier.

We can summarize the main points of the preceding discussions as follows.

♦**Theorem 11.5.1. Generating continuous-time finite duration orthonormal wavelets.** Suppose $\Psi(\omega)$ and $\Phi(\omega)$ are defined as in (11.5.1).

1. If $G_s(z)$ and $H_s(z)$ are causal FIR filters of order N, then $\phi(t)$ and $\psi(t)$ are causal with duration equal to N. So the functions defined according to $\psi_{k\ell}(t) \triangleq 2^{-k/2}\psi(2^{-k}t - \ell)$ have finite duration.
2. Suppose the FIR filter pair $[G_s(z), H_s(z)]$ is paraunitary i.e., (11.4.13) holds (i.e., the polyphase matrix $\mathbf{R}(z)$ is paraunitary). If in addition $|G_s(e^{j0})| = \sqrt{2}$ and $G_s(e^{j\omega}) \neq 0$ for $|\omega| \leq \pi/2$, the wavelet basis functions $\psi_{k\ell}(t)$ are orthonormal. ◊

Deeper Significance of the Nyquist(1) Condition (11.5.42)

Let $R(z)$ denote the z-transform of the sampled autocorrelation function $r_\phi(nT)$ with $T = 1$, that is, $R(z) = \sum_n r_\phi(n)z^{-n}$. It can be shown (Problem 11.19) that $R(z)$ satisfies the equation

$$\widetilde{G}_s(z)G_s(z)S(z)\Big|_{\downarrow 2} = S(z). \qquad (11.5.48)$$

In other words if we substitute $S(z) = R(z)$ into the left side of (11.5.48), it reduces to $R(z)$. In the paraunitary case, we have $\widetilde{G}_s(z)G_s(z)\Big|_{\downarrow 2} = 1$ so that the function $S(z) = constant$ is a solution to (11.5.48). If it turns out that the *only* solution to (11.5.48) is a constant, then this implies in particular that $R(z)$ is a constant, i.e., that $r_\phi(\tau)$ is Nyquist(1).

11.5.4 Regularity Considerations

In Fig. 11.5-11 we show the response $|G_s(e^{j\omega})|$ and the corresponding wavelet function $\psi(t)$ generated from the filter pair $[G_s(z), H_s(z)]$, for two cases. In both the examples, $G_s(z)$ and $H_s(z)$ are fifth order FIR filters, designed to satisfy the conditions of Theorem 11.5.1. So the basis functions $\psi_{k\ell}(t)$ are orthonormal.

We see that $\psi(t)$ is much more 'smooth' or 'regular' in part (a) than in part (b) of the figure. Qualitatively speaking, the latter function has strong 'high-frequency' components, so that it is less 'regular'. Smoothness of $\psi(t)$ can be obtained if $G_s(e^{j\omega})$ has a sufficient number of zeros (say K zeros) at $\omega = \pi$ (i.e., $z = -1$). In this case $G_s(e^{j2^{-m}\omega})$ has K zeros at each of the frequencies

$$\omega = 2^m\pi + 2^m 2\pi\ell, \quad m = 2,3,4\ldots \quad \ell = 0,1,2\ldots \qquad (11.5.49)$$

Using this it can be shown (Problem 11.23) that $\Psi(\omega)$ has K zeros at each of the frequencies $4\pi n$, for $n = 1, 2, 3 \ldots$ If $\Psi(\omega)$ is plotted with a logarithmic scale for the frequency axis (Bode plot), these zero locations get more and more crowded as the frequency increases.

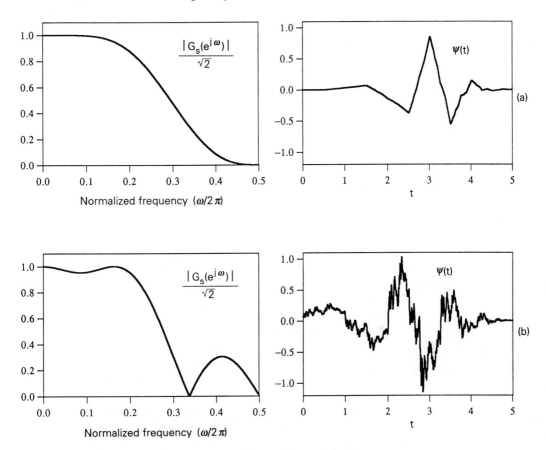

Figure 11.5-11 Two examples of plots of $|G_s(e^{j\omega})|$ and $\psi(t)$. (a) $\psi(t)$ looks regular or smooth and (b) $\psi(t)$ does not look smooth.

The smoothness or regularity of $\psi(t)$ improves with the number of zeros of $G_s(e^{j\omega})$ at $\omega = \pi$. For a more quantitative statement of this, see Daubechies [1988] where the author defines regularity in terms of the asymptotic decrease of the product (11.5.1) as ω increases, and relates this decrease to the number of zeros of $G_s(e^{j\omega})$ at $\omega = \pi$.

Let $G_s(z)$ and $H_s(z)$ be FIR with order N. For orthonormality of $\psi_{k\ell}(t)$, the filters have to satisfy (11.4.13). In particular, $\widetilde{G}_s(z)G_s(z)|_{\downarrow 2} = 1$, which means that $\widetilde{G}_s(z)G_s(z)$ is a half-band filter. In other words, $G_s(z)$ should be a spectral factor of the half-band filter $F(z) \triangleq \widetilde{G}_s(z)G_s(z)$. Subject to this constraint we will show that, no more than $(N+1)/2$ zeros of $G_s(z)$ can be located at $\omega = \pi$. We will also show how to find the coefficients of $G_s(z)$ such that it has $(N+1)/2$ zeros at $\omega = \pi$. These results are based on the theory of maximally flat FIR filters, developed in [Herrmann, 1971].

Maximally Flat FIR Filters

In Chapter 3 we outlined a number of techniques for FIR filter design, but left out maximally flat (linear-phase) FIR filters. These are the FIR counterparts, in some sense, of IIR Butterworth filters. If we design a maximally flat FIR half band filter and take $G_s(z)$ to be one of its spectral factors, then the function $\psi(t)$ designed as in the previous section has a much smoother plot, as pointed out by Daubechies.

Meaning of maximal flatness. Refer to Fig. 11.5-12, which shows the response $F(e^{j\omega})$ of a zero-phase lowpass filter $F(z)$, with $F(e^{j\pi}) = 0$. Suppose the derivative $dF(e^{j\omega})/d\omega$ has N_0 zeros at $\omega = 0$ and N_π zeros at $\omega = \pi$. We say that the (degree of) flatness is N_0 at $\omega = 0$ and N_π at $\omega = \pi$. Let N_d be the largest possible number of zeros of the derivative in the range $0 \leq \omega \leq \pi$. [This number is determined by the order of $F(z)$.] If $N_d = N_0 + N_\pi$, we say that $F(e^{j\omega})$ is maximally flat. This means that, for a given flatness at $\omega = 0$, the flatness at $\omega = \pi$ has been maximized (or vice versa). Notice that unlike an IIR Butterworth filter, the flatness need not be the same at $\omega = 0$ and π.

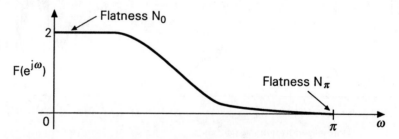

Figure 11.5-12 Pertaining to maximally flat FIR filters.

Closed form expression for the filter coefficients. We now show how to find the coefficients of maximally flat FIR filters. We discuss the problem in terms of a polynomial $P(y)$, and then make the change of variables $y \rightarrow \sin^2(\omega/2)$ to obtain the filter response $F(e^{j\omega})$. Here $P(y) = \sum_{n=0}^{N} p_n y^n$ (i.e., polynomial with order N). Let $P(0) = 2$, and let $K - 1$ denote the degree of flatness at $y = 1$ (Fig. 11.5-13). This means $P(y)$ has K zeros at $y = 1$. Let the degree of flatness at $y = 0$ be $L - 1$. Thus the derivative $dP(y)/dy$ has $K + L - 2$ zeros. Maximal flatness implies that $dP(y)/dy$ has no other zeros, so that the order of $P(y)$ is

$$N = K + L - 1. \tag{11.5.50}$$

Since $P(y)$ has K zeros at $y = 1$, we can write

$$P(y) = Q(y)(1 - y)^K, \tag{11.5.51}$$

where $Q(y)$ is a polynomial with order $L - 1$, that is,

$$Q(y) = \sum_{\ell=0}^{L-1} q_\ell y^\ell. \tag{11.5.52}$$

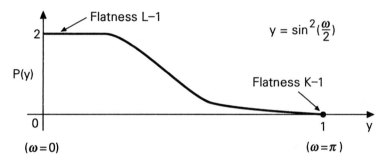

Figure 11.5-13 The maximally flat polynomial.

Our aim now is to find the L unknowns q_ℓ by imposing the following L constraints on $P(y)$:

$$P(0) = 2, \quad P^{(n)}(0) = 0, \ 1 \le n \le L - 1, \tag{11.5.53}$$

where $P^{(n)}(0)$ denotes the nth derivative evaluated at $y = 0$. The process is enormously simplified by rewriting (11.5.51) as

$$Q(y) = P(y)(1 - y)^{-K}. \tag{11.5.54}$$

By applying Leibnitz's rule for the derivative of a product (e.g., see p. 147 of Apostol [1961]), we obtain

$$Q^{(\ell)}(0) = \sum_{n=0}^{\ell} \binom{\ell}{n} P^{(n)}(0) \left(\frac{d^{\ell-n}(1-y)^{-K}}{dy^{\ell-n}} \bigg|_{y=0} \right). \tag{11.5.55}$$

In view of (11.5.53), only the term with $n = 0$ survives, and we obtain

$$Q^{(\ell)}(0) = 2 \frac{d^\ell (1-y)^{-K}}{dy^\ell} \bigg|_{y=0} = 2K(K+1)\ldots(K+\ell-1) = \frac{2(K+\ell-1)!}{(K-1)!} \tag{11.5.56}$$

Since $q_\ell = \left(Q^{(\ell)}(0)/\ell!\right)$, we finally obtain

$$q_\ell = 2\binom{K+\ell-1}{\ell}, \qquad (11.5.57)$$

so that

$$P(y) = (1-y)^K \sum_{\ell=0}^{L-1} 2\binom{K+\ell-1}{\ell} y^\ell. \qquad (11.5.58)$$

Summarizing, this function $P(y)$ represents the polynomial with smallest order ($= K+L-1$) having the following properties: (a) $P(0) = 2, P(1) = 0$, (b) degree of flatness $L-1$ at $y = 0$, and (c) degree of flatness $K-1$ at $y = 1$. Substituting $y = \sin^2(\omega/2)$, we obtain the zero-phase maximally flat FIR filter

$$F(e^{j\omega}) = \cos^{2K}(\omega/2) \sum_{\ell=0}^{L-1} 2\binom{K+\ell-1}{\ell} \sin^{2\ell}(\omega/2). \qquad (11.5.59)$$

The filter $F(z)$ has order $2N = 2(K+L-1)$. The transfer function $F(z)$ is verified to be (Problem 11.18)

$$F(z) = z^K \left(\frac{1+z^{-1}}{2}\right)^{2K} \underbrace{\sum_{\ell=0}^{L-1} 2(-z)^\ell \binom{K+\ell-1}{\ell} \left(\frac{1-z^{-1}}{2}\right)^{2\ell}}_{\widehat{F}(z)}. \qquad (11.5.60)$$

The filter indicated as $\widehat{F}(z)$ does not have zeros on the unit circle. Its purpose is to provide the passband flatness.

Figure 11.5-14 Effect of changing K, for fixed order $2(K+L-1)$.

Wavelets with Regularity, from Half-band Maximally Flat Filters

The effect of relative values of K and L is clearly demonstrated in Fig. 11.5-14. For fixed order $2(K+L-1)$, increase of K results in wider stopband.

If $K = L$, we obtain a response with symmetry around $\pi/2$. More precisely, in this case we have
$$F(z) + F(-z) = 2. \qquad (11.5.61)$$
(See below for proof.) In other words, $F(z)$ is a half band filter.

Design procedure. With $F(z)$ so chosen if we define $G_s(z)$ to be a spectral factor of $F(z)$ and take $H_s(z)$ in the usual manner, that is, $H_s(z) = -z^{-N}\widetilde{G}_s(-z)$, the paraunitary conditions (11.4.13) are satisfied. Furthermore $G_s(e^{j\omega}) \neq 0$ for $|\omega| \leq \pi/2$ so that the conditions of Theorem 11.5.1 are satisfied. If the function $\Psi(\omega)$ is now constructed as in (11.5.1b), then $\{\psi_{k\ell}(t)\}$ is a set of finite duration orthonormal functions with maximum regularity. Notice that even though $G_s(z)$ is a spectral factor of $F(z)$, only a spectral factor $S(z)$ of the function $\widehat{F}(z)$ [indicated in (11.5.60)] has to be computed. We then have
$$G_s(z) = \left(\frac{1+z^{-1}}{2}\right)^K S(z). \qquad (11.5.62)$$

Proof that (11.5.61) holds when $K = L$. When $K = L$, the polynomial $P(y)$ has flatness $K - 1$ at each of the points $y = 0$ and $y = 1$. Defining $R(y) = P(1-y)$, we see that this polynomial has the same flatness at these two points, but satisfies $R(0) = 0$ and $R(1) = 2$. Thus $P(y) + R(y) = 2$ at $y = 0$ as well as $y = 1$. This sum has the same flatness $K - 1$ at $y = 0$ and $y = 1$, i.e., its derivative has a total of $2K - 2$ zeros if we count the points $y = 0$ and $y = 1$. Since the order of $P(y) + R(y)$ is $2K - 1$, the derivative cannot have any further zeros in $0 < y < 1$. This shows that $P(y) + R(y) = 2$ for all y. Substituting $y = \sin^2(\omega/2)$, this gives $F(e^{j\omega}) + F(-e^{j\omega}) = 2$ which is equivalent to (11.5.61). $\qquad \triangledown\triangledown\triangledown$

From the above discussion it is also clear that the maximum possible number of zeros of $G_s(e^{j\omega})$ at the frequency $\omega = \pi$ is given by $K = (N+1)/2$.

Order estimation. In traditional signal processing applications, maximally flat FIR filters are not used as commonly as equiripple filters. This is because, for a given set of specifications (Fig. 3.1-1), the filter order is much higher. For example if we let $2\delta_1 = \delta_2 = 0.05$, then the filter order for a given transition width Δf is estimated to be [Kaiser, 1979]
$$2N \approx \frac{1}{2(\Delta f)^2}. \qquad (11.5.63)$$

The order grows as $1/(\Delta f)^2$, and not as $1/\Delta f$ as in the equiripple case!

Design Example 11.5.2: Wavelet Regularity

Let $K = L = 3$ so that the maximally flat filter $F(z)$ has order $2(K + L - 1) = 10$. The spectral factor $G_s(z)$ has order $N = 5$, with three zeros at $\omega = \pi$ (since $K = 3$). The response $|G_s(e^{j\omega})|/\sqrt{2}$ for this example was shown in Fig. 11.5-11(a), along with the function $\psi(t)$ derived from this filter.

Table 11.5.1 provides a summary of the key concepts and equations, pertaining to the generation of continuous time wavelets.

11.6 CONCLUDING REMARKS

Wavelets were thoroughly studied in the mathematics literature only after the mid 1980s. Most of the analysis was confined to the continuous time case, which has a wider scope for deeper mathematial issues. Discrete time wavelet transforms, on the other hand, are equivalent to tree structured digital filter banks, and can be understood with the help of elementary signal processing theory, and a fair amount of matrix theory.

Even before the development of wavelets and paraunitary filter banks, nonuniform filter banks have been used in speech processing literature [Nelson, et al., 1972], [Schafer, et al., 1975]. The motivation at that time was that, nonuniform bandwidths could be used to exploit the nonuniform frequency resolution of the human ear [Flanagan, 1972]. The more recent results on wavelets in the mathematical and signal processing literature enhance our understanding, and enable us to perform orthonormal decomposition.

As we saw in this chapter, discrete time orthonormal wavelet transforms are very easy to implement, simply by designing a two channel paraunitary QMF bank and then building the tree structure. If we use the cascaded lattice structure (Sec. 6.4), then the paraunitary property is retained in spite of multiplier quantization. This means that both the perfect reconstruction and the wavelet orthonormality properties can be retained in spite of multiplier quantization. Furthermore, the lattice structure generates the complete class of FIR orthonormal basis functions.

It does not appear to be appropriate to obtain discrete time wavelets by discretizing the continuous time version. Such an approach would typically result in the loss of many of the desirable properties such as orthonormality and perfect reconstruction. Furthermore, as we found in Sec. 11.5, continous time wavelets are often generated by starting from discrete time filter banks anyway.

Thus, if the wavelet application is already in the digital domain, it is really not necessary to understand the deeper results [e.g., the fundamental functions $\phi(t), \psi(t)$, self similarity, and so forth] developed in Sec. 11.5. In this case it is sufficient to understand the results of Sec. 11.4; the nonuniform nature of the digital filter responses (as well as the nonuniform decimation) already provides the nonuniform time-frequency grid [Fig. 11.3-3(b)], which is the key to many of the advantages of wavelet transformation.

Symmetric wavelet basis. In Chap. 7 we saw that two-channel FIR perfect reconstruction QMF banks cannot simultaneously have linear phase and paraunitary properties, unless the filters have fairly trivial forms. As a

TABLE 11.5.1 Generation of continuous-time wavelet basis

Definition of the infinite products:

$$\Phi(\omega) = \frac{1}{\sqrt{2}} G_s(e^{j\omega/2}) \prod_{m=2}^{\infty} \frac{1}{\sqrt{2}} G_s(e^{j2^{-m}\omega}), \quad (11.5.1a)$$

$$\Psi(\omega) = \frac{1}{\sqrt{2}} H_s(e^{j\omega/2}) \prod_{m=2}^{\infty} \frac{1}{\sqrt{2}} G_s(e^{j2^{-m}\omega}), \quad (11.5.1b)$$

where $G_s(z) = \sum_n g_s(n) z^{-n}$ and $H_s(z) = \sum_n h_s(n) z^{-n}$. Then

$$\Phi(\omega) = \frac{1}{\sqrt{2}} G_s(e^{j\omega/2}) \Phi(\omega/2), \quad \Psi(\omega) = \frac{1}{\sqrt{2}} H_s(e^{j\omega/2}) \Phi(\omega/2). \quad (11.5.15)$$

Equivalently

$$\phi(t) = \sqrt{2} \sum_{n=-\infty}^{\infty} g_s(n) \phi(2t - n), \quad \psi(t) = \sqrt{2} \sum_{n=-\infty}^{\infty} h_s(n) \phi(2t - n). \quad (11.5.18)$$

If $G_s(z)$ and $H_s(z)$ are causal FIR (order N), the functions $\psi(t)$ and $\phi(t)$ are causal, and of finite duration N.

Definition of the basis functions: $\quad \psi_{k\ell}(t) = 2^{-k/2} \psi(2^{-k} t - \ell)$

Orthonormality

Suppose $G_s(z)$ and $H_s(z)$ are causal and FIR, satisfying:
1. $|G_s(e^{j\omega})|^2 + |H_s(e^{j\omega})|^2 = 2$
2. $H_s(z) = z^{-N} \widetilde{G}_s(-z)$ for some odd N.

Equivalently,

$$G_s(z)\widetilde{G}_s(z)\Big|_{\downarrow 2} = 1, \quad H_s(z)\widetilde{H}_s(z)\Big|_{\downarrow 2} = 1, \quad G_s(z)\widetilde{H}_s(z)\Big|_{\downarrow 2} = 0. \quad (11.4.13)$$

In other words, $[G_s(z), H_s(z)]$ is an FIR paraunitary pair, i.e., the polyphase matrix $\mathbf{R}(z)$ satisfies $\widetilde{\mathbf{R}}(z)\mathbf{R}(z) = \mathbf{I}$. If we further have

$$|G_s(e^{j0})| = \sqrt{2}, \quad \text{and} \quad G_s(e^{j\omega}) \neq 0, \ |\omega| \leq \pi/2,$$

then $\{\psi_{k\ell}(t)\}$ forms an orthonormal set. In particular, each function $\psi_{k\ell}(t)$ has unit energy. The Haar basis (Example 11.5.5) is a familiar example of this. Also see Lemma 11.5.1 and Theorem 11.5.1.

Regularity. If the above $G_s(z)$ is designed to be a spectral factor of a maximally flat FIR half-band filter $F(z)$ [i.e., $F(z)$ as in (11.5.60) with $K = L$] then $\psi(t)$ exhibits very smooth behavior. See Fig. 11.5-11 for demonstration. Thus $\{\psi_{k\ell}(t)\}$ are *regular, orthonormal,* and of *finite duration* for finite k.

result, if we confine the orthonormal wavelet basis functions $\eta_{km}(n)$ in Sec. 11.4.2 to have linear phase, then the functions $F_k(z)$ are severely restricted. If we give up the paraunitary property (hence wavelet orthonormality), it is possible to design linear phase perfect reconstruction systems (i.e., symmetric wavelet basis functions) with greater flexibility of coefficients. This was demonstrated in Chap. 7. Further discussion of symmetric nonorthonormal wavelets can be found in Vetterli and Herley [1992]. For the case of *symmetric orthonormal* wavelets, see [Soman, Vaidyanathan and Nguyen, 1992]. For the case of nonbinary tree strutures (as used in wavelet packets) the simultaneous imposition of paraunitary and linear phase properties does not severely restrict the filter responses, and it is possible to obtain nontrivial, symmetric, orthonormal wavelet basis functions. Also see Problem 11.17.

PROBLEMS

11.1. Consider the following system where $H(e^{j\omega})$ is an ideal filter, with center frequency ω_0.

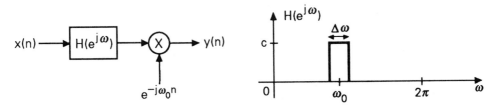

Figure P11-1

Let $c > 0$ be the height of the passband response. Assume that the Fourier transform $X(e^{j\omega})$ is constant in the passband of the filter. We know that $y(n)$ is a slowly varying signal for small $\Delta\omega$ (slower for smaller $\Delta\omega$).
  a) Find an expression for $y(n)$.
  b) For what value of n is $|y(n)|$ maximum?
  c) Let $|y(n)|$ be maximum for $n = n_0$. Suppose we wish to have $X(e^{j\omega_0}) = y(n_0)$. Show then that $c = 2\pi/\Delta\omega$.

This shows that, in order for the outputs of a filter bank to deliver a snap-shot of the (time-varying) Fourier transform, the filter *heights* should be inversely proportional to the bandwidths.

11.2. Consider the STFT system in (11.2.2) where $v(n)$ is of finite duration N. Suppose we implement this as a uniform DFT system (11.2.5), where $H_0(z)$ is as in (11.2.6). If we wish to have an FIR inversion system, show that $v(n)$ cannot have more than M nonzero coefficients.

11.3. Consider the discrete-time STFT and its inverse given in (11.2.1) and (11.2.9) respectively.
  a) Show that if the relation (11.2.1) is used in (11.2.9) this results in $x(n) = x(n) \sum_m |v(m)|^2$.
  b) Let z_0 be a zero of $\sum_k v^*(k)z^{-k}$. Show that if we substitute z_0^m in place of $X_{STFT}(e^{j\omega}, m)$ on the right hand side of (11.2.9), it reduces to zero. This shows that if $X_{STFT}(e^{j\omega}, m)$ satisfies the inverse relation (11.2.9), then so does $X_{STFT}(e^{j\omega}, m) + cz_0^m$ for any constant c.

11.4. Consider the sampled STFT given in (11.2.27). Here the STFT is evaluated on a two-dimensional grid with samples located at $\Omega = k\Omega_0$ and $\tau = nT$. Define the quantities

$$Y(\tau, \Omega) = \frac{1}{T} \sum_{k=-\infty}^{\infty} \sum_{n=-\infty}^{\infty} \mathcal{X}(k, n) e^{-j\Omega(nT)} e^{j\tau(k\Omega_0)}, \qquad (P11.4a)$$

$$X_1(\tau, \Omega) = \sum_{m=-\infty}^{\infty} x(\tau + mT) e^{-j\Omega(mT)}, \qquad (P11.4b)$$

$$V_1(\tau, \Omega) = \sum_{m=-\infty}^{\infty} v^*(\tau + mT) e^{-j\Omega(mT)}. \qquad (P11.4c)$$

Assume that the sample spacings Ω_0 and T are related as $\Omega_0 T = 2\pi$.
a) Verify that $Y(\tau, \Omega)$ is periodic in τ with period T, and periodic in Ω with period Ω_0. Also verify that $X_1(\tau, \Omega)$ and $V_1(\tau, \Omega)$ are periodic in Ω with period Ω_0.
b) Assuming that summations and integrals can be interchanged when necessary, show that

$$Y(\tau, \Omega) = X_1(\tau, \Omega) V_1^*(\tau, \Omega). \qquad (P11.4d)$$

Thus, given the sampled STFT $\mathcal{X}(k, n)$ and the window $v(t)$, we can recover $x(t)$ as follows:
a) Compute $Y(\tau, \Omega)$ for $0 \le \tau < T$ and $0 \le \Omega < \Omega_0$.
b) Compute $V_1(\tau, \Omega)$ for $0 \le \tau < T$ and $0 \le \Omega < \Omega_0$.
c) Using (P11.4d), compute $X_1(\tau, \Omega)$ for $0 \le \tau < T$ and $0 \le \Omega < \Omega_0$. (This assumes $V_1(\tau, \Omega) \ne 0$ in this range).
d) Find $x(\tau + mT)$ by using inverse Fourier transform relation corresponding to (P11.4b). This gives $x(\tau + mT)$ for any integer m and for any τ in the range $0 \le \tau < T$. So $x(t)$ can be recovered for any t.

11.5. For the signals given in Fig. 11.2-11 (a), (b), and (c), verify the indicated expressions for the RMS duration D_t.

11.6. Recall the RMS durations defined in (11.2.23). We now show $D_t D_f \ge 0.5$. For simplicity, assume $x(t)$ is a real function of the real variable t. Let $X(j\Omega)$ denote its Fourier transform. For the purpose of this problem, it is useful to review Cauchy-Schwartz inequality [Appendix C, Problem C.7(c)]. All integrals are in the range $-\infty$ to ∞. Assume that all relevant integrals exist, and that $x(t)$ has 'sufficient decay' so that $tx^2(t) \to 0$ as $t \to \pm\infty$.
a) Prove the inequality

$$\left(\int tx(t) \frac{dx(t)}{dt} dt \right)^2 \le \int t^2 x^2(t) dt \int \left(\frac{dx(t)}{dt} \right)^2 dt. \qquad (P11.6)$$

b) Using the fact that $j\Omega X(j\Omega)$ is the Fourier transform of $dx(t)/dt$, prove that the right hand side of the above inequality is equal to $E^2 D_t^2 D_f^2$, where $E = \int x^2(t) dt$.
c) Show that $\int tx(t) \frac{dx(t)}{dt} dt = -0.5E$. (Use integration by parts, i.e., $\int u\,dv = uv - \int v\,du$.)
d) Combine these results to prove $D_t D_f \ge 0.5$. From Cauchy-Schwartz inequality we know that equality occurs if and only if $tx(t) = cdx(t)/dt$ for some constant c. By integrating this, show that $D_t D_f = 0.5$ if and only if $x(t) = Ae^{-\alpha t^2}$ for some real A and $\alpha > 0$. In other words, $x(t)$ is a Gaussian pulse.

11.7. The continuous-time Fourier transform pair is given by (2.1.20) and (2.1.21). From this one can verify the Fourier transform relations: (i) $dx(t)/dt \leftrightarrow j\Omega X(j\Omega)$, and (ii) $tx(t) \leftrightarrow jdX(j\Omega)/d\Omega$. Now suppose $x(t) = e^{-t^2/(2\sigma^2)}$. (Note that the subscripts a have been dropped for simplicity.)

a) Using the above Fourier transform properties, show that $X(j\Omega)$ satisfies the differential equation $[dX(j\Omega)/d\Omega] = -\sigma^2 \Omega X(j\Omega)$.
b) By appropriate integration show that $X(j\Omega) = ce^{-\sigma^2 \Omega^2/2}$.
c) By using the following facts:

$$c = X(0) = \int e^{-t^2/(2\sigma^2)} dt, \quad x(0) = 1 = \frac{c}{2\pi} \int e^{-\sigma^2 \Omega^2/2} d\Omega,$$

show that $c = \sigma\sqrt{2\pi}$. Thus $X(j\Omega) = \sigma\sqrt{2\pi} e^{-\sigma^2 \Omega^2/2}$, a well-known result.

11.8. Let $x(t) = e^{-t^2/(2\sigma^2)}$ where $\sigma > 0$. We know from Problem 11.7 that $X(j\Omega) = \sigma\sqrt{2\pi} e^{-\sigma^2 \Omega^2/2}$. Denote the RMS time duration by $D_t(\sigma)$.
a) Show that the RMS frequency duration is given by

$$D_f(\sigma) = \frac{D_t(\sigma)}{\sigma^2} \qquad (P11.8)$$

b) By using the fact that $D_t(\sigma)D_f(\sigma) = 0.5$ for Gaussian $x(t)$, show that $D_t(\sigma) = \sigma/\sqrt{2}$.

11.9. Let $A(z)$ and $B(z)$ be rational transfer functions, and let m_1 and m_2 be positive integers. Show that

$$\Big(A(z^{m_1})B(z)\Big)\Big|_{\downarrow m_1 m_2} = \Big(A(z)(B(z)|_{\downarrow m_1})\Big)\Big|_{\downarrow m_2}. \qquad (P11.9)$$

11.10. Show that the Type 2 polyphase matrix $\mathbf{R}(z)$ of a synthesis bank is related to the filter matrix (11.4.24) according to (11.4.25).

11.11. Consider the synthesis bank shown below.

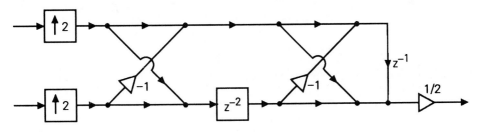

Figure P11-11

We know how to associate a polyphase matrix $\mathbf{R}(z)$ with this system. Verify that $\mathbf{R}(z)$ satisfies $\widetilde{\mathbf{R}}(z)\mathbf{R}(z) = \mathbf{I}$. From Sec. 11.4.1 we know how to generate an orthonormal set of discrete-time wavelet basis functions $\eta_{km}(n)$ from this filter bank. List the four sequences $\eta_{km}(n)$ for $k = 0, 1$, and $m = 0, 1$. Verify that these are indeed orthonormal.

11.12. Consider the two-level tree structured synthesis bank shown in Fig. P11-12.

Let $G_s(z) = c(1 + z^{-1})$ and $H_s(z) = c(1 - z^{-1})$, where c is some positive constant. From Sec. 11.3.3 we know how to generate discrete-time wavelet basis functions $\eta_{km}(n)$ from this filter bank.

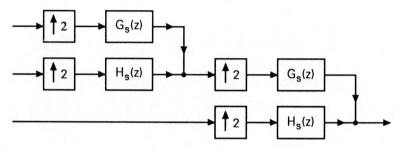

Figure P11-12

a) List the six sequences $\eta_{km}(n)$ for $k = 0, 1,$ and 2, and $m = 0, 1$.
b) Pick any two sequences from the above list as you wish, and verify that they are orthonormal for appropriate choice of c.

11.13. Show that the condition (11.4.14) can be rewritten as (11.4.15). Clearly explain where the inequality $2^{n_k} \geq 2^{n_\ell}$ is used in this rewriting.

11.14. Consider the infinite product $\prod_{m=1}^{\infty} a^{b^m}$, where $|b| < 1$. We show that this converges to the quantity $S = a^{b/(1-b)}$. Note that, since $|b| < 1$, we can write $S = a^{(\sum_{m=1}^{\infty} b^m)}$. Define the partial product $S_L = \prod_{m=1}^{L} a^{b^m}$. Show that $S_L - S = S \times (a^{-b^{L+1}/(1-b)} - 1)$. Hence show that $S_L \to S$ as $L \to \infty$.

11.15. Let $G_s(e^{j\omega/2})$ and $H_s(e^{j\omega/2})$ be as in Fig. 11.5-1. Then $\Psi(\omega)$ is the ideal bandpass function shown in Fig. 11.5-2. Prove then that the wavelet basis functions $\psi_{k\ell}(t)$ [defined in (11.5.3)] form an orthonormal set. [*Hint.* It might help to note that $\int_0^{2\pi} e^{j\omega n} d\omega = 2\pi \delta(n)$.]

11.16. Consider the Harr basis generated in Example 11.5.5. Sketch $\psi_{3,0}(t)$. Also sketch the functions $\psi_{1,\ell}(t)$ for $0 \leq \ell \leq 3$, and verify that these are orthogonal to $\psi_{3,0}(t)$.

11.17. In Sec. 11.5 we generated the functions $\phi(t)$ and $\psi(t)$ starting from the discrete-time transfer functions $G_s(z)$ and $H_s(z)$. Suppose the filters $G_s(z)$ and $H_s(z)$ are as in (11.5.23), that is, causal FIR. From Chap. 7 we know that $G_s(z)$ and $H_s(z)$ cannot have linear phase unless we give up the paraunitary property of the polyphase matrix. (The only exception to this gives trivial frequency responses, as seen in Chap. 7). In Chap. 7 we also found that we can obtain two-channel FIR perfect reconstruction systems with linear phase analysis and synthesis filters if we give up the paraunitary property. Under this condition, the functions $\phi(t)$ and $\psi(t)$ exhibit some kind of symmetry, which you prove in this problem. We say that a function $f(t)$ is symmetric if $f(t) = f(2t_0 - t)$ for some finite t_0 [antisymmetric if $f(t) = -f(2t_0 - t)$] and t_0 is said to be the center of symmetry.

a) Assume $g_s(n) = g_s(N - n)$ and $h_s(n) = -h_s(N - n)$ (which implies linear phase in the real coefficient case). Does $\phi(t)$ exhibit symmetry or antisymmetry? If so, what is the center of symmetry or antisymmetry?

b) Repeat part (a) with $\psi(t)$ instead of $\phi(t)$.

By using the FIR perfect reconstruction QMF bank with the above synthesis filters, it is possible to obtain the so-called bi-orthonormal symmetric wavelet functions with finite duration. For further details, see Vetterli and Herley [1992], and Soman, Vaidyanathan and Nguyen [1992].

11.18. Starting from the frequency response (11.5.59), verify that the transfer function $F(z)$ is indeed given by (11.5.60).

11.19. Let $r_\phi(\tau)$ be the autocorrelation function of the scaling function $\phi(t)$, defined as in (11.5.40). Define the z-transform $R(z) = \sum_n r_\phi(n) z^{-n}$. It can then be shown that $R(z)$ satisfies

$$\left. \widetilde{G}_s(z) G_s(z) R(z) \right|_{\downarrow 2} = R(z). \qquad (P11.19)$$

a) Verify the above equation for the function $\phi(t)$ in Fig. 11.5-4. Repeat the same for Fig. 11.5-9.

b) More generally, give a proof of (P11.19).

11.20. Consider the functions $\Phi(\omega)$ and $\Psi(\omega)$ defined in Sec. 11.5, and assume that $[G_s(z), H_s(z)]$ is a paraunitary pair. Show then that

$$|\Psi(\omega)|^2 + |\Psi(2\omega)|^2 + |\Psi(4\omega)|^2 + \ldots + |\Psi(2^L\omega)|^2 + |\Phi(2^L\omega)|^2 = |\Phi(\omega/2)|^2,$$

for any integer $L \geq 0$. This is pictorially demonstrated as shown in Fig. P11-20. if the "squeezed" bandpass filters generated from $|\Psi(\omega)|^2$ and the 'squeezed' lowpass filter $|\Phi(2^L\omega)|^2$ are added, the result is precisely the *stretched* lowpass filter $|\Phi(\omega/2)|^2$ (The figure assumes magnitude symmetry with respect to $\omega = 0$.)

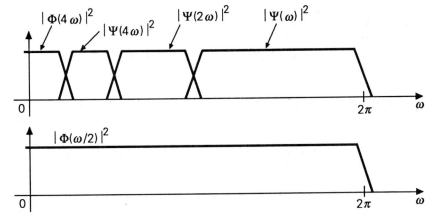

Figure P11-20

11.21. Consider the maximally decimated filter bank of Fig. 5.4-1, with a wide sense stationary input $x(n)$. Suppose the analysis bank is FIR and paraunitary (i.e., the polyphase matrix $\mathbf{E}(z)$ (Fig. 5.5-3(b)) is paraunitary). Let $x(n)$ be zero-mean and white. Show that the decimated subband output signals $y_k(n)$ are uncorrelated, i.e., $E[y_k(n)y_m^*(i)] = 0$ for $k \neq m$. (For consistency, we have used $y_k(n)$ in place of the notation $v_k(n)$ used in Fig. 5.4-1.) Show also that all signals $y_k(n)$ are white, with equal variance for all k. (*Note.* These properties would not, in general, be true if $\mathbf{E}(z)$ were not paraunitary.)

11.22. *Most general, nonuniform, orthonormal, discrete-time wavelets.* In Problem 5.32 we introduced the filter bank with nonuniform integer decimators n_k. Assume that this system is maximally decimated, that is, $\sum 1/n_k = 1$. Let $y_k(n)$, $0 \leq k \leq M-1$ be the decimated subband outputs. Assuming perfect reconstruction (i.e., $\widehat{x}(n) = x(n)$) show that we can express $x(n)$ in terms of the synthesis filters as

$$x(n) = \sum_{k=0}^{M-1} \sum_m y_k(m) f_k(n - n_k m). \qquad (P11.22a)$$

Thus, $x(n)$ has been expanded in terms of the basis functions $\eta_{km}(n) = f_k(n - n_k m)$, and $y_k(m)$ can be regarded as the generalized wavelet coefficients.

a) Recall that the basis $\eta_{km}(n)$ is orthonormal if the synthesis filters satisfy (11.2.19). Show that this orthonormality condition can be rewritten as

$$\sum_{n=-\infty}^{\infty} f_{k_1}^*(n) f_{k_2}(n - g_{k_1 k_2} i) = \delta(k_1 - k_2)\delta(i), \qquad (P11.22b)$$

where $g_{k_1 k_2}$ is the greatest common divisor (gcd) of n_{k_1} and n_{k_2}.

b) How would you extend the result of Problem 11.21 to this nonuniform orthonormal case?

11.23. Assume that $G_s(e^{j\omega})$ has K zeros at the frequency $\omega = \pi$, and that the infinite product defining $\Psi(\omega)$ converges. Show that $\Psi(\omega)$ has K zeros at each frequency of the form $4\pi n$, $n =$ positive integer.

12

Multidimensional Multirate Systems

12.0 INTRODUCTION

In this chapter we extend the basic concepts of multirate signal processing to the case of multidimensional (MD) signals, for example, two dimensional (2D) signals, and three dimensional (3D) signals. An example of a 2D signal is an image, whose intensity is a function of two variables (horizontal and vertical coordinates) [Jain, 1989]. A popular example of a 3D signal is an image which varies in time (such as a movie), "time" being the third dimension. An example of a 4D signal is the temperature at a point in space, as a function of time. Multidimensional multirate systems find applications in image coding [Woods, 1991], sampling format conversion between various video standards, and high-definition television systems, to mention a few.

Some of the earliest contributions to multidimensional filter banks were due to Vetterli [1984], Woods and O'Neil [1986], and Wackershruther [1986a]. The idea of "sampling" a multidimensional signal is fundamental to many of our discussions here. This is a nontrivial concept, and involves the mathematical concept of *lattices* [Cassels, 1959]. The use of lattices in sampling has been recognized for several decades, for example, see Petersen and Middleton [1962], and references therein. A detailed treatment of 2D sampling can be found in Dudgeon and Mersereau [1984]. Theoretical aspects of multidimensional systems are treated in Bose [1982].

The use of lattices has also been extended to the case of discrete time decimation and interpolation [Mersereau and Speake, 1983], [Dubois, 1985]. More recently, lattices have been used for the description and analysis of multidimensional filter banks [Ansari and Lee, 1988], [Ramstad, 1988], [Viscito and Allebach, 1988b], [Bamberger, 1990], [Karlsson and Vetterli, 1990], [Simoncelli and Adelson, 1990], [Vaidyanathan, 1990c, 1991b], [Chen and Vaidyanathan, 1991,1992a,b]. In particular the perfect reconstruction problem has been addressed in Viscito and Allebach [1988b and 1991], Ansari and Guillemot [1990], Karlsson and Vetterli [1990], and Kovačević and Vetterli [1992]. Design techniques for 2D filter banks can be found in Ansari and Guillemot [1990], Smith and Eddins [1990], Fettweis, et al. [1990] and Bamberger and Smith [1992]. The last two references show how to design filters which can be used for "directional decomposition" of images.

An article on MD filter banks, with excellent tutorial value can be found in Viscito and Allebach [1991]. Some of the notations to be developed in Sec. 12.4–12.9 are based on this reference and on [Vaidyanathan, 1990c].

Chapter Outline

Some of the basic concepts on MD signals are reviewed in Sec. 12.1–12.3. An attempt to do full justice to the basics will take up several chapters. The exposition here, accordingly, is brief (but technically complete). It is by no means a substitute for complete texts such as Dudgeon and Mersereau [1984], Jain [1989], and Lim [1990].

Sec. 12.4 to 12.10 are, however, self contained. They deal with multi-dimensional decimation, interpolation, polyphase decomposition, and filter banks. Four Tables in the end summarize key notations and concepts.

12.1 MULTIDIMENSIONAL SIGNALS

A D-dimensional signal $x_a(t_0, t_1, \ldots, t_{D-1})$ is a function of D real variables $t_0, t_1, \ldots, t_{D-1}$. We define the column vector

$$\mathbf{t} = \begin{bmatrix} t_0 & t_1 & \ldots & t_{D-1} \end{bmatrix}^T, \qquad (12.1.1)$$

and abbreviate the signal as $x_a(\mathbf{t})$. The subscript a indicates 'analog' which actually means that t_i are continuous variables. Even though these variables do not (necessarily) represent time, it is customary to call them so, and refer to $x_a(\mathbf{t})$ as a continuous 'time' signal.

12.1.1 The Fourier Transform

The Fourier transform of $x_a(\mathbf{t})$ (if it exists) is defined to be the D-dimensional integral

$$\begin{aligned} X_a(j\boldsymbol{\Omega}) = \int_{t_{D-1}=-\infty}^{\infty} & e^{-j\Omega_{D-1}t_{D-1}} \ldots \\ \ldots \int_{t_1=-\infty}^{\infty} & e^{-j\Omega_1 t_1} \int_{t_0=-\infty}^{\infty} x_a(\mathbf{t}) e^{-j\Omega_0 t_0} \, dt_0 dt_1 \ldots dt_{D-1}. \end{aligned} \qquad (12.1.2)$$

Thus, the Fourier transform is a function of D real variables (frequencies) $\Omega_0, \Omega_1, \ldots, \Omega_{D-1}$. (Both $x_a(\mathbf{t})$ and $X_a(j\boldsymbol{\Omega})$ are *scalar functions* of *vectors*). The inverse transform relation analogous to (2.1.21) is

$$\begin{aligned} x_a(\mathbf{t}) = \frac{1}{(2\pi)^D} \int_{\Omega_{D-1}=-\infty}^{\infty} & e^{j\Omega_{D-1}t_{D-1}} \ldots \\ \ldots \int_{\Omega_1=-\infty}^{\infty} & e^{j\Omega_1 t_1} \int_{\Omega_0=-\infty}^{\infty} X_a(j\boldsymbol{\Omega}) e^{j\Omega_0 t_0} \, d\Omega_0 d\Omega_1 \ldots d\Omega_{D-1}. \end{aligned} \qquad (12.1.3)$$

Defining the column vector of frequencies

$$\mathbf{\Omega} = [\,\Omega_0 \quad \Omega_1 \quad \ldots \quad \Omega_{D-1}\,]^T, \qquad (12.1.4)$$

we can abbreviate these relations as

$$X_a(j\mathbf{\Omega}) = \int_{-\infty}^{\infty} x_a(\mathbf{t}) e^{-j\mathbf{\Omega}^T \mathbf{t}} d\mathbf{t}, \qquad (12.1.5)$$

$$x_a(\mathbf{t}) = \frac{1}{(2\pi)^D} \int_{-\infty}^{\infty} X_a(j\mathbf{\Omega}) e^{j\mathbf{\Omega}^T \mathbf{t}} d\mathbf{\Omega}. \qquad (12.1.6)$$

The notation $\int_a^b$ indicates that the MD signal is integrated with respect to all variables, with *each* variable ranging from a to b. The above Fourier transform relations resemble the 1D versions [eqns. (2.1.20) and (2.1.21)] with the exceptions that (a) Ωt is replaced with the inner product $\mathbf{\Omega}^T \mathbf{t}$, and (b) there is a scale factor $(1/2\pi)^D$ rather than $1/2\pi$ in the inverse transform relation.

Bandlimited signals. A multidimensional signal is said to be band limited if the Fourier transform $X_a(j\mathbf{\Omega})$ is identically zero everywhere except in a designated finite region. The region where $X_a(j\mathbf{\Omega})$ is allowed to be nonzero is said to be the *support* of $X_a(j\mathbf{\Omega})$. Figure 12.1-1 shows several examples, with gray areas indicating the support. The examples in parts (a) and (b) are lowpass signals. If the quantity $|X_a(j\mathbf{\Omega})|$ is a nonzero constant in the gray areas, it can be considered to be a lowpass filter [i.e., $x_a(\mathbf{t})$ is the impulse response of a 2D lowpass filter]. Part (c), on the other hand, would represent a bandpass filter.

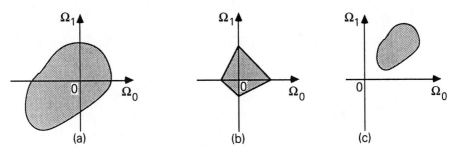

Figure 12.1-1 Demonstration of supports for $X_a(j\mathbf{\Omega})$.

Discrete-Time Signals

Fourier transform relations for discrete-time 1D signals were defined in (2.1.4). In the D-dimensional case, the signal is denoted as $x(\mathbf{n})$, where the 'time' index $\mathbf{n}$ is a column vector

$$\mathbf{n} = [\,n_0 \quad n_1 \quad \ldots \quad n_{D-1}\,]^T. \qquad (12.1.7)$$

Thus, $x(\mathbf{n})$ is a function of D integer variables $n_k, 0 \le k \le D-1$. We refer to $\mathbf{n}$ as an integer vector (or just "integer"). The Fourier transform, if it exists, is defined analogous to the 1D case, and is a function of D frequency variables $\omega_k, 0 \le k \le D-1$. With

$$\boldsymbol{\omega} = [\omega_0 \quad \omega_1 \quad \ldots \quad \omega_{D-1}]^T, \qquad (12.1.8)$$

the Fourier transform is defined by

$$X(\boldsymbol{\omega}) = \sum_{\mathbf{n} \in \mathcal{N}} x(\mathbf{n}) e^{-j\boldsymbol{\omega}^T \mathbf{n}}, \qquad (12.1.9)$$

where $\mathcal{N}$ denotes the set of all D-component integer vectors. The inverse Fourier transform relation is

$$x(\mathbf{n}) = \frac{1}{(2\pi)^D} \int_0^{2\pi} X(\boldsymbol{\omega}) e^{j\boldsymbol{\omega}^T \mathbf{n}} d\boldsymbol{\omega}. \qquad (12.1.10)$$

Since

$$e^{-j\boldsymbol{\omega}^T \mathbf{n}} = e^{-j\omega_0 n_0} e^{-j\omega_1 n_1} \ldots e^{-j\omega_{D-1} n_{D-1}}, \qquad (12.1.11)$$

we see that $X(\boldsymbol{\omega})$ is periodic in each variable ω_i with period 2π.

12.1.2 The z-Transform

Consider the summation

$$X_z(\mathbf{z}) = \sum_{\mathbf{n} \in \mathcal{N}} x(\mathbf{n}) z_0^{-n_0} z_1^{-n_1} \ldots z_{D-1}^{-n_{D-1}}, \qquad (12.1.12)$$

which is called the z-transform of $x(\mathbf{n})$. Here $\mathbf{z}$ denotes the vector

$$\mathbf{z} = [z_0 \quad z_1 \quad \ldots \quad z_{D-1}]^T.$$

This summation does not, in general, converge for arbitrary $\mathbf{z}$. For details about convergence, see Lim [1990]. If it converges for all z_k of the form

$$z_k = e^{j\omega_k}, 0 \le k \le D-1, \qquad (12.1.13)$$

(ω_k real) it reduces to the Fourier transform $X(\boldsymbol{\omega})$.

Notice the subscript z on X in (12.1.12). This is meant to distinguish the above from $X(\boldsymbol{\omega})$. This is necessary because we have to substitute $z_k = e^{j\omega_k}$ into $X_z(\mathbf{z})$ (rather than substitute $\mathbf{z} = \boldsymbol{\omega}$) to obtain $X(\boldsymbol{\omega})$. Introducing the notation

$$\mathcal{Z}(\mathbf{n}) \triangleq z_0^{n_0} z_1^{n_1} \ldots z_{D-1}^{n_{D-1}}, \qquad (12.1.14)$$

the z-transform becomes

$$X_z(\mathbf{z}) = \sum_{\mathbf{n} \in \mathcal{N}} x(\mathbf{n}) \mathcal{Z}(-\mathbf{n}). \qquad (12.1.15)$$

The notation $\mathcal{Z}(\mathbf{n})$ helps us to write the multidimensional z-transform in a manner which resembles the 1D z-transform.[†] Notice that when $z_i = e^{j\omega_i}$, the quantity $e^{-j\boldsymbol{\omega}^T \mathbf{n}}$ takes the place of $\mathcal{Z}(-\mathbf{n})$.

Special case of 2D signals. For the special case of 2D signals, we shall often use notations such as $x(n_0, n_1)$, $X_z(z_0, z_1)$, and $X(\omega_0, \omega_1)$ in place of $x(\mathbf{n})$, $X_z(\mathbf{z})$ and $X(\boldsymbol{\omega})$.

Key Properties of the Fourier and z-Transforms

We now state a number of properties of Fourier and z-transforms. Proofs for some of these are requested in Problem 12.1.

1. Linearity. From the definition we can see that if $X_{z,1}(\mathbf{z})$ and $X_{z,2}(\mathbf{z})$ are the transforms of $x_1(\mathbf{n})$ and $x_2(\mathbf{n})$, then the transform of $a_1 x_1(\mathbf{n}) + a_2 x_2(\mathbf{n})$ is given by $a_1 X_{z,1}(\mathbf{z}) + a_2 X_{z,2}(\mathbf{z})$ for arbitrary a_1, a_2.

2. Shifting a sequence. The sequence $y(\mathbf{n}) \triangleq x(\mathbf{n} - \mathbf{k})$ is said to be the shifted version of $x(\mathbf{n})$, with $\mathbf{k}$ denoting the vector-shift. To demonstrate this consider a 2D sequence $x(n_0, n_1)$ which is equal to unity at the sample locations indicated in Fig. 12.1-2(a), and zero elsewhere. The shifted version is shown in Fig. 12.1-2(b) for $\mathbf{k} = \begin{bmatrix} 2 \\ 1 \end{bmatrix}$.

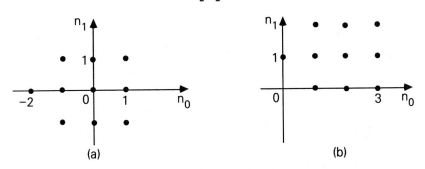

Figure 12.1-2 (a) A sequence $x(\mathbf{n})$ and (b) its shifted version $x(\mathbf{n} - \mathbf{k})$.

It is easily verified that the z-transform of the shifted sequence $x(\mathbf{n} - \mathbf{k})$ is given by $\mathcal{Z}(-\mathbf{k}) X_z(\mathbf{z})$. So $\mathcal{Z}(-\mathbf{k})$ can be regarded as a delay (or shift) operator, which introduces $\mathbf{k}$ 'units' of delay. It can also be shown that (a) $\mathcal{Z}(-\mathbf{k}) = 1/\mathcal{Z}(\mathbf{k})$, and (b) $\mathcal{Z}(\mathbf{k}_1 + \mathbf{k}_2) = \mathcal{Z}(\mathbf{k}_1) \mathcal{Z}(\mathbf{k}_2)$.

[†] An alternative, $\mathbf{z}^{(\mathbf{n})}$, has some advantages; see comments after Example 12.4.2.

3. *Convolution theorem.* Given two MD sequences $x(\mathbf{n})$ and $h(\mathbf{n})$ define the convolution sum $y(\mathbf{n}) = \sum_{\mathbf{m} \in \mathcal{N}} h(\mathbf{m})x(\mathbf{n}-\mathbf{m})$. Then, the transforms are related as $Y_z(\mathbf{z}) = H_z(\mathbf{z})X_z(\mathbf{z})$. In terms of Fourier transforms, this becomes $Y(\omega) = H(\omega)X(\omega)$.

Modulation and Frequency-Shift

Consider the sequence $y(\mathbf{n}) = x(\mathbf{n})e^{\mathbf{a}^T \mathbf{n}}$, where $\mathbf{a}$ is a column vector. Then $Y_z(\mathbf{z})$ is obtained from $X_z(\mathbf{z})$ by replacing each variable z_i with $z_i e^{-a_i}$ where a_i is the ith element of $\mathbf{a}$. Defining the diagonal matrix

$$\Lambda_{\mathbf{a}} = \text{diag}\,[\,e^{-a_0} \quad e^{-a_1} \quad \ldots \quad e^{-a_{D-1}}\,]. \qquad (12.1.16)$$

we can write $Y_z(\mathbf{z}) = X_z(\Lambda_{\mathbf{a}}\mathbf{z})$. Thus, in the 2D case,

$$Y_z(z_0, z_1) = X_z(e^{-a_0}z_0, e^{-a_1}z_1).$$

In particular suppose $\mathbf{a}$ is purely imaginary, that is, $\mathbf{a} = j\mathbf{b}$ for real $\mathbf{b}$. Then we can verify that $Y(\omega) = X(\omega - \mathbf{b})$. So multiplication of $x(\mathbf{n})$ with $e^{j\mathbf{b}^T\mathbf{n}}$ for real $\mathbf{b}$ amounts to a frequency shift by the amount $\mathbf{b}$.

12.1.3 Multidimensional Digital Filters

Figure 12.1-3 shows the schematic for a multidimensional digital filter. This is an LTI system characterized by a transfer function

$$H_z(\mathbf{z}) = \sum_{\mathbf{n} \in \mathcal{N}} h(\mathbf{n})\mathcal{Z}(-\mathbf{n}), \qquad (12.1.17a)$$

where $h(\mathbf{n})$ is the impulse response of the filter. The input-output relation of the system is given by the generalized convolution,

$$y(\mathbf{n}) = \sum_{\mathbf{k} \in \mathcal{N}} x(\mathbf{k})h(\mathbf{n}-\mathbf{k}) = \sum_{\mathbf{k} \in \mathcal{N}} h(\mathbf{k})x(\mathbf{n}-\mathbf{k}). \qquad (12.1.17b)$$

or, in terms of z-transforms by $Y_z(\mathbf{z}) = H_z(\mathbf{z})X_z(\mathbf{z})$. Given a two dimensional image $x(\mathbf{n})$ with finite size, the convolution operation in (12.1.17b) produces a larger image $y(\mathbf{n})$. In fact $y(\mathbf{n})$ has infinite extent if the filter is IIR. There exist techniques to cut down the size of the image by appropriate processing of the boundaries.

The frequency response $H(\omega)$ of the filter is given by

$$H(\omega) = \sum_{\mathbf{n} \in \mathcal{N}} h(\mathbf{n})e^{-j\omega^T \mathbf{n}}. \qquad (12.1.18)$$

This is $H_z(\mathbf{z})$ evaluated at $z_k = e^{j\omega_k}, 0 \le k \le D-1$. Figure 12.1-4 shows some typical frequency responses for the 2D case. The passband region (i.e., region where $|H(\omega)| \approx 1$) is said to be the support of the filter.

Figure 12.1-3 Schematics for multidimensional digital filters.

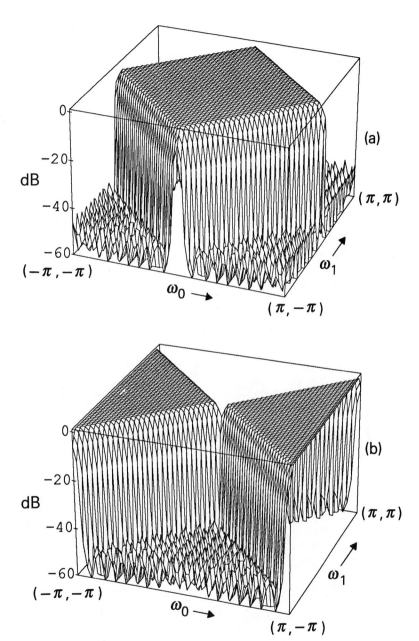

Figure 12.1-4 Typical two-dimensional frequency response plots. The magnitude in dB is plotted. (a) Diamond filter, and (b) fan filter.

Sec. 12.1 Multidimensional signals

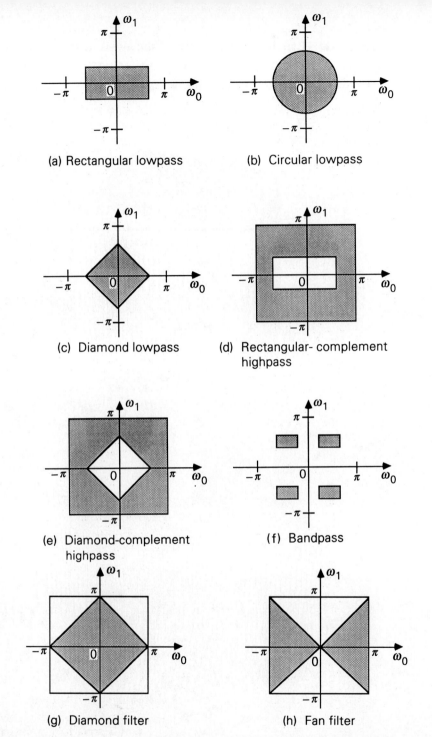

Figure 12.1-5 Several possible supports for 2D filters. Gray areas denote passbands. If $H(\omega)$ is constant in gray areas, then (a) and (f) are separable filters.

It is convenient to show the responses of filters in terms of their support. Figure 12.1-5 shows several examples of this kind. Note in particular that a lowpass response can have many possible supports, such as rectangular, circular, diamond, and so on. The supports shown in Fig. 12.1-5(g) and (h) correspond to the filters in Fig. 12.1-4. These are, respectively, called diamond and fan filters, because of the shape of the support.

The supports in the figure are shown only for the region $-\pi \leq \omega_i \leq \pi$, but the response is periodic (with period 2π) in each direction. This is demonstrated in Fig. 12.1-6 for lowpass and highpass filters.

Separable filters. A multidimensional filter is said to be separable if

$$H(\boldsymbol{\omega}) = H_0(e^{j\omega_0})H_1(e^{j\omega_1})\ldots H_{D-1}(e^{j\omega_{D-1}}). \qquad (12.1.19)$$

In other words, the transfer function is a product of D one-dimensional transfer functions. Separability implies, in particular, that the plot of $H(\boldsymbol{\omega})$ with respect to any one frequency variable ω_i is unchanged (except for a scale factor) if the remaining frequency variables are changed. To demonstrate nonseparability, consider the 2D filter response in Fig. 12.1-7(a). The plot of $H(\omega_0, \omega_1)$ with respect to ω_0 is shown in parts (b) and (c), for two values of ω_1. These plots are not scaled versions of each other so that the response is not separable. On the other hand, Fig. 12.1-5 (a) and (f) represent separable filters (if $H(\boldsymbol{\omega})$ is constant in the gray regions). Here are some analytical examples: (a) $z_0^{-1}z_1^{-1}$ (separable), and (b) $1 - z_0^{-1}z_1^{-1}$ (nonseparable).

The advantage of separable filters is that they can be designed easily from 1D filters. However, there exist many nonseparable filters which can be designed from 1D versions by clever mappings. A simple example is $1-H(\boldsymbol{\omega})$, which is not necessarily separable even if $H(\boldsymbol{\omega})$ is. [If $H(\boldsymbol{\omega})$ is unity in the gray area of Fig. 12.1-5(a), then $1 - H(\boldsymbol{\omega})$ is as in Fig. 12.1-5(d).] More nontrivial examples of these will be mentioned in Sec. 12.7. (For a preview the reader can see Fig. 12.7-2!) Also see Problem 12.36.

Linear phase and zero-phase filters. A multidimensional filter $H(\boldsymbol{\omega})$ is said to have linear phase if

$$H(\boldsymbol{\omega}) = ce^{-j\mathbf{k}^T\boldsymbol{\omega}}H_R(\boldsymbol{\omega}), \qquad (12.1.20)$$

where $\mathbf{k}$ is a real constant vector, $H_R(\boldsymbol{\omega})$ is a real function, and c is a (possibly complex) constant. (Notice that it is possible for $H_R(\boldsymbol{\omega})$ to be negative, but in most cases this happens only in the stopbands). Linear phase systems do not introduce phase distortion (in the same sense as for the 1D case, Sec. 2.4.2). If $H(\boldsymbol{\omega})$ has linear phase, then so does the frequency-shifted version $H(\boldsymbol{\omega} - \mathbf{b})$ and the time shifted-version $h(\mathbf{n} - \mathbf{m})$.

A zero-phase filter is a linear phase filter whose frequency response is real for all $\boldsymbol{\omega}$. In other words, the response can be written as in (12.1.20) with $\mathbf{k} = \mathbf{0}$, and $c = 1$. It can be shown (Problem 12.3) that a filter has zero-phase if, and only if, the impulse response satisfies

$$h(\mathbf{n}) = h^*(-\mathbf{n}). \qquad (12.1.21)$$

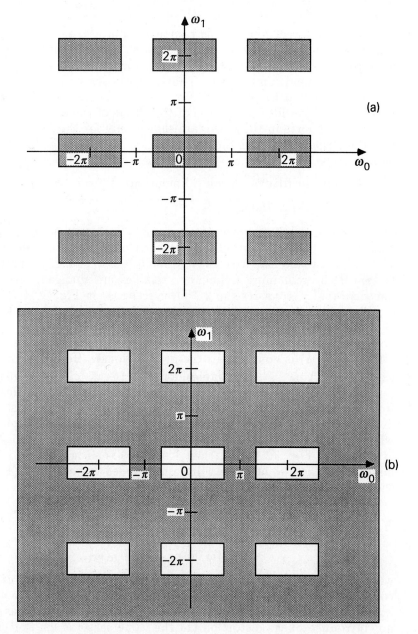

Figure 12.1-6 Demonstration of two-dimensional periodicity of the 2D digital filter response. (a) Rectangular lowpass filter and (b) highpass filter.

In the 2D real coefficient case, the condition

$$h(n_0, n_1) = h(-n_0, -n_1), \qquad (12.1.22)$$

is therefore necessary and sufficient to ensure zero phase.

Linearity of phase is important in image processing (unlike in many speech applications where some amount of phase nonlinearity is acceptable). Examples demonstrating this point can be found in Lim [1990] and references therein.

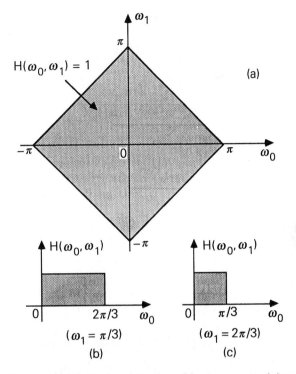

Figure 12.1-7 Demonstration of a nonseparable response. (a) Support of the filter, (b) and (c) responses for two fixed values of ω_1.

12.2 SAMPLING A MULTIDIMENSIONAL SIGNAL

The theory behind sampling of a multidimensional signal is fundamentally more complicated because of the many different ways to choose the sampling geometry. To illustrate, we first consider the 2D case. Given a 2D signal $x_a(t_0, t_1)$, the simplest sampled version we can visualize is

$$x(n_0, n_1) = x_a(n_0 T_0, n_1 T_1), \qquad (12.2.1)$$

where n_0 and n_1 are integers. So all the values of t_0 and t_1, which are integer multiples of T_0 and T_1, are included by the sampler. The sample points are thus arranged in a rectangular pattern, as demonstrated in Fig. 12.2-1 for $T_0 = 2, T_1 = 1$. This scheme is therefore termed *rectangular sampling*. We now demonstrate a more general sampling pattern. The motivation for considering such generalizations arises from the fact that rectangular sampling is often not the most efficient way to sample a bandlimited signal (as justified in Sec. 12.3.2).

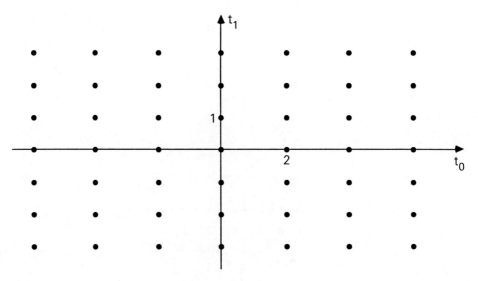

Figure 12.2-1 Demonstration of rectangular sampling. Here $T_0 = 2$, and $T_1 = 1$.

Example 12.2.1

Consider the sequence generated as

$$x(n_0, n_1) = x_a(n_0 - n_1, n_0 + 2n_1). \qquad (12.2.2)$$

The sample points are shown in Fig. 12.2-2. To see how this figure has been obtained, note that the sample locations, which are $t_0 = n_0 - n_1$ and $t_1 = n_0 + 2n_1$, can be written compactly as

$$\mathbf{t} \triangleq \begin{bmatrix} t_0 \\ t_1 \end{bmatrix} = \underbrace{\begin{bmatrix} 1 & -1 \\ 1 & 2 \end{bmatrix}}_{\mathbf{V}} \begin{bmatrix} n_0 \\ n_1 \end{bmatrix}. \qquad (12.2.3)$$

The matrix $\mathbf{V}$ is called the *sampling matrix*. Every sample location $\mathbf{t}$ is of the form

$$\mathbf{t} = n_0 \mathbf{v}_0 + n_1 \mathbf{v}_1, \qquad (12.2.4)$$

where $\mathbf{v}_0 = \begin{bmatrix} 1 \\ 1 \end{bmatrix}$ and $\mathbf{v}_1 = \begin{bmatrix} -1 \\ 2 \end{bmatrix}$. In other words, the sample locations are vectors $\mathbf{t}$ that are *integer linear combinations* of the columns $\mathbf{v}_0$ and $\mathbf{v}_1$ of the sampling matrix $\mathbf{V}$.

Given the matrix $\mathbf{V}$, the sample points can be graphically obtained as follows: first sketch the vectors $\mathbf{v}_0$ and $\mathbf{v}_1$ (heavy lines in the figure). Then draw two sets of equispaced parallel lines such that the two vectors form two sides of a parallelogram generated by these lines. The sample points are then located at the intersections of these sets of lines.

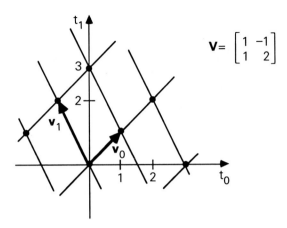

Figure 12.2-2 The set of sample points generated by **V** in Example 12.2.1.

In general **V** need not be an integer matrix (see next example), but it must be real and nonsingular. The *sampling geometry* in Fig. 12.2-2 is only one of an infinite number of possible ones. As we will see, the choice of sampling geometry affects the efficiency (in terms of required number of samples per unit volume) with which bandlimited signals can be represented.

Example 12.2.2

Suppose we sample a 2D signal $x_a(\mathbf{t})$ with the sampling matrix

$$\mathbf{V} = \begin{bmatrix} \cos\theta & -\sin\theta \\ \sin\theta & \cos\theta \end{bmatrix}. \tag{12.2.5}$$

This is a Givens rotation matrix [Eq. (6.1.9), with $-\theta$ in place of θ_m]. This means that the sample points

$$\mathbf{t} = \mathbf{V}\mathbf{n} = \begin{bmatrix} \cos\theta & -\sin\theta \\ \sin\theta & \cos\theta \end{bmatrix} \begin{bmatrix} n_0 \\ n_1 \end{bmatrix}, \tag{12.2.6}$$

are obtained by rotating the sample points corresponding to

$$\mathbf{n} = \begin{bmatrix} 1 & 0 \\ 0 & 1 \end{bmatrix} \begin{bmatrix} n_0 \\ n_1 \end{bmatrix}. \tag{12.2.7}$$

by an amount θ in the counterclockwise direction. Equation (12.2.7) in turn represents rectangular sampling with unit spacing between adjacent samples in both directions. See Fig. 12.2-3 which demonstrates the above rotation for $\theta = \pi/4$. We can consider the sample points to be located at the corners and centers of appropriate hexagons, which fill the 2D plane [see Fig 12.2-3(c)]. This is a special case of *hexagonal* sampling to be defined at the end of Sec. 12.3.2.

From Fig. 12.2-3(b) we also see that the samples can be considered to be located at equidistant points with spacing $s \ (=\sqrt{2})$ along equi spaced horizontal lines (spacing $s/2$). The samples on any horizontal line are, however, displaced with respect to those on the adjacent horizontal line by the amount $s/2$. These comments continue to hold if the word 'horizontal' is replaced with 'vertical' everywhere.

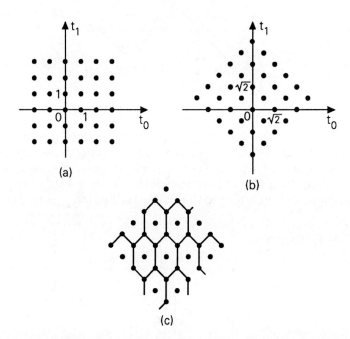

Figure 12.2-3 (a) Sample points for rectangular sampling with $T_0 = T_1 = 1$, (b) rotated sample points obtained by using **V** with $\theta = 45°$, and (c) demonstration that the rotated sample points are located at the center and corners of hexagonal figures.

12.2.1 Multidimensional Sampling

The sampled version of a D-dimensional signal $x_a(\mathbf{t})$ is defined as

$$x(\mathbf{n}) = x_a(\mathbf{Vn}), \qquad (12.2.8)$$

where **V** is a real $D \times D$ nonsingular matrix

$$\mathbf{V} = [\mathbf{v}_0 \quad \mathbf{v}_1 \quad \ldots \quad \mathbf{v}_{D-1}], \qquad (12.2.9)$$

and $\mathbf{n} \in \mathcal{N}$. The set of all sample points is the set

$$\mathbf{t} = \mathbf{V}\mathbf{n}, \quad \mathbf{n} \in \mathcal{N}, \qquad (12.2.10)$$

that is, the set of vectors $\sum_{k=0}^{D-1} n_k \mathbf{v}_k$. This is the set of all integer linear combinations of the columns $\mathbf{v}_0, \ldots, \mathbf{v}_{D-1}$ of $\mathbf{V}$. This set, denoted $LAT(\mathbf{V})$, is called the *lattice* generated by the matrix $\mathbf{V}$. The matrix $\mathbf{V}$ is called the *sampling matrix*, and is also said to be the *basis* which generates $LAT(\mathbf{V})$.

Evidently $LAT(\mathbf{V})$ is a discrete set, but has infinite number of elements. For the 2D examples above $\mathbf{V}$ was indicated in (12.2.3) and (12.2.5) respectively. For the rectangular case (12.2.1), we have

$$\mathbf{V} = \begin{bmatrix} T_0 & 0 \\ 0 & T_1 \end{bmatrix}. \qquad (12.2.11)$$

More generally rectangular sampling is *defined* to be a sampling scheme for which $\mathbf{V}$ is diagonal.

Nonuniqueness of the Lattice-Generator V

The matrix $\mathbf{V}$ that generates a lattice is not unique. Thus the lattice generated in Example 12.2.1 could also have been generated by the sampling matrix

$$\widehat{\mathbf{V}} = \begin{bmatrix} 0 & -1 \\ 3 & 2 \end{bmatrix}. \qquad (12.2.12)$$

Similarly, the rectangular lattice with $T_0 = T_1 = 1$, which is generated by $\mathbf{V} = \mathbf{I}$ could also have been generated by

$$\widehat{\mathbf{V}} = \begin{bmatrix} 1 & 1 \\ 1 & 2 \end{bmatrix}. \qquad (12.2.13)$$

In Problem 12.5 the reader is requested to verify both of these statements graphically. The theory underlying these, though, is given in Lemma 12.2.1. But first we need a definition.

Unimodular integer matrix. An integer matrix $\mathbf{E}$ is said to be unimodular if $[\det \mathbf{E}] = \pm 1$. For such a matrix, the inverse $\mathbf{E}^{-1}$ is also an integer (since $\mathbf{E}^{-1}$ is equal to the matrix of cofactors divided by the determinant; Appendix A). Conversely, suppose the inverse $\mathbf{F}$ of an integer matrix $\mathbf{E}$ is integer. We have $\mathbf{EF} = \mathbf{I}$ so that $[\det \mathbf{E}][\det \mathbf{F}] = 1$. Since each determinant must be an integer, this shows that both the determinants have unit magnitude, that is, both the matrices are unimodular. Summarizing, an integer matrix is unimodular if, and only if, its inverse is an integer matrix.

♠**Lemma 12.2.1.** Let $\mathbf{V}$ be a $D \times D$ real nonsingular matrix generating the lattice $LAT(\mathbf{V})$ [i.e., $\mathbf{V}$ is a basis for $LAT(\mathbf{V})$].

(a) Define $\widehat{\mathbf{V}} = \mathbf{VE}$ where $\mathbf{E}$ is some $D \times D$ integer matrix satisfying

$$\det \mathbf{E} = \pm 1. \qquad (12.2.14)$$

Then $\widehat{\mathbf{V}}$ is another basis for $LAT(\mathbf{V})$. [In other words, $LAT(\mathbf{V}) = LAT(\mathbf{VE})$.]

(b) Let $\widehat{\mathbf{V}}$ be another basis for $LAT(\mathbf{V})$. Then there exists integer matrix $\mathbf{E}$ with $[\det \mathbf{E}] = \pm 1$, such that $\widehat{\mathbf{V}} = \mathbf{VE}$. ◊

Proof. First consider part (a). Let $\mathbf{x}$ belong to $LAT(\mathbf{V})$ so that $\mathbf{x} = \mathbf{Vn}$ for some integer $\mathbf{n}$. So we can write

$$\mathbf{x} = \mathbf{VEE}^{-1}\mathbf{n} = \mathbf{VEm} = \widehat{\mathbf{V}}\mathbf{m} \qquad (12.2.15)$$

where $\mathbf{m} = \mathbf{E}^{-1}\mathbf{n}$. Since $\mathbf{E}^{-1}$ is also integer, $\mathbf{m}$ is an integer. This proves that every vector in $LAT(\mathbf{V})$ can also be written as $\widehat{\mathbf{V}}\mathbf{m}$ for integer $\mathbf{m}$. Conversely, consider any vector of the form $\mathbf{y} = \widehat{\mathbf{V}}\mathbf{m}$. We can write $\mathbf{y} = \mathbf{VEm} = \mathbf{Vn}$ for integer $\mathbf{n}$, so that $\mathbf{y}$ belongs to $LAT(\mathbf{V})$. Thus, $\mathbf{V}$ and $\widehat{\mathbf{V}}$ generate the same lattice.

Now consider proving part (b) which is more fun. The columns $\mathbf{v}_i$ of $\mathbf{V}$ are evidently points belonging to the lattice $LAT(\mathbf{V})$. Since $\widehat{\mathbf{V}}$ is also a basis, we can express these columns as $\widehat{\mathbf{V}}\mathbf{n}_i$ for integer $\mathbf{n}_i$. So we have

$$\mathbf{V} = \widehat{\mathbf{V}}\underbrace{[\mathbf{n}_0 \quad \mathbf{n}_1 \quad \ldots \quad \mathbf{n}_{D-1}]}_{\mathbf{F}}, \qquad (12.2.16)$$

for integer vectors $\mathbf{n}_i$. By the same argument we can express each column of $\widehat{\mathbf{V}}$ as $\mathbf{Vm}_i$ for integers $\mathbf{m}_i$, that is,

$$\widehat{\mathbf{V}} = \mathbf{V}\underbrace{[\mathbf{m}_0 \quad \mathbf{m}_1 \quad \ldots \quad \mathbf{m}_{D-1}]}_{\mathbf{E}}. \qquad (12.2.17)$$

Combining these two expressions we obtain $\mathbf{V} = \mathbf{VEF}$. Since $\mathbf{V}$ is nonsingular, this implies $\mathbf{EF} = \mathbf{I}$ so that

$$\det \mathbf{E} \det \mathbf{F} = 1. \qquad (12.2.18)$$

But the determinants of $\mathbf{E}$ and $\mathbf{F}$ are evidently integers, so that the above equation implies $\det \mathbf{E} = \pm 1$ indeed. ▽▽▽

Comments. Part (a) of the lemma should be carefully interpreted. Consider the two sequences $x(\mathbf{n}) = x_a(\mathbf{Vn})$ and $y(\mathbf{n}) = x_a(\widehat{\mathbf{V}}\mathbf{n})$. Then it is not necessarily true that $x(\mathbf{n}) = y(\mathbf{n})$. Part (a) merely says that the samples of $y(\mathbf{n})$ are obtained by a rearrangement (permutation) of the samples of $x(\mathbf{n})$.

12.2.2 Sampling Density and Determinant of V

The determinant of the sampling matrix **V** plays a fundamental role in the sampling of a MD signal. We will see that $1/|\det \mathbf{V}|$ gives the sampling density, that is, number of samples per unit volume (area in the 2D case). For the special case of 1D signals we have $\mathbf{V} = T$ where T is the sample spacing. In this case $1/|\det \mathbf{V}| = 1/T$, which *is* the sampling density (number of samples per unit length).

The Fundamental Parallelepiped (FPD)

Given a sampling matrix **V**, suppose we sketch all the basis vectors $\mathbf{v}_k$ (columns of **V**) in a D-dimensional coordinate space. This is demonstrated in Fig. 12.2-4 for the sampling matrix **V** used in Example 12.2.1. We can now complete a parallelepiped with these vectors as edges as shown. (This can also be done for 3D and higher dimensional cases.) This is called the fundamental parallelepiped [denoted $FPD(\mathbf{V})$] of the sampling matrix **V**. In the 1D case, $FPD(V)$ is the segment $0 \le x < V$ of the real line.

The points which fall inside $FPD(\mathbf{V})$ (gray area in the figure) can be represented as the set $\mathbf{Vx}$ where $\mathbf{x} = \begin{bmatrix} x_0 \\ x_1 \end{bmatrix}$, with $0 \le x_0, x_1 < 1$.

More general definition. Let $[a,b)^D$ denote the set of D-component vectors **x** such that the components satisfy $a \le x_i < b$. Let **V** be a $D \times D$ real nonsingular matrix. Then the fundamental parallelepiped generated by **V** is defined as

$$FPD(\mathbf{V}) = \text{set of all points } \mathbf{Vx} \text{ with } \mathbf{x} \in [0,1)^D. \qquad (12.2.19)$$

According to this definition the set $\mathbf{x} \in [0,1)^D$ itself can be denoted as $FPD(\mathbf{I})$.

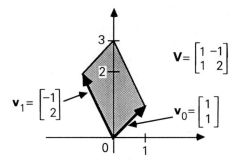

Figure 12.2-4 The fundamental parallelepiped $FPD(\mathbf{V})$ for a specific example of **V**.

The Role of the Determinant

It is well known in geometry that the volume of the fundamental parallelepiped $FPD(\mathbf{V})$ is equal to $|\det \mathbf{V}|$. (The term "volume," which we use for convenience, degenerates to "area" in the 2D case and 'length' in the 1D case.) This is particularly easy to verify in the 2D case. Thus, consider Fig. 12.2-5 which shows a parallelepiped with one edge aligned to the horizontal axis. (There is no loss of generality in such alignment since we can always rotate a parallelepiped without changing its area). With a and b denoting the lengths of the sides, the area is given by $ab|\sin\theta|$. On the other hand, the matrix $\mathbf{V}$ generating $FPD(\mathbf{V})$ has columns $\mathbf{v}_0 = \begin{bmatrix} a \\ 0 \end{bmatrix}$, and $\mathbf{v}_1 = \begin{bmatrix} b\cos\theta \\ b\sin\theta \end{bmatrix}$ so that

$$\det \mathbf{V} = \det \begin{bmatrix} a & b\cos\theta \\ 0 & b\sin\theta \end{bmatrix} = ab\sin\theta, \qquad (12.2.20)$$

verifying that the area equals $|\det \mathbf{V}|$.

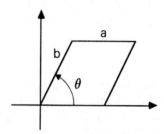

Figure 12.2-5 The area of the parallelogram shown above is equal to $ab|\sin\theta|$.

Representation of Arbitrary Points in Terms of Lattice Points

Given any real number y, we can always express it as $y = x + n$ where n is a unique integer and x is a unique real number with $0 \le x < 1$ [i.e., $x \in [0,1)$]. Here n is merely the "integer" part (and x the fraction part) of y. Similarly, given a real vector $\mathbf{y}$ we can express it as

$$\mathbf{y} = \mathbf{x} + \mathbf{n}, \quad \mathbf{x} \in [0,1)^D, \text{ and } \mathbf{n} \in \mathcal{N}, \qquad (12.2.21)$$

simply by expressing each component as above. Evidently $\mathbf{x}$ and $\mathbf{n}$ are unique.

In the scalar case, a generalization of such decomposition is this: given a fixed number $v > 0$, we can express an arbitrary real u as

$$u = u_P + vn, \quad 0 \le u_P < v, \quad n = \text{integer}. \qquad (12.2.22a)$$

Moreover u_P and n are unique for a given u. Note that vn belongs to the 'lattice' generated by the real number v (which is the set of all points of the form vm, m = integer). So we can write $u = u_P + u_L$, with $u_P \in FPD(v)$ and $u_L \in LAT(v)$.

Decomposing a vector u with respect to matrix V. Here is an easy generalization, to the MD case, of the above decomposition: Let **u** be any real D-component vector, and **V** any $D \times D$ real nonsingular matrix. We can then write

$$\mathbf{u} = \mathbf{u}_P + \mathbf{u}_L, \qquad (12.2.22b)$$

where $\mathbf{u}_P$ and $\mathbf{u}_L$ are unique vectors such that

$$\mathbf{u}_P \in FPD(\mathbf{V}), \quad \text{and} \quad \mathbf{u}_L \in LAT(\mathbf{V}). \qquad (12.2.22c)$$

In other words, any real vector **u** can be uniquely written as a sum of a vector $\mathbf{u}_L$ in the discrete set $LAT(\mathbf{V})$ and a vector $\mathbf{u}_P$ in the continuous set $FPD(\mathbf{V})$. One often says that $\mathbf{u}_P$ is the remainder of **u** modulo the matrix **V**.

The above decomposition is justified as follows: define $\mathbf{y} = \mathbf{V}^{-1}\mathbf{u}$. Then **y** is a real vector and can be decomposed as in (12.2.21). So $\mathbf{u} = \mathbf{V}\mathbf{y} = \mathbf{V}\mathbf{x} + \mathbf{V}\mathbf{n}$. So we have (12.2.22b), with $\mathbf{u}_P = \mathbf{V}\mathbf{x}$ and $\mathbf{u}_L = \mathbf{V}\mathbf{n}$. Since **x** and **n** are restricted as in (12.2.21), we have $\mathbf{u}_P \in FPD(\mathbf{V})$ and $\mathbf{u}_L \in LAT(\mathbf{V})$ indeed. The uniqueness of $\mathbf{u}_P$ and $\mathbf{u}_L$ follow from that of **x** and **n**.

Geometric interpretation. The set of points $\mathbf{u}_P + \mathbf{u}_L$ where $\mathbf{u}_L$ is a fixed point in the lattice and $\mathbf{u}_P$ varies over $FPD(\mathbf{V})$ can be considered to be a shifted version of $FPD(\mathbf{V})$. Thus consider the example

$$\mathbf{V} = \begin{bmatrix} 1 & -1 \\ 1 & 2 \end{bmatrix}. \qquad (12.2.23)$$

Figure 12.2-6 demonstrates $FPD(\mathbf{V})$ as well as shifted versions for some values of $\mathbf{u}_L$. According to the above decomposition, we can cover the entire D-dimensional Euclidean space by taking the union of the set $FPD(\mathbf{V})$ and all these shifted copies. Referring to Fig. 12.2-6, the gray area contains all vectors **u** with $\mathbf{u}_L = \mathbf{0}$ in (12.2.22b). By shifting this gray area to a lattice point (and repeating this for each lattice point), we can cover all points in the two dimensional plane.

Sampling Density

By definition of $FPD(\mathbf{V})$, only one lattice point falls *inside* it, namely the origin. Similarly, there is one lattice point in each shifted version. The number of sample points per unit volume is equal to the number of $FPD(\mathbf{V})$'s which we can fit into unit volume. Since $|\det \mathbf{V}|$ is the volume of $FPD(\mathbf{V})$, we conclude that the number of lattice points per unit volume (i.e., sampling densitiy) is equal to $1/|\det \mathbf{V}|$. So we have

$$\text{Sampling density } \rho = \frac{1}{|\det \mathbf{V}|}. \qquad (12.2.24)$$

For the sampling matrix in Example 12.2.1 we have $\rho = 1/3$ whereas for Example 12.2.2, we have $\rho = 1$.

The sampling density is an important concept. Intuitively we can see that for an MD signal which is bandlimited in a certain way, a certain minimum sampling density is required, if we desire to retain all information (and avoid aliasing). We will make this intuition more precise in the succeeding sections.

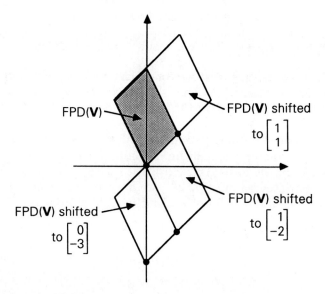

Figure 12.2-6 $FPD(\mathbf{V})$ and copies obtained by shifting it to various points on $LAT(\mathbf{V})$ (shown by heavy dots).

12.2.3 Aliasing Effects Created by Sampling

Given a signal $x_a(t)$ with Fourier transform $X_a(j\Omega)$, suppose we sample it with sampling matrix $\mathbf{V}$ and obtain $x(\mathbf{n}) = x_a(\mathbf{Vn})$. How is the Fourier transform $X(\omega)$ of $x(\mathbf{n})$ related to $X_a(j\Omega)$? Recall what happens in the 1D case: if we define $x(n) = x_a(nT)$, then

$$X(e^{j\omega}) = \frac{1}{T} \sum_{k=-\infty}^{\infty} X_a\left(j\frac{1}{T}(\omega - 2\pi k)\right). \qquad (12.2.25)$$

In other words, $X(e^{j\omega})$ is formed by adding an infinite number of uniformly shifted copies of $X_a(j\Omega)$ (shifted by integer multiples of $2\pi/T$), replacing Ω with ω/T, and scaling the result by $1/T$. So in the time domain the samples are spaced apart by T whereas in the frequency domain there is a periodicity of $2\pi/T$.

Effect of Sampling in the MD Case

In the MD case there is a similar effect. Let $x_a(\mathbf{t})$ be a D-dimensional signal with Fourier transform $X_a(j\Omega)$. Define the sequence $x(\mathbf{n}) = x_a(\mathbf{Vn})$.

Then its Fourier transform is given by

$$X(\omega) = \frac{1}{|\det \mathbf{V}|} \sum_{\mathbf{k} \in \mathcal{N}} X_a\bigl(j\mathbf{V}^{-T}(\omega - 2\pi\mathbf{k})\bigr), \qquad (12.2.26)$$

where $\mathcal{N}$ is the set of all D-component integers.† With $\mathbf{V} = T$ (1D case), this reduces to (12.2.25).

The reciprocal lattice. Before proving this, some discussion is in order. The matrix $2\pi\mathbf{V}^{-T}$ will play a crucial role in future discussions, and deserves a special notation

$$\mathbf{U} \triangleq 2\pi\mathbf{V}^{-T}. \qquad (12.2.27)$$

The lattice $LAT(\mathbf{V}^{-T})$ is said to be the reciprocal lattice of $LAT(\mathbf{V})$. The lattice $LAT(2\pi\mathbf{V}^{-T})$ is called the *scaled reciprocal lattice*. For example, let $\mathbf{V}$ be as in (12.2.23). For this case

$$\mathbf{V} = \begin{bmatrix} 1 & -1 \\ 1 & 2 \end{bmatrix}, \quad \mathbf{V}^{-T} = \frac{1}{3}\begin{bmatrix} 2 & -1 \\ 1 & 1 \end{bmatrix}, \quad \mathbf{U} = 2\pi\mathbf{V}^{-T} = \frac{2\pi}{3}\begin{bmatrix} 2 & -1 \\ 1 & 1 \end{bmatrix}. \qquad (12.2.28)$$

Figure 12.2-7 shows the lattice and the scaled reciprocal lattice.

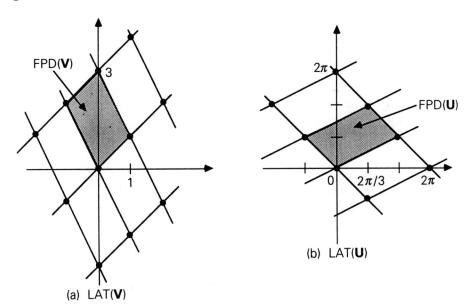

Figure 12.2-7 (a) The lattice generated by $\mathbf{V}$, and (b) the scaled reciprocal lattice, generated by $\mathbf{U} = 2\pi\mathbf{V}^{-T}$. The matrices are as in (12.2.28).

---

† The notation $\mathbf{V}^{-T}$ stands for $(\mathbf{V}^{-1})^T$.

Pictorial interpretation of (12.2.26). We now illustrate the procedure for constructing $X(\omega)$ from $X_a(j\Omega)$. For this we first make the change of variables

$$\Omega = \mathbf{V}^{-T}\omega, \qquad (12.2.29)$$

(analogous to $\Omega \to \omega/T$) to rewrite (12.2.26) as

$$X(\mathbf{V}^T\Omega) = \frac{1}{|\det \mathbf{V}|} \sum_{\mathbf{k}\in\mathcal{N}} X_a\big(j(\Omega - \mathbf{U}\mathbf{k})\big). \qquad (12.2.30)$$

Consider a signal $x_a(\mathbf{t})$ bandlimited to the gray region shown in Fig. 12.2-8(a). According to the above expression, we produce an infinite number of copies of $X_a(j\Omega)$ by shifting its origin to the points on $LAT(\mathbf{U})$. Thus $X(\mathbf{V}^T\Omega)$ has support equal to the union of gray regions in Fig. 12.2-8(b). Finally by making the change of variables $\Omega = \mathbf{V}^{-T}\omega$ we obtain the Fourier transform $X(\omega)$ of $x(\mathbf{n})$. This change of variables maps the gray region in Fig. 12.2-8(b) to the gray region in Fig. 12.2-8(c) so that $X(\omega)$ is indeed periodic (period $= 2\pi$) in each variable ω_i.

The shifted versions of $X_a(j\Omega)$ mentioned above are called *alias components*. If any of these has overlap with the unshifted version, we cannot recover $X_a(j\Omega)$ from $X(\omega)$. This is similar to the aliasing effect in the 1D case. If the signal $X_a(j\Omega)$ is "appropriately bandlimited", and if the sampling matrix $\mathbf{V}$ is appropriately chosen, this overlap can be avoided. We will return to this very important point later.

Proof of the relation (12.2.26). From (12.1.6) we have

$$x(\mathbf{n}) = x_a(\mathbf{V}\mathbf{n}) = \frac{1}{(2\pi)^D} \int_{\Omega=-\infty}^{\infty} X_a(j\Omega) e^{j\Omega^T \mathbf{V}\mathbf{n}} d\Omega. \qquad (12.2.31)$$

With the change of variables $\Omega = \mathbf{V}^{-T}\omega$ this becomes

$$x(\mathbf{n}) = \frac{1}{(2\pi)^D |\det \mathbf{V}|} \int_{\omega=-\infty}^{\infty} X_a(j\mathbf{V}^{-T}\omega) e^{j\omega^T \mathbf{n}} d\omega. \qquad (12.2.32)$$

(Note that $d\Omega$ should be replaced with $d\omega/|\det \mathbf{V}|$.) Now we can decompose the vector ω as in (12.2.22b), with $\mathbf{V} = 2\pi \mathbf{I}$. In other words, we can write $\omega = \omega_P - 2\pi \mathbf{k}$ where $\omega_P \in [0, 2\pi)^D$ and $\mathbf{k}$ is an integer vector. By substituting this into (12.2.32) (and dropping the subscript P on ω) we obtain

$$x(\mathbf{n}) = \frac{1}{(2\pi)^D |\det \mathbf{V}|} \int_{\omega \in FPD(2\pi\mathbf{I})} \sum_{\mathbf{k}\in\mathcal{N}} X_a\big(j\mathbf{V}^{-T}(\omega - 2\pi\mathbf{k})\big) e^{j\omega^T \mathbf{n}} d\omega, \qquad (12.2.33)$$

which, by comparison with (12.1.10), yields the desired result.

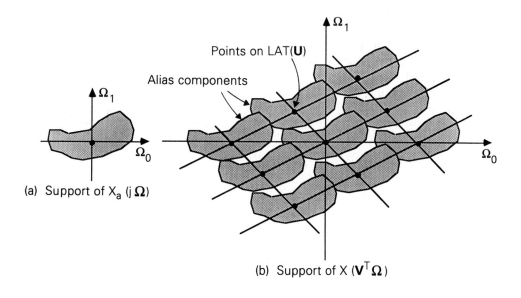

(a) Support of $X_a(j\Omega)$

(b) Support of $X(\mathbf{V}^T\Omega)$

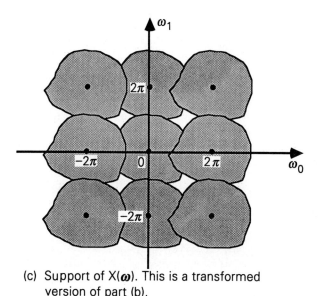

(c) Support of $X(\omega)$. This is a transformed version of part (b).

Figure 12.2-8 Demonstrating frequency domain effect of sampling of a 2D signal.

The Periodicity Matrix U

In the 1D case, we say that a function $f(\Omega)$ of a real variable Ω is periodic if there exists $U > 0$ such that $f(\Omega) = f(\Omega - Uk)$ for all Ω, for all integers k. Graphically this means that we can shift its origin by any integer multiple of U, without changing the function. In the D-dimensional case the idea is similar. We say that $f(\mathbf{\Omega})$ is periodic if

$$f(\mathbf{\Omega}) = f(\mathbf{\Omega} - \mathbf{U}\mathbf{k}), \quad \text{for all } \mathbf{\Omega}, \tag{12.2.34}$$

for all integer vectors $\mathbf{k}$. The matrix $\mathbf{U}$ (which is required to be real and nonsingular), is said to be a "period" or "periodicity matrix." A periodicity matrix for which the determinant has smallest magnitude is said to be a fundamental periodicity matrix. Now recall that the set of vectors of the form $\mathbf{U}\mathbf{k}$ is the lattice generated by $\mathbf{U}$. The right hand side of (12.2.34) is obtained by shifting the origin of the plot for $f(\mathbf{\Omega})$ so that its new origin is at the lattice point $\mathbf{U}\mathbf{k}$. So $f(\mathbf{\Omega})$ is periodic (with periodicity matrix $\mathbf{U}$) if it does not change as a result of relocating the origin to an arbitrary lattice point generated by $\mathbf{U}$.

It is now clear that the function in Fig. 12.2-8(b) is periodic in $\mathbf{\Omega}$ with periodicity matrix $\mathbf{U}$. This is consistent with the fact that $X(\boldsymbol{\omega})$ is periodic in each of the variables ω_i, i.e., it has periodicity matrix $2\pi\mathbf{I}$ [Fig. 12.2-8(c)].

Summary and tables. Much of the theory of multidimensional sampling, decimation and interpolation can be understood without difficulty, as long as the above matrix notations are clearly understood (and efficiently used). As an aid to easy reading, we will summarize the main fundamentals in Tables 12.10.1 and 12.10.2 at the end of Sec. 12.10.

12.3 MINIMUM SAMPLING DENSITY

Suppose we sample a bandlimited lowpass signal $x_a(\mathbf{t})$ using the sampling matrix $\mathbf{V}$. We know that the Fourier transform of the sampled version is periodic in $\mathbf{\Omega}$ with periodicity matrix $\mathbf{U} = 2\pi\mathbf{V}^{-T}$. The fundamental parallelepiped generated by $\mathbf{U}$ has area

$$|\det \mathbf{U}| = \frac{(2\pi)^D}{|\det \mathbf{V}|} = (2\pi)^D \rho, \tag{12.3.1}$$

where ρ is the sampling density $1/|\det \mathbf{V}|$. So the separation between 'copies' of the basic spectrum $X_a(j\mathbf{\Omega})$ increases as the sampling density ρ increases. This suggests that if we sample a bandlimited signal at sufficiently high density, then we can avoid overlap between any two terms in (12.2.30). In this manner aliasing can been avoided. We can recover $x_a(\mathbf{t})$ from such sampled version by use of a lowpass filter whose passband includes the region of support of $X_a(j\mathbf{\Omega})$, and stopband includes the regions of support of $X_a(j(\mathbf{\Omega} - 2\pi\mathbf{V}^{-T}\mathbf{k}))$ for all $\mathbf{k} \neq \mathbf{0}$.

It is clear that for a given region of support and a particular sampling geometry, ρ has to exceed a minimum sampling density ρ_{min}. For a given region of support, the quantity ρ_{min} itself may depend on the sampling geometry (as demonstrated below). So a judicious choice of sampling geometry can help to reduce the number of samples per unit volume required to represent a given bandlimited signal. We will consider a specific class of bandlimited signals, namely those with circular support, and compare two sampling geometries.

12.3.1 Rectangular Sampling Under Circular Support

Consider a 2D lowpass signal $x_a(t)$ with circular support for $X_a(j\Omega)$ (Fig. 12.3-1(a)). The circle is centered at the origin $(0,0)$, and has radius σ. Suppose we wish to use rectangular sampling. So we have

$$\mathbf{V} = \begin{bmatrix} v_0 & 0 \\ 0 & v_1 \end{bmatrix}, \quad \mathbf{U} = 2\pi \mathbf{V}^{-T} = 2\pi \begin{bmatrix} v_0^{-1} & 0 \\ 0 & v_1^{-1} \end{bmatrix}. \quad (12.3.2)$$

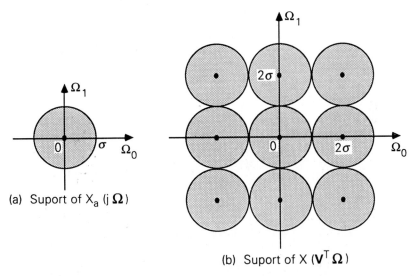

(a) Suport of $X_a(j\Omega)$

(b) Suport of $X(\mathbf{V}^T\Omega)$

Figure 12.3-1 Effect of rectangular sampling (circular support).

The region of support of (12.2.30) is the union of all circles of radius σ with centers at the lattice points generated by $\mathbf{U}$. It is clear [Fig. 12.3-1(b)] that there is no overlap between $X_a(j\Omega)$ and any of the alias components if, and only if,

$$\frac{2\pi}{v_k} \geq 2\sigma. \quad (12.3.3)$$

So the choice

$$v_0 = v_1 = \frac{\pi}{\sigma} \quad (12.3.4)$$

results in minimum sampling density for rectangular sampling. This minimum density is

$$\rho_{min} = \frac{1}{|\det \mathbf{V}|} = \frac{1}{v_0 v_1} = \frac{\sigma^2}{\pi^2} \quad \text{(rectangular sampling)}. \quad (12.3.5)$$

12.3.2 Hexagonal Sampling Under Circular Support

Consider again the same circular lowpass support as above. Rectangular sampling at minimal density will result in the support region shown in Fig. 12.3-1(b). The gaps in this figure (white areas) indicate regions where the sampled signal has no energy. Is it possible to reduce these gaps by changing the geometry of the lattice points? In other words, can we obtain a closer packing of these circles?

Consider Fig. 12.3-2(a) which shows a different arrangement of the circles. Once again the circles are arranged in rows, but there is a subtle difference: the centers of the circles in any given row are *midway* between the centers in the previous row.

It is already clear from the figure that the white areas are smaller compared to Fig. 12.3-1(b). We will show below that the circles are packed tighter. A second way to obtain this arrangement is as follows: (a) fill the 2D frequency plane with hexagons having *equal sides* [one of these hexagons being centered at $(0,0)$], and (b) draw a circle (with radius σ) at the center and around each corner of each hexagon. If the sides of the hexagons are chosen large enough, we can avoid overlap between circles.†

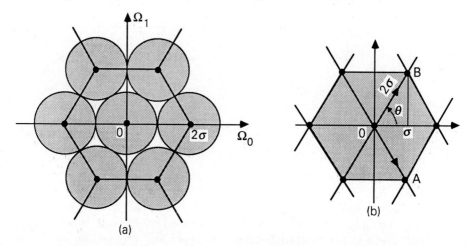

Figure 12.3-2 Pertaining to hexagonal sampling.

The transform of the sampled signal will have such a support if the lattice points generated by **U** are located at the corners and centers of the hexagonal figures. Such lattice is indicated separately in Fig. 12.3-2(b), and can be obtained by taking the columns of **U** to be the vectors OA and OB indicated in the figure.

By symmetry of the hexagonal figure it is clear that the angle $\theta = 2\pi/6$.

---

† Notice that, in this arrangement, each circle has six "nearest neighbours," whereas there are only *four* in the rectangular arrangement of Fig. 12.3-1(b).

(This is also clear from the requirement $2\sigma \cos \theta = \sigma$ which we can see from the figure.) So the vectors OA and OB can be written in regular notation as

$$OA : \begin{bmatrix} \sigma \\ -2\sigma \sin \theta \end{bmatrix} = \begin{bmatrix} \sigma \\ -\sqrt{3}\sigma \end{bmatrix} \qquad OB : \begin{bmatrix} \sigma \\ 2\sigma \sin \theta \end{bmatrix} = \begin{bmatrix} \sigma \\ \sqrt{3}\sigma \end{bmatrix}. \qquad (12.3.6)$$

The matrix $\mathbf{U}$ should, therefore, be chosen as

$$\mathbf{U} = \sigma \begin{bmatrix} 1 & 0 \\ 0 & \sqrt{3} \end{bmatrix} \begin{bmatrix} 1 & 1 \\ -1 & 1 \end{bmatrix}. \qquad (12.3.7)$$

The corresponding sampling matrix $\mathbf{V} = 2\pi \mathbf{U}^{-T}$ can be simplified to

$$\mathbf{V} = \frac{\sqrt{2\pi}}{\sqrt{3}\sigma} \begin{bmatrix} \sqrt{3} & 0 \\ 0 & 1 \end{bmatrix} \begin{bmatrix} \cos \theta & -\sin \theta \\ \sin \theta & \cos \theta \end{bmatrix}, \qquad (12.3.8)$$

with $\theta = -45°$. To calculate the minimum sampling density note that $[\det \mathbf{V}] = 2\pi^2/\sqrt{3}\sigma^2$, so that

$$\rho_{min} = \frac{\sqrt{3}\sigma^2}{2\pi^2} \approx \frac{0.866\sigma^2}{\pi^2} \approx \frac{\sigma^2}{1.15\pi^2} \quad \text{(hexagonal sampling)}. \qquad (12.3.9)$$

Summarizing, if we wish to sample a bandlimited signal with circular support (radius σ), then the minimum sampling density to avoid aliasing is given by (12.3.9) for hexagonal sampling, and by (12.3.5) for rectangular sampling. So hexagonal sampling is *more* efficient by a factor of about 1.15; it packs the circles more tightly in the frequency domain. The sample points generated by $\mathbf{V}$ are depicted in Fig. 12.3-3, and we see that we can fit hexagons into this sampling geometry, that is, the samples are located at the centers and corners of hexagons.

We conclude this section by defining hexagonal sampling (for the 2D case) to be one for which the sampling matrix has the form

$$\mathbf{V} = \begin{bmatrix} s_0 & 0 \\ 0 & s_1 \end{bmatrix} \begin{bmatrix} \cos \theta & -\sin \theta \\ \sin \theta & \cos \theta \end{bmatrix} \mathbf{E} \qquad (12.3.10)$$

where s_0 and s_1 are nonzero real numbers (stretching factors), $\theta = \pm 45°$, and $\mathbf{E}$ is a unimodular integer matrix. Evidently (12.3.8) is a special case of this. The sampling matrix used in (12.2.6) (with $\theta = 45°$) is another special case. Notice that the sides of the hexagons are not necessarily equal, and depend on the scale factors s_0, s_1.

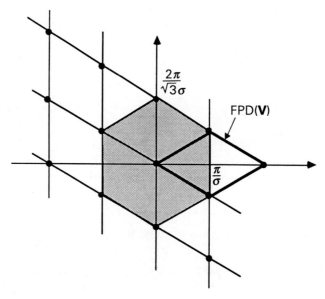

Figure 12.3-3 Sample points generated by **V** (hexagonal sampling).

12.4 MULTIRATE FUNDAMENTALS

Basic multirate ideas for 1D discrete-time signals and systems were introduced in Chapter 4. These include decimation, interpolation, polyphase decompostion, Nyquist filtering, and filter bank structures. We now extend these for the multidimensional case. Some of the notations to be introduced in this section (e.g., $\mathbf{z}^{(\mathbf{M})}$) were first introduced in [Viscito and Allebach, 1988b], and are very useful in many of the developments.

12.4.1 Basic Building Blocks and Tools

Multidimensional Decimators

Given a 1D sequence $x(n)$, the M-fold decimated version was defined as $y(n) = x(Mn)$. In an analogous manner, given the D-dimensional sequence $x(\mathbf{n})$, we define the M-fold decimated version as

$$y(\mathbf{n}) = x(\mathbf{Mn}), \qquad (12.4.1)$$

where **M** is a $D \times D$ nonsingular matrix of integers. Now we know that the set of vectors **Mn** is the lattice $LAT(\mathbf{M})$ generated by **M**. Since **M** itself is an integer, this lattice contains only integer vectors. Basically the decimated version $y(\mathbf{n})$ contains only those samples of $x(\mathbf{n})$ which fall at the points belonging to $LAT(\mathbf{M})$. Figure 12.4-1(a)-(e) demonstrates this for some examples of **M**. In this figure, black dots represent the lattice points, that is, sample points retained by the decimator. White circles represent

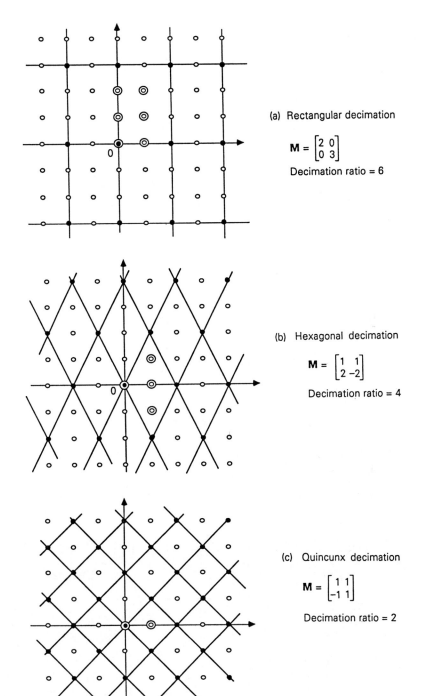

Figure 12.4-1 (a)–(c) Demonstration of two-dimensional decimation. *Continued* →

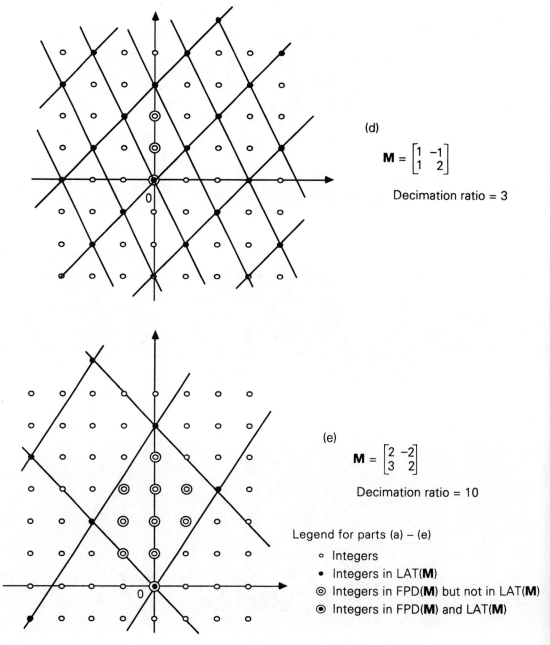

Figure 12.4-1 *Continuation.* (d), (e) Demonstration of two-dimensional decimation.

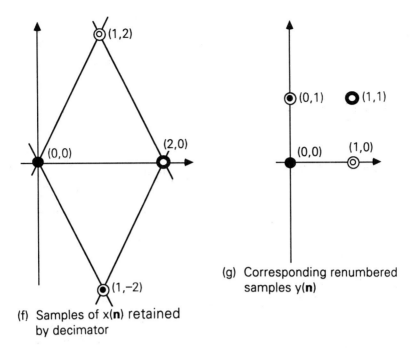

Figure 12.4-1 *Continuation.* (f), (g) Demonstration of how a decimator renumbers the samples. Here **M** is the hexagonal decimator (part (b)).

integers which are sample points discarded by the decimator. The black dots and white circles which fall inside $FPD(\mathbf{M})$ are highlighted by the surrounding outer circles.

The decimator in part (a) is rectangular. More generally, a rectangular decimator is defined to be one for which **M** is diagonal. The decimator in (b) is said to be hexagonal. The one in (c) is called a quincunx decimator. See end of Sec. 12.4.5 for further comments on the term "hexagonal."

It is important to notice that the decimator which retains a subset of samples of $x(\mathbf{n})$, also re-numbers them. This is demonstrated in parts (f) and (g) for the hexagonal decimator. The samples $x(0,0), x(1,2), x(1,-2)$ and $x(2,0)$, which are among the samples retained by the decimator, are mapped as follows:

$$x(0,0) \to y(0,0) \quad x(1,2) \to y(1,0) \quad x(1,-2) \to y(0,1) \quad x(2,0) \to y(1,1).$$

Decimation ratio. We know that $FPD(\mathbf{M})$ contains a finite number of integer vectors. Denote this set of integers as $\mathcal{N}(\mathbf{M})$, that is,

$$\begin{aligned}\mathcal{N}(\mathbf{M}) &= \text{set of integer vectors in } FPD(\mathbf{M})\\ &= \text{set of integers of the form } \mathbf{Mx}, \quad \mathbf{x} \in [0,1)^D.\end{aligned} \quad (12.4.2)$$

Sec. 12.4 Multirate fundamentals

The number of integers $J(\mathbf{M})$ in this set is given by [†]

$$J(\mathbf{M}) = |\det \mathbf{M}| \qquad (12.4.3)$$

The decimator retains one sample point in the $FPD(\mathbf{M})$ (namely the origin) and rejects the rest. So it retains one out of $|\det \mathbf{M}|$ samples. In other words the $\mathbf{M}$-fold decimator has decimation ratio $|\det \mathbf{M}|$.

The decimation ratio is indicated in Fig. 12.4-1 for each of the five examples. Thus consider part (d) where we have $|\det \mathbf{M}| = 3$. Accordingly, we have three integers in $FPD(\mathbf{M})$, indicated by double circles. (One of these has a black inner circle indicating that it is a lattice point.) We, therefore, retain one out of three points, and the sampling density is reduced by three indeed. Similarly in example (e) of this figure, we have $|\det \mathbf{M}| = 10$ so that one out of ten samples are retained by the decimator. Indeed, there are ten double circles in the FPD, one of which has a black inner circle.

Reasons for defining generalized decimators. The advantage of defining generalized (non rectangular) decimators is that for certain types of sequences, proper choice of $\mathbf{M}$ permits higher decimation ratio. For example, we will see later that if $x(\mathbf{n})$ is bandlimited to the region shown in Fig. 12.5-2(a), rectangular decimation cannot reduce the sample density by more than a factor of two (without creating aliasing), whereas hexagonal decimation can reduce the density by *four* with no aliasing. Non rectangular decimation has also been found to be more efficient than rectangular, in image coding practice.

Multidimensional Expanders

In the 1D case, we defined the M-fold expander to be a device with input output relation

$$y(n) = \begin{cases} x(n/M), & n = \text{mul. of } M \\ 0 & \text{otherwise.} \end{cases} \qquad (12.4.4)$$

A multidimensional expander is defined similarly, but in terms of the expander matrix $\mathbf{M}$. This is a $D \times D$ nonsingular integer matrix, and generates a lattice $LAT(\mathbf{M})$. The 'expanded' output $y(\mathbf{n})$ is zero unless $\mathbf{n}$ belongs to this lattice. More precisely,

$$y(\mathbf{n}) = \begin{cases} x(\mathbf{M}^{-1}\mathbf{n}) & \text{if } \mathbf{n} \in LAT(\mathbf{M}) \\ 0 & \text{otherwise.} \end{cases} \qquad (12.4.5)$$

Thus $y(\mathbf{n}) = 0$ for all $\mathbf{n}$, with the exception that

$$y(\mathbf{M}\mathbf{m}) = x(\mathbf{m}). \qquad (12.4.6)$$

[†] To prove this, note that the volume of $FPD(\mathbf{M})$ is equal to $|\det \mathbf{M}|$. This means that we can fit $|\det \mathbf{M}|$ hypercubes with unit-length sides into this parallelepiped. We have one integer per hypercube, so that there are $|\det \mathbf{M}|$ integers in $FPD(\mathbf{M})$.

Referring again to Fig. 12.4-1, the output of the M-fold expander is zero at the points indicated by white circles, and typically nonzero at other places.

In all discussions to follow, the symbol M is used to denote nonsingular integer matrices (unless stated otherwise).

12.4.2 Polyphase Decomposition

For 1D sequences this was introduced in Sec. 4.3. Given a sequence $x(n)$, we defined its Type 1 polyphase components as $x_k(n) = x(nM+k), 0 \leq k \leq M-1$. We could then express $X(z) = \sum_{k=0}^{M-1} z^{-k} X_k(z^M)$. This decomposition uses the fact that any integer n can be represented as

$$n = k + Mn_0, \qquad (12.4.7)$$

where n_0 and k are unique integers such that $0 \leq k \leq M-1$. (This the well-known *division theorem* for integers.) Evidently k is the remainder (and n_0 the quotient) when we divide n by M.

For a D-dimensional sequence $x(\mathbf{n})$ a similar polyphase representation is possible, and is useful when we deal with M-fold decimators and expanders. The representation is based on the following fact.

♠**Division theorem for integer vectors.** Let M be a $D \times D$ nonsingular integer matrix, and let **n** be some D-dimensional integer. We can express **n** as

$$\mathbf{n} = \mathbf{k} + \mathbf{M}\mathbf{n}_0, \qquad (12.4.8)$$

where $\mathbf{k} \in \mathcal{N}(\mathbf{M})$ and $\mathbf{n}_0$ is an integer vector. Moreover **k** and $\mathbf{n}_0$ are unique for given **n**. ◇

Proof. We know we can write **n** as in (12.2.22b), that is,

$$\mathbf{n} = \mathbf{n}_P + \mathbf{n}_L, \qquad (12.4.9)$$

where $\mathbf{n}_P$ and $\mathbf{n}_L$ are unique vectors with

$$\mathbf{n}_P \in FPD(\mathbf{M}), \quad \text{and} \quad \mathbf{n}_L \in LAT(\mathbf{M}). \qquad (12.4.10)$$

Since **M** is an integer, $\mathbf{n}_L$ is an integer. So it follows from (12.4.9) that $\mathbf{n}_P$ is integer as well. Since $\mathbf{n}_P \in FPD(\mathbf{M})$, it follows that $\mathbf{n}_P \in \mathcal{N}(\mathbf{M})$. By letting $\mathbf{k} = \mathbf{n}_P$ and $\mathbf{n}_L = \mathbf{M}\mathbf{n}_0$, we therefore obtain (12.4.8). ▽▽▽

Modulo notation. We say that $\mathbf{n}_0$ is the quotient and **k** the remainder obtained by dividing **n** with **M**. In many applications the remainder **k** is more important than the quotient. We often use modulo notation to indicate this remainder. Thus:

$$\mathbf{k} = \mathbf{n} \bmod \mathbf{M}, \quad \text{or } \mathbf{k} = ((\mathbf{n}))_\mathbf{M} \quad \text{or just } \mathbf{k} = ((\mathbf{n})). \qquad (12.4.11)$$

These notations imply that **k** is the unique integer in $\mathcal{N}(\mathbf{M})$ satisfying (12.4.8).

Polyphase decomposition

The number of possible values of $\mathbf{k}$ in (12.4.8) is equal to $J(\mathbf{M})$. Since any integer can be written as (12.4.8), we can classify the set of all integers into $J(\mathbf{M})$ sets, depending on the value of the remainder $\mathbf{k}$.[†] Thus, given any sequence $x(\mathbf{n})$ and the matrix $\mathbf{M}$, we can identify $J(\mathbf{M})$ unique subsequences,

$$x_{\mathbf{k}}(\mathbf{n}) = x(\mathbf{Mn} + \mathbf{k}), \quad \mathbf{k} \in \mathcal{N}(\mathbf{M}). \qquad (12.4.12)$$

The signals $x_{\mathbf{k}}(\mathbf{n})$ defined as above are called the Type 1 polyphase components of $x(\mathbf{n})$. Notice that $x_{\mathbf{k}}(\mathbf{n})$ is merely the $\mathbf{M}$-fold decimated version of $x(\mathbf{n} + \mathbf{k})$.

Defining $y_{\mathbf{k}}(\mathbf{n})$ to be the $\mathbf{M}$-fold expanded version of $x_{\mathbf{k}}(\mathbf{n})$, that is,

$$y_{\mathbf{k}}(\mathbf{n}) = \begin{cases} x_{\mathbf{k}}(\mathbf{M}^{-1}\mathbf{n}), & \mathbf{n} \in LAT(\mathbf{M}) \\ 0 & \text{otherwise,} \end{cases} \qquad (12.4.13)$$

we can express each sample of $x(\mathbf{n})$ in terms of one of the $y_{\mathbf{k}}(\mathbf{n})$'s as follows: compute $\mathbf{k} = \mathbf{n} \mod \mathbf{M}$. This determines the unique integer $\mathbf{k} \in \mathcal{N}(\mathbf{M})$. Then

$$x(\mathbf{n}) = y_{\mathbf{k}}(\mathbf{n} - \mathbf{k}). \qquad (12.4.14)$$

Example 12.4.1

Consider the matrix

$$\mathbf{M} = \begin{bmatrix} 1 & 1 \\ 2 & -2 \end{bmatrix} \quad \text{(hexagonal } \mathbf{M}\text{)}. \qquad (12.4.15)$$

We have $[\det \mathbf{M}] = -4$ so that $J(\mathbf{M}) = |\det \mathbf{M}| = 4$. The set $\mathcal{N}(\mathbf{M})$, therefore, has four elements (i.e., there are 4 polyphase components in this case). These four elements are the integer vectors in $FPD(\mathbf{M})$. Fig. 12.4-2(a) shows $FPD(\mathbf{M})$ as well as these four integer vectors inside it. The set $\mathcal{N}(\mathbf{M})$ is thus seen to be

$$\mathbf{k}_0 = \begin{bmatrix} 0 \\ 0 \end{bmatrix}, \ \mathbf{k}_1 = \begin{bmatrix} 1 \\ 0 \end{bmatrix}, \ \mathbf{k}_2 = \begin{bmatrix} 1 \\ 1 \end{bmatrix}, \ \mathbf{k}_3 = \begin{bmatrix} 1 \\ -1 \end{bmatrix} \quad \text{(hexagonal } \mathbf{M}\text{)}.$$
$$(12.4.16a)$$

The set of all two-component integers can therefore be partitioned into the four subsets shown in Fig. 12.4-2(b). We have used four different

---

[†] In set theoretic language, the set of all integer vectors $\mathbf{n}$ can be partitioned to $J(\mathbf{M})$ equivalence classes. Also see end of this section for terms such as *cosets* and *sublattices*.

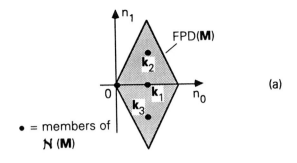

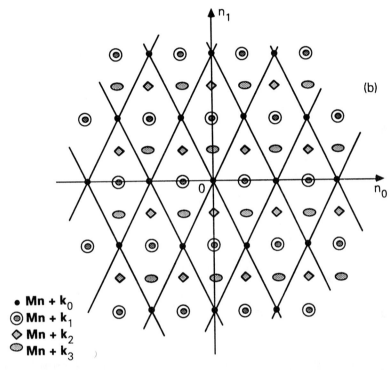

Figure 12.4-2 Demonstrating the four polyphase components generated by hexagonal **M** in (12.4.15). (a) Identifying the set $\mathcal{N}(\mathbf{M})$. (b) The partition of the two dimensional integer index **n**, which identifies the four polyphase components. (© Adopted from Sadhana, 1990 [Vaidyanathan, 1990c].)

symbols to indicate the sample points in the four subsets. For example black dots • indicate sample locations of the form $\mathbf{Mn} + \mathbf{k}_0$.

By definition, vectors $\mathbf{k}_i$ have the form $\mathbf{Mx}_i$ with $\mathbf{x}_i \in [0,1)^2$. The vectors $\mathbf{x}_i$ can be verified to be

$$\mathbf{x}_0 = \begin{bmatrix} 0 \\ 0 \end{bmatrix}, \quad \mathbf{x}_1 = \begin{bmatrix} 0.5 \\ 0.5 \end{bmatrix}, \quad \mathbf{x}_2 = \begin{bmatrix} 0.75 \\ 0.25 \end{bmatrix}, \quad \mathbf{x}_3 = \begin{bmatrix} 0.25 \\ 0.75 \end{bmatrix}. \quad (12.4.16b)$$

Sec. 12.4 Multirate fundamentals

12.4.3 Transform Domain Expressions

Linear Transformation of Polygons

We now study the effects of the decimator and expander in the frequency domain. In order to visualize these effects pictorially, we will use several 2D examples which use polygonal support for $X(\omega)$. In many of the discussions and examples, we are required to perform transformations of the form $\mathbf{q} = \mathbf{Vp}$ where $\mathbf{p} = \begin{bmatrix} p_0 \\ p_1 \end{bmatrix}$, $\mathbf{q} = \begin{bmatrix} q_0 \\ q_1 \end{bmatrix}$, and $\mathbf{V}$ is a 2×2 real nonsingular matrix.

Suppose $\mathcal{P}$ is some polygon in the plane (p_0, p_1) (Fig. 12.4-3(a)). How does this map into the (q_0, q_1) plane? The answer is that the mapped region continues to remain a polygon with the same number of edges, and moreover the mapped versions of pairs of touching-edges continue to remain touching-edges. This is demonstrated in Fig. 12.4-3(b) where edges are numbered for convenience. Finally, for every interior point in the (p_0, p_1) polygon, there exists a unique point interior to the (q_0, q_1) polygon and vice versa. The procedure to find the mapped polygon is as follows: simply map each vertex using the transformation $\mathbf{q} = \mathbf{Vp}$, and then join these vertices by straightlines. Proofs of these statements are developed in Problem 12.6.

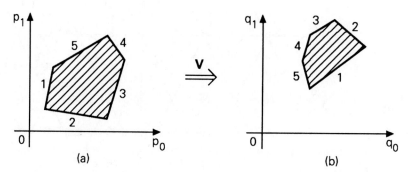

Figure 12.4-3 Mappping of a polygon under linear transformation.

Multidimensional Expander (Transform Domain Analysis)

From (12.4.5) we have

$$Y(\omega) = \sum_{\mathbf{n} \in LAT(\mathbf{M})} x(\mathbf{M}^{-1}\mathbf{n})e^{-j\omega^T \mathbf{n}}$$
$$= \sum_{\mathbf{m} \in \mathcal{N}} x(\mathbf{m})e^{-j\omega^T \mathbf{Mm}} = X(\mathbf{M}^T \omega). \qquad (12.4.17)$$

Thus the effect of the expander is merely to perform a (invertible) transformation of the frequency vector ω. It is clear that the periodicity matrix for $Y(\omega)$ is $2\pi \mathbf{M}^{-T}$ rather than $2\pi \mathbf{I}$. So $|\det \mathbf{M}|$ copies of the basic spectrum $X(\omega)$ are squeezed into the region $[0, 2\pi)^D$.

This effect is graphically demonstrated in Fig. 12.4-4 for the expander matrix

$$\mathbf{M} = \begin{bmatrix} 1 & -1 \\ 1 & 2 \end{bmatrix}. \qquad (12.4.18)$$

The support of $X(\omega)$ which is shown in Fig. 12.4-4(a) is mapped into the polygon indicated in Fig. 12.4-4(b), using the rules indicated above. Since the periodicity matrix for $Y(\omega)$ is given by $2\pi \mathbf{M}^{-T}$, we have additional copies of this polygon, centered at the points of $LAT(2\pi \mathbf{M}^{-T})$. Of these, those that are *not* centered around integer multiples of 2π are the *images* created by the expander. In the figure, images are shown in light gray. If these are eliminated by filtering, only the dark gray areas will remain, and will constitute the 'interpolated' signal. In Fig. 12.4-4(b), the parallelogram in heavy lines indicates $FPD(2\pi \mathbf{M}^{-T})$ for reference purposes.

Evidently we can recover $X(\omega)$ from $Y(\omega)$ by doing a change of variables $\omega \to \mathbf{M}^{-T}\omega$. This is achieved trivially by passing $y(\mathbf{n})$ through an M-fold decimator.

We can express (12.4.17) in terms of z-domain quantities. The z-transform of the signal $y(\mathbf{n})$ is given by

$$\begin{aligned} Y_z(\mathbf{z}) &= \sum_{\mathbf{n} \in \mathcal{N}} y(\mathbf{n}) \mathcal{Z}(-\mathbf{n}) \\ &= \sum_{\mathbf{m} \in \mathcal{N}} x(\mathbf{m}) \mathcal{Z}(-\mathbf{Mm}) \quad \text{by (12.4.5)}. \end{aligned} \qquad (12.4.19)$$

Definition of $\mathbf{z}^{(\mathbf{M})}$

We now introduce a new notation, viz., $\mathbf{z}^{(\mathbf{M})}$. The quantity $\mathbf{z}^{(\mathbf{M})}$ is a column vector with D components, obtained from the column vector $\mathbf{z} = [z_0 \ldots z_{D-1}]^T$. The kth component of $\mathbf{z}^{(\mathbf{M})}$ is obtained as follows:

$$(\mathbf{z}^{(\mathbf{M})})_k = z_0^{M_{0,k}} z_1^{M_{1,k}} \ldots z_{D-1}^{M_{D-1,k}}. \qquad (12.4.20)$$

With this notation it is easily verified that

$$Y_z(\mathbf{z}) = X_z(\mathbf{z}^{(\mathbf{M})}) \quad \text{(M-fold expander)}, \qquad (12.4.21)$$

which has an appealing resemblance to the corresponding 1D relation $Y(z) = X(z^M)$.

For diagonal $\mathbf{M}$ (i.e., rectangular interpolators), the relation (12.4.21) is easy to interpret. For example if $\mathbf{M} = \begin{bmatrix} 2 & 0 \\ 0 & 3 \end{bmatrix}$, then $Y_z(z_0, z_1) = X_z(z_0^2, z_1^3)$. To appreciate the meaning of the notation $\mathbf{z}^{(\mathbf{M})}$ further, we consider a non rectangular example.

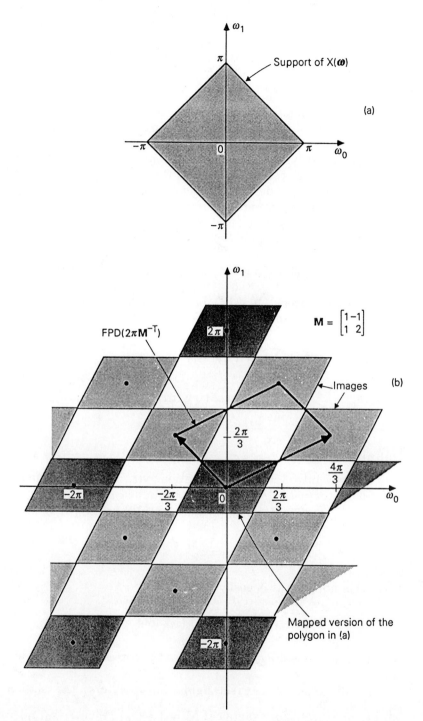

Figure 12.4-4 Frequency domain effect (imaging) caused by an expander. (a) Support of $X(\boldsymbol{\omega})$ and (b) support of $Y(\boldsymbol{\omega}) = X(\mathbf{M}^T\boldsymbol{\omega})$. The white regions imply $Y(\boldsymbol{\omega}) = 0$; the light gray regions are images.

Example 12.4.2

Consider the hexagonal matrix M in (12.4.15). The kth element of $\mathbf{z^{(M)}}$ is given by
$$z_0^{M_{0,k}} z_1^{M_{1,k}}, \quad k = 0, 1,$$
so that we have
$$\mathbf{z^{(M)}} = \begin{bmatrix} z_0 z_1^2 \\ z_0 z_1^{-2} \end{bmatrix} \quad \text{(hexagonal M)}.$$

So for the hexagonal expander, the output can be rewritten as
$$Y_z(z_0, z_1) = X_z(z_0 z_1^2, z_0 z_1^{-2}).$$

Mathematically, the above notation can also be used for rectangular matrices. In particular, consider $\mathbf{z^{(n)}}$ where $\mathbf{n}$ is a $D \times 1$ column vector as in (12.1.7) [Viscito and Allebach, 1988b]. Then the above definition yields $\mathbf{z^{(n)}} = z_0^{n_0} z_1^{n_1} \ldots z_{D-1}^{n_{D-1}}$. This is consistent with the notation $\mathcal{Z}(\mathbf{n})$ introduced in (12.1.14), i.e., $\mathbf{z^{(n)}} = \mathcal{Z}(\mathbf{n})$. The notation $\mathbf{z^{(n)}}$ has the advantage that it reveals the dependence on $\mathbf{z}$ as well as $\mathbf{n}$. It can be verified that $[\mathbf{z^{(M)}}]^{(\mathbf{n})} = \mathbf{z^{(Mn)}}$, which is similar to the relation $[z^M]^n = z^{Mn}$ in the scalar (1D) case. Furthermore, $[\mathbf{z^{(L)}}]^{(M)} = \mathbf{z^{(LM)}}$.

Multidimensional Decimator (Transform Domain Analysis)

Consider the decimated version $y(\mathbf{n}) = x(\mathbf{Mn})$. If $X(\boldsymbol{\omega})$ and $Y(\boldsymbol{\omega})$ denote the Fourier transforms of the sequences $x(\mathbf{n})$ and $y(\mathbf{n})$, how are these related? Recall that in the 1D case, the transform of $y(n)$ is given by (4.1.4). This means that $Y(e^{j\omega})$ is a sum of the stretched version $X(e^{j\omega/M})/M$ with $M - 1$ uniformly shifted copies of this stretched version. In the MD case a very similar result holds, except that $\boldsymbol{\omega}$ is now a vector so that 'stretching' and 'shifting' are more sophisticated matrix operations. When M is diagonal the result is easy to see; we merely apply the 1D formula for each frequency variable separately.

For the M-fold decimator, we will establish the relation
$$Y(\boldsymbol{\omega}) = \frac{1}{J(\mathbf{M})} \sum_{\mathbf{k} \in \mathcal{N}(\mathbf{M}^T)} X\left(\mathbf{M}^{-T}(\boldsymbol{\omega} - 2\pi \mathbf{k})\right), \qquad (12.4.22)$$

where $\mathcal{N}(\mathbf{M}^T)$ is the set of integers of the form $\mathbf{M}^T \mathbf{x}$, with $\mathbf{x} \in [0, 1)^D$ [consistent with the notation $\mathcal{N}(.)$ defined in (12.4.2)]. Once again in the 1D case this reduces to (4.1.4). The number of terms in the above summation is equal to $J(\mathbf{M})$. The term $X(\mathbf{M}^{-T}\boldsymbol{\omega})$ is the 'stretched' version, and the terms with $\mathbf{k} \neq \mathbf{0}$ are the shifted versions. Even though the stretched version does not necessarily have period 2π in each frequency variable, the sum (12.4.22) does (Problem 12.7). If we replace $\mathbf{k} \in \mathcal{N}(\mathbf{M}^T)$ in (12.4.22) with $\mathbf{k} \in \mathcal{N}(\mathbf{M}^T\mathbf{E})$ for some unimodular integer matrix $\mathbf{E}$, the result of the summation remains unchanged (Problem 12.33).

Before proving the above relation, we will consider some examples.

Example 12.4.3

Let
$$\mathbf{M} = \begin{bmatrix} 2 & 0 \\ 0 & 3 \end{bmatrix}. \qquad (12.4.23)$$

Here $J(\mathbf{M}) = 6$ so that there are six terms in $Y(\omega)$. The elements $\mathbf{k}_i$ in the set $\mathbf{k} \in \mathcal{N}(\mathbf{M}^T)$ are

$$\begin{bmatrix} 0 \\ 0 \end{bmatrix}, \begin{bmatrix} 1 \\ 0 \end{bmatrix}, \begin{bmatrix} 0 \\ 1 \end{bmatrix}, \begin{bmatrix} 1 \\ 1 \end{bmatrix}, \begin{bmatrix} 0 \\ 2 \end{bmatrix}, \begin{bmatrix} 1 \\ 2 \end{bmatrix}. \qquad (12.4.24)$$

Figure 12.4-5 demonstrates the supports of $X(\omega)$ and the "stretched version" $X(\mathbf{M}^{-T}\omega)$, for bandlimited $x(\mathbf{n})$ (with circular support). Notice that the stretched version in general has different "size" and "shape," compared to $X(\omega)$. The figure also shows some of the shifted versions, which are closest to $X(\mathbf{M}^{-T}\omega)$.

Note that, in this example, the shifted versions which take part in the summation have overlap with the stretched version so that there is aliasing.

Beauty of the RHS in (12.4.22): It can be shown that any shifted version $X(\mathbf{M}^{-T}(\omega - 2\pi\mathbf{m}))$ for arbitrary integer $\mathbf{m}$ can be rewritten in the form $X(\mathbf{M}^{-T}(\omega - 2\pi\mathbf{k}))$ for some $\mathbf{k} \in \mathcal{N}(\mathbf{M}^T)$. So the RHS contains every possible distinct shifted copy of $X(\mathbf{M}^{-T}(\omega))$. To see this note that $\mathbf{m}$ can be expressed as $\mathbf{m} = \mathbf{k} + \mathbf{M}^T\mathbf{n}$ for integer $\mathbf{n}$, with $\mathbf{k} \in \mathcal{N}(\mathbf{M}^T)$. Thus

$$X(\mathbf{M}^{-T}(\omega - 2\pi\mathbf{m})) = X(\mathbf{M}^{-T}(\omega - 2\pi\mathbf{k}) - 2\pi\mathbf{n}), \qquad (12.4.25)$$

which reduces to $X(\mathbf{M}^{-T}(\omega - 2\pi\mathbf{k}))$ in view of the periodicity of $X(.)$.

Derivation of (12.4.22). We will derive this expression by assuming that $x(\mathbf{n})$ has been obtained by sampling an analog signal $x_a(\mathbf{t})$ with some sampling matrix $\mathbf{V}$. There is no loss of generality here, but a reader wishing to obtain an independent derivation can go through Problem 12.30. We know (Sec. 12.2.3) that $X(\omega)$ and $X_a(j\Omega)$ are related as

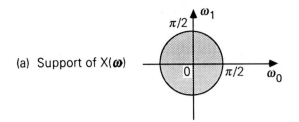

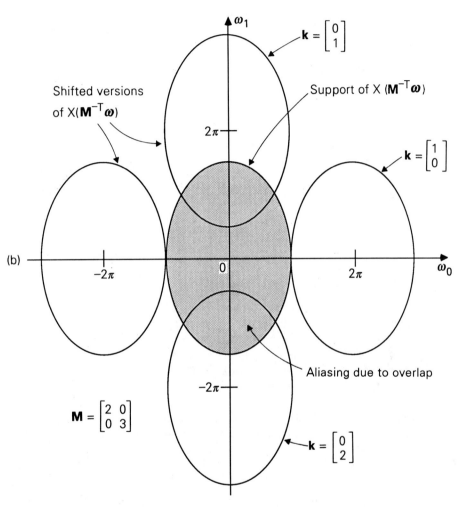

Figure 12.4-5 Illustrating the frequency domain effects of decimation. (a) Support of the signal before decimation, and (b) supports of the stretched and various shifted versions.

$$X(\omega) = \frac{1}{|\det \mathbf{V}|} \sum_{\mathbf{n}_0 \in \mathcal{N}} X_a\Big(j\mathbf{V}^{-T}(\omega - 2\pi \mathbf{n}_0)\Big), \qquad (12.4.26)$$

Sec. 12.4 Multirate fundamentals   585

where $\mathcal{N}$ is the set of D-dimensional integers. Evidently $y(\mathbf{n})$ is obtainable directly from $x_a(\mathbf{t})$ as

$$y(\mathbf{n}) = x_a(\mathbf{VMn}). \qquad (12.4.27)$$

So $Y(\omega)$ is given by

$$Y(\omega) = \frac{1}{|\det \mathbf{VM}|} \sum_{\mathbf{n} \in \mathcal{N}} X_a\big(j\mathbf{V}^{-T}\mathbf{M}^{-T}(\omega - 2\pi\mathbf{n})\big). \qquad (12.4.28)$$

We now use the division theorem [similar to (12.4.8)] to write the summation index $\mathbf{n}$ as

$$\mathbf{n} = \mathbf{M}^T \mathbf{n}_0 + \mathbf{k}, \qquad (12.4.29)$$

with $\mathbf{k} \in \mathcal{N}(\mathbf{M}^T)$ and $\mathbf{n}_0 \in \mathcal{N}$. Thus, $Y(\omega)$ can be rearranged as

$$Y(\omega) = \frac{1}{|\det \mathbf{VM}|} \sum_{\mathbf{k} \in \mathcal{N}(\mathbf{M}^T)} \sum_{\mathbf{n}_0 \in \mathcal{N}} X_a\big(j\mathbf{V}^{-T}(\mathbf{M}^{-T}\omega - 2\pi\mathbf{M}^{-T}\mathbf{k} - 2\pi\mathbf{n}_0)\big),$$

$$= \frac{1}{J(\mathbf{M})} \sum_{\mathbf{k} \in \mathcal{N}(\mathbf{M}^T)} \frac{1}{|\det \mathbf{V}|} \sum_{\mathbf{n}_0 \in \mathcal{N}} X_a\big(j\mathbf{V}^{-T}(\widehat{\omega} - 2\pi\mathbf{n}_0)\big),$$

$$(12.4.30)$$

where $\widehat{\omega} = \mathbf{M}^{-T}(\omega - 2\pi\mathbf{k})$. The inner summation can be written in terms of $X(\omega)$ so that $Y(\omega)$ reduces to (12.4.22) indeed!

Polyphase decomposition (Transform Domain Expression)

We wish to write the multidimensional polyphase decomposition in terms of z-domain quantities so that we can obtain an expression analogous to (4.3.7). Recall that the k-th polyphase component $x_\mathbf{k}(\mathbf{n})$ of $x(\mathbf{n})$ is defined as in (12.4.12), and that $x(\mathbf{n})$ can be expressed uniquely in terms of the polyphase components $y_\mathbf{k}(\mathbf{n})$ as in (12.4.14).

In terms of z-domain block diagrams we can express $x_\mathbf{k}(\mathbf{n})$ and $y_\mathbf{k}(\mathbf{n})$ as in Fig. 12.4-6. The relation (12.4.14) shows that we can obtain $x(\mathbf{n})$ from the set of signals $y_\mathbf{k}(\mathbf{n})$ as shown in the figure. Combining these we arrive at Fig. 12.4-7 which is merely a decomposition of $x(\mathbf{n})$ into its polyphase components and re-synthesis of $x(\mathbf{n})$ from these components. From this it is clear that we can express $X(\omega)$ as

$$X(\omega) = \sum_{\mathbf{k} \in \mathcal{N}(\mathbf{M})} e^{-j\omega^T \mathbf{k}} X_\mathbf{k}(\mathbf{M}^T \omega) \quad \text{(Type 1 polyphase)}, \qquad (12.4.31)$$

or in the z-domain as

$$X_z(\mathbf{z}) = \sum_{\mathbf{k} \in \mathcal{N}(\mathbf{M})} \mathcal{Z}(-\mathbf{k}) X_{z,\mathbf{k}}(\mathbf{z}^{(\mathbf{M})}), \quad \text{(Type 1 polyphase)}. \qquad (12.4.32)$$

Figure 12.4-7 can also be looked upon as a generalization of the delay chain structure (Fig. 5.6-2), which represented a simple perfect reconstruction

QMF bank! In Sec. 12.9 this will be used as a starting point for design of multidimensional QMF banks.

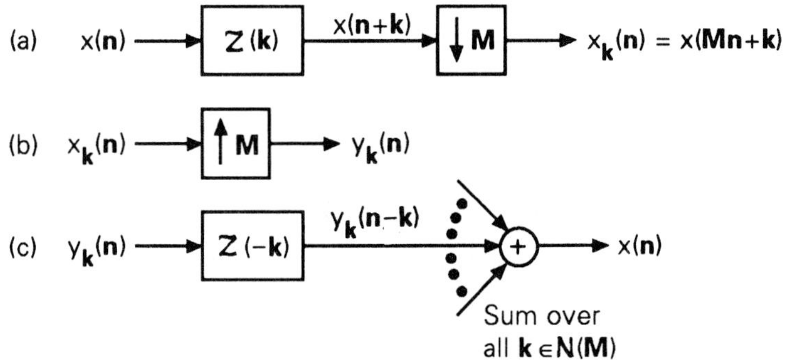

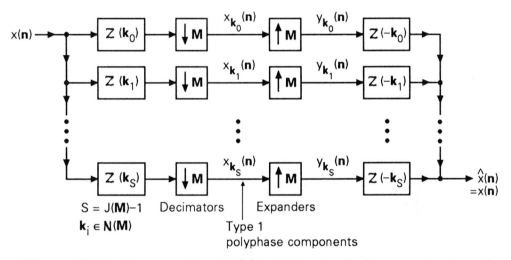

Figure 12.4-6 Block-diagram representations of the relations among the signals $x(\mathbf{n})$, $x_k(\mathbf{n})$, and $y_k(\mathbf{n})$.

Figure 12.4-7 Decomposition of $x(\mathbf{n})$ into Type 1 polyphase components, and re-synthesis of $x(\mathbf{n})$.

Example 12.4.4

Consider the rectangular decimation matrix

$$\mathbf{M} = \begin{bmatrix} 2 & 0 \\ 0 & 2 \end{bmatrix}.$$

Here $J(\mathbf{M}) = 4$ and the integers in $\mathcal{N}(\mathbf{M})$ are

$$\begin{bmatrix} 0 \\ 0 \end{bmatrix}, \begin{bmatrix} 1 \\ 0 \end{bmatrix}, \begin{bmatrix} 0 \\ 1 \end{bmatrix}, \text{ and } \begin{bmatrix} 1 \\ 1 \end{bmatrix}.$$

Sec. 12.4   Multirate fundamentals   587

So the quantities $\mathcal{Z}(-\mathbf{k})$ in (12.4.32) are

$$\mathcal{Z}(-\mathbf{k}_0) = 1, \ \mathcal{Z}(-\mathbf{k}_1) = z_0^{-1}, \ \mathcal{Z}(-\mathbf{k}_2) = z_1^{-1}, \ \mathcal{Z}(-\mathbf{k}_3) = z_0^{-1}z_1^{-1}.$$

We now have $\mathbf{z}^{(\mathbf{M})} = \begin{bmatrix} z_0^2 \\ z_1^2 \end{bmatrix}$ so that (12.4.32) becomes

$$\begin{aligned}X_z(z_0, z_1) = {}& X_{z,\mathbf{k}_0}(z_0^2, z_1^2) + z_0^{-1} X_{z,\mathbf{k}_1}(z_0^2, z_1^2) \\ &+ z_1^{-1} X_{z,\mathbf{k}_2}(z_0^2, z_1^2) + z_0^{-1}z_1^{-1} X_{z,\mathbf{k}_3}(z_0^2, z_1^2).\end{aligned}$$

Example 12.4.5. Polyphase Decomposition for Hexagonal M

Consider the hexagonal matrix (12.4.15) again. Here $J(\mathbf{M}) = 4$, and the set of integers $\mathcal{N}(\mathbf{M})$ is given in (12.4.16a). We can now identify the elements $\mathcal{Z}(-\mathbf{k})$ as

$$\mathcal{Z}(-\mathbf{k}_0) = 1, \ \mathcal{Z}(-\mathbf{k}_1) = z_0^{-1}, \ \mathcal{Z}(-\mathbf{k}_2) = z_0^{-1}z_1^{-1}, \ \mathcal{Z}(-\mathbf{k}_3) = z_0^{-1}z_1.$$

In Example 12.4.2 we identified the meaning of $\mathbf{z}^{(\mathbf{M})}$ for this M. By using this we can rewrite the expression (12.4.32) as

$$\begin{aligned}X_z(z_0, z_1) = {}& X_{z,\mathbf{k}_0}(z_0 z_1^2, z_0 z_1^{-2}) + z_0^{-1} X_{z,\mathbf{k}_1}(z_0 z_1^2, z_0 z_1^{-2}) \\ &+ z_0^{-1} z_1^{-1} X_{z,\mathbf{k}_2}(z_0 z_1^2, z_0 z_1^{-2}) + z_0^{-1} z_1 X_{z,\mathbf{k}_3}(z_0 z_1^2, z_0 z_1^{-2}).\end{aligned}$$

which is the Type 1 polyphase decomposition.

Type 2 polyphase decomposition. Given any integer $\mathbf{n}$, the decomposition (12.4.8) results in Type 1 polyphase decomposition. Now it is also possible to obtain a decomposition of the form $\mathbf{n} = -\mathbf{k} + \mathbf{M}\mathbf{n}_0$, where $\mathbf{k} \in \mathcal{N}(\mathbf{M})$ and $\mathbf{n}_0$ is integer. We can use this to obtain the Type 2 polyphase decomposition given by

$$X_z(\mathbf{z}) = \sum_{\mathbf{k} \in \mathcal{N}(\mathbf{M})} \mathcal{Z}(\mathbf{k}) X'_{z,\mathbf{k}}(\mathbf{z}^{(\mathbf{M})}), \quad \text{(Type 2 polyphase).} \quad (12.4.33)$$

where $x'_\mathbf{k}(\mathbf{n})$ are called Type 2 polyphase components. Notice that Type 2 decomposition involves advance operators $\mathcal{Z}(\mathbf{k})$. We can avoid these by use of an overall fixed delay operator in the definition (as we did in the 1D case), but we will not do so. In most 2D applications, advance operators are realizable (because we would often work with pictures rather than real-time signals).

We can perform polyphase decompositions on a transfer function $H_z(\mathbf{z})$ precisely as in (12.4.32) and (12.4.33). Figs. 12.4-8(a),(b) show these polyphase implementations. The quantities $E_{z,\mathbf{k}}(\mathbf{z})$ and $R_{z,\mathbf{k}}(\mathbf{z})$ are the polyphase components. Figure 12.4-9 shows the details for the hexagonal matrix of Example 12.4.5.

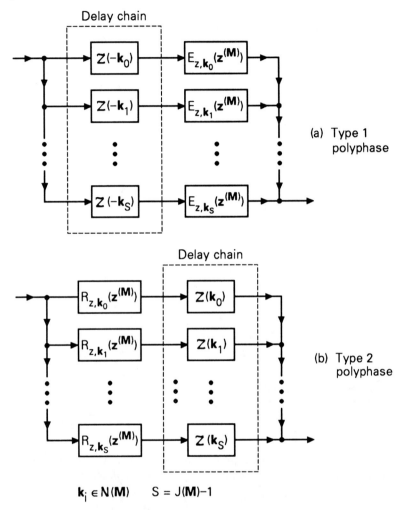

Figure 12.4-8 Polyphase implementations of a transfer function.

12.4.4 Generalized orthogonal exponentials, and DFT

From Chapter 4 we know that the $M \times M$ DFT matrix plays an important role in filter bank design. For example, the uniform DFT bank (Sec. 4.3.2) is a very efficient way to implement M filters at the cost of (almost) one filter. The entries of $\mathbf{W}$ have the form $e^{-j2\pi k\ell/M}$, so that $\mathbf{W}^\dagger \mathbf{W} = M\mathbf{I}$.

More explicitly,
$$\sum_{\ell=0}^{M-1} e^{-j2\pi k\ell/M} = \begin{cases} M & k = 0, \\ 0 & 1 \le k \le M-1. \end{cases} \quad (12.4.34)$$

In other words, the columns of **W** are pairwise orthogonal.

For the 1D case, we know that many important properties, such as Mth band property (Sec. 4.6), and the relation between AC matrix and polyphase matrix (Sec. 5.5) can be expressed in terms of the elements of the DFT matrix. Such valuable relations also exist in the MD case, provided the generalized-DFT matrix is defined properly. Such a definition can be found in Dudgeon and Merseareau [1984]. We will arrive at this definition in a different way, by applying polyphase decomposition.

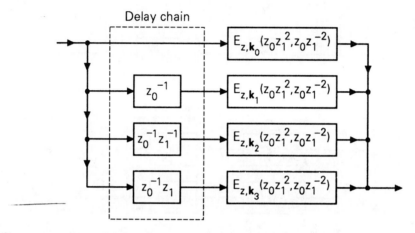

Figure 12.4-9 Type 1 polyphase implementation of a filter, with hexagonal **M** as in (12.4.15).

Generalized Exponentials

The polyphase component $x_\mathbf{k}(\mathbf{n})$ is the decimated version of $x(\mathbf{n}+\mathbf{k})$. Using the relation (12.4.22) we therefore arrive at

$$X_\mathbf{k}(\boldsymbol{\omega}) = \frac{1}{J(\mathbf{M})} \sum_{\mathbf{m} \in \mathcal{N}(\mathbf{M}^T)} e^{j\left(\mathbf{M}^{-T}(\boldsymbol{\omega}-2\pi\mathbf{m})\right)^T \mathbf{k}} X\left(\mathbf{M}^{-T}(\boldsymbol{\omega}-2\pi\mathbf{m})\right). \quad (12.4.35)$$

We know that $X(\boldsymbol{\omega})$ can be expressed in terms of $X_\mathbf{k}(\boldsymbol{\omega})$ as indicated in (12.4.31). Substituting the above into this and rearranging, we obtain

$$X(\boldsymbol{\omega}) = \frac{1}{J(\mathbf{M})} \sum_{\mathbf{m} \in \mathcal{N}(\mathbf{M}^T)} X(\boldsymbol{\omega} - 2\pi\mathbf{M}^{-T}\mathbf{m}) \sum_{\mathbf{k} \in \mathcal{N}(\mathbf{M})} e^{-j2\pi\mathbf{m}^T \mathbf{M}^{-1}\mathbf{k}}. \quad (12.4.36)$$

Since this relation holds for all possible functions $X(\omega)$, we conclude (see Problem 12.9)

$$\sum_{\mathbf{k} \in \mathcal{N}(\mathbf{M})} e^{-j2\pi \mathbf{m}^T \mathbf{M}^{-1} \mathbf{k}} = \begin{cases} J(\mathbf{M}), & \mathbf{m} = \mathbf{0}, \\ 0, & \mathbf{m} \in \mathcal{N}(\mathbf{M}^T), \ \mathbf{m} \neq \mathbf{0}. \end{cases} \quad (12.4.37a)$$

This is the MD extension of the orthogonality relation (12.4.34). Here $\mathbf{m}$ and $\mathbf{k}$ are D-component integers, and $\mathbf{M}$ is a $D \times D$ integer matrix. The reader can verify that the above equation can also be rearranged as

$$\sum_{\mathbf{m} \in \mathcal{N}(\mathbf{M}^T)} e^{-j2\pi \mathbf{m}^T \mathbf{M}^{-1} \mathbf{k}} = \begin{cases} J(\mathbf{M}) & \mathbf{k} = \mathbf{0}, \\ 0 & \mathbf{k} \in \mathcal{N}(\mathbf{M}), \ \mathbf{k} \neq \mathbf{0}. \end{cases} \quad (12.4.37b)$$

Furthermore, by conjugating both sides, we see that these equations hold if all the minus signs on the exponentials are dropped.

The set of sequences $e^{-j2\pi \mathbf{m}^T \mathbf{M}^{-1} \mathbf{n}}$ is called the *generalized orthogonal exponentials*. In Problem 12.29 a second derivation of (12.4.37) is presented, based on the fact that these are single-frequency sequences. This derivation will not make use of results such as (12.4.22) and is therefore more independent.

Generalized DFT Matrix

We define the generalized DFT matrix to be the $J(\mathbf{M}) \times J(\mathbf{M})$ matrix $\mathbf{W}^{(g)}$ whose elements are given by

$$[\mathbf{W}^{(g)}]_{\mathbf{m},\mathbf{k}} = e^{-j2\pi \mathbf{m}^T \mathbf{M}^{-1} \mathbf{k}}, \quad \mathbf{m} \in \mathcal{N}(\mathbf{M}^T), \mathbf{k} \in \mathcal{N}(\mathbf{M}). \quad (12.4.38)$$

Notice that the row and column indices for the matrix are taken as *vectors* $\mathbf{m}$ and $\mathbf{k}$, which might appear to be strange. However, imagine that the sets of integers $\mathbf{k}$ and $\mathbf{m}$ are ordered in some way (the ordering itself being immaterial). Then there is no difficulty in identifying the meaning of $[\mathbf{W}^{(g)}]_{\mathbf{m},\mathbf{k}}$. Thus, given the vector $\mathbf{k}$, suppose it is the ith vector in the set $\mathcal{N}(\mathbf{M})$. Similarly, let $\mathbf{m}$ be the ℓth vector in the set $\mathcal{N}(\mathbf{M}^T)$. Then the notation $[\mathbf{W}^{(g)}]_{\mathbf{m},\mathbf{k}}$ actually means the (i, ℓ) element $[\mathbf{W}^{(g)}]_{i,\ell}$. We continue to use vector subscripts on matrices and vectors, as it is more convenient.

In view of (12.4.37), $\mathbf{W}^{(g)}$ is a unitary matrix satisfying

$$[\mathbf{W}^{(g)}]^\dagger \mathbf{W}^{(g)} = J(\mathbf{M})\mathbf{I}, \quad (12.4.39)$$

so that

$$[\mathbf{W}^{(g)}]^{-1} = [\mathbf{W}^{(g)}]^\dagger / J(\mathbf{M}). \quad (12.4.40)$$

In Sec. 12.9 we indicate further applications of the generalized DFT matrix in MD multirate systems. Notice that $\mathbf{W}^{(g)}_{\mathbf{M}}$ would be a more appropriate notation, but we omit the subscript for simplicity.

Example 12.4.6: Generalized DFT Matrix for Hexagonal M

Consider again $\mathbf{M}$ as in (12.4.15). To construct the matrix $\mathbf{W}^{(g)}$, the sets $\mathcal{N}(\mathbf{M})$ and $\mathcal{N}(\mathbf{M}^T)$ should first be identified. Since $[\det \mathbf{M}] = -4$, we have $J(\mathbf{M}) = 4$ so that each of these sets has four members. $\mathcal{N}(\mathbf{M})$ is the set of integers in $FPD(\mathbf{M})$, and is given in (12.4.16a). Similarly $\mathcal{N}(\mathbf{M}^T)$ is the set of integers in $FPD(\mathbf{M}^T)$. From Fig. 12.4-10(b) we identify the elements of $\mathcal{N}(\mathbf{M}^T)$ as

$$\mathbf{m}_0 = \begin{bmatrix} 0 \\ 0 \end{bmatrix}, \quad \mathbf{m}_1 = \begin{bmatrix} 1 \\ 0 \end{bmatrix}, \quad \mathbf{m}_2 = \begin{bmatrix} 1 \\ -1 \end{bmatrix}, \quad \mathbf{m}_3 = \begin{bmatrix} 2 \\ -1 \end{bmatrix}. \qquad (12.4.41)$$

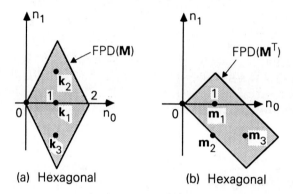

(a) Hexagonal   (b) Hexagonal

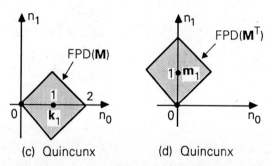

(c) Quincunx   (d) Quincunx

Figure 12.4-10 The fundamental parallelepiped for various cases. Such figures help us to identify the integer sets $\mathcal{N}(\mathbf{M})$ and $\mathcal{N}(\mathbf{M}^T)$. (© Adopted from Sadhana, 1990 [Vaidyanathan, 1990c].)

We now compute the sixteen products $\mathbf{m}_i^T \mathbf{M}^{-1} \mathbf{k}_\ell = \mathbf{m}_i^T \mathbf{x}_\ell$, where $\mathbf{x}_\ell$ are given in (12.4.16b). The sixteen computed products are arranged in matrix form below:

$$\frac{1}{8} \begin{bmatrix} 0 & 0 & 0 & 0 \\ 0 & 4 & 6 & 2 \\ 0 & 0 & 4 & -4 \\ 0 & 4 & 10 & -2 \end{bmatrix}. \qquad (12.4.42)$$

So the matrix $\mathbf{W}^{(g)}$ is

$$\mathbf{W}^{(g)} = \begin{bmatrix} 1 & 1 & 1 & 1 \\ 1 & W^2 & W^3 & W \\ 1 & W^4 & W^6 & W^2 \\ 1 & W^6 & W^9 & W^3 \end{bmatrix} \quad \text{(for hexagonal } \mathbf{M}\text{)}, \qquad (12.4.43)$$

where $W \triangleq e^{-j2\pi/4}$. Therefore, $\mathbf{W}^{(g)}$ is a column-permuted version of the traditional (1D) 4×4 DFT matrix! This, however, is not a general fact. For example, try the 2×2 matrix $\mathbf{M} = 2\mathbf{I}$. Also see Problem 12.13. The exact relation between $\mathbf{W}^{(g)}$ and 'traditional' DFT matrices is given in Sec. 12.10.

12.4.5 Fine Points About the Decimation Matrix

Recall that the samples of $x(\mathbf{n})$ which are retained by the decimator $\mathbf{M}$, are assigned appropriate locations to form $y(\mathbf{n})$. [See Fig. 12.4-1(f),(g)]. Recall also that $LAT(\mathbf{M})$ does not change if we replace $\mathbf{M}$ with $\mathbf{ME}$ where $\mathbf{E}$ is an integer unimodular matrix. We can think up an infinite number of $\mathbf{E}$-matrices like this. For example in the 2D case, any integer matrix of the form $\mathbf{E} = \begin{bmatrix} 1 & 0 \\ x & 1 \end{bmatrix}$ works. As a result, there exist infinite number of decimation matrices which retain the same subset of samples from $x(\mathbf{n})$. However, the ordering or "arrangement" of these retained samples is different, depending on $\mathbf{E}$. In other words, the samples in $x(\mathbf{Mn})$ are permuted versions of the samples in $x(\mathbf{MEn})$.

For example, with $\mathbf{E} = \begin{bmatrix} 1 & 0 \\ 1 & 1 \end{bmatrix}$, we can obtain simpler equivalent forms for the hexagonal and qunicunx matrices in Fig. 12.4-1. Thus, we have the equivalent forms

$$\begin{bmatrix} 1 & 1 \\ 2 & -2 \end{bmatrix}, \begin{bmatrix} 2 & 1 \\ 0 & -2 \end{bmatrix} \quad \text{(hexagonal } \mathbf{M}\text{)}, \qquad (12.4.44)$$

and

$$\begin{bmatrix} 1 & 1 \\ -1 & 1 \end{bmatrix}, \begin{bmatrix} 2 & 1 \\ 0 & 1 \end{bmatrix} \quad \text{(quincunx } \mathbf{M}\text{)}. \qquad (12.4.45)$$

Comment on the term "hexagonal"

The reader should be warned that the jargon "hexagonal" is somewhat misleading. For, we can fit hexagonal patterns into several other kinds of decimators, which are not given that label. In Fig. 12.4-11 we demonstrate this for hexagonal, quincunx and rectangular(!) decimators. Thus, even the samples retained by a rectangular decimator are located at the centers and corners of a *tilted* hexagon.

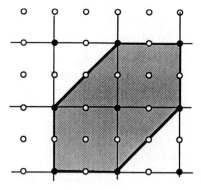

(a) Rectangular decimation

$$\mathbf{M} = \begin{bmatrix} 2 & 0 \\ 0 & 2 \end{bmatrix}$$

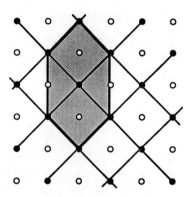

(b) Quincunx decimation

$$\mathbf{M} = \begin{bmatrix} 1 & 1 \\ -1 & 1 \end{bmatrix}$$

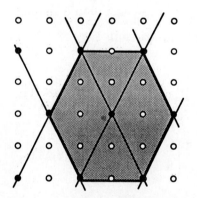

(c) Hexagonal decimation

$$\mathbf{M} = \begin{bmatrix} 1 & 1 \\ 2 & -2 \end{bmatrix}$$

Figure 12.4-11 Fitting hexagons into the set of samples retained by decimators.

Unimodular Decimation Matrices

What happens when the decimator matrix itself has determinant $= \pm 1$? This means that the decimation ratio is unity, and the result is a mere rearrangement of samples. No samples are lost in the decimation process. We can verify in this case that $Y(\omega) = X(\mathbf{E}^{-T}\omega)$ (as the set $\mathcal{N}(\mathbf{E}^T)$ has only one element, viz., the zero-vector). Thus, with $[\det \mathbf{E}] = \pm 1$ there is no aliasing, and the frequency variable undergoes an invertible linear transformation.

Figure 12.4-12 demonstrates this for

$$\mathbf{E} = \begin{bmatrix} 1 & 1 \\ 1 & 2 \end{bmatrix}. \qquad (12.4.46)$$

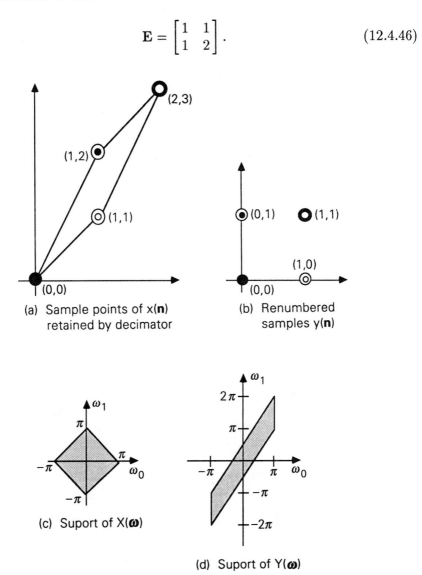

Figure 12.4-12 Demonstrating the effect of a 'decimation' matrix with unit determinant (unimodular decimator).

The sample points of $x(\mathbf{n})$ get mapped into $y(\mathbf{n})$ as follows:

$$x(0,0) \to y(0,0) \quad x(1,1) \to y(1,0) \quad x(1,2) \to y(0,1) \quad x(2,3) \to y(1,1),$$

and so on. If the support of $X(\boldsymbol{\omega})$ is a diamond as in part (c), the transformed support [support of $Y(\boldsymbol{\omega})$] is as in part (d). Notice that these supports actually repeat with periodicity 2π in each direction.

It can be verified from the definitions that when $\mathbf{E}$ is unimodular, the $\mathbf{E}$-fold decimator is equivalent to the $\mathbf{E}^{-1}$-fold expander.

Some Algebraic Language

Much of our discussions (such as, for example, polyphase decomposition) were based on the division theorem. As mentioned earlier, this theorem is a tool by which we can partition the set $\mathcal{N}$ of all $D \times 1$ integers into $J(\mathbf{M})$ disjoint subsets $\mathcal{N}_\mathbf{k}$. The subset $\mathcal{N}_\mathbf{k}$ is the set of all integers $\mathbf{n}$ for which $(\mathbf{n} \bmod \mathbf{M}) = \mathbf{k}$, that is, division by $\mathbf{M}$ would yield the remainder $\mathbf{k} \in \mathcal{N}(\mathbf{M})$. From the division theorem one can infer that no two subsets have a common member, and that the union of these $J(\mathbf{M})$ subsets is the set of all integers $\mathcal{N}$.

It is sometimes useful to express these in terms of *cosets* and *sublattices*. We now state a few facts in this connection, for which proofs can be found in Cassels, [1959]. Also see Problem 12.31. (We will not use these, and present them only for completeness.) Let $\mathbf{V}_1$ and $\mathbf{V}_2$ be $D \times D$ real nonsingular matrices (not necessarily integers). We say that $LAT(\mathbf{V}_2)$ is a sublattice of $LAT(\mathbf{V}_1)$ if it is a subset of $LAT(\mathbf{V}_1)$. Under this subset relation, one can verify that $\mathbf{L} \stackrel{\Delta}{=} \mathbf{V}_1^{-1} \mathbf{V}_2$ is an integer matrix. So $|\det \mathbf{L}| = |\det \mathbf{V}_2|/|\det \mathbf{V}_1|$, and is guaranteed to be an integer. This integer is denoted as ρ, that is, $\rho = |\det \mathbf{V}_2|/|\det \mathbf{V}_1|$.

Given a fixed vector $\mathbf{v}_1 \in LAT(\mathbf{V}_1)$, consider the set of all vectors of the form $\mathbf{v}_1 + \mathbf{v}_2$ where $\mathbf{v}_2$ varies over the set $LAT(\mathbf{V}_2)$. This is essentially the lattice $LAT(\mathbf{V}_2)$, shifted by $\mathbf{v}_1$. This shifted set (often indicated as $\mathbf{v}_1 + LAT(\mathbf{V}_2)$) is called a *coset* of $LAT(\mathbf{V}_2)$ in $LAT(\mathbf{V}_1)$. It can be shown that the number of distinct cosets is equal to ρ defined above. No two non identical cosets have a common element, and the union of all the ρ non identical cosets is the bigger lattice $LAT(\mathbf{V}_1)$. In set theoretic language, the cosets therefore *partition* the bigger lattice into ρ subsets (called *equivalence classes*).

Cosets and polyphase decomposition. The most familiar example we have seen in this section is the case where $\mathbf{V}_1 = \mathbf{I}$ and $\mathbf{V}_2 = $ integer matrix $\mathbf{M}$. In this case, the lattice $LAT(\mathbf{M})$ is a sublattice of $LAT(\mathbf{I})$ (which in turn is merely the set $\mathcal{N}$ of all integers). A coset of $LAT(\mathbf{M})$ in $LAT(\mathbf{I})$ is the set of all integers of the form (12.4.8), where $\mathbf{k}$ is some fixed integer. There are precisely $|\det \mathbf{M}|$ distinct cosets, one for each $\mathbf{k} \in \mathcal{N}(\mathbf{M})$. The reader will now recognize that the kth coset gives rise to the kth polyphase component $x_\mathbf{k}(\mathbf{n})$ of a sequence $x(\mathbf{n})$.

Tables 12.10.1 and 12.10.2 at the end of Sec. 12.10 contain summaries of many of the notations and definitions introduce so far.

12.5 ALIAS-FREE DECIMATION

In the 1D case we know that if the support of $X(\omega)$ is limited to $-\pi/M \leq \omega < \pi/M$, then decimation by M does not create aliasing. Such bandlimiting can be accomplished by use of decimation filters. In the MD case, we can similarly restrict the support of $X(\omega)$ so that decimation by the matrix **M** does not cause aliasing. We will describe the appropriate bandlimiting support. This information can be used to design decimation filters (i.e., anti-alias filters which precede a decimator).

12.5.1 The Symmetric Parallelepiped

Given real nonsingular **V**, we know how to associate an $FPD(\mathbf{V})$ with it. (Sec. 12.2.2). The symmetric parallelepiped $SPD(\mathbf{V})$, on the other hand, is defined as the set of vectors of the form

$$\mathbf{V}\mathbf{x}, \quad \mathbf{x} \in [-1,1)^D. \tag{12.5.1}$$

So the only difference from the set $FPD(\mathbf{V})$ is that, for $SPD(\mathbf{V})$ each component x_i is allowed to be negative as well, that is, $-1 \leq x_i < 1$. Notice that in the 1D case, $SPD(V)$ is the segment $-V \leq x < V$ of the real line.

In Fig. 12.5-1(a), the set $FPD(\mathbf{V})$ is shown for

$$\mathbf{V} = \begin{bmatrix} 1 & -1 \\ 1 & 2 \end{bmatrix}. \tag{12.5.2}$$

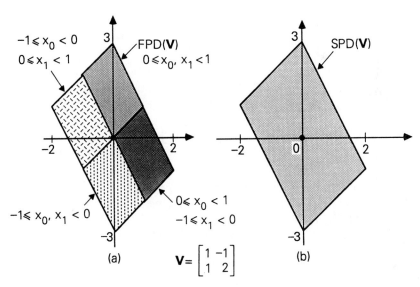

Figure 12.5-1 Demonstrating the relation between $FPD(\mathbf{V})$ and $SPD(\mathbf{V})$.

This is the set of points $\mathbf{Vx}$, with $0 \leq x_0, x_1 < 1$. This figure also shows the set of points $\mathbf{Vx}$ for three other cases, viz., (a) $-1 \leq x_0 < 0, 0 \leq x_1 < 1$, (b)

$0 \leq x_0 < 1, -1 \leq x_1 < 0$, and (c) $-1 \leq x_0, x_1 < 0$. The union of the four sets in Fig. 12.5-1(a) is equal to $SPD(\mathbf{V})$ shown in Fig. 12.5-1(b). One can verify that $SPD(\mathbf{V})$ can be obtained by appropriately shifting and scaling $FPD(\mathbf{V})$. More specifically, we have

$$SPD(\mathbf{V}) = FPD(2\mathbf{V}) - \mathbf{V}\begin{bmatrix}1\\1\\\vdots\\1\end{bmatrix}. \qquad (12.5.3)$$

This is clearly seen in the above example.

The Region $SPD(\pi \mathbf{M}^{-T})$

The set of frequencies defined by $SPD(\pi \mathbf{M}^{-T})$ will play a crucial role in decimation and interpolation. This set reduces to

$$\frac{-\pi}{M} \leq \omega < \frac{\pi}{M}$$

for the 1D case. The gray areas in Fig. 12.5-2 show $SPD(\pi \mathbf{M}^{-T})$ for three cases, viz.,

$$\mathbf{M} = \begin{bmatrix} 1 & 1 \\ 2 & -2 \end{bmatrix}, \quad \mathbf{M}^{-T} = \frac{1}{4}\begin{bmatrix} 2 & 2 \\ 1 & -1 \end{bmatrix}, \quad \text{(hexagonal M)}, \qquad (12.5.4)$$

$$\mathbf{M} = \begin{bmatrix} 1 & 1 \\ -1 & 1 \end{bmatrix}, \quad \mathbf{M}^{-T} = \frac{1}{2}\begin{bmatrix} 1 & 1 \\ -1 & 1 \end{bmatrix}, \quad \text{(quincunx M)}, \qquad (12.5.5)$$

and

$$\mathbf{M} = \begin{bmatrix} 1 & 1 \\ 1 & 2 \end{bmatrix}, \quad \mathbf{M}^{-T} = \begin{bmatrix} 2 & -1 \\ -1 & 1 \end{bmatrix}, \quad ([\det M] = 1 \text{ here.}) \qquad (12.5.6)$$

Notice that the volume of $SPD(\mathbf{M}^{-T})$ is equal to that of $[-\pi, \pi)^D$ divided by $J(\mathbf{M})$. In Fig. 12.5-2(c) we have $J(\mathbf{M}) = 1$ (so that an M-fold "decimator" merely rearranges the samples, rather than discarding any sample), and $SPD(\pi \mathbf{M}^{-T})$ has the same volume as $[-\pi, \pi)^2$. But since $SPD(\pi \mathbf{M}^{-T})$ is not rectangular, it does not fit into the region $[-\pi, \pi)^2$.

12.5.2 Alias-Free Decimation

Since $X(\omega)$ has periodicity matrix $2\pi \mathbf{I}$, the region of support given by $SPD(\pi \mathbf{M}^{-T})$ should be interpreted as the region

$$\omega = \pi \mathbf{M}^{-T}\mathbf{x} + 2\pi \mathbf{m}, \quad \mathbf{x} \in [-1, 1)^D, \mathbf{m} \in \mathcal{N}. \qquad (12.5.7)$$

Suppose the quantity $X(\omega)$ is bandlimited to the above region. We will show that M-fold decimation will not cause aliasing. In fact we will prove a slightly stronger result, namely, if the support of $X(\omega)$ is restricted to

$$\boldsymbol{\omega} = \mathbf{c} + \pi \mathbf{M}^{-T}\mathbf{x} + 2\pi \mathbf{m}, \quad \mathbf{x} \in [-1,1)^D, \mathbf{m} \in \mathcal{N}, \tag{12.5.8}$$

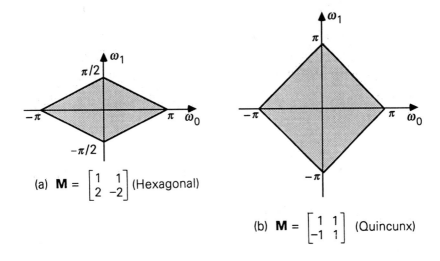

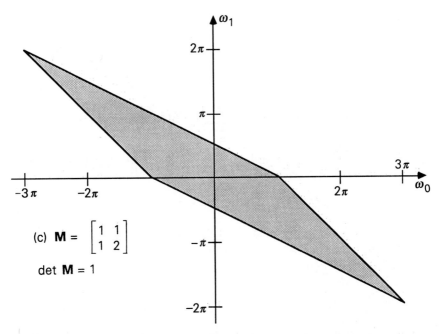

Figure 12.5-2 Demonstration of $SPD(\pi\mathbf{M}^{-T})$ (gray region) for three cases.

for arbitrary **c**, there is no aliasing due to M-fold decimation. In particular, by picking the vector **c** appropriately, we can obtain the alias-free support

$$\boldsymbol{\omega} = 2\pi \mathbf{M}^{-T}\mathbf{x} + 2\pi \mathbf{m}, \quad \mathbf{x} \in [0,1)^D, \, \mathbf{m} \in \mathcal{N}. \quad (12.5.9)$$

Proof. We have to show that there is no overlap between any two terms in the summation of (12.4.22). If $X(\boldsymbol{\omega})$ has support as in (12.5.8), then the stretched and shifted version $X(\mathbf{M}^{-T}(\boldsymbol{\omega} - 2\pi\mathbf{k}))$ has support

$$\boldsymbol{\omega} = \mathbf{M}^T\mathbf{c} + \pi\mathbf{x} + 2\pi\mathbf{M}^T\mathbf{m} + 2\pi\mathbf{k}, \quad \mathbf{x} \in [-1,1)^D, \mathbf{m} \in \mathcal{N}. \quad (12.5.10)$$

Suppose there is overlap between two terms. This implies

$$\mathbf{M}^T\mathbf{c} + \pi\mathbf{x}_1 + 2\pi\mathbf{M}^T\mathbf{m}_1 + 2\pi\mathbf{k}_1 = \mathbf{M}^T\mathbf{c} + \pi\mathbf{x}_2 + 2\pi\mathbf{M}^T\mathbf{m}_2 + 2\pi\mathbf{k}_2, \quad (12.5.11)$$

where $\mathbf{x}_i \in [-1,1)^D, \mathbf{k}_i \in \mathcal{N}(\mathbf{M}^T), \mathbf{m}_i \in \mathcal{N}$. In other words,

$$\mathbf{x} + \mathbf{M}^T\mathbf{m} = \mathbf{k}, \quad (12.5.12)$$

where $\mathbf{x} = 0.5(\mathbf{x}_1 - \mathbf{x}_2) \in (-1,1)^D$, $\mathbf{m} = \mathbf{m}_1 - \mathbf{m}_2 \in \mathcal{N}$, and $\mathbf{k} = \mathbf{k}_2 - \mathbf{k}_1$. Since $\mathbf{M}^T\mathbf{m}$ and $\mathbf{k}$ are integers, the above equation implies $\mathbf{x} = \mathbf{0}$. Since $\mathbf{k}_i = \mathbf{M}^T\mathbf{y}_i$ for $\mathbf{y}_i \in [0,1)^D$, we, therefore, have $\mathbf{m} \in (-1,1)^D$. This means $\mathbf{m} = \mathbf{0}$, i.e., $\mathbf{k}_1$ and $\mathbf{k}_2$ are not distinct. Summarizing, if $\mathbf{k}_1$ and $\mathbf{k}_2$ *are* distinct, there cannot be an overlap between the two terms!

Special Case of Two-Dimensional Signals

In the 2D case

$$\mathbf{M} = \begin{bmatrix} M_{00} & M_{01} \\ M_{10} & M_{11} \end{bmatrix}, \quad (12.5.13)$$

so that

$$\mathbf{M}^{-T} = \frac{1}{\det \mathbf{M}} \begin{bmatrix} M_{11} & -M_{10} \\ -M_{01} & M_{00} \end{bmatrix}. \quad (12.5.14)$$

The region $\pi\mathbf{M}^{-T}\mathbf{x}$, $\mathbf{x} \in [-1,1)^2$ can then be expressed in terms of the frequencies ω_0 and ω_1 as

$$-\pi \le M_{00}\omega_0 + M_{10}\omega_1 < \pi, \quad -\pi \le M_{01}\omega_0 + M_{11}\omega_1 < \pi. \quad (12.5.15)$$

To demonstrate, consider Fig. 12.5-3(a), which shows the above region of support for quincunx M given in (12.5.5). For clarity we have shown a larger region than $[-\pi,\pi)^2$. The gray region in Fig. 12.5-3(b) is the stretched version $X(\mathbf{M}^{-T}\boldsymbol{\omega})$. Since $|\det \mathbf{M}| = 2$, there is only one shifted version $X(\mathbf{M}^{-T}(\boldsymbol{\omega} - 2\pi\mathbf{k}))$ of interest. This occupies the white areas in the figure.

It should be emphasized that if $X(\boldsymbol{\omega})$ has support (12.5.7), then the stretched version $X(\mathbf{M}^{-T}\boldsymbol{\omega})$ has support $\boldsymbol{\omega} = \pi\mathbf{x} + 2\pi\mathbf{M}^T\mathbf{m}$, which therefore fills the region $[-\pi,\pi)^D$ completely. Figure 12.5-3(b) clearly demonstrates this.

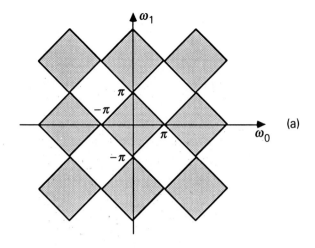

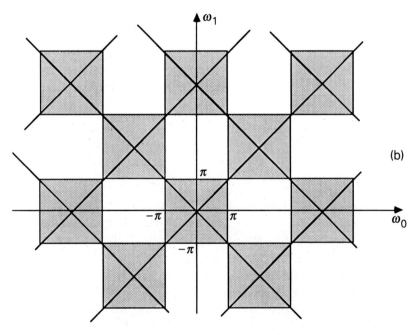

Figure 12.5-3 Supports of (a) $X(\omega)$, and (b) $X(\mathbf{M}^{-T}\omega)$, for quincunx $\mathbf{M}$.

Subtle Things About the Support (12.5.7)

Ignoring the term $2\pi\mathbf{m}$ in (12.5.7) (which just represents the periodicity), the support of $X(\omega)$ is given by

$$\omega = \pi\mathbf{M}^{-T}\mathbf{x}, \quad \mathbf{x} \in [-1,1)^D, \qquad (12.5.16)$$

which of course is $SPD(\pi\mathbf{M}^{-T})$. In the 1D case we know that, as long as $M > 1$, this region is a strict subset of $-\pi \leq \omega < \pi$. In the 2D (or MD)

Sec. 12.5 Alias-free decimation 601

case, such a result does not hold! That is, $SPD(\pi \mathbf{M}^{-T})$ is not necessarily a subset of $[-\pi, \pi)^D$. This was demonstrated in Fig. 12.5-2(c) for the matrix $\mathbf{M}$ in (12.5.6). (In fact any integer unimodular matrix which is not a trivial modification of $\mathbf{I}$ is such an example.) This strange feature, however, does not violate any fundamental properties of the Fourier transform $X(\boldsymbol{\omega})$, as it continues to be periodic in each variable ω_i with period 2π [because the complete support is given by (12.5.7)].

Deeper Nonuniqueness of Support of a Decimation Filter

Since the vector $\mathbf{c}$ in (12.5.8) is arbitrary, the desired support for the decimation filter is non unique, for example, it can be a bandpass rather than lowpass filter. There is a deeper reason for nonuniqueness which actually permits more complicated modifications of the support, than mere shifts. For this recall that the lattices generated by $\mathbf{M}$ and $\mathbf{M}_1 \triangleq \mathbf{ME}$ are the same if $\mathbf{E}$ is an integer unimodular matrix. So we can use a modified decimation filter with support

$$\boldsymbol{\omega} = \pi \mathbf{M}_1^{-T}\mathbf{x} + 2\pi \mathbf{m}, \quad \mathbf{x} \in [-1, 1)^D, \ \mathbf{m} \in \mathcal{N}, \qquad (12.5.17)$$

and decimate the filter output by $\mathbf{M}$, without causing aliasing. (Problem 12.11) And since $\mathbf{E}$ is arbitrary (except for unimodularity), we can generate an infinite number of permissible decimation filters. As an example, with quincunx $\mathbf{M}$ we can obtain

$$\mathbf{M}_1 = \underbrace{\begin{bmatrix} 1 & 1 \\ -1 & 1 \end{bmatrix}}_{\mathbf{M}} \underbrace{\begin{bmatrix} 1 & 0 \\ 1 & -1 \end{bmatrix}}_{\mathbf{E}} = \begin{bmatrix} 2 & -1 \\ 0 & -1 \end{bmatrix}. \qquad (12.5.18)$$

Figs. 12.5-4(a),(b) show $SPD(\pi \mathbf{M}^{-T})$ and $SPD(\pi \mathbf{M}_1^{-T})$, either of which can be used as the passband region of the decimation filter. In view of periodicity of frequency response, the support in Fig. 12.5-4(b), in reality, degenerates to the infinite vertical strip shown in Fig. 12.5-4(c).

Most general alias-free support. What is the most general region of support, which will result in alias-free decimation? In the 1D case this support can be described as a union of an arbitrary number of supports such that (a) no two of them overlap modulo $2\pi/M$, and (b) the total support width is no greater than $2\pi/M$ [Sathe and Vaidyanathan, 1992]. Some discussions for the multidimensional case can be found in Chap. 2 of Bamberger [1990].

12.5.3 Frequency Partitioning

Consider the frequency region

$$\boldsymbol{\omega} = \pi \mathbf{M}^{-T}\mathbf{x} + 2\pi \mathbf{m} + 2\pi \mathbf{M}^{-T}\mathbf{k}, \quad \mathbf{x} \in [-1, 1)^D, \mathbf{m} \in \mathcal{N}, \mathbf{k} \in \mathcal{N}(\mathbf{M}^T). \qquad (12.5.19)$$

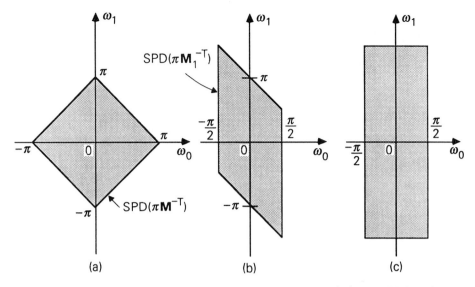

Figure 12.5-4 (a) and (b): two possible supports of $X(\omega)$ for which quincunx decimation does not cause aliasing. The support in (c) is equivalent to (b) because of periodicity with respect to ω_1.

A simple modification of the preceding discussions shows that these regions are distinct for two different values of $\mathbf{k} \in \mathcal{N}(\mathbf{M}^T)$. Furthermore, the union of all the above regions as $\mathbf{k}$ varies over $\mathcal{N}(\mathbf{M}^T)$ covers the entire frequency domain. [This can be proved using the vector decomposition result (12.2.22b)]. The number of distinct regions is equal to the number of distinct values of $\mathbf{k}$ which, in turn, is equal to $J(\mathbf{M})$.

Summarizing, the set $\mathcal{N}(\mathbf{M}^T)$ can be used to *partition* the frequency region into $J(\mathbf{M})$ regions. The volume of each of the regions (12.5.19), which falls inside the fundamental region $[-\pi, \pi)^D$, is equal to the volume of $[-\pi, \pi)^D$ divided by $J(\mathbf{M})$.

For a preview of such a partitioning, see Fig. 12.8-4 which is obtained for the case where $\mathbf{M}$ is hexagonal. Such a partitioning is useful in filter bank theory. For example, each of the $J(\mathbf{M})$ regions can be used as a subband, and we can decimate each subband signal by $\mathbf{M}$, to obtain a maximally decimated filter bank. We will return to this topic in Sec. 12.9.

12.6 CASCADE CONNECTIONS

As in the 1D case, it is of interest to study cascaded arrangements of decimators, expanders and transfer functions. Fig. 12.6-1(a) shows a cascade of two decimators. Here $\mathbf{M}_1$ and $\mathbf{M}_2$ are $D \times D$ nonsingular integer matrices. This structure is equivalent to a single decimator with decimation matrix $\mathbf{M} = \mathbf{M}_1 \mathbf{M}_2$. To see this note that $y_1(\mathbf{n}) = x(\mathbf{M}_1 \mathbf{n})$ so that

$$y_2(\mathbf{n}) = y_1(\mathbf{M}_2 \mathbf{n}) = x(\mathbf{M}_1 \mathbf{M}_2 \mathbf{n}) = x(\mathbf{M} \mathbf{n}), \qquad (12.6.1)$$

which is M-fold decimated version. Notice that the hexagonal decimator can be regarded as a cascade of a rectangular and the quincunx decimators:

$$\underbrace{\begin{bmatrix} 1 & 1 \\ 2 & -2 \end{bmatrix}}_{\text{hexagonal}} = \begin{bmatrix} 1 & 0 \\ 0 & -2 \end{bmatrix} \underbrace{\begin{bmatrix} 1 & 1 \\ -1 & 1 \end{bmatrix}}_{\text{quincunx}}$$

Next, Fig. 12.6-1(b) shows a cascade of two expanders, and the equivalent circuit. This equivalence is easier to verify in the frequency domain. Thus, $Y_1(\omega) = x(\mathbf{M}_2^T \omega)$, so that

$$Y_2(\omega) = Y_1(\mathbf{M}_1^T \omega) = X(\mathbf{M}_2^T \mathbf{M}_1^T \omega) = X(\mathbf{M}^T \omega), \qquad (12.6.2)$$

which indeed is the M-fold expander output.

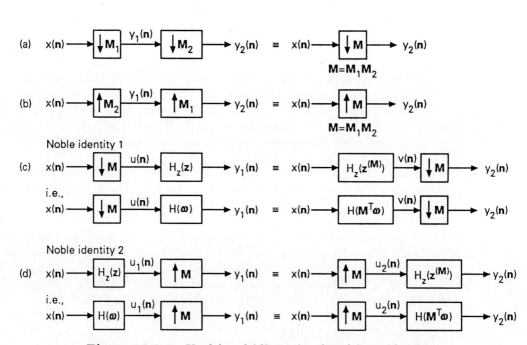

Figure 12.6-1 Useful multidimensional multirate identities.

Noble Identities

The noble identities, which were introduced in Fig. 4.2-3 for the 1D case, can be extended to the multidimensional case. Figs. 12.6-1(c) and (d) show these identities. (Note that in block diagrams, filters are labeled either as $H_z(\mathbf{z})$ or as $H(\omega)$ interchangeably.) Fig. 12.6-2 demonstrates these identities for the case of rectangular decimators and expanders. See Example 12.6.1 below for a more sophisticated case.

The second noble identity follows immediately from the frequency domain relation (12.4.17). To prove the first noble identity, denote $G(\omega) = H(\mathbf{M}^T\omega)$ so that $g(\mathbf{n})$ is the M-fold expanded version, that is, $g(\mathbf{Mn}) = h(\mathbf{n})$ and $g(\mathbf{k}) = 0$ otherwise. So

$$v(\mathbf{n}) = \sum_{\mathbf{m}} g(\mathbf{m})x(\mathbf{n}-\mathbf{m}) = \sum_{\mathbf{k}} h(\mathbf{k})x(\mathbf{n}-\mathbf{Mk}). \quad (12.6.3)$$

Thus $y_2(n) = v(\mathbf{Mn}) = \sum_{\mathbf{k}} h(\mathbf{k})x(\mathbf{Mn}-\mathbf{Mk})$. From the figure we also see that

$$y_1(n) = \sum_{\mathbf{k}} h(\mathbf{k})u(\mathbf{n}-\mathbf{k}). \quad (12.6.4)$$

Using $u(\mathbf{n}) = x(\mathbf{Mn})$, we therefore obtain $y_1(n) = \sum_{\mathbf{k}} h(\mathbf{k})x(\mathbf{Mn}-\mathbf{Mk}) = y_2(n)$ indeed.

The Polyphase Identity

In Sec. 4.3 we considered the cascade arrangement shown in Fig. 4.3-13(a) and showed that this is equivalent to a linear time invariant system with transfer function $E_0(z)$ [0th polyphase component of $H(z)$]. A similar result is true for the D-dimensional case, and is shown in Fig. 12.6-3(a). Here $E_0(\omega)$ represents the 0th Type 1 polyphase component of $H(\omega)$. An application of this is indicated in Fig. 12.6-3(b). Proofs of these are requested in Problem 12.10.

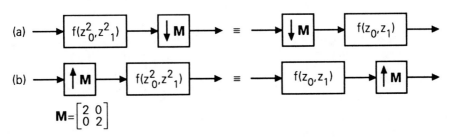

Figure 12.6-2 Noble identitites for (a) rectangular decimator, and (b) rectangular expander.

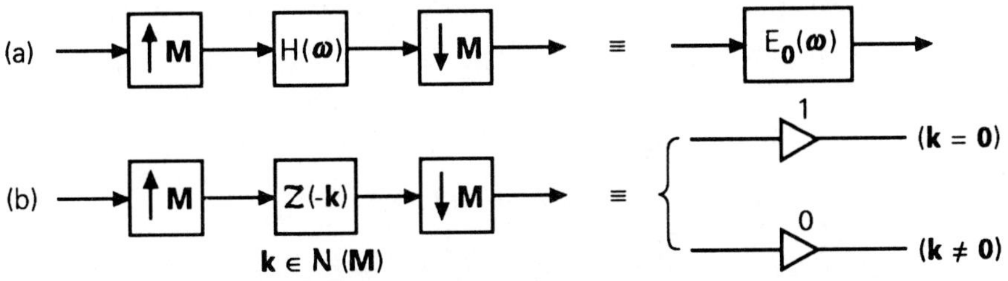

Figure 12.6-3 (a) The polyphase identity, and (b) an application.

Decimation and Interpolation Filters

Recall from Sec. 12.5 that decimation by **M** does not result in aliasing errors if the transform of the signal is bandlimited to $SPD(\pi \mathbf{M}^{-T})$. This bandlimiting is achieved by use of a decimation filter $H(\omega)$ as shown in Fig. 12.6-4(a).

Similarly, Fig. 12.6-4(b) shows an expander followed by a lowpass filter $H(\omega)$, whose purpose is to suppress the images created by the expander. Recall that there are $J(\mathbf{M}) - 1$ images in $[-\pi, \pi)^D$. These can be suppressed if $H(\omega)$ is chosen to have a passband support $SPD(\pi \mathbf{M}^{-T})$. If the signal $x(n)$ is itself bandlimited in some way, then other choices of $H(\omega)$ can be used to suppress the images. For example if $X(\omega)$ and **M** are as in Fig. 12.4-4, then the images [light-gray regions in Fig. 12.4-4(b)] can be suppressed if $H(\omega)$ has support equal to the dark-gray areas.

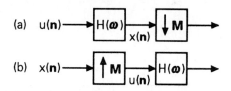

Figure 12.6-4 (a) Decimation filter, and (b) interpolation filter.

Example 12.6.1: Polyphase Implementation and Noble Identities

Let **M** be the hexagonal decimator (eqn. (12.4.15)). For this case $\mathbf{z}^{(\mathbf{M})}$ was computed in Example 12.4.2. The noble identities can therefore be redrawn in this case as shown in Figs. 12.6-5(a) and (b). Figs. 12.6-5(c)-(e) show this for some specific examples of $f(z_0, z_1)$.

For this **M** we have already obtained the Type 1 polyphase implementation (Fig. 12.4-9). Now consider a decimation filter with this **M**. The decimator **M**, which follows the structure of Fig. 12.4-9 can be moved using the noble identity so that the result is as shown in Fig. 12.6-5(f).

To obtain an improved feeling for these strange rules for moving decimators, consider the specific case shown in Fig. 12.6-5(c): this says that decimation of $x(n_0 - 1, n_1 - 2)$ is the same as first decimating $x(\mathbf{n})$ by **M** and delaying only in the horizontal direction, that is,

$$x\left(\mathbf{Mn} - \begin{bmatrix} 1 \\ 2 \end{bmatrix}\right) = x\left(\mathbf{M}\left(n - \begin{bmatrix} 1 \\ 0 \end{bmatrix}\right)\right). \qquad (12.6.5)$$

This is true for all inputs if and only if

$$\mathbf{M} \begin{bmatrix} 1 \\ 0 \end{bmatrix} = \begin{bmatrix} 1 \\ 2 \end{bmatrix}. \qquad (12.6.6)$$

This equation merely says that the 0th column of **M** should be $\begin{bmatrix} 1 \\ 2 \end{bmatrix}$, which is true by choice of **M**! Similarly the rule in Fig. 12.6-5(d) is another way of saying that the 1st column of **M** is equal to $\begin{bmatrix} 1 \\ -2 \end{bmatrix}$.

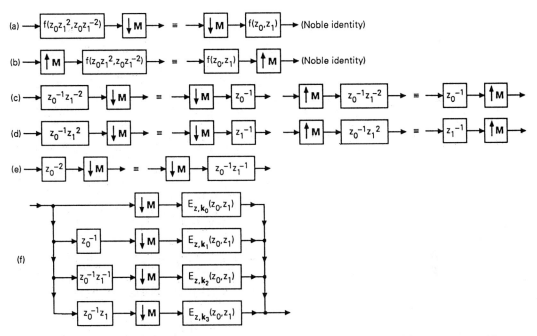

Figure 12.6-5 Some details for hexagonal **M**. Parts (a), (b) indicate noble identities. Parts (c) – (e) are specific examples. Part (f) shows polyphase implementation of a decimation filter.

More generally in the D-dimensional case, the noble identity for moving a decimator can be broken into D components. The kth component says that if we delay the decimated sequence $x(\mathbf{Mn})$ in the kth direction only, it is the same as 'first delaying $x(\mathbf{n})$ by the kth column $\mathbf{m}_k$ of **M**, and then decimating'. In other words, noble identity says

$$\mathbf{m}_k = \mathbf{M} \begin{bmatrix} 0 \\ 1 \\ 0 \end{bmatrix}, \qquad (12.6.7)$$

where the '1' on the RHS occurs in the kth position. This of course is an "obvious identity."

Tables 12.10.1–12.10-4 at the end of Sec. 12.10 serve as a quick reference guide to many of the notations and concepts introduced so far.

12.7 MULTIRATE FILTER DESIGN

The design of decimation and interpolation filters for general **M** is a nontrivial problem. The most common frequency response of the filter $H(\omega)$ in Fig. 12.6-4 is lowpass with passband region $SPD(\pi \mathbf{M}^{-T})$ (Sec. 12.5). More specifically, the ideal frequency response has the form

$$H(\omega) = \begin{cases} 1 & \text{if } \omega = \pi \mathbf{M}^{-T}\mathbf{x} + 2\pi \mathbf{m}, \text{ for some } \mathbf{x} \in [-1,1)^D, \mathbf{m} \in \mathcal{N}, \\ 0 & \text{otherwise.} \end{cases}$$

(12.7.1)

This support was demonstrated in Fig. 12.5-2 for various choices of **M**. In practice, we approximate the above ideal response. With FIR filters it is possible to constrain the filter to have zero-phase, so that only the magnitude response is approximated. In the IIR case, $H(\omega)$ introduces phase distortion (which is not acceptable in many image processing applications). We now consider some examples of multirate filter design.

A number of techniques for 2D digital filter design have been reported in the past [Rabiner and Gold, 1975], [Dudgeon and Mersereau, 1984] and [Lim, 1990]. One of these techniques [McClellan, 1973] designs the 2D filter by transformation of a 1D filter. In the next few subsections, we will describe more recent techniques [Ansari and Lau, 1987], [Chen and Vaidyanathan, 1991] for decimation filter design. Further results can be found in Smith and Eddins [1990], and Bamberger and Smith [1992].

Example 12.7.1

Suppose we wish to decimate a 2D signal $x(\mathbf{n})$ using the quincunx matrix

$$\mathbf{M} = \begin{bmatrix} 1 & 1 \\ -1 & 1 \end{bmatrix}. \quad (12.7.2)$$

The black dots in Fig. 12.7-1(a) (which are the lattice points generated by **M**) are retained by the decimator. We have $J(\mathbf{M}) = |\det \mathbf{M}| = 2$ so that the decimator reduces the sample density by a factor of two. Figure 12.7-1(b) is a reminder that we can fit a hexagon to the sample points.

The region $SPD(\pi \mathbf{M}^{-T})$ for the above matrix **M** is shown in Fig. 12.7-1(c), and the lowpass decimation filter $H(\omega)$ should have this support. In the region $[-\pi, \pi)^2$ we should, therefore, have

$$H(\omega) = \begin{cases} 1 & \text{if } \omega \text{ is in the diamond region} \\ 0 & \text{otherwise,} \end{cases} \quad (12.7.3)$$

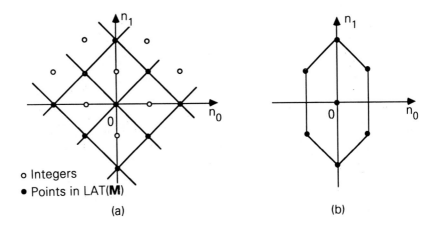
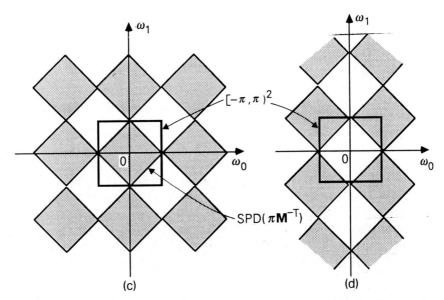

Figure 12.7-1 (a) Lattice generated by quincunx $\mathbf{M}$, (b) fitting hexagons to lattice points, (c) lowpass decimation filter, and (d) highpass decimation filter.

for the ideal decimation filter. As indicated, the above response repeats periodically with period 2π in each frequency variable.

For $\mathbf{M}$ as above, we now identify the set $\mathcal{N}(\mathbf{M}^T)$ which appears in the decimation formula (12.4.22). This is the set of integers of the form

$$\mathbf{k} = \begin{bmatrix} 1 & -1 \\ 1 & 1 \end{bmatrix} \begin{bmatrix} x_0 \\ x_1 \end{bmatrix}, \tag{12.7.4}$$

with $0 \le x_i < 1$. With x_i in this range, the only combinations which result in integer valued $\mathbf{k}$ are: (a) $x_0 = x_1 = 0$ and (b) $x_0 = x_1 = 0.5$.

Sec. 12.7 Multirate filter design

The corresponding values of **k** are, respectively,

$$\mathbf{k}_0 = \begin{bmatrix} 0 \\ 0 \end{bmatrix}, \quad \mathbf{k}_1 = \begin{bmatrix} 0 \\ 1 \end{bmatrix}. \qquad (12.7.5)$$

With the decimation filter having support as in Fig. 12.7-1(c), the stretched version $X(\mathbf{M}^{-T}\boldsymbol{\omega})$ has the support shown earlier in Fig. 12.5-3(b) (gray area). The only shifted version appearing in (12.4.22) is $X(\mathbf{M}^{-T}(\boldsymbol{\omega} - 2\pi\mathbf{k}_1))$, and its support is the white area in Fig. 12.5-3(b).

Figure 12.7-1(d) shows a different type of decimation filter, which is highpass. (This is obtained by shifting the lowpass response by π in both directions.) Thus, in the region $[-\pi, \pi)^2$, the ideal response can be taken as

$$H(\boldsymbol{\omega}) = \begin{cases} 0 & \text{if } \boldsymbol{\omega} \text{ is in the diamond region} \\ 1 & \text{otherwise.} \end{cases} \qquad (12.7.6)$$

The decimated signal is still free from aliasing. The stretched version now has support identical to the *white areas* in Fig. 12.5-3(b) with the shifted versions occupying the gray areas.

12.7.1 Design of Diamond- and Fan-shaped Filters

From the previous discussions we see that the diamond response is applicable in some multirate designs (e.g., quincunx decimation). We now present an elegant method [Ansari and Lau, 1987] for designing such a response, starting from a 1D filter. Consider an ideal 1D filter $G(e^{j\omega})$ with magnitude response as shown in Fig. 12.7-2(a). Suppose we define a 2D transfer function $G(z_0 z_1)$. This has frequency response $G(e^{j(\omega_0+\omega_1)})$. We will sketch this response in the 2D plane for the region $[-\pi, \pi]^2$. For this note that as ω_0 and ω_1 span the range $-\pi \leq \omega_i \leq \pi$, the sum $\omega_0 + \omega_1$ spans the range $-2\pi \leq \omega \leq 2\pi$. Thus, in the 2D region $[-\pi, \pi]^2$ we have

$$|G(e^{j(\omega_0+\omega_1)})| = \begin{cases} 1 & 0 \leq |\omega_0 + \omega_1| \leq \pi/2, \\ 1 & 3\pi/2 \leq |\omega_0 + \omega_1| \leq 2\pi, \\ 0 & \text{otherwise.} \end{cases} \qquad (12.7.7)$$

This response is shown in Fig. 12.7-2(b). Next consider the 2D transfer function $G(z_0 z_1^{-1})$. This has frequency response $G(e^{j(\omega_0-\omega_1)})$ so that its support is obtained by flipping Fig. 12.7-2(b) with respect to the horizontal axis. If we now consider the product

$$K_z(z_0, z_1) \triangleq G(z_0 z_1) G(z_0 z_1^{-1}), \qquad (12.7.8)$$

the magnitude response is as in Fig. 12.7-2(c), which does look like a diamond, but has the wrong size. Moreover, there are passbands around the four corner frequencies $(\pm \pi, \pm \pi)$.

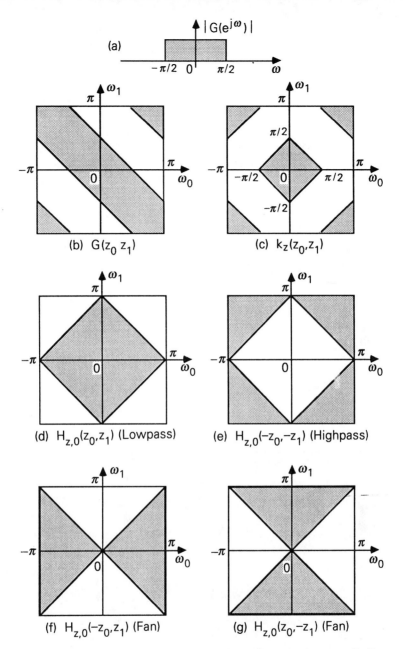

Figure 12.7-2 Nonseparable 2D filters designed from a 1D filter.

Let $k(n_0, n_1)$ denote the impulse response of $K_z(z_0, z_1)$. Consider the decimated impulse response $k(2n_0, 2n_1)$. We will show that this has the desired support shown in Fig. 12.7-2(d). For this note that the the transfer function of $k(2n_0, 2n_1)$ is given by

$$\frac{1}{4}[K_z(z_0, z_1) + K_z(-z_0, z_1) + K_z(z_0, -z_1) + K_z(-z_0, -z_1)]\Big|_{z_i \to z_i^{1/2}}$$

which, in view of (12.7.8), reduces to

$$\frac{1}{2}[K_z(z_0, z_1) + K_z(-z_0, z_1)]\Big|_{z_i \to z_i^{1/2}} \qquad (12.7.9)$$

Now the frequency response of $K_z(-z_0, z_1)$ is obtained by shifting the response of $K_z(z_0, z_1)$ to the right by π. If we add this to the response of $K_z(z_0, z_1)$ and then perform $z_i \to z_i^{1/2}$ (i.e., stretch the axes by doing $\omega_i \to 0.5\omega_i$), the result has the desired diamond support [Fig. 12.7-2(d)] indeed.

We now express the above 2D filter in terms of polyphase components of $G(z)$. Thus, let

$$G(z) = E_0(z^2) + z^{-1}E_1(z^2). \qquad (12.7.10)$$

Substituting this into (12.7.9), the decimation filter takes the form

$$H_{z,0}(z_0, z_1) = E_0(z_0 z_1)E_0(z_0 z_1^{-1}) + z_0^{-1}E_1(z_0 z_1)E_1(z_0 z_1^{-1}). \qquad (12.7.11)$$

Summarizing, we first design a 1D lowpass filter $G(z)$ whose magnitude response approximates Fig. 12.7-2(a). We then take the 2D filter to be as in (12.7.11) where $E_0(z)$ and $E_1(z)$ are the polyphase components of $G(z)$. Then the response of the 2D filter is an approximation of Fig. 12.7-2(d). In a similar manner, the 2D transfer function

$$H_{z,1}(z_0, z_1) = E_0(z_0 z_1)E_0(z_0 z_1^{-1}) - z_0^{-1}E_1(z_0 z_1)E_1(z_0 z_1^{-1}) \qquad (12.7.12)$$

results in an approximation of the highpass response shown in Fig. 12.7-2(e). To see this note that $H_{z,1}(z_0, z_1) = H_{z,0}(-z_0, -z_1)$ so that its frequency response is obtained by shifting by π in both directions.

Figs. 12.7-2(f),(g) show responses $H_{z,0}(-z_0, z_1)$ and $H_{z,0}(z_0, -z_1)$ respectively. Filters with such responses are said to be *fan* filters. So we see that a number of useful nonseparable responses can be designed by starting from a one dimensional prototype, and performing clever manipulations. All of the above mentioned filters will serve as decimation filters which avoid aliasing with quincunx **M**.

12.7.2 Multistage Systems

In Chap. 4 we saw examples of filters, expanders and decimators in cascade, such as Fig. 4.4-5 (multistage decimators), and Fig. 4.1-10(b) (fractional decimation circuit). We now consider some multidimensional examples of multistage designs.

Example 12.7.2: Rectangular to Hexagonal Conversion

Figure 12.7-3 shows a multistage structure in which a sequence $x(\mathbf{n})$ is first interpolated and then decimated, to accomplish some purpose. We now indicate one example of such a "purpose," for the 2D case. Suppose

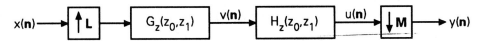

Figure 12.7-3 Technique for conversion from rectangular to hexagonal sampling.

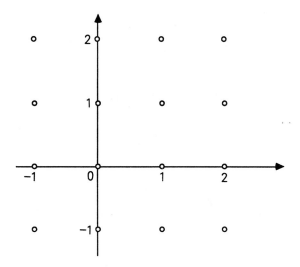

(a) Sample locations for x(**n**)

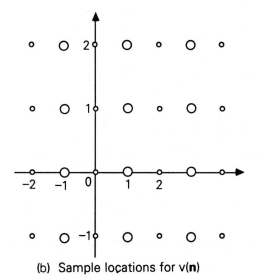

(b) Sample locations for v(**n**)

Figure 12.7-4 (a), (b)   *Figure continued* →

Sec. 12.7   Multirate filter design   613

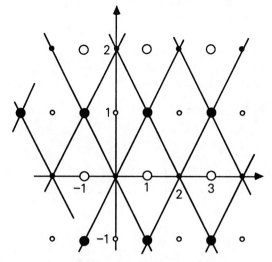

(c) The decimation lattice working on v(**n**)

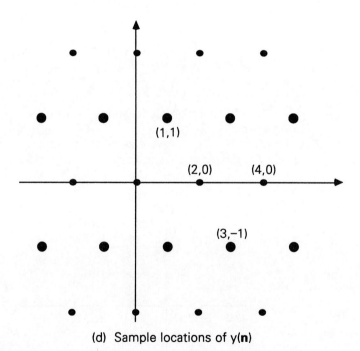

(d) Sample locations of y(**n**)

Figure 12.7-4 (c), (d). Sample points for various sequences, in the process of converting from rectangular to hexagonal sampling.

the signal $x(\mathbf{n})$ is a rectangularly sampled version of some analog signal $x_a(\mathbf{t})$. We wish to convert it to the hexagonally sampled version $y(\mathbf{n})$, without changing the sampling density. One technique for this would

614   Chap. 12. Multidimensional multirate systems

be to reconvert $x(\mathbf{n})$ into the analog version, and then resample using a hexagonal lattice to obtain $y(\mathbf{n})$.

A more elegant technique would be to perform the conversion directly in the discrete domain, using the above structure. Here $x(\mathbf{n})$ is first interpolated to produce $v(\mathbf{n})$, which is a rectangular sampled version with two times higher density in the horizontal direction. Figures 12.7-4(a),(b) demonstrate the sample locations before and after interpolation. (The big circles are the newly generated sample locations.) Such interpolation is accomplished by use of the expander matrix

$$\mathbf{L} = \begin{bmatrix} 2 & 0 \\ 0 & 1 \end{bmatrix}, \qquad (12.7.13)$$

followed by lowpass filtering with $G_z(z_0, z_1)$. This filtering merely removes the image-support.

Next we *decimate* $v(\mathbf{n})$ using the quincunx matrix (12.7.2). Figure 12.7-4(c) shows the lattice generated by this matrix. The sample points retained by this decimator are shown as black circles (with big circles again denoting the new samples generated by the interpolator). Figure 12.7-4(d) separately shows the sample locations of the resulting sequence $y(n)$, and we clearly see the hexagonal pattern. For clarity, the figure also indicates the coordinates of some samples of $y(n)$.

It should be noticed that the above scheme does not alter the sampling density [i.e., $x(n)$ and $y(n)$ have same density] because [det M] = [det L] here.

How should we design the filters $G_z(z_0, z_1)$ and $H_z(z_0, z_1)$ in the figure? Since $G_z(z_0, z_1)$ merely removes the image [light-gray region in Fig. 12.7-5(b)], its response should be zero here, [and unity in the dark-gray region]. The decimation filter $H_z(z_0, z_1)$ precedes M. From Sec. 12.5 we know that if its passband coincides with the gray area in Fig. 12.5-4(c), aliasing is avoided. But this is same as the passband of the interpolation filter, so that we really have no need for $H_z(z_0, z_1)$! So we set $H_z(z_0, z_1) = 1$.

Example 12.7.3: Two Stage Decimation

Consider the structure of Fig. 12.7-6(a) in which a signal is decimated in two stages, resulting in overall decimation by $\mathbf{M} = \mathbf{M}_1 \mathbf{M}_2$. The associated filtering is accomplished in two stages. The filters $H_1(\omega)$ and $H_2(\omega)$ are required to ensure that $x_1(\mathbf{n})$ and $x_2(\mathbf{n})$ are appropriately bandlimited to avoid aliasing. For this we design $H_i(\omega)$ to be lowpass with appropriate support as described earlier. For analysis purposes we can rearrange the structure using noble identities as shown in Fig. 12.7-6(b). The overall decimation filter is therefore $H_1(\omega)H_2(\mathbf{M}_1^T \omega)$. If this filter is lowpass with response as in (12.7.1), then decimation by M does not cause aliasing.

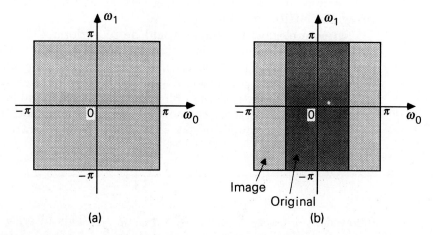

Figure 12.7-5 (a) Support of $X(\omega)$, and (b) support of the output of the expander (**L** as in (12.7.13)).

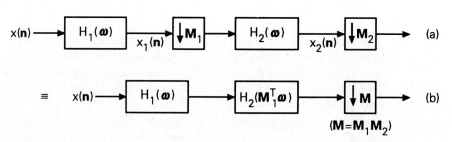

Figure 12.7-6 (a) A two-stage decimator, and (b) its equivalent structure.

As a specific example, let

$$\mathbf{M}_1 = \begin{bmatrix} 1 & -1 \\ 1 & 2 \end{bmatrix}, \quad \mathbf{M}_2 = 2\begin{bmatrix} 1 & -1 \\ 0 & 1 \end{bmatrix}, \qquad (12.7.14)$$

so that

$$\mathbf{M} = 2\begin{bmatrix} 1 & -2 \\ 1 & 1 \end{bmatrix}. \qquad (12.7.15)$$

From here we have

$$\mathbf{M}_1^{-T} = \frac{1}{3}\begin{bmatrix} 2 & -1 \\ 1 & 1 \end{bmatrix}, \ \mathbf{M}_2^{-T} = \frac{1}{2}\begin{bmatrix} 1 & 0 \\ 1 & 1 \end{bmatrix}, \ \mathbf{M}^{-T} = \frac{1}{3}\begin{bmatrix} 0.5 & -0.5 \\ 1 & 0.5 \end{bmatrix} \qquad (12.7.16)$$

The regions $SPD(\pi\mathbf{M}_1^{-T})$ and $SPD(\pi\mathbf{M}_2^{-T})$ are sketched in Fig. 12.7-7. If the filters $H_i(\omega)$ are designed to have passband coinciding with $SPD(\pi\mathbf{M}_i^{-T})$, then neither of the decimators creates aliasing.

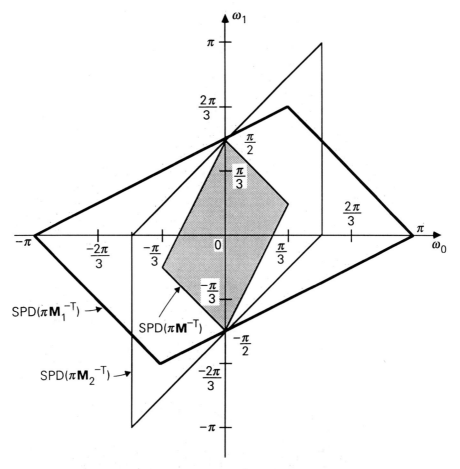

Figure 12.7-7 Pertaining to example 12.7.3.

With the support of $H_2(\omega)$ chosen as above, the reader can verify that the support of $H_2(\mathbf{M}_1^T\boldsymbol{\omega})$ is

$$\pi\mathbf{M}^{-T}\mathbf{x} + 2\pi\mathbf{M}_1^{-T}\mathbf{m}, \quad \mathbf{x} \in [-1,1)^D, \mathbf{m} \in \mathcal{N}. \qquad (12.7.17a)$$

This is the same as the desired support for single stage decimation with the exception that the second term is $2\pi\mathbf{M}_1^{-T}\mathbf{m}$ rather than $2\pi\mathbf{m}$. In other words, the support $SPD(\pi\mathbf{M}^{-T})$ repeats with periodicity matrix $2\pi\mathbf{M}_1^{-T}$ (rather than $2\pi\mathbf{I}$) so that there are additional images of $SPD(\pi\mathbf{M}^{-T})$. To understand this better, let us decompose the integer $\mathbf{m}$ (using division theorem, Sec. 12.4.2) as $\mathbf{m} = \mathbf{M}_1^T\mathbf{m}_1 + \mathbf{m}_0$, with $\mathbf{m}_0 \in \mathcal{N}(\mathbf{M}_1^T)$ and $\mathbf{m}_1 \in \mathcal{N}$. Then (12.7.17a) becomes

$$\pi\mathbf{M}^{-T}\mathbf{x} + 2\pi\mathbf{M}_1^{-T}\mathbf{m}_0 + 2\pi\mathbf{m}_1, \quad \mathbf{x} \in [-1,1)^D, \mathbf{m}_0 \in \mathcal{N}(\mathbf{M}_1^T), \mathbf{m}_1 \in \mathcal{N}. \qquad (12.7.17b)$$

Since $\mathbf{m}_0$ can take $J(\mathbf{M}_1) - 1$ nonzero values from the set $\mathcal{N}(\mathbf{M}_1^T)$, there are $J(\mathbf{M}_1) - 1$ unwanted image-passbands in $H_2(\mathbf{M}_1^T \boldsymbol{\omega})$. In Problem 12.12 the reader is requested to show that none of these extra images overlaps with $SPD(\pi \mathbf{M}^{-T})$, and that these images are eliminated by $H_1(\boldsymbol{\omega})$ in the cascaded filter $H_1(\boldsymbol{\omega}) H_2(\mathbf{M}_1^T \boldsymbol{\omega})$.

Summarizing, the cascaded filter $H_1(\boldsymbol{\omega}) H_2(\mathbf{M}_1^T \boldsymbol{\omega})$ has passband support given by (12.5.7), consistent with the decimation matrix $\mathbf{M}$.

Subtle Things About Multistage Decimation

We now consider a different example which brings up a strange situation for which there is no 1D analogy. Thus consider the decimation matrix

$$\mathbf{M} = \begin{bmatrix} 1 & -4 \\ 1 & 2 \end{bmatrix} = \underbrace{\begin{bmatrix} 1 & -1 \\ 1 & 2 \end{bmatrix}}_{\mathbf{M}_1} \underbrace{\begin{bmatrix} 1 & -2 \\ 0 & 2 \end{bmatrix}}_{\mathbf{M}_2}. \qquad (12.7.18)$$

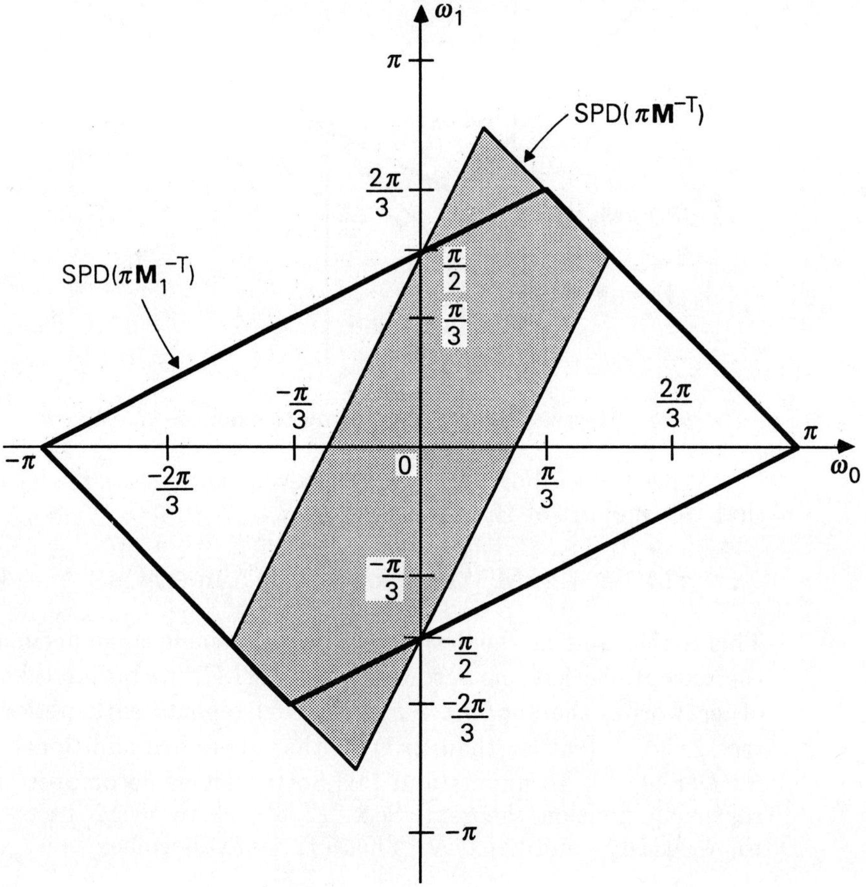

Figure 12.7-8 Demonstration that $SPD(\pi \mathbf{M}^{-T})$ may not be contained in $SPD(\pi \mathbf{M}_1^{-T})$. Matrices are as in (12.7.18).

Figure 12.7-8 shows the region $SPD(\pi \mathbf{M}^{-T})$, which can be taken as the support for the decimation filter. Suppose we wish to implement the decimation in two stages as in Fig. 12.7-6(a), and take the support of $H_1(\boldsymbol{\omega})$ in the usual way that is, as $SPD(\pi \mathbf{M}_1^{-T})$ (Fig. 12.7-8). From the figure we see that $SPD(\pi \mathbf{M}^{-T})$ does *not* fit inside $SPD(\pi \mathbf{M}_1^{-T})$. (Such a situation would *never* arise in the 1D case because of the condition $M > M_1$.) This means that the filter $H_1(\boldsymbol{\omega})$ would cut off a portion of the signal that would normally be passed by $H(\boldsymbol{\omega})$. In other words, the product $H_1(\boldsymbol{\omega})H_2(\mathbf{M}_1^T\boldsymbol{\omega})$ has a support which leaves out part of the desired support $SPD(\pi\mathbf{M}^{-T})$.

Summarizing, the idea of performing decimation in two or more stages does not work with the above choice of $H_1(\boldsymbol{\omega})$. In the multidimensional case, one has to be careful in choosing the factors $\mathbf{M}_1$ and $\mathbf{M}_2$ and the supports of $H_1(\boldsymbol{\omega})$ and $H_2(\boldsymbol{\omega})$. In Problem 12.34 we request the reader to show that the region $SPD(\pi \mathbf{M}^{-T})$ will be contained within $SPD(\pi \mathbf{M}_1^{-T})$ as long as $SPD(\mathbf{M}_2^{-T})$ is contained within $[-1,1)^D$. In Example 12.7.3 above, the choice of factors happened to satisfy this condition, as evidenced from Fig. 12.7-7.

12.7.3. Multidimensional Filters From 1D Filters

In the previous section we saw that a certain class of 2D filters can be designed efficiently starting from 1D filters. These are useful in quincunx decimation. Now consider the more general problem where we wish to approximate the decimation filter response (12.7.1) for arbitrary $\mathbf{M}$, and for arbitrary number of dimensions D (i.e., $\mathbf{M}$ is $D \times D$). Such design is normally complicated; the design as well as implementation of general (non separable) multidimensional filters has much higher complexity that 1D filters. Both of these complexities grow exponentially with the number of dimensions D [Dudgeon and Mersereau, 1984].

We will now outline a more efficient procedure for approximating the response (12.7.1) [Chen and Vaidyanathan, 1991]. The design as well as implementation complexity of this procedure will grow *linearly* with the number of dimensions D (rather than exponentially). The method works for arbitrary $\mathbf{M}$ and arbitrary number of dimensions D. Further advantages and properties will be summarized at the end of this section. To explain the method, we first look at the impulse response of $H(\boldsymbol{\omega})$.

The Ideal Impulse Response

Let $H(\boldsymbol{\omega}) = \sum_{\mathbf{n} \in \mathcal{N}} h(\mathbf{n}) e^{-j \boldsymbol{\omega}^T \mathbf{n}}$, i.e., let $h(\mathbf{n})$ be the impulse response of $H(\boldsymbol{\omega})$. With $H(\boldsymbol{\omega})$ given as in (12.7.1), we obtain

$$h(\mathbf{n}) = \frac{1}{(2\pi)^D} \int_{\boldsymbol{\omega} \in SPD(\pi \mathbf{M}^{-T})} e^{j\boldsymbol{\omega}^T \mathbf{n}} d\boldsymbol{\omega}$$

$$= \frac{1}{2^D J(\mathbf{M})} \int_{\mathbf{x} \in [-1,1)^D} e^{j\pi \mathbf{x}^T \mathbf{M}^{-1} \mathbf{n}} d\mathbf{x}$$

$$= \frac{1}{2^D J(\mathbf{M})} \int_{\mathbf{x} \in [-1,1)^D} e^{j\pi \mathbf{x}^T \mathbf{m}} d\mathbf{x}, \quad (\mathbf{m} = \mathbf{M}^{-1}\mathbf{n})$$

$$= \frac{1}{2^D J(\mathbf{M})} \prod_{i=0}^{D-1} \int_{x_i=-1}^{1} e^{j\pi m_i x_i} dx_i. \quad (12.7.19)$$

Note that m_i are the components of the D-vector $\mathbf{m} = \mathbf{M}^{-1}\mathbf{n}$. We know $\mathbf{M}^{-1} = [\text{Adj } \mathbf{M}]/[\det \mathbf{M}]$. Since $J(\mathbf{M}) = |\det \mathbf{M}|$, we can write

$$\mathbf{m} = \mathbf{M}^{-1}\mathbf{n} = \frac{\widehat{\mathbf{M}}\mathbf{n}}{J(\mathbf{M})}, \quad (12.7.20)$$

with $\widehat{\mathbf{M}} = J(\mathbf{M})\mathbf{M}^{-1} = \pm[\text{Adj } \mathbf{M}]$. Note that $\mathbf{m}$ is not necessarily an integer vector. From the above we obtain

$$h(\mathbf{n}) = \frac{1}{J(\mathbf{M})} \prod_{i=0}^{D-1} \frac{\sin(\pi m_i)}{\pi m_i}. \quad (12.7.21)$$

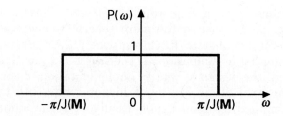

Figure 12.7-9 The one-dimensional lowpass prototype.

Relation to 1D filters

Consider a one dimensional ideal filter with frequency response $P(\omega)$ as shown in Fig. 12.7-9. Its impulse response is given by

$$p(n) = \frac{\sin(\frac{\pi n}{J(\mathbf{M})})}{\pi n}. \quad (12.7.22)$$

Starting from this $P(\omega)$, suppose we define the D-dimensional filter

$$H^{(s)}(\boldsymbol{\omega}) = P(\omega_0)P(\omega_1)\ldots P(\omega_{D-1}). \quad (12.7.23)$$

This is a separable lowpass filter, with passband support $SPD(\pi\mathbf{I}/J(\mathbf{M}))$. Its impulse response is

$$h^{(s)}(\mathbf{n}) = p(n_0)p(n_1)\ldots p(n_{D-1}), \quad (12.7.24)$$

that is,
$$h^{(s)}(\mathbf{n}) = \prod_{i=0}^{D-1} \frac{\sin(\frac{\pi n_i}{J(\mathbf{M})})}{\pi n_i} \qquad (12.7.25)$$

Now consider the quantity $h^{(s)}(\widehat{\mathbf{M}}\mathbf{n})$, which is the $\widehat{\mathbf{M}}$-fold decimated version of $h^{(s)}(\mathbf{n})$. Since

$$\widehat{\mathbf{M}}\mathbf{n} = J(\mathbf{M})\mathbf{M}^{-1}\mathbf{n} = J(\mathbf{M})\mathbf{m}, \qquad (12.7.26)$$

we obtain the following very simple relation between the sequence $h(\mathbf{n})$ in (12.7.21), and the sequence $h^{(s)}(\widehat{\mathbf{M}}\mathbf{n})$:

$$h(\mathbf{n}) = c h^{(s)}(\widehat{\mathbf{M}}\mathbf{n}), \qquad (12.7.27)$$

where $c = [J(\mathbf{M})]^{D-1}$. In other words, $h(\mathbf{n})$ is obtained simply by $\widehat{\mathbf{M}}$-fold decimation of the D-dimensional separable sequence $h^{(s)}(\mathbf{n})$, followed by scaling with c.

Design Procedure

The above derivation gives us the following procedure for designing the decimation filter $H(\omega)$ for M-fold decimator.

1. First design a 1D lowpass filter $P(\omega)$ which approximates the response of Fig. 12.7-9. Let $p(n)$ denote its impulse response.
2. Define the D-dimensional separable filter $h^{(s)}(\mathbf{n})$ as in (12.7.24).
3. Finally obtain the impulse response $h(\mathbf{n})$ of $H(\omega)$ by decimating $h^{(s)}(\mathbf{n})$ with the matrix $\widehat{\mathbf{M}}$ and scaling, that is, as in (12.7.27).

Note that the method can be applied for any decimation matrix **M**, and for any number of dimensions. We will now demonstrate the idea for hexagonal **M**. In this case, we wish the ideal passband of $H(\omega)$ to be as in Fig. 12.5-2(a). Figure 12.7-10 shows the various frequency responses. The filter $P(\omega)$ is a linear phase equiripple filter of order 66 with response as in Fig. 12.7-10(a). The 2D separable filter $h^{(s)}(\mathbf{n})$ has the response shown in Fig. 12.7-10(b), and finally Fig. 12.7-10(c) shows the response of the decimation filter $H(\omega)$. The minimum stopband attenuation of the 1D filter is about 60 dB and that of the non separable 2D filter is about 53 dB.

Further Comments About the Method

A number of further details about this technique can be found in Chen and Vaidyanathan [1991]. It is shown in particular that

1. The resulting filter $H(\omega)$ can be *implemented* with computational complexity proportional to $N \times D$ rather than N^D, where N is the order of the 1D filter $P(\omega)$.
2. If δ_1 and δ_2 are the peak passband and stopband ripples of the 1D filter $P(\omega)$, then the peak passband and stopband ripples of the multidimensional filter $H(\omega)$ are upper bounded by

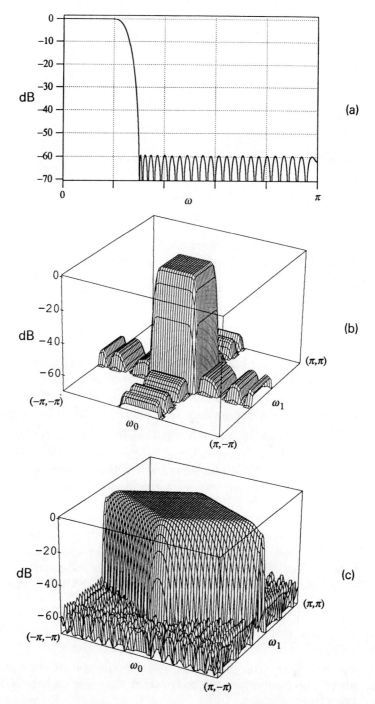

Figure 12.7-10 Magnitude response plots in dB for (a) one dimensional filter $P(\omega)$, (b) two-dimensional separable filter $H^{(s)}(\boldsymbol{\omega})$ and (c) two-dimensional nonseparable filter $H(\boldsymbol{\omega})$.

$$\widehat{\delta}_1 \leq \left(J(\widehat{\mathbf{M}}) - 1\right)\delta_2 + D\delta_1, \quad \widehat{\delta}_2 \leq J(\widehat{\mathbf{M}})\delta_2. \tag{12.7.28}$$

3. If we wish $H(\omega)$ to be a zero-phase filter, we can achieve this by designing the 1D filter $P(\omega)$ to have zero phase.
4. If we wish $H(\omega)$ to be a Mth band filter (see Sec. 12.8.1), we can achieve this by forcing $P(\omega)$ to be a $J(\mathbf{M})$th band filter.

In the above filter design method, we have restricted the elements of $\mathbf{M}$ to be integers. The method can easily be extended to the case of lowpass filters with passband supports of the form $SPD(\pi \mathbf{P}^{-T})$ where $\mathbf{P}$ is a nonsingular *rational* matrix (i.e., whose elements are rational numbers).

12.8 SPECIAL FILTERS AND FILTER BANKS

In Sec. 4.6 several special classes of filters and filter banks were discussed, with applications in multirate systems. These include Nyquist filters, uniform DFT filter banks, and several types of complementary functions. We now indicate the multidimensional extensions of some of these. Some of these have appeared in Renfors [1989].

12.8.1 Nyquist(M) Filters or M-th Band Filters

In the 1D case, there are several equivalent ways to define the 'Nyquist' property. The same is true in multidimensions. A discrete time filter with impulse response $h(\mathbf{n})$ is said to have the Mth band property if

$$h(\mathbf{Mn}) = 0, \quad \mathbf{n} \neq \mathbf{0}, \tag{12.8.1}$$

where $\mathbf{M}$ is a nonsingular integer matrix. In analogy with the 1D case, such a filter is also called a Nyquist($\mathbf{M}$) filter. If an interpolation filter (for M-fold interpolation) has this property, then the original set of samples is preserved in the interpolated version, as we saw in the 1D case.

We know that any sequence $x(\mathbf{n})$ can be expressed in terms of the polyphase components $x_\mathbf{k}(\mathbf{n})$ as in (12.4.32). The quantity $h(\mathbf{Mn})$ is in fact the 0th polyphase component $e_0(\mathbf{n})$ of the sequence $h(\mathbf{n})$. So the M-th band property (12.8.1) is equivalent to the condition that the Fourier transform $E_0(\omega)$ of the 0th polyphase component $e_0(\mathbf{n})$ be a constant.

Since $e_0(\mathbf{n})$ is the decimated version of $h(\mathbf{n})$, we have

$$E_0(\omega) = \frac{1}{J(\mathbf{M})} \sum_{\mathbf{m} \in \mathcal{N}(\mathbf{M}^T)} H\left(\mathbf{M}^{-T}(\omega - 2\pi \mathbf{m})\right). \tag{12.8.2}$$

The fact that this is constant implies that the summation above is constant. Equivalently,

$$\sum_{\mathbf{m} \in \mathcal{N}(\mathbf{M}^T)} H(\omega_0 - 2\pi \mathbf{M}^{-T}\mathbf{m}) = constant, \tag{12.8.3}$$

for all $\boldsymbol{\omega}_0$. In other words, if we make copies of $H(\boldsymbol{\omega})$ by shifting the origin to the points $2\pi\mathbf{M}^{-T}\mathbf{m}$ (which are points on the lattice $LAT(2\pi\mathbf{M}^{-T})$), and add the copies for all $\mathbf{m} \in \mathcal{N}(\mathbf{M}^T)$, the result is constant! This is an extension of the 1D property (4.6.6).

Summarizing, the definition of **M**-th band (or Nyquist(**M**)) property can be taken to be any one of the following: (a) condition (12.8.1), or (b) condition (12.8.3), or (c) the property that $E_0(\boldsymbol{\omega})$ be constant. †

Ideal Decimation Filter: Example of an Mth Band Filter

Recall that in the 1D case, an ideal lowpass filter with passband edge π/M has the Mth band property. Now consider an ideal multidimensional filter $H(\boldsymbol{\omega})$ with passband support $SPD(\pi\mathbf{M}^{-T})$. (Fig. 12.5-2 demonstrated this for several values of **M**). We now show that such an ideal filter has the Mth band property.

From Sec. 12.7.3 we know that the impulse response has the form (12.7.27) where $h^{(s)}(\mathbf{n})$ is a separable filter, formed from the 1D filter $p(n)$. We see that $p(J(\mathbf{M})n) = 0$ for $n \neq 0$. This implies $h^{(s)}(J(\mathbf{M})\mathbf{n}) = 0, \mathbf{n} \neq \mathbf{0}$. From (12.7.27) we conclude that

$$h(\mathbf{Mn}) = ch^{(s)}(\widehat{\mathbf{M}}\mathbf{Mn}) = ch^{(s)}(J(\mathbf{M})\mathbf{n}) = 0, \quad \mathbf{n} \neq \mathbf{0}.$$

proving that $H(\boldsymbol{\omega})$ is an Mth band filter indeed.

12.8.2 Uniform DFT Filter Banks

One dimensional uniform DFT filter banks were studied in Chap. 4. In these systems, a set of M filters was derived from a prototype $H_0(z)$ as shown in Fig. 4.3-5 where **W** is the $M \times M$ DFT matrix. The M filters satisfy the shift relation $H_k(z) = H_0(zW^k)$. The quantities $E_k(z)$ are polyphase components of the prototype $H_0(z)$. To generalize this for the MD case, we first consider an example.

Example 12.8.1

Consider the system of Fig. 12.8-1. Here the input $x(\mathbf{n})$ is passed through a set of delays

$$\mathcal{Z}(-\mathbf{k}), \quad \mathbf{k} \in \mathcal{N}(\mathbf{M}). \tapleftarrow \quad\quad (12.8.4)$$

---

† One can make the definition more general by saying that the filter $h(\mathbf{n})$ is Nyquist(**M**) if, for some integer **k**, $h(\mathbf{n} - \mathbf{k})$ satisfies any of the above conditions.

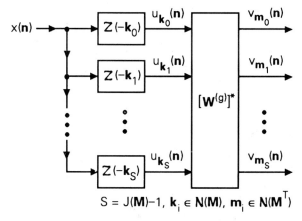

Figure 12.8-1 Multidimensional extension of the uniform DFT analysis bank.

Since there are $J(\mathbf{M})$ integers in the set $\mathcal{N}(\mathbf{M})$, the result is a vector of the $J(\mathbf{M})$ signals $u_\mathbf{k}(\mathbf{n})$, $\mathbf{k} \in \mathcal{N}(\mathbf{M})$. This vector is then passed through the matrix $[\mathbf{W}^{(g)}]^*$, to produce the set of $J(\mathbf{M})$ signals $v_\mathbf{m}(\mathbf{n})$, $\mathbf{m} \in \mathcal{N}(\mathbf{M}^T)$. This system is a multidimensional LTI system with one input and $J(\mathbf{M})$ outputs, and is characterized by the $J(\mathbf{M})$ transfer functions

$$H_\mathbf{m}(\omega) = \frac{V_\mathbf{m}(\omega)}{X(\omega)}, \quad \mathbf{m} \in \mathcal{N}(\mathbf{M}^T). \tag{12.8.5}$$

By using the expression (12.4.38) for the entries of $\mathbf{W}^{(g)}$, we obtain

$$\begin{aligned} H_\mathbf{m}(\omega) &= \sum_{\mathbf{k} \in \mathcal{N}(\mathbf{M})} e^{-j\omega^T \mathbf{k}} e^{j2\pi \mathbf{m}^T \mathbf{M}^{-1}\mathbf{k}} \\ &= \sum_{\mathbf{k} \in \mathcal{N}(\mathbf{M})} e^{-j(\omega - 2\pi \mathbf{M}^{-T}\mathbf{m})^T \mathbf{k}}. \end{aligned} \tag{12.8.6}$$

Thus

$$H_\mathbf{m}(\omega) = H_0(\omega - 2\pi \mathbf{M}^{-T}\mathbf{m}), \quad \mathbf{m} \in \mathcal{N}(\mathbf{M}^T), \tag{12.8.7}$$

where the proptotype response is

$$H_0(\omega) = \sum_{\mathbf{k} \in \mathcal{N}(\mathbf{M})} e^{-j\omega^T \mathbf{k}}. \tag{12.8.8}$$

(Note that $\mathbf{m}_0 = \mathbf{k}_0 = \mathbf{0}$.) Equation (12.8.7) shows that the responses $H_\mathbf{m}(\omega)$ are shifted versions of the response $H_0(\omega)$. The shifted locations are the points on the scaled reciprocal lattice corresponding to $\mathbf{M}$. So the system of Fig. 12.8-1 represents a multidimensional uniform-DFT analysis bank.

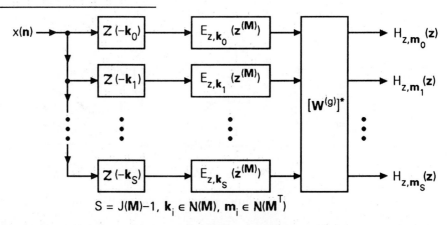

Figure 12.8-2 More general multidimensional uniform DFT analysis bank. (© Adopted from Sadhana, 1990 [Vaidyanathan, 1990c].)

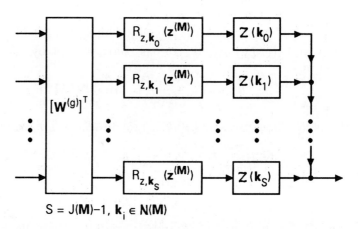

Figure 12.8-3 The multidimensional uniform DFT synthesis bank. (© Adopted from 1990 Sadhana.)

Figure 12.8-2 shows a further generalization of this system. The analysis filters are again related by (12.8.7) (Problem 12.17), with the prototype now given by

$$H_0(\omega) = \sum_{\mathbf{k} \in \mathcal{N}(\mathbf{M})} e^{-j\omega^T \mathbf{k}} E_{\mathbf{k}}(\mathbf{M}^T \omega). \qquad (12.8.9)$$

[Remember that according to our notations the quantity $E_{z,\mathbf{k}}(\mathbf{z}^{(\mathbf{M})})$ translates to $E_{\mathbf{k}}(\mathbf{M}^T \omega)$, if written in terms of ω.] Next, Fig. 12.8-3 represents a uniform DFT *synthesis bank*. In Problem 12.18 the reader is requested to show that the synthesis filters $F_{\mathbf{m}}(\omega)$ are related exactly as in (12.8.7) (with

H replaced by F everywhere), where the prototype is now given by

$$F_0(\omega) = \sum_{\mathbf{k}\in\mathcal{N}(M)} e^{j\omega^T \mathbf{k}} R_{\mathbf{k}}(M^T\omega). \qquad (12.8.10)$$

Example 12.8.2

Consider an uniform DFT analysis bank, with hexagonal M [given in (12.4.15)]. In Example 12.4.6 we calculated the elements of the set $\mathcal{N}(M^T)$. From that we can verify the shift vectors $2\pi M^{-T}\mathbf{m}$ in (12.8.7) to be

$$\begin{bmatrix}0\\0\end{bmatrix}, \pi\begin{bmatrix}1\\0.5\end{bmatrix}, \pi\begin{bmatrix}0\\1\end{bmatrix}, \pi\begin{bmatrix}1\\1.5\end{bmatrix}. \qquad (12.8.11)$$

Assume that the prototype filter $H_0(\omega)$ [i.e., $H_{\mathbf{m}_0}(\omega)$] is lowpass with passband support as in Fig. 12.8-4 [which is $SPD(\pi M^{-T})$]. Then, the three shifted filters have supports as shown in the same figure.

If we keep in mind the fact that each of the responses in this figure is periodic (periodicity matrix $2\pi I$), we can verify that the four passbands (which are disjoint), fill the entire frequency plane. If the prototype filter had some other kind of support (inconsistent with our choice of M), this would not be true.

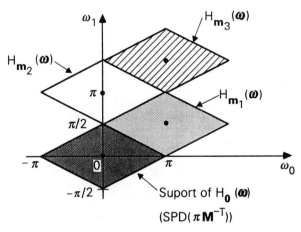

Figure 12.8-4 Example 12.8.2. Filter supports in the uniform DFT bank.

12.9 MAXIMALLY DECIMATED FILTER BANKS

Figure 12.9-1 shows the maximally decimated filter bank structure (QMF bank) for the multidimensional case. The analysis filters $H_{z,\mathbf{k}}(\mathbf{z})$ essentially divide the signal $x(\mathbf{n})$ into subbands in the D-dimensional freqency domain.

The output of each analysis filter is decimated by **M**. Since the decimator reduces the sample density by $J(\mathbf{M})$ ($= |\det \mathbf{M}|$), we use a total of $J(\mathbf{M})$ analysis filters so that the sampling density is preserved by the complete analysis bank. The decimated subband signals $v_\mathbf{k}(\mathbf{n})$ are then expanded **M**-fold, and passed through the $J(\mathbf{M})$ synthesis filters $F_{z,\mathbf{k}}(\mathbf{z})$. The outputs of these filters are added to obtain the reconstructed signal $\hat{x}(\mathbf{n})$. Notice that in Fig. 12.9-1, the index **k** which identifies the filter-number is a D-dimensional integer belonging to $\mathcal{N}(\mathbf{M}^T)$. The reader can regard this convention to be a matter of convenience. [One could equally have chosen to use a scalar index i (as in $H_i(\mathbf{z})$), and let i run from 0 to $J(\mathbf{M}) - 1$.]

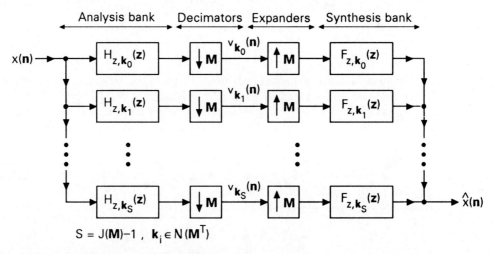

Figure 12.9-1 The multidimensional maximally decimated filter bank.

12.9.1 Analysis of the Structure

By using the transform domain relations for decimators and expanders (Sec. 12.4.3), one can analyze the filter bank system to obtain the expression

$$\widehat{X}(\omega) = \frac{1}{J(\mathbf{M})} \sum_{\mathbf{m}\in\mathcal{N}(\mathbf{M}^T)} X(\omega - 2\pi\mathbf{M}^{-T}\mathbf{m}) \sum_{\mathbf{k}\in\mathcal{N}(\mathbf{M}^T)} F_\mathbf{k}(\omega) H_\mathbf{k}(\omega - 2\pi\mathbf{M}^{-T}\mathbf{m})$$

(12.9.1)

for the reconstructed signal. This shows that $\widehat{X}(\omega)$ is a linear combination of $X(\omega)$ and the shifted versions $X(\omega - 2\pi\mathbf{M}^{-T}\mathbf{m})$ (which are the alias components). These aliasing terms can be eliminated if the filters are chosen such that

$$\sum_{\mathbf{k}\in\mathcal{N}(\mathbf{M}^T)} F_\mathbf{k}(\omega) H_\mathbf{k}(\omega - 2\pi\mathbf{M}^{-T}\mathbf{m}) = 0, \quad \mathbf{m} \neq \mathbf{0},\ \mathbf{m} \in \mathcal{N}(\mathbf{M}^T). \quad (12.9.2)$$

Under this alias-free condition we have

$$\widehat{X}(\omega) = \underbrace{\frac{1}{J(\mathbf{M})} \sum_{\mathbf{k} \in \mathcal{N}(\mathbf{M}^T)} F_{\mathbf{k}}(\omega) H_{\mathbf{k}}(\omega)}_{T(\omega)} X(\omega), \qquad (12.9.3)$$

with $T(\omega)$ representing the overall distortion function. If $|T(\omega)|$ is a nonzero constant for all ω (i.e., $T_z(z)$ is allpass) the system is free from amplitude distortion; if $T(\omega) = c e^{-j\omega^T \mathbf{n}_0}$ for some integer $\mathbf{n}_0$, the system has perfect reconstruction property, i.e., $\widehat{x}(\mathbf{n}) = c x(\mathbf{n} - \mathbf{n}_0)$. The quantities

$$A_{\mathbf{m}}(\omega) \triangleq \frac{1}{J(\mathbf{M})} \sum_{\mathbf{k} \in \mathcal{N}(\mathbf{M}^T)} F_{\mathbf{k}}(\omega) H_{\mathbf{k}}(\omega - 2\pi \mathbf{M}^{-T} \mathbf{m}), \quad \mathbf{m} \in \mathcal{N}(\mathbf{M}^T) \quad (12.9.4)$$

which are the weighting functions for $X(\omega - 2\pi \mathbf{M}^{-T} \mathbf{m})$ in (12.9.1), can be collected into a vector $\mathbf{A}(\omega)$, and expressed neatly in matrix vector notation as

$$\mathbf{A}(\omega) = \frac{\mathbf{H}(\omega) \mathbf{f}(\omega)}{J(\mathbf{M})}, \qquad (12.9.5)$$

where $\mathbf{H}(\omega)$ is the $J(\mathbf{M}) \times J(\mathbf{M})$ alias component (AC) matrix with elements

$$[\mathbf{H}(\omega)]_{\mathbf{m},\mathbf{k}} = H_{\mathbf{k}}(\omega - 2\pi \mathbf{M}^{-T} \mathbf{m}), \qquad (12.9.6)$$

and $\mathbf{f}(\omega)$ is the synthesis filter vector with elements $[\mathbf{f}(\omega)]_{\mathbf{k}} = F_{\mathbf{k}}(\omega)$. The condition for alias cancelation is equivalent to setting $\mathbf{A}(\omega)$ in (12.9.5) to be of the form

$$\begin{bmatrix} X \\ 0 \\ \vdots \\ 0 \end{bmatrix}. \qquad (12.9.7)$$

Given the set of analysis filters (and hence the matrix $\mathbf{H}(\omega)$), we can therefore find the synthesis filters satisfying alias-free condition by solving the equation (12.9.5), subject only to invertibility of the AC matrix.

We can rewrite the above formulation in terms of z-transform variables easily. For this recall (Sec. 12.1) that if the frequency variable ω is replaced with $\omega - \mathbf{b}$ for a real vector

$$\mathbf{b} = [b_0 \quad b_1 \quad \ldots \quad b_{D-1}]^T,$$

then the vector $\mathbf{z}$ is replaced with $e^{-j\mathbf{\Lambda}_\mathbf{b}} \mathbf{z}$. In other words, the scalar variable z_i is replaced with $z_i e^{-j b_i}$. An example will make this idea clearer.

Example 12.9.1

Consider the hexagonal decimator (12.4.15) again. We know that the elements of $\mathcal{N}(\mathbf{M}^T)$ are as in (12.4.41), so that the four vectors $\mathbf{M}^{-T}\mathbf{m}$ involved in (12.9.4) are

$$\mathbf{x}_0 = \begin{bmatrix} 0 \\ 0 \end{bmatrix}, \mathbf{x}_1 = \begin{bmatrix} 0.5 \\ 0.25 \end{bmatrix}, \mathbf{x}_2 = \begin{bmatrix} 0 \\ 0.5 \end{bmatrix}, \mathbf{x}_3 = \begin{bmatrix} 0.5 \\ 0.75 \end{bmatrix}.$$

The AC matrix has four rows, one for each $\mathbf{x}_i$. The 0th row corresponds to $\mathbf{x}_0$, and is given by

$$[H_{\mathbf{k}_0}(z_0, z_1) \quad H_{\mathbf{k}_1}(z_0, z_1) \quad H_{\mathbf{k}_2}(z_0, z_1) \quad H_{\mathbf{k}_3}(z_0, z_1)],$$

where the subscript z on H is omitted for simplicity. For other values of $\mathbf{x}_i$, we obtain the corresponding row by replacing z_0 with $z_0 e^{-j2\pi x_{i,0}}$ and z_1 with $z_1 e^{-j2\pi x_{i,1}}$. Thus the entire AC matrix is given by

$$\begin{bmatrix} H_{\mathbf{k}_0}(z_0, z_1) & H_{\mathbf{k}_1}(z_0, z_1) & H_{\mathbf{k}_2}(z_0, z_1) & H_{\mathbf{k}_3}(z_0, z_1) \\ H_{\mathbf{k}_0}(-z_0, -jz_1) & H_{\mathbf{k}_1}(-z_0, -jz_1) & H_{\mathbf{k}_2}(-z_0, -jz_1) & H_{\mathbf{k}_3}(-z_0, -jz_1) \\ H_{\mathbf{k}_0}(z_0, -z_1) & H_{\mathbf{k}_1}(z_0, -z_1) & H_{\mathbf{k}_2}(z_0, -z_1) & H_{\mathbf{k}_3}(z_0, -z_1) \\ H_{\mathbf{k}_0}(-z_0, jz_1) & H_{\mathbf{k}_1}(-z_0, jz_1) & H_{\mathbf{k}_2}(-z_0, jz_1) & H_{\mathbf{k}_3}(-z_0, jz_1) \end{bmatrix}.$$

Next consider the rectangular decimator $\mathbf{M} = \begin{bmatrix} 2 & 0 \\ 0 & 2 \end{bmatrix}$. In this case the AC matrix is verified to be

$$\begin{bmatrix} H_{\mathbf{k}_0}(z_0, z_1) & H_{\mathbf{k}_1}(z_0, z_1) & H_{\mathbf{k}_2}(z_0, z_1) & H_{\mathbf{k}_3}(z_0, z_1) \\ H_{\mathbf{k}_0}(-z_0, z_1) & H_{\mathbf{k}_1}(-z_0, z_1) & H_{\mathbf{k}_2}(-z_0, z_1) & H_{\mathbf{k}_3}(-z_0, z_1) \\ H_{\mathbf{k}_0}(z_0, -z_1) & H_{\mathbf{k}_1}(z_0, -z_1) & H_{\mathbf{k}_2}(z_0, -z_1) & H_{\mathbf{k}_3}(z_0, -z_1) \\ H_{\mathbf{k}_0}(-z_0, -z_1) & H_{\mathbf{k}_1}(-z_0, -z_1) & H_{\mathbf{k}_2}(-z_0, -z_1) & H_{\mathbf{k}_3}(-z_0, -z_1) \end{bmatrix}.$$

Choice of Frequency Responses

Recall that if a decimation filter has passband support restricted to $SPD(\pi\mathbf{M}^{-T})$ (or an arbitrarily shifted version of this), then there is no aliasing. We can therefore choose the analysis filters in Fig. 12.9-1 according to this criteria. Since filters are never ideal, we still have some aliasing at the outputs of the decimators. This aliasing should be canceled by appropriate choice of synthesis filters.

For example suppose we use the hexagonal decimator (12.4.15). Since $|\det \mathbf{M}| = 4$, we have four analysis filters. Figure 12.5-2(a) shows the region $SPD(\pi\mathbf{M}^{-T})$. Suppose we take the passband of $H_{z,0}(z_0, z_1)$ to be this region. This region, of course repeats with period 2π in both directions. By using four properly shifted copies of this region, we can fill the entire 2D frequency plane as demonstrated in Fig. 12.9-2. (Only the region $[-\pi, \pi)^2$ is shown fully). We can therefore take the passbands of the four analysis filters to be these four regions. Again, in practice, we have nonzero transition bands, so that there is overlap between passbands. So the passbands have areas exceeding the ideal areas. The decimators therefore cause aliasing which should eventually be canceled.

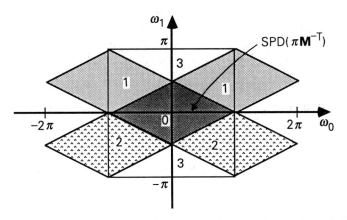

Figure 12.9-2 A possible choice of the passband regions for the four analysis filters, when hexagonal decimators are used in the QMF bank.

It is often desirable in practice to limit the filter coefficients to be real. This can be accomplished if the passband supports are chosen to have appropriate symmetry in the frequency domain.

12.9.2 Polyphase Representation

As in the 1D case, the theory and design of maximally decimated filter banks is considerably simplified by use of the polyphase representation. This has been observed in Viscito and Allebach [1988b], and used in Karlsson and Vetterli [1990] and Vaidyanathan [1991b]. To obtain this, recall that each analysis filter can be represented in the form

$$H_{z,\mathbf{k}}(\mathbf{z}) = \sum_{\mathbf{m} \in \mathcal{N}(\mathbf{M})} \mathcal{Z}(-\mathbf{m}) E_{z,\mathbf{k},\mathbf{m}}(\mathbf{z^{(M)}}), \quad \mathbf{k} \in \mathcal{N}(\mathbf{M}^T), \qquad (12.9.8)$$

where $E_{z,\mathbf{k},\mathbf{m}}(\mathbf{z})$ are the Type 1 polyphase components. The $J(\mathbf{M}) \times J(\mathbf{M})$ matrix $\mathbf{E}_z(\mathbf{z})$ with elements

$$[\mathbf{E}_z(\mathbf{z})]_{\mathbf{k},\mathbf{m}} \triangleq E_{z,\mathbf{k},\mathbf{m}}(\mathbf{z}), \qquad (12.9.9)$$

is the polyphase component matrix (or polyphase matrix) for the analysis bank. With this definition we can redraw the analysis bank as indicated in Fig. 12.9-3.

For the synthesis filters we will use the Type 2 polyphase decomposition

$$F_{z,\mathbf{k}}(\mathbf{z}) = \sum_{\mathbf{m}\in\mathcal{N}(\mathbf{M})} \mathcal{Z}(\mathbf{m}) R_{z,\mathbf{m},\mathbf{k}}(\mathbf{z}^{(\mathbf{M})}), \quad \mathbf{k}\in\mathcal{N}(\mathbf{M}^T). \tag{12.9.10}$$

The $J(\mathbf{M}) \times J(\mathbf{M})$ matrix $\mathbf{R}_z(\mathbf{z})$ with elements

$$[\mathbf{R}_z(\mathbf{z})]_{\mathbf{m},\mathbf{k}} \triangleq R_{z,\mathbf{m},\mathbf{k}}(\mathbf{z}), \tag{12.9.11}$$

is the (Type 2) polphase component matrix (or polyphase matrix) for the synthesis bank. We can thus redraw the synthesis bank as shown in the figure.

By invoking the noble identities (Fig. 12.6-1) we now move the decimators and expanders to arrive at the equivalent structure of Fig. 12.9-4, where $\mathbf{P}_z(\mathbf{z}) = \mathbf{R}_z(\mathbf{z})\mathbf{E}_z(\mathbf{z})$. Any multidimensional QMF bank can be represented like this.

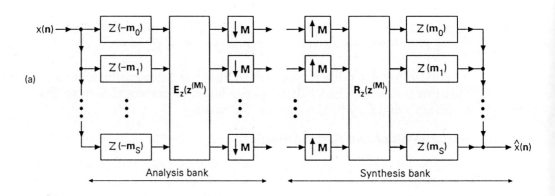

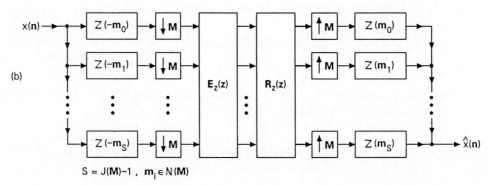

Figure 12.9-3 Representation of analysis and synthesis banks in terms of polyphase matrices.

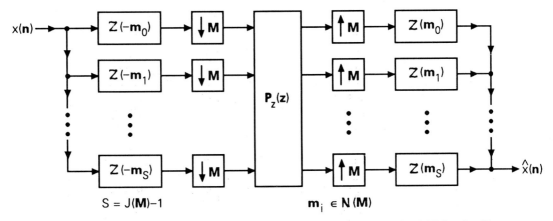

Figure 12.9-4 Equivalent structure for the multidimensional QMF bank. Here $\mathbf{P}_z(z) = \mathbf{R}_z(z)\mathbf{E}_z(z)$.

12.9.3 Perfect Reconstruction (PR) Systems

The theory of perfect reconstruction systems (Chap. 5) is easily extended to the MD case. For the 1D case, we started from the delay chain structure of Fig. 5.6-2 (which is a simple PR system) and obtained more general systems by insertion of the polyphase matrices $\mathbf{E}(z)$ and $\mathbf{R}(z)$. A similar approach works in the MD case. The extension of the delay chain structure is shown in Fig. 12.9-5(a). This system is essentially the QMF bank of Fig. 12.9-1 with the filters taken as

$$H_{z,\mathbf{k}_i}(z) = \mathcal{Z}(-\mathbf{m}_i), \quad F_{z,\mathbf{k}_i}(z) = \mathcal{Z}(\mathbf{m}_i), \quad \mathbf{k}_i \in \mathcal{N}(\mathbf{M}^T), \mathbf{m}_i \in \mathcal{N}(\mathbf{M}). \tag{12.9.12}$$

Thus the $\mathbf{k}_i$th analysis filter is a delay in the direction $\mathbf{m}_i$, whereas the synthesis filter is an advance operator in the same direction. It can be verified (precisely as we did in Sec. 12.4.3 for Fig. 12.4-7) that this is a perfect reconstruction system, and satisfies $\widehat{x}(\mathbf{n}) = x(\mathbf{n})$. For the special case where $\mathbf{M}$ is the rectangular decimator in Example 12.4.4, this reduces to Fig. 12.9-5(b).

Now consider an arbirary QMF bank (Fig. 12.9-1). This can always be redrawn as in Fig. 12.9-3. If the product $\mathbf{R}_z(z)\mathbf{E}_z(z) = \mathbf{I}$, this reduces to the perfect reconstruction system Fig. 12.9-5(a).

Example 12.9.2: QMF Bank with Hexagonal Decimation

Let $\mathbf{M}$ be the hexagonal decimation matrix (12.4.15). The delay elements $\mathcal{Z}(-\mathbf{m}_i)$ for this case were identified in Example 12.4.5. Using this, the perfect reconstruction system of Fig. 12.9-5(a) therefore reduces to Fig. 12.9-6(a). If we insert matrices $[\mathbf{W}^{(g)}]^*$ and $[\mathbf{W}^{(g)}]^T$ as

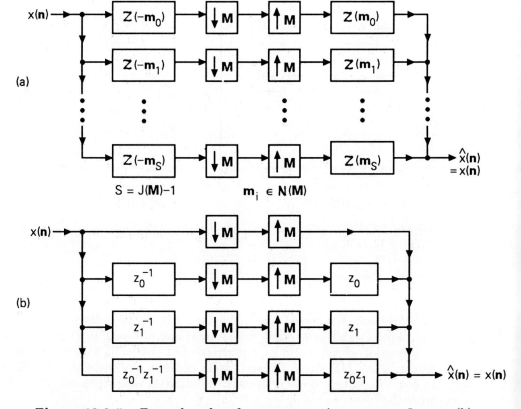

Figure 12.9-5 Examples of perfect reconstruction systems. In part (b), $M = 2I$.

shown in part (b), the perfect reconstruction property continues to hold (because the product of the two matrices equals $J(M)I$). We therefore obtain a perfect reconstruction system in which the analysis filters form a uniform-DFT bank (Sec. 12.8.2). The matrix $W^{(g)}$ was calculated in (12.4.43). Using this we obtain

$$H_{z,k_0}(z_0, z_1) = 1 + z_0^{-1} + z_0^{-1}z_1^{-1} + z_0^{-1}z_1,$$
$$H_{z,k_1}(z_0, z_1) = 1 - z_0^{-1} + jz_0^{-1}z_1^{-1} - jz_0^{-1}z_1,$$
$$H_{z,k_2}(z_0, z_1) = 1 + z_0^{-1} - z_0^{-1}z_1^{-1} - z_0^{-1}z_1,$$
$$H_{z,k_3}(z_0, z_1) = 1 - z_0^{-1} - jz_0^{-1}z_1^{-1} + jz_0^{-1}z_1.$$

Figure 12.9-6(c) shows an example in which the filters have higher order. Here the polyphase matrix for the analysis bank is

$$\mathbf{E}_z(\mathbf{z}) = \mathbf{R}_1 \mathbf{\Lambda}_z(\mathbf{z}) \mathbf{R}_0, \qquad (12.9.13)$$

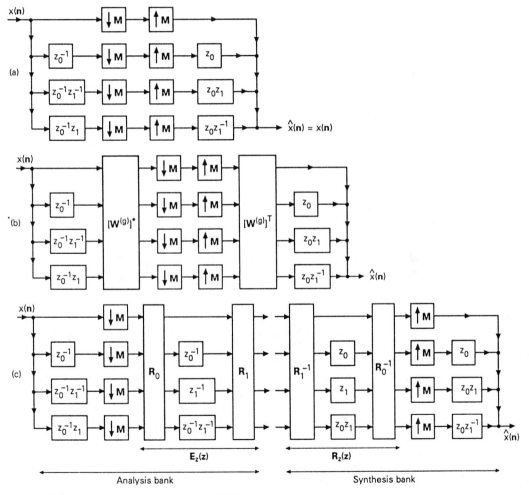

Figure 12.9-6 Examples of FIR perfect reconstruction systems, when M is the hexagonal decimator.

where

$$\Lambda_z(\mathbf{z}) = \begin{bmatrix} 1 & 0 & 0 & 0 \\ 0 & z_0^{-1} & 0 & 0 \\ 0 & 0 & z_1^{-1} & 0 \\ 0 & 0 & 0 & z_0^{-1}z_1^{-1} \end{bmatrix}, \qquad (12.9.14)$$

and $\mathbf{R}_m$ are nonsingular. With the synthesis bank chosen as indicated in the figure, the complete QMF bank is equivalent to the perfect reconstruction system of Fig 12.9-6(a). In fact this idea can be generalized by extending the cascade (12.9.13) to $\mathbf{R}_K \Lambda_z(\mathbf{z}) \mathbf{R}_{K-1} \ldots \Lambda_z(\mathbf{z}) \mathbf{R}_0$.

If the matrices $\mathbf{R}_m$ are unitary then $\mathbf{E}_z(\mathbf{z})$ is paraunitary, that is, satisfies the property

$$\widetilde{\mathbf{E}}_z(z_0, z_1)\mathbf{E}_z(z_0, z_1) = c\mathbf{I}, \quad c > 0. \qquad (12.9.15)$$

where "tilde" notation implies the following: (a) transpose the matrix, (b) replace z_i with z_i^{-1}, and (c) conjugate the coefficients. When $\mathbf{R}_m$ is unitary, we can replace $\mathbf{R}_m^{-1}$ with $\mathbf{R}_m^\dagger$ and obtain perfect reconstruction. With this choice the Type 2 polyphase matrix for the synthesis bank becomes $\mathbf{R}_z(z_0, z_1) = \widetilde{\mathbf{E}}_z(z_0, z_1)$. The unitary matrices can be parameterized as explained in Sec. 14.6, and these parameters optimized to obtain good designs for the analysis filters.

Further properties of multidimensional paraunitary filter banks can be found in Problems 12.19 and 12.27.

The ideas used in the above example can be easily applied for the case of rectangular decimators. For example we can insert unitary matrices and delays at appropriate places in Fig. 12.9-5(b) and so on ... It should be noticed that the rectangular decimator structure *does not* necessarily restrict the analysis filters to be separable. See Problem 12.20.

12.9.4 QMF Bank with Quincunx Decimators

We now consider a 2D QMF bank with the decimators and expanders taken to be the quincunx matrix. Since $J(\mathbf{M}) = 2$, the QMF bank has two channels. The quincunx matrix [Fig. 12.4-1(c)] was discussed earlier on many occassions. In Table 12.10.4 we collect many of the facts pertaining to this matrix. Based on this we can infer that the perfect reconstruction system of Fig. 12.9-5(a) reduces to Fig. 12.9-7. Moreover, the noble identities take the simplified form shown in Fig. 12.9-8. Finally the polyphase decomposition for the two analysis filters can be written as

$$H_z(z_0, z_1) = E_{z,0}(z_0 z_1^{-1}, z_0 z_1) + z_0^{-1} E_{z,1}(z_0 z_1^{-1}, z_0 z_1) \quad \text{(Type 1)}$$
$$H_z(z_0, z_1) = R_{z,0}(z_0 z_1^{-1}, z_0 z_1) + z_0 R_{z,1}(z_0 z_1^{-1}, z_0 z_1) \quad \text{(Type 2)}.$$
$$(12.9.16)$$

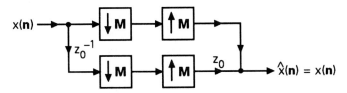

Figure 12.9-7 The simple perfect reconstruction system with quincunx decimator.

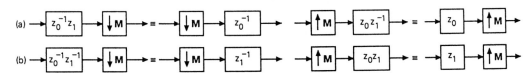

Figure 12.9-8 Noble identities for quincunx **M**.

Choice of Analysis Filters

In Example 12.7.1 and Sec. 12.7.1 we considered filters for the quincunx decimator. From these we see that if the analysis filter $H_{z,0}(z_0, z_1)$ is ideal lowpass with a diamond shaped passband support [Fig. 12.7-1(c)], then aliasing can be avoided. Similarly if $H_{z,1}(z_0, z_1)$ is ideal highpass with support as in Fig. 12.7-1(d), this also avoids aliasing.

Filters with such shapes can be approximated by starting from an ideal 1D filter with magnitude response as in Fig. 12.7-2(a). If this 1D filter has polyphase components $E_0(z)$ and $E_1(z)$, then the above lowpass and highpass filters can be obtained as in (12.7.11) and (12.7.12) respectively. These can be used as the analysis filters in the quincunx QMF bank. Since practical filters are not ideal, there is aliasing, which has to be canceled by the synthesis bank.

Example 12.9.3

Figure 12.9-9 shows a QMF bank with quincunx decimators. The signal $x(n)$ is split into two subbands [typically with supports approximating Figs. 12.7-1(c),(d)] and then decimated by **M**. This creates aliasing. Our aim is to choose the synthesis filters such that aliasing is canceled. (After this we have to worry about amplitude and phase distortions too.) We now concentrate on a specific technique which completely eliminates aliasing and amplitude distortion.

Using the expressions (12.7.11) and (12.7.12), the analysis bank can be redrawn as in Fig. 12.9-9(b). Note that the analysis filters are related as

$$H_{z,1}(z_0, z_1) = H_{z,0}(-z_0, -z_1). \qquad (12.9.17)$$

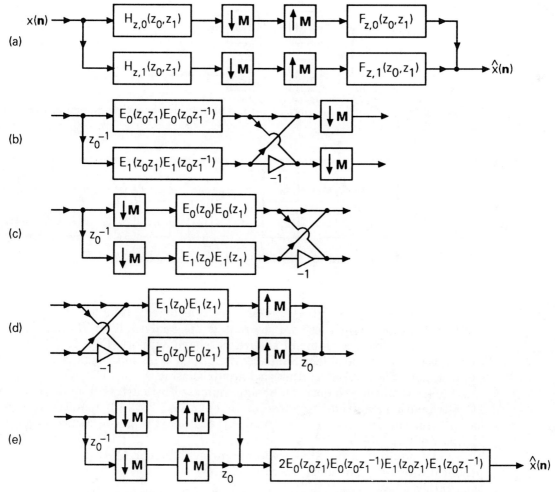

Figure 12.9-9 (a) The quincunx QMF bank, (b) polyphase implementation of analysis bank, (c) equivalent circuit for analysis bank, (d) proposed structure for synthesis bank, and (e) equivalent structure for complete QMF bank.

By invoking the noble identities we can move the decimators to obtain the equivalent of Fig. 12.9-9(c). If we now choose the synthesis bank as in Fig. 12.9-9(d), then the complete system can be redrawn (again with the help of noble identities) as in Fig. 12.9-9(e). By comparing with the perfect reconstruction system of Fig. 12.9-7 we conclude that this is an alias free system with transfer function (or distortion function)

$$T_z(z_0, z_1) = 2E_0(z_0z_1)E_0(z_0z_1^{-1})E_1(z_0z_1)E_1(z_0z_1^{-1}). \qquad (12.9.18)$$

With synthesis bank as in Fig. 12.9-9(d), the expressions for the

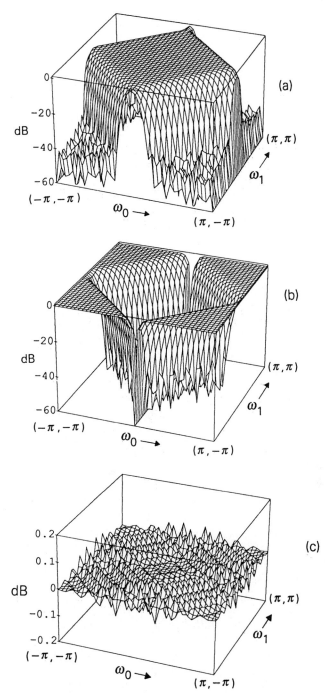

Figure 12.9-10 Alias-free quincunx QMF bank, starting from Johnston's 32D filters. The plots show magnitude in dB. (a) $H_{z,0}(z_0, z_1)$, (b) $H_{z,1}(z_0, z_1)$ and (c) $T_z(z_0, z_1)$.

synthesis filters are given by

$$F_{z,0}(z_0, z_1) = z_0 H_{z,0}(z_0, z_1), \quad F_{z,1}(z_0, z_1) = -z_0 H_{z,1}(z_0, z_1). \quad (12.9.19)$$

Thus, by designing a 1D prototype filter $G(z)$ to approximate the response of Fig. 12.7-2(a), we can identify the analysis filter $H_{z,0}(z_0, z_1)$ as in (12.7.11), and the remaining three filters according to (12.9.17) and (12.9.19). In particular if the 1D prototype is FIR, then the 2D analysis and synthesis filters are FIR as well.

Note that this QMF bank has correspondence with the 1D QMF bank studied in Sec. 5.2.2 where the filters were related as in (5.2.1) and (5.2.2). In Problem 12.26 we show that $T_z(z_0, z_1)$ is allpass if the distortion function $T(z)$ of the 1D system is allpass. Fig. 12.9-10 shows a design example. Here the 1D filter $G(z)$ is taken to be the lowpass filter $H_0(z)$ in Johnston's 1D QMF bank design (Sec. 5.2.2). More specifically, $G(z)$ is taken to be Johnston's 32D filter (tabulated in Crochiere and Rabiner [1983]). Figure 12.9-10(a) and (b) show the magnitude responses (dB) of the lowpass and highpass filters $H_{z,0}(z_0, z_1)$ and $H_{z,1}(z_0, z_1)$, with peak magnitude normalized to zero dB.

Figure 12.9-10(c) shows the magnitude of the distortion function $T_z(z_0, z_1)$ in dB, with nominal distortion normalized to zero dB. The peak distortion function is about 0.05dB. This is two times the corresponding value in the 1D Johnston design.

The reader is requested to find the AC matrix for the quincunx QMF bank in Problem 12.24. Further discussions on hexagonal and quincunx filter banks can be found in Ansari and Guillemot [1990], and Karlsson and Vetterli [1990].

12.9.5 Tree-Structured QMF Banks with Separable Filters

One of the simplest approaches for the design of 2D (and higher dimensional) QMF banks is to start from 1D QMF banks and connect them in a tree structure. Fig. 12.9-11 shows such a system. Here, the filters $H_0(z), H_1(z)$ are analysis filters of a 1D QMF bank, and $F_0(z), F_1(z)$ are the corresponding synthesis filters. Similarly $G_0(z), G_1(z)$ and $P_0(z), P_1(z)$ are the analysis and synthesis filters of a 1D QMF bank. The decimators $\mathbf{M}_0$ and $\mathbf{M}_1$ are given by

$$\mathbf{M}_0 = \begin{bmatrix} 2 & 0 \\ 0 & 1 \end{bmatrix}, \quad \mathbf{M}_1 = \begin{bmatrix} 1 & 0 \\ 0 & 2 \end{bmatrix}. \quad (12.9.20)$$

In other words, $\mathbf{M}_0$ decimates by two only in the horizontal direction (and $\mathbf{M}_1$ in vertical direction). The overall behavior of this 2D QMF bank (including aliasing and other distortions), evidently depends on the 1D QMF banks from which it is derived.

The 2D analysis filters which result from this approach are restricted to be separable, and can therefore be designed easily by designing 1D filters.

Many properties of 1D filter banks (such as, for example, perfect reconstruction) are automatically inherited by the 2D system. See Problems 12.22 and 12.23 where we study this in greater detail.

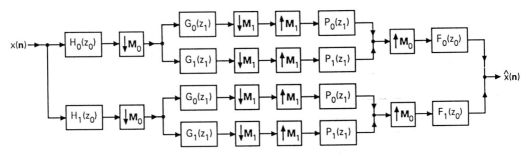

Figure 12.9-11 A tree-structured arrangement of one-dimensional QMF banks, which produces a separable two-dimensional QMF bank.

12.10 CONCLUDING REMARKS

In this chapter we have presented several fundamental aspects of multidimensional multirate systems. Many of the 1D concepts have been extended, by using the theory of lattices as the starting point. Even though our coverage was limited, it is our hope that the reader will be able to make further extensions of 1D results, with this chapter as the starting point. For example, one can extend many of the wavelet concepts (Chap. 11), by using the ideas in Sec. 12.9 [Kovačević and Vetterli, 1992]. Tables 12.10.1–12.10.4 at the end of this section summarize the key concepts, which can be used as a constant reference while reading the chapter.

12.10.1 Diagonalization of the Decimator Matrix

The main reason why the multidimensional ideas are more complicated is because the decimation matrix $\mathbf{M}$ is not diagonal. If it were diagonal, we could do most of our work with separable filters, which in turn can be designed starting from 1D filters. In this chapter we found that some multirate filters (even with nondiagonal $\mathbf{M}$) can be designed starting from 1D techniques (e.g., Sec. 12.7.1 and 12.7.3).

Any $D \times D$ integer matrix $\mathbf{M}$ can be factorized as [Smith, 1861]

$$\mathbf{M} = \mathbf{U}\mathbf{\Lambda}\mathbf{V} \quad \text{(Smith form)} \quad (12.10.1)$$

where $\mathbf{U}$ and $\mathbf{V}$ are integer unimodular matrices and $\mathbf{\Lambda}$ is a diagonal integer matrix. In this decompostion, we can always ensure that the diagonal elements λ_i of $\mathbf{\Lambda}$ are positive (assuming $\mathbf{M}$ nonsingular) and such that

$$\lambda_i | \lambda_{i+1}, \quad (12.10.2)$$

that is, λ_i is a factor of λ_{i+1}. Under these restrictions, $\boldsymbol{\Lambda}$ is unique for a given $\mathbf{M}$, even though $\mathbf{U}$ and $\mathbf{V}$ are not. Furthermore we can find the elements λ_i as $\lambda_i = \Delta_{i+1}/\Delta_i$ where Δ_i is the greatest common divisor of all the $i \times i$ minors of $\mathbf{M}$. (For $i = 0$ take $\Delta_0 = 1$.) The proof of this is very similar to the proof of the same result for polynomial matrices (Smith form decomposition, Theorem 13.5.1).† Here are some examples:

$$\underbrace{\begin{bmatrix} 1 & 1 \\ 2 & -2 \end{bmatrix}}_{\text{hexagonal } \mathbf{M}} = \underbrace{\begin{bmatrix} 1 & 0 \\ 2 & -1 \end{bmatrix}}_{\mathbf{U}} \underbrace{\begin{bmatrix} 1 & 0 \\ 0 & 4 \end{bmatrix}}_{\boldsymbol{\Lambda}} \underbrace{\begin{bmatrix} 1 & 1 \\ 0 & 1 \end{bmatrix}}_{\mathbf{V}}. \quad (12.10.3)$$

$$\underbrace{\begin{bmatrix} 1 & 1 \\ -1 & 1 \end{bmatrix}}_{\text{quincunx } \mathbf{M}} = \underbrace{\begin{bmatrix} 1 & 0 \\ -1 & 1 \end{bmatrix}}_{\mathbf{U}} \underbrace{\begin{bmatrix} 1 & 0 \\ 0 & 2 \end{bmatrix}}_{\boldsymbol{\Lambda}} \underbrace{\begin{bmatrix} 1 & 1 \\ 0 & 1 \end{bmatrix}}_{\mathbf{V}}. \quad (12.10.4)$$

This decomposition brings the multidimensional problems much closer to the 1D case, since decimation by $\boldsymbol{\Lambda}$ is equivalent to independent decimation in each dimension. We now mention a few applications of this decomposition. More detailed discussions and applications can be found in Vaidyanathan [1991b].

Characterization of Decimators

The $\mathbf{M}$-fold decimator can now be redrawn as in Fig. 12.10-1. $\mathbf{U}$ permutes the samples of $x(\mathbf{n})$. The matrix $\boldsymbol{\Lambda}$ performs independent decimation by the factor λ_i in each dimension i. The matrix $\mathbf{V}$ then permutes the decimated samples. Any $\mathbf{M}$-fold decimator is a succession of these three operations. Similar comment holds for expanders.

Figure 12.10-1 Three stage drawing of a decimator.

Since $\mathbf{V}$ is unimodular, $LAT(\mathbf{U\Lambda V}) = LAT(\mathbf{U\Lambda})$ (Sec. 12.2.1). Thus the samples retained by a decimator can be represented by $LAT(\mathbf{U\Lambda})$, i.e., a permutation of samples (due to $\mathbf{U}$) followed by independent decimation by λ_i in ith dimension.

Consider the set of all decimators with decimation ratio 2. These can be characterized as $\mathbf{U\Lambda}$. Since $|\det \mathbf{M}| = 2$, we have $|\det \boldsymbol{\Lambda}| = 2$, and the

---

† In fact such a factorization holds for any matrix whose elements belong to a so-called "principal ideal domain" (pid) [Forney, 1970], [Vidyasagar, 1985]. The set of integers (as well as the set of polynomials with coefficients belonging to a field) just happen to be pid's. Also see Sec. 13.5.1.

only possibility is
$$\Lambda = \text{diag} \begin{bmatrix} 1 & 1 & \ldots & 1 & 2 \end{bmatrix}, \qquad (12.10.5)$$

since $\lambda_i | \lambda_{i+1}$. As a further example, every 2D decimation matrix with decimation ratio 4 can be characterized by $\mathbf{U}\Lambda$ where Λ has one of the following two forms

$$\begin{bmatrix} 1 & 0 \\ 0 & 4 \end{bmatrix}, \begin{bmatrix} 2 & 0 \\ 0 & 2 \end{bmatrix}. \qquad (12.10.6)$$

We therefore have a very satisfactory way to characterize all decimators of a given ratio.

As a further theoretical application, one can rederive the frequency domain relation (12.4.22) in a more straightforward manner, by applying the 1D relation (4.1.4) repeatedly for each dimension. Another application is that, by factorizing the elements λ_i into primes, we can identify many possible multistage implementations of a given decimator.

Further Significance of the Property $\lambda_i | \lambda_{i+1}$

Since λ_i is a factor of λ_{i+1}, we can write

$$\lambda_0 = \alpha_0, \ \lambda_1 = \alpha_0 \alpha_1, \ \lambda_2 = \alpha_0 \alpha_1 \alpha_2 \ldots \qquad (12.10.7)$$

where α_i are integers. For example, in the 3D case,

$$\Lambda = \begin{bmatrix} \alpha_0 & 0 & 0 \\ 0 & \alpha_0 & 0 \\ 0 & 0 & \alpha_0 \end{bmatrix} \begin{bmatrix} 1 & 0 & 0 \\ 0 & \alpha_1 & 0 \\ 0 & 0 & \alpha_1 \end{bmatrix} \begin{bmatrix} 1 & 0 & 0 \\ 0 & 1 & 0 \\ 0 & 0 & \alpha_2 \end{bmatrix}. \qquad (12.10.8)$$

Thus, decimation by $\mathbf{M} = \mathbf{U}\Lambda\mathbf{V}$ can be interpreted as a succession of these steps: (a) permute samples by $\mathbf{U}$, (b) decimate in all directions by α_0, (c) decimate in all but the 0th direction by α_1, (d) decimate in the last direction by α_2, (e) and finally permute by $\mathbf{V}$. This interpretation can be generalized for arbitrary dimensions.

Revisiting the Polyphase Implementation using Smith Form

Figure 12.10-2(a) shows a decimation filter, with decimator $\mathbf{M} = \mathbf{U}\Lambda\mathbf{V}$. By using the noble identity along with the fact that the unimodular decimator $\mathbf{U}$ is equivalent to a $\mathbf{U}^{-1}$-fold expander, we can redraw this as in Fig.12.10-2(b). Here $G(\omega) = H(\mathbf{U}^{-T}\omega)$. Since Λ is diagonal, we can perform a traditional (i.e., 1D) polyphase decomposition of $G(\omega)$ in each dimension independently (e.g., as in Example 12.4.4), and then apply the 1D noble identities in each dimension individually. This results in Fig. 12.10-3.

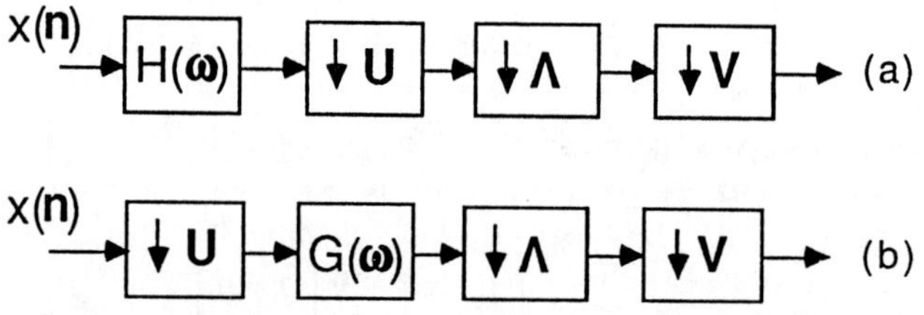

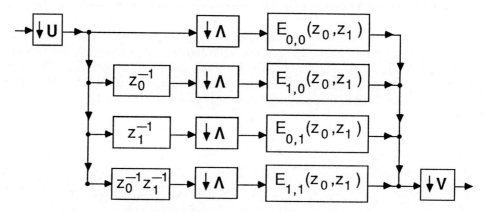

Figure 12.10-2 The decimation filter, redrawn.

Figure 12.10-3 Polyphase implementation after diagonalization.

Revisiting the Generalized DFT Matrix using Smith Form

In Sec. 12.4.4, we defined the generalized DFT matrix $\mathbf{W}^{(g)}$ for arbitrary $\mathbf{M}$. Let $\mathbf{M} = \mathbf{U}\mathbf{\Lambda}\mathbf{V}$ and let $\mathbf{W}_{\lambda_i}$ be the traditional $\lambda_i \times \lambda_i$ DFT matrix, i.e., $[\mathbf{W}_{\lambda_i}]_{km} = e^{-j2\pi km/\lambda_i}$. It can be shown that $\mathbf{W}^{(g)}$ is a 'Kronecker product' of all the $\mathbf{W}_{\Lambda_i}$'s. More accurately,

$$\mathbf{W}^{(g)} = \mathbf{P}_1[\mathbf{W}_{\lambda_0} \otimes \mathbf{W}_{\lambda_1} \otimes \ldots \otimes \mathbf{W}_{\lambda_{D-1}}]\mathbf{P}_2, \qquad (12.10.9)$$

where $\mathbf{P}_1$ and $\mathbf{P}_2$ are appropriate permutation-matrices [depending on $\mathbf{U}, \mathbf{V}$ and the ordering of the integer vectors $\mathbf{k}$ and $\mathbf{m}$ in (12.4.38)]. Here $\otimes$ is the Kronecker product defined as

$$\underbrace{\mathbf{A}}_{I \times K} \otimes \underbrace{\mathbf{B}}_{J \times L} = \underbrace{\begin{bmatrix} a_{0,0}\mathbf{B} & \cdots & a_{0,K-1}\mathbf{B} \\ \vdots & \cdots & \vdots \\ a_{I-1,0}\mathbf{B} & \cdots & a_{I-1,K-1}\mathbf{B} \end{bmatrix}}_{IJ \times KL}. \qquad (12.10.10)$$

This was also observed in Problem 2.20 of Dudgeon and Mersereau [1984].

For the hexagonal matrix $\mathbf{M}$ we have $\lambda_0 = 1$ and $\lambda_1 = 4$ so that $\mathbf{W}^{(g)}$ reduces to a permutation of the 1D DFT matrix $\mathbf{W}_4$, as shown by (12.4.43). On the other hand, if $\mathbf{M} = \begin{bmatrix} 2 & 0 \\ 0 & 2 \end{bmatrix}$, then $\mathbf{W}^{(g)}$ (which is still 4×4) is a Kronecker product of the 1D DFT matrix $\mathbf{W}_2$ with itself because, in this case, $\lambda_0 = \lambda_1 = 2$. So, in general, the form of $\mathbf{W}^{(g)}$ for a given $\mathbf{M}$ is governed essentially by the diagonal matrix $\boldsymbol{\Lambda}$.

12.10.2 Commutation of Decimators and Expanders

One of the interesting results in the 1D case is that an M-fold decimator and L-fold expander can be commuted in a cascade whenever L and M are relatively prime. This result was used in Sec. 4.3.3 to obtain polyphase structures for fractional decimation. The equivalents of these results in the multidimensional case are quite complicated. They have been developed by Kovačević and Vetterli [1991b], for the 2D case. A problem of considerable interest is the generalization of this for the MD case. Many authors have worked on this, and a number of results (with a wide variety of viewpoints as well as details) can be found in Chen and Vaidyanathan [1992a,b], [1993], [Gopinath and Burrus, 1991], [Evans, et al., 1992], and [Kalker, 1992].

TABLE 12.10.1 Lattice related notions in D-dimensional multirate systems

Unless context dictates otherwise, 'vector' stands for $D \times 1$ column vector. Integer vectors are often referred to as 'integers'. 'MD' stands for 'Multidimensional,' and 2D for 'two-dimensional'.

Notations and terminology.
1. $[a, b)^D$: set of $D \times 1$ real vectors $\mathbf{x}$ with elements x_i in the range $a \leq x_i < b$.
2. $\mathcal{N}$: set of all $D \times 1$ integer vectors.
3. Unless mentioned otherwise: $\mathbf{V} = D \times D$ nonsingular real matrix, and $\mathbf{M} = D \times D$ nonsingular integer matrix. An integer matrix $\mathbf{E}$ is said to be unimodular if $[\det \mathbf{E}] = \pm 1$.
4. $LAT(\mathbf{V})$: lattice generated by $\mathbf{V}$, i.e., set of all vectors of the form $\mathbf{Vn}, \mathbf{n} \in \mathcal{N}$.
5. $LAT(\mathbf{V}^{-T})$: reciprocal lattice. $LAT(2\pi \mathbf{V}^{-T})$: scaled reciprocal lattice. Here $\mathbf{V}^{-T}$ denotes $(\mathbf{V}^{-1})^T$.
6. $FPD(\mathbf{V})$: fundamental parallelepiped generated by $\mathbf{V}$. This is the set of all vectors of the form $\mathbf{Vx}, \mathbf{x} \in [0, 1)^D$. See Fig. 12.2-4. *Note:* $FPD(\mathbf{I})$ is same as $[0, 1)^D$.
7. $SPD(\mathbf{V})$: symmetric parallelepiped generated by $\mathbf{V}$. This is the set of all vectors of the form $\mathbf{Vx}, \mathbf{x} \in [-1, 1)^D$. See Fig. 12.5-1. *Note:* $SPD(\mathbf{I})$ is same as $[-1, 1)^D$.
8. $LAT(\mathbf{M})$, $FPD(\mathbf{M})$ and $SPD(\mathbf{M})$ defined similarly.
9. $J(\mathbf{M}) = |\det \mathbf{M}|$.
10. $\mathcal{N}(\mathbf{M})$: set of all integer vectors in $FPD(\mathbf{M})$. Has $J(\mathbf{M})$ elements.
11. $\mathcal{N}(\mathbf{M}^T)$: set of all integer vectors in $FPD(\mathbf{M}^T)$. Has $J(\mathbf{M})$ elements.
12. $\mathbf{z} = [z_0 \ldots z_{D-1}]^T$.
13. $\mathbf{z}^{(\mathbf{M})}$: $D \times 1$ vector with kth element $(\mathbf{z}^{(\mathbf{M})})_k = z_0^{M_{0,k}} z_1^{M_{1,k}} \ldots z_{D-1}^{M_{D-1,k}}$.
14. Let $\mathbf{n} = [n_0 \ldots n_{D-1}]^T$. Then $\mathbf{z}^{(\mathbf{n})} = \mathcal{Z}(\mathbf{n}) = z_0^{n_0} z_1^{n_1} \ldots z_{D-1}^{n_{D-1}}$.

Important properties.
1. The volume of $FPD(\mathbf{V})$ is equal to $|\det \mathbf{V}|$.
2. $J(\mathbf{M})$ (i.e., $|\det \mathbf{M}|$) = number of integers in $\mathcal{N}(\mathbf{M})$ (hence in $\mathcal{N}(\mathbf{M}^T)$).
3. Any real vector $\mathbf{u}$ can be expressed as $\mathbf{u} = \mathbf{u}_P + \mathbf{u}_L$, where $\mathbf{u}_P \in FPD(\mathbf{V})$ and $\mathbf{u}_L \in LAT(\mathbf{V})$. Here $\mathbf{u}_P$ and $\mathbf{u}_L$ are unique for given $\mathbf{V}$.
4. In particular, therefore, any integer vector $\mathbf{n}$ can be expressed as $\mathbf{n} = \mathbf{n}_P + \mathbf{n}_L$, where $\mathbf{n}_P \in \mathcal{N}(\mathbf{M})$ and $\mathbf{n}_L \in LAT(\mathbf{M})$ (and the integers $\mathbf{n}_P$ and $\mathbf{n}_L$ are unique for given $\mathbf{M}$). In other words, $\mathbf{n} = \mathbf{n}_P + \mathbf{Mn}_0$ (division theorom for integers; $\mathbf{n}_0$ is the quotient and $\mathbf{n}_P$ the remainder).
5. $LAT(\mathbf{V}_1) = LAT(\mathbf{V}_2)$ iff $\mathbf{V}_1 = \mathbf{V}_2 \mathbf{E}$ for unimodular integer $\mathbf{E}$.

TABLE 12.10.2 Fundamentals at a glance

Sampling a continuous 'time' signal.
1. Let $x(\mathbf{n}) = x_a(\mathbf{Vn})$. Then $X(\omega) = \sum_{\mathbf{k} \in \mathcal{N}} X_a(j\mathbf{V}^{-T}(\omega - 2\pi\mathbf{k}))/|\det \mathbf{V}|$.
2. Sampling density $\rho = 1/|\det \mathbf{V}|$. (Number of samples/volume.)
3. The sampling is said to be rectangular if $\mathbf{V}$ is diagonal.
4. Minimum sampling density required for alias-free sampling depends on choice of sampling geometry. If $X_a(j\Omega)$ has circular support centered at origin, then hexagonal sampling is more efficient than rectangular.

Decimation of discrete 'time' signal.
1. $y(\mathbf{n}) = x(\mathbf{Mn})$ = M-fold decimated version of $x(\mathbf{n})$. Decimation ratio = $|\det \mathbf{M}|$. (Fig. 12.4-1.)
2. $Y(\omega) = \sum_{\mathbf{k} \in \mathcal{N}(\mathbf{M}^T)} X(\mathbf{M}^{-T}(\omega - 2\pi\mathbf{k}))/|\det \mathbf{M}|$.
3. Decimation does not cause aliasing if $X(\omega)$ has support $SPD(\pi\mathbf{M}^{-T})$ or any frequency-shifted version of this. The support $SPD(\pi(\mathbf{ME})^{-T})$ also works for any unimodular integer matrix $\mathbf{E}$.

'Expander' for the discrete 'time' signal. $y(\mathbf{n})$ defined in (12.4.5) is the M-fold expanded version of $x(\mathbf{n})$. Here $Y(\omega) = X(\mathbf{M}^T\omega)$. (Equivalently $Y_z(\mathbf{z}) = X_z(\mathbf{z}^{(\mathbf{M})})$.) $Y(\omega)$ has $J(\mathbf{M}) - 1$ images, centered at lattice points of $2\pi\mathbf{M}^{-T}$ (scaled reciprocal lattice of $\mathbf{M}$).

Polyphase decomposition. The signal $x_\mathbf{k}(\mathbf{n}) \stackrel{\Delta}{=} x(\mathbf{Mn} + \mathbf{k})$ ($\mathbf{k} \in \mathcal{N}(\mathbf{M})$) is the kth Type 1 polyphase component of $x(\mathbf{n})$. For given $\mathbf{M}$, there are $J(\mathbf{M})$ such components. See Fig. 12.4-2. Similarly Type 2 polyphase components are $x'_\mathbf{k}(\mathbf{n}) = x(\mathbf{Mn} - \mathbf{k})$. In the transform domain $X_z(\mathbf{z}) =$

$$\sum_{\mathbf{k} \in \mathcal{N}(\mathbf{M})} \mathcal{Z}(-\mathbf{k}) X_{z,\mathbf{k}}(\mathbf{z}^{(\mathbf{M})}) \text{ (Type 1)} = \sum_{\mathbf{k} \in \mathcal{N}(\mathbf{M})} \mathcal{Z}(\mathbf{k}) X'_{z,\mathbf{k}}(\mathbf{z}^{(\mathbf{M})}) \text{ (Type 2)}$$

The notation $\mathbf{z}^{(\mathbf{k})}$ is also used instead of $\mathcal{Z}(\mathbf{k})$.

Generalized DFT matrix. $J(\mathbf{M}) \times J(\mathbf{M})$ matrix $\mathbf{W}^{(g)}$ with elements defined as $[\mathbf{W}^{(g)}]_{\mathbf{m},\mathbf{k}} = e^{-j2\pi\mathbf{m}^T\mathbf{M}^{-1}\mathbf{k}}$, with $\mathbf{m} \in \mathcal{N}(\mathbf{M}^T)$ and $\mathbf{k} \in \mathcal{N}(\mathbf{M})$. For fixed row index $\mathbf{m}$, the entries $e^{-j2\pi\mathbf{m}^T\mathbf{M}^{-1}\mathbf{k}}$ can be considered to be one period of the sequence $e^{-j2\pi\mathbf{m}^T\mathbf{M}^{-1}\mathbf{n}}$. (Single-frequency sequence with frequency $-2\pi\mathbf{M}^{-T}\mathbf{m}$; periodicity matrix $\mathbf{M}$.) $\mathbf{W}^{(g)}$ is unitary; $[\mathbf{W}^{(g)}]^\dagger \mathbf{W}^{(g)} = J(\mathbf{M})\mathbf{I}$. This is equivalent to the conditions (12.4.37a,b).

Noble identities. See Fig. 12.6-1(c), (d).

TABLE 12.10.3 Summary of properties of Hexagonal M

The hexagonal decimator matrix is $\mathbf{M} = \begin{bmatrix} 1 & 1 \\ 2 & -2 \end{bmatrix}$. The lattice generated by this is shown in Fig. 12.4-1(b). We have $J(\mathbf{M}) = 4$ so that the decimation ratio is four. Regions such as $FPD(\mathbf{M})$, $FPD(\mathbf{M}^T)$ and $SPD(\pi\mathbf{M}^{-T})$ are shown in Figures 12.4-10 and 12.5-2(a). The elements of $\mathcal{N}(\mathbf{M})$ are

$$\mathbf{k}_0 = \begin{bmatrix} 0 \\ 0 \end{bmatrix}, \ \mathbf{k}_1 = \begin{bmatrix} 1 \\ 0 \end{bmatrix}, \ \mathbf{k}_2 = \begin{bmatrix} 1 \\ 1 \end{bmatrix}, \ \mathbf{k}_3 = \begin{bmatrix} 1 \\ -1 \end{bmatrix}.$$

and those of $\mathcal{N}(\mathbf{M}^T)$ are

$$\mathbf{m}_0 = \begin{bmatrix} 0 \\ 0 \end{bmatrix}, \ \mathbf{m}_1 = \begin{bmatrix} 1 \\ 0 \end{bmatrix}, \ \mathbf{m}_2 = \begin{bmatrix} 1 \\ -1 \end{bmatrix}, \ \mathbf{m}_3 = \begin{bmatrix} 2 \\ -1 \end{bmatrix}.$$

(Warning: The letters $\mathbf{k}$ and $\mathbf{m}$ are not standard, and are sometimes interchanged.) The delay elements $\mathcal{Z}(-\mathbf{k})$ in the Type 1 polyphase decomposition (12.4.32) are

$$\mathcal{Z}(-\mathbf{k}_0) = 1, \ \mathcal{Z}(-\mathbf{k}_1) = z_0^{-1}, \ \mathcal{Z}(-\mathbf{k}_2) = z_0^{-1}z_1^{-1}, \ \mathcal{Z}(-\mathbf{k}_3) = z_0^{-1}z_1.$$

The vector $\mathbf{z}^{(\mathbf{M})}$ is given by

$$\mathbf{z}^{(\mathbf{M})} = \begin{bmatrix} z_0 z_1^2 \\ z_0 z_1^{-2} \end{bmatrix}.$$

Using these, the Type 1 polyphase decomposition can be written as

$$X_z(z_0, z_1) = X_{z,\mathbf{k}_0}(z_0 z_1^2, z_0 z_1^{-2}) + z_0^{-1} X_{z,\mathbf{k}_1}(z_0 z_1^2, z_0 z_1^{-2})$$
$$+ z_0^{-1} z_1^{-1} X_{z,\mathbf{k}_2}(z_0 z_1^2, z_0 z_1^{-2}) + z_0^{-1} z_1 X_{z,\mathbf{k}_3}(z_0 z_1^2, z_0 z_1^{-2}).$$

The generalized DFT matrix $\mathbf{W}^{(g)}$ is given by

$$\mathbf{W}^{(g)} = \begin{bmatrix} 1 & 1 & 1 & 1 \\ 1 & W^2 & W^3 & W \\ 1 & W^4 & W^6 & W^2 \\ 1 & W^6 & W^9 & W^3 \end{bmatrix}, \quad W = e^{-j2\pi/4}.$$

The noble identities. See Fig. 12.6-5.
The AC matrix. See Example 12.9.1.

TABLE 12.10.4 Summary of properties of quincunx M

The quincunx decimator matrix is $\mathbf{M} = \begin{bmatrix} 1 & 1 \\ -1 & 1 \end{bmatrix}$. The lattice generated by this is shown in Fig. 12.4-1(c). We have $J(\mathbf{M}) = 2$ so that the decimation ratio is two. Regions such as $FPD(\mathbf{M})$, $FPD(\mathbf{M}^T)$ and $SPD(\pi \mathbf{M}^{-T})$ are shown in Figures 12.4-10 and 12.5-2(b). The elements of $\mathcal{N}(\mathbf{M})$ are

$$\mathbf{k}_0 = \begin{bmatrix} 0 \\ 0 \end{bmatrix}, \ \mathbf{k}_1 = \begin{bmatrix} 1 \\ 0 \end{bmatrix},$$

and those of $\mathcal{N}(\mathbf{M}^T)$ are

$$\mathbf{m}_0 = \begin{bmatrix} 0 \\ 0 \end{bmatrix}, \ \mathbf{m}_1 = \begin{bmatrix} 0 \\ 1 \end{bmatrix}.$$

(Warning: The letters $\mathbf{k}$ and $\mathbf{m}$ are not standard, and are sometimes interchanged.) The delay elements $\mathcal{Z}(-\mathbf{k})$ in the Type 1 polyphase decomposition (12.4.32) are

$$\mathcal{Z}(-\mathbf{k}_0) = 1, \ \mathcal{Z}(-\mathbf{k}_1) = z_0^{-1}.$$

The vector $\mathbf{z}^{(\mathbf{M})}$ is given by

$$\mathbf{z}^{(\mathbf{M})} = \begin{bmatrix} z_0 z_1^{-1} \\ z_0 z_1 \end{bmatrix}.$$

Using these, the Type 1 polyphase decomposition can be written as

$$X_z(z_0, z_1) = X_{z,\mathbf{k}_0}(z_0 z_1^{-1}, z_0 z_1) + z_0^{-1} X_{z,\mathbf{k}_1}(z_0 z_1^{-1}, z_0 z_1).$$

The generalized DFT matrix $\mathbf{W}^{(g)}$ is given by

$$\mathbf{W}^{(g)} = \begin{bmatrix} 1 & 1 \\ 1 & -1 \end{bmatrix}.$$

The noble identities. See Fig. 12.9-8.

PROBLEMS

12.1. Let $X_z(\mathbf{z})$ and $H_z(\mathbf{z})$ denote the z-transforms of the D-dimensional sequences $x(\mathbf{n})$ and $h(\mathbf{n})$.

a) Show that the z-transform of $x(\mathbf{n} - \mathbf{k})$ is $\mathcal{Z}(-\mathbf{k})X_z(\mathbf{z})$.

b) With $\mathbf{a} = [a_0 \ \ldots \ a_{D-1}]^T$, show that the z-transform of the sequence $y(\mathbf{n}) = x(\mathbf{n})e^{\mathbf{a}^T \mathbf{n}}$ is given by $X_z(\mathbf{\Lambda_a z})$, where $\mathbf{\Lambda_a}$ is as in (12.1.16).

c) In part (b) consider the 2D case with $\mathbf{a} = \begin{bmatrix} 2 \\ 3 \end{bmatrix}$. Express $Y_z(z_0, z_1)$ in terms of $X_z(z_0, z_1)$.

d) Establish that the z-transform of the convolution sum (12.1.17b) is $Y_z(\mathbf{z}) = H_z(\mathbf{z})X_z(\mathbf{z})$.

12.2. Shown below are several 2D filter responses. (Only the region $[-\pi, \pi)^2$ shown). Assume that $H(\omega) = 0$ in the white (unshaded) areas. The values of $H(\omega)$ in the shaded areas are indicated ajdacent to them. Which of these responses is separable?

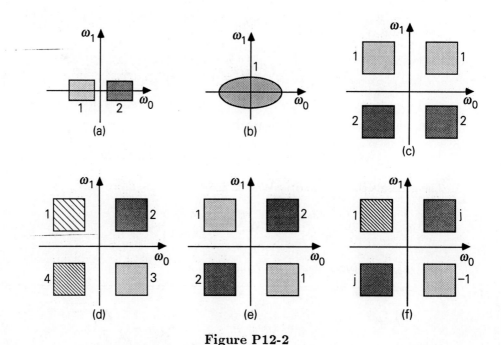

Figure P12-2

12.3. Let $h(\mathbf{n})$ be a D-dimensional digital filter. If the frequency response $H(\omega)$ is real for all ω, the filter is said to have zero-phase. Show that the zero-phase property holds if and only if $h(\mathbf{n}) = h^*(-\mathbf{n})$.

12.4. Consider the following two matrices.

$$\mathbf{M}_1 = \begin{bmatrix} 2 & -1 \\ -1 & 3 \end{bmatrix}, \quad \mathbf{M}_2 = \begin{bmatrix} 4 & -2 \\ -2 & 2 \end{bmatrix}. \quad (P12.4)$$

For each of these:
a) Sketch the lattice $LAT(\mathbf{M}_i)$.
b) Clearly indicate the fundamental parallelepiped $FPD(\mathbf{M}_i)$, and highlight the integer points which belong to $FPD(\mathbf{M}_i)$.
c) Verify that $FPD(\mathbf{M}_i)$ contains exactly $|\det \mathbf{M}_i|$ integer points.
d) Sketch the reciprocal lattice.

12.5. (a) Sketch the lattice generated by the matrix $\widehat{\mathbf{V}}$ in (12.2.12), and verify that this is the same as that generated by $\mathbf{V}$ in Example 12.2.1. (b) Sketch the lattice generated by $\widehat{\mathbf{V}}$ in (12.2.13) and verify that it is the set of *all* integers.

12.6. Consider the linear transformation

$$\begin{bmatrix} q_0 \\ q_1 \end{bmatrix} = \begin{bmatrix} v_{00} & v_{01} \\ v_{10} & v_{11} \end{bmatrix} \begin{bmatrix} p_0 \\ p_1 \end{bmatrix}, \quad (P12.6)$$

where the 2×2 matrix is real nonsingular. This maps a point from the coordinate system (p_0, p_1) to the system (q_0, q_1) (see figure below).

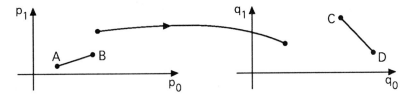

Figure P12-6

Show that a straight line AB gets mapped into a straight line CD. Hence argue that a polygon gets mapped into a polygon with the same number of sides. (*Hint.* A vertex in (p_0, p_1) plane is mapped into a unique vertex in (q_0, q_1) plane.)

12.7. Consider the function $Y(\omega) = \sum_{\mathbf{k}} X(\mathbf{M}^{-T}(\omega - 2\pi \mathbf{k}))$ where $\mathbf{M}$ is a nonsingular integer matrix, and $\mathbf{k}$ ranges over all integers in $\mathcal{N}(\mathbf{M}^T)$. Show that $Y(\omega)$ is periodic in ω with periodicity matrix $2\pi \mathbf{I}$.

12.8. Consider a bandlimited $x(\mathbf{n})$ with support for $X(\omega)$ shown below.

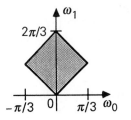

Figure P12-8

Suppose we decimate $x(\mathbf{n})$ by the matrix $\mathbf{M} = \begin{bmatrix} 1 & -1 \\ 1 & 2 \end{bmatrix}$. Show the support of the stretched version $X(\mathbf{M}^{-T}\omega)$ and all the shifted versions which enter the formula (12.4.22). Is there aliasing?

12.9. Consider the equality

$$X(\omega) = \sum_{\mathbf{m} \in \mathcal{N}(\mathbf{M}^T)} f(\mathbf{m}) X(\omega - 2\pi \mathbf{M}^{-T}\mathbf{m}), \qquad (P12.9)$$

where $\mathbf{M}$ is a nonsingular integer matrix. Suppose $f(\mathbf{m})$ is such that this holds for all choice of $X(\omega)$. Show then that $f(\mathbf{m})$ is zero for all nonzero $\mathbf{m} \in \mathcal{N}(\mathbf{M}^T)$, and that $f(\mathbf{0}) = 1$.

12.10. Consider the following multirate system, where $\mathbf{M}$ is $D \times D$ integer nonsingular matrix, and $J(\mathbf{M}) = |\det \mathbf{M}|$.

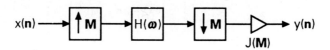

Figure P12-10

a) Show that this is a (linear and) shift invariant system. What is the transfer function?
b) Suppose $H(\omega)$ has the response given in (12.7.1). Show then that $y(\mathbf{n}) = x(\mathbf{n})$.
c) In part (b), suppose we replace $H(\omega)$ with $H(\omega - \mathbf{v})$ where $\mathbf{v}$ is some real vector. Show that the relation $y(\mathbf{n}) = x(\mathbf{n})$ continues to hold.

12.11. Consider the decimation filter circuit of Fig. 12.6-4(a) for $\mathbf{M}$-fold decimation.
a) Define $\mathbf{M}_1 = \mathbf{ME}$ where $\mathbf{E}$ is an arbitrary unimodular integer matrix. Suppose the support of $H(\omega)$ is equal to $SPD(\pi\mathbf{M}_1^{-T})$ (and of course repeats with periodicity matrix $2\pi\mathbf{I}$). Show then that decimation does not result in aliasing [i.e., there is no overlap between the stretched version and shifted versions in (12.4.22)].
b) Instead of using the decimation filter in part (a), suppose we use the filter $H(\omega - \mathbf{v})$ where $\mathbf{v}$ is an arbitrary vector. Show that decimation still does not result in aliasing.

12.12. This is by way of completing Example 12.7.3, which the reader should first review. Recall that the support of $H_2(\mathbf{M}_1^T\omega)$ is as in (12.7.17a) and is basically $SPD(\pi\mathbf{M}^{-T})$ repeated with periodicity matrix $2\pi\mathbf{M}_1^{-T}$ rather than $2\pi\mathbf{I}$. So the support has additional images of $SPD(\pi\mathbf{M}^{-T})$ represented by the nonzero values of $\mathbf{m}_0$ in (12.7.17b).
a) Consider the regions

$$\pi\mathbf{M}^{-T}\mathbf{x} + 2\pi\mathbf{M}_1^{-T}\mathbf{m}_0 + 2\pi\mathbf{m}_1, \qquad (P12.12a)$$

and

$$\pi\mathbf{M}^{-T}\mathbf{x} + 2\pi\mathbf{M}_1^{-T}\mathbf{m}_0' + 2\pi\mathbf{m}_1, \qquad (P12.12b)$$

where $\mathbf{m}_0$ and $\mathbf{m}'_0$ are fixed unequal integers in $\mathcal{N}(\mathbf{M}_1^T)$, $\mathbf{x}$ varies over $[-1,1)^D$ and $\mathbf{m}_1$ varies over $\mathcal{N}$. Show that these two regions are disjoint.

b) Show that with $H_1(\omega)$ chosen to have passband region $SPD(\pi \mathbf{M}_1^{-T})$, the support of $H_1(\omega)H_2(\mathbf{M}_1^T \omega)$ has passband region $SPD(\pi \mathbf{M}^{-T})$ (i.e., the $\mathbf{m}_0$-term is removed in (12.7.17b)). Thus $H_1(\omega)$ eliminates the images of passband (which were generated by letting $\mathbf{m}_0$ be nonzero).

12.13. For the following matrices, compute the generalized DFT matrix $\mathbf{W}^{(g)}$ defined in Section 12.4.4.

$$\mathbf{M}_1 = \begin{bmatrix} 2 & 0 \\ 0 & 2 \end{bmatrix}, \quad \mathbf{M}_2 = \begin{bmatrix} 2 & 2 \\ 2 & 4 \end{bmatrix}, \quad \mathbf{M}_3 = \begin{bmatrix} 1 & 0 \\ 5 & 4 \end{bmatrix}. \tag{P12.13}$$

Which of these DFT matrices are obtainable by *row and column permutations* of the traditional 1D DFT matrix of corresponding size?

12.14. Consider the matrix

$$\mathbf{M} = \begin{bmatrix} 1 & -1 \\ 1 & 3 \end{bmatrix}. \tag{P12.14}$$

a) Sketch the lattice generated by $\mathbf{M}$. What is the decimation ratio $J(\mathbf{M})$ for $\mathbf{M}$-fold decimation?
b) Give sketches of $FPD(\mathbf{M})$, $FPD(\mathbf{M}^T)$, and $SPD(\mathbf{M}^T)$.
c) Identify the elements in the sets $\mathcal{N}(\mathbf{M})$ and $\mathcal{N}(\mathbf{M}^T)$.
d) Identify the elements $\mathcal{Z}(-\mathbf{k})$ which would appear in the Type 1 polyphase decomposition. Hence write down this decomposition (similar to what we wrote in Examples 12.4.4, 12.4.5).
e) Identify all the elements of the generalized DFT matrix $\mathbf{W}^{(g)}$.

12.15. Let $\mathbf{M}$ be as in Problem 12.14.
a) Give a sketch of $SPD(\pi \mathbf{M}^{-T})$.
b) Give three possible sketches for the support of a (ideal) decimation filter for $\mathbf{M}$-fold decimation, which would prevent aliasing. (Make sure not *all* of them are shifted versions of the same thing. And avoid the example $H(\omega) \equiv 0$, of course!)

12.16. Let $\mathbf{M}$ be as in Problem 12.14.
a) Express the elements of the vector $\mathbf{z}^{(\mathbf{M})}$ in terms of z_0, z_1 and other known quantities.
b) Since this is a 2D system, there are two noble identities for decimators and two for expanders (analogous to those in Fig. 12.6-5(c),(d)). Draw these explicitly.
c) Give a schematic of the polyphase implementation of a decimation filter (analogous to Fig. 12.6-5(f)).

12.17. Consider the analysis bank shown in Fig. 12.8-2. Show that the filter $H_0(\omega)$ (i.e. $H_{\mathbf{m}_0}(\omega)$) is given by (12.8.9), and that the remaining filters are related to $H_0(\omega)$ as in (12.8.7).

12.18. For the synthesis bank shown in Fig. 12.8-3 show that the filter $F_0(\omega)$ is given by (12.8.10) and that the remaining filters are given by $F_\mathbf{m}(\omega) = F_0(\omega - 2\pi \mathbf{M}^{-T}\mathbf{m})$.

12.19. In this problem we consider the 2D quincunx QMF bank (Section 12.9.4). Assume all filters are FIR with real coefficients.

  a) Redraw the QMF bank in the polyphase form (Fig. 12.9-3(b)), explicitly indicating the elements $\mathcal{Z}(-\mathbf{m}_i)$ and $\mathcal{Z}(\mathbf{m}_i)$ in terms of z_0 and z_1.

  b) Suppose $\mathbf{E}_z(\mathbf{z})$ (i.e., $\mathbf{E}_z(z_0, z_1)$, which is a more convenient notation in this case) has the form

  $$\mathbf{E}_z(z_0, z_1) = \begin{bmatrix} E_{z,0}(z_0, z_1) & E_{z,1}(z_0^{-1}, z_1^{-1}) \\ E_{z,1}(z_0, z_1) & -E_{z,0}(z_0^{-1}, z_1^{-1}) \end{bmatrix}. \qquad (P12.19a)$$

  Express the analysis filters $H_{z,0}(z_0, z_1)$ and $H_{z,1}(z_0, z_1)$ in terms of the components $E_{z,0}(z_0, z_1)$ and $E_{z,1}(z_0, z_1)$, and show that

  $$H_{z,1}(z_0, z_1) = -z_0^{-1} H_{z,0}(-z_0^{-1}, -z_1^{-1}). \qquad (P12.19b)$$

  c) In part (b) if the analysis filter $H_{z,0}(z_0, z_1)$ has the diamond shaped passband as in Fig. 12.7-2(d), what is the passband region of $H_{z,1}(z_0, z_1)$?

  d) Suppose the matrix in (P12.19a) is such that the first column is power complementary, that is,

  $$E_{z,0}(z_0^{-1}, z_1^{-1}) E_{z,0}(z_0, z_1) + E_{z,1}(z_0^{-1}, z_1^{-1}) E_{z,1}(z_0, z_1) = c, \qquad (P12.19c)$$

  for some constant $c > 0$. Show then that $\mathbf{E}_z(z_0, z_1)$ is paraunitary, that is, $\mathbf{E}_z^T(z_0^{-1}, z_1^{-1}) \mathbf{E}_z(z_0, z_1) = c\mathbf{I}$.

  e) With $\mathbf{E}_z(z_0, z_1)$ assumed to be paraunitary, show that the analysis filters form a power complementary pair.

  f) Assume $\mathbf{E}_z(z_0, z_1)$ is paraunitary. With $\mathbf{R}_z(z_0, z_1)$ chosen as

  $$\mathbf{R}_z(z_0, z_1) = \mathbf{E}_z^T(z_0^{-1}, z_1^{-1}), \qquad (P12.19d)$$

  we now have a perfect reconstruction system. Express the synthesis filters $F_{z,0}(z_0, z_1)$ and $F_{z,1}(z_0, z_1)$ in terms of $E_{z,0}(z_0, z_1)$ and $E_{z,1}(z_0, z_1)$.

  g) With $\mathbf{R}_z(z_0, z_1)$ chosen as above, show that

  $$F_{z,0}(z_0, z_1) = H_{z,0}(z_0^{-1}, z_1^{-1}), \quad F_{z,1}(z_0, z_1) = H_{z,1}(z_0^{-1}, z_1^{-1}). \qquad (P12.19e)$$

  h) Re-express the relations (P12.19b) and (P12.19e) in terms of the impulse response coefficients $h_0(n_0, n_1), h_1(n_0, n_1), f_0(n_0, n_1)$, and $f_1(n_0, n_1)$.

12.20. Consider a QMF bank with $\mathbf{M} = \begin{bmatrix} 2 & 0 \\ 0 & 2 \end{bmatrix}$. Evidently there are four channels since $[\det \mathbf{M}] = 4$. Suppose the analysis filters have the form

$$\begin{bmatrix} H_{z,0}(z_0, z_1) \\ H_{z,1}(z_0, z_1) \\ H_{z,2}(z_0, z_1) \\ H_{z,3}(z_0, z_1) \end{bmatrix} = \mathbf{T} \begin{bmatrix} 1 \\ z_0^{-1} \\ z_1^{-1} \\ z_0^{-1} z_1^{-1} \end{bmatrix}, \qquad (P12.20)$$

where **T** is nonsingular.
- a) Suppose the 0th row of **T** is $[1 \ 1 \ 1 \ 2]$. Construct the remaining three rows (obviously not unique) so that **T** comes out to be nonsingular. With this **T**, find a set of FIR synthesis filters for perfect reconstruction.
- b) With **T** chosen as in (a), show that $H_{z,0}(z_0, z_1)$ is nonseparable [i.e., cannot be written as $H_0(z_0)H_1(z_1)$].

12.21. In Section 12.9 we formulated the multidimensional QMF problem in terms of the AC matrix $\mathbf{H}_z(\mathbf{z})$, as well as in terms of the polyphase matrix $\mathbf{E}_z(\mathbf{z})$. In the 1D case, we know that these matrices are related as in (5.5.8).
- a) In the multidimensional case, show that the relation is

$$\mathbf{H}_z(\mathbf{z}) = [\mathbf{W}^{(g)}]^* \mathbf{\Lambda}_z(\mathbf{z}) \mathbf{E}_z^T(\mathbf{z}^{(\mathbf{M})}), \qquad (P12.21a)$$

where $\mathbf{W}^{(g)}$ is the generalized DFT matrix for the chosen **M**, and

$$\mathbf{\Lambda}_z(\mathbf{z}) = \text{diag} \ [\mathcal{Z}(-\mathbf{k}_0) \ \ \mathcal{Z}(-\mathbf{k}_1) \ \ \ldots \ \ \mathcal{Z}(-\mathbf{k}_s)], \ \ s = J(\mathbf{M}) - 1.$$
$$(P12.21b)$$

Here $\mathbf{k}_i$ are the distinct integers in the set $\mathcal{N}(\mathbf{M})$, appropriately numbered. (It might be notationally easier to obtain the relation in terms of the frequency vector ω first. In any case, also give the the relation using ω rather than **z**.)
- b) With **M** taken as the hexagonal decimator, explicitly fill the details of the above relation.

12.22. Consider the tree structure of Fig. 12.9-11, with $\mathbf{M}_0$ and $\mathbf{M}_1$ as in (12.9.20). This is equivalent to a four channel two-dimensional QMF bank (Fig. 12.9-1).
- a) Express the four analysis filters and four synthesis filters in terms of the one dimensional filters in the figure. Are these two dimensional filters separable?
- b) What is the decimation matrix **M**?

12.23. Consider again the tree structure of Fig. 12.9-11, with $\mathbf{M}_0$ and $\mathbf{M}_1$ as in (12.9.20). This is designed starting from two one-dimensional QMF banks, viz.,

$$\{H_0(z), H_1(z), F_0(z), F_1(z)\} \quad \text{and} \quad \{G_0(z), G_1(z), P_0(z), P_1(z)\}.$$

- a) Suppose the 1D QMF banks are alias-free. Show that the 2D QMF bank is alias-free.
- b) Repeat part (a) with "alias-free" replaced by "alias-free, and free from amplitude distortion."
- c) Repeat part (a) with "alias-free" replaced by "alias-free, and free from phase distortion."
- d) Repeat part (a) with "alias-free" replaced by "perfect reconstruction."

12.24. Write down the entries of the AC matrix for maximally decimated QMF banks with the following decimation matrices.
- a) Decimator as in Fig. 12.4-1(d).
- b) The quincunx decimator.

12.25. Let $T_z(\mathbf{z})$ be a linear phase function, and define $S_z(\mathbf{z}) = T_z(\mathbf{z}^{(\mathbf{M})})$. Show that $S_z(\mathbf{z})$ has linear phase as well.

12.26. Consider the quincunx QMF bank of Fig. 12.9-9(a) again. Let the filter $H_{z,0}(z_0, z_1)$ be designed as in (12.7.11) starting from the 1D prototype $G(z) = E_0(z^2) + z^{-1}E_1(z^2)$. Assuming that the remaining filters are chosen as in (12.9.17) and (12.9.19), we know that the system is alias-free, and can be redrawn as in Fig. 12.9-9(e). The QMF bank now has distortion function $T_z(z_0, z_1)$ given in (12.9.18).

We shall now compare this with the 1D QMF bank with filters

$$H_0(z) = G(z), H_1(z) = H_0(-z), F_0(z) = H_0(z), F_1(z) = -H_1(z). \quad (P12.26)$$

From Chapter 5 we know this system is alias-free with distortion function $T(z) = 2z^{-1}E_0(z^2)E_1(z^2)$.

a) Suppose the 1D prototype $G(z)$ is of the form $\sum_{n=0}^{N} g(n)z^{-n}$ where N is odd, and $g(n)$ is real with $g(n) = g(N-n)$. From Sec. 5.2.2 we know that $T(z)$ has linear phase. Prove that for the 2D QMF bank $T_z(z_0, z_1)$ has linear phase as well.

b) Instead, suppose the 1D prototype $G(z)$ is such that $T(z)$ is allpass. Show then that $T_z(z_0, z_1)$ is allpass as well.

c) Finally suppose $G(z)$ is such that $T(z)$ is a delay z^{-K}. So the 1D system has perfect reconstruction. Show that the 2D QMF bank also has perfect reconstruction.

12.27. Recall that any QMF bank (Fig. 12.9-1) can be drawn in polyphase form as in Fig. 12.9-3(a). Suppose the matrix $\mathbf{E}_z(\mathbf{z})$ is FIR and paraunitary, that is, $\widetilde{\mathbf{E}}_z(\mathbf{z})\mathbf{E}_z(\mathbf{z}) = c\mathbf{I}, c > 0$. (See Example 12.9.2 for definition of the tilde notation.)

a) Show that the analysis filters forms a power complementary set, that is, $\sum_{\mathbf{k}} |H_{\mathbf{k}}(\omega)|^2 = $ constant.

b) Show that the AC matrix $\mathbf{H}_z(\mathbf{z})$ is paraunitary (*Hint.* Use Problem 12.21).

c) Suppose the matrix $\mathbf{R}_z(\mathbf{z})$ is chosen as $\widetilde{\mathbf{E}}_z(\mathbf{z})$. Express $\widehat{x}(\mathbf{n})$ in terms of $x(\mathbf{n})$ and show that we have perfect reconstruction.

d) With $\mathbf{R}_z(\mathbf{z}) = \widetilde{\mathbf{E}}_z(\mathbf{z})$, show that the synthesis filters are given by the relation $F_{z,\mathbf{k}}(\mathbf{z}) = \widetilde{H}_{z,\mathbf{k}}(\mathbf{z})$.

12.28. Consider the multirate system shown below, where $\mathbf{M}$ is some nonsingular integer matrix, $S = J(\mathbf{M}) - 1$, and $\mathbf{m}_i \in \mathcal{N}(\mathbf{M})$.

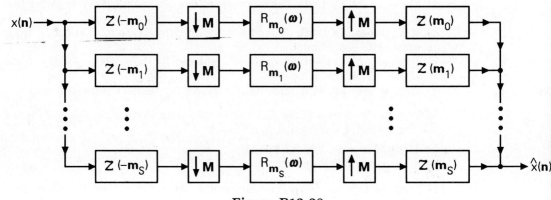

Figure P12-28

This is essentially the system of Fig. 12.9-5(a) with the additional transfer functions $R_{\mathbf{m}_i}(\omega)$ inserted. Find a set of necessary and sufficient conditions on these transfer functions such that the system is alias-free. Under this condition, what is the transfer function $\widehat{X}(\mathbf{z})/X(\mathbf{z})$?

12.29. In this chapter we obtained the frequency domain formula (12.4.22) for a decimator based on the view that $x(\mathbf{n})$ is the sampled version of some underlying 'analog' signal $x_a(\mathbf{t})$. We then used the formula (12.4.22) in conjunction with our knowledge of polyphase decomposition, and obtained the generalized DFT and the orthogonality relation (12.4.37a).

We shall now take a reverse approach, which may be more appealing to some readers. In this Problem we shall derive the orthogonality relation (12.4.37a) independently, based on first principles, and then use it in the next Problem to get the decimation formula (12.4.22). This is closer to the approach taken in Chapter 4 for the one-dimensional case.

a) Let $x(\mathbf{n}) = e^{j\boldsymbol{\omega}_0^T \mathbf{n}}$, where $\boldsymbol{\omega}_0$ is some real $D \times 1$ vector. This is called a single-frequency sequence, and the Fourier transform is zero except at a 'single frequency' $\boldsymbol{\omega}_0$ (and repeats periodically). More specifically, show that the transform is

$$X(\boldsymbol{\omega}) = (2\pi)^D \sum_{\mathbf{i} \in \mathcal{N}} \delta(\boldsymbol{\omega} - \boldsymbol{\omega}_0 + 2\pi\mathbf{i}), \qquad (P12.29a)$$

by verifying that it satisfies (12.1.10).

b) Let $x_0(\mathbf{n}) = e^{j\boldsymbol{\omega}_0^T \mathbf{n}}$, and $x_1(\mathbf{n}) = e^{j\boldsymbol{\omega}_1^T \mathbf{n}}$. Assume $\boldsymbol{\omega}_0$ and $\boldsymbol{\omega}_1$ are two distinct frequencies, i.e., $\boldsymbol{\omega}_0 - \boldsymbol{\omega}_1 \neq 2\pi\mathbf{i}$ for integer $\mathbf{i}$. This means the Fourier transforms of the sequences $x_0(\mathbf{n})$ and $x_1(\mathbf{n})$ do not overlap, that is, $X_0(\boldsymbol{\omega})X_1(\boldsymbol{\omega}) = 0$ for all $\boldsymbol{\omega}$. Using this, prove that

$$\sum_{\mathbf{n} \in \mathcal{N}} x_0(\mathbf{n}) x_1(-\mathbf{n}) = 0. \qquad (P12.29b)$$

(*Hint.* Convolution in "time" is multiplication in frequency.)

c) Now consider the special case of (b) where the frequency vectors are $\boldsymbol{\omega}_0 = 2\pi \mathbf{M}^{-T}\mathbf{m}_0$ and $\boldsymbol{\omega}_1 = 2\pi \mathbf{M}^{-T}\mathbf{m}_1$ for integer $\mathbf{m}_0, \mathbf{m}_1$. Here $\mathbf{M}$ is a nonsingular integer matrix. Now the sequences are given by

$$x_0(\mathbf{n}) = e^{j 2\pi \mathbf{m}_0^T \mathbf{M}^{-1} \mathbf{n}}, \quad x_1(\mathbf{n}) = e^{j 2\pi \mathbf{m}_1^T \mathbf{M}^{-1} \mathbf{n}}. \qquad (P12.29c)$$

Verify that these are periodic with periodicity matrix $\mathbf{M}$.

d) Let $\boldsymbol{\omega}_0$ and $\boldsymbol{\omega}_1$ be distinct in (c). Using (P12.29b) verify that

$$\sum_{\mathbf{k} \in \mathcal{N}(\mathbf{M})} e^{j 2\pi (\mathbf{m}_0 - \mathbf{m}_1)^T \mathbf{M}^{-1} \mathbf{k}} = 0. \qquad (P12.29d)$$

(*Hint.* Decompose appropriate integer-vectors using appropriate division theorems.)

e) Hence prove the relation (12.4.37a).

12.30. We shall now derive the relation (12.4.22) for a decimator, using the orthogonality condition (12.4.37a). The latter can be derived independently as in the previous problem. The M-fold decimated sequence $y(\mathbf{n}) = x(\mathbf{Mn})$ can be considered to be generated by a two stage process: first define

$$v(\mathbf{n}) = \begin{cases} x(\mathbf{n}) & \text{if } \mathbf{n} \in LAT(\mathbf{M}), \\ 0 & \text{otherwise,} \end{cases} \qquad (P12.30a)$$

and then define
$$y(\mathbf{n}) = v(\mathbf{Mn}). \qquad (P12.30b)$$

a) Show that $v(\mathbf{n})$ can be expressed as

$$v(\mathbf{n}) = \frac{1}{J(\mathbf{M})} \sum_{\mathbf{m} \in \mathcal{N}(\mathbf{M}^T)} x(\mathbf{n}) e^{j2\pi \mathbf{m}^T \mathbf{M}^{-1}\mathbf{n}}, \qquad (P12.30c)$$

where $J(\mathbf{M}) = |\det \mathbf{M}|$. Hence derive a relation between $V(\omega)$ and $X(\omega)$.
b) Show that $Y(\omega) = V(\mathbf{M}^{-T}\omega)$.
c) By combining the results of parts (a) and (b), obtain the relation (12.4.22).

Note. In this problem be careful not to employ any results which have been derived *based* on (12.4.22), as it would imply a 'circular' proof!

12.31. At the end of Section 12.4.5 we introduced sublattices and cosets. Let $\mathbf{V}_1$ and $\mathbf{V}_2$ be $D \times D$ real nonsingular matrices and let $LAT(\mathbf{V}_2)$ be a sublattice of $LAT(\mathbf{V}_1)$.

a) Prove that $\mathbf{L} \triangleq \mathbf{V}_1^{-1}\mathbf{V}_2$ is an integer matrix. (*Hint.* Given any $D \times 1$ integer $\mathbf{n}$, we can find $D \times 1$ integer $\mathbf{m}$ such that $\mathbf{V}_2\mathbf{n} = \mathbf{V}_1\mathbf{m}$.)
b) Let $\mathcal{C}_a$ and $\mathcal{C}_b$ be cosets of $LAT(\mathbf{V}_2)$ in $LAT(\mathbf{V}_1)$. Prove that these are either identical or disjoint.
c) Let $\rho = |\det \mathbf{L}|$. (Evidently this is an integer.) Prove that there are precisely ρ distinct cosets of $LAT(\mathbf{V}_2)$ in $LAT(\mathbf{V}_1)$, and that the union of these cosets is $LAT(\mathbf{V}_1)$.

12.32. Recall how we obtained the 2D filter $H_z(z_0, z_1)$ of Fig. 12.7-2(d) starting from the 1D lowpass response $G(e^{j\omega})$. Now suppose that $G(e^{j\omega})$ is highpass as shown below.

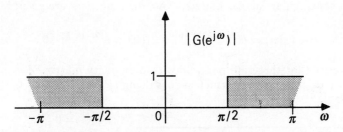

Figure P12-32

Sketch the support of $H(\omega_0, \omega_1)$.

12.33. In (12.4.22) suppose we replace the summation index $\mathbf{k} \in \mathcal{N}(\mathbf{M}^T)$ with $\mathbf{k} \in \mathcal{N}(\mathbf{M}^T\mathbf{E})$ for some unimodular integer matrix $\mathbf{E}$. Show that the summation remains unchanged.

12.34. Let $\mathbf{M}_1$ and $\mathbf{M}_2$ be nonsingular integer matrices and let $\mathbf{M} = \mathbf{M}_1\mathbf{M}_2$. Suppose the region $SPD(\mathbf{M}_2^{-T})$ is contained within $[-1,1)^D$. Show then that $SPD(\pi\mathbf{M}^{-T})$ will be contained within $SPD(\pi\mathbf{M}_1^{-T})$.

12.35. Consider two zero-phase 2D digital filters $H_z(z_0, z_1)$ and $G_z(z_0, z_1)$ with frequency response supports as shown below.

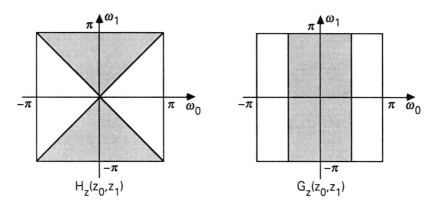

Figure P12-35

The response is equal to unity in the shaded region and zero everywhere else in $[-\pi, \pi)^2$. What are the frequency responses of the filters (i) $H_z(z_0^2, z_1)G_z(z_0, z_1)$ and (ii) $H_z(z_0, z_1) - H_z(z_0^2, z_1)G_z(z_0, z_1)$. *Note.* Using such combinations and clever extensions of these, one can design so-called "directional filter banks." See Bamberger and Smith [1990 and 1992] and Fettweis, et al. [1990] for further details.

12.36. Let $G_z(z_0, z_1)$ be a separable filter. Show that $1 + G_z(z_0, z_1)$ is necessarily nonseparable (unless $G_z(z_0, z_1)$ has the degenerate form $G_0(z_0)$ or $G_1(z_1)$).

PART 4 Multivariable and Lossless Systems

13

Review of Discrete-Time Multi-Input Multi-Output LTI Systems

13.0 INTRODUCTION

In this chapter we review several important concepts from the theory of multi-input multi-output linear time invariant systems (abbreviated MIMO LTI systems). MIMO systems are sometimes also called *multivariable* systems, and should not be confused with multidimensional systems discussed in Chap. 12. There are excellent texts on this topic, for example, Kailath [1980] and Chen [1970 and 1984]. Other references for this chapter include Desoer and Schulman [1974], Gantmacher [1959], Rosenbrock [1970], and Vidyasagar [1985]. While many of these references emphasize continuous-time systems, our presentation here is entirely for the discrete-time case. Some of the results, for example, those in Sec. 13.9 and 13.10 have not appeared in text books before.

Because of the "reference" nature of this chapter, the writing style is unlike any other chapter in this text. We have chosen to present various facts and results in the form of lemmas and theorems. This has enabled us to review a large number of advanced results in an economic manner, while at the same time ensuring completeness. In this chapter we will frequently use matrix theory results, particularly concepts such as rank, nonsingularity, orthogonal complements, and so on (summarized in Appendix A).

13.1 MULTI-INPUT MULTI-OUTPUT SYSTEMS

In Sec. 2.2 we introduced $p \times r$ LTI systems, that is, LTI systems with r inputs and p outputs. The kth output in response to all the inputs is given by (2.2.1), where $H_{km}(z)$ is the transfer function from the input $u_m(n)$ to the output $y_k(n)$. The MIMO system is characterized by the $p \times r$ transfer matrix $\mathbf{H}(z)$ or, equivalently, by the impulse response matrix $\mathbf{h}(n)$. In Fig. 2.2-2 we indicated two ways of representing the system. The output sequence $\mathbf{y}(n)$ is related to the input $\mathbf{u}(n)$ in terms of the matrix convolution (2.2.7) (equivalently (2.2.4) in the z-domain). We request the reader to review Sec. 2.2, because the same notations will be used here.

A system with $p = r = 1$ is said to be a single input single output (SISO) system, or a *scalar* system. Many of the concepts which follow easily in the SISO case turn out to be complicated for MIMO systems because matrices are involved. Development of basic ideas such as irreducible rational functions, minimal realizations, and transmission zeros require more careful thought in the MIMO case. In this chapter we will review many of these concepts.

Outline

Section 13.2 introduces matrix polynomials and their properties. In Sec. 13.3 we study the matrix fraction description which is useful for describing MIMO transfer functions. Section 13.4 gives a detailed exposure to state space description of MIMO LTI systems. The Smith-McMillan form is introduced in Sec. 13.5. Sections 13.6 and 13.7 deal with poles and zeros of MIMO LTI systems. The degree or McMillan degree of a system is studied in Sec. 13.8. Finally Sec. 13.9 and 13.10 present some results for the case of FIR MIMO systems, which are useful in filter bank research.

Polynomials and integers. While many of the results of this chapter are developed for polynomials and polynomial matrices, they hold equally well for integers and integer matrices. This is because, the set of polynomials as well as the set of integers belong to a common algebraic structure called the *principal ideal domain* [Forney, 1970], [Vidyasagar, 1985]. In Sec. 12.10.1 we have already applied one of these results (e.g., Smith decomposition of integer matrices) for the implementation of multidimensional decimators. Further applications (to multidimensional multirate systems) of the integer counterpart of the results of this chapter can be found in Chen and Vaidyanathan [1992a,b], [1993], [Gopinath and Burrus, 1991], [Evans, et al., 1992], and [Kalker, 1992].

13.2 MATRIX POLYNOMIALS

Matrix polynomials (or polynomial matrices) play a very crucial role in the description and understanding of multivariable systems. Both FIR and IIR systems can be represented in terms of matrix polynomials. In this section we study their properties.

A $p \times r$ polynomial matrix $\mathbf{P}(z)$ in the variable z is simply a $p \times r$ matrix whose entries are polynomials in z. The matrix can be expressed as

$$\mathbf{P}(z) = \sum_{n=0}^{K} \mathbf{p}(n) z^n. \qquad (13.2.1a)$$

If $\mathbf{p}(K) \neq \mathbf{0}$ the quantity K is called the *order* of the polynomial (not to be confused with degree which has a subtler meaning; Sec. 13.3, 13.8). When there is no room for confusion, we use the term 'polynomial' rather than 'polynomial matrix'.

A polynomial in z^{-1} with order K, that is, a system of the form

$$\mathbf{H}(z) = \sum_{n=0}^{K} \mathbf{h}(n) z^{-n}, \qquad (13.2.1b)$$

is said to be a *causal FIR system* for obvious reasons. The system in (2.2.9) is such an example.

Rank and "Normal Rank" of a Polynomial Matrix

Given a polynomial matrix $\mathbf{P}(z)$, it is clear that the rank usually depends on the value of z. In any case the rank $\rho(z)$ cannot exceed $\min(p, r)$. As an example, for the system given by

$$\begin{bmatrix} 1+z & 1+2z \\ 2+2z & 2+4z \end{bmatrix}, \qquad (13.2.2)$$

it is clear that $\rho(z) < 2$ for all z because the second row is proportional to the first for all z. On the other hand, the matrix

$$\begin{bmatrix} 2+z & 1+z \\ 1+z & 2+z \end{bmatrix} \qquad (13.2.3)$$

has $\rho(z) = 2$ for all z except $z = -3/2$. To see this note that the determinant is equal to $3 + 2z$ which is nonzero (so that the matrix has full rank) unless $z = -3/2$. For the example in (13.2.2), on the other hand, the determinant is identically zero so that $\rho(z) < 2$ for all z.

Normal rank. Let $\rho(z)$ denote the rank of a $p \times r$ polynomial at z. Then the normal rank is defined to be the maximum value of $\rho(z)$ in the entire z plane. Matrices for which the normal rank is the maximum possible $[= \min(p, r)]$ are said to have full normal rank.

Thus (13.2.3) has full normal rank whereas (13.2.2) does not. For the special case of a $M \times M$ (i.e., square) matrix $\mathbf{P}(z)$, the normal rank is full (i.e., equal to M) if and only if the determinant (which is a polynomial in z) is not identically zero for all z. In this case the rank is less than M only at a finite number of points in the z plane (viz., at the locations of the zeros of $[\det \mathbf{P}(z)]$).

Unimodular Matrices

A unimodular matrix $\mathbf{U}(z)$ in the variable z is a square polynomial matrix in z with constant nonzero determinant. Here are some examples:

$$\begin{bmatrix} 1 & 0 \\ 1 + z^3 & 1 \end{bmatrix}, \quad \begin{bmatrix} 1 + z^2 & z \\ 2z & 2 \end{bmatrix}.$$

It is readily verified that the product of unimodular matrices is unimodular. Using this, one can generate more complicated examples.

For $M \times M$ unimodular $\mathbf{U}(z)$ the normal rank is M. In fact the rank $\rho(z) = M$ for all z. Moreoever, since

$$\mathbf{U}^{-1}(z) = \frac{\text{Adj } \mathbf{U}(z)}{\det \mathbf{U}(z)} \qquad (13.2.4)$$

(where 'Adj' denotes the adjugate; Appendix A) the inverse is also a polynomial in z. Since $\mathbf{U}(z)\mathbf{U}^{-1}(z) = \mathbf{I}$, the determinant of the inverse is a (nonzero) constant. So the inverse of unimodular $\mathbf{U}(z)$ is again a unimodular polynomial matrix.

Conversely, if $\mathbf{U}(z)$ is a polynomial matrix with polynomial inverse $\mathbf{V}(z)$ then $\mathbf{U}(z)\mathbf{V}(z) = \mathbf{I}$ so that $[\det \mathbf{U}(z)][\det \mathbf{V}(z)] = 1$. Since $[\det \mathbf{U}(z)]$ and $[\det \mathbf{V}(z)]$ are polynomials, this means that these determinants are constants so that $\mathbf{U}(z)$ is unimodular. Summarizing, a polynomial matrix $\mathbf{U}(z)$ is unimodular if and only if its inverse is a polynomial matrix.

Common Factors, Glcd, and Coprimeness

Two scalar polynomials $P(x)$ and $Q(x)$ are said to be coprime (or relatively prime) if they do not have common factors (i.e., common divisors) with order ≥ 1. For example $(1+x)$ and $(2+x)(3+x)$ are coprime, whereas $(2+x)$ and $(2+x)(3+x)$ are not. A scalar transfer function $H(z) = P(z)/Q(z)$ is said to be in *irreducible* form if the polynomials $P(z)$ and $Q(z)$ are relatively prime.

In the next section we will express rational transfer matrices in irreducible form. This is algebraically more complicated because matrices are involved. We will require the concept of coprimeness of matrix polynomials; and we have to distinguish between *left* divisors and *right* divisors. Given a $p \times r$ polynomial $\mathbf{Q}(z)$ we say that the $p \times p$ polynomial $\mathbf{L}(z)$ is a left divisor (or factor) of $\mathbf{Q}(z)$ if

$$\mathbf{Q}(z) = \mathbf{L}(z)\mathbf{Q}_1(z) \qquad (13.2.5)$$

for some *polynomial* $\mathbf{Q}_1(z)$. Suppose $\mathbf{L}(z)$ is a left divisor of two polynomials $\mathbf{P}(z)$ and $\mathbf{Q}(z)$ (with the same number of rows p). That is

$$\mathbf{Q}(z) = \mathbf{L}(z)\mathbf{Q}_1(z), \quad \mathbf{P}(z) = \mathbf{L}(z)\mathbf{P}_1(z), \qquad (13.2.6)$$

for some polynomials $\mathbf{Q}_1(z)$ and $\mathbf{P}_1(z)$. Then $\mathbf{L}(z)$ is said to be a left common divisor (abbreviated lcd) of $\mathbf{Q}(z)$ and $\mathbf{P}(z)$. For example, let

$$\mathbf{P}(z) = \begin{bmatrix} 1 & -1+z+z^2 \\ z+z^2 & z+z^2 \end{bmatrix}, \quad \text{and} \quad \mathbf{Q}(z) = \begin{bmatrix} z & 2+z \\ 2z & 0 \end{bmatrix}. \quad (13.2.7)$$

We can rewrite these as

$$\mathbf{P}(z) = \underbrace{\begin{bmatrix} 1+z & -1 \\ z & z \end{bmatrix}}_{\mathbf{L}(z)} \begin{bmatrix} 1 & z \\ z & 1 \end{bmatrix}, \quad \mathbf{Q}(z) = \underbrace{\begin{bmatrix} 1+z & -1 \\ z & z \end{bmatrix}}_{\mathbf{L}(z)} \begin{bmatrix} 1 & 1 \\ 1 & -1 \end{bmatrix}, \quad (13.2.8)$$

so that $\mathbf{L}(z)$ is an lcd.

Notice that any $p \times p$ unimodular matrix $\mathbf{U}(z)$ is an lcd of $\mathbf{P}(z)$ and $\mathbf{Q}(z)$ because $\mathbf{U}^{-1}(z)$ is polynomial and we can write

$$\mathbf{P}(z) = \mathbf{U}(z)\underbrace{\mathbf{U}^{-1}(z)\mathbf{P}(z)}_{\text{polynomial}}, \quad \mathbf{Q}(z) = \mathbf{U}(z)\underbrace{\mathbf{U}^{-1}(z)\mathbf{Q}(z)}_{\text{polynomial}}. \quad (13.2.9)$$

The greatest left common divisor (glcd). Let the polynomial $\mathbf{L}(z)$ be an lcd of the two polynomials $\mathbf{P}(z)$ and $\mathbf{Q}(z)$. From (13.2.6) we see that every left factor of $\mathbf{L}(z)$ is also an lcd. We say that $\mathbf{L}(z)$ is a greatest lcd (abbreviated glcd) if *every other lcd* is a left factor of $\mathbf{L}(z)$, that is, if $\mathbf{L}_1(z)$ is any lcd then

$$\mathbf{L}(z) = \mathbf{L}_1(z)\mathbf{R}_1(z) \quad (13.2.10)$$

for some polynomial $\mathbf{R}_1(z)$. So every lcd is a left factor of the glcd, and conversely every left factor of the glcd is an lcd, as one can readily verify. (In a similar way one can define right common divisors (rcd), and grcd. We will skip details, as we do not intend to use them.)

Glcd is not unique. The glcd of two polynomial matrices is not unique. For example if $\mathbf{L}(z)$ is a glcd of $\mathbf{Q}(z)$ and $\mathbf{P}(z)$, then so is $\mathbf{L}(z)\mathbf{W}(z)$ for unimodular $\mathbf{W}(z)$ (Problem 13.2).

Coprimeness. Two polynomial matrices $\mathbf{Q}(z)$ and $\mathbf{P}(z)$ are said to be left coprime if every lcd (hence the glcd) is unimodular. For the scalar case this is equivalent to saying that there are no common factors of order > 0. The matrices $\mathbf{P}(z)$ and $\mathbf{Q}(z)$ in the example (13.2.7) are not left coprime because there is an lcd $\mathbf{L}(z)$ [shown in (13.2.8)], which is not unimodular.

From the above definitions and the properties of unimodular matrices, one can obtain the following two results which we use later:

♠**Fact 13.2.1.** Two polynomial matrices $\mathbf{P}(z)$ and $\mathbf{Q}(z)$ are left coprime if and only if $\mathbf{P}(z)\mathbf{W}(z)$ and $\mathbf{Q}(z)\mathbf{V}(z)$ are left coprime for every pair of unimodular $\mathbf{W}(z)$ and $\mathbf{V}(z)$. ◊

Proof. See Problem 13.2. ▽▽▽

♠**Fact 13.2.2. Inverses of square polynomial matrices.** Let $\mathbf{P}(z)$ be a $p \times p$ polynomial matrix in z with normal rank p. Let α be a zero of its determinant, i.e., $(z - \alpha)$ a factor of $[\det \mathbf{P}(z)]$. Then $\mathbf{P}^{-1}(z)$ has a 'pole' at $z = \alpha$. In other words, at least one element of $\mathbf{P}^{-1}(z)$ has the factor $(z - \alpha)$ in its denominator. ◇

Proof. We know

$$\mathbf{P}^{-1}(z) = \frac{\text{Adj } \mathbf{P}(z)}{\det \mathbf{P}(z)} = \frac{\text{Adj } \mathbf{P}(z)}{(z - \alpha)^L B(z)}, \qquad (13.2.11)$$

where L is a positive integer and $B(z)$ is a polynomial with $B(\alpha) \neq 0$. It is sufficient to prove that the factor $(z - \alpha)^L$ in the denominator is not completely canceled by the adjugate. We can rewrite (13.2.11) as

$$\mathbf{P}(z)[\text{Adj } \mathbf{P}(z)] = (z - \alpha)^L B(z)\mathbf{I}. \qquad (13.2.12)$$

Suppose the adjugate completely cancels the factor $(z - \alpha)^L$. After such cancelation, the determinant of the left hand side above is still zero for $z = \alpha$ [because of $\mathbf{P}(z)$], but that of the right hand side is nonzero. This is evidently a contradiction! ▽▽▽

13.3 MATRIX FRACTION DESCRIPTIONS

We know that any scalar rational transfer function $H(z)$ can be written as $P(z)/Q(z)$, where $P(z)$ and $Q(z)$ are polynomials in z. Similarly, a causal $p \times r$ transfer matrix $\mathbf{H}(z)$ with rational entries $H_{km}(z)$ (i.e., a causal rational system) can be written as

$$\underbrace{\mathbf{H}(z)}_{p \times r} = \underbrace{\mathbf{Q}^{-1}(z)}_{p \times p} \underbrace{\mathbf{P}(z)}_{p \times r}, \qquad (13.3.1)$$

where $\mathbf{Q}(z)$ and $\mathbf{P}(z)$ are matrix polynomials in z with indicated sizes, and $\mathbf{Q}(z)$ has full normal rank $(= p)$. This is called a left matrix fraction description (abbreviated as *left* MFD). Such a description helps to study the properties of the system more compactly, and also gives rise to implementations which are more efficient than directly implementing each element $H_{km}(z)$ of $\mathbf{H}(z)$.

The simplest way to obtain the MFD would be as follows: (a) write each element $H_{km}(z)$ in the form $H_{km}(z) = A_{km}(z)/B_{km}(z)$ where $A_{km}(z)$ and $B_{km}(z)$ are polynomials in z (and *not* in z^{-1}, which is more standard in digital filtering theory), (b) compute the least common multiple (LCM) $D(z)$ of the denominators $B_{km}(z)$, (c) re-express $H_{km}(z)$ as $H_{km}(z) = P_{km}(z)/D(z)$ (which in general is not in irreducible form), and (d) define $\mathbf{P}(z) = [P_{km}(z)]$, $\mathbf{Q}(z) = D(z)\mathbf{I}$.

In a similar way, a right MFD is defined to be of the form $\mathbf{P}_1(z)\mathbf{Q}_1^{-1}(z)$. In this chapter we will not have much occassion to use this form. Unless mentioned otherwise, the term MFD stands for left MFD.

Example 13.3.1

Suppose
$$\mathbf{H}(z) = \begin{bmatrix} 1 \\ \frac{1}{1+az^{-1}} \end{bmatrix}. \qquad (13.3.2)$$

This can be rewritten as
$$\mathbf{H}(z) = \begin{bmatrix} 1 \\ \frac{z}{z+a} \end{bmatrix} = \frac{1}{z+a}\begin{bmatrix} z+a \\ z \end{bmatrix} = \underbrace{[(z+a)\mathbf{I}]^{-1}}_{\mathbf{Q}^{-1}(z)} \underbrace{\begin{bmatrix} z+a \\ z \end{bmatrix}}_{\mathbf{P}(z)}. \qquad (13.3.3)$$

which is a valid MFD. It can be verified that the following choice:
$$\mathbf{Q}(z) = \frac{1}{a}\begin{bmatrix} z+a & -(z+a) \\ -z & z+a \end{bmatrix}, \quad \mathbf{P}(z) = \begin{bmatrix} 1 \\ 0 \end{bmatrix}, \qquad (13.3.4)$$

results in a second possible MFD for this system.

The matrices $\mathbf{Q}(z)$ and $\mathbf{P}(z)$ need not be unique (i.e., the MFD is not unique) as the above example shows. For every $\mathbf{H}(z)$ we can always find a so-called irreducible MFD, as we will describe later. Broadly speaking, this is analogous to scalar systems expressed in the form $P(z)/Q(z)$ where $P(z)$ and $Q(z)$ have no common factors. Irreducible MFDs, again, are not unique.

Degree or McMillan Degree of a System

A very fundamental concept in the study of MIMO systems is the degree μ of a system (also called the *McMillan degree*). The degree μ of a $p \times r$ causal rational system $\mathbf{H}(z)$ is the minimum number of delay units (i.e., z^{-1} elements) required to implement $\mathbf{H}(z)$. We often use "deg" as an abbreviation for degree (as in $[\deg \mathbf{H}(z)]$). The degree is not defined for noncausal systems because they cannot be implemented with delays alone.

For scalar (i.e., SISO) systems the degree is an easily understood concept. Thus consider an Nth order causal FIR filter $\sum_{n=0}^{N} h(n)z^{-n}$, with $h(N) \neq 0$. The degree is then N. Similarly an Nth order IIR filter (Chap. 2) has degree N. For MIMO systems, on the other hand, the situation is more complicated. Thus, let
$$\mathbf{H}(z) = \sum_{n=0}^{N} \mathbf{h}(n)z^{-n},$$

with $\mathbf{h}(N) \neq \mathbf{0}$. The order is N (which is merely the highest power appearing in the expression for $\mathbf{H}(z)$). But the degree is in general $\geq N$. For example consider

$$\mathbf{H}(z) = z^{-1}\mathbf{I} = \begin{bmatrix} z^{-1} & 0 \\ 0 & z^{-1} \end{bmatrix}. \quad (13.3.5a)$$

It is clear that this requires at least two delays for its implementation (Fig. 13.3-1) so that the degree is two, whereas the order $N = 1$. A more advanced tool is therefore necessary to tell what the degree is. We will deal with this issue in Sec. 13.8.

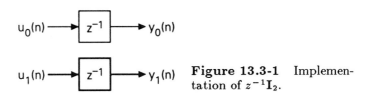

Figure 13.3-1 Implementation of $z^{-1}\mathbf{I}_2$.

Example 13.3.2

Consider the FIR system

$$\mathbf{H}(z) = z^{-1}\begin{bmatrix} 1 & 2 \\ 1 & 2 \end{bmatrix}. \quad (13.3.5b)$$

We can rewrite $\mathbf{H}(z)$ as

$$\mathbf{H}(z) = z^{-1}\begin{bmatrix} 1 \\ 1 \end{bmatrix}\begin{bmatrix} 1 & 2 \end{bmatrix} = \begin{bmatrix} 1 \\ 1 \end{bmatrix}z^{-1}\begin{bmatrix} 1 & 2 \end{bmatrix}.$$

So the system can be implemented as in Fig. 13.3-2 with only one delay, proving that the degree is unity. More generally, consider the example $\mathbf{H}(z) = \mathbf{R}z^{-1}$ where $\mathbf{R}$ is $M \times M$ with rank ρ. This means that we can write

$$\mathbf{R} = \underbrace{\mathbf{T}}_{M \times \rho} \underbrace{\mathbf{S}}_{\rho \times M} \quad (13.3.6)$$

so that $\mathbf{H}(z) = z^{-1}\mathbf{R} = \mathbf{T}[z^{-1}\mathbf{I}]\mathbf{S}$. This shows that we can implement the system using ρ delays (Fig. 13.3-3). So the system $\mathbf{H}(z)$ has degree $\leq \rho$.

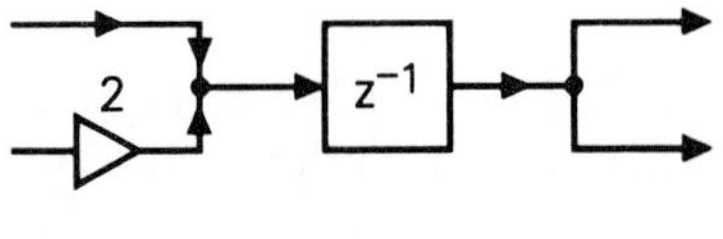

Figure 13.3-2 The system in Example 13.3.2.

It turns out that the degree of $\mathbf{H}(z)$ above is precisely equal to the rank ρ (Problem 13.8). This is readily verified to be true for the examples (13.3.5a) and (13.3.5b).

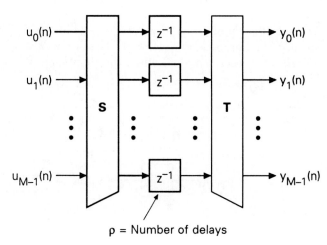

Figure 13.3-3 Pertaining to Example 13.3.2.

Irreducible MFD, and Order of [det Q(z)]

For SISO systems the expression $H(z) = A(z)/B(z)$ is said to be irreducible if $A(z)$ and $B(z)$ are coprime. For the matrix case if $\mathbf{L}(z)$ is a glcd of $\mathbf{Q}(z)$ and $\mathbf{P}(z)$, then we can obtain a MFD $\mathbf{H}(z) = \mathbf{Q}_1^{-1}(z)\mathbf{P}_1(z)$ where $\mathbf{Q}_1(z)$ and $\mathbf{P}_1(z)$ are as in (13.2.6). Having done so, the matrices $\mathbf{Q}_1(z)$ and $\mathbf{P}_1(z)$ are left coprime, and the MFD $\mathbf{Q}_1^{-1}(z)\mathbf{P}_1(z)$ is said to be irreducible. Now

$$\det \mathbf{Q}(z) = [\det \mathbf{L}(z)]\det \mathbf{Q}_1(z). \qquad (13.3.7)$$

If $\mathbf{P}(z)$ and $\mathbf{Q}(z)$ are not left coprime, then $\mathbf{L}(z)$ is not unimodular so that the order of $[\det \mathbf{Q}(z)]$ exceeds that of $[\det \mathbf{Q}_1(z)]$. Thus given a reducible MFD, we can always find an irreducible MFD for which the determinant of $\mathbf{Q}_1(z)$ has reduced order.

In Sec. 13.5.2 we show that all irreducible MFDs of a given system $\mathbf{H}(z)$ have same order for $[\det \mathbf{Q}(z)]$. From above discussions it follows that this is also the smallest order of $[\det \mathbf{Q}(z)]$, among all possible MFDs.

Order versus degree. It is worth emphasizing the distinction between order and degree. The function $1 + z^{-1}$ is a polynomial in z^{-1} with order = 1. Moreover it can also be viewed as a causal system with degree = 1, because it can be implemented with one delay. The function $1 + z$ is a polynomial in z with order = 1, but we cannot say that its degree = 1. This is because this system cannot be implemented with one *delay* element! It is noncausal, and its degree is simply undefined! For this reason, we have used the term "order" rather than "degree" when discussing the determinant of $\mathbf{Q}(z)$ which is a polynomial in the advance operator z.

13.4 STATE SPACE DESCRIPTIONS

We know that a transfer function $H(z)$ can be implemented in many ways such as the direct form, cascade form and so on. Given any such implementation, the outputs of the delay elements are called state variables, and the system behavior can be completely described in the time domain by a set of equations which involve the input, state variables and output. This is called the *state space description* of the structure.

To demonstrate state space descriptions, consider the example of a direct form structure shown in Fig. 13.4-1. This is a causal system with transfer function $H(z) = P(z)/Q(z)$, where

$$P(z) = \sum_{n=0}^{N} p_n z^{-n}, \quad Q(z) = 1 + \sum_{n=1}^{N} q_n z^{-n}. \qquad (13.4.1)$$

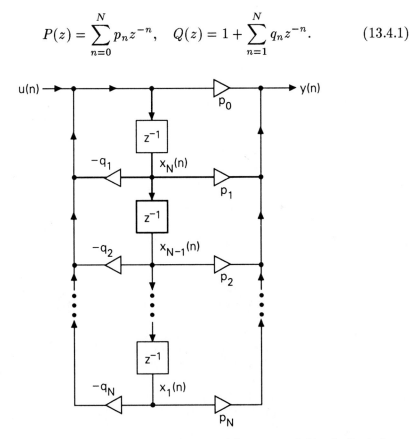

Figure 13.4-1 The direct form structure, with state variables indicated.

Since there are N delays, we have N state variables $x_1(n), \ldots, x_N(n)$ as shown. The input and output signals are $u(n)$ and $y(n)$, respectively. The state space description is

$$x_1(n+1) = x_2(n)$$
$$x_2(n+1) = x_3(n)$$
$$\vdots$$
$$x_N(n+1) = -\sum_{k=1}^{N} q_{N+1-k} x_k(n) + u(n), \qquad (13.4.2)$$
$$y(n) = \sum_{k=1}^{N} (p_{N+1-k} - p_0 q_{N+1-k}) x_k(n) + p_0 u(n).$$

This can be expressed compactly in matrix-vector notation as

$$\mathbf{x}(n+1) = \mathbf{A}\mathbf{x}(n) + \mathbf{B}\mathbf{u}(n), \quad \text{(state equation)}, \qquad (13.4.3)$$

$$\mathbf{y}(n) = \mathbf{C}\mathbf{x}(n) + \mathbf{D}\mathbf{u}(n), \quad \text{(output equation)}, \qquad (13.4.4)$$

where

$$\mathbf{A} = \begin{bmatrix} 0 & 1 & 0 & \cdots & 0 \\ 0 & 0 & 1 & \cdots & 0 \\ \vdots & \vdots & \vdots & \ddots & \vdots \\ 0 & 0 & 0 & \cdots & 1 \\ -q_N & -q_{N-1} & \cdots & \cdots & -q_1 \end{bmatrix}, \qquad (13.4.5)$$

$$\mathbf{B} = [\,0 \quad 0 \quad \cdots \quad 0 \quad 1\,]^T,$$
$$\mathbf{C} = [\,p_N - p_0 q_N \quad p_{N-1} - p_0 q_{N-1} \quad \cdots \quad p_1 - p_0 q_1\,] \qquad (13.4.6)$$
$$\mathbf{D} = p_0,$$

$$\mathbf{x}(n) = [\,x_1(n) \quad x_2(n) \quad \cdots \quad x_N(n)\,], \qquad (13.4.7)$$

$\mathbf{y}(n) = y(n)$, and $\mathbf{u}(n) = u(n)$. We have used bold-faced $\mathbf{u}(n)$ and $\mathbf{y}(n)$ because we will use these equations for the MIMO case also.

We can write down the state space description in the above manner for any structure representing a $p \times r$ causal LTI system $\mathbf{H}(z)$. (See examples below.) With N denoting the number of delays, the state equation (13.4.3) is a set of N equations. The output equation is a set of p equations. The sizes of the matrices are:

$$\mathbf{A}: N \times N; \quad \mathbf{B}: N \times r; \quad \mathbf{C}: p \times N; \quad \mathbf{D}: p \times r. \qquad (13.4.8)$$

Figure 13.4-2 shows a schematic of the state space description.

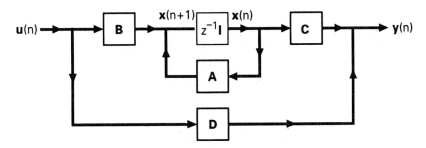

Figure 13.4-2 Schematic representation of the state space structure.

Main Points About State Space Descriptions

1. **State variables, state vector and state transition matrix.** The outputs $x_i(n)$ of the delay elements are the state variables, and the vector $\mathbf{x}(n)$ of state variables is called the state vector (or just the 'state' at time n). The matrix $\mathbf{A}$ is called the *state transition matrix* (STM). One can verify that if the state $\mathbf{x}(n_0)$ is given then the quantities $\mathbf{x}(n), n > n_0$ and $\mathbf{y}(n), n \geq n_0$ can be found by knowing $\mathbf{u}(m)$ for $m \geq n_0$. [See (13.4.11) later.]

2. **State space description.** The state equations represent the storage part (or the part of the system having the memory elements z^{-1}), and can be considered to be the recursive part, since $\mathbf{x}(n+1)$ is computed from $\mathbf{x}(n)$. (This does not necessarily imply the existence of feedback; see FIR example below). The output equation is the nonrecursive part. The quadruple $(\mathbf{A}, \mathbf{B}, \mathbf{C}, \mathbf{D})$ is said to be the state space description of the structure.

3. **Meaning of D.** Notice from Fig. 13.4-2 that the quantity $\mathbf{D}$ is given by setting $z = \infty$, that is, $\mathbf{D} = \mathbf{H}(\infty)$. This means that we can find the value of $\mathbf{D}$ by wiping out the delay elements in the structure (i.e., set $z^{-1} = 0$). Since $\mathbf{H}(\infty) = \mathbf{h}(0)$ for a causal system, we conclude

$$\mathbf{D} = \mathbf{H}(\infty) = \mathbf{h}(0). \qquad (13.4.9)$$

Essentially $\mathbf{D}$ represents the 'direct path' (i.e., delay-free path) from the input to the output.

4. **Implicit causality.** It should be borne in mind that (13.4.3) and (13.4.4) are implicitly restricted to causal systems. The memory part [eqn. (13.4.3)] is clearly executed in a causal manner. The output at time n is completely determined by the input $\mathbf{u}(m), m \leq n$ (equivalently by the input $\mathbf{u}(m)$ for $n_0 \leq m \leq n$ and the state vector $\mathbf{x}(n_0)$ for arbitrary n_0). Unless mentioned otherwise, all results in this chapter are restricted to causal systems.

Example 13.4.1: An FIR System

Consider the FIR system of Fig. 13.4-3 with two inputs, two outputs and two state variables. The state equations are easily verified to be

$$\underbrace{\begin{bmatrix} x_1(n+1) \\ x_2(n+1) \end{bmatrix}}_{\mathbf{x}(n+1)} = \begin{bmatrix} 0 & 0 \\ 1 & 0 \end{bmatrix} \underbrace{\begin{bmatrix} x_1(n) \\ x_2(n) \end{bmatrix}}_{\mathbf{x}(n)} + \begin{bmatrix} 0 & 1 \\ k_1 & 0 \end{bmatrix} \underbrace{\begin{bmatrix} u_0(n) \\ u_1(n) \end{bmatrix}}_{\mathbf{u}(n)},$$

and the output equations are

$$\underbrace{\begin{bmatrix} y_0(n) \\ y_1(n) \end{bmatrix}}_{\mathbf{y}(n)} = \begin{bmatrix} k_1 & k_2 \\ k_1 k_2 & 1 \end{bmatrix} \underbrace{\begin{bmatrix} x_1(n) \\ x_2(n) \end{bmatrix}}_{\mathbf{x}(n)} + \begin{bmatrix} 1 & 0 \\ k_2 & 0 \end{bmatrix} \underbrace{\begin{bmatrix} u_0(n) \\ u_1(n) \end{bmatrix}}_{\mathbf{u}(n)}.$$

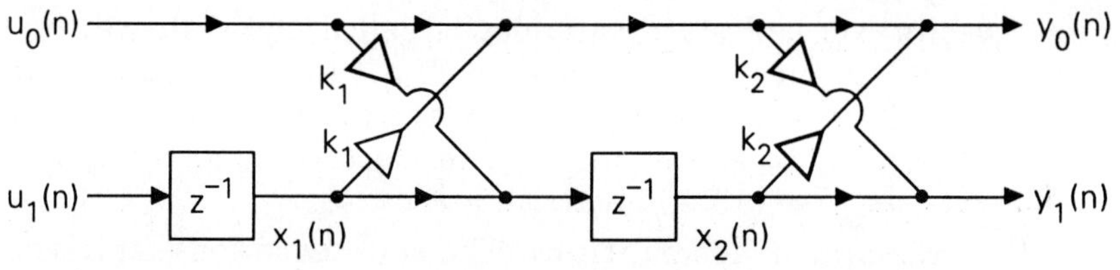

Figure 13.4-3 FIR example for state space description.

So the state space description of the structure is

$$\mathbf{A} = \begin{bmatrix} 0 & 0 \\ 1 & 0 \end{bmatrix}, \quad \mathbf{B} = \begin{bmatrix} 0 & 1 \\ k_1 & 0 \end{bmatrix}, \quad \mathbf{C} = \begin{bmatrix} k_1 & k_2 \\ k_1 k_2 & 1 \end{bmatrix}, \quad \mathbf{D} = \begin{bmatrix} 1 & 0 \\ k_2 & 0 \end{bmatrix}.$$

Notice that $\mathbf{D}$ could also be obtained by replacing the delay elements in the figure with zero.

Since $\mathbf{A} \neq \mathbf{0}$, the vector $\mathbf{x}(n+1)$ depends on $\mathbf{x}(n)$. This is represented by the feedback path in Fig. 13.4-2. However, from Fig. 13.4-3 we note that there is no feedback connection in the structure!

Example 13.4.2: An IIR System (the Coupled Form Structure)

Consider the IIR filter structure of Fig. 13.4-4 with one input, two outputs and two state variables, as labeled. This is called the *coupled form structure* in digital filtering literature [Oppenheim and Schafer, 1989]. The state space description is verified to be

$$\mathbf{A} = R \begin{bmatrix} \cos\theta & -\sin\theta \\ \sin\theta & \cos\theta \end{bmatrix}, \quad \mathbf{B} = \begin{bmatrix} 1 \\ 0 \end{bmatrix}, \quad \mathbf{C} = \begin{bmatrix} R\sin\theta & R\cos\theta \\ 0 & 1 \end{bmatrix}, \quad \mathbf{D} = \mathbf{0}.$$

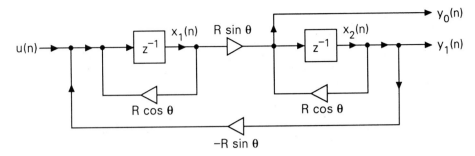

Figure 13.4-4 The coupled form IIR structure.

The fact that $\mathbf{D} = \mathbf{0}$ (i.e., $\mathbf{H}(\infty) = \mathbf{0}$) implies that there is no direct (delay-free) path between the input and any of the outputs, as can be verified from the figure.

State space "structures," "descriptions," and "realizations." It is important to distinguish between state space structures and state space descriptions. This is best explained with the above examples. For the structure of Example 13.4.2 we found that the elements of the matrices $(\mathbf{A}, \mathbf{B}, \mathbf{C}, \mathbf{D})$ are equal to the values of the multipliers in the structure so that Fig. 13.4-4 is an *implementation* of (or a *structure* for) $(\mathbf{A}, \mathbf{B}, \mathbf{C}, \mathbf{D})$. In Example 13.4.1 on the other hand, this is not true because of the appearance of the product $k_1 k_2$ in $\mathbf{C}$. So $(\mathbf{A}, \mathbf{B}, \mathbf{C}, \mathbf{D})$ is merely the state space description of the structure in Fig. 13.4-3 but the figure is *not* a structure for $(\mathbf{A}, \mathbf{B}, \mathbf{C}, \mathbf{D})$. Whenever we say 'the realization $(\mathbf{A}, \mathbf{B}, \mathbf{C}, \mathbf{D})$' we just mean a structure whose state space *description* agrees with $(\mathbf{A}, \mathbf{B}, \mathbf{C}, \mathbf{D})$.

13.4.1 Properties of State Space Descriptions

Transfer Function

The state space description $(\mathbf{A}, \mathbf{B}, \mathbf{C}, \mathbf{D})$ completely determines the input output behavior of the system. We now proceed to substantiate this by writing down the transfer function as well as the time domain input-output relations in terms of $\mathbf{A}$, $\mathbf{B}$, $\mathbf{C}$, and $\mathbf{D}$.

Taking z-transforms of both sides of (13.4.3) and (13.4.4) we obtain

$$z\mathbf{X}(z) = \mathbf{A}\mathbf{X}(z) + \mathbf{B}\mathbf{U}(z)$$
$$\mathbf{Y}(z) = \mathbf{C}\mathbf{X}(z) + \mathbf{D}\mathbf{U}(z). \qquad (13.4.10a)$$

From these we obtain $\mathbf{Y}(z) = \mathbf{H}(z)\mathbf{U}(z)$ where

$$\mathbf{H}(z) = \mathbf{D} + \mathbf{C}(z\mathbf{I} - \mathbf{A})^{-1}\mathbf{B}. \qquad (13.4.10b)$$

Time Domain Description

In order to express the output $\mathbf{y}(n)$ in terms of $(\mathbf{A},\mathbf{B},\mathbf{C},\mathbf{D})$ and the input, we first write $\mathbf{y}(n)$ in terms of the initial state vector (or 'initial condition') $\mathbf{x}(0)$ and the input $\mathbf{u}(m)$, $0 \leq m \leq n$. By repeated application of (13.4.3) we find

$$\mathbf{x}(n) = \mathbf{A}^n \mathbf{x}(0) + \sum_{m=0}^{n-1} \mathbf{A}^m \mathbf{B} \mathbf{u}(n-m-1), \quad n \geq 1, \qquad (13.4.11a)$$

so that from (13.4.4)

$$\mathbf{y}(n) = \mathbf{C}\mathbf{A}^n \mathbf{x}(0) + \sum_{m=0}^{n-1} \mathbf{C}\mathbf{A}^m \mathbf{B} \mathbf{u}(n-m-1) + \mathbf{D}\mathbf{u}(n), \quad n \geq 1. \qquad (13.4.11b)$$

In particular if we have $\mathbf{x}(0) = \mathbf{0}$ then

$$\mathbf{y}(n) = \mathbf{D}\mathbf{u}(n) + \sum_{k=1}^{n} \mathbf{C}\mathbf{A}^{k-1}\mathbf{B}\mathbf{u}(n-k). \qquad (13.4.12)$$

Impulse response. By comparing the above equation with (2.2.7), we can write the the impulse response as

$$\mathbf{h}(k) = \begin{cases} \mathbf{D} & k=0 \\ \mathbf{C}\mathbf{A}^{k-1}\mathbf{B}, & k>0. \end{cases} \qquad (13.4.13)$$

Thus, all coefficients of the impulse response can be calculated from the matrices $(\mathbf{A},\mathbf{B},\mathbf{C},\mathbf{D})$. A second way to obtain this based on power series expansion of (13.4.10b) is addressed in Problem 13.18.

Poles of H(z) and Eigenvalues of A

We say that z_p is a pole of $\mathbf{H}(z)$ if it is a pole of at least one of the elements $H_{km}(z)$ of the matrix $\mathbf{H}(z)$. Now from (13.4.10b) we see that

$$\mathbf{H}(z) = \mathbf{D} + \frac{\mathbf{C}\mathbf{R}(z)\mathbf{B}}{\det(z\mathbf{I}-\mathbf{A})}, \qquad (13.4.14)$$

where $\mathbf{R}(z)$ is the adjugate of $(z\mathbf{I}-\mathbf{A})$. Thus, if z_p is a pole then it is a zero of the determinant appearing in (13.4.14). So

$$\begin{aligned} z_p = pole &\Rightarrow \det(z_p\mathbf{I}-\mathbf{A}) = 0 \\ &\Rightarrow (z_p\mathbf{I}-\mathbf{A})\mathbf{v} = \mathbf{0} \quad \text{(for some } \mathbf{v} \neq \mathbf{0}) \\ &\Rightarrow \mathbf{A}\mathbf{v} = z_p\mathbf{v}, \quad \mathbf{v} \neq \mathbf{0}, \end{aligned} \qquad (13.4.15)$$

which shows that poles of $\mathbf{H}(z)$ are eigenvalues of $\mathbf{A}$. So the system $\mathbf{H}(z)$ is stable if all the eigenvalues satisfy $|\lambda_i| < 1$.

But we have to be careful with the converse. In general, not all eigenvalues λ of $\mathbf{A}$ are poles of $\mathbf{H}(z)$, unless $(\mathbf{A}, \mathbf{B}, \mathbf{C}, \mathbf{D})$ is minimal (to be defined in Section 13.4.2). For the moment just note that even if the determinant in the denominator of (13.4.14) is zero at $z = \lambda$, this can, in principle, cancel with a factor $(z - \lambda)$ in every entry of $\mathbf{CR}(z)\mathbf{B}$, so that λ may not in reality be a pole.

It is possible for all eigenvalues of $\mathbf{A}$ to be zero (as in Example 13.4.1). This implies that $\mathbf{H}(z)$ is FIR (i.e., has all poles at $z = 0$).

Similarity Transformations

Given a state space description $(\mathbf{A}, \mathbf{B}, \mathbf{C}, \mathbf{D})$ for a system $\mathbf{H}(z)$, suppose we define a new set of matrices

$$\mathbf{A}_1 = \mathbf{T}^{-1}\mathbf{AT}, \quad \mathbf{B}_1 = \mathbf{T}^{-1}\mathbf{B}, \quad \mathbf{C}_1 = \mathbf{CT}, \quad \mathbf{D}_1 = \mathbf{D}, \qquad (13.4.16a)$$

where $\mathbf{T}$ is any nonsingular matrix. We then see that the transfer function corresponding to this description is given by $\mathbf{D}_1 + \mathbf{C}_1(z\mathbf{I} - \mathbf{A}_1)^{-1}\mathbf{B}_1$ which upon substitution from (13.4.16a) reduces to $\mathbf{H}(z)$ given by (13.4.10b). In other words, $(\mathbf{A}_1, \mathbf{B}_1, \mathbf{C}_1, \mathbf{D})$ is an equivalent state space description for the system $\mathbf{H}(z)$. In this way we can find an infinite number of equivalent descriptions because $\mathbf{T}$ is arbitrary. The matrix $\mathbf{T}$ is called a *similarity transformer* (or transformation).

We can rewrite the state equation (13.4.3) as

$$\mathbf{T}^{-1}\mathbf{x}(n+1) = \underbrace{\mathbf{T}^{-1}\mathbf{AT}}_{\mathbf{A}_1} \mathbf{T}^{-1}\mathbf{x}(n) + \underbrace{\mathbf{T}^{-1}\mathbf{B}}_{\mathbf{B}_1} \mathbf{u}(n),$$

which shows that the new system has the *transformed* state vector $\mathbf{T}^{-1}\mathbf{x}(n)$. So, the similarity transformation changes the internal state vector while retaining the same input-output relation. In the next subsection we will use this idea to generate the so-called *minimal* realizations.

Notice that the eigenvalues of $\mathbf{A}$ are the same as those of $\mathbf{A}_1$ (Appendix A). From Cayley-Hamilton theorem (the same Appendix) we conclude that the quantities $\mathbf{A}^N$ and $\mathbf{A}_1^N$ can be expressed as

$$\mathbf{A}^N = \sum_{k=0}^{N-1} \alpha_k \mathbf{A}^k, \quad \mathbf{A}_1^N = \sum_{k=0}^{N-1} \alpha_k \mathbf{A}_1^k, \qquad (13.4.16b)$$

by using the *same* set of scalars α_k. This is a useful fact.

13.4.2 Minimal Realizations

The number N of delay elements in a structure is equal to the size of the state vector $\mathbf{x}(n)$. So N is said to be the *dimension of the state space*. This integer N also governs the size of $\mathbf{A}$ (which is $N \times N$).

A structure (or implementation or realization) for a transfer function is said to be *minimal* if the number of delay elements N is the smallest possible, viz., the degree μ of $\mathbf{H}(z)$. So, a structure is minimal if and only if the state space has smallest dimension (i.e., size of $\mathbf{A}$ is smallest). In this section we introduce two properties called reachability and observability, and show that a structure is minimal if and only if it satisfies these two properties.

Example 13.4.3

To motivate, consider Fig. 13.4-5 which shows a nonminimal structure. The system transfer function is $H(z) = z^{-1}$, which has degree one. But there are four delays in the structure. The first is unconnected to the output $y(n)$, and its output is not observable; the second delay is unconnected to the input $u(n)$ and is not reachable, that is, its output cannot be changed by choice of input $u(n)$. The third delay represents an unreachable and unobservable component. The fourth delay is the only useful one, as it is both reachable and observable.

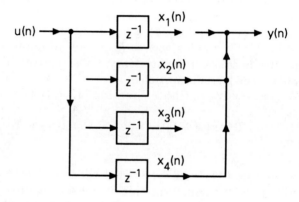

Figure 13.4-5 A simple example of a structure with unreachable and unobservable states.

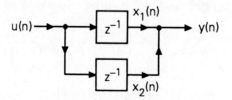

Figure 13.4-6 A more subtle example of a structure with unreachable states.

Not all structures are as simple to analyze and understand as the above one. Thus consider Fig. 13.4-6, where $H(z) = 2z^{-1}$. Evidently this has degree one, but the structure has two delays. Both delays are connected to the input as well as to the output. However, this structure is not 'reachable' in the sense that we cannot find an input sequence to achieve arbitrary values for the state variables [as they are constrained by $x_1(n) = x_2(n) = u(n-1)$].

Decoupling unreachable and unobservable state variables. As the above example shows, there are situations where we cannot identify a particular state variable to be unreachable, even though the system has more delays than necessary. We will show that such structures can be transformed (using the similarity transformation) in such a manner that the transformed state vector has two sets of state variables. One of these is the set of unreachable variables and can be discarded, thereby reducing the number of delay elements. In a similar way, we can perform a transformation that separates or "decouples" the unobservable state variables, which can then be discarded. The result of these manipulations will be a structure with minimum number of delays.

It is now time to make our definitions and analysis mathematically more precise.

Reachability

Consider a structure for $\mathbf{H}(z)$ with N delays. Let $(\mathbf{A}, \mathbf{B}, \mathbf{C}, \mathbf{D})$ be the state space description so that $\mathbf{A}$ is $N \times N$. The structure for $\mathbf{H}(z)$ is said to be reachable (or *completely reachable*, often abbreviated *cr*) if we can reach any specified final state $\mathbf{x}_f$ starting from any arbitrary inital state $\mathbf{x}_i$ by application of an appropriate *finite length* input sequence. Because of the shift invariant nature of the system, we will assume that the initial time is zero, that is, $\mathbf{x}(0) = \mathbf{x}_i$.

To study the conditions for reachability, recall that (13.4.11a) gives us the general relation between any state $\mathbf{x}(n), n > 0$ and the initial state $\mathbf{x}(0)$. Given arbitrary initial state $\mathbf{x}(0)$, we can reach any specified state $\mathbf{x}_f$ at time n (i.e., force $\mathbf{x}(n) = \mathbf{x}_f$) by choice of the inputs $\mathbf{u}(m)$ in the summation of (13.4.11a), provided the matrix

$$[\mathbf{B} \quad \mathbf{AB} \quad \ldots \quad \mathbf{A}^{n-1}\mathbf{B}] \qquad (13.4.17)$$

has full row rank N (Appendix A). If this is not the case, we can try to reach the desired state $\mathbf{x}_f$ at time $n+1$. Now suppose that we have not been able to reach the desired state up until time N. This means that the matrix

$$\mathcal{R}_{\mathbf{A},\mathbf{B}} \stackrel{\Delta}{=} \underbrace{[\mathbf{B} \quad \mathbf{AB} \quad \ldots \quad \mathbf{A}^{N-1}\mathbf{B}]}_{N \times Nr} \qquad (13.4.18)$$

does not have rank N. If this is the case, then any further waiting will prove to be unfruitful. In other words, it will not be possible to reach the state

$\mathbf{x}_f$ for any $n > N$. The reason for this is that the additional columns of the matrix (13.4.17), which are of the form

$$\mathbf{A}^N\mathbf{B}, \ \mathbf{A}^{N+1}\mathbf{B}, \ \ldots \qquad (13.4.19)$$

can be expressed as linear combinations of the columns of $\mathcal{R}_{\mathbf{A},\mathbf{B}}$ (Cayley-Hamilton theorem) so that we do not obtain additional linearly independent columns after time N. In other words, the rank of (13.4.17) does not increase as n is increased beyond N. Summarizing, we have proved the following important result.

♦**Lemma 13.4.1. Reachability.** A structure with N delays (i.e., N state variables) is reachable if and only if the matrices $\mathbf{A}$ and $\mathbf{B}$ are such that the matrix $\mathcal{R}_{\mathbf{A},\mathbf{B}}$ has full rank N. ◊

Remarks

1. Consider the $N \times N$ matrix $\mathcal{R}_{\mathbf{A},\mathbf{B}}\mathcal{R}_{\mathbf{A},\mathbf{B}}^\dagger$. We know this is positive semi definite for any $\mathcal{R}_{\mathbf{A},\mathbf{B}}$. The full-rank condition on $\mathcal{R}_{\mathbf{A},\mathbf{B}}$ is equivalent to the condition $\mathcal{R}_{\mathbf{A},\mathbf{B}}\mathcal{R}_{\mathbf{A},\mathbf{B}}^\dagger > 0$ (i.e., positive definite).
2. Since reachability is governed by the two matrices $\mathbf{A}$ and $\mathbf{B}$, we often say "$(\mathbf{A}, \mathbf{B})$ is reachable" instead of "the structure is reachable."
3. Evidently reachability is a property of the *structure* implementing $\mathbf{H}(z)$, rather than a property of $\mathbf{H}(z)$.
4. The requirement of reachability is stronger than *controllability* which is a notion we will not use (see Problem 13.7).

What if a structure is not reachable? This means that we can perform a transformation on the structure so that the new state vector reveals the unreachable state variables explicitly. These variables can then be eliminated to obtain an equivalent structure for $\mathbf{H}(z)$ with fewer delays. More formally we will prove:

♦**Lemma 13.4.2. Reduction to reachable form.** Suppose $(\mathbf{A},\mathbf{B})$ is such that $\mathcal{R}_{\mathbf{A},\mathbf{B}}$ does not have full rank N. Then there exists a similarity transformation $\mathbf{T}$ such that the equivalent system (13.4.16a) has matrices $\mathbf{A}_1$ and $\mathbf{B}_1$ of the form

$$\mathbf{A}_1 = \begin{matrix} \rho \\ N-\rho \end{matrix} \overset{\rho \quad N-\rho}{\begin{pmatrix} \mathbf{A}_{11} & \mathbf{A}_{12} \\ 0 & \mathbf{A}_{22} \end{pmatrix}}, \quad \mathbf{B}_1 = \begin{matrix} \rho \\ N-\rho \end{matrix} \overset{r}{\begin{pmatrix} \mathbf{B}_{11} \\ 0 \end{pmatrix}}. \qquad (13.4.20)$$

where ρ denotes the rank of $\mathcal{R}_{\mathbf{A},\mathbf{B}}$. ◊

Consequence of the Lemma. If we have a structure for which the matrices $\mathbf{A}_1$ and $\mathbf{B}_1$ have the form (13.4.20), then the state equation can be partitioned as

$$\begin{aligned}\mathbf{x}_1(n+1) &= \mathbf{A}_{11}\mathbf{x}_1(n) + \mathbf{A}_{12}\mathbf{x}_2(n) + \mathbf{B}_{11}\mathbf{u}(n), \\ \mathbf{x}_2(n+1) &= \mathbf{A}_{22}\mathbf{x}_2(n).\end{aligned} \qquad (13.4.21)$$

So the subset of states $\mathbf{x}_2(n)$ is independent of the input $\mathbf{u}(n)$. Moreover its evolution in time is independent of $\mathbf{x}_1(n)$. This means that $\mathbf{x}_2(n)$ is the *unreachable component* of the state (as it cannot be changed in any way by changing the input). Assuming that the initial state is zero, the quantity $\mathbf{x}_2(n)$ is zero for all future time, and does not affect the input-output behavior [and hence the transfer function $\mathbf{H}(z)$].

The reduced system $\mathbf{A}_{11}, \mathbf{B}_{11}, \mathbf{C}_{11}, \mathbf{D}$ (with state vector $\mathbf{x}_1(n)$ of reduced dimension ρ) has the same input-output behavior as the original system. Here $\mathbf{C}_{11}$ represents the leftmost $p \times \rho$ submatrix of $\mathbf{C}_1 \triangleq \mathbf{CT}$.

Proof of Lemma 13.4.2. Let $\mathbf{t}_0, \mathbf{t}_1, \ldots, \mathbf{t}_{\rho-1}$ denote a set of ρ independent columns of $\mathcal{R}_{\mathbf{A},\mathbf{B}}$. Evidently all columns of $\mathcal{R}_{\mathbf{A},\mathbf{B}}$ (which are N-vectors) are linear combinations of these. Let $\mathbf{t}_\rho, \ldots, \mathbf{t}_{N-1}$ be a basis for the orthogonal complement of the column space of $\mathcal{R}_{\mathbf{A},\mathbf{B}}$. Defining the $N \times N$ nonsingular matrix

$$\mathbf{T} = [\mathbf{t}_0 \quad \ldots \quad \mathbf{t}_{\rho-1} \quad \mathbf{t}_\rho \quad \ldots \quad \mathbf{t}_{N-1}], \qquad (13.4.22)$$

we then have

$$\mathcal{R}_{\mathbf{A},\mathbf{B}} = \mathbf{T} \begin{bmatrix} \mathbf{P} \\ \mathbf{0} \end{bmatrix} \qquad (13.4.23)$$

where the $\mathbf{0}$ on the RHS is $(N - \rho) \times Nr$. From this and from the definition (13.4.18) of $\mathcal{R}_{\mathbf{A},\mathbf{B}}$ we deduce, in particular, that $\mathbf{B}$ has the form

$$\mathbf{B} = \mathbf{T} \begin{bmatrix} \mathbf{B}_{11} \\ \mathbf{0} \end{bmatrix}. \qquad (13.4.24)$$

Since $\mathbf{A}^N$ can be expressed as a linear combination of $\mathbf{A}^k$, $0 \leq k \leq N - 1$ (Cayley-Hamilton theorem), Eq. (13.4.23) also implies

$$\mathbf{A} \underbrace{[\mathbf{B} \quad \mathbf{AB} \quad \ldots \quad \mathbf{A}^{N-1}\mathbf{B}]}_{\mathcal{R}_{\mathbf{A},\mathbf{B}}} = \mathbf{T} \begin{bmatrix} \times \\ \mathbf{0} \end{bmatrix}, \qquad (13.4.25)$$

where $\times$ denotes possibly nonzero entries. Since the first ρ columns of $\mathbf{T}$ span the column space of $\mathcal{R}_{\mathbf{A},\mathbf{B}}$, (13.4.25) implies

$$\mathbf{AT} = \mathbf{T} \begin{bmatrix} \mathbf{A}_{11} & \mathbf{A}_{12} \\ \mathbf{0} & \mathbf{A}_{22} \end{bmatrix}, \quad \text{i.e.,} \quad \mathbf{A} = \mathbf{T} \begin{bmatrix} \mathbf{A}_{11} & \mathbf{A}_{12} \\ \mathbf{0} & \mathbf{A}_{22} \end{bmatrix} \mathbf{T}^{-1}, \qquad (13.4.26)$$

where $\mathbf{A}_{11}$ is $\rho \times \rho$, and $\mathbf{A}_{12}$ and $\mathbf{A}_{22}$ are of appropriate sizes. The results (13.4.24) and (13.4.26) establish the existence of a similarity transformer $\mathbf{T}$ such that the new state space description has the form (13.4.20). $\triangledown\triangledown\triangledown$

Observability

A structure for $\mathbf{H}(z)$ is said to be observable (or *completely observable,* often abbreviated *co*) if the state $\mathbf{x}(m)$ at time m can be uniquely determined by observing a *finite-length segment* of the output sequence starting from time m, and knowing the input sequence for the corresponding set of sample values. In view of shift invariance we shall set $m = 0$ in our discussions. Like reachability, observability is a property of the structure and not of the transfer function $\mathbf{H}(z)$. Once again, if the structure is not observable, then measurement of more than N output samples does not help to identify the initial state [where N is the size of $\mathbf{x}(n)$]. So the structure is observable if the knowledge of $\mathbf{u}(n), \mathbf{y}(n), 0 \leq n \leq N - 1$ can be used to find $\mathbf{x}(0)$ uniquely.

To determine the conditions for observability, recall that $\mathbf{y}(n)$ can be expressed as in (13.4.11b). In terms of the impulse response $\mathbf{h}(k)$ defined in (13.4.13) we therefore get

$$\underbrace{\begin{bmatrix} \mathbf{y}(0) \\ \mathbf{y}(1) \\ \vdots \\ \mathbf{y}(N-1) \end{bmatrix}}_{\text{call this } \mathbf{Y}} = \underbrace{\begin{bmatrix} \mathbf{C} \\ \mathbf{CA} \\ \vdots \\ \mathbf{CA}^{N-1} \end{bmatrix}}_{\mathcal{S}_{\mathbf{C},\mathbf{A}}} \mathbf{x}(0) + \underbrace{\begin{bmatrix} \mathbf{h}(0) & 0 & \cdots & 0 \\ \mathbf{h}(1) & \mathbf{h}(0) & \cdots & 0 \\ \vdots & \vdots & \ddots & \vdots \\ \mathbf{h}(N-1) & \mathbf{h}(N-2) & \cdots & \mathbf{h}(0) \end{bmatrix} \begin{bmatrix} \mathbf{u}(0) \\ \mathbf{u}(1) \\ \vdots \\ \mathbf{u}(N-1) \end{bmatrix}}_{\text{call this } \mathbf{f}}.$$
(13.4.27)

Given the quantities $\mathbf{y}(n)$ and $\mathbf{u}(n)$ for $0 \leq n \leq N - 1$, that is, given $\mathbf{Y}$ and $\mathbf{f}$, there exists $\mathbf{x}(0)$ satisfying (13.4.27) because, by definition, $\mathbf{y}(n)$ satisfies (13.4.27). It is also clear (Appendix A) that we can find a *unique* value of $\mathbf{x}(0)$ if and only if

$$\mathcal{S}_{\mathbf{C},\mathbf{A}} \triangleq \underbrace{\begin{bmatrix} \mathbf{C} \\ \mathbf{CA} \\ \vdots \\ \mathbf{CA}^{N-1} \end{bmatrix}}_{Np \times N} \qquad (13.4.28)$$

has full column rank N. Under this condition, the unique solution is

$$\mathbf{x}(0) = [\mathcal{S}_{\mathbf{C},\mathbf{A}}^{\dagger} \mathcal{S}_{\mathbf{C},\mathbf{A}}]^{-1} \mathcal{S}_{\mathbf{C},\mathbf{A}}^{\dagger} (\mathbf{Y} - \mathbf{f}). \qquad (13.4.29)$$

Summarizing, we have proved:

◆**Lemma 13.4.3. Observability.** A structure is observable if and only if the matrices $\mathbf{C}$ and $\mathbf{A}$ are such that $\mathcal{S}_{\mathbf{C},\mathbf{A}}$ has full rank N. ◇

Remarks. (a) Once again, the full-rank condition on $\mathcal{S}_{C,A}$ is equivalent to the positive definite condition $\mathcal{S}_{C,A}^{\dagger}\mathcal{S}_{C,A} > 0$. (b) We often say that (C,A) is observable instead of "the structure is observable."

If (C,A) is not observable, it is possible to apply a similarity transformation such that the transformed state reveals the unobservable state variables. These variables can then be eliminated, resulting in a system with fewer delays. This is the consequence of the following lemma.

♠**Lemma 13.4.4. Reduction to observable form.** Suppose (C,A) is such that $\mathcal{S}_{C,A}$ does not have full rank N. Then there exists a similarity transformation T such that the equivalent system (13.4.16a) has matrices C_1 and A_1 of the form

$$C_1 = p\ \begin{pmatrix} \overset{\rho}{C_{11}} & \overset{N-\rho}{0} \end{pmatrix}, \quad A_1 = \begin{matrix} \rho \\ N-\rho \end{matrix}\begin{pmatrix} \overset{\rho}{A_{11}} & \overset{N-\rho}{0} \\ A_{21} & A_{22} \end{pmatrix} \quad (13.4.30)$$

where ρ denotes the rank of $\mathcal{S}_{C,A}$. ◊

Proof. Very similar to proof of Lemma 13.4.2. ▽▽▽

If the state space description has C_1 and A_1 of the form (13.4.30), then the state vector can be partitioned into $\begin{bmatrix} x_1(n) \\ x_2(n) \end{bmatrix}$ where $x_1(n)$ has ρ elements. The part $x_2(n)$ does not affect $x_1(n)$ or the output and is therefore the nonobservable part. The reduced system $A_{11}, B_{11}, C_{11}, D$ (with state vector $x_1(n)$ of dimension ρ) has same input-output behavior as the original system. Here B_{11} represents the first ρ rows of $B_1 = T^{-1}B$.

A Beautiful Significance of $\mathcal{S}_{C,A}$ and $\mathcal{R}_{A,B}$

Suppose we form the product $\mathcal{S}_{C,A}\mathcal{R}_{A,B}$. The result is

$$\mathcal{S}_{C,A}\mathcal{R}_{A,B} = \begin{bmatrix} h(1) & h(2) & \cdots & h(N) \\ h(2) & h(3) & \cdots & h(N+1) \\ \vdots & \vdots & \cdots & \vdots \\ h(N) & h(N+1) & \cdots & h(2N-1) \end{bmatrix} \quad (13.4.31)$$

This is an $Np \times Nr$ matrix. It can also be considered as an $N \times N$ block matrix, with each block having size $p \times r$. So the product of the matrices which arise in the reachability and observability conditions is merely a matrix of the impulse response coefficients $h(n)$ for $1 \le n \le 2N - 1$.

Reachability and Observability Imply Minimality

It is clear from the above discussions that if a structure is minimal (i.e., N is as small as possible, viz., $N =$ McMillan degree μ) then it has to be reachable and observable. For, otherwise, we can find a smaller matrix

A_{11} and a corresponding set of matrices B_{11} and C_{11} resulting in the same transfer function $H(z)$. The converse of this result is provided by:

♠**Lemma 13.4.5.** Suppose a realization (A, B, C, D) is reachable and observable. Then it is minimal, that is, there does not exist an equivalent structure for the same transfer function with fewer delays. ◊

Proof. We prove this by contradiction. Let N denote the state space dimension for (A, B, C, D) and let (a, b, c, D) be an equivalent realization with smaller dimension $\rho < N$ (i.e., a is $\rho \times \rho$, and so on). Since the structures represent the same transfer function, the impulse response coefficients $h(n)$ are the same for both. From (13.4.13) we therefore have

$$h(n) = CA^{n-1}B = ca^{n-1}b, \quad n > 0. \qquad (13.4.32)$$

By using the result (13.4.31) we immediately arrive at

$$\underbrace{\begin{bmatrix} C \\ CA \\ \vdots \\ CA^{N-1} \end{bmatrix}}_{\mathcal{S}_{C,A}} \underbrace{[B \ AB \ \ldots \ A^{N-1}B]}_{\mathcal{R}_{A,B}} = \underbrace{\begin{bmatrix} c \\ ca \\ \vdots \\ ca^{N-1} \end{bmatrix}}_{\mathcal{S}} \underbrace{[b \ ab \ \ldots \ a^{N-1}b]}_{\mathcal{R}}$$

(13.4.33)

from which we obtain

$$\underbrace{\mathcal{S}_{C,A}^\dagger \mathcal{S}_{C,A}}_{N \times N} \underbrace{\mathcal{R}_{A,B} \mathcal{R}_{A,B}^\dagger}_{N \times N} = \underbrace{\mathcal{S}_{C,A}^\dagger \mathcal{S}}_{N \times \rho} \underbrace{\mathcal{R} \mathcal{R}_{A,B}^\dagger}_{\rho \times N}. \qquad (13.4.34)$$

As (A, B, C, D) is reachable and observable, the $N \times N$ matrices $\mathcal{S}_{C,A}^\dagger \mathcal{S}_{C,A}$ and $\mathcal{R}_{A,B} \mathcal{R}_{A,B}^\dagger$ are nonsingular. So, the LHS of (13.4.34) has rank N. But the rank of the RHS is at most $\rho < N$. This is a contradiction. ▽▽▽

The results of the above three Lemmas can be summarized as follows.

♠**Theorem 13.4.1. Minimality, reachability, and observability.** A realization (A, B, C, D) of a transfer function $H(z)$ is minimal if and only if it is reachable and observable (i.e., (A, B) reachable and (C, A) observable). ◊

In all future discussions, the word "minimal" is therefore synonymous to the condition "reachable as well as observable." This is also abbreviated as *crco* (i.e., completely reachable and completely observable).

Minimal Realizations are Related by Similarity Transforms

A nice property of minimal realizations is that all minimal realizations of a system $H(z)$ are related by similarity transformations:

♦**Lemma 13.4.6.** Let $(\mathbf{A},\mathbf{B},\mathbf{C},\mathbf{D})$ and $(\mathbf{a},\mathbf{b},\mathbf{c},\mathbf{D})$ be two minimal realizations of a $p \times r$ system $\mathbf{H}(z)$. Then there exists a unique similarity transformation $\mathbf{T}$ which transforms $(\mathbf{A},\mathbf{B},\mathbf{C},\mathbf{D})$ to $(\mathbf{a},\mathbf{b},\mathbf{c},\mathbf{D})$. ◊

Proof. Let N denote the state space dimension so that $\mathbf{A}$ and $\mathbf{a}$ are both $N \times N$. Our aim is to prove the existence of a unique $N \times N$ nonsingular $\mathbf{T}$ such that

$$\mathbf{a} = \mathbf{T}^{-1}\mathbf{A}\mathbf{T}, \quad \mathbf{b} = \mathbf{T}^{-1}\mathbf{B}, \quad \mathbf{c} = \mathbf{C}\mathbf{T}. \qquad (13.4.35)$$

Assuming $\mathbf{T}$ exists, uniqueness is established as follows: Eq. (13.4.35) implies $\mathbf{c}\mathbf{a}^n = \mathbf{C}\mathbf{A}^n\mathbf{T}$ for all integers $n \geq 0$ so that $\mathcal{S}_{\mathbf{c},\mathbf{a}} = \mathcal{S}_{\mathbf{C},\mathbf{A}}\mathbf{T}$. This implies $\mathcal{S}_{\mathbf{C},\mathbf{A}}^\dagger \mathcal{S}_{\mathbf{c},\mathbf{a}} = \mathcal{S}_{\mathbf{C},\mathbf{A}}^\dagger \mathcal{S}_{\mathbf{C},\mathbf{A}}\mathbf{T}$. Since minimality implies observability, $\mathcal{S}_{\mathbf{C},\mathbf{A}}^\dagger \mathcal{S}_{\mathbf{C},\mathbf{A}}$ is $N \times N$ nonsingular. So $\mathbf{T}$ is uniquely determined as

$$\mathbf{T} = [\mathcal{S}_{\mathbf{C},\mathbf{A}}^\dagger \mathcal{S}_{\mathbf{C},\mathbf{A}}]^{-1} \mathcal{S}_{\mathbf{C},\mathbf{A}}^\dagger \mathcal{S}_{\mathbf{c},\mathbf{a}}. \qquad (13.4.36)$$

Using (13.4.33) one can verify that $\mathbf{T}$ can also be expressed as

$$\mathbf{T} = \mathcal{R}_{\mathbf{A},\mathbf{B}} \mathcal{R}_{\mathbf{a},\mathbf{b}}^\dagger [\mathcal{R}_{\mathbf{a},\mathbf{b}} \mathcal{R}_{\mathbf{a},\mathbf{b}}^\dagger]^{-1} \qquad (13.4.37)$$

We now prove existence of $\mathbf{T}$. We know that a relation similar to (13.4.33) holds, from which we have

$$\mathcal{R}_{\mathbf{A},\mathbf{B}} = \underbrace{[\mathcal{S}_{\mathbf{C},\mathbf{A}}^\dagger \mathcal{S}_{\mathbf{C},\mathbf{A}}]^{-1} \mathcal{S}_{\mathbf{C},\mathbf{A}}^\dagger \mathcal{S}_{\mathbf{c},\mathbf{a}}}_{N \times N} \mathcal{R}_{\mathbf{a},\mathbf{b}}. \qquad (13.4.38a)$$

This step has been possible because the required inverse exists (by observability). For clarity, let us rewrite this more explicitly:

$$[\mathbf{B} \quad \mathbf{A}\mathbf{B} \quad \ldots \quad \mathbf{A}^{N-1}\mathbf{B}] = \mathbf{T}[\mathbf{b} \quad \mathbf{a}\mathbf{b} \quad \ldots \quad \mathbf{a}^{N-1}\mathbf{b}], \qquad (13.4.38b)$$

where $\mathbf{T}$ is the $N \times N$ matrix indicated in (13.4.38a). Note that this choice of $\mathbf{T}$ agrees with (13.4.36) as expected. From this we obtain $\mathbf{B} = \mathbf{T}\mathbf{b}$, which is one of the three relations in (13.4.35). Now from (13.4.32) we have

$$\mathcal{S}_{\mathbf{C},\mathbf{A}}\mathbf{A}^N\mathbf{B} = \mathcal{S}_{\mathbf{c},\mathbf{a}}\mathbf{a}^N\mathbf{b}. \qquad (13.4.39)$$

So if we append the columns $\mathbf{A}^N\mathbf{B}$ and $\mathbf{a}^N\mathbf{b}$ to the matrices $\mathcal{R}_{\mathbf{A},\mathbf{B}}$ and $\mathcal{R}_{\mathbf{a},\mathbf{b}}$ in (13.4.33) respectively, the equality continues to hold. In a way similar to (13.4.38b) we then obtain

$$\mathbf{A}\underbrace{[\mathbf{B} \quad \mathbf{A}\mathbf{B} \quad \ldots \quad \mathbf{A}^{N-1}\mathbf{B}]}_{\mathcal{R}_{\mathbf{A},\mathbf{B}}} = \mathbf{T}\mathbf{a}\underbrace{[\mathbf{b} \quad \mathbf{a}\mathbf{b} \quad \ldots \quad \mathbf{a}^{N-1}\mathbf{b}]}_{\mathcal{R}_{\mathbf{a},\mathbf{b}}}. \qquad (13.4.40)$$

Postmultiplying both sides of (13.4.40) by $\mathcal{R}_{\mathbf{a,b}}^{\dagger}$ and rearranging we obtain $\mathbf{AT} = \mathbf{Ta}$, which again is one of the relations in (13.4.35). Finally, from (13.4.33) we have

$$\mathbf{C}\mathcal{R}_{\mathbf{A,B}} = \mathbf{c}\mathcal{R}_{\mathbf{a,b}}. \qquad (13.4.41)$$

Postmultiplying with $\mathcal{R}_{\mathbf{a,b}}^{\dagger}$ and rerranging, we obtain $\mathbf{CT} = \mathbf{c}$. Summarizing, we have shown that if $\mathbf{T}$ is defined as in (13.4.36) [equivalently (13.4.37)], then all the three relations in (13.4.35) are satisfied. This establishes the existence of $\mathbf{T}$ with the advertised properties. Notice that the minimality properties are crucial in the proof because we use the inverses of $\mathcal{R}_{\mathbf{A,B}}\mathcal{R}_{\mathbf{A,B}}^{\dagger}$ and $\mathcal{S}_{\mathbf{C,A}}^{\dagger}\mathcal{S}_{\mathbf{C,A}}$. ▽▽▽

Real coefficient systems. Suppose $\mathbf{H}(z)$ is a real transfer matrix (i.e., all elements $H_{km}(z)$ have real coefficients). It is clear that there exists a structure with real valued multipliers (because we can realize each $H_{km}(z)$ individually in direct form!). What is perhaps not so obvious is the fact that there also exist *minimal* realizations with real multiplier values. A proof is requested in Problem 13.9.

The PBH Test for Reachability and Observability

The rank conditions on the matrices $\mathcal{R}_{\mathbf{A,B}}$ and $\mathcal{S}_{\mathbf{C,A}}$ can be expressed in an elegant form, in terms of the eigenvectors of $\mathbf{A}$. This result is of considerable theoretical value (as it simplifies many proofs). It is called the *PBH condition* (or test) because it was invented by Popov, Belevitch, and Hautus.

To motivate the basic idea, suppose $\mathbf{v}$ is an eigenvector of $\mathbf{A}$ which is at the same time orthogonal to all rows of $\mathbf{C}$, i.e.,

$$\mathbf{Av} = \lambda \mathbf{v}, \quad \text{and} \quad \mathbf{Cv} = \mathbf{0}, \quad \mathbf{v} \neq \mathbf{0}. \qquad (13.4.42)$$

We then have $\mathbf{CAv} = \lambda \mathbf{Cv} = \mathbf{0}$. More generally, we can verify $\mathbf{CA}^n\mathbf{v} = \mathbf{0}$ for any integer $n \geq 0$ so that $\mathcal{S}_{\mathbf{C,A}}\mathbf{v} = \mathbf{0}$. This implies that the rank of $\mathcal{S}_{\mathbf{C,A}}$ is less than N, so that the system is not observable. The PBH test asserts an even stronger result, summarized as follows:

♦**Theorem 13.4.2. The PBH test.** This can be stated in two parts.
1. The pair $(\mathbf{C}, \mathbf{A})$ is observable if and only if there does *not* exist a nonzero vector $\mathbf{v}$ such that $\mathbf{Av} = \lambda \mathbf{v}$ and $\mathbf{Cv} = \mathbf{0}$.
2. The pair $(\mathbf{A}, \mathbf{B})$ is reachable if and only if there does *not* exist a nonzero vector $\mathbf{w}$ such that $\mathbf{w}^T\mathbf{A} = \lambda \mathbf{w}^T$ and $\mathbf{w}^T\mathbf{B} = \mathbf{0}$. ◊

Proof. We prove only part 1, as the other part is similar. We already showed that (13.4.42) implies that $(\mathbf{C}, \mathbf{A})$ is not observable. Conversely, suppose $(\mathbf{C}, \mathbf{A})$ is not observable. Then we can apply a similarity transform $\mathbf{T}$ and obtain the equivalent form (13.4.30). Let $\mathbf{w}$ be an eigenvector of $\mathbf{A}_{22}$.

We then have

$$A_1 \begin{bmatrix} 0 \\ \mathbf{w} \end{bmatrix} = \begin{bmatrix} A_{11} & 0 \\ A_{21} & A_{22} \end{bmatrix} \begin{bmatrix} 0 \\ \mathbf{w} \end{bmatrix} = \begin{bmatrix} 0 \\ A_{22}\mathbf{w} \end{bmatrix} = \begin{bmatrix} 0 \\ \lambda \mathbf{w} \end{bmatrix} = \lambda \begin{bmatrix} 0 \\ \mathbf{w} \end{bmatrix},$$

so that $\mathbf{u} \triangleq \begin{bmatrix} 0 \\ \mathbf{w} \end{bmatrix}$ is an eigenvector of A_1. Furthermore

$$C_1 \mathbf{u} = [C_{11} \quad 0] \begin{bmatrix} 0 \\ \mathbf{w} \end{bmatrix} = 0.$$

This implies that the nonzero vector $\mathbf{v} \triangleq T\mathbf{u}$ satisfies (13.4.42). ▽▽▽

We conclude this subsection with the following easily proved result.

♠**Fact 13.4.1. Observability and reachability preserved under similarity transform.** Let (C, A) be observable. Then the pair (C_1, A_1) obtained by a similarity transformation is also observable. Same holds for reachability. ◇

13.4.3 Stability and Lyapunov Lemma

Recall from Sec. 2.2 that $H(z)$ is said to be stable if each element $H_{km}(z)$ is stable (i.e., has all poles inside the unit circle). Given any minimal realization (A, B, C, D) for $H(z)$, it turns out (see Theorem 13.6.1 later) that λ is a pole of $H(z)$ if and only if it is an eigenvalue of A.

So stability is equivalent to the condition that all eigenvalues λ_i of A satisfy $|\lambda_i| < 1$. If A satisfies this, we say that A *is stable*. The following property of stable matrices is very valuable in system theoretic work.

♠**Lemma 13.4.7.** Let A be stable. Then $A^n \to 0$ as $n \to \infty$. ◇

Proof. If A has distinct eigenvalues, we can write $A = T\Lambda T^{-1}$ where Λ is a diagonal matrix with diagonal elements representing the eigenvalues of A, and the columns of T are the eigenvectors. So $A^n = T\Lambda^n T^{-1}$. Since $|\lambda_i| < 1$, the quantity λ_i^n goes to zero as $n \to \infty$, so that $A^n \to 0$.

For the case where eigenvalues are *not distinct*, we cannot in general diagonalize A. In this case we can still apply the unitary triangularization result (Sec. A.7, Appendix A). According to this, *any* square matrix A can be written as $A = U\Delta U^\dagger$, where U and Δ are square matrices with $U^\dagger U = I$, and $\Delta =$ lower triangular. The diagonal elements of Δ are equal to the eigenvalues of A.

We have $A^n = U\Delta^n U^\dagger$ (using $U^\dagger U = I$) so that proving $A^n \to 0$ is equivalent to proving $\Delta^n \to 0$. To prove this we adopt the following simple trick [Franklin, 1968]: define a lower triangular matrix $\widehat{\Delta}$ as follows:

1. The diagonal elements $[\widehat{\Delta}]_{ii}$ (i.e., eigenvalues of $\widehat{\Delta}$) are distinct, with $|[\Delta]_{ii}| \leq [\widehat{\Delta}]_{ii} < 1$. This is always possible since stability of $\mathbf{A}$ assures us that $|[\Delta]_{ii}| < 1$ for all i.
2. $[\widehat{\Delta}]_{ij} \geq |[\Delta]_{ij}|$ for all (i,j).

As the eigenvalues of $\widehat{\Delta}$ are distinct and bounded by unity, we conclude $\widehat{\Delta}^n \to \mathbf{0}$. Since all the elements of $\widehat{\Delta}$ are nonnegative and satisfy $[\widehat{\Delta}]_{ij} \geq |[\Delta]_{ij}|$, it is clear that $[\widehat{\Delta}^n]_{ij} \geq [\Delta^n]_{ij}$ for all integers $n > 0$ so that $\Delta^n \to \mathbf{0}$ indeed! ▽▽▽

The stability of $\mathbf{A}$ can be expressed elegantly in terms of a simple algebraic equation. This is called the discrete-time Lyapunov Lemma, even though it is a blending of results due to many authors [Anderson and Moore, 1979].

♠**Lemma 13.4.8. Discrete-time Lyapunov lemma.** We will state this in two parts for convenience.
1. Let $\mathbf{A}$ be $N \times N$, $\mathbf{C}$ be $p \times N$ and let $(\mathbf{C}, \mathbf{A})$ be observable. Let the equation

$$\mathbf{A}^\dagger \mathbf{P} \mathbf{A} + \mathbf{C}^\dagger \mathbf{C} = \mathbf{P}, \qquad (13.4.43)$$

be satisfied for some Hermitian positive definite $\mathbf{P}$. Then $\mathbf{A}$ is stable.
2. Let $\mathbf{Q}$ be some $N \times N$ Hermitian positive semidefinite matrix so that it can be written as $\mathbf{Q} = \mathbf{C}^\dagger \mathbf{C}$ for some (possibly rectangular) $\mathbf{C}$. Let $\mathbf{A}$ be $N \times N$ stable with $(\mathbf{C}, \mathbf{A})$ observable. Then the algebraic equation (13.4.43) has a unique solution $\mathbf{P}$, and the solution is Hermitian positive definite. ◊

Proof. First consider part 1. Since $\mathbf{P}$ is Hermitian and positive definite, there exists nonsingular $\mathbf{T}$ such that $\mathbf{P} = [\mathbf{T}\mathbf{T}^\dagger]^{-1}$. So we can rearrange (13.4.43) as

$$\mathbf{T}^\dagger \mathbf{A}^\dagger \mathbf{T}^{-\dagger} \mathbf{T}^{-1} \mathbf{A} \mathbf{T} + \mathbf{T}^\dagger \mathbf{C}^\dagger \mathbf{C} \mathbf{T} = \mathbf{I}. \qquad (13.4.44)$$

Define $\mathbf{A}_1 = \mathbf{T}^{-1} \mathbf{A} \mathbf{T}$ and $\mathbf{C}_1 = \mathbf{C}\mathbf{T}$. Then (13.4.44) becomes

$$\mathbf{A}_1^\dagger \mathbf{A}_1 + \mathbf{C}_1^\dagger \mathbf{C}_1 = \mathbf{I}. \qquad (13.4.45)$$

Now suppose λ is an eigenvalue of $\mathbf{A}$ (hence that of $\mathbf{A}_1$) so that $\mathbf{A}_1 \mathbf{v} = \lambda \mathbf{v}$ for some $\mathbf{v} \neq \mathbf{0}$. From (13.4.45) we have $\mathbf{v}^\dagger \mathbf{A}_1^\dagger \mathbf{A}_1 \mathbf{v} + \mathbf{v}^\dagger \mathbf{C}_1^\dagger \mathbf{C}_1 \mathbf{v} = \mathbf{v}^\dagger \mathbf{v}$, that is, $(1 - |\lambda|^2)\mathbf{v}^\dagger \mathbf{v} = [\mathbf{C}_1 \mathbf{v}]^\dagger [\mathbf{C}_1 \mathbf{v}]$. Since $[\mathbf{C}_1 \mathbf{v}]^\dagger [\mathbf{C}_1 \mathbf{v}] \geq 0$, this proves that $|\lambda| < 1$ unless $\mathbf{C}_1 \mathbf{v} = \mathbf{0}$. But by PBH test (Theorem 13.4.2) this cannot happen because observability of $(\mathbf{C}, \mathbf{A})$ implies that of $(\mathbf{C}_1, \mathbf{A}_1)$ (by Fact 13.4.1). This proves Part 1.

Now consider part 2. Define

$$\mathbf{P} = \sum_{n=0}^{\infty} [\mathbf{A}^\dagger]^n \mathbf{C}^\dagger \mathbf{C} \mathbf{A}^n. \qquad (13.4.46)$$

Since **A** is stable, this summation converges [p. 64, Anderson and Moore, 1979]. Evidently **P** is Hermitian and positive semidefinite. Moreover, the sum of the first N terms is precisely equal to $\mathcal{S}_{C,A}^\dagger \mathcal{S}_{C,A}$, which is positive definite (by observability), so that **P** is positive definite. It is readily verified by substitution that this **P** satisfies (13.4.43).

To prove that **P** is the only solution, suppose $\mathbf{P}_1$ and $\mathbf{P}_2$ are solutions to (13.4.43). Then $\mathbf{A}^\dagger \mathbf{R} \mathbf{A} = \mathbf{R}$, where $\mathbf{R} = \mathbf{P}_1 - \mathbf{P}_2$. By repeated use of this, one obtains $[\mathbf{A}^\dagger]^n \mathbf{R} \mathbf{A}^n = \mathbf{R}$, for all $n \geq 0$. By Lemma 13.4.7, $\mathbf{A}^n$ goes to zero as $n \to \infty$ so that this implies $\mathbf{R} = \mathbf{0}$. So **P** is unique. $\triangledown\triangledown\triangledown$

13.5 THE SMITH-McMILLAN FORM

For transfer matrices $\mathbf{H}(z)$ and their structures, a study of advanced concepts such as transmission zeros, minimality, degree, and related things is greatly facilitated by a diagonalization result known as the Smith-McMillan decomposition. In this section we study these results.

13.5.1 The Smith Form of a Polynomial Matrix

Given a $p \times r$ polynomial matrix $\mathbf{P}(z)$, it is possible to obtain simpler forms such as triangular and diagonal forms [Smith, 1861], by performing certain operations called elementary operations on the matrix.

Elementary Operations on Polynomial Matrices

An elementary row operation on $\mathbf{P}(z)$ is defined to be any one of the following:

Type 1. Interchange two rows.
Type 2. Multiply a row with a nonzero constant c.
Type 3. Add a polynomial multiple of a row to another row.

Elementary column operations are defined in a similar way. The above row operations can be performed by premultiplying $\mathbf{P}(z)$ with an appropriate square matrix, called an elementary matrix. There are three types of elementary matrices corresponding to the above three operations. Examples of 3×3 elementary matrices are shown below.

$$\underbrace{\begin{bmatrix} 0 & 0 & 1 \\ 0 & 1 & 0 \\ 1 & 0 & 0 \end{bmatrix}}_{Type\ 1}, \underbrace{\begin{bmatrix} 1 & 0 & 0 \\ 0 & c & 0 \\ 0 & 0 & 1 \end{bmatrix}}_{Type\ 2}, \underbrace{\begin{bmatrix} 1 & 0 & 0 \\ 0 & 1 & 0 \\ \alpha(z) & 0 & 1 \end{bmatrix}}_{Type\ 3}. \qquad (13.5.1)$$

The first matrix interchanges row 0 with row 2. (Remember that rows and columns are numbered starting from zero.) The second matrix multiplies all elements of row 1 with $c \neq 0$. The third matrix replaces row 2 with

$$(\text{row 2}) + (\alpha(z) \times \text{row 0}), \qquad (13.5.2)$$

where $\alpha(z)$ is a polynomial in z.

Notice that all the three types of elementary matrices are unimodular. Repeated application of various row operations amounts to premultiplication of $\mathbf{P}(z)$ with a $p \times p$ unimodular matrix. (We will see later that *any* unimodular matrix is a product of the three types of elementary matrices.)

Elementary column operations are equivalent to *post* multiplication of $\mathbf{P}(z)$ with one of the three types of elementary matrices. Repeated elementary column operations amounts to post multiplication by an $r \times r$ unimodular matrix.

Elementary operations can perform wonders. For example, by repeated row and column operations, one can reduce $\mathbf{P}(z)$ into a diagonal matrix, whose diagonal entries are polynomials. This result is theoretically extremely powerful, and helps us to obtain many conclusions which would otherwise be difficult to derive.

The division theorem. The key property which enables us to obtain the diagonalization is the *division theorem* for polynomials. This states that if $P_1(z)$ and $P_2(z)$ are scalar polynomials in z with order of $P_1(z) \geq$ order of $P_2(z)$, then we can find unique polynomials $Q(z)$ (quotient polynomial) and $R(z)$ (reminder polynomial) such that

$$P_1(z) = Q(z)P_2(z) + R(z), \qquad (13.5.3)$$

and such that the order of $R(z) <$ order of $P_2(z)$. (This includes the case $R(z) = 0$.)

Example 13.5.1

Let

$$\mathbf{P}(z) = \begin{matrix} & 0 & 1 \\ 0 & \begin{pmatrix} z+1 & z \\ 2z^2 + 3 & 2(z+1)^2 \end{pmatrix} \\ 1 & & \end{matrix}. \qquad (13.5.4)$$

If we divide the $(1,0)$ element $2z^2 + 3$ by the $(0,0)$ element $z + 1$ we obtain

$$2z^2 + 3 = \underbrace{2(z-1)}_{Q(z)}(z+1) + \underbrace{5}_{R(z)}. \qquad (13.5.5)$$

By using this fact we can perform an elementary row operation of Type 3 to reduce the $(1,0)$ element to a constant $(= R(z) = 5)$. Thus

$$\begin{bmatrix} 1 & 0 \\ -2(z-1) & 1 \end{bmatrix} \underbrace{\begin{bmatrix} z+1 & z \\ 2z^2+3 & 2(z+1)^2 \end{bmatrix}}_{\mathbf{P}(z)} = \begin{bmatrix} z+1 & z \\ 5 & 2(3z+1) \end{bmatrix}. \qquad (13.5.6)$$

We can now reduce the (0,0) element to a constant by replacing row 0 with

$$(\text{row } 0) - z/5 \times (\text{row } 1) \quad (\text{Type 3 operation}).$$

The result is

$$\begin{bmatrix} 1 & -z/5 \\ 0 & 1 \end{bmatrix} \begin{bmatrix} z+1 & z \\ 5 & 2(3z+1) \end{bmatrix} = \begin{bmatrix} 1 & 3z(1-2z)/5 \\ 5 & 2(3z+1) \end{bmatrix}. \quad (13.5.7)$$

Next the $(0,1)$ element can be forced to zero by an elementary column operation as follows:

$$\begin{bmatrix} 1 & 3z(1-2z)/5 \\ 5 & 2(3z+1) \end{bmatrix} \begin{bmatrix} 1 & -3z(1-2z)/5 \\ 0 & 1 \end{bmatrix} = \begin{bmatrix} 1 & 0 \\ 5 & 6z^2 + 3z + 2 \end{bmatrix}. \quad (13.5.8)$$

Finally the $(1,0)$ element can be forced to be zero by an elementary row operation:

$$\begin{bmatrix} 1 & 0 \\ -5 & 1 \end{bmatrix} \begin{bmatrix} 1 & 0 \\ 5 & 6z^2 + 3z + 2 \end{bmatrix} = \begin{bmatrix} 1 & 0 \\ 0 & 6z^2 + 3z + 2 \end{bmatrix} = \mathbf{\Gamma}(z), \quad (13.5.9)$$

eventually resulting in a diagonal matrix. The sequence of row operations can be represented as

$$\begin{bmatrix} 1 & 0 \\ -5 & 1 \end{bmatrix} \begin{bmatrix} 1 & -z/5 \\ 0 & 1 \end{bmatrix} \begin{bmatrix} 1 & 0 \\ -2(z-1) & 1 \end{bmatrix} = \begin{bmatrix} (2z^2 - 2z + 5)/5 & -z/5 \\ -2z^2 - 3 & z+1 \end{bmatrix},$$

which we denote $\mathbf{W}_1(z)$. The only column operation is represented by

$$\mathbf{V}_1(z) = \begin{bmatrix} 1 & 3z(2z-1)/5 \\ 0 & 1 \end{bmatrix}, \quad (13.5.10)$$

so that

$$\mathbf{W}_1(z) \mathbf{P}(z) \mathbf{V}_1(z) = \mathbf{\Gamma}(z). \quad (13.5.11)$$

Since $\mathbf{W}_1(z)$ and $\mathbf{V}_1(z)$ are unimodular, their inverses are also unimodular polynomials in z. So we have the decomposition

$$\mathbf{P}(z) = \mathbf{W}(z) \mathbf{\Gamma}(z) \mathbf{V}(z), \quad (13.5.12)$$

where $\mathbf{\Gamma}(z)$ is diagonal and $\mathbf{W}(z), \mathbf{V}(z)$ are unimodular. If desired, the diagonal entries of $\mathbf{\Gamma}(z)$ can further be forced to be monic polynomials (i.e., highest power has coefficient unity), by use of Type 2 operations.

The above example is a demonstration of a general diagonalization theorem which can be stated as follows.

♦**Theorem 13.5.1. The Smith form** [Smith, 1861]. Let $\mathbf{P}(z)$ be a $p \times r$ matrix polynomial in z. Then there exists finite number of elementary row and column operations which reduce $\mathbf{P}(z)$ into a diagonal polynomial matrix. So we can write

$$\mathbf{P}(z) = \underbrace{\mathbf{W}(z)}_{p \times p} \underbrace{\mathbf{\Gamma}(z)}_{p \times r} \underbrace{\mathbf{V}(z)}_{r \times r} \quad \text{(Smith form decomposition)}. \tag{13.5.13}$$

where $\mathbf{W}(z)$ and $\mathbf{V}(z)$ are unimodular matrix polynomials in the variable z and

$$\mathbf{\Gamma}(z) = \begin{bmatrix} \gamma_0(z) & 0 & \cdots & 0 & 0 & \cdots & 0 \\ 0 & \gamma_1(z) & \cdots & 0 & 0 & \cdots & 0 \\ \vdots & \vdots & \ddots & \vdots & \vdots & & \vdots \\ 0 & 0 & \cdots & \gamma_{\rho-1}(z) & 0 & \cdots & 0 \\ 0 & 0 & \cdots & 0 & 0 & \cdots & 0 \\ \vdots & \vdots & & \vdots & \vdots & & \vdots \\ 0 & 0 & \cdots & 0 & 0 & \cdots & 0 \end{bmatrix}. \tag{13.5.14}$$

Here ρ is the normal rank of $\mathbf{P}(z)$. Moreover the unimodular matrices can be so chosen that the polynomials $\gamma_i(z)$ are monic (i.e., highest power has coefficient unity), and $\gamma_i(z)$ is a factor of $\gamma_{i+1}(z)$, that is,

$$\gamma_i(z) \big| \gamma_{i+1}(z), \quad 0 \leq i \leq \rho - 2. \tag{13.5.15a}$$

Finally, for a given $\mathbf{P}(z)$, the matrix $\mathbf{\Gamma}(z)$ is unique, and the elements $\gamma_i(z)$ are given by

$$\gamma_i(z) = \Delta_{i+1}(z)/\Delta_i(z), \tag{13.5.15b}$$

where $\Delta_i(z), i > 0$, is the greatest common divisor of all the $i \times i$ minors of $\mathbf{P}(z)$, and $\Delta_0(z) = 1$. $\mathbf{\Gamma}(z)$ is called the *Smith form* of $\mathbf{P}(z)$. ◊

Sketch of proof. We will assume that $P_{00}(z)$ is nonzero and has the smallest order among all nonzero elements. [This can be arranged by use of Type 1 row and column operations, which will be part of $\mathbf{W}(z)$ and $\mathbf{V}(z)$.] Suppose there is a nonzero element $P_{0k}(z)$ in the 0th row. Let $Q(z)$ and $R(z)$ denote the quotient and remainder when we attempt to divide $P_{0k}(z)$ with $P_{00}(z)$, that is,

$$P_{0k}(z) = Q(z)P_{00}(z) + R(z). \tag{13.5.16a}$$

We can now perform a Type 3 column operation so that the resulting matrix has the element $R(z)$ in place of $P_{0k}(z)$. (Only the kth column is affected by this operation.) Since $R(z)$ is the remainder, its order is smaller than that of $P_{00}(z)$. By performing another Type 1 operation we can bring $R(z)$ to the $(0,0)$ location. In this way we can keep reducing the order of the

$(0,0)$ element. Since this can proceed only a finite number of times, all the remaining elements in the 0th row eventually become zero. In a similar way, by performing elementary *row* operations, we can convert all the elements in the 0th column (except the $(0,0)$ element) to zero. At this point, the polynomial matrix takes the form

$$\begin{bmatrix} \gamma_0(z) & 0 \\ 0 & \mathbf{S}(z) \end{bmatrix}, \qquad (13.5.16b)$$

where $\gamma_0(z)$ is a scalar polynomial and $\mathbf{S}(z)$ is a matrix polynomial.

Suppose $\gamma_0(z)$ is *not* a factor of all the elements of $\mathbf{S}(z)$. For example let the $(1,1)$ element of (13.5.16b) be one such. We can then add the 1st row to the 0th row (Type 3 operation), and create a situation whereby the order of the $(0,0)$ element can be reduced further. Since the order reduction cannot proceed indefinitely, we eventually obtain the form (13.5.16b) where $\gamma_0(z)$ is a factor of all the elements in $\mathbf{S}(z)$.

We can repeat the entire set of operations on the smaller matrix $\mathbf{S}(z)$. Continuing in this way, we eventually arrive at a diagonal matrix $\mathbf{\Gamma}(z)$ whose elements satisfy (13.5.15a). The monic nature of the polynomials $\gamma_i(z)$ can be ensured trivially by use of Type 2 operations.

It remains to prove (13.5.15b). For this first consider the case where $\mathbf{W}(z) = \mathbf{I}_p$ and $\mathbf{V}(z) = \mathbf{I}_r$, so that $\mathbf{P}(z) = \mathbf{\Gamma}(z)$. In view of the divisibility condition (13.5.15a), we can write

$$\gamma_0(z) = \alpha_0(z), \quad \gamma_1(z) = \alpha_0(z)\alpha_1(z), \quad \gamma_2(z) = \alpha_0(z)\alpha_1(z)\alpha_2(z), \ldots$$

where $\alpha_i(z)$ are polynomials in z. The 1×1 minors of $\mathbf{\Gamma}(z)$ are

$$0, \; \alpha_0(z), \; \alpha_0(z)\alpha_1(z), \; \ldots$$

so that their gcd $\Delta_1(z) = \alpha_0(z)$. So $\gamma_0(z) = \Delta_1(z)/\Delta_0(z)$, since $\Delta_0(z) = 1$. Similarly we can verify $\gamma_1(z) = \Delta_2(z)/\Delta_1(z)$ and so on. This proves (13.5.15b) when $\mathbf{P}(z) = \mathbf{\Gamma}(z)$. Finally consider the case when $\mathbf{W}(z)$ and $\mathbf{V}(z)$ are not identity. These matrices perform elementary operations on the rows and columns of $\mathbf{\Gamma}(z)$. It can be shown by use of the so-called Binet-Cauchy theorem [Gantmacher, 1959] that the gcd of $i \times i$ minors is unaffected (except for scale factors) by these operations. So (13.5.15b) continues to hold. ▽▽▽

Notice that (13.5.13) implies, in particular,

$$\det \mathbf{P}(z) = c \times \det \mathbf{\Gamma}(z). \qquad (13.5.17)$$

The matrix $\mathbf{\Gamma}(z)$ is called the Smith form of $\mathbf{P}(z)$, and (13.5.13) the Smith decomposition. Here is an example of $\mathbf{\Gamma}(z)$ that satisfies all conditions stated in the Theorem:

$$\begin{bmatrix} (2+z) & 0 & 0 \\ 0 & (2+z)(3-z^2) & 0 \\ 0 & 0 & (2+z)^2(3-z^2) \\ 0 & 0 & 0 \end{bmatrix}.$$

Another example is $\Gamma(z) = (a+z)\mathbf{I}$.

Example 13.5.2

Even though the Smith form $\Gamma(z)$ is unique, the matrices $\mathbf{W}(z)$ and $\mathbf{V}(z)$ are not. For example,

$$\begin{bmatrix} 1 & 0 \\ z^2 & z \end{bmatrix} = \begin{bmatrix} 1 & 0 \\ z^2 & 1 \end{bmatrix}\begin{bmatrix} 1 & 0 \\ 0 & z \end{bmatrix}\begin{bmatrix} 1 & 0 \\ 0 & 1 \end{bmatrix} = \begin{bmatrix} 1 & 0 \\ 0 & 1 \end{bmatrix}\begin{bmatrix} 1 & 0 \\ 0 & z \end{bmatrix}\begin{bmatrix} 1 & 0 \\ z & 1 \end{bmatrix},$$

which shows that there are (at least) two possible choices for $\mathbf{W}(z)$ and $\mathbf{V}(z)$.

Application to unimodular matrices. Now suppose that $\mathbf{P}(z)$ is itself a unimodular matrix, i.e., a square matrix with nonzero constant determinant. In view of (13.5.17), the Smith form has to be $\Gamma(z) = \mathbf{I}$ in this case. Since $\mathbf{W}(z)$ and $\mathbf{V}(z)$ are products of elementary matrices, this proves that *any* unimodular matrix is a product of elementary matrices. This important result can be summarized as

♦**Corollary 13.5.1. Factorization of unimodular matrices.** A square polynomial matrix $\mathbf{P}(z)$ is unimodular if and *only if* it is a product of finite number of elementary matrices. ◇

13.5.2 Application in Glcd Extraction

Given the polynomial matrices $\mathbf{P}(z)$ and $\mathbf{Q}(z)$ in the MFD $\mathbf{Q}^{-1}(z)\mathbf{P}(z)$, suppose we wish to extract a greatest left common divisor (glcd). This can be done by applying the above decomposition theory. For this let

$$\mathbf{S}(z) \triangleq p\;(\overset{p}{\mathbf{Q}(z)}\;\overset{r}{\mathbf{P}(z)}). \qquad (13.5.18)$$

Performing the Smith decomposition we get

$$\mathbf{W}_1(z)[\mathbf{Q}(z)\;\mathbf{P}(z)]\mathbf{V}_1(z) = [\times\;\;\mathbf{0}], \qquad (13.5.19)$$

where × is $p \times p$ denoting possibly nonzero polynomial entries. The fact that × is diagonal is irrelevant in this application. Since $\mathbf{W}_1(z)$ is unimodular, its inverse is a polynomial in z and we can rewrite (13.5.19) as

$$[\mathbf{Q}(z)\;\mathbf{P}(z)]\mathbf{V}_1(z) = [\mathbf{R}(z)\;\;\mathbf{0}] \qquad (13.5.20)$$

where $\mathbf{R}(z)$ is a $p \times p$ polynomial matrix. Now $\mathbf{V}(z) \triangleq \mathbf{V}_1^{-1}(z)$ is a polynomial in z (since $\mathbf{V}_1(z)$ is unimodular), so we can write (13.5.20) as

$$[\mathbf{Q}(z)\;\mathbf{P}(z)] = [\mathbf{R}(z)\;\;\mathbf{0}]\begin{bmatrix} \mathbf{V}_{00}(z) & \mathbf{V}_{01}(z) \\ \mathbf{V}_{10}(z) & \mathbf{V}_{11}(z) \end{bmatrix} \qquad (13.5.21)$$

where $\mathbf{V}_{ij}(z)$ are polynomials of appropriate sizes. So

$$\mathbf{Q}(z) = \mathbf{R}(z)\mathbf{V}_{00}(z), \quad \mathbf{P}(z) = \mathbf{R}(z)\mathbf{V}_{01}(z) \tag{13.5.22}$$

which shows that $\mathbf{R}(z)$ is an lcd of $\mathbf{P}(z)$ and $\mathbf{Q}(z)$.

We now show that $\mathbf{R}(z)$ is in fact a glcd of $\mathbf{Q}(z)$ and $\mathbf{P}(z)$. By partitioning $\mathbf{V}_1(z)$ in an obvious manner we can write (13.5.20) as

$$[\mathbf{Q}(z) \quad \mathbf{P}(z)] \begin{bmatrix} \mathbf{V}_q(z) & \times \\ \mathbf{V}_p(z) & \times \end{bmatrix} = [\mathbf{R}(z) \quad \mathbf{0}] \tag{13.5.23}$$

where $\times$ denotes entries whose details are irrelevant. From this equation we obtain

$$\mathbf{Q}(z)\mathbf{V}_q(z) + \mathbf{P}(z)\mathbf{V}_p(z) = \mathbf{R}(z). \tag{13.5.24}$$

This shows that any lcd of $\mathbf{P}(z)$ and $\mathbf{Q}(z)$ is also a left divisor of $\mathbf{R}(z)$. So $\mathbf{R}(z)$ is in fact a glcd of $\mathbf{Q}(z)$ and $\mathbf{P}(z)$. Summarizing, we have established:

♦**Lemma 13.5.1. Extension of Euclid's theorem.** Let $\mathbf{Q}(z)$ and $\mathbf{P}(z)$ be matrix polynomials in z with sizes $p \times p$ and $p \times r$, respectively, and let $\mathbf{R}(z)$ be a glcd. Then there exist matrix polynomials $\mathbf{V}_p(z)$ and $\mathbf{V}_q(z)$ such that $\mathbf{Q}(z)\mathbf{V}_q(z) + \mathbf{P}(z)\mathbf{V}_p(z) = \mathbf{R}(z)$. ◊

Comments

1. When $\mathbf{Q}(z)$ and $\mathbf{P}(z)$ are scalar polynomials, this reduces to the well-known Euclid's theorem [Sec. 2.3, Bose, 1985]. This says that when $Q(z)$ and $P(z)$ are polynomials in z with greatest common factor $R(z)$, there exist polynomials $V_p(z)$ and $V_q(z)$ such that $Q(z)V_q(z) + P(z)V_p(z) = R(z)$.

2. When $\mathbf{Q}(z)$ and $\mathbf{P}(z)$ are left coprime, we can take $\mathbf{R}(z) = \mathbf{I}$. So there exist polynomial matrices $\mathbf{V}_q(z)$ and $\mathbf{V}_p(z)$ such that

$$\mathbf{Q}(z)\mathbf{V}_q(z) + \mathbf{P}(z)\mathbf{V}_p(z) = \mathbf{I}_p. \tag{13.5.25}$$

This result is called Bezout's identity. Conversely, (13.5.25) implies that $\mathbf{Q}(z)$ and $\mathbf{P}(z)$ are left coprime (why?).

Obtaining deeper results about irreducible MFDs

Using (13.5.25) we can develop deeper results on irreducible matrix fraction descriptions.

♦**Lemma 13.5.2.** Let $\mathbf{Q}^{-1}(z)\mathbf{P}(z)$ and $\mathbf{Q}_2^{-1}(z)\mathbf{P}_2(z)$ be two MFDs representing $\mathbf{H}(z)$ and let $\mathbf{Q}^{-1}(z)\mathbf{P}(z)$ be irreducible. Then $\mathbf{Q}_2(z)\mathbf{Q}^{-1}(z)$ is a polynomial matrix. ◊

Proof. Since $\mathbf{Q}^{-1}(z)\mathbf{P}(z)$ is irreducible, (13.5.25) holds for some polynomials $\mathbf{V}_p(z)$ and $\mathbf{V}_q(z)$. But $\mathbf{Q}^{-1}(z)\mathbf{P}(z) = \mathbf{Q}_2^{-1}(z)\mathbf{P}_2(z)$ so that

$$\mathbf{Q}(z)\mathbf{V}_q(z) + \underbrace{\mathbf{Q}(z)\mathbf{Q}_2^{-1}(z)\mathbf{P}_2(z)}_{\mathbf{P}(z)}\mathbf{V}_p(z) = \mathbf{I}.$$

This can be rearranged as

$$\mathbf{Q}_2(z)\mathbf{V}_q(z) + \mathbf{P}_2(z)\mathbf{V}_p(z) = \mathbf{Q}_2(z)\mathbf{Q}^{-1}(z).$$

Since the left hand side is a polynomial, this proves the desired result. ▽▽▽

The above lemma shows that we can write

$$\mathbf{Q}_2(z) = \mathbf{L}(z)\mathbf{Q}(z), \quad \mathbf{P}_2(z) = \mathbf{L}(z)\mathbf{P}(z),$$

where $\mathbf{L}(z) = \mathbf{Q}_2(z)\mathbf{Q}^{-1}(z)$ is a polynomial (an lcd of $\mathbf{Q}_2(z)$ and $\mathbf{P}_2(z)$). In the above lemma if both the MFDs are irreducible, then we can interchange their roles to argue that $\mathbf{Q}(z)\mathbf{Q}_2^{-1}(z)$ is a polynomial (i.e., $\mathbf{L}^{-1}(z)$ is a polynomial. In other words, $\mathbf{L}(z)$ is not only a polynomial, but is unimodular. Summarizing, we have proved:

♠**Corollary 13.5.2.** Let $\mathbf{Q}_1^{-1}(z)\mathbf{P}_1(z)$ and $\mathbf{Q}_2^{-1}(z)\mathbf{P}_2(z)$ be irreducible MFDs for the same system $\mathbf{H}(z)$. Then $\mathbf{Q}_2(z)\mathbf{Q}_1^{-1}(z)$ is a *unimodular* polynomial matrix. So we can relate the MFDs by

$$\mathbf{Q}_2(z) = \mathbf{V}(z)\mathbf{Q}_1(z), \quad \mathbf{P}_2(z) = \mathbf{V}(z)\mathbf{P}_1(z),$$

where $\mathbf{V}(z) = \mathbf{Q}_2(z)\mathbf{Q}_1^{-1}(z)$ is a unimodular polynomial matrix. ◇

This gives rise to a few other very valuable conclusions. First

$$\det \mathbf{Q}_2(z) = c[\det \mathbf{Q}_1(z)],$$

where $c = \det \mathbf{V}(z) =$ constant, so that the determinant of $\mathbf{Q}(z)$ is the same (except for scale factor) for all irreducible MFDs of a given system. In particular the *order* of this determinant is same. From Sec. 13.3 we also know that this order is strictly *smaller* for irreducible MFDs than for reducible ones. Summarizing we have:

♠**Corollary 13.5.3.** The quantity $[\det \mathbf{Q}(z)]$ is the same (except for constant scale factor) for all irreducible MFDs of a fixed system $\mathbf{H}(z)$. This means in particular that the order of $[\det \mathbf{Q}(z)]$ is the same for all irreducible MFDs of $\mathbf{H}(z)$. Furthermore, the order of $[\det \mathbf{Q}(z)]$ is strictly smaller for irreducible MFDs, than for any reducible MFD. ◇

13.5.3 The Smith-McMillan Form for Transfer Matrices

Let $\mathbf{H}(z)$ be a $p \times r$ rational transfer matrix representing a causal LTI system. Assume each element $H_{km}(z)$ has been expressed as $H_{km}(z) = P_{km}(z)/D(z)$ where $D(z)$ is the least common multiple of the denominators of $H_{km}(z)$. Here $P_{km}(z)$ and $D(z)$ are polynomials in the variable z. Define the $p \times r$ matrix $\mathbf{P}(z) = [P_{km}(z)]$, and let (13.5.12) be its Smith decomposition. We can then write

$$\mathbf{H}(z) = \underbrace{\mathbf{W}(z)}_{p \times p} \underbrace{\mathbf{\Lambda}(z)}_{p \times r} \underbrace{\mathbf{V}(z)}_{r \times r}, \qquad (13.5.26)$$

where

$$\Lambda(z) = \begin{bmatrix} \lambda_0(z) & 0 & \cdots & 0 & 0 & \cdots & 0 \\ 0 & \lambda_1(z) & \cdots & 0 & 0 & \cdots & 0 \\ \vdots & \vdots & \ddots & \vdots & \vdots & & \vdots \\ 0 & 0 & \cdots & \lambda_{\rho-1}(z) & 0 & \cdots & 0 \\ 0 & 0 & \cdots & 0 & 0 & \cdots & 0 \\ \vdots & \vdots & & \vdots & \vdots & \cdots & \vdots \\ 0 & 0 & \cdots & 0 & 0 & \cdots & 0 \end{bmatrix}. \quad (13.5.27)$$

Evidently $\lambda_i(z) = \gamma_i(z)/D(z)$. By canceling off the common factors between $\gamma_i(z)$ and $D(z)$ we can write this in irreducible form as

$$\lambda_i(z) = \frac{\alpha_i(z)}{\beta_i(z)}, \quad 0 \leq i \leq \rho - 1. \quad (13.5.28)$$

In view of the divisibility property (13.5.15a), the polynomials $\alpha_i(z)$ and $\beta_i(z)$ satisfy

$$\alpha_i(z)|\alpha_{i+1}(z), \quad \beta_{i+1}(z)|\beta_i(z), \quad 0 \leq i \leq \rho - 2. \quad (13.5.29)$$

Equation (13.5.26) is called the *Smith-McMillan decomposition* of $\mathbf{H}(z)$, and $\Lambda(z)$ the Smith-McMillan form of $\mathbf{H}(z)$. We summarize these as

♠**Theorem 13.5.2. Smith-McMillan form.** Let $\mathbf{H}(z)$ be a $p \times r$ rational transfer matrix representing a causal discrete-time system. Then it can be decomposed into the form (13.5.26) where

1. $\mathbf{W}(z)$ and $\mathbf{V}(z)$ are unimodular matrix polynomials in z.
2. $\Lambda(z)$ is a $p \times r$ diagonal matrix as in (13.5.27) where ρ is the normal rank of $\mathbf{H}(z)$.
3. The diagonal elements of $\Lambda(z)$ can be expressed as in (13.5.28) where $\alpha_i(z)$ and $\beta_i(z)$ are relatively prime polynomials in z satisfying (13.5.29). Furthermore the polynomials $\alpha_i(z)$ and $\beta_i(z)$ are unique up to a constant scale factor. ◊

It is worth emphasizing here that the Smith-form (13.5.14) is developed for a polynomial matrix $\mathbf{P}(z)$, whereas the Smith-McMillan form (13.5.27) is developed for a *causal* rational LTI system $\mathbf{H}(z)$.

Example 13.5.3

Let

$$\mathbf{H}(z) = \begin{bmatrix} \dfrac{z^{-3}}{(1+2z^{-1})^2(1+3z^{-1})^2} & \dfrac{-z^{-1}}{(1+3z^{-1})^2} \\ \dfrac{z^{-1}}{(1+3z^{-1})^2} & \dfrac{-z^{-1}}{(1+3z^{-1})^2} \end{bmatrix} \quad (13.5.30)$$

$$= \dfrac{1}{(z+2)^2(z+3)^2} \underbrace{\begin{bmatrix} z & -z(z+2)^2 \\ z(z+2)^2 & -z(z+2)^2 \end{bmatrix}}_{\mathbf{P}(z)}.$$

The following is a Smith-decomposition of $\mathbf{P}(z)$:

$$\begin{bmatrix} 1 & 0 \\ (z+2)^2 & 1 \end{bmatrix} \begin{bmatrix} z & 0 \\ 0 & z(z+2)^2(z^2+4z+3) \end{bmatrix} \begin{bmatrix} 1 & -(z+2)^2 \\ 0 & 1 \end{bmatrix}, \quad (13.5.31)$$

so that $\mathbf{H}(z)$ can be written as

$$\underbrace{\begin{bmatrix} 1 & 0 \\ (z+2)^2 & 1 \end{bmatrix}}_{\mathbf{W}(z)} \underbrace{\begin{bmatrix} \dfrac{z}{(z+2)^2(z+3)^2} & 0 \\ 0 & \dfrac{z(z^2+4z+3)}{(z+3)^2} \end{bmatrix}}_{\mathbf{\Lambda}(z)} \underbrace{\begin{bmatrix} 1 & -(z+2)^2 \\ 0 & 1 \end{bmatrix}}_{\mathbf{V}(z)}.$$

(13.5.32)

We can now identify $\alpha_i(z)$ and $\beta_i(z)$ as

$$\begin{aligned} \alpha_0(z) &= z, \quad \alpha_1(z) = z(z^2+4z+3) \\ \beta_0(z) &= (z+2)^2(z+3)^2, \quad \beta_1(z) = (z+3)^2. \end{aligned} \quad (13.5.33)$$

It is clear that the divisibility conditions (13.5.29) are satisfied. This example demonstrates a very important point: the Smith-McMillan form $\mathbf{\Lambda}(z)$ is not necessarily causal even though $\mathbf{H}(z)$ is causal. We can rewrite $\mathbf{\Lambda}(z)$ in (13.5.32) as

$$\begin{bmatrix} \dfrac{z^{-3}}{(1+2z^{-1})^2(1+3z^{-1})^2} & 0 \\ 0 & \dfrac{z(1+4z^{-1}+3z^{-2})}{(1+3z^{-1})^2} \end{bmatrix}. \quad (13.5.34)$$

So $\lambda_0(z)$ has a causal inverse-transform but $\lambda_1(z)$ does not!

Example 13.5.4

We now demonstrate the theory for a rectangular system with $p = 2, r = 1$. Let

$$\mathbf{H}(z) = \begin{bmatrix} 1 + bz^{-1} \\ \dfrac{1}{1 + az^{-1}} \end{bmatrix}. \tag{13.5.35}$$

The reader can readily verify that this can be decomposed as

$$\mathbf{H}(z) = \underbrace{\begin{bmatrix} (z+a)(z+b) & z + \dfrac{a^2 + b^2 + ab}{a+b} \\ z^2 & z - \dfrac{ab}{a+b} \end{bmatrix}}_{\mathbf{W}(z)} \underbrace{\begin{bmatrix} \dfrac{1}{z(z+a)} \\ 0 \end{bmatrix}}_{\mathbf{\Lambda}(z)}. \tag{13.5.36}$$

Since $r = 1$, $\mathbf{V}(z)$ is a scalar and $\mathbf{V}(z) = 1$ here. Notice that for $p \times 1$ systems, $\mathbf{\Lambda}(z)$ is also $p \times 1$ and since only its 'diagonal' elements can be nonzero, only the 0th element is nonzero. In this example, $\lambda_0(z) = z^{-2}/(1 + az^{-1})$ which happens to be causal.

Irreducible MFDs From Smith-McMillan Decomposition

From the decomposition (13.5.26), it is possible to obtain an irreducible MFD for $\mathbf{H}(z)$. For this define two diagonal matrices

$$\begin{aligned} \mathbf{\Lambda}_\beta(z) &= \text{diag }[\beta_0(z), \ldots, \beta_{p-1}(z), 1, \ldots, 1], \quad (p \times p) \\ \mathbf{\Lambda}_\alpha(z) &= \text{diag }[\alpha_0(z), \ldots, \alpha_{p-1}(z), 0, \ldots, 0], \quad (p \times r) \end{aligned} \tag{13.5.37}$$

so that

$$\mathbf{H}(z) = \mathbf{W}(z)\mathbf{\Lambda}(z)\mathbf{V}(z) = \underbrace{\mathbf{W}(z)\mathbf{\Lambda}_\beta^{-1}(z)}_{\mathbf{Q}_1^{-1}(z)} \underbrace{\mathbf{\Lambda}_\alpha(z)\mathbf{V}(z)}_{\mathbf{P}_1(z)}. \tag{13.5.38}$$

Since $\alpha_i(z)$ and $\beta_i(z)$ are relatively prime, the matrices $\mathbf{\Lambda}_\beta(z)$ and $\mathbf{\Lambda}_\alpha(z)$ are left coprime (Problem 13.13). So by Fact 13.2.1, the matrices defined by $\mathbf{Q}_1(z) \triangleq \mathbf{\Lambda}_\beta(z)\mathbf{W}^{-1}(z)$ and $\mathbf{P}_1(z) \triangleq \mathbf{\Lambda}_\alpha(z)\mathbf{V}(z)$ are left coprime proving that the MFD given by $\mathbf{Q}_1^{-1}(z)\mathbf{P}_1(z)$ is irreducible!

♦**Corollary 13.5.4.** If $\mathbf{Q}^{-1}(z)\mathbf{P}(z)$ is an irreducible MFD of $\mathbf{H}(z)$, then $[\det \mathbf{Q}(z)] = c_1 \prod_{i=0}^{p-1} \beta_i(z)$ for some constant c_1. This follows by combining Corollary 13.5.2 with the fact that $\mathbf{Q}_1^{-1}(z)\mathbf{P}_1(z)$ in (13.5.38) is irreducible. ◊

13.5.4 Structural Interpretations of Unimodular Matrices and Smith-McMillan Forms

FIR system with FIR inverse. Let $\mathbf{V}(z)$ be an $r \times r$ unimodular polynomial in z. This represents a (noncausal) FIR system with FIR inverse (Fig. 13.5-1). The FIR nature of $\mathbf{V}(z)$ means that a finite length input $\mathbf{u}(n)$ produces a finite length output $\mathbf{a}(n)$. And since $\mathbf{V}^{-1}(z)$ is also FIR, every finite length output of $\mathbf{V}(z)$ is produced by a *unique, finite length,* input! This statement is evidently not true for arbitrary FIR $\mathbf{V}(z)$ (Problem 13.20). Moreover since $\mathbf{V}(z)$ and $\mathbf{V}^{-1}(z)$ have only positive powers of z, it follows that if the input is zero after a certain time m, then so is the output, and vice versa. That is,

$$\mathbf{u}(n) = \mathbf{0}, \forall n > m \iff \mathbf{a}(n) = \mathbf{0}, \forall n > m. \tag{13.5.39}$$

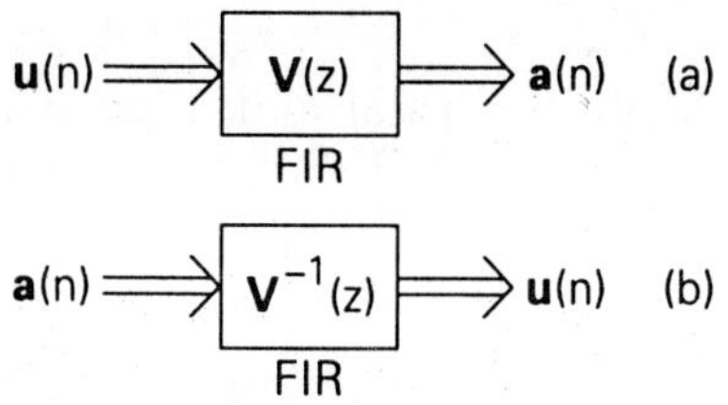

Figure 13.5-1 A unimodular system and its inverse.

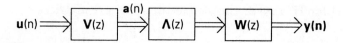

Figure 13.5-2 Implementation of $\mathbf{H}(z)$ in terms of quantitites involved in Smith-McMillan decomposition.

"Watching" the Poles and Zeros

Now consider the Smith-McMillan decomposition, which is pictorially represented in Fig. 13.5-2. Here $\mathbf{V}(z)$ is an FIR unimodular prefilter and $\mathbf{W}(z)$ an FIR unimodular post filter. This representation allows us to 'observe' the poles and zeros of the elements $\lambda_i(z)$ from 'outside' just by measuring the output $\mathbf{y}(n)$ in response to a cleverly chosen *finite length* input sequence $\mathbf{u}(n)$. †

---

† For the case of continuous time systems, we cannot attach such a simple and elegant interpretation for the decomposition $\mathbf{W}(s)\mathbf{\Gamma}(s)\mathbf{V}(s)$. This is due to the fact that the matrices $\mathbf{W}(s)$ and $\mathbf{V}(s)$, which are unimodular polynomials in s, are unrealizable. This in turn has to do with the fact that the transfer function s^k (kth order differentiator) represents an unstable system for $k > 0$. In contrast, for the discrete time case, $\mathbf{W}(z)$ and $\mathbf{V}(z)$ are not

For example, suppose we wish to 'observe' the properties of $\lambda_0(z)$. This can be done by applying an input $\mathbf{u}(n)$ such that the intermediate signal $\mathbf{a}(n)$ has z-transform

$$\mathbf{A}(z) = \begin{bmatrix} 1 \\ 0 \end{bmatrix}. \qquad (13.5.40)$$

The appropriate input is

$$\mathbf{U}(z) = \mathbf{V}^{-1}(z)\begin{bmatrix} 1 \\ 0 \end{bmatrix} = \mathbf{R}(z), \qquad (13.5.41)$$

where $\mathbf{R}(z)$ is merely the 0th column of $\mathbf{V}^{-1}(z)$. Since $\mathbf{V}(z)$ is unimodular, the input (13.5.41) is a polynomial in z, that is, $\mathbf{u}(n)$ is FIR. The output of the system $\mathbf{H}(z)$ is

$$\begin{aligned}
\mathbf{Y}(z) &= \mathbf{W}(z)\mathbf{\Lambda}(z)\mathbf{V}(z)\mathbf{U}(z) \\
&= \mathbf{W}(z)\mathbf{\Lambda}(z)\begin{bmatrix} 1 \\ 0 \end{bmatrix} \\
&= \mathbf{W}(z)\begin{bmatrix} \lambda_0(z) \\ 0 \end{bmatrix} \\
&= \lambda_0(z)\mathbf{W}_0(z) = \frac{\alpha_0(z)\mathbf{W}_0(z)}{\beta_0(z)}.
\end{aligned} \qquad (13.5.42)$$

In (13.5.42) the quantity $\mathbf{W}_0(z)$, which is the 0th column of $\mathbf{W}(z)$, is a polynomial in z. Since $\mathbf{W}(z)$ is unimodular, this column cannot have a factor of the form $(z - z_p)$ common to all its elements because this would imply that $[\det \mathbf{W}(z)]$ has a factor $(z - z_p)$ violating unimodularity. So none of the factors of $\beta_0(z)$ is canceled in (13.5.42).

Summarizing, we can say that z_p is a pole of $\mathbf{Y}(z)$ if and only if it is a pole of $\lambda_0(z)$. And z_0 is a zero of $\mathbf{Y}(z)$ if and only if it is a zero of $\lambda_0(z)$. So all the crucial dynamical properties of each of the elements $\lambda_i(z)$ can be communicated to the output $\mathbf{y}(n)$ in this manner in *finite time* since an FIR input $\mathbf{u}(n)$ will convey this information to the output terminal! The relation between poles, zeros and the polynomials $\alpha_i(z), \beta_i(z)$ will made more precise in the next few sections.

13.6 POLES OF TRANSFER MATRICES

The point z_p is said to be a pole of $\mathbf{H}(z)$ if it is a pole of some element $H_{km}(z)$ in $\mathbf{H}(z)$. Recall from Sec. 2.2 that a causal system cannot have any pole at $z = \infty$, and that $\mathbf{H}(\infty) = \mathbf{h}(0)$ where $\mathbf{h}(n)$ is the causal impulse response.

---

only realizable but also FIR (which is the best type of systems we can hope for!)

In terms of time domain significance, poles are related to the solutions of the homogeneous difference equation describing the system (Problem 13.22). In this section we will present several manifestations of a pole, both in the z-domain and time domain. The reader interested only in the main result can proceed to Theorem 13.6.1 and Lemma 13.6.4.

♠**Lemma 13.6.1. Poles of H(z) and zeros of [det Q(z)].** Let $\mathbf{H}(z)$ be a $p \times r$ rational LTI system with irreducible MFD $\mathbf{Q}^{-1}(z)\mathbf{P}(z)$. Then z_p is a pole of $\mathbf{H}(z)$ if and only if $[\det \mathbf{Q}(z_p)] = 0$.   ◊

Proof. We have

$$\mathbf{H}(z) = \mathbf{Q}^{-1}(z)\mathbf{P}(z) = \mathbf{S}(z)\mathbf{P}(z)/[\det \mathbf{Q}(z)], \qquad (13.6.1)$$

where $\mathbf{S}(z)$ is the adjugate of $\mathbf{Q}(z)$. Since $\mathbf{S}(z)$ and $\mathbf{P}(z)$ are polynomials in z, we see that any factor which occurs in the denominator of any $H_{km}(z)$ must also be a factor of $[\det \mathbf{Q}(z)]$. So if z_p is a pole of $\mathbf{H}(z)$ then it is a zero of $[\det \mathbf{Q}(z)]$.

Conversely suppose z_p is a zero of $[\det \mathbf{Q}(z)]$. From (13.6.1) we see that, in general, this can cancel with every element in the numerator matrix $\mathbf{S}(z)\mathbf{P}(z)$. So it is not obvious that z_p is a pole. The reasoning is somewhat subtler, and depends on the fact that the MFD is irreducible. Irreducibility means that $\mathbf{Q}(z)$ and $\mathbf{P}(z)$ are left coprime so that (13.5.25) holds for some polynomials $\mathbf{V}_q(z)$ and $\mathbf{V}_p(z)$, that is,

$$\mathbf{V}_q(z) + \mathbf{H}(z)\mathbf{V}_p(z) = \frac{\text{Adj } \mathbf{Q}(z)}{[\det \mathbf{Q}(z)]} \qquad (13.6.2)$$

Now if z_p is a zero of $[\det \mathbf{Q}(z)]$, then $[\text{Adj } \mathbf{Q}(z)]$ cannot completely cancel this zero (Fact 13.2.2). So z_p is a pole of the RHS of (13.6.2), and hence of the LHS. But since $\mathbf{V}_q(z)$ and $\mathbf{V}_p(z)$ are polynomial matrices, this implies that z_p is a pole of $\mathbf{H}(z)$ indeed.   ▽▽▽

♠**Lemma 13.6.2. Poles of H(z) and zeros of $\beta_0(z)$.** The point z_p is a pole of $\mathbf{H}(z)$ if and only if it is a zero of the polynomial $\beta_0(z)$ in the Smith-McMillan form.   ◊

Proof. We know from the previous section that (13.5.38) is an irreducible MFD obtained from the Smith-McMillan decomposition. In this MFD,

$$\mathbf{Q}_1(z) = \mathbf{\Lambda}_\beta(z)\mathbf{W}^{-1}(z), \qquad (13.6.3)$$

where $\mathbf{W}(z)$ (hence $\mathbf{W}^{-1}(z)$) is unimodular, that is, has constant non zero determinant. Clearly, $[\det \mathbf{Q}_1(z_p)] = 0$ if and only if $[\det \mathbf{\Lambda}_\beta(z_p)] = 0$. But $[\det \mathbf{\Lambda}_\beta(z)]$ is the product of all $\beta_i(z)$, and moreover, $\beta_0(z)$ contains all other $\beta_i(z)$ as factors [by (13.5.29)]. This shows that $[\det \mathbf{Q}_1(z_p)] = 0$ if and only if $\beta_0(z_p) = 0$. By Lemma 13.6.1 we, therefore, conclude that z_p is a pole if and only if $\beta_0(z_p) = 0$.   ▽▽▽

Time Domain Dynamical Interpretation of a Pole

Qualitatively speaking, z_p is a pole if $\mathbf{H}(z_p)$ "blows up." But this statement hardly provides any physical insight about the meaning of a pole.

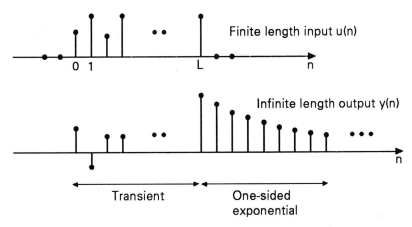

Figure 13.6-1 Time domain interpretation of a pole of a scalar LTI system.

In Problem 2.4 we saw that the pole of a scalar causal system $H(z)$ has a nice time domain interpretation in terms of causal inputs and outputs. According to this, z_p is a pole if and only if there exists a finite length input $u(n)$ such that the system output takes the form z_p^n for all n greater than some finite integer (Fig. 13.6-1).

For MIMO systems a similar interpretation can be given, as elaborated in the next Lemma, which is a discrete-time version of the result presented in Desoer and Schulman [1974].

♦**Lemma 13.6.3. Time domain meaning of a pole.** Let $\mathbf{H}(z)$ be a $p \times r$ rational transfer function representing a causal system. Let $z_p \neq 0$. Then z_p is a pole if and only if there exists a finite length input (i.e., FIR input) $\mathbf{u}(n)$ such that the output takes the form

$$\mathbf{y}(n) = z_p^n \mathbf{v}, \quad \text{for all } n > K, \tag{13.6.4}$$

for some vector $\mathbf{v} \neq \mathbf{0}$ and for some finite integer K. ◊

Comments. Why exclude $z_p = 0$? For any rational $\mathbf{H}(z)$ we can trivially find FIR input such that output is FIR. For example, if $H(z) = P(z)/Q(z)$, just pick $U(z) = Q(z)$ so that $Y(z) = P(z) =$ FIR. So there exists FIR input such that (13.6.4) holds with $z_p = 0$. This does not necessarily mean that there is a pole at the origin because the above argument holds for any $P(z)/Q(z)$.

Proof of Lemma 13.6.3. First suppose that there exists an FIR input

$\mathbf{U}(z)$ producing output $\mathbf{y}(n)$ of the above form. Then

$$\mathbf{Y}(z) = \frac{\mathbf{v}}{1 - z_p z^{-1}} + \mathbf{Y}_1(z), \qquad (13.6.5)$$

where $\mathbf{Y}_1(z)$ is FIR. So $\mathbf{Y}(z)$ has the pole z_p. Since $\mathbf{Y}(z) = \mathbf{H}(z)\mathbf{U}(z)$ and $\mathbf{U}(z)$ is FIR, it is clear that $\mathbf{H}(z)$ has a pole at z_p. (This argument uses the assumption $z_p \neq 0$.)

Now consider the converse. If z_p is a pole then an element, say $H_{00}(z)$ has this pole. If we apply the input $[\,S(z) \; 0 \; \ldots \; 0\,]^T$ for some FIR $S(z)$, we obtain $\mathbf{Y}(z) = S(z)\mathbf{H}_0(z)$ where $\mathbf{H}_0(z)$ is the 0th column of $\mathbf{H}(z)$. We can always choose $S(z)$ to cancel all the poles of $\mathbf{H}_0(z)$ except z_p. We then have $\mathbf{Y}(z) = \mathbf{N}(z)/(1 - z^{-1}z_p)$ for FIR $\mathbf{N}(z)$. This can be written in the form (13.6.5) for FIR $\mathbf{Y}_1(z)$ and nonzero $\mathbf{v}$, so that (13.6.4) follows. ▽▽▽

Summary of Various Manifestations of a Pole

♠**Theorem 13.6.1.** Let $\mathbf{H}(z)$ be a $p \times r$ causal rational transfer function, with irreducible MFD $\mathbf{Q}^{-1}(z)\mathbf{P}(z)$. Also let $\mathbf{A}$ be the state transition matrix of some minimal realization of $\mathbf{H}(z)$. Finally let the Smith-McMillan form of $\mathbf{H}(z)$ be as in (13.5.27), with $\alpha_i(z)$ and $\beta_i(z)$ relatively prime. Then the first four statements below are equivalent. If $z_p \neq 0$, then all the five statements are equivalent.

1. z_p is a pole of $\mathbf{H}(z)$.
2. z_p is a zero of $[\det \mathbf{Q}(z)]$.
3. z_p is a zero of $\beta_0(z)$.
4. z_p is an eigenvalue of $\mathbf{A}$.
5. There exists an FIR input such that the output takes the form $z_p^n \mathbf{v}$ for all n greater than a finite integer, for some $\mathbf{v} \neq \mathbf{0}$. ◇

Proof. It only remains to prove that statement 4 is equivalent to the others. From (13.4.14) we already know that if z_p is a pole it is an eigenvalue of $\mathbf{A}$. It only remains to prove the converse under the condition that $(\mathbf{A}, \mathbf{B}, \mathbf{C}, \mathbf{D})$ is minimal. Now minimality implies reachability. This means that we can apply a finite length input $\mathbf{u}(0), \ldots, \mathbf{u}(N-1)$ (with zero initial state) and reach the state $\mathbf{x}(N) = \mathbf{v}$, where $\mathbf{v}$ is an eigenvector of $\mathbf{A}$ corresponding to eigenvalue λ. We then have

$$\mathbf{x}(N+1) = \mathbf{A}\mathbf{x}(N) = \mathbf{A}\mathbf{v} = \lambda\mathbf{v}, \quad \mathbf{x}(N+2) = \mathbf{A}\mathbf{x}(N+1) = \lambda\mathbf{A}\mathbf{v} = \lambda^2\mathbf{v},$$

and so on, so that $\mathbf{x}(N+k) = \lambda^k \mathbf{v}, k \geq 1$. Thus the output is

$$\mathbf{y}(N+k) = \lambda^k \mathbf{C}\mathbf{v} = \lambda^{N+k}[\mathbf{C}\mathbf{v}/\lambda^N], \quad k \geq 1, \qquad (13.6.6)$$

for the case where $\lambda \neq 0$. Since $(\mathbf{C}, \mathbf{A})$ is observable, $\mathbf{C}\mathbf{v} \neq \mathbf{0}$ (by PBH test). So $\mathbf{y}(n) = \lambda^n \mathbf{w}$ ($\mathbf{w} \neq \mathbf{0}$) for $n > N$. Thus λ is a pole (by Lemma 13.6.3). If $\lambda = 0$, the result is still true. See Problem 13.23. ▽▽▽

Order of a pole

We say that a scalar system $H(z)$ has a pole z_p of order K if $(z-z_p)^K$ appears in the denominator. For MIMO systems, we use the Smith-McMillan form to define pole order. Recall that the functions $\lambda_i(z)$ have the irreducible form (13.5.28) where the $\alpha_i(z)$ and $\beta_i(z)$ satisfy (13.5.29). This means in particular that $\beta_0(z)$ contains all other $\beta_i(z)$ as factors. If z_p is a zero of $\beta_0(z)$ with order K, we say that $\mathbf{H}(z)$ has a pole of order K at $z = z_p$.

The reason for this definition can be seen from the structural meaning of the Smith-McMillan decomposition: we can always find an FIR input (13.5.41) such that the output has the form (13.5.42). Now let us write $\beta_0(z) = \beta_0'(z)(z-z_p)^K$ so that $\beta_0'(z)$ is a polynomial in z which does not vanish at $z = z_p$. If we replace the above FIR input with

$$\mathbf{U}(z) = \beta_0'(z)\mathbf{R}(z), \tag{13.6.7}$$

it is clear that the output is replaced with

$$\mathbf{Y}(z) = \frac{\alpha_0(z)\mathbf{W}_0(z)}{(z-z_p)^K}. \tag{13.6.8}$$

And since $\alpha_0(z)\mathbf{W}_0(z)$ is nonzero for $z = z_p$, there is no cancelation of factors in (13.6.8).

Conversely, suppose an FIR input $\mathbf{U}(z)$ produces output of the form (13.6.8). We know $\mathbf{Y}(z) = \mathbf{W}(z)\mathbf{\Lambda}(z)\mathbf{V}(z)\mathbf{U}(z)$. Since $\mathbf{U}(z)$, $\mathbf{V}(z)$, and $\mathbf{W}(z)$ are FIR, it is clear that at least one element of $\mathbf{\Lambda}(z)$ has the factor $(z-z_p)^K$ in the denominator. This implies, in particular, that $\beta_0(z)$ has the factor $(z-z_p)^K$. We summarize these discussions as follows:

♠**Lemma 13.6.4. Dynamical meaning of order of a pole.** Let $\mathbf{H}(z)$ be some $p \times r$ causal rational discrete-time system and let $z_p \neq 0$. Then z_p is a pole of order $\geq K$ if and only if there exists an FIR input $\mathbf{U}(z)$ such that the output takes the form $\mathbf{Y}(z) = \mathbf{Y}_1(z)/(z-z_p)^K$ where $\mathbf{Y}_1(z)$ is FIR with $\mathbf{Y}_1(z_p) \neq \mathbf{0}$. ◇

13.7 ZEROS OF TRANSFER MATRICES

The definition of poles of a transfer matrix $\mathbf{H}(z)$ is really very simple; we say that z_p is a pole if it is a pole of some element $H_{km}(z)$. This gives rise to several equivalent meanings for a pole as summarized above. But the definition of a zero is more complicated as elaborated next.

Fine Points in the Definition of Zeros

One can define a zero z_0 to be such that $\mathbf{H}(z_0) = \mathbf{0}$, but this is too restricted. It means that every element $H_{km}(z)$ has a zero at z_0. Many practical systems will simply fail to satisfy this definition for any z. For example, the system in (13.5.35) is not zero for any z.

A second possibility would be to take a hint from Theorem 13.6.1 for poles, and define z_0 to be a zero if the determinant of $\mathbf{P}(z_0)$ is zero (where $\mathbf{Q}^{-1}(z)\mathbf{P}(z)$ is an irreducible MFD). This again is meaningless if $\mathbf{P}(z)$ is not square (i.e., if $p \neq r$.) A natural extension of this idea, however, is to define z_0 to be a zero if $\mathbf{P}(z_0)\mathbf{v} = \mathbf{0}$ for some vector $\mathbf{v} \neq \mathbf{0}$. In terms of time domain, this means that there exists an input of the form

$$\mathbf{u}(n) = z_0^n \mathbf{v},$$

which results in zero output. To see this note that (2.2.13) implies

$$\mathbf{y}(n) = \mathbf{H}(z_0)\mathbf{v}z_0^n = \mathbf{Q}^{-1}(z_0)\mathbf{P}(z_0)\mathbf{v}z_0^n = \mathbf{0}.$$

In other words, an exponential input aligned in the direction of $\mathbf{v}$ produces zero output. This behavior is very similar to the scalar case (where $\mathbf{v}$ would just be a nonzero scalar).

Upon deeper thinking this definition has some triviality associated with it. For example suppose $\mathbf{H}(z) = [\,1 \quad z^{-1}\,]$. Given any point $z = z_0$, consider the input $z_0^n \mathbf{v}$ where $\mathbf{v} = \begin{bmatrix} 1 \\ -z_0 \end{bmatrix}$. Since $\mathbf{H}(z_0)\mathbf{v} = \mathbf{0}$, this will result in zero output for all n. More generally, if the normal rank ρ of $\mathbf{P}(z)$ is less than r, then according to this definition any z_0 is a zero of the system!

An *improved definition*, and the most commonly accepted one, says that z_0 is a zero if the rank of $\mathbf{P}(z_0)$ is less than the normal rank of $\mathbf{P}(z)$. We will rephrase this in terms of the Smith-McMillan form, so that we can also explain the meaning of 'order' of a zero.

Zeros in terms of Smith-McMillan form

We say that z_0 is a zero of $\mathbf{H}(z)$ if it is a zero of $\alpha_i(z)$ for some i. In view of the divisibility property (13.5.29), this is equivalent to the condition $\alpha_{\rho-1}(z_0) = 0$. We say that z_0 is a zero of order K if $\alpha_{\rho-1}(z) = (z-z_0)^K \alpha'(z)$ where $\alpha'(z)$ is a polynomial in z with $\alpha'(z_0) \neq 0$. So if z_0 is a zero then the rank of $\mathbf{P}(z_0)$ falls below the normal rank ρ.

Zeros at infinity? In Sec. 2.2 we saw that a causal system cannot have a pole at $z = \infty$. However, a causal system might have a zero at ∞, as in $H(z) = z^{-1}$. The definitions in the above paragraph exclude this situation. This is because $\alpha_{\rho-1}(z)$, being a nonzero polynomial in z, cannot be zero at ∞. Furthermore, we do not obtain meaningful answer for the rank of the polynomial $\mathbf{P}(z)$ if we set $z = \infty$. In order to study the behavior at $z = \infty$, one usually performs the mapping $z \to 1/z$ and studies the behavior at the origin. The properties of zeros to be presented in this section hold only for zeros at finite points.

As explained above, we cannot attach a nontrivial time domain significance to zeros, unless the normal rank is full [i.e., unless $\rho = \min(p, r)$]. Under the full normal-rank assumption, some results in this connection have been derived in Desoer and Schulman [1974], for continuous-time systems.

These are restated below in discrete-time language. We show that if z_0 is a zero then there exists an IIR input with a pole at z_0, such that the output is FIR. In other words, the zero of $\mathbf{H}(z)$ cancels the 'pole' of the input. More precisely we have:

♠**Theorem 13.7.1. Time domain dynamics of zeros.** Let $\mathbf{H}(z)$ be a $p \times r$ rational causal LTI system with $p \geq r$, and let the normal rank be $\rho = r$. Let $z_0 \neq 0$. Then z_0 is a zero of order $\geq K$ (as defined above in terms of $\alpha_{\rho-1}(z)$), if and only if there exists an input of the form

$$\mathbf{U}(z) = \frac{\mathbf{G}(z)}{(z-z_0)^K}, \quad \mathbf{G}(z) \text{ FIR}, \ \mathbf{G}(z_0) \neq \mathbf{0}, \qquad (13.7.1)$$

such that the output $\mathbf{Y}(z)$ is FIR. ◇

Remarks. If we assume $K = 1$ for a moment, we can express the above input in the time domain as

$$\mathbf{u}(n) = \mathbf{c} z_0^n \mathcal{U}(n) + \mathbf{f}(n)$$

where $\mathbf{c}$ is a constant vector, $\mathcal{U}(n)$ is the unit step function, and $\mathbf{f}(n)$ is a finite length sequence. Thus the input looks like the exponential sequence $\mathbf{c} z_0^n$ for sufficiently large n. For sufficiently large n the output is, however, zero (because it is FIR according to the theorem)! This is the time domain interpretation of a zero, and is based only on *one sided* input and output sequences. This is unlike the traditional interpretation which depends on a *two sided* exponential input $z_0^n \mathbf{v}$. The point $z_0 = 0$ is excluded in the above result for reasons similar to those discussed after Lemma 13.6.3.

Proof of Theorem 13.7.1. First assume that z_0 is a zero of order $\geq K$ that is, $\alpha_{\rho-1}(z) = (z-z_0)^K \alpha'_{\rho-1}(z)$, where $\alpha'_{\rho-1}(z)$ is a polynomial in z. We have to prove the existence of an input with above form such that the output is FIR. From the Smith-McMillan decomposition we know

$$\mathbf{Y}(z) = \mathbf{W}(z)\mathbf{\Lambda}(z)\mathbf{V}(z)\mathbf{U}(z). \qquad (13.7.2)$$

Since $p \geq r = \rho$,

$$\mathbf{\Lambda}(z) = \begin{bmatrix} \lambda_0(z) & 0 & 0 & \cdots & 0 \\ 0 & \lambda_1(z) & 0 & \cdots & 0 \\ \vdots & \vdots & \vdots & \ddots & \vdots \\ 0 & 0 & 0 & \cdots & \lambda_{\rho-1}(z) \\ 0 & 0 & 0 & \cdots & 0 \\ \vdots & \vdots & \vdots & \ddots & \vdots \\ 0 & 0 & 0 & \cdots & 0 \end{bmatrix}. \qquad (13.7.3)$$

Suppose we pick $\mathbf{U}(z)$ such that

$$\mathbf{V}(z)\mathbf{U}(z) = \begin{bmatrix} 0 \\ 0 \\ \vdots \\ 0 \\ \dfrac{\beta_{\rho-1}(z)}{(z-z_0)^K} \end{bmatrix}. \qquad (13.7.4)$$

Then

$$\mathbf{Y}(z) = \frac{\mathbf{W}_{\rho-1}(z)\alpha_{\rho-1}(z)\beta_{\rho-1}(z)}{\beta_{\rho-1}(z)(z-z_0)^K} = \mathbf{W}_{\rho-1}(z)\alpha'_{\rho-1}(z), \qquad (13.7.5)$$

which is FIR since $\mathbf{W}(z)$ is unimodular. (Here $\mathbf{W}_{\rho-1}(z)$ is the $(\rho-1)$th column of $\mathbf{W}(z)$. The input satisfying (13.7.4) is of the form (13.7.1) where

$$\mathbf{G}(z) = \mathbf{T}_{\rho-1}(z)\beta_{\rho-1}(z). \qquad (13.7.6)$$

The quantity $\mathbf{T}_{\rho-1}(z)$ is the $(\rho-1)$th column of the inverse of $\mathbf{V}(z)$. Since $\mathbf{V}(z)$ is unimodular, its inverse and hence $\mathbf{T}_{\rho-1}(z)$ are FIR, and furthermore $\mathbf{T}_{\rho-1}(z_0) \neq 0$. Also $\beta_{\rho-1}(z)$ is relatively prime to $\alpha_{\rho-1}(z)$ so that $\beta_{\rho-1}(z_0) \neq 0$. So $\mathbf{G}(z)$ is FIR with $\mathbf{G}(z_0) \neq \mathbf{0}$, proving the desired result.

Conversely, suppose an input of the form (13.7.1) produces output $\mathbf{Y}(z)$. Now

$$\mathbf{Y}(z) \text{ is FIR} \Rightarrow \mathbf{W}^{-1}(z)\mathbf{Y}(z) \text{ is FIR} \quad (\text{since } \mathbf{W}(z) \text{ unimodular})$$
$$\Rightarrow \frac{\mathbf{\Lambda}(z)\mathbf{V}(z)\mathbf{G}(z)}{(z-z_0)^K} \text{ is FIR.} \qquad (13.7.7)$$

Since $\mathbf{G}(z)$ is FIR with $\mathbf{G}(z_0) \neq \mathbf{0}$, we have $\mathbf{V}(z_0)\mathbf{G}(z_0) \neq \mathbf{0}$ (because $\mathbf{V}(z)$ is unimodular). The FIR nature of $\mathbf{Y}(z)$ implies then that there is at least one diagonal element $\lambda_i(z)$ in $\mathbf{\Lambda}(z)$, having the factor $(z-z_0)^K$ in the numerator. (This reasoning uses the assumption $z_0 \neq 0$). This means that z_0 is a zero of order (at least) K. $\qquad \triangledown\triangledown\triangledown$

For the $p < r$ case the above result does not hold because for every z, there exists some nonzero vector $\mathbf{v}$ such that $\mathbf{P}(z)\mathbf{v} = \mathbf{0}$. For this case, a more useful dynamical significance is developed in Problem 13.21.

Case of Square Matrices: the Cleanest Case

Suppose $\mathbf{H}(z)$ is $p \times p$ so that $\mathbf{P}(z)$ in an irreducible MFD $\mathbf{Q}^{-1}(z)\mathbf{P}(z)$ is $p \times p$. If in addition $\mathbf{P}(z)$ has normal rank $\rho = p$, then z_0 is a zero if and only if $[\det \mathbf{P}(z_0)] = 0$. The dynamical significance offered by Theorem 13.7.1 continues to hold in this case.

So this is the cleanest case in the sense that the meaning of zeros and poles in terms of the MFD quantities are very similar: z_p is a pole if, and only if, $[\det \mathbf{Q}(z_p)] = 0$ whereas z_0 is a zero if, and only if, $[\det \mathbf{P}(z_0)] = 0$. Also remember that z_p and z_0 are zeros of $\beta_0(z)$ and $\alpha_{\rho-1}(z)$, respectively.

Poles and zeros at same place. For an SISO system a pole z_p and a zero z_0 cannot be at the same place (i.e., we cannot have $z_p = z_0$) because these would simply cancel. In the MIMO case, a pole and a zero *can* exist at the same point without canceling. An example is

$$\mathbf{H}(z) = \text{diag} \left[1 + az^{-1} \quad \frac{1}{1 + az^{-1}} \right]. \tag{13.7.8}$$

In such cases, $\prod \alpha_i(z)$ and $\prod \beta_i(z)$ are not coprime even though $\alpha_i(z)$ and $\beta_i(z)$ are coprime for each i.

13.8 DEGREE OF A TRANSFER MATRIX

Let $\mathbf{H}(z)$ be a $p \times r$ causal rational system. In Sec. 13.3 we defined the degree μ of $\mathbf{H}(z)$ to be the minimum number of delay elements required for its implementation. Letting N denote the size of the state vector $\mathbf{x}(n)$ in any realization of $\mathbf{H}(z)$, we showed that $N = \mu$ (i.e., the realization is minimal) if and only if $(\mathbf{C}, \mathbf{A})$ is observable and $(\mathbf{A}, \mathbf{B})$ is reachable.

There is an extremely important result in realization theory [Kalman, 1965] which expresses the degree μ of $\mathbf{H}(z)$ in terms of the quantities $\beta_i(z)$ in the Smith-McMillan decomposition. Recall that $\beta_i(z)$ and $\alpha_i(z)$ are polynomials in z. With μ_i denoting the order of $\beta_i(z)$, the degree of $\mathbf{H}(z)$ is

$$\mu = \sum_{i=0}^{\rho-1} \mu_i. \tag{13.8.1}$$

See Kalman [1965] or Forney [1970] for a proof.

Warning. It is important to notice that this result applies only under the following conditions:

1. $\mathbf{H}(z)$ is causal (otherwise it cannot be realized with delay elements alone, and our definition of degree is meaningless).
2. The matrices $\mathbf{W}(z)$ and $\mathbf{V}(z)$ are polynomials in z, and so are the quantities $\alpha_i(z)$ and $\beta_i(z)$. If these are re-expressed as polynomials in z^{-1}, the above result does not hold.

Relation to $[\det \mathbf{Q}(z)]$. An immediate consequence of (13.8.1) can be obtained from (13.5.38), which gives an irreducible MFD for $\mathbf{H}(z)$. The order of $[\det \mathbf{Q}_1(z)]$ is equal to μ (as the determinant of $\mathbf{W}(z)$ is constant). But the order of $[\det \mathbf{Q}(z)]$ is the same in all irreducible MFDs of $\mathbf{H}(z)$ (Corollary 13.5.3). So we conclude that $\mu =$ order of $[\det \mathbf{Q}(z)]$. We now summarize all results pertaining to degree.

♠**Theorem 13.8.1. On degree of H(z).** Let $\mathbf{H}(z)$ be a causal rational $p \times r$ discrete-time system. Then the following are true.

1. Let (13.5.27) be the Smith-McMillan form. Then the degree μ is given by (13.8.1) where μ_i is the order of $\beta_i(z)$.
2. Let $\mathbf{Q}^{-1}(z)\mathbf{P}(z)$ be an irreducible MFD for $\mathbf{H}(z)$, where $\mathbf{P}(z)$ and $\mathbf{Q}(z)$ are polynomials in z. Then, $\mu =$ order of $[\det \mathbf{Q}(z)]$.
3. Finally a realization $(\mathbf{A}, \mathbf{B}, \mathbf{C}, \mathbf{D})$ for $\mathbf{H}(z)$ is minimal (i.e., has the smallest possible size $\mu \times \mu$ for $\mathbf{A}$) if, and only if, $(\mathbf{C}, \mathbf{A})$ is observable and $(\mathbf{A}, \mathbf{B})$ is reachable. ◇

Degree of H(z) Versus Degree of [det H(z)]

Suppose $\mathbf{H}(z)$ is $p \times p$ with irreducible MFD $\mathbf{Q}^{-1}(z)\mathbf{P}(z)$. We know that the order of $[\det \mathbf{Q}(z)]$ is equal to the degree μ of $\mathbf{H}(z)$. Does the degree of $[\det \mathbf{H}(z)]$ have any role in this connection? Can we relate it to μ? We now provide an answer.

For example if $\mathbf{H}(z) = \begin{bmatrix} 1 & 0 \\ f(z) & 1 \end{bmatrix}$ where $f(z)$ is causal, the degree of $\mathbf{H}(z)$ is equal to that of $f(z)$ whereas the degree of $[\det \mathbf{H}(z)]$ is zero regardless of $f(z)$. So the degree of $[\det \mathbf{H}(z)]$ can be as small as zero. Next, how large can it be? It turns out that it cannot be greater than μ. To see this note that (13.5.26) yields

$$\det \mathbf{H}(z) = c\alpha(z)/\beta(z) \qquad (13.8.2)$$

where c is constant, $\alpha(z) = \prod \alpha_i(z)$ and $\beta(z) = \prod \beta_i(z)$. We know the order of $\beta(z)$ equals μ. The order of $\alpha(z)$ is $\leq \mu$ because of causality (remember that $\alpha(z)$ and $\beta(z)$ are polynomials in z). So, the degree of (13.8.2) is at most μ. It is less than μ if there are cancelations in the ratio (13.8.2). Summarizing, we have proved

$$0 \leq \deg [\det \mathbf{H}(z)] \leq \mu \qquad (13.8.3)$$

where μ is the McMillan degree of $\mathbf{H}(z)$.

13.9 FIR TRANSFER MATRICES

The above results apply for FIR as well as IIR systems. In this text we frequently deal with MIMO FIR systems. These arise, for example, in multirate filter banks in the form of polyphase matrices for the analysis and synthesis banks. If the filter bank is FIR then these polyphase matrices are FIR and it is important to understand their system-theoretic properties.

Degree of FIR Transfer Matrices

Suppose $\mathbf{H}(z)$ is $p \times r$ causal FIR so that

$$\mathbf{H}(z) = \mathbf{h}(0) + \mathbf{h}(1)z^{-1} + \ldots + \mathbf{h}(K)z^{-K}, \qquad (13.9.1)$$

with $\mathbf{h}(K) \neq 0$. The quantity K is called the order of $\mathbf{H}(z)$, and $K+1$ the length. If μ is the degree of $\mathbf{H}(z)$ then $\mu \geq K$ because we require at least K delays to realize this system.

Remarks

1. Examples of the form $\mathbf{H}(z) = z^{-1}\mathbf{I}$ demonstrate that μ can be larger than K.
2. We also know from Example 13.3.2 that if $\mathbf{H}(z) = \mathbf{h}(K)z^{-K}$ then the degree is equal to sK where s is the rank of the (possibly rectangular) matrix $\mathbf{h}(K)$.
3. For the special case of $p \times 1$ FIR systems the transfer matrix is a column vector. This arises, for example, when $\mathbf{H}(z)$ is an analysis bank with p analysis filters. In this case the degree μ is equal to K because $\mathbf{h}(K)$ has rank one. $\mathbf{H}(z)$ can then be implemented as in Fig. 13.9-1 requiring K delays. Similarly if $\mathbf{H}(z)$ is a row vector (as in a synthesis bank) then $\mu = K$.

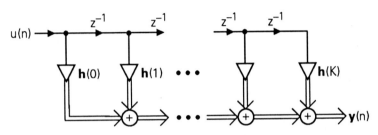

Figure 13.9-1 Implementation of $p \times 1$ FIR $\mathbf{H}(z)$.

State Space Description of FIR Systems

Given a structure for a $p \times r$ causal FIR transfer matrix, one can find a state space description $(\mathbf{A}, \mathbf{B}, \mathbf{C}, \mathbf{D})$ in the usual manner. Assuming that the realization is minimal, the eigenvalues of $\mathbf{A}$ are the poles, all of which lie at the origin of the z plane. So, all eigenvalues of $\mathbf{A}$ are zero. Some of the other properties are summarized in the next lemma.

♦**Lemma 13.9.1.** *On FIR state space description.* Consider a minimal realization $(\mathbf{A}, \mathbf{B}, \mathbf{C}, \mathbf{D})$ of a $p \times r$ causal FIR system with order K [eqn. (13.9.1)]. Then, $\mathbf{A}$ has all eigenvalues equal to zero. Moreover, (a) $\mathbf{A}^K\mathbf{B} = \mathbf{0}$, (b) $\mathbf{C}\mathbf{A}^K = \mathbf{0}$, and (c) $\mathbf{A}^N = \mathbf{0}$, where N is the state space dimension (i.e., $\mathbf{A}$ is $N \times N$). Conversely, if $\mathbf{A}^N = \mathbf{0}$, then $(\mathbf{A}, \mathbf{B}, \mathbf{C}, \mathbf{D})$ represents an FIR system. ◊

Proof. We know $\mathbf{h}(n) = \mathbf{C}\mathbf{A}^{n-1}\mathbf{B}$ for $n > 0$. Since $\mathbf{h}(n) = \mathbf{0}$ for $n > K$, we have $\mathbf{C}\mathbf{A}^i\mathbf{B} = \mathbf{0}$, for $i \geq K$. So we can write, in particular,

$$\underbrace{\begin{bmatrix} \mathbf{C} \\ \mathbf{C}\mathbf{A} \\ \vdots \\ \mathbf{C}\mathbf{A}^{N-1} \end{bmatrix}}_{\mathcal{S}_{\mathbf{C},\mathbf{A}}} \mathbf{A}^K\mathbf{B} = \mathbf{0}. \qquad (13.9.2)$$

Since $(\mathbf{A}, \mathbf{B}, \mathbf{C}, \mathbf{D})$ is minimal, $\mathcal{S}_{\mathbf{C},\mathbf{A}}$ has full column rank N, so this implies $\mathbf{A}^K \mathbf{B} = \mathbf{0}$. The proof of $\mathbf{C}\mathbf{A}^K = \mathbf{0}$ is similar. Next $\mathbf{A}^K \mathbf{B} = \mathbf{0}$ also implies $\mathbf{A}^N \mathbf{B} = \mathbf{0}$, since $N \geq K$. This means

$$\mathbf{A}^N \underbrace{[\mathbf{B} \quad \mathbf{AB} \quad \ldots \quad \mathbf{A}^{N-1}\mathbf{B}]}_{\mathcal{R}_{\mathbf{A},\mathbf{B}}} = \mathbf{0}. \tag{13.9.3}$$

Since $\mathcal{R}_{\mathbf{A},\mathbf{B}}$ has full row rank N, this implies $\mathbf{A}^N = \mathbf{0}$. A second way to prove $\mathbf{A}^N = \mathbf{0}$ is to note that the characteristic equation is $\lambda^N = 0$, and $\mathbf{A}$ has to satisfy this equation (Cayley-Hamilton theorem; Appendix A).

Conversely, suppose $\mathbf{A}^N = \mathbf{0}$. We know that the eigenvalues of $\mathbf{A}^N$ are given by λ_i^N, so that we have $\lambda_i^N = 0$, that is, $\lambda_i = 0$. So all eigenvalues of $\mathbf{A}$ are equal to zero [i.e., poles of $\mathbf{H}(z)$ are at the origin] proving that $\mathbf{H}(z)$ is FIR. $\triangledown\triangledown\triangledown$

Smith-McMillan Form for the FIR Case

We now present some examples of FIR Smith-McMillan forms, highlighting the main features.

Example 13.9.1

Consider
$$\mathbf{H}(z) = \begin{bmatrix} z^{-1} + z^{-2} & z^{-1} \\ 2 + 3z^{-2} & 2(1 + z^{-1})^2 \end{bmatrix}$$

which is 2×2 causal FIR. To find the Smith-McMillan form we first express every entry as a ratio of polynomials in z. Thus

$$\mathbf{H}(z) = \frac{1}{z^2} \underbrace{\begin{bmatrix} 1 + z & z \\ 3 + 2z^2 & 2(1+z)^2 \end{bmatrix}}_{\mathbf{P}(z)} \tag{13.9.4}$$

The Smith-form for this $\mathbf{P}(z)$ has been worked out in Example 13.5.1. The Smith-McMillan form for $\mathbf{H}(z)$ is

$$\mathbf{\Lambda}(z) = \begin{bmatrix} \dfrac{1}{z^2} & 0 \\ 0 & \dfrac{2 + 3z + 6z^2}{z^2} \end{bmatrix} \tag{13.9.5}$$

so that
$$\begin{aligned} \alpha_0(z) &= 1, \ \alpha_1(z) = 2 + 3z + 6z^2, \\ \beta_0(z) &= z^2, \ \beta_1(z) = z^2. \end{aligned} \tag{13.9.6}$$

From these we have $\mu_0 = 2, \mu_1 = 2$ so that the degree is $\mu = 2 + 2 = 4$.

Example 13.9.2.

Consider the causal FIR system

$$\mathbf{H}(z) = \begin{bmatrix} 1 & 0 \\ z^{-1} & 1 \end{bmatrix}. \qquad (13.9.7)$$

This obviously has degree one. It is easily verified that

$$\mathbf{H}(z) = \begin{bmatrix} z & -1 \\ 1 & 0 \end{bmatrix} \begin{bmatrix} z^{-1} & 0 \\ 0 & z \end{bmatrix} \begin{bmatrix} 1 & z \\ 0 & 1 \end{bmatrix}, \qquad (13.9.8)$$

so that

$$\begin{aligned} \alpha_0(z) &= 1, \ \alpha_1(z) = z, \\ \beta_0(z) &= z, \ \beta_1(z) = 1. \end{aligned} \qquad (13.9.9)$$

So $\mu = 1 + 0 = 1$ as expected. The main point to notice in this example is that all elements $\lambda_i(z)$ are powers of z. From (13.9.8) we see that the determinant of $\mathbf{H}(z)$ is constant, which is also obvious by inspection of (13.9.7). This $\mathbf{H}(z)$ therefore happens to be a unimodular polynomial in the variable z^{-1}. Note that we have both positive and negative powers of z in $\lambda_i(z)$. This is fine because $\Lambda(z)$ can be noncausal for causal $\mathbf{H}(z)$.

Since $\alpha_1(z) = \beta_0(z) = z$, we see that the system has a pole as well as a zero at $z = 0$. If we write $\mathbf{H}(z)$ in irreducible MFD form we expect the rank of $\mathbf{P}(z)$ to drop at $z = 0$. This is clearly verified from the following MFD for $\mathbf{H}(z)$:

$$\mathbf{H}(z) = \begin{bmatrix} 1 & 0 \\ z^{-1} & 1 \end{bmatrix} = \underbrace{\begin{bmatrix} 1 & 0 \\ 0 & z \end{bmatrix}^{-1}}_{\mathbf{Q}^{-1}(z)} \underbrace{\begin{bmatrix} 1 & 0 \\ 1 & z \end{bmatrix}}_{\mathbf{P}(z)}$$

The reader should notice that for the FIR case the polynomials $\beta_i(z)$ are of the form z^{μ_i}. This is consistent with the fact that for causal FIR systems all poles are at $z = 0$. In view of (13.5.29) we also have $\mu_0 \geq \mu_1 \geq \ldots \geq \mu_{p-1}$.

13.10 CAUSAL INVERSES OF CAUSAL SYSTEMS

13.10.1 Some Basic Results (FIR and IIR Systems)

Let $\mathbf{H}(z)$ be a causal $p \times p$ system (FIR or IIR), and let there exist a causal $p \times p$ inverse $\mathbf{G}(z)$, so that

$$\mathbf{G}(z)\mathbf{H}(z) = \mathbf{I}_p. \qquad (13.10.1)$$

Denoting the impulse response coefficients of $\mathbf{H}(z)$ and $\mathbf{G}(z)$ by $\mathbf{h}(n)$ and $\mathbf{g}(n)$ respectively, we have from (13.10.1),

$$\mathbf{g}(0)\mathbf{h}(0) = \mathbf{I}_p \qquad (13.10.2)$$

so that $\mathbf{g}(0)$ and $\mathbf{h}(0)$ are nonsingular.

State Space Description of the Inverse System

Let $(\mathbf{A}, \mathbf{B}, \mathbf{C}, \mathbf{D})$ be a minimal realization of $\mathbf{H}(z)$ so that

$$\mathbf{H}(z) = \mathbf{D} + \mathbf{C}(z\mathbf{I} - \mathbf{A})^{-1}\mathbf{B}. \qquad (13.10.3)$$

If $\mathbf{D}$ is nonsingular we can apply the matrix-inversion lemma (A.4.7) given in Appendix A, to obtain a causal $p \times p$ inverse system

$$\mathbf{G}(z) = \mathbf{D}_1 + \mathbf{C}_1(z\mathbf{I} - \mathbf{A}_1)^{-1}\mathbf{B}_1, \qquad (13.10.4)$$

where

$$\begin{aligned} \mathbf{A}_1 &= \mathbf{A} - \mathbf{B}\mathbf{D}^{-1}\mathbf{C}, \\ \mathbf{B}_1 &= \mathbf{B}\mathbf{D}^{-1}, \\ \mathbf{C}_1 &= -\mathbf{D}^{-1}\mathbf{C}, \\ \mathbf{D}_1 &= \mathbf{D}^{-1}. \end{aligned} \qquad (13.10.5)$$

This result is true as long as $\mathbf{D}$ is nonsigular. Since $\mathbf{h}(0) = \mathbf{D}$, we conclude that a $p \times p$ causal system has a causal inverse if and only if $\mathbf{h}(0)$ is nonsingular.

It can be shown (Problem 13.10) that minimality of $(\mathbf{A}, \mathbf{B}, \mathbf{C}, \mathbf{D})$ implies that of $(\mathbf{A}_1, \mathbf{B}_1, \mathbf{C}_1, \mathbf{D})$. This means in particular that if a $p \times p$ causal system $\mathbf{H}(z)$ has a causal inverse $\mathbf{G}(z)$, then $\mathbf{H}(z)$ and $\mathbf{G}(z)$ have the same degree. The same is not true of rectangular systems. For example if

$$\mathbf{H}(z) = \begin{bmatrix} 1 - z^{-1} \\ z^{-1} \end{bmatrix}, \qquad (13.10.6)$$

then

$$\mathbf{G}(z) = \begin{bmatrix} 1 & 1 \end{bmatrix}, \qquad (13.10.7)$$

is a valid left inverse (i.e., $\mathbf{G}(z)\mathbf{H}(z) = 1$). But the degrees of $\mathbf{H}(z)$ and $\mathbf{G}(z)$ are, respectively, one and zero.

13.10.2 Causal unimodular systems

In Sec. 5.6.2 we mentioned that the polyphase matrix $\mathbf{E}(z)$ of any causal FIR perfect reconstruction QMF bank can be written as a product of a causal paraunitary system and a causal unimodular system. In Chap. 6 we elaborated on paraunitary systems (which are further studied in Chap.

14). In this section we will present some details about causal unimodular systems.

The phrase 'causal unimodular system' stands for a square matrix polynomial $\mathbf{H}(z)$ in z^{-1}, which is unimodular. This implies that $\mathbf{H}(z)$ is causal FIR, and has a causal FIR inverse. Conversely, suppose $\mathbf{H}(z)$ is a $p \times p$ causal FIR system with a causal FIR inverse $\mathbf{G}(z)$. Then $\mathbf{H}(z)$ and $\mathbf{G}(z)$ are unimodular polynomials (in z^{-1}). To see this note that (13.10.1) implies that the product of determinants is unity, and since each determinant can at best be a polynomial in z^{-1}, it has to be a constant. Summarizing, the phrase "causal unimodular system" is synonymous to "casual FIR system with causal FIR inverse."

State Space Description for Causal Unimodular Systems

With $(\mathbf{A}, \mathbf{B}, \mathbf{C}, \mathbf{D})$ denoting a state space description of $\mathbf{H}(z)$, the causal unimodular inverse $\mathbf{G}(z)$ has a realization $\mathbf{A}_1, \mathbf{B}_1, \mathbf{C}_1, \mathbf{D}_1$ given by (13.10.5). Since both $\mathbf{G}(z)$ and $\mathbf{H}(z)$ are FIR, it follows that all the eigenvalues of $\mathbf{A}$ and $\mathbf{A}_1$ are equal to zero. Conversely, suppose $\mathbf{H}(z)$ is some $p \times p$ causal system with nonsingular $\mathbf{h}(0)$ so that the causal inverse $\mathbf{G}(z)$ exists. Suppose $(\mathbf{A}, \mathbf{B}, \mathbf{C}, \mathbf{D})$ is a realization such that $\mathbf{A}$ and $\mathbf{A} - \mathbf{BD}^{-1}\mathbf{C}$ have all eigenvalues equal to 0. Then, $\mathbf{H}(z)$ as well as its inverse $\mathbf{G}(z)$ are (causal) FIR. In other words $\mathbf{H}(z)$ is causal unimodular. We now summarize these results as follows:

♦**Theorem 13.10.1.** Let $\mathbf{H}(z)$ be some $p \times p$ causal rational system, and let $\mathbf{h}(0)$ be nonsingular.

1. With $(\mathbf{A}, \mathbf{B}, \mathbf{C}, \mathbf{D})$ denoting a minimal realization of $\mathbf{H}(z)$, the matrix $\mathbf{D}$ is therefore nonsingular, and a $p \times p$ causal rational inverse $\mathbf{G}(z)$ exists. The matrices (13.10.5) give a minimal realization of $\mathbf{G}(z)$, and $\mathbf{G}(z)$ has the same McMillan degree as $\mathbf{H}(z)$.
2. If in addition $\mathbf{H}(z)$ is unimodular then so is $\mathbf{G}(z)$, and both $\mathbf{A}$ and $\mathbf{A}_1$ have all eigenvalues equal to zero.
3. Finally if $(\mathbf{A}, \mathbf{B}, \mathbf{C}, \mathbf{D})$ is a minimal realization such that $\mathbf{A}$ and $\mathbf{A}_1$ have all eigenvalues equal to zero, then $\mathbf{H}(z)$ is causal unimodular (that is, causal FIR with a causal FIR inverse). ◊

Example 13.10.1

Let

$$\mathbf{H}(z) = \begin{bmatrix} 1 & 0 & 0 \\ z^{-1} & 1 & 0 \\ 1 & z^{-1} & 1 \end{bmatrix}. \qquad (13.10.8)$$

This happens to be a triangular matrix, and $[\det [\mathbf{H}(z)] = 1$ by inspection. So this is causal unimodular. The inverse is verified to be

$$\mathbf{G}(z) = \begin{bmatrix} 1 & 0 & 0 \\ -z^{-1} & 1 & 0 \\ -1 + z^{-2} & -z^{-1} & 1 \end{bmatrix}, \qquad (13.10.9)$$

which is also causal unimodular as expected.

Let us see if we can guess the degree of $\mathbf{H}(z)$ by inspection. From (13.10.8) it is obvious that at most two delays are required to implement $\mathbf{H}(z)$ (since z^{-1} occurs only twice), i.e., the degree ≤ 2. From (13.10.9) we see that $\mathbf{G}(z)$ has degree ≥ 2 (since z^{-2} appears). But the degrees of $\mathbf{H}(z)$ and $\mathbf{G}(z)$ are the same according to the above theorem, so we conclude that the degree is precisely two!

Even though the degrees of $\mathbf{H}(z)$ and $\mathbf{G}(z)$ are equal, their orders (highest power of z^{-1} appearing in the transfer function) need not be the same. In our example, the order of $\mathbf{H}(z)$ is unity whereas that of $\mathbf{G}(z)$ is two.

Smith-McMillan Form for Causal FIR Unimodular Systems

Suppose $\mathbf{H}(z)$ is a $p \times p$ unimodular polynomial in z^{-1}. What does the Smith-McMillan form $\Lambda(z)$ look like? Since $\mathbf{H}(z)$ is FIR, $\beta_i(z) = z^{\mu_i}$. So

$$\det \mathbf{H}(z) = constant \times z^{-\mu} \prod_{i=0}^{p-1} \alpha_i(z). \qquad (13.10.10)$$

Since the determinant is required to be constant by unimodularity, $\alpha_i(z)$ must also have the form $\alpha_i(z) = z^{m_i}$, with $\sum_i m_i = \sum_i \mu_i$. This fact can be verified to be true in Example 13.9.2. Summarizing, we have

$$\Lambda(z) = \begin{bmatrix} z^{n_0} & 0 & 0 & \cdots & 0 \\ 0 & z^{n_1} & 0 & \cdots & 0 \\ \vdots & \vdots & \vdots & \ddots & \vdots \\ 0 & 0 & 0 & \cdots & z^{n_{p-1}} \end{bmatrix}, \qquad (13.10.11)$$

with $\sum_{i=0}^{p-1} n_i = 0$. The degree μ is equal to $\sum_i |n_i|$ with the summation carried over negative n_i only.

PROBLEMS

13.1. Verify Parseval's relation (2.2.12) for vector signals.

13.2. Let $\mathbf{P}(z)$ and $\mathbf{Q}(z)$ be matrix polynomials in z, with the same number of rows.

a) Let $\mathbf{L}(z)$ be a greatest left common divisor (glcd) of $\mathbf{P}(z)$ and $\mathbf{Q}(z)$. Show then that $\mathbf{L}(z)\mathbf{W}(z)$ is also a glcd, for any unimodular $\mathbf{W}(z)$.

b) Let $\mathbf{L}_1(z)$ and $\mathbf{L}_2(z)$ be two glcds of $\mathbf{P}(z)$ and $\mathbf{Q}(z)$. Show then that $\mathbf{L}_1(z) = \mathbf{L}_2(z)\mathbf{V}(z)$ where $\mathbf{V}(z)$ is an appropriate unimodular matrix.

c) Supply a proof for Fact 13.2.1.

13.3. For the following system:

$$\mathbf{H}(z) = \begin{bmatrix} \frac{1+2z^{-1}}{3+z^{-1}} & \frac{1}{3+z^{-1}} \\ \frac{1+2z^{-1}}{(2+z^{-1})(3+z^{-1})} & 1 \end{bmatrix},$$

find a left MFD.

13.4. Consider the following MFD.

$$\mathbf{H}(z) = \underbrace{\begin{bmatrix} 1+z & z \\ z & 1+z \end{bmatrix}^{-1}}_{\mathbf{Q}^{-1}(z)} \underbrace{\begin{bmatrix} 2+z & 1+z \\ z & z \end{bmatrix}}_{\mathbf{P}(z)}$$

Find another MFD $\widehat{\mathbf{Q}}^{-1}(z)\widehat{\mathbf{P}}(z)$ for this same system. To avoid trivial answers, make sure that $\mathbf{Q}(z) \neq f(z)\widehat{\mathbf{Q}}(z)$ for scalar $f(z)$.

13.5. For each of the structures shown in Fig. P13-5, write down the state space description $(\mathbf{A}, \mathbf{B}, \mathbf{C}, \mathbf{D})$. For each case answer the following: (a) Is $(\mathbf{A}, \mathbf{B})$ reachable? (b) Is $(\mathbf{C}, \mathbf{A})$ observable? (c) Is $(\mathbf{A}, \mathbf{B}, \mathbf{C}, \mathbf{D})$ minimal?

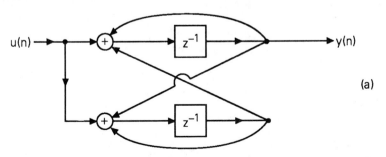

Figure P13-5(a)

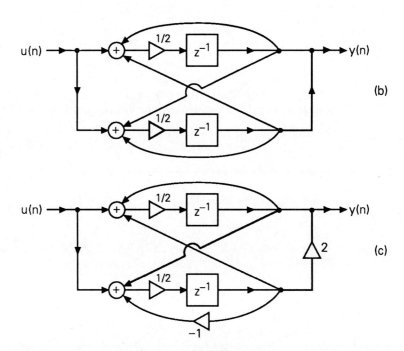

Figure P13-5(b), (c)

13.6. For each structure in Fig. P13-5 find out whether **A** is stable.

13.7. Consider the state space equations (13.4.3), where **A** is $N \times N$. The pair $(\mathbf{A}, \mathbf{B})$ is said to be controllable [Anderson and Moore, 1979] if we can start from any arbitrary initial state $\mathbf{x}(0)$ and force $\mathbf{x}(N) = \mathbf{0}$, by appropriate choice of $\mathbf{u}(0), \ldots, \mathbf{u}(N-1)$. Evidently if $(\mathbf{A}, \mathbf{B})$ is reachable (Sec. 13.4.2) it is also controllable. However, the converse is not true.

    a) Show that the structure shown in Fig. P13-5(a) is controllable but not reachable.

    b) More generally, show that $(\mathbf{A}, \mathbf{B})$ is controllable as long as every column of $\mathbf{A}^N$ can be expressed as a linear combination of the columns of $\mathcal{R}_{\mathbf{A},\mathbf{B}}$ defined in (13.4.18).

13.8. Consider the system $\mathbf{H}(z) = z^{-1}\mathbf{R}$ where $\mathbf{R}$ is $M \times M$ with rank ρ. Then we can write $\mathbf{H}(z) = \mathbf{TS}z^{-1}$ where $\mathbf{T}$ and $\mathbf{S}$ are $M \times \rho$ and $\rho \times M$ respectively, with rank ρ. This leads to the implementation of Fig. 13.3-3.

    a) Find the state space description $(\mathbf{A}, \mathbf{B}, \mathbf{C}, \mathbf{D})$ of this structure.

    b) Show that this structure for $\mathbf{H}(z)$ is minimal, that is, we cannot find a structure with fewer than ρ delays.

13.9. Let $\mathbf{H}(z)$ be a $p \times r$ causal rational system with real impulse response $\mathbf{h}(n)$. Show that there exists a *minimal* structure with real valued matrices **A**, **B**, **C** and **D**. (*Hint.* If you prefer, find a nonminimal realization with real multipliers, and apply results of Sec. 13.4.2).

13.10. Let $\mathbf{H}(z)$ represent an $M \times M$ causal system and let $(\mathbf{A}, \mathbf{B}, \mathbf{C}, \mathbf{D})$ be a valid state space description. Assume $\mathbf{h}(0)$ is nonsingular. Then (13.10.5) gives the state space description of the inverse system $\mathbf{G}(z)$. Show that if $(\mathbf{A}, \mathbf{B}, \mathbf{C}, \mathbf{D})$ is minimal, then so is the system (13.10.5). (You can use PBH test if you wish.)

13.11. For
$$\mathbf{P}(z) = \begin{bmatrix} 1+z & 2+z \\ 2+z & 3+z \end{bmatrix},$$
find the Smith form $\mathbf{\Gamma}(z)$ as well as diagonalizing unimodular matrices $\mathbf{W}(z)$ and $\mathbf{V}(z)$. Now consider the causal FIR system
$$\mathbf{H}(z) = \begin{bmatrix} z^{-1}+1 & 2z^{-1}+1 \\ 2z^{-1}+1 & 3z^{-1}+1 \end{bmatrix}.$$

a) Find the Smith-McMillan form $\mathbf{\Lambda}(z)$, as well as diagonalizing unimodular matrices $\mathbf{W}(z)$ and $\mathbf{V}(z)$.
b) What is the McMillan degree of this system?
c) Find an irreducible MFD $\mathbf{Q}^{-1}(z)\mathbf{P}(z)$ for $\mathbf{H}(z)$, and evaluate the determinant of $\mathbf{Q}(z)$. Verify that this determinant has order equal to the McMillan degree.

13.12. Consider the causal unimodular system
$$\mathbf{H}(z) = \begin{bmatrix} 1+3z^{-1} & 3 \\ z^{-1} & 1 \end{bmatrix}.$$

a) Find the Smith-McMillan form $\mathbf{\Lambda}(z)$.
b) What is the McMillan degree of this system?
c) Find an irreducible MFD $\mathbf{Q}^{-1}(z)\mathbf{P}(z)$ for the system. Evaluate the determinant of $\mathbf{Q}(z)$ and verify that it has order equal to the McMillan degree.

13.13. Consider the diagonal matrices $\mathbf{\Lambda}_\beta(z)$ and $\mathbf{\Lambda}_\alpha(z)$ in (13.5.37). Assume that $\alpha_i(z)$ and $\beta_i(z)$ (which are polynomials in z) are relatively prime for each i. Show that the matrices $\mathbf{\Lambda}_\beta(z)$ and $\mathbf{\Lambda}_\alpha(z)$ are left coprime.

13.14. Let $\mathbf{H}(z) = \mathbf{I} + z^{-1}\mathbf{h}_1$.
a) Let $\mathbf{h}_1^2 = \mathbf{0}$. Show that $\mathbf{H}(z)$ has the inverse $\mathbf{I} - z^{-1}\mathbf{h}_1$, and hence that $\mathbf{H}(z)$ is causal unimodular.
b) More generally let $\mathbf{h}_1^L = \mathbf{0}$ for some integer $L > 0$. Show that $\mathbf{H}(z)$ is still unimodular, and find the causal FIR inverse!

13.15. The purpose of this problem is to get acquainted with practically *all* the concepts introduced in this chapter, using a simple example. Consider the following left MFD:
$$\mathbf{H}(z) = \underbrace{\begin{bmatrix} 1+z & z \\ z & 1+z \end{bmatrix}^{-1}}_{\mathbf{Q}^{-1}(z)} \underbrace{\begin{bmatrix} 2+z & 1+z \\ z & z \end{bmatrix}}_{\mathbf{P}(z)}$$

a) What is the normal rank of $\mathbf{P}(z)$?
b) Work out the four elements $H_{km}(z), 0 \leq k, m \leq 1$ of the transfer matrix $\mathbf{H}(z)$.

c) Find the causal impulse response matrix $\mathbf{h}(n)$ corresponding to $\mathbf{H}(z)$. Express all the entries $h_{km}(n)$ in closed form.

d) Compute the coefficients $\mathbf{h}(0)$ and $\mathbf{h}(1)$.

e) Is the causal system stable?

f) Find $[\det \mathbf{Q}(z)]$. Hence argue that the system $\mathbf{H}(z)$ has degree one.

g) Argue that the above MFD is irreducible.

h) Find a minimal structure (i.e., structure with only one delay) for $\mathbf{H}(z)$. Write down the state space description $(\mathbf{A}, \mathbf{B}, \mathbf{C}, \mathbf{D})$ for the structure.

i) Find the Smith-form $\mathbf{\Gamma}(z)$ for $\mathbf{P}(z)$ by first computing all appropriate minors, and then using (13.5.15b). Also find the Smith-McMillan form $\mathbf{\Lambda}(z)$ for $\mathbf{H}(z)$. Using the formula (13.8.1), verify again that the degree is one.

j) Using the Smith-McMillan form identify the poles and zeros of $\mathbf{H}(z)$. These should agree with the zeros of $[\det \mathbf{Q}(z)]$ and $[\det \mathbf{P}(z)]$ respectively (since the MFD is irreducible). Verify that this is so.

13.16. Let $\mathbf{P}$ be $p \times r$ with rank r. We know that $\mathbf{P}^\dagger \mathbf{P} > 0$. However, in general we cannot claim $\mathbf{P}^T \mathbf{P} > 0$ even if it turns out to be Hermitian (unless, of course, $\mathbf{P}$ is real). Demonstrate this with an example.

13.17. Consider an r input p output system with input and output denoted as $\mathbf{x}(n)$ and $\mathbf{y}(n)$, respectively. Suppose the following properties are satisfied:

a) The response to a shifted input $\mathbf{x}(n-K)$ is equal to $\mathbf{y}(n-K)$, and this is true for all K and all possible input sequences $\mathbf{x}(n)$.

b) If the responses to $\mathbf{x}_a(n)$ and $\mathbf{x}_b(n)$ are equal to $\mathbf{y}_a(n)$ and $\mathbf{y}_b(n)$, then the response to $\alpha \mathbf{x}_a(n) + \beta \mathbf{x}_b(n)$ is equal to $\alpha \mathbf{y}_a(n) + \beta \mathbf{y}_b(n)$. And this is true for all scalars α, β and for all possible pairs of inputs $\mathbf{x}_a(n)$ and $\mathbf{x}_b(n)$.

Then show that the input output behavior of the system can be characterized by the convolution relation (2.2.7). Conversely, if (2.2.7) holds prove that the above two conditions are satisfied. This shows that the above two conditions can be taken as the defining properties of a MIMO LTI system.

13.18. Recall the relation between the impulse response $\mathbf{h}(k)$ and state space descriptions, given in (13.4.13). This was derived by first obtaining the expression (13.4.12) and then comparing with the convolution sum. A second procedure would be to start from (13.4.10b) and express $(z\mathbf{I} - \mathbf{A})^{-1}$ as a power series. Rederive (13.4.13) using this idea. (Hint: Note that $\sum_{k=0}^{L} \mathbf{A}^k = (\mathbf{I} - \mathbf{A})^{-1}(\mathbf{I} - \mathbf{A}^{L+1})$. Assume $\mathbf{A}$ is stable so that $\mathbf{A}^n \to \mathbf{0}$ as $n \to \infty$.)

13.19. Let $\mathbf{H}(z)$ be a $M \times 1$ rational causal system (i.e., an M-channel analysis bank). Show that it can always be rewritten as

$$\mathbf{H}(z) = \mathbf{W}(z) \begin{bmatrix} \lambda_0(z) \\ 0 \\ \vdots \\ 0 \end{bmatrix},$$

where $\mathbf{W}(z)$ is an $M \times M$ unimodular polynomial in z, and $\lambda_0(z)$ is a rational transfer function.

13.20. Consider the FIR filter $H(z) = (z+a)(z+b)$. Suppose we wish the output $y(n)$ to be such that $Y(z) = (z+b)$, i.e., FIR. Find an appropriate input. Show that there does *not* exist an FIR input which would result in this output.

13.21. Let $\mathbf{H}(z) = \mathbf{Q}^{-1}(z)\mathbf{P}(z)$ be a $p \times r$ causal rational system. In Sec. 13.7 we studied the meaning of a transmission zero, and its dynamical interpretation (Theorem 13.7.1) for the case $p \geq r$. We now consider the case when $p < r$. In this case the normal rank of $\mathbf{P}(z) < r$ so that for any value of z_0 we can find $\mathbf{v} \neq \mathbf{0}$ such that $\mathbf{P}(z_0)\mathbf{v} = \mathbf{0}$. For this reason, it is tricky to find a useful time domain interpretation of zeros. In what follows, we provide a useful and nontrivial time domain interpretation.

Let z_0 be a zero. Let the normal rank of $\mathbf{P}(z)$ be equal to p. Let $\mathbf{c}$ be any arbitrary $r \times 1$ vector. Then show that there exist finite length sequences $\mathbf{s}(n)$ and $\mathbf{t}(n)$ with $\mathbf{T}(z) \neq \mathbf{0}$ for any z, such that the input

$$\mathbf{U}(z) = \frac{\mathbf{c}}{1 - z_0 z^{-1}} + \mathbf{S}(z) \quad (P13.21)$$

produces an output $\mathbf{Y}(z)$ for which $\mathbf{T}^T(z)\mathbf{Y}(z)$ is a finite length sequence. [Here $\mathbf{S}(z)$ and $\mathbf{T}(z)$ represent the z-transforms of $\mathbf{s}(n)$ and $\mathbf{t}(n)$.]

Note. If the normal rank of $\mathbf{P}(z)$ were less than p, this would be true whether z_0 is a zero or not. Also note that in the above statement $\mathbf{c}$ is an arbitrary vector; this is what makes the statement nontrivial!

Hints. Use the fact that $(z - z_0)$ is a factor of $\alpha_{p-1}(z)$ in the Smith-McMillan form. The trick is to find a way to cancel the IIR components of the output, introduced by the first term in (P13.21), and by the quantity $\beta_{p-1}(z)$ in the Smith-McMillan form. It helps to remember that $\alpha_{p-1}(z)$ and $\beta_{p-1}(z)$ are relatively prime.

13.22. Consider the MFD $\mathbf{H}(z) = \mathbf{Q}^{-1}(z)\mathbf{P}(z)$, and let the polynomials $\mathbf{Q}(z)$ and $\mathbf{P}(z)$ be given by

$$\mathbf{Q}(z) = \sum_{n=0}^{L} \mathbf{q}_n z^n, \quad \mathbf{P}(z) = \sum_{n=0}^{L} \mathbf{p}_n z^n. \quad (P13.22a)$$

The input-output relation $\mathbf{Y}(z) = \mathbf{H}(z)\mathbf{U}(z)$ [i.e., $\mathbf{Q}(z)\mathbf{Y}(z) = \mathbf{P}(z)\mathbf{U}(z)$], can now be expressed in the time domain as follows:

$$\sum_{m=0}^{L} \mathbf{q}_m \mathbf{y}(n+m) = \sum_{m=0}^{L} \mathbf{p}_m \mathbf{u}(n+m). \quad (P13.22b)$$

This difference equation completely describes the input-output behavior of the system. The difference equation with $\mathbf{u}(n) = \mathbf{0}$, that is,

$$\sum_{m=0}^{L} \mathbf{q}_m \mathbf{y}(n+m) = \mathbf{0}, \quad (P13.22c)$$

is said to be the homogeneous equation.

a) Let z_p be a pole of $\mathbf{H}(z)$. Show that the homogeneous equation has a solution of the form $\mathbf{y}(n) = \mathbf{v} z_p^n$, for some constant non zero vector $\mathbf{v}$.

b) Conversely, let $\mathbf{v} z_p^n$ be a solution to the homogeneous equation ($\mathbf{v} \neq \mathbf{0}$, $z_p \neq 0$). Assuming that the MFD is irreducible, show that z_p is a pole of $\mathbf{H}(z)$.

13.23. One of the results claimed by Theorem 13.6.1 is this: If $(\mathbf{A}, \mathbf{B}, \mathbf{C}, \mathbf{D})$ is a minimal realization of a causal rational system $\mathbf{H}(z)$, then z_p is a pole of $\mathbf{H}(z)$ if and only if it is an eigenvalue of $\mathbf{A}$. The proof that an eigenvalue λ of $\mathbf{A}$ is a pole of $\mathbf{H}(z)$ assumed that $\lambda \neq 0$. We now develop a proof (suggested to the author by Prof. John Doyle, Caltech), which works even when $\lambda = 0$. This is based on the fact that given any $N \times N$ matrix, we can always apply a similarity transform to obtain a special form called the Jordan form [Chen, 1970,1984]. Since the similarity transform leaves the transfer function unchanged, we will develop the proof based on this form. The Jordan form is a block-diagonal matrix, given by

$$\mathbf{A} = \begin{bmatrix} \mathbf{A}_0 & 0 & \cdots & 0 \\ 0 & \mathbf{A}_1 & \cdots & 0 \\ \vdots & \vdots & \ddots & \vdots \\ 0 & 0 & \cdots & \mathbf{A}_{L-1} \end{bmatrix} \quad (P13.23a)$$

where $\mathbf{A}_i$ are $n_i \times n_i$ matrices of the special lower triangular form

$$\mathbf{A}_i = \begin{bmatrix} \lambda_i & 0 & \cdots & 0 & 0 \\ \times & \lambda_i & \cdots & 0 & 0 \\ \vdots & \vdots & \ddots & \vdots & \vdots \\ 0 & 0 & \cdots & \times & \lambda_i \end{bmatrix} \quad (P13.23b)$$

Here λ_i is an eigenvalue of $\mathbf{A}$ of multiplicity n_i, and $\times$ denotes possibly nonzero entries (with values = 0 or 1, but that is irrelevant here). Let us assume that the minimal realization $(\mathbf{A}, \mathbf{B}, \mathbf{C}, \mathbf{D})$ has $\mathbf{A}$ in the above form.

a) By partitioning $\mathbf{B}$ and $\mathbf{C}$ as

$$\mathbf{B} = \begin{bmatrix} \mathbf{B}_0 \\ \mathbf{B}_1 \\ \vdots \\ \mathbf{B}_{L-1} \end{bmatrix}, \quad \text{and} \quad \mathbf{C} = [\mathbf{C}_0 \ \mathbf{C}_1 \ \cdots \ \mathbf{C}_{L-1}] \quad (P13.23c)$$

show that we can express $\mathbf{H}(z)$ as

$$\mathbf{H}(z) = \sum_{k=0}^{L-1} \underbrace{\mathbf{C}_k (z\mathbf{I} - \mathbf{A}_k)^{-1} \mathbf{B}_k}_{\mathbf{H}_k(z)} + \mathbf{D} \quad (P13.23d)$$

b) Let $\lambda_0 = 0$, that is, $\mathbf{A}$ has the eigenvalue zero, with multiplicity $n_0 > 0$. By definition, $\lambda_i \neq 0$ for $i > 0$. Show that, all the poles of $\mathbf{H}_0(z)$ are at $z = 0$. Show also that none of $\mathbf{H}_k(z), k > 0$ has a pole at $z = 0$.

c) Thus, unless $\mathbf{H}_0(z)$ is a constant, $\mathbf{H}(z)$ has a pole at $z = 0$. We now show that $\mathbf{H}_0(z)$ cannot be constant, by using the minimality of $(\mathbf{A}, \mathbf{B}, \mathbf{C}, \mathbf{D})$.

(i) First show that minimality of $(\mathbf{A},\mathbf{B},\mathbf{C},\mathbf{D})$ implies $\mathbf{C}_0 \neq 0$ as well as $\mathbf{B}_0 \neq 0$. (ii) Then show that if $\mathbf{H}_0(z)$ is constant, then

$$\mathcal{S}_{\mathbf{C},\mathbf{A}} \begin{bmatrix} \mathbf{B}_0 \\ 0 \\ \vdots \\ 0 \end{bmatrix} = \mathbf{0}, \qquad (P13.23e)$$

thus violating complete observability, and hence minimality of the realization $(\mathbf{A},\mathbf{B},\mathbf{C},\mathbf{D})$.

Summarizing, we have proved that $\mathbf{H}_0(z)$ is not a constant, and therefore has pole(s) at $z=0$. Thus the eigenvlaues $\lambda_0 = 0$ of the matrix $\mathbf{A}$ imply that $\mathbf{H}(z)$ has pole(s) at $z = 0$.

13.24. Let $x(n)$ and $y(n)$ be the input and output of an LTI system $H(z)$, and define the blocked versions $\mathbf{x}_B(n)$ and $\mathbf{y}_B(n)$ as in (10.1.1), Chap. 10. Show that the M-input M-output system which produces $\mathbf{y}_B(n)$ in response to $\mathbf{x}_B(n)$ is an LTI system. In other words, verify that the defining property of a multi-input multi-output LTI system (given in Problem 13.17) is satisfied. (In Chap. 10 we just *assumed* this, and proceeded to show that the transfer matrix is pseudocirculant.)

13.25. Let $(\mathbf{A},\mathbf{B},\mathbf{C},D)$ be a minimal state space representation of a causal scalar transfer function $H(z)$ with degree N. Let $\mathbf{H}(z)$ be the blocked version of $H(z)$ with block length M (Sec. 10.1.1). Find a state space description $\widehat{\mathbf{A}}, \widehat{\mathbf{B}}, \widehat{\mathbf{C}}, \widehat{\mathbf{D}}$ for $\mathbf{H}(z)$ in terms of $(\mathbf{A},\mathbf{B},\mathbf{C},D)$ and M. Make sure your answer is such that $\widehat{\mathbf{A}}$ has size no larger than $N \times N$. Can you show that the blocked version has McMillan degree N?

14

Paraunitary and Lossless Systems

14.0 INTRODUCTION

In Sec. 3.4 we studied allpass functions. An allpass function $H(z)$ satisfies the relation $|H(e^{j\omega})| = c \neq 0$ for all ω. When $c = 1$ this implies that the input signal $u(n)$ has the same energy as the output signal $y(n)$. For this reason a stable allpass function is said to be a (single-input single-output) lossless system. The quantity c does not have a fundamental role, even though it is convenient to leave it there in the definition. The term 'lossless' is used even if $c \neq 1$.

In Sec. 6.1 we defined multi-input multi-output (MIMO) lossless and paraunitary systems. For these, the transfer matrix $\mathbf{H}(z)$ (not necessarily square) is unitary on the unit circle. We applied these for the design of perfect reconstruction systems (Chap. 6 and 8). In this chapter we present a self-contained treatment of MIMO lossless systems, and develop useful structures for these. All the structures are based on factorization of the transfer matrix. Some of the structures developed here have already been used in Sec. 6.4 (two channel QMF banks), Sec. 6.5 (M-channel filter banks), and Sec. 8.5 (cosine modulated filter banks), but in these earlier chapters we did not provide a formal development of these structures.

Outline

After a brief review of history, we provide a self contained discrete-time development of lossless systems. The basics are reviewed in Sec. 14.2. We develop structures for lossless systems in Sec. 14.3 and 14.4. Sec. 14.5 presents the state-space manifestation of lossless property. In Sec. 14.6 we present structures for unitary matrices (which are of importance in implementation of lossless systems). Sec. 14.7 and 14.8 deal with advanced results such as the Smith-McMillan form, pole-zero patterns, and modulus properties of lossless systems. Sec. 14.9 deals with structures for the IIR case, and Sec. 14.10 presents further modified structures which have certain advantages. In Sec. 14.11 we consider quantization of the parameters in these structures, and show how the lossless property can be retained in spite of quantization.

14.1 A BRIEF HISTORY

Classical electrical networks based on inductors and capacitors (LC networks) can be considered to be the source from which the concept of losslessness originated. LC networks are lossless, that is, do not generate or dissipate energy. We now state their connection to the discrete-time case, even though our developments in the rest of the chapter do not use this connection. Our presentation in this chapter will be self-contained and entirely in the discrete-time domain.

An electrical network with M ports is associated with M port currents and voltages. With such an electrical multiport we can associate an $M \times M$ impedance matrix $\mathbf{Z}(s)$. This relates the M voltages and currents as $\mathbf{V}(s) = \mathbf{Z}(s)\mathbf{I}(s)$. Also defined in the classical literature is the scattering matrix

$$\mathbf{S}(s) \triangleq \Big(\mathbf{Z}(s) - \mathbf{I}\Big)\Big(\mathbf{Z}(s) + \mathbf{I}\Big)^{-1}. \qquad (14.1.1)$$

If the multiport is lossless (e.g., an LC network) then the impedance matrix $\mathbf{Z}(s)$ satisfies the property that $\mathbf{Z}(j\Omega)$ has real part equal to zero for all Ω. This is equivalent to the property that $\mathbf{S}(j\Omega)$ is normalized-unitary. If we now use the bilinear transformation to obtain

$$\mathbf{H}(z) = \mathbf{S}(s)\Big|_{s=\frac{1-z^{-1}}{1+z^{-1}}} \qquad (14.1.2)$$

then $\mathbf{H}(z)$ is normalized-unitary on the unit circle of the z-plane. In other words, $\mathbf{H}(z)$ is paraunitary in the sense defined in Sec. 6.1. In this text $\mathbf{H}(z)$ is $p \times r$ (i.e., not necessarily square). Recall, in contrast, that $\mathbf{Z}(s)$ is square because it relates M currents to M voltages. And $\mathbf{S}(s)$ is square by its definition in terms of $\mathbf{Z}(s)$.

References on passivity and losslessness include Brune [1931], Darlington [1939], Guillemin [1957], Potapov [1960], Belevitch [1968], Balabanian and Bickart [1969], and [Anderson and Vongpanitlerd, 1973]. Many advanced results can be found in [Belevitch, 1968].

Passivity and losslessness have been applied for the design of robust digital filter structures [Fettweis, 1971], [Deprettere and Dewilde, 1980], [Rao and Kailath, 1984], [Vaidyanathan and Mitra, 1984]. Paraunitary transfer matrices were applied to the design of perfect reconstruction filter banks in Vaidyanathan [1987a], and subsequent references cited in Vaidyanathan [1990a]. They have also been used in the design of two-dimensional perfect reconstruction filter banks [Karlsson and Vetterli, 1990]. Results on discrete time lossless systems and their factorizations can be found in Potapov [1960],

Deprettere and Dewilde [1980], Vaidyanathan, et al. [1989], and Doğanata and Vaidyanathan [1990]. See also references at the beginning of Chap. 6.

Our presentation will be entirely in the discrete-time domain. We believe that this widens the cross section of readers who can appreciate the value of these concepts.

14.2 FUNDAMENTALS OF LOSSLESS SYSTEMS

At this point, the reader should review Sec. 6.1, where we introduced lossless systems first. In what follows, we will summarize the main points of Sec. 6.1, for quick reference.

1. *Definition.* Let $\mathbf{H}(z)$ be a $p \times r$ transfer matrix representing a causal system. It is said to be lossless if (a) each entry $H_{km}(z)$ is stable and (b) $\mathbf{H}(e^{j\omega})$ is unitary on the unit circle, that is,

$$\mathbf{H}^\dagger(e^{j\omega})\mathbf{H}(e^{j\omega}) = c^2 \mathbf{I}, \quad \forall\, \omega, \qquad (14.2.1a)$$

for some real constant $c \neq 0$. [The matrix $\mathbf{I}$ above has size $r \times r$, that is, it is $\mathbf{I}_r$]. If in addition the coefficients of $\mathbf{H}(z)$ are real [i.e., $\mathbf{H}(z)$ real for real z], we say that $\mathbf{H}(z)$ is *lossless bounded real* (abbreviated LBR). In order to satisfy (14.2.1a) we require $p \geq r$.

2. *Paraunitary property.* For rational transfer functions, (14.2.1a) implies

$$\widetilde{\mathbf{H}}(z)\mathbf{H}(z) = c^2 \mathbf{I}, \quad \forall z, \qquad (14.2.1b)$$

which is termed the *paraunitary* property. Conversely, (14.2.1b) implies (14.2.1a). We can therefore define a lossless system to be a causal stable paraunitary system. For square matrices, (14.2.1b) implies $\mathbf{H}^{-1}(z) = \widetilde{\mathbf{H}}(z)/c^2$. So the inverse is obtained by use of "tilde" operation.

3. *Normalized systems.* If a lossless system has $c^2 = 1$ in (14.2.1a) we say that it is normalized-lossless. Correspondingly, (14.2.1a) and (14.2.1b) are termed *normalized-unitary* and *normalized-paraunitary*.

Whenever causality and stability are clear from the context, we will use "paraunitary" and "lossless" interchangeably. In signal processing literature, the term "paraunitary" is somewhat more popular than "lossless" because of the more common use of FIR systems (which are stable automatically).

14.2.1 Properties Induced by Losslessness

The paraunitary property induces a number of other properties. We now summarize these, but elaborate only those not already presented in Sec. 6.1. More advanced properties will be presented throughout this chapter. In what follows, $\mathbf{H}(z)$ is $p \times r$ causal lossless and $\mathbf{H}(z) = \sum_{n=0}^{\infty} \mathbf{h}(n)z^{-n}$.

1. *Energy balance property.* From the input-output relation $\mathbf{Y}(e^{j\omega}) = \mathbf{H}(e^{j\omega})\mathbf{U}(e^{j\omega})$ we obtain

$$\mathbf{Y}^\dagger(e^{j\omega})\mathbf{Y}(e^{j\omega}) = c^2 \mathbf{U}^\dagger(e^{j\omega})\mathbf{U}(e^{j\omega}) \quad \text{(using (14.2.1a))}. \qquad (14.2.2)$$

We can integrate both sides of this equation in the range $0 \le \omega \le 2\pi$ and then apply the vector version of Parseval's theorem (2.2.12) to arrive at

$$\sum_{n=-\infty}^{\infty} \mathbf{y}^\dagger(n)\mathbf{y}(n) = c^2 \sum_{n=-\infty}^{\infty} \mathbf{u}^\dagger(n)\mathbf{u}(n). \qquad (14.2.3)$$

The summations above represent the energies of the sequences $\mathbf{y}(n)$ and $\mathbf{u}(n)$. We denote them as E_y and E_u respectively. Thus

Output energy $E_y = c^2 \times$ (input energy E_u) (for lossless systems). $\qquad (14.2.4)$

Note that for $c^2 = 1$ we have $E_y = E_u$, justifying the name "lossless" (which is employed even when $c^2 \neq 1$). Conversely, suppose (14.2.3) holds for every input sequence $\mathbf{u}(n)$. Then the system satisfies (14.2.1a). A proof of this is developed in Problem 14.18.

2. The impulse response $\mathbf{h}(n)$ satisfies

$$\sum_{n=0}^{\infty} \mathbf{h}^\dagger(n)\mathbf{h}(n+k) = c^2 \mathbf{I}\delta(k). \qquad (14.2.5)$$

This follows by substituting $\mathbf{H}(z) = \sum_{n=0}^{\infty} \mathbf{h}(n)z^{-n}$ into (14.2.1b) and equating like powers of z. For FIR case, letting $\mathbf{H}(z) = \sum_{n=0}^{L} \mathbf{h}(n)z^{-n}$ with $L > 0$, this property implies, in particular,

$$\mathbf{h}^\dagger(0)\mathbf{h}(L) = \mathbf{0}. \qquad (14.2.6)$$

When $p = r$, this means that $\mathbf{h}(0)$ and $\mathbf{h}(L)$ are singular matrices.

3. *Determinant is allpass.* If $p = r$ then the determinant of $\mathbf{H}(z)$ is a stable allpass function. In particular, if $\mathbf{H}(z)$ is FIR then $[\det \mathbf{H}(z)]$ is of the form az^{-K}, where K is a nonnegative integer and $a \neq 0$.

4. *Power complementary (PC) property.* For a $M \times 1$ causal stable transfer matrix it is clear that the lossless property is equivalent to the power complementary property.

5. *Submatrices of lossless $\mathbf{H}(z)$.* Every column of a lossless transfer matrix is itself lossless. Any $p \times L$ submatrix of $\mathbf{H}(z)$ is lossless.

In Sec. 6.1 several examples and interconnections of lossless systems were presented, which should be reviewed at this time. Notable among these is the Givens rotation matrix $\mathbf{R}_m$. This is given in (6.1.9), and has the flowgraph shown in Fig. 6.1-2(a). We also defined the diagonal lossless system $\mathbf{\Lambda}(z)$ as in (6.1.10). We used cascades of the above two types of building blocks to generate more general lossless systems (e.g., Fig. 6.1-4).

14.2.2 Structures for Lossless Systems

Basic Philosophy

In this chapter we will derive structures for lossless systems, both FIR and IIR. Losslessness implies $\tilde{\mathbf{H}}(z)\mathbf{H}(z) = c^2\mathbf{I}$. So the coefficients of $\mathbf{H}(z)$ are not arbitrary, but are constrained in some manner. As a result, the number of 'degrees' of freedom for a lossless system is less than the number of coefficients. If we pick an arbitrary coefficient of $\mathbf{H}(z)$, and vary it in an arbitrary manner, the lossless property is not retained. The same is true of constant unitary matrices. For example, consider $M \times M$ real $\mathbf{R}$ with $\mathbf{R}^T\mathbf{R} = \mathbf{I}$. Even though $\mathbf{R}$ has M^2 real elements, there are only $M(M-1)/2$ degrees of freedom (Sec. 14.6).

We will obtain structures for lossless systems such that they have the minimum number of free parameters. (For example an $M \times M$ real $\mathbf{R}$ with $\mathbf{R}^T\mathbf{R} = \mathbf{I}$ will be implemented with exactly $M(M-1)/2$ parameters.) An arbitrarily chosen parameter in the structure can be varied in an arbitrary manner (except for mild constraints which will become obvious later), and yet the transfer matrix will retain its lossless or unitary property. So, unlike the original matrix, the structure has *fewer parameters*, and these are essentially *unconstrained*. The lossless property is *implicitly imposed* by the structural interconnection. Thus, a good structure, in most cases, is an economic way to express a well-defined family of systems. This is the fundamental philosophy behind all structures we derive.

An example of a lossless structure is the cascaded lattice derived in Chap. 3 (see Fig. 3.4-8). In this structure, the transfer function $G_N(z)$ is lossless (stable and allpass) as long as $|k_m| < 1$ for all m.

Factorization of Lossless Matrices

Each structure we obtain is a cascade of simple building blocks. We arrive at this product-form by performing a factorization of the lossless matrix. For this reason, the terms "structure" and "factorization" are often used interchangeably. All factorizations will have the following features.

1. Each building block has a simple "standard" form. It is easy to preserve its lossless property in spite of parameter variation, as long as the parameters satisfy some simple constraint (e.g., realness, or unit-norm property, etc.) For example, we can always factorize 2×2 real coefficient FIR lossless systems in terms of Givens rotations (Fig. 14.3-2). The lossless property is preserved regardless of the choice of the real valued parameters θ_m.
2. The factorization is *complete*, that is, covers a well defined family of lossless systems. For example, we will show that *every* $M \times M$ FIR lossless system can be factorized as in (14.4.10).
3. The factorization is *minimal*. This means that the number of delays in the resulting structure is equal to the (McMillan) degree of $\mathbf{H}_N(z)$.

Value of these structures. Perhaps the greatest advantage of these factorizations is that they give rise to structures which cover a *complete* family of lossless systems, and at the same time the parameters of the structures can be varied *independently* and *arbitrarily*, without impairing the lossless property. In Chap. 6 and 8, where we had to optimize filter bank responses under lossless constraint, such structural representations were useful. The lossless lattice structures (Sec. 6.4) can also be used to generate orthonormal wavelets (Sec. 11.4).

14.3 LOSSLESS SYSTEMS WITH TWO OUTPUTS

In this section we restrict attention to real coefficient FIR lossless systems with two outputs, that is, 2×2 and 2×1 systems. (These restrictions will be removed in the next section). Here are the main results we wish to establish:

1. For the 2×2 case we will show that the transfer matrix can be implemented as in Fig. 14.3-2.
2. For the 2×1 case we will show that the transfer matrix can be implemented as in Fig. 14.3-3.

14.3.1 2×2 Real-Coefficient FIR Lossless Systems

Suppose we are given the 2×2 FIR lossless system

$$\mathbf{H}_N(z) = \sum_{n=0}^{L} \mathbf{h}(n) z^{-n}, \quad \mathbf{h}(L) \neq \mathbf{0}, \qquad (14.3.1)$$

where $\mathbf{h}(n)$ are real. If $L = 0$, then $\mathbf{h}(0)$ is a real unitary matrix and can always be represented as (Problem 14.12)

$$\alpha \begin{bmatrix} \cos\theta_0 & \sin\theta_0 \\ -\sin\theta_0 & \cos\theta_0 \end{bmatrix} \begin{bmatrix} 1 & 0 \\ 0 & \pm 1 \end{bmatrix}, \qquad (14.3.2)$$

so that the discussion is trivial. So we will assume $L > 0$.

The meaning of subscript N in (14.3.1) will be clear in a moment. We will first show how $\mathbf{H}_N(z)$ can be expressed in terms of a FIR lossless system $\mathbf{H}_{N-1}(z)$ which is a "reduced system" in some sense to be described. Repeated application of this operation will result in the desired factorization.

The Degree-Reduction Step

From (14.2.6) we know that $\mathbf{h}(0)$ is singular so that there exists a real nonzero vector $\mathbf{v}$ such that $\mathbf{v}^T \mathbf{h}(0) = \mathbf{0}$. We can always scale $\mathbf{v}$ to have unit norm and write $\mathbf{v}^T = [\sin\theta_N \quad \cos\theta_N]$ for real θ_N, so that

$$[\sin\theta_N \quad \cos\theta_N] \mathbf{h}(0) = \mathbf{0}. \qquad (14.3.3)$$

Now consider the product

$$\underbrace{\begin{bmatrix} \cos\theta_N & -\sin\theta_N \\ \sin\theta_N & \cos\theta_N \end{bmatrix}}_{\text{call this } \mathbf{R}_N^T} \underbrace{[\mathbf{h}(0) + \mathbf{h}(1)z^{-1} + \ldots + \mathbf{h}(L)z^{-L}]}_{\mathbf{H}_N(z)}. \qquad (14.3.4)$$

Here $\mathbf{R}_N$ is the Givens rotation operator (as in (6.1.9)), and is therefore unitary, with $\mathbf{R}_N^T \mathbf{R}_N = \mathbf{I}$. In view of (14.3.3) we have

$$\begin{bmatrix} \cos\theta_N & -\sin\theta_N \\ \sin\theta_N & \cos\theta_N \end{bmatrix} \mathbf{h}(0) = \begin{bmatrix} \times & \times \\ 0 & 0 \end{bmatrix},$$

where $\times$ denotes possibly nonzero entries, so that (14.3.4) becomes

$$\underbrace{\begin{bmatrix} \cos\theta_N & -\sin\theta_N \\ \sin\theta_N & \cos\theta_N \end{bmatrix}}_{\mathbf{R}_N^T} \mathbf{H}_N(z) = \begin{bmatrix} \mathbf{G}_0(z) \\ z^{-1}\mathbf{G}_1(z) \end{bmatrix} = \underbrace{\begin{bmatrix} 1 & 0 \\ 0 & z^{-1} \end{bmatrix}}_{\mathbf{\Lambda}(z)} \underbrace{\begin{bmatrix} \mathbf{G}_0(z) \\ \mathbf{G}_1(z) \end{bmatrix}}_{\mathbf{H}_{N-1}(z)},$$

(14.3.5)

where $\mathbf{G}_0(z)$ and $\mathbf{G}_1(z)$ are *causal* FIR row-vectors. We can rewrite this as

$$\mathbf{H}_N(z) = \underbrace{\begin{bmatrix} \cos\theta_N & \sin\theta_N \\ -\sin\theta_N & \cos\theta_N \end{bmatrix}}_{\mathbf{R}_N} \mathbf{\Lambda}(z) \mathbf{H}_{N-1}(z). \quad (14.3.6)$$

From this it is easily verified that $\widetilde{\mathbf{H}}_{N-1}(z)\mathbf{H}_{N-1}(z) = \widetilde{\mathbf{H}}_N(z)\mathbf{H}_N(z)$. Since $\widetilde{\mathbf{H}}_N(z)\mathbf{H}_N(z) = c^2\mathbf{I}$, we conclude that $\mathbf{H}_{N-1}(z)$ is paraunitary. Furthermore, by construction, $\mathbf{H}_{N-1}(z)$ is causal and FIR.

We know that the determinant of $\mathbf{H}_N(z)$ is a delay, that is,

$$[\det \mathbf{H}_N(z)] = \beta z^{-N}.$$

From (14.3.6) we see that

$$[\det \mathbf{H}_{N-1}(z)] = z [\det \mathbf{H}_N(z)] = \beta z^{-(N-1)}. \quad (14.3.7)$$

The matrix $\mathbf{H}_{N-1}(z)$ is thus a reduced system in the sense that its determinant has lower degree.

Summarizing, we have expressed $\mathbf{H}_N(z)$ in terms of another 2×2 causal real coefficient FIR lossless matrix $\mathbf{H}_{N-1}(z)$ (see Fig. 14.3-1). In this process, the *degree of determinant* has been reduced by unity. Essentially, we have extracted a lossless degree-one building block to obtain a reduced remainder $\mathbf{H}_{N-1}(z)$.

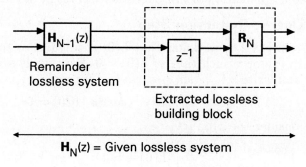

Figure 14.3-1 The degree reduction step in the factorization of 2×2 lossless $\mathbf{H}_N(z)$.

Complete factorization of $\mathbf{H}_N(z)$

Given $\mathbf{H}_N(z)$, we can express it as in (14.3.6) and repeat this process by expressing $\mathbf{H}_{N-1}(z)$ in terms of $\mathbf{H}_{N-2}(z)$ and so on. After N repetitions we arrive at the matrix $\mathbf{H}_0(z)$ with $[\det \mathbf{H}_0(z)] = \beta$. If $\mathbf{H}_0(z)$ is not a constant we can perform further reduction to obtain a causal system $\mathbf{H}_{-1}(z)$, whose determinant is βz (from (14.3.7)). This being impossible, we conclude that $\mathbf{H}_0(z)$ is a constant unitary matrix, which can therefore be expressed as in (14.3.2). Summarizing, we have been able to express $\mathbf{H}_N(z)$ as

$$\mathbf{H}_N(z) = \alpha \mathbf{R}_N \mathbf{\Lambda}(z) \mathbf{R}_{N-1} \ldots \mathbf{\Lambda}(z) \mathbf{R}_0 \begin{bmatrix} 1 & 0 \\ 0 & \pm 1 \end{bmatrix}, \qquad (14.3.8)$$

where $\alpha = \sqrt{|\beta|}$,

$$\mathbf{R}_m = \begin{bmatrix} \cos \theta_m & \sin \theta_m \\ -\sin \theta_m & \cos \theta_m \end{bmatrix}, \quad \theta_m \text{ real} \quad \text{(Givens rotation)}, \qquad (14.3.9a)$$

and

$$\mathbf{\Lambda}(z) = \begin{bmatrix} 1 & 0 \\ 0 & z^{-1} \end{bmatrix}. \qquad (14.3.9b)$$

This gives the structure of Fig. 14.3-2.

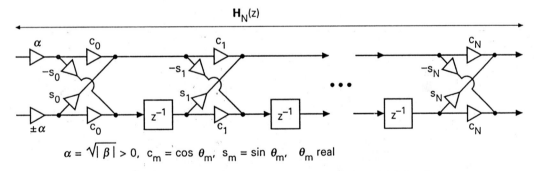

$\alpha = \sqrt{|\beta|} > 0$, $c_m = \cos \theta_m$, $s_m = \sin \theta_m$, θ_m real

Figure 14.3-2 Lattice structure for 2×2 real coefficient FIR lossless system.

Minimality. Since $\mathbf{H}_N(z)$ can be implemented as in Fig. 14.3-2 which has N delays, we have

$$[\deg \mathbf{H}_N(z)] \leq N. \qquad (14.3.10)$$

From (14.3.8) we also have $[\deg \det \mathbf{H}_N(z)] = N$. But from Sec. 13.8 we know $[\deg \mathbf{H}_N(z)] \geq [\deg \det \mathbf{H}_N(z)]$ so that

$$[\deg \mathbf{H}_N(z)] \geq N. \qquad (14.3.11)$$

From the above two inequalities we conclude [deg $\mathbf{H}_N(z)$] = N. This proves that the structure is minimial.

Comment on degree of determinant. It follows from the above result, that the degree (i.e., McMillan degree) of the lossless system $\mathbf{H}_N(z)$ is equal to the degree of its determinant. This fact, proved above for a very specific case, is true for *any* lossless system as justified in Sec. 14.7. Each degree-reduction step described above, therefore, reduces the McMillan degree of the system by unity. Notice that the degree-reduction step does not necessarily reduce the highest power of z^{-1} in the expression for $\mathbf{H}_N(z)$. As seen in Chap. 13, this highest power does not indicate the degree, and is not directly involved in the reduction process.

Summarizing, we have shown the following:

♠**Theorem 14.3.1. Factorizing 2×2 FIR paraunitary systems.** Let $\mathbf{H}_N(z)$ be 2×2 causal real-coefficient FIR lossless, with [det $\mathbf{H}_N(z)$] = βz^{-N}. Then, we can implement it in terms of Givens rotations, as shown in Fig. 14.3-2 where θ_m are real, and $\alpha = \sqrt{|\beta|}$. Equivalently, $\mathbf{H}_N(z)$ can be factorized as in (14.3.8), where $\mathbf{R}_m$ and $\mathbf{\Lambda}(z)$ are lossless building blocks as in (14.3.9). Furthermore the structure is minimal, that is, the number of delays N equals the degree of $\mathbf{H}_N(z)$. There are $N+2$ parameters in the structure, viz., the $N+1$ angles θ_m and the real number $\alpha \neq 0$. *Conversely, it is also clear that the transfer matrix of the above structure is lossless, regardless* of the values of θ_m and α. ◇

14.3.2 2×1 real-coefficient FIR lossless systems

A lossless system with two outputs ($p = 2$) can be either 2×2 or 2×1 (since $p \geq r$.) Suppose we are given a 2×1 lossless system

$$\mathbf{P}_N(z) = \begin{bmatrix} H_0(z) \\ H_1(z) \end{bmatrix}. \quad (14.3.12)$$

Losslessness implies the power complementary property

$$\widetilde{H}_0(z)H_0(z) + \widetilde{H}_1(z)H_1(z) = c^2, \quad c > 0. \quad (14.3.13)$$

If $\mathbf{P}_N(z)$ is (causal and) FIR, we can write

$$H_0(z) = \sum_{n=0}^{N} h_0(n)z^{-n}, \quad H_1(z) = \sum_{n=0}^{N} h_1(n)z^{-n}. \quad (14.3.14)$$

Our aim is to find a lattice structure for this, similar to Fig. 14.3-2. For this define

$$\mathbf{H}_N(z) = \begin{bmatrix} H_0(z) & -z^{-N}\widetilde{H}_1(z) \\ H_1(z) & z^{-N}\widetilde{H}_0(z) \end{bmatrix}. \quad (14.3.15)$$

This is evidently causal (because of the use of z^{-N} at appropriate places) and FIR, and moreover satisfies $\widetilde{\mathbf{H}}_N(z)\mathbf{H}_N(z) = c^2\mathbf{I}$.

The matrix $\mathbf{H}_N(z)$ is therefore a lossless system. Furthermore it has real coefficients, so that it can be realized as in Fig. 14.3-2. By ignoring the lower input terminal, we obtain the structure of Fig. 14.3-3, for the original 2×1 system $\mathbf{P}_N(z)$. Without loss of generality, assume that at least one of $h_0(N), h_1(N)$ is nonzero so that the degree of $\mathbf{P}_N(z)$ is N. The structure, which evidently has N delays, is therefore minimal!

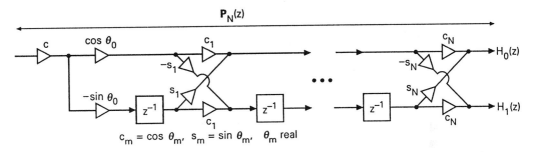

Figure 14.3-3 Lattice structure for 2×1 real coefficient FIR lossless system.

Given $\mathbf{P}_N(z)$, the above structure can be synthesized by identifying θ_m. This is done by defining $\mathbf{H}_N(z)$ as above, and peforming the degree reduction step described earlier in Sec. 14.3.1. Thus, given a power-complementary pair of FIR filters $H_0(z)$ and $H_1(z)$, we can always find the structure of Fig. 14.3-3. Summarizing, we have proved the following [Vaidyanathan, 1986]:

♦**Theorem 14.3.2. Factorizing 2×1 FIR paraunitary systems.** Let $\mathbf{P}_N(z)$ be a 2×1 causal real-coefficient FIR lossless system with degree N. Then it can be factorized as

$$\mathbf{P}_N(z) = c\mathbf{R}_N \mathbf{\Lambda}(z) \mathbf{R}_{N-1} \ldots \mathbf{\Lambda}(z) \begin{bmatrix} \cos\theta_0 \\ -\sin\theta_0 \end{bmatrix}, \qquad (14.3.16)$$

where c, θ_0 are real, and $\mathbf{R}_m$ and $\mathbf{\Lambda}(z)$ are as in (14.3.9). Thus $\mathbf{P}_N(z)$ can be realized as in Fig. 14.3-3, and the structure is minimal. Conversely, the transfer matrix of the above structure is lossless, regardless of the values of the $N+1$ real parameters θ_m and the parameter $c > 0$. ◇

Notice that if $c < 0$, we can replace it with $|c|$ and replace θ_0 with $\theta_0 + \pi$ to get the same transfer functions $H_0(z)$ and $H_1(z)$.

14.4 STRUCTURES FOR M × M and M × 1 FIR LOSSLESS SYSTEMS

We now generalize the results of the previous section. The generalization will take place in two respects. First, we remove the restriction $p = 2$, and consider $M \times M$ FIR lossless systems. Second the coefficients are not restricted to be real. At the end of this section we also consider factorization of power complementary filter banks (i.e., $M \times 1$ FIR lossless systems).

Since the structures for 2×2 systems were based on planar rotations, one would expect that $M \times M$ lossless systems can also be factorized using generalized rotations (i.e., $M \times M$ unitary matrices). This indeed is possible as we show in Sec. 14.10.2. In this section, however, we establish a much simpler factorization based on diadic matrices (to be defined). In addition to its simplicity, this also offers some practical advantages under finite wordlength conditions. To be more specific, *rotation based* structures lose the paraunitary property under coefficient quantization, whereas *diadic based* structures do not (as proved in Sec. 14.11).

♠**Main points of this section.** The structures introduced in this section are based on a fundamentally different lossless building block. In this section we do the following:

1. We first introduce the degree-one lossless building block shown in Fig. 14.4-1.
2. We then show (Sec. 14.4.2) that any $M \times M$ FIR lossless system $\mathbf{H}_N(z)$ (with degree N) can be expressed as a product of N of these building blocks, and a constant unitary matrix $\mathbf{H}_0$ [as in (14.4.10)]. This results in the structure of Fig. 14.4-3. When $\mathbf{H}_N(z)$ has real coefficients, the multipliers in the structure are real.
3. Finally we show (Sec. 14.4.3) that any $M \times 1$ FIR lossless system (i.e., an FIR power complementary filter bank) can be factorized as in (14.4.12), resulting in the structure of Fig. 14.4-4, where $\mathbf{P}_0$ is a constant nonzero vector. ◇

14.4.1 A Fundamental Degree-One Building Block

The structures we develop in this section are based on the $M \times M$ transfer matrix given by

$$\mathbf{V}_m(z) = \mathbf{I} - \mathbf{v}_m \mathbf{v}_m^\dagger + z^{-1} \mathbf{v}_m \mathbf{v}_m^\dagger, \qquad (14.4.1)$$

where $\mathbf{v}_m$ is a (possibly complex) $M \times 1$ vector with unit norm, that is,

$$\mathbf{v}_m^\dagger \mathbf{v}_m = 1. \qquad (14.4.2)$$

Note the appearance of the *diadic form* $\mathbf{v}_m \mathbf{v}_m^\dagger$ in (14.4.1). The resulting structures are said to be *diadic-based*. It is readily verified that $\mathbf{V}_m(z)$ can be implemented as in Fig. 14.4-1. Since there is only one delay in the figure, $\mathbf{V}_m(z)$ has degree = 1.

The system $\mathbf{V}_m(z)$ is normalized-lossless, that is, $\widetilde{\mathbf{V}}_m(z) \mathbf{V}_m(z) = \mathbf{I}$. To prove this, define for convenience $\mathbf{P} = \mathbf{v}_m \mathbf{v}_m^\dagger$ and $\mathbf{Q} = \mathbf{I} - \mathbf{v}_m \mathbf{v}_m^\dagger$. Note that $\mathbf{P}$ and $\mathbf{Q}$ are Hermitian. Using $\mathbf{v}_m^\dagger \mathbf{v}_m = 1$, the following identities are easily

verified:
$$\mathbf{P}^2 = \mathbf{P}, \mathbf{Q}^2 = \mathbf{Q}, \mathbf{PQ} = \mathbf{QP} = \mathbf{0}. \qquad (14.4.3)$$

As a result
$$\widetilde{\mathbf{V}}_m(z)\mathbf{V}_m(z) = \Big[\mathbf{Q} + z\mathbf{P}\Big]\Big[\mathbf{Q} + z^{-1}\mathbf{P}\Big]$$
$$= \mathbf{Q}^2 + \mathbf{P}^2 + z\mathbf{PQ} + z^{-1}\mathbf{QP} = \mathbf{Q} + \mathbf{P} = \mathbf{I}.$$

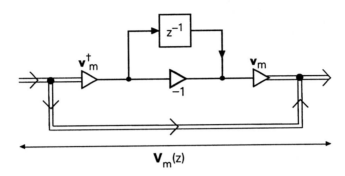

Figure 14.4-1 Implementation of $\mathbf{V}_m(z)$ using one delay.

Determinant of $\mathbf{V}_m(z)$. In the next section, where we obtain a structure for FIR lossless systems based on degree reduction, a knowledge of the determinant of $\mathbf{V}_m(z)$ is required. We now prove

$$\det \mathbf{V}_m(z) = z^{-1}. \qquad (14.4.4)$$

Proof. We have $\mathbf{V}_m(z)\mathbf{v}_m = z^{-1}\mathbf{v}_m$ (using $\mathbf{v}_m^\dagger \mathbf{v}_m = 1$). So $\mathbf{v}_m$ is an eigenvector with eigenvalue $= z^{-1}$. Next if $\mathbf{u}$ is any vector orthogonal to $\mathbf{v}_m$ (i.e., $\mathbf{v}_m^\dagger \mathbf{u} = 0$), then $\mathbf{V}_m(z)\mathbf{u} = \mathbf{u}$. Suppose now that $\mathbf{u}_k, 0 \le k \le M-1$ is a set of M mutually orthogonal unit-norm vectors (with $\mathbf{u}_{M-1} \stackrel{\Delta}{=} \mathbf{v}_m$). Defining the $M \times M$ unitary matrix $\mathbf{U} = [\mathbf{u}_0 \ldots \mathbf{u}_{M-1}]$, we then have

$$\mathbf{V}_m(z)\mathbf{U} = \mathbf{U}\begin{bmatrix} \mathbf{I} & \mathbf{0} \\ \mathbf{0} & z^{-1} \end{bmatrix}. \qquad (14.4.5)$$

Taking determinants we arrive at (14.4.4). ▽▽▽

Summarizing, $\mathbf{V}_m(z)$ has the following properties:

1. It is a normalized lossless system, that is, $\widetilde{\mathbf{V}}_m(z)\mathbf{V}_m(z) = \mathbf{I}$.
2. It has degree $= 1$.
3. $[\det \mathbf{V}_m(z)] = z^{-1}$.
4. $\mathbf{V}_m(1) = \mathbf{I}$ (follows from definition).

Example 14.4.1: Higher-degree Lossless Systems

To get familiar with the above properties of $\mathbf{V}_m(z)$, consider the product of two such building blocks, viz., $\mathbf{V}_1(z)\mathbf{V}_2(z)$. This is lossless. Assume, for sake of this example, that $\mathbf{v}_1$ is orthogonal to $\mathbf{v}_2$ (i.e., $\mathbf{v}_1^\dagger \mathbf{v}_2 = 0$). Then, we can verify that $\mathbf{V}_1(z)\mathbf{V}_2(z) = \mathbf{I} - \mathbf{DD}^\dagger + z^{-1}\mathbf{DD}^\dagger$ where $\mathbf{D} = [\mathbf{v}_1 \; \mathbf{v}_2]$.

Similarly, consider the product $\mathbf{V}_0(z)\ldots\mathbf{V}_{M-1}(z)$, where each factor is as in (14.4.1). Let the M vectors $\mathbf{v}_m$, $0 \le m \le M-1$ be pairwise orthogonal unit-norm vectors. Then it can be verified (Problem 6.10) that this product reduces to $\mathbf{I} - \mathbf{EE}^\dagger + z^{-1}\mathbf{EE}^\dagger$ where $\mathbf{E} = [\mathbf{v}_0 \; \mathbf{v}_1 \; \ldots \; \mathbf{v}_{M-1}]$. Because of our choice of the vectors $\mathbf{v}_m$, $\mathbf{E}$ is normalized unitary (i.e., $\mathbf{EE}^\dagger = \mathbf{I}$) so that the above product reduces to $z^{-1}\mathbf{I}$.

This shows that trivial lossless systems such as $z^{-1}\mathbf{I}$ (more generally $z^{-K}\mathbf{I}$) can be written as products of degree-one systems of the form (14.4.1). We will see later that any $M \times M$ FIR lossless system is essentially a product of these building blocks.

Most General Degree-One FIR Lossless System

The reader might wonder how the strange-looking building block $\mathbf{V}_m(z)$ was invented. We now present a logical development to show that it arises naturally when we try to identify the *most general* degree-one FIR lossless system.

More formally, let $\mathbf{H}(z)$ be a $p \times r$ causal FIR lossless system with degree $= 1$. We show that it can be written as

$$\mathbf{H}(z) = \underbrace{[\mathbf{I} - \mathbf{vv}^\dagger + z^{-1}\mathbf{vv}^\dagger]}_{p \times p} \underbrace{\mathbf{H}(1)}_{p \times r}, \qquad (14.4.6)$$

where $\mathbf{v}$ is $p \times 1$ with $\mathbf{v}^\dagger\mathbf{v} = 1$. This is proved as follows:

Proof. We can write $\mathbf{H}(z)$ as $\mathbf{H}(z) = \mathbf{h}(0) + \mathbf{h}(1)z^{-1}$. This has degree $= 1$ if and only if $\mathbf{h}(1)$ has rank $= 1$ (Problem 13.8). Now consider the difference $\mathbf{H}(z) - \mathbf{H}(1)$. This is equal to zero for $z = 1$ so that we can write $\mathbf{H}(z) - \mathbf{H}(1) = (1 - z^{-1})\mathbf{S}$. Here $\mathbf{S}$ is a constant because the degree-one restriction forbids higher powers of z. So we can write

$$\mathbf{H}(z) = (1 - z^{-1})\mathbf{S} + \mathbf{H}(1)$$
$$= \Big((1 - z^{-1})\mathbf{SR} + \mathbf{I}\Big)\mathbf{H}(1),$$
$$= \underbrace{\Big(\mathbf{I} + \mathbf{SR} - z^{-1}\mathbf{SR}\Big)}_{p \times p} \underbrace{\mathbf{H}(1)}_{p \times r},$$

where $\mathbf{R} = \mathbf{H}^\dagger(1)/c^2$. Here we have used the fact that $\mathbf{H}^\dagger(1)\mathbf{H}(1) = c^2\mathbf{I}$. Since $\mathbf{H}(z)$ is paraunitary and $\mathbf{H}(1)$ unitary, it is clear from above that the quantity $\mathbf{V}(z) \triangleq (\mathbf{I} + \mathbf{SR} - z^{-1}\mathbf{SR})$ satisfies $\widetilde{\mathbf{V}}(z)\mathbf{V}(z) = \mathbf{I}$. Because of the degree-one restriction, $\mathbf{SR}$ has rank one, and can be expressed as $\mathbf{wu}^\dagger$ where $\mathbf{u}$ and $\mathbf{w}$ are nonzero $p \times 1$ vectors. By setting $\widetilde{\mathbf{V}}(z)\mathbf{V}(z) = \mathbf{I}$ and equating the coefficients of z on both sides we get $\mathbf{wu}^\dagger = -\|\mathbf{w}\|^2 \mathbf{uu}^\dagger$, that is, $\mathbf{wu}^\dagger$ can be written as $-\mathbf{vv}^\dagger$. So we can rewrite $\mathbf{V}(z) = \mathbf{I} - \mathbf{vv}^\dagger + z^{-1}\mathbf{vv}^\dagger$. By using the fact that $\mathbf{V}(-1)$ satisfies $\mathbf{V}^\dagger(-1)\mathbf{V}(-1) = \mathbf{I}$, we obtain $\mathbf{vv}^\dagger(\mathbf{v}^\dagger\mathbf{v} - 1) = 0$. Since $\mathbf{v} \neq \mathbf{0}$, this implies $\mathbf{v}^\dagger\mathbf{v} = 1$ indeed. This completes the proof that $\mathbf{H}(z)$ has the form (14.4.6). ▽▽▽

14.4.2 Factorization of $M \times M$ FIR Lossless Matrices

Our next aim is to show that an arbitrary $M \times M$ FIR lossless system can be factorized in terms of $\mathbf{V}_m(z)$. This is based on a degree-reduction procedure as before, and will again involve the reduction of degree of the determinant.

Degree-Reduction Procedure

Let $\mathbf{H}_N(z)$ be an $M \times M$ causal FIR lossless system with determinant $= \beta z^{-N}$. We first show how we can express $\mathbf{H}_N(z)$ in terms of the building block $\mathbf{V}_m(z)$, and a "reduced" lossless system $\mathbf{H}_{N-1}(z)$ with determinant $\beta z^{-(N-1)}$.

First recall that the FIR nature of $\mathbf{H}_N(z)$ implies (14.2.6), so that $\mathbf{h}(0)$ is singular [unless $\mathbf{h}(L) = \mathbf{0}$ in which case no reduction is necessary]. So, there exists a vector $\mathbf{v}_N \neq \mathbf{0}$ such that

$$\mathbf{v}_N^\dagger \mathbf{h}(0) = \mathbf{0}. \tag{14.4.7}$$

Without loss of generality, assume $\mathbf{v}_N$ has unit norm. Consider the product

$$\mathbf{H}_{N-1}(z) \triangleq \underbrace{[\mathbf{I} - \mathbf{v}_N\mathbf{v}_N^\dagger + z\mathbf{v}_N\mathbf{v}_N^\dagger]}_{\widetilde{\mathbf{V}}_N(z)} \underbrace{[\mathbf{h}(0) + \ldots + z^{-L}\mathbf{h}(L)]}_{\mathbf{H}_N(z)}. \tag{14.4.8}$$

In view of (14.4.7), the term $z\mathbf{v}_N\mathbf{v}_N^\dagger \mathbf{h}(0) = \mathbf{0}$ so that the above product remains causal. Since $\mathbf{V}_N(z)\widetilde{\mathbf{V}}_N(z) = \mathbf{I}$, we can rewrite (14.4.8) as

$$\mathbf{H}_N(z) = \mathbf{V}_N(z)\mathbf{H}_{N-1}(z). \tag{14.4.9}$$

Since $\mathbf{H}_N(z)$ and $\mathbf{V}_N(z)$ are paraunitary, it is easily verified that $\mathbf{H}_{N-1}(z)$ is paraunitary. From (14.4.9) we obtain

$$[\det \mathbf{H}_N(z)] = z^{-1}[\det \mathbf{H}_{N-1}(z)] \quad (\text{since } [\det \mathbf{V}_N(z)] = z^{-1}).$$

So we conclude $[\det \mathbf{H}_{N-1}(z)] = \beta z^{-(N-1)}$, since $[\det \mathbf{H}_N(z)] = \beta z^{-N}$.

Summarizing, we have expressed $\mathbf{H}_N(z)$ as in Fig. 14.4-2, where $\mathbf{V}_N(z)$ is a degree-one lossless system and $\mathbf{H}_{N-1}(z)$ is another causal FIR lossless system. The system $\mathbf{H}_{N-1}(z)$ has reduced determinantal degree $N - 1$. Essentially we have *extracted* the degree-one lossless building block $\mathbf{V}_N(z)$ to obtain a reduced lossless *remainder* $\mathbf{H}_{N-1}(z)$.

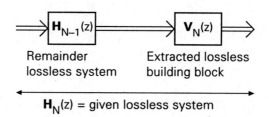

Figure 14.4-2 The degree reduction step in the factorization of $M \times M$ lossless $\mathbf{H}_N(z)$.

Complete Factorization of the M×M FIR Lossless System

The lossless system $\mathbf{H}_{N-1}(z)$ satisfies all the properties of $\mathbf{H}_N(z)$ except for reduced degree of determinant. If we repeat this reduction N times, we eventually obtain the remainder $\mathbf{H}_0(z)$. This is a causal FIR lossless system with $[\det \mathbf{H}_0(z)] = \beta$. If $\mathbf{H}_0(z)$ is not a constant matrix, we can repeat the reduction process to obtain a causal remainder $\mathbf{H}_{-1}(z)$ with determinant βz, which is an impossible situation. The conclusion is that $\mathbf{H}_0$ is a constant unitary matrix. Thus, the structure of Fig. 14.4-3 follows. As in Sec. 14.3, we can also show that the structure is minimal. The overall factorization is given by

$$\mathbf{H}_N(z) = \mathbf{V}_N(z)\mathbf{V}_{N-1}(z)\ldots\mathbf{V}_1(z)\mathbf{H}_0, \qquad (14.4.10)$$

where $\mathbf{V}_m(z)$ is as in (14.4.1) with the $M \times 1$ vectors $\mathbf{v}_m$ satisfying $\mathbf{v}_m^\dagger \mathbf{v}_m = 1$, and $\mathbf{H}_0$ is $M \times M$ unitary.

Notice, once again, that the highest power of z^{-1} is not necessarily reduced at each step of the reduction process. For example, if this reduction is performed on the matrix $\mathbf{H}_N(z) = z^{-1}\mathbf{I}$, it will require M steps before the coefficient of z^{-1} becomes zero!

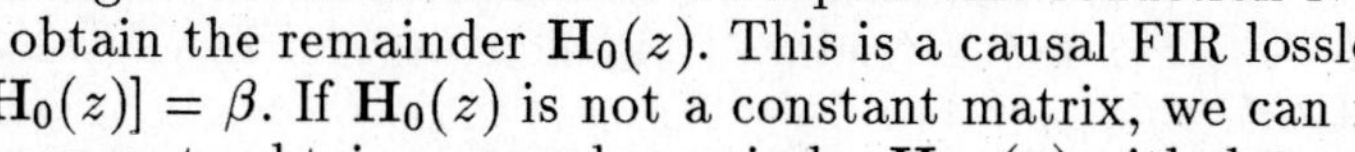

Figure 14.4-3 Factorization of $\mathbf{H}_N(z)$. Building blocks are as in Fig. 14.4-1.

Real-coefficient case. Suppose $\mathbf{H}_N(z)$ has real coefficients. The vector $\mathbf{v}_N$, which is required to satisfy (14.4.7) can therefore be taken to be real. So the remainder $\mathbf{H}_{N-1}(z)$ has real coefficients. Continuing this argument we see that all the vectors $\mathbf{v}_m$ and the matrix $\mathbf{H}_0$ in Fig. 14.4-3 will come out to be real. So there exists a structure of the said form with real multipliers.

Degree of the determinant of a lossless system. We now consider a by-product of the above result. There are N delays in the structure, since each building block $\mathbf{V}_m(z)$ requires one delay. Since the structure is minimal, it follows that N is the degree of $\mathbf{H}_N(z)$. In other words, for an $M \times M$ causal FIR lossless system with degree N, we have

$$[\deg \mathbf{H}_N(z)] = [\deg \det \mathbf{H}_N(z)], \qquad (14.4.11)$$

where deg [.] denotes the McMillan degree. (It turns out, as we show in Sec. 14.7, that the FIR restriction in the above statement is not necessary.) The special relation (14.4.11) is not necessarily true for *arbitrary* (non lossless) transfer functions. For example, if $\mathbf{H}_N(z)$ is unimodular (Sec. 13.10.2), then its determinant has degree = 0 regardless of degree of $\mathbf{H}_N(z)$!

Summarizing this subsection, we have proved the following factorization result:

♠**Theorem 14.4.1. Factorization of $M \times M$ FIR paraunitary system.** Let $\mathbf{H}_N(z)$ be $M \times M$ causal FIR lossless, with $[\det \mathbf{H}_N(z)] = \beta z^{-N}$. We can then factorize it as in (14.4.10). Furthermore, the structure resulting from the factorization (shown in Fig. 14.4-3) is minimal. (In the real coefficient case, we can ensure that $\mathbf{v}_m$ and $\mathbf{H}_0$ are real.) *Conversely*, the structure represents a degree N lossless system whenever $\mathbf{v}_m$ has unit norm and $\mathbf{H}_0$ is unitary. ◇

Note. Since $\mathbf{V}_m(1) = \mathbf{I}$, we see from (14.4.10) that $\mathbf{H}_N(1) = \mathbf{H}_0$.

14.4.3 Factorization of $M \times 1$ FIR Lossless Matrices

Let $\mathbf{P}_N(z)$ be an $M \times 1$ causal FIR lossless system with degree N. (For example, $\mathbf{P}_N(z)$ could represent an analysis bank with M filters.) We will show that it can be expressed as

$$\mathbf{P}_N(z) = \mathbf{V}_N(z)\mathbf{V}_{N-1}(z)\ldots\mathbf{V}_1(z)\mathbf{P}_0, \qquad (14.4.12)$$

where $\mathbf{V}_m(z)$ are as in (14.4.1) with the $M \times 1$ vectors $\mathbf{v}_m$ satisfying $\mathbf{v}_m^\dagger \mathbf{v}_m = 1$, and $\mathbf{P}_0$ is a nonzero vector. As a result, we can implement the system as in Fig. 14.4-4 requiring N delays, so that the structure is minimal.

Conversely the above structure represents a lossless system $\mathbf{P}_N(z)$ as long as $\mathbf{P}_0 \neq \mathbf{0}$ and $\mathbf{v}_m^\dagger \mathbf{v}_m = 1$. (This is obvious because a product of lossless systems is lossless.)

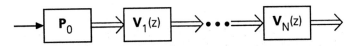

Figure 14.4-4 Factorization of $\mathbf{P}_N(z)$.

Degree-Reduction Step

The proof of the above factorization is again based on a degree reduction step. Given $\mathbf{P}_N(z)$ with properties stated above, we can write it as

$$\mathbf{P}_N(z) = \sum_{n=0}^{N} \mathbf{p}(n) z^{-n}. \qquad (14.4.13)$$

Since this is a column vector, the degree-N property is equivalent to the condition $\mathbf{p}(N) \neq \mathbf{0}$. From (14.2.6) it follows that

$$\mathbf{p}^\dagger(N)\mathbf{p}(0) = 0. \qquad (14.4.14)$$

Consider the new $M \times 1$ transfer matrix

$$\mathbf{P}_{N-1}(z) \triangleq \underbrace{\left(\mathbf{I} - \mathbf{v}_N \mathbf{v}_N^\dagger + z\mathbf{v}_N \mathbf{v}_N^\dagger\right)}_{\widetilde{\mathbf{V}}_N(z)} \underbrace{\left(\mathbf{p}(0) + \ldots + z^{-N}\mathbf{p}(N)\right)}_{\mathbf{P}_N(z)}. \qquad (14.4.15)$$

If we choose the unit-norm vector $\mathbf{v}_N$ as

$$\mathbf{v}_N = \mathbf{p}(N)/\|\mathbf{p}(N)\|, \qquad (14.4.16)$$

then in view of (14.4.14), the noncausal term in (14.4.15) becomes

$$z\mathbf{v}_N \mathbf{v}_N^\dagger \mathbf{p}(0) = \mathbf{0}. \qquad (14.4.17)$$

Moreover, the coefficient of z^{-N} becomes

$$\left(\mathbf{I} - \mathbf{v}_N \mathbf{v}_N^\dagger\right) \mathbf{p}(N) = \|\mathbf{p}(N)\| \left(\mathbf{I} - \mathbf{v}_N \mathbf{v}_N^\dagger\right) \mathbf{v}_N = \|\mathbf{p}(N)\| \left(\mathbf{v}_N - \mathbf{v}_N\right) = \mathbf{0},$$

since $\mathbf{v}_N^\dagger \mathbf{v}_N = 1$. So $\mathbf{P}_{N-1}(z)$ is causal and FIR with degree $< N$. We can rewrite (14.4.15) as

$$\mathbf{P}_N(z) = \mathbf{V}_N(z) \mathbf{P}_{N-1}(z), \qquad (14.4.18)$$

which shows that $\mathbf{P}_{N-1}(z)$ cannot have degree smaller than $N-1$ [since $\mathbf{P}_N(z)$ has degree N]. So the degree of $\mathbf{P}_{N-1}(z)$ is precisely $N-1$.

It is clear from (14.4.18) that $\mathbf{P}_{N-1}(z)$ is paraunitary. Summarizing, we have expressed $\mathbf{P}_N(z)$ in terms of the $M \times M$ degree-one building block $\mathbf{V}_N(z)$ and the reduced degree FIR lossless system $\mathbf{P}_{N-1}(z)$. See Fig. 14.4-5.

Completion of the factorization. $\mathbf{P}_{N-1}(z)$ has all the properties of $\mathbf{P}_N(z)$, except that its degree is $N-1$. Repeating the degree-reduction process N times, we obtain the structure of Fig. 14.4-4 where $\mathbf{P}_0$ is a nonzero vector. When $\mathbf{P}_N(z)$ has real coefficients, all the vectors $\mathbf{v}_m$ and $\mathbf{P}_0$ will be real.

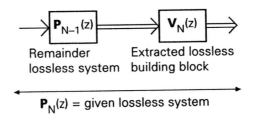

Figure 14.4-5 The degree reduction step in the factorization of $M \times 1$ lossless $\mathbf{P}_N(z)$.

14.4.4 Uniqueness of the Factorizations

We have obtained two major factorization results in the preceding subsections. For a given lossless system, are these factorizations unique? The answer depends on the case under consideration, and is best itemized as follows.

$M \times M$ **case.**

1. Since $\mathbf{V}_m(1) = \mathbf{I}$, we have $\mathbf{H}_0 = \mathbf{H}_N(1)$, so that $\mathbf{H}_0$ is unique.
2. $\mathbf{V}_m(z)$ **are not unique.** The unit-norm vector $\mathbf{v}_N$ is required to be such that (14.4.7) holds. No other condition is imposed on $\mathbf{v}_N$. As a result, it is not unique [unless $\mathbf{h}(0)$ happens to have rank $M-1$ in which case $\mathbf{v}_N$ is unique upto a scale factor of unit magnitude]. Thus the vectors $\mathbf{v}_m$ are, in general, not unique. So, $\mathbf{V}_m(z)$ are not unique.
3. Now consider the product $\mathbf{V}_N(z)\ldots\mathbf{V}_1(z)$. Since $\mathbf{H}_0$ is unitary, its inverse exists and this product has value $\mathbf{H}_N(z)\mathbf{H}_0^{-1}$ (see (14.4.10)). From the uniqueness of $\mathbf{H}_0$ we conclude that this product *is* unique, even though individual factors $\mathbf{V}_m(z)$ are not.

$M \times 1$ **case**

1. Once again $\mathbf{P}_0 = \mathbf{P}_N(1) =$ unique.
2. $\mathbf{V}_m(z)$ **are unique!** Recall that $\mathbf{v}_N$ is required to satisfy the condition

$$(\mathbf{I} - \mathbf{v}_N\mathbf{v}_N^{\dagger})\mathbf{p}(N) = \mathbf{0}. \qquad (14.4.19)$$

This implies $\mathbf{p}(N) = (\mathbf{v}_N^{\dagger}\mathbf{p}(N))\mathbf{v}_N$, which proves $\mathbf{v}_N = s \times \mathbf{p}(N)$ for scalar s. Thus $\mathbf{v}_N$ is unique upto a unit-magnitude scale factor (since its norm is also constrained to be unity). Since this scale factor does not affect the diadic $\mathbf{v}_N\mathbf{v}_N^{\dagger}$, the matrix $\mathbf{V}_N(z)$ is unique, and so is the remainder $\mathbf{P}_{N-1}(z)$. Repeating this argument we conclude that, for given $\mathbf{P}_N(z)$, the vectors $\mathbf{v}_m$ are unique upto unit-magnitude scale factor, and that $\mathbf{V}_m(z)$ are unique.

Warning. The above uniqueness result is valid only when the factorization of $\mathbf{P}_N(z)$ is minimal, that is, the number of factors $\mathbf{V}_m(z)$ equals N as in the above method. It is possible to obtain, using other means, an infinite number of nonminimal factorizations as shown in Problem 14.11.

14.5 STATE SPACE MANIFESTATION OF LOSSLESS PROPERTY

Given a structure for a transfer matrix $\mathbf{H}(z)$, we can always write down its state space description $(\mathbf{A},\mathbf{B},\mathbf{C},\mathbf{D})$ as described in Sec. 13.4. Now consider the so-called *realization matrix*

$$\mathbf{R} \triangleq \begin{bmatrix} \mathbf{A} & \mathbf{B} \\ \mathbf{C} & \mathbf{D} \end{bmatrix} \quad \text{(realization matrix)}. \quad (14.5.1)$$

It turns out that the lossless property of $\mathbf{H}(z)$ can be related to this matrix in a elegant way. In this section we will do the following.

1. We first show that whenever $\mathbf{R}$ is normalized-unitary (i.e., $\mathbf{R}^\dagger \mathbf{R} = \mathbf{I}$), and $\mathbf{A}$ is stable, the transfer function is lossless.
2. This does not imply that *any* structure for a lossless system has unitary $\mathbf{R}$. We show, however, that the FIR structures developed in the previous section have unitary $\mathbf{R}$.
3. We then show that for any normalized lossless system (FIR or IIR), we can find a structure such that $\mathbf{R}^\dagger \mathbf{R} = \mathbf{I}$ (Theorem 14.5.1).
4. We finally show that if a structure has stable $\mathbf{A}$ and unitary $\mathbf{R}$, it is necessarily minimal (i.e., number of delays is equal to degree of the system).

Thus the unitary property of the function $\mathbf{H}_N(e^{j\omega})$ for each ω is equivalent to the unitary property of a single *constant* matrix $\mathbf{R}$.

Brief historical notes. In the continuous-time world, the properties of state space descriptions of lossless electrical networks are well understood. Letting $\mathbf{Z}(s)$ denote the impedance matrix and $\mathbf{S}(s)$ the scattering matrix (Sec. 14.1) of a lossless electrical network, one can analyze the effects of lossless property on their state space descriptions. The results are known in various forms, e.g., the Kalman Yakubovic lemma, LBR lemma, LPR Lemma and so forth. An excellent comprehensive study of this topic can be found in Anderson and Vongpanitlerd [1973]. The discrete-time version of one of these results is particularly applicable in digital signal processing, and was presented in Vaidyanathan [1985b], and further reviewed in Prabhakara Rao and Dewilde [1987]. The results presented in this section are elaborations of these two discrete-time references.

Proof That $\mathbf{R}^\dagger \mathbf{R} = \mathbf{I}$ Implies $\mathbf{H}(z)$ is Lossless

Let $(\mathbf{A},\mathbf{B},\mathbf{C},\mathbf{D})$ be the state-space description of some structure for a causal (possibly IIR, but stable) $p \times r$ transfer function $\mathbf{H}(z)$. Let $\mathbf{R}$ defined in (14.5.1) be normalized-unitary, that is, $\mathbf{R}^\dagger \mathbf{R} = \mathbf{I}$. We now prove that $\mathbf{H}(z)$ is normalized-lossless.

Since $\mathbf{H}(z)$ is already given to be stable, it only remains to prove $\widetilde{\mathbf{H}}(z)\mathbf{H}(z) = \mathbf{I}$. For this recall (Sec. 13.4.1) that $\mathbf{H}(z) = \mathbf{D} + \mathbf{C}(z\mathbf{I} - \mathbf{A})^{-1}\mathbf{B}$, so that $\widetilde{\mathbf{H}}(z)\mathbf{H}(z)$ is equal to

$$\mathbf{D}^\dagger \mathbf{D} + \mathbf{B}^\dagger(z^{-1}\mathbf{I} - \mathbf{A}^\dagger)^{-1}\mathbf{C}^\dagger\mathbf{C}(z\mathbf{I} - \mathbf{A})^{-1}\mathbf{B} \\ + \mathbf{D}^\dagger \mathbf{C}(z\mathbf{I} - \mathbf{A})^{-1}\mathbf{B} + \mathbf{B}^\dagger(z^{-1}\mathbf{I} - \mathbf{A}^\dagger)^{-1}\mathbf{C}^\dagger\mathbf{D}. \tag{14.5.2}$$

Since $\mathbf{R}$ is normalized unitary, we have

$$\mathbf{A}^\dagger \mathbf{A} + \mathbf{C}^\dagger \mathbf{C} = \mathbf{I}, \tag{14.5.3}$$

$$\mathbf{A}^\dagger \mathbf{B} + \mathbf{C}^\dagger \mathbf{D} = \mathbf{0}, \tag{14.5.4}$$

$$\mathbf{B}^\dagger \mathbf{B} + \mathbf{D}^\dagger \mathbf{D} = \mathbf{I}. \tag{14.5.5}$$

By using these we can show that proving $\widetilde{\mathbf{H}}(z)\mathbf{H}(z) = \mathbf{I}$ is equivalent to proving $\mathbf{B}^\dagger(z^{-1}\mathbf{I} - \mathbf{A}^\dagger)^{-1}\mathbf{F}(z)(z\mathbf{I} - \mathbf{A})^{-1}\mathbf{B} = \mathbf{0}$, where

$$\mathbf{F}(z) = (z^{-1}\mathbf{I} - \mathbf{A}^\dagger)(z\mathbf{I} - \mathbf{A}) + (z^{-1}\mathbf{I} - \mathbf{A}^\dagger)\mathbf{A} + \mathbf{A}^\dagger(z\mathbf{I} - \mathbf{A}) + \mathbf{A}^\dagger \mathbf{A} - \mathbf{I}. \tag{14.5.6}$$

Upon simplification $\mathbf{F}(z)$ reduces to $\mathbf{0}$ so that $\widetilde{\mathbf{H}}(z)\mathbf{H}(z) = \mathbf{I}$ indeed. This completes the proof.

Energy Balance Property

From the definition of $(\mathbf{A}, \mathbf{B}, \mathbf{C}, \mathbf{D})$ we know that the realization matrix $\mathbf{R}$ relates two vector sequences in the following way:

$$\begin{bmatrix} \mathbf{x}(n+1) \\ \mathbf{y}(n) \end{bmatrix} = \mathbf{R} \begin{bmatrix} \mathbf{x}(n) \\ \mathbf{u}(n) \end{bmatrix}. \tag{14.5.7}$$

Thus $\mathbf{R}^\dagger \mathbf{R} = \mathbf{I}$ implies

$$\|\mathbf{x}(n+1)\|^2 + \|\mathbf{y}(n)\|^2 = \|\mathbf{x}(n)\|^2 + \|\mathbf{u}(n)\|^2, \tag{14.5.8}$$

where $\|\mathbf{v}\|^2$ is the energy $\mathbf{v}^\dagger \mathbf{v}$ of the vector $\mathbf{v}$. With $E_x(n) = \|\mathbf{x}(n)\|^2$, $E_u(n) = \|\mathbf{u}(n)\|^2$, and $E_y(n) = \|\mathbf{y}(n)\|^2$, this becomes

$$E_x(n+1) - E_x(n) = E_u(n) - E_y(n), \tag{14.5.9}$$

which implies a beautiful energy balance property. The increase in the state-vector energy at time n is equal to the instantaneous input energy minus instantaneous output energy. In other words, a portion of the input energy E_u goes to the output and the rest goes to increase the internal energy.

There is no dissipation. This is a consequence of losslessness of $\mathbf{H}(z)$. For nonparaunitary systems, there are no structures with $\mathbf{R}^\dagger \mathbf{R} = \mathbf{I}$.

14.5.1 Finding Structures with Unitary Realization Matrix R

Not every structure for a normalized-lossless system has unitary $\mathbf{R}$. (example: try the direct form structure for a second order allpass function). The next question, therefore, is how to find structures for lossless systems such that $\mathbf{R}$ is normalized unitary. For FIR $\mathbf{H}(z)$, the structures in Figs. 14.4-3 and 14.4-4 already satisfy this property. Before proving this, we first consider an example.

Example 14.5.1

Consider the degree-one lossless system $\mathbf{I} - \mathbf{vv}^\dagger + z^{-1}\mathbf{vv}^\dagger$ with $\|\mathbf{v}\| = 1$. Figure 14.5-1 reproduces the structure, with various signals labeled. By comparing this with the standard state-space equations (Sec. 13.4), we conclude $\mathbf{A} = 0$, $\mathbf{B} = \mathbf{v}^\dagger$, $\mathbf{C} = \mathbf{v}$ and $\mathbf{D} = \mathbf{I} - \mathbf{vv}^\dagger$. Computing $\mathbf{R}^\dagger \mathbf{R}$ we find

$$\mathbf{R}^\dagger \mathbf{R} = \begin{bmatrix} \mathbf{v}^\dagger \mathbf{v} & \mathbf{v}^\dagger (\mathbf{I} - \mathbf{vv}^\dagger) \\ (\mathbf{I} - \mathbf{vv}^\dagger)\mathbf{v} & (\mathbf{I} - \mathbf{vv}^\dagger)^2 + \mathbf{vv}^\dagger \end{bmatrix} = \mathbf{I} \quad (\text{using } \mathbf{v}^\dagger \mathbf{v} = 1),$$

(14.5.10)

so that $\mathbf{R}$ normalized unitary.

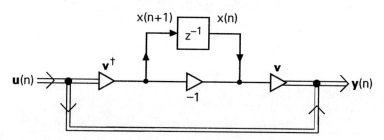

Figure 14.5-1 The degree-one FIR lossless building block.

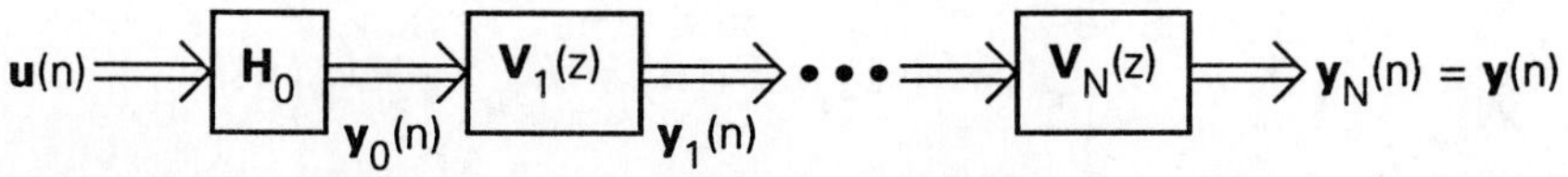

Figure 14.5-2 The cascade of Fig. 14.4-3 reproduced, with internal signals labeled.

FIR Structures with Unitary Realization Matrix R

More generally, we claim this: let $\mathbf{H}_N(z)$ be an $M \times r$ causal FIR matrix realized as in Fig. 14.5-2 with $M \times M$ building blocks $\mathbf{V}_m(z)$ as in Fig. 14.4-1. Let the $M \times r$ matrix $\mathbf{H}_0$ satisfy $\mathbf{H}_0^\dagger \mathbf{H}_0 = \mathbf{I}$ so that $\mathbf{H}_N(z)$ is normalized lossless. Then $\mathbf{R}$ is normalized unitary. (Note that structures with $r = 1$ and $r = M$, which were shown in Fig. 14.4-3 and 14.4-4, are special cases.)

To prove this, let $\mathbf{R}_m$ denote the $\mathbf{R}$-matrix for the mth section $\mathbf{V}_m(z)$. From Example 14.5.1 we know that $\mathbf{R}_m^\dagger \mathbf{R}_m = \mathbf{I}$, so that

$$|x_m(n+1)|^2 + \|\mathbf{y}_m(n)\|^2 = |x_m(n)|^2 + \|\mathbf{y}_{m-1}(n)\|^2, \quad \forall n. \qquad (14.5.11)$$

This holds for $1 \leq m \leq N$. By adding these (and using $\|\mathbf{u}(n)\| = \|\mathbf{y}_0(n)\|$ and $\mathbf{y}(n) = \mathbf{y}_N(n)$) we obtain

$$\|\mathbf{x}(n+1)\|^2 + \|\mathbf{y}(n)\|^2 = \|\mathbf{x}(n)\|^2 + \|\mathbf{u}(n)\|^2, \quad \forall n.$$

Summarizing, the vectors $\begin{bmatrix} \mathbf{x}(n+1) \\ \mathbf{y}(n) \end{bmatrix}$ and $\begin{bmatrix} \mathbf{x}(n) \\ \mathbf{u}(n) \end{bmatrix}$ have the same norm for all n. But these vectors are related as in (14.5.7). Since the above argument holds regardless of the value of the vector $\begin{bmatrix} \mathbf{x}(n) \\ \mathbf{u}(n) \end{bmatrix}$, $\mathbf{R}$ must be normalized unitary (Problem A.17, Appendix A).

Existence of Structures with Unitary Realization Matrix R

From previous sections we know that any $M \times 1$ or $M \times M$ causal FIR lossless system can be realized as in Fig. 14.5-2. As a result, these FIR systems always have an implementation with unitary $\mathbf{R}$. Furthermore, structures based on planar rotations (as given by Figs. 14.3-2 and 14.3-3) also have unitary $\mathbf{R}$ (see Problem 14.13). We now prove a stronger result.

♦**Theorem 14.5.1. State space manifestation of losslessness.** Let $\mathbf{H}(z)$ be a $p \times r$ causal stable system. Then it is normalized lossless if and only if there exists a structure with state space description $(\mathbf{A}, \mathbf{B}, \mathbf{C}, \mathbf{D})$ such that the realization matrix satisfies $\mathbf{R}^\dagger \mathbf{R} = \mathbf{I}$. ◊

Proof. The 'if' part has already been proved. For the 'only if' part, assume $\mathbf{H}(z)$ is normalized lossless. Let $(\mathbf{A}_1, \mathbf{B}_1, \mathbf{C}_1, \mathbf{D})$ be some minimal realization. We will construct an equivalent minimal realization $(\mathbf{A}, \mathbf{B}, \mathbf{C}, \mathbf{D})$ such that $\mathbf{R}^\dagger \mathbf{R} = \mathbf{I}$. Since $\mathbf{A}_1$ is stable and $(\mathbf{C}_1, \mathbf{A}_1)$ observable (due to minimality), we know that there exists Hermitian positive definite $\mathbf{P}$ such that $\mathbf{A}_1^\dagger \mathbf{P} \mathbf{A}_1 + \mathbf{C}_1^\dagger \mathbf{C}_1 = \mathbf{P}$. Since $\mathbf{P}$ is positive definite, we can find a nonsingular $\mathbf{T}$ such that $\mathbf{P} = \mathbf{T}^{-\dagger} \mathbf{T}^{-1}$ (Appendix A). So we have $\mathbf{A}_1^\dagger \mathbf{T}^{-\dagger} \mathbf{T}^{-1} \mathbf{A}_1 + \mathbf{C}_1^\dagger \mathbf{C}_1 = \mathbf{T}^{-\dagger} \mathbf{T}^{-1}$, that is,

$$\mathbf{A}^\dagger \mathbf{A} + \mathbf{C}^\dagger \mathbf{C} = \mathbf{I}, \qquad (14.5.12)$$

where $\mathbf{A} \triangleq \mathbf{T}^{-1}\mathbf{A}_1\mathbf{T}$ and $\mathbf{C} \triangleq \mathbf{C}_1\mathbf{T}$. By using $\mathbf{T}$ as a similarity transformation, we can now define an equivalent *minimal* state-space structure $(\mathbf{A}, \mathbf{B}, \mathbf{C}, \mathbf{D})$ for $\mathbf{H}(z)$.

So far we have not used the lossless property. We will now show that when $\mathbf{H}(z)$ is normalized lossless, the condition (14.5.12) implies $\mathbf{R}^\dagger \mathbf{R} = \mathbf{I}$. That is (14.5.3) automatically implies (14.5.4) and (14.5.5).

Suppose that we apply a finite-energy input $\mathbf{u}(n)$ such that $\mathbf{u}(n) = \mathbf{0}, n \geq M$. Then

$$\begin{bmatrix} \mathbf{x}(n+1) \\ \mathbf{y}(n) \end{bmatrix} = \begin{bmatrix} \mathbf{A} \\ \mathbf{C} \end{bmatrix} \mathbf{x}(n), \quad n \geq M,$$

that is,

$$\|\mathbf{x}(n+1)\|^2 + \|\mathbf{y}(n)\|^2 = \|\mathbf{x}(n)\|^2, \quad n \geq M \quad \text{[using (14.5.12)]}.$$

Adding this for $n \geq M$, we get

$$\sum_{n=M}^{\infty} \|\mathbf{y}(n)\|^2 = \|\mathbf{x}(M)\|^2. \tag{14.5.13}$$

Note that as $n \to \infty$, $\mathbf{x}(n) \to \mathbf{0}$ and $\mathbf{y}(n) \to \mathbf{0}$ since $\mathbf{A}$ is stable. The lossless property $\widetilde{\mathbf{H}}(z)\mathbf{H}(z) = \mathbf{I}$ implies $\sum_{n=-\infty}^{\infty} \|\mathbf{y}(n)\|^2 = \sum_{n=-\infty}^{\infty} \|\mathbf{u}(n)\|^2$, that is,

$$\|\mathbf{x}(M)\|^2 + \sum_{n=-\infty}^{M-1} \|\mathbf{y}(n)\|^2 = \sum_{n=-\infty}^{M-1} \|\mathbf{u}(n)\|^2, \tag{14.5.14}$$

by using (14.5.13). By causality, the quantities $\mathbf{x}(M)$ and $\mathbf{y}(n), n \leq M-1$ do not depend on $\mathbf{u}(n), n \geq M$, and hence on the assumption that $\mathbf{u}(n)$ is zero for $n \geq M$. So (14.5.14) continues to hold if M is replaced by $M+1$, i.e.,

$$\|\mathbf{x}(M+1)\|^2 + \sum_{n=-\infty}^{M} \|\mathbf{y}(n)\|^2 = \sum_{n=-\infty}^{M} \|\mathbf{u}(n)\|^2. \tag{14.5.15}$$

Subtracting (14.5.14) from (14.5.15), we obtain

$$\|\mathbf{x}(M+1)\|^2 + \|\mathbf{y}(M)\|^2 = \|\mathbf{x}(M)\|^2 + \|\mathbf{u}(M)\|^2. \tag{14.5.16}$$

This is true for all possible $\mathbf{u}(M)$. More subtle is the fact that this is true for all possible $\mathbf{x}(M)$. To see this note that the minimality of $(\mathbf{A}, \mathbf{B}, \mathbf{C}, \mathbf{D})$ assures us that $(\mathbf{A}, \mathbf{B})$ is reachable so that we can obtain any value for $\mathbf{x}(M)$ by choosing $\mathbf{u}(M-1), \mathbf{u}(M-2), \ldots, \mathbf{u}(M-N)$ appropriately where N is the degree of $\mathbf{H}(z)$.

Since (14.5.16) holds for all $\mathbf{x}(M)$ and $\mathbf{u}(M)$, we conclude in view of (14.5.7) that $\mathbf{R}^\dagger \mathbf{R} = \mathbf{I}$. ▽▽▽

Remarks. Note that the result holds for FIR as well as IIR systems. Examples of IIR lossless structures for which $\mathbf{R}^\dagger \mathbf{R} = \mathbf{I}$, can be found in Problem 14.28. For the case of real-coefficient lossless $\mathbf{H}(z)$, one can verify the existence of *real* $(\mathbf{A}, \mathbf{B}, \mathbf{C}, \mathbf{D})$ such that $\mathbf{R}^T \mathbf{R} = \mathbf{I}$.

14.5.2 Minimality of Structure and Unitariness of R

Let $(\mathbf{A}, \mathbf{B}, \mathbf{C}, \mathbf{D})$ be the state space description of some structure for an $M \times M$ transfer function $\mathbf{H}(z)$ with the following two properties: (a) $\mathbf{A}$ is stable, and (b) $\mathbf{R}$ defined in (14.5.1) satisfies $\mathbf{R}^\dagger \mathbf{R} = \mathbf{I}$. Then the structure is a minimal realization of $\mathbf{H}(z)$.

We now prove this result. From Sec. 13.4.2 we know that proving minimality of a structure is equivalent to proving that $(\mathbf{A}, \mathbf{B})$ is completely reachable and $(\mathbf{C}, \mathbf{A})$ is completely observable. The property $\mathbf{R}^\dagger \mathbf{R} = \mathbf{I}$ implies in particular (14.5.3). Using this and the fact that $\mathbf{A}$ is stable, we show that $(\mathbf{C}, \mathbf{A})$ is completely observable. For this assume that $(\mathbf{C}, \mathbf{A})$ is not observable. Then there exists $\mathbf{v} \neq \mathbf{0}$ such that $\mathbf{Cv} = \mathbf{0}$ and $\mathbf{Av} = \lambda \mathbf{v}$. (This follows from Theorem 13.4.2). From (14.5.3) we then obtain $|\lambda|^2 \mathbf{v}^\dagger \mathbf{v} = \mathbf{v}^\dagger \mathbf{v}$, that is, $|\lambda| = 1$, violating stability of $\mathbf{A}$. This proves that $(\mathbf{C}, \mathbf{A})$ is completely observable.

For $M \times M$ matrix $\mathbf{H}(z)$, $\mathbf{R}$ is square so that $\mathbf{R}^\dagger \mathbf{R} = \mathbf{I} \Rightarrow \mathbf{RR}^\dagger = \mathbf{I}$. If we write out the condition $\mathbf{RR}^\dagger = \mathbf{I}$ in terms of $\mathbf{A}, \mathbf{B}, \mathbf{C}$ and $\mathbf{D}$, one of the three equations is $\mathbf{AA}^\dagger + \mathbf{BB}^\dagger = \mathbf{I}$. By using this along with the fact that $\mathbf{A}$ is stable, we can prove (similar to the observability proof) that $(\mathbf{A}, \mathbf{B})$ is completely reachable. These together establish that the structure is minimal.

Remarks

1. This result is *not* restricted to FIR $\mathbf{H}(z)$. However, it does not in general hold for $p \times r$ case with $p \neq r$ (Problem 14.15).
2. In the above, we assumed that $\mathbf{A}$ is stable, and proved that $\mathbf{R}^\dagger \mathbf{R} = \mathbf{I}$ implies minimality. As a complement of this, if one assumes that $(\mathbf{A}, \mathbf{B}, \mathbf{C}, \mathbf{D})$ is minimal and that $\mathbf{R}^\dagger \mathbf{R} = \mathbf{I}$, it turns out that $\mathbf{A}$ is stable. See Problem 14.16.
3. In the next section we show how unitary matrices can be factorized into fundamental unitary building blocks. If such factorization is applied to the $\mathbf{R}$-matrix in (14.5.1), we obtain new structures for lossless systems.

14.6 FACTORIZATION OF UNITARY MATRICES

Unitary matrices play an important role in the theory and implementation of lossless systems. For example:

1. The lossless factorization in Fig. 14.4-3 involves a unitary matrix $\mathbf{H}_0$.
2. The factorization in Fig. 14.3-2 is in terms of 2×2 real unitary matrices.
3. For a lossless system we can always find a structure such that the realization matrix $\mathbf{R}$ defined in (14.5.1) is a constant unitary matrix.

In view of their frequent occurence, it will be very useful to develop cascade structures for unitary matrices, which is the purpose of this section. An $M \times M$ matrix $\mathbf{R}$ is unitary if

$$\mathbf{R}^{\dagger}\mathbf{R} = d\mathbf{I}, \quad d > 0. \tag{14.6.1}$$

When $d = 1$ we say that $\mathbf{R}$ is normalized-unitary. Unitary matrices are evidently lossless. From (14.6.1) we have $\|\mathbf{R}\mathbf{v}\| = \sqrt{d}\|\mathbf{v}\|$ for any $M \times 1$ vector $\mathbf{v}$. Conversely (Problem A.17 in Appendix A) if $\mathbf{A}$ is some $M \times M$ matrix with $\|\mathbf{A}\mathbf{v}\| = \sqrt{d}\|\mathbf{v}\|$ for all $\mathbf{v}$ (for fixed d) then $\mathbf{A}^{\dagger}\mathbf{A} = d\mathbf{I}$.

♠Main points of this section. It is clear that a product of unitary matrices is unitary. In this section we show that *any* $M \times M$ unitary $\mathbf{R}$ is a product of finite number of standard unitary building blocks. To be more specific, the following results will be presented.

1. **Rotation based structure.** We show that when $\mathbf{R}$ is real, it can be factorized into a product of $M(M-1)/2$ simpler real unitary matrices, each of which represents a *planar* rotation operator (defined below). This will result in the structure of Fig. 14.6-3. Equation (14.6.4) below shows a 3×3 example.
2. **Diadic based structure.** For complex $\mathbf{R}$ a similar result is possible with the planar rotations replaced by complex versions. We will not prove this result (which can be found in Murnaghan [1962]). Instead, we prove a more convenient factorization called Householder (or diadic-based) factorization. According to this, $\mathbf{R}$ can be factorized in terms of the very elegant building block (14.6.11), resulting in the structure of Fig. 14.6-4. This also holds for real $\mathbf{R}$, in which case $\mathbf{u}_k$ are real. An advantage of diadic-based structures over rotation-based ones is that we can quantize the multipliers without losing unitariness (Sec. 14.11).
3. **Degress of freedom.** Even though $\mathbf{R}$ has M^2 complex elements, the unitary constraint imposes a relation among these. So the number of 'free parameters' is less that M^2. Thus for complex unitary $\mathbf{R}$ we have only about M^2 *real-valued* free parameters. And for real unitary $\mathbf{R}$ we have about $M(M-1)/2$ real-valued free parameters (see end of Sec. 14.6.2 for precise statements). The factorizations to be presented are in terms of these free parameters. ◊

The results presented in this section can be deduced easily from a number of texts and papers [Givens, 1958], [Murnaghan, 1962], [Golub and Van Loan, 1989].

14.6.1 Factorization of Real Unitary Matrices Using Givens Rotations

Recall the Givens or planar rotation matrix given in (14.3.9a) and repeated below:

$$\mathbf{G} = \begin{bmatrix} c & s \\ -s & c \end{bmatrix}, \quad c = \cos\theta, s = \sin\theta, \quad \theta \text{ real}. \quad (14.6.2)$$

$\mathbf{G}$ satisfies $\mathbf{G}^T\mathbf{G} = \mathbf{I}$. It can be verified (Problem 14.12) that any 2×2 real unitary matrix $\mathbf{R}$ can be expressed in terms of Givens rotation as

$$\sqrt{d}\begin{bmatrix} c & s \\ -s & c \end{bmatrix}\begin{bmatrix} \mu_0 & 0 \\ 0 & \mu_1 \end{bmatrix}, \quad \mu_i = \pm 1, d > 0. \quad (14.6.3)$$

Less obvious is the fact that any 3×3 real unitary $\mathbf{R}$ can be expressed as

$$\sqrt{d}\underbrace{\begin{bmatrix} 1 & 0 & 0 \\ 0 & c_0 & s_0 \\ 0 & -s_0 & c_0 \end{bmatrix}}_{(1,2) \text{ rotation}}\underbrace{\begin{bmatrix} c_1 & 0 & s_1 \\ 0 & 1 & 0 \\ -s_1 & 0 & c_1 \end{bmatrix}}_{(0,2) \text{ rotation}}\underbrace{\begin{bmatrix} c_2 & s_2 & 0 \\ -s_2 & c_2 & 0 \\ 0 & 0 & 1 \end{bmatrix}}_{(0,1) \text{ rotation}}\underbrace{\begin{bmatrix} \mu_0 & 0 & 0 \\ 0 & \mu_1 & 0 \\ 0 & 0 & \mu_2 \end{bmatrix}}_{\text{sign matrix}},$$
$$(14.6.4)$$

where $c_k = \cos\theta_k$, $s_k = \sin\theta_k$ (θ_k being real) and $\mu_k = \pm 1$. Notice also that, given any set of real numbers $\{\theta_k\}$, matrices of the form (14.6.4) are unitary. There are three rotation matrices in (14.6.4). Each of these performs rotation in a plane (two-dimensional subspace of the three dimensional space). For example the leftmost matrix above performs a planar rotation in the $(1,2)$ plane.[†] So we say that $\mathbf{R}$ has been factorized in terms of planar rotations or Givens rotations.

More generally, it will be shown that any $M \times M$ real unitary $\mathbf{R}$ can be expressed in terms of $M(M-1)/2$ planar rotation operators. This is based on the following size-reduction step.

Size-reduction step.

Let $\mathbf{R}$ be $M \times M$ real with $\mathbf{R}^T\mathbf{R} = \mathbf{I}$. We show that it can be written as

$$\mathbf{R} = \Theta \begin{bmatrix} \mathbf{S} & 0 \\ 0 & \mu \end{bmatrix} \quad (14.6.5)$$

where $\mathbf{S}$ is a real $(M-1) \times (M-1)$ matrix with $\mathbf{S}^T\mathbf{S} = \mathbf{I}$, $\mu = \pm 1$, and $\Theta = \Theta_{M-2}\Theta_{M-3}\ldots\Theta_0$ where Θ_m is the building block shown in Fig. 14.6-1, with θ_m real. Evidently $\Theta_m^T\Theta_m = \mathbf{I}$, so that $\Theta^T\Theta = \mathbf{I}$. The matrix Θ_m performs a Givens rotation in the plane defined by the indices $(m, M-1)$ in the M dimensional space.

[†] As always, row and column indices start from zero.

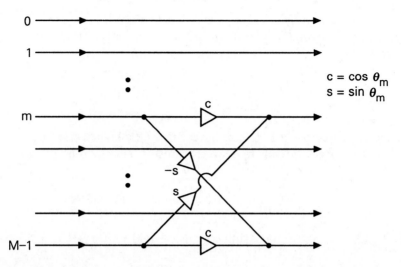

Figure 14.6-1 The building block Θ_m which is a planar Givens rotation in M-dimensional space.

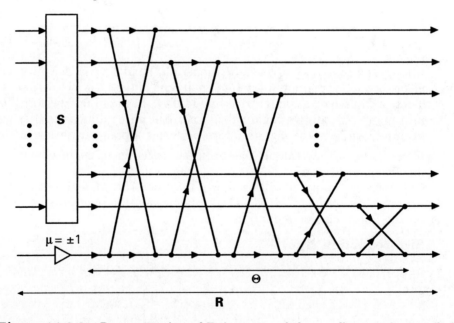

Figure 14.6-2 Representation of **R** in terms of the smaller matrix **S** and the sequence of rotations Θ. Each criss-cross here represents a Givens rotation.

Justifying the factorization (14.6.5). Let us first understand the product

$$\begin{bmatrix} \mathbf{I} & \mathbf{0} & \mathbf{0} \\ \mathbf{0} & c & -s \\ \mathbf{0} & s & c \end{bmatrix} \mathbf{R}, \qquad (14.6.6)$$

748   Chap. 14. Paraunitary and lossless systems

where $c = \cos\theta$ and $s = \sin\theta$. We will show how to find real θ such that this product has the form

$$\begin{bmatrix} \times & \times & \cdots & \times & \times \\ \vdots & \vdots & \ddots & \vdots & \vdots \\ \times & \times & \cdots & \times & \times \\ \times & \times & \cdots & \times & 0 \\ \times & \times & \cdots & \times & \times \end{bmatrix}, \qquad (14.6.7)$$

where $\times$ denotes possibly nonzero entries. The premultiplying matrix in (14.6.6) is Θ_{M-2}^T. This premultiplication has the effect of replacing the last two rows of $\mathbf{R}$ by linear combinations of these rows. *No other rows are affected.* Our aim is to choose θ so that the last element of the $(M-2)$th row, that is, the $(M-2, M-1)$ element, is forced to be zero. This element is given in terms of the elements of $\mathbf{R}$ by

$$R_{M-2,M-1}\cos\theta - R_{M-1,M-1}\sin\theta. \qquad (14.6.8)$$

To force this to be zero we choose θ such that

$$\tan\theta = \frac{R_{M-2,M-1}}{R_{M-1,M-1}}. \qquad (14.6.9)$$

This gives a unique value of θ in the range $-\pi/2 < \theta \leq \pi/2$. If $R_{M-1,M-1} = 0$, then (14.6.9) gives $\theta = \pi/2$, that is, we simply take $c = 0, s = 1$.

It is clear that if the premultiplying matrix in (14.6.6) is chosen to be Θ_k^T for some other k, then we can force the last entry of the kth row to be zero. So by successive premultiplication with Θ_m^T for $m = M-2, M-3, \ldots 0$, we can obtain a matrix of the form

$$\Theta_0^T \Theta_1^T \ldots \Theta_{M-2}^T \mathbf{R} = \begin{bmatrix} \times & \times & \cdots & \times & 0 \\ \vdots & \vdots & \ddots & \vdots & \vdots \\ \times & \times & \cdots & \times & 0 \\ \times & \times & \cdots & \times & \alpha \end{bmatrix}. \qquad (14.6.10)$$

Since every matrix on the left hand side is normalized-unitary, so is the right hand side matrix. So each column and each row of right hand side has unit norm. By looking at the last column we therefore conclude $\alpha = \pm 1$. Since the last row has to have unit norm, the remaining elements in the last *row* are also zero! Summarizing, the right hand side above has the form $\begin{bmatrix} \mathbf{S} & \mathbf{0} \\ \mathbf{0} & \mu \end{bmatrix}$. Since $\Theta_m^T = \Theta_m^{-1}$, we can rewrite (14.6.10) to arrive at the claimed factorization (14.6.5).

Completing the Factorization by Repeated Size-Reduction

Pictorially we can represent the above factorization as in Fig. 14.6-2, where each criss-cross represents a Givens rotation. It is clear that we can

Sec. 14.6 Factorization of unitary matrices

repeat the factorization recursively. Thus, we can replace the $(M-1) \times (M-1)$ matrix $\mathbf{S}$ with $\mathbf{S} = \mathbf{\Phi}\begin{bmatrix} \mathbf{T} & \mathbf{0} \\ \mathbf{0} & \pm 1 \end{bmatrix}$ where $\mathbf{T}$ is $(M-2) \times (M-2)$ with $\mathbf{T}^T \mathbf{T} = \mathbf{I}$, and $\mathbf{\Phi}$ is a product of $M-2$ planar rotations. If we keep repeating this we eventually arrive at a factorization of $\mathbf{R}$ as in Fig. 14.6-3.

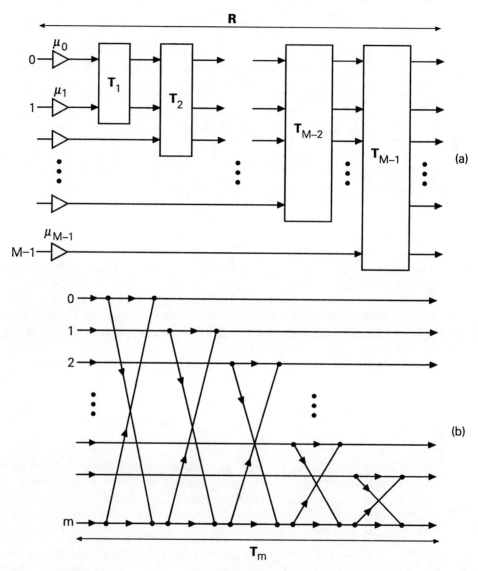

Figure 14.6-3 (a) Complete factorization of $M \times M$ real $\mathbf{R}$ with $\mathbf{R}^T \mathbf{R} = \mathbf{I}$. (b) Internal details of $\mathbf{T}_m$. Each criss-cross is a Givens rotation.

This factorization is entirely in terms of Givens rotations. The matrix $\mathbf{T}_m$ contains m rotations. The total number of rotation operators is $(M -$

$1) + (M-2) + \ldots + 1 = M(M-1)/2$.

Summary. We have proved this: Let $\mathbf{R}$ be $M \times M$ real with $\mathbf{R}^T\mathbf{R} = \mathbf{I}$. We can then factorize $\mathbf{R}$ as in Fig. 14.6-3 in terms of $M(M-1)/2$ planar rotations, and M sign parameters μ_i. Conversely, the transfer matrix $\mathbf{R}$ of the above structure necessarily satisfies $\mathbf{R}^T\mathbf{R} = \mathbf{I}$ for all possible choices of the rotation angles.

14.6.2 Householder Factorization of Unitary Matrices

We now show how to express unitary matrices in terms of what are called *Householder* building blocks. This factorization holds (and is equally simple) for real as well as complex matrices.

An $M \times M$ Householder unitary matrix has the form [†]

$$\mathbf{H} = \mathbf{I} - 2\mathbf{u}\mathbf{u}^\dagger, \quad \mathbf{u} \text{ is } M \times 1, \mathbf{u}^\dagger\mathbf{u} = 1 \quad \text{(Householder matrix).} \quad (14.6.11)$$

This is unitary, and $\mathbf{H}^\dagger\mathbf{H} = \mathbf{I}$. Evidently $\mathbf{H}$ is Hermitian so that $\mathbf{H}^{-1} = \mathbf{H}$. It is interesting to note that if we set $z = -1$ in the lossless system $\mathbf{V}_m(z)$ in (14.4.1), we obtain the Householder matrix.

Turning Arbitrary Vectors into Fixed Form

Much of the value of Householder matrices comes from the fact that they can be used to 'turn' a vector $\mathbf{v}$ into a vector of the form $[0 \ldots 0 \ X \ 0 \ldots 0]^T$.

More formally, we claim this: let $\mathbf{v} = [v_0 \ v_1 \ \ldots \ v_{M-1}]^T$ be some $M \times 1$ nonzero vector and let $\theta = \arg v_i$ so that $v_i = |v_i|e^{j\theta}$. Define

$$\mathbf{w} = \mathbf{v} + s\|\mathbf{v}\|e^{j\theta}\mathbf{e}_i, \quad (14.6.12)$$

where $\mathbf{e}_i = [0 \ldots 0 \ 1 \ 0 \ldots 0]^T$ with the '1' in the ith position, and $s = \pm 1$. Then $\mathbf{w} \neq \mathbf{0}$ for at least one choice of s. Assuming $\mathbf{w} \neq \mathbf{0}$ we then have

$$\left(\mathbf{I} - 2\mathbf{u}\mathbf{u}^\dagger\right)\mathbf{v} = -s\|\mathbf{v}\|e^{j\theta}\mathbf{e}_i, \quad (14.6.13)$$

where $\mathbf{u} = e^{j\phi}\mathbf{w}/\|\mathbf{w}\|$ and ϕ is an arbitrary real number.

We now prove this. If $\mathbf{w} = \mathbf{0}$ for $s = 1$ as well as $s = -1$, this implies $\mathbf{v} = \mathbf{0}$, violating the stated conditions. So $\mathbf{w} \neq \mathbf{0}$ for at least one choice of s indeed. Assume this choice is made. By explicit computation,

$$\mathbf{w}^\dagger\mathbf{w} = 2\|\mathbf{v}\|\Big(\|\mathbf{v}\| + s|v_i|\Big), \quad \mathbf{w}^\dagger\mathbf{v} = \|\mathbf{v}\|\Big(\|\mathbf{v}\| + s|v_i|\Big). \quad (14.6.14)$$

[†] A more general definition permits a $M \times N$ unitary matrix $\mathbf{A}$ in place of $\mathbf{u}$. We will not require it here.

The left hand side of (14.6.13) is equal to

$$\mathbf{v} - 2\frac{\mathbf{w}\mathbf{w}^\dagger \mathbf{v}}{\mathbf{w}^\dagger \mathbf{w}}. \qquad (14.6.15)$$

Substituting from (14.6.14), this simplifies to $\mathbf{v} - \mathbf{w}$, which by the definition of $\mathbf{w}$, reduces to the right hand side of (14.6.13) indeed.

The above fact is useful for factorizing a $M \times M$ unitary matrix in terms of Householder matrices, as stated next.

Householder Factorization

Let $\mathbf{R}$ be $M \times M$ unitary, that is, $\mathbf{R}^\dagger \mathbf{R} = d\mathbf{I}, d > 0$. We will show that $\mathbf{R}$ can be written as

$$\mathbf{R} = \sqrt{d}\mathbf{H}_1 \mathbf{H}_2 \ldots \mathbf{H}_{M-1} \mathbf{D} \qquad (14.6.16)$$

where $\mathbf{D}$ is diagonal with $d_{ii} = e^{j\theta_i}$ (θ_i real), and $\mathbf{H}_k$ are Householder matrices, i.e., $\mathbf{H}_k = \mathbf{I} - 2\mathbf{u}_k \mathbf{u}_k^\dagger$ with $\mathbf{u}_k^\dagger \mathbf{u}_k = 1$. Figure 14.6-4 shows the resulting structure for $\mathbf{R}$. Conversely, it is clear that the transfer matrix of the above structure is always unitary as long as $\mathbf{u}_k^\dagger \mathbf{u}_k = 1$, and θ_i are real.

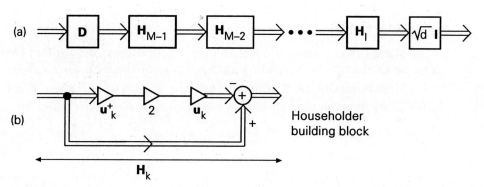

Figure 14.6-4 (a) Cascaded structure for unitary $\mathbf{R}$, and (b) Householder building blocks.

To prove the above factorization, we will work with $\mathbf{S} \triangleq \mathbf{R}/\sqrt{d}$ which satisfies

$$\mathbf{S}^\dagger \mathbf{S} = \mathbf{I}. \qquad (14.6.17)$$

Each column of $\mathbf{S}$ has unit norm. Consider the 0th column $\mathbf{s}_0$ of $\mathbf{S}$. We know there exists a unit-norm vector $\mathbf{u}_1$ such that $(\mathbf{I} - 2\mathbf{u}_1 \mathbf{u}_1^\dagger)\mathbf{s}_0 = e^{j\theta_0}\mathbf{e}_0$. As a result we have

$$\underbrace{\left(\mathbf{I} - 2\mathbf{u}_1 \mathbf{u}_1^\dagger\right)}_{\mathbf{H}_1} \mathbf{S} = \begin{bmatrix} e^{j\theta_0} & \times & \ldots & \times \\ 0 & \times & \ldots & \times \\ \vdots & \vdots & \ddots & \vdots \\ 0 & \times & \ldots & \times \end{bmatrix} \qquad (14.6.18)$$

where × denotes possibly nonzero entries. Since $\mathbf{S}$ and $\mathbf{H}_1$ are normalized unitary, the right hand side of (14.6.18) is also normalized-unitary. In particular this implies that the 0th *row* is equal to $[e^{j\theta_0} \; 0 \; \ldots \; 0]$. So

$$\mathbf{H}_1 \mathbf{S} = \begin{bmatrix} e^{j\theta_0} & \mathbf{0} \\ \mathbf{0} & \mathbf{T} \end{bmatrix}, \quad \text{where } \mathbf{T}^\dagger \mathbf{T} = \mathbf{I}. \qquad (14.6.19)$$

We can now repeat the process on $\mathbf{T}$ to reduce it to $\begin{bmatrix} e^{j\theta_1} & \mathbf{0} \\ \mathbf{0} & \mathbf{U} \end{bmatrix}$. Repeating this $M-1$ times we get

$$\mathbf{H}_{M-1} \mathbf{H}_{M-2} \ldots \mathbf{H}_1 \mathbf{S} = \mathbf{D}, \qquad (14.6.20)$$

where $\mathbf{D}$ and $\mathbf{H}_k$ are as defined above. Since $\mathbf{H}_k = \mathbf{H}_k^{-1}$, (14.6.20) implies $\mathbf{S} = \mathbf{H}_1 \mathbf{H}_2 \ldots \mathbf{H}_{M-1} \mathbf{D}$ so that (14.6.16) follows.

Comments

1. Note that the factorization (14.6.16) is not restricted to real $\mathbf{R}$. If $\mathbf{R}$ is real, the quantities $\mathbf{u}_k$ and $\mathbf{D}$ are real. In particular the diagonal elements of $\mathbf{D}$ satisfy $d_{ii} = \pm 1$.
2. *Special nature of* $\mathbf{u}_k$. By construction, the unit norm vectors $\mathbf{u}_k$ have a certain restricted form. For example, $\mathbf{u}_2$ has 0th element equal to zero; $\mathbf{u}_3$ has the top two elements equal to zero and so on. We thus have

$$[\mathbf{u}_1 \; \mathbf{u}_2 \; \ldots \; \mathbf{u}_{M-1}] = \begin{bmatrix} \times & 0 & 0 & \ldots & 0 \\ \times & \times & 0 & \ldots & 0 \\ \vdots & \vdots & \ddots & \vdots & \vdots \\ \times & \times & \times & \ldots & 0 \\ \times & \times & \times & \ldots & \times \\ \times & \times & \times & \ldots & \times \end{bmatrix} \qquad (14.6.21)$$

where × denotes possibly nonzero entries. It should also be noticed that $\mathbf{u}_k$ appears only in the form $\mathbf{u}_k \mathbf{u}_k^\dagger$ (diadic form) in $\mathbf{H}_k$. So if we multiply $\mathbf{u}_k$ with a scalar $e^{j\alpha_k}$, this does not affect $\mathbf{H}_k$. We can exploit this to reduce a nonzero entry of $\mathbf{u}_k$ to be real valued.

Number of Free Parameters in the Factorization

We can now count the number of parameters used in the expression (14.6.20) to characterize the unitary matrix. We will use the results of this counting later on, in order to characterize the cost of implementing lossless systems.

The vector $\mathbf{u}_k$ has $M - k + 1$ possible nonzero complex-valued entries. The $M-1$ vectors $\mathbf{u}_k$ are therefore associated with $\left(\sum_{i=2}^{M} i \right)$ complex-valued parameters, which is equivalent to $M(M+1) - 2$ real valued parameters.

However, each $\mathbf{u}_k$ is constrained to have unit-norm, that is, $\mathbf{u}_k^\dagger \mathbf{u}_k = 1$. This constraint takes away one degree of freedom from $\mathbf{u}_k$. Moreover, one component of each $\mathbf{u}_k$ can be taken to be real without loss of generality. This takes away another degree of freedom. Summarizing, the total number of real degrees of freedom associated with the set of vectors $\{\mathbf{u}_k\}$ is then $M(M+1) - 2 - 2(M-1) = M^2 - M$. Add to this the real valued parameters θ_i (associated with $\mathbf{D}$) and $\sqrt{d}$. This results in $M^2 + 1$ real valued freedoms. This, of course, reduces to M^2 if the matrix is *normalized*-unitary.

For real unitary matrices, the diagonal elements of $\mathbf{D}$ are ± 1, and are not counted as degrees of freedom. The kth unit-norm vector $\mathbf{u}_k$ has $M - k$ real-valued freedoms. So the total number of freedoms is $\sum_{k=1}^{M-1}(M-k)$ which simplifies to $M(M-1)/2$. This is consistent with Sec. 14.6-1. Counting $\sqrt{d}$, this becomes $M(M-1)/2 + 1$.

One can arrive at this counting-result in a direct way, even without using the factorization result (14.6.20). This is done in Problem 14.17. We summarize the above discussions as:

Degrees of freedom in a unitary matrix. An $M \times M$ unitary matrix is completely characterized by $M^2 + 1$ real-valued degrees of freedom (M^2 for the normalized case). A real $M \times M$ unitary matrix is completely characterized by $M(M-1)/2 + 1$ degrees of freedom ($M(M-1)/2$ for the normalized case).

14.7 SMITH-McMILLAN FORM AND POLE-ZERO PATTERN

In this and the next few sections we study a number of deeper properties of lossless systems. This includes the Smith McMillan form, pole-zero pattern (this section), factorization of IIR lossless systems (Sec. 14.9), new diadic based structures, and structures based on planar rotations (Sec. 14.10). While these concepts are not used anywhere else in this text, they do find applications in advanced research on multirate filter banks.

Reminder About Smith-McMillan Form

In Sec. 13.5 we saw that every causal rational transfer matrix $\mathbf{H}(z)$ can be written in the form (13.5.26) which is called the *Smith-McMillan decomposition*. Assume that $\mathbf{H}(z)$ is $M \times M$. Then we can write

$$\mathbf{H}(z) = \mathbf{W}(z)\mathbf{\Lambda}(z)\mathbf{V}(z), \qquad (14.7.1)$$

where $\mathbf{W}(z)$ and $\mathbf{V}(z)$ are $M \times M$ unimodular matrices (polynomials in z with *constant* nonzero determinants) and $\mathbf{\Lambda}(z)$ is diagonal. The ith diagonal element $\lambda_i(z)$ of $\mathbf{\Lambda}(z)$ has the form

$$\lambda_i(z) = \frac{\alpha_i(z)}{\beta_i(z)}, \qquad 0 \le i \le M-1, \qquad (14.7.2)$$

where $\alpha_i(z)$ and $\beta_i(z)$ are polynomials in z with *no common factors*, and satisfy the following divisibility condition

$$\alpha_i(z) | \alpha_{i+1}(z), \quad \beta_{i+1}(z) | \beta_i(z) \qquad 0 \le i \le M-2. \qquad (14.7.3)$$

Losslessness implies that $\mathbf{H}(e^{j\omega})$ has full rank M so that none of the $\alpha_i(z)$'s is identically zero.

Recall that the zeros of $\beta_i(z)$ and $\alpha_i(z)$ are the poles and zeros of $\mathbf{H}(z)$ respectively. Because of the causality of $\mathbf{H}(z)$, the order μ of $\prod_i \beta_i(z)$ (as a polynomial in z) is at least as large as that of $\prod_i \alpha_i(z)$. Note that $\alpha_i(z)$ and $\beta_j(z)$ are polynomials in z and *not* z^{-1}. Under these conditions, if μ_i denotes the degree of $\beta_i(z)$, the degree of $\mathbf{H}(z)$ is

$$\mu = \sum_{i=0}^{M-1} \mu_i \quad \text{(McMillan degree)}. \tag{14.7.4}$$

That is, $\mathbf{H}(z)$ can be implemented with μ memory elements, but not less.

♠**Main points of this section.** Let $\mathbf{H}(z)$ be an $M \times M$ causal lossless system. We will study its Smith-form matrix $\Lambda(z)$ and obtain the following conclusions.

1. If $\mathbf{H}(z)$ is FIR, the elements $\lambda_i(z)$ are pure delays, that is, $\lambda_i(z) = z^{-n_i}$, with $n_i \geq 0$.
2. If p is a pole of $\mathbf{H}(z)$, then $1/p^*$ is a zero. If q is a zero, then $1/q^*$ is a pole. Since all poles are inside the unit circle, the zeros lie outside. So the polynomials $\alpha_i(z)$ and $\beta_j(z)$ do not have common factors for any choice of i and j.
3. The degree of $\mathbf{H}(z)$ is equal to the degree of its determinant (which in turn is allpass). ◊

Example 14.7.1

As always, it is best to begin with examples. Consider the FIR lossless system

$$\mathbf{H}(z) = \begin{bmatrix} 1+z^{-1} & 1-z^{-1} \\ 1-z^{-1} & 1+z^{-1} \end{bmatrix}. \tag{14.7.5}$$

The Smith-McMillan decomposition is

$$\begin{bmatrix} 1+z^{-1} & 1-z^{-1} \\ 1-z^{-1} & 1+z^{-1} \end{bmatrix} = \underbrace{\begin{bmatrix} -1 & 1 \\ 1 & 1 \end{bmatrix}}_{\mathbf{W}(z)} \underbrace{\begin{bmatrix} z^{-1} & 0 \\ 0 & 1 \end{bmatrix}}_{\Lambda(z)} \underbrace{\begin{bmatrix} -1 & 1 \\ 1 & 1 \end{bmatrix}}_{\mathbf{V}(z)}. \tag{14.7.6}$$

In this example $\mathbf{W}(z)$ and $\mathbf{V}(z)$ turn out to be constant nonsingular matrices. Also

$$\alpha_0(z) = \alpha_1(z) = 1, \ \beta_0(z) = z, \ \text{and} \ \beta_1(z) = 1. \tag{14.7.7}$$

So $\mu_0 = 1, \mu_1 = 0$ and $\mu = 1$ so that $\mathbf{H}(z)$ has degree one. Notice that $\Lambda(z)$ is causal.

From Example 13.9.2 we know that the Smith-McMillan form $\Lambda(z)$ is not necessarily causal, even though $\mathbf{H}(z)$ is. However, for FIR lossless systems, $\Lambda(z)$ is always causal as shown below.

Smith-McMillan Form for FIR Lossless Systems

Suppose $\mathbf{H}(z)$ is FIR. Then, we know from Sec. 13.9 that $\beta_i(z)$ have the form z^{n_i}, $n_i \geq 0$, so that the determinant of $\mathbf{H}(z)$ is $a \prod_{i=0}^{M-1} z^{-n_i} \alpha_i(z)$ where a is constant. If this determinant is a delay [which is the case, for instance, when $\mathbf{H}(z)$ is lossless] then $\alpha_i(z)$ must also have the form z^{m_i}, $m_i \geq 0$ so that $[\det \mathbf{H}(z)] = az^{-K}$. Thus,

$$\det \mathbf{H}(z) = az^{-K}, \qquad K = \mu - \sum_{i=0}^{M-1} m_i, \quad m_i \geq 0. \qquad (14.7.8)$$

From Sec. 14.4 we know that if $\mathbf{H}(z)$ is causal FIR lossless, then $[\det \mathbf{H}(z)] = az^{-\mu}$. From (14.7.8) it therefore follows that $m_i = 0$ for all i. So we have $\lambda_i(z) = z^{-n_i}$, with $\sum_{i=0}^{M-1} n_i = \mu$. Furthermore $n_0 \geq n_1 \geq \ldots \geq n_{M-1} \geq 0$ in view of (14.7.3).

Summarizing, we have proved the following: Let $\mathbf{H}(z)$ be $M \times M$ causal FIR lossless. Then, the Smith-McMillan form is

$$\Lambda(z) = \begin{bmatrix} z^{-n_0} & 0 & \ldots & 0 \\ 0 & z^{-n_1} & \ldots & 0 \\ \vdots & \vdots & \ddots & \vdots \\ 0 & 0 & \ldots & z^{-n_{M-1}} \end{bmatrix},$$

where $n_0 \geq n_1 \geq \ldots \geq n_{M-1} \geq 0$. In particular, therefore, $\Lambda(z)$ is causal. Also, the degree $\mu = \sum_{i=0}^{M-1} n_i$.

Poles and zeros

Let p be a pole of an $M \times M$ causal lossless system $\mathbf{H}(z)$. We will prove that $1/p^*$ is a zero. This is a beautiful extension of a similar property of stable allpass functions, proved earlier in Sec. 3.4.1.

Proof. With α_i and β_i defined as in (14.7.2), define

$$\mathbf{A}(z) = \text{diag}[\alpha_0(z) \quad \ldots \quad \alpha_{M-1}(z)], \quad \mathbf{B}(z) = \text{diag}[\beta_0(z) \quad \ldots \quad \beta_{M-1}(z)].$$

Then $\Lambda(z) = \mathbf{B}^{-1}(z)\mathbf{A}(z)$, so that $\mathbf{H}(z) = \mathbf{W}(z)\mathbf{B}^{-1}(z)\mathbf{A}(z)\mathbf{V}(z)$. Defining $\mathbf{W}_1(z) = \mathbf{W}^{-1}(z)$ this becomes

$$\mathbf{H}(z) = [\mathbf{B}(z)\mathbf{W}_1(z)]^{-1}\mathbf{A}(z)\mathbf{V}(z).$$

Using the paraunitary property $\mathbf{H}(z)\widetilde{\mathbf{H}}(z) = c^2\mathbf{I}$, we, therefore, obtain

$$c^2\mathbf{B}(z)\mathbf{P}(z)\widetilde{\mathbf{B}}(z) = \mathbf{A}(z)\mathbf{Q}(z)\widetilde{\mathbf{A}}(z), \qquad (14.7.9)$$

where $\mathbf{P}(z) \triangleq \mathbf{W}_1(z)\widetilde{\mathbf{W}}_1(z), \mathbf{Q}(z) \triangleq \mathbf{V}(z)\widetilde{\mathbf{V}}(z)$. Since $\mathbf{W}(z)$ is unimodular, so is $\mathbf{W}^{-1}(z)$. So, the $M \times M$ matrices $\mathbf{P}(z)$ and $\mathbf{Q}(z)$ have constant nonzero determinants, hence full rank for all z. Now let p be a pole of $\mathbf{H}(z)$, that is, $\beta_i(p) = 0$ for some i, so that the ith row of the left hand side of (14.7.9) is zero for $z = p$. So the ith row of the right hand side is also zero for $z = p$, that is,

$$\alpha_i(p)[Q_{i,0}(p)\tilde{\alpha}_0(p) \quad Q_{i,1}(p)\tilde{\alpha}_1(p) \quad \ldots \quad Q_{i,M-1}(p)\tilde{\alpha}_{M-1}(p)] = \mathbf{0}. \qquad (14.7.10)$$

But we know $\alpha_i(p) \neq 0$ since $\alpha_i(z)$ and $\beta_i(z)$ are relatively prime. Moreover, the ith row of $\mathbf{Q}(p)$ cannot have all elements equal to zero since $\mathbf{Q}(z)$ has full rank for all z. So $\tilde{\alpha}_j(p) = 0$ for some j, that is, $\alpha_j(1/p^*) = 0$ for some j so that $1/p^*$ is a zero indeed. $\triangledown\triangledown\triangledown$

Conversely, if q is a zero, one can show that $1/q^*$ is a pole.

All zeros are outside. As a consequence of the above result, all zeros of the lossless system are strictly *outside* the unit circle (since poles are inside)! So $\alpha_i(z)$ and $\beta_j(z)$ are relatively prime even if $i \neq j$. This would *not*, in general, be true if the system were just stable but not lossless.

We now turn to another important result which can be deduced from the Smith-McMillan form.

♠**Theorem 14.7.1. Degree of a lossless system.** Let $\mathbf{H}(z)$ be $M \times M$ causal lossless. Then $A(z) \triangleq \det \mathbf{H}(z)$ is stable allpass, and

$$[\deg [\det \mathbf{H}(z)]] = [\deg \mathbf{H}(z)]. \qquad (14.7.11)$$

Also p is a pole of $A(z)$ if, and only if, it is a pole of $\mathbf{H}(z)$ and q is a zero of $A(z)$ if and only if it is a zero of $\mathbf{H}(z)$. ◇

Proof. The stable allpass property of $A(z)$ follows from $\widetilde{\mathbf{H}}(z)\mathbf{H}(z) = c^2\mathbf{I}$. Next, from the Smith-McMillan decomposition we have

$$A(z) = a \prod_{i=0}^{M-1} \frac{\alpha_i(z)}{\beta_i(z)}. \qquad (14.7.12)$$

Since $\alpha_i(z)$ and $\beta_j(z)$ are relatively prime for any i and j, there are no uncanceled factors in (14.7.12). Thus the degree of $A(z)$ is equal to that of the product $\prod_{i=0}^{M-1} \beta_i(z)$. This, in turn is equal to μ, which is the degree of $\mathbf{H}(z)$.

Since there are no uncanceled factors in (14.7.12), it is clear that q is a zero of $\mathbf{H}(z)$ [i.e., a zero of some $\alpha_i(z)$] if, and only if, it is a zero of $A(z)$; The same holds for poles. $\triangledown\triangledown\triangledown$

14.8 THE MODULUS PROPERTY

In Sec. 3.4.1 we proved a property called the *modulus* property, for stable allpass functions. We now establish the following extension of this result: Any $M \times M$ causal normalized-lossless $\mathbf{H}(z)$ satisfies

$$\mathbf{H}^\dagger(z)\mathbf{H}(z) \begin{cases} = \mathbf{I}, & \text{for } |z| = 1 \\ \leq \mathbf{I}, & \text{for } |z| > 1 \\ \geq \mathbf{I}, & \text{for } |z| < 1. \end{cases} \qquad (14.8.1)$$

The property '$\mathbf{H}^\dagger(z)\mathbf{H}(z) = \mathbf{I}$' for $|z| = 1$ follows from definition. We will now prove

$$\mathbf{H}^\dagger(z)\mathbf{H}(z) \leq \mathbf{I}, \quad |z| > 1. \qquad (14.8.2)$$

As in Sec. 3.4.1, one could try to invoke a matrix version of the maximum modulus theorem (which is proved in Problem 14.30). We will, instead, use a direct energy-balance argument here. We apply a specific type of input, and bring out the inequality $\mathbf{H}^\dagger(z)\mathbf{H}(z) \leq \mathbf{I}$ in a simple way. Thus let the input to $\mathbf{H}(z)$ be

$$\mathbf{u}(n) = \begin{cases} a^n \mathbf{v}, & n \leq L \\ \mathbf{0} & n > L, \end{cases} \qquad (14.8.3)$$

for some integer L. Here $\mathbf{v}$ is an arbitrary nonzero vector. So $\mathbf{u}(n)$ is a truncated exponential. Assume $|a| > 1$ so that $\mathbf{u}(n)$ is bounded as $n \to -\infty$. With $\mathbf{h}(n)$ denoting the impulse response of $\mathbf{H}(z)$, the output is

$$\mathbf{y}(n) = \sum_{m=-\infty}^{n} a^m \mathbf{h}(n-m)\mathbf{v}, \quad n \leq L. \qquad (14.8.4)$$

This can be simplified to yield

$$\mathbf{y}(n) = a^n \mathbf{H}(a)\mathbf{v}, \quad n \leq L. \qquad (14.8.5)$$

(The value of $\mathbf{y}(n)$ for $n > L$ will not enter our reasoning). The z-transform $\mathbf{H}(z)$ converges for $|z| > 1$ because $\mathbf{H}(z)$ is causal and stable (Chap. 2). So the quantity $\mathbf{H}(a)$ above is well defined. We now have the following sequence of inequalities:

$$\left(\sum_{n=-\infty}^{L} |a|^{2n} \right) \mathbf{v}^\dagger \mathbf{H}^\dagger(a)\mathbf{H}(a)\mathbf{v} = \sum_{n=-\infty}^{L} \mathbf{y}^\dagger(n)\mathbf{y}(n) \quad \text{(by (14.8.5))}$$

$$\leq \sum_{n=-\infty}^{\infty} \mathbf{y}^\dagger(n)\mathbf{y}(n)$$

$$= \sum_{n=-\infty}^{\infty} \mathbf{u}^\dagger(n)\mathbf{u}(n) \quad \text{(since } \widetilde{\mathbf{H}}(z)\mathbf{H}(z) = \mathbf{I})$$

$$= \left(\sum_{n=-\infty}^{L} |a|^{2n} \right) \mathbf{v}^\dagger \mathbf{v}. \quad \text{[by (14.8.3)]}$$

$$(14.8.6)$$

Consider the leftmost and rightmost quantities above. Since $|a| > 1$, the summation $\sum_{n=-\infty}^{L} |a|^{2n}$ converges to a finite value, and can be canceled. This results in

$$\mathbf{v}^\dagger \mathbf{H}^\dagger(a)\mathbf{H}(a)\mathbf{v} \leq \mathbf{v}^\dagger \mathbf{v}, \qquad (14.8.7)$$

for any $\mathbf{v}$, so that $\mathbf{H}^\dagger(a)\mathbf{H}(a) \leq \mathbf{I}$ for any a with $|a| > 1$.

Now consider proving $\mathbf{H}^\dagger(z)\mathbf{H}(z) \geq \mathbf{I}$ for $|z| < 1$. By using the property $\widetilde{\mathbf{H}}(z)\mathbf{H}(z) = \mathbf{I}$ in (14.8.2), we can rearrange it as

$$\left[\widetilde{\mathbf{H}}(z)\widetilde{\mathbf{H}}^\dagger(z)\right]^{-1} \leq \mathbf{I}, \qquad |z| > 1. \qquad (14.8.8)$$

By invoking the following facts (Problem 14.25): (a) $\mathbf{P} \leq \mathbf{I} \iff \mathbf{P}^{-1} \geq \mathbf{I}$, (b) $\mathbf{P} \leq \mathbf{I} \iff \mathbf{P}^T \leq \mathbf{I}$, and (c) $\mathbf{P} \leq \mathbf{I} \iff \mathbf{P}^* \leq \mathbf{I}$ for any Hermitian positive definite $\mathbf{P}$, we can verify that (14.8.8) implies $\mathbf{H}^\dagger(z)\mathbf{H}(z) \geq \mathbf{I}$ for $|z| < 1$ indeed.

For $p \times r$ lossless systems a similar property is true.

Are the Inequalities Strict?

In Sec. 3.4.1 we found that the above inequalities were strict, unless the allpass function was just a constant. In the matrix case, however, this is not true. For the $M \times M$ case, we now explore the significance of the equality

$$\mathbf{H}^\dagger(a)\mathbf{H}(a) = \mathbf{I}, \qquad |a| > 1. \qquad (14.8.9)$$

This is possible if and only if the inequality in (14.8.6) holds with equality. This means, in particular, that $\mathbf{y}(n) = \mathbf{0}$ for all $n > L$. In other words, the output $\mathbf{y}(n)$ ceases as soon as the input $\mathbf{u}(n)$ becomes zero!

This is equivalent to saying that if $\mathbf{h}(n)$ denotes the (causal) impulse response of $\mathbf{H}(z)$ then $\mathbf{h}(n)\mathbf{v} = \mathbf{0}$ for $n \neq 0$ (Problem 14.26). In other words, $\mathbf{H}(z)\mathbf{v} = \mathbf{c}$ for all z. That is, the system $\mathbf{H}(z)$ acts like a memoryless system (i.e., just a constant) for inputs in the direction $\mathbf{v}$ [i.e., for inputs of the form $f(n)\mathbf{v}$ where $f(n)$ is scalar]. If $\mathbf{H}(z)$ is not memoryless in any direction, then (14.8.2) holds with strict inequality. We then say that $\mathbf{H}(z)$ is a full-memory lossless system.

14.9 STRUCTURES FOR IIR LOSSLESS SYSTEMS

All the structures developed so far were for FIR lossless systems. We now consider the IIR case. It turns out that the main factorization results of Sec. 14.4 can be extended to the IIR case simply by replacing the fundamental building block $\mathbf{V}_m(z)$ [eqn. (14.4.1)], with

$$\mathbf{V}_m(z) = \mathbf{I} - \mathbf{v}_m \mathbf{v}_m^\dagger + \left(\frac{-a_m^* + z^{-1}}{1 - a_m z^{-1}}\right)\mathbf{v}_m \mathbf{v}_m^\dagger, \qquad (14.9.1)$$

where $|a_m| < 1$ [Doğanata and Vaidyanathan, 1990]. The main results of this section depend on the properties of this building block, which are summarized next.

Properties of $\mathbf{V}_m(z)$

1. $\mathbf{V}_m(z)$ has been obtained by replacing z^{-1} in (14.4.1) with the stable allpass function $G(z) = (-a_m^* + z^{-1})/(1 - a_m z^{-1})$. The structure for $\mathbf{V}_m(z)$ is shown in Fig. 14.9-1.
2. The pole of $\mathbf{V}_m(z)$ is at a_m so that it is stable.
3. $\mathbf{V}_m(z)$ is a degree-one lossless system satisfying (a) $\widetilde{\mathbf{V}}_m(z)\mathbf{V}_m(z) = \mathbf{I}$, and (b) $[\det \mathbf{V}_m(z)] = (-a_m^* + z^{-1})/(1 - a_m z^{-1})$. These can be proved precisely as in Sec. 14.4.1.

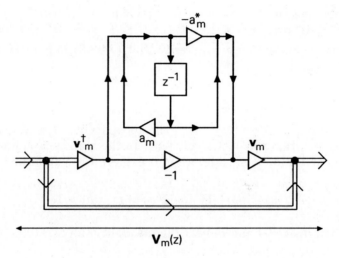

Figure 14.9-1 Implementation of the degree-one IIR lossless building block $\mathbf{V}_m(z)$.

14.9.1 Structures for $M \times M$ IIR Lossless Systems

We now show that any $M \times M$ causal lossless transfer matrix $\mathbf{H}_N(z)$ with degree N can be factorized as

$$\mathbf{H}_N(z) = \mathbf{V}_N(z)\mathbf{V}_{N-1}(z)\ldots\mathbf{V}_1(z)\mathbf{H}_0, \qquad (14.9.2)$$

where $\mathbf{H}_0$ is $M \times M$ unitary, and $\mathbf{V}_m(z)$ are as in (14.9.1) with $|a_m| < 1$. This results in the structure of Fig. 14.4-3 with building blocks as in Fig. 14.9-1. Conversely, it is clear that the above structure represents a lossless system of degree N as long as $\mathbf{v}_m^\dagger \mathbf{v}_m = 1$ and $|a_m| < 1$ for all m.

Remarks. (a) The resulting structure has N delays (one per $\mathbf{V}_m(z)$) so that it is minimal. (b) The above result also covers the FIR case, since we can set $a_m = 0$ for all m.

The Degree Reduction Step

The proof of the above factorization is based on a successive degree-reduction technique. In all discussions we will freely use the fact that the degree of $\mathbf{H}_N(z)$ is equal to the degree of the determinant (Theorem 14.7.1). Consider the function

$$\mathbf{H}_{N-1}(z) = \underbrace{\left(\mathbf{I} - \mathbf{v}_N\mathbf{v}_N^\dagger + \frac{1 - a_N z^{-1}}{-a_N^* + z^{-1}}\mathbf{v}_N\mathbf{v}_N^\dagger\right)}_{\widetilde{\mathbf{V}}_N(z)} \mathbf{H}_N(z). \qquad (14.9.3)$$

Evidently this is paraunitary. Suppose we choose a_N to be a pole of $\mathbf{H}_N(z)$. From Sec. 14.7 it then follows that $\mathbf{H}_N(z)$ has a zero at $1/a_N^*$. But any zero of $\mathbf{H}_N(z)$ is also a zero of its determinant (Theorem 14.7.1). This means that there exists a column vector $\mathbf{v} \neq \mathbf{0}$ such that $\mathbf{v}^\dagger \mathbf{H}_N(1/a_N^*) = \mathbf{0}^\dagger$. Choosing $\mathbf{v}_N = \mathbf{v}/\|\mathbf{v}\|$ we see that $\mathbf{v}_N^\dagger \mathbf{H}_N(z)$ will then have the factor $(-a_N^* + z^{-1})$ which, therefore, cancels the "unstable pole" of $\widetilde{\mathbf{V}}_N(z)$. With the only unstable pole canceled, $\mathbf{H}_{N-1}(z)$ is stable and hence lossless.

From Theorem 14.7.1 we know that the determinant of $\mathbf{H}_N(z)$ has the factor $(-a_N^* + z^{-1})$ in its numerator and the factor $(1 - a_N z^{-1})$ in the denominator. Since $\mathbf{V}_N(z)$ is lossless, these same factors appear in the determinant of $\mathbf{V}_N(z)$. When we take determinants of (14.9.3), these factors therefore cancel on the right hand side. So $[\det \mathbf{H}_{N-1}(z)]$ has degree $N-1$. And since $\mathbf{H}_{N-1}(z)$ is also lossless, Theorem 14.7.1 can be applied again to conclude that its degree is equal to $N-1$.

Summarizing, we have extracted the degree-one lossless system $\mathbf{V}_N(z)$ from $\mathbf{H}_N(z)$ to obtained the 'remainder' lossless system $\mathbf{H}_{N-1}(z)$ with degree $N-1$.

Obtaining the complete factorization. By repeated application of the degree reduction step we eventually obtain the remainder $\mathbf{H}_0(z)$. This is a lossless system with constant determinant. By Theorem 14.7.1 this is therefore a constant. This establishes the factorization (14.9.2).

14.9.2 $M \times 1$ IIR lossless systems

$M \times 1$ lossless systems find application as power complementary filter banks. An $M \times 1$ IIR system can be represented as $\mathbf{H}_N(z) = \mathbf{P}_N(z)/D_N(z)$ where $\mathbf{P}_N(z)$ is $M \times 1$ causal FIR and

$$D_N(z) = \prod_{k=1}^{N}(1 - z_k z^{-1}). \qquad (14.9.4)$$

To avoid redundancies we assume $z_k \neq 0$ and that $\mathbf{P}_N(z)/D_N(z)$ is an irreducible representation, that is, the greatest common factor of $D_N(z)$

and all the elements of $\mathbf{P}_N(z)$ is a constant. As a result the system $\mathbf{H}_N(z)$ has exactly N nonzero poles ($z_k \neq 0$). It should be kept in mind, however, that the entries of the numerator $\mathbf{P}_N(z)$ can have degree $\geq N$. So there may be an arbitrary number of poles at the origin, representing the FIR part.

Degree-Reduction Step

We will first establish the following: Let $\mathbf{H}_m(z)$ be $M \times 1$ causal lossless with irreducible representation $\mathbf{H}_m(z) = \mathbf{P}_m(z)/D_m(z)$ where $D_m(z) = \prod_{k=1}^{m}(1 - z_k z^{-1})$. Let $a_m = z_m$, and

$$\mathbf{v}_m = \mathbf{P}_m(z_m)/\|\mathbf{P}_m(z_m)\|, \qquad (14.9.5a)$$

so that $\mathbf{V}_m(z)$ in (14.9.1) is completely defined. We can then write

$$\mathbf{H}_m(z) = \mathbf{V}_m(z)\mathbf{H}_{m-1}(z), \qquad (14.9.5b)$$

where $\mathbf{H}_{m-1}(z)$ is $M \times 1$ causal lossless with irreducible representation $\mathbf{H}_{m-1}(z) = \mathbf{P}_{m-1}(z)/D_{m-1}(z)$ where $D_{m-1}(z) = \prod_{k=1}^{m-1}(1 - z_k z^{-1})$.

Proof: Since $\mathbf{V}_m^{-1}(z) = \widetilde{\mathbf{V}}_m(z)$, the relation (14.9.5b) can be rewritten as $\mathbf{H}_{m-1}(z) = \widetilde{\mathbf{V}}_m(z)\mathbf{H}_m(z)$, that is,

$$\frac{\mathbf{P}_{m-1}(z)}{D_{m-1}(z)} = \left(\mathbf{I} - \mathbf{v}_m\mathbf{v}_m^\dagger + \left(\frac{1 - z_m z^{-1}}{-z_m^* + z^{-1}}\right)\mathbf{v}_m\mathbf{v}_m^\dagger\right)\frac{\mathbf{P}_m(z)}{D_m(z)}, \qquad (14.9.6)$$

since $a_m = z_m$. With $\mathbf{v}_m$ chosen as in (14.9.5a) we have

$$(\mathbf{I} - \mathbf{v}_m\mathbf{v}_m^\dagger)\mathbf{P}_m(z_m) = \|\mathbf{P}_m(z_m)\|(\mathbf{I} - \mathbf{v}_m\mathbf{v}_m^\dagger)\mathbf{v}_m = \mathbf{0}. \qquad (14.9.7)$$

This means that $(\mathbf{I} - \mathbf{v}_m\mathbf{v}_m^\dagger)\mathbf{P}_m(z)$ has the factor $(1 - z_m z^{-1})$. So the factor $(1 - z_m z^{-1})$, which is also present in $D_m(z)$, gets canceled in the right hand side of (14.9.6).

We now face the second main issue: the quantity $(-z_m^* + z^{-1})$ in the denominator of (14.9.6) represents an unstable pole at $z = 1/z_m^*$, and must be canceled. It turns out, however, that the choice (14.9.5a) also results in such a cancelation – a consequence of losslessness!

To show this, we will prove that $\mathbf{v}_m^\dagger\mathbf{P}_m(z)$ has the factor $(-z_m^* + z^{-1})$. Losslessness of $\mathbf{H}_m(z)$ implies

$$\widetilde{\mathbf{P}}_m(z)\mathbf{P}_m(z) = c^2\widetilde{D}_m(z)D_m(z). \qquad (14.9.8)$$

The right hand side is zero for $z = z_m$ (due to $D_m(z)$) and for $z = 1/z_m^*$ [due to $\widetilde{D}_m(z)$]. So $\widetilde{\mathbf{P}}_m(1/z_m^*)\mathbf{P}_m(1/z_m^*) = 0$. By rearranging this, one can conclude that

$$\mathbf{v}_m^\dagger \mathbf{P}_m(1/z_m^*) = 0,$$

so that $\mathbf{v}_m^\dagger \mathbf{P}_m(z)$ does have the factor $(-z_m^* + z^{-1})$.

Summarizing, $\mathbf{H}_{m-1}(z)$ is $M \times 1$ causal, and is stable because the poles are z_k, $1 \le k \le m-1$, $|z_k| < 1$. Evidently, $\mathbf{H}_{m-1}(z)$ is paraunitary (from (14.9.5b)). Moreover $\mathbf{P}_{m-1}(z)/D_{m-1}(z)$ is irreducible (otherwise, as seen from (14.9.5b), $\mathbf{P}_m(z)/D_m(z)$ cannot be irreducible). So the remainder $\mathbf{H}_{m-1}(z)$ has all the claimed properties. ▽▽▽

Remarks. (a) The remainder $\mathbf{H}_{m-1}(z)$ therefore has only $m-1$ poles $z_k \neq 0$ (subset of the original set of m poles). Thus the number of nonzero poles has been reduced by one. (b) Since $D_m(z_m) = 0$, we are assured that $\mathbf{P}_m(z_m) \neq 0$ (by irreducibility of $\mathbf{P}_m(z)/D_m(z)$. So the above expression (14.9.5a) for $\mathbf{v}_m$ is well-defined.

Obtaining the complete factorization. By repeated use of this reduction-step, we obtain the complete factorization, which is summarized as follows: Let $\mathbf{H}_N(z)$ be $M \times 1$ causal IIR lossless having N poles z_k, with $|z_k| > 0$. Then it can be expressed as

$$\mathbf{H}_N(z) = \mathbf{V}_N(z) \ldots \mathbf{V}_1(z) \mathbf{G}(z), \qquad (14.9.9)$$

where $\mathbf{V}_m(z)$ are as in (14.9.1) with $a_m = z_m$, $\|\mathbf{v}_m\| = 1$, and $\mathbf{G}(z)$ is $M \times 1$ causal FIR lossless.

The ordering of the N poles z_k in (14.9.4) is arbitrary. So we obtain a number of equivalent structures simply by renumbering the poles.

Real Coefficient Systems

Recall the FIR versions of the above results, which were given in Sec. 14.4. In the FIR case, whenever the coefficients of $\mathbf{H}_N(z)$ were real, the vectors $\mathbf{v}_m$ turned out to be real so that we could obtain a structure with real multipliers.

In the IIR case, this is not necessarily so because the poles z_m are not necessarily real. It is, however, true that the poles occur in complex conjugate pairs. The corresponding pairs of $\mathbf{V}_m(z)$ can be combined to obtain real coefficient structures. But the structures lose the beautiful simplicity of the individual building blocks $\mathbf{V}_m(z)$.

14.10 MODIFIED LOSSLESS STRUCTURES

In this section we obtain two new structures, by modifying the structures obtained earlier. The main results are summarized below.

♠**Main points of this section.** We start with the familiar cascaded lossless structure of Fig. 14.10-1 with building blocks $\mathbf{V}_m(z)$ as in Fig. 14.9-1. By performing some simple matrix manipulations, we will arrive at two new structures. These are cascaded interconnections of fundamental unitary matrices (e.g., Householder matrices), separated by a diagonal matrix of the form

$$\Phi_m(z) = \begin{bmatrix} \mathbf{I} & \mathbf{0} \\ \mathbf{0} & x_m \end{bmatrix}, \qquad (14.10.1)$$

where

$$x_m = \begin{cases} z^{-1}, & \text{(FIR case)}, \\ (-a_m^* + z^{-1})/(1 - a_m z^{-1}), & \text{(IIR case)}. \end{cases} \quad (14.10.2)$$

1. **The first structure** (which, we believe, has not been reported before) has the form shown in Fig. 14.10-4, were $\mathbf{W}_m$ are Householder unitary matrices (defined in Sec. 14.6.2). This shows that any lossless system can be implemented as a cascade of Householder matrices separated by $\mathbf{\Phi}_m(z)$. This structure has the advantage that it requires fewer additions compared to the earlier structures. It is also conceptually simpler, and is given entirely in terms of the very well-known Householder matrix.
2. **The second structure** has the form shown in Fig. 14.10-5. Once again it is a cascade of unitary matrices $\mathbf{\Theta}_m$ separated by $\mathbf{\Phi}_m(z)$, but the unitary matrices are not of Householder type. Instead they are based on planar rotations as shown in Fig. 14.10-5(b). This structure will be derived only for the FIR real coefficient case, and is the precise generalization (to the $M \times M$ case) of the lattice structure developed in Sec. 14.3.1 for the 2×2 FIR case (Fig. 14.3-2). This structure was derived in [Doğanata, et al., 1988] using a different procedure.

14.10.1 Lossless Structures Based on Householder Matrix

In earlier sections, we developed structures for FIR and IIR lossless systems. Both $M \times M$ and $M \times 1$ cases were covered. We saw that the basic building block in all these cases had the same form, viz.,

$$\mathbf{V}_m(z) = \mathbf{I} - \mathbf{v}_m \mathbf{v}_m^\dagger + x_m \mathbf{v}_m \mathbf{v}_m^\dagger, \quad (14.10.3)$$

where x_m is as in (14.10.2), and $\mathbf{v}_m^\dagger \mathbf{v}_m = 1$. All the structures had the common form reproduced in Fig. 14.10-1, where N is the degree of the transfer matrix $\mathbf{H}(z)$. The size of $\mathbf{H}_0$ is equal to that of $\mathbf{H}(z)$.

The matrix $\mathbf{V}_m(z)$ has all the properties studied in Sec. 14.4.1 (with z^{-1} replaced by x_m). In particular, therefore eqn. (14.4.5) holds, that is, there exists an $M \times M$ matrix $\mathbf{U}_m$ with $\mathbf{U}_m^\dagger \mathbf{U}_m = \mathbf{I}$, such that

$$\mathbf{V}_m(z) = \mathbf{U}_m \underbrace{\begin{bmatrix} \mathbf{I} & 0 \\ 0 & x_m \end{bmatrix}}_{\mathbf{\Phi}_m(z)} \mathbf{U}_m^\dagger. \quad (14.10.4)$$

We can thus redraw Fig. 14.10-1 as in Fig. 14.10-2, where $\mathbf{R}_m \triangleq \mathbf{U}_{m+1}^\dagger \mathbf{U}_m$ for $1 \le m \le N-1$, $\mathbf{R}_N = \mathbf{U}_N$, and $\mathbf{R}_0 = \mathbf{U}_1^\dagger \mathbf{H}_0$. Clearly $\mathbf{R}_m$ are normalized unitary. Even though each of these matrices has M^2 degrees of freedom (Sec.

14.6), they can be replaced with much simpler unitary matrices (Householder matrices), as we now show.

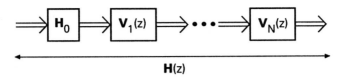

Figure 14.10-1 General form of all lossless structures. $\mathbf{V}_m(z)$ can be FIR or IIR. $\mathbf{H}_0$ is $M \times M$ or $M \times 1$.

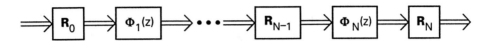

Figure 14.10-2 An equivalent structure for Fig. 14.10-1.

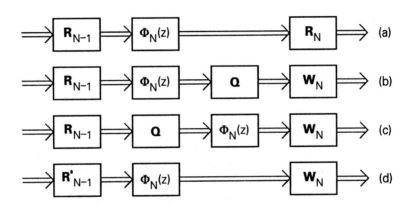

Figure 14.10-3 Further simplification of Fig. 14.10-2.

Figure 14.10-4 The simplified equivalent structure for Fig. 14.10-1.

Given any normalized unitary matrix $\mathbf{S}$, we can find unit norm $\mathbf{u}_1$ such that (14.6.18) holds. We will use this idea, with $\mathbf{R}_N$ in place of $\mathbf{S}$. By a slight modification, we can show that there exists unit norm $\mathbf{w}_N$ such that

$$\underbrace{\left(\mathbf{I} - 2\mathbf{w}_N \mathbf{w}_N^\dagger\right)}_{\mathbf{W}_N} \mathbf{R}_N = \begin{bmatrix} \mathbf{T} & 0 \\ 0 & e^{j\theta_N} \end{bmatrix}. \qquad (14.10.5)$$

Since $\mathbf{W}_N^{-1} = \mathbf{W}_N$, we can write (14.10.5) as

$$\mathbf{R}_N = \mathbf{W}_N \underbrace{\begin{bmatrix} \mathbf{T} & 0 \\ 0 & e^{j\theta_N} \end{bmatrix}}_{\mathbf{Q}} \qquad (14.10.6)$$

The cascade of the building blocks $\mathbf{R}_{N-1}, \mathbf{\Phi}_N(z)$, and $\mathbf{R}_N$ which appears in Fig. 14.10-2 can now be manipulated as follows: (a) replace $\mathbf{R}_N$ with (14.10.6) (Fig. 14.10-3(b)), (b) interchange $\mathbf{\Phi}_N(z)$ with $\mathbf{Q}$ (Fig. 14.10-3(c)), and (c) combine $\mathbf{R}_{N-1}$ with $\mathbf{Q}$ and obtain a new normalized unitary matrix $\mathbf{R}'_{N-1} = \mathbf{Q}\mathbf{R}_{N-1}$ [Fig. 14.10-3(d)]. The second step is possible because

$$\mathbf{Q}\mathbf{\Phi}_N(z) = \begin{bmatrix} \mathbf{T} & 0 \\ 0 & e^{j\theta_N} \end{bmatrix} \begin{bmatrix} \mathbf{I} & 0 \\ 0 & x_N \end{bmatrix} = \mathbf{\Phi}_N(z)\mathbf{Q}. \qquad (14.10.7)$$

Since $\mathbf{R}'_{N-1}$ is just another normalized unitary matrix, we can repeat the above three operations on the cascade of $\mathbf{R}_{N-2}, \mathbf{\Phi}_{N-1}(z)$, and $\mathbf{R}'_{N-1}$. This process can be repeated until we obtain the cascaded structure of Fig. 14.10-4 where the building blocks $\mathbf{W}_m$ have the form

$$\mathbf{W}_m = \left(\mathbf{I} - 2\mathbf{w}_m \mathbf{w}_m^\dagger\right), \quad \mathbf{w}_m^\dagger \mathbf{w}_m = 1 \quad \text{(Householder matrix)}. \qquad (14.10.8)$$

The leftmost matrix $\mathbf{G}_0$ continues to be a unitary matrix. In general this is different from $\mathbf{H}_0$ or $\mathbf{R}_0$, but has the same size viz., $M \times M$ or $M \times 1$ as the case may be.

Summarizing, we have proved the following: Any $M \times M$ causal lossless system $\mathbf{H}(z)$ can be implemented as in Fig. 14.10-4 where $\mathbf{W}_m$ are Householder matrices, $\mathbf{\Phi}_m(z)$ is as in (14.10.1) and $\mathbf{G}_0$ is an $M \times M$ unitary matrix. The quantity x_m is as in (14.10.2) with $|a_m| < 1$. (For the FIR case just set $a_m = 0$ for all m.) The same holds for the $M \times 1$ case, but now $\mathbf{G}_0$ is $M \times 1$.

A second (more direct) proof of this result without the above manipulations is developed in Problem 14.23. It is clear that each building block $\mathbf{W}_m$ requires one less addition than $\mathbf{V}_m(z)$ used in Fig. 14.10-1. Also, the structure is conceptually simpler.

The above manipulations do not affect the number of delays so that the structure continues to be minimal. Also the number of parameters in the structures are unchanged, viz., N unit norm vectors and a unitary matrix.

14.10.2 Structures Based on Givens Rotations

For the FIR real coefficient case, a slight modification of the preceding arguments results in a different structure, based on Givens rotations. Recall that Fig. 14.10-1 is equivalent to Fig. 14.10-2. When the lossless system $\mathbf{H}(z)$ is FIR with real coefficients, the unitary matrices $\mathbf{R}_m$ are real. We know that a real normalized-unitary matrix can be factorized into the form (14.6.5) with $\mathbf{S}^\dagger \mathbf{S} = \mathbf{I}$, and $\mu = \pm 1$, where Θ is a sequence of $M - 1$ Givens rotations (Fig. 14.6-2). Suppose we factorize $\mathbf{R}_N$ like this, that is, as

$$\mathbf{R}_N = \Theta_N \underbrace{\begin{bmatrix} \mathbf{S} & 0 \\ 0 & \mu \end{bmatrix}}_{\text{call this } \mathbf{Q}} \qquad (14.10.9)$$

Once again we can perform the above type of manipulations: (a) replace $\mathbf{R}_N$ with (14.10.9), (b) interchange $\Phi_N(z)$ with $\mathbf{Q}$ and (c) combine $\mathbf{R}_{N-1}$ with $\mathbf{Q}$. Repeating this N times, we arrive at the equivalent structure shown in Fig. 14.10-5(a) where each Θ_m is a sequence of $M - 1$ Givens rotations as shown in Fig. 14.10-5(b). The matrix $\mathbf{E}_0$ is $M \times M$ (or $M \times 1$) unitary. (In the $M \times M$ case it can further be expressed as a sequence of $M(M - 1)/2$ Givens rotations.)

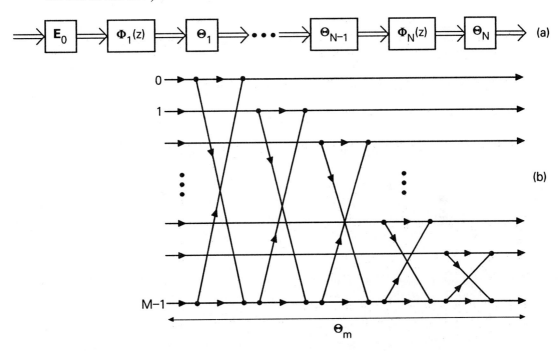

Figure 14.10-5 (a) The equivalent structure for Fig. 14.10-1, when the lossless system is real coefficient FIR. (b) Details of the building block Θ_m.

Summarizing, we have proved this: Let $\mathbf{H}(z)$ be an $M \times M$ FIR real coefficient lossless system. It can then be implemented as in Fig. 14.10-5(a) where

$$\Phi_m(z) = \begin{bmatrix} \mathbf{I} & 0 \\ 0 & z^{-1} \end{bmatrix},$$

Sec. 14.10 Modified lossless structures

and where $\boldsymbol{\Theta}_m$ represents a sequence of $M-1$ Givens rotations as shown in Fig. 14.10-5(b). The leftmost matrix $\mathbf{E}_0$ is $M \times M$ unitary, that is, a sequence of $M(M-1)/2$ Givens rotations as in Fig. 14.6-3 (possibly followed by a fixed scale factor α on all lines). For the $M \times 1$ case, the same result is true except that $\mathbf{E}_0$ is now a nonzero column vector.

Remarks

1. Once again, the structure of Fig. 14.10-5(a) is minimal. Notice also that the number of parameters involved in the new structure (for $M \times M$ case) is equal to $N(M-1) + M(M-1)/2$ plus a possible scale factor. This is the same as the number of degrees of freedom in Fig. 14.10-1 with real $\mathbf{v}_m$ and $\mathbf{H}_0$.
2. For the real-coefficient *IIR case*, the above unitary building blocks are not guaranteed to be real. The above results can however be easily extended to the case where the matrices are complex. The only significant additional complication is that the Givens rotation should be replaced by its complex version.

14.11 PRESERVING LOSSLESS PROPERTY UNDER QUANTIZATION

In Chap. 9 we discussed finite wordlength (FWL) effects in the digital implementation of filters. We considered three effects, namely (a) roundoff noise, (b) limit cycles, and (c) coefficient quantization. When implementing filter banks, another important FWL effect comes to play. This is the effect of coefficient quantization on such properties as alias cancelation and perfect reconstruction (PR). It turns out that by careful implementation of the filter bank system, one can ensure that the alias cancelation and/or PR property are not affected by coefficient quantization. We saw examples of this in Chap. 5 and 6. Thus, in Sec. 5.3.5 we saw that the two channel IIR power symmetric QMF bank can be implemented such that aliasing as well as amplitude distortion are eliminated in spite of coefficient quantization. In Sec. 6.4.1 we saw that the two channel FIR QMF lattice has the PR property even when the coefficients are quantized.

We now show that these "robustness properties" hold also for the M channel case. We know from Sec. 6.2 that an M channel FIR filter bank can be designed to have the PR property by forcing $\mathbf{E}(z)$ to be paraunitary and taking the synthesis filters to be $f_k(n) = h_k^*(N-n)$. We show how the paraunitary (or lossless) property can be preserved in presence of quantization. This in turn shows how to perserve the PR property of the filter bank in presence of coefficient quantization.

14.11.1 Quantization of Unitary Matrices

An $M \times M$ unitary matrix $\mathbf{R}$ satisfies (14.6.1). Let the columns of $\mathbf{R}$ be denoted as $\mathbf{r}_k$, $0 \le k \le M-1$. Then unitariness implies

$$\mathbf{r}_k^\dagger \mathbf{r}_m = 0, \quad k \ne m. \tag{14.11.1}$$

Moreover, all columns have same norm, that is,

$$\mathbf{r}_m^\dagger \mathbf{r}_m = d > 0, \quad 0 \le m \le M-1. \qquad (14.11.2)$$

These two properties completely summarize the features of unitary matrices. If we quantize the elements of the matrix $\mathbf{R}$ *directly* then it is not possible to satisfy unitariness. For example, if the elements of the column $\mathbf{r}_0$ are quantized independently of $\mathbf{r}_1$, the mutual orthogonality is not preserved.

Limitations of Planar Rotation Building Blocks

First consider real unitary matrices. From Sec. 14.6.1 we know that these can be completely characterized by $M(M-1)/2$ angles θ_k and a real scalar $c \ne 0$. For example, a 3×3 real unitary matrix $\mathbf{R}$ can be represented as in (14.6.4).

If we are interested only in storing the matrix in quantized form, we can store the angles θ_k. Then an approximate version of $\mathbf{R}$ can be recovered by computing (using higher precision) the cosines and sines of these angles and evaluating the above product, which will remain unitary to the extent that the effect of quantizations of sines and cosines can be neglected.

But if $\mathbf{R}$ is *implemented* as a cascade as in (14.6.4), then c_k and s_k are the *multipliers* in this structure, and their quantization will result in loss of unitariness. To see this note that if we multiply two matrices satisfying (14.11.1) and (14.11.2), the result also satisfies (14.11.1) and (14.11.2). But if we multiply two matrices satisfying only (14.11.1), the result does not satisfy even (14.11.1) (try it!). So the property (14.11.2) (equality of column-norms) is crucial for the building blocks. However the columns in the 3×3 matrix factors in (14.6.4) have unequal norms after the c_k and s_k are quantized because

$$\{Q[\cos \theta_k]\}^2 + \{Q[\sin \theta_k]\}^2 \ne 1, \qquad (14.11.3)$$

in general. The value of the left hand side above depends on θ_k, and on the details of quantization. [If we attempt to replace those entries in (14.6.4) which have value '1' by the norms of the quantized columns, this would require infinite precision because norms involve square-roots.] Because of this, after the three 3×3 matrix factors are multiplied, the result does not satisfy *either* (14.11.1) or (14.11.2).

A lucky exception to this difficulty is the $M=2$ case. In this case the real orthogonal matrix after quantization becomes

$$\begin{bmatrix} Q[\cos \theta] & Q[\sin \theta] \\ -Q[\sin \theta] & Q[\cos \theta] \end{bmatrix} \qquad (14.11.4)$$

and remains orthogonal. The two columns have same norm after quantization, even though this norm is not unity.

In the more general $M \times M$ case, the implementation of unitary matrices in terms of Given's rotations is such that unitariness is not preserved under coefficient quantization. For the same reason, structures for lossless transfer matrices based on Givens rotations (Fig. 14.10-5) do not preserve losslessness when the coefficients are quantized (except for the 2 × 1 and 2 × 2 cases).

Preserving Unitariness with Householder Building Blocks

Instead of using Givens rotations, suppose we implement $\mathbf{R}$ using the Householder factorization (14.6.16). Here the vectors $\mathbf{u}_k$ and the parameters d_{ii} (diagonal elements of $\mathbf{D}$) can be varied (say, quantized) completely independent of each other and yet the form (14.6.16) represents a unitary matrix as long as $\mathbf{u}_k$ have unit norm, and $|d_{ii}|$ are same for all i.

The unit norm requirement on $\mathbf{u}_k$ cannot be satisfied when the components are quantized to a given precision. To overcome this difficulty, we rewrite (14.6.16) as

$$\mathbf{R} = \widehat{\mathbf{G}}_1 \widehat{\mathbf{G}}_2 \ldots \widehat{\mathbf{G}}_{M-1} \widehat{\mathbf{D}} \qquad (14.11.5)$$

where

$$\widehat{\mathbf{G}}_k = \|\mathbf{u}_k\|^2 \mathbf{I} - 2\mathbf{u}_k \mathbf{u}_k^\dagger, \quad \widehat{\mathbf{D}} = \frac{\sqrt{d}}{\alpha} \mathbf{D}, \quad \alpha = \prod_{k=1}^{M-1} \|\mathbf{u}_k\|^2. \qquad (14.11.6)$$

In (14.11.6) we do not require $\mathbf{u}_k$ to have unit-norm any more. The matrix $\widehat{\mathbf{G}}_k$ is a scaled version of a unitary matrix and is unitary for arbitrary nonzero $\mathbf{u}_k$. So the product (14.11.5) is unitary even after quantization *as long as* the diagonal elements $\widehat{d}_{ii}$ of $\widehat{\mathbf{D}}$ have same magnitude after quantization. If this is not the case, then the product (14.11.5) does not satisfy (14.11.2) even though (14.11.1) holds. For the real case we have $|\widehat{d}_{ii}| = \sqrt{d}/\alpha$ for all i so that even this minor difficulty does not arise.

Summarizing, the form (14.11.5) permits us to generate every unitary matrix by independently varying the complex parameters $\mathbf{u}_k$, and the diagonal entries of $\widehat{\mathbf{D}}$. In particular if these are quantized independently, the property (14.11.1) still holds. (Notice that $\|\mathbf{u}_k\|^2$ in (14.11.6) requires higher precision, but it is still finite). For real $\mathbf{R}$, the unitary property [i.e., *both* (14.11.1) and (14.11.2)] is itself preserved. So (14.11.5) leads to a structurally unitary system.

14.11.2 Preserving Losslessness under Quantization

The simplest structure which we can discuss under this topic is the lattice structure of Figs. 14.3-2 and 14.3-3 for the real-coefficient FIR lossless case. Recall that each $\mathbf{R}_m$ is the Givens rotation (14.3.9a). When this is quantized as in (14.11.4), it remains unitary with equal norm for both columns. So the cascaded structure continues to be lossless, and the QMF bank of Fig. 6.4-1 preserves the PR property, as already seen in Sec. 6.4.1.

For the $M \times M$ case, rotation based structures do not preserve lossless property, as explained earlier. Consider however the diadic based FIR building block in Fig. 14.4-1. This represents a degree-one lossless system if and only if the column vector $\mathbf{v}_m$ has unit-norm. Now the unit-norm property is usually lost when the elements of $\mathbf{v}_m$ are quantized, and $\mathbf{V}_m(z)$ does not remain lossless. To overcome this difficulty, note that $\mathbf{V}_m(z)$ can be rewritten as

$$\mathbf{V}_m(z) = \mathbf{I} - \frac{\mathbf{v}_m \mathbf{v}_m^\dagger}{\|\mathbf{v}_m\|^2} + z^{-1} \frac{\mathbf{v}_m \mathbf{v}_m^\dagger}{\|\mathbf{v}_m\|^2}, \qquad (14.11.7)$$

where $\mathbf{v}_m$ is an arbitrary nonzero vector. Define $\mathbf{U}_m(z) = \|\mathbf{v}_m\|^2 \mathbf{V}_m(z)$, that is,

$$\mathbf{U}_m(z) = \|\mathbf{v}_m\|^2 \mathbf{I} - \mathbf{v}_m \mathbf{v}_m^\dagger + z^{-1} \mathbf{v}_m \mathbf{v}_m^\dagger, \qquad (14.11.8)$$

so that

$$\widetilde{\mathbf{U}}_m(z) \mathbf{U}_m(z) = \|\mathbf{v}_m\|^4 \mathbf{I}. \qquad (14.11.9)$$

Evidently $\mathbf{U}_m(z)$ is lossless regardless of the value of the (nonzero) vector $\mathbf{v}_m$. The unit-norm constraint on $\mathbf{v}_m$ has now been dropped so that its components can be quantized to any precision without violating losslessness of $\mathbf{U}_m(z)$. Of course, the quantities $\|\mathbf{v}_m\|^2$ and $\mathbf{v}_m \mathbf{v}_m^\dagger$ require higher precisions, but these are finite precisions.

Now rewrite the cascade (14.4.10) as

$$\mathbf{H}_N(z) = \mathbf{U}_N(z) \mathbf{U}_{N-1}(z) \ldots \mathbf{U}_1(z) \mathbf{H}_0. \qquad (14.11.10)$$

If we implement the unitary matrix $\mathbf{H}_0$ using Householder building blocks as explained above, then $\widetilde{\mathbf{H}}_0 \mathbf{H}_0$ remains a constant diagonal matrix, inspite of coefficient quantization. Consequently $\widetilde{\mathbf{H}}_N(z) \mathbf{H}_N(z)$ is a constant diagonal matrix. This is as good as saying $\widetilde{\mathbf{H}}_N(z) \mathbf{H}_N(z) = c^2 \mathbf{I}$, for most applications.

IIR Lossless Systems

The above discussions can be generalized for the IIR lossless structures developed in Sec. 14.9. We will now recall a particularly simple system studied in Sec. 5.3, viz., IIR power symmetric QMF banks (Fig. 5.2-5). Here the polyphase matrix of the analysis bank is

$$\mathbf{E}(z) = 0.5 \begin{bmatrix} 1 & 1 \\ 1 & -1 \end{bmatrix} \begin{bmatrix} a_0(z) & 0 \\ 0 & a_1(z) \end{bmatrix} \qquad (14.11.11)$$

where $a_0(z)$ and $a_1(z)$ are unit-magnitude allpass. So $\mathbf{E}(z)$ is paraunitary. The allpass filters $a_i(z)$ can be written as products of first order filters (5.3.25) with *real* coefficients $\alpha_{j,i}$. These filters remain unit-magnitude allpass in spite of quantization of $\alpha_{j,i}$. So $\mathbf{E}(z)$ remains lossless. As a result, the system of Fig. 5.2-5 is free from aliasing and amplitude distortion in spite of multiplier quantization.

14.12 SUMMARY AND TABLES

The main points of the chapter are summarized in Tables 14.12.1, 14.12.2 and 14.12.3. References on lossless systems were cited in Sec. 14.1, and a list can be found in the Bibliography at the end of the text.

TABLE 14.12.1 Losslessness at a glance.

In what follows, $\mathbf{H}(z)$ is a $p \times r$ causal transfer matrix:
$\mathbf{H}(z) = \sum_{n=0}^{\infty} \mathbf{h}(n) z^{-n}$ (IIR); $\quad \mathbf{H}(z) = \sum_{n=0}^{L} \mathbf{h}(n) z^{-n}, \mathbf{h}(L) \neq \mathbf{0}$ (FIR).

Definitions.

1. *Paraunitary property.* $\mathbf{H}(z)$ is paraunitary if $\widetilde{\mathbf{H}}(z)\mathbf{H}(z) = c^2 \mathbf{I}_r, \forall z$, where $c^2 > 0$. (Note: this requires $p \geq r$.) In particular this means $\mathbf{H}^\dagger(e^{j\omega}) \mathbf{H}(e^{j\omega}) = c^2 \mathbf{I}_r, \forall \omega$.
2. *Losslessness.* $\mathbf{H}(z)$ is lossless if it is stable and paraunitary.
3. *LBR* (Lossless Bounded Real) property: implies lossless with real coefficients.
4. *Normalized-lossless.* Implies lossless with $c^2 = 1$, that is $\widetilde{\mathbf{H}}(z)\mathbf{H}(z) = \mathbf{I}_r$.
5. *Unitariness.* $\mathbf{R}$ is unitary if $\mathbf{R}^\dagger \mathbf{R} = d\mathbf{I}, d > 0$ (*normalized* if $d = 1$).

Basic properties of lossless systems.

1. Output energy = $c^2 \times$ Input energy, i.e.,
$$\sum_n \mathbf{y}^\dagger(n)\mathbf{y}(n) = c^2 \sum_n \mathbf{u}^\dagger(n)\mathbf{u}(n), \quad \forall \; \mathbf{u}(n).$$
2. $\sum_{n=0}^{\infty} \mathbf{h}^\dagger(n) \mathbf{h}(n+k) = c^2 \mathbf{I}_r \delta(k)$.
3. $\mathbf{h}^\dagger(0) \mathbf{h}(L) = \mathbf{0}$ (FIR case), assuming $L > 0$.
4. For $p = r$ case, $[\det \mathbf{H}(z)] = A(z) = allpass\; (= az^{-\mu}, \mu \geq 0$, for FIR case).
5. Let $p = r$. The degree of $\mathbf{H}(z)$ (i.e., the minimum number of delays required to implement $\mathbf{H}(z)$) is equal to the degree of determinant, i.e.,
$$[\deg \mathbf{H}(z)] = [\deg [\det \mathbf{H}(z)]].$$

This follows from Section 14.4.2 for FIR case, and is proved in Section 14.7 for the general case.

State-space manifestation.

Recall state space equations: $\begin{cases} \mathbf{x}(n+1) = \mathbf{Ax}(n) + \mathbf{Bu}(n) \\ \mathbf{y}(n) = \mathbf{Cx}(n) + \mathbf{Du}(n) \end{cases}$

1. $\mathbf{H}(z)$ normalized lossless $\Rightarrow$ there exists minimal realization for which the matrix $\mathbf{R} \triangleq \begin{bmatrix} \mathbf{A} & \mathbf{B} \\ \mathbf{C} & \mathbf{D} \end{bmatrix}$ (*realization matrix*) is normalized unitary.
2. Let $(\mathbf{A}, \mathbf{B}, \mathbf{C}, \mathbf{D})$ describe some structure of $\mathbf{H}(z)$. Let $\mathbf{A}$ be stable and $\mathbf{R}$ defined above be normalized unitary. Then (i) the structure is minimal if $p = r$, and (ii) $\mathbf{H}(z)$ is normalized-lossless (even if $p \neq r$).

For summary of further properties see Table 14.12.3.

TABLE 14.12.2 Lossless systems: factorizations and structures.

In what follows $\mathbf{H}_N(z)$ denotes a causal lossless system.

Basic building blocks.
1. $\mathbf{V}_m(z) = \mathbf{I} - \mathbf{v}_m\mathbf{v}_m^\dagger + z^{-1}\mathbf{v}_m\mathbf{v}_m^\dagger$, where $\mathbf{v}_m$ is a unit norm vector. Fig. 14.4-1 shows the structure. $\mathbf{V}_m(z)$ is degree-one lossless, and $[\det \mathbf{V}_m(z)] = z^{-1}$.
2. Householder matrix $\mathbf{H}_k = \mathbf{I} - 2\mathbf{u}_k\mathbf{u}_k^\dagger$, where $\mathbf{u}_k$ is a unit norm vector. Fig. 14.6-4(b) shows the structure.

Structures for FIR lossless system $\mathbf{H}_N(z)$.
1. *Rotation based.* If $\mathbf{H}_N(z)$ is 2×2 FIR with real coefficients, it can be implemented using planar rotations as in Fig. 14.3-2. Corresponding 2×1 lossless structure is shown in Fig. 14.3-3. Generalizations of rotation-based structures for $M \times M$ (and $M \times 1$) real coefficient case is given in Section 14.10 (Fig. 14.10-5).
2. *Diadic-based.* If $\mathbf{H}_N(z)$ is $M \times M$ or $M \times 1$ degree-N FIR, it can be factorized into the form
$$\mathbf{H}_N(z) = \mathbf{V}_N(z)\mathbf{V}_{N-1}(z)\ldots\mathbf{V}_1(z)\mathbf{H}_0$$
where $\mathbf{H}_0$ is $M \times M$ unitary or $M \times 1$ nonzero as the case may be. If coefficients of $\mathbf{H}_N(z)$ are real, the vectors $\mathbf{v}_m$ in $\mathbf{V}_m(z)$ and the matrix $\mathbf{H}_0$ are real.
3. *Minimality.* All structures mentioned above are minimal in delays.

Structures for unitary matrices.
1. *Rotation based.* Any real $M \times M$ matrix $\mathbf{R}$ with $\mathbf{R}^T\mathbf{R} = \mathbf{I}$ can be factorized into the form shown in Fig. 14.6-3(a), involving $M(M-1)/2$ planar rotations.
2. *Diadic based.* Any $M \times M$ matrix $\mathbf{R}$ with $\mathbf{R}^\dagger\mathbf{R} = \mathbf{I}$ can be factorized into the form
$$\mathbf{R} = \mathbf{H}_1\mathbf{H}_2\ldots\mathbf{H}_{M-1}\mathbf{D} \qquad \text{(Householder factorization)},$$
where $\mathbf{H}_k$ are Householder matrices, and $\mathbf{D}$ is diagonal with $|d_{ii}| = 1$. If $\mathbf{R}$ is real, then $\mathbf{H}_k$ and $\mathbf{D}$ are real.

Variations. Variations of the above structures are given in Section 14.10, along with structures for the IIR case. See Table 14.12.3 for summary.

TABLE 14.12.3 Lossless systems: further properties.

In what follows, $\mathbf{H}(z)$ is a causal transfer matrix:
$\mathbf{H}(z) = \sum_{n=0}^{\infty} \mathbf{h}(n)z^{-n}$ (IIR); $\quad \mathbf{H}(z) = \sum_{n=0}^{L} \mathbf{h}(n)z^{-n}, \mathbf{h}(L) \neq 0$ (FIR).

Smith McMillan form

Let $\mathbf{H}(z)$ be $M \times M$ with Smith-McMillan form $\mathbf{\Lambda}(z)$. The diagonal elements of $\mathbf{\Lambda}(z)$ are of the form $\lambda_i(z) = \alpha_i(z)/\beta_i(z)$, where $\alpha_i(z)$ and $\beta_i(z)$ are relatively prime polynomials in z. Assume further that $\mathbf{H}(z)$ is lossless. Then the following are true.

1. When $\mathbf{H}(z)$ is FIR, $\lambda_i = z^{-n_i}$, where n_i are integers with $n_0 \geq n_1 \geq \ldots \geq n_{M-1} \geq 0$.
2. In general if p is a pole of $\mathbf{H}(z)$ then $1/p^*$ is a zero, and vice versa. So all zeros are outside unit circle. The polynomials $\alpha_i(z)$ and $\beta_j(z)$ are relatively prime for all pairs of i, j.
3. The degree of $\mathbf{H}(z)$ is the same as the degree of $[\det \mathbf{H}(z)]$. Every pole of $\mathbf{H}(z)$ is a pole of $[\det \mathbf{H}(z)]$. The same holds for zeros.

Modulus property

Let $\mathbf{H}(z)$ be $M \times M$ lossless with $\widetilde{\mathbf{H}}(z)\mathbf{H}(z) = \mathbf{I}$. Then $\mathbf{H}^\dagger(z)\mathbf{H}(z) \leq \mathbf{I}$ for $|z| > 1$. And $\mathbf{H}^\dagger(z)\mathbf{H}(z) \geq \mathbf{I}$ for $|z| < 1$.

Structures.

1. Let $\mathbf{H}(z)$ be $M \times M$ lossless, with degree N. Then it can be factorized as

$$\mathbf{H}(z) = \mathbf{V}_N(z)\mathbf{V}_{N-1}(z)\ldots\mathbf{V}_1(z)\mathbf{H}_0$$
$$= \mathbf{W}_N\mathbf{\Phi}_N(z)\mathbf{W}_{N-1}\mathbf{\Phi}_{N-1}(z)\ldots\mathbf{W}_1\mathbf{\Phi}_1(z)\mathbf{G}_0$$

where (i) $\mathbf{V}_m(z)$ is the degree-one building block (14.9.1), (ii) $\mathbf{W}_m$ is the Householder unitary matrix (14.10.8), (iii) $\mathbf{\Phi}_m(z)$ are diagonal matrices as in (14.10.1), and (iv) $\mathbf{H}_0$ and $\mathbf{G}_0$ are unitary matrices. In the real coefficient FIR case, $\mathbf{H}(z)$ can further be written as

$$\mathbf{H}(z) = \mathbf{\Theta}_N\mathbf{\Phi}_N(z)\mathbf{\Theta}_{N-1}\mathbf{\Phi}_{N-1}(z)\ldots\mathbf{\Theta}_1\mathbf{\Phi}_1(z)\mathbf{E}_0$$

where each $\mathbf{\Theta}_m$ is a sequence of $M-1$ planar rotations (Fig. 14.10-5(b)).

2. Let $\mathbf{H}(z)$ be $M \times 1$ lossless. Then it can be factorized as

$$\mathbf{H}(z) = \mathbf{V}_N(z)\mathbf{V}_{N-1}(z)\ldots\mathbf{V}_1(z)\mathbf{G}(z),$$

where $\mathbf{V}_m(z)$ are as in (14.9.1) and $\mathbf{G}(z)$ is causal FIR $M \times 1$ lossless.

PROBLEMS

14.1. Let $H(z)$ be a rational transfer function satisfying (14.2.1a). Prove then that (14.2.1b) holds.

14.2. Show that any $p \times L$ submatrix of a $p \times r$ lossless matrix is lossless.

14.3. Let $\mathbf{H}(z) = \sum_{n=0}^{\infty} \mathbf{h}(n) z^{-n}$, $\mathbf{h}(0) \neq \mathbf{0}$, be $M \times M$ lossless with degree > 0. For FIR $\mathbf{H}(z)$ we know that $\mathbf{h}(0)$ is singular. For IIR $\mathbf{H}(z)$, $\mathbf{h}(0)$ may or may not be singular. Give examples of both types.

14.4. Consider the following structure where $\mathbf{T}(z)$ is normalized lossless.

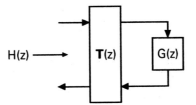

Figure P14-4

Let $G(z)$ be stable with $|G(e^{j\omega})| < 1$ for all ω. Prove that (i) $H(z)$ is stable and (ii) $|H(e^{j\omega})| \leq 1$ for all ω.

14.5. Consider the following structure where $|\alpha| < 1$.

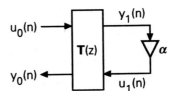

Figure P14-5

Define $H(z) \triangleq Y_1(z)/U_0(z)$. What is the maximum possible value of $|H(e^{j\omega})|$ subject to the constraint that $\mathbf{T}(z)$ is normalized lossless?

14.6. Shown below is a feedback-cascade of 2×2 transfer matrices. The cascade has N building blocks, and is terminated at the right with a multiplier G_0. An example of this type was presented in Chap. 3, viz., the lattice structures for allpass functions (e.g., see Fig. 3.4-8).

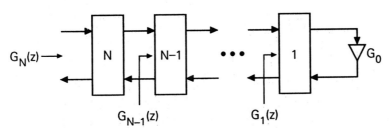

Figure P14-6

We assume that the cascaded feedback connection does not generate delay free loops. This assumption is valid for the lattice structures. Assume that each 2×2 building block above is normalized-lossless, and that the transfer function $G_N(z)$ is stable. Show that $G_N(z)$ is allpass when $|G_0| = 1$ and that $|G_N(e^{j\omega})| \leq 1$ when $|G_0| < 1$.

14.7. Consider the 2×2 lossless system

$$\mathbf{H}(z) = \begin{bmatrix} 1 + 2z^{-1} - z^{-2} & 1 + z^{-2} \\ 1 + z^{-2} & 1 - 2z^{-1} - z^{-2} \end{bmatrix}. \qquad (P14.7)$$

Clearly its degree ≥ 2.

a) Verify that the determinant of $\mathbf{H}(z)$ is αz^{-2} for some $\alpha \neq 0$ so that $[\deg \mathbf{H}(z)] = 2$ exactly.
b) Obtain the cascaded structure based on planar rotations (Sec. 14.3.1).
c) Obtain the diadic-based cascaded structure (Sec. 14.4.2).

14.8. Consider the lossless system

$$\mathbf{P}_2(z) = \begin{bmatrix} 1 \\ z^{-1} \\ z^{-2} \end{bmatrix}. \qquad (P14.8)$$

a) Find the vectors $\mathbf{v}_m$ and $\mathbf{P}_0$ in the diadic based factorization given by (14.4.12).
b) Suppose we are given a $M \times 1$ causal FIR lossless system $\mathbf{P}_N(z)$. We know this can be factorized as in (14.4.12). Suppose we construct a unitary matrix $\mathbf{H}_0$ whose 0th column is $\mathbf{P}_0$. Consider now the product (14.4.10). This is evidently lossless, with 0th column equal to $\mathbf{P}_N(z)$. So we have *embedded* the $M \times 1$ lossless system $\mathbf{P}_N(z)$ into an $M \times M$ lossless system $\mathbf{H}_N(z)$. As a demonstration of this, let $\mathbf{P}_N(z)$ be as in (P14.8). Find a 3×3 lossless matrix $\mathbf{H}_N(z)$ with 0th column equal to $\mathbf{P}_N(z)$.

14.9. Let $\mathbf{H}(z) = \begin{bmatrix} z^{-1} & 0 \\ 0 & z^{-1} \end{bmatrix}$. This is evidently lossless with degree two. Formally perform the degree reduction steps to arrive at the cascaded structure in terms of planar rotations (Sec. 14.3.1).

14.10. Show that any 2×2 causal lossless system (not necessarily FIR) has the form

$$\begin{bmatrix} H_0(z) & F(z)\widetilde{H}_1(z) \\ H_1(z) & -F(z)\widetilde{H}_0(z) \end{bmatrix}, \qquad (P14.10)$$

where (i) the 0th column is lossless, and (ii) $\widetilde{F}(z)F(z) = 1$. Note that (14.3.15) is a special case of this.

14.11. Let $\mathbf{P}(z)$ be $M \times 1$ causal FIR lossless, factorized as $\mathbf{P}(z) = \mathbf{S}^{(p)}(z)\mathbf{P}_0$ where $\mathbf{S}^{(p)}(z)$ is a product of finite number of building blocks of the form (14.4.1). We know that if the factorization is minimal, then $\mathbf{S}^{(p)}(z)$ is unique (Sec. 14.4.4). However if the degree of $\mathbf{S}^{(p)}(z)$ exceeds that of $\mathbf{P}(z)$ (nonminimal factorization) then there exist infinite number of matrices $\mathbf{S}^{(p)}(z)$ satisfying

$\mathbf{P}(z) = \mathbf{S}^{(p)}(z)\mathbf{P}_0$. Prove this! [*Hint.* $(\mathbf{I} - \mathbf{vv}^\dagger + z^{-1}\mathbf{vv}^\dagger)\mathbf{P}_0 = \mathbf{P}_0$ for *any* $\mathbf{v}$ orthogonal to $\mathbf{P}_0$.]

14.12. Show that any 2×2 real unitary matrix has the form (14.3.2).

14.13. Consider the lossless structure of Fig. 14.3-2 based on planar rotations. Assume $N = 1$ and $\alpha = 1$. Find the matrices $(\mathbf{A}, \mathbf{B}, \mathbf{C}, \mathbf{D})$ which occur in the state space description. Verify that the realization matrix (14.5.1) satisfies $\mathbf{R}^\dagger \mathbf{R} = \mathbf{I}$.

14.14. As a generalization of Problem 14.13, consider the following cascaded structure

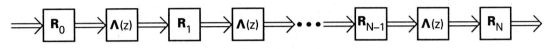

Figure P14-14

where $\mathbf{R}_m$ are $M \times M$ with $\mathbf{R}_m^\dagger \mathbf{R}_m = \mathbf{I}$, and

$$\boldsymbol{\Lambda}(z) = \begin{bmatrix} \mathbf{I}_{M-1} & 0 \\ 0 & z^{-1} \end{bmatrix}. \qquad (P14.14)$$

With $(\mathbf{A}, \mathbf{B}, \mathbf{C}, \mathbf{D})$ denoting the state space description, show that the realization matrix $\mathbf{R}$ defined in (14.5.1) satisfies $\mathbf{R}^\dagger \mathbf{R} = \mathbf{I}$.

14.15. Give example of a structure for an $M \times 1$ causal lossless system such that (i) the realization matrix defined in (14.5.1) satisfies $\mathbf{R}^\dagger \mathbf{R} = \mathbf{I}$, and (ii) the structure is *not* minimal. This demonstrates that the result of Sec. 14.5.2 does not hold hold unless $\mathbf{H}(z)$ is $M \times M$.

14.16. Suppose $(\mathbf{A}, \mathbf{B}, \mathbf{C}, \mathbf{D})$ is the state-space description of a minimal structure for some system. Let the $\mathbf{R}$-matrix (14.5.1) be such that $\mathbf{R}^\dagger \mathbf{R} = \mathbf{I}$. Prove that $\mathbf{A}$ is stable.

14.17. Let $\mathbf{R}$ be $M \times M$ so that it has M^2 entries, that is, $2M^2$ real degrees of freedom in absence of further constraints. Suppose we impose the constraint $\mathbf{R}^\dagger \mathbf{R} = \mathbf{I}$. So the columns $\mathbf{R}_k$ satisfy (i) $\mathbf{R}_k^\dagger \mathbf{R}_m = 0$ for $k \neq m$ and (ii) $\|\mathbf{R}_k\| = 1$ for all k. Since the M^2 complex entries are constrained by these two conditions, the number of freedoms in choosing these M^2 elements is less that $2M^2$. Show by direct counting (i.e., without using any factorization result) that the number of freedoms is M^2.

14.18. Let $\mathbf{H}(z)$ be a stable rational transfer function such that (14.2.3) holds for every input sequence $\mathbf{u}(n)$. Prove that $\mathbf{H}(z)$ is lossless. (*Hint.* Use Sec. 14.5.)

14.19. Consider the unitary matrix

$$\begin{bmatrix} 1 & 1 \\ -1 & 1 \end{bmatrix}. \qquad (P14.19)$$

Identify the vectors $\mathbf{u}_k$, and the quantities d and $\mathbf{D}$ in the Householder factorization (14.6.16).

14.20. Repeat Problem 14.19 for the unitary matrix

$$\mathbf{R} = \begin{bmatrix} 1 & 1 & 0 \\ -1 & 1 & 0 \\ 0 & 0 & \sqrt{2} \end{bmatrix} \qquad (P14.20)$$

14.21. Consider the two-input two-output structure shown below, where $0 < \sigma < 1$.

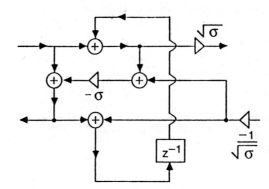

Figure P14-21

a) Show that the transfer matrix is given by

$$\mathbf{T}(z) = \frac{\begin{bmatrix} 1-\sigma & \sqrt{\sigma}(1+z^{-1}) \\ \sqrt{\sigma}(1+z^{-1}) & -(1-\sigma)z^{-1} \end{bmatrix}}{1+\sigma z^{-1}}, \qquad (P14.21)$$

and that it is lossless.

b) Find the Smith-McMillan form $\mathbf{\Lambda}(z)$. Also find appropriate unimodular $\mathbf{W}(z)$ and $\mathbf{V}(z)$ such that $\mathbf{T}(z) = \mathbf{W}(z)\mathbf{\Lambda}(z)\mathbf{V}(z)$.

14.22. We know that the IIR lossless example in Problem 14.21 can be expressed as $\mathbf{V}_m(z)\mathbf{H}_0$ where $\mathbf{V}_m(z)$ is as in (14.9.1). Identify a_m, $\mathbf{H}_0$, and the unit norm vector $\mathbf{v}_m$.

14.23. Let $\mathbf{H}_m(z) = \sum_{n=0}^{L} \mathbf{h}(n) z^{-n}$ with $\mathbf{h}(0) \neq 0$ (i.e., $\mathbf{H}_m(\infty) \neq 0$) be an $M \times M$ lossless system with degree $= m$.

a) We wish to find an $M \times 1$ vector $\mathbf{w}_m$ with $\|\mathbf{w}_m\| = 1$ such that

$$\mathbf{H}_m(z) = [\mathbf{I} - 2\mathbf{w}_m \mathbf{w}_m^\dagger] \begin{bmatrix} \mathbf{I} & 0 \\ 0 & z^{-1} \end{bmatrix} \mathbf{H}_{m-1}(z), \qquad (P14.23)$$

where $\mathbf{H}_{m-1}(z)$ is causal FIR lossless with $\mathbf{H}_{m-1}(\infty) \neq 0$ and with degree $m-1$. Show that there exists a $\mathbf{w}_m$ satisfying all these requirements. Explain how this $\mathbf{w}_m$ can be found.

b) Using this result show that any $M \times M$ causal FIR lossless system with degree N can be factorized as in Fig. 14.10-4 where (i) $\mathbf{\Phi}_m(z) = \begin{bmatrix} \mathbf{I} & 0 \\ 0 & z^{-1} \end{bmatrix}$, (ii) $\mathbf{W}_m = \mathbf{I} - 2\mathbf{w}_m \mathbf{w}_m^\dagger$ with $\|\mathbf{w}_m\| = 1$, and (iii) $\mathbf{G}_0$ is $M \times M$ unitary.

14.24. Consider the lossless system

$$\mathbf{H}(z) = \begin{bmatrix} 1 + z^{-1} & 1 - z^{-1} \\ 1 - z^{-1} & 1 + z^{-1} \end{bmatrix}. \qquad (P14.24a)$$

a) Verify that $[\det \mathbf{H}(z)] = \alpha z^{-1}$ for some $\alpha \neq 0$, so that $\mathbf{H}(z)$ has degree one.

b) We know from Sec. 14.10 that $\mathbf{H}(z)$ can be factorized as

$$\mathbf{H}(z) = \mathbf{W}_1 \begin{bmatrix} 1 & 0 \\ 0 & z^{-1} \end{bmatrix} \mathbf{G}_0 \qquad (P14.24b)$$

where $\mathbf{G}_0$ is unitary and $\mathbf{W}_1 = \mathbf{I} - 2\mathbf{w}_1\mathbf{w}_1^\dagger$, with $\|\mathbf{w}_1\| = 1$. Find $\mathbf{w}_1$ and $\mathbf{G}_0$.

c) We know from Sec. 14.4 that $\mathbf{H}(z)$ can also be factorized as $\mathbf{H}(z) = \mathbf{V}_1(z)\mathbf{H}_0$ where $\mathbf{V}_1(z)$ is as in (14.4.1) with $\|\mathbf{v}_1\| = 1$, and $\mathbf{H}_0$ is unitary. Find $\mathbf{v}_1$ and $\mathbf{H}_0$.

14.25. Let $\mathbf{P}$ be $M \times M$ Hermitian positive definite, and let λ_i be its eigenvalues. We know that $\lambda_i > 0$. Prove the following:

a) $\mathbf{P} \leq \mathbf{I} \iff \lambda_i \leq 1, \forall i$.
b) $\mathbf{P} \leq \mathbf{I} \iff \mathbf{P}^{-1} \geq \mathbf{I}$.
c) $\mathbf{P} \leq \mathbf{I} \iff \mathbf{P}^T \leq \mathbf{I}$.
d) Prove that (14.8.8) implies $\mathbf{H}^\dagger(z)\mathbf{H}(z) \geq \mathbf{I}$ for $|z| < 1$.
e) Now let $\mathbf{k}$ be a column vector with $\mathbf{k}^\dagger \mathbf{k} < 1$. Show that $\mathbf{k}\mathbf{k}^\dagger < \mathbf{I}$, that is, $\mathbf{I} - \mathbf{k}\mathbf{k}^\dagger > 0$.

Note that since $\mathbf{P}$ is Hermitian, $\mathbf{P}^T = \mathbf{P}^*$ so that (c) also holds with $\mathbf{P}^T$ replaced by $\mathbf{P}^*$.

14.26. Let $\mathbf{H}(z)$ be a causal stable transfer function. Let the input (14.8.3) ($|a| > 1$) produce an output which satisfies $\mathbf{y}(n) = \mathbf{0}$ for $n > L$. Prove then that the impulse response $\mathbf{h}(n)$ of the system $\mathbf{H}(z)$ satisfies $\mathbf{h}(n)\mathbf{v} = \mathbf{0}$ for $n > 0$.

14.27. We now consider the state space description $(\mathbf{A}, \mathbf{B}, \mathbf{C}, \mathbf{D})$, and the matrix

$$\mathbf{R} \triangleq \begin{bmatrix} \mathbf{A} & \mathbf{B} \\ \mathbf{C} & \mathbf{D} \end{bmatrix} \qquad (P14.27)$$

for some IIR lossless structures.

a) Write down $\mathbf{R}$ for the example in Problem 14.21. Is $\mathbf{R}$ unitary?
b) Consider Fig. 14.9-1. Assume $\|\mathbf{v}_m\| = 1$ and $|a_m| < 1$. Write down the state space matrices $(\mathbf{A}, \mathbf{B}, \mathbf{C}, \mathbf{D})$ and hence $\mathbf{R}$. Is $\mathbf{R}$ unitary?

14.28. Consider the structure shown below.

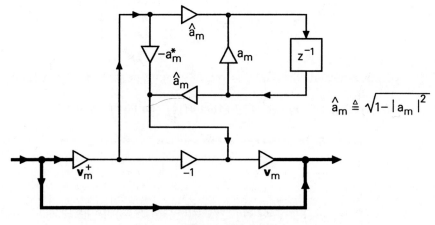

Figure P14-28

This is the *normalized* IIR lossless building block.

a) Write down the realization matrix $\mathbf{R}$ and *show* that $\mathbf{R}^\dagger \mathbf{R} = \mathbf{I}$.

b) Suppose we use this building block in place of $\mathbf{V}_m(z)$ in Fig. 14.10-1. Assume $\mathbf{H}_0$ is $p \times r$ with $\mathbf{H}_0^\dagger \mathbf{H}_0 = \mathbf{I}$. Show that the realization matrix for the entire system satisfies $\mathbf{R}^\dagger \mathbf{R} = \mathbf{I}$.

14.29. Consider Problem 14.6 again. In particular let each 2×2 building block be as in Problem 14.21. There are N building blocks, and each is characterized by a parameter σ_m. Each building block requires 3 multipliers. Now suppose we replace each building block with the following modified version.

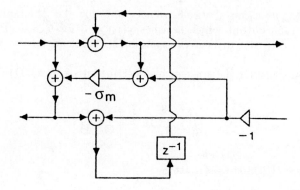

Figure P14-29

This requires only one multiplier (and four adders). Show that its transfer matrix is given by

$$\mathbf{T}_m(z) = \frac{\begin{bmatrix} 1-\sigma_m & \sigma_m(1+z^{-1}) \\ 1+z^{-1} & -(1-\sigma_m)z^{-1} \end{bmatrix}}{1+\sigma_m z^{-1}}, \qquad (P14.29)$$

and that $G_N(z)$ is unaffected (as long as the values of σ_m and G_0 are unchanged). This structure requires only $N+1$ multipliers. Evidently the inequality $|G_N(e^{j\omega})| \le 1$ continues to be true as long as $0 \le \sigma_m < 1$ and $|G_0| < 1$. In other words, we have a *strcturally bounded* implementation, which therefore enjoys low passband sensitivity (Sec. 9.7).

Note: Low sensitivity structures of the above form can be used to implement a restricted class of filters for example, Butterworth filters. More general filters (such as elliptic) can be obtained by introduction of degree-two lossless building blocks; see Swamy and Thyagarajan [1975] for details. These are closely related to wave digital filters, as elaborated in Vaidyanathan [1985a]. They are, however, of restricted use because of the lack of pipelineability in VLSI implementations; see Kung, et al. [1985].

14.30. We now give another proof of the matrix version of the maximum modulus theorem (suggested to the author by Prof. John Doyle at Caltech). This is perhaps more direct, and works for any closed contour, not just the unit circle.

a) Let $\mathbf{A}$ be any $M \times M$ matrix. Let μ be the maximum magnitude of the quantity $\mathbf{x}^\dagger \mathbf{A} \mathbf{y}$, as $\mathbf{x}$ and $\mathbf{y}$ are varied over all unit-norm vectors. Let λ denote the maximum eigenvalue of $\mathbf{A}^\dagger \mathbf{A}$. Show that $\mu^2 = \lambda$.

b) Let $\mathbf{H}(z)$ be an $M \times M$ matrix function of the complex variable z. Assume that each entry $H_{km}(z)$ is analytic on and inside a closed contour $\mathcal{C}$, and let $\mathbf{H}^\dagger(z)\mathbf{H}(z) \le \mathbf{I}$ on the contour. Show then that $\mathbf{H}^\dagger(z)\mathbf{H}(z) \le \mathbf{I}$ everywhere inside the contour. (You can use the scalar version of the maximum modulus theorem, Sec. 3.4.1).

Appendix A

Review of Matrices

A.0 INTRODUCTION

There are several excellent references on matrices, e.g., MacDuffee [1946], Gantmacher [1959], Bellman [1960], Franklin [1968], Halmos [1974], Horn and Johnson [1985], and Golub and Van Loan [1989], to name a few. Our aim here is to review those results from matrix theory which have direct relevance to this text. Most proofs can be found, or deduced from, the above texts. The material here is somewhat dense, as it is primarily meant to be a reference. Most of the deeper results mentioned here are required only in Chap. 13 and 14.

A.1 DEFINITIONS AND EXAMPLES

A $p \times r$ matrix $\mathbf{A}$ is a collection of pr elements (real or complex numbers) arranged in p rows and r columns. Thus

$$\mathbf{A} = \begin{bmatrix} A_{00} & A_{01} & \cdots & A_{0,r-1} \\ A_{10} & A_{11} & \cdots & A_{1,r-1} \\ \vdots & \vdots & \ddots & \vdots \\ A_{p-1,0} & A_{p-1,1} & \cdots & A_{p-1,r-1} \end{bmatrix}. \tag{A.1.1}$$

One often writes $\mathbf{A} = [A_{ij}]$ (or with a comma, as in $[A_{i,j}]$), and $A_{ij} = [\mathbf{A}]_{ij}$. So i is the row index and j the column index, both starting at zero. A $p \times 1$ matrix is said to be a p-vector or *column vector* (or just a vector). A $1 \times r$ matrix is called a *row vector*. Thus,

$$\underbrace{\begin{bmatrix} 1+j & -j \\ 2 & 3 \end{bmatrix}}_{2 \times 2\ matrix}, \quad \underbrace{\begin{bmatrix} 1 & 2 \end{bmatrix}}_{row\ vector}, \quad \underbrace{\begin{bmatrix} 1 \\ 2 \end{bmatrix}}_{column\ vector}. \tag{A.1.2}$$

To save space, one often writes a column vector as the 'transpose' of a row vector. Thus $[1 \quad 2j]^T$ stands for the column vector

$$\begin{bmatrix} 1 \\ 2j \end{bmatrix}. \tag{A.1.3}$$

(See below for definition of transpose). The elements of a vector **v** are commonly denoted as v_i or $v(i)$. Whether **v** is a row or a column is usually clear from the context.

A square matrix is a matrix with $p = r$. Thus the first matrix in (A.1.2) is square. If a matrix is not square, it is said to be rectangular. A 1×1 matrix (i.e., just a single element) is said to be a scalar. A $p \times r$ null matrix, denoted **0**, has all elements equal to zero. If $\mathbf{A} = \mathbf{0}$, **A** is said to be zero or null.

If two matrices have the same number of rows p and same number of columns r, they have the same size.

Diagonal matrices. The elements A_{ii} of a matrix **A** are called its diagonal elements. A matrix for which all elements are zero except possibly the diagonal elements is called a diagonal matrix. Examples are

$$\begin{bmatrix} 2 & 0 \\ 0 & 1 \end{bmatrix}, \begin{bmatrix} 1 & 0 & 0 \\ 0 & j & 0 \end{bmatrix}. \qquad (A.1.4)$$

The set of diagonal elements is sometimes called the main diagonal. Note that a diagonal matrix need not be square. A square diagonal matrix with all diagonal elements equal to unity is said to be the *identity matrix*, denoted as **I**. Examples are

$$\begin{bmatrix} 1 & 0 & 0 \\ 0 & 1 & 0 \\ 0 & 0 & 1 \end{bmatrix}, \begin{bmatrix} 1 & 0 \\ 0 & 1 \end{bmatrix}, \qquad (A.1.5)$$

which are 3×3 and 2×2 respectively. If the size of **I** is not clear from the context, a subscript will be used. The above examples represent, respectively, $\mathbf{I}_3$ and $\mathbf{I}_2$.

The notation

$$\text{diag}\,[d_0 \quad d_1 \quad \ldots \quad d_{N-1}]$$

stands for a $N \times N$ diagonal matrix **A** with diagonal elements $A_{ii} = d_i$.

Triangular matrices. A lower triangular matrix is one for which the elements above the main diagonal are equal to zero. An upper triangular matrix is one for which the elements below the main diagonal are equal to zero. Examples are:

$$\underbrace{\begin{bmatrix} 1 & 0 & 0 \\ 2 & 3 & 0 \\ 5 & 1 & 2 \\ 1 & 4 & 1 \end{bmatrix}}_{lower\ triangular}, \underbrace{\begin{bmatrix} 1 & 2 & 4 \\ 0 & 3 & 1 \\ 0 & 0 & 1 \end{bmatrix}}_{upper\ triangular}. \qquad (A.1.6)$$

A.2 BASIC OPERATIONS

A number of operations, including arithmetic, can be performed with matrices.

Transpose and transpose-conjugate. Given $\mathbf{A} = [A_{ij}]$, we denote its transpose as $\mathbf{A}^T$. It is defined by $[\mathbf{A}^T]_{ij} = A_{ji}$. In other words, the (i,j) entry of the transpose is same as the (j,i) entry of **A**. The transpose-conjugate of **A**, denoted

$\mathbf{A}^\dagger$, is obtained by conjugating every element of $\mathbf{A}^T$. For example with $\mathbf{A}$ equal to the 2×2 matrix in (A.1.2), we have

$$\mathbf{A}^T = \begin{bmatrix} 1+j & 2 \\ -j & 3 \end{bmatrix}, \quad \mathbf{A}^\dagger = \begin{bmatrix} 1-j & 2 \\ j & 3 \end{bmatrix}. \qquad (A.2.1)$$

Note that if $\mathbf{A}$ is $p \times r$ then $\mathbf{A}^T$ as well as $\mathbf{A}^\dagger$ are $r \times p$. Thus the transpose of a column vector is a row vector, and vice versa.

For a square matrix $\mathbf{A}$, the notation $\mathbf{A}^{-1}$ stands for its inverse (defined and discussed in Sec. A.4). The notation $\mathbf{A}^{-T}$ stands for $(\mathbf{A}^{-1})^T$. Similarly, $\mathbf{A}^{-\dagger}$ stands for $(\mathbf{A}^{-1})^\dagger$.

Submatrices. A submatrix of $\mathbf{A}$ is any matrix formed by deleting an arbitrary set of rows and an arbitrary set of columns.

Arithmetic Operations

Addition and scalar multiplication. Two matrices with the same size can be added or subtracted by adding or subtracting corresponding elements. The notation $c\mathbf{A}$ stands for a matrix which is obtained from $\mathbf{A}$ by multiplying each element A_{ij} with c. This operation is called scalar multiplication. Thus, $c\mathbf{A} = [cA_{ij}]$. Given two matrices $\mathbf{A}$ and $\mathbf{B}$ of the same size, the matrix $\mathbf{P} = c\mathbf{A} + d\mathbf{B}$ has elements $P_{ij} = cA_{ij} + dB_{ij}$. Matrix addition is evidently commutative, that is, $\mathbf{A} + \mathbf{B} = \mathbf{B} + \mathbf{A}$. Note also that $[\mathbf{A} + \mathbf{B}]^T = \mathbf{A}^T + \mathbf{B}^T$.

Matrix multiplication. Given two matrices $\mathbf{A}$ and $\mathbf{B}$ with sizes $p \times m$ and $m \times r$ the product $\mathbf{C} = \mathbf{AB}$ is defined by defining the elements of $\mathbf{C}$ as

$$C_{ij} = \sum_{k=0}^{m-1} A_{ik} B_{kj}, \quad 0 \leq i \leq p-1, \ 0 \leq j \leq r-1. \qquad (A.2.2)$$

Schematically,

$$\underbrace{\mathbf{C}}_{p \times r} = \underbrace{\mathbf{A}}_{p \times m} \underbrace{\mathbf{B}}_{m \times r}. \qquad (A.2.3)$$

Note that the number of columns of $\mathbf{A}$ has to be the same as the number of rows of $\mathbf{B}$ (this is called compatibility requirement), and that $\mathbf{C}$ is $p \times r$. For example,

$$\begin{bmatrix} 1 & 1 \\ 1 & 2 \\ 1 & 1 \end{bmatrix} \begin{bmatrix} 2 & 3 \\ 1 & 4 \end{bmatrix} = \begin{bmatrix} 3 & 7 \\ 4 & 11 \\ 3 & 7 \end{bmatrix}.$$

Whenever we write $\mathbf{AB}$, the sizes of $\mathbf{A}$ and $\mathbf{B}$ are understood to be appropriate to make the product valid. The product $\mathbf{PQR}$ of three matrices is defined as $(\mathbf{PQ})\mathbf{R}$ provided that the sizes of the matrices are compatible. Note that matrix multiplications is associative, that is, $(\mathbf{PQ})\mathbf{R} = \mathbf{P}(\mathbf{QR})$. However, it is not commutative, i.e., in general $\mathbf{AB} \neq \mathbf{BA}$. For example if $\mathbf{A}$ is 2×3 and $\mathbf{B}$ is 3×4, then $\mathbf{AB}$ is well-defined but $\mathbf{BA}$ is not defined at all. It can be shown that $(\mathbf{AB})^T = \mathbf{B}^T \mathbf{A}^T$ and $(\mathbf{AB})^\dagger = \mathbf{B}^\dagger \mathbf{A}^\dagger$. The notation $\mathbf{A}^n$ stands for the product $\mathbf{AAA}\ldots\mathbf{A}$ (n times).

The reversal matrix. Matrices of the form

$$\begin{bmatrix} 0 & 1 \\ 1 & 0 \end{bmatrix}, \quad \begin{bmatrix} 0 & 0 & 1 \\ 0 & 1 & 0 \\ 1 & 0 & 0 \end{bmatrix}$$

are said to be reversal matrices. The general notation for an $M \times M$ reversal matrix is $\mathbf{J}_M$ with subscript omitted when obvious. The matrix $\mathbf{JA}$ is obtained from $\mathbf{A}$ by renumbering the rows in reverse order. Similarly $\mathbf{AJ}$ is obtained by renumbering the columns in reverse order. Given a diagonal matrix $\mathbf{\Lambda}$, the product $\mathbf{J\Lambda J}$ represents a new diagonal matrix with diagonal elements in reverse order. For example

$$\mathbf{J} \begin{bmatrix} a & 0 & 0 \\ 0 & b & 0 \\ 0 & 0 & c \end{bmatrix} \mathbf{J} = \begin{bmatrix} c & 0 & 0 \\ 0 & b & 0 \\ 0 & 0 & a \end{bmatrix}.$$

Trace of a matrix. The trace of a square matrix $\mathbf{A}$, denoted $\text{Tr}(\mathbf{A})$ is defined to be the sum of the diagonal elements, i.e., $\sum_i A_{ii}$. It can be shown that $\text{Tr}(\mathbf{AB}) = \text{Tr}(\mathbf{BA})$ as long as both products are meaningful.

Norms, Inner Products and Outer Products

Given two N-vectors $\mathbf{u}$ and $\mathbf{v}$, consider $\alpha = \mathbf{u}^\dagger \mathbf{v}$. This is a scalar quantity and is called the *inner product* of $\mathbf{u}$ with $\mathbf{v}$. The vectors are said to be (mutually) orthogonal if $\mathbf{u}^\dagger \mathbf{v} = 0$.

The inner product of $\mathbf{u}$ with $\mathbf{u}$, that is, $\mathbf{u}^\dagger \mathbf{u}$ is called the *energy* of $\mathbf{u}$. Denoting the elements of $\mathbf{u}$ as u_i, we see that $\mathbf{u}^\dagger \mathbf{u} = \sum_{i=0}^{N-1} |u_i|^2 \geq 0$. For example

$$\mathbf{u} = \begin{bmatrix} 1 \\ 2j \end{bmatrix} \Rightarrow \mathbf{u}^\dagger \mathbf{u} = 5.$$

The quantity $\mathbf{u}^\dagger \mathbf{u}$ is nonzero (hence positive) unless $\mathbf{u} = \mathbf{0}$.

The norm $\|\mathbf{u}\|$ of $\mathbf{u}$ is defined as the positive square root of its energy, i.e.,

$$\|\mathbf{u}\| = \sqrt{\mathbf{u}^\dagger \mathbf{u}} \quad \text{(norm of } \mathbf{u}\text{)}. \qquad (A.2.4)$$

Sometimes this is also called the $\mathcal{L}_2$ *norm*, and denoted as $\|\mathbf{u}\|_2$.

Given a p-vector $\mathbf{u}$ and an r-vector $\mathbf{v}$, the quantity $\mathbf{A} = \mathbf{u}\mathbf{v}^\dagger$ is a $p \times r$ matrix and is called the *outer product* of $\mathbf{u}$ with $\mathbf{v}$. This quantity is also called a *diadic* matrix. Example:

$$\mathbf{u} = \begin{bmatrix} u_0 \\ u_1 \end{bmatrix}, \quad \mathbf{v} = \begin{bmatrix} v_0 \\ v_1 \\ v_2 \end{bmatrix}, \quad \mathbf{uv}^\dagger = \begin{bmatrix} u_0 v_0^* & u_0 v_1^* & u_0 v_2^* \\ u_1 v_0^* & u_1 v_1^* & u_1 v_2^* \end{bmatrix}. \qquad (A.2.5)$$

Cauchy-Schwartz inequality. Given two column vectors $\mathbf{u}$ and $\mathbf{v}$, it can be shown that

$$|\mathbf{u}^\dagger \mathbf{v}| \leq \|\mathbf{u}\| \|\mathbf{v}\|, \qquad (A.2.6)$$

with equality if and only if $\mathbf{v} = c\mathbf{u}$ for some scalar c. For example,

$$\mathbf{u} = \begin{bmatrix} 1 \\ 1 \end{bmatrix}, \mathbf{v} = \begin{bmatrix} 1 \\ 2 \end{bmatrix} \Rightarrow \mathbf{u}^\dagger \mathbf{u} = 2, \mathbf{v}^\dagger \mathbf{v} = 5, \mathbf{u}^\dagger \mathbf{v} = 3,$$

and (A.2.6) holds with strict inequality.

A.3 DETERMINANTS

There are several equivalent definitions of the determinant of a $p \times p$ matrix $\mathbf{A}$. We will conveniently define it recursively as follows:

$$\det \mathbf{A} = \sum_{k=0}^{p-1} (-1)^{k+m} a_{km} M_{km}, \qquad (A.3.1)$$

where m is a fixed integer in $0 \leq m \leq p-1$. Here M_{km} is the determinant of the $(p-1) \times (p-1)$ submatrix obtained by deleting the kth row and mth column of $\mathbf{A}$.

Minors and cofactors. The quantity M_{km} is said to be the *minor* of the element a_{km}. The quantity $(-1)^{k+m} M_{km}$ is said to be the *cofactor* of a_{km}. In (A.3.1), the fixed column-index m is arbitrary.

In the above formula, the determinant has been computed by working with the mth column. Similarly, one can work with the mth row and obtain the determinant as

$$\det \mathbf{A} = \sum_{k=0}^{p-1} (-1)^{k+m} a_{mk} M_{mk}. \qquad (A.3.2)$$

Whenever we write [det $\mathbf{A}$], it is implicit that $\mathbf{A}$ is square. The determinant of $\mathbf{A}$ is denoted either as [det $\mathbf{A}$] or as $|\mathbf{A}|$.

For 2×2 matrices the above formula is simplified to

$$\begin{vmatrix} a & b \\ c & d \end{vmatrix} = ad - bc$$

Here are some examples:

$$\begin{vmatrix} 1 & 1 \\ 1 & 2 \end{vmatrix} = 1.$$

$$\begin{vmatrix} 1 & 4 & 5 \\ 2 & 1 & 1 \\ 3 & 1 & 2 \end{vmatrix} = 1 \times \begin{vmatrix} 1 & 1 \\ 1 & 2 \end{vmatrix} - 2 \times \begin{vmatrix} 4 & 5 \\ 1 & 2 \end{vmatrix} + 3 \times \begin{vmatrix} 4 & 5 \\ 1 & 1 \end{vmatrix} = -8.$$

Determinants of block-diagonal matrices. Let $\mathbf{A}$ be a square matrix of the form

$$\mathbf{A} = \begin{bmatrix} \mathbf{P} & \mathbf{0} \\ \mathbf{0} & \mathbf{Q} \end{bmatrix}.$$

Then we can show that [det $\mathbf{A}$] = [det $\mathbf{P}$][det $\mathbf{Q}$].

Principal and leading-principal minors. In general, the determinant of any square submatrix of $\mathbf{A}$ is said to be a minor of $\mathbf{A}$. Let $\mathbf{A}$ be square. A principal minor is any minor whose diagonal elements are also the diagonal elements of **A**. Thus for a 3×3 matrix **A** the principal minors are

$$a_{00}, a_{11}, a_{22}, \quad \begin{vmatrix} a_{00} & a_{01} \\ a_{10} & a_{11} \end{vmatrix}, \begin{vmatrix} a_{00} & a_{02} \\ a_{20} & a_{22} \end{vmatrix}, \begin{vmatrix} a_{11} & a_{12} \\ a_{21} & a_{22} \end{vmatrix}, \text{and } [\det \mathbf{A}].$$

Next, a leading principal minor of **A** is a principal minor such that if a_{kk} is an element, then so is a_{ii} for all $i < k$. Thus, the leading principal minors for a 3×3 matrix are:

$$a_{00}, \quad \begin{vmatrix} a_{00} & a_{01} \\ a_{10} & a_{11} \end{vmatrix}, \text{ and } [\det \mathbf{A}]. \tag{A.3.3}$$

Properties of Determinants

a) Let $\mathbf{C} = \mathbf{AB}$. Then $[\det \mathbf{C}] = [\det \mathbf{A}][\det \mathbf{B}]$.
b) If **B** is obtained from **A** by interchanging two rows (or two columns), then $[\det \mathbf{B}] = -[\det \mathbf{A}]$.
c) $[\det \mathbf{A}^T] = [\det \mathbf{A}]$.
d) For a $p \times p$ matrix **A**, $[\det c\mathbf{A}] = c^p [\det \mathbf{A}]$, for any scalar c.
e) The determinant of a diagonal matrix is the product of its diagonal elements. The same is true for lower or upper triangular square matrices.
f) If any row is a scalar multiple of another row, the determinant is zero. If any row is zero, the determinant is zero. These statements also hold if 'row' is replaced with 'column' everywhere.

Singular and nonsingular matrices. A square matrix is said to be singular if $[\det \mathbf{A}] = 0$, and nonsingular if $[\det \mathbf{A}] \neq 0$. The product **AB** of two square matrices is nonsingular if and only if each of **A** and **B** is nonsingular (since the determinant of **AB** is the product of individual determinants).

A.4 LINEAR INDEPENDENCE, RANK, AND RELATED ISSUES

Let $\mathbf{v}_k, 0 \leq k \leq m-1$ be a set of m column vectors. A linear combination of these vectors is an expression of the form $\sum_{k=0}^{m-1} \alpha_k \mathbf{v}_k$, and is clearly a vector of the same size. (Here α_k are, in general, complex numbers.) The set $\mathcal{S}$ of all linear combinations of these vectors is said to be the *space* or *vector space* spanned by these vectors.

Another way to define a vector space in our context is this: a vector space is a collection of vectors of a given size such that every possible linear combination from this collection also belongs to this collection. In particular, the null vector (the vector with all components equal to zero) is a member of the vector space.

Linear independence. We say that the vectors $\mathbf{v}_k$, $0 \leq k \leq m-1$ are linearly dependent if there exists a set of m scalars α_k, not all zero, such that $\sum_{k=0}^{m-1} \alpha_k \mathbf{v}_k = 0$. The set of vectors is linearly independent if they are not linearly dependent. For row vectors, we have an identical definition.

Basis vectors. The set of all vectors of the form $\sum_{k=0}^{m-1} \alpha_k \mathbf{v}_k$ is said to be the space spanned by the m column-vectors $\mathbf{v}_k$. If $\mathcal{S}$ is the space spanned by a set of *linearly independent* vectors, then these vectors are said to form a basis for this space. The minimum number of basis vectors required to span the space under question is called the *dimension* of the space. The basis set is not unique. For example, either of the following two sets of vectors spans the entire space of two-component vectors:

$$\underbrace{\begin{bmatrix}1\\0\end{bmatrix}, \begin{bmatrix}0\\1\end{bmatrix}}_{set\ 1}, \quad \underbrace{\begin{bmatrix}1\\1\end{bmatrix}, \begin{bmatrix}0\\1\end{bmatrix}}_{set\ 2} \qquad (A.4.1)$$

Rank of a Matrix

There are several equivalent ways to define rank (even though the equivalence is not obvious). We define the rank of a $p \times r$ matrix $\mathbf{A}$ to be an integer $\rho(\mathbf{A})$ (denoted just ρ when there is no confusion) such that there exists a $\rho \times \rho$ nonsingular submatrix of $\mathbf{A}$ but there does not exist a larger nonsingular submatrix. If ρ_1 denotes the largest number of columns of $\mathbf{A}$ that form a linearly independent set, we say that ρ_1 is the column rank of $\mathbf{A}$. Evidently $\rho_1 \leq r$, and if $\rho_1 = r =$ number of columns, we say that $\mathbf{A}$ has full column rank. Similarly we can define row rank ρ_2, and full row-rank matrices. It turns out that, for any matrix, $\rho = \rho_1 = \rho_2$. It is also clear from the definition that a $p \times p$ matrix is nonsingular if and only if it has full rank p.

♦**Fact A.4.1. Important properties of rank.** We now list the key features of rank.
a) $\rho = \rho_1 = \rho_2$, as stated above.
b) A $p \times p$ matrix is nonsingular if and only if the rank $\rho = p$.
c) $\rho(\mathbf{AB}) \leq \min(\rho(\mathbf{A}), \rho(\mathbf{B}))$.
d) If $\rho(\mathbf{A}) = 0$ then $\mathbf{A} = \mathbf{0}$.
e) Suppose $\mathbf{A}$ is $N \times N$ with rank $\rho < N$. This means that there are ρ linearly independent columns, from which all columns can be generated by linear combination. As a result, $\mathbf{A}$ can be written as

$$\mathbf{A} = \underbrace{\mathbf{B}}_{N \times \rho} \underbrace{\mathbf{C}}_{\rho \times N}. \qquad (A.4.2)$$

Any $N \times N$ matrix $\mathbf{A}$ with rank ρ can be factorized like this.

f) *Sylvester's inequality.* Let $\mathbf{P}$ and $\mathbf{Q}$ be $M \times N$ and $N \times K$ matrices with ranks ρ_p and ρ_q. Let ρ_{pq} be the rank of $\mathbf{PQ}$. Then,

$$\rho_p + \rho_q - N \leq \rho_{pq} \leq \min(\rho_p, \rho_q).$$

g) Given two square matrices $\mathbf{A}$ and $\mathbf{B}$, the products $\mathbf{AB}$ and $\mathbf{BA}$ may not have the same rank. However the matrices $\mathbf{I} - \mathbf{AB}$ and $\mathbf{I} - \mathbf{BA}$ do have the same rank. See Problem A.13. ◊

Diadic matrices. If $\rho(\mathbf{A}) = 1$, then every column of $\mathbf{A}$ is a scalar multiple of every other column. The same is true of the rows. In this case we can write $\mathbf{A} = \mathbf{uv}^\dagger$ where $\mathbf{u}$ and $\mathbf{v}$ are column vectors (so that $\mathbf{v}^\dagger$ is a row vector). In other words, any rank-one matrix is an outer product of two column vectors, and is sometimes called a *diadic* matrix. Here is an example:

$$\mathbf{A} = \begin{bmatrix} 1 & 2 & 3j \\ 2 & 4 & 6j \\ 1 & 2 & 3j \\ 4 & 8 & 12j \end{bmatrix} = \underbrace{\begin{bmatrix} 1 \\ 2 \\ 1 \\ 4 \end{bmatrix}}_{\mathbf{u}} \underbrace{[1 \quad 2 \quad 3j]}_{\mathbf{v}^\dagger}, \qquad (A.4.3)$$

so that

$$\mathbf{v} = \begin{bmatrix} 1 \\ 2 \\ -3j \end{bmatrix} \qquad (A.4.4)$$

in this example.

Range space and null space. Given a $p \times r$ matrix $\mathbf{A}$, the space spanned by its columns is said to be the range space (or *column* space) of $\mathbf{A}$. This space is the set of all vectors of the form $\mathbf{Ax}$ where $\mathbf{x}$ is any r-component column vector. The dimension of the range space is equal to the rank of $\mathbf{A}$. Note that the elements of the range space are p-vectors. The null space of $\mathbf{A}$ is the set of all r-vectors $\mathbf{y}$ such that $\mathbf{Ay} = 0$. It can be shown that the set of all linear combinations from the range space of $\mathbf{A}$ and the null space of $\mathbf{A}^\dagger$ is equal to the complete space of all vectors of size p.

Orthogonal complements. Let $\mathbf{t}_0, \mathbf{t}_1 \ldots \mathbf{t}_{\rho-1}$ be a set of linearly independent N-vectors and let $\mathcal{V}$ be the vector space spanned by them. (Clearly, $\rho \leq N$). Now consider the set $\mathcal{V}^\perp$ of all N-vectors orthogonal to all the vectors in $\mathcal{V}$ (i.e., orthogonal to all the above $\mathbf{t}_i$). The set $\mathcal{V}^\perp$ is itself a vector space, and is called the *orthogonal complement* of $\mathcal{V}$. It has dimension $N - \rho$. Any N-vector $\mathbf{x}$ can be expressed as a linear combination of one vector in $\mathcal{V}$ and one in $\mathcal{V}^\perp$. That is,

$$\mathbf{x} = \mathbf{x}_0 + \mathbf{x}_1, \quad \mathbf{x}_0 \in \mathcal{V}, \ \mathbf{x}_1 \in \mathcal{V}^\perp$$

Moreover, for a given $\mathbf{x}$, the components $\mathbf{x}_0$ and $\mathbf{x}_1$ are unique. Letting $\mathbf{t}_\rho \ldots \mathbf{t}_{N-1}$ denote a basis for $\mathcal{V}^\perp$, the matrix

$$\begin{bmatrix} \mathbf{t}_0 & \mathbf{t}_1 & \ldots & \mathbf{t}_{N-1} \end{bmatrix}$$

is $N \times N$ nonsingular. Its columns span the space of all N-vectors.

The annihilating vector. Given a matrix $\mathbf{A}$, any vector $\mathbf{y}$ such that $\mathbf{Ay} = 0$ is called an annihilating vector for $\mathbf{A}$. Evidently, the set of all annihilating vectors is equal to the null space of $\mathbf{A}$ defined above. A nonzero annihilating vector exists whenever the column rank of $\mathbf{A}$ is less than full, so that a linear combination of the columns can be made zero. For square matrices, this is equivalent to the condition that the determinant be zero.

Inverse of a Square Matrix

Given a $p \times r$ matrix $\mathbf{A}$, we say that the $r \times p$ matrix $\mathbf{L}$ is a left inverse if $\mathbf{LA} = \mathbf{I}_r$. We say that the $r \times p$ matrix $\mathbf{R}$ is a right inverse if $\mathbf{AR} = \mathbf{I}_p$. Inverses may or may not exist, and in general are not unique. For the case of square matrices, however the following are true: (a) An inverse exists if and only if $\mathbf{A}$ has full rank, i.e., $\mathbf{A}$ is nonsingular, and (b) when they exist, the left and right inverses are the same, and unique.

There is a closed form expression for the inverse of a nonsingular square matrix, given by

$$\mathbf{A}^{-1} = \frac{\text{Adj } \mathbf{A}}{\det \mathbf{A}}, \qquad (A.4.5)$$

where $[\text{Adj } \mathbf{A}]$ is the *adjugate* of $\mathbf{A}$ (referred to as adjoint in some texts), defined as

$$[\text{Adj } \mathbf{A}]_{ij} = \text{cofactor of } A_{ji}. \qquad (A.4.6)$$

In other words, the (i,j) element of the adjugate is equal to the cofactor of the (j,i) element of $\mathbf{A}$.

The matrix-inversion lemma. The following inversion formula, which holds whenever $\mathbf{P}$ and $\mathbf{R}$ are nonsingular, is very useful in system theoretic work. ($\mathbf{Q}$ and $\mathbf{S}$ need not be square).

$$(\mathbf{P} + \mathbf{QRS})^{-1} = \mathbf{P}^{-1} - \mathbf{P}^{-1}\mathbf{Q}(\mathbf{SP}^{-1}\mathbf{Q} + \mathbf{R}^{-1})^{-1}\mathbf{SP}^{-1}, \qquad (A.4.7)$$

Linear Equations

Consider an equation of the form $\mathbf{Ax} = \mathbf{b}$, where $\mathbf{A}$ is $N \times N$, and $\mathbf{b}$ and $\mathbf{x}$ are N-vectors. Given the quantities $\mathbf{A}$ and $\mathbf{b}$, we wish to find $\mathbf{x}$ satisfying this equation. Basically we have a set of N linear equations in N unknowns (the elements of $\mathbf{x}$). Given $\mathbf{A}$ and $\mathbf{b}$, there are several possibilities:

1. If $\mathbf{A}$ is nonsingular, there exists a unique solution given by $\mathbf{x} = \mathbf{A}^{-1}\mathbf{b}$.
2. If $\mathbf{A}$ is singular, then there are two possibilities: either there does not exist a solution, or there is an infinite number of solutions. (Since $\mathbf{A}$ is singular, $\mathbf{Av} = \mathbf{0}$ for some $\mathbf{v} \neq \mathbf{0}$ which shows that if $\mathbf{x}$ is a solution, then $\mathbf{x} + c\mathbf{v}$ is also a solution for *any* scalar c.)

More generally, consider the equation

$$\mathbf{Ax} = \mathbf{b}, \qquad (A.4.8)$$

where $\mathbf{A}$ is $N \times M$. The following are true.

1. If $\mathbf{A}$ has rank N (which implies $N \leq M$), then there exists $\mathbf{x}$ such that (A.4.8) holds.
2. If the rank of $\mathbf{A}$ is less than N, then depending on $\mathbf{b}$ there may or may not exist $\mathbf{x}$ satisfying (A.4.8).
3. In any case, if a solution exists, it is *unique* if and only if the rank of the matrix $\mathbf{A}$ equals the number of columns M.

A.5 EIGENVALUES AND EIGENVECTORS

Given a $N \times N$ square matrix $\mathbf{A}$, consider $D(s) = \det[s\mathbf{I} - \mathbf{A}]$. This is a polynomial in s with order N, called the *characteristic polynomial*. The N roots of $D(s)$ are said to be the N eigenvalues of $\mathbf{A}$. If a particular root λ_0 has multiplicity K, that is, if $D(s)$ has the factor $(s - \lambda_0)^K$, then the eigenvalue λ_0 is said to have multiplicity K. From the above definition, one can verify that λ is an eigenvalue of $\mathbf{A}$ if and only if there exists a nonzero vector $\mathbf{v}$ such that

$$\mathbf{Av} = \lambda \mathbf{v}. \qquad (A.5.1)$$

The vector $\mathbf{v}$ is said to be an eigenvector of $\mathbf{A}$ corresponding to the eigenvalue λ.

For example, let $\mathbf{A} = \begin{bmatrix} 3 & 1 \\ 1 & 3 \end{bmatrix}$. Then

$$D(s) = \begin{vmatrix} s-3 & -1 \\ -1 & s-3 \end{vmatrix} = s^2 - 6s + 8 = (s-4)(s-2),$$

so that the eigenvalues are $\lambda_0 = 4$ and $\lambda_1 = 2$. Furthermore

$$\begin{bmatrix} 3 & 1 \\ 1 & 3 \end{bmatrix} \begin{bmatrix} 1 \\ 1 \end{bmatrix} = 4 \begin{bmatrix} 1 \\ 1 \end{bmatrix}, \quad \begin{bmatrix} 3 & 1 \\ 1 & 3 \end{bmatrix} \begin{bmatrix} 1 \\ -1 \end{bmatrix} = 2 \begin{bmatrix} 1 \\ -1 \end{bmatrix},$$

so that the corresponding eigenvectors are $\begin{bmatrix} 1 \\ 1 \end{bmatrix}$ and $\begin{bmatrix} 1 \\ -1 \end{bmatrix}$.

Properties of Eigenvalues and Eigenvectors

a) If $\mathbf{v}$ is an eigenvector of $\mathbf{A}$, then so is $c\mathbf{v}$ for any scalar $c \neq 0$.

b) If $\mathbf{v}_1$ and $\mathbf{v}_2$ are eigenvectors corresponding to distinct eigenvalues λ_1 and λ_2, then $\mathbf{v}_1$ and $\mathbf{v}_2$ are "distinct". More accurately, we cannot write $\mathbf{v}_1 = \alpha \mathbf{v}_2$ for any scalar α.

c) Let $\mathbf{A}$ be $N \times N$ with N distinct eigenvalues λ_k, $0 \le k \le N-1$. In other words, none of the eigenvalues is a multiple root of $D(s)$. Then the corresponding eigenvectors $\mathbf{v}_k$, $0 \le k \le N-1$ are linearly independent. Also, each eigenvector $\mathbf{v}_k$ is unique (except, of course, for a scale factor). Notice, in general, that if $\mathbf{A}$ has less than N distinct eigenvalues, then there may or may not exist a set of N linearly independent eigenvectors.

d) Suppose λ is a complex eigenvalue of a real matrix $\mathbf{A}$. Then its conjugate λ^* is also an eigenvalue. Also if an eigenvalue of a real matrix is complex then the corresponding eigenvector is necessarily complex.

e) $\mathbf{A}$ has an eigenvalue equal to zero if and only if it is singular.

f) The eigenvalues of $\mathbf{A}$ are same as those of $\mathbf{A}^T$.

g) For a square (lower or upper) triangular matrix, the eigenvalues are equal to the diagonal elements. (The same is true of a diagonal matrix, which is a special case.) This follows by noting that, in these cases, $[\det(s\mathbf{I} - \mathbf{A})] = \prod(s - a_{ii})$.

h) Let $\mathbf{A}$ be $N \times N$ with eigenvalues λ_i. Then the determinant and trace can be expressed as

$$\det \mathbf{A} = \prod_{i=0}^{N-1} \lambda_i, \quad \text{Tr}(\mathbf{A}) = \sum_{i=0}^{N-1} \lambda_i.$$

i) For nonsingular $\mathbf{A}$, the eigenvalues of $\mathbf{A}^{-1}$ are reciprocals of those of $\mathbf{A}$.

j) If λ_k are the eigenvalues of $\mathbf{A}$, the eigenvalues of $\mathbf{A} + \sigma \mathbf{I}$ are equal to $\lambda_k + \sigma$. *Proof.* Let $\mathbf{A}\mathbf{v} = \lambda \mathbf{v}$, then $(\mathbf{A} + \sigma \mathbf{I})\mathbf{v} = \mathbf{A}\mathbf{v} + \sigma \mathbf{v} = (\lambda + \sigma)\mathbf{v}$.

It is possible for all eigenvalues to be equal to zero, even if $\mathbf{A} \neq \mathbf{0}$. An example is a triangular matrix with all diagonal elements equal to zero.

This appears to be a good place to summarize the various ways in which singularity of a matrix can manifest:

♠**Fact A.5.1. On singularity.** Let $\mathbf{A}$ be $N \times N$. Then the following statements are equivalent:

a) $\mathbf{A}$ is singular.
b) $[\det \mathbf{A}] = 0$.
c) There exists an eigenvalue of $\mathbf{A}$ equal to zero.
d) There exists a nonzero vector $\mathbf{v}$ such that $\mathbf{A}\mathbf{v} = \mathbf{0}$.
e) The rank of $\mathbf{A}$ is less than N.
f) The N columns (and rows) of $\mathbf{A}$ are not linearly independent.
g) $\mathbf{A}$ has no inverse.

h) The equation $\mathbf{Ax} = \mathbf{b}$ has no unique solution $\mathbf{x}$ (i.e., it has either no solutions, or an infinite number of them).   ◊

Eigenspaces. Suppose the N eigenvalues of $\mathbf{A}$ are not distinct. It is then conceivable that an eigenvalue, say λ_0, has more than one eigenvector. Suppose, for example, that $\{\mathbf{v}_0, \mathbf{v}_1, \mathbf{v}_2\}$ is a set of three linearly independent eigenvectors corresponding to λ_0, i.e., $\mathbf{Av}_k = \lambda_0 \mathbf{v}_k$ for $k = 0, 1, 2$. Then any linear combination of $\mathbf{v}_0, \mathbf{v}_1$ and $\mathbf{v}_2$ (i.e., any vector in the space spanned by $\mathbf{v}_0, \mathbf{v}_1$ and $\mathbf{v}_2$) is an eigenvector for λ_0. This vector space spanned by $\mathbf{v}_0, \mathbf{v}_1$, and $\mathbf{v}_2$ is the *eigenspace* corresponding to λ_0.

Similarity transformations. Given a square matrix $\mathbf{A}$, suppose we define $\mathbf{A}_1 = \mathbf{T}^{-1}\mathbf{AT}$ where $\mathbf{T}$ is some nonsingular matrix. It turns out that $\mathbf{A}_1$ and $\mathbf{A}$ have the same set of eigenvalues. (*Proof:* $\mathbf{Av} = \lambda\mathbf{v} \Rightarrow \mathbf{T}^{-1}\mathbf{AT}(\mathbf{T}^{-1}\mathbf{v}) = \lambda(\mathbf{T}^{-1}\mathbf{v})$). This transformation of $\mathbf{A}$ to $\mathbf{A}_1$ is said to be a similarity transformation.

Diagonalization

Suppose $\mathbf{A}$ is $N \times N$, and assume that it has N linearly independent eigenvectors $\mathbf{t}_k$. (This does not necessarily mean that there are N distinct eigenvalues.) We can write $\mathbf{At}_k = \lambda_k \mathbf{t}_k$, $0 \le k \le N-1$. We can compactly write these as one matrix equation:

$$\mathbf{A} \underbrace{[\mathbf{t}_0 \quad \mathbf{t}_1 \quad \ldots \quad \mathbf{t}_{N-1}]}_{\mathbf{T}} = \underbrace{[\mathbf{t}_0 \quad \mathbf{t}_1 \quad \ldots \quad \mathbf{t}_{N-1}]}_{\mathbf{T}} \mathbf{\Lambda}, \qquad (A.5.2)$$

where $\mathbf{\Lambda}$ is an $N \times N$ diagonal matrix with kth diagonal element equal to λ_k. We can rearrange this as

$$\mathbf{T}^{-1}\mathbf{AT} = \mathbf{\Lambda}, \quad i.e., \quad \mathbf{A} = \mathbf{T}\mathbf{\Lambda}\mathbf{T}^{-1}. \qquad (A.5.3)$$

This shows that if there exist N linearly independent eigenvectors, then we can diagonalize $\mathbf{A}$ by applying a similarity transformation. Conversely, whenever we can find $\mathbf{T}$ such that $\mathbf{T}^{-1}\mathbf{AT}$ is diagonal, the columns of $\mathbf{T}$ are eigenvectors of $\mathbf{A}$ with corresponding eigenvalues appearing on the diagonals of $\mathbf{T}^{-1}\mathbf{AT}$. For example, let $\mathbf{A} = \begin{bmatrix} 3 & 1 \\ 1 & 3 \end{bmatrix}$. We have already computed the eigenvalues and eigenvectors above. From these we obtain

$$\underbrace{\begin{bmatrix} 1 & 1 \\ 1 & -1 \end{bmatrix}^{-1}}_{\mathbf{T}^{-1}} \underbrace{\begin{bmatrix} 3 & 1 \\ 1 & 3 \end{bmatrix}}_{\mathbf{A}} \underbrace{\begin{bmatrix} 1 & 1 \\ 1 & -1 \end{bmatrix}}_{\mathbf{T}} = \underbrace{\begin{bmatrix} 4 & 0 \\ 0 & 2 \end{bmatrix}}_{\mathbf{\Lambda}}.$$

An $N \times N$ matrix $\mathbf{A}$ is said to be *diagonalizable* if it can be written as in (A.5.3) for some diagonal $\mathbf{\Lambda}$ and nonsingular $\mathbf{T}$. The following points are worth noting.

1. Not every $N \times N$ matrix is diagonalizable. For example suppose $\mathbf{A}$ is a nonzero matrix such that the only possible eigenvalue is $\lambda = 0$. (An example is $\mathbf{A} = \begin{bmatrix} 0 & 0 \\ 1 & 0 \end{bmatrix}$.) If $\mathbf{A}$ is diagonalizable, then $\mathbf{A} = \mathbf{T}\mathbf{\Lambda}\mathbf{T}^{-1}$ with $\mathbf{\Lambda} = \mathbf{0}$ so that $\mathbf{A} = \mathbf{0}$, which is a contradiction.

2. Every matrix with N distinct eigenvalues is diagonalizable because there are N linearly independent eigenvectors.

3. There is a class of matrices called normal matrices (defined below), which are diagonalizable even if the eigenvalues are not distinct.

Cayley-Hamilton Theorem

Recall that the characteristic polynomial $D(s)$ is defined as $[\det(s\mathbf{I} - \mathbf{A})]$ and has the form $D(s) = s^N + d_{N-1}s^{N-1} + \ldots + d_0$. The equation $D(s) = 0$ is called the *characteristic equation* (its solutions being the eigenvalues). It turns out that the $N \times N$ matrix $\mathbf{A}$ satisfies this equation, that is,

$$\mathbf{A}^N + d_{N-1}\mathbf{A}^{N-1} + \ldots + d_1\mathbf{A} + d_0\mathbf{I} = 0. \qquad (A.5.4)$$

This result is called the *Cayley-Hamilton theorem*. This says that the matrix $\mathbf{A}^N$ can be expressed as a linear combination of the lower powers.

A.6 SPECIAL TYPES OF MATRICES

We now discuss a number of special types of matrices which arise in our discussions throughout the text.

Hermitian matrix. $\mathbf{H}$ is said to be Hermitian if $\mathbf{H}^\dagger = \mathbf{H}$. This implies $H_{ij} = H_{ji}^*$. Note that a Hermitian matrix, by definition, is square. A real Hermitian matrix is *symmetric* (i.e., $\mathbf{H}^T = \mathbf{H}$). Variations of this class are the skew-Hermitian matrix ($\mathbf{H}^\dagger = -\mathbf{H}$), and antisymmetric matrix ($\mathbf{H}^T = -\mathbf{H}$).

Any matrix $\mathbf{A}$ can be written as

$$\mathbf{A} = \mathbf{A}_h + \mathbf{A}_s$$

where $\mathbf{A}_h$ is Hermitian and $\mathbf{A}_s$ is skew-Hermitian. For this, just define

$$\mathbf{A}_h = \frac{\mathbf{A} + \mathbf{A}^\dagger}{2}, \quad \mathbf{A}_s = \frac{\mathbf{A} - \mathbf{A}^\dagger}{2}$$

Unitary matrix. $\mathbf{U}$ is said to be unitary if $\mathbf{U}^\dagger \mathbf{U} = c\mathbf{I}$ for some $c > 0$. This means that every pair of columns is mutually orthogonal, and that all columns have the same norm $\sqrt{c}$. If the unitary matrix is square, then $\mathbf{U}^T$ as well as $\mathbf{U}^\dagger$ are unitary. Thus, for a square unitary matrix $\mathbf{U}\mathbf{U}^\dagger = \mathbf{U}^\dagger\mathbf{U} = c\mathbf{I}$. If $c = 1$, $\mathbf{U}$ is *normalized* unitary. A real unitary matrix is usually said to be an orthogonal matrix (ortho*normal* if $\mathbf{U}^T\mathbf{U} = \mathbf{I}$). In Chap. 14 one can find more details about unitary matrices, planar rotations, Householder forms, and factorizations.

Let $\mathbf{y} = \mathbf{U}\mathbf{x}$. If $\mathbf{U}$ is unitary, it is clear that $\mathbf{y}^\dagger \mathbf{y} = c\mathbf{x}^\dagger \mathbf{x}$, for any choice of $\mathbf{x}$. So a unitary matrix changes the norms of all vectors by the same factor $\sqrt{c}$. Conversely, suppose a matrix $\mathbf{U}$ is such that $\mathbf{y}^\dagger \mathbf{y} = c\mathbf{x}^\dagger \mathbf{x}$ for all vectors $\mathbf{x}$. Then $\mathbf{U}$ is unitary (Problem A.17).

Circulant matrices. A square matrix is right circulant if each row is obtained by a right circular shift of the previous row. Example:

$$\mathbf{A} = \begin{bmatrix} a_0 & a_1 & a_2 \\ a_2 & a_0 & a_1 \\ a_1 & a_2 & a_0 \end{bmatrix}. \qquad (A.6.1)$$

The left-circulant property is similarly defined. Unless mentioned otherwise, 'circulant' denotes right circulants.

Normal matrix. $\mathbf{A}$ is said to be normal if $\mathbf{A}\mathbf{A}^\dagger = \mathbf{A}^\dagger\mathbf{A}$. By definition, $\mathbf{A}$ has to be a square matrix. It can be verified that the following matrices are normal: (a) Hermitian and skew-Hermitian matrices, (b) square unitary matrices and (c) circulants.

The DFT and IDFT Matrices

A matrix of special interest in digital signal processing is the Discrete Fourier Transform (DFT) matrix. This is a $N \times N$ matrix defined as $\mathbf{W}_N = [W_N^{km}]$ where $W_N = e^{-j2\pi/N}$. In other words, the entry at the kth row and mth column is equal to $e^{-j2\pi km/N}$. Evidently this is a symmetric (but complex) matrix. Examples are:

$$\mathbf{W}_2 = \begin{bmatrix} 1 & 1 \\ 1 & -1 \end{bmatrix}, \quad \mathbf{W}_4 = \begin{bmatrix} 1 & 1 & 1 & 1 \\ 1 & -j & -1 & j \\ 1 & -1 & 1 & -1 \\ 1 & j & -1 & -j \end{bmatrix}.$$

The subscripts N on $\mathbf{W}$ and W are usually omitted if they are clear from the context. The matrix $\mathbf{W}$ satisfies the property $\mathbf{W}^\dagger \mathbf{W} = N\mathbf{I}$ so that it is unitary. Given a finite length sequence $x(n)$, $0 \le n \le N-1$, suppose we define the vector $\mathbf{x} = [x(0) \; x(1) \ldots \; x(N-1)]^T$, and compute the vector $\mathbf{X} = \mathbf{W}\mathbf{x}$. Then the components of $\mathbf{X}$, viz., $X(k), 0 \le k \le N-1$ are said to form the DFT coefficients of the sequence $x(n)$. The sequence $x(n)$ is the inverse DFT (abbreviated IDFT) of the sequence $X(k)$. The matrix $\mathbf{W}^{-1}$ (which is equal to $\mathbf{W}^\dagger/N$) is called the *IDFT matrix*. Notice that $\mathbf{W}$ is symmetric, that is, $\mathbf{W}^T = \mathbf{W}$ so that $\mathbf{W}^\dagger = \mathbf{W}^*$.

The DFT and IDFT relations are more commonly written as

$$X(k) = \sum_{m=0}^{N-1} x(m) W^{km}, \quad x(m) = \frac{1}{N} \sum_{k=0}^{N-1} X(k) W^{-km}. \quad (A.6.2)$$

Toeplitz matrices. An $N \times N$ matrix $\mathbf{A}$ is said to be Toeplitz if the elements A_{ij} are determined completely by the difference $i - j$. For example,

$$\mathbf{A} = \begin{bmatrix} a_0 & a_1 & a_2 \\ a_3 & a_0 & a_1 \\ a_4 & a_3 & a_0 \end{bmatrix}, \quad (A.6.3)$$

is Toeplitz. Pictorially, if we draw a line parallel to the main diagonal, then all elements on this line are equal. Thus, a Toeplitz matrix is completely determined by the 0th row and 0th column, that is, by $2N - 1$ elements. Notice that circulants are Toeplitz.

If we replace each of the $2N - 1$ elements in a Toeplitz matrix by a (possibly rectangular) matrix, we obtain a block-Toeplitz matrix. An example is

$$\mathbf{A} = \begin{bmatrix} \mathbf{a}_0 & \mathbf{a}_1 & \mathbf{a}_2 \\ \mathbf{a}_3 & \mathbf{a}_0 & \mathbf{a}_1 \\ \mathbf{a}_4 & \mathbf{a}_3 & \mathbf{a}_0 \end{bmatrix}, \quad (A.6.4)$$

where $\mathbf{a}_i$ are themselves 2×2 matrices.

Vandermonde matrices. An $N \times N$ matrix, each of whose rows has the form

$$[1 \quad a_i \quad a_i^2 \quad \ldots \quad a_i^{N-1}] \tag{A.6.5}$$

is a Vandermonde matrix. Example:

$$\mathbf{A} = \begin{bmatrix} 1 & a_0 & a_0^2 \\ 1 & a_1 & a_1^2 \\ 1 & a_2 & a_2^2 \end{bmatrix}. \tag{A.6.6}$$

The transponse of a Vandermonde matrix, for example,

$$\mathbf{A} = \begin{bmatrix} 1 & 1 & 1 \\ a_0 & a_1 & a_2 \\ a_0^2 & a_1^2 & a_2^2 \end{bmatrix}, \tag{A.6.7}$$

is also said to be a Vandermonde matrix. Note that the DFT matrix is Vandermonde. The determinant of a Vandermonde matrix is given by

$$\det \mathbf{A} = \prod_{i>j}(a_i - a_j). \tag{A.6.8}$$

For example, if $\mathbf{A}$ is as in (A.6.6),

$$\det \mathbf{A} = (a_1 - a_0)(a_2 - a_0)(a_2 - a_1). \tag{A.6.9}$$

It follows that a Vandermonde matrix is nonsingular if and only if the a_i's are distinct.

Eigenstructures of Special Matrices

Some of the above mentioned special matrices satisfy special properties related to eigenvalues and eigenvectors.

♠**Fact A.6.1. Normal matrices.** The $N \times N$ matrix $\mathbf{A}$ is normal if and only if we can find $N \times N$ unitary $\mathbf{U}$ such that $\mathbf{U}^{-1}\mathbf{A}\mathbf{U}$ is diagonal, that is, if and only if we can write

$$\mathbf{A}\mathbf{U} = \mathbf{U}\mathbf{\Lambda} \tag{A.6.10}$$

for diagonal $\mathbf{\Lambda}$ and unitary $\mathbf{U}$. This means that normal matrices are precisely those for which there exists a complete set of *mutually orthogonal eigenvectors* (i.e., unitary diagonalization is possible). Without loss of generality we can assume the columns of $\mathbf{U}$ to have unit norm. Then (A.6.10) is the same as

$$\mathbf{A} = \mathbf{U}\mathbf{\Lambda}\mathbf{U}^\dagger, \quad \text{where} \quad \mathbf{U}^\dagger\mathbf{U} = \mathbf{U}\mathbf{U}^\dagger = \mathbf{I}. \tag{A.6.11}$$

This is identical to (A.5.3) with $\mathbf{T} = \mathbf{U}$. ◇

Notice, as a corollary, that if all the eigenvalues of a normal matrix are identical, then it has the form $\mathbf{A} = \lambda \mathbf{I}$ (where λ is this common eigenvalue). The same is not true for arbitary matrices, for example, $\mathbf{A} = \begin{bmatrix} 1 & 0 \\ 2 & 1 \end{bmatrix}$.

♠**Fact A.6.2. Special normal matrices.** Since Hermitian, unitary and circulant square matrices are normal, they can be written as in (A.6.11). In addition, the following are true.

a) If $\mathbf{A}$ is Hermitian, all eigenvalues are real. Moreover, $\mathbf{v}^\dagger \mathbf{A} \mathbf{v}$ is real for all vectors $\mathbf{v}$.
b) If $\mathbf{A}$ is unitary ($\mathbf{A}^\dagger \mathbf{A} = c\mathbf{I}, c > 0$), then all the eigenvalues have magnitude $\sqrt{c}$.
c) If $\mathbf{A}$ is $M \times M$ circulant, then we can write (A.6.10) with $\mathbf{U} = \mathbf{W}/\sqrt{M}$, where $\mathbf{W}$ is the DFT matrix. The eigenvalues of $\mathbf{A}$ are the DFT coefficients of the 0th row of $\mathbf{A}$. The eigenvectors are the columns of $\mathbf{W}/\sqrt{M}$.

Quadratic Forms and Positive Definite Matrices

For any $N \times N$ matrix $\mathbf{P}$, the scalar $\mathbf{v}^\dagger \mathbf{P} \mathbf{v}$ is said to be a quadratic form. In particular when $\mathbf{P}$ is Hermitian we know that $\mathbf{v}^\dagger \mathbf{P} \mathbf{v}$ is real. If this is positive for *all nonzero* $\mathbf{v}$, we say that $\mathbf{P}$ is positive definite. Notice that this property is defined only for Hermitian matrices.

Based on the properties of $\mathbf{v}^\dagger \mathbf{P} \mathbf{v}$ we can in fact identify a number of definitions as follows:

$$\mathbf{v}^\dagger \mathbf{P} \mathbf{v} \begin{cases} > 0, \ \forall \mathbf{v} \neq \mathbf{0} & \text{(positive definite)} \\ \geq 0, \ \forall \mathbf{v} & \text{(positive semidefinite or nonnegative definite)} \\ < 0, \ \forall \mathbf{v} \neq \mathbf{0} & \text{(negative definite)} \\ \leq 0, \ \forall \mathbf{v} & \text{(negative semidefinite)}. \end{cases} \quad (A.6.12)$$

Here are some examples:

$$\underbrace{\begin{bmatrix} 2 & 1 \\ 1 & 2 \end{bmatrix}}_{\text{positive definite}} \quad \underbrace{\begin{bmatrix} 2 & j \\ -j & 2 \end{bmatrix}}_{\text{positive definite}} \quad \underbrace{\begin{bmatrix} 1 & 1 \\ 1 & 1 \end{bmatrix}}_{\text{positive semidefinite}}$$

(See test for positive (semi)definiteness below). By definition, a positive definite matrix is also positive semidefinite. Notice that $\mathbf{P}$ is positive definite (semidefinite) if and only if $-\mathbf{P}$ is negative definite (semidefinite). It is possible that $\mathbf{P}$ does not belong to any of these categories. (Example: $\mathbf{P} = \begin{bmatrix} 1 & 0 \\ 0 & -1 \end{bmatrix}$.) In other words, it is possible that $\mathbf{v}^\dagger \mathbf{P} \mathbf{v}$ has different signs for different $\mathbf{v}$. In that case, $\mathbf{P}$ is said to be indefinite.

Matrix inequalities. If $\mathbf{P}$ is positive definite, we indicate it as $\mathbf{P} > 0$ ($\mathbf{P} \geq 0$ for semidefinite). Given two Hermitian matrices $\mathbf{P}$ and $\mathbf{Q}$ of the same size, we write $\mathbf{P} > \mathbf{Q}$ if $\mathbf{P} - \mathbf{Q}$ is positive definite ($\mathbf{P} \geq \mathbf{Q}$ if $\mathbf{P} - \mathbf{Q}$ is positive semidefinite). Notice however that, in general, the difference matrix $\mathbf{P} - \mathbf{Q}$ can be indefinite, even though $\mathbf{P}$ and $\mathbf{Q}$ are definite.

Properties of Positive Definite Matrices

For convenience of reference we now list a number of properties of positive (semi)definite matrices. We encourage the reader to verify these for the examples shown above.

a) All diagonal elements of a positive definite (semidefinite) matrix are positive (nonnegative).

b) The Hermitian matrix **P** is positive definite (semidefinite) if and only if all the eigenvalues are positive (nonnegative).
c) *Test for positive definiteness.* The Hermitian matrix **P** is positive definite if and only if all leading principal minors of **P** are positive, and positive semidefinite if and only if all principal minors are nonnegative.
d) If $\mathbf{P} \geq 0$ and $\mathbf{Q} \geq 0$, then $\mathbf{P} + \mathbf{Q} \geq 0$. If in addition $\mathbf{Q} > 0$ (or $\mathbf{P} > 0$), then $\mathbf{P} + \mathbf{Q} > 0$.
e) *Square roots.* Given a positive number a, we know that we can find a real square root. The beauty of positive definite matrices is that we can define square roots in a similar way. Given a Hermitian matrix **P**, if we can factorize it as $\mathbf{P} = \mathbf{Q}^\dagger \mathbf{Q}$ for some **Q** (possibly rectangular), we say that **Q** is a square root. For example, consider $\mathbf{A} = \begin{bmatrix} 3 & 1 \\ 1 & 3 \end{bmatrix}$. This is Hermitian, and we have already calculated its eigenvalues to be 4 and 2, so that it is positive definite. By using the diagonalization result for this, we obtain

$$\begin{bmatrix} 3 & 1 \\ 1 & 3 \end{bmatrix} = \frac{1}{2} \begin{bmatrix} 1 & 1 \\ 1 & -1 \end{bmatrix} \begin{bmatrix} 4 & 0 \\ 0 & 2 \end{bmatrix} \begin{bmatrix} 1 & 1 \\ 1 & -1 \end{bmatrix}$$

$$= \underbrace{\frac{1}{\sqrt{2}} \begin{bmatrix} 1 & 1 \\ 1 & -1 \end{bmatrix} \begin{bmatrix} 2 & 0 \\ 0 & \sqrt{2} \end{bmatrix}}_{\mathbf{Q}^\dagger} \underbrace{\frac{1}{\sqrt{2}} \begin{bmatrix} 2 & 0 \\ 0 & \sqrt{2} \end{bmatrix} \begin{bmatrix} 1 & 1 \\ 1 & -1 \end{bmatrix}}_{\mathbf{Q}}.$$

Square roots are not unique. For example, suppose **Q** is a square root, then **UQ** is also a square root for any normalized unitary **U**. It is clear that if there exists a square root for **P**, then **P** is nonnegative definite because $\mathbf{v}^\dagger \mathbf{P} \mathbf{v} = \mathbf{v}^\dagger \mathbf{Q}^\dagger \mathbf{Q} \mathbf{v} = \mathbf{w}^\dagger \mathbf{w} \geq 0$. Conversely, it can be shown that any $N \times N$ nonnegative definite **P** with rank ρ can be factorized as $\mathbf{Q}^\dagger \mathbf{Q}$ where **Q** is $\rho \times N$. One technique to find such a factor **Q** is called *Cholesky decomposition* [Golub and Van Loan, 1989], which produces a lower triangular square root.
f) Suppose **Q** is $p \times r$, with $p \geq r$. Evidently the rank of $\mathbf{Q} \leq r$. Define the $r \times r$ positive semidefinite matrix $\mathbf{P} = \mathbf{Q}^\dagger \mathbf{Q}$. This is nonsingular (hence positive definite) if and only if **Q** has full rank r.
g) *Determinant and diagonal elements.* Let **P** be $N \times N$ Hermitian positive definite, and let P_{ii} denote its diagonal elements. Then

$$\det \mathbf{P} \leq \prod_{i=0}^{N-1} P_{ii},$$

with equality if and only if **P** is diagonal. See Problem A.19 for a proof.

♠**Fact A.6.3. Positive definite matrices.** Let **P** be $N \times N$ Hermitian. Then the following statements are equivalent.
a) **P** is positive definite (i.e., $\mathbf{v}^\dagger \mathbf{P} \mathbf{v} > 0$ for all vectors $\mathbf{v} \neq \mathbf{0}$.)
b) All eigenvalues of **P** are positive.
c) There exists an $N \times N$ nonsingular square root **Q**.
d) There exists an $N \times N$ nonsingular *lower triangular* square root $\mathbf{\Delta}_\ell$.
e) There exists an $N \times N$ nonsingular *upper triangular* square root $\mathbf{\Delta}_u$.

f) All leading principal minors of **P** are positive. ◇

A.7 UNITARY TRIANGULARIZATION

It should be noticed that an arbitrary square matrix may not be diagonalizable [i.e., expressible in the form (A.5.3)]. The class of matrices which can be diagonalized by unitary matrices [i.e., which can be expressed as in (A.6.11)] is even smaller (namely *normal* matrices). However, *every* square matrix can be *triangularized* by a unitary transformation, as stated next.

♠**Fact A.7.1.** Let **A** be an arbitrary $N \times N$ matrix. Then, we can always write it in the form

$$\mathbf{A} = \mathbf{U}\boldsymbol{\Delta}\mathbf{U}^\dagger, \quad (A.7.1)$$

where **U** is $N \times N$ normalized unitary (i.e., $\mathbf{U}^\dagger\mathbf{U} = \mathbf{I}$), and $\boldsymbol{\Delta}$ is lower triangular. (A.7.1) can be regarded as a similarity transformation of **A** into $\boldsymbol{\Delta}$. So the eigenvalues of **A** are the same as those of $\boldsymbol{\Delta}$, which in turn are the diagonal elements of $\boldsymbol{\Delta}$. Note that the columns of **U** are not necessarily the eigenvectors of **A**, unlike in diagonalization. ◇

This result, due to Schur, is of great importance. As an example,

$$\begin{bmatrix} 2+j & j \\ -j & 2-j \end{bmatrix} = \underbrace{\frac{1}{\sqrt{2}}\begin{bmatrix} 1 & 1 \\ 1 & -1 \end{bmatrix}}_{\mathbf{U}} \underbrace{\begin{bmatrix} 2 & 0 \\ 2j & 2 \end{bmatrix}}_{\boldsymbol{\Delta}} \underbrace{\frac{1}{\sqrt{2}}\begin{bmatrix} 1 & 1 \\ 1 & -1 \end{bmatrix}}_{\mathbf{U}^\dagger}.$$

Since triangular matrices play an important role in many applications, it is useful to summarize some of their properties.

♠**Fact. A.7.2. Properties of triangular matrices.** Let **A** and **B** be $N \times N$ lower triangular. Then,

a) The product **AB** is lower triangular.
b) [det **A**] is equal to the product of diagonal elements A_{ii}.
c) The eigenvalues of **A** are equal to the diagonal elements A_{ii}.
d) If all diagonal elements are such that $|A_{ii}| < 1$, then $\mathbf{A}^n \to \mathbf{0}$ as $n \to \infty$.
e) If all diagonal elements are equal to zero then $\mathbf{A}^N = \mathbf{0}$. ◇

Property (d) above finds application in stability analysis (Chap. 13). Property (e) is useful in the study of FIR systems. The above results hold if "lower triangular" is replaced with "upper triangular" everywhere.

A.8 MAXIMIZATION AND MINIMIZATION

Suppose **A** is Hermitian. Consider the quadratic form $\mathbf{v}^\dagger\mathbf{A}\mathbf{v}$. If we constrain **v** to be a unit norm vector, then this quadratic form cannot take arbitrarily large or small values. The extreme values are determined by the eigenvalues of **A** as summarized in the following result.

♠**Fact A.8.1. Rayleigh's principle.** Let **A** be $N \times N$ Hermitian. We know that the eigenvalues are real. Let λ_{min} and λ_{max} be the smallest and largest eigenvalues. Then the maximum value of $\mathbf{v}^\dagger\mathbf{A}\mathbf{v}$, as we vary **v** over all unit norm vectors, is equal to λ_{max} and occurs if and only if $\mathbf{v}$ is an eigenvector corresponding to λ_{max}. Similarly the minimum value of $\mathbf{v}^\dagger \mathbf{A} \mathbf{v}$ over unit norm $\mathbf{v}$ is equal to λ_{min} and occurs if, and only if, $\mathbf{v}$ is an eigenvector corresponding to λ_{min}. ◊

Note that λ_{max} may have multiplicity > 1, in which case the eigenvector which maximizes the quadratic form is any vector from the corresponding eigenspace.

The "power method" for computing λ_{max} and its eigenvector

Let $\mathbf{A}$ be Hermitian positive semidefinite. So $\lambda_{min} \geq 0$ and $\lambda_{max} \geq 0$. Suppose we perform the following iteration

$$\mathbf{v}_k = \mathbf{A} \mathbf{v}_{k-1}, \qquad (A.8.1)$$

with the initial vector $\mathbf{v}_0$ chosen arbitrarily. Unless $\mathbf{v}_0$ is orthogonal to every vector in the eigenspace of λ_{max}, this iteration eventually converges to an eigenvector corresponding to λ_{max}. This technique is called the *power method*. Once an eigenvector $\mathbf{v}$ is so computed, we compute λ_{max} from $\mathbf{A}\mathbf{v} = \lambda_{max}\mathbf{v}$.

If we are interested in computing λ_{min} and a corresponding eigenvector, there are several tricks we can use. If $\mathbf{A}$ is nonsingular, we can invert it (so that $1/\lambda_{min}$ is its largest eigenvalue. If we wish to avoid inversion (which is time consuming) we can first compute λ_{max} and define $\mathbf{B} = \lambda_{max}\mathbf{I} - \mathbf{A}$. This is positive semidefinite with largest eigenvalue $\lambda_{max} - \lambda_{min}$ which can be computed by the power method. So λ_{min} can be found.

A.9 PROPERTIES PRESERVED IN MATRIX PRODUCTS

Let $\mathbf{A}$ and $\mathbf{B}$ be $N \times N$ matrices satisfying a given property. It is often important to know whether the product $\mathbf{C} = \mathbf{AB}$ satisfies the same property. For example, the product of two unitary matrices is unitary, but the product of two Hermitian matrices is not necessarily Hermitian.

Properties preserved under matrix multiplication. (a) Unitariness, (b) circulant property, (c) nonsingularity, (d) lower (hence upper) triangular property.

Properties not necessarily preserved. (a) Hermitian property, (b) positive definiteness, (c) Toeplitz property, (d) Vandermonde property, (e) *normal* property, and (f) stability (i.e., all eigenvalues inside the unit circle). See Problem A.16.

PROBLEMS

A.1. We know that matrix products do not commute, that is, in general $\mathbf{AB} \neq \mathbf{BA}$.
  a) Demonstrate this with an example when $\mathbf{A}$ and $\mathbf{B}$ are (i) 2×2 and (ii) 3×3.
  b) Find examples of 2×2 matrices $\mathbf{A}$ and $\mathbf{B}$ such that $\mathbf{AB} = \mathbf{BA}$. (To avoid trivial answers, make sure the matrices are non diagonal.)

A.2. Which of the following matrices is diagonalizable?

$$\begin{bmatrix} 0 & j \\ j & 0 \end{bmatrix}, \begin{bmatrix} 0 & j \\ -j & 0 \end{bmatrix}, \begin{bmatrix} 0 & 1 \\ 0 & 0 \end{bmatrix}.$$

A.3. Consider the Toeplitz matrix

$$\mathbf{A} = \begin{bmatrix} 3 & 2 & 1 \\ 2 & 3 & 2 \\ 1 & 2 & 3 \end{bmatrix}.$$

  a) Compute the quantity $\mathbf{A}^2$ and verify that it is not Toeplitz. This shows that the product of Toeplitz matrices may not be Toeplitz.
  b) Compute the determinant and verify that this is nonsingular. Find the inverse.

A.4. Verify Cayley-Hamilton theorem (A.5.4), for $\mathbf{A} = \begin{bmatrix} 1 & 1 \\ 1 & 1 \end{bmatrix}$.

A.5. Let $\mathbf{A}$ and $\mathbf{B}$ be $N \times N$ lower triangular matrices. Thus, $A_{ij} = 0$ for $j > i$ and $B_{jk} = 0$ for $k > j$. Using these prove that $\mathbf{AB}$ is lower triangular.

A.6. Check whether each of the following matrices has any of these properties: (a) Hermitian, (b) positive definiteness, (c) unitariness, and (d) normal property.

$$\begin{bmatrix} 2 & 1 \\ 1 & 2 \end{bmatrix}, \begin{bmatrix} 2 & 2 \\ -2 & 2 \end{bmatrix}, \begin{bmatrix} 0 & j \\ j & 0 \end{bmatrix}, \begin{bmatrix} 2 & 2 \\ 2 & 1 \end{bmatrix}.$$

A.7. Find the eigenvalues and eigenvectors of the matrix $\begin{bmatrix} 1 & 1 \\ 1 & -1 \end{bmatrix}$. Is the matrix positive definite?

A.8. Evaluate all the leading principal minors of the matrices

$$\begin{bmatrix} 3 & 2 & 1 \\ 2 & 3 & 2 \\ 1 & 2 & 3 \end{bmatrix}, \begin{bmatrix} 3 & 2 & 0 \\ 2 & 3 & 0 \\ 0 & 0 & 1 \end{bmatrix}, \begin{bmatrix} 3 & 2 & 1 \\ 2 & 1 & 2 \\ 1 & 2 & 3 \end{bmatrix}.$$

Which of these matrices are positive definite?

A.9. Let $\mathbf{A} = \mathbf{v}\mathbf{v}^\dagger$ where $\mathbf{v}$ is an $N \times 1$ matrix, i.e., a column vector.
  a) Show that $\mathbf{v}^\dagger \mathbf{v}$ is an eigenvalue of $\mathbf{A}$, with corresponding eigenvector $\mathbf{v}$.

b) Show that the remaining $N - 1$ eigenvalues are equal to zero. Find a set of N independent eigenvectors for $\mathbf{A}$.

A.10. Let $\mathbf{H}$ be Hermitian and $\mathbf{S}$ skew-Hermitian. Show that $\mathbf{v}^\dagger \mathbf{H} \mathbf{v}$ is real and $\mathbf{v}^\dagger \mathbf{S} \mathbf{v}$ imaginary for any choice of $\mathbf{v}$.

A.11.
a) It is possible for a nonzero matrix $\mathbf{A}$ to be such that $\mathbf{v}^T \mathbf{A} \mathbf{v} = 0$ for all $\mathbf{v}$. Show that $\mathbf{A} = \begin{bmatrix} 0 & 1 \\ -1 & 0 \end{bmatrix}$ is such an example.
b) Next show that if $\mathbf{v}^\dagger \mathbf{A} \mathbf{v} = 0$ for all vectors $\mathbf{v}$, then $\mathbf{A} = \mathbf{0}$. (*Hint*: Write $\mathbf{A} = \mathbf{A}_h + \mathbf{A}_s$ where $\mathbf{A}_h$ is Hermitian and $\mathbf{A}_s$ is skew Hermitian.)

A.12. Show that a lower triangular matrix cannot be unitary unless it is diagonal.

A.13. Let $\mathbf{A} = \begin{bmatrix} 1 & 1 \\ 1 & 1 \end{bmatrix}$, $\mathbf{B} = \begin{bmatrix} 1 & 1 \\ -1 & -1 \end{bmatrix}$.
a) Compute $\mathbf{AB}$ and $\mathbf{BA}$ and verify that they do not have the same rank.
b) Compute $\mathbf{I} - \mathbf{AB}$ and $\mathbf{I} - \mathbf{BA}$ and verify that these have the same rank.
c) (This is tricky.) More generally, let $\mathbf{A}$ and $\mathbf{B}$ be arbitrary square matrices of the same size. Show that $\mathbf{I} - \mathbf{AB}$ and $\mathbf{I} - \mathbf{BA}$ have the same rank.

A.14. Find examples of Hermitian positive definite matrices $\mathbf{P}$ and $\mathbf{Q}$ such that $\mathbf{P} - \mathbf{Q}$ is indefinite. (Avoid trivial answers by finding *non diagonal* examples!)

A.15. The product of two right-circulant matrices is right-circulant. Prove this for the 3×3 case.

A.16. Let $\mathbf{A}$ and $\mathbf{B}$ be square matrices with a certain property in common (e.g., unitary, circulant, and so on.) If the product $\mathbf{AB}$ also has this property, we say that the property is preserved under multiplication. Prove by examples that the following properties are not necessarily preserved under multiplication: (i) Hermitian, (ii) Vandermonde, (iii) normal property, and (iv) stability (i.e., all eigenvalues of $\mathbf{A}$ have magnitude less than unity).

A.17. Let $\mathbf{U}$ be a square matrix, and let $\mathbf{y} = \mathbf{U}\mathbf{x}$. Let $\mathbf{U}$ be such that $\mathbf{y}^\dagger \mathbf{y} = \mathbf{x}^\dagger \mathbf{x}$ for all vectors $\mathbf{x}$. Show that $\mathbf{U}^\dagger \mathbf{U} = \mathbf{I}$.

A.18. Prove the following matrix identity

$$\underbrace{\begin{bmatrix} \mathbf{A} & \mathbf{B} \\ \mathbf{C} & \mathbf{D} \end{bmatrix}}_{\mathbf{R}} = \begin{bmatrix} \mathbf{I} & \mathbf{B}\mathbf{D}^{-1} \\ \mathbf{0} & \mathbf{I} \end{bmatrix} \begin{bmatrix} \mathbf{A} - \mathbf{B}\mathbf{D}^{-1}\mathbf{C} & \mathbf{0} \\ \mathbf{0} & \mathbf{D} \end{bmatrix} \begin{bmatrix} \mathbf{I} & \mathbf{0} \\ \mathbf{D}^{-1}\mathbf{C} & \mathbf{I} \end{bmatrix},$$

where $(\mathbf{A}, \mathbf{B}, \mathbf{C}, \mathbf{D})$ are matrices of appropriate dimensions, $\mathbf{A}$ and $\mathbf{D}$ are square (hence $\mathbf{R}$ is square), and $\mathbf{D}$ is nonsingular. Hence prove that

$$\det \mathbf{R} = [\det \mathbf{D}] \times \det [\mathbf{A} - \mathbf{B}\mathbf{D}^{-1}\mathbf{C}].$$

A.19. Let $\mathbf{P}$ be $N \times N$ Hermitian positive definite. Partition it as

$$\mathbf{P} = \begin{bmatrix} P_{00} & \mathbf{p}_{10}^\dagger \\ \mathbf{p}_{10} & \mathbf{P}_{11} \end{bmatrix}. \tag{$PA.19a$}$$

Here P_{00} is scalar, whereas $\mathbf{P}_{11}$ is $(N-1) \times (N-1)$. Evidently $\mathbf{p}_{10}$ is a column vector.

a) Using the definition of positive definiteness, show that P_{00} is real and positive. Also show that $\mathbf{P}_{11}$ is Hermitian positive definite.

b) Using the previous problem, show that

$$\det \mathbf{P} = \left(P_{00} - \mathbf{p}_{10}^{\dagger} \mathbf{P}_{11}^{-1} \mathbf{p}_{10}\right) \det \mathbf{P}_{11}. \qquad (PA.19b)$$

c) Using some or all of the results proved above, show that

$$\det \mathbf{P} \leq P_{00}[\det \mathbf{P}_{11}], \qquad (PA.19c)$$

with equality if and only if $\mathbf{p}_{10} = \mathbf{0}$.

d) Let P_{ii} denote the diagonal elements of $\mathbf{P}$. By repeated application of the above result, prove that

$$\det \mathbf{P} \leq \prod_{i=0}^{N-1} P_{ii}, \qquad (PA.19d)$$

with equality if and only if $\mathbf{P}$ is diagonal.

A.20. Let $\mathbf{P}$ be as in Problem A.19. We now provide a second proof of the inequality in (PA.19d).

a) Prove that $\mathbf{P}$ can be written as $\mathbf{P} = \mathbf{D}\mathbf{Q}\mathbf{D}^{\dagger}$ where $\mathbf{D}$ is a diagonal matrix of positive elements, and $Q_{ii} = 1$ for all i.

b) Hence show that $[\det \mathbf{P}] = [\det \mathbf{Q}] \times [\prod_{i=0}^{N-1} P_{ii}]$.

c) Now show that $\det \mathbf{Q} \leq 1$, with equality if and only if $\mathbf{Q} = \mathbf{I}$. This establishes the desired result (PA.19d). [*Hint.* The determinant is the product of eigenvalues and the trace is the sum of eigenvalues. Use this along with the arithmetic/geometric mean inequality, which can be found in Appendix C; see discussion around Eq. (C.2.3).]

Appendix B

Review of random processes

B.0 INTRODUCTION

This appendix reviews some results from the theory of random processes, which will be essential when analyzing roundoff noise in filter banks (Chap. 9). These results will also be used in studying the effects of subband quantization in Appendix C. Detailed treatements of random processes can be found in a number of references, e.g., Papoulis [1965], Davenport [1970], Oppenheim and Schafer [1975], Peebles [1987], and Therrien [1992].

B.1 REAL RANDOM VARIABLES

We assume familiarity with the notions of probability and random variables, which are basic to all further discussions. Let X be a real *random variable* (abbreviated as rv or r.v.).[†] We will deal with *continuous random variables*. This means that X can take any value in a continuous range such as $a \leq X \leq b$, where a and b may not be finite. In this case, we have to describe the r.v. by a probability density function (rather than just probability), denoted as $f_X(x)$. This function is defined such that the integral

$$\int_{x_1}^{x_2} f_X(x) dx, \qquad (B.1.1)$$

represents the probability that X is in the range $x_1 \leq x \leq x_2$. It satisfies the following properties: (a) $f_X(x) \geq 0$ and (b) $\int_{-\infty}^{\infty} f_X(x) dx = 1$.

Figure B.1-1 shows an example of $f_X(x)$ called the uniform density function. It is easy to see that this satisfies the property $\int_{-\infty}^{\infty} f_X(x) dx = 1$. The density function $f_X(x)$ can exceed unity, since it is not a probability by itself. In the example shown, if $(b-a) < 1$, the value of $f_X(x)$ exceeds unity. Fig. B.1-2 shows another popular

---

[†] Formally, an r.v. is defined to be a mapping from a *sample space* to the real line (more generally the complex plane); see Papoulis [1965].

density called the Gaussian density. This is analytically given by

$$f_X(x) = \frac{1}{\sigma_X \sqrt{2\pi}} \exp\left(\frac{-(x-m_X)^2}{2\sigma_X^2}\right). \quad (B.1.2)$$

The meanings of m_X and σ_X will be explained below. For the moment note that as m_X decreases, the center of the plot shifts to the left, whereas as σ_X decreases the plot gets narrower and taller. †

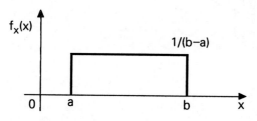

Figure B.1-1 The uniform probability density function.

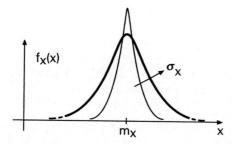

Figure B.1-2 The Gaussian density function.

Expected Values

Let $g(X)$ be a function of an r.v. X (e.g., $g(X) = X^2, \sin X$ etc.). In general $g(X)$ is itself a random variable. The expected value of $g(X)$ is defined to be

$$E[g(X)] = \int_{-\infty}^{\infty} g(x) f_X(x) dx. \quad (B.1.3)$$

Here are some standard examples: $m_X \triangleq E[X]$ is the expected value of the r.v. X. The quantity $E[X^2]$ is called the mean square value of X. The quantity $E[(X-m_X)^2]$ which is the mean square value of the r.v. $(X - m_X)$ is called the variance of X, denoted σ_X^2. Summarizing,

$$\text{Expected value } m_X = E[X]$$
$$\text{Mean square value} = E[X^2] \quad (B.1.4)$$
$$\text{Variance } \sigma_X^2 = E[(X-m_X)^2].$$

---

† In this section we have used upper case letters such as X for the random variables, and lower case letters such as x, x_1 for the values assume by the random variables. This is a useful notation, but sometimes becomes infeasible because of conflicting notational conventions, and other demands for upper case letters. In many cases we will leave it to the reader's judgement to make the distinction between an r.v. and the value assigned to it.

804 App. B. Review of random processes

These are related as
$$E[X^2] = \sigma_X^2 + m_X^2. \quad (B.1.5)$$

m_X is also called the mean value or *mean* of X. The quantity σ_X (square root of variance) is called the *standard deviation* of X. From these definitions one can verify that for the Gaussian density, the mean and variance are indeed given by m_X and σ_X^2 appearing in (B.1.2). Refering to Fig. B.1-2 we see that the mean m_X agrees with the center of the plot. (This is not in general true for arbitrary density functions.) For the uniform density function one can verify that

$$m_X = \frac{b+a}{2}, \quad \sigma_X^2 = \frac{(b-a)^2}{12}. \quad (B.1.6)$$

In general, the function $f_X(x)$ can have impulses (Dirac delta functions) in it. These are required to take care of the discrete nature of the density function in some applications; for example, if the probability for X to take the value 7 is a half, then $f_X(x)$ must have a term $0.5\delta_a(x-7)$. In this text, we have no need to accomodate such impulses, so we assume $f_X(x)$ to be impulse-free.

Collection of Random Variables

Suppose X and Y are two random variables. These are jointly described by the *joint probability density* function $f_{XY}(x,y)$. This is defined such that

$$\int_{x=x_1}^{x_2} \int_{y=y_1}^{y_2} f_{XY}(x,y)dxdy, \quad (B.1.7)$$

is equal to the probability that X and Y are in the range $x_1 \le X \le x_2$ and $y_1 \le Y \le y_2$. The joint density is nonnegative and such that if we set $x_1 = y_1 = -\infty$ and $x_2 = y_2 = \infty$, the above integral reduces to unity. Moreover, the density function of one of the random variables, say that of X, can be recovered from $f_{XY}(x,y)$ by integrating over the other r.v. Y, that is,

$$f_X(x) = \int_{y=-\infty}^{\infty} f_{XY}(x,y)dy. \quad (B.1.8)$$

$f_Y(y)$ can be obtained similarly. $f_X(x)$ and $f_Y(y)$ are said to be *marginal density* functions.

Given a function $g(X,Y)$ of the two random variables, we define its expected value as

$$E[g(X,Y)] = \int_{x=-\infty}^{\infty} \int_{y=-\infty}^{\infty} g(x,y) f_{XY}(x,y)dxdy. \quad (B.1.9)$$

The quantity
$$R_{XY} \triangleq E[XY], \quad (B.1.10)$$

is called the *cross correlation* between the real random variables X and Y. The quantity
$$C_{XY} \triangleq E[(X-m_X)(Y-m_Y)], \quad (B.1.11)$$

is called the *cross covariance* between X and Y. These are related as

$$R_{XY} = C_{XY} + m_X m_Y, \quad (B.1.12)$$

a generalization of (B.1.5).

Depending on the behavior of $f_{XY}(x, y)$, R_{XY} and C_{XY}, two random variables are often classified into useful types. We say that X and Y are

1. *statistically independent* if $f_{XY}(x, y) = f_X(x) f_Y(y)$.
2. *uncorrelated* if $E[XY] = E[X]E[Y]$, that is, $R_{XY} = m_X m_Y$ i.e., $C_{XY} = 0$.
3. *orthogonal* if $E[XY] = 0$.

Note that uncorrelatedness and orthogonality are identical, if one of the random variables has zero mean. It can be shown that statistical independence implies uncorrelatedness but the converse is in general not true. For example, let $X = \cos\theta$ and $Y = \sin\theta$ where θ is a real r.v., with uniform density function in the range $0 \leq \theta < 2\pi$. Then X and Y are uncorrelated but not statistically independent. If $f_{XY}(x, y)$ is Gaussian (Sec. B.5), then it can be shown that the uncorrelated property implies independence.

In a manner identical to the above, if we are given several random variables $X_0, X_1, \ldots X_{M-1}$, we can define a joint density function and various expected values. See Sec. B.5.

B.2 REAL RANDOM PROCESSES

Let $\{X(n)\}$ be a real sequence such that each sample $X(n)$ is a random variable. We say that $\{X(n)\}$ is a random process (with braces ususally omitted). (Once again a formal definition based on a mapping from a "sample space" to the space of functions can be found in the references mentioned before). We use $X(n)$ to indicate the random process and $x(n)$ to denote a particular realization, i.e., $x(n)$ is the value taken by the random variable $X(n)$ in a particular measurement.

When we attempt to characterize the random process, several practical difficulties arise. Each sample $X(n)$ is characterized by a density function, and pairs of samples such as $X(n-1), X(n)$ are characterized by two dimensional joint density functions. In fact, given any set of samples such as $X(n_1), X(n_2), \ldots X(n_M)$, we have to characterize them by a M dimensional joint density. One can see that the complete characterization becomes very complicated. Note also that in general $x(n)$ does not have finite energy, and its z-transform defined in the usual way may not converge anywhere. So we require a different tool to understand and characterize these waveforms.

However, as we will see, a partial characterization based on expected values is sufficient for many applications, for example, noise analysis in digital filter banks. We can define various kinds of expected values for any random process. The quantity $E[X(n)]$ is the mean of the random process, and in general depends on n. The quantity

$$R(m, n) = E[X(m)X(n)], \qquad (B.2.1)$$

is the cross correlation between the real random variables $X(m)$ and $X(n)$.

Wide sense stationary (WSS) processes

A wide sense stationary (WSS) random process is one for which $E[X(n)]$ is independent of n and $R(m, n)$ depends only on the difference $m - n$. We can therefore define the functions

$$\begin{aligned} m_X &= E[X(n)] = \text{mean value}, \\ R_{XX}(k) &= E[X(n)X(n-k)] = \text{autocorrelation sequence}. \end{aligned} \qquad (B.2.2)$$

The quantity k in $R_{XX}(k)$ is called the lag variable (as it describes a difference in time index). We say that $R_{XX}(k)$ is the autocorrelation of the WSS process at lag k.

For applications such as noise analysis in linear systems, the noise source (such as the output noise of a roundoff quantizer), can often be modeled satisfactorily as a wide sense stationary random process [Oppenheim and Schafer, 1975]. In these applications, the quantity of main interest is *noise variance* at the system output. This can be calculated from the autocorrelation $R_{XX}(k)$ of the noise source. It is not necessary to know the higher dimensional density functions (or even the two dimensional density function) in order to perform this analysis. This simplifies our study of random processes, as well as noise analysis to a large extent. From this point on, we will concentrate entirely on WSS random processes.

From $R_{XX}(k)$ we can define several useful quantities: the energy of the random process is

$$E[X^2(n)] = R_{XX}(0), \qquad (B.2.3)$$

and the variance of the process is

$$\sigma_X^2 = E[(X(n) - m_X)^2]. \qquad (B.2.4)$$

These quantities are related as

$$\sigma_X^2 = R_{XX}(0) - m_X^2. \qquad (B.2.5)$$

The covariance squence $C_{XX}(k)$ is defined as

$$C_{XX}(k) = E[(X(n) - m_X)(X(n-k) - m_X)], \qquad (B.2.6)$$

and leads to

$$R_{XX}(k) = C_{XX}(k) + m_X^2. \qquad (B.2.7)$$

$C_{XX}(k)$ is the autocorrelation of the zero mean process $X(n) - m_X$.

Power Spectral Density

The power spectrum (or power spectral density) of the WSS process is defined as the Fourier transform of $R_{XX}(k)$, that is,

$$S_{XX}(e^{j\omega}) = \sum_{k=-\infty}^{\infty} R_{XX}(k) e^{-j\omega k}. \qquad (B.2.8)$$

For a real process, $R_{XX}(k)$ is symmetric (that is, $R_{XX}(-k) = R_{XX}(k)$), so that $S_{XX}(e^{j\omega})$ is real-valued. In fact it can be shown that $S_{XX}(e^{j\omega}) \geq 0$ (whether the process is real or complex). Note that we can write $S_{XX}(e^{j\omega})$ as a sum of two terms:

$$S_{XX}(e^{j\omega}) = \sum_{k=-\infty}^{\infty} C_{XX}(k) e^{-j\omega k} + m_X^2 2\pi \sum_{m=-\infty}^{\infty} \delta(\omega + 2\pi m). \qquad (B.2.9)$$

From the relation (B.2.8), we can express $R_{XX}(0)$ as

$$R_{XX}(0) = \int_{-\pi}^{\pi} S_{XX}(e^{j\omega}) \frac{d\omega}{2\pi}. \qquad (B.2.10)$$

In particular, for a zero mean WSS process, the variance is equal to the above integral.

In many cases, the correlation between the samples $x(n)$ and $x(n-k)$ becomes weaker as k grows. (This is not always true, for example if $R_{XX}(k)$ has periodic components). This means that $C_{XX}(k) \to 0$ as $k \to \infty$, i.e., $R_{XX}(k) \to m_X^2$ as $k \to \infty$. So $S_{XX}(e^{j\omega})$ has an impulse component $2\pi m_X^2 \delta_a(\omega)$, and a nonimpulsive component due to $C_{XX}(k)$ (Fig. B.2-1).

For our applications, $S_{XX}(e^{j\omega})$ will be assumed to be free from impulse functions except possibly at $\omega = 0$. This means that $R_{XX}(k)$ does not have any periodic components except the m_X^2 term which has period one. In this case, Fig. B.2-1(a) is a typical plot of $R_{xx}(k)$.

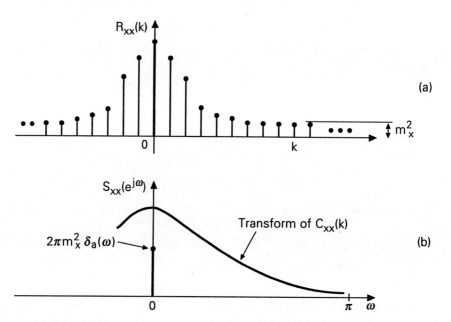

Figure B.2-1 (a) A typical autocorrelation sequence, asymptotically tending to m_x^2, and (b) the corresponding power spectrum, revealing an impulse at $\omega = 0$, representing nonzero mean.

White Random Process

A random process is said to be white if any pair of samples are uncorrelated, i.e., $E[X(n)X(m)] = E[X(n)]E[X(m)]$ for $m \neq n$. If a white process is WSS, then

$$R_{XX}(k) = \begin{cases} m_X^2 & k \neq 0 \\ \sigma_X^2 + m_X^2 & k = 0. \end{cases} \quad (B.2.11)$$

This can be compactly expressed as

$$R_{XX}(k) = \sigma_X^2 \delta(k) + m_X^2. \quad (B.2.12)$$

Evidently the covariance sequence is $C_{XX}(k) = \sigma_X^2 \delta(k)$. The power spectrum of a white WSS process is given by

$$S_{XX}(e^{j\omega}) = \sigma_X^2 + 2\pi m_X^2 \delta(\omega), \qquad (B.2.13)$$

in $-\pi \leq \omega < \pi$, and repeats with period 2π. In other words, it has a constant component with height σ_X^2 representing the variance of $X(n)$, and an impulse component representing m_X^2. A zero-mean WSS white random process is therefore characterized by a flat power spectrum (height σ_X^2) and an autocorrelation $R_{XX}(k) = \sigma_X^2 \delta(k)$. Fig. B.2-2 summarizes these. If a WSS process is not white, it is said to be *colored*.

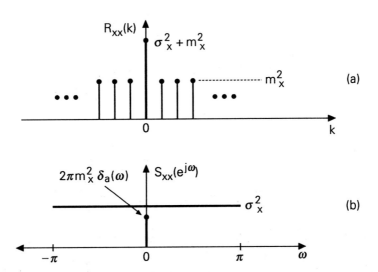

Figure B.2-2 (a) The autocorrelation sequence of a white WSS process, and (b) corresponding power spectral density.

Joint Description of Two or More Random Processes

In many applications, it is necessary to deal with the effect of several random signals at a time. In a digital filter structure, for example, there may exist a number of noise waveforms, generated by a number of quantizers, and it is necessary to find the effective noise power at the filter output.

Suppose $X(n)$ and $Y(n)$ are two real random processes. These are said to be *jointly WSS* if each process is WSS and if $E[X(n)Y(n-k)]$ is a function only of the lag k. In this case we define

$$R_{XY}(k) = E[X(n)Y(n-k)], \qquad (B.2.14)$$

which is called the cross-correlation between the processes. The two processes are said to be uncorrelated if $R_{XY}(k) = E[X(n)]E[Y(n-k)] = m_X m_Y$ for all k. If one of the processes has zero mean, then uncorrelatedness implies $R_{XY}(k) = 0$ for all k.

Ergodicity. A WSS process is said to be ergodic if the statistical averages (such as $E[X(n)]$, $E[X^2(n)]$ etc.), are equal to the corresponding time averages over any single realization of the process. The ergodicity assumption enables us to estimate these expected values by using time averages.

B.3 PASSAGE THROUGH LTI SYSTEMS

Consider Fig. B.3-1 where $H(z) = \sum_n h(n)z^{-k}$ represents the transfer function of a (stable) LTI system with real impulse response $h(n)$. Suppose the input $X(n)$ is a real WSS random process with mean m_X, autocorrelation $R_{XX}(k)$, and power spectral density $S_{XX}(e^{j\omega})$. Then $Y(n)$ is a WSS random process. Letting m_Y, $R_{YY}(k)$ and $S_{YY}(e^{j\omega})$ be its mean, autocorrelation and power spectrum, we can make the following statements.

a) $m_Y = m_X \sum_n h(n)$.
b) $S_{YY}(e^{j\omega}) = S_{XX}(e^{j\omega})|H(e^{j\omega})|^2$.
c) $R_{YY}(k) = \sum_n R_{hh}(n) R_{XX}(k-n)$ where $R_{hh}(k) = \sum_n h(n)h(n-k)$.

$R_{hh}(k)$ is called the (deterministic) autocorrelation of $h(n)$. All summations above are from $-\infty$ to ∞. Thus $R_{YY}(k)$ is the convolution of $R_{hh}(k)$ with $R_{XX}(k)$.

As an application of this, suppose $x(n)$ is zero mean WSS white with variance σ_X^2. Then

$$R_{XX}(k) = \sigma_X^2 \delta(k), \quad \text{and} \quad S_{XX}(e^{j\omega}) = \sigma_X^2, \qquad (B.3.1)$$

so that

$$R_{YY}(k) = \sigma_X^2 R_{hh}(k), \quad \text{and} \quad S_{YY}(e^{j\omega}) = \sigma_X^2 |H(e^{j\omega})|^2. \qquad (B.3.2)$$

In other words, the output power spectrum is precisely equal to the magnitude-squared response of $H(z)$ (scaled by σ_X^2). The output process $Y(n)$ is therefore not white but becomes 'colored' by the LTI system $H(z)$ (unless $H(z)$ is allpass). The variance of the output process is

$$R_{YY}(0) = \sigma_X^2 R_{hh}(0) = \sigma_X^2 \sum_n h^2(n). \qquad (B.3.3)$$

In other words, the variance gets amplified by the energy in the impulse response $h(n)$.

Figure B.3-1 Passing a random process through an LTI system.

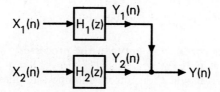

Figure B.3-2 Two uncorrelated WSS processes going through two systems.

Fig. B.3-2 shows another application of the concepts introduced above. Here $H_1(z) = \sum_n h_1(n) z^{-n}$ and $H_2(z) = \sum_n h_2(n) z^{-n}$ are two stable LTI systems with real impulse responses. Suppose the inputs $X_1(n)$ and $X_2(n)$ are zero mean, white, real, jointly WSS random processes. Assume further that the two processes are uncorrelated as defined in the previous section. It can then be shown that $Y_1(n)$ and $Y_2(n)$ are zero mean, (possibly colored), real, jointly WSS random processes, *uncorrelated* to each other. The variances of $Y_1(n)$ and $Y_2(n)$ are, respectively,

$$\sigma_{Y_1}^2 = \sigma_{X_1}^2 \sum_n h_1^2(n), \quad \sigma_{Y_2}^2 = \sigma_{X_2}^2 \sum_n h_2^2(n). \qquad (B.3.4)$$

Because of the fact that $Y_1(n)$ and $Y_2(n)$ are uncorrelated, it can be shown that the variance of $Y(n)$ in Fig. B.3-2 is the sum of variances of $Y_1(n)$ and $Y_2(n)$. So $Y(n)$ is a zero mean WSS random process with variance

$$\sigma_Y^2 = \sigma_{Y_1}^2 + \sigma_{Y_2}^2. \qquad (B.3.5)$$

This kind of an analysis finds applications in studying roundoff noise in filter banks (Chap. 9).

B.4 THE COMPLEX CASE

A complex r.v. $X = X_r + jX_i$ is a complex quantity whose real and imaginary parts are random variables, possibly correlated. We can define the joint density function $f_{X_r, X_i}(x_r, x_i)$ to describe the pair of random variables (X_r, X_i). In defining the expected value of functions of X (such as $E[X]$, $E[|X|^2]$ and so on), we merely use this joint density in the appropriate integral. Quantities such as mean and variance generalize nicely. For example the variance is defined now as $E[|X - m_X|^2]$. To make the presentation more efficient, we shall discuss the joint properties of two complex random variables X and Y (such as cross correlation etc.), and set $X = Y$ to obtain the single complex r.v. case. The following is a summary of definitions and properties.

Definitions

1. $m_X \triangleq E[X]$ = mean value of X.
2. $R_{XY} \triangleq E[XY^*]$ = cross correlation between X and Y.
3. $C_{XY} \triangleq E[(X - m_X)(Y - m_Y)^*]$ = cross covariance between X and Y.
4. $\sigma_X^2 \triangleq C_{XX}$ = variance of X.
5. X, Y uncorrelated if $R_{XY} = m_X m_Y^*$.
6. X, Y orthogonal if $R_{XY} = 0$.
7. X, Y (statistically) independent if $f_{XY}(x, y) = f_X(x) f_Y(y)$.

Properties

1. $R_{XX} = \sigma_X^2 + |m_X|^2$.
2. $R_{XY} = R_{YX}^*$ and $C_{XY} = C_{YX}^*$.
3. $R_{XY} = C_{XY} + m_X m_Y^*$.
4. $R_{XX} = E[XX^*] = E[|X|^2]$ = mean square value.
5. X, Y uncorrelated if and only if $C_{XY} = 0$. Statistical independence implies uncorrelatedness, but converse is not true. Uncorrelatedness is same as orthogonality if m_X or m_Y is zero.

Complex Random Processes

A complex random process is defined in the same way as a real process except that $X(n)$ is now complex. We define the mean to be $E[X(n)]$ and autocorrelation to be $E[X(n)X^*(n-k)]$. The process is called WSS if the mean is constant and the autocorrelation is independent of n. Two processes $X(n)$ and $Y(n)$ are jointly WSS, if each of them is individually WSS, and the cross correlation defined as $E[X(n)Y^*(n-k)]$ is independent of n. For the jointly WSS case, we summarize various definitions and properties next.

Definitions

1. $m_X = E[X(n)] =$ mean value.
2. $R_{XY}(k) = E[X(n)Y^*(n-k)] =$ cross correlation between $X(n)$ and $Y(n)$.
3. $C_{XY}(k) = E[(X(n) - m_X)(Y(n-k) - m_Y)^*] =$ cross covariance between $X(n)$ and $Y(n)$.
4. $S_{XY}(e^{j\omega}) = \sum_{k=-\infty}^{\infty} R_{XY}(k)e^{-j\omega k} =$ cross power spectral density.

Properties

1. $R_{XY}(k) = C_{XY}(k) + m_X m_Y^*$.
2. $R_{YX}(k) = R_{XY}^*(-k)$.
3. $S_{YX}(e^{j\omega}) = S_{XY}^*(e^{j\omega})$.
4. In particular,
   $R_{XX}(k) = C_{XX}(k) + |m_X|^2$.
   $R_{XX}(k) = R_{XX}^*(-k)$ (Hermitian symmetric sequence).
   $S_{XX}(e^{j\omega})$ is real (in fact nonnegative).

Passage Through LTI Systems

Suppose $X(n)$ is a complex WSS process. Let this be input to a stable LTI system $H(z) = \sum_n h(n)z^{-n}$, to generate output $Y(n)$. Then $Y(n)$ is WSS. We will now state further properties of the process $Y(n)$. For this define $R_{hh}(k)$ to be the deterministic autocorrelation of $h(n)$, that is,

$$R_{hh}(k) = \sum_n h(n)h^*(n-k).$$

Or equivalently, $R_{hh}(k) = \sum_n h^*(n)h(n+k)$. Taking Fourier transforms on both sides, we get $\sum_k R_{hh}(k)e^{-j\omega k} = |H(e^{j\omega})|^2$. The following properties are true.

1. $m_Y = m_X \sum_n h(n)$.
2. $R_{YY}(k) = \sum_n R_{XX}(n)R_{hh}(k-n)$. (convolution)
3. $S_{YY}(e^{j\omega}) = S_{XX}(e^{j\omega})|H(e^{j\omega})|^2$.
4. $R_{YX}(k) = \sum_m h(m)R_{XX}(k-m)$. (convolution)
5. $S_{YX}(e^{j\omega}) = H(e^{j\omega})S_{XX}(e^{j\omega})$.

We can obtain the case of real processes (with real LTI systems) by dropping the conjugate signs.

B.5. THE VECTOR CASE

At the end of Sec. B.1 it was mentioned that a collection of random variables $X_0 \ldots X_{M-1}$ can be described by a joint density function. We denote this as $f_{\mathbf{X}}(\mathbf{x})$

where $\mathbf{X} = [X_0 \ldots X_{M-1}]^T$. The vector $\mathbf{X}$ is said to be a random vector, or just r.v. for simplicity. $f_{\mathbf{X}}(\mathbf{x})$ is a nonnegative scalar function of the vector $\mathbf{x}$ and has properties similar to $f_{XY}(x,y)$ in Sec. B.1.

The 'mean' or expectation $E[\mathbf{X}]$ is the vector $[E[X_0] \ldots E[X_{M-1}]]^T$. The vector $\mathbf{X}$ has zero mean if this expectation is the null vector. We define the *cross correlation matrix* between the vectors $\mathbf{X}$ and $\mathbf{Y}$ as $E[\mathbf{X}\mathbf{Y}^\dagger]$.

Note in particular that $\mathbf{R} = E[\mathbf{X}\mathbf{X}^\dagger]$ is the *autocorrelation matrix* of $\mathbf{X}$. The autocorrelation of $(\mathbf{X} - E[\mathbf{X}])$ is said to be the *covariance* matrix of $\mathbf{X}$, and is denoted $\mathbf{C}$. Both $\mathbf{R}$ and $\mathbf{C}$ are $M \times M$ Hermitian positive semidefinite matrices. We sometimes use subscripts (e.g., $\mathbf{R}_{\mathbf{xx}}$) when it is necessary to distinguish between more than one random vector. Notice that the k,m element of $\mathbf{C}$ is the cross covariance between X_k and X_m. If every pair of components of $\mathbf{X}$ are uncorrelated, then $\mathbf{C}$ becomes a diagonal matrix. If in addition all components have same variance σ^2 then $\mathbf{C} = \sigma^2 \mathbf{I}$.

Gaussian random vector. An $M \times 1$ real random vector $\mathbf{X}$ with mean $\mathbf{m}$ and covariance $\mathbf{C}$ is said to be Gaussian if the joint density function is given by

$$f_{\mathbf{X}}(\mathbf{x}) = \frac{1}{(2\pi)^{M/2}[\det \mathbf{C}]^{1/2}} \exp\left(-\frac{1}{2}(\mathbf{x}-\mathbf{m})^T \mathbf{C}^{-1}(\mathbf{x}-\mathbf{m})\right). \qquad (B.5.1)$$

In this case we also say that the set of random variables X_k (components of $\mathbf{X}$) are jointly Gaussian. The implicit assumption in the above definition is that $\mathbf{C}$ is nonsingular. If the components of $\mathbf{X}$ are uncorrelated, then $\mathbf{C}$ is diagonal, and the above function can be written as the product $f_{X_0}(x_0) \ldots f_{X_{M-1}}(x_{M-1})$ where $f_{X_i}(x_i)$ are one-variable Gaussian density functions. This shows, in particular, that uncorrelated Gaussain random variables are also statistically independent.

Vector Random Processes

An $M \times 1$ vector random process $\{\mathbf{X}(n)\}$ (with curly braces often omitted) is a sequence where all the samples $\mathbf{X}(n)$ are random vectors. This is said to be a wide sense stationary (WSS) process if the vector $E[\mathbf{X}(n)]$ and the matrix $E[\mathbf{X}(n)\mathbf{X}^\dagger(n-k)]$ are independent of n. In this case, we define the autocorrelation to be

$$\mathbf{R}_{\mathbf{xx}}(k) = E[\mathbf{X}(n)\mathbf{X}^\dagger(n-k)]. \qquad (B.5.2)$$

This is an $M \times M$ matrix sequence. The power spectrum of the WSS process $\mathbf{X}(n)$ is defined as

$$\mathbf{S}_{\mathbf{xx}}(z) = \sum_k \mathbf{R}_{\mathbf{xx}}(k) z^{-k}. \qquad (B.5.3)$$

Thus, $\mathbf{S}_{\mathbf{xx}}(e^{j\omega})$ is the Fourier transform of $\mathbf{R}_{\mathbf{xx}}(k)$. Some of the key properties of these quantities are listed below.

1. $\mathbf{R}_{\mathbf{xx}}(k) = \mathbf{R}_{\mathbf{xx}}^\dagger(-k)$.
2. $\mathbf{S}_{\mathbf{xx}}(e^{j\omega}) = \mathbf{S}_{\mathbf{xx}}^\dagger(e^{j\omega})$.
3. The matrix $\mathbf{R}_{\mathbf{xx}}(0)$ is positive (semi)definite, and so is $\mathbf{S}_{\mathbf{xx}}(e^{j\omega})$ for *all* ω.

Two random processes $\mathbf{X}(n)$ and $\mathbf{Y}(n)$ are said to be jointly WSS if (i) each of them is WSS, and (ii) $E[\mathbf{X}(n)\mathbf{Y}^\dagger(n-k)]$ is independent of n. In this case, we define the cross correlation $\mathbf{R_{xy}}(k)$ and cross power spectrum $\mathbf{S_{xy}}(z)$ as

$$\mathbf{R_{xy}}(k) = E[\mathbf{X}(n)\mathbf{Y}^\dagger(n-k)] \qquad (B.5.4)$$

$$\mathbf{S_{xy}}(z) = \sum_k \mathbf{R_{xy}}(k) z^{-k}. \qquad (B.5.5)$$

Passage through LTI systems. Let $\mathbf{X}(n)$ be a $r \times 1$ WSS random process. Let this be the input to a $p \times r$ LTI system $\mathbf{H}(z)$. Then the output $\mathbf{Y}(n)$ is a $p \times 1$ vector WSS process. The power spectra of the processes are related as

$$\mathbf{S_{yy}}(z) = \mathbf{H}(z)\mathbf{S_{xx}}(z)\widetilde{\mathbf{H}}(z). \qquad (B.5.6)$$

An interesting example occurs when $\mathbf{X}(n)$ is a zero-mean process with

$$\mathbf{R_{xx}}(k) = \begin{cases} 0 & k \neq 0 \\ \sigma^2 \mathbf{I}, & k = 0 \end{cases} \qquad (B.5.7)$$

This means that (a) any two samples $\mathbf{X}(k)$ and $\mathbf{X}(m)$ are uncorrelated, (b) any two components of $\mathbf{X}(k)$ (for any fixed k) are uncorrelated, and (c) each component of $\mathbf{X}(k)$ has the same variance σ^2. We can abbreviate (B.5.7) as $\mathbf{R_{xx}}(k) = \sigma^2 \delta(k)\mathbf{I}$. In this case, $\mathbf{S_{xx}}(z) = \sigma^2 \mathbf{I}$, so that

$$\mathbf{S_{yy}}(z) = \sigma^2 \mathbf{H}(z)\widetilde{\mathbf{H}}(z). \qquad (B.5.8)$$

Vector Processes from Scalar Processes

Suppose $X(n)$ is a scalar process and we form the vector process $\mathbf{X}(n)$ by partitioning $X(n)$ into successive blocks of M samples, that is,

$$\mathbf{X}(n) = [X(nM) \quad X(nM-1) \quad \ldots \quad X(nM-M+1)]^T \qquad (B.5.9)$$

According to the notations of Chap. 10, this is the 'blocked version' of $X(n-M+1)$. If $X(n)$ is WSS, then so is the blocked version $\mathbf{X}(n)$. Furthermore it can be shown that if $X(n)$ is WSS, the power spectrum matrix $\mathbf{S_{xx}}(z)$ of $\mathbf{X}(n)$ has the pseudocirculant property (defined in Sec. 5.7.2).

Let $X(n)$ be real WSS, with autocorrelation $R(k)$. Let $\mathbf{R}_M(k)$ be the autocorrelation sequence of the blocked version $\mathbf{X}(n)$. (The subscript M is a reminder that $\mathbf{X}(n)$ is $M \times 1$.) Then $\mathbf{R}_M(0)$ is a (real) symmetric Toeplitz matrix (Appendix A). For example, with $M = 3$, we have

$$\mathbf{R}_3(0) = \begin{bmatrix} R(0) & R(1) & R(2) \\ R(1) & R(0) & R(1) \\ R(2) & R(1) & R(0) \end{bmatrix}. \qquad (B.5.10)$$

$\mathbf{R}_M(0)$ is said to be the $M \times M$ autocorrelation matrix associated with the scalar process $X(n)$. This matrix plays a fundamental role in several problems, for example, optimal linear prediction [Jayant and Noll, 1984]. Clearly $\mathbf{R}_0(0) = R(0) > 0$ and is nonsingular. However, $\mathbf{R}_M(0)$ may become singular for sufficiently large M.

If $\mathbf{R}_M(0)$ is singular then so is $\mathbf{R}_n(0)$ for any $n > M$ (since $\mathbf{R}_M(0)$ is the upper left submatrix of $\mathbf{R}_n(0)$).

Harmonic processes. A scalar WSS process is Harmonic(N) if its power spectrum $S_{xx}(e^{j\omega})$ is zero everywhere except at N frequencies ω_m in the range $0 \leq \omega < 2\pi$. In other words,

$$S_{xx}(e^{j\omega}) = 2\pi \sum_{m=0}^{N-1} c_m \delta_a(\omega - \omega_m), \quad 0 \leq \omega < 2\pi, \quad (B.5.11)$$

with $c_m \neq 0$ for any m. Thus the autocorrelation is a sum of 'single frequency terms', that is,

$$R_{xx}(k) = \sum_{m=0}^{N-1} c_m e^{j\omega_m k}. \quad (B.5.12)$$

It can be shown that $X(n)$ is Harmonic(N) if $\mathbf{R}_{N+1}$ is singular but $\mathbf{R}_N$ nonsingular.

Appendix C

Quantization of Subband Signals

C.0 INTRODUCTION

In this appendix we study the quantization in the subbands of subband coders (Sec. 4.5) and transform coders (Sec. 6.6). In both of the above coders, the main aim is to quantize a set of "derived" signals (subband signals) rather than the original signal directly. The benefits to be gained by such quantization can be quantified by using the concept of "coding gain." In this appendix we will study the coding gain. We will also consider the problem of optimal bit allocation among the subbands, and obtain expressions for the optimal coding gains for subband coding as well as orthogonal transform coding. We also briefly review a coding technique called differential PCM, and state the connections to the so-called rate-distortion function in information theory.

Some of the early references on these topics are Huang and Schultheiss [1963], and Segall [1976]. More recent references include Ramstad [1982], Woods and O'Neil [1986], Westerink, et al. [1988], Jain [1989], Mallat [1989a], Akansu and Liu [1991], Pearlman in Woods [1991], and Gersho and Gray [1992]. A detailed and systematic treatment can be found in Chap. 11 and 12 of Jayant and Noll [1984], with many practical examples and illustrations. Our aim here is to provide a brief review.

We will assume that the input $x(n)$ is a real zero-mean wide sense stationary (WSS) random process. So all the filtered and decimated signals will share this property. The results reviewed in Appendices A and B will be freely used here. The abbreviation r.v. stands for "random variable."

C.1 QUANTIZER NOISE VARIANCE

First consider Fig. C.1-1 where a real r.v. x is quantized to b bits, to obtain the signal v. Let the most significant (leftmost) bit have weight 2^ℓ. So the permissible range for x is $-2^{\ell+1} < x < 2^{\ell+1}$ (there is a sign bit which we will never show explicitly).

The rightmost bit has weight $2^{-(b-1)}2^\ell$. The quantization step size is therefore $\Delta = 2^{-(b-1)}2^\ell$, that is, $\Delta = 2^{(\ell+1)}2^{-b}$. Note that this agrees with (9.2.1) where we considered the special case of $\ell = -1$ (fixed point fraction). Essentially ℓ determines the allowed size of the number whereas b determines the accuracy or precision used for its representation. In Fig. C.1-1 we do not indicate the binary point any more as it will be irrelevant for our discussion.

The quantization error (or noise) is $q = v - x$. We assume that this is an r.v., uniformly distributed in $-\Delta/2 \leq q \leq \Delta/2$, as in a roundoff quantizer. (This assumption is reasonable only when b is sufficiently large.) Then, its variance is $\sigma^2 = \Delta^2/12$, that is,

$$\sigma^2 = \frac{2^{2(\ell+1)}2^{-2b}}{12} \quad \text{(quantizer } error \text{ or } noise \text{ variance)}. \qquad (C.1.1)$$

Fig. C.1-2 shows a 2-bit example, for two choices of ℓ. There are seven levels because of the unshown sign bit. Thus for $\ell = 1$, the 2-bit number can take the values $0, \pm 1, \pm 2$ and ± 3.

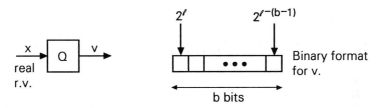

Figure C.1-1 Quantization of a real random variable to b bits.

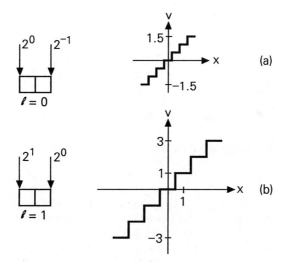

Figure C.1-2 Two cases of quantization with $b = 2$. (a) $\ell = 0$ and (b) $\ell = 1$.

We see that the choice of ℓ governs the range over which x can vary without causing overflow (i.e., this range is the *dynamic range*). If we increase this range by increasing ℓ, the quantizer noise variance σ^2 increases.

There is an elegant way in which we can mathematically relate the quantizer noise variance, the signal variance σ_x^2 (that is, variance of x) and b. This is done by relating ℓ to σ_x^2, and then eliminating ℓ from (C.1.1). The details of this depends on the statistics of x. We now demonstrate this with two examples.

Uniform input. First suppose that the input x is a uniformly distributed r.v., in the permissible range $-2^{\ell+1} < x < 2^{\ell+1}$. Then its variance is

$$\sigma_x^2 = \frac{2^{2(\ell+1)}}{3} \qquad (C.1.2a)$$

so that the quantization error variance is

$$\sigma^2 = 0.25 \times 2^{-2b}\sigma_x^2 \qquad \text{(for uniform r.v. } x\text{)} \qquad (C.1.2b)$$

Gaussian input. As a second example, let x be a zero mean Gaussian r.v. with variance σ_x^2. In this case its range of values is unlimited. Since the *permissible* range is only $-2^{\ell+1} < x < 2^{\ell+1}$, there is a nonzero probability of overflow. (The probability of overflow in the preceding example was zero.) Let ℓ be chosen such that $2^{(\ell+1)} = 3\sigma_x$. This means that the probability that x will exceed the permitted range is equal to 0.0026 [pp. 314, Peebles, 1987]. Using the value $\sigma_x = 2^{(\ell+1)}/3$ and the expression (C.1.1) we find the quantization error variance to be

$$\sigma^2 = 0.75 \times 2^{-2b}\sigma_x^2 \qquad \text{(for Gaussian r.v. } x\text{)}. \qquad (C.1.3)$$

In both the above examples, we see that, for a given number of bits b, the quantization error variance is proportional to signal variance, and to the factor 2^{-2b}, that is,

$$\sigma^2 = c \times 2^{-2b}\sigma_x^2 \qquad (C.1.4)$$

This relation is in fact true for a broader class of signals than the above examples [Chap. 4, Jayant and Noll, 1984]. The constant of proportionality c depends on the statistics of x and on the allowed overflow probability. If we increase ℓ to reduce the overflow probability, then the noise variance σ^2 increases. This is said to be the *noise/dynamic range* interaction. We will repeatedly use the expression (C.1.4) in this appendix. We will implicitly assume that c is the same for all the random variables entering a particular discussion. This is reasonable if all the density functions have the same form, for example, Gaussian.

C.2 THE IDEAL SUBBAND CODER

Consider the M-channel QMF bank reproduced in Fig. C.2-1. Here the decimated subband signals $v_k(n)$ are quantized as in any practical system. Let the quantizer Q_k in the kth channel be a b_k bit quantizer (in the same sense as in Fig. C.1-1). Let 2^{ℓ_k} denote the weight on the left most bit. We can replace the quantizer with a noise source $q_k(n)$ as shown by the broken lines. We will assume that $q_k(n)$ is wide sense stationary (WSS) and white, and is uniformly distributed with zero mean and variance $\sigma_{q_k}^2$. Also, any two noise sources are assumed uncorrelated[†].

Imagine that the filters are ideal and nonoverlapping, with equal bandwidths (Fig. C.2-2). [‡] We consider the real coefficient case for simplicity, so that only the

---

[†] These assumptions are easy to criticize, because b_k is often small. However, without such assumptions, it is not possible to gain much theoretical insight into the operations of the system.

[‡] This implies, in particular, that the subband signals are uncorrelated with each other – recall Problem 9.12. For further comments on this see Sec. C.4.1.

frequency region $0 \leq \omega \leq \pi$ is shown. This system will be called the *ideal* subband coding (SBC) system. We assume $H_k(e^{j\omega}) = F_k(e^{j\omega})$ for all k, and that $H_k(e^{j\omega}) = \sqrt{M}$ in the passband and $H_k(e^{j\omega}) = 0$ in the stopband. The ideal nature of filters ensures that there is no aliasing due to decimation. The distortion function $T(z)$ [Eq. (5.4.14)] is unity so that we would have $\hat{x}(n) = x(n)$ (perfect reconstruction) if there were no quantizers. We will now analyze the effect of quantizers on the reconstruction error.

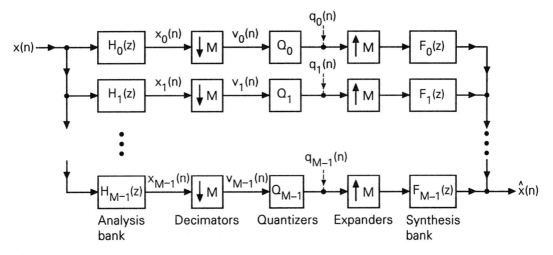

Figure C.2-1 The M-channel filter bank with quantizers in the subbands.

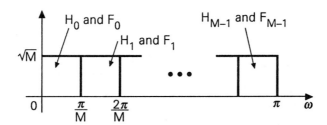

Figure C.2-2 Magnitude responses of filters for the ideal subband coding system.

Even though $q_k(n)$ is WSS, the output of the expander is not (Sec. 9.3). It can be shown, however, that if the bandpass filters $F_k(e^{j\omega})$ are ideal as above, then their outputs *are* WSS, and have power spectral density $\sigma_{q_k}^2 |F_k(e^{j\omega})|^2/M$ [Sathe and Vaidyanathan, 1993]. This therefore represents bandlimited white noise, confined entirely to the passband of $F_k(z)$. Its variance is $\sigma_{q_k}^2/M$ so that the total noise variance at the node of $\hat{x}(n)$ is

$$\sigma_q^2 = \frac{1}{M} \sum_{k=0}^{M-1} \sigma_{q_k}^2 \quad \text{(output noise variance)}. \qquad (C.2.1)$$

C.2.1 Optimum Bit Allocation

With b_k denoting the number of bits per sample of the kth subband signal, the quantity

$$b \triangleq \frac{1}{M} \sum_{k=0}^{M-1} b_k \qquad (C.2.2)$$

represents the *average bit rate*, that is, average number of bits per subband. Essentially b represents the average number of bits per sample of $x(n)$, transmitted by the decimated analysis bank.

Suppose the average bit rate b is fixed. How should we allocate the bits b_k to the individual subbands so that the total output noise variance (C.2.1) is minimized? Qualitatively speaking, we have to allocate fewer bits ($b_k < b$) for subbands with low energy, and more bits ($b_k > b$) for dominant subbands. An extreme case occurs when most of the energy is in the lowpass subband, in which case we set $b_0 = Mb$ and $b_k = 0$ otherwise. It is clear that this is much more efficient than assigning b bits per sample of the original signal $x(n)$. To address the bit allocation problem more quantitatively, we use the following standard result.

The AM-GM inequality. [Beckenbach and Bellman, 1961]. Given a set of M nonnegative numbers a_k, define the arithmetic mean (AM) and geometric mean (GM) to be

$$\begin{aligned}\text{Arithmetic mean:} \quad & \frac{1}{M} \sum_{k=0}^{M-1} a_k, \\ \text{Geometric mean:} \quad & \left(\prod_{k=0}^{M-1} a_k\right)^{1/M}\end{aligned} \qquad (C.2.3)$$

We then have $AM \geq GM$ with equality if and only if $a_0 = a_1 = \ldots = a_{M-1}$. See Problem C.2 for a proof.

Assuming that the filter bank input $x(n)$ is zero mean WSS with variance σ_x^2, the signal $x_k(n)$ [output of $H_k(z)$] is WSS with zero mean and variance

$$\sigma_{x_k}^2 = \frac{1}{2\pi} \int_{-\pi}^{\pi} |H_k(e^{j\omega})|^2 S_{xx}(e^{j\omega}) \, d\omega, \qquad (C.2.4)$$

where $S_{xx}(e^{j\omega})$ is the power spectrum of $x(n)$. The decimated signal $v_k(n)$ is WSS with the same variance $\sigma_{x_k}^2$. With $|H_k(e^{j\omega})|$ as in Fig. C.2-2, one can verify that $\sigma_x^2 = \sum_k \sigma_{x_k}^2 / M$ where σ_x^2 is the variance of the input signal $x(n)$. The quantizer variance $\sigma_{q_k}^2$ is related to the variance of $v_k(n)$ in a way similar to (C.1.4), so that

$$\sigma_{q_k}^2 = c \times 2^{-2b_k} \sigma_{x_k}^2. \qquad (C.2.5)$$

Here we have assumed that c is the same for all quantizers. This is valid if the quantizer inputs have identical statistical distribution (e.g., all of them Gaussian).

According to (C.2.1) the output noise variance σ_q^2 is the AM of $\sigma_{q_k}^2$ and so we have

$$\sigma_q^2 \geq \left(\prod_{k=0}^{M-1} \sigma_{q_k}^2\right)^{1/M} \quad \text{(AM-GM inequality)}$$

$$= c\left(\prod_{k=0}^{M-1} 2^{-2b_k} \sigma_{x_k}^2\right)^{1/M} \quad \text{[using (C.2.5)]}$$

$$= c \times \left(2^{-2\sum b_k/M}\right)\left(\prod_{k=0}^{M-1} \sigma_{x_k}^2\right)^{1/M} \quad (C.2.6)$$

$$= c \times 2^{-2b}\left(\prod_{k=0}^{M-1} \sigma_{x_k}^2\right)^{1/M} \quad \text{[using (C.2.2)]}.$$

So the smallest possible value of σ_q^2 is given by the last line of the above equation. This quantity depends on the input signal $x(n)$, and the average number of bits b. This lower bound is achieved if and only if the AM-GM inequality on the first line becomes an equality, which happens if and only if $\sigma_{q_k}^2$ is the same for all k, that is,

$$\sigma_{q_k}^2 = \sigma_q^2 \quad 0 \leq k \leq M-1. \quad (C.2.7)$$

So the optimal bit allocation strategy is such that *variances of all quantizer noise sources are equalized*. We then have

$$2^{2b_k} = \frac{c \times \sigma_{x_k}^2}{\sigma_q^2}, \quad (C.2.8a)$$

so that b_k is proportional to $\log \sigma_{x_k}^2$. Under this condition, the minimized output noise variance is given by

$$\sigma_{q,SBC}^2 = c \times 2^{-2b}\left(\prod_{k=0}^{M-1} \sigma_{x_k}^2\right)^{1/M}. \quad (C.2.8b)$$

The subscript SBC is used because we will soon compare this with other schemes.

Align the least-significant bits! If $\sigma_{q_k}^2$ is the same for all quantizers, the step sizes are identical. The optimal bit allocation strategy (C.2.7) therefore tells us that we must align the rightmost bits (least-significant bits) of the binary words representing the subbands. This is demonstrated in Fig. C.2-3.

More Explicit Expression for Bit Allocation

The optimal bit allocation scheme (C.2.8a) indicates that *more* bits are required for subband signals having *higher* variance. The reconstruction error variance σ_q^2 has the optimum value given by (C.2.8b). Using this we can rearrange (C.2.8a) as

$$b_k = b + 0.5 \log_2 \frac{\sigma_{x_k}^2}{\left(\prod_{i=0}^{M-1} \sigma_{x_i}^2\right)^{1/M}} \quad \text{(optimal bit allocation)}. \quad (C.2.9)$$

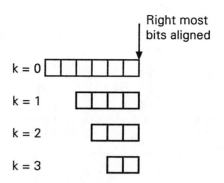

Figure C.2-3 Demonstration of optimal bit allocation. Here $\sigma_{x_0} = 4\sigma_{x_1} = 8\sigma_{x_2} = 16\sigma_{x_3}$.

Some numerical examples will be considered in Problem C.9. Several comments are now in order.

1. The above bit-allocation strategy may not lead to integer valued solutions for b_k. We can either round b_k to an integer, or use a (periodically) time varying allocation which results in an average value of nearly b_k bits.
2. From the above expression we see that if b is too small, then some of the b_k might turn out to be negative. However, for a given variance distribution $\{\sigma_{x_k}^2\}$, we can find a sufficiently large b so that $b_k > 0$ for all k.
3. If some of the b_k are very small, then the standard noise model assumptions (white, uncorrelated, uniform-distribution, etc.) will not hold, and the above analysis will not apply. This happens, for example, if some of the subband variances $\sigma_{x_k}^2$ are very small. Notice, in particular, that if $\sigma_{x_k}^2 = 0$ for some k, then the noise model assumptions are not meaningful any more. Many of the conclusions of this Appendix will be incorrect or meaningless under this situation. For example, the righthand side of (C.2.8b) becomes zero.
4. Assume that b_k are large enough so that the above analysis is meaningful. If we now increment b by one bit, then according to (C.2.9) each of the subband bits b_k is increased by one.

C.2.2 Coding Gain for the Ideal SBC System

In order to evaluate the usefulness of subband coding for a particular $x(n)$, we compare it with the direct quantization scheme shown in Fig. C.1-1 (imagine now that x in this figure is replaced with the sequence $x(n)$). This is often called the PCM (pulse code modulation) scheme. Let the quantizer Q use b bits so that the average transmission rate is the same as the above SBC system. The quantization noise is given by (C.1.4) so that

$$\sigma_{q,PCM}^2 = c \times 2^{-2b} \sigma_x^2. \qquad (C.2.10)$$

The subband coding gain is defined as

$$G_{SBC}(M) = \frac{\sigma_{q,PCM}^2}{\sigma_{q,SBC}^2} \qquad (C.2.11)$$

Note that this is a function of M. In the optimal case

$$G_{SBC}(M) = \frac{\sigma_x^2}{\left(\prod_{k=0}^{M-1} \sigma_{x_k}^2\right)^{1/M}} = \frac{\frac{1}{M}\sum_{k=0}^{M-1} \sigma_{x_k}^2}{\left(\prod_{k=0}^{M-1} \sigma_{x_k}^2\right)^{1/M}} \qquad (C.2.12)$$

So the coding gain can be interpreted in two ways: first it is the ratio of the input variance σ_x^2 to the geometric mean of the subband signal variances $\sigma_{x_k}^2$. Second, it is the ratio of the arithmetic to geometric mean of the subband signal variances. By the AM-GM inequality, we have

$$G_{SBC}(M) \geq 1$$

so that subband coding can never be worse than PCM (under the above assumptions on filters and statistics). As explained earlier, we will assume that $\sigma_{x_k}^2$ is nonzero for all k, so that the denominator in (C.2.12) is nonzero.

We have the least coding gain (i.e., $G_{SBC}(M) = 1$) if and only if all the subband signals have the same variance (i.e., the integral in (C.2.4) is the same for all subbands). This happens, for example, when $x(n)$ is white (i.e., when $S_{xx}(e^{j\omega})$ is constant); in this case $G_{SBC}(M) = 1$ for all M.

Coding Gain in dB, and the "6 dB per bit" Rule

The coding gain is a measure of improvement obtained by subband coding over traditional quantization, for fixed bit rate. Instead of fixing the bits, suppose we fix the reconstruction error variance. Thus, suppose we use b bits per sample for the subband coder, and b' bits per sample for the direct (PCM) quantizer. If we wish to make $\sigma_{q,PCM}^2 = \sigma_{q,SBC}^2$, then the coding gain $G_{SBC}(M)$ is related to b and b' as

$$G_{SBC}(M) = 2^{2(b'-b)}, \quad \text{i.e.,} \quad b' - b = 0.5 \log_2 G_{SBC}(M).$$

For example if $G_{SBC}(M) = 4$, then $b' - b = 1$, that is, we gain *one bit per sample* by using subband coding.

It is often convenient to express the coding gain in dB. This is given by

$$10 \log_{10} G_{SBC}(M) = 6.02(b' - b).$$

Thus a coding gain $\approx$ 6 dB implies that we have gained one bit per sample. In Sec. 4.5.2 we mentioned that there exist speech coding systems which operate at rates such as 2 to 4 bits per sample. Thus, an improvement in the coding gain by about 6 dB is considered to be relatively significant in these applications.

Behavior of $G_{SBC}(M)$ as M Increases

Is $G_{SBC}(M)$ a **nondecreasing function** of M? In most practical applications, the coding gain increases (at least, it does not decrease) as the number of subbands M increases. However, there is no 'theorem' which asserts that this is always the case. We can in fact construct a simple counter example as follows. (Also see Problem C.4).

Example C.2.1: $G_{SBC}(M)$ Can Decrease with Increasing M

Consider an input process $x(n)$ with power spectrum as in Fig. C.2-4. Suppose we apply a three-band ideal SBC system to this input. In this case, the area of $S_{xx}(e^{j\omega})$ in each of the subbands is the same, that is, $\sigma_{x_0}^2 = \sigma_{x_1}^2 = \sigma_{x_2}^2 = 3$. So $G_{SBC}(3) = 1$. If we now apply the same input to a two-band ideal SBC, then we find $\sigma_{x_0}^2 = 11/3$ and $\sigma_{x_1}^2 = 7/3$ so that $G_{SBC}(2) > 1$. In fact $G_{SBC}(2) \approx 1.0256$. In other words, $G_{SBC}(3) < G_{SBC}(2)$ showing that the subband coding gain can *decrease* as M increases!

It can, however, be verified that if the number of bands is increased in integer multiples, then the gain can never decrease, that is, $G_{SBC}(nM) \geq G_{SBC}(M)$ for any positive integer n (Problem C.3).

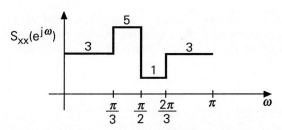

Figure C.2-4 An input power spectrum for which $G_{SBC}(3) < G_{SBC}(2)$.

Behavior as M tends to infinity. It will be interesting to see what happens to $G_{SBC}(M)$ as $M \to \infty$. We can study this by rewriting the product in the denominator of (C.2.12) as follows:

$$\begin{aligned}\Big(\prod_{k=0}^{M-1} \sigma_{x_k}^2\Big)^{1/M} &= \exp\Big(\log_e \Big(\prod_{k=0}^{M-1} \sigma_{x_k}^2\Big)^{1/M}\Big) \\ &= \exp\Big(\frac{1}{M}\log_e \Big(\prod_{k=0}^{M-1} \sigma_{x_k}^2\Big)\Big) \quad (C.2.13)\\ &= \exp\Big(\frac{1}{M}\sum_{k=0}^{M-1} \log_e \sigma_{x_k}^2\Big)\end{aligned}$$

The quantity $\sigma_{x_k}^2$ in the above expression was defined in (C.2.4). (*Note.* we have to integrate only the region $0 \leq \omega \leq \pi$ since all time-domain quantities are assumed real). If M is sufficiently large then the passband of $H_k(e^{j\omega})$ is sufficiently narrow and we can assume $S_{xx}(e^{j\omega})$ is nearly constant in this passband. This gives $\sigma_{x_k}^2 \approx S_{xx}(e^{j\omega_k})$ where ω_k is the center frequency for $H_k(e^{j\omega})$. So the above equation simplifies to

$$\left(\prod_{k=0}^{M-1} \sigma_{x_k}^2\right)^{1/M} \approx \exp\left(\frac{1}{M} \sum_{k=0}^{M-1} \log_e S_{xx}(e^{j\omega_k})\right) \quad (C.2.14)$$

$$\approx \exp\left(\frac{1}{2\pi} \int_{-\pi}^{\pi} \log_e S_{xx}(e^{j\omega}) \, d\omega\right)$$

for sufficiently large M. Summarizing, we have

$$\lim_{M \to \infty} G_{SBC}(M) = \frac{\sigma_x^2}{\exp\left(\frac{1}{2\pi} \int_{-\pi}^{\pi} \log_e S_{xx}(e^{j\omega}) \, d\omega\right)} \quad (C.2.15)$$

where $S_{xx}(e^{j\omega})$ is the power spectrum (and σ_x^2 the variance) of $x(n)$. Note that we can rewrite this as

$$\lim_{M \to \infty} G_{SBC}(M) = \frac{\frac{1}{2\pi} \int_{-\pi}^{\pi} S_{xx}(e^{j\omega}) \, d\omega}{\exp\left(\frac{1}{2\pi} \int_{-\pi}^{\pi} \log_e S_{xx}(e^{j\omega}) \, d\omega\right)} \quad (C.2.16)$$

which is called the asymptotic coding gain. Thus, in the limit as $M \to \infty$, the AM-GM ratio (C.2.12) is replaced with the above integral version! In obtaining this expression we have assumed that $S_{xx}(e^{j\omega})$ is nonzero, i.e., $S_{xx}(e^{j\omega}) > 0$, for all ω [so that $\log_e S_{xx}(e^{j\omega})$ does not blow up].

The reciprocal of the quantity in (C.2.15), i.e.,

$$\gamma_x^2 \triangleq \frac{\exp\left(\frac{1}{2\pi} \int_{-\pi}^{\pi} \log_e S_{xx}(e^{j\omega}) \, d\omega\right)}{\sigma_x^2} \quad (C.2.17)$$

is said to be the *spectral flatness measure* for the process $x(n)$. Notice that $0 \le \gamma_x \le 1$ (since $1 \le G_{SBC}(M) \le \infty$). γ_x is largest (i.e., $\gamma_x = 1$) for a flat spectrum i.e., if $x(n)$ is white. As γ_x gets smaller and smaller, the spectrum is considered to be more and more "nonflat." Thus the asymptotic coding gain *increases* as the spectrum gets more and more 'nonflat'.

Deeper details about behavior of $G_{SBC}(M)$.

We now state further properties of the coding gain of the ideal subband coder. All of these are justified in Problems C.3–C.5.

1. First, it can be shown that $G_{SBC}(M) \le G_{SBC}(nM)$ for any integer $n > 0$. By letting $n \to \infty$ we see that $G_{SBC}(M) \le G_{SBC}(\infty)$ for any M, even though $G_{SBC}(M)$ may increase or decrease with M.

2. Suppose we have a power spectrum with piecewise constant behavior as shown in Fig. C.2-5. In other words, $S_{xx}(e^{j\omega})$ is constant within each passband of the M-channel subband coder. Then, for this value of M, we have $G_{SBC}(M) = G_{SBC}(nM)$ for any integer $n > 0$ (Problem C.3). Letting $n \to \infty$ we conclude $G_{SBC}(M) = G_{SBC}(\infty)$. Notice, however, that this does *not* imply $G_{SBC}(L) = G_{SBC}(\infty)$ for *all* $L > M$ (Problem C.4). In any case, this example shows that the maximum coding gain $G_{SBC}(\infty)$ can be achieved with *finite* number of subbands M, when $S_{xx}(e^{j\omega})$ has certain properties [even if $x(n)$ is not white].

3. Consider now the converse situation: suppose the power spectrum of $x(n)$ is such that $G_{SBC}(nM)$ is the same for all integers $n > 0$, for some fixed M. Then $S_{xx}(e^{j\omega})$ has the piecewise constant behavior as demonstrated in Fig. C.2-5 (Problem C.3). As a corollary, if $G_{SBC}(M) = G_{SBC}(\infty)$ for some finite M, then $G_{SBC}(nM) = G_{SBC}(\infty)$ and $S_{xx}(e^{j\omega})$ has this behavior.

4. Finally suppose that $G_{SBC}(i) = G_{SBC}(M)$ for all $i \geq M$. Then $S_{xx}(e^{j\omega})$ is constant [i.e., $x(n)$ is white] (Problem C.5). This means, in fact, that $G_{SBC}(i) = 1$ for all i.

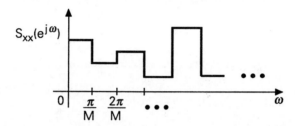

Figure C.2-5 A piecewise constant power spectrum.

C.3 THE ORTHOGONAL TRANSFORM CODER

Transform coding was reviewed in Sec. 6.6. This is a special case of subband coding where the analysis and synthesis filters are FIR with length $\leq M$. The responses of these filters are far from the ideal responses considered in the previous section.

We will consider only real signals and matrices for simplicity of discussions. Fig. C.3-1 shows the system with quantizers inserted in the subbands (i.e., the transform domain coefficients $y_k(n)$ are quantized). Since the matrix $\mathbf{T}$ is orthogonal with $\mathbf{TT}^T = \mathbf{T}^T\mathbf{T} = \mathbf{I}$, we have replaced $\mathbf{T}^{-1}$ with $\mathbf{T}^T$. In absence of quantizers this is a perfect reconstruction system with $\widehat{x}(n) = x(n - M + 1)$.

C.3.1 Optimum Bit Allocation for Fixed Orthogonal T

We will assume that the quantizers are as in the previous section (i.e., b_k bit quantizer for the kth subband, etc.). We also use the same assumptions about the noise model as in the ideal subband coding case. With the average bit rate b (defined as in (C.2.2)) fixed, what is the best allocation of bits among the subbands that minimizes the noise variance at the output node [i.e., node of $\widehat{x}(n)$]? It turns out that the answer does *not* depend on the orthogonal matrix $\mathbf{T}$ as we show next.

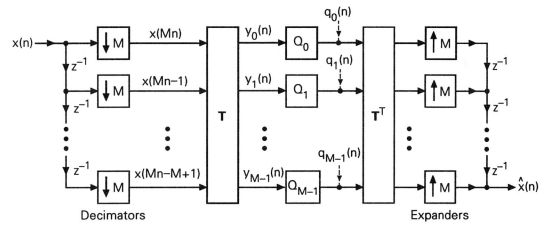

Figure C.3-1 The orthogonal transform coding system.

As in the previous section, the quantizers can be replaced with noise sources $q_k(n)$ as shown in broken lines. These sources are at the input of the matrix $\mathbf{T}^T$, and generate a set of noise components, say $\widehat{q}_k(n)$, at the output of $\mathbf{T}^T$ (This system is similar to Fig. 9.3-2, analyzed in Chap. 9). The noise component $q(n)$ at the node of $\widehat{x}(n)$ is the *interlaced* version of the noise components $\widehat{q}_k(n)$. So, the average output noise variance over a period of length M is given by

$$\sigma_q^2 = \frac{1}{M} \sum_{k=0}^{M-1} \widehat{\sigma}_{q_k}^2, \qquad (C.3.1)$$

where $\widehat{\sigma}_{q_k}^2$ is the variance of $\widehat{q}_k(n)$. Since $\mathbf{T}^T\mathbf{T} = \mathbf{T}\mathbf{T}^T = \mathbf{I}$, we can rewrite the above expression as

$$\sigma_q^2 = \frac{1}{M} \sum_{k=0}^{M-1} \sigma_{q_k}^2, \qquad (C.3.2)$$

where $\sigma_{q_k}^2$ is the quantizer noise variance (i.e., variance of $q_k(n)$). To see this note that if we define the vectors

$$\mathbf{q}(n) = [\,q_0(n) \quad \ldots \quad q_{M-1}(n)\,]^T, \quad \widehat{\mathbf{q}}(n) = [\,\widehat{q}_0(n) \quad \ldots \quad \widehat{q}_{M-1}(n)\,]^T, \qquad (C.3.3)$$

then $\widehat{\mathbf{q}}(n) = \mathbf{T}^T \mathbf{q}(n)$, and

$$\widehat{\mathbf{q}}^T(n)\widehat{\mathbf{q}}(n) = \mathbf{q}^T(n)\mathbf{T}\mathbf{T}^T \mathbf{q}(n) = \mathbf{q}^T(n)\mathbf{q}(n). \qquad (C.3.4)$$

Taking expected values, we get $\sum \sigma_{q_k}^2 = \sum \widehat{\sigma}_{q_k}^2$. †

We can now proceed exactly as in (C.2.6) to show that the output noise variance σ_q^2 is minimized if and only if all the quantizer noise variances $\sigma_{q_k}^2$ are equalized [i.e., (C.2.7) holds]. Equivalently, the rightmost bits of the binary representations

---

† Notice that the above derivation *does not* assume that the quantizer noise sources are either white or uncorrelated.

for the transform coefficients $y_k(n)$ should be aligned (similar to Fig. C.2-3). This in turn is accomplished if b_k are chosen according to $2^{2b_k} = c \times \sigma_{y_k}^2/\sigma_q^2$ where $\sigma_{y_k}^2$ is the variance of $y_k(n)$ (the kth output of $\mathbf{T}$ in Fig. C.3-1). Proceeding as in the previous section, we obtain the expression

$$\sigma_{q,TC}^2 = c \times 2^{-2b} \left(\prod_{k=0}^{M-1} \sigma_{y_k}^2 \right)^{1/M}. \qquad (C.3.5)$$

for the minimized average noise variance at the output node [i.e., node of $\hat{x}(n)$]. Notice that the above derivations make use of the orthogonal nature of $\mathbf{T}$; the results are therefore not necessarily valid for nonorthogonal $\mathbf{T}$.

C.3.2 An Optimal Transform: the KLT

Note that the product of variances $\sigma_{y_k}^2$ depends on $x(n)$ and on the orthogonal matrix $\mathbf{T}$. Given the WSS input $x(n)$, how should we choose $\mathbf{T}$ so that this product is minimized? We now answer this.

Defining the vectors

$$\begin{aligned} \mathbf{y}(n) &= [\, y_0(n) \quad \ldots \quad y_{M-1}(n)\,]^T, \\ \mathbf{x}(n) &= [\, x(Mn) \quad \ldots \quad x(Mn-M+1)\,]^T, \end{aligned} \qquad (C.3.6)$$

we see that $\mathbf{y}(n) = \mathbf{T}\mathbf{x}(n)$ so that

$$\mathbf{y}(n)\mathbf{y}^T(n) = \mathbf{T}\mathbf{x}(n)\mathbf{x}^T(n)\mathbf{T}^T. \qquad (C.3.7)$$

If we take expected values on both sides, we obtain

$$\mathbf{R_{yy}}(0) = \mathbf{T}\mathbf{R_{xx}}(0)\mathbf{T}^T, \qquad (C.3.8)$$

where $\mathbf{R_{xx}}(0) = E[\mathbf{x}(n)\mathbf{x}^T(n)]$ and $\mathbf{R_{yy}}(0) = E[\mathbf{y}(n)\mathbf{y}^T(n)]$. Note that $\mathbf{R_{xx}}(0)$ and $\mathbf{R_{yy}}(0)$ are Hermitian positive (semi) definite matrices. [$\mathbf{R_{xx}}(0)$ is also Toeplitz but that is irrelevant here.]

The diagonal elements of $\mathbf{R_{yy}}(0)$ are, by definition,

$$[\mathbf{R_{yy}}(0)]_{kk} = E[y_k^2(n)] = \sigma_{y_k}^2 \qquad (C.3.9)$$

which represents the variance of $y_k(n)$ (since we assume zero mean). So the product in (C.3.5) is equal to the product of diagonal elements of $\mathbf{R_{yy}}(0)$, i.e., diagonal elements of $\mathbf{T}\mathbf{R_{xx}}(0)\mathbf{T}^T$. Our aim therefore is to choose $\mathbf{T}$ (for a given $\mathbf{R_{xx}}(0)$) such that this product of diagonal elements is minimized.

For this we will use an inequality which has to do with the product of diagonal elements of a positive definite matrix. First recall that $\mathbf{R_{yy}}(0)$ is positive semidefinite by definition. From (C.3.8) we see that $\mathbf{R_{yy}}(0)$ is nonsingular as long as $\mathbf{R_{xx}}(0)$ is nonsingular. We will assume that $\mathbf{R_{xx}}(0)$ is nonsingular (otherwise $x(n)$ would be a harmonic process; Appendix B). So $\mathbf{R_{yy}}(0)$ is a positive definite matrix, and we can apply the following result from Appendix A (Sec. A.6).

$$\prod_{k=0}^{M-1} [\mathbf{R_{yy}}(0)]_{kk} \geq \det \mathbf{R_{yy}}(0), \qquad (C.3.10)$$

with equality if and only if $\mathbf{R_{yy}}(0)$ is diagonal. We will also use the fact that

$$\det \mathbf{R_{yy}}(0) = \det \mathbf{R_{xx}}(0), \qquad (C.3.11)$$

which follows from (C.3.8), using $[\det \mathbf{T}] = \pm 1$. Summarizing the above discussions, we have

$$\begin{aligned}\prod_{k=0}^{M-1} \sigma_{y_k}^2 &= \prod_{k=0}^{M-1} [\mathbf{R_{yy}}(0)]_{kk} \\ &\geq \det \mathbf{R_{yy}}(0) \\ &= \det \mathbf{R_{xx}}(0).\end{aligned} \qquad (C.3.12)$$

Note that $\mathbf{R_{xx}}(0)$ is completely determined by the random process $x(n)$, whereas $\mathbf{R_{yy}}(0)$ depends on $\mathbf{T}$. Equality holds on the second line above if and only if $\mathbf{R_{yy}}(0)$ is diagonal (i.e., $y_k(n)$ uncorrelated with $y_m(n)$ for $k \neq m$). So we have to choose $\mathbf{T}$ such that it diagonalizes $\mathbf{R_{xx}}(0)$ [see (C.3.8)]. By comparing this with the discussion on eigenvectors (Sec. A.5) we see that $\mathbf{T}^T$ should be chosen such that its columns are the orthonormal eigenvectors of the Hermitian matrix $\mathbf{R_{xx}}(0)$.

This $\mathbf{T}$ is called the *Karhunen-Loeve* Transform (KLT) for the signal $x(n)$. Clearly it depends on the block size M of the transform. This choice of $\mathbf{T}$, along with the bit allocation scheme described above results in the smallest possible average output noise variance $\sigma_{q,TC}^2$. This variance is obtained by substituting the minimum value of $\prod_{k=0}^{M-1} \sigma_{y_k}^2$ into (C.3.5):

$$\sigma_{q,KLT}^2 = c \times 2^{-2b} \left(\det \mathbf{R_{xx}}(0)\right)^{1/M} \qquad \text{(min. output noise var.)} \qquad (C.3.13)$$

Even though this minimized variance is unique, the KLT matrix itself is not, since the set of orthonormal eigenvectors is not necessarily unique.

Suboptimal transforms. In practice, the KLT is inconvenient to implement for several reasons. First, the KLT matrix depends on the input statistics, and has to be estimated by numerical means. If the input statistics varies (i.e., input is not quite WSS, but slowly varying), then the KLT has to be somehow recomputed or updated. Finally, the KLT matrix in general does not have any specific structure (other than unitariness), and cannot be implemented in a 'fast manner'. In practice, there exist several suboptimal substitutes, which overcome these difficulties. For example, in speech applications, the DCT matrix (Sec. 8.4) provides good coding gain. There also exist fast techniques to implement this matrix [Yip and Rao, 1987]. For further details, see Sec. 12.6 of Jayant and Noll [1984].

Toeplitz Property

Since $x(n)$ is WSS, the $M \times M$ matrix $\mathbf{R_{xx}}(0)$ has the Toeplitz property (Appendix A). For example if $M = 4$, one can verify that

$$\mathbf{R_{xx}}(0) = \begin{bmatrix} R(0) & R(1) & R(2) & R(3) \\ R(1) & R(0) & R(1) & R(2) \\ R(2) & R(1) & R(0) & R(1) \\ R(3) & R(2) & R(1) & R(0) \end{bmatrix}$$

where $R(k) = E[x(n)x(n-k)]$, that is, $R(k)$ is the autocorrelation sequence of the real WSS process $x(n)$. Note that $R(0) = \sigma_x^2$ = variance of $x(n)$. The signal $x(n)$ is white if and only if $\mathbf{R_{xx}}(0) = \sigma_x^2 \mathbf{I}_M$ for any choice of the size M.

C.3.3 Coding Gain for the Transform-Coder

The transform coding gain $G_{TC}(M)$ is defined as the ratio of $\sigma_{q,PCM}^2$ in (C.2.10) to $\sigma_{q,TC}^2$. Using (C.3.5)

$$G_{TC}(M) = \frac{\sigma_x^2}{\left(\prod_{k=0}^{M-1} \sigma_{y_k}^2\right)^{1/M}} \qquad (C.3.14)$$

As for the ideal SBC case, we can express this as an AM/GM ratio. For this note that the relation $\mathbf{y}(n) = \mathbf{T}\mathbf{x}(n)$ implies $\mathbf{y}^T(n)\mathbf{y}(n) = \mathbf{x}^T(n)\mathbf{x}(n)$. Taking expected values we get $\sum_k \sigma_{y_k}^2 = M\sigma_x^2$ so that

$$G_{TC}(M) = \frac{\sum_{k=0}^{M-1} \sigma_{y_k}^2}{M\left(\prod_{k=0}^{M-1} \sigma_{y_k}^2\right)^{1/M}} \qquad (C.3.15)$$

Thus, the coding gain is the ratio of the AM to GM of the variances of the subband outputs. This situation is precisely as in the ideal SBC case. This is interesting because the ideal SBC case and the orthogonal transform case use totally different assumptions on filters. In the former, we use ideal bandpass filters, whereas in the latter we use FIR filters of length $\leq M$, having orthogonal polyphase matrix $\mathbf{T}$.

Coding Gain in the Optimal Case

From Sec. C.3.2 we know that the maximum value of the coding gain is achieved when $\mathbf{T}$ is the KLT matrix. In this case we can rewrite (C.3.14) as

$$G_{KLT}(M) = \frac{\sigma_x^2}{\left(\det \mathbf{R_{yy}}(0)\right)^{1/M}}$$

$$= \frac{\sigma_x^2}{\left(\det \mathbf{R_{xx}}(0)\right)^{1/M}} = \frac{\sum_{k=0}^{M-1} \sigma_{y_k}^2}{M\left(\det \mathbf{R_{xx}}(0)\right)^{1/M}} \qquad (C.3.16)$$

Since the diagonal elements of $\mathbf{R_{yy}}(0)$ are the variances $\sigma_{y_k}^2$, the summation on the numerator is the *trace* of $\mathbf{R_{yy}}(0)$. By using the fact (Appendix A) that the trace of $\mathbf{AB}$ is same as that of $\mathbf{BA}$ (whenever the two products are meaningful) we can show from (C.3.8) that the trace of $\mathbf{R_{yy}}(0)$ is same as that of $\mathbf{R_{xx}}(0)$. As a result, we can rewrite (C.3.16) as

$$G_{KLT}(M) = \frac{\text{Trace of } \mathbf{R_{xx}}(0)}{M\left(\det \mathbf{R_{xx}}(0)\right)^{1/M}} \qquad (C.3.17)$$

But the trace of a matrix is the sum of its eigenvalues, whereas the determinant is the product of eigenvalues. Denoting the eigenvalues of $\mathbf{R_{xx}}(0)$ as λ_k we therefore have

$$G_{KLT}(M) = \frac{\sum_{k=0}^{M-1} \lambda_k}{M \left(\prod_{k=0}^{M-1} \lambda_k \right)^{1/M}} \qquad (C.3.18)$$

Thus, the optimal (maximized) transform coding gain is the ratio of the AM to GM of the eigenvalues of the input autocorrelation matrix $\mathbf{R_{xx}}(0)$ (i.e., the matrix $E[\mathbf{x}(n)\mathbf{x}^T(n)]$). So this coding gain satisfies $G_{KLT}(M) \geq 1$.

The worst optimal coding gain ($G_{KLT}(M) = 1$) occurs if and only if all eigenvalues of $\mathbf{R_{xx}}(0)$ are equal. Since $\mathbf{R_{xx}}(0)$ is Hermitian, this will happen if and only if $\mathbf{R_{xx}}(0)$ itself is diagonal, that is, $\mathbf{R_{xx}}(0) = \sigma_x^2 \mathbf{I}$. [This does not imply that $x(n)$ is white.]

At the end of Appendix B we saw that for a Harmonic process, the autocorrelation matrix of appropriate size is singular. For a 'nearly' harmonic process (i.e., process with many sharp peaks in the power spectrum), the determinant of the autocorrelation matrix is very small, and the above coding gain can be quite large.

$G_{KLT}(M)$ is a Monotone Function of M

In the previous section we saw that the subband coding gain $G_{SBC}(M)$ (with ideal filters) may increase or decrease with M. It turns out, however, that the optimal transform coding gain $G_{KLT}(M)$ is monotone, that is,

$$G_{KLT}(M+1) \geq G_{KLT}(M). \qquad (C.3.19)$$

We will now proceed to prove this. [The reader wishing to skip the proof can go to the next subtitle without loss of continuity.]

Proof that $G_{KLT}(M)$ is monotone. First let us replace the notation $\mathbf{R_{xx}}(0)$ with $\mathbf{R}_M$ where the subscript indicates that this matrix has size $M \times M$. Note that $\mathbf{R}_M$ is completely determined by the WSS process $x(n)$. Proving (C.3.19) is equivalent to proving

$$\left(\det \mathbf{R}_{M+1} \right)^{1/(M+1)} \leq \left(\det \mathbf{R}_M \right)^{1/M}, \qquad (C.3.20)$$

that is,

$$\det \mathbf{R}_{M+1} \leq \det \mathbf{R}_M \left(\det \mathbf{R}_M \right)^{1/M} \qquad (C.3.21)$$

In order to prove this, we will need a result from the theory of linear prediction. (We use the result without proof, but will cite a precise reference.) In linear prediction theory, one tries to predict the sample $x(n)$ using a linear combination of past values:

$$x_p(n) = -a_1 x(n-1) - a_2 x(n-2) - \ldots - a_k x(n-k). \qquad (C.3.22)$$

The parameter k above is the prediction order. The quantity $e_k(n) = x(n) - x_p(n)$ is called the prediction error. Here $x(n)$ is a real WSS process and a_i are real valued prediction coefficients. In optimal linear prediction, these coefficients are chosen such that the mean-squared prediction error $E[e_k^2(n)]$ is minimized. We use $\mathcal{E}_k$ to denote the minimized mean-squared prediction error for the kth order predictor. As one would intuitively expect, it can be shown that $\mathcal{E}_{k+1} \leq \mathcal{E}_k$ for any integer $k \geq 0$. Also $\mathcal{E}_0 = E[x^2(n)] = \sigma_x^2$.

There is a fundamental connection between $\mathcal{E}_k$ and the $M \times M$ autocorrelation matrices $\mathbf{R}_M$ associated with $x(n)$. It can be shown that

$$\det \mathbf{R}_M = \mathcal{E}_{M-1}\mathcal{E}_{M-2}\ldots\mathcal{E}_0, \quad (M \geq 1). \qquad (C.3.23)$$

This can be obtained, for example, by repeated use of Eq. (6.48), Jayant and Noll [1984]. From this we deduce

$$\frac{\det \mathbf{R}_{M+1}}{\det \mathbf{R}_M} = \mathcal{E}_M, \quad (M \geq 1). \qquad (C.3.24)$$

From the above two equations we see that proving (C.3.21) is equivalent to proving

$$\left(\mathcal{E}_M\right)^M \leq \mathcal{E}_{M-1}\mathcal{E}_{M-2}\ldots\mathcal{E}_0. \qquad (C.3.25)$$

But this follows immediately from the fact that $\mathcal{E}_{k+1} \leq \mathcal{E}_k$ for any $k \geq 0$. Summarizing, we have proved that $G_{KLT}(M)$ is a monotone (nondecreasing) function of M.

Behavior as M tends to infinity. The behavior as $M \to \infty$ can be studied by studying the behavior of $[\det \mathbf{R}_M]$. Since $x(n)$ is WSS, the matrix $\mathbf{R}_M$ is Toeplitz (in addition to being Hermitian and positive definite). There is a theorem for such matrices [Grenander and Szego, 1958], [Gray, 1972], [Jayant and Noll, 1984] which says that, under fairly general conditions,

$$\lim_{M \to \infty} \left(\det \mathbf{R}_M\right)^{1/M} = \exp\left(\frac{1}{2\pi}\int_{-\pi}^{\pi} \log_e S_{xx}(e^{j\omega})\,d\omega\right). \qquad (C.3.26)$$

As a result, we have

$$\lim_{M \to \infty} G_{KLT}(M) = \frac{\sigma_x^2}{\exp\left(\dfrac{1}{2\pi}\displaystyle\int_{-\pi}^{\pi} \log_e S_{xx}(e^{j\omega})\,d\omega\right)} \qquad (C.3.27)$$

This is the same as $\lim_{M \to \infty} G_{SBC}(M)$ so that $G_{SBC}(\infty) = G_{KLT}(\infty)$. So for sufficiently large M, the coding gain obtainable with ideal SBC is the same as the coding gain offered by the KLT!

Relation to linear prediction. By using (C.3.26) in (C.3.24) we see that the minimized prediction error $\mathcal{E}_M$ has the following limit:

$$\mathcal{E}_\infty = \lim_{M \to \infty} \mathcal{E}_M = \exp\left(\frac{1}{2\pi}\int_{-\pi}^{\pi} \log_e S_{xx}(e^{j\omega})\,d\omega\right).$$

Since $\mathcal{E}_{k+1} \leq \mathcal{E}_k$, the quantity $\mathcal{E}_\infty$ represents the smallest possible prediction error for a given $S_{xx}(e^{j\omega})$. We can now write

$$G_{SBC}(\infty) = G_{KLT}(\infty) = \sigma_x^2/\mathcal{E}_\infty = 1/\gamma_x^2 \qquad (C.3.28)$$

where $\mathcal{E}_\infty$ is the variance of the prediction error for infinite predictor order. Thus the asymptotic coding gains are the same for the ideal subband coder and the KLT. Recall that γ_x^2 stands for the spectral flatness measure, and that $0 \leq \gamma_x \leq 1$. The quantity $\sigma_x^2/\mathcal{E}_\infty$ is called the *prediction gain*. Its significance is discussed in Sec. C.5.1 later, in the context of differential PCM (DPCM) techniques. Also see Problem C.8.

Autoregressive (AR) process. There exists a class of random processes for which $e_k(n)$ becomes white for some finite k, say $k = N$. When this happens, the error variance cannot be decreased further by increasing the prediction order. In other words, $\mathcal{E}_N = \mathcal{E}_\infty$, given by the expression presented above. Such a process $x(n)$ is called an autoregressive process of order N, abbreviated as $AR(N)$. Many real-life waveforms (e.g., speech) have been successfully approximated as AR processes [Jayant and Noll, 1984].

Deeper Details About Behavior of $G_{KLT}(M)$

We know that $G_{KLT}(M)$ is monotone in M, and that $G_{KLT}(M) \geq 1$ with equality if, and only if, $\mathbf{R}_M = \sigma_x^2 \mathbf{I}$. Also, $G_{KLT}(M) \leq G_{KLT}(\infty)$. We specifically proved $G_{KLT}(M+1) \geq G_{KLT}(M)$. It turns out (Problem C.6) that if $G_{KLT}(M+1) = G_{KLT}(M)$ then we will in fact have

$$G_{KLT}(M+1) = G_{KLT}(M) = G_{KLT}(M-1) = \ldots = G_{KLT}(1) = 1. \qquad (C.3.29)$$

This has some interesting corollaries.

1. If $G_{KLT}(i) > 1$ then $G_{KLT}(M) > G_{KLT}(M-1)$ for any $M \geq i$. Similarly if $G_{KLT}(i+1) > G_{KLT}(i)$ for some i, then $G_{KLT}(M+1) > G_{KLT}(M)$ for any $M \geq i$. In other words, once $G_{KLT}(i)$ starts increasing, it is a *strictly* increasing function of i.

2. Unless $x(n)$ is white, $G_{KLT}(M)$ *cannot* achieve the maximum value $G_{KLT}(\infty)$ for finite M. (For, if it does, then $G_{KLT}(\infty) = G_{KLT}(i+1) = G_{KLT}(i)$ for arbitrarily large i. Using a relation like (C.3.29) we can then conclude that $G_{KLT}(i) = 1$ for all i. So $\mathbf{R}_M$ is diagonal for all M, that is, $x(n)$ is white). This is unlike the ideal subband coder where we can achieve the maximum coding gain for finite M, with certain nonwhite inputs [e.g., if $S_{xx}(e^{j\omega})$ is constant within each of the M subbands, then $G_{SBC}(M) = G_{SBC}(\infty)$.]

C.4 SIMILARITIES AND DIFFERENCES

It is often stated that transform coding (TC) is a special case of subband coding (SBC). This statement has its origin in the fact that Fig. C.3-1 essentially represents a subband coder in which the filters are FIR with length $\leq M$. However, when we compare an ideal SBC system (system with ideal bandpass filters as in Fig. C.2-2) with the optimal TC system (KLT), we notice some subtle differences.

For example, consider the coding gains $G_{TC}(M)$ and $G_{SBC}(M)$. We have shown that $G_{TC}(M)$ (with KLT) can never decrease with increasing M, whereas a similar statement is not true for $G_{SBC}(M)$. It is important to realize that the optimal TC system should be redesigned as the input statistics changes. The ideal SBC system, on the other hand, does not exploit input statistics.

Notice that in both of the above systems, we are essentially transforming the signal in such a way that the energy (or variance) tends to be concentrated (or compacted or packed) into a few subbands. This *energy compaction* is the key to subband coding as well as transform coding.

The De-correlating Property

The ideal SBC system has the property that the subband signals $v_k(n)$ and $v_\ell(m)$ are uncorrelated for $k \neq \ell$. This is true for arbitrary m, n. [This is a consequence of Problem 9.12.] The transform coding system with KLT is such that $y_k(n)$ and $y_\ell(n)$ are uncorrelated for $k \neq \ell$, since $\mathbf{R_{yy}}(0)$ is diagonal. However, we cannot in general say that $y_k(n)$ and $y_\ell(m)$ are uncorrelated for $n \neq m$). In other words, for ideal SBC the random *processes* $\{v_k(n)\}$ and $\{v_\ell(m)\}$ are uncorrelated. For KLT, only the random *variables* $y_k(n)$ and $y_\ell(n)$ are uncorrelated.

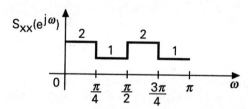

Figure C.4-1 Power spectrum for the example where $G_{TC}(2) > G_{SBC}(2)$.

Example C.4.1: Situation Where $G_{KLT}(M) > G_{SBC}(M)$

Consider a WSS input with power spectrum as in Fig. C.4-1, and let $M = 2$. The areas of $S_{xx}(e^{j\omega})$ in the two subbands are equal, so that $G_{SBC}(2) = 1$. However, we can show that $G_{KLT}(2) > 1$. For this it is sufficient to show that the 2×2 matrix $\mathbf{R_{xx}}(0)$ is not diagonal. Now

$$\mathbf{R_{xx}}(0) = \begin{bmatrix} E[x^2(n)] & E[x(n)x(n-1)] \\ E[x(n)x(n-1)] & E[x^2(n)] \end{bmatrix}. \qquad (C.4.1)$$

The quantity $E[x(n)x(n-1)]$ is the autocorrelation sequence of $x(n)$ evaluated at lag $= 1$, and is equal to

$$\frac{1}{2\pi}\int_{-\pi}^{\pi} S_{xx}(e^{j\omega})e^{j\omega}\,d\omega = \frac{\sqrt{2}-1}{\pi} \neq 0 \qquad (C.4.2)$$

showing that $\mathbf{R_{xx}}(0)$ is not diagonal. So $G_{KLT}(2) > 1 = G_{SBC}(2)$.

C.4.1 Summary of Main Points

The subband coding system is shown in Fig. C.2-1 and the transform coding system in Fig. C.3-1. Here $x(n)$ is a real zero mean, nonharmonic, WSS process. The matrix $\mathbf{T}$, as well as the filter coefficients in subband coding case are assumed to be real. The filters $H_k(z)$ and $F_k(z)$ in subband coding are ideal (hence we call it the ideal SBC) with responses as in Fig. C.2-2. The transform coding matrix $\mathbf{T}$ is an orthogonal matrix with $\mathbf{T}^T\mathbf{T} = \mathbf{T}\mathbf{T}^T = \mathbf{I}$.

The quantizers Q_k are b_k bit quantizers (as in Fig. C.1-1). These can be replaced with noise sources $q_k(n)$ which are assumed to be white, pairwise uncorrelated, and zero-mean with variance $\sigma_{q_k}^2$.

1. For both ideal SBC and TC, the (average) noise variance at the output node [node of $\widehat{x}(n)$] is given by (C.3.2). If the average number of bits per sample of $x(n)$ is fixed, then this variance is minimized by allocating the bits to quantizers such that the quantizer noise variances $\sigma_{q_k}^2$ are equalized.

2. With bits allocated as above, the output noise variance is given by (C.2.8b) for ideal SBC and by (C.3.5) for TC. These expressions are identical, and have the common form $c \times 2^{-2b} \times$ (geometric mean of subband signal variances).

3. The *coding gain* $G_{SBC}(M)$ for the ideal SBC system is given by the ratio (C.2.12) which is the arithmetic to geometric mean of the subband signal variances. We have $G_{SBC}(M) \geq 1$, with $G_{SBC}(M) = 1$ if and only if all the M subband signals have same variance. $G_{SBC}(M)$ usually increases as M increases, but there are examples where it can decrease. For deeper details about variation of $G_{SBC}(M)$, see end of Sec. C.2.

4. The *coding gain* $G_{TC}(M)$ can also be expressed as the ratio of arithmetic to geometric means of the transform coefficient variances (eqn. (C.3.15)). So $G_{TC}(M) \geq 1$. For a given process $x(n)$, this gain is maximized when $\mathbf{T}$ is chosen to be the KLT. In this case, $G_{TC}(M)$ is denoted as $G_{KLT}(M)$. This is given by (C.3.18) which is the ratio of the arithmetic to geometric means of the *eigenvalues* of $\mathbf{R_{xx}}(0)$ (which is the $M \times M$ autocorrelation matrix formed from $x(n)$). The maximized coding gain satisfies $G_{KLT}(M) \geq 1$ and is equal to unity if and only if $\mathbf{R_{xx}}(0)$ is diagonal with $\mathbf{R_{xx}}(0) = \sigma_x^2 \mathbf{I}$. When KLT is used, $G_{KLT}(M)$ can never decrease with increasing M. For deeper details, see end of Sec. C.3.

5. For ideal SBC, the subband signals $v_k(n)$ and $v_\ell(m)$ are uncorrelated for $k \neq \ell$, for any n, m. For transform coding with KLT, the random variables $y_k(n)$ and $y_\ell(n)$ are uncorrelated for $k \neq \ell$ for each n.

6. For ideal SBC, the above uncorrelating property is true for any (WSS) input $x(n)$. For the KLT case, it is true only for a specific input signal because $\mathbf{T}$ was computed from the autocorrelation matrix of $x(n)$. Thus the KLT is a system optimized according to the input statistics. If the statistics of the input changes, the matrix $\mathbf{T}$ has to be readjusted to obtain the optimal coding gain. In many practical applications, the KLT is replaced with a fixed matrix such as the DCT, which is independent of the input. This results in a suboptimal transform coder.

7. If $x(n)$ is white then $G_{SBC}(M) = G_{KLT}(M) = 1$, for all M.

8. As the number of subbands M grows indefinitely, the coding gains $G_{SBC}(M)$ and $G_{KLT}(M)$ approach each other, and

$$G_{SBC}(\infty) = G_{KLT}(\infty) = \frac{\sigma_x^2}{\exp\left(\dfrac{1}{2\pi} \displaystyle\int_{-\pi}^{\pi} \log_e S_{xx}(e^{j\omega})\, d\omega\right)} \quad (C.4.3)$$

that is, $G_{SBC}(\infty) = G_{KLT}(\infty) = \sigma_x^2/\mathcal{E}_\infty$ where $\mathcal{E}_\infty$ is the variance of the infinite order linear-prediction error and σ_x^2 is the variance of $x(n)$. Thus the asymptotic coding gains for ideal subband coding and KLT coding are identical. The reciprocal of this asymptotic coding gain is the spectral flatness measure γ_x^2 defined in (C.2.17). Moreover, the above coding gain is equal to the asymptotic coding gain of the ideal DPCM system (Sec. C.5.1). All of these, in turn, are equal to the information theoretic bound on the 'achievable coding gain' for a Gaussian process (see Sec. C.5.2).

9. It is possible for $G_{SBC}(M)$ to attain the maximum value $G_{SBC}(\infty)$ for finite M. This happens when $S_{xx}(e^{j\omega})$ is constant in each of the M subbands. (This does not necessarily mean that $x(n)$ is white.) Such a situation is not possible in transform coding; if $G_{KLT}(M) = G_{KLT}(\infty)$ for finite M then $G_{KLT}(M) = 1$ for *all* M, and $x(n)$ is white.

C.4.2 Relation to Paraunitary Filter Banks

It turns out that the ideal SBC and the orthogonal TC systems are two extreme cases of paraunitary filter banks. Recall that a paraunitary filter bank is a system having the form in Fig. 5.5-3(b), where $\mathbf{E}(z)$ is paraunitary (Chap. 6). If the matrix $\mathbf{R}(z)$ is chosen as $\widetilde{\mathbf{E}}(z)$, the system has perfect reconstruction.

For the ideal SBC case (filters as in Fig. C.2-2) it is easily verified that $\mathbf{E}(z)$ is paraunitary. For this recall (Sec. 6.2.2) that $\mathbf{E}(z)$ is paraunitary if and only if the AC matrix $\mathbf{H}(z)$ is paraunitary. By using the nonoverlapping nature of the filters, it can be shown that $\mathbf{H}(z)$ is indeed paraunitary. Since ideal filters have infinite number of coefficients, the paraunitary matrix $\mathbf{E}(z)$ however has infinite 'order' and moreover is noncausal.

As another extreme, if we choose $\mathbf{E}(z) = \mathbf{T}$ (i.e., a paraunitary system of degree *zero*) we obtain the orthogonal TC system. So we see that the ideal SBC system and orthogonal TC system are two extreme cases of paraunitary perfect reconstruction systems.

Maximizing the Paraunitary Coding Gain

For paraunitary perfect reconstruction systems, an analysis of coding gain can be found in Soman and Vaidyanathan [1991]. The fact that *optimal bit allocation* equalizes quantizer noise variances is a direct consequence of the *paraunitary* property. Ideal SBC and orthogonal transform coding therefore share this property. It turns out that the problem of determining the optimal FIR paraunitary $\mathbf{E}(z)$ of a given degree (in the sense of maximizing the coding gain) is much more complicated than the problem of finding the best constant unitary $\mathbf{T}$. A discussion for the special case of degree-one paraunitary $\mathbf{E}(z)$ can be found in Malvar [1990]. In what follows, we give a flavor for the general maximization problem.

Figure C.4-2 shows the polyphase representation of an M channel maximally decimated filter bank, with quantizers in the subbands. We have represented the quantizers with noise sources $q_k(n)$. We will use the same statistical noise model assumptions as in Sec. C.2. Let $\mathbf{E}(z)$ be paraunitary with $\widetilde{\mathbf{E}}(z)\mathbf{E}(z) = \mathbf{I}$. Let $\mathbf{R}(z) = \widetilde{\mathbf{E}}(z)$ so that $\widehat{x}(n) = x(n - M + 1)$. Clearly $\widetilde{\mathbf{R}}(z)\mathbf{R}(z) = \mathbf{I}$ as well.

The study of coding gain here is similar to that in Sec. C.3.1 for unitary transform coding systems. Thus let $\mathbf{q}(n)$ and $\widehat{\mathbf{q}}(n)$ be as in (C.3.3), where $\widehat{q}_k(n)$ are the noise components at the outputs of $\mathbf{R}(z)$ caused by the noise sources $q_k(n)$. The noise reaching the output node $\widehat{x}(n)$ is the interlaced version of $\widehat{q}_k(n)$ so that the average output noise variance over a duration of M samples is again as in (C.3.1).

We now show that the summation (C.3.1) again reduces to (C.3.2), by using the paraunitary property of $\mathbf{R}(z)$. For this first note that the summation in (C.3.2) is the trace of $\mathbf{R}_{qq}(0)$, where $\mathbf{R}_{qq}(0)$ is equal to $E[\mathbf{q}(n)\mathbf{q}^T(n)]$. Since $\mathbf{R}_{qq}(k)$ is the inverse Fourier transform of the power spectral matrix $\mathbf{S}_{qq}(e^{j\omega})$, we have $\mathbf{R}_{qq}(0) = \int_{-\pi}^{\pi} \mathbf{S}_{qq}(e^{j\omega}) d\omega/2\pi$.

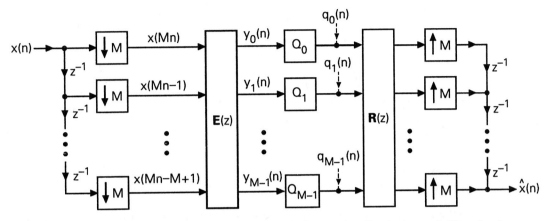

Figure C.4-2 Polyphase representation of the maximally decimated filter bank with subband quantizers.

Defining $\widehat{\mathbf{q}}(n)$ similar to (C.3.3), its power spectrum is (Appendix B)

$$\widehat{\mathbf{S}}_{qq}(e^{j\omega}) = \mathbf{R}(e^{j\omega})\mathbf{S}_{qq}(e^{j\omega})\mathbf{R}^{\dagger}(e^{j\omega}). \qquad (C.4.4)$$

Letting Tr ($\mathbf{A}$) denote the trace of the matrix $\mathbf{A}$, we have

$$\begin{aligned}
\text{Tr }[\widehat{\mathbf{S}}_{qq}(e^{j\omega})] &= \text{Tr }[\mathbf{R}(e^{j\omega})\mathbf{S}_{qq}(e^{j\omega})\mathbf{R}^{\dagger}(e^{j\omega})] \\
&= \text{Tr }[\mathbf{R}^{\dagger}(e^{j\omega})\mathbf{R}(e^{j\omega})\mathbf{S}_{qq}(e^{j\omega})] \quad \text{(since Tr }[\mathbf{AB}]=\text{Tr }[\mathbf{BA}]) \quad (C.4.5) \\
&= \text{Tr }[\mathbf{S}_{qq}(e^{j\omega})] \quad \text{(since } \mathbf{R}^{\dagger}(e^{j\omega})\mathbf{R}(e^{j\omega}) = \mathbf{I}).
\end{aligned}$$

Thus Tr $[\widehat{\mathbf{S}}_{qq}(e^{j\omega})] = $ Tr $[\mathbf{S}_{qq}(e^{j\omega})]$. Integrating this, we obtain

$$\text{Tr }[\widehat{\mathbf{R}}_{qq}(0)] = \text{Tr }[\mathbf{R}_{qq}(0)],$$

so that the average output noise variance is given by (C.3.2), as in orthogonal transform coding.†

With the number of bits per sample fixed as before, the quantizer variance $\sigma_{q_k}^2$ must again be the same for all k (in order to minimize σ_q^2). The minimized variance

---

† This analysis does not assume that the quantizer noise sources are either white or uncorrelated.

σ_q^2 is then given by the righthand side of (C.3.5), where $\sigma_{y_k}^2$ is the variance of the signal $y_k(n)$ [kth output of $\mathbf{E}(z)$].

Recall that the variance $\sigma_{y_k}^2$ is the kth diagonal element of $\mathbf{R_{yy}}(0)$. In view of the inequality (C.3.10), we once again conclude that the paraunitary $\mathbf{E}(z)$ which minimizes σ_q^2 is such that $\mathbf{R_{yy}}(0)$ is diagonal. (However, unlike in transform coding, this diagonal property is not sufficient for optimality; see below). The coding gain is still given by the first line of (C.3.16), that is,

$$G_{PU}(M,N) = \frac{\sigma_x^2}{\left(\det \mathbf{R_{yy}}(0)\right)^{1/M}} \qquad (C.4.6)$$

The subscript PU stands for "paraunitary." This coding gain is a function of the number of subbands M, and the degree N of the paraunitary system. If $N=0$ we obtain the case of traditional transform coding.

Notice that we *cannot* rewrite this as in the second line of (C.3.16) because $[\det \mathbf{R_{yy}}(0)] \neq [\det \mathbf{R_{xx}}(0)]$ in this case. This is because the relation (C.3.8) is not true any more, but only the transform domain version

$$\mathbf{S_{yy}}(e^{j\omega}) = \mathbf{E}(e^{j\omega})\mathbf{S_{xx}}(e^{j\omega})\mathbf{E}^\dagger(e^{j\omega}) \qquad (C.4.7)$$

holds. The relation $[\det \mathbf{R_{yy}}(0)] = [\det \mathbf{R_{xx}}(0)]$ is, therefore, replaced with

$$\det \mathbf{S_{yy}}(e^{j\omega}) = \det \mathbf{S_{xx}}(e^{j\omega}), \qquad \forall \omega. \qquad (C.4.8)$$

As a result, $[\det \mathbf{R_{yy}}(0)]$ is not independent of $\mathbf{E}(z)$ (unlike in orthogonal transform coding). The purpose of $\mathbf{E}(z)$ is therefore two fold. First it has to minimize $[\det \mathbf{R_{yy}}(0)]$ for a given $\mathbf{S_{xx}}(e^{j\omega})$. Second, it has to diagonalize $\mathbf{R_{yy}}(0)$. Finding such a theoretical solution for $\mathbf{E}(z)$ (subject to the degree-N paraunitary constraint) is not an easy task. This theoretical problem (i.e., the extension of the KLT solution for the case of degree N paraunitary $\mathbf{E}(z)$) is still an open problem.

To obtain more insight, recall that an FIR paraunitary $\mathbf{E}(z)$ can be expressed as in Fig. C.4-3 where $\mathbf{W}_m$ are householder (unitary) matrices, $\mathbf{\Lambda}(z)$ is diagonal with diagonal elements $[1,1,\ldots 1, z^{-1}]$, and $\mathbf{T}$ is a constant unitary matrix. (This structure is obtained by transposition of Fig. 14.10-4). The matrices $\mathbf{W}_m$ should be chosen so that $[\det \mathbf{R_{vv}}(0)]$ (which equals $[\det \mathbf{R_{yy}}(0)]$) is minimized. The rightmost matrix $\mathbf{T}$ should then be chosen such that the autocorrelation matrix $\mathbf{R_{yy}}(0) = \mathbf{T}\mathbf{R_{vv}}(0)\mathbf{T}^T$ is diagonalized. That is, $\mathbf{T}$ is the KLT matrix for its input vector $\mathbf{v}(n)$. Notice that there is no change of determinant *across* the matrices $\mathbf{W}_m$, but only across the matrices $\mathbf{\Lambda}(z)$. However, the choice of $\mathbf{W}_m$ affects the extent to which the determinants change across the succeeding $\mathbf{\Lambda}(z)$ blocks.

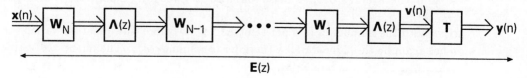

Figure C.4-3 Expressing the paraunitary matrix $\mathbf{E}(z)$ as a cascaded structure.

C.5 RELATION TO OTHER METHODS

We now present the relation between the above coding techniques and a technique called differential PCM. The connection to information theoretic bounds will also be discussed.

C.5.1 The DPCM Technique

Consider the system shown in Fig. C.5-1 where $x(n)$ is a real WSS process, and Q is a b-bit roundoff quantizer as before. The filter $C(z)$ is a linear predictor, which predicts the sample $x_q(n)$ based on its past values $x_q(n-k), k > 0$. That is,

$$\widehat{x}_q(n) = -a_1 x_q(n-1) - a_2 x_q(n-2) - \ldots - a_N x_q(n-N), \qquad (C.5.1)$$

where N is the predictor order. Note that with this notation we have $C(z) = -\sum_{k=1}^{N} a_k z^{-k}$. The quantity $\widehat{x}_q(n)$ is the predicted value. This system is called a *differential pulse code modulation* (DPCM) system. At first sight this might appear to be a strange circuit, but the following simple observations reveal its operation as well as ingenuity.

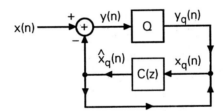

Figure C.5-1 The differential PCM technique.

Observation 1

The signals in the structure satisfy the following equations:

$$y(n) = x(n) - \widehat{x}_q(n)$$
$$x_q(n) = y_q(n) + \widehat{x}_q(n). \qquad (C.5.2)$$

If we add these two equations, then $\widehat{x}_q(n)$ is eliminated, and we get

$$x_q(n) - x(n) = \underbrace{y_q(n) - y(n)}_{\text{quantizer error } q(n)}. \qquad (C.5.3)$$

Thus the difference between $x_q(n)$ and the input signal $x(n)$ is equal to the error introduced by the quantizer. If the quantizer error is small, then $x(n) \approx x_q(n)$.

Observation 2

Let $e_N(n)$ be the prediction error, that is, $e_N(n) = x_q(n) - \widehat{x}_q(n)$. It can then be verified that the quantizer input is

$$y(n) = e_N(n) - q(n).$$

Since $q(n)$ is the error which results due to quantization of $y(n)$, it is reasonable to assume that the variance of $q(n)$ is negligible compared to the variance σ_y^2 of $y(n)$. Thus, from the above equation

$$\sigma_y^2 \approx \text{variance of the prediction error } e_N(n).$$

If the process $x_q(n)$ is highly predictable (i.e., its flatness measure (C.2.17) is small), it is therefore reasonable to assume

$$\text{variance of the prediction error } \ll \sigma_{x_q}^2.$$

Combining these results, we conclude $\sigma_y^2 \ll \sigma_x^2$.

The quantizer error σ_q^2 is given by

$$\sigma_q^2 = c \times 2^{-2b} \sigma_y^2, \qquad (C.5.4)$$

according to Sec. C.1. If $x(n)$ were directly quantized as in PCM using b bits, the quantization error would be

$$\sigma_{q,PCM}^2 = c \times 2^{-2b} \sigma_x^2, \qquad (C.5.5)$$

which is much larger (since $\sigma_y^2 \ll \sigma_x^2$). The quantity (C.5.4) is also the variance of the error $x_q(n) - x(n)$, in view of (C.5.3). Thus if we regard $x_q(n)$ as the reconstructed version of $x(n)$, the reconstruction error has variance

$$\sigma_{q,DPCM}^2 = c \times 2^{-2b} \sigma_y^2, \qquad (C.5.6)$$

so that $\sigma_{q,DPCM}^2 \ll \sigma_{q,PCM}^2$.

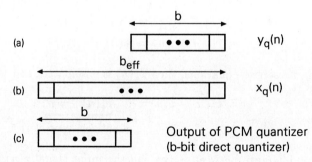

Figure C.5-2 Binary patterns for (a) $y_q(n)$, (b) $x_q(n)$, and (c) output of PCM quantizer.

Figure C.5-2 depicts the situation in terms of binary representations. We have assumed that there is a quantizer at the output of $C(z)$ to avoid infinite bit accumulation in the feedback loop. This new quantizer can be chosen to have its rightmost bit aligned with that of Q. So $y_q(n)$ and $x_q(n)$ have the same step size. However, since $x(n)$ has larger variance than $y(n)$, the signal $x_q(n)$ uses more bits to the left than $y_q(n)$.

The binary pattern which would result if $x(n)$ were directly quantized using b bits is shown in Fig. C.5-2(c). Since the stepsize here is larger than that in Fig. C.5-2(b), there is more quantization error with PCM, for a fixed number of bits b. Thus, even though $x_q(n)$ is an approximation of $x(n)$, obtained by using a b-bit quantizer in a feedback loop, it is a better approximation than a direct b bit quantized version. In other words, the system provides an effective number of bits $b_{eff} > b$. To see this quantitatively, we rewrite (C.5.6) as

$$\sigma^2_{q,DPCM} = c \times \frac{2^{-2b}\sigma^2_y}{\sigma^2_x} \times \sigma^2_x = c \times 2^{-2b_{eff}}\sigma^2_x.$$

Thus $\sigma^2_{q,PCM}$ with b_{eff} bits is identical to $\sigma^2_{q,DPCM}$ with b bits, where

$$b_{eff} = b + \log_2(\sigma_x/\sigma_y). \qquad (C.5.7)$$

Observation 3

We see that $y_q(n)$ requires fewer bits per sample than $x_q(n)$ because of its smaller variance. So it is more economical to transmit $y_q(n)$, and then recover $x_q(n)$. The circuit which recovers $x_q(n)$ from $y_q(n)$ is shown in Fig. C.5-3 (a). (This follows directly from Fig. C.5-1). This can be redrawn as in Fig. C.5-3(b) where $A(z) = 1 - C(z)$. This figure represents an allpole IIR filter $1/A(z)$. We can regard Fig. C.5-1 as the *coder* (or transmitter) and Fig. C.5-3(b) as the *decoder* or receiver.

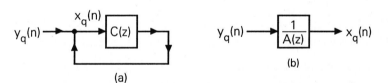

Figure C.5-3 (a) Generation of $x_q(n)$ from $y_q(n)$, and (b) equivalent circuit with $A(z) = 1 - C(z)$.

Essentially we have incorporated a b bit quantizer Q in a feedback loop, to obtain an effective accuracy of b_{eff} bits. A variation of this idea has also been used in the design of high precision A/D converters, by properly incorporating a low precision A/D converter, a linear predictor, and a high precision D/A converter in the feedback loop.

The above heuristic explanation assumes that the signal $x(n)$ has a small spectral flatness measure γ^2_x (i.e., it is highly predictable). We can make these discussions more quantitative; see Jayant and Noll [1984]. In this section we will be content with presenting the coding gain which is

$$G_{DPCM}(N) \triangleq \frac{\sigma^2_{q,PCM}}{\sigma^2_{q,DPCM}} = \frac{\sigma^2_x}{\sigma^2_y}. \qquad (C.5.8)$$

The argument N denotes the predictor order. From the above discussion we see that σ_y^2 is nearly equal to the variance $\mathcal{E}_N$ of the Nth order optimal prediction error $(x(n) - \hat{x}(n))$. So the coding gain is

$$G_{DPCM}(N) \approx \frac{\sigma_x^2}{\mathcal{E}_N}. \qquad (C.5.9)$$

This is precisely the prediction gain of an Nth order predictor. As $N \to \infty$ this approaches the asymptotic coding gains $G_{SBC}(\infty)$ and $G_{TC}(\infty)$ [see (C.3.28)]. Thus, the coding gains of all the three schemes have identical asymptotic value! The coding gain is unity if $x(n)$ is white, since $\mathcal{E}_N = \sigma_x^2$ in that case (Sec. C.3.3). The gain increases as the spectral flatness measure (C.2.17) decreases.

Thus, with the incorporation of an optimal linear predictor in a feedback loop, it is possible to achieve the same coding gain as obtained by ideal subband coding and KLT coding. It turns out that if an optimal predictor is used in an *open* loop, then there is nothing to be gained, that is, the coding gain is unity. This statement is clarified in Problems C.7 and C.8.

C.5.2 Information Theoretic Performance Bound

We have studied three techniques, viz., subband coding, transform coding, and DPCM. Each of these systems can be considered to be a "sophisticated quantizer." For example the Transform coding system of Fig. C.3-1 "quantizes" $x(n)$ into $\hat{x}(n)$. The DPCM system "quantizes" $x(n)$ into $x_q(n)$. In each case, we have embedded traditional roundoff quantizer(s) Q into a sophisticated structure made of filters, decimators, linear predictors and so on.

It is intriguing to note that each of the above techniques gives the same asymptotic coding gain under ideal conditions. It is, therefore, reasonable to expect that this asymptotic gain has a fundamental significance. This indeed is the case, and has its foundations in the results of Shannon in communication theory [Shannon, 1949]. We will now state this connection. For further details, the reader should see Appendix D of Jayant and Noll [1984], and references therein.

Rate Distortion Bound

The specific theory which is applicable in this context is the so-called rate-distortion theory [Berger, 1971]. Given a WSS process $x(n)$, suppose we wish to use an average of b bits per sample to transmit it. The technique which encodes the signal $x(n)$ into a b bit version is the 'coding scheme'. Examples are subband coding, DPCM, and so on. We can conceive of an unlimited variety of such coding schemes, for example, a subband coder in which each subband quantizer itself is replaced with a sophisticated device such as a DPCM coder.

In any case, assume that we have obtained a reconstructed version $x_{rec}(n)$ of the signal $x(n)$, from its coded version. (For example, $x_{rec}(n) = \hat{x}(n)$ in the scheme of Fig. C.3-1, and $x_{rec}(n) = x_q(n)$ in Fig. C.5-1.) The reconstruction error is $e(n) = x_{rec}(n) - x(n)$. Rate distortion theory provides a lower bound on the variance of $e(n)$, as a function of b (average number of bits per sample). This bound is denoted as $D_{min}(b)$. This is the minimum possible distortion at bit rate b; there exists no coding technique which can result in a smaller distortion.

A plot of $D_{min}(b)$ versus b therefore provides a yardstick for the performance of practical coding systems. If a practical system has a $D(b)$ plot well above $D_{min}(b)$, then there is much room for improvement in terms of coder design. (Fig. C.5-4 shows some qualitative plots). On the other hand, if the $D(b)$ curve is very close to this bound, then any further improvement will usually require excessive cost for coder design. That is, the additional cost for coder design is excessive in relation to the reduction in the cost of transmission. (The latter cost refers to the bits per second resulting after coding.)

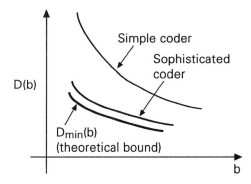

Figure C.5-4 Qualitative plots of rate distortion functions.

The case of a Gaussian process. We now give the functional form of $D(b)$ for a specific case. Let $x(n)$ be a zero mean real WSS Gaussian process, with power spectral density $S_{xx}(e^{j\omega})$. For 'sufficiently small' distortions (i.e., assuming that b is 'sufficiently' large), we have

$$D_{min}(b) = 2^{-2b} \exp\left(\frac{1}{2\pi} \int_{-\pi}^{\pi} \log_e S_{xx}(e^{j\omega})\, d\omega\right). \qquad (C.5.10)$$

This quantity appeared several times during our discussions of subband and transform coding gains. It also appeared in the spectral flatness measure (C.2.17). The maximum coding gain obtainable (using any scheme whatsoever) is theoretically bounded by

$$G_{max} = \frac{\sigma_{q,PCM}^2}{D_{min}(b)} = \frac{c \times \sigma_x^2}{\exp\left(\dfrac{1}{2\pi} \int_{-\pi}^{\pi} \log_e S_{xx}(e^{j\omega})\, d\omega\right)} \qquad (C.5.11)$$

which agrees with $G_{SBC}(\infty), G_{KLT}(\infty)$ and $G_{DPCM}(\infty)$. In other words, the benefits to be gained from each of these coding systems approach the theoretical rate distortion limit as the number of bands (or the predictor order in the DPCM case) approaches infinity. Notice that the bound G_{max} has the form c/γ_x^2 where γ_x^2 is the spectral flatness measure (C.2.17).

The reader will notice that the theoretical bound (C.5.11) does not fully agree with (C.3.28) because of the constant c. We will not specify c more exactly, since it has to do with the probability of overflow (Sec. C.1). Our main aim is to draw attention to the fact that the formula (C.5.11) has the same functional dependencies on σ_x^2 and $S_{xx}(e^{j\omega})$, as do the asymptotic coding gains $G_{KLT}(\infty), G_{SBC}(\infty)$, and $G_{DPCM}(\infty)$.

Rate distortion theory tells us that if we are allowed to use b bits per sample then the mean squared distortion cannot be smaller than $D_{min}(b)$. Now consider a subband coding system, where we are allowed to use b_k bits per sample in the kth subband. The overall bit rate b is related to b_k as in (C.2.2). Instead of using a simple roundoff quantizer in each subband, suppose we use a DPCM coder or an even more sophisticated device. No matter how we perform this coding of each subband, the overall reconstruction error will never be smaller than $D_{min}(b)$, even if the quantization error in each subband is as low as its own rate distortion bound $D_{min}(b_k)$.

In particular, if we are allowed to use quantizers of arbitrary degree of sophistication, then a PCM coder with such an 'advanced' quantizer can even be better than a subband coder with similar quantizers in each subband. (Of course, we will not think of it as a PCM coder any more, according to traditional language.) Techniques such as subband coding, transform coding, and DPCM are superior to PCM, if we perform the comparison under the condition that all the boxes labeled Q in their flow diagrams are simple roundoff quantizers.

PROBLEMS

Note. Unless mentioned otherwise, the input $x(n)$ is real WSS with zero mean, and all filters have real coefficients.

C.1. Consider the expression (C.1.4) which relates the quantizer noise variance σ^2 to the signal variance σ_x^2. We know that c depends on statistics of x, and we evaluated c for a number of statistical distributions.

a) Now assume x has the following density function:

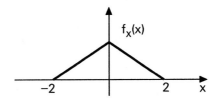

Figure PC-1

How would you choose ℓ to prevent overflow? Evaluate c.

b) Next let x be Gaussian, and suppose we choose ℓ such that $2^{\ell+1} = 4\sigma_x$. What is the probability of overflow? What is the value of c?

C.2. The AM-GM inequality says that if a_k are nonnegative numbers then

$$\frac{1}{M}\sum_{k=0}^{M-1} a_k \geq \Big(\prod_{k=0}^{M-1} a_k\Big)^{1/M} \qquad (PC.2a)$$

with equality if and only if $a_0 = a_1 = \ldots = a_{M-1}$. We now outline a proof. (This appears to be Cauchy's original proof! [Bellman, 1960]).

a) Let $x_0, x_1 \geq 0$. Prove that $0.5(x_0 + x_1) \geq \sqrt{x_0 x_1}$. (*Hint.* Use $(\sqrt{x_0} - \sqrt{x_1})^2 \geq 0$.) When does equality occur?

b) Prove AM-GM inequality for $M = 4$. [*Hint.* Set $x_0 = (a_0 + a_1)/2$, $x_1 = (a_2 + a_3)/2$ in part (a).]

c) Hence prove, by induction, that such an inequality holds for $M = 2^m$, for all positive integers m.

d) It remains to complete the proof for arbitrary positive integers M. For this it is sufficient to show that it holds for $M - 1$ if it holds for M. Thus, let $b_0, \ldots b_{M-2} \geq 0$. Define $a_i = b_i$ for $0 \leq i \leq M - 2$ and

$$a_{M-1} = \frac{b_0 + \ldots + b_{M-2}}{M-1}. \qquad (PC.2b)$$

Assuming (PC.2a) to be true, show that

$$\frac{1}{M-1}\sum_{i=0}^{M-2} b_i \geq \Big(\prod_{i=0}^{M-2} b_i\Big)^{1/(M-1)}. \qquad (PC.2c)$$

Also prove that equality holds if and only if all b_i are equal. (This technique is called the *backward induction.*)

C.3. Consider the subband coder with ideal filters (as in Sec. C.2).
   a) Show that the coding gain satisfies $G_{SBC}(nM) \geq G_{SBC}(M)$.
   b) Suppose the power spectrum $S_{xx}(e^{j\omega})$ of the input $x(n)$ is constant in each of the M subbands (e.g., as in Fig. C.2-5). Show that $G_{SBC}(nM) = G_{SBC}(M)$ for all positive integers n.
   c) Conversely, suppose $G_{SBC}(nM) = G_{SBC}(M)$ for all positive integers n. Argue that $S_{xx}(e^{j\omega})$ is constant in each of the M subbands.

C.4. Let the power spectrum of $x(n)$ be as shown below.

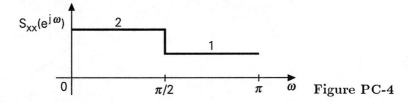

Figure PC-4

With ideal subband coding filters as in Sec. C.2, explicitly evaluate the coding gains $G_{SBC}(2)$ and $G_{SBC}(\infty)$, and verify that they are the same. Argue (without explicit evaluation) that $G_{SBC}(3) < G_{SBC}(2)$.

C.5. Let the input $x(n)$ to the ideal subband coder of Sec. C.2 be such that $G_{SBC}(i) = G_{SBC}(M)$ for all $i \geq M$ for some integer $M > 1$. Show that $x(n)$ is white!

C.6. For the KLT transform coder, suppose $G_{TC}(M+1) = G_{TC}(M)$ for some $M > 0$. Show then that (C.3.29) holds. *Hint.* Use (C.3.25).

C.7. Consider the system shown below where Q is a roundoff quantizer satisfying the standard assumptions of Sec. C.1. Let $x(n)$ be real, zero mean, and WSS with power spectrum $S_{xx}(e^{j\omega})$. Let σ_q^2 denote the quantizer noise variance.

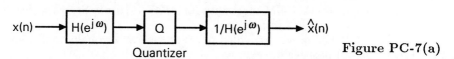

Figure PC-7(a)

   a) Find an expression for the variance of the reconstruction error $\hat{x}(n) - x(n)$.
   b) Find an expression for the coding gain in terms of $S_{xx}(e^{j\omega})$ and $H(e^{j\omega})$.
   c) For fixed $S_{xx}(e^{j\omega})$ find a filter $H(e^{j\omega})$ which maximizes the coding gain. *Hint.* Use Cauchy Schwartz inequality, that is,

$$\left(\int u(\alpha)v(\alpha)d\alpha\right)^2 \leq \int u^2(\alpha)d\alpha \int v^2(\alpha)d\alpha \qquad (PC.7)$$

for real $u(\alpha)$ and $v(\alpha)$, with equality if and only if $u(\alpha) = Kv(\alpha)$ for some constant K.

   d) What is the maximized coding gain? Evaluate it for the example where $S_{xx}(e^{j\omega})$ has the form shown in Fig. PC-7(b).

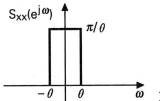

Figure PC-7(b)

Plot this coding gain as a function of θ for $0 < \theta < \pi$. You will see that the gain is very large for a 'peaky' spectrum and small for a relatively 'flat' spectrum.

Note. In general, the optimal $H(e^{j\omega})$ and its inverse may not be realizable filters. One has to replace these filters with practical (causal, stable) approximations, which result in suboptimal coding gains.

C.8. Consider the linear prediction equation (C.3.22). With $e_k(n)$ denoting the prediction error $x(n) - x_p(n)$, we see that $e_k(n)$ and $x(n)$ can be generated from each other as shown below, where $A_k(z) = 1 + \sum_{i=1}^{k} a_i z^{-i}$.

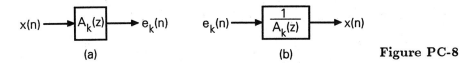

(a)   (b)   **Figure PC-8**

Consider the case where the predictor coefficients are optimal, that is, $E[e_k^2(n)]$ is minimized. It is then well known in linear prediction theory that $A_k(z)$ has all zeros inside the unit circle (unless $x(n)$ is a special process called 'harmonic process'). So $1/A_k(z)$ is stable IIR, and the structure of Fig. PC-7(a) is realizable with $H(z) = A_k(z)$. Let $\mathcal{E}_k$ denote the minimized prediction error variance $E[e_k^2(n)]$. It is known that the variance σ_x^2 of $x(n)$ and the quantity $\mathcal{E}_k$ are related as

$$\frac{\sigma_x^2}{\mathcal{E}_k} = \frac{1}{2\pi} \int_{-\pi}^{\pi} \frac{1}{|A(e^{j\omega})|^2} \, d\omega \qquad (PC.8)$$

(This ratio is called the 'prediction gain'). Assuming the truth of (PC.8), show that if $H(z) = A_k(z)$ in Fig. PC-7(a), then the coding gain is unity.

Comment. Thus, if the linear prediction system is used in an open loop configuration, we do not gain anything. On the other hand, if we use it in a closed loop as in DPCM, then the coding gain is nearly equal to the prediction gain (PC.8), as we saw in Sec. C.5.

Note. The result (PC.8) is obvious when $e_k(n)$ is white [from Fig. PC-8(b)], but is rather subtle in the general case. It can be proved by using some of the deeper properties (e.g., the autocorrelation matching property, [Makhoul, 1975]) of the optimal predictor.

C.9. Consider the optimum bit allocation formula (C.2.9). With $M = 2$, $\sigma_{x_0}^2 = 64$ and $\sigma_{x_1}^2 = 1$, what is the minimum b that ensures $b_0 \geq 1$ and $b_1 \geq 1$? Assume that b is chosen to be two times this minimum value. What are optimal values of b_0 and b_1? Are these integers?

C.10. Consider the ideal subband coder. Let the coding gain be such that $G_{SBC}(M) = G_{SBC}(N) = G_{SBC}(\infty)$ for two relatively prime integers M and N where $M, N > 1$. Show then that the input $x(n)$ is white.

Appendix D

Spectral factorization techniques

D.0 INTRODUCTION

In Sec. 3.2.5 we defined the spectral factor $H_0(z)$ of a FIR transfer function $H(z)$ with nonnegative frequency response $H(e^{j\omega})$. This section should be reviewed at this time. The spectral factor $H_0(z)$ is not unique, and we can find all possible solutions by finding the zeros of $H(z)$ and grouping them appropriately. We will now describe an algorithm, reported in Mian and Nainer [1982], for computing a spectral factor $H_0(z)$ *without finding the zeros* of $H(z)$. The spectral factor given by this algorithm has minimum phase (and is therefore unique). Moreover, the method works even if $H(z)$ has zeros *on the unit circle,* unlike some other techniques that have been reported in the literature.

D.1 THE COMPLEX CEPSTRUM

The technique to be described here uses some fundamental properties of the so-called complex cepstrum of a sequence. A detailed treatment of this topic, along with references to poineering work in this area, can be found in Chap. 12 of Oppenheim and Schafer [1989]. Here we will present the definition and only those properties that are relevant for the spectral factorization algorithm.

Let $x(n)$ be a sequence with z-transform $X(z)$. In general this converges in some annulus in the z-plane. We shall specify this region explicitly only when it becomes relevant later on. Consider the function $\widehat{X}(z) \stackrel{\Delta}{=} [\ln X(z)]$. This is in general complex valued [as is $X(z)$]. Assuming that this has a nontrivial region of convergence, we can find its inverse z-transform $\widehat{x}(n)$. This sequence $\widehat{x}(n)$ is said to be the *complex cepstrum* of $x(n)$.[†] We also say "complex cepstrum of $X(z)$" when that is more convenient.

We will now consider some examples to bring out a few key properties.

[†] $\widehat{x}(n)$ is not necessarily a complex sequence; the term 'complex' is used to emphasize the fact that a complex logarithm is involved.

Example D.1.1

Let $X(z) = 1 - \alpha_k z^{-1}$ which is the z-transform of the causal FIR sequence $x(n)$ whose nonzero samples are $\{1, -\alpha_k\}$. Then $\widehat{X}(z) = \ln(1 - \alpha_k z^{-1})$. By using the expansion

$$\ln(1+a) = -\sum_{n=1}^{\infty} \frac{(-a)^n}{n}, \qquad |a| < 1, \tag{D.1.1}$$

we obtain the power series expansion

$$\widehat{X}(z) = -\sum_{n=1}^{\infty} \frac{\alpha_k^n z^{-n}}{n}, \qquad |z| > |\alpha_k|. \tag{D.1.2}$$

Note that the region of convergence is $|z| > |\alpha_k|$. The complex cepstrum $\widehat{x}(n)$ of the sequence $x(n)$ is the inverse z-transform of $\widehat{X}(z)$ i.e.,

$$\widehat{x}(n) = \frac{-\alpha_k^n}{n} \mathcal{U}(n-1), \tag{D.1.3}$$

where $\mathcal{U}(n)$ denotes the unit-step function. In this example $\widehat{x}(n)$ is a *causal* sequence. If $|\alpha_k| \leq 1$, then this sequence is bounded, and approaches zero for $n \to \infty$. Note that $\widehat{x}(n)$ is infinitely long, even though $x(n)$ has finite duration.

Next consider the example $Y(z) = 1 - \beta_k z$. This is the z-transform of the noncausal FIR sequence $y(n)$ whose nonzero sample values are $\{-\beta_k, 1\}$. The quantity $\widehat{Y}(z) \triangleq [\ln Y(z)]$ can be expressed as

$$\widehat{Y}(z) = \sum_{n=-\infty}^{-1} \frac{\beta_k^{-n} z^{-n}}{n}, \qquad |z| < \frac{1}{|\beta_k|}. \tag{D.1.4}$$

The region of convergence is $|z| < |\beta_k^{-1}|$, and the complex cepstrum of $y(n)$ is

$$\widehat{y}(n) = \frac{\beta_k^{-n}}{n} \mathcal{U}(-n-1), \tag{D.1.5}$$

which is anticausal. If $|\beta_k| \leq 1$, $\widehat{y}(n)$ is bounded, and approaches zero for $n \to -\infty$.

D.1.1 Minimum and Maximum Phase Filters

Let $H_0(z) = \sum_{n=0}^{N} h_0(n) z^{-n}$ be a causal minimum phase FIR filter (i.e., no zeros outside the unit circle). Let a_k, $1 \leq k \leq n_a$ be the zeros on the unit circle, and let b_k, $1 \leq k \leq n_b$ be those inside. Then,

$$H_0(z) = \prod_{k=1}^{n_a}(1 - a_k z^{-1}) \prod_{k=1}^{n_b}(1 - b_k z^{-1}), \qquad |a_k| = 1, |b_k| < 1, \tag{D.1.6}$$

where we have assumed $h_0(0) = 1$ for simplicity. Since $\ln XY = \ln X + \ln Y$, the quantity $[\ln H_0(z)]$ is the sum of the logarithms of the first order factors in (D.1.6).

Using the ideas in Example D.1.1, we see that the function $\widehat{H}_0(z) \triangleq [\ln H_0(z)]$ converges in the region $|z| > 1$, and that the complex cepstrum of $h_0(n)$ is the causal sequence

$$\widehat{h}_0(n) = \underbrace{-\sum_{k=1}^{n_a} \frac{a_k^n}{n} \mathcal{U}(n-1)}_{c_a(n)} \underbrace{- \sum_{k=1}^{n_b} \frac{b_k^n}{n} \mathcal{U}(n-1)}_{c_b(n)}. \qquad (D.1.7)$$

If $H_0(z)$ has real coefficients, then complex zeros are accompanied by their conjugates so that the summations above are real.

As a second example, consider the filter

$$H_1(z) = \prod_{k=1}^{n_a}(1 - a_k z^{-1}) \prod_{k=1}^{n_b}(1 - b_k z), \quad |a_k| = 1, |b_k| < 1. \qquad (D.1.8)$$

This has the zeros a_k on unit circle, and the zeros $1/b_k$ *outside* the unit circle. This is a maximum phase filter. Its logarithm can be expressed as

$$\ln H_1(z) = \sum_{k=1}^{n_a} \ln(1 - a_k z^{-1}) + \sum_{k=1}^{n_b} \ln(1 - b_k z). \qquad (D.1.9)$$

Using Example D.1.1 we see that the first summation converges for $|z| > 1$ and the second summation for $|z| < \min_k |b_k^{-1}|$. So the above z-transform has region of convergence $1 < |z| < \min_k |b_k^{-1}|$. Its inverse transform, that is, the cepstrum of $h_1(n)$ is given by

$$\widehat{h}_1(n) = \underbrace{-\sum_{k=1}^{n_a} \frac{a_k^n}{n} \mathcal{U}(n-1)}_{c_a(n)} + \underbrace{\sum_{k=1}^{n_b} \frac{b_k^{-n}}{n} \mathcal{U}(-n-1)}_{c_b(-n)}. \qquad (D.1.10)$$

Thus the complex cepstrum of $h_0(n)$ is causal (see (D.1.7)) whereas that of $h_1(n)$ has an anticausal part $c_b(-n)$ (contributed by the zeros $1/b_k$ outside the unit circle).

An intriguing relation. Since $\ln XY = \ln X + \ln Y$, the complex cepstrum of $G(z) \triangleq H_1(z)H_0(z)$ is given by

$$\widehat{g}(n) = 2c_a(n) + c_b(n) + c_b(-n). \qquad (D.1.11)$$

On the other hand, the complex cepstrum of $H_0^2(z)$ is equal to $2\widehat{h}_0(n)$. From (D.1.7) we have

$$2\widehat{h}_0(n) = 2c_a(n) + 2c_b(n) \qquad (D.1.12)$$

We can therefore obtain $\widehat{h}_0(n)$ from $\widehat{g}(n)$ as follows: simply fold the anticausal part $c_b(-n)$, add to the causal part, and divide by two. We will use this idea later.

D.1.2 Subtleties in the Computation

There are some subtle issues which must be carefully considered when computing the complex cepstrum. Consider $\widehat{X}(z) = [\ln X(z)]$. If the region of convergence of this quantity includes the unit circle, we can evaluate $\widehat{X}(e^{j\omega}) = \ln X(e^{j\omega})$, and then find $\widehat{x}(n)$ by performing an inverse Fourier transformation.

If, on the other hand, $\widehat{X}(z)$ does not converge on the unit circle, we can always take care of this as follows: define $x_1(n) = \rho^{-n}x(n)$ so that $X_1(z) = X(\rho z)$. For appropriate choice of ρ, the quantity $\widehat{X}_1(z) \triangleq [\ln X_1(z)]$ will converge on the unit circle. Having computed its inverse Fourier transform $\widehat{x}_1(n)$, we can then obtain the complex cepstrum of the original sequence $x(n)$ using the relation $\widehat{x}(n) = \rho^n \widehat{x}_1(n)$. In what follows, we shall therefore assume that $X(z)$ converges on the unit circle, that is, that $X(e^{j\omega})$ exists.

Subtleties About the Phase Response

Let $X(e^{j\omega}) = |X(e^{j\omega})|e^{j\phi(\omega)}$. Then its complex logarithm can be expressed as

$$\widehat{X}(e^{j\omega}) = \ln X(e^{j\omega}) = \ln|X(e^{j\omega})| + j\phi(\omega). \qquad (D.1.13)$$

We know that the phase function $\phi(\omega)$ is not uniquely determined by $X(e^{j\omega})$ since we can replace $\phi(\omega)$ with $\phi(\omega) + 2\pi k$ for integer k, without changing $X(e^{j\omega})$. However, $[\ln X(e^{j\omega})]$ is affected by the choice of k.

In Sec. 2.4.1 we stated that there are two kinds of phase responses, unwrapped and wrapped, denoted respectively as $\phi_u(\omega)$ and $\phi_w(\omega)$. For many applications, we need not distinguish between these, and the subscript is omitted. But in the computation of $[\ln X(e^{j\omega})]$, one uses the unwrapped phase $\phi_u(\omega)$ (which is free from discontinuities in $\phi_w(\omega)$ caused by the modulo 2π operation on phase). The reason for this is the following: whenever we consider the complex cepstrum of a product [e.g., equation (D.1.6)], we would like the result to be the sum of individual complex cepstra (e.g., as in (D.1.7)). From (D.1.13) we see that this will happen only if we consistently use unwrapped phases everywhere. (The sum of unwrapped phases gives the total unwrapped phase.)

The unwrapped phase $\phi_u(\omega)$ can be computed using the algorithm in Tribolet [1977] (which we will not repeat here). Furthermore, a computer program for computation of the complex cepstrum, which incorporates phase unwrapping, is available in Tribolet and Quatieri [1979].

When is the unwrapped phase periodic? The wrapped phase response $\phi_w(\omega)$ is periodic with period 2π whereas the unwrapped phase response $\phi_u(\omega)$ is not necessarily so. (Example: let $H(z) = z^{-K}$, then $\phi_u(\omega) = -K\omega$). This will create some difficulties in (D.1.13) because the Fourier transform $\widehat{X}(e^{j\omega})$ of the sequence $\widehat{x}(n)$ is supposed to be periodic.

It can be shown (Problems D.1 and D.2) that if we have a transfer function of the form $(1 - \alpha z^{-1})$ where α is a possibly complex number with $|\alpha| < 1$, then the unwrapped phase is still periodic with period 2π. This is also true of $(1 - \alpha z)$ for $|\alpha| < 1$. More generally, this is true for products of the form

$$\prod(1 - \alpha_k z^{-1})\prod(1 - \beta_k z), \quad |\alpha_k|, |\beta_k| < 1. \qquad (D.1.14)$$

Thus, the factors corresponding to the zeros inside the unit circle should be represented as $(1 - \alpha z^{-1}), |\alpha| < 1$ rather than as $(1 - \alpha' z), |\alpha'| > 1$. Similarly the zeros outside the unit circle must be represented by $(1 - \beta z), |\beta| < 1$ rather than as $(1 - \beta' z^{-1}), |\beta'| > 1$. This will ensure that the unwrapped phase is periodic.

Elimination of the nonperiodic component. Notice that any FIR transfer function with *no zeros on the unit circle* can be written as

$$H(z) = cz^{-K} \prod(1 - \alpha_k z^{-1}) \prod(1 - \beta_k z), \quad |\alpha_k|, |\beta_k| < 1 \qquad (D.1.15)$$

where K is an integer, α_k are the zeros inside the unit circle and $(1/\beta_k)$ are the zeros outside. The unwrapped phase $\phi_u(\omega)$ of this system is not periodic, because of the factor z^{-K}. So the Fourier transform $[\ln H(e^{j\omega})]$ is not periodic if we use the unwrapped phase. On the other hand, if we use the wrapped phase, then the imaginary part of $[\ln H(e^{j\omega})]$ has discontinuities (jumps of 2π).

In practice, this situation is avoided by estimating K and then replacing $H(z)$ with $z^K H(z)$. This eliminates the nonperiodic component of the unwrapped phase. Given the quantity $H(z)$, the algorithm in Tribolet and Quatieri [1979] first obtains the unwrapped phase, and then estimates K, and actually computes the complex cepstrum of $z^K H(z)$. As a result, the output of this algorithm is unaffected if the input to the algorithm [i.e., $H(z)$] is replaced with $z^{-L} H(z)$ for arbitrary integer L.

The computation of K in (D.1.15) is performed as follows: suppose the unwrapped phase $\phi_u(\omega)$ has been found. Compute the difference $\phi_u(\pi) - \phi_u(-\pi)$. Since the nonperiodicity comes only from $e^{-j\omega K}$, this difference is equal to $-2K\pi$. From this K can be found.

D.2 A CEPSTRAL INVERSION ALGORITHM

It is possible to recover $x(n)$ uniquely from $\widehat{x}(n)$. Techniques for such inversion can be found in Chap. 12 of Oppenheim and Schafer [1989]. We now present a procedure for the case where $X(z)$ is rational, and $x(n)$ satisfies the following properties.

1. $x(n)$ is real, causal, and stable (i.e., all poles of $X(z)$ are inside the unit circle).
2. $X(z)$ has no zeros outside the unit circle (i.e., it is a mimimum phase function).
3. $x(0) > 0$.

In this case the cepstral sequence $\widehat{x}(n)$ has the form (D.1.7), and is causal, as well as real. (The poles of $X(z)$ give rise to terms similar to those in (D.1.7), except for sign). Both $X(z)$ and $\widehat{X}(z)$ converge for $|z| > 1$. In this region, we can write

$$X(z) = e^{\widehat{X}(z)} = 1 + \widehat{X}(z) + \frac{\widehat{X}^2(z)}{2!} + \ldots \qquad (D.2.1)$$

where $\widehat{X}(z) = \widehat{x}(0) + \widehat{x}(1)z^{-1} + \widehat{x}(2)z^{-2} + \ldots$ Equating the constant coefficients on both sides of (D.2.1), we get

$$x(0) = 1 + \widehat{x}(0) + \frac{\widehat{x}^2(0)}{2!} \ldots = e^{\widehat{x}(0)}. \qquad (D.2.2)$$

Suppose we are given the causal sequence $\widehat{x}(n)$. Then we can compute $x(0)$ from the above equation, and obtain the remaining samples of $x(n)$ by using the recursion

$$x(n) = \sum_{k=1}^{n} \frac{k}{n} \widehat{x}(k) x(n-k), \quad n > 0. \qquad (D.2.3)$$

To see this note that the equation $X(z) = e^{\widehat{X}(z)}$ implies $X'(z) = X(z)\widehat{X}'(z)$ (where prime denotes derivative with respect to z). Using the fact that $zX'(z)$ is the

z-transform of $-nx(n)$, we obtain $nx(n) = x(n) * n\hat{x}(n)$, where $*$ stands for convolution. Eqn. (D.2.3) follows from this. This equation also shows that if $x(n)$ has finite duration (say L), we can compute it from the first L samples of $\hat{x}(n)$ even though $\hat{x}(n)$ itself does not have finite duration.

D.3 A SPECTRAL FACTORIZATION ALGORITHM

Suppose we are given the coefficients of the FIR filter $H(z)$, with frequency response $H(e^{j\omega}) \geq 0$ (which, in particular, has zero-phase). We will assume that the coefficients are real i.e., $H(e^{j\omega})$ is even (as in most filter bank applications). Thus if z_k is a zero of $H(z)$, then so is z_k^*. Recall that the minimum-phase spectral factor $H_0(z)$ is obtained by retaining all the zeros inside the unit circle, and one out of every double zero on the unit circle. (See Sec. 3.2.5.) So $H_0(z)$ has real zeros and complex-conjugate pairs of zeros, and therefore has real coefficients. We now present a method to extract this real-coefficient minimum-phase spectral factor $H_0(z)$.

$H(z)$ is a zero phase filter, and has the form $H(z) = \sum_{n=-N}^{N} h(n) z^{-n}$. We can write this function as

$$H(z) = A z^K \prod_{k=1}^{n_a} (1 - a_k z^{-1})^2 \prod_{k=1}^{n_b} (1 - b_k z^{-1}) \prod_{k=1}^{n_b} (1 - b_k z), \qquad (D.3.1)$$

with $|a_k| = 1$ and $|b_k| < 1$. In writing (D.3.1) we have used the following facts: (a) if α is a zero of $H(z)$ then so is $1/\alpha^*$, (b) the zeros on the unit circle have even multiplicity (since $H(e^{j\omega}) \geq 0$), and (c) if α is a zero then so is α^* (since the coefficients are real).

The minimum phase spectral factor has the form

$$H_0(z) = B \prod_{k=1}^{n_a} (1 - a_k z^{-1}) \prod_{k=1}^{n_b} (1 - b_k z^{-1}). \qquad (D.3.2)$$

From this we have $h_0(0) = B$ so that B is real. It can be shown that $A = B^2 > 0$. (*Proof.* Evaluate $H(z)$ and $H_0(z)$ at $z = 1$ and use $H(1) = H_0^2(1)$. This gives $A = B^2$ as long as $H(e^{j0}) \neq 0$, which is a valid assumption for the lowpass case).

Our aim is to compute $H_0(z)$ [i.e., its coefficients $h_0(n)$] from $H(z)$, by first computing the complex cepstrum of $h(n)$. The program in Tribolet and Quatieri [1979] can be adapted for this purpose. As explained at the end of Sec. D.1, this program will automatically estimate and eliminate z^K. In this sense the program is insensitive to a time-shift of the input data. So we will focus our attention only on the effective input function

$$F(z) = A \prod_{k=1}^{n_a} (1 - a_k z^{-1})^2 \prod_{k=1}^{n_b} (1 - b_k z^{-1}) \prod_{k=1}^{n_b} (1 - b_k z). \qquad (D.3.3)$$

From this we first compute the complex cepstrum $\hat{f}(n)$. This can be done by inverse transformation of $[\ln F(z)]$. But we have to be careful: $[\ln F(z)]$ does not converge on the unit circle (since $F(z)$ has zeros there). Fortunately, however, $F(z)$ does not have any zeros in the region $1 < |z| < \min_k \frac{1}{|b_k|}$, so that we can choose

this to be the region of convergence of $[\ln F(z)]$. Then, the function $F_1(z) \triangleq F(\rho z)$ which is the z-transform of
$$f_1(n) \triangleq \rho^{-n} f(n), \qquad (D.3.4)$$
converges on the unit circle, for ρ in the range
$$1 < \rho < \min_k \frac{1}{|b_k|}$$

We can therefore work with $[\ln F_1(e^{j\omega})]$ and compute the inverse Fourier transform $\widehat{f}_1(n)$, and then obtain $\widehat{f}(n) = \rho^n \widehat{f}_1(n)$. This has the form
$$\widehat{f}(n) = (\ln A)\delta(n) + 2c_a(n) + c_b(n) + c_b(-n), \qquad (D.3.5)$$
where $c_a(n)$ and $c_b(n)$ are real valued causal sequences defined in (D.1.7) and $c_b(-n)$ is anticausal. (Clearly $\widehat{f}(n)$ is real as long as $A > 0$. See comments below.) By folding the anticausal part $c_b(-n)$, adding to the causal part, and dividing by two, we therefore obtain
$$\Big(0.5 \ln A\Big)\delta(n) + c_a(n) + c_b(n). \qquad (D.3.6)$$

But the complex cepstrum $\widehat{h}_0(n)$ of the minimum phase spectral factor $H_0(z)$ [eqn. (D.3.2)] is also given by this expression (using $A = B^2$). Thus if we use the causal sequence (D.3.6) as $\widehat{x}(n)$ in (D.2.3), then the inverted sequence $x(n)$ will be equal to $h_0(n)$. In other words, the spectral factor $H_0(z)$ has been determined.

D.3.1 Computational Issues

The program in Tribolet and Quatieri [1979] computes the complex cepstrum $\widehat{f}_1(n)$ of $f_1(n)$. The user has the responsibility of choosing ρ and supplying $f_1(n)$ defined in (D.3.4). The program first evaluates $F_1(e^{j\omega})$ for a finite number (say L) of values of ω. This is done by computing the DFT of the sequence $f_1(n)$. This gives the sequence
$$F_1(e^{j2\pi k/L}), \quad 0 \leq k \leq L-1. \qquad (D.3.7)$$
The inverse DFT of $\big[\ln F_1(e^{j2\pi k/L})\big]$ is then obtained [by using the unwrapped phase of $F_1(e^{j\omega})$ for the purpose]. This gives an approximation of $\widehat{f}_1(n)$ [complex cepstrum of $f_1(n)$]. The accuracy of this approximation depends on L. Once $\widehat{f}_1(n)$ is thus obtained, the user proceeds as described above, in order to complete the spectral factorization.

The DFT and IDFT are typically done by using the fast Fourier transform (FFT). For further discussion on the choice of L and ρ, see [Mian and Nainer, 1982]. Usually the choice $L = 8N$ gives acceptable results. The choice of ρ is tricky because the zeros b_k are not given. In practice, this choice might require trial and error. Once the spectral factor $H_0(z)$ is identified, one can plot $|H_0(e^{j\omega})|^2$ and compare with the plot of $H(e^{j\omega})$, to check the accuracy as well as to make sure that there have been no fatal errors.

D.3.2 Summary of the Spectral-Factorization Procedure

In the above derivation, we have provided several computational details, so that the user is well-informed about these. However, the implementaion itself is very simple, and does not require many of the above details. Here is a summary of the procedure.

1. Given the function $H(z) = \sum_{n=-N}^{N} h(n)z^{-n}$ with $H(e^{j\omega}) \geq 0$, define the causal sequence $g(n) = \rho^{-n}h(n-N)$, $0 \leq n \leq 2N$, where $1 < \rho < \min_k 1/|b_k|$. Since b_k are unknown, this choice of ρ requires some guess work. In the author's experience, the value $\rho = 1.02$ produced excellent results for most of half band filters $H(z)$ used in the two-channel QMF problem. Note that the user need not know the quantity K in (D.3.1), since the program in Tribolet and Quatieri [1979] is insensitive to any time-shift of its input.

2. With $g(n)$ used as the input to the program in Tribolet and Quatieri [1979], the output is $\widehat{g}(n)$, which is the complex cepstrum of $g(n)$. Notice that the length of $\widehat{g}(n)$ is typically much longer than $2N + 1$, since a much longer DFT is used in the computation. (Unlike $g(n)$, the sequence $\widehat{g}(n)$ is noncausal, and the values for negative n should be carefully identified from the output; see instruction in the program listing of Tribolet and Quatieri [1979].)

3. Fold the anticausal part of $\widehat{g}(n)$, add it to the causal part, and divide by two. This gives the causal sequence (D.3.6). By using this causal sequence as $\widehat{x}(n)$ in (D.2.2) and (D.2.3), evaluate the sequence $x(n)$. This sequence is equal to $h_0(n)$. The spectral factor $H_0(z) = \sum_{n=0}^{N} h_0(n)z^{-n}$ has therefore been determined.

Design Example D.3.1.

We now consider an example where $H(z)$ is a zero phase FIR equiripple halfband filter with order $2N = 178$. The response $H(e^{j\omega})$ has been ensured to be nonnegative by first designing a causal linear phase filter $G(e^{j\omega})$ by use of the McClellan-Parks program, and then defining $H(z) = z^N G(z) + \delta$ where δ is the peak stopband ripple of $G(e^{j\omega})$. Figure D.3-1 shows the response $H(e^{j\omega})$. The stopband edge is at $\omega_S = 0.527\pi$. The figure also shows the response $|H_0(e^{j\omega})|$ of the minimum phase spectral factor $H_0(z)$, computed using the above technique. The stopband attenuation of $H(e^{j\omega})$ is about 80 dB and that of $H_0(e^{j\omega})$ is about 40 dB. The value $\rho = 1.02$ was used in the computation.

There have been other approaches for spectral factorization of FIR filters. See for example Friedlander [1983]. While these have some advantages over the one described above, they work only if there are no zeros on the unit circle.

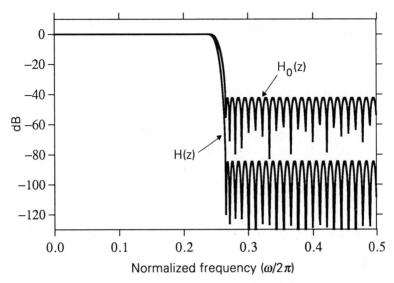

Figure D.3-1 Design example D.3.1. Magnitude responses of $H(z)$ and its spectral factor $H_0(z)$.

PROBLEMS

D.1. Consider the transfer function $H(z) = 1 - \alpha z$, with $0 < \alpha < 1$. This can be written as $H(z) = \alpha(\frac{1}{\alpha} - z)$ and the phase response $\phi(\omega)$ is equal to the phase of
$$G(e^{j\omega}) = \frac{1}{\alpha} - e^{j\omega}.$$
It is convenient to use a vector diagram to trace the behavior of $\phi(\omega)$. This is shown below.

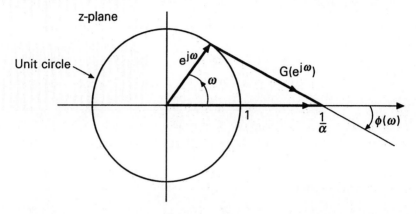

Figure PD-1

Similar diagrams can be drawn for arbitray rational transfer functions. (For example, see pp. 220–221 of Oppenheim and Schafer [1989].) As ω varies from 0 to 2π, the vector $G(e^{j\omega})$ changes both in magnitude and angular orientation. Its angle $\phi(\omega)$ can span a range exceeding 2π. So, the unwrapped phase $\phi_u(\omega)$ can be conceptually obtained by tracing the variation of $\phi(\omega)$, but carefully avoiding any "modulo 2π reduction." (Thus, if $\phi(\omega)$ is plotted with respect to ω, there will be no abrupt jumps of 2π.) In what follows, we therefore have $\phi(\omega) = \phi_u(\omega)$. By using this diagram, we can also find out whether the unwrapped phase is periodic. For this note that $\phi(0) - \phi(2\pi) = 2\pi m$ for some integer m, since $G(e^{j\omega})$ is periodic. $\phi(\omega)$ is periodic with period 2π if and only if $m = 0$.

a) For $H(z) = 1 - \alpha z$, with $0 < \alpha < 1$, show that $\phi_u(\omega)$ is periodic with period 2π.
b) Hence show that the same is true for $H(z) = 1 - \alpha z$, where α is possibly complex with $|\alpha| < 1$. (*Hint.* Try a frequency shift.)
c) Hence show that the same is true for $H(z) = 1 - \alpha z^{-1}$, where α is possibly complex with $|\alpha| < 1$.

Note. Part (b) takes care of zeros outside the unit circle, whereas part (c) takes care of zeros inside.

D.2. Let $H(z) = 1 - \alpha z$ with $\alpha > 1$. Show that the unwrapped phase does not satisfy $\phi_u(0) = \phi_u(2\pi)$. By a simple argument, extend this to the case where α is complex with $|\alpha| > 1$. Show finally that the same is true for $H(z) = 1 - \alpha z^{-1}$, $|\alpha| > 1$. (*Note.* It helps to first read Problem D.1).

Appendix E

Mason's Gain Formula

Mason's gain formula is a very convenient tool to obtain transfer functions of complicated structures. It can be applied to any signal flow graph (see below), for example, the coupled form structure reproduced in Fig. E-1(a). We now state this result, and present some examples of its use. Further examples and details can be found in B. C. Kuo [1975]. For orginal work and theoretical developments see S. J. Mason [1953,1956].

Terminology

A signal flow graph is a collection of *nodes* interconnected by *directed branches.* In Fig. E-1(a) the nodes are indicated by circled numbers. Each node is associated with a signal, for example, node 1 is associated with

$$s_1(n) = u(n) - (R\sin\theta)y(n) + (R\cos\theta)s_2(n),$$

where $s_2(n)$ is the signal associated with node 2. Each branch has a gain, for example, the branch from node 1 to node 2 in the above example has gain z^{-1}. The gain is indicated adjacent to the branch. Whenever a branch is unlabeled, its gain is assumed to be unity.

The input and output nodes of the graph are typically labeled as $u(n)$ and $y(n)$ respectively. Notice that the signal associated with node 4 is also the output, i.e., $s_4(n) = y(n)$.

Paths. A path is merely a succession of branches, directed the same way. The *path gain* is the product of individual branch gains. (This explains why the gains are always indicated in terms of z-transforms rather than in time domain). The symbol $P_k(z)$ denotes the gain for the kth path. In Figs. E-1(b),(c) we have indicated some paths of the flow graph of Fig. E-1(a). These paths have gains

$$P_1(z) = -R\sin\theta, \quad P_2(z) = z^{-2}R\sin\theta. \qquad (E.1)$$

If a path starts from the input node and ends at the output node *with no node encountered more than once*, it is said to be a *forward path*. In the above example $P_2(z)$ is a forward path. A signal flow graph can have many forward paths.

Deleting a path. The operatin of 'path deletion' is central to Mason's gain formula. Deleting a path $P_k(z)$ means (i) delete all branches in the path, and (ii)

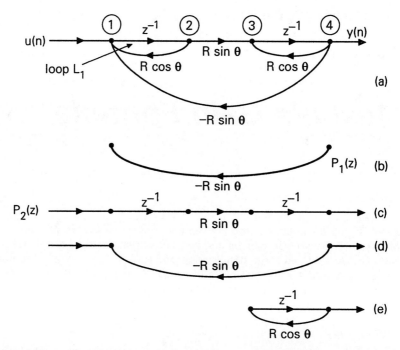

Figure E-1 (a) Signal flow graph of the coupled form structure, (b), (c) two examples of 'paths', (d) graph that remains after deleting the path '$R\sin\theta$.' (e) graph that remains after deleting the path z^{-1} which connects nodes 1 and 2.

delete all branches that touch any node on the path. Figs. E-1(d),(e) show examples of the remaining graph after deleting certain paths from Fig. E-1(a). Note that if we delete the horizontal path which connects nodes 1 and 4 in Fig. E-1(a), the resulting graph is empty!

Loops. A path which starts and ends at the same node with no other node encountered more than once is said to be a *loop*. The product of the branch gains is said to be the *loop gain*. In the above example, the loop labeled L_1 has loop gain $z^{-1}R\cos\theta$. A signal flow graph can have any number of loops. Two loops are said to be *touching* if they share a common node, and *nontouching* if they do not. Figure E-2 shows examples of touching and nontouching loops.

Determinant of a Signal Flow Graph

The determinant of a signal flow graph is defined as

$$\Delta(z) = 1 - \text{sum of all loop gains}$$
$$+ \text{sum of products of gains of pairs of nontouching loops}$$
$$- \text{sum of products of gains of triples of nontouching loops} \quad (E.2)$$
$$+ \ldots$$

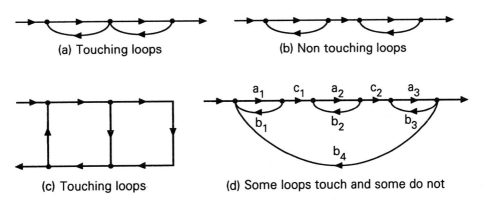

Figure E-2 Examples of touching and nontouching loops. (a) Touching, (b) nontouching, (c) touching, and (d) both types.

To demonstrate this, consider Fig. E-2(d). We have

Loops: $a_1 b_1$, $a_2 b_2$, $a_3 b_3$, $a_1 c_1 a_2 c_2 a_3 b_4$.

Nontouching pairs: $(a_1 b_1, a_2 b_2)$, $(a_1 b_1, a_3 b_3)$, $(a_2 b_2, a_3 b_3)$.

Nontouching triples: $(a_1 b_1, a_2 b_2, a_3 b_3)$.

So the determinant is

$$\Delta(z) = 1 - (a_1 b_1 + a_2 b_2 + a_3 b_3 + a_1 c_1 a_2 c_2 a_3 b_4)$$
$$+ (a_1 b_1 a_2 b_2 + a_1 b_1 a_3 b_3 + a_2 b_2 a_3 b_3) \qquad (E.3)$$
$$- a_1 b_1 a_2 b_2 a_3 b_3.$$

Notice that a graph with no loops has $\Delta = 1$. This is true, in particular, for the "empty graph," that is, graph with no nodes.

Mason's formula

Let $H(z)$ denote the transfer function $Y(z)/U(z)$ of a signal flow graph. Define the following notations:

1. $\Delta(z)$ = determinant of the graph.
2. L = number of forward paths, with $P_k(z), 1 \leq k \leq L$ denoting the forward path gains.
3. $\Delta_k(z)$ = determinant of the graph that remains after deleting the kth forward path $P_k(z)$.

Then the transfer function is

$$H(z) = \frac{\sum_{k=1}^{L} P_k(z) \Delta_k(z)}{\Delta(z)} \qquad \text{(Mason's formula)}. \qquad (E.4)$$

Example E.1

Consider the direct form structure of Fig. E-3. We have only one loop, with gain az^{-1}. So the determinant $\Delta(z) = 1 - az^{-1}$. There are two forward paths, with gains $P_1(z) = b_0$ and $P_2(z) = b_1 z^{-1}$. If either of these paths is deleted, the result is the empty graph, so that $\Delta_1(z) = \Delta_2(z) = 1$. Applying (E.4) we obtain

$$H(z) = \frac{Y(z)}{U(z)} = \frac{b_0 + b_1 z^{-1}}{1 - az^{-1}}. \qquad (E.5)$$

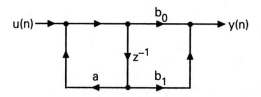

Figure E-3 First order direct form structure.

Example E.2.

Consider the coupled form structure shown in Fig. E-1(a). For this we have:

Loops: $z^{-1} R \cos\theta$, $z^{-1} R \cos\theta$, and $-z^{-2} R^2 \sin^2\theta$.

Nontouching pairs: $z^{-1} R \cos\theta, z^{-1} R \cos\theta$.

There are no nontouching triples and so on. Using (E.2) the determinant of the structure is

$$\Delta(z) = 1 - 2z^{-1} R \cos\theta + z^{-2} R^2. \qquad (E.6)$$

There is only one forward path, with $P_1(z) = z^{-2} R \sin\theta$. If this is deleted, the result is an empty graph, and we obtain $\Delta_1(z) = 1$. Using (E.4) we therefore obtain

$$H(z) = \frac{Y(z)}{U(z)} = \frac{z^{-2} R \sin\theta}{1 - 2z^{-1} R \cos\theta + z^{-2} R^2}. \qquad (E.7)$$

Example E.3.

We now consider the lattice structure of Fig. E-4, which provides an exampe where $\Delta_k(z)$ are nontrivial. Here $\widehat{k}_i$ stands for $\sqrt{(1 - k_i^2)}$, and $-1 < k_i < 1$. First let us list the loops:

Loops: $-k_1 k_2 z^{-1}$, $-k_1 z^{-1}$, and $-k_2 \widehat{k}_1^2 z^{-2}$.

Nontouching pairs: $(-k_1 k_2 z^{-1}, -k_1 z^{-1})$.

Using (E.2) we then obtain the determinant

$$\Delta(z) = 1 + k_1(k_2 + 1)z^{-1} + k_2 z^{-2}, \quad \text{using } \widehat{k}_i = \sqrt{1 - k_i^2}. \qquad (E.8)$$

Next, we have three forward paths. The gains are

$$P_1(z) = k_2, \quad P_2(z) = k_1 \widehat{k}_2^2 z^{-1}, \quad P_3(z) = \widehat{k}_1^2 \widehat{k}_2^2 z^{-2}.$$

The graph that remains after deletion of the forward path $P_k(z)$ is shown in Fig. E-5 for $k = 1, 2$. (If $P_3(z)$ is deleted, the remaining graph empty.) The determinants of these 'remaining' graphs are

$$\Delta_1(z) = \Delta(z), \quad \Delta_2(z) = 1 + k_1 z^{-1}, \quad \Delta_3(z) = 1. \qquad (E.9)$$

Substituting these into (E.4) we arrive at the transfer function

$$H(z) = \frac{Y(z)}{U(z)} = \frac{k_2 + k_1(k_2 + 1)z^{-1} + z^{-2}}{1 + k_1(k_2 + 1)z^{-1} + k_2 z^{-2}}. \qquad (E.10)$$

which is an allpass function, since k_i are real.

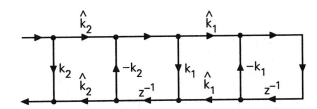

Figure E-4 The lattice structure example.

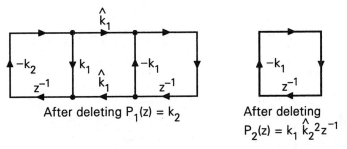

Figure E-5 Graphs that remain after deletion of certain paths from Fig. E-4.

PROBLEMS

E.1. Find the transfer function of the following structure using Mason's formula, assuming that $y(n)$ is the output.

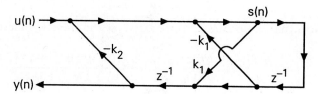

Figure PE-1

If $s(n)$ is regarded as the output instead, what will be the transfer function?

E.2. Find the transfer function of the following structure using Mason's formula.

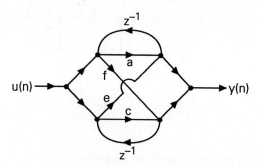

Figure PE-2

E.3. Find the transfer function of the following structure, and show that it is allpass, whenever b_2 and b_3 are real.

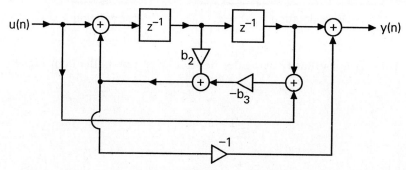

Figure PE-3

E.4. Consider the two-input two-output system shown in Fig. PE-4. This is characterized by a 2×2 transfer matrix, $\mathbf{H}(z)$, that is,

$$\begin{bmatrix} Y_0(z) \\ Y_1(z) \end{bmatrix} = \underbrace{\begin{bmatrix} H_{00}(z) & H_{01}(z) \\ H_{10}(z) & H_{11}(z) \end{bmatrix}}_{\mathbf{H}(z)} \begin{bmatrix} U_0(z) \\ U_1(z) \end{bmatrix}.$$

Finds the elements $H_{km}(z)$ of this transfer matrix using Mason's formula.

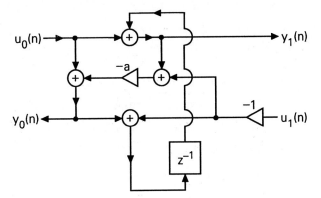

Figure PE-4

GLOSSARY OF SYMBOLS

Bold faced quantities denote matrices and vectors.

$|a|$ denotes the absolute value.

det $\mathbf{A}$ denotes the determinant of $\mathbf{A}$.

$\deg[\mathbf{H}(z)]$ denotes the McMillan degree of $\mathbf{H}(z)$.

$\text{Tr}(\mathbf{A})$ denotes the Trace of $\mathbf{A}$.

$\mathbf{A}^T$ denotes the transpose of $\mathbf{A}$.

$\mathbf{A}^\dagger$ denotes the transpose conjugate of $\mathbf{A}$.

$\mathbf{A}^*$ denotes the conjugate of $\mathbf{A}$.

$\mathbf{A}^{-1}$ denotes the inverse of $\mathbf{A}$.

$\mathbf{A}^{-T}$ denotes the inverse transpose of $\mathbf{A}$.

$\mathbf{A}^{-\dagger}$ denotes the inverse transpose conjugate of $\mathbf{A}$.

$\widetilde{\mathbf{H}}(z) = \mathbf{H}^\dagger(1/z^*)$. Thus $\widetilde{\mathbf{H}}(e^{j\omega}) = \mathbf{H}^\dagger(e^{j\omega})$. Page 30.

$\|\mathbf{a}\|$ denotes the $\mathcal{L}_2$ norm of the vector $\mathbf{a}$. Page 785.

$a(n)\big|_{\downarrow M}$ denotes the decimated version $a(Mn)$, and $A(z)\big|_{\downarrow M}$ denotes its z-transform.

$W_M = e^{-j2\pi/M}$. Subscript M is often omitted.

$\mathbf{W}$ denotes the $M \times M$ DFT matrix with $[\mathbf{W}]_{km} = W^{km}$.

$\mathbf{E}(z)$ denotes the Polyphase matrix of the analysis bank. Page 231.

$\mathbf{R}(z)$ denotes the Polyphase matrix of the synthesis bank. Page 232.

$\delta(n)$ denotes the unit pulse or impulse function. Page 12.

$\delta_a(t)$ denotes the Dirac delta function or impulse function. Page 12.

$\mathcal{U}(n)$ denotes the unit step function. Page 12.

$\triangleq$ denotes "defined as".

Further details on notations

General. Chapter 2, pages 28–31

Multidimensional multirate. Chapter 12, pages 646, 647.

Abbreviations and acronyms. Please look in the main index.

LIST OF IMPORTANT SUMMARIES (TABLES)

- Four types of linear phase filters. Page 35.
- Comparison of filter design techniques. Page 91.
- Design of power symmetric elliptic filters. Page 212.
- Optimal IIR power symmetric filters. Page 217.
- Two-channel QMF bank. Page 267.
- IIR allpass based QMF bank. Page 268.
- FIR power symmetric QMF bank. Page 269.
- The M-channel filter bank. Page 270.
- Matrix notations in filter bank theory. Page 271.
- Perfect reconstruction paraunitary filter banks. Page 327.
- Comparison of filter banks. Pages 330, 331.
- Comparison of linear phase filter banks. Page 346.
- Comparison of cosine modulated filter banks. Pages 386, 387.
- Roundoff noise in filter banks. Page 417.
- Unifying block filtering, LPTV systems, and filter banks. Page 437.
- The short-time Fourier transform. Page 477.
- The continuous-time wavelet transform. Page 490.
- The discrete-time wavelet transform. Page 511.
- Generation of continuous-time wavelet basis. Page 537.
- Lattices in multidimensional multirate systems. Page 646.
- Multidimensional multirate fundamentals. Page 647.
- Properties of the hexagonal matrix M. Page 648.
- Properties of the quincunx matrix M. Page 649.
- Lossless systems at a glance. Page 772.
- Lossless systems: factorization and structures. Page 773.
- Lossless systems: further properties. Page 774.

LIST OF IMPORTANT SUMMARIES (FIGURES)

- Signal flow graph notations. Page 19.
- Filter design specifications. Page 43.
- Multirate system interconnections. Pages 118, 119.
- The noble identities. Page 119.
- The polyphase identity. Page 133.
- Multidimensional multirate identities. Page 604.
- Multidimensional polyphase identity. Page 605.

Bibliography

This is an alphabetical list of all the references cited in the text. For specific topics, see the listing *Some references by topic,* which follows this listing.

Ahmed, N., Natarajan, T., and Rao, K.R. "Discrete cosine transform," *IEEE Trans. Computers,* vol. C-23, pp. 90-93, 1974.

Akansu, A.N., and Haddad, R.A., *Multiresolution signal decomposition: transforms, subbands, and wavelets,* Academic Press, Inc., 1992.

Akansu, A. N., and Liu, Y. "On signal decomposition techniques," *Optical engr.,* vol. 30, pp. 912-920, July 1991.

Akansu, A. N., Haddad, R. A., and Caglar, H. "The binomial QMF-wavelet transform for multiresolution signal decomposition," *IEEE Trans. on Signal Processing,* vol. SP-40, 1992.

Allen, J. B., and Rabiner, L. R. "A unified theory of short-time spectrum analysis and synthesis," Proc. IEEE, vol. 65, pp. 1558–1564, Nov. 1977.

Anderson, B. D. O., and Vongpanitlerd, S. *Network analysis and synthesis,* Prentice Hall, Inc., Englewood Cliffs, NJ, 1973.

Anderson, B. D. O., and Moore, J. B. *Optimal filtering,* Prentice Hall, Inc., Englewood Cliffs, NJ, 1979.

Ansari, R., and Liu, Bede "Interpolators and decimators as periodically time varying filters," *IEEE Int. Symp. on Circuits and Systems,* pp. 447-450, 1981.

Ansari, R., and Liu, Bede "A class of low-noise computationally efficient recursive digital filters with applications to sampling rate alterations," *IEEE Trans. on Acoust. Speech and Signal Proc.,* vol. ASSP-33, pp. 90-97, Feb. 1985.

Ansari, R., and Lau C.-L. "Two-dimensional IIR filters for exact reconstruction in tree-structured subband decomposition," *Electronics letters,* vol. 23, pp. 633-634, June 1987.

Ansari, R., and Lee, S. H. "Two-dimensional nonrectangular interpolation, decimation, and filter banks," Proc. Int. Conf. on Acoust. Speech and Signal Proc., New York, 1988.

Ansari, R., and Guillemot, C. "Exact reconstruction filter banks using diamond FIR filters," Proc. Int. Conf. on new trends in Comm. Control, and Signal Proc., Elsevier, Ankara, Turkey, July, 1990.

Antoniou, A. *Digital filters: analysis and design* McGraw Hill Book Co., 1979.

Apostol, T. M. *Calculus,* vol. 1, Blaisdell Publ. Co., New York, 1961.

Apostol, T. M. *Mathematical analysis,* Addison-Wesley Publ. Co., 1974

Balabanian, N., and Bickart, T. A. *Electrical network theory,* John Wiley & Sons, Inc., 1969.

Bamberger, R. H. *The directional filter bank: a multirate filter bank for the directional decomposition of images,* Doctoral dissertation, Georgia Inst. of Tech., Nov. 1990.

Bamberger, R. H., and Smith, M. J. T. "A filter bank for the directional decomposition of images: theory and design," *IEEE Trans. on Signal Processing,* vol. SP-40, April 1992.

Barnes, C. W., and Shinnaka, S. "Block-shift invariance and block implementation of discrete-time filters," *IEEE Trans. Circuits and Sys.,* vol. CAS-27, pp. 667-672, Aug. 1980.

Barnes, C. W., Tran, B. N., and Leung, S. H. "On the statistics of fixed-point roundoff error," *IEEE Trans. on Acoust. Speech and Signal Processing,* pp. 595-606, vol. ASSP-33, June 1985.

Barnwell, III, T. P. "Sub-band coder design incorporating recursive quadrature filters and optimum ADPCM coders," *IEEE Trans. on Acoust. Speech and Signal Proc.,* vol. ASSP-30, pp. 751-765, Oct. 1982.

Beckenbach, E., and Bellman, R. *An introduction to inequalities,* Random House, 1961.

Belevitch, V. *Classical network theory,* Holden Day, Inc., San Francisco, 1968.

Bellanger, M., Bonnerot, G., and Coudreuse, M. "Digital filtering by polyphase network: application to sample rate alteration and filter banks," *IEEE Trans. on Acoust. Speech and Signal Proc.,* vol. ASSP-24, pp. 109-114, April 1976.

Bellanger, M. "On computational complexity in digital transmultiplexer filters," *IEEE Trans. on Comm.,* vol. COM-30, pp. 1461-1465, July 1982.

Bellanger, M. *Digital processing of signals: theory and practice,* John Wiley and Sons, 1989.

Bellman, R. *Introduction to matrix analysis,* McGraw-Hill Book Co., Inc., 1960.

Berger, T. *Rate distortion theory,* Prentice Hall, Inc., Englewood Cliffs, NJ, 1971.

Bitmead, R., and Anderson, B. D. O. "Adaptive frequency sampling filters," *IEEE Trans. on Acoustics, Speech and Signal Proc.,* vol. ASSP-29, pp. 684-693, June 1981.

Black, H. S. *Modulation theory,* New York, Van Nostrand, 1953.

Blahut, R. E. *Fast algorithms for digital signal processing,* Addison-Wesley Publ. Co., 1985.

Bloom, P.J. "High-quality digital audio in the entertainment industry: an overview of achievements and challenges," *IEEE ASSP magazine,* vol. 2, pp. 2-25, Oct. 1985.

Bose, N. K. *Applied multidimensional systems theory,* Van Nostrand Reinhold, 1982.

Bose, N. K. *Digital filters,* Elsevier Science Publ. Co. Inc., 1985.

Brandt, A. "Multilevel adaptive solutions to boundary value problems," *Mathematics of Computation,* vol. 31., pp. 333-390, 1977.

Briggs, W. L. *A multigrid tutorial,* SIAM, 1987.

Brown, Jr., J. L. "Multi-channel sampling of lowpass signals," *IEEE Trans. on Circuits and Systems,* vol. CAS-28, pp. 101-106, Feb. 1981.

Brune, O. "Synthesis of a finite two terminal network whose driving point impedance is a prescribed function of frequency," *J. Math., and Phys.,* vol. 10, pp. 191-235, 1931.

Bruton, L. T., and Vaughan-Pope, D. A. "Synthesis of digital ladder filters from LC filters," *IEEE Trans. on Circuits and Systems,* vol. CAS-23, pp. 395-402, June 1976.

Bullock III, R. M. "A new three-way allpass crossover network design," *J. of Audio Engr. Soc.,* pp. 315-322, May 1986.

Burrus, C. S. "Block implementation of digital filters," *IEEE Trans. on Circuit theory,* vol. CT-18, pp. 697-701, Nov. 1971.

Burt, P. J., and Adelson, E. H. "The Laplacian pyramid as a compact image code," *IEEE Trans. Comm.,* vol. COM-31, pp. 532-540, April 1983.

Caglar, H., Liu, Y., and Akansu, A.N. "Statistically optimized PR-QMF design," Proc. SPIE Visual Comm. Image Proc., pp. 86-94, Boston, 1991.

Cassels, J. W. S. *An introduction to the geometry of numbers,* Springer-Verlag, Berlin, 1959.

Cassereau, P. "A new class of orthogonal transforms for image processing," S. M. thesis, Dept. EECS, Mass. Inst. Technol., Cambridge, May 1985.

Chen, C.-T. *Introduction to linear system theory,* Holt, Rinehart and Winston, Inc., 1970.

Chen, C.-T. *Linear system theory and design,* Holt, Rinehart and Winston, Inc., 1984.

Chen, T., and Vaidyanathan, P. P. "Multidimensional multirate filters derived from one dimensional filters," *Electronics letters,* vol. 27, pp. 225-228, Jan. 1991.

Chen, T., and Vaidyanathan, P. P. "Commutativity of D-dimensional decimation and interpolation matrices, and application to rational decimation systems," Proc. IEEE Int. Conf. Acoust. Speech, and Signal Proc., San Francisco, CA, March 1992a.

Chen, T., and Vaidyanathan, P. P. "Least common right/left multiples of integer matrices and applications to multidimensional multirate systems," Proc. IEEE Int. Symp. Circuits and Sys., San Diego, CA, pp. 935-938, May 1992b.

Chen, T., and Vaidyanathan, P. P. "The role of integer matrices in multidimensional multirate systems," *IEEE Trans. Signal Proc.,* vol. SP-41, May 1993.

Chu, P. L. "Quadrature mirror filter design for an arbitrary number of equal bandwidth channels," *IEEE Trans. Acoust., Speech, Signal Proc.,* vol. ASSP-33, pp. 203-218, Feb. 1985.

Chui, C. K. *An introduction to wavelets,* Academic Press, 1992.

Chui, C. K. (edited), *Wavelets: a tutorial in theory and applications,* Academic Press, 1992.

Churchill, R. V., and Brown, J. W. *Introduction to complex variables and applications,* McGraw-Hill Book Co., New York, 1984.

Clark, G., Mitra, S. K., and Parker, S. "Block implementation of adaptive digital filters," *IEEE Trans. Acoust., Speech, Signal Proc.,* vol. ASSP-29, pp. 744-752,

June 1981.

Coifmann, R., Meyer, Y., Quake, S., and Wickerhauser, V. "Signal processing with wavelet packets," Numerical Algorithm Research Group, Yale Univ., 1990.

Constantinides, A. G., and Valenzuela, R. A. "A class of efficient interpolators and decimators with applications in transmultiplexers," Proc. IEEE Int. Symp. on Circuits and Sys., pp. 260-263, Rome, May 1982.

Cox, R. V. "The design of uniformly and nonuniformly spaced pseudo QMF," *IEEE Trans. Acoust., Speech, Signal Proc.*, vol. ASSP-34, pp. 1090-1096, Oct. 1986.

Cox, R. V., Bock, D. E., Bauer, K. B., Johnston, J. D. and Snyder, J. H. "The analog voice privacy system," *AT&T Tech. J.*, vol. 66, pp. 119-131, Jan-Feb 1987.

Crochiere, R. E., Webber, S. A., and Flanagan, J. L. "Digital coding of speech in subbands," *Bell Syst. Tech. J.*, vol. 55, pp. 1069-1085, Oct. 1976.

Crochiere, R. E. "On the design of subband coders for low bit rate speech communication," *Bell System. Tech. J.*, vol. 56, pp. 747-771, May-June 1977.

Crochiere, R. E., and Rabiner, L. R. "Interpolation and decimation of digital signals: a tutorial review," *Proc. IEEE* vol. 69, pp. 300-331, March 1981.

Crochiere, R. E. "Subband coding," *Bell System Tech. J.*, vol. 60, pp. 1633-1654, Sept. 1981.

Crochiere, R. E., and Rabiner, L. R. *Multirate digital signal processing*, Englewood Cliffs, NJ: Prentice Hall, 1983.

Croisier, A., Esteban, D., and Galand, C. "Perfect channel splitting by use of interpolation/decimation/tree decomposition techniques," Int. Symp. on Info., Circuits and Systems, Patras, Greece, 1976.

Darlington, S. "Synthesis of reactance four-poles," *J. Math. Phys.*, vol. 18, pp. 257-353, Sept., 1939.

Daubechies, I. "Orthonormal bases of compactly supported wavelets," *Comm. on Pure and Appl. Math.*, vol. 4, pp. 909-996, Nov. 1988.

Daubechies, I. "The wavelet transform, time-frequency localization and signal analysis," *IEEE Trans. on Info. Theory*, vol. IT-36, pp. 961-1005, Sept. 1990.

Daubechies, I. *Ten lectures on wavelets*, SIAM, CBMS series, April 1992.

Davenport, Jr., W. B. *Probability and random processes, An introduction for applied scientists and engineers*, McGraw-Hill, Inc., New York, 1970.

Davis, P. J. *Circulant matrices*, New York, Wiley, 1979.

Dentino, M., McCool, J., and Widrow, B. "Adaptive filtering in the frequency domain," *Proc. IEEE*, vol. 66, pp. 1658-1659, Dec. 1978.

Deprettere, E., Dewilde, P. "Orthogonal cascade realization of real multiport digital filters," *Int. J. Circuit Theory and Appl.*, vol. 8, pp. 245-277, 1980.

Desoer, C. A., and Schulman, J. D. "Zeros and poles of matrix transfer functions and their dynamic interpretation," *IEEE Trans. on Circuits and Systems*, vol. CAS-21, pp. 1-8, 1974.

Digital Audio, IEEE ASSP magazine, Special issue, vol. 2, Oct. 1985.

Doğanata, Z., and Vaidyanathan, P. P. "On one-multiplier implementations of FIR lattice structures," *IEEE Trans. on Circuits and Systems,* vol. CAS-34, pp. 1608-1609, Dec. 1987.

Doğanata, Z., Vaidyanathan, P. P., and Nguyen, T. Q. "General synthesis procedures for FIR lossless transfer matrices, for perfect-reconstruction multirate filter bank applications," *IEEE Trans. on Acoustics, Speech and Signal Proc.,* vol. ASSP-36, pp. 1561-1574, Oct. 1988.

Doğanata, Z., and Vaidyanathan, P. P. "Minimal structures for the implementation of digital rational lossless systems," IEEE Trans. Acoustics, Speech and Signal Proc., vol. ASSP-38, pp 2058-2074, Dec. 1990.

Dubois, E. "The sampling and reconstruction of time-varying imagery with application in video systems," *Proc. of the IEEE,* vol. 73, pp. 502-522, April 1985.

Dudgeon, D. E., and Mersereau, R. M. "Multidimensional digital signal processing," Prentice Hall, Inc., Englewoods Cliffs, N.J., 1984.

Esteban, D., and Galand, C. "Application of quadrature mirror filters to split band voice coding schemes, Proc. IEEE Int. Conf. Acoust. Speech, and Signal Proc., pp. 191-195, May 1977.

Evangelista, G. "Orthogonal wavelet transforms and filter banks," Proc. of the 23rd Annual Asilomar Conference on Signals, Systems and Computers, 1989.

Evans, B. L., McClellan, J. H., and Bamberger, R. H. "A symbolic algebra for linear multidimensional multirate systems," Proc. Int. Conf. Info. Sciences and Systems, Princeton, N. J., March 1992.

Fettweis, A. "Digital filter structures related to classical filter networks," *AEU,* vol. 25, pp. 79-89, Feb. 1971.

Fettweis, A. "Wave digital lattice filters," *Int. J. Circuit Theory and Appl.,* vol. 2, pp. 203-211, June 1974.

Fettweis, A., Nossek, J. A., and Meerkotter, K. "Reconstruction of signals after filtering and sampling rate reduction," IEEE Trans. Acoust. Speech and Signal Proc., vol. ASSP-33, pp. 893-902, August 1985.

Fettweis, A., Leickel, T., Bolle, M., Sauvagerd, U. "Realization of filter banks by means of wave digital filters," Proc. IEEE Int. Symp. Circuits and Sys., pp. 2013-2016, New Orleans, May 1990.

Flanagan, J. L. *Speech analysis, synthesis and perception,* Springer-Verlag, New York, 1972.

Forney, Jr., G. D. "Convolutional codes I: algebraic structure," *IEEE Trans. on Info. Theory,* vol. IT-16, pp. 720-738, Nov. 1970.

Franklin, J. N. *Matrix theory,* Prentice Hall, Englewood Cliffs, NJ., 1968.

Friedlander, B. "A lattice algorithm for factoring the spectrum of a moving average process," *IEEE Trans. on Automatic Control,* vol. AC-28, pp. 1051-1055, Nov. 1983.

Gabor, D. "Theory of communications," *J. Inst. Elec. Eng.,* (London), vol. 93, pp. 429-457, 1946.

Galand, C., and Esteban, D. "16 Kbps real-time QMF subband coding implementation," Proc. Int. Conf. on Acoust. Speech and Signal Proc., pp. 332-335, Denver, CO, April 1980.

Galand, C. R., and Nussbaumer, H. J. "New quadrature mirror filter structures," *IEEE Trans. Acoustics, Speech and Signal Proc.*, vol. ASSP-32, pp. 522-531, June 1984.

Galand, C. R. and Nussbaumer, H. J. "Quadrature mirror filters with perfect reconstruction and reduced computational complexity," Proc. of the IEEE Int. Conf. on Acoust. Speech and Signal Proc., pp. 525-528, April 1985.

Gantmacher, F. R. *The theory of matrices*, vol. 1,2, Chelsa Publishing Co., N.Y., 1959.

Gaszi, L. "Explicit formulas for lattice wave digital filters," *IEEE Trans. on Circuits and Systems*, vol. CAS-32, pp. 68-88, Jan. 1985.

Gersho, A., and Gray, R. M. *Vector quantization and signal compression*, Kluwer Academic Press, 1992.

Gharavi, H., and Tabatabai, A. "Subband coding of digital images using two dimensional quadrature mirror filtering," Proc. SPIE Conf. on Visual Comm., and Image Proc., vol. 707, pp. 51-61, Cambridge, Sept. 1986.

Gharavi, H., and Tabatabai, A. "Subband coding of monochrome and color images," *IEEE Trans. on Circuits and Systems.*, vol. CAS-35, pp. 207-214, Feb. 1988.

Gilloire, A. "Experiments with subband acoustic echo cancelers for teleconferencing," Proc. IEEE Int. Conf. on Acoust. Speech and Signal Proc., pp. 2141-2144, Dallas, April 1987.

Gilloire, A., and Vetterli, M. "Adaptive filtering in subbands with critical sampling: analysis, experiments, and applications to acoustic echo cancelation," *IEEE Trans. on Signal Processing*, vol. SP-40, pp. 1862-1875, Aug. 1992.

Givens, W. "Computation of plane unitary rotations transforming a general matrix to triangular form," *SIAM J. Appl. Math.*, vol. 6, pp. 26-50, 1958.

Golub, G. H., and Van Loan, C. F. *Matrix computations*, Johns Hopkins Univ. Press, 1989.

Gopinath, R. A., and Burrus, C. S. "Wavelet transforms and filter banks," In *Wavelets and applications*, edited by C. H. Chui, Academic Press, 1991.

Gopinath, R. A., and Burrus, C. S. "On upsampling, downsampling and rational sampling filter banks," Technical report, Rice University, Nov. 1991.

Gradshteyn, I. S., and Ryzhik, I. M. *Tables of integrals, series and products*, Academic Press, Inc., 1980.

Gray, Jr., A. H., and Markel, J. D. "Digital lattice and ladder filter synthesis," *IEEE Trans. Audio Electroacoust.*, vol. AU-21, pp. 491-500, Dec. 1973.

Gray, Jr., A. H., and Markel, J. D. "A normalized digital filter structure," *IEEE Trans. Acoust. Speech and Signal Proc.*, vol. ASSP-23, pp. 268-277, June 1975.

Gray, Jr., A. H. "Passive cascaded lattice digital filters," *IEEE Trans. on Circuits and Systems*, vol. CAS-27, pp. 337-344, May 1980.

Gray, R. M. "On asymptotic eigenvalue distribution of Toeplitz matrices," *IEEE Trans. on Info. Theory,* vol. IT-18, pp. 725-730, Nov. 1972.

Gray, R. M. "Vector quantization," *IEEE ASSP magazine,* vol. 1, pp. 4-29, April 1984.

Grenander, U., and Szego, G. *Toeplitz forms and their applications,* University of California Press, Berkeley, CA, 1958.

Grenez, F. "Chebyshev design of filters for subband coders," *IEEE Trans. Acoust. Speech and Signal Proc.,* vol. ASSP-36, pp. 182-185, Feb. 1988.

Grossman, A., and Morlet, J. *Decomposition of Hardy functions into square integrable wavelets of constant shape, SIAM J. Math. Anal.,* vol. 15, pp. 723-736, 1984.

Guessoum, A., and Mersereau, R. M. "Fast algorithms for the multidimensional discrete Fourier transform," *IEEE Trans. Acoust. Speech and Signal Proc.,* vol. ASSP-34, pp. 937-943, August 1986.

Guillemin, E. A. *Synthesis of passive networks,* Wiley, New York, 1957.

Halmos, P. R. *Finite dimensional vector spaces,* Springer-Verlag, Inc., New York, 1974.

Haykin, S. *Adaptive filter theory,* Prentice Hall, Englewood Cliffs, NJ., 1991.

Helms, H. D. "Digital filters with equiripple or minimax responses," *IEEE Trans. Audio Electroacoust.* vol. AU-19, pp. 87-94, March 1971.

Herrmann, O., and Schüssler, W. "Design of nonrecursive digital filters with minimum phase," *Electronics letters* vol. 6, pp. 329-330, May 1970.

Herrmann, O. "On the approximation problem in nonrecursive digital filter design," *IEEE Trans. Circuit theory,* vol. CT-18, pp. 411-413, May 1971.

Hlawatsch, F., and Boudreaux-Bartels, G. F. "Linear and quadratic time-frequency signal representations," *IEEE Signal Processing Magazine,* vol. 9, pp. 21-67, April 1992.

Hoang, P. -Q., and Vaidyanathan, P. P. "Nonuniform multirate filter banks: theory and design," Proc. IEEE Int. Symp. on Circuits and Systems, pp. 371-374, Portland, Oregon, May 1989.

Honda, M., and Itakura, F. "Bit allocation in time and frequency domains for predictive coding of speech," *IEEE Trans. on Acoust. Speech and Signal Proc.,* vol. ASSP-32, pp. 465-473, June 1984.

Horn, R. A., and Johnson, C. R. *Matrix analysis,* Cambridge Univ. Press, 1985.

Horng, B-R., Samueli, H., and Willson, A. N., Jr. "The design of low-complexity linear phase FIR filter banks using powers-of-two coefficients with an application to subband image coding," *IEEE Trans. Circuits and Syst. for Video Technology,* vol. 1, pp. 318-324, Dec. 1991.

Horng, B-R., and Willson, A. N., Jr. "Lagrange multiplier approaches to the design of two-channel perfect reconstruction linear phase FIR filter banks," *IEEE Trans. Signal Processing,* vol. 40, pp. 364-374, Feb. 1992.

Hsiao, C.-C. "Polyphase filter matrix for rational sampling rate conversions," Proc. IEEE Int. Conf. on ASSP, pp. 2173-2176, Dallas, April 1987.

Huang, Y., and Schultheiss, P. M. "Block quantization of correlated Gaussian random variables," *IEEE Trans. Comm. Syst.* pp. 289-296, Sept. 1963.

IMSL, Fortran subroutines for mathematical applications, Houston, Texas, April 1987.

ICASSP, Set of papers on digital audio, Proc. of the IEEE Conf. Acoust. Speech and Signal Proc., pages 3597-3620, Toronto, Canada, May 1991.

Jackson, L. B. "On the interaction of roundoff noise and dynamic range in digital filters," *Bell Syst. Tech. J.*, vol. 49, pp. 159-184, Feb. 1970.

Jackson, L. B. *Digital filters and signal processing*, Kluwer Academic Publishers, 1989.

Jackson, L. B. *Signals, systems and transforms*, Addison-Wesley Publ. Co., 1991.

Jain, A. K., *Fundamentals of digital image processing*, Prentice Hall, Englewood Cliffs, NJ., 1989.

Jain, V. K., and Crochiere, R. E. "Quadrature mirror filter design in the time domain," *IEEE Trans. on Acoust. Speech and Signal Proc.*, vol. ASSP-32, pp. 353-361, April 1984.

Jayant, N. S., and Noll, P. *Digital coding of waveforms*, Prentice Hall, Inc., Englewood Cliffs, 1984.

Jerri, A. J. "The Shannon Sampling theorem – its various extensions and applications: a tutorial review" *Proc. of the IEEE*, pp. 1565-1596, Nov. 1977.

Johnston, J. D. "A filter family designed for use in quadrature mirror filter banks," Proc. IEEE Int. Conf. Acoust. Speech and Signal Proc., pp. 291-294, April 1980.

Kailath, T. *Linear Systems*, Prentice Hall, Inc., Englewood Cliffs, N.J., 1980.

Kaiser, J. F., "Nonrecursive digital filter design using the I_0-sinh window function," Proc. IEEE Int. Symp. on Circuits and Sys., San Francisco, pp, 20-23 April 1974.

Kaiser, J. F. "Design subroutine (MXFLAT) for symmetric FIR lowpass digital filters with maximally flat pass and stop bands," in *Programs for digital signal processing*, IEEE Press, N. Y., 1979.

Kalman, R. E. "Irreducible realizations and the degree of a rational matrix," *SIAM J. Appl. Math.*, vol. 13, pp. 520-544, June 1965.

Kalker, A. A. C. M., "Commutativity of up/down sampling," *Electronics Letters*, vol. 28, pp. 567-569, March 1992.

Karlsson, G., and Vetterli, M. "Theory of two-dimensional multirate filter banks," *IEEE Trans. on Acoust. Speech and Signal Proc.*, vol. ASSP-38, pp. 925-937, June 1990.

Koilpillai, R. D. and Vaidyanathan, P. P. "New results on cosine-modulated FIR filter banks satisfying perfect reconstruction," Proc. IEEE Int. Conf. Acoust. Speech and Signal Proc., pp. 1793-1796, Toronto, Canada, May 1991a.

Koilpillai, R. D. and Vaidyanathan, P. P. "A spectral factorization approach to pseudo-QMF design," Proc. IEEE Int. Symp. Circuits and Sys., pp. 160-163, Singapore, June 1991b.

Koilpillai, R. D., Nguyen, T. Q., and Vaidyanathan, P. P. "Some results in the theory of cross-talk free transmultiplexers," *IEEE Trans. on Signal Processing*, vol. ASSP-39, pp. 2174-2183, Oct. 1991.

Koilpillai, R. D. and Vaidyanathan, P. P. "Cosine-modulated FIR filter banks satisfying perfect reconstruction," *IEEE Trans. on Signal Processing*, vol. SP-40, pp. 770–783, April 1992.

Kovačević, J., and Vetterli, M. "Perfect reconstruction filter banks with rational sampling rate changes," Proc. IEEE Int. Conf. Acoust. Speech and Signal Proc. pp. 1785-1788, Toronto, Canada, May 1991a.

Kovačević, J., and Vetterli, M. "The commutativity of up/downsampling in two dimensions," *IEEE Trans. on Info. Theory*, vol. IT-37, pp. 695-698, May 1991b.

Kovačević, J., and Vetterli, M. "Nonseparable multidimensional perfect reconstruction filter banks and wavelet bases for R^n," *IEEE Trans. on Info. Theory*, vol. IT-38, Feb. 1992.

Kreyszig, E. *Advanced Engineering Mathematics*, John Wiley and Sons, Inc., 1972.

Kung, S. Y., Whitehouse, H. J., and Kailath, T. *VLSI and modern signal processing*, Prentice Hall, Inc., Englewood Cliffs, NJ, 1985.

Kuo, B. C. *Automatic control systems*, Prentice Hall, Inc., Englewood Cliffs, NJ, 1975.

Lagadec, R., Pelloni, D., and Weiss, D. "A 2-channel, 16-bit digital sampling frequency converter for professional digital audio," Proc. IEEE Int. Conf. Acoust. Speech and Signal Proc., Paris, France, May 1982.

Lim, J. S. *Two-dimensional signal and image processing*, Englewood Cliffs, NJ: Prentice Hall, 1990.

Lim, Y. C. "Frequency response masking approach for the synthesis of sharp linear phase digital filters," *IEEE Trans. on Circuits and Systems*, vol. CAS-33, pp. 357-364, April 1986.

Liu, V. C., and Vaidyanathan, P. P. "Compression of two-dimensional band-limited signals using sub-sampling theorems," *Journal of the Institution of Electronics and Telecomm. Engineers*, India, vol. 34, No. 5, pp. 416-422, 1988.

Liu, V. C., and Vaidyanathan, P. P. "On the factorization of a subclass of 2-D digital FIR lossless matrices for 2-D QMF bank applications," *IEEE Trans. on Circuits and Systems*, vol. CAS-37, pp. 852-854, June 1990.

MacDuffee, C. C. *The theory of matrices*, Chelsa Publ. Co., New York, 1946.

Makhoul, J. "Linear prediction: a tutorial review," *Proc. IEEE*, vol. 63, pp. 561-580, 1975.

Mallat, S. "A theory for multiresolution signal decomposition: the wavelet representation," *IEEE Trans. on Pattern Anal., and Machine Intell.*, vol. 11, pp. 674-693, July 1989a.

Mallat, S. "Multiresolution approximations and wavelet orthonormal bases of $L^2(R)$," *Trans. of Amer. Math. Soc.*, vol. 315, pp. 69-87, Sept. 1989b.

Malvar, H. S., and Staelin, D. H. "The LOT: Transform coding without blocking effects," *IEEE Trans. Acoust., Speech, Signal Proc.*, vol. ASSP-37, pp. 553-559, April, 1989.

Malvar, H. S. "Lapped transforms for efficient transform/subband coding," *IEEE Trans. Acoust., Speech, Signal Proc.*, vol. ASSP-38, pp. 969-978, June 1990a.

Malvar, H. S. "Modulated QMF filter banks with perfect reconstruction," *Electronics letters,* vol. 26, pp. 906-907, June 1990b.

Malvar, H. S. "Extended lapped transforms: fast algorithms and applications," Proc. IEEE Int. Conf. on Acoustics, Speech and Signal Proc., pp. 1797-1800, Toronto, Canada, May, 1991.

Malvar, H. S. *Signal processing with lapped transforms,* Artech House, Norwood, MA, 1992.

Markel, J. D., and Gray, A. H., Jr. *Linear prediction of speech,* Springer-Verlag, New York, 1976.

Marshall, Jr., T. G. "Structures for digital filter banks," Proc. IEEE Int. Conf. on Acoustics, Speech and Signal Proc., pp. 315-318, Paris, April 1982.

Mason, S. J. *Feedback theory – some properties of signal flow graphs, Proc. IRE,* vol. 41, pp. 1144-1156, Sept. 1953.

Mason, S. J. *Feedback theory – further properties of signal flow graphs, Proc. IRE,* vol. 44, pp. 920-926, July 1956.

Masson, J., and Picel, Z. "Flexible design of computationally efficient nearly perfect QMF filter banks, " Proc. of the IEEE Int. Conf. on ASSP, pp. 14.7.1-14.7.4, Tampa, FL, March, 1985.

Mau, J. "Perfect reconstruction modulated filter banks," Proc. of the IEEE Int. Conf. on ASSP, vol. IV, pp. 273–276, March 1992.

McClellan, J. H. "The design of two-dimensional digital filters by transformation," Proc. 7th Annual Princeton Conf. Info. Sci., and Syst., pp. 247-251, 1973.

McClellan, J. H., and Parks, T. W. "A unified approach to the design of optimum FIR linear-phase digital filters," *IEEE Trans. Circuit Theory,* vol. CT-20, pp. 697-701, Nov. 1973.

McGee, W. F., and Zhang, G. "Logarithmic filter banks," Proc. IEEE Int. Symp. Circuits and Sys., pp. 661-664, New Orleans, May 1990.

Mersereau, R. M. "The processing of hexagonally sampled two-dimensional signals," *Proc. of the IEEE,* vol. 67, pp. 930-949, June 1979.

Mersereau, R. M., and Speake, T. C. "The processing of periodically sampled multidimensional signals," *IEEE Trans. on Acoust., Speech and Signal Proc.,* vol. ASSP-31, pp. 188-194, Feb. 1983.

Meyer, R. A., and Burrus, C. S. "A unified analysis of multirate and periodically time varying digital filters," *IEEE Trans. on Circuits and Systems,* vol. CAS-22, pp. 162-168, Mar. 1975.

Meyer, Y. *Ondelettes et fonctions splines* Seminaire EDP, Ecole Polytechnique, Paris, 1986.

Mian, G. A., and Nainer, A. P. "A fast procedure to design equiripple minimum-phase FIR filters," *IEEE Trans. on Circuits and Sys.,* vol. CAS-29, pp. 327-331, May 1982.

Mintzer, F. "On half-band, third-band and Nth band FIR filters and their design," *IEEE Trans. on Acoustics, Speech and Signal Proc.,* vol. ASSP-30, pp. 734-738, Oct. 1982.

Mintzer, F. "Filters for distortion-free two-band multirate filter banks," *IEEE Trans. on Acoustics, Speech and Signal Proc.,* vol. ASSP-33, pp. 626-630,

June 1985.

Mitra, S. K., and Hirano, K. "Digital allpass networks," *IEEE Trans. on Circuits and Syst.*, vol. CAS-21, pp. 688-700, Sept. 1974.

Mitra, S. K., and Gnanasekaran, R. "Block implementation of recursive digital filters: new structures and properties," *IEEE Trans. Circuits and Sys.*, vol. CAS-25, pp. 200-207, April 1978.

Mitra, S. K., Damonte, A. J., Fujii, N., and Neuvo, Y. "Tunable active crossover networks," *J. Audio Eng. Soc.*, vol. 31, pp. 762-769, Oct. 1985.

Mou, Z. J., and Duhamel, P. "Fast FIR filtering: algorithms and implementations," *Signal Processing*, vol. 13, pp. 377-384, Dec. 1987.

Mueller, K.H., "A new approach to optimum pulse shaping in sampled systems using time-domain filtering," *Bell Sys. Tech. J.*, vol. 52, pp. 723-729, May-June 1973.

Murnaghan, F. D. *The unitary and rotation groups*, Spartan Books, Washington, D.C., 1962.

NAG Fortran Library, *The numerical algorithms group*, Downers Grove, IL, July 1988.

Nashashibi, A., and Charalambous, C. "2-D FIR eigenfilters," *Proc. Int. Symp. on Circuits and Sys.* Espoo, Finland, pp. 1037-1040, June 1988.

Nayebi, K., Barnwell, III, T. P. and Smith, M. J. T. "A general time domain analysis and design framework for exact reconstruction FIR analysis/synthesis filter banks," *Proc. IEEE Int. Symp. Circuits and Sys.*, pp. 2022-2025, New Orleans, May 1990.

Nayebi, K., Barnwell, III, T. P. and Smith, M. J. T. "The design of perfect reconstruction nonuniform band filter banks," *Proc. IEEE Int. Conf. Acoust. Speech and Signal Proc.*, pp. 1781-1784, Toronto, Canada, May 1991a.

Nayebi, K., Barnwell, III, T. P. and Smith, M. J. T. "Design of low delay FIR analysis-synthesis filter bank systems," *Proc. 25th Annual Conf. on Information Sciences and systems*, The Johns Hopkins Univ., pp. 233-238, 1991b.

Nayebi, K., Barnwell, III, T. P. and Smith, M. J. T. "Nonuniform filter banks: a reconstruction and design theory," *IEEE Trans. on Signal Proc.*, vol. SP-41, June 1993.

Nelson, G. A., Pfeifer, L. L., and Wood, R. C. "High speed octave band digital filtering," *IEEE Trans. on Audio and Electroacoust.*, vol. AU-20, pp. 8-65, March 1972.

Neuvo, Y., Dong, C.-Y., and Mitra, S. K. "Interpolated finite impulse response filters," *IEEE Trans. on Acoust. Speech and Signal Proc.*, vol. ASSP-32, pp. 563-570, June 1984.

Nguyen, T. Q., and Vaidyanathan, P. P. "Maximally decimated perfect-reconstruction FIR filter banks with pairwise mirror-image analysis (and synthesis) frequency responses," *IEEE Trans. on Acoust. Speech and Signal Proc.*, vol. ASSP-36, pp. 693-706, May 1988.

Nguyen, T. Q., and Vaidyanathan, P. P. "Two-channel perfect reconstruction FIR QMF structures which yield linear phase FIR analysis and synthesis filters,"

IEEE Trans. on Acoustics, Speech and Signal Proc., vol. ASSP-37, pp. 676-690, May 1989.

Nguyen, T. Q., and Vaidyanathan, P. P. "Structures for M-channel perfect reconstruction FIR QMF banks which yield linear-phase analysis filters," *IEEE Trans. on Acoustics, Speech and Signal Processing*, vol. ASSP-38, pp. 433-446, March 1990.

Nguyen, T. Q. "A quadratic constrained least-squares approach to the design of digital filter banks," *Proc. IEEE Int. Symp., Circuits and Sys.*, pp. 1344–1347, San Diego, CA, May 1992a.

Nguyen, T. Q. "A class of generalized cosine-modulated filter bank," *Proc. IEEE Int. Symp., Circuits and Sys.*, pp. 943–946, San Diego, CA, May 1992b.

Nussbaumer, H. J. "Pseudo QMF filter bank," *IBM Tech. disclosure Bulletin*, vol. 24, pp. 3081-3087, Nov. 1981.

Nyquist, H. "Certain topics in telegraph transmission theory," *AIEE Trans.*, vol. 47, pp. 617-644, 1928.

Oetken, G., Parks, T. W., and Schüssler, W. "New results in the design of digital interpolators," *IEEE Trans. on Acoustics, Speech and Signal Proc.*, vol. ASSP-23, pp. 301-309, June 1975.

Oppenheim, A. V., and Schafer, R. W. *Digital signal processing*, Prentice Hall, Inc., Englewood Cliffs, NJ, 1975.

Oppenheim, A. V., Willsky, A. S., and Young, I. T. *Signals and systems*, Prentice Hall, Inc., 1983.

Oppenheim, A. V., and Schafer, R. W. *Discrete-time signal processing*, Prentice Hall, Inc., Englewood Cliffs, NJ, 1989.

Padmanabhan, M., and Martin, K. "Some further results on modulated/extended lapped transforms," *Proc. IEEE Int. Conf. Acoust. Speech and Signal Proc.*, vol. IV, pp. 265-268, San Francisco, CA, March 1992.

Papoulis, A. *Probability, random variables, and stochastic processes*, McGraw-Hill, Inc., 1965.

Papoulis, A. *Signal analysis* McGraw-Hill, Inc., 1977a.

Papoulis, A. "Generalized sampling expansion," *IEEE Trans. on Circuits and Sys.*, pp. 652-654, Nov. 1977b.

Parks, T. W., and McClellan, J. H. "Chebyshev approximation for nonrecursive digital filters with linear phase" *IEEE Trans. on Circuit Theory*, vol. CT-19, pp. 189-194, March 1972.

Peebles, Jr., P. Z. *Probability, random variables, and random signal principles*, McGraw-Hill, Inc., 1987.

Pei, S.-C., and Shyu, J.-J. "A computer program for designing linear phase FIR digital filters by eigen-approach," *Proc. Int. Symp. on Circuits and Sys.*, pp. 367-370, Portland, Oregon, May 1989.

Pei, S.-C., and Shyu, J.-J. "2-D FIR eigenfilters: a least squares approach," *IEEE Trans. on Circuits and Sys.*, pp. 24-34, Jan. 1990.

Petersen, D. P., and Middleton, D. "Sampling and reconstruction of wave-number limited functions in N-dimensional Euclidean spaces," *Information and Control,* vol. 5, pp. 279-323, 1962.

Petraglia, A., and Mitra, S. K. "High speed A/D conversion using QMF banks," Proc. IEEE Int. Symp. on Circuits and Syst., pp. 2797-2800, New Orleans, May 1990.

Portnoff, M. R. "Time-frequency representation of digital signals and systems based on short-time Fourier Analysis," *IEEE Trans. on Acoust. Speech and Signal Proc.,* vol. ASSP-28, pp. 55-69, Feb. 1980.

Potapov, V. P. "The multiplicative structure of J-contractive matrix functions," *Amer. Math. Soc., Translation series 2,* vol. 15, pp. 131-243, 1960.

Prabhakara Rao, C. V. K., and Dewilde, P. "On lossless transfer functions and orthogonal realizations," *IEEE Trans. on Circuits and Systems,* vol. CAS-34, pp. 677-678, June 1987.

Press, W. H., Flannery, B. P., Teukolsky, S. A. and Vetterling, W. T., *Numerical recipes,* Cambridge Univ. Press, 1989.

Princen, J. P., and Bradley, A. P. "Analysis/synthesis filter bank design based on time domain aliasing cancelation," *IEEE Trans. on Acoustics, Speech and Signal Processing,* vol. ASSP-34, pp. 1153-1161, Oct. 1986.

Rabiner, L. R., and Gold, B. *Theory and application of digital signal processing,* Prentice Hall, Inc., Englewood Cliffs, 1975.

Rabiner, L. R., McClellan, J. H., and Parks, T. W. "FIR digital filter design techniques using weighted Chebyshev approximation," *Proc. IEEE,* vol. 63, pp. 595-610, April 1975.

Rabiner, L. R., and Schafer, R. W. *Digital processing of speech signals,* Prentice Hall, Inc., Englewood Cliffs, NJ, 1978.

Ramesh, R. "A multirate signal processing approach to block decision feedback equalization," IEEE Globecom '90, San Diego, CA, pp. 1685-1690, Dec. 1990.

Ramstad, T. A. "Subband coder with a simple bit allocation algorithm: a possible candidate for digital mobile telephony," Proc. IEEE Int. Conf. Acoust. Speech and Signal Proc., pp. 203-207, May 1982.

Ramstad, T. A. "Digital methods for conversion between arbitrary sampling frequencies," *IEEE Trans. on Acoustics, Speech and Signal Processing,* vol. ASSP-32, pp. 577-591, June 1984a.

Ramstad, T. A. "Analysis/synthesis filter banks with critical sampling," International Conf. on Digital Signal Processing, Florence, Sept. 1984b.

Ramstad, T. A. "IIR filterbank for subband coding of images," Proc. of the IEEE Int. Symp. on Circuits and Systems, pp. 827-830, Espoo, Finland, June 1988.

Ramstad, T. A. "Cosine modulated analysis-synthesis filter bank with critical sampling and perfect reconstruction," Proc. IEEE Int. Conf. Acoust. Speech and Signal Proc., pp. 1789-1792, Toronto, Canada, May 1991.

Rao, S. K., and Kailath, T. 'Orthogonal digital lattice filters for VLSI implementation," *IEEE Trans. on Circuits and Systems,* vol. CAS-31, pp. 933-945, Nov. 1984.

Rao, R. P., and Pearlman, W. A. "On entropy of pyramid structures," *IEEE Trans. on Information Theory,* vol. 37, pp. 407-413, March, 1991.

Regalia, P. A., and Mitra, S. K. "A class of magnitude complementary loudspeaker crossovers," *IEEE Trans. ASSP,* vol. ASSP-35, pp. 1509-1516, Nov. 1987.

Regalia, P. A., Mitra, S. K., and Vaidyanathan, P. P. "The digital allpass filter: a versatile signal processing building block," *Proc. IEEE,* vol. 76, pp. 19-37, Jan. 1988.

Renfors, M., and Saramäki, T. "Recursive Nth band digital filters, Parts I and II," *IEEE Trans. on Circuits and Systems,* vol. CAS-34, pp. 24-51, Jan. 1987.

Renfors, M. "Multidimensional sampling structure conversion with one dimensional Nth band filters," Proc. of the IEEE Int. Symp. on Circuits and Systems, pp. 1502-1507, Portland, OR, May 1989.

Rioul, O., and Vetterli, M. "Wavelets and signal processing," *IEEE Signal Processing magazine,* pp. 14-38, Oct. 1991.

Roberts, R. A., and Mullis, C. T. *Digital signal processing,* Addison-Wesley Publ. Co., 1987.

Rosenbrock, H. H. *State space and multivariable theory,* Wiley-Interscience, 1970.

Rothweiler, J. H. "Polyphase quadrature filters, a new subband coding technique," Proc. of the IEEE Int. Conf. on ASSP, pp. 1980-1983, Boston, MA, April 1983.

Safranek, R. J., MacKay, K., Jayant, N. S., and Kim, T. "Image coding based on selective quantization of the reconstruction noise in the dominant subband," Proc. of the IEEE Int. Conf. on ASSP, pp. 765-768, April 1988.

Saramäki, T. "On the design of digital filters as a sum of two allpass filters," *IEEE Trans. on Circuits and Systems,* vol. CAS-32, pp. 1191-1193, Nov. 1985.

Saramäki, T. "A class of window functions with nearly minimum sidelobe energy for designing FIR filters," Proc. IEEE Int. Symp. on Circuits and Sys., Portland, pp.359-362, May 1989.

Sathe, V., and Vaidyanathan, P. P. "Efficient adaptive identification and equalization of bandlimited channels using multirate/multistage FIR filters," Proc. of the Asilomar Conference on Signals, Systems and Computers, Nov. 1990.

Sathe, V., and Vaidyanathan, P. P. "Analysis of the effects of multirate filters on stationary random inputs, with applications in adaptive filtering," Proc. IEEE Int. Conf. Acoust. Speech and Signal Proc., pp. 1681-1684, Toronto, Canada, May 1991.

Sathe, V., and Vaidyanathan, P. P. "Effects of multirate systems on the statistical properties of random signals," *IEEE Trans. on Signal Processing,* vol. ASSP-41, pp. 131-146, Jan 1993.

Schafer, R. W., and Rabiner, L. R. "A digital signal processing approach to interpolation," *Proc. of the IEEE,* vol. 61, pp. 692-702, June 1973.

Schafer, R. W., Rabiner, L. R., and Herrmann, O. "FIR digital filter banks for speech analysis," *Bell Syst. Tech. J.,* vol. 54, pp. 531-544, Mar. 1975.

Scheuermann, H., and Gockler, H. "A comprehensive survey of digital transmultiplexing methods," *Proc. of the IEEE,* vol. 69, pp. 1419-1450, Nov. 1981.

Schüssler, W. "A stability theorem for discrete-time systems," *IEEE Trans. on Acoust., Speech and Signal Proc.,* vol., ASSP-24, pp. 87-89, Feb. 1976.

Sedra, A. S., and Brackett, P. O. *Filter theory and design, active and passive,* Matrix publishers, Inc., Beaverton, OR, 1978.

Segall, A. "Bit allocation and encoding for vector sources," *IEEE Trans. on Info. Theory,* pp. 162-169, March 1976.

Shannon, C. E. "Communications in the presence of noise," *Proc. of the IRE,* vol. 37, pp. 10-21, Jan. 1949.

Shynk, J. J. "Frequency-domain and multirate adaptive filtering," *IEEE Signal Processing Magazine,* vol. 9, Jan. 1992.

Simoncelli, E. P., and Adelson, E. H. "Nonseparable extensions of quadrature mirror filters to multiple dimensions," *Proc. IEEE,* vol. 78, pp. 652-664, April, 1990.

Slepian, D. "Prolate spheroidal wave functions, Fourier analysis, and uncertainty – V: the discrete case," *Bell Sys. Tech. J.,* vol. 57, pp. 1371-1430, May-June 1978.

Smith, H. J. S. "On systems of linear indeterminate equations and congruences," Philos. Trans. Royal Soc. London, vol. 151, pp. 293-326, 1861.

Smith, M. J. T., and Barnwell III, T. P. "A procedure for designing exact reconstruction filter banks for tree structured subband coders," Proc. IEEE Int. Conf. Acoust. Speech, and Signal Proc., pp. 27.1.1-27.1.4, San Diego, CA, March 1984.

Smith, M. J. T., and Barnwell III, T. P. "A unifying framework for analysis/synthesis systems based on maximally decimated filter banks," Proc. IEEE Int. Conf. Acoust. Speech, and Signal Proc., pp. 521-524, Tampa, FL, March 1985.

Smith, M. J. T., and Barnwell III, T. P. "Exact reconstruction techniques for tree-structured subband coders," *IEEE Trans. on Acoustics, Speech and Signal Proc.,* pp. 434-441, June 1986.

Smith, M. J. T., and Barnwell, III, T. P. "A new filter-bank theory for time-frequency representation," *IEEE Trans. on Acoustics, Speech and Signal Proc.,* vol. ASSP-35, pp. 314-327, March 1987.

Smith, M. J. T., and Eddins, S. L. "Analysis/synthesis techniques for subband image coding," *IEEE Trans. on Acoustics, Speech and Signal Proc.,* vol. ASSP-38, pp. 1446-1456, Aug. 1990.

Soman, A. K., and Vaidyanathan, P. P. "On optimal bit allocation strategies for paraunitary subband coders," Proc. of the 25rd Annual Asilomar Conference on Signals, Systems and Computers, 1991.

Soman, A. K., and Vaidyanathan, P. P. "Paraunitary filter banks and wavelet packets" Proc. IEEE Int. Conf. Acoust. Speech, and Signal Proc., San Francisco, March 1992a.

Soman, A. K., and Vaidyanathan, P. P. "Coding gain in multirate paraunitary filter banks," Proc. IEEE Int. Symp. Circuits and Sys., San Diego, CA, May 1992b.

Soman, A. K., Vaidyanathan, P. P., and Nguyen, T. Q. "Linear phase paraunitary filter banks: theory, factorizations and applications," *IEEE Trans. on Signal Processing,* vol. SP-41, Dec. 1993.

Soman, A. K., and Vaidyanathan, P. P. "On orthonormal wavelets and paraunitary filter banks," *IEEE Trans. on Signal Processing*, vol. SP-41, March 1993.

Strang, G. "Wavelets and dilation equations: a brief introduction," *SIAM review*, vol. 31, pp. 614-627, Dec. 1989.

Swaminathan, K., and Vaidyanathan, P.P. "Theory and design of uniform DFT, parallel, quadrature mirror filter banks," *IEEE Trans. on Circuits and Systems*, vol. CAS-33, pp. 1170-1191, Dec. 1986.

Swamy, M. N. S., and Thyagarajan, K. S. "A new type of wave digital filters," *J. Franklin Inst.*, vol. 300, pp. 41-58, July 1975.

Szczupak, J., Mitra, S. K., and Fadavi-Ardekani, J. "A computer-based method of realization of structurally LBR digital allpass networks," *IEEE Trans. on Circuits and Systems*, vol. CAS-35, pp. 755-760, June 1988.

Tan, S., and Vandewalle, J. "Fundamental factorization theorems for rational matrices over complex or real fields," *Proc. IEEE Int. Symp. on Circuits and Syst.*, pp. 1183-1186, Espoo, Finland, June 1988.

Tewfik, A. H., Sinha, D., and Jorgensen, P. E. "On the optimal choice of a wavelet for signal representation," *IEEE Trans. Info. Theory*, pp. 747-765, March 1992.

Tewfik, A. H., and Kim, M. "Correlation structure of the discrete wavelet coefficients of fractional brownian motion," *IEEE Trans. Info. Theory*, vol. IT-38, pp. 904-909, March 1992.

Thong, T. "Practical consideration for a continuous-time digital spectrum analyzer," *Proc. of the IEEE Int. Symp. on Circuits and Systems*, pp. 1047-1050, Portland, May 1989.

Tribolet, J. M. "A new phase unwrapping algorithm," *IEEE Trans. on Acoustics, Speech and Signal Proc.*, vol. ASSP-25, pp. 170-177, April 1977.

Tribolet, J. M., and Quatieri, T. F. "Computation of the complex cepstrum," In *Programs for digital signal processing*, IEEE Press, 1979.

Vaidyanathan, P. P., and Mitra, S. K. "Low passband sensitivity digital filters: A generalized viewpoint and synthesis procedures," *Proc. of the IEEE*, vol. 72, pp. 404-423, April 1984.

Vaidyanathan, P. P. "A unified approach to orthogonal digital filters and wave digital filters, based on LBR two-pair extraction", *IEEE. Trans. on Circuits and Systems*, vol. CAS-32, pp. 673-686, July 1985a.

Vaidyanathan, P. P. "The discrete-time bounded-real lemma in digital filtering," *IEEE. Trans. on Circuits and Systems*, vol. CAS-32, pp. 918-924, Sept. 1985b.

Vaidyanathan, P. P., and Mitra, S. K. "A general family of multivariable digital lattice filters," *IEEE. Trans. on Circuits and Systems*, vol. CAS-32, pp. 1234-1245, Dec. 1985.

Vaidyanathan, P. P., Mitra, S. K., and Neuvo, Y. "A new approach to the realization of low sensitivity IIR digital filters," *IEEE Trans. on Acoustics, Speech and Signal Processing*, vol. ASSP-34, pp. 350-361, April 1986.

Vaidyanathan, P. P. "Passive cascaded lattice structures for low sensitivity FIR filter design, with applications to filter banks," *IEEE Trans. on Circuits and Systems,* Vol. CAS-33, pp. 1045-1064, Nov. 1986.

Vaidyanathan, P. P., and Nguyen, T. Q. "Eigenfilters: a new approach to least-squares FIR filter design and applications including Nyquist filters," *IEEE Trans. on Circuits and Systems,* vol. CAS-34, pp. 11-23, January 1987a.

Vaidyanathan, P. P., and Nguyen, T. Q. "A trick for the design of FIR half-band filters," *IEEE Trans. on Circuits and Systems,* vol. CAS-34, pp. 297-300, Mar. 1987b.

Vaidyanathan, P. P., Regalia, P., and Mitra, S. K. "Design of doubly complementary IIR digital filters using a single complex allpass filter, with multirate applications," *IEEE Trans. on Circuits and Systems,* vol. CAS-34, pp. 378-389 April 1987.

Vaidyanathan, P. P., and Mitra, S. K. "A unified structural interpretation of some well-known stability-test procedures for linear systems," *Proc. of the IEEE,* vol. 75, pp. 478-497, April 1987.

Vaidyanathan, P. P. "Theory and design of M-channel maximally decimated quadrature mirror filters with arbitrary M, having perfect reconstruction property," *IEEE Trans. on Acoustics, Speech and Signal Processing,* vol. ASSP-35, pp. 476-492, April 1987a.

Vaidyanathan, P. P. "Quadrature mirror filter banks, M-band extensions and perfect-reconstruction techniques," *IEEE ASSP magazine,* vol. 4, pp. 4-20, July 1987b.

Vaidyanathan, P. P. "Design and implementation of digital FIR filters," in *Handbook on Digital Signal Processing,* edited by D. F. Elliott, Academic Press Inc., pp. 55-172, 1987c.

Vaidyanathan, P. P., and Hoang, P.-Q. "Lattice structures for optimal design and robust implementation of two-channel perfect reconstruction QMF banks," *IEEE Trans. on Acoustics, Speech and Signal Processing,* vol. ASSP-36, pp. 81-94, January 1988.

Vaidyanathan, P. P., and Mitra, S. K. "Polyphase networks, block digital filtering, LPTV systems, and alias-free QMF banks: a unified approach based on pseudocirculants," *IEEE Trans. Acoust., Speech, Signal Proc.,* vol. ASSP-36, pp. 381-391, March 1988.

Vaidyanathan, P. P., and Liu, Vincent C. "Classical sampling theorems in the context of multirate and polyphase digital filter bank structures," *IEEE Trans. on Acoustics, Speech and Signal Proc.,* vol. ASSP-36, pp. 1480-1495, Sept. 1988.

Vaidyanathan, P. P., Nguyen, T. Q., Doğanata, Z., and Saramäki, T. "Improved technique for design of perfect reconstruction FIR QMF banks with lossless polyphase matrices," *IEEE Trans. on Acoustics, Speech and Signal Proc.,* vol. ASSP-37, pp. 1042-1056, July 1989.

Vaidyanathan, P. P., and Doğanata, Z. "The role of lossless systems in modern digital signal processing: a tutorial," Special issue on Circuits and Systems, *IEEE Trans. on Education,* pp. 181-197, August 1989.

Vaidyanathan, P. P. "Multirate digital filters, filter banks, polyphase networks, and applications: a tutorial," *Proc. of the IEEE,* vol. 78, pp. 56-93, Jan.

1990a.

Vaidyanathan, P. P. "How to capture all FIR perfect reconstruction QMF banks with unimodular matrices?" Proc. IEEE Int. Symp. Circuits and Sys., pp. 2030-2033, New Orleans, May 1990b.

Vaidyanathan, P. P. "Fundamentals of multidimensional multirate digital signal processing," *Sadhana,* vol. 15, pp. 157-176, Nov. 1990c.

Vaidyanathan, P. P., and Liu, V. C. "Efficient reconstruction of bandlimited sequences from nonuniformly decimated versions by use of polyphase filter banks," *IEEE Trans. on Acoust. Speech and Signal Proc.,* vol. ASSP-38, pp. 1927-1936, Nov. 1990.

Vaidyanathan, P. P. "On coefficient-quantization and computational roundoff effects in lossless multirate filter banks," *IEEE Trans. on Signal Proc.,* vol. SP-39, pp. 1006-1008, April, 1991a.

Vaidyanathan, P. P. "The role of Smith-form decomposition of integer-matrices, in multidimensional multirate systems," Proc. IEEE Int. Conf. Acoust. Speech and Signal Proc., pp. 1777-1780, Toronto, Canada, May 1991b.

Vaidyanathan, P. P. "Lossless systems in wavelet transforms," Proc. of the IEEE Int. Symp. on Circuits and Systems, pp. 116-119, Singapore, June 1991c.

Vary, P. "On the design of digital filter banks based on a modified principle of polyphase," *AEU,* vol. 33, pp. 293-300, 1979.

Veldhuis, R. N. J., Breeuwer, M., and Van der wall, R. G. "Subband coding of digital audio signals" *Philips J. Res.,* vol. 44, pp. 329-343, 1989.

Vetterli, M. "Multidimensional sub-band coding: some theory and algorithms," *Signal Processing,* vol. 6, pp. 97-112, April 1984.

Vetterli, M. "Splitting a signal into subsampled channels allowing perfect reconstruction," Proc. IASTED Conf. Appl. Signal Proc., and digital filtering, Paris, France, June 1985.

Vetterli, M. "Filter banks allowing for perfect reconstruction," *Signal Processing,* vol. 10., pp. 219-244, April 1986a.

Vetterli, M. "Perfect transmultiplexers," Proc. IEEE Int. Conf. Acoust. Speech and Signal Proc., pp. 2567-2570 Tokyo, Japan, April 1986b.

Vetterli, M. "A theory of multirate filter banks," *IEEE Trans. Acoust. Speech and Signal Proc.,* vol. ASSP-35, pp. 356-372, March 1987.

Vetterli, M. "Running FIR and IIR filtering using multirate filter banks," *IEEE Trans. Acoust. Speech and Signal Proc.,* vol. ASSP-36, pp. 730-738, May 1988.

Vetterli, M., and Le Gall, D. "Analysis and design of perfect reconstruction filter banks satisfying symmetry constraints," Proc. Princeton Conf. Inform. Sci. Syst., pp. 670–675, Mar 1988.

Vetterli, M., and Le Gall, D. "Perfect reconstruction FIR filter banks: some properties and factorizations," *IEEE Trans. on Acoustics, Speech and Signal Processing,* vol. ASSP-37, 1057-1071, July 1989.

Vetterli, M., Kovačević, J., and Le Gall, D. "Perfect reconstruction filter banks for HDTV representation and coding," *Image Communication,* vol. 2, pp. 349-364, Oct. 1990.

Vetterli, M., and Herley, C. "Wavelets and filter banks," *IEEE Trans. on Signal Processing*, vol. SP-40, 1992.

Vidyasagar, M. *Control system synthesis:* a factorization approach, MIT Press, 1985.

Viscito, E., and Allebach, J. "The design of tree-structured M-channel filter banks using perfect reconstruction filter blocks," Proc. of the IEEE Int. Conf. on ASSP, pp. 1475-1478, New York, April 1988a.

Viscito, E., and Allebach, J. "Design of perfect reconstruction multidimensional filter banks using cascaded Smith form matrices," Proc. of the IEEE Int. Symp. on Circuits and Systems, Espoo, Finland, pp. 831-834, June 1988b.

Viscito, E., and Allebach, J. P. "The analysis and design of multidimensional FIR perfect reconstruction filter banks for arbitrary sampling lattices," *IEEE. Trans. on Circuits and Systems*, vol. CAS-38, pp. 29-41, Jan. 1991.

Von Neumann, J. *Mathematical foundations of quantum mechanics*, Princeton Univ., 1955.

Wackersreuther, G. "On two-dimensional polyphase filter banks," *IEEE Trans. on Acoustics, Speech and Signal Proc.*, vol. ASSP-34, pp. 192-199, Feb. 1986a.

Wackersreuther, G. "Some new aspects of filters for filter banks," *IEEE Trans. on Acoustics, Speech and Signal Proc.*, vol. ASSP-34, pp. 1182-1200, Oct. 1986b.

Westerink, P. H., Biemond, J., Boekee, D. E. "An optimal allocation algorithm for subband coding," Proc. IEEE Int. Conf. Acoust. Speech and Signal Proc., pp. 757-760, New York, April 1988.

Whittaker, J. M. "The Fourier theory of the cardinal functions," Proc. Math. Soc. Edinburgh, vol. 1, pp. 169-176, 1929.

Widrow, B., and Stearns, S. D. *Adaptive signal processing*, Prentice Hall, Englewood Cliffs, NJ., 1985.

Woods, J. W., and O'Neil, S. D. "Subband coding of images," *IEEE Trans. on Acoust. Speech and Signal Proc.*, vol. 34, pp. 1278-1288, Oct. 1986.

Woods, J. W. *Subband image coding*, Kluwer Academic Publishers, Inc., 1991.

Wornell, G. W. "A Karhunen-Loeve-Like expansion of $1/f$ processes via wavelets," *IEEE Trans. on Info. Theory*, vol. IT-36, July 1990.

Wornell, G. W., and Oppenheim, A. V. "Wavelet based representations for a class of self-similar signals with application to fractal modulation," *IEEE Trans. Info. Theory*, vol. IT-38, March 1992.

Yip, P., and Rao, K. R. "Fast discrete transforms," in *Handbook of digital signal processing*, edited by D. F. Elliott, Academic Press, San Diego, CA, 1987.

Zou, H., and Tewfik. A. H., "Design and parameterization of M-band orthonormal wavelets," Proc. IEEE Int. Symp. Circuits and Sys., pp. 983–986, San Diego, CA, 1992.

Some references by topic

Books (general DSP, filters, etc.) [1]
Antoniou, 1979
Bellanger, 1989
Bose, 1985
Haykin, 1991
Jackson, 1989, 1991
Oppenheim and Schafer, 1975 and 1989
Oppenheim, Willsky and Young, 1983
Rabiner and Gold, 1975
Roberts and Mullis, 1987

Books (multirate-related)
Akansu and Haddad, 1992
Briggs, 1987
Chui, 1992
Crochiere and Rabiner, 1983
Daubechies, 1992
Malvar, 1992
Meyer, 1986
Woods, 1991

Books (multidimensional signals)
Bose, 1982
Dudgeon and Mersereau, 1984
Jain, 1989
Lim, J. S., 1990

Filter Banks (two channel and tree-structured)
Croisier, Esteban, and Galand, 1976
Fettweis, Nossek, and Meerkötter, 1985
Galand and Nussbaumer, 1984, 1985
Grenez, 1988
Horng and Willson, Jr., 1992

---

[1] In these few pages we give a partial list of references for special topics, intended to be a starting point. Many excellent articles cited in the main bibliography are not listed here. Further citations to specific topics can be found at the beginnings of the appropriate chapters and appendices.

Jain and Crochiere, 1984
Johnston, 1980
Mintzer, 1985
Nguyen and Vaidyanathan, 1989
Smith and Barnwell, 1984, 1986
Vaidyanathan, Regalia and Mitra, 1987
Vaidyanathan and Hoang, 1988

Filter Banks (M-channel)
Akansu, 1991
Bellanger, 1976
Caglar, Liu and Akansu, 1991
Chu, 1985
Cox, 1986
Koilpillai and Vaidyanathan, 1991, 1992
Kovačević and Vetterli, 1991a
Malvar, 1990b, 1991
Masson and Picel, 1985
McGee and Zhang, 1990
Nayebi, Barnwell, and Smith, 1990, 1991a,b
Nelson, Pfeifer, and Wood, 1972
Nguyen and Vaidyanathan, 1988, 89, 90
Nussbaumer, 1981
Princen and Bradley, 1986
Ramstad, 1984b
Rothweiler, 1983
Smith and Barnwell, 1985, 1987
Swaminathan and Vaidyanathan, 1986
Vaidyanathan, 1987a, 1987b, 1990a
Vaidyanathan and Mitra, 1988
Vaidyanathan, Nguyen, Doğanata, and Saramäki, 1989
Vetterli, 1984, 1985, 1987
Vetterli and Le Gall, 1989
Viscito and Allebach, 1988a
Wackersreuther, 1986b

Filter Banks (multidimensional)
Ansari and Lau, 1987
Ansari and Lee, 1988

Ansari and Guillemot, 1990
Bamberger and Smith, 1992
Chen and Vaidyanathan, 1991, 92, 93
Karlsson and Vetterli, 1990
Kovačević and Vetterli, 1992
Simoncelli and Adelson, 1990
Smith and Eddins, 1990
Vaidyanathan, 1990c, 1991b
Vetterli, 1984
Viscito and Allebach, 1991
Wackersreuther, 1986a
Woods and O'Neil, 1986

Paraunitary and Lossless Systems

Anderson and Vongpanitlerd, 1973
Balabanian and Bickart, 1969
Belevitch, 1968
Doğanata, Vaidyanathan and Nguyen, 1988
Doğanata and Vaidyanathan, 1990
Deprettere and Dewilde, 1980
Liu and Vaidyanathan, 1990
Prabhakara Rao and Dewilde, 1987
Vaidyanathan, 1985b
Vaidyanathan and Doğanata, 1989
Vaidyanathan, Nguyen, Doğanata, and Saramäki, 1989
Soman, Vaidyanathan, and Nguyen, 1992
Vetterli, and Le Gall, 1989

Sampling (general)

Brown, J. L., Jr., 1981.
Jerri, 1977
Nyquist, 1928
Papoulis, 1977a,b
Petersen and Middleton, 1962
Shannon, 1949
Vaidyanathan and Liu, 1988, 1990
Whittaker, 1929

Sampling (multidimensional)

Dubois, 1985
Dudgeon and Mersereau, 1984
Mersereau, 1979
Mersereau and Speake, 1983
Petersen and Middleton, 1962
Liu and Vaidyanathan, 1988

Short-time Fourier Transform

Allen and Rabiner, 1977
Gabor, 1946
Hlawatsch and Boudreaux-Bartels, 1992
Oppenheim and Schafer, 1989
Portnoff, 1980
Rabiner and Schafer, 1978

Subband Coding (one dimensional)

Akansu and Haddad, 1992
Barnwell, 1982
Crochiere, Webber, and Flanagan, 1976
Crochiere, 1977, 1981
Esteban and Galand, 1977
Galand and Esteban, 1980
Jayant and Noll, 1984

Subband Coding (multidimensional)

Gharavi and Tabatabai, 1986, 1988
Ramstad, 1988
Safranek, MacKay, Jayant, and Kim, 1988
Smith and Eddins, 1990
Vetterli, 1984
Woods and O'Neil, 1986
Woods, 1991

Transform Coding and Bit Allocation

Akansu, Haddad, and Caglar, 1992
Cassereau, 1985
Gersho and Gray, 1992
Honda and Itakura, 1984
Huang and Schultheiss, 1963
Jain, 1989
Jayant and Noll, 1984
Malvar and Staelin, 1989
Pearlman (in Woods, 1991)
Ramstad, 1982
Segall, 1976
Soman and Vaidyanathan, 1991, 1992b
Westerink, Biemond and Boekee, 1988
Woods, 1991

Tutorial-valued papers

Allen and Rabiner, 1977
Crochiere and Rabiner, 1981
Dubois, 1985
Hlawatsch and Boudreaux-Bartels, 1992

Jerri, 1977
Portnoff, 1980
Rioul and Vetterli, 1991
Vaidyanathan, 1987b, 1990a, 1990c
Vaidyanathan and Doğanata, 1989
Vetterli, 1987
Viscito and Allebach, 1991

Wavelets and Multiresolution

Chui, 1992
Coifmann, Meyer, Quake, and Wickerhauser, 1990
Daubechies, 1988, 1990
Evangelista, 1989
Gopinath and Burrus, 1991
Grossman and Morlet, 1984
Hlawatsch and Boudreaux-Bartels, 1992
Kovačević and Vetterli, 1992
Mallat, 1989a, 1989b
Rioul and Vetterli, 1991
Soman and Vaidyanathan, 1991, 1992a,b
Strang, 1989
Tewfik and Kim, 1992
Vetterli and Herley, 1992
Wornell, 1990
Wornell and Oppenheim, 1992

Index

Absolutely summable, 17
AC matrix (*see* Alias component matrix)
Adaptive multirate filtering, 151
A/D converters (*see* Analog to digital conversion)
Addition rate (*see* APU)
Adjoint, 789
Adjugate of a matrix, 789
Adjustable multilevel filters, 164
Alias cancelation, 193-194 (*see also* Filter banks)
Alias component (AC) matrix, 193
   circulant, 278
   singularity, 278
   multidimensional filter banks, 629
      hexagonal M, 630
      rectangular M, 630
Alias components, 193
Alias-free decimation (multidimensional), 597-602
Alias-free filter banks, 193 (*see also* Filter banks)
Aliasing, 23
   due to decimation, 105
   in filter banks, 191 (*see also* Filter banks)
   peak distortion, 367
Aliasing in multidimensional sampling, 564, 566 (*see also* Multidimensional sampling)
Allpass complementary (AC), 158
Allpass decomposition, 84-90
   Butterworth, Chebyshev and elliptic filters, 87-91
   efficiency of, 89
   even order case, 90
   generalizations, 99
   pole interlace property, 89
   sufficient conditions for, 85
   theorem, 85
Allpass filters (functions), 71-83 (*see also* Allpass structures)
   continuous-time, 98, 99
   polyphase components, 181

   properties, 71-77
      autocorrelation of, 74
      general form, 72,73
      lossless property, 74
      modulus property, 74
      monotone phase, 76
      poles and zeros, 72
      time domain meaning, 74
Allpass structures, 77-83
   cascade-form, 78
   direct-form, 77, 78
   lattice structures, 79-83
   normalized, 82
   one-multiplier, 83
   stability, 81
   by multiplier extraction, 79
All-pole systems, 17, 64, 66
AM (Arithmetic mean), 820
Amplitude response, 34
Amplitude distortion (AMD), 195 (*see also* Filter banks)
Analog filters, 63-68 (*see also* Optimal filters)
Analog signals, 23
Analog to digital conversion, 24
   with filter banks, 163, 164
   with hybrid QMF banks, 163, 164
   with oversampling, 143-145
Analog QMF banks, 162-163, 274
Analog voice privacy systems, 148
Analysis filter bank, 113, 188
   vector notation, 224
Analysis filters, 113, 188
Analysis/synthesis system, 146, 188
Analytic continuation, 37
Annihilating vector, 789
Antialiasing filters, 2, 144
Anticausality, 16
Antisymmetric matrix, 793
Antisymmetric polynomials (filters), 31
Applications of multirate systems, 143-151
   adaptive filtering, 151

digital audio, 143–145
multilevel filtering, 164
subband coding, 145–148
transmultiplexers, 148
tunable filters, 164
voice privacy, 148
APU (Additions per unit time), 123
AR (Autoregressive process), 833
Arithmetic mean-geometric mean (AM-GM) inequality, 820
proof, 845
Asymptotic coding gain ($M \to \infty$):
ideal subband coder (SBC), 824–825
information theoretic bound, 842
orthogonal transform coder, 832
Attenuation characteristics of filters, 44
maximum passband, 44
minimum stopband, 43
Attenuation characteristics of subband filters, 6
Audio, digital (*see* Digital audio)
Autocorrelation:
of allpass functions, 74
deterministic, 41, 810
of a random process, 806
Autocorrelation matrix, 813
of a scalar random process, 814
Autoregressive (AR) process, 833
Average bit rate, 820

b-bit number, 397
Backward difference approach (IIR filter design), 96
Backward induction, 845
Bandedge, 43
Bandlimited signals (σ-BL)
continuous-time, 23, 41
discrete-time, 454
Bandlimited sequence and causality, 454
Basis vectors, 787
Bessel function in FIR design, 48
Bezout's identity, 693
BIBO stability, 17
Bilinear transformation, 62
Binary numbers, 397
Bit allocation (*see also* Optimum bit allocation)
in subband coding, 146, 147
Bit rate, average, 820
Blocked sequence, 428
Blocked version of a transfer function, 428
LTI property, 721

state space description, 721
Block filters, 427, 428
and alias-free filter banks, 430
in multirate notation, 428–429
and pseudocirculants, 430–432
Blocking effects in signal coding, 324
Block length, 428
Block Toeplitz matrix, 794
Bounded input bounded output (BIBO) stability, 17
Bounded real (BR) functions, 83
Bounded sequences, 13
Bounded transfer functions, 83
BR functions (*see* Bounded real functions)
Butterworth filters, 63–66 (*see also* Optimal filters)
allpass decomposition, 87
design, 65
properties, 64

Cascade form structures, 18
real coefficient case, 20
Cauchy-Schwartz inequality
integral version, 846
summation version, 785
Causality, 16, 22
Causal inverses of causal (MIMO) systems, 711–714
Causal unimodular systems, 712, 713
state space descriptions, 713
Smith-McMillan form, 714
Cayley-Hamilton theorem, 793
Cepstrum (*see* Complex cepstrum)
Channel distortion in filter banks, 279
Characteristic equation, 793
Characteristic polynomial, 790
Chebyshev filters, 66, 96–98 (*see also* Optimal filters)
Chebyshev polynomial, 66, 96–97
Cholesky decomposition, 797
Circulant matrix, 249, 793
alias cancelation (AC) matrix, 278
eigenvalues and eigenvectors, 796
pseudocirculant, 249
right-circulant, 250
Circularly bandlimited signal, 569
rectangular sampling, 569
hexagonal sampling, 570
co (*see* Complete observability)
Coding gain of ideal subband coder (SBC), 822–826
asymptotic ($M \to \infty$), 824–825
information theoretic bound, 842
nonmonotonicity, 824

paraunitary filter banks, 836, 838
white input, 848
Coding gain of orthogonal transform coder, 830–833
   in terms of eigenvalues, 831
   monotone behavior, 831
   optimal, 830
   asymptotic ($M \to \infty$), 832
Coefficient quantization effects, 418–423 (see also Quantization effects)
   in filter banks, 421–423
      FIR QMF lattice, 421
      power symmetric IIR, 421
   low sensitivity, 419
   magnitude response of filter, 419
   stability, 418, 419
   and lattice, 81, 419
Cofactors of a matrix, 786
Colored random process, 809
Comb sequence, 104, 118
Commutativity of decimator and expander, 119
   application, 128, 129
   multidimensional case, 645
Commutator model, 130–131
Compaction of energy by coding, 834
Compatible sets in filter bank theory, 285
Complementary transfer functions, 157–159
   allpass complementary (AC), 158
   doubly complementary (DC), 158
      application in digital audio, 161
      design, 159
   Euclidean complementary (EC), 159
   power complementary (PC), 158
   strictly complementary (SC), 157
Complete observability (co), 680 (see Observability)
Complete reachability (cr), 677 (see Reachability)
Complex cepstrum, 849
   inversion, 853
   minimum phase FIR, 850
   and unwrapped phase, 852
Complexity (see Computational complexity, and also Filter banks)
Complex quantizer, 405
Complex random processes, 812
Complex random variables, 811
Compressor, 101 (see also Decimator)
Computational complexity (see also 'Complexity' under 'Filter banks')

decimation and interpolation filters, 123
DFT bank, 126
fractional decimator, 130
multistage decimation and interpolation, 139, 141
Computational overflow, 398
Constant Q system, 483
Continuous-time filter banks, 275–277
Continuous-time signals, 22–24
Continuous-time systems, 22–24
Controllable, 716
Convergence of z-transform, 14 (see also Region of convergence)
Convolution, 15
   MIMO systems, 25
Coprime polynomials, 663
Coprime polynomial matrices, 664
Cosets, 596, 658
Cosine modulated filter banks, 353–393
   complexity comparisons, 385–391
   using DCT and DST, 388, 389
   definition, 363
   paraunitary perfect reconstruction, 377–391
      advantages, 390
      complexity, 382
      design, 380–383
      Theorem for, 378
   polyphase structures, 370–373, 388, 389
   pseudo QMF, 354 (see also under Pseudo QMF)
Cosine modulated perfect reconstruction systems (see Cosine modulated filter banks)
Cosine modulation and interpolation, 179
cr (see Complete reachability)
crco (Completely reachable and completely observable), 682
Coupled form structure, 672–673, 860
   and Mason's formula, 862
Covariance matrix, 813
Covariance sequence of a random process, 807
Cross correlation
   deterministic signals, 41
   random processes, 809
   random variables, 805
   vector random processes, 814
   vector random variables, 813
Cross correlation matrix, 813
Cross covariance between random variables, 805

Crossover network, 161
Cross power spectrum, 814
Cross talk, 149 (*see also* Transmultiplexers)
Cutoff frequency, 45, 49

D/A conversion (Digital to analog conversion), 24
D/A (Digital to analog) conversion
  with hybrid QMF banks, 162
  in digital audio, 162–163
DCT (*see* Discrete cosine transform)
Decibel (dB) plots, 33
Decimation filters, 105-107
  for boundary value problems, 175, 176 (*see also* Multigrid methods)
  fractional, 109–111
  linear phase, 125
  polyphase structures for, 122-125
  time domain description, 117
Decimation of periodic sequences, 180
Decimator, 100 (*see also* Decimation)
  aliasing due to, 105
  causality, 111-113
  fractional, 109–111
  nonuniform, 446, 452
  physical time scale, 111
  time varying property, 105
  transform domain analysis, 102–105
Decorrelating property of coders, 834
Degree of determinant
  general, 708
  lossless systems, 729, 737, 757
Degree of flatness, 532
Degree reduction step, 727, 735
Degree of a system, 27
Degree (McMillan) of a transfer matrix, 666–668, 707–708
  equivalent definitions, 707–708
  lossless systems, 757
Delay chain, 124
  in filter banks, 235
  perfect reconstruction system, 235, 236
  vector notation, 224
Delay-free loops, 21
Delta function, 12
Density functions
  probability, 803
  power spectrum, 807
Derivative sampling theorem, 277, 439
Design specifications (*see* Digital filters)

Determinant of autocorrelation, and prediction error, 832
Determinant, degree of (*see* Degree of determinant)
Determinant of the denominator matrix, 694
Determinant inequality for positive definite matrices, 797
Determinant of a lossless system, degree of, 757
Determinant of a matrix, 786
  block diagonal, 786
  block partitioned, 801
  and eigenvalues, 791
  properties, 787
Determinant of the sampling matrix, 562
Determinant of a signal flow graph, 860
Deterministic autocorrelation, 41, 810
  of allpass functions, 74
Deterministic cross correlation, 41
DFT (*see also* Discrete Fourier Transform), 794
DFT filter banks, 113, 182, 184
  decimated, 127
  polyphase implementation, 125
  and short-time Fourier transform, 468, 469
DFT matrix, 794
Diadic matrix, 785, 788
Diadic based structures, 732
Diagonalization of matrices, 792
  with unitary matrices, 795
Diagonal matrix, 783
Difference equations, 18
  MIMO systems, 719
Difference-sampling theorems, 440-444
  and filter banks, 441, 442
Differential pulse code modulation (DPCM), 839–844
  coder and decoder, 841
  coding gain, 841, 842
  effective number of bits, 841
Digital to analog conversion, 24
Digital audio (music), 2, 143-145
  doubly complementary (DC) filters in, 161
  sampling rate alteration, 143
  subband coding, 147
Digital filter banks (*see* Filter banks)
Digital filters, 31, 42-99
  (*see also* FIR filter design, IIR filter design, Optimal filters, Time domain constraints)
  design specifications, 42-45

FIR, 17, 34–37
    transmission zeros, 32
    types of (lowpass, etc.), 32
Digital filter structures (*see* Structures for transfer functions)
Digital signal, 23
Dilation, 488
Dimension of state space, 675
Dimension of vector space, 787
Dirac delta function $\delta_a(t)$, 12
Direct-form structure, 18
Discrete cosine transform (DCT), 373
    orthonormality, 374
    in transform coding, 829
Discrete Fourier Transform (DFT), 794
    DFT filter bank, 113
    generalized to multidimensions, 590, 591
Discrete sine transform (DST), 373
Discrete-time signals, 12
Discrete-time systems, 12–41
Discrete-time wavelet transform (DTWT), 491
    definition, 495
    FIR basis, 500
    inversion, 495
    orthonormal, 500–510
    summary, 511
    and tree-structured filter banks, 493
Discrete wavelet transform (DWT), 485
Discretizing differential equations, 168
Disk partitioning diagram, 500
Distortion transfer function, 195 (*see also* Filter banks)
Division theorem for integer vectors, 577, 596
    and polyphase decomposition, 578
Division theorem for polynomials, 688
Dolph-Chebyshev window, 50 (*see also* FIR filter design)
Doubly complementary (DC), 158
Downsampler, 100 (*see also* Decimator)
DPCM (*see* Differential pulse code modulation)
DST (*see* Discrete sine transform)
DTWT (*see* Discrete-time wavelet transform)
Duration, RMS, 478
    and uncertainty principle, 478
DWT (*see* Discrete wavelet transform)
Dynamic range, 397, 400, 817

Ear, resolution, 457
Eigenfilters, 53–56 (*see also* Optimal filters)
    for design of Nyquist filters, 155
Eigenfunctions of LTI systems, 16, 39
Eigenspaces, 792
Eigenstructures of special matrices, 795–796
Eigenvalues and eigenvectors of matrices, 790
    computation (extremum values), 799
    and determinants, 791
    properties, 791
    and trace, 791
Eigenvalues and poles, 674, 720–721
Elementary matrix operations, 687
Elliptic filters 66–70 (*see also* Optimal filters)
    allpass decomposition, 87
    design of, 66
    family of, 69, 70
    power symmetric, 204, 205, 213
    properties, 68–71
    transfer function, 67
    uniqueness, 70
Electrical multiport network, 723
Energy balance in state equations, 741
Energy compaction by coding, 834
Energy of a random process, 807
Energy of a sequence, 15
Energy of a vector, 785
Equiripple filters (*see also* Optimal filters)
    FIR, 56, 57
        Estimation of order, 57
        McClellan-Parks algorithm, 57
    IIR, 66 (*see also* Elliptic filters)
Ergodicity, 809
Error vector 170–177 (*see also* Multigrid methods)
Euclidean complementary (EC), 159
Euclid's theorem, 129, 159
Euclid's theorem, matrix version, 693
Expander, 101 (*see also* Interpolation)
    causality, 111–113
    imaging due to, 102
    physical time scale, 111
    time varying property, 105
    transform domain analysis, 102
Expected value of a random variable, 804
Exponential sequences, 13, 27
    one sided, 13

multidimensional, orthogonal, 590, 591
Extraction of lossless building blocks, 728, 736

Factorization of lossless (paraunitary) matrices
   2×1 real FIR, 730–731
   2×2 real FIR, 302, 303, 727–730
   $M \times 1$ FIR, 737
   $M \times M$ FIR, 314, 735–737
   completeness, 726
   general FIR, 731–740
   Givens rotation based, 766, 768
   Householder matrix based, 765, 766
   IIR case, 759–763
   minimality, 726, 729
   philosophy, 726
   real coefficient case, 736, 763
   uniqueness, 739
Factorization of unitary matrices, 315, 745–754 (*see also* under Unitary matrix)
Fan filters, 551, 611, 612 (*see also* Multidimensional filters)
FDM (*see* Frequency division multiplexing)
Filter banks, 113
   alias cancelation, 193, 196
   alias component (AC) matrix, 228
      inversion, 229
      singularity, 230
   alias-free filter banks, 6, 193
      examples, 245–249
      most general, 249–254
      necessary and sufficient conditions, 249, 250
      perfect reconstruction, 253 (*see also* under Perfect reconstruction)
      pseudocirculant condition, 250
   aliasing, 191, 225
   alias terms, 225, 226
      gain of, 226
   amplitude distortion (AMD), 195, 226
      elimination, 201-203
      minimization, 199
   analysis, 113
   with channel distortion, 278
   compatible sets, 285
   complexity
      comparisons, 329–332, 385–391
      cosine modulated, 385
      Johnston's FIR design, 201

   paraunitary (general), 329
   paraunitary (cosine modulated), 385
   power symmetric (allpass based IIR), 215, 217
   power symmetric FIR, 221–223
   pseudo QMF, 373
   QMF lattice, 311
   continuous-time, 275–277
   cosine modulated (*see* Cosine modulated filter banks)
   DFT (or uniform DFT) 113, 116
   distortion transfer function, 195
   errors in, 188, 225
   history, 189–190
   with ideal filters, 226
   linear-phase filters (approximate reconstruction), 198–201
   linear-phase filters (perfect reconstruction), 337–352
   maximally decimated, 188, 224
   noise analysis (*see* under Noise analysis)
   noise comparison 412–416
   nonuniform bandwidth, decimation, 190, 256, 284, 482
   nonmaximally decimated, 283
   octave spacing, 5
   overall transfer function, 195
   paraunitary (*see* Paraunitary filter banks)
   perfect reconstruction, 196, 228 (*see also* under Perfect reconstruction)
   phase distortion (PHD), 196, 226
      elimination, 198
   polyphase representation, 197, 230–234
      IIR (special design), 201–218
   power symmetric
      and allpass filters, 207
      Butterworth, 211
      definitions, 205, 206
      design (FIR), 219
      design (IIR) 210–213, 215
      elliptic, 208, 211, 213
      FIR (perfect reconstruction), 218-220
      IIR (allpass based), 204–218
      IIR complex, 275
      IIR even ordered, 273
      poles of, 208
   pseudo QMF (*see* Pseudo QMF)
   quantization effects (*see* under Quantization effects)
   robustness to quantization, 218

spectrum analyzers, 5–6
summary (Tables), 267–271
time varying property, 193–195
transmultiplexing (*see* Transmultiplexers)
tree-structured, 254–259, 280–282, 335
  binary, 254
  and Laplacian pyramid, 258
  and multiresolution, 256, 257
  and nonuniform decimation, 280–282
  unconventional modifications, 277, 279, 285
  and wavelets, 457–538
Filter bank transform, 475
Filter design techniques, comparison, 67, 91
Filters (*see* Analog filters, and Digital filters)
Finite impulse response (FIR) filters, 45–60 (*see also* Time domain constraints, Nyquist filters)
  four types of, 35
  length of, 17
  nonlinear (minimum) phase, 59
  optimal FIR filters (*see* Equiripple filters, and Eigenfilters)
  order of, 17, 28
  order estimation, 48, 57
  phase response
    linear phase, 34–37
    maximum phase, 28
    minimum phase, 28
    mixed phase, 28
  window technique, 46
    Dolph-Chebyshev window, 50
    Kaiser window, 47–49
    limitations, 50, 53
    minimax window, 50
    optimal windows, 50–53
    prolate spheroidal windows, 50–53
FIR filter design (*see* Finite impulse response filters)
FIR lattice structures, 302
FIR sequences, 17
FIR transfer matrices, 708–711
  state space description, 709
  Smith-McMillan form, 710
Filter structures (*see* Structures for transfer functions)
Fixed point binary fraction, 397
Flatness constraints, 45
Flatness, degree of, 532
Fourier transform (FT)
  continuous-time, 22
  discrete-time, 13–14
  and filter banks, 465
  inverse of, 14, 22
FPD (*see* Fundamental parallelepiped)
Fractional sampling rate alteration, 109–111
  in digital audio, 143–145
  polyphase implementation, 127
Frequency, 22
Frequency division multiplexing (FDM), 148, 149
Frequency response, 31
Frequency response matrix, 26
FT (*see* Fourier transform)
Full memory lossless system, 759
Fundamental parallelepiped, 561

Gabor lattice, 481
Gabor transform, 476
Gaussian random process, 843
Gaussian random variable
  density function, 804, 806
  uncorrelatedness and independence, 806, 813
  vector case, 813
Gaussian elimination, 170, 172
Gaussian window, 476
  duration (RMS), 541
  Fourier transform, 541
  optimality (Gabor transform), 478
Gcd (Greatest common divisor), 511, 544
General degree-one paraunitary (lossless) building block $\mathbf{V}_m(z)$, 734
Generalized DFT (multidimensional), 590, 591 (*see also* Multidimensional DFT matrix)
Generalized orthogonal exponentials (multidimensional), 591
Generalized STFT (*see* Short-time Fourier transform)
Gibbs phenomenon, 46
Givens rotation, 290, 291, 747
  in paraunitary factorization, 728
  limitations of, 769
Glcd, 664 (*see* Greatest left common divisor)
GM (Geometric mean), 820
Gray and Markel structure, 79 (*see also* Allpass structures)
Grcd, 664 (*see* Greatest right common divisor)
Greatest left common divisor (glcd), 664

Index 897

identification of, 692
Greatest right common divisor (grcd), 664
Group delay, 32, 34

Haar basis, 517, 520, 521
Half-band filters 153–156 (*see also* Nyquist(M) filters)
    definitions, 153, 154
    design, 155, 185–186
    in perfect reconstruction filter bank design, 220–221
Harmonic process, 815, 831
Height conventions for wavelet filter banks, 483, 497, 539
Hermitian matrix, 793
    eigenvalues, 796
Hermitian polynomials, and generalizations, 30
Hexagonal decimator M573
    AC matrix, 630
    noble identity, 606, 607
    perfect reconstruction filter bank, 633–636
    polyphase decomposition, 588
    summary, 648
    uniform DFT filter bank, 627
Hexagonal sampling, 558
    general definition, 571
    circular support, 570
Hilbert transformers, 35, 447
Homogeneous difference equations, 719, 720
Householder unitary matrices, 315, 751
    in lossless factorization, 765, 766
Hurwitz polynomial, 28
Hybrid QMF banks, 162–164

Ideal lowpass filters, 45
Ideal subband coder, 818, 819 (*see also* Optimum bit allocation, and Coding gain)
Identity matrix, 783
IDFT (*see* Inverse DFT)
IDFT matrix, 794
IIR (*see* Infinite impulse response)
IIR filter design (*see* Infinite impulse response filter design)
IFIR approach (*see* Interpolated FIR)
IIR filters based on allpass filters, 84–90 (*see also* Allpass decomposition)
Image compression (coding), 7, 147
Images, 102 (*see also* Expander)
Image suppressor, 137

Imaging due to expanders, 102
Impedance matrix, 723
Implementations of transfer functions, 18 (*see also* Structures for transfer functions)
Impulse response, 15
    finite (FIR), 17
    infinite (IIR), 17
    matrix, 25
    and state space description, 674
Impulse function, 12
Indefinite matrices, 796
Infinite impulse response (IIR) filter design, 60–70 (*see also* Optimal filters)
    bilinear transform, design using, 62
    pole crowding, 62
    poles of sharp-cutoff filters, 62
    working principle, 60, 61
Infinite impulse response (IIR) systems, 17
    linear phase, 37
Infinite products, 512, 514
    convergence, 514, 516
Information theoretic bound on coders, 842
    Gaussian processes, 843
Inner product, 785
Interconnections of multirate building blocks, 118–119
Interlacing with expanders and delay chain, 129
Interpolated FIR (IFIR) approach, 134–138
    for decimation filters, 138–143
    extensions, 137, 138, 183
    for interpolation filters, 184
    for narrowband filters, 135
    for wideband filters, 183
Interpolation, 100–113
    fractional, 111
    polynomial fitting, 111
    interpolation filters, 105–109
    and modulation, 109, 178–179
    time domain description, 117
    structures for, 123–125
Interpolation filters for boundary value problems, 175, 176 (*see also* Multigrid methods)
Interpolation and cosine modulation, 179
Interpolator, 101 (*see also* Expander, and Interpolation)
Inverse Chebyshev filters, 98
Inverse DFT (IDFT), 115, 794

Inverse of a matrix, 789–790
Irreducible transfer functions, 16, 663
Irreducible matrix fraction descriptions, 668
    deeper results, 693, 694
    Smith-McMillan form, 697
Iterative techniques for solving equations, 170 (see also Multigrid methods)

Johnston's technique, 199, 200
Joint probability density functions, 805, 806
Jointly wide sense stationary process, 809
Jordan form, 720

Kaiser window, 48–50 (see also FIR filter design)
Kalman Yacubovic lemma, 740
Karhunen-Loeve Transform (KLT), 324, 828–829
KLT (see Karhunen-Loeve Transform)
Kronecker product of matrices, 644

$\mathcal{L}_2$-norm
    of a function, 402
    of a vector, 785
$\mathcal{L}_2$ scaling, 402, 403
$\mathcal{L}_p$-norm, 402
$\mathcal{L}_p$ scaling, 401–402, 424–425
Lag variable, 41, 807
Laplace's transform, 22
Laplacian pyramid, 258
Lapped orthogonal transform (LOT), 322–326
$LAT(\mathbf{V})$, 559
Lattice $LAT(\mathbf{V})$, 559
Lattice sampling (multidimensional), 559
Lattice structures (see QMF lattice, Allpass structures, and FIR lattice)
LBR lemma, 740
LBR transfer functions, 83
Lcd (see Left common divisor)
LC networks, 723
Leading principal minors, 786
Least squares design, 46, 54
Left common divisor (lcd), 664
    greatest (glcd), 664
Left coprimeness, 664
Left divisor, 663
Left matrix fraction description, 665
Leibnitz's rule, 533
Limit cycles, 416
    granular or roundoff, 417
    overflow, 417
    suppression, 418
Linear combination, 787
Linear equations, 790
    solvability, 790
    uniqueness, 790
Linear independence, 787
Linear periodically time varying (LPTV) systems
    in filter banks, 132, 193, 435
    formal definition, 433
    and pseudocirculants, 435
Linear phase filter banks (approximate reconstruction), 198–201 (see also under Filter banks)
Linear phase filter banks (perfect reconstruction), 337–352
    comparison with Johnston's design, 352
    complexity, 343–345
    design using lattice, 342, 343
    initialization, 345
    lattice structures for, 339
    one-multiplier version, 351
    lattice synthesis, 347–350
    paraunitariness, 338
    power complementarity, 338
Linear phase filters 34–37 (see also Digital filters)
    advantages of, 36
    amplitude response, 34
    conditions for, 35, 37
    efficient structures for, 36
    four types of, 35
    polyphase components, 125, 182, 183
    zero locations of, 37
    zero-phase response, 34
Linear prediction, 831
Linear prediction error and determinant of autocorrelation matrix, 832
Linear predictive coder
    closed loop, 839, 842
    open loop, 842, 847
Linear shift invariant (LSI) systems, 15
Linear systems, 15
Linear time invariant (LTI) systems
    definition, 15
    causality, 16
    eigenfunctions of, 16
    MIMO case, definition, 718
Linear time varying (LTV) property, 105,

decimators and interpolators, 178
definition, 433
Localization
   in time, 117, 460
   optimal, 478
Lossless bounded real (LBR), 83, 288
Lossless building block $\mathbf{V}_m(z)$, 732
   (*see also* under Paraunitary, degree-one, building block)
Lossless property, 288 (*see also* entries beginning Paraunitary)
   under quantization, 768–771
   state space manifestation, 740
   theorem, 743
Lossless transfer functions, 83
Lossless transfer matrices, 288
Lossless (paraunitary) systems
   degree of, 757
   definition, 724
   determinant, 725
   energy balance, 724
   factorizations, 726 (*see* under Factorizations)
   history, 723
   impulse response, 725
   modulus property, 758
   most general, 2 × 2, 776
   poles and zeros, 755, 756
   Smith-McMillan form, 754–757
      FIR case, 756
   structures, 726
   summary and tables, 772–774
   zero location, 757
LOT (*see* Lapped orthogonal transform)
Low sensitivity structures, 419–423
LPR lemma, 740
LPTV (*see* Linear periodically time varying system)
LSI systems (*see* Linear shift invariant systems)
LTI systems (*see* Linear time invariant systems)
LTV systems (*see* Linear time varying property)
Lyapunov Lemma, 686

Mth band filters, 151-156 (*see also* Nyquist filters)
M-channel QMF (M-channel filter banks), 223
Mth root of unity W_M, 104
MacClellan-Parks algorithm, 57
MacMillan degree, 27 (*see also* Degree of a system)
Magnitude response, 32
   in dB, 44
   normalized, 43
Magnitude truncation, 398
Main diagonal of a matrix, 783
Marginal density functions, 805
Mason's gain formula, 859, 861
   for MIMO systems (example), 864
Matrices, review of, 782–802
Matrices, 29–30 (*see also* under specific entries, e.g., Positive definite matrices)
   addition, 784
   antisymmetric, 793
   circulant, 793
   diadic, 785, 788
   diagonal, 783
   DFT, 794
   eigenvalues and eigenvectors, 790–793
   functions, 29–30
   Hermitian, 793
   identity, 783
   inverse, 789–790
   as lowpass filters, 174, 176, 187
   multiplication, 784
   nonsingular, 787
   normal, 794
   null, 783
   null space, 789
   paraconjugation, 30
   positive definite, 796
   range space, 789
   rank of, 788
   rectangular, 783
   singular, 787
   skew-Hermitian, 793
   square, 783
   stability, 685, 799
   symmetric, 793
   tilde operation, 29
   Toeplitz, 794
   trace of, 785
   transpose, 783
   transpose conjugation, 783
   transposition, 29
   triangular, 783
      determinants, 787
   unitary, 793
   Vandermonde, 795
Matrix adjugate, 789
Matrix convolution, 25
Matrix diagonalization, 792
Matrix fraction descriptions (MFD), 665
   irreducible, 668
Matrix inequalities, 796

Matrix inverse, 789–790
Matrix inversion lemma, 790
    application, 712
Matrix polynomials, 661 (*see* Polynomial matrices)
Matrix products
    properties preserved, 799, 801
Matrix triangularization, 798
Maximally decimated filter banks, 5, 188–285 (*see also* under Filter banks)
Maximally flat filters
    FIR, 532
        half-band, 535
            order estimation, 535
    IIR (Butterworth) 63–65
Maximization of quadratic forms, 798
Maximum eigenvalue, computation, 799
Maximum modulus theorem, 75
    matrix version, 758, 781
Maximum phase, 28
McClellan-Parks algorithm, 137
McMillan degree, 666 (*see* Degree of a system)
MD (*see* under Multidimensional)
Mean opinion score (MOS), 147
Mean of a random variable, 805
Mean square value of a random variable, 804
Memoryless system, 759
MFD (*see* Matrix fraction descriptions)
MIMO (*see* Multi-input multi-output systems)
MIMO LTI system, definition, 718
Minimal structures, 21, 676
        relation to reachability and observability, 681, 682
        and similarity transform, 682, 683
        real coefficient systems, 684, 716
        and unitariness of realization matrix, 745
Minimax property, 44, 97
Minimum phase, 28
Minors of a matrix, 786
Mirror image polynomials, 31
Mixed phase, 28
Modulo notation (integer vectors), 577
Modulus property of allpass filters, 74
Modulus property of lossless systems, 758
Monotone phase response of allpass filters, 76
MOS (Mean Opinion Score), 147
Mother wavelet $\psi(t)$, 488

MPU (Multiplications per unit time), 123
Multidimensional complex exponential, 657
Multidimensional decimation filters, 606–607
    alias-free, 598–600
    design, 608–623
    diamond shape, 610–612
    fan shape, 610–612
    multistage, 615–619
    nonuniqueness, 602
    from one-dimensional filters, 610, 619
Multidimensional decimators, 572
    aliasing due to, 584, 585
    decimation ratio, 575
    hexagonal **M**, 573, 593
    nonuniqueness, 593
    quincunx **M**, 573
    rectangular **M**, 573, 575
    transform domain analysis, 583–586
    unimodular, 594
Multidimensional DFT matrix, 590, 591
    hexagonal **M**, 592
Multidimensional expanders, 576
    images due to, 581
    transform domain analysis, 580
Multidimensional frequency partitioning, 602, 603
Multidimensional (generalized) DFT
    and Smith decomposition, 644
    as a Kronecker product, 644
Multidimensional filter banks, 627–641
    AC matrix, 629
        hexagonal **M**, 630
    AC matrix and relation to polyphase matrix, 655
    analysis of, 628
    delay chain system, 633
    paraunitary, 654, 656
    perfect reconstruction, 633
        hexagonal **M**, 633
    polyphase representation, 631–632
    quincunx, 654, 656
    tree-structured, 640–641
Multidimensional filters
    diamond filters, 551
    fan filters, 551, 553
    frequency responses, 550, 551
    frequency supports, 552
    linear phase, 553

from one-dimensional filters, 610, 619
separable, 553
zero phase, 553, 650
Multidimensional interpolation filters, 606
Multidimensional Mth band filters, 623
Multidimensional multirate building blocks
cascade connections, 603
Multidimensional multirate identities, 603–605
Multidimensional multirate systems, summary
hexagonal M, 648
multirate fundamentals, 647
notations and lattice concepts, 646
quincunx M, 649
Multidimensional multistage designs, 612
Multidimensional noble identity, 604
hexagonal M, 606-607
quincunx M, 637
rectangular M, 605
Multidimensional Nyquist(M) filters, 623
Multidimensional paraunitary filter banks, 636
Multidimensional polyphase decomposition, 578
and cosets, 596
decimation filters, 606, 607
hexagonal M, 578–579
transform domain, 586, 588
type 2, 588
Multidimensional sampling, 555, 558
aliasing, 564–566, 568
and Fourier transform, 565, 567
general, 558
hexagonal, 558, 570
definition, 571
rectangular, 555, 559
sampling density, 561, 563
minimum, 568
Multidimensional signals, 546
bandlimited, 547
convolution, 550
discrete-time, 547
Fourier transform, 546–548
modulation, 550
z-transform, 548–549
Multidimensional subband coding, 7
Multidimensional (uniform) DFT filter banks, 624–627

Multigrid methods, 9, 168–177
for boundary value problems, 168, 169
dense and coarse grids, 177
information exchange between grids, 176-177
Multi-input multi-output (MIMO) systems, 10, 24–27, 660–714
causality, 26
frequency response matrix, 26
impulse response matrix, 24
poles, 27
stability, 27
transfer matrix, 24
Multilevel filters, 164
in nonuniform sampling and reconstruction, 447
Multiple use of a filter, 93, 94
Multiplexing, 123 (see also TDM, FDM, and Transmultiplexers)
Multiplication rate (see MPU)
Multiport, electrical, 723
Multirate adaptive filtering, 151
Multirate building blocks, 100-118
interconnections, 118–119
Multirate identities, 118–120
basic ones, 118
noble identities, 119
polyphase identity, 133
multidimensional, 603–605
Multirate system, 1
Multiresolution methods, 256, 257, 496
Multistage implementations, 134–143 (see also IFIR approach)
decimation filters, 134, 138
interpolation filters, 143
complexity, 139, 141
Multivariable systems, 10, 660
Music, digital, 2, 4
subband coding, 147

Negative definite matrices, 796
Negative frequency, 14
Negative number representation, 397
Noble identities, 119
multidimensional, 604 (see also Multidimensional noble identity)
Nodes to be scaled, 401
Noise analysis, 405
allpass cascade, 404
cascaded structure, 404
cosine modulated filter bank, 425, 426
decimators, 405

direct form (FIR), 403
expanders, 405
filter banks, 408
filter bank output noise, 412
  comparisons, 412–416,
  summary, 417
fractional sampling rate changer, 425
lattice (IIR), 424
multistage (IFIR) filter designs, 425
paraunitary (lossless) systems, 408
paraunitary filter banks, 409, 412
transmultiplexers, 425
Noise gain, 399
Noise model assumptions, 398
Noise source, 395, 398
Noise transfer function, 396
Noise variance, 398, 399
Nonmaximally decimated filter banks, 283
Nonrectangular decimation, 576 (see also Multidimensional decimators)
Nonsingular matrix, 787
Nontouching loops, 860
Nonuniform decimator, 446, 452
Nonuniform filter banks (see also Filter banks)
  uses, 457
  and wavelets, 482, 498
Nonuniform sampling, 440, 444–453
  filter bank model, 446
Normalization of frequency Ω, 66
Normalized frequency $\omega/2\pi$, 43
Normalized paraunitary, 289
Normalized specifications, 43, 44
Normalized lattice structure, 82 (see also Allpass structures)
Normalized unitary, 793
Normal matrix, 794
  and diagonalization, 795
  eigenvectors, 795
Normal rank, 662
Norms, 785
Notations
  general 28–31
  digital filter building blocks, 18
Null matrix, 783
Null space, 789
Nyquist(M) filters, 54, 151–156
  continuous-time, 153
  definition, 152
  design, 155–157
  eigenfilter approach, 155

frequency domain manifestation, 153
in interpolation filtering, 151–152
multidimensional, 623
in multilevel filter design, 165
and power-complementary filters, 159
Nyquist property and wavelet orthonormality, 529, 530
Nyquist rate, 23, 276
Nyquist sampling theorem, 23
  generalizations, 275, 277
  derivative sampling, 277, 439

Observability, 680
  condition for, 680
  reduction to observable form, 681
  test for, 684
Octave-band filter banks, 8
One-multiplier lattice, 83 (see also Allpass structures)
Optimal filters, 44 (see also Analog filters, and Digital filters)
  criteria for, 44
  eigenfilters, 53–56
  equiripple, 44
    FIR, 56, 57
    IIR Chebyshev, 96–98
    IIR elliptic, 66–70
  least squares, 45
  maximally flat filters
    FIR, 532
    IIR (Butterworth), 63, 64
Optimal windows (see FIR filter design)
Optimal window for short-time Fourier transform, 478
Optimum bit allocation
  ideal subband coder, 820
  orthogonal transform coder, 826–828
  paraunitary filter banks, 836
Order estimation for filters
  FIR filters, 134
    equiripple design, 57
    maximally flat, 535
    window design, 48
  IIR filters
    Butterworth, 65
    Chebyshev, 97
    Elliptic, 66
Order of a FIR transfer matrix, 666–668, 708
Order of a pole
  MIMO case, 703
Order of a system

definition, 16
FIR case, 17
MIMO case, 708
Order of a zero, 704
Orthogonal complement, 789
Orthogonal exponentials, generalized (multidimensional) 590, 591
Orthogonal matrix, 793
Orthogonal random variables, 806
Orthogonal transform coder, 826 (*see also* Transform coders)
    suboptimal, 829
Orthogonal vectors, 785
Orthonormal basis, 94
    and least squares, 95
    and optimality of rectangular window, 95
    and binary tree structure, 504
Orthonormal basis, definition
    wavelets (continuous-time), 489
    wavelets (discrete-time), 496
Orthonormality expressed in z-domain, 502, 509
Orthonormality and uncorrelatedness, 544
Orthonormality of wavelet basis (continuous-time) 510–530
    completeness, 524
    finite duration, 524
    generation, 522
    Haar basis, 517, 520, 521
    and paraunitariness, 512, 523
    sufficient conditions (lemma), 529
    sufficient conditions (theorem), 530
Orthonormality of wavelet basis (discrete-time) 500–510
    generation (design) 503
    and paraunitary property, 503–510
    M-channel, 507–510
Orthonormal matrix, 793
Outer product, 785
Outline of the book, 8–11
Output equation, 670
Overflow, 398
    probability, 403, 818
Oversampling, 2, 143–145

Paraconjugation, 28–30
Paraunitary, degree-one, building block $\mathbf{V}_m(z)$, 732
    generality, 734
    IIR case, 759, 780
    properties, 733
Paraunitary matrix or system (*see also* under Lossless)
    definition, 288
    determinant, 289
    examples, 290–294
    history, 287
    IIR, 293
    inteconnections, 290
    multidimensional, 636
    normalized, 289
    properties, 289
    and quantization, 770–771
Paraunitary matrix factorization (*see under* Factorization)
Paraunitary (perfect reconstruction) filter banks, 286–336
    alias-component (AC) matrix, 297
    advantages, 286, 328, 329
    bit allocation and coding gain, 836–838
    complexity,
        cosine modulated, 385–389
        general, 329
        QMF lattice, 311
    cosine modulated, 377, 378
    definition, 294
    energy of analysis filters, 297
    and ideal subband filters, 297
    IIR, 295
    M-channel, design 314–322
    and Nyquist property, 297
    perfect reconstruction, 294–296
    power complementary property, 296
    properties, 294–298
    quantization and perfect reconstruction, 307
    two-channel case, 298–314
        design, 301
        lattice structure (QMF lattice), 302
        power symmetric property, 298
        relation among analysis filters, 299
    summary, 327
    synthesis filters, 295
Paraunitary polyphase matrix, 238, 294
Paraunitary property and uncorrelatedness, 408, 544
Paraunitary pseudocirculant matrix, 431
Parseval's relation, 14
    for vector signals, 27
Passband, 43
Passband edge, 43
Passband ripple, 43
Passive quantizers, 418

Passivity, 723
PBH test, 684
PC (*see* Power complementary)
PCM scheme, 822
Peak ripple in filter design, 43, 93, 94
Peak aliasing distortion, 367
Perfect reconstruction (PR) systems, 132
    definition in maximally decimated systems, 196, 228
    delay-chain systems, 235, 236
    DFT filter banks, 132–133, 241
    FIR case, 238
    general theory, 234–238
    linear phase (*see* Linear phase filter banks)
    necessary and suffcient conditions, 237, 253
    summary, 326–332
    Venn diagram, 326, 328
Periodically time varying, 130
Periodicity matrix, 567
    and multidimensional sampling, 568
Periodic sequences and decimation, 180
Phase distortion (PHD), 2
    in digital audio, 143
    in filter banks, 196 (*see also* under Filter banks)
Phase response, 32 (*see also* Wrapped and Unwrapped phase)
Pid (Principal ideal domain), 661
Planar rotation, 291
Pole crowding, 62
Pole interlace property, 89
Poles, 16
    dynamical meaning, 39
    and stability, 17, 27
    time domain meaning, 39
Poles of a transfer matrix, 27, 674
    and homogeneous difference equation, 719, 720
    and eigenvalues of $\mathbf{A}$, 674, 720–721
    order of, 703
    summary of various manifestations, 702
    time domain (dynamical) meaning, 701
    with zeros at the same place, 707
    and zeros of det $\mathbf{Q}(z)$, 700
    and zeros of $\beta_0(z)$, 700
Poles and zeros of lossless systems, 755, 756 (*see also* Lossless systems)

Polygons, linear transformation, 580
Polynomials, 28 (*see also* Finite impulse response (FIR) functions)
Polynomial matrices, 661
    normal rank, 662
    order, 662
    rank, 662
Polyphase component, 121
    linear phase filters, 125, 182, 183
    matrix, 231
Polyphase decomposition, 6, 120–134
    allpass filters, 181
    in filter banks, 230–234
    justification of name, 166
    linear phase filters, 125, 182, 183
    multidimensional filter banks, 631
    origin of the term, 166
    for transmultiplexers, 262, 263
    Type 1, 121
    Type 2, 122
Polyphase identity, 133, 261
    multidimensional, 605
Polyphase implementation
    and commutator models, 130–131
    decimation filters, 122
    linear phase, 125
    DFT filter bank, 125
    fractional decimation circuit, 127–130
    interpolation filters, 123
Polyphase matrix
    and AC matrix, 234
    definition, 231
Polyphase representation (*see* Polyphase decomposition)
Positive definite matrices, 796
    determinant inequality, 797
    application, 828–829
    proofs, 801, 802
    properties, 796
    square roots of, 797
    test for, 797
Power complementary (PC) functions, 83, 85, 88, 90, 158
    and Nyquist(M) filters, 159
Power method (eigenvalue computation), 799
Power spectral density, 807 (*see also* Power spectrum)
Power spectrum matrix of a vector random process, 813
Power symmetric filters, filter banks, 204-223 (*see also* under Filter banks)

Index   905

Power symmetry and paraunitariness, 298, 301
Prediction coefficients, 831
Prediction error, 831
Prediction gain, 833, 847
Prediction order, 831
Principal ideal domain (pid), 642, 661
Principal minors, 786
Probability density function, 803
    uniform, 803, 804
    Gaussian, 804
Probability of overflow, 403, 818
Prolate spheroidal sequences, 50–53
    (*see also* FIR filter design)
    zeros of, 52, 53, 95
    symmetry of, 95
PR (Perfect reconstruction) QMF (*see* under Perfect reconstruction, and under Filter banks)
PR systems (*see* under Perfect reconstruction, and under Filter banks)
Pseudocirculant matrix, 249
    and alias-free filter banks, 250
    and block filters, 430
    and LPTV systems, 435, 436
    paraunitary property, 431
    product of, 454
    summary, 437
    and vector random processes, 814
Pseudo QMF, 7, 354–373
    complexity, 373
    design, 363–370
    polyphase implementation, 370–373
    theory, 354–363
        alias cancelation (approximate), 357, 358
        amplitude distortion reduction, 363–365
        phase distortion elimination, 359

Q (Quality factor), 483
QMF (*see* Quadrature mirror filter)
QMF lattice, linear-phase, 339, 351
QMF lattice (paraunitary), 302–314
    advantages, 313
    complexity, 311
    design based on, 309
    hierarchical property, 306
    properties, 304–307
    recursions for, 306–307, 334
    robustness to quantization, 307
    two-multiplier version, 304
Quadratic forms, 796
    maximization and minimization, 798
Quadrature mirror filter (QMF) bank, 147, 188, 189, 223
    analog/digital hybrid, 162, 163
    reason for the name, 196
    use of the misnomer, 223
Quantization effects (*see also* Coefficient quantization effects, and Noise analysis)
    filter bank coefficients, 218
    multiplier quantization (*see* Coefficient quantization)
    robustness of filter banks to, 218
    signal quantization, 395
    subband signals, 189, 191 (*see also* under Subband signals)
    types of, 394
Quantization of paraunitary matrices, 770–771
Quantization step size, 397, 816
Quantization of subband signals, 816–848
Quantization of unitary matrices, 769, 770
Quantizer, 397
    complex, 405
    error, 395
    noise source, 395, 398
    passive, 418
    types of, 398
Quantizer noise (error) variance, 816–818
    equalization, and bit allocation, 821
Quincunx **M**, 573
    filter banks, 636–640, 654
    noble identities 637
    polyphase decomposition, 636
    summary, 649
Quotient (integer vector division), 577

Random processes, 806
    harmonic, 815
    and LTI systems, 810
    and multirate building blocks, 406–408
    vector, 813
    white, 808, 809
    wide sense stationary (WSS), 806, 813
Random variables (r.v.) 803
Random vectors, 812
    Gaussian, 813
Range space, 789
Rank of a matrix, 788

Rate distortion theory, 842–844
Rational decimation (*see* Fractional sampling rate alteration)
Rational transfer functions, 16
Rayleigh's principle, 52, 798
Rcd (Right common divisor), 664
Reachability, 677
    conditions for, 678
    reduction to reachable form, 678
    test for, 684
Reactance functions, 98
    and allpass filters, 98
    and LC networks, 98
Real coefficient systems, 17
    cascade form, 20
Realization matrix **R**, 740, 777, 779
    and losslessness, 740
    and minimality, 745
    unitariness, 742
Reciprocal conjugate pairs, 37, 72
Reciprocal lattice, 565
Rectangular decimator, 573, 575
Rectangular to hexagonal conversion, 612–615
Rectangular sampling, 555, 559
Rectangular window
    in filter design, 46
    and Gibbs phenomenon, 45
    and optimality, 95
Recursive difference equation, 18
Reflection zeros, 69, 98, 209
Region of convergence (ROC)
    z-transform, 14
    Laplace's transform, 22
    MIMO systems, 27
Regularity, 530 (*see also* Wavelets with regularity)
Relatively prime
    integers, 118, 128, 179
    polynomials, 16
        Euclid's theorem, 159
Remainder (vector division), 577
Residual, 170
Residual vector, 170–177
    decimation of, 175
    smoothing of, 173–174
Resolution (in frequency), 460
Reversal matrix, 785
Right common divisor (rcd), 664
Right matrix fraction description (MFD), 666
Ripple (in filter response), 43
RMS (root mean square) duration of a signal, 478
Robust (to quantization) filter banks, 218

ROC (*see* Region of convergence)
Root of unity, 104
Rotation, planar (*see* Givens rotation)
Roundoff arithmetic, 398
Roundoff noise model, 396, 398 (*see also* Quantization effects)
Roundoff noise (*see* Noise analysis)
    and dynamic range, 400
r.v. (random variable), 803

Sampling (*see also* under Multidimensional sampling)
    and aliasing, 23
    definition, 22
    frequency, 22
    at Nyquist rate, 23
    period, 22
    rate, 22
Sampling density, 561, 563 (*see also* under Multidimensional sampling)
    and determinant, 563
    minimum, 568
Sampling geometry, 557
Sampling matrix, 556, 559
Sampling rate alteration, 2
    in digital audio, 145
    fractional, 109-111
Sampling rate compressor, 101 (*see also* Decimator)
Sampling rate expander, 101 (*see also* Expander)
Sampling theorems
    derivative based, 277, 439
    difference based, 440
    nonuniform, 440
    traditional, 436
    unconventional, 436–453
Scalar systems (*see* SISO systems)
SBC (*see* Subband coding)
Scaling a structure, 400, 401
    $\mathcal{L}_2$ scaling, 402
    $\mathcal{L}_p$ scaling, 401
    lattice structure (IIR), 424
    sum-scaling, 401
Scaling function $\phi(t)$, 512
    and Nyquist property, 528, 529
    recursion for, 518
    self similarity, 519
Scattering matrix, 723
Self similarity, 519
Sequences, 12–15
    bounded, 13
    energy of, 15
    exponential, 13

single-frequency, 13
sinusoidal, 13
unit-pulse, 12
unit-step, 12
Shift-invariant systems, 15
Short-time Fourier Transform (STFT)
   basis functions, 475
   continuous-time case, 476, 478
   definitions, 463, 464
   discrete-time, 463–476
   and filter banks, 464, 468
   generalizations, 474
   inversion, 472–474
   motivation, 459
   orthonormal basis, 475, 476
   summary, 477
   time-frequency tradeoff, 469
   and uniform DFT, 468
   window, choice of, 469
   window, step size, 470
Sigma-bandlimited (σ-BL), 23, 276
Signal to noise ratio, 401, 402
Signal flowgraph, 859
Sign-bit, 397
Sign-magnitude convention, 397
Similarity transform, 675, 792
   and Jordan form, 720
   and minimality, 682, 683
Single-frequency sequence, 13
   Fourier transform of, 14
Single-input single-output (SISO) system, 25, 30
Single rate system, 1
Singular matrices, 787
   summary, 791
Sinusoids, 13
SISO (*see* Single-input single-output system)
Skew Hermitian matrix, 793
Skew Hermitian polynomials, 30
Sliding windows, 117
   and the DFT, 126
Smith form decomposition for integer matrices, 641
   in multidimensional decimators, 641–643
   in multidimensional polyphase decomposition, 643
   in generalized multidimensional DFT, 644
Smith form decomposition for polynomial matrices, 687, 690
Smith-McMillan decomposition, 695
Smith-McMillan form for rational transfer matrices, 694–699
   and irreducible MFD, 697

lossless systems, 754–757 (*see also* Lossless systems)
noncausality, 696
structural meaning, 698, 699
FIR case, 710
Smoothing of residual vector, 173, 174, 186 (*see also* Multigrid methods)
SPD (*see* Symmetric parallelepiped)
Space (vector space), 787
Special sequences, 12–13 (*see also* Sequences)
Special types of filters, 83, 84 (*see also* Nyquist(M) filters, Complementary transfer functions)
   applications of, 161–164
Spectral factorization, 57–60
   phase of factors, 59
Spectral factorization techniques, 220, 849–856
   algorithm, 855, 856
   theory, 854, 855
Spectral flatness measure, 825
Spectrum analyzers, 116, 474-475 (*see also* Filter banks)
   and sliding windows, 117
Speech coding, 4–5
Square roots of matrices, 797
Stability
   conditions for, 40
   definition, 17
   and lattice structures, 81, 99
   and eigenvalues of $\mathbf{A}$, 685
   second order case, 41
   triangle, 41
Stable matrix, 685, 799
Stalling phenomenon, 171–172, 186 (*see also* Multigrid methods)
Standard deviation, 805
State equations, 670
State space descriptions, 669
   FIR, 709
   impulse response, 674
   transfer functions, 673
State space manifestation of lossless property, 740, 743
State space realizations, 673
State space structures, 673
State transition matrix (STM), 671
State variables, 669, 671
State vector, 671
Statistically independent random variables, 806
Step size, 397
STFT (*see* Short time Fourier Transform)

STM (State transition matrix), 671
Stopband, 43
    attenuation, 43
    edge, 43
    ripple, 43
Strictly complementary (SC), 157
Structurally bounded (passive) implementation, 781
Structural passivity, 419
    and low sensitivity, 420
Structures for allpass filters (see Allpass structures)
Structures for lossless systems (see under Factorization)
Structures for transfer functions,
    direct form, 18
    cascade form, 18
    minimal, 21
Subband coding (SBC), 4
    bit allocation, 146, 821
    signal compression, 145–148
Subband coder, ideal (see Ideal subband coder)
Subband decomposition, 3
Subband quantization, 816–848
Subband signals, 3
    quantization of, 189, 816–844
Subband and transform coders
    similarities and differences, 833–838
    summary, 834–836
Sublattices, 596, 658
Submatrices, 784
Suboptimal orthogonal transforms, 829
Subsampler, 101 (see also Decimator)
Sum-scaling, 401
Support of a bandlimited signal, 547
Symmetric matrix, 793
Symmetric parallelepiped [$SPD(\mathbf{V})$], 597
    in alias-free decimation filtering, 598
Symmetric polynomials (filters), and their variations, 31
Symmetric wavelet basis, 536, 542, 543
Symmetry conditions in filter bank design, 318
Synthesis bank, 113
    vector notation, 224
Synthesis filters, 3, 113, 188

TDM (see Time division multiplexing)
Tilde notation, 28–30
Time division multiplexing (TDM), 148, 149
Time-domain constraints in filter design, 54
    Nyquist constraint, 54 (see also Nyquist filters)
Time-domain description of multirate filters, 117
Time-frequency grid
    short-time Fourier transform, 471
    wavelet transform, 485, 486
Time-frequency representation
    motivation, 459–461
Time-invariant systems, 15
Time-localization (see Localization)
Time-multiplexing, 123
Toeplitz matrix, 794
    block, 794
    determinant, 832
    product, 800
Tolerance regions in filter design, 43, 44
Touching loops, 860
Trace of a matrix, 785
    applications, 830–831
    and eigenvalues, 791
Transfer functions, 15
    poles of, 16
    rational, 16
    real, 16
    and state space descriptions, 673
    zeros of, 16
Transfer matrices (see Multi-input multi-output systems)
    row and column vectors, 26
Transform coders, 322 (see also KLT, LOT)
    analysis of, 826–833
    optimum bit allocation, 826
    coding gain, 830
    decorrelation property, 834
Transform and subband coders
    similarities and differences, 833–838
    summary, 834–836
Transition band, 42
Transition bandwidth, 43
Transmission zeros, 32, 35, 69, 209 (see also Digital filters)
Transmultiplexers, 148–151, 259–266
    cross-talk, 149, 262
    cross-talk free, 265
    guard bands, 259
    perfect reconstruction, 264
    polyphase representation, 262–265
    relation to QMF bank, 264

Transpose of a matrix, 783
Transposed direct form, 18
Tree-structured filter banks (*see* under Filter banks)
    multidimensional, 640–641
Triangularization, unitary, 798
Triangular matrices, 783
    determinant, 787
    eigenvalues, 791
    properties and summary, 798
Truncation arithmetic, 398
Tunable filters, 166
Two's complement arithmetic, 397
    and scaling, 401
Type 1–4 FIR filters, 35

Uncertainty principle:
    demonstration, 459
    proof of, 540
    quantitative, 478
Uncorrelatedness
    Gaussian case, 806, 813
    random processes, 809
    random variables, 806
Uniform density, 803, 804
Uniform DFT filter banks, 116 (*see also* DFT filter banks)
Unimodular integer matrices, 559
    in multidimensional decimation, 594–596
Unimodular polynomial matrices, 663
    causal systems, 712, 713
    factorization, 692
    in perfect reconstruction, 238
Uniqueness of paraunitary factorizations, 739
Unitary diagonalization, 795
Unitary matrix
    definition, 746, 793
    degrees of freedom, 754, 777
    eigenvalues, 796
    factorization, 316, 745–754
        Givens rotation based, 747–751
        Householder (diadic) based, 751–753
    general 2×2, 727
    Givens rotation, 290, 291
    normalized, 289, 793
    quantization, 769, 770
Unitary triangularization, 798
Unit circle, 14
    zeros on, 32
Unit-pulse sequence $\delta(n)$, 12
Unit-step sequence, 12
Unreachable state component, 679

Unwrapped phase, 33
    in complex cepstrum computation, 852
    periodicity, 852, 858
Upsampler, 101 (*see also* Expander)

Vandermonde matrix, 795
Variance of a random variable, 804
Vector random processes, 813
    wide sense stationary, 813
    and LTI systems, 814
Vector random variables, 812
Vector space, 787
Voice privacy systems, 148
Von Neumann lattice, 481

Wavelet basis, finite duration, 521
Wavelet basis, symmetric, 536, 542, 543
Wavelet coefficients
    continuous-time, 487
    discrete-time, 494
Wavelet function $\psi(t)$, 488, 512
    Haar basis, 517, 520, 521
    recursion for, 518
    zeros in the frequency domain, 531, 544
Wavelet orthonormality (*see under* Orthonormal, and Orthonormality)
    and uncorrelatedness, 544
Wavelet packets, 510
Wavelet transforms, 457–544
    basis functions, 487, 488
    and filters, 488
    definitions
        discrete, 485
        discrete-time, 491 (*see also* Discrete-time wavelet transform)
        general, 487
    introduction, 461, 482–486
    inversion, 487–489
        using synthesis bank, 488
    and nonuniform decimation, 485
    and nonuniform filter banks, 483–484
    orthonormal basis, 489 (*see under* Orthonormal)
    summary, 490, 491, 497–500
    time-frequency grid, 485, 486
Wavelets, summary of
    continuous-time wavelets, 490
    discrete-time wavelets, 511
    generation of continuous-time orthonormal wavelets, 537
    short-time Fourier transform, 477

Wavelets with regularity, 530, 531
    design, 535
        and maximally flat FIR filters, 532, 534, 535
        and zeros of QMF filter, 531
White random processes, 808, 809
Wide sense stationary (WSS) process, 806
    passage through LTI systems, 810
    vector case, 813
Windows (*see* FIR filter design)
Windows, optimal for STFT, 478 (*see also* Short-time Fourier transform)
Wordlength, 397
Wrapped phase, 33
WSS (*see* Wide sense stationary)
WUZE process, 407

Zero-phase filter, 34, 45
Zero-phase response, 34
Zeros of a transfer function, 16
    time domain (dynamical) meaning, 39, 40
Zeros of a transfer matrix, 703, 704
    definition, 704
    time domain (dynamical) interpretation, 705, 719
z-transform, 13–14
    convergence, 14
    definition, 13
    multidimensional, 548